HANDBOOK OF MASSIVE DATA SETS

Handbook of Massive Data Sets

Edited by

James Abello
AT&T Labs Research

Panos M. Pardalos
University of Florida

and

Mauricio G. C. Resende
AT&T Labs Research

KLUWER ACADEMIC PUBLISHERS
DORDRECHT / BOSTON / LONDON

A C.I.P. Catalogue record for this book is available from the Library of Congress.

```
QA
76.9
.D3
H3474
2002
```

ISBN 1-4020-0489-3

Published by Kluwer Academic Publishers,
P.O. Box 17, 3300 AA Dordrecht, The Netherlands.

Sold and distributed in North, Central and South America
by Kluwer Academic Publishers,
101 Philip Drive, Norwell, MA 02061, U.S.A.

In all other countries, sold and distributed
by Kluwer Academic Publishers,
P.O. Box 322, 3300 AH Dordrecht, The Netherlands.

Printed on acid-free paper

All Rights Reserved
© 2002 Kluwer Academic Publishers
No part of the material protected by this copyright notice may be reproduced or
utilized in any form or by any means, electronic or mechanical,
including photocopying, recording or by any information storage and
retrieval system, without written permission from the copyright owner.

Printed in the Netherlands.

To our wives and children:
>Sandra and Andrea,
>Rosemary and Akis, and
>Lucia, Sasha, and Alec

CONTENTS

PREFACE xi

PART I.   INTERNET AND THE WORLD WIDE WEB

1. Algorithmic Aspects of Information Retrieval on the Web
   3
   A. Broder and M. Henzinger

2. High-Performance Web Crawling 25
   M. Najork and A. Heydon

3. Internet Growth: Is There a "Moore's Law" for Data Traffic? 47
   K. G. Coffman and A. M. Odlyzko

PART II.   MASSIVE GRAPHS

4. Random Evolution in Massive Graphs 97
   W. Aiello, F. Chung and L. Lu

5. Property Testing in Massive Graphs 123
   O. Goldreich

PART III.   STRING PROCESSING AND DATA COMPRESSION

6. String Pattern Matching for a Deluge Survival Kit ... 151
   A. Apostolico and M. Crochemore

7. Searching Large Text Collections 195
   R. Baeza-Yates, A. Moffat and G. Navarro

8. Data Compression 245
   D. Salomon

PART IV. EXTERNAL MEMORY ALGORITHMS AND DATA STRUCTURES

9. External Memory Data Structures 313
   L. Arge

10. External Memory Algorithms 359
    J. S. Vitter

PART V. OPTIMIZATION

11. Data Envelopment Analysis (DEA) in Massive Data Sets
    419
    J. H. Dulá and F. J. López

12. Optimization Methods in Massive Data Sets 439
    P. S. Bradley, O. L. Mangasarian and D. R. Musicant

13. Wavelets and Multiscale Transforms in Astronomical Image Processing 473
    F. Murtagh and J.L. Starck

14. Clustering in Massive Data Sets 501
    F. Murtagh

PART VI. DATA MANAGEMENT

15. Managing and Analyzing Massive Data Sets with Data Cubes 547
    M. Riedewald, D. Agrawal and A. El Abbadi

16. Data Squashing: Constructing Summary Data Sets .. 579
    W. DuMouchel

CONTENTS

17. Mining and Monitoring Evolving Data 593
    V. Ganti and R. Ramakrishnan

18. Data Quality in Massive Data Sets 643
    M. F. Goodchild and K. C. Clarke

19. Data Warehousing 661
    T. Johnson

20. Aggregate View Management in Data Warehouses ... 711
    Y. Kotidis

21. Semistructured Data and XML 743
    D. Suciu

PART VII. ARCHITECTURE ISSUES

22. Overview of High Performance Computers 791
    A. J. van der Steen and J. Dongarra

23. The National Scalable Cluster Project 853
    R. Grossman and R. Hollebeek

24. Sorting and Selection on Parallel Disk Models 875
    S. Rajasekaran

PART VIII. APPLICATIONS

25. Billing in the Large 895
    A. Hume

26. Detecting Fraud in the Real World 911
    M. H. Cahill, D. Lambert, J. C. Pinheiro and D. X. Sun

27. Massive Datasets in Astronomy 931
    R. J. Brunner, S. G. Djorgovski, T. A. Prince and A. S. Szalay

28. Data Management in Environmental Information Systems 981
    O. Günther

29. Massive Data Sets Issues in Earth Observing 1093
    R. Yang and M. Kafatos

30. Mining Biomolecular Data Using Background Knowledge and Artificial Neural Networks 1141
    Q. Ma, J. T. L. Wang and J. R. Gattiker

31. Massive Data Set Issues in Air Pollution Modelling   1169
    Z. Zlatev

INDEX 1221

Preface

The proliferation of massive data sets brings with it a series of special computational challenges. This "data avalanche" arises in a wide range of scientific and commercial applications. With advances in computer and information technologies, many of these challenges are beginning to be addressed by diverse inter-disciplinary groups, that include computer scientists, mathematicians, statisticians and engineers, working in close cooperation with application domain experts. High profile applications include astrophysics, bio-technology, demographics, finance, geographical information systems, government, medicine, telecommunications, the environment and the internet. John R. Tucker of the Board on Mathematical Sciences has stated:

> "My interest in this problem (Massive Data Sets) is that I see it as the most important cross-cutting problem for the mathematical sciences in practical problem solving for the next decade, because it is so pervasive."

The Handbook of Massive Data Sets is comprised of articles written by experts on selected topics that deal with some major aspect of massive data sets. It contains chapters on information retrieval both in the internet and in the traditional sense, web crawlers, massive graphs, string processing, data compression, clustering methods, wavelets, optimization, external memory algorithms and data structures, the US national cluster project, high performance computing, data warehouses, data cubes, semi-structured data, data squashing, data quality, billing in the large, fraud detection, and data processing in astrophysics, air pollution, biomolecular data, earth observation and the environment.

The handbook is addressed to academic and industrial researchers, graduate students and practitioners interested in the theoretical and technological approaches being pursued in the manipulation and analysis of very large data sets. The highly interdisciplinary nature of research in this area is providing a fruitful ground where a variety of ideas and methods come together. This volume is a sample of some of the major techniques currently in use in massive data set processing. We trust that the included material is helpful and stimulating. It provides a good indication of the vitality of of this cross-cutting field.

We would like to take the opportunity to thank the authors of the chapters, the anonymous referees, AT&T Labs Research, and the Center for Applied Optimization, University of Florida for supporting this effort. The first editor wants to express his appreciation to Dave Belanger for his continued support. Special thanks and appreciation go to

Dr. Hong-Xuan Huang for assisting us with LaTeX and other issues in the preparation of the camera-ready copy of this handbook. Finally, we would like to thank the Kluwer Academic Publishers for their assistance.

J. Abello, P.M. Pardalos, and M.G.C. Resende
July 2001

I

INTERNET AND THE WORLD WIDE WEB

Chapter 1

ALGORITHMIC ASPECTS OF INFORMATION RETRIEVAL ON THE WEB

Andrei Broder
AltaVista Company, San Mateo, California, USA
andrei.broder@av.com

Monika Henzinger
Google Incorporated, Mountain View, California, USA
monika@google.com

Abstract   The Web explosion offers a bonanza of novel problems. In particular, information retrieval in the Web context requires methods and ideas that have not been addressed in the classic information retrieval literature. This chapter will survey emerging techniques for information retrieval in the Web context and discuss some of the pertinent open problems.

Keywords: Web, Internet, Information retrieval.

1.   Introduction

In this chapter we discuss algorithmic issues specific to information retrieval on the Web. We intentionally avoid issues pertaining to information retrieval in general (e.g. stemming, query expansion) and algorithmic issues related to the Web, but not to information retrieval (e.g. Web caching or e-commerce). This paper is loosely based on a tutorial talk that we presented at the 39th Annual Symposium on Foundations of Computer Science (FOCS 98) in Palo Alto, California.

We will first give a short introduction to pre-Web information retrieval, which we denote *classical information retrieval* (Section 2). Section 3 discusses information retrieval on the Web in general. In section 4 algorithmic issues in general purpose search engines are presented. Section 5 describes further Web information retrieval tools. In section 6 we will conclude.

Throughout the chapter we will use the term *document* and *(Web) page* interchangeably.

2. Classical Information Retrieval

In the field of Information Retrieval the following classic problem setting is studied: A user tries to satisfy an information need in a given collection of documents. For that purpose the user inputs a request into the information retrieval system containing the collection. The goal of the system is to retrieve documents with information content that is relevant to the user's information need. See (Sparck-Jones and Willet 1997, Baeza-Yates and Ribeiro-Neto 1999) for an in-depth discussion of information retrieval.

An information retrieval system has to handle two tasks: Firstly, processing the collection to build an internal data structure that allows efficient access to the relevant documents, and secondly processing queries, i.e., searching the collection.

There are various strategies, called *models* to determine which documents to return. The *basic model* simply returns all documents that contain at least one of the query terms or all of the query terms. The *logic model* extends the basic model with the logical operators AND, OR, and NOT.

The *vector space model* introduced by Salton and his associates (Salton 1971; 1981) considers a high-dimensional vector space with one dimension per term. Each document or query is represented as a vector in this vector space, called *term vector*. The entry for a term in the term vector, also called the *weight* of the term, is positive if the term is contained in the document and it is zero otherwise. Entries of terms occurring in the document are positive. More specifically, the weight of the term is usually a function that increases with the frequency of the term within the document and that decreases with the number of documents in the collection containing the term. The idea is that (a) the more documents the term appears in, the less "characteristic" the term is for the document, and (b) the more often the term appears in the document, the more "characteristic" the term is for the document (assuming that some content-less words, called *stop-words*, like "the", "a", etc. were removed). The term vectors of documents might be normalized to one to account for different document lengths. The similarity between a document and a query is usually computed by the dot-product of their term vectors. For a given query this assigns a non-negative score to each document. To answer a query the documents with positive score are returned in decreasing order of score. For a detailed description of these algorithms and their implementation see the chapter on Searching Large Text Collections.

There are further models, for example, probabilistic models (Maron and Kuhns 1960, Robertson and Jones 1976, Bookstein and Swanson 1974, van Rijsbergen 1979, Fuhr 1989), and cognitive models (Salton 1980, Ellis 1992), which we will not discuss in this survey.

3. Information Retrieval on the Web

Information Retrieval on the Web in a variant of classical information retrieval. As in classical information retrieval, a user tries to satisfy an information need in a collection in documents. In this case the collection of documents consists of all the Web pages in the publicly accessible Web. Given a user query the goal is to retrieve *high quality* Web pages that are relevant to the user's need. Finding high quality documents is an additional requirement that arises in the Web context as discussed below.

3.1. Comparison with Classical Information Retrieval

To explain the difference between classical information retrieval and information retrieval on the Web we compare the two. Basically the dissemblances can be partitioned into two parts, namely dissemblances in the documents and dissemblances in the users.

We first discuss the dissemblances in the documents.

1. *Hypertext:* Documents on the Web are different from standard, text-only documents since they are connected by hyperlinks. On the average, there are roughly 10 links per document. Additionally Web pages are presented in HTML, a language that exposes various structures of the document. This can be exploited for information retrieval purposes.

2. *Heterogenity of documents:* In classical information retrieval the documents in a collection are usually quite homogeneous and in particular on the same level of quality. For example, the collection might exist of Wall Street Journal articles. In contrast, documents on the Web are very heterogenity. They vary in the types of documents, in their quality, and in their language.

(a) *Different types of documents:* In addition to text, Web pages might contain multimedia content like audio and images. They also might be dynamically generated when a request for the document is issued. The latter fact makes processing and representing the whole collection challenging if not impossible.

(b) *Quality Variance:* There is a huge variance in the quality of documents on the Web. From scientific articles to meaningless sequences of words, everything can be found on the Web. This poses an additional requirement on search engines: they not only have to find relevant documents, but relevant documents of high quality.

(c) *Variance in Languages*: There are more than 100 languages represented on the Web. Since certain information retrieval techniques, such as reducing a word to its stem, depend on the language of the word, this poses additional requirements on a Web information retrieval system. For example, to determine what kind of stemming is appropriate the system has to identify the underlying language. This is made difficult by the fact that many Web documents are short and might consist of a mixture of languages.

3 *Number of documents:* The size of the Web, currently estimated at over 1 billion documents, is much larger than any collection of documents processed by an information retrieval system before. It is estimated that the Web currently grows by 10% per month.

4 *Lack of stability:* While the collections for which commercial IR systems exist are relatively stable, the Web changes constantly. It grows continuously, and additionally some of the existing Web pages are frequently modified. For example, in a recent study by Cho and Garcia-Molina (1999) more than 20% of the Web pages in the study changed per day, with over 40% changing in the dot com domain. Additionally, the livespan of 10% of the Web pages was under 1 week.

5 *Duplication:* Over 20% of the documents on the Web are near or exact duplicates of other documents on the Web (Broder et al. 1997). It is not known how many semantic duplicates there are.

6 *Non-running text:* Many pages contain links and the corresponding anchor text, but no or not much running text. For example, a typical home page of a company contains contains images, some headlines and some links with anchor text, but no running text. The

traditional information retrieval techniques assumed that the documents contain running text. Their performance on non-running text has not been researched extensively.

The users on the Web behave differently than the users of classical information retrieval systems. The users of the latter systems are mostly trained librarians. To fulfill an information need they ask complicated queries containing a large number of query terms and operators to the system. They carefully scan all the answers and, if necessary, reformulate the query based on the answers they received.

To analyze the user behavior, Silverstein et al. (1998) analyzed a six-week long AltaVista query log. We cite their numbers below.

1. *Poor queries.* The queries in the log are usually short, 2.35 term in the average and the terms are often imprecise. They contain sub-optimal syntax. For example, 80% of the queries in the log use no operators. All this seems to imply that the users do not put much effort into formulating the query.

2. *Reaction to results.* When receiving the answers the users do not evaluate them all. This is very understandable since there are frequently thousands of answers to broad queries. Instead, for 85% of the queries in the log only the first result screen, consisting of the first 10 answers, is requested. Most queries, namely 78% of them, are not modified. Frequently, however, users follow the links that are on an answer page, i.e., they explore some of the pages that can be reached from answer pages.

3. *Heterogenity of users.* There is wide variance in education and Web experience between Web users.

Thus, the main challenge of information retrieval on the Web is how to meet the user needs given the heterogeneity of the Web pages and the poorly made queries.

However some of the above differences and some additional facts make information retrieval on the Web actually simpler. From the point of view of the collection, these are:

1. Duplication/Redundancy: It suffices to retrieve one of the duplicates. So even if a page is not returned, there is still hope that a copy of it is returned.

2. Hyperlinks: Information in the hyperlinks can be used to improve the ranking algorithm.

3 HTML-Structure: Also information in the HTML-structure, like for example that two links belong to the same list, can be exploited to improve the ranking algorithm.

4 Statistics: Statistics about page or query result popularity can also help to improve the ranking algorithm.

From the point of view of the users the following simplifies the retrieval problem:

1 Plurality of tools: While using a search engine, the user can use a multitude of other tools, including other search engines, online dictionaries and thesauri, etc.

2 Timeliness: Publication and indexing on the Web happens much faster than classical methods. This enables users to retrieve very recent information, including real-time information such as news, or weather data.

3 Interactivity: If a user wants to refine a query, he can do so and get the answer very quickly – much faster than in a library. This gives the user the option to try many different refinements and modifications. As discussed above, however, not many users take advantage of this feature.

4 Following links: The user's information need is often satisfied if a page pointing to the page that contains the information is returned.

3.2. Quantifying the Quality of Results

As we discussed above there are various differences between classic information retrieval and information retrieval on the Web. The obvious question is, then, whether the measures to evaluate classic information retrieval systems can be applied to the Web as well.

The prevalent effectiveness measures in classic IR are recall and precision. *Recall* is the percentage of relevant pages that are returned. *Precision* is the percentage of returned pages that are relevant. *Precision at rank* X is the percentage of top X pages that are relevant.

In the Web context the same ratios are interesting, but instead of using "relevance" alone to determine whether a page is a good answer, "relevance" and "high-quality" have to be used. We simply say a page is *valuable* for a query if it is a high-quality page on the topic of the query. Thus, precision, for example, is the number of valuable pages that are returned.

3.3. Web IR Tools

Web IR tools can be roughly classified into the following categories:

1 *General-purpose search engines:* There are many direct search engines, like AltaVista (AltaVista 2000), Excite (Excite 2000), Google (Google 2000), HotBot (Hotbot 2000), etc. Each of them has its own set of Web pages which they search to answer a query. None of them can deal with dynamic pages at the time of this writing. There are also some indirect search engines, also called meta-search engines, like MetaCrawler, DogPile, AskJeeves, etc. They typically have no or only a small set of their own Web pages. Instead, they send the query to a multitude of direct search engines and then compile an answer set in some way from the answers of the individual search engines.

2 *Hierarchical directories:* A different approach is taken by hierarchical directories, like Yahoo! (Yahoo! 2000) or the dmoz open directory project (dmoz 2000). At each level the user is required to choose one of a given set of categories to get to the next level. The last level, but potentially also some levels before, contains links to Web pages on the selected topic.

3 *Specialized search engines:* There also exist search engines that are specialized either on a topic, e.g. PubMed (PubMed 2000), a search engine specialized on medical publications, or on functions, like the Ahoy home page finder (Ahoy 2000) or various applet finders. Finally there are various shoping robots, which are basically search engines for commercially available products. Shopping robots are usually able to access pages with dynamic content.

4 *Other search paradigms:* There are various other search paradigms. A Search-by-example feature exists in various incarnations. Also various collaborative filtering approaches and notification systems exist on the Web.

In the following we discuss the techniques for some of the above tools with an emphasis on direct search engines.

4. General Purpose Search Engines

General purpose search engines attempt to index a sizeable portion of the Web across all topics and domains. Each such engine consists of three major components:

- A *spider* or *crawler* or *robot* collects documents by recursively following links from a large set of starting pages. This collection of pages is called *corpus*. The corpus is typically augmented with pages obtained from direct submissions to search engines and various other sources. Each crawler has different policies with respect to which links are followed, how deep various sites are explored, etc. As a result, there is surprisingly little correlation among the corpora of various engines (Bharat and Broder 1998, Lawrence and Giles 1998).

- The *indexer* processes the data and represents it usually in the form of fully inverted files. However, each major search engine uses different representation schemes and has different policies with respect to which words are indexed, capitalization, support for unicode, stemming, whether locations within documents are stored, etc. As a result, the query capabilities of various engines vary considerably.

- The *query processor* is the system that accepts input queries and returns matching answers, in an order determined by a *ranking* algorithm. It consists of a *front end* that transforms the input and brings it to a standard format (e.g. via word stemming, capitalization rules, parallelization of boolean expressions, compound identification [e.g. "San Francisco" is interpreted as the equivalent of a single word], synonym expansion, natural language processing, etc.) and a *back end* that finds the matching documents and ranks them.

Clearly the construction of large scale search engines involves numerous algorithmic issues. We discuss below some of them in more detail.

4.1. Ranking

The goal of the ranking algorithm is to order the answers to a query in decreasing order of value. For this purpose a numerical score is assigned to each document and the documents are output in decreasing order of the score. This score is typically a combination of query-independent and query-dependent criteria. A *query-independent* criteria assigns an intrinsic value to a document, regardless of the actual query. Typical examples are the length, the vocabulary, publication data (like the site to which it belongs, the date of the last change, etc.), the number of citations (indegree), anti-porn heuristics, human annotations, etc. A *query-dependent* criteria is a score which is determined only with respect to a particular query. Typical examples are the cosine measure of similarity used in the vector space model, the query-dependent connectivity-based technique described below, and statistics on which answers previous users selected for the same query.

The algorithmically most interesting of these techniques are query-independent and query-dependent connectivity-based techniques which we describe below. The first assumption behind both techniques is that a link from page A to page B means that the author of page A recommends page B. A second assumption is that a link often connects related pages.

Of course these assumptions do not always hold. A hyperlink should not be considered a recommendation if page A and B have the same author or if the hyperlink was generated for example by a Web-authoring tool.

The idea of studying "referrals" is not new. In classic IR there was a subfield, called bibliometrics, where citations were analyzed. See, e.g., (Kessler 1963, Garfield 1972, Small 1973). The field of sociometry analyzes *prestige* of members of a group to estimate the interpersonal relationships within a group. They developed algorithms (Katz 1953, Mizruchi et al. 1986) very similar to the two ranking algorithms that we present below. Furthermore, other Web related research exploit the hyperlink structure of the Web (Pirolli et al. 1996, Arocena et al. 1997, Spertus 1997, Carriere and Kazman 1997, Kleinberg 1998, Brin and Page 1998).

4.1.1 A Graph Representation for the Web.

There are various ways the Web can be represented as a graph. In this subsection we assume the most straightforward representation: The graph contains a node for each page u and there exists a directed edge (u, v) if and only if page u contains a hyperlink to page v.

A complete depiction of this graph in computer memory requires careful compression and/or complex graph algorithms. (See Bharat et al. (1998) that report a construction involving over 1 billion edges.) Fortunately, some of the algorithms below can be completed in batch mode. Others might be implemented using efficient external-memory graph algorithms. This seems a challenging open problem.

4.1.2 Query-independent Connectivity-based Ranking.

The first assumption of connectivity based techniques immediately leads to a simple query-independent criteria: The larger the number of hyperlinks pointing to a page, also called *inlinks*, the better the page (Carriere and Kazman 1997). The main drawback of this criteria is that each link is equally weighted. Thus, it cannot distinguish the quality of a page that gets pointed to by i low-quality pages from the quality of a page that gets pointed to by i high-quality pages. Obviously it is therefore easy to make a page appear to be high-quality – just create many other pages that point to it.

To remedy this problem, Brin and Page (1998) invented the PageRank measure. Their basic idea is to weight each hyperlink proportionally to the quality of the page containing the hyperlink. To determine the quality of a page, they use its PageRank, which leads to the following recursive definition of the PageRank $R(p)$ of a page p:

$$R(p) = \epsilon/n + (1 - \epsilon) \cdot \sum_{(q,p) \text{ exists}} R(q)/outdegree(q),$$

where

- ϵ is a dampening factor usually set between 0.1 and 0.2;
- n is the number of pages on the Web;
- $outdegree(q)$ is the number of hyperlinks on page q.

Alternatively, the PageRank can be defined by the stationary distribution of the following infinite, random walk $p_1, p_2, p_3, \ldots$, where each p_i is a node in the graph: The walk starts at each node with equal probability. To determine node p_{i+1} a biased coin is flipped: With probability ϵ node p_{i+1} is chosen uniformly at random from all nodes in the graph, with probability $1 - \epsilon$ it is chosen uniformly at random from all nodes q such that edge (p_i, q) exists in the graph.

The PageRank measure is used in the Google search engine.

4.1.3 Query-dependent Connectivity-based Ranking.
A query-dependent connectivity-based ranking algorithm was developed by Kleinberg (1998). We describe it next.

Given a user query, the algorithm first constructs a query specific graph which is a subgraph of the graph described in subsection 4.1.1. Then it iteratively computes a *hub* score and an *authority* score for each node in the graph. The documents are then ranked by hub and authority scores, respectively.

Nodes, i.e., documents that have high authority scores are expected to have relevant content, whereas documents with high hub scores are expected to contain *hyperlinks* to relevant content. The intuition is as follows. A document which points to many others is a good hub, and a document that many documents point to is a good authority. Recursively, a document that points to many good authorities is an even better hub, and similarly a document pointed to by many good hubs is an even better authority.

The query-dependent *neighborhood graph* is constructed as follows. A *start set* of documents matching the query is fetched from a search engine (say the top 200 matches). This set is augmented by its *neighborhood*, which is the set of documents that either point to or are pointed to by documents in the start set. Since the indegree of nodes can be very large, in practice a limited number of predecessors (say 50) of a document are included. The documents in the start set and its neighborhood together form the nodes of the neighborhood graph. Hyperlinks between documents *not on the same host* form the directed edges. Hyperlinks within the same host are assumed to be by the same author and hence are not considered to be a recommendation. The computation of hub and authority scores is done as follows.

(1) Let N be the set of nodes in the neighborhood graph.
(2) For every node n in N, let $H[n]$ be its hub score and $A[n]$ its authority score.
(3) Initialize $H[n]$ to 1 for all n in N.
(4) While the vectors H and A have not converged:
(5)   For all n in N, $A[n] := \sum_{(n',n) \in N} H[n']$
(6)   For all n in N, $H[n] := \sum_{(n,n') \in N} A[n']$
(7)   Normalize the H and A vectors.

Kleinberg proved that the H and A vectors will eventually converge, i.e., that termination is guaranteed when finite precision computation is used.

Note that the algorithm does not claim to find *all* valuable pages for a query, since there may be some that have good content but have not been linked to by many authors or that do not belong to the neighborhood graph.

There are two types of problems with this approach: First, since it only considers a relatively small part of the Web graph, adding a few edges can potentially change the resulting hubs and authority scores considerably. Thus it is easier for authors of Web pages to manipulate the hubs and authority scores than it is to manipulate the PageRank score. A second problem is that if the neighborhood graph contains more pages on a topic different from the query, then it can happen that the top authority and hub pages are on this different topic. This problem was called *topic drift*. Various papers (Chakrabarti et al. 1998a;c, Bharat and Henzinger 1998) suggest the use of edge weights and content analysis of either the documents or the anchor text to deal with these problems. In a user study (Bharat and Henzinger 1998) it was shown that this considerably improves the precision at rank 10.

4.1.4 Open Problems.
There are many open problems related to connectivity-based ranking techniques. Both PageRank and Kleinberg's algorithm compute the principal eigenvector. As mentioned by Kleinberg, non-principal eigenvectors might also convey useful information.

It would also be interesting to compare the performance of the query-dependent and query-independent techniques and determine their weaknesses and strengths.

Finally other graph structures, like the graph of co-citations (there exists an undirected edge (u,v) iff u and v are co-cited), could be derived from the Web and used for ranking.

4.2. Duplicate Filtering

The Web has undergone an exponential growth since its birth, and this has lead to the proliferation of documents that are identical or near identical. As mentioned earlier experiments indicate that over 20% of the publicly available documents on the Web are duplicates or near-duplicates. These documents arise innocently (e.g. local copies of popular documents, mirroring), maliciously (e.g., "spammers" and "robot traps"), and erroneously (spider mistakes). In any case they represent a serious problem for indexing software for two main reasons: first, indexing of duplicates wastes expensive resources; and second, users are seldom interested in seeing documents that are "roughly the same" in response to their queries.

This informal concept does not seem to be well captured by any of the standard distances defined on strings (Hamming, Levenshtein, etc.). Furthermore the computation of these distances usually requires the pairwise comparison of entire documents. For a very large collection of documents this is not feasible, and a sampling mechanism per document is necessary. Furthermore, for efficiency it is preferable to store only a short sketch for each document and processing n documents should require $O(n)$ or $O(n \log n)$.

We describe here the solution presented in (Broder 1997, Broder et al. 1997), based on the notion of *resemblance*, a mathematical concept that captures well the informal notion of syntactic similarity. Other similarity detection mechanisms are presented in (Manber 1994, Brin et al. 1995, Shivakumar and García-Molina 1995, Heintze 1996). Techniques for detecting duplicated sites (i.e., collections of web pages on the same host) are given in (Bharat et al. 1999, Cho et al. 2000).

The basic approach in (Broder 1997) for computing resemblance has two aspects: First, resemblance is expressed as a set intersection problem, and second, the relative size of intersections is evaluated by a process of random sampling that can be done independently for each document.

The reduction to a set intersection problem is done via a process called *shingling*, which was previously used by Brin et al. (1995). We view each document as a sequence of tokens. We can take tokens to be letters, or words, or lines. We assume that we have a parser program that takes an arbitrary document and reduces it to a canonical sequence of tokens. (Here "canonical" means that any two documents that differ only in formatting or other information that we chose to ignore, for instance punctuation, formatting commands, capitalization, and so on, will be reduced to the same sequence.) For the remainder of this section, document means a canonical sequence of tokens.

A contiguous subsequence of w tokens contained in D is called a *shingle*. Given a document D, we can associate to it its w-*shingling* defined as the set of all shingles of size w contained in D. So for instance the 4-shingling of

$$(\texttt{a,rose,is,a,rose,is,a,rose})$$

is the set

$$\{(\texttt{a,rose,is,a}), (\texttt{rose,is,a,rose}), (\texttt{is,a,rose,is})\}$$

(It is possible to use alternative definitions, based on multisets. See (Broder 1997) for details.)

Rather than dealing with shingles directly, it is more convenient to associate to each shingle a numeric uid (unique id). This is done by *fingerprinting* the shingle. (Fingerprints are short tags for larger objects. They have the property that if two fingerprints are different, then the corresponding objects are certainly different and there is only a small probability that two different objects have the same fingerprint. This probability is typically exponentially small in the length of the fingerprint.) For reasons explained in (Broder 1997), it is particularly advantageous to use Rabin fingerprints (Rabin 1981) that have a very fast software implementation (Broder 1993).

Via shingling, each document D gets an associated set S_D. For the purpose of the discussion here we can view S_D as a set of natural numbers. (The size of S_D is about equal to the number of words in D.) The *resemblance* $r(A, B)$ of two documents, A and B, is defined as

$$r(A, B) = \frac{|S_A \cap S_B|}{|S_A \cup S_B|}.$$

Experiments seem to indicate that high resemblance (that is, close to 1) captures well the informal notion of "near-duplicate" or "roughly the same".

To compute the resemblance of two documents it suffices to keep for each document a relatively small, fixed size *sketch*. The sketches can be computed fairly fast (linear in the size of the documents) and given two sketches the resemblance of the corresponding documents can be computed in linear time in the size of the sketches.

This is done as follows. Assume that for all documents of interest $S_D \subseteq \{1, \ldots, n\}$. (In practice $n = 2^{64}$.) Let π be chosen uniformly at random over S_n, the set of permutations of $[n]$. Then

$$\Pr(\min\{\pi(S_A)\} = \min\{\pi(S_B)\}) = \frac{|S_A \cap S_B|}{|S_A \cup S_B|} = r(A, B). \qquad (1.1)$$

Hence, we can choose, say, 100 independent random permutations $\pi_1, \ldots, \pi_{100}$. For each document D, we store the list

$$\bar{S}_A = (\min\{\pi_1(S_A)\}, \min\{\pi_2(S_A)\}, \ldots, \min\{\pi_{100}(S_A)\}).$$

Then we can readily estimate the resemblance of A and B by computing how many corresponding elements in $\bar{S}_A$ and $\bar{S}_B$ are common. (For a set of documents, we avoid quadratic processing time, because a particular value for any coordinate is usually shared by only a few documents. For details see (Broder 1997, Broder et al. 1997).)

In practice, it is impossible to choose π uniformly at random from S_n. We are thus led to consider smaller families of permutations that still satisfy (or approximately satisfy) equation (1.1). Such families are called *min-wise independent*. The theory of min-wise independent permutations is discussed in (Broder et al. 1998).

For large scale Web indexing it is not necessary to determine the actual resemblance value: it suffices to determine whether newly encountered documents are duplicates or near-duplicates of documents already indexed; this determination can be made using a sample of only a few tens of bytes per document, called *features*.

The goal is that with high probability, two documents share more than a certain number of features if and only if their resemblance is very high.

Consider two documents, A and B, that have resemblance ρ. If ρ is close to 1, then almost all the elements of $\bar{S}_A$ and $\bar{S}_B$ will be pairwise equal. The idea of duplicate filtering is to divide every sketch into k groups of s elements each, called *features*. The probability that all the elements of a group are pair-wise equal is simply ρ^s and the probability that two sketches have r or more equal groups is

$$P_{k,s,r} = \sum_{r \leq i \leq k} \binom{k}{i} \rho^{s \cdot i} (1 - \rho^s)^{k-i}.$$

The remarkable fact is that for suitable choices of $[k, s, r]$ the polynomial $P_{k,s,r}$ behaves as a very sharp high-band pass filter even for small values of k. For instance, for 6 features

- The probability that two documents that have resemblance greater than 97.5% *do not share* at least two features is less than 0.01. The probability that two documents that have resemblance greater than 99% *do not share* at least two features is less than 0.00022.

- The probability that two documents that have resemblance less than 77% *do share* two or more features is less than 0.01 The probability that two documents that have resemblance less than 50% share two or more features is less than 0.6×10^{-7}.

Note that using 6 features of 8 bytes we store on 48 bytes/document.

4.3. Collecting Documents

Another challenging issue related to information retrieval on the Web is creating the collection, i.e., collecting all pages on the Web. There is no search engine that claims to index all pages on the Web. Most search engines use a program to find and download the pages. This program usually maintains a priority queue of pages that have yet to be downloaded. Whenever a pages is downloaded its hyperlinks are checked to see whether they point to undiscovered pages and if so these pages are added to the queue. They are either added to the beginning or to the end of the queue depending on whether a breadth-first or depth-first graph exploration is intended.

However, since most search engines cannot store all Web pages in their index, their goal is actually to visit the highest-quality pages first. To achieve this goal Cho, Garcia-Molina, and Page suggested to order the pages in the queue according to the highest potential PageRank (Cho et al. 1998).

Web pages can become outdated. Thus, another important goal is to keep the index "fresh", i.e., to refetch outdated pages and remove deleted pages from the index. Coffman, Liu, and Weber suggested for this purpose to order the documents in the queue according to their rate of change (Coffman et al. 1997).

A third interesting question is how to evenly distribute the load internally and externally. Internally means how to parallelize the crawler efficiently considering that the response time to each fetch request as well as the size of the answer are unpredictable and considering that there are additional system constraints (for example on the number of threads, the number of open connections, etc.) External load balancing means that a crawler should not overload any server or connection (See Koster for general guidelines (Koster 1993).) This is important since a well-connected crawler can saturate the entire outside bandwidth of some small countries. Thus, the queuing discipline used must be acceptable to the community.

5. Search-by-example

Web search engines take a query as input and produce a set of (hopefully) relevant pages that match the query terms. While useful in many circumstances, search engines have the disadvantage that users have to formulate queries that specify their information need, which is prone to errors. Search-by-example is a different approach to Web searching. The input to the search process is not a set of query terms, but the URL (Universal Resource Locator) of a page, and the output is a set of related Web pages. A related Web page is one that addresses the same topic as the original page, but is not necessarily semantically identical. For example, given www.nytimes.com, other newspapers and news organizations on the Web would be considered related pages.

Of course, in contrast to search engines, this approach requires that the user has already found a page of interest. If this page was found by Google (Google 2000), then a link on the result page, executes a search by example. Otherwise, there exists a button that provides related pages on Alexa's toolbar (alexa 2000). Not much is published about what algorithms these services use. Alexa says that its service exploits usage data, the hyperlink structure, and the content of the pages.

Kleinberg (1998) suggested that a modification of his query-dependent connectivity-based technique could be used for a search-by-example feature. Dean and Henzinger (1999) extended this approach to consider the order of hyperlinks on a page. They also showed that returning the most co-cited pages as the best related pages (suggested already in (Spertus 1997)) works surprisingly well.

6. Final Conclusions

In this chapter, we mostly discussed Web information retrieval methods and tools that take advantage of the Web particularities to mitigate some of the difficulties that Web information retrieval encounters.

There are many interesting research topics in Web information retrieval that we did not discuss in this chapter. We list a few here. Categorizing Web pages into a given hierarchical category (see, e.g., (Chakrabarti et al. 1998b)), clustering query results (see, e.g., (Zamir and Etzioni 1998)), and summarization of Web pages (see, e.g., (Tombros and Sanderson 1998)) are interesting research topics. Various collaborative filtering approaches were applied to the Web (see, e.g., (Terveen et al. 1997, Konstan et al. 1997)). Finally there exist new approaches to learn how to access dynamic pages (Doorenbos et al. 1997).

Bibliography

Ahoy. http://centauri-prime.cs.washington.edu:6060, 2000.

alexa. http://alexa.com, 2000.

AltaVista. http://altavista.com, 2000.

G. O. Arocena, A. O. Mendelzon, and G. A. Mihaila. Applications of a web query language. In *WWW6*, pages 587–595, 1997.

R. Baeza-Yates and B. Ribeiro-Neto. *Modern Information Retrieval*. Addison-Wesley, 1999.

K. Bharat, A. Broder, J. Dean, and M. Henzinger, 1999. Workshop on Organizing Webspace at the Fourth ACM Conference on Digital Libraries.

K. Bharat and A. Z. Broder. A technique for measuring the relative size and overlap of public web search engines. In *WWW7*, pages 379–388, 1998.

K. Bharat, A.Z. Broder, M. Henzinger, P. Kumar, and S. Venkatasubramanian. The connectivity server: Fast access to linkage information on the web. In *WWW7*, pages 469–477, 1998.

K. Bharat and M. Henzinger. Improved algorithms for topic distillation in hyperlinked environments. In *Proceedings of the 21st International ACM SIGIR Conference on Research and Development in Information Retrieval (SIGIR'98)*, pages 111–104, 1998.

A. Bookstein and D. Swanson. Probabilistic models for automatic indexing. *Journal of the American Society for Information Science*, 25: 312–318, 1974.

S. Brin, J. Davis, and H. García-Molina. Copy detection mechanisms for digital documents. In M.J. Carey and D.A. Schneider, editors,

Proceedings of the 1995 ACM SIGMOD International Conference on Management of Data, pages 398–409, 1995.

S. Brin and L. Page. The anatomy of a large-scale hypertextual web search engine. In *WWW7*, pages 107–117, 1998.

A.Z. Broder. Some applications of Rabin's fingerprinting method. In R. Capocelli, A. De Santis, and U. Vaccaro, editors, *Sequences II: Methods in Communications, Security, and Computer Science*, pages 143–152. Springer-Verlag, 1993.

A.Z. Broder. On the resemblance and containment of documents. In *Proceedings of Compression and Complexity of Sequences*, pages 21–29. IEEE Computer Society, 1997.

A.Z. Broder, M. Charikar, A. Frieze, and M. Mitzenmacher. Min-wise independent permutations. In *Proceedings of the 30th Annual ACM Symposium on Theory of Computing (STOC-98)*, pages 327–336. ACM Press, 1998.

A.Z. Broder, S.C. Glassman, M.S. Manasse, and G. Zweig. Syntactic clustering of the Web. In *WWW6*, pages 391–404, 1997.

J. Carriere and R. Kazman. Webquery: Searching and visualizing the web through connectivity. In *WWW6*, pages 701–711, 1997.

S. Chakrabarti, B. Dom, D. Gibson, S. Kumar, P. Raghavan, S. Rajagopalan, and A. Tomkins. Experiments in topic distillation, 1998a. ACM-SIGIR'98 Post-Conference Workshop on Hypertext Information Retrieval for the Web.

S. Chakrabarti, B. Dom, and P. Indyk. Enhanced hypertext categorization using hyperlinks. In L.M. Haas and A. A. Tiwary, editors, *SIGMOD 1998, Proceedings ACM SIGMOD International Conference on Management of Data*. ACM Press, 1998b.

S. Chakrabarti, B.P.R. Dom, S. Rajagopalan, D. Gibson, and J. Kleinberg. Automatic resource compilation by analyzing hyperlink structure and associated text. In *WWW7*, pages 65–74, 1998c.

J. Cho and H. Garcia-Molina. The evolution of the web and implications for an incremental crawler. Technical report, Stanford University, Stanford, California, 1999.

J. Cho, H. García-Molina, and L. Page. Efficient crawling through URL ordering. In *WWW7*, pages 161–172, 1998.

J. Cho, N. Shivakumar, and H. Garcia-Molina. Finding replicated web collections. In *Proceedings of the 2000 ACM Internation Conference on Management of Data (SIGMOD)*, 2000.

E.G. Coffman, Z. Liu, and R.R. Weber. Optimal robot scheduling for web search engines. Technical Report 3317, INRIA, 1997.

J. Dean and M.R. Henzinger. Finding related web pages in the world wide web. In *Proceedings of the Eighth International World Wide Web Conference*, pages 389–401, 1999.

dmoz, 2000. http://dmoz.org/.

R.B. Doorenbos, O. Etzioni, and D.S. Weld. A scalable comparison-shopping agent for the World-Wide Web. In W.L. Johnson and B. Hayes-Roth, editors, *Proceedings of the 1st International Conference on Autonomous Agents*, pages 39–48. ACM Press, 1997.

D. Ellis. The physical and cognitive paradigms in information retrieval research. *Journal of Documentation*, 48:45–64, 1992.

Excite, 2000. http://excite.com/.

N. Fuhr. Models for retrieval with probabilistic indexing. *Information Precessing and Management*, 25:55–72, 1989.

E. Garfield. Citation analysis as a tool in journal evaluation. *Science*, 178:471–479, 1972.

Google, 2000. http://google.com/.

N. Heintze. Scalable document fingerprinting. In *Second USENIX Workshop on Electronic Commerce*, pages 191–200, 1996.

Hotbot, 2000. http://hotbot.com/.

L. Katz. A new status index derived from sociometric analysis. *Psychometrika*, 18:39–43, 1953.

M.M. Kessler. Bibliographic coupling between scientific papers. *American Documentation*, 14:10–25, 1963.

J. Kleinberg. Authoritative sources in a hyperlinked environment. In *Proceedings of the 9th Annual ACM-SIAM Symposium on Discrete Algorithms*, pages 668–677, 1998.

J. Konstan, B. Miller, D. Maltz, J. Herlocker, L. Gordon, and J. Riedl. Grouplens: Collaborative filtering for usenet news. *Communications of the ACM*, 40:77–87, 1997.

M. Koster, 1993. http://info.webcrawler.com/mak/projects/robots/guidelines.html.

S. Lawrence and C.L. Giles. Searching the World Wide Web. *Science*, 280:98, 1998.

U. Manber. Finding similar files in a large file system. In *Proceedings of the Winter 1994 USENIX Conference*, pages 1–10, 1994.

M. Maron and J. Kuhns. On relevance, probabilistic indexing and information retrieval. *Journal of the Association for Computing Machinery*, 7:216–244, 1960.

M.S. Mizruchi, P. Mariolis, M. Schwartz, and B. Mintz. Techniques for disaggregating centrality scores in social networks. In N.B. Tuma, editor, *Sociological Methodology*, pages 26–48. Jossey-Bass, 1986.

P. Pirolli, J. Pitkow, and R. Rao. Silk from a sow's ear: Extracting usable structures from the web. In *Proceedings of the Conference on Human Factors in Computing Systems (CHI 96)*, pages 118–125, 1996.

PubMed, 2000. http://ncbi.nlm.nih.gov/.

M.O. Rabin. Fingerprinting by random polynomials. Technical Report TR-15-81, Center for Research in Computing Technology, Harvard University, 1981.

S.E. Robertson and K.S. Jones. Relevance weighting of search terms. *Journal of the American Society for Information Science*, 27:129–146, 1976.

G. Salton. *The SMART System – Experiments in Automatic Document Processing*. Prentice Hall, 1971.

G. Salton. The relevance of the cognitive paradigm for information science. In O. Harbo and L. Kajberg, editors, *Theory and Application of Information Research. Proccedings of the 2nd International Research Forum on Information Science*, pages 49–61. Mansell, 1980.

G. Salton. The smart environment for retrieval system evaluation – Advantages and problem areas. In K.S. Jones, editor, *Information Retrieval Experiment*, pages 316–329. Butterworths, 1981.

N. Shivakumar and H. García-Molina. SCAM: A copy detection mechanism for digital documents. In *DL '95 Proceedings*, pages 0–, 1995.

C. Silverstein, M. Henzinger, J. Marais, and M. Moricz. Analysis of a very large AltaVista query log. Technical Report 1998-014, Compaq Systems Research Center, Palo Alto, California, 1998.

H. Small. Co-citation in the scientific literature: A new measure of the relationship between two documents. *Journal of the American Society for Information Science*, 24:265–269, 1973.

K. Sparck-Jones and P. Willet, editors. *Readings in Information Retrieval*. Morgan Kaufmann, 1997.

E. Spertus. Parasite: Mining structural information on the web. In *WWW6*, pages 587–595, 1997.

L. Terveen, W. Hill, B. Amento, D. McDonald, and J. Creter. Phoaks: A system for sharing recommendations. *Communications of the ACM*, 40:59–62, 1997.

A. Tombros and M. Sanderson. Advantages of query biased summaries in information retrieval. In *Proceedings of the 21st International ACM SIGIR Conference on Research and Development in Information Retrieval (SIGIR'98)*, pages 2–10, 1998.

C.J. van Rijsbergen. *Information Retrieval*. Butterworths, 1979.

Yahoo!, 2000. http://yahoo.com/.

O. Zamir and O. Etzioni. Web document clustering: A feasibility demonstration. In *Proceedings of the 21st International ACM SIGIR Conference on Research and Development in Information Retrieval (SIGIR'98)*, pages 46–54, 1998.

Chapter 2

HIGH-PERFORMANCE WEB CRAWLING

Marc Najork
Compaq Computer Corporation Systems Research Center, Palo Alto, CA 94301, USA
marc.najork@compaq.com

Allan Heydon*
Model N, Inc., South San Francisco, CA 94080, USA
aheydon@modeln.com

Abstract   High-performance web crawlers are an important component of many web services. For example, search services use web crawlers to populate their indices, comparison shopping engines use them to collect product and pricing information from online vendors, and the Internet Archive uses them to record a history of the Internet. The design of a high-performance crawler poses many challenges, both technical and social, primarily due to the large scale of the web. The web crawler must be able to download pages at a very high rate, yet it must not overwhelm any particular web server. Moreover, it must maintain data structures far too large to fit in main memory, yet it must be able to access and update them efficiently. This chapter describes our experience building and operating such a high-performance crawler.

Keywords: Web crawling, Internet archive, Search engines, Java, HTTP, Checkpointing, Link extractor, Breadth first traversal, Name resolution, Fingerprinting, Mercator.

1.   Introduction

A web crawler (also known as a web robot or spider) is a program for downloading web pages. Given a set *s* of "seed" Uniform Resource Loca-

---
*This work was performed while the author was a member of the research staff at Compaq's Systems Research Center.

tors (URLs), the crawler repeatedly removes one URL from s, downloads the corresponding page, extracts all the URLs contained in it, and adds any previously unknown URLs to s.

Although the web crawling algorithm is conceptually simple, designing a high-performance web crawler comparable to the ones used by the major search engines is a complex endeavor. All the challenges inherent in building such a high-performance crawler are ultimately due to the scale of the web. In order to crawl a billion pages in a month, a crawler must download about 400 pages every second. Moreover, the crawler must store several data structures (such as the set s of URLs remaining to be downloaded) that must all scale gracefully beyond the limits of main memory.

We have built just such a high-performance web crawler, called Mercator, which has the following characteristics:

Distributed. A Mercator crawl can be distributed in a symmetric fashion across multiple machines for better performance.

Scalable. Mercator is scalable in two respects. First, due to its distributed architecture, Mercator's performance can be scaled by adding extra machines to the crawling cluster. Second, Mercator has been designed to be able to cope with a rapidly growing web. In particular, its data structures use a bounded amount of main memory, regardless of the size of the web being crawled. This is achieved by storing the vast majority of data on disk.

High performance. During our most recent crawl, which ran on four Compaq DS20E 666 MHz Alpha servers and which saturated our 160 Mbit/sec Internet connection, Mercator downloaded about 50 million documents per day over a period of 17 days.

Polite. Despite the need for speed, anyone running a web crawler that overloads web servers soon learns that such behavior is considered unacceptable. At the very least, a web crawler should not attempt to download multiple pages from the same web server simultaneously; better, it should impose a limit on the portion of a web server's resources it consumes. Mercator can be configured to obey either of these politeness policies.

Continuous. There are many crawling applications (such as maintaining a fresh search engine index) where it is desirable to continuously refetch previously downloaded pages. This naturally raises the question of how to interleave the downloading of old pages with newly discovered ones. Mercator solves this problem by providing a priority-based mechanism for scheduling URL downloads.

Extensible. No two crawling tasks are the same. Ideally, a crawler should be designed in a modular way, where new functionality can be added by third parties. Mercator achieves this ideal through a component-based architecture. Each of Mercator's main components is specified by an abstract interface. We have written numerous implementations of each component, and third parties can write new implementations from scratch or extend ours through object-oriented subclassing. To configure Mercator for a particular crawling task, users supply a configuration file that causes the appropriate components to be loaded dynamically.

Portable. Mercator is written entirely in Java, and thus runs on any platform for which there exists a Java virtual machine. In particular, it is known to run on Windows NT, Linux, Tru64 Unix, Solaris, and AIX.

There is a natural tension between the high performance requirement on the one hand, and the scalability, politeness, extensibility, and portability requirements on the other. Simultaneously supporting all of these features is a significant design and engineering challenge.

This chapter describes Mercator's design and implementation, the lessons we've learned in the process of building it, and our experiences in performing large crawls.

2. A Survey of Web Crawlers

Web crawlers are almost as old as the web itself (Koster 2001). The first crawler, Matthew Gray's Wanderer, was written in the spring of 1993, roughly coinciding with the first release of NCSA Mosaic (Gray 2001). Several papers about web crawling were presented at the first two World Wide Web conferences (Eichmann 1994, McBryan 1994, Pinkerton 1994). However, at the time, the web was three to four orders of magnitude smaller than it is today, so those systems did not address the scaling problems inherent in a crawl of today's web.

Obviously, all of the popular search engines use crawlers that must scale up to substantial portions of the web. However, due to the competitive nature of the search engine business, the designs of these crawlers have not been publicly described. There are two notable exceptions: the Google crawler and the Internet Archive crawler. Unfortunately, the descriptions of these crawlers in the literature are too terse to enable reproducibility.

The original Google crawler (Brin and Page 1998) (developed at Stanford) consisted of five functional components running in different processes. A *URL server process* read URLs out of a file and forwarded them to multiple crawler processes. Each *crawler process* ran on a different machine, was single-threaded, and used asynchronous I/O to fetch data from up to 300 web servers in parallel. The crawlers transmitted downloaded pages to a single *StoreServer process*, which compressed the pages and stored them to disk. The pages were then read back from disk by an *indexer process*, which extracted links from HTML pages and saved them to a different disk file. A *URL resolver process* read the link file, derelativized the URLs contained therein, and saved the absolute URLs to the disk file that was read by the URL server. Typically, three to four crawler machines were used, so the entire system required between four and eight machines.

Research on web crawling continues at Stanford even after Google has been transformed into a commercial effort. The Stanford WebBase project has implemented a high-performance distributed crawler, capable of downloading 50 to 100 documents per second (Hirai et al. 2000). Cho and others have also developed models of document update frequencies to inform the download schedule of incremental crawlers (Cho et al. 1998).

The Internet Archive also used multiple machines to crawl the web (Burner 1997, Int 2001). Each crawler process was assigned up to 64 sites to crawl, and no site was assigned to more than one crawler. Each single-threaded crawler process read a list of seed URLs for its assigned sites from disk into per-site queues, and then used asynchronous I/O to fetch pages from these queues in parallel. Once a page was downloaded, the crawler extracted the links contained in it. If a link referred to the site of the page it was contained in, it was added to the appropriate site queue; otherwise it was logged to disk. Periodically, a batch process merged these logged "cross-site" URLs into the site-specific seed sets, filtering out duplicates in the process.

The WebFountain crawler shares several of Mercator's characteristics: it is distributed, continuous (the authors use the term "incremental"), polite, and configurable (Edwards et al. 2001). Unfortunately, as of this writing, WebFountain is in the early stages of its development, and data about its performance is not yet available.

3. Mercator's Architecture

The basic algorithm executed by any web crawler takes a list of *seed* URLs as its input and repeatedly executes the following steps:

> Remove a URL from the URL list, determine the IP address of its host name, download the corresponding document, and extract any links contained in it. For each of the extracted links, ensure that it is an absolute URL (derelativizing it if necessary), and add it to the list of

URLs to download, provided it has not been encountered before. If desired, process the downloaded document in other ways (e.g., index its content).

This basic algorithm requires a number of functional components:

- a component (called the *URL frontier*) for storing the list of URLs to download;
- a component for resolving host names into IP addresses;
- a component for downloading documents using the HTTP protocol;
- a component for extracting links from HTML documents; and
- a component for determining whether a URL has been encountered before.

The remainder of this section describes how Mercator refines this basic algorithm.

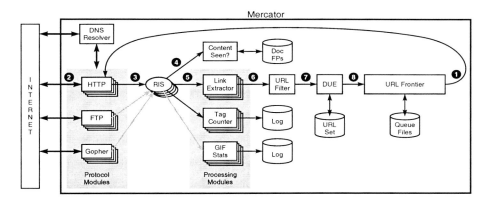

Figure 2.1. Mercator's main components.

Figure 2.1 shows Mercator's main components. Crawling is performed by multiple worker threads, typically numbering in the hundreds. Each worker repeatedly performs the steps needed to download and process a document. The first step of this loop ❶ is to remove an absolute URL from the shared URL frontier for downloading.

An absolute URL begins with a *scheme* (e.g., "http"), which identifies the network protocol that should be used to download it. In Mercator, these network protocols are implemented by *protocol modules*. The protocol modules to be used in a crawl are specified in a user-supplied configuration file, and are dynamically loaded at the start of the crawl.

The default configuration includes protocol modules for HTTP, FTP, and Gopher.

Based on the URL's scheme, the worker selects the appropriate protocol module for downloading the document. It then invokes the protocol module's *fetch* method, which downloads the document from the Internet ❷ into a per-thread *RewindInputStream* ❸ (or RIS for short). A RIS is an I/O abstraction that is initialized from an arbitrary input stream, and that subsequently allows that stream's contents to be re-read multiple times.

Courteous web crawlers implement the Robots Exclusion Protocol, which allows web masters to declare parts of their sites off limits to crawlers (Koster 1996). The Robots Exclusion Protocol requires a web crawler to fetch a resource named "/robots.txt" containing these declarations from a web site before downloading any real content from it. To avoid downloading this resource on every request, Mercator's HTTP protocol module maintains a fixed-sized cache mapping host names to their robots exclusion rules. By default, the cache is limited to 2^{18} entries, and uses an LRU replacement strategy.

Once the document has been written to the RIS, the worker thread invokes the *content-seen test* to determine whether this document with the same content, but a different URL, has been seen before ❹. If so, the document is not processed any further, and the worker thread goes back to step ❶.

Every downloaded document has a *content type*. In addition to associating schemes with protocol modules, a Mercator configuration file also associates content types with one or more *processing modules*. A processing module is an abstraction for processing downloaded documents, for instance extracting links from HTML pages, counting the tags found in HTML pages, or collecting statistics about GIF images. In general, processing modules may have side-effects on the state of the crawler, as well as on their own internal state.

Based on the downloaded document's content type, the worker invokes the *process* method of each processing module associated with that content type ❺. For example, the Link Extractor and Tag Counter processing modules in Figure 2.1 are used for text/html documents, and the GIF Stats module is used for image/gif documents.

By default, a processing module for extracting links is associated with the content type text/html. The *process* method of this module extracts all links from an HTML page. Each link is converted into an absolute URL and tested against a user-supplied *URL filter* to determine if it should be downloaded ❻. If the URL passes the filter, it is submitted to the *duplicate URL eliminator* (DUE) ❼, which checks if the URL has been seen before, namely, if it is in the URL frontier or has already been downloaded. If the URL is new, it is added to the frontier ❽.

Finally, in the case of continuous crawling, the URL of the document that was just downloaded is also added back to the URL frontier. As noted earlier, a mechanism is required in the continuous crawling case for interleaving the downloading of new and old URLs. Mercator uses a randomized priority-based scheme for this purpose. A standard configuration for continuous crawling typically uses a frontier implementation that attaches priorities to URLs based on their download history, and whose dequeue method is biased towards higher priority URLs. Both the degree of bias and the algorithm for computing URL priorities are pluggable components. In one of our configurations, the priority of documents that do not change from one download to the next decreases over time, thereby causing them to be downloaded less frequently than documents that change often.

In addition to the numerous worker threads that download and process documents, every Mercator crawl also has a single background thread that performs a variety of tasks. The background thread wakes up periodically (by default, every 10 seconds), logs summary statistics about the crawl's progress, checks if the crawl should be terminated (either because the frontier is empty or because a user-specified time limit has been exceeded), and checks to see if it is time to checkpoint the crawl's state to stable storage.

Checkpointing is an important part of any long-running process such as a web crawl. By *checkpointing* we mean writing a representation of the crawler's state to stable storage that, in the event of a failure, is sufficient to allow the crawler to recover its state by reading the checkpoint and to resume crawling from the exact state it was in at the time of the checkpoint. By this definition, in the event of a failure, any work performed after the most recent checkpoint is lost, but none of the work up to the most recent checkpoint. In Mercator, the frequency with which the background thread performs a checkpoint is user-configurable; we typically checkpoint anywhere from 1 to 4 times per day.

The description so far assumed the case in which all Mercator threads are run in a single process. However, Mercator can be configured as a multi-process distributed system. In this configuration, one process is designated the *queen*, and the others are *drones*.[1] Both the queen and the

---

[1] This terminology was inspired by the common practice of referring to web crawlers as spiders. In fact, our internal name for the distributed version of Mercator is Atrax, after *atrax robustus*, also known as the Sydney Funnel Web Spider, one of the few spider species that lives in colonies.

drones run worker threads, but only the queen runs a background thread responsible for logging statistics, terminating the crawl, and initiating checkpoints.

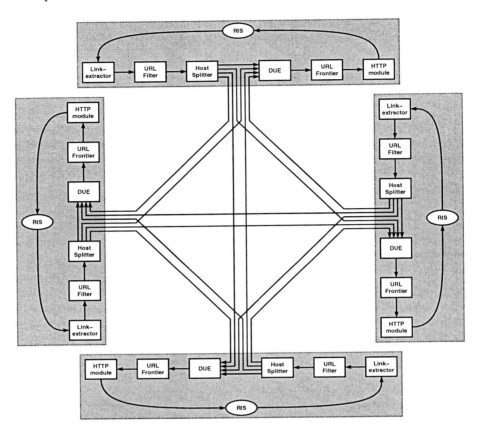

Figure 2.2. A four-node distributed crawling hive

In its distributed configuration, the space of host names is partitioned among the queen and drone processes. Each process is responsible only for the subset of host names assigned to it. Hence, the central data structures of each crawling process — the URL frontier, the URL set maintained by the DUE, the DNS cache, etc. — contain data only for its hosts. Put differently, the state of a Mercator crawl is fully partitioned across the queen and drone processes; there is no replication of data.

In a distributed crawl, when a Link Extractor extracts a URL from a downloaded page, that URL is passed through the URL Filter, into a *host splitter* component. This component checks if the URL's host name is assigned to this process or not. Those that are assigned to this process are passed on to the DUE; the others are routed to the appropriate peer process, where it is then passed to that process's DUE component. Since about 80% of links are relative, the vast majority of discovered URLs remain local to the crawling process that discovered them. Moreover, Mercator buffers the outbound URLs so that they may be transmitted in batches for efficiency. Figure 2.2 illustrates this design.

The above description omits several important implementation details. Designing data structures that can efficiently handle hundreds of millions of entries poses many engineering challenges. Central to these concerns are the need to balance memory use and performance. The following subsections provide additional details about the URL frontier, the DUE, and the DNS resolver components. A more detailed description of the architecture and implementation is available elsewhere (Heydon and Najork 1999, Najork and Heydon 2001).

3.1. A Polite and Prioritizing URL Frontier

Every web crawler must keep track of the URLs to be downloaded. We call the data structure for storing these URLs the *URL frontier*. Despite the name, Mercator's URL frontier actually stores objects that encapsulate both a URL and the download history of the corresponding document.

Abstractly speaking, a frontier is a URL repository that provides two major methods to its clients: one for adding a URL to the repository, and one for obtaining a URL from it. Note that the clients control in what order URLs are added, while the frontier controls in what order they are handed back out. In other words, the URL frontier controls the crawler's download schedule.

Like so many other parts of Mercator, the URL frontier is a pluggable component. We have implemented about half a dozen versions of this component. The main difference between the different versions lies in their scheduling policies. The policies differ both in complexity and in the degree of "politeness" (i.e., rate-limiting) they provide to the crawled web servers.

Most crawlers work by performing a breadth-first traversal of the web, starting from the pages in the seed set. Such traversals are easily implemented by using a first-in/first-out (FIFO) queue. However, the prevalence of relative URLs on web pages causes a high degree of host locality within the FIFO queue; that is, the queue contains runs of URLs with the same host name. If all of the crawler's threads dequeue URLs from a single FIFO queue, many of them will issue HTTP requests to the same web server simultaneously, thereby overloading it. Such behavior is considered socially unacceptable (in fact, it has the potential to crash some web servers).

Such overloads can be avoided by limiting the number of outstanding HTTP requests to any given web server. One way to achieve this is by ensuring that at any given time, only one thread is allowed to contact a particular web server. We call this the *weak politeness guarantee*.

We implemented a frontier that met the weak politeness guarantee and used it to perform several crawls, each of which fetched tens of millions of documents. During each crawl, we received a handful of complaints from various web server administrators. It became clear that our weak politeness guarantee was still considered too rude by some. The problem is that the weak politeness guarantee does not prevent a stream of requests from being issued to the same host without any pauses between them.

Figure 2.3 shows our most sophisticated frontier implementation. In addition to providing a stronger politeness guarantee that rate-limits the stream of HTTP requests issued to any given host, it also distributes the work among the crawling threads as evenly as possible (subject to the politeness requirement), and it provides a priority-based scheme for scheduling URL downloads. The frontier consists of a front-end (the top part of the figure) that is responsible for prioritizing URLs, and a back-end (the bottom part of the figure) that is responsible for ensuring strong politeness.

When a URL u is added to the frontier, a pluggable *prioritizer* component computes a priority value p between 1 and k based on the URL and its download history (e.g. whether the document has changed since the last download), and inserts u into front-end FIFO queue p.

The back-end maintains n FIFO queues, each of which is guaranteed to be non-empty and to contain URLs of only a single host, and a table T that maintains a map from hosts to back-end queues. Moreover, it maintains a heap data structure that contains a handle to each FIFO queue, and that is indexed by a timestamp indicating when the web server corresponding to the queue may be contacted again. Obtaining a URL from the frontier involves the following steps: First, the calling thread removes the root item from the heap (blocking, if necessary, until its timestamp is in the past). It then returns the head URL u from the corresponding back-end queue q. The calling thread will subsequently download the corresponding document.

Once the download has completed, u is removed from q. If q becomes empty, the calling thread refills q from the front end. This is done by choosing a front-end queue at random with a bias towards "high-priority" queues, and removing a URL u' from it. If one of the other

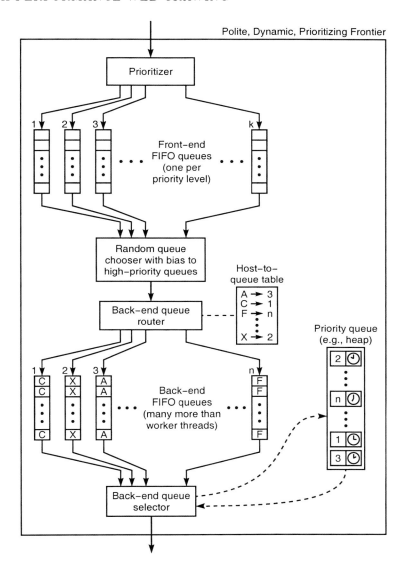

Figure 2.3. Our best URL frontier implementation

back-end queues contains URLs with the same host component as u', u' is inserted into that queue and process of refilling q goes on. Otherwise, u' is added to q, and T is updated accordingly. Also, the calling thread computes the time at which u's host may be contacted again, and reinserts the handle to q with that timestamp back into the heap.

Note that the number n of back-end queues and the degree of rate-limiting go hand in hand: The larger the degree of rate-limiting, the more back-end queues are required to keep all the crawling threads busy. In our production crawls, we typically use 3 times as many back-end queues as crawling threads, and we wait 10 times as long as it took us to download a URL from a host before contacting that host again. These values are sufficient to keep all threads busy and to keep the rate of complaints to a bare minimum.

In a crawl of the entire web, the URL frontier soon outgrows the available memory of even the largest machines. It is therefore necessary to store most of the frontier's URLs on disk. In Mercator, each of the FIFO queues stores the bulk of its URLs on disk, and buffers only a fixed number of the URLs at its head and tail in memory.

3.2. Efficient Duplicate URL Eliminators

In the course of extracting links, any web crawler will encounter multiple links to the same document. To avoid downloading and processing a document multiple times, a duplicate URL eliminator (DUE) guards the URL frontier. Extracted links are *submitted* to the DUE, which passes new ones to the frontier while ignoring those that have been submitted to it before.

One of our implementations of the DUE maintains an in-memory hash table of all URLs that have been encountered before. To save space, the table stores 8-byte checksums of the URLs rather than the URLs themselves. We compute the checksums using Rabin's fingerprinting algorithm (Broder 1993, Rabin 1981), which has good spectral properties and gives exponentially small probabilistic bounds on the likelihood of a collision. The high-order 3 bytes of the checksum are used to index into the hash table spine. Since all checksums in the same overflow bucket would have the same high-order 3 bytes, we actually store only the 5 low-order bytes. Taking pointer and counter overhead into account, storing the checksums of 1 billion URLs in such a hash table requires slightly over 5 GB.

This implementation is very efficient, but it requires a substantial hardware investment; moreover, the memory requirements are proportional to the size of the crawl. A disk-based approach avoids these problems, but is difficult to implement efficiently. Our first disk-based implementation essentially stored the URL fingerprint hash table on disk. By caching popular fingerprints in memory, only one in every six DUE submissions required a disk access. However, due to the spectral properties of the fingerprinting function, there is very little locality in the stream of fingerprints that miss on the in-memory cache, so virtually every disk access required a disk seek. On state-of-the-art disks, the average seek requires about 8 ms, which would enable us to perform 125 seeks or 750 DUE submissions per second. Since the average web page contains about 10 links, this would limit the crawling rate to 75 downloads per second. Initially, this bottleneck is masked by the operating system's file buffer cache, but once the disk-based hash table grows larger than the file buffer cache, the seek operations become the performance bottleneck.

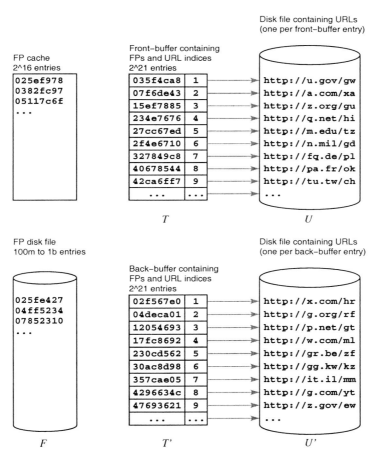

Figure 2.4. Our most efficient disk-based DUE implementation

Both designs described so far add never-before-seen URLs to the frontier *immediately*. By buffering URLs, we can amortize the access to the disk, thereby increasing the throughput of the DUE. Figure 2.4 shows the main data structures of our most efficient disk-based DUE implementation. This implementation's largest data structure is a file F of sorted URL fingerprints.

When a URL u is submitted to the DUE, its fingerprint fp is computed. Next, fp is checked against a cache of popular URLs and an in-memory hash table T. If fp is contained in either, no further action is required. Otherwise, u is appended to a URL disk file U, and a mapping from fp to u's ordinal in U is added to T.

Once T's size exceeds a predefined limit, the thread that "broke the camel's back" atomically copies T's content to a table T' consisting of fingerprint/ordinal pairs, empties T, and renames U to U'. Once this very short atomic operation completes, other crawling threads are free to submit URLs to the DUE, while the back-breaking thread adds the new content of T' and U' into F and the frontier, respectively. It first sorts T' by fingerprint value, and then performs a linear merge of T' and F, *marking* every row in T' whose fingerprint was added to F. Next, it sorts T' by ordinal value. Finally, it scans both T' and U' sequentially, and adds all URLs in U' that are marked in T' to the frontier.

In real crawls, we found that this DUE implementation performs at least twice as well as the one that is seek-limited; however, its throughput still deteriorates over time, because the time needed to merge T' into F eventually exceeds the time required to fill T.

3.3. The Trouble with DNS

Hosts on the Internet are identified by Internet Protocol (IP) addresses, which are 32-bit numbers. IP addresses are not mnemonic. This problem is avoided by the use of symbolic host names, such as *cnn.com*, which identify one or more IP addresses. Any program that contacts sites on the internet whose identities are provided in the form of symbolic host names must resolve those names into IP addresses. This process is known as *host name resolution*, and it is supported by the *domain name service* (DNS). DNS is a globally distributed service in which name servers refer requests to more authoritative name servers until an answer is found. Therefore, a single DNS request may take seconds or even tens of seconds to complete, since it may require many round-trips across the globe.

DNS name resolution is a well-documented bottleneck of most web crawlers. We tried to alleviate this bottleneck by caching DNS results, but that was only partially effective. After some probing, we discovered that the Java interface to DNS lookups is synchronized. Further investigation revealed that the DNS interface on most flavors of Unix (i.e., the `gethostbyname` function provided as part of the Berkeley Internet Name

Domain (BIND) distribution (BIND 2001)) is also synchronized. This meant that only one DNS request per address space on an uncached name could be outstanding at once. The cache miss rate is high enough that this limitation causes a severe bottleneck.

To work around these problems, we made DNS resolution one of Mercator's pluggable components. We implemented a multi-threaded DNS resolver component that does not use the resolver provided by the host operating system, but rather directly forwards DNS requests to a local name server, which does the actual work of contacting the authoritative server for each query. Because multiple requests can be made in parallel, our resolver can resolve host names much more rapidly than either the Java or Unix resolvers.

This change led to a significant crawling speedup. Before making the change, performing DNS lookups accounted for 70% of each thread's elapsed time. Using our custom resolver reduced that elapsed time to 14%. (Note that the actual number of CPU cycles spent on DNS resolution is extremely low. Most of the elapsed time is spent waiting for remote DNS servers.) Moreover, because our resolver can perform resolutions in parallel, DNS is no longer a bottleneck; if it were, we would simply increase the number of worker threads.

4. Experiences from a Large Crawl

This section describes the results of our crawling experiments. Our crawling cluster consists of four Compaq DS20E AlphaServers, each one equipped with 4 GB of main memory, 650 GB of disk, and a 100 Mbit/sec Ethernet card. The cluster is located close to the Internet backbone. Our ISP rate-limits our bandwidth to 160 Mbits/sec.

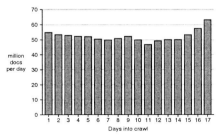
(a) Documents downloaded per day

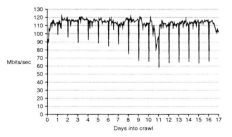
(b) Megabits downloaded per second

Figure 2.5. Mercator's performance over a 17-day crawl

In December 2000, we performed a crawl that processed 891 million URLs over the course of 17 days.[2] Figure 2.5a shows the number of URLs processed per day of the crawl; Figure 2.5b shows the bandwidth consumption over the life of the crawl. The periodic downspikes are caused by the crawler checkpointing its state once a day. The crawl was network-limited over its entire life; CPU load was below 50%, and disk activity was low as well.

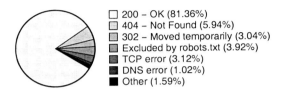

Figure 2.6. Outcome of download attempts

As any web user knows, not all download attempts are successful. During our crawl, we collected statistics about the outcome of each download attempt. Figure 2.6 shows the outcome percentages. Of the 891 million processed URLs, 35 million were excluded from download by robots.txt files, and 9 million referred to a nonexistent web server; in other words, the crawler performed 847 million HTTP requests. 725 million of these requests returned an HTTP status code of 200 (i.e., were successful), 94 million returned an HTTP status code other than 200, and 28 million encountered a TCP failure.

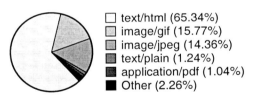

Figure 2.7. Distribution of content types

There are many different types of content on the internet, such as HTML pages, GIF and JPEG images, MP3 audio files, and PDF documents. The MIME (Multipurpose Internet Mail Extensions) standard defines a naming scheme for these content types (Freed and Borenstein 1996). We have collected statistics about the distribution of content

---

[2] As a point of comparison, the current Google index contains about 700 million fully-indexed pages (the index size claimed on the Google home page – 1.35 billion — includes URLs that have been discovered, but not yet downloaded).

HIGH-PERFORMANCE WEB CRAWING

types of the successfully downloaded documents. Overall, our crawl discovered 3,173 different content types (many of which are misspellings of common content types). Figure 2.7 shows the percentages of the the most common types. HTML pages (of type *text/html*) account for nearly two-thirds of all documents; images (in both GIF and JPEG formats) account for another 30%; all other content types combined account for less than 5%.

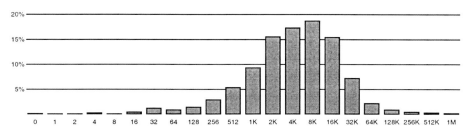

Figure 2.8.   Distribution of document sizes

Figure 2.8 is a histogram showing the document size distribution. In this figure, the documents are distributed over 22 bins labeled with exponentially increasing document sizes; a document of size n is placed in the rightmost bin with a label not greater than n. Of the 725 million documents that were successfully downloaded, 67% were between 2K and 32K bytes in size, corresponding to the four tallest bars in the figure.

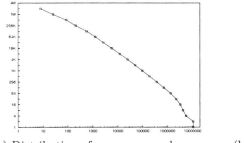 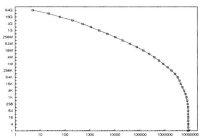

(a) Distribution of pages over web servers   (b) Distribution of bytes over web servers

Figure 2.9.   Document and web server size distributions

Figure 2.9 shows the distribution of content across web servers. Figure 2.9a measures the content using a granularity of whole pages, while Figure 2.9b measures content in bytes. Both figures are plotted on a log-log scale, and in both, a point (x, y) indicates that x web servers had at least y pages/bytes. The near-linear shape of the plot in Figure 2.9a indicates that the distribution of pages over web servers is Zipfian.

Finally, Figure 2.10 shows the distributions of web servers and web pages across top-level domains. About half of the servers and pages fall into the .com domain. For the most part, the numbers of hosts and pages in a top-level domain are well-correlated. However, there are some interesting wrinkles. For example, the .edu domain contains only about 1.53% of the hosts, but 5.56% of the total pages. In other words, the average university web server contains almost four times as many pages as the average server on the web at large.

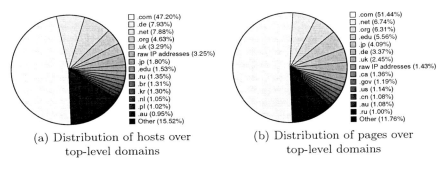

(a) Distribution of hosts over top-level domains

(b) Distribution of pages over top-level domains

Figure 2.10. Distribution of hosts and pages over top-level domains

5. Conclusion

High-performance web crawlers are an important component of many web services. Building a high-performance crawler is a non-trivial endeavor: the data manipulated by the crawler is too big to fit entirely in memory, so there are performance issues related to how to balance the use of disk and memory. This chapter has enumerated the main components required in any crawler, and it has discussed design alternatives for some of those components. In particular, the chapter described Mercator, an extensible, distributed, high-performance crawler written entirely in Java.

Mercator's design features a crawler core for handling the main crawling tasks, and extensibility through a component-based architecture that allows users to supply new modules at run-time for performing customized crawling tasks. These extensibility features have been quite successful. We were able to adapt Mercator to a variety of crawling tasks, and the new code was typically quite small (tens to hundreds of lines). Moreover, the flexibility afforded by the component model encouraged us to experiment with different implementations of the same functional components, and thus enabled us to discover new and efficient data structures. In our experience, these innovations produced larger performance gains than low-level tuning of our user-space code (Heydon and Najork 2000).

Mercator's scalability design has also worked well. It is easy to configure the crawler for varying memory footprints. For example, we have run it on machines with memory sizes ranging from 128 MB to 2 GB. The ability to configure Mercator for a wide variety of hardware platforms makes it possible to select the most cost-effective platform for any given crawling task.

Although our use of Java as an implementation language was met with considerable scepticism when we began the project, we have not regretted the choice. Java's combination of features — including threads, garbage collection, objects, and exceptions — made our implementation easier and more elegant. Moreover, on I/O-intensive applications, Java has little negative impact on performance. Profiling Mercator running on Compaq AlphaServers reveals that over 60% of the cycles are spent in the kernel and in C libraries; less than 40% are spent executing (JIT-compiled) Java bytecode. In fact, Mercator is faster than any other web crawler for which performance numbers have been published.

Mercator has proven to be extremely popular. It has been incorporated into AltaVista's Search Engine 3 product, and it is being used as the web crawler for AltaVista's American and European search sites. Our colleague Raymie Stata has performed Mercator crawls that collected over 12 terabytes of web content, which he has contributed to the Internet Archive's web page collection. Raymie also performed a continuous crawl on behalf of the Library of Congress to monitor coverage of the 2000 U.S. Presidential election. That crawl took daily snapshots of about 200 election-related web sites during the five months preceding the inauguration. Finally, Mercator has been an enabler for other web research within Compaq. For example, we have configured Mercator to perform random walks instead of breadth-first search crawls, using the crawl traces to estimate the quality and sizes of major search engine indices (Henzinger et al. 1999; 2000).

Bibliography

BIND. Berkeley internet name domain, 2001.

S. Brin and L. Page. The anatomy of a large-scale hypertextual web search engine. In *Proceedings of the Seventh International World Wide Web Conference*, pages 107–117, April 1998.

A. Broder. Some applications of rabin's fingerprinting method. In R. Capocelli, A. De Santis, and U. Vaccaro, editors, *Sequences II: Methods in Communications, Security, and Computer Science*, pages 143–152. Springer-Verlag, 1993.

M. Burner. Crawling towards eternity: Building an archive of the world wide web. *Web Techniques Magazine*, 2(5), May 1997.

J. Cho, H. G. Molina, and L. Page. Efficient crawling through url ordering. In *Proceedings of the Seventh International World Wide Web Conference*, pages 161–172, April 1998.

J. Edwards, K. McCurley, and J. Tomlin. An adaptive model for optimizing performance of an incremental web crawler. In *Proceedings of the Tenth International World Wide Web Conference*, pages 106–113, May 2001.

D. Eichmann. The rbse spider – balancing effective search against web load. In *Proceedings of the First International World Wide Web Conference*, pages 113–120, 1994.

N. Freed and N. Borenstein. Multipurpose internet mail extensions part two: Media types, November 1996.

M. Gray. Internet growth and statistics: Credits and background, 2001.

M. Henzinger, A. Heydon, M. Mitzenmacher, and M. A. Najork. Measuring index quality using random walks on the web. In *Proceedings of the Eighth International World Wide Web Conference*, pages 213–225, May 1999.

M. Henzinger, A. Heydon, M. Mitzenmacher, and M. A. Najork. On near-uniform url sampling. In *Proceedings of the Ninth International World Wide Web Conference*, pages 295–308, May 2000.

A. Heydon and M. Najork. Mercator: A scalable, extensible web crawler. *World Wide Web*, 2:219–229, December 1999.

A. Heydon and M. Najork. Performance limitations of the java core libraries. *Concurrency: Practice and Experience*, 12:363–373, May 2000.

J. Hirai, S. Raghavan, H. G. Molina, and A. Paepcke. Webbase: A repository of web pages. In *Proceedings of the Ninth International World Wide Web Conference*, pages 277–293, May 2000.

The internet archive, 2001.

M. Koster. A method for web robots control, December 1996.

M. Koster. The web robots pages, 2001.

O. A. McBryan. Genvl and wwww: Tools for taming the web. In *Proceedings of the First International World Wide Web Conference*, pages 79–90, 1994.

M. Najork and A. Heydon. On high-performance web crawling. Technical report, Compaq Systems Research Center, 2001. SRC Research Report, forthcoming.

B. Pinkerton. Finding what people want: Experiences with the webcrawler. In *Proceedings of the Second International World Wide Web Conference*, 1994.

M. O. Rabin. Fingerprinting by random polynomials. Technical report, Center for Research in Computing Technology, Harvard University, 1981. Report TR-15-81.

Chapter 3

INTERNET GROWTH: IS THERE A "MOORE'S LAW" FOR DATA TRAFFIC?

K. G. Coffman
AT&T Labs - Research, Florham Park, NJ 07932, USA
kgc@research.att.com

A. M. Odlyzko
AT&T Labs - Research, Florham Park, NJ 07932, USA
amo@research.att.com

Abstract   Internet traffic is approximately doubling each year. This growth rate applies not only to the entire Internet, but to a large range of individual institutions. For a few places we have records going back several years that exhibit this regular rate of growth. Even when there are no obvious bottlenecks, traffic tends not to grow much faster. This reflects complicated interactions of technology, economics, and sociology, similar to, but more delicate than those that have produced "Moore's Law" in semiconductors.

A doubling of traffic each year represents extremely fast growth, much faster than the increases in other communication services. If it continues, data traffic will surpass voice traffic around the year 2002. However, this rate of growth is slower than the frequently heard claims of a doubling of traffic every three or four months. Such spectacular growth rates apparently did prevail over a two-year period 1995-6. Ever since, though, growth appears to have reverted to the Internet's historical pattern of a single doubling each year.

Progress in transmission technology appears sufficient to double network capacity each year for about the next decade. However, traffic growth faster than a tripling each year could probably not be sustained for more than a few years. Since computing and storage capacities will also be growing, as predicted by the versions of "Moore's Law" appropriate for those technologies, we can expect demand for data transmission to continue increasing. A doubling in Internet traffic each year appears a likely outcome.

If Internet traffic continues to double each year, we will have yet another form of "Moore's Law." Such a growth rate would have several important implications. In the intermediate run, there would be neither a clear "bandwidth glut" nor a "bandwidth scarcity," but a more balanced situation, with supply and demand growing at comparable rates. Also, computer and network architectures would be strongly affected, since most data would stay local. Programs such as Napster would play an increasingly important role. Transmission would likely continue to be dominated by file transfers, not by real time streaming media.

Keywords: Internet traffic, Moore's law, Bandwidth.

1. Introduction

An earlier paper (Coffman and Odlyzko 1998) estimated the sizes of telecommunications networks in the U.S. and the traffic they carried. (We concentrated on the U.S. because of lack of data about other countries, and because the disparate development stages of their communications infrastructures make cross-country comparisons difficult.) Our conclusion was that by year-end 1997, the voice network was still the largest in terms of bandwidth, although data networks were almost as large. On the other hand, the traffic carried by the voice network was far larger than that of the data networks. These estimates are summarized in Table 3.1, with the units of measurement being terabytes per month. (A voice call for the purposes of this measurement was counted as two 64 Kb/s streams of data, and fax and modem calls carried on the switched long distance network were counted as voice. For details, see Coffman and Odlyzko (1998).)

Table 3.1. Traffic on U.S. long distance networks, year-end 1997

network	traffic (TB/month)
US voice	40,000
Internet	2,500 - 4,000
other public data networks	500
private lines	3,000 - 5,000

Table 3.2 is an update of Table 3.1, showing our estimates for traffic on U.S. long distance networks around December 2000. As before, the estimates for the voice, private line, and other public data networks (primarily ATM and Frame Relay) are uncontroversial, consistent with other publicly available sources. On the other hand, the estimates for

the public Internet are much less certain, less certain than those we made for year-end 1997. We explain this in greater detail below.

Table 3.2. Traffic on U.S. long distance networks, year-end 2000.

network	traffic (TB/month)
US voice	53,000
Internet	20,000 - 35,000
other public data networks	3,000
private line	6,000 - 11,000

In the earlier paper (Coffman and Odlyzko 1998) we also considered the growth rates of various networks. While reliable information had long been available for most networks (with increases on the order of 10% a year for the voice network), very little was known about the Internet. There were many claims of huge growth rates, usually of a doubling of traffic every three or four months, which corresponds to annual growth of around 1,000%. We confirmed that such growth rates did hold during 1995 and 1996. (More precisely, the traffic carried by the Internet backbones grew by a cumulative factor of 100 between year-end 1994 and year-end 1996. We did not obtain precise estimates for traffic during those two years, so do not know how growth was distributed within those two years.) However, we also showed that growth had slowed down to about 100% during 1997. Remarkably enough, that was almost exactly the growth rate registered by the Internet backbone in the early 1990s (see Coffman and Odlyzko (1998) for data), and is close to other measures of growth rates during earlier periods (cf. Coffman and Odlyzko (1998), Paxson (1994)). In fact, if we assume that the traffic on the initial ARPANET links that were activated in the summer of 1969 amounted to a few thousand bytes per month, we find that the average growth rate in traffic has been very close to 100% a year for the entire 30-year history of the Internet and its predecessors. However, growth rates varied substantially over that period as the initially small research networks evolved. The steady annual doubling of traffic only appeared in the early 1990s, when the NSFNet backbone grew to a significant size in both traffic and number of participants.

Table 3.3 shows our estimates for traffic on U.S. Internet backbones. The data through the end of 1994 are based on careful measurements of the NSFNet backbone (and neglect what were thought to be much smaller private backbones and research networks). For the late 1990s, the figures are less certain. As we explain in Section 3, there is less information today than was available a couple of years ago about different carriers' IP networks, either about their sizes or the traffic they carry. Therefore our current estimates are based less on solid data than those of Coffman and Odlyzko (1998), and more on extrapolating from the circumstantial evidence we have accumulated.

Table 3.3. Traffic on Internet backbones in U.S. For each year, shows estimated traffic in terabytes during December of that year.

year	TB/month
1990	1.0
1991	2.0
1992	4.4
1993	8.3
1994	16.3
1995	?
1996	1,500
1997	2,500 - 4,000
1998	5,000 - 8,000
1999	10,000 - 16,000
2000	20,000 - 35,000

Much of this chapter is devoted to a study of historical growth rates in Internet traffic. We find that for a variety of institutions, some with abundant bandwidth, others with congested links, traffic tends to about double each year. Even when the bandwidth of congested links is increased, in most cases traffic does not explode in the much-feared "tragedy of the commons" phenomenon (Gupta et al. 1995). Instead, it continues growing at about 100% per year. Malicious behavior is a smaller problem than is often feared. What dominates is the time it takes for new applications and processes to be widely adopted. That, and technological progress, combine to produce the "Moore's Law" of data traffic, in a process similar to that operating in other areas.

In considering prospects for future growth, one approach is to simply look at the historical record. Since the Internet has been growing at about 100% a year for its entire history, one can then extrapolate this growth rate into the future, and predict that traffic will continue to double each year. We prefer to do more than that, and consider how fast supply and demand are likely to grow. Those considerations lead us to the same conclusion as the simple extrapolation, namely that traffic is likely to about double each year. (By an approximate doubling we mean growth rates of between 70% and 150% per year.)

Much is written about the near infinite capacity of fiber, and of the upcoming bandwidth glut (or at least the overabundance of available fiber). For example, the introduction to Gilder (1997) talks of "the coming world of cheap, unlimited bandwidth." In the US alone there are now over half a dozen long haul carriers that either have or will have very substantial national fiber networks. The conventional wisdom is that the exploding increase in Internet traffic is the main driver for the expansion of these networks. It also seems to be implied that the "ever increasing" capacities of WDM (wavelength division multiplexing) systems (both in terms of the number of channels and the individual channel rates) coupled with this forecasted "fiber glut" will result in the national networks being easily able to accommodate whatever growth rate the Internet throws at it. We do not think the carrying capacity of the network, at least the long haul national backbone network(s), can or will grow to accommodate arbitrary traffic growth rates. In fact we believe that if traffic grows by factors of more than two or three a year for any sustained period, the transport backbones are likely to become a very serious bottleneck. However, 100% annual growth rates appear to be realizable without unusually large new investments for the foreseeable future (which means at least five years). We explore this issue in detail in Section 5.

The demand for data transmission is potentially insatiable. As we show in Section 6, there is already so much data stored on hard disks (and much more on magnetic tapes and optical drives) that only a small fraction of it can traverse the long distance networks without saturating them. Furthermore, data storage capacities are about doubling each year. Hence the evolution of computer and network architectures will come from the subtle interaction of supply and demand, which will be mediated by economics and sociology. It appears that an approximate doubling of Internet traffic each year is a reasonable course of evolution for the Internet, even though there may be brief periods of higher or lower growth rates.

A doubling of Internet traffic each year is not the doubling every 100 days that is often claimed. However, it is extremely rapid by comparison with older communication technologies. Still, since other technologies (storage and processing power) will also be growing, many expectations for radical changes will not be realized. For example, there have been predictions that the growing bandwidth of long distance networks would lead to dramatic changes (Gilder 1997).

> [The] change in the relationship between the bandwidth of networks and the bandwidth of computers will transform the architecture of information technology. As Robert Lucky of Bellcore [recently renamed

Telcordia] puts it, "Perhaps we should transmit signals thousands of miles to avoid even the simplest processing functions."

The projections of this chapter show that this is very unlikely. Bandwidth will continue to be in short supply compared to storage. Further, the delays forced by speed of light limitations as well as communications overheads mean that most data will be processed locally. We will have to take processing to the data, not vice versa.

For decades, the most frequent predictions for data networks were that eventually they would be dominated by traffic such as voice and video. That has not happened yet. If the doubling of traffic each year continues, it will not be true in the long run either, simply since there will be far more data traffic than needed to handle real time streaming transmissions.

Section 2 outlines the wide range of commonly expressed opinions about the growth rate of the Internet. Section 3, the bulk of this chapter, is then devoted to an examination of the historical data we have been able to find. Section 4 discusses the disruptive innovations of the past, such as browsers, and of the present (primarily Napster and related programs), and the effects they have on traffic statistics. Section 5 outlines the history and projections for advances in photonic transmission. Section 6 discusses the potential demand for data bandwidth. Section 7 concludes that growth in data traffic is determined by factors similar to those that produce the standard "Moore's laws" in other areas. Section 8 has some final concluding remarks.

2. Skeptics and Cheerleaders

There continue to be voices skeptical of the growth prospects of the Internet. In particular, A. Michael Noll has been a persistent naysayer. Back in 1991, he estimated what the maximal feasible volume of data transfers could be (pp. 171–175 of Noll (1991)). Since that time, the volume of data traffic has surged far beyond his original prediction. The main reason for this development is that he did not foresee the arrival of graphics-rich content, such as Web pages. His estimates assumed only text would be transmitted, and all of it would be processed by people. While that prediction has turned out to be wrong, he continues predicting that data traffic will not exceed voice traffic unless multimedia begins dominating the Internet (Noll 1999). While our estimates do confirm that there is still more voice traffic than data traffic, Noll's estimates for data traffic are far too conservative. As an example, we examined the publicly available statistics for data traffic at the University of Southern California, where Noll is a professor. These statistics were available at

⟨http://foo.usc.edu/netstats⟩, and in early 1999 showed a considerably higher volume of data flow than Noll estimated in Noll (1999). (By early 2000, the volume of data flow at USC had grown by about 70%, from an average of about 20 Mb/s to the campus and 10 Mb/s out from the campus to about 30 Mb/s in and 20 Mb/s out. Unfortunately, between late 2000 and early 2001, those traffic statistics stopped being updated, so we do not have data about more recent growth patterns.)

While there are some skeptics about the prospects for the Internet, there are vastly larger ranks of people who claim astronomical growth rates. However, they invariably talk only of rates of increase, and never cite precise verifiable figures. The most common claim one hears is that "Internet traffic is doubling every three or four months." As was pointed out in Coffman and Odlyzko (1998), many of these claims appear to trace back to statements of John Sidgmore of MCI WorldCom's UUNet or his colleagues. A March 2000 news report (Howe 2000), for example, cites MCI WorldCom president Bernard J. Ebbers as saying that his company "has recently had to add capacity to its global network at a rate of 800 percent annually to keep up with soaring demand for Net traffic." Yet the February 10, 2000 press release by MCI WorldCom that accompanied the earnings report for the fourth quarter of 1999 refers to "[g]ains in data services ... measured by an 87 percent increase in Voice Grade Equivalents (VGEs), which capture the volume of local data circuits." The two statements may refer to different parts of the MCI WorldCom data network. However, eventually capacities of long distance links are unlikely to grow much faster than those of local ones. Hence we are inclined to believe that the "87 percent increase" of the official press release describes overall growth more accurately. Revenue increases from data services for MCI WorldCom (reported in their audited financial statements) are also far more consistent with annual growth rates of 100 percent than 800 percent.

It could be that the phrase "doubling of Internet traffic every three [or four] months" has lost its literal meaning. Perhaps it is being used as a figure of speech, in the way that many people use "exponential growth" (which in mathematics has a precise meaning) to describe any fast growth. There are just too many examples where such statements are either implausible or even demonstrably incorrect. For example, Keith Mitchell, executive chairman of LINX, the London Internet Exchange, Ltd., is quoted in Jander (2000) as saying in March, 2000, that "[LINX] traffic doubles every hundred days or so." This rate of growth would increase traffic over a year by a factor of 12. Yet an examination of the publicly available statistics for LINX as well as discussions with technical staff at LINX showed that traffic had grown by a factor of about 4 between March 1999 and March 2000. That is certainly fast, but corresponds to a doubling of traffic every 180 days, not every 100 days. (Current traffic statistics for LINX are available at ⟨http://ochre.linx.net/⟩. There is further discussion of LINX in the next section.)

Whether Internet traffic doubles every three months or just once a year has huge consequences for network design as well as the telecommunications industry. Much of the excitement about and funding for novel technologies appear to be based on expectations of unrealistically high growth rates (cf. Bruno (2000)). Yet it should have been obvious that such growth rates cannot be sustained for long, and in particular could not have been going on for long. A doubling of Internet traffic every three months would produce an increase by a factor of 16 in one year. Hence, from the end of 1994 to the end of 2000, it would have grown by a factor of almost 17 million. Until the end of 1994, the Internet backbone was funded by the National Science Foundation, and was well instrumented. Hence we know that it carried about 15 TB (terabytes) of traffic each month. Had that traffic grown by a factor of 16 million in the intervening 6 years, we would now have about 240 exabytes (exabyte is 10^{18} bytes) of traffic on Internet backbones each month. If we generously assume that there are 500 million Internet users in the world today, that volume of traffic would translate into about 1.5 Mb/s of U.S. backbone traffic for each user around the clock! This is enough for reasonably high quality video (if one uses appropriate compression). Yet most Internet users have access only to modems that transmit at best at 28 Kb/s. Moreover, those modems are in use typically for less than an hour per day, and on average transmit about 5 Kb/s while they are connected to the Internet. Even the vast majority of enterprises as well as some universities have links to the Internet that run no faster than T1 speeds (i.e., maximal rates of 1.5 Mb/s). The bottom line is that current user behavior falls well short of the usage levels one would expect had Internet traffic been doubling every three months since the end of 1994.

Assuming a doubling of Internet traffic every four months produces traffic estimates that are only slightly less absurd. Actual traffic at well-wired institutions in the U.S. (primarily corporations and some universities) tends to average out to something between one and three thousand bits per second per person (averaged over a complete week).

3. Historical Record

The online data for LINX in April 2001 showed growth of about 300% from early 2000 to early 2001. Earlier versions of those same online statistics as well as conversations with LINX technical personnel show that LINX has been experiencing those growth rates for several years. One can find other examples of such high growth rates, and sometimes even higher ones. However, there are also numerous examples of much more slowly growing links. In this section we present growth rates from a variety of sources, and attempt to put them into context. Our general conclusion is that Internet traffic appears to be growing at about 100% a year. By this we mean that the growth rate appears to be between 70% and 150% per year, as we cannot be more precise given the limitations of the data.

Although our paper (Coffman and Odlyzko 1998) appears to have been the first one to point out the slowing down of Internet traffic growth, others observed the same phenomenon soon afterwards. For example, there is an article by Peter Sevcik (Sevcik 1999) from early 1999, as well as reports from market researcher firms such as Probe that also pointed out that the claims of a doubling every three or four months were not correct.

An important example that supports our thesis of Internet traffic doubling about once each year is that of Telstra, the dominant Australian telecommunications carrier. A January 15, 2001 news story (Cochrane 2001) cited official Telstra figures as follows:

> Australia's biggest ISP, Telstra Big Pond, says total daily traffic grew from about 4 TB a day in November 1999 to more than 9 TB a day at the same time last year. It continued the 1998-99 growth rate of 225 per cent, when traffic demand rose from 1.8 TB to about 4 TB in November. Consumption first broke the 1 TB a day barrier in 1997.

Disregarding the obvious mistake (an increase from 4 TB to 9 TB represents growth by 125%, not 225%), we find annual growth rate for Telstra falling in the range we derive for each of several consecutive years. (The printed version of the story, but not the online one, has a detailed chart, showing daily Telstra Internet traffic from beginning of 1997 to November 2000. It shows that traffic grew at about 100% a year over that full 4-year period.) There is further discussion of Telstra later in this section.

Our presentation in this section follows the pattern of the above paragraph for Telstra, and is largely in terms of a narrative. It would be much more effective to have a table or chart combining data for a variety of institutions. Unfortunately the statistics we have collected are not sufficiently systematic to do this. They come from a variety of sources in many formats and for different periods. We concentrate on large and stable institutions for which we have more than a year's worth of data. For some links, one can obtain detailed statistics going back some years from the Web page we indicate. For most, though, the public Web page shows only the graphs produced by the MRTG (multi-router traffic grapher) tool (Oetiker and Rand 2001). (At some institutions, MRTG is beginning to be partially displaced by a more modern program, RRDtool (Oetiker 2001).) This excellent program displays the exact averages of in and out traffic over the previous day, week, month, and year, and also produces graphs with the traffic profiles for those periods. This means that by downloading one of the MRTG pages, one can estimate the average traffic a year earlier. It is in principle possible to decode the .gif files produced by MRTG to obtain more precise values, but we have not done so, and have relied on "eye-balling" the graphs to estimate traffic. As an example, consider Onvoy, the main ISP in Minnesota. Its growth trends are discussed later in this section. We have exact MRTG readings and graphs from October 1999 and June 2000. That means we have exact traffic data for those two months, and can estimate traffic back to October 1998, but no further. In a few cases we have obtained more precise statistics from network administrators.

As a brief note on conversion factors, traffic that averages 100 Mb/s is equivalent to about 30 TB/month. (It is 32.4 TB for a 30-day month, but such precision is excessive given the uncertainties in the data we have.)

When the NSF Internet backbone was phased out in early 1995, it was widely claimed that most of the Internet backbone traffic was going through the Network Access Points or NAPs, which tended to provide decent statistics on their traffic. Currently it is thought that only a small fraction of backbone traffic goes through the NAPs, while most goes through private peering connections. Furthermore, NAP statistics are either no longer available, or not as reliable. Here we just mention a few cases. The data for the Chicago NAP from the summer of 1996 through May of 1999 is available at ⟨http://nap.aads.net/~napstat/⟩. From August of 1996 to very early in 1997 (February) the traffic profile was moderately flat. However, from February 1997 until May 1999 there was a fairly consistent growth that resulted in about a 12 fold increase in traffic, to a final level of about 1.2 Gb/s. A twelve-fold increase in a little over 2 years implies an annual growth rate of around 3.5. On the other hand, a different picture emerges when we examine the statistics for MAE-East (which a few years ago was often claimed to handle a third of all Internet traffic). They are available at ⟨http://www.mae.net/east/stats.html⟩. In March of 2000, the average traffic there was about 1.5 Gb/s. That is the same traffic as this NAP handled in March 1998 (and twice what it handled in March 1997). Thus practically nothing can be concluded about current growth rates

INTERNET GROWTH: "MOORE'S LAW"

of Internet traffic by examining the statistics of the public NAPs in the U.S.

LINX, the London Internet exchange, was mentioned already in Section 2. A July 1999 LINX press release announced that it had achieved a traffic level of 1.0 Gb/s, up from 180 Mb/s a year earlier, for a growth rate of 455%. However, no definition was offered how these "traffic levels" were defined. An examination of the MRTG graphs obtained from ⟨http://www2.linx.net/info/⟩ (which are unreliable towards the end of the period they cover due to counter overflows in the routers and the more recent statistics available at ⟨http://ochre.linx.net⟩) showed that the average traffic (averaged over a whole month) went from about 200 Mb/s in September 1998 to 360 Mb/s in March 1999, to 1.1 Gb/s by the end of 1999, and 4.5 Gb/s by May 2001. Thus the growth rate has been fairly steady at about 300% per year.

AMS-IX, the Amsterdam Internet Exchange, ⟨http://www.ams-ix.net/⟩, has shown growth by a factor of 4.6 from June 1999 to May 2000, to a level of about 800 Mb/s, then a further growth by a factor of 2.5 to April 2001, to a level of 2 Gb/s. Several smaller exchanges, especially in countries that used to lag in Internet penetration, sometimes show similarly high growth rates. For example, the Slovak Internet eXchange, ⟨http://www.six.sk/mrtg/switch.six.sk.b.html⟩, increased its traffic approximately 4-fold in the year ending June 2000, to an average level of about 40 Mb/s. However, by May 2001, traffic grew only about 80% during the intervening 11 months, to a level of 73 Mb/s. HKIX, a commercial exchange created by The Chinese University of Hong Kong, ⟨http://www.hkix.net⟩ (with aggregate statistics available more directly at ⟨http://www.cuhk.edu.hk/hkix/stat/aggt/hkix-aggregate.html⟩), doubled its traffic in 1999, and then tripled it in the first six months of 2000 to an average of about 250 Mb/s. By May 2001, traffic had grown to about 600 Mb/s, for a growth rate of about 150% per year. (Total Internet bandwidth from Hong Kong to other countries tripled between September 1999 and September 2000, and then doubled in the next six months, according to statistics compiled by Hong Kong telecommunications regulators.) Even BNIX, ⟨http://www.belnet.be/bnix/⟩, located in Belgium, a country that already has extensive Internet deployment, experienced a 5-fold rise from middle of 1999 to the middle of 2000, to a level of 120 Mb/s, and a further rise to about 300 Mb/s by May 2001. INEX, an Irish exchange, ⟨http://www.inex.ie⟩, saw its traffic increase from about 3 Mb/s in June 1999 to about 5 Mb/s in June 2000, and 10 Mb/s in May 2001. SIX, the Seattle Internet Exchange, ⟨http://www.altopia.com/six⟩, saw approximately 150% growth in the 12 months to June 2000, to a level of about 50 Mb/s, and a further approximately 100% growth in the 12 months to May 2001, to a level of about 100 Mb/s. FICIX, the Finnish exchange, ⟨http://www.ficix.fi/⟩, tripled its traffic in a bit less than two years, from an average of about 70 Mb/s in September 1998 to about 210 Mb/s in June 2000, for an annual growth rate of about 70%. By May 2001, though, its traffic had grown to about 450 Mb/s, for a growth rate of slightly more than 100%. (The statistics of traffic for individual FICIX members tend to show similar growth rates to that of the exchange aggregate.) VIX, the Vienna Internet Exchange, ⟨http://www.vix.at/⟩, approximately doubled its traffic between May 2000 and May 2001, to a level of about 450 Mb/s.

Traffic interchange statistics are hard to interpret, unless one has data for most exchanges, which we do not. Much of the growth one sees can come from ISPs moving from one exchange to another, moving their traffic from one exchange to another, or else coming to an exchange in preference to buying transit from another ISP. At LINX, a large part of its growth is almost surely caused by more ISPs exchanging their traffic there. Between March 1999, and March 2000, the ranks of ISPs that are members of LINX have grown by about two thirds, based on the data on the LINX home page. Hence the average per-member traffic through LINX may have increased only around 120% during that year. On the other hand, FICIX membership appears to have been much more stable.

Unfortunately the largest ISPs do not release reliable statistics. This situation was better even a couple of years ago. For example, MCI used to publish precise data about the traffic volumes on their Internet backbone. Even though they were among the first ISPs to stop providing official network maps, one could obtain good estimates of the MCI Internet backbone capacity from Vint Cerf's presentations. These sources dried up when MCI was acquired by WorldCom, and the backbone was sold to Cable & Wireless. As was noted in Coffman and Odlyzko (1998), the traffic growth rate for that backbone had been in the range of 100% a year before the change.

Today, one can obtain some idea of the sizes (but not traffic) of various ISP networks through the backbone maps available at Boardwatch (2001). They may not be too reliable, but provide some indication of capacity and capacity growth. Although networks appear to have been growing at faster rates than the doubling of traffic we estimate, even they have not been growing at anything like the mythological doubling every three months.

The only large U.S. ISP to provide detailed network statistics is AboveNet, at ⟨http://www.above.net/traffic/⟩. We have recorded the MRTG data for AboveNet for March 1999, June 1999, February 2000, June 2000, November 2000, and April 2001. The average utilizations of the links in the AboveNet long-haul backbone during those 6 months were 18%, 16%, 29%, 12%, 11%, and 10%, respectively. (The large drop between February and June 2000 was caused by deployment of massive new capacity, including four OC48s. By the middle of 2001, there were even some OC192 links. One of the reasons we concentrate on traffic and not network sizes in this chapter is that extensive new capacity is being deployed at an irregular schedule, and is often lightly utilized. Thus it is hard to obtain an accurate picture of the evolution of network capacity.) If we just add up the volumes for each link separately, we find that traffic quadrupled in 15 months between March 1999 and June 2000. (Similarly, it grew by a factor of slightly over 3 between June 2000 and April 2001.) These increases represent annual growth rates of about 200%. However, this figure has to be treated with caution, as actual traffic almost surely increased somewhat more slowly. During these periods, AboveNet expanded geographically, with links to Japan and Europe, so that at the end it probably carried packets over more hops than before. In this chapter we count only bytes that are delivered to customers, and count them once, no matter how many backbone links they traverse. Hence the sum of the traffic figures for the AboveNet links has to be deflated by the average number of hops that a packet makes over the backbones. (In particular, the sum of the volumes over all the links in the AboveNet network in June 2000, which comes to 1,400 TB, has to be deflated by this average number of hops if one desires to compare it to the volumes in Table 3.2.)

Even when we do have data for a single carrier, such as AboveNet, some of the growth we see there may be coming from gains in market share, both from gains within a geographical region, and from greater geographical reach, and not from general growth in the market. More interesting examples are those of Telstra and Onvoy, since their geographical reach did not change. We discuss them next.

Telstra, the dominant Australian carrier, has operated within the same geographical region, and its traffic growth rate may be an approximation to that of the entire Australian market. The only traffic statistics that Telstra has provided are those in Cochrane (2001), cited at the beginning of this section. (In addition to an approximate doubling of Internet traffic over several years, that reference also provides some data about Telstra ATM and Frame Relay traffic, which just about doubled during the year 2000.) In addition, Telstra does present network maps at ⟨http://www.telstra.net⟩. In January 1998, the total bandwidth to the U.S., including some provided over a satellite link, was 146 Mb/s. (The bandwidth to other countries has historically been almost negligible compared to that to the U.S..) In March 2000, that bandwidth had grown to 592 Mb/s. By late June 2000, it had shrunk to 515 Mb/s. However, by September 2000, it was up to 980 Mb/s, and by April 2001, it was 1245 Mb/s. Thus the growth rate of international bandwidth was close to 100% per year over three consecutive years, closely paralleling the growth of Internet traffic, as disclosed by Telstra in Cochrane (2001). On the other hand, domestic Australian Internet bandwidth grew much faster. Whereas in September 1999, the highest capacity domestic links were 155 Mb/s (OC3), during 2000 several were upgraded to 2488 Mb/s (OC48).

The Telstra data provides a rough check on our estimates for Internet backbone traffic. According to public statements by Telstra officials (Taggart 1999), 60% of Telstra Internet traffic is with the U.S. Let us assume that the links from the U.S. to Australia are run at an average of 60% of capacity. (This is rather heavy loading, but not unprecedented on expensive links, cf. Table 3.6.) Let us assume that the reverse direction is operated at an average of 20% of capacity, which seems reasonable by comparison with the data in Table 3.5. Then we find that at year-end 2000, Telstra was probably receiving about 200 TB/month from the U.S., and transmiting about 70 TB/month. If this represents 60% of Telstra's Internet traffic (and this fits reasonably well with the data in Cochrane (2001)), then (without making allowances for other carriers or for the likelihood that some of the traffic on the link to the U.S. is destined for or comes from other countries) we obtain an estimate of about 500 TB/month for all of Australia's Internet traffic. Since the U.S. has about 15 times as many inhabitants as Australia, is somewhat richer on a per capita basis, and has better developed Internet infrastructure and lower prices, the estimates in Table 3.2 appear in the right range.

Onvoy, ⟨http://www.onvoy.net⟩ has evolved from the research and educational MRNET to the largest commercial ISP in Minnesota. Its traffic statistics were available at ⟨http://graphs.onvoy.com/infrastructure⟩ and over the two years from June 1998 to June 2000 showed an annual growth rate in traffic of only about 50%, to a level in June 2000 of 155 Mb/s from the Internet and 100 Mb/s towards the Internet. (More recent statistics are not available any more.)

IP-Plus, the ISP operated by Swisscom, the dominant carrier in Switzerland, has extensive statistics about their network at ⟨http://www.ip-plus.net⟩. As of May 2001, they showed only very moderate growth over the preceding year, almost surely under 100%, for the domestic links.

For the general market, the growth in usage by residential customers, at least in the U.S., has slowed down. Their ranks are growing at only about 20% per year, and the time spent online is growing at under 20% per year (cf. Odlyzko (2000a)). Thus the traffic they generate (about 5 Kb/s from the Internet towards their PCs when they are online) is increasing less than 50% a year. In other countries, the pattern is different. For example, in France in 1999, the number of residential Internet users grew by almost 150%, just as it had done the previous year, while the average time online stayed constant (Odlyzko 2000a). Therefore it is likely that French residential traffic grew by a factor of 2.5 in 1999. Around the world, the number of residential customers appears to be growing at about 50% per year, but their usage tends to be static as a result of per-minute charging (cf. Odlyzko (2000a)).

The traffic from residential U.S. customers may very well increase at a faster rate in the near future. The growth in the number of users is likely to diminish, as we reach saturation. (You cannot double the ranks of subscribers if more than half the people are already signed up!) However, broadband access, in the shape of cable modems and DSL (digital subscriber line), and to a lesser extent fixed wireless links, will stimulate usage. The evidence so far is that users who switch to cable modem or DSL access increase their time online by 50 to 100%, and the total volume of data they download per month by factors of 5 to 10. A 5 or 10-fold growth in data traffic would correspond to a doubling of traffic every four months if everyone were to switch to such broadband access in a year. However, that is not going to happen. At the end of 1999, there were about 3 million households in the U.S. with broadband access. At the end of 2000, it is estimated there were about 6 million, and as of May 2001, the estimates for year-end 2001 were for about 11 million. That is approximately a doubling each year. (Not all of this growth will be even. The ranks of DSL subscribers apparently grew about four-fold in 2000, but in early 2001, with the bankruptcy of several DSL providers, growth appears to have slowed down dramatically.) The traffic from a typical residential broadband customer is likely to grow beyond the level we see today, as more content becomes available, and especially as more content that requires high bandwidth is produced, and as people learn to exploit high bandwidth links for their own communication needs. Still, it is hard to see average traffic per customer among those with broadband connections growing at more than 50% a year. Together with a doubling in the ranks of such customers, this might produce a tripling of traffic from this source. Since the ranks of customers with regular modems are unlikely to decrease much, if any, and since their traffic dominates, it appears that we most likely will see total residential customer traffic growing no faster than 200% per year, and probably closer to 100% per year. (Access from information appliances, which are forecast to proliferate, is unlikely to have a major impact on total traffic, since the mobile radio link will continue to have small bandwidth compared to wired connections. There may be much greater traffic to mobile gateways, but it appears unlikely that such traffic will be huge.)

Growth in traffic can be broken down into growth in the number of traffic sources, and growth in traffic per source. For LINX, much of the increase in traffic may be coming from an increase in member ISPs and increased peering among those ISPs. For individual ISPs, much of the increase in traffic may also be coming from new customers. Yet in the end, that kind of growth is limited, as the market gets saturated. We will concentrate in the rest of this section on rates of growth in traffic from stable sources. Now nothing is completely stable, as the number of devices per person is likely to continue growing, especially with the advent of information appliances and wireless data transmission. Hence we will consider growth in traffic from large institutions that are already well wired, such as corporations and universities. Most corporations do not publicize information about their network traffic, and many do not even collect it. However, there are some exceptions. For example, Lew Platt, the former CEO of Hewlett-Packard, used to regularly cite the HP Intranet in his presentations. The last such report, dated September 7, 1998, and available at ⟨http://www.hp.com/financials/textonly/personnel/ceo/rules.html⟩, stated that this network carried 20 TB/month, and a comparison with previous reports shows that this volume of traffic had been doubling each year for at least the previous two years. (As an interesting point of comparison, the entire NSFNet Internet backbone carried 15 TB/month at its peak at the end of 1994.) Several other corporations have provided us with data showing similar rates of growth for their Intranet traffic, although some indicated their growth has slowed down, and a few have had practically no growth at all recently.

Internal corporate traffic appears to be growing more slowly than the public Internet traffic. Data for retail private lines (i.e., those sold to corporate and government entities for their own internal use, not for connecting to the Internet, and not for use by ISPs) as well as for Frame Relay and ATM services show aggregate growth in bandwidth (and therefore most likely also traffic) in a a range of 30 to 40% per year. (Growth is slow for retail private lines, and faster for Frame Relay and ATM.) This is remarkably close to the growth rate observed in the late 1970s in the U.S., which was around 30% per year (de Sola Pool et al. 1984), as well as to the growth rate of total private line data bandwidth, local and long distance, through most of the 1990s (Galbi 2000). Thus it is the corporate traffic to the public Internet that is growing at 100% per year. (Currently over two thirds of the volume on the public Internet appears to be business to business.) Thus the acceleration in the overall growth rate of data traffic to the range of 100% per year from the old

INTERNET GROWTH: "MOORE'S LAW"

30% or so a year appears to be a reflection of the advantages of the Internet, with its open standards, and any-to-any connectivity.

In the rest of this section we concentrate on publicly available information, primarily about academic, research, and government networks. These might be thought of as unrepresentative of the corporate or private residential users. Our view is just the opposite, that these are the institutions that are worth studying the most, since they normally already have broadband access to the Internet, tend to be populated by technically sophisticated users, and tend to try out new technologies first. The spread of Napster through universities is a good example of the last point. We suggest (see Section 6 for more detail) that Napster and related tools, such as Gnutella and Wrapster, are just the forerunners of other programs for sharing of general information, and not just for disseminating pirated MP3 files. As we explain in Section 6, there is already much more digital data on hard disks alone than shows up on today's Internet. Further, this situation is likely to continue.

Table 3.4. Growth in data traffic at Library of Congress. For each year, shows total traffic in gigabytes during February of that year and the rate of increase over the previous year.

year	GB/month	increase
1995	14.0	
1996	31.2	123%
1997	109.4	251%
1998	282.0	158%
1999	535.0	88%
2000	741.1	39%
2001	1202.6	62%

The growth rates we observe among the institutions that make traffic statistics publicly available vary tremendously. For example, Table 3.4 presents data for the Library of Congress, taken from the online statistics at ⟨http://lcweb.loc.gov/stats/⟩. There was a pronounced slowdown in the growth rate of traffic, followed by a noticeable by not huge increase. On the other hand, other sources show no such effect. The AT&T Labs - Research public Web server has experienced a consi ent growth rate in the volume of downloads of about 50% a year ur il the beginning of 2000, and then more than a doubling of traffic d :ing 2000. This growth continued even though this server contains prin rily high quality technical material, that is of interest primarily to people who are already well-connected.

Table 3.5 shows statistics for the transatlantic link of the JANET network, which provides connectivity to British academic institutions. (More complete data is available at ⟨http://bill.ja.net/⟩.) There are several interesting features of this data. One is that for a long time, there was increasing asymmetry of the traffic, with the preponderance of traffic from the U.S. over that to the U.S. growing. (There are some signs of a reversal of this trend, starting in 2001.) Another is that traffic with the U.S. is increasing faster than with the LINX exchange. Perhaps even more interesting is that the growth rate of this traffic shows no signs of exploding, even though the link capacity has grown, and so the average utilization has decreased. In March 1999, JANET had two T3s across the Atlantic, for an aggregate capacity of 90 Mb/s. By March 2000, these were replaced by two OC3s, providing 310 Mb/s. In January 2001, a third OC3 was added. Hence the utilization of the U.S. to U.K. link decreased from 64.8% in March 1999 to 47.0% in March 2000 and remained at about that level in March 2001. The increased capacity was not filled up immediately. (JANET links within the U.K. are about to be upgraded to OC48 or OC192, and it will be interesting to see what effect this has on the load on the transatlantic link.)

Table 3.5. Growth in JANET traffic. Shows terabytes transmitted on the link from the U.S. to the British JANET academic network in March of each year, and the rate of increase from previous year.

year	US to UK TB/month	UK to US TB/month	increase in US to UK traffic
1997	3.73	2.95	
1998	8.79	4.44	136%
1999	19.52	9.51	122
2000	48.76	14.90	150
2001	75.18	28.94	54

The prevalent opinion appears to be that in data networks, "if you build it, they will fill it." Our evidence supports this, but with the important qualification that "they" will not fill it immediately. That certainly has been the experience in local area networks, LANs. The prevalence of lightly utilized long distance corporate links was noted in Odlyzko (1998). That paper also discussed the vBNS research network, which was extremely lightly loaded. Here we cite another example of a large network with low utilizations and moderate growth rates. Abilene is the network created by the Internet2 consortium of U.S. universities (Dunn 1999). Its backbone consists of 13 OC48 (2.4 Gb/s) links. The average utilization in June 2000 was about 1.5%, and by April 2001 it had grown to 4.1%, for an annual growth rate of somewhat under 300%. (That also appears to have been the growth rate over the year ending June 2000.) Yet most members had OC3 or OC12 links to the Abilene backbone. Thus in spite of the uncongested access and backbone links, traffic did not explode. (Moreover, the 300% growth rate in traffic may be partially a reflection of some of the growth that would normally have gone over the commercial ISP links being redirected over Abilene instead. A substantial fraction of the traffic growth appears to have come from additional members joining the consortium.) Access to vBNS was restricted to certain research projects. On the other hand, Abilene is open to any traffic between the member universities, and thus it does not have the same limits to growth.

The research networks cited above have low utilizations. It should be emphasized that this is not a sign of inefficiency. Many novel applications require high bandwith to be effective. That (together with some additional factors, such as high growth rates, lumpy capacity, and pricing structure) contributes to the generally much lower utilization of data networks than of the long distance voice network (Odlyzko 1998).

Even on more congested links, it often happens that an increase in capacity does not lead to a dramatic increase in traffic. This is illustrated by several examples. Figure 3.1 shows statistics for the traffic from the public Internet to the University of Waterloo over a period of 7 years, through the end of 1999. (This is the longest such time series that we have been able to obtain.) Detailed statistics for the Waterloo network are available at ⟨http://www.ist.uwaterloo.ca/cn/#Stats⟩, but Fig. 3.1 is based on additional historical data provided to us by this institution.) Just as for the JANET network discussed above, for the SWITCH network to be discussed later, as well as for most access links, there is much more traffic from the public Internet to the institution than in the other direction. Hence we concentrate on this more congested link, since it offers more of a barrier. We see that even substantial jumps in link capacity did not affect the growth rate much. Traffic during most of that period kept about doubling each year. This growth rate slowed down substantially in the last two years, to about 55% from early 1999 to early 2000, and about 33% from then to early 2001. (We do not include most of that period in our graph, since it is hard to provide comparable data, as new connections to research networks were opened up, which, however, are not available for general Internet access.) This was primarily the result of a budget-driven limitation on the capacity of the public link. It led to an imposition of official limits on individual users, limits we will discuss later. Even with those limits, usage has been rising, and the link is completely saturated for large parts of the day. (There has been growth in connections to research networks, but those links are much more lightly utilized. Hence it is hard to say precisely what the growth rate of the entire Internet traffic at Waterloo has been recently. This is also a problem at many other institutions. That is why we do not show traffic statistics for 2000 and early 2001 in Figure 3.1.)

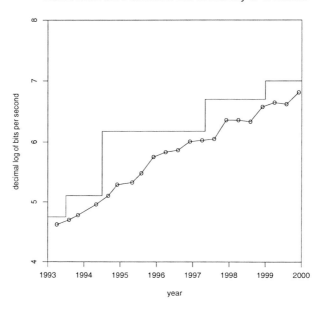

Figure 3.1. Traffic on the link from the public Internet to the University of Waterloo. The line with circles shows average traffic during the month of heaviest traffic in each school term. The step function is the full capacity of the link.

The same phenomenon of traffic that just about doubles each year, no matter what happens to capacity, can be observed in the statistics for the SWITCH network, which provides connectivity for Swiss academic and research institutions. The history and operations of this network are described in Harms (1994), Reichl et al. (1999), and extensive current and historical data is available at ⟨http://www.switch.ch/lan/stat/⟩. The data used to prepare Table 3.6 was provided to us by SWITCH. As is noted in Reichl et al. (1999), the transatlantic link has historically been the most expensive part of the SWITCH infrastructure, and at times was more expensive than the entire network within Switzerland. It is therefore not surprising that this link tends to be the most congested in the SWITCH network. Even so, increasing its capacity did not lead to a dramatic change in the growth rate of traffic. If we compare increases in volume of data received between November of one year and January of the following year, there was an unusually high jump from Nov. 1998 to Jan. 1999, by 42%. This was in response to extreme congestion experienced at the end of 1998, congestion that produced extremely poor service, with packet loss rates during peak periods exceeding 20%. However, over longer periods of time, the growth rate has been rather steady at close to 100% per year and independent of the capacity of the link. Especially noteworthy was the large increase in capacity in the year 2000, caused partially by the dramatic declines in prices of transatlantic transmission. It allowed SWITCH to have two OC3 links, providing redundancy. Utilization dropped noticeably, to a low level of about 10% (and a level of about 3% in the less heavily utilized direction from Switzerland to the U.S.), but traffic continued growing at about the same pace as before.

More detailed data about other types of SWITCH traffic can be found at ⟨http://www.switch.ch/lan/stat/⟩, through the "Public access" link. The listings available there as of early 2001, as well as those from previous years, show that various transmissions tended to grow at 100 to 150% per year. (Some of the growth has come from growth in the number of institutions. For example, for the largest SWITCH customer, traffic between June 1992 and June 2000 grew by a factor of 90, for an annual growth rate of 75%.) Occasionally there have been bigger jumps, such as the explosion in the category of traffic "leaving SWITCHlan" in early 2000, caused by the installation of an Akamai server that provides data to many European educational and research institutions.

The NORDUNet network connects research and educational institutions in the Nordic countries. It has detailed traffic statistics online, at ⟨http://www.nordu.net/stats/⟩, that go back to November 1996. Over this period, adding up the traffic over all the interfaces shown in the data, we find that total traffic has been growing at about 130% a year. The link to the U.S. went from 56 Kb/s in 1990 to 1.4 Gb/s in January 2001, for a compound annual growth rate of bandwidth over 10 years of about 150%. Over the last few years, the bandwidth to the U.S. has been growing at about 100% a year; the first OC3 (155 Mb/s) link was installed in February 1998, the second in January 1999, and by February 2000 there were four OC3 links. The fifth OC3 was put into service in June 2000, and in January 2001, four more were installed. Traffic on the main U.S. links (ignoring the 45 Mb/s connection from Iceland) grew

Table 3.6. Growth in SWITCH traffic. Average traffic flow from the U.S. to SWITCH, the Swiss academic and research network.

month	year	traffic flow Mb/s	link capacity Mb/s	average utilization(%)
May	1996	1.51	3	50.4
Jul	1996	1.90	3	63.3
Sep	1996	1.99	3	66.3
Nov	1996	2.21	3	73.6
Jan	1997	2.37	3	67.6
Mar	1997	2.62	4	65.4
May	1997	2.86	4	71.4
Jul	1997	3.17	8	39.7
Sep	1997	2.87	8	35.9
Nov	1997	3.24	8	40.5
Jan	1998	3.88	8	48.5
Mar	1998	4.20	8	52.5
May	1998	5.05	8	63.1
Jul	1998	5.14	8	64.3
Sep	1998	5.66	8	70.7
Nov	1998	6.20	8	77.5
Jan	1999	8.78	24	36.6
Mar	1999	9.41	24	39.2
May	1999	10.63	32	33.2
Jul	1999	10.03	32	31.3
Sep	1999	11.62	32	36.3
Nov	1999	13.26	32	41.4
Jan	2000	15.52	56	27.7
Mar	2000	17.81	56	31.8
May	2000	15.92	64	24.9
Jul	2000	19.94	155	12.9
Sep	2000	24.86	155	16.0
Nov	2000	28.37	155	18.3
Jan	2001	28.75	310	9.3
Mar	2001	32.00	310	10.3

from 54 and 30 Mb/s in March 1998 (54 Mb/s from the U.S. to NOR-DUNet, and 30 Mb/s in the reverse direction) to 237 and 122 Mb/s in March 2000, almost exactly a 100% growth rate over those two years.

The traffic statistics for the European TEN-155 research network (which consists largely of OC3, 155 Mb/s, links) are available at ⟨http://stats.dante.org.uk/mystere/⟩. Some of the links were heavily congested in the middle of 2000, while overall utilization has been moderate. Some of the historical traffic data for TEN-155 has been lost. However, DANTE (the organization that runs it) has provided us with data for the access link to the German national research network DFN, one of the largest contributors to TEN-155 traffic. This data, covering the period from the end of January 1999 to the beginning of July 2000, shows annual growth rates of about 70 and 90% (for the two directions of traffic).

Merit Network is a non-profit ISP that serves primarily Michigan educational institutions. It has some statistics available online at ⟨http://www.merit.net/michnet/statistics/direct.html⟩ that goes back to January 1993. This data was used to construct the graph in Figure 3.2. The information for January 1993 through June 1998 shows only the number of inbound IP packets. The data for months since July 1998 is more complete, but it is so complete, with details of so many interfaces, that we have not yet figured out how to utilize it fully and obtain figures comparable to those for the earlier periods. Hence we have used only the earlier information for January 1993 through June 1998. The resulting time series is a reasonable although imperfect representation of a straight line, modulated by the periodic variations introduced by the academic calendar. The growth rate is almost exactly 100% per year.

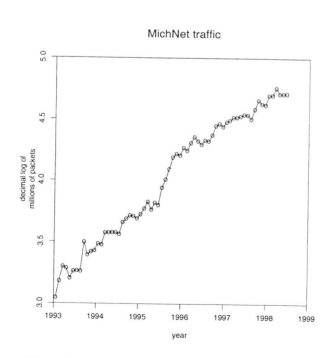

Figure 3.2. Traffic from Merit Network to customers.

We conclude by presenting data traffic growth rates for three universities. The University of Toronto in the spring of 1998 had an 8 Mb/s connection to the Internet. By the spring of 2000, the bandwidth had been increased to 25 Mb/s. At both times, the link was run at about a 50% utilization (average of both directions), so it was very congested. (For current data and that of the previous year, see ⟨http://www.noc.utoronto.ca/netstats/index.html⟩.) Therefore, the compound annual growth rate for both traffic and bandwidth was about 70%. By the spring of 2001, the link capacity had only been increased to 30 Mb/s, and the average utilization was 65%. Thus even in the face of deteriorating quality of transmission, users had increased their traffic by about 55%.

Princeton University had two standard connections to the Internet in the springs of both 1998 and 2000. Their combined bandwidth was 31 Mb/s all this time. (They were lightly utilized in 1998, and only moderately heavily in 2000.) By 2000, Princeton also had a 155 Mb/s connection to the vBNS research network. (MRTG data for all these links is available at ⟨http://wwwnet.princeton.edu/monitoring.html⟩.) The combined traffic increased from 4.2 and 1.9 Mb/s in 1998 to 14.7 and 11.4 Mb/s in 2000. If we combine the data rates for the two directions, we find a growth rate of about 100% per year. By spring 2001, the bandwidth of the two commercial connections to the Internet had been increased to 41 Mb/s, while the vBNS connection remained at 155 Mb/s. Combined traffic had grown to 27.0 and 20.4 Mb/s, for an almost exact doubling since spring of 2000.

The connection between the University of California at Santa Cruz and the Internet is currently dominated by traffic from student residences. Thus it shows very strong seasonal effects. If we consider total Internet traffic during the first 7 days in June (before the start of the summer vacation), we find 143% growth from 1998 to 1999, and 205% growth from 1999 to 2000, to a final average level of 22.7 Mb/s. Much of the recent increase was caused by Napster (and will be discussed in the next section).

Many more examples can be cited, such as that of USC that was mentioned in Section 2, or that of Utah State University, with traffic statistics available at ⟨http://thingy.usu.edu/network-stats/uen-ds3.html⟩. At ⟨http://www.calren2.net/router-stats/⟩ there is data about some of the campuses of the University of California (CalREN2 provides external connectivity to most campuses in this system.) The article (McCredie 2000) mentions that the Berkeley campus has seen its traffic to the outside grow by a factor of 40 in 7 years, for a compound annual growth rate of 70%. Some of the other average growth rates appear higher.

An interested reader can find pointers to other universities, corporations, as well as exchanges that make their traffic statistics available in the listing of MRTG users at Oetiker and Rand (2001). In addition, there is an increasing number of gigapops that are being formed, and they usually show their traffic statistics on the Web. For URLs, see (Dunn 1999). (It should be mentioned, though, that most of the institutions connecting to the gigapops continue to have other connections to the Internet. Thus growth rates of their traffic at the gigapops by themselves may sometimes be misleading.)

The general conclusion we draw from the examples listed above, as well as from numerous others, is that data traffic has a remarkable tendency to double each year. There are slower and faster growth rates. Overall, though, they tend to cluster in the vicinity of 100% a year. We have not seen any large institutions with traffic doubling anywhere close to three or even four months.

The growth rates noted above are often affected strongly by restrictions imposed at various levels. We will discuss this question further in Section 7, and at this stage just remark on some of the explicit limits imposed by network administrators. We noted that at the University of Waterloo, the growth rate slowed down to 55% from early 1999 to early 2000, and to 33% over the following year. This was probably caused largely by the congestion on the Internet link, and the explicit limits on individual student download rates, described at ⟨http://www.ist.uwaterloo.ca/cn/#Stats⟩. The arrival of Napster (discussed in the next section) led many institutions to either ban its use, or limit traffic rates to some parts of the campus (typically student dormitories), or else to limit rates of individual flows. Push technologies were stifled at least partially because enterprise network administrators blocked them at their firewalls. Email often has size restrictions that block large attachments (and in some cases all attachments are still banned). Teleconferencing is only slowly being experimented with on corporate intranets, and even packetized voice sees very limited (although growing) use.

Even spammers exercise some control. One of us has been collecting all the spam messages that have made it through the AT&T Labs - Research spam filter to his email account. They total several thousand over the last four years. An interesting observation is that the average size of these spam messages has increased over this period, but not rapidly, namely from 4900 bytes to 7600 bytes, for a compound annual growth rate of only 12%. Spammers want to send many messages from their connections, so have an interest in keeping them short (especially if they are attempting to avoid being recognized as spammers and shut down). At the same time, most of their customers are probably not ready to process complicated attachments in any case, and, connected by slow modems, would not have the patience to download large files. (Hardly any spam messages contain Microsoft Office attachments, the main reason corporate email messages are much larger.) This provides an informal but apparently effective limitation on how big spam messages get. There appears to be an increase in large spam messages in html format, but this process started in a serious way only in 2000. Even now, in early 2001, relatively few html spam messages show up, although the Web has been prominent for many years.

Similar constraints apply to most of the content seen on the Web. As long as a large fraction of potential users have limited bandwidth, such as through dial modems, managers of Web servers will have an incentive to keep individual pages moderate in size.

The general conclusion from the above discussion is that Internet traffic is subject to a variety of constraints and feedback loops, at different levels and operating on different time scale. Some are applied by network managers, others by individual users. The interaction of these constraints with rising demands is what produces the growth rates we see.

To sustain the high growth rate of Internet traffic will require the creation of new applications that will generate huge traffic volumes. We estimate that as of year-end 1999, U.S. Internet backbone traffic was about a quarter to a third of voice traffic. At current growth rates (100% per year for the Internet, 10% for voice), by year-end 2004 there will be 8 times as much Internet as voice traffic. If voice is packetized at that stage, it will likely be compressed as well, and even at very moderate 4:1 compression, would then amount to just 3% of Internet traffic. Thus voice will not fill the pipes that are likely to exist, and neither will traditional Web surfing. Thus we have the dilemma of service providers, network administrators, and equipment suppliers: to sustain the growth rates that the industry has come to depend on, and to accommodate the progress in technology (to be discussed more extensively later), we need new applications. Such applications are likely to appear disruptive to network operations today, and so often have to be controlled. In the long run, though, they have to be encouraged.

4. Disruptive Innovations: Browsers, Napster, ...

It is often said that everything changes so rapidly on the Internet that it is impossible to forecast far into the future. The next "killer app" could disrupt any plans that one makes. Yet there have been just two "killer apps" in the history of the Internet: email and the Web (or, more precisely, Web browsers, which made the Web usable by the masses). Many other technologies that had been widely touted as the next "killer app," such as push technology, have fizzled. Furthermore, only the Web can be said to have been truly disruptive. From the first release of the Mosaic visual browser around the middle of 1993, it apparently took under 18 months before Web traffic became dominant on Internet backbones. It appears overwhelmingly likely that it was the appearance of browsers that then led, in combination with other developments, to that abnormal spurt of a doubling of Internet traffic every three or four months in 1995 and 1996.

What were the causes of the 100-fold explosion in Internet backbone traffic over the two-year period of 1995 and 1996? We do not have precise data, but it appears that there were four main factors, all interrelated. Browsers passed some magic threshold of usability, so many more people were willing to use computers and online information services. Users of the established online services, primarily AOL, CompuServe, and Prodigy, started using the Internet. The text-based transmissions of those services, which probably averaged only a few hundred bits per second per connected user, were replaced by the graphics-rich content of the Web, so transmission rates increased to a few thousand bits per second. Finally, flat rate access plans led to a tripling of the time that individual users spent online (Odlyzko 2000a), as well as faster growth in number of users.

The Internet was able to support this explosion in traffic because it was utilizing the existing infrastructure of the telephone network. At that time, the Internet was tiny compared to the voice network. It is likely that the data network that handles control and billing for the AT&T long distance voice services by itself was carrying more traffic than the NSF Internet backbone did at its peak at the end of 1994. Today, by contrast, the public Internet is rapidly moving towards being the main network, so quantum jumps in traffic cannot be tolerated so easily.

In late 1999, a new application appeared that attracted extensive attention and led to many predictions that network traffic would see a major impact. It was Napster. Numerous articles in the press have cited Napster's ability to "overwhelm Internet lines", and have claimed that it has forced numerous universities to ban or limit its use. The impression one got from these press reports was that Napster was causing a quantum jump in Internet traffic, and was driving the traffic growth rates well beyond the normal range. However, upon close examination this does not appear to be completely accurate, and the use of Napster has not increased growth rates much beyond the annual doubling or tripling rates, even within university environments, where Napster is most popular. That is not to say that is has not resulted in huge amounts of traffic, nor that it has not had serious impact on several major networks.

Napster provides software that enables users connected to the Internet to exchange and/or download MP3 music files. The Napster (web) site matches users seeking certain music files with other users who have those files on their computer. The Napster system preferentially uses as sources of files machines that have high bandwidth connections. This means that universities are the primary sources, since other organizations with fast dedicated links, mainly corporations, do not allow such traffic. The result is that although college students are often cited as the greatest users of MP3 files, it is the traffic from universities that gets boosted the most. (Since that direction of traffic is typically much less heavily used than the reverse one, the impact of Napster is much less severe than if the dominant direction of traffic were reversed.) Regular modem users are usually not affected, since their connections are too slow and evanescent. However, the proliferation of cable modems and DSL connections that have "always-on" high bandwidth connectivity is leading to problems for some residential users and their ISPs, especially since the uplink is the one that invariably has the more limited bandwidth.

Napster has attracted huge attention because of its perceived potential to facilitate violations of copyright. This threat has led to litigation, and several universities have blocked access to the central Napster server as a result. (Whether such bans can be effective is questionable, as there are ways to bypass them. Some universities have adopted an attitude of watchful waiting, cf. Plonka (2001).) While the legal aspects of Napster and their implications for the music business are important questions, we will not deal with them in this chapter.

A key reason that Napster is of great interest to us is that similar types of sharing applications effectively turn consumers of information into providers of information. (The World Wide Web was designed for such information sharing, but for some types of files Napster and its kin are preferable.) These applications will effectively turn the traditional consumer PCs into Internet servers which will output large amounts of traffic to other users. In Napster's case this has been predominantly MP3 music files, but other programs, such as Gnutella, work with more general data. It is highly probable that such applications could be one of the key factors that fuel the continued annual doubling or tripling of data traffic.

INTERNET GROWTH: "MOORE'S LAW"

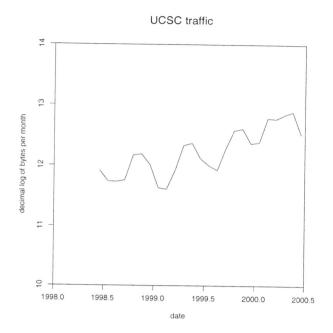

Figure 3.3. Traffic between UCSC and the Internet.

Napster first became noticeable in the summer of 1999. Its share of the total Internet traffic on many of the university networks has grown from essentially nothing to around 25% of the total traffic. The amount of Napster traffic that is reported by several university networks (such as UC Santa Cruz, University of Michigan, University of Michigan, Indiana, UC Berkeley, Northwestern University, and Oregon State University to name a few) range from around 20% at some as high as 50%. However, the reported numbers are often very preliminary, and in some cases they compare Napster traffic to total traffic, while in others it appears that the high values may represent a comparison only to the out traffic. In any event this is a phenomenal growth rate for any single application.

With the caveat that the numbers are approximate and preliminary, we did a quick estimate of the impact of Napster on growth rates for some of these university networks. For example, prior to the introduction of Napster, UC Berkeley's network traffic appeared to be growing at roughly 70% annually (i.e., doubling every 15 to 16 months) (McCredie 2000). It was reported that by the spring of 2000, Napster traffic had grown to 50% of the total. (It is not clear whether this is a percentage of the total traffic or only the out traffic). If we assume that the non-Napster traffic continued to grow at this rate, and assume that Napster traffic is now 50% of the total, then the overall annual growth rate (since the introduction of Napster) is around 4× per year. If, however, Napster only makes up 30% of the total then it works out to an annual growth rate of about 3.2× per year.

We have detailed data for the University of California at Santa Cruz (UCSC). Its traffic reporting system is at ⟨http://noc.ucsc.edu/mrtg/-data/routers/to_campus:_commcat-comm-g.html⟩. In Figure 3.3 we show the monthly traffic between UCSC and the Internet during the crucial period when Napster made its appearance. There were no controls on this traffic in 1999, other than email warnings to owners of machines with large data transfers. There are huge variations from month to month, caused by the academic calendar, but one can see that traffic basically doubled from the spring semester of 1999 to the fall one. By early 2000, Napster traffic was about 50% of outgoing transmissions, and 10% of incoming ones. In March 2000, a rate limitation was imposed on the traffic from the dormitories, which has limited the impact of Napster. Napster is the obvious culprit in the increase in outgoing traffic to about 60% of that of incoming traffic in February 2000. The limits did result in the drop in the ratio of outgoing to incoming traffic to about 40% in April 2000, but then this ratio crept up to over 50% in May and June.

The University of Wisconsin-Madison has done the most to carefully monitor Napster and its kin, has high bandwidth on campus and to the Internet, and as of mid-2000 has not as yet taken any steps to limit Napster traffic (Plonka 2001). (For current analysis of their traffic by protocol, application, and so on, see ⟨http://wwwstats.net.wisc.edu/⟩. Several other universities are using this software, developed by Dave Plonka at the University of Wisconsin, for analyzing their traffic as well.) Figure 3.4 shows the effect of Napster. This campus is very unusual in that even before Napster made its appearance, there was about as much outgoing as incoming traffic. Napster has led to a disproportionate increase in outgoing traffic. (At some times it apparently led to this traffic reaching capacity limit for the link.) Figure 3.4 compares the average rates for data traffic during the week starting May 7, 1999 to those for the week starting May 12, 2000. Both were exam weeks at the end of the regular school year. (Transmissions dropped dramatically in the following weeks.) Outgoing traffic increased 157%, and total traffic 138%. If we exclude Napster, the other traffic increased by 81%. Thus Napster has had a noticeable effect on the growth rate of traffic on this campus, but not an outlandish one.

Several networks that report Napster traffic of as much as 30% are not doing anything to limit Napster since they claim that they still have plenty of bandwidth. Others have imposed limits on the total bandwidth

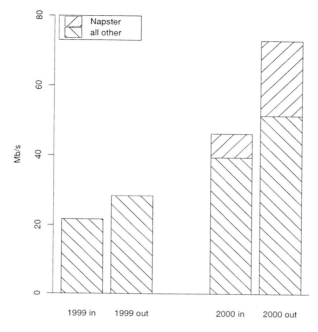

Figure 3.4. Change in volume and composition of traffic between the University of Wisconsin in Madison and the Internet. Average transmission rates in spring 1999 and spring 2000.

available to the dormitories, or else are limiting rates of individual flows. Note that if your traffic doubles each year, a one-time 30% increment is noticeable when it occurs, but becomes minor in a couple of years. A much more serious problem arises if a new application continues growing. Even if Napster does not grow much more, either because of legal action or because of music demand being met by other sources. video files are likely to become increasingly common, and the question is how fast that will drive the growth of total traffic.

Aside from Napster, occasionally even a large institution will experience a local perturbation in its data traffic patterns caused by one particular application. For example, the SETI@home distributed computing project, ⟨http://setiathome.ssl.berkeley.edu⟩, uses idle time on about three million PCs (as of mid-2001) to search for signs of extraterrestrial intelligence in signals collected by radio telescopes. This project is run out of the Space Sciences Institute at the University of California at Berkeley, and within a year of inception accounted for about a third of the outgoing campus traffic (McCredie 2000). (Moreover, this was extremely asymmetrical traffic, with large sets of data to be analyzed going out to the participating PCs, and small final results coming back. That most of the data went away from campus made this application less disruptive than it would have been otherwise.) Its disruptive effect is moderated by limiting its transmission rate to about 20 Mb/s.

At the University of California at Santa Cruz, a complete copy of the available genome sequence was made available for public download in early July 2000. This, combined with coverage in the popular press and on the popular Slashdot online discussion list, led to an immediate surge in traffic, far exceeding the effects of Napster. If the interest in this database continues, it will require reengineering of the campus network.

The SETI@home project is interesting for several reasons. It is cited in McCredie (2000) as a major new disruptive influence. Yet it contributes only about 20 Mb/s to the outgoing traffic. An increasing number of PCs and workstations are connected at 100 Mb/s, and even Gigabit Ethernet (1,000 Mb/s) is coming to the desktop. This means that for the foreseeable future, a handful of workstations will in principle be capable of saturating any Internet link. Given the projections for bandwidth (discussed in Section 5), a few thousand machines will continue to be capable of saturating all the links in the entire Internet. Thus control on user traffic will have to be exercised to prevent accidental as well as malicious disruptions of service. However, it seems likely that such control could be limited to the edges of the network. In fact, such control will pretty much have to be exercised at the edges of the network. QoS will not help by itself, since a malicious attacker who takes over control of a machine will be able to subvert any automatic controls.

Finally, after considering current disruptions from SETI@home and Napster, we go back and consider browsers and the Web again. They were cited as disruptive back in 1994 and 1995. (Mosaic was first released unofficially around the middle of 1993, officially in the fall of 1993, and took off in 1994.) However, when we consider the growth rates for the University of Waterloo (Fig. 3.1), for MichNet (Figure 3.2), or for SWITCH (for which Table 3.6 only covers the period since early 1996, but which apparently had regular growth throughout the 1990s, according to Harms (1994)), we do not see anything anomalous, just the steady doubling of traffic each year or so. If we consider the composition of the traffic, there were major changes. For example, Figure 3.5 shows the evolution of traffic between the University of Waterloo and the Internet. (It is based on analysis of traffic during the third week in each March, and more complete results are available at http://www.ist.uwaterloo.ca/cn/Stats/ext-prot.html.) The Web

INTERNET GROWTH: "MOORE'S LAW"

did take over, but much more slowly than on Internet backbones. There are no good data sets, but it has been claimed that by the end of 1994, Web traffic was more than half of the volume of the commercial backbones. On the other hand, the data for the NSFNet backbone, available at ⟨http://www.merit.edu/merit/archive/nsfnet/statistics/.index.html⟩, show that Web traffic was only approaching 20% there by the end of 1994, a level similar to that for the University of Waterloo. Thus at well-wired academic institutions such as the University of Waterloo and others that dominated NSFNet traffic, the impact of the Web was muted.

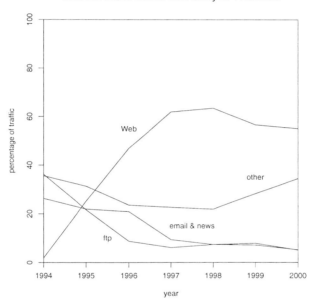

Figure 3.5. Composition of traffic between the University of Waterloo and the Internet. Based on data collected in March of each year.

Perhaps the main lesson to be drawn from the discussion in this section is that the most disruptive factor is simply rapid growth by itself. A doubling of traffic each year is very rapid, much more rapid than in other communication services. Figure 3.5 shows email and netnews shrinking as fractions of the traffic at the University of Waterloo, from a quarter to about 5%. Yet the byte volume of these two applications grew by a factor of 12 during the 6 years covered by the graph, for a growth rate of over 50% per year, which is very rapid by most standards. If we are to continue the doubling of traffic each year, new applications will have to keep appearing and assuming dominant roles. An interesting data point is that even at the University of Wisconsin in Madison, which analyzes its data traffic very carefully, about 40% of the transmissions escape classification. That is consistent with information from a few corporate networks, where the managers report that upwards of half of their traffic is of unknown types. (A vast majority of network managers do not even attempt to perform such analyses.) This shows how difficult coping with rapid growth is.

5. Technology Trends: Growth in Bandwidth

Bandwidth is growing rapidly, primarily through the introduction of WDM systems of increasing capacity, and to a much lesser extent through installation of additional fiber. However, this increase is not as fast as some would have us believe. For example, George Gilder has predicted that "bandwidth will triple each year for the next 25" (Gilder 1997). While he was right in predicting rapid growth, the actual rate of increase has so far been somewhat more modest. In Table 3.7 we show the progress that has occurred so far, and is likely to occur in the next few years. This table was generated by using a variety of vendor release dates along with our detailed technical understanding of such high-speed transport systems. We are considering commercial systems and not research hero experiments. Historically, there was a lag of three years or more between the lab demonstration of a given transmission capacity, and its introduction into the commercial environment. Recently, however, this time lag appears to have diminished, but this is very debatable. There is much hype about commercial transmission capacities, and it is still a bit unclear as to how early a given high capacity transmission system will become widely deployed.

Even the rate of increase for the maximum capacity achieved within lab experiments has slowed down considerably. For example, the first 1 Tb/s experiments were done in late 1995 to early 1996, and it appears that at year's end 1999 the maximum rate was only around 3 Tb/s. This is a growth of "only" three times in roughly 4 years, a rate much less than what was demonstrated in the previous several years.

Even when a system does become commercially widely available, it takes a while for it to affect the available bandwidth of a network. It has to be installed and tested, which all takes time. Further, not all old systems get replaced right away. This is similar to the PC situation, in which machines tend to be replaced on a three or four year cycle.

The projections in Table 3.7 only go out to 2007. Beyond that point there appear to be serious barriers to improvements in the current tech-

Table 3.7. Widespread deployment of WDM systems.

system description	fiber capacity	wide deployment
8 × 2.5 Gb/s	20 Gb/s	1996
16 × 2.5 Gb/s	40 Gb/s	1997
32 × 2.5 Gb/s	80 Gb/s	1999
80 × 2.5 Gb/s	200 Gb/s	2000
40 × 10 Gb/s	400 Gb/s	mid to late 2000
160 × 2.5 Gb/s	400 Gb/s	mid to late 2000
80 × 10 Gb/s	800 Gb/s	late 2001
160 × 10 Gb/s	1.6 Tb/s	late 2002
40 × 40 Gb/s	1.6 Tb/s	late 2002
80 × 40 Gb/s	3.2 Tb/s	late 2003 to early 2004
100 × 40 Gb/s	4 Tb/s	2005
160 × 40 Gb/s	6.4 Tb/s	2007

nologies. For example, the erbium-doped fiber amplifiers that are crucial for the current systems cover only a limited range of wavelengths. It could be that those barriers will be overcome in time, just as similar barriers to Moore's Law for semiconductors have repeatedly been overcome. However, that does not affect our arguments. Given the time lags between research experiments and deployment, it appears that for the next half a dozen years at least we will be limited to the progress shown in Table 3.7, which corresponds to approximately a doubling of capacity each year for each fiber.

The effective transport capacity can be increased by measures other than boosting the capacity of each fiber. One can "light up" existing fibers that are presently not used. (Much of the capacity, especially among the new carriers, is "dark," not used for any transmission yet.) One can also "light up" more channels on installed DWDM systems. (Today it is rare for all channels to be in use.) The currently dominant SONET technologies can be abandoned in favor of service restoration at the IP level. (SONET rings typically use fiber with three times the transport capacity of the circuits that provide actual service.) Private lines, which are used at low fractions of their capacity, can be replaced by virtual private networks (VPNs) over the public Internet, which is run at higher average utilization. All these measures together, combined with advances in DWDM, could allow an increase in the traffic carried by long distance network by factors of three or so each year for the next decade. However, substantially faster growth rates would require deploying new fiber. This is increasingly easy to do, as much of the new construction provides empty conduits in addition to the one carrying the fiber that is installed. However, there are limits to how fast new fiber could be installed. If network capacity were to grow by 200% per year, and the DWDM allowed the capacity of each fiber to grow only 100%, eventually we would need to increase the volume of fiber by 50% a year. Compound interest is very powerful, and a 50% annual growth rate results in growth by a factor of almost 60 over a decade. As a result, we do not expect that growth rates of even 300% per year in traffic could be sustained over many years.

Although technology will be providing increasing transport capacity, and also ways to use that capacity more efficiently, there will also be countervailing tendencies that are hard to estimate. One is that of the distance dependence of traffic. This subject is treated in Coffman and Odlyzko (1998) and in somewhat more detail in Odlyzko (2000a). Voice telephone and private line data traffic are strongly local. On the other hand, it appears that Internet traffic at present is not. Will that change? We do not know enough to predict this yet, although there are some signs of increasing locality. Caching and content distribution networks will be pushing content towards the edges. However, only a fraction (usually estimated at well under half) of the Web traffic is cachable. There is also likely to be a growth in non-Web traffic, as we explain later. This, combined with the continuing growth in the volume of data on the Internet (to be discussed later) may be such that these measures will have little effect. In addition, it is desire for low transaction latency that is driving the development of data networks, and this is what underlies the low average utilizations we observe. It is possible that as transmission capacity becomes less expensive, users will demand lower utilizations. This would be analogous to what appears to have happened, and to still be happening, in LANs. A decade ago, average utilizations appear to have been perhaps 10 times as high as those we see today. (We do not have solid data on this subject, unfortunately.) Yet there is a rapid movement towards gigabit Ethernet, and soon towards ten gigabit Ethernet, although average utilizations are low, on the order of 1%.

As a final remark, we should note that the bandwidth we discuss here is not directly comparable to the bandwidths of data and voice networks discussed in Coffman and Odlyzko (1998). That paper considered only circuits that are actually used to carry customer traffic. It ignored all the redundancy that is present to provide fault tolerance, as well as the dark fiber. It also ignored the differences between air distance and actual distance along fibers, the loss of capacity from various framing schemes, and many other factors.

6. Technology Trends: Growth in Demand

The approximate doubling of transmission capacity of each fiber that is shown in Table 3.7 is analogous to the famous "Moore's Law" in the semiconductor industry. In 1965, Gordon E. Moore, then in charge of R&D at Fairchild Semiconductor, made a simple extrapolation from three data points in his company's product history. He predicted that the number of transistors per chip would about double each year for the next 10 years. This prediction was fulfilled, but when Moore revisited the subject in 1975, he modified his projection for further progress by predicting that the doubling period would be closer to 18 months. (For the history and fuller discussion of "Moore's Law", see (Schaller 1997).) Remarkably enough, this growth rate has been sustained over the following 25 years. There have been many predictions that progress was about to come to a screeching halt (including some recent ones), but the most that can be said is that there may have been some slight slowdown recently. (For example, according to the calculations of (Eldering et al. 1999), the number of transistors in leading-edge microprocessors doubles every 2.2 years. On the other hand, the doubling period is lower for commodity memories.) Experts in the semiconductor area are confident that Moore's 1975 prediction for rate of improvement can be fulfilled for at least most of the next decade.

Predictions similar to Moore's had been made before in other areas, and in (Licklider 1965) (which was written before Moore made his famous prediction) they were made for the entire spectrum of computing and communications. However, it is Moore's Law that has entered the vernacular as a description of the steady and predictable progress of technology that improves at an exponential rate (in the precise mathematical sense).

Moore's Law results from a complex interaction of technology, sociology, and economics. No new laws of nature had to be discovered, and there have been no dramatic breakthroughs. On the other hand, an enormous amount of research had to be carried out to overcome the numerous obstacles that were encountered. It may have been incremental research, but it required increasing ranks of very clever people to undertake it. Further, huge investments in manufacturing capacity had to be made to produce the hardware. Perhaps even more important, the resulting products had to be integrated into work and life styles of the institutions and individuals using them. For further discussions of the genesis, operations, and prospects of Moore's Law, see Eldering et al. (1999), Schaller (1997). The key point is that Moore's Law is not a natural law, but depends on a variety of factors. Still, it has held with remarkable regularity over many decades.

While Moore's Law does apply to a wide variety of technologies, the actual rates of progress vary tremendously among different areas. For example, battery storage is progressing at a snail's pace, compared to microprocessor improvements. This has significant implications for mobile Internet access, limiting processor power and display quality. Display advances are more rapid than those in power storage, but nowhere near fast enough to replace paper as the preferred technology for general reading, at least not at any time in the next decade. (This implies, in particular, that the bandwidth required for a single video transmission will be growing slowly.) DRAMs are growing in size in accordance with Moore's Law, but their speeds are improving slowly. Microprocessors are rapidly increasing their speed and size (which allows for faster execution through parallelism and other clever techniques), but memory buses are improving slowly. For some quantitative figures on recent progress, see Gray and Shenoy (2000). From the standpoint of a decade ago, we have had tidal waves of just about everything, processing power, main memory, disk storage, and so on. For a typical user, the details of the PC on the desktop (MHz rating of the processor, disk capacity) do not matter too much. It is generally assumed that in a couple of years a new and much more powerful machine will be required to run the new applications, and that it will be bought for about the same price as the current one. In the meantime, the average utilization of the processor is low (since it is provided for peak performance only), compression is not used, and wasteful encodings of information (such as 200 KB Word documents conveying a simple message of a few lines) are used. The stress is not on optimizing the utilization of the PC's resources, but on making life easy for the user.

To make life easy for the end user, though, clever engineering is employed. Because the tidal waves of different technologies are advancing at different rates, optimizing user experience requires careful architectural decisions (Gray and Shenoy 2000, Hennessy and Patterson 1990). In particular, since processing power and storage capacity are growing the fastest, while communication within a PC is improving much more slowly, elaborate memory hierarchies are built. They start with magnetic hard disks, and proceed through several levels of caches, invisibly to the user. The resulting architecture has several interesting implications, explored in Gray and Shenoy (2000). For example, mirroring disks is becoming preferable to RAID fault tolerant schemes that are far more efficient but slower.

Table 3.8. Worldwide hard disk drive market. (Based on Sept. 1998 and Aug. 2000 IDC reports.)

year	revenues (billions)	storage capacity (terabytes)
1995	$21.593	76,243
1996	24.655	147,200
1997	27.339	334,791
1998	26.969	695,140
1999	29.143	1,463,109
2000	32.519	3,222,153
2001	36.219	7,239,972
2002	40.683	15,424,824
2003		30,239,756
2004		56,558,700

The density of magnetic disk storage increased at about 30% per year from 1956 to 1991, doubling every two and a half years (Eco 1997). (Total deployed storage capacity increased faster, as the number of disks shipped grew.) In the 1990s, the growth rate accelerated, and in the late 1990s increased yet again. By some accounts, the densities in disk drives are about doubling each year. For our purposes, the most relevant figure will be the total storage capacity of disk drives. Table 3.8 shows data from IDC reports, which shows capacity shipped each year slightly more than doubling through the year 2003, and then slowing down somewhat. An interesting comment is that in the earlier 1998 report, the slowing of the growth rate was expected to occur already in 1999. Similar projections from Disk/Trend (⟨http://www.disktrend.com/⟩) also suggest that the total capacity of disk drives shipped will continue doubling through at least the year 2002. Given the advances in research on magnetic storage, it seems that a doubling each year until the year 2010 might be achievable (with some contribution from higher revenues, as shown in Table 3.8, but most coming from better technology). After about 2010, it appears that magnetic storage progress will be facing serious limits, but by then more exotic storage technologies may become competitive.

It seems safest to assume that total magnetic disk storage capacity will be doubling each year for the next decade. However, even if there is a slowdown, say to a 70% annual growth rate, this will not affect our arguments too much. The key point is that storage capacity is likely to grow at rates not much slower than those of network capacity. Furthermore, total installed storage is already immense. Table 3.8 shows that at the beginning of the year 2000, there were about 3,000,000 TB of magnetic disk storage. If we compare that with the estimates of

Table 3.2 for network traffic, we see that it would take between 250 and 400 months to transmit all the bits on existing disks over the Internet backbones. This comparison is meant as just a thought exercise. The backbones considered in Table 3.2 are just those in the U.S., whereas disks counted in Table 3.8 are spread around the world. A large fraction of the disk space is empty, and much of the content is duplicated (such as those hundreds of millions of copies of Windows 98), so nobody would want to send them over the Internet. Still, this thought exercise is useful in showing that there is a huge amount of digital data that could potentially be sent over the Internet. Further, this pool of digital data is about doubling each year.

An interesting estimate of the volume of information in the world is presented in Lesk (1997). (For a recent update of Lesk's study, with more detail as well as with more current data, see the report prepared at University of California at Berkeley under the leadership of Peter Lyman and Hal Varian (Lyman and Varian 2000).) It shows that already in the year 1997 we were on the threshold of being able to store all data that has ever been generated (meaning books, movies, music, and so on) in digital format on hard disks. By now we are well past that threshold, so future growth in disk capacities will have be devoted to other types of data that we have not dealt with before. Some of that capacity will surely be devoted to duplicate storage (such as a separate copy of an increasingly bloated operating system on each machine). Most of the storage, though, will have to be filled by new types of data. The same process that is yielding faster processors and larger memories is also leading to improved cameras and sensors. These will yield huge amounts of new data, that had not been available before. It appears impossible to predict precisely what type of data this will be. Much is likely to be video storage, from cameras set up as security measures, or else ones that record our every movement. There could also be huge amounts of data from medical sensors on our bodies. What is clear, though, is that "[t]he typical piece of information will *never* be looked at by a human being" (Lesk 1997). There will simply not be enough of the traditional "content" (books, movies, music), nor even of the less formal type of "content" that individuals will be generating on their own.

Huge amounts of data that is generated by machines use by other machines suggests that data networks will also be dominated by transfers of such data. This was already predicted in de Sola Pool et al. (1984), and more recently in Odlyzko (2000b), St. Arnaud (1997), St. Arnaud et al. (1998). Given an exponential growth rate in volume of data transfers, it was clear that at some point in the future most of the data flying through the networks would be neither seen nor heard by any human being. Thus we can expect that streaming media with real-time quality requirements will be a decreasing fraction of total traffic at some point within the next decade.

There will surely be an increase in the raw volume of streaming real-time traffic, as applications such as videoconferencing move onto the Internet. However, as a fraction of total traffic, such transmissions will not only decrease eventually, but may not grow much at all even in the intermediate future. (Recall that at the University of Waterloo over the last 6 years, the volume of email grew about 50% a year, but as a fraction of total traffic it is almost negligible now.) The huge imbalance in volume of storage and capacities of long distance data networks strongly suggests that even the majority of traditional "content" will be transmitted as files, and not in streaming form. For more detailed arguments supporting this prediction, see Odlyzko (2000b). This development (in which "content" is sent around as files for local storage and playback) is already making its appearance with MP3, Napster, and related programs.

The huge hard disk storage volumes also mean that most data will have to be generated locally. There will surely also be much duplication (such as operating systems, movies, and so on that would be stored on millions of computers). Aside from that, there will likely be huge volumes of locally generated data (such as from security cameras and medical sensors) that will be used (if at all) only in highly digested form.

7. Is There a "Moore's Law" for Data Traffic?

The examples in Section 3 support the notion that there is a "Moore's Law" for data traffic, with transmission volumes doubling each year. Even at large institutions that already have access to state-of-the art technology, data traffic to the public Internet tends to follow this rule of doubling each year. This is not a natural law, but, like all other versions of "Moore's Law," reflects a complicated process, the interaction of technology and the speed with which new technologies are absorbed.

A "Moore's Law" for data traffic is different from those in other areas, since it depends in a much more direct way on user behavior. In semiconductors, consumer willingness to pay drives the research, development, and investment decisions of the industry, but the effects are indirect. In data traffic, though, changes can potentially be much faster. A residential customer with dial modem access to the Internet could increase the volume of data transfer by a factor of about five very quickly. All it would take would be installation of one of the software packages that prefetch Web sites that are of potential interest, and which fill in the slack between transmissions initiated by the user. Similarly, a university's T3 connection to the Internet could potentially be filled by a single workstation sending data to another institution. Thus any "Moore's Law" for data traffic is by nature much more fragile than the standard "Moore's Law" for semiconductors, for example. Thus it is remarkable that we see so much regularity in growth rates of data transfers.

Links to the public Internet are usually the most expensive parts of a network, and are regarded as key choke points. They are where congestion is seen most frequently at institutional networks. Yet the "mere" annual doubling of data traffic even at institutions that have plenty of spare capacity on their Internet links means that there are other barriers that matter. The obvious one is the public Internet itself. It is often (some would say usually) congested. A terabit pipe does not help if it is hooked up to a megabit link, and so providing a lightly utilized link to the Internet does not guarantee good end-to-end performance. Yet that is not the entire explanation either, since corporate Intranets, which tend to have adequate bandwidth, and seldom run into congestion, tend to grow no faster than a doubling of traffic each year. There are other obstructions, such as servers, middleware, and, perhaps most important, services and user interfaces. People do not care about getting many bits. What they care about are applications. However, applications take time to be developed, deployed, and adopted. To quote J. Licklider (who deserves to be called "the grandfather of the Internet" for his role in setting up the research program that led to the Internet's creation),

> A modern maxim says: "People tend to overestimate what can be done in one year and to underestimate what can be done in five or ten years."
>
> (footnote on p. 17 of Licklider (1965))

"Internet time," where everything changes in 18 months, has a grain of truth, but is largely a myth. Except for the ascendancy of browsers, most substantial changes take 5 to 10 years. As an example, it is at least five years since voice over IP was first acclaimed as the "next big thing." Yet its impact so far has been surprisingly modest. It is coming, but it is not here today, and it won't be here tomorrow. People take time to absorb new technologies.

What is perhaps most remarkable is that even at institutions with congested links to the Internet, traffic doubles or almost doubles each year. Users appear to find the Internet attractive enough that they exert pressure on their administration to increase the capacity of the connection. Existing constraints, such as those on email attachments, or on packetized voice, or video, as well as the basic constraint of limited bandwidth, are gradually loosened. Note that this is similar to the process that produces the standard Moore's Law for PCs. Intel, Micron, Toshiba, and the rest of the computer industry would surely produce faster advances if users bought new PCs every year. Instead, a typical

PC is used for three to four years. On one hand there is pressure to keep expenditures on new equipment and software under control, and also to minimize the complexity of the computing and communications support job. On the other hand, there is pressure to upgrade, either to better support existing applications, or to introduce new ones. Over the last three decades, the conflict between these two pressures has produced a steady progress in computers. Similar pressures appear to be in operation in data networking.

In conclusion, we cannot be certain that Internet traffic will continue doubling each year. All we can say is that historically it has tended to double each year. Still, trends in both transmission and in other information technologies appear to provide both the demand and the supply that will allow a continuing doubling each year. Since betting against such "Moore's laws" in other areas has been a loser's game for the last few decades, it appears safest to assume that data traffic will indeed follow the same pattern, and grow at close to 100% per year.

8. Conclusions and Speculations

The main conclusion we draw from the data and arguments of the previous sections is that Internet traffic is likely to continue doubling each year for the next decade or so. We next discuss the likely implications of such a growth rate.

There have been many predictions that data traffic would grow rapidly, and that this would produce a quantum change in our information environment. That has happened, and is continuing to happen. However, some of the predictions have been too optimistic. Bill Gates predicted in 1994 that we would have "unlimited bandwidth" within a decade (see Gilder (1994)). That has not happened and will not happen. We are indeed experiencing a "tidal wave of bandwidth," as George Gilder forecast (Gilder 1994). However, that tidal wave is accompanied by other tidal waves, of processing power and especially storage. In particular, there is far more stored data than transport capacity, and this will not change materially. At today's Internet traffic rates, it would take over 20 years to transmit all the bits that are on magnetic hard drives. The precedents of Moore's laws suggest that we should expect total capacity of magnetic storage to continue doubling each year, as it has been doing for a while. If it does, and data traffic also doubles each year, as we have shown is likely, then the relation of storage capacity and transport will not change. Even if data traffic triples each year (as it might for a few years, given the spare fiber capacity that exists, and other factors we have discussed in Section 5), at the end of this decade it would still take four months to transmit all the hard disk contents over the Internet. Therefore most data will stay local.

Locality of data, as well as speed of light limitations and communication overheads will mean that information architectures will not change radically. Today it already makes sense to cache practically everything (Gray and Shenoy 2000). This will be even more true in the future. Caching and content distribution will play major roles. Yet the value of the Net is largely in the mass of data out there, most of which has little value to most people. Thus there will be tremendous value in crawlers and other aids to finding and organizing all that content.

Another general conclusion is that there will be neither a "bandwidth glut" nor a "bandwidth shortage." It appears that supply and demand will be growing at comparable rates. Thus pricing is likely to play a major role in the evolution of traffic. (As was noted in Coffman and Odlyzko (1998), Odlyzko (2000a), data transmission prices have been increasing through most of the 1990s, and have only recently showed signs of decrease. Once they start declining rapidly, many of the constraints on usage that we see today are likely to be relaxed.)

Streaming real time transmission is bound to grow in absolute volume. As a fraction of total traffic, it may increase for a while. However, eventually it is very likely to decline, as demand for this type of traffic will not be growing as fast as network capacity. Repeating what we said in Section 4, if the doubling of traffic each year is to continue, new applications will have to keep appearing and assuming dominant roles.

A doubling of traffic each year will mean that network operators will continue to scramble to meet disruptive demands of new applications. We will not have the smooth and predictable growth that has been common in other communication services.

Acknowledgements.. We thank Mark Boolootian, Michael Bruce, Doug Carson, Kim Claffy, Jim Gray, Jose-Marie Griffiths, Willi Huber, Cheng Jin, Don King, Michael Lesk, Alan McCord, Daniel McRobb, Jeff Ogden, Saverio Pangoli, Jim Pepin, Dave Plonka, Gil Press, Larry Rabiner, Roger Watt, and Dan Werthimer for comments and useful information.

Bibliography

Boardwatch magazine, 2001. `http://www.boardwatch.com`.

L. Bruno. Fiber optimism: Nortel, Lucent, and Cisco are battling to win the high-stakes fiber-optics game. *Red Herring*, June 2000. Available at `http://www.herring.com/mag/issue79/mag-fiber-79.html`.

N. Cochrane. We're insatiable: Now it's 20 million million bytes a day. *Melbourne Age*, Jan. 2001. Available at http://www.it.fairfax.com.au/networking/20010115/A13694--2001Jan15.html.

K.G. Coffman and A.M. Odlyzko. The size and growth rate of the internet. *First Monday*, Oct. 1998. http://firstmonday.org/, Also available at http://www.research.att.com/~amo.

I. de Sola Pool, H. Inose, N. Takasaki, and R. Hurwitz. *Communications Flows: A Census in the United States and Japan.* North-Holland, 1984.

L. Dunn. The Internet2 project. *The Internet Protocol Journal*, 2, Dec. 1999. Available at http://www.cisco.com/warp/public/-759/ipj_issues.html.

Not Moore's Law. *The Economist*, July 1997.

C.A. Eldering, M. L. Sylla, and J. A. Eisenach. Is there a Moore's Law for bandwidth? *IEEE Communications Magazine*, pages 2–7, Oct. 1999.

D. Galbi. Bandwidth use and pricing trends in the U.S. *Telecommunications Policy*, 24, Dec. 2000. Available at http://www.-galbithink.org.

G. Gilder. The bandwidth tidal wave. *Forbes ASAP*, Dec. 1994. Available at http://www.forbes.com/asap/gilder/telecosm10a.htm.

G. Gilder. Fiber keeps its promise: Get ready, bandwidth will triple each year for the next 25. *Forbes*, April 1997. Available at http://www.forbes.com/asap/97/0407/090.htm.

J. Gray and P. Shenoy. Rules of thumb in data engineering. In *Proc. 16th Intern. Conf. on Data Engineering*, 2000. Also available at http://research.microsoft.com/~gray.

A. Gupta, D.O. Stahl, and A.B. Whinston. The Internet: A future tragedy of the commons?, 1995. Available at http://cism.-bus.utexas.edu/res/wp.html.

J. Harms. From SWITCH to SWITCH* – Extrapolating from a case study. In *Proc. INET'94*, pages 341–1–341–6, 1994. Available at http://info.isoc.org/isoc/whatis/conferences/inet/94/papers/index.html.

J.L. Hennessy and D.A. Patterson. *Computer Architecture: A Quantitative Approach.* Morgan Kaufmann, 1990.

P.J. Howe. MCI chief sees big outlays to handle net traffic: Ebbers estimates $100B to upgrade network. *Boston Globe*, March 2000.

M. Jander. LINX to Cisco: "Good Riddance". *Light Reading*, March 2000. Available at http://www.lightreading.com/document.asp?doc_id=266.

M. Lesk. How much information is there in the world?, 1997. Available at http://www.lesk.com/mlesk/diglib.html.

J.C.R. Licklider. *Libraries of the Future.* MIT Press, 1965.

P. Lyman and H. R. Varian. How much information?, 2000. Available at http://www.sims.berkeley.edu/how-much-info/.

J. McCredie. UC Berkeley must manage campus network growth. *The Daily Californian*, 2000. Available at http://www.dailycal.org/article.asp?id=1912&ref=news.

A.M. Noll. *Introduction to Telephones and Telephone Traffic.* Artech House, 2nd edition, 1991.

A.M. Noll. Technical opinion: Does data traffic exceed voice traffic? *Comm. ACM*, 42:121–124, June 1999.

A.M. Odlyzko. Data networks are lightly utilized, and will stay that way, 1998. Available at http://www.research.att.com/~amo.

A.M. Odlyzko. The history of communications and its implications for the Internet, 2000a. Available at http://www.research.att.com/~amo.

A.M. Odlyzko. The Internet and other networks: Utilization rates and their implications. *Information Economics & Policy*, 12:341–365, 2000b.

T. Oetiker. RRDtool, 2001. http://ee-staff.ethz.ch/~oetiker/webtools/rrdtool/.

T. Oetiker and D. Rand. The Multi Router Traffic Grapher, 2001. See http://ee-staff.ethz.ch/~oetiker/webtools/mrtg/mrtg.html.

V. Paxson. Growth trends in wide-area TCP connections. *IEEE Network*, pages 8–17, 1994.

D. Plonka. UW-Madison Napster traffic measurement, 2001. Available at http://net.doit.wisc.edu/data/Napster.

P. Reichl, S. Leinen, and B. Stiller. A practical review of pricing and cost recovery for Internet services. In *Proc. 2nd Internet Economics Workshop Berlin (IEW'99)*, 1999. Available at http://www.tik.ee.ethz.ch/~cati/.

R.R. Schaller. Moore's law: Past, present, and future. *IEEE Spectrum*, 34:52–59, 1997. Available through Spectrum online search at http://www.spectrum.ieee.org.

P. Sevcik. The myth of Internet growth. *Business Communications Review*, pages 12–14, 1999. Available at http://www.bcr.com/bcrmag/01/99p12.htm.

B. St. Arnaud. The future of the Internet is NOT multimedia. *Network World*, Nov. 1997. Available at http://www.canarie.ca/~bstarn/publications.html.

B. St. Arnaud, J. Coulter, J. Fitchett, and S. Mokbel. Architectural and engineering issues for building an optical Internet, 1998. Full version available at http://www.canarie.ca/~bstarn/optical-internet.html.

S. Taggart. Telstra: The prices fight. *Wired News*, December 1999. http://www.wired.com/news/politics/0,1283,32961,00.html.

II

MASSIVE GRAPHS

Chapter 4

RANDOM EVOLUTION IN MASSIVE GRAPHS

William Aiello
AT&T Labs Research, Florham Park, NJ 07932, USA
aiello@research.att.com

Fan Chung
University of California, San Diego, La Jolla, CA 92093, USA
fan@ucsd.edu

Linyuan Lu
University of California, San Diego, La Jolla, CA 92093, USA
llu@euclid.ucsd.edu

Abstract   Many massive graphs (such as WWW graphs and Call graphs) share certain universal characteristics which can be described by the so-called the "power law". In this paper, we first briefly survey the history and previous work on power law graphs. Then we give four evolution models for generating power law graphs by adding one node/edge at a time. We show that for any given edge density and desired distributions for in-degrees and out-degrees (not necessarily the same, but adhered to certain general conditions), the resulting graph almost surely satisfy the power law and the in/out-degree conditions. We show that our most general directed and undirected models include nearly all known models as special cases. In addition, we consider another crucial aspect of massive graphs that is called "scale-free" in the sense that the frequency of sampling (w.r.t. the growth rate) is independent of the parameter of the resulting power law graphs. We show that our evolution models generate scale-free power law graphs.

Keywords: Massive graphs, Power laws, Scale invariance, Random evolution.

1. Introduction

The number of Internet hosts as of January 2000 topped 70 million and is estimated to be growing at 63% per year (NGI 2000). The number of web pages indexed by large search engines now exceeds 500 million and it is estimated that over 4,000 web sites are created everyday. Is it possible to determine simple structural properties for such massive and dynamic graphs as the Internet and the World Wide Web? For example, are these graphs connected? If not, what is the size and diameter of the largest component? Are there interesting structural properties which govern or influence the development and use of these physical and virtual networks?

Of course, answering these questions exactly is quite likely not possible. However, in many other areas of the physical, biological, and social sciences and in engineering where the size and dynamic nature of the data sets similarly do not allow for exact answers, progress in understanding has nonetheless been achieved through an iterative interplay between experimental data and modeling, where both the data and the modeling often have a random or statistical basis. Such an interplay is in its early stages for the study of several massive, dynamic graphs such as the World Wide Web. The starting point of this interplay began when several groups independently made an important observation: the degree distributions of several different massive graphs, including the WWW graph, follow a power law (Barabási and Albert 1999, Barabási et al. 1999, Kumar et al. 1999b, Broder et al. 2000). In a power law degree distribution, the fraction of nodes with degree d is proportional to $1/d^\alpha$ for some constant $\alpha \geq 0$. In this paper we present and analyze a general random graph evolution model which yields graphs with power law degree distributions. Below we first review the empirical findings for graphs with power law degree distributions followed by an overview of previous modeling work for such graphs. Then we discuss the models and results presented in this paper. In particular, we examine three important aspects of power law graphs, (1) analyzing their evolution, (2) their asymmetry of in-degrees and out-degrees, and (3) their "scale invariance."

2. History of Power Law Graphs

2.1. Early History

The history of power laws can be traced back to statistical analysis in a variety of fields, including linguistics, academic citation, physical sciences, or even in nature or economy. In 1926, Lotka (1926) plotted the distribution of authors in the decennial index of Chemical Abstracts (1907-1916), and he found that the number of authors is inversely proportional to the square of the number of papers published by those authors (which is often called *Lotka's law* or *inverse square law* and *Yule's law* (Yule 1944)). Zipf (1949) observed that the frequency of English words follows a power law function. That is, the word frequency that has rank i among all word frequencies is proportional to $1/i^a$ where a is close to 1. This is called *Zipf's law* or *Zipf's distribution*. As Simon (1957) noted in an influential paper in 1957, this distribution is also common to various phenomena, such as word frequencies in large samples of prose, city sizes and income distributions. There has been a large number of research papers on power laws in natural language (Miller et al. 1958, Rousseau and Zhang 1992, Tuldava 1996), bibliometrics (Egghe and Rousseau 1990, Fairthore 1969, Gilbert 1997, Koenig and Harrell 1995, Silagadze 1997), social sciences (Murphy 1973, Hill 1970, Makse et al. 1995) and nature (Mandelbrot 1977, Martindale and Konopka 1996, Schroeder 1991).

2.2. Empirical Power Laws

Power laws in massive graphs have recently been reported in a variety of contexts. In 1999, Kumar et al. (1999b) reported that a web crawl of a pruned data set from 1997 containing about 40 million pages revealed that the in-degree and out-degree distributions of the web followed a power law. Barabási and Albert (1999) and Barabási et al. (1999) independently reported the same phenomenon on the approximately 325 thousand node nd.edu subset of the web. Both reported a power of approximately 2.1 for the in-degree power law and 2.7 for the out-degree (although the degree sequence for the out-degree deviates from the power law for small degrees). More recently, these figures have been confirmed for a Web crawl of approximately 200 million nodes (Broder et al. 2000). Thus, the power law fit of the degree distribution of the Web appears to be remarkably stable over time and scale.

Faloutsos et al. (1999) have also observed a power law for the degree distribution of the Internet network. They reported that the distribution of the out-degree for the interdomain routing tables fits a power law with a power of approximately 2.2 and that this power remained the same over several different snapshots of the network. At the router level, the out-degree distribution for a single snapshot in 1995 followed a power law with a power of approximately 2.6.

In addition to the Web graph and the Internet graph, several other massive graphs exhibit a power law for the degree distribution. The graph derived from telephone calls during a period of time over one or more carriers' networks is called a call graph. Using data collected by Abello et al. (1998), Aiello et al. (2000) observe that their call graphs are power law graphs. Both the in-degrees and the out-degrees have a power of 2.1. The graphs derived from the U.S. power grid and from the co-stars graph of actors (where there is an edge between two actors if they have appeared together in a movie) also obey a power law (Barabási and Albert 1999). Thus, a power law fit for the degree distribution appears to be a ubiquitous and robust property for many massive real-world graphs.

2.3. Modeling Power Law Graphs

As discussed above, many of the graphs above are so large and dynamic that answering simple structural questions exactly by empirical means is very difficult or infeasible. It is important, therefore, to develop models which match empirically observed behavior and yet are themselves amenable to structural analysis. Good models often guide further empirical analysis which often subsequently requires the models to be refined, and so on.

To begin our discussion of modeling power law graphs, first note that the standard random graph models, $\mathcal{G}(n,p)$, $\mathcal{G}(n,|E|)$, and $\tilde{\mathcal{G}}^n$, will not suffice (see, for example, Bollobás (1998)). In these models, the choice of edges have a high degree of independence. Hence, the distribution of degrees decays exponentially from the expected or average degree.

In order for a power law degree distribution to emerge, the choice of edges must be correlated. To achieve this correlation, two basic approaches have been taken thus far. We review them in turn. The first basic approach is exemplified in Aiello et al. (2000). They do not attempt to explain how graphs with a power law degree distribution arise. Rather, they focus on classes of graphs with a power law degree distribution and they derive the structures and properties (such as connected components (Aiello et al. 2000), diameters (Lu 2001), etc.) as a function of the power. Newman et al. (2001) take a similar approach but use different methods of analysis. Certain questions are likely to prove more amenable to analysis using the later approach than the former and vice versa. Thus, the two approaches are complementary.

The second approach to modeling power law graphs attempts to model the evolution of such graphs and the manner in which the power law degree distribution arises. We briefly overview the history along the following three aspects of power law graphs, (1) their evolution, (2) their asymmetry of in-degrees and out-degrees, and (3) the "scale-free" phenomenon.

2.3.1 The evolution of power law graphs.

Barabási and Albert (1999) describe the following graph evolution process. They start with a small initial graph. At each time step they add a new node and an edge between the new node and each of m random nodes in the existing graph, where m is a parameter of the model. The random nodes are not chosen uniformly. Instead, the probability of picking a node is weighted according to its existing degree (the edges are assumed to be undirected). That is, if there are e_t edges at time t and node v has degree $\delta_{v,t}$ at time t, then the probability of picking node v is $\delta_{v,t}/2e_t$. Using heuristic analysis (e.g., the analysis assumes that the discrete degree distribution is differentiable) they derive a power law for the degree distribution with a power of 3, regardless of m. Clearly, the fact that the power is 3 regardless of the parameter m is a drawback of the model. Moreover, it can easily be shown that all of edges (except, perhaps, those of the small initial graph) of a resulting graph can be decomposed into m disjoint forests (i.e., the graph has arboricity m). Presumably, most massive real-world graphs with power law degree distributions have a richer structure than this.

The main intuition behind the development of a power law degree distribution for this model is as follows. Nodes which acquire a relatively large degree early on in the process have an "advantage" and continue to accumulate added degree because of the preferential selection of nodes with high degree. Barabasi and Albert show that if the preferential selection of high degree nodes is replaced by a uniform selection of nodes then the power law behavior of the degree distribution does not result. Moreover, if the number of nodes is fixed, as opposed to constantly increasing, then the power law degree distribution again fails to occur.

Kumar et al. also describe a random graph evolution process (Kumar et al. 1999a). Unlike that of Barabási and Albert (1999), their random graphs are directed. Their model has the advantage that the power in the power-law is a function of a parameter of the model. Their model is as follows. A node and an edge are added at every time step. With probability $1 - \alpha$, a directed self-loop is added to the new node. With probability α, an edge is added from the new node to a randomly selected node. The node is selected in proportion to its current in-degrees. That is, if there are t edges at time t, the probability of picking node v at time t is $\delta^{in}_{v,t}/t$ where $\delta^{in}_{v,t}$ is the in-degree of v at time t. They analyze this evolution process with a heuristic analysis and they derive a power law for the degree distribution with a power of $1/\alpha$. As we will see, this model is a special case of our general model for which our analysis yields a power of $1 + 1/\alpha$. The above model has a similar drawback as that of Barabási and Albert (1999): the resulting random graph is a tree.

2.3.2 Asymmetry of in-degrees and out-degrees.

Kumar et al. (1999a) provide a general model which they call the (α, β) model which has the advantage that the in-degree and the out-degree both follow a power law. The powers in the power law for the in-degree and out-degree need not be the same; they can be controlled independently by α and β. As before a node and an edge are added at every time step. Let w_t be the node added at step t. At each time step, two nodes are chosen from the existing graph. Node u is selected according to its out degree, i.e., the probability that u is chosen is $\delta_{u,t}^{\text{out}}/t$. Node v is selected according to its in degree, i.e., the probability that v is chosen is $\delta_{v,t}^{\text{in}}/t$. Then two coins are tossed. The "origin" coin is "u" with probability α and "w_t" with probability $1 - \alpha$. The "destination" coin is "v" with probability β and "w_t" with probability $1 - \beta$. The new edge is added from the outcome of the origin coin to the outcome of the destination coin. That is, an edge is added from: u to v with probability $\alpha\beta$; from u to w_t with probability $\alpha(1 - \beta)$; from w_t to v with probability $(1 - \alpha)\beta$, and from w_t to w_t with probability $(1 - \alpha)(1 - \beta)$. They claim an out-degree power law with a power of $1/\alpha$ and an in-degree power law with a power of $1/\beta$. (As with their first model, the (α, β) model is a special case of our model. Our analysis yields power laws with powers $1 + 1/\alpha$ and $1 + 1/\beta$ for the out-degree and in-degree, respectively.)

While the above model allows for different powers laws for the in-degree and out-degree and yields graphs which which do not have small arboricity, it has the following restrictive property. Suppose that at time step t, the origin and destination coins are w_t and v, respectively. In this case, w_t will have out-degree 1 and in-degree 0 at time $t + 1$. Hence, w_t cannot be chosen as node v in time step $t + 1$ and thus its in-degree will be 0 at time $t + 2$. Continuing in this manner, w_t will always have in-degree 0. Thus, with high probability, a constant fraction (approximately $(1 - \alpha)\beta$) of the nodes will have in-degree 0. Likewise, with high probability, a constant fraction (approximately $\alpha(1-\beta)$) of the nodes will have out-degree 0. While some real-world power law graphs may have this property, it is likely that some, e.g., the Web, do not. Also note that this model is restricted to graphs with density 1 since one node and one edge are added at every time step.

Recently, Kumar et al. (2000) proposed three evolution models — "linear growth copying", "exponential growth copying", and "linear growth variants". The *Linear growth coping* model adds one new vertex with

d out-links at a time. The destination of the i-th out-link of the new vertex is either copied from the corresponding out-link of a "prototype" vertex (chosen randomly) or a random vertex. They showed that the in-degree sequence follows the power law. These models were designed explicitly to model the World Wide Web. Indeed, they show that their model has a large number of complete bipartite subgraphs, as has been observed in the WWW graph, whereas several other models, including that of Aiello et al. (2000), do not. But this (and the linear growth variants model) has the similar drawback as the first model in Kumar et al. (1999a). The out-degree of every vertex is always a constant. Edges and vertices in the *exponential growth copying* model increase exponentially. This exponential growth copying model does not have the same drawback as the other two models have. However, it is not clear whether its out-degrees satisfy the power law distribution.

2.3.3 Scale-free property for power law graphs. Power-laws or heavy tailed distributions are often associated with self-similarity and scaling laws. Indeed, by comparing the web crawls of Barabási and Albert (1999), Barabási et al. (1999), Broder et al. (2000), and Kumar et al. (1999b), we see that the same power law appears to govern various subgraphs of the web as well as the whole. However, while some subgraphs obey the same power law and appear to be self-similar, clearly, there exists subgraphs of the web which would not obey the power law (e.g., the subgraph defined by all nodes with out-degree 100). The natural problem is thus: formally define and analyze a scale-free property for power law graphs. While there may be several types of scaling behavior exhibited by power law graphs, to the best of our knowledge, we give the first such definition and show that our model exhibits this scale-free property.

3. Our Results

Below we will describe a sequence of graph evolution models. The first three, Models A, B, and C, are for directed graphs and are increasingly more general. The first two are primarily illustrative although they may have merits as models in their own right due to their parsimony. Model C incompasses all of the directed graph models above, except that of Kumar et al. (2000). We also describe a fourth model, Model D, which is the natural analogue of Model C for undirected graphs.

Consider the following simple model which we call model A. At each time step, a new node is added with probability $1 - \alpha$. The node starts with in-weight 1 and out-weight 1. Whenever the node is the origin (destination) of a new edge, the out-weight (in-weight) is increased by

1. That is, the in-weight (out-weight) of a node u at time t is just $w_{u,t}^{\text{in}} = 1 + \delta_{u,t}^{\text{in}}$ ($w_{u,t}^{\text{out}} = 1 + \delta_{u,t}^{\text{out}}$). With probability α a random edge is added to the existing nodes. The origin (destination) of the new edge is chosen proportional to the current in-weights (out-weights) of the nodes. That is, u (v) is chosen as the origin (destination) of the new edge at time t with probability $w_{u,t}^{\text{out}}/t$ ($w_{v,t}^{\text{in}}/t$). Note that the expected number of edges in the graph is α and the expected number of nodes is $1 - \alpha$. Call the ratio of the former to the latter $\Delta = \alpha/(1-\alpha)$ as it is a measure of the density of the graph. As a corollary to our general result, we will show that this model yields a power law with power $2 + 1/\Delta$ for both the in-degrees and the out-degrees. Thus, this model allows for graphs of varying density. For this model we also derive the *joint* distribution for the in-degrees and out-degrees. We show that the number of nodes with in-degree i and out-degree j is proportional to $(i+j)^{3+1/\Delta}$.

Note that when an edge is added among existing nodes, the probabilities concerning which edge is added are functions of the current degree distribution. Thus, the probability distribution of the new degree distribution is a function of the current degree distribution. This is difficult to solve recursively since the current degree distribution, itself, has a probability distribution. However, this means that the expected value of the new degree distribution is a function of the current degree distribution. Moreover, as we will see, the change in the degree distribution from step to step is bounded. Thus, we observe that the evolution of the degree distribution is a semi martingale where deviation from the expected value of the final degree distribution occurs with exponentially small tails. Due to linearity of expectation, we are able to solve for the expected value of the final degree distribution recursively. These recursive equations and their solutions are non-standard, to the best of our knowledge, and may be of independent interest.

One drawback of model A is that the density parameter Δ and the power in the power law cannot be controlled independently. They are both functions of the parameter α. Moreover, the in and out degree have the same power. A simple modification to model A yields model B which overcomes both drawbacks. When a new node is added with probability $1 - \alpha$ at a time step, it will be given in-weight γ^{in} and out-weight γ^{out}. Thus, the in-weight (out-weight) of a node u at time t is just $w_{u,t}^{\text{in}} = \gamma^{\text{in}} + \delta_{u,t}^{\text{in}}$ ($w_{u,t}^{\text{out}} = \gamma^{\text{out}} + \delta_{u,t}^{\text{out}}$). As before, when an edge is added with probability α, the origin of the edge is chosen with probability proportional to the current out-weights and the destination is chosen with probability proportional to the current in-weights. We show that this graph evolution process yields graphs with power law degree distributions with powers $2 + \gamma^{\text{in}}/\Delta$, and $2 + \gamma^{\text{out}}/\Delta$ for the in-degrees and out-degrees, respectively. Note that the powers for the in-degrees and out-degrees and the density can all be controlled separately. This is the simplest model of which we are aware for which this is the case. Moreover, the model does not suffer from any of the other drawbacks mentioned above such as small arboricity or a constant fraction of nodes with no incoming edges.

While the above model may indeed be the simplest with which to model a real-world power law graph on the basis of measurements of the density of the graph and the powers for the in-degrees and out-degrees, it may not capture other features of the graph which are measurable. Hence, we would also like a more general model which, for example, would include the above model as well as that of Kumar et al. (1999a). Consider now model C. Suppose that at each time step four numbers $m^{e,e}, m^{n,e}, m^{e,n}, m^{n,n}$ are drawn according to some probability distribution to be specified later. We assume that the four random variables are bounded. These four random variables need not be independent. In this time step $m^{e,e}$ edges are added between existing nodes in the graph. Of course, as before, the origin and destination of these edges are chosen independently according to the current out-degrees and in-degrees, respectively. Likewise, $m^{n,e}$ edges are added from the new node to existing nodes chosen independently according to the current in-degrees. Likewise, $m^{e,n}$ edges are added from existing nodes (chosen independently according to the current out-degrees) to the new node. Finally, $m^{n,n}$ directed self loops are added to the new node. Of course, each of these random variables has a well-defined expectation which we denote $\mu^{e,e}, \mu^{n,e}, \mu^{e,n}, \mu^{n,n}$, respectively. We show that this general process still yields a power law degree distribution. We derive a power of $2+(\mu^{n,n}+\mu^{n,e})/(\mu^{e,n}+\mu^{e,e})$ for the out-degree. Consider the rightmost ratio in this expression. By definition, the first element of a superscript refers to the origination of the random edges. Notice that both terms in the numerator have the new node as the origination and both terms in the denominator have existing nodes as the origination. We also derive a power of $2+(\mu^{n,n}+\mu^{e,n})/(\mu^{n,e}+\mu^{e,e})$ for the in-degree. Analogously to the above expression, notice that in the rightmost ratio of this expression both terms in the numerator represent the new node being the destination of the random edges and both terms of the denominator represent existing nodes being the destination of random edges.

Note that the first simple model of Kumar et al. (1999a) has $\mu^{n,e} = \alpha, \mu^{n,n} = 1-\alpha$ and $\mu^{e,e} = \mu^{e,n} = 0$. Substituting this into our result gives an in-degree power of $2+(1-\alpha)/\alpha = 1+1/\alpha$. The (α, β) model of Kumar et al. (1999a) gives $\mu^{e,e} = \alpha\beta, \mu^{n,e} = (1-\alpha)\beta, \mu^{e,n} = \alpha(1-\beta), \mu^{n,n} = (1-\alpha)(1-\beta)$. Using our general results this gives an out-degree power of $1 + 1/\alpha$ and an in-degree power of $1 + /\beta$. Also note that our model A has $\mu^{e,e} = \alpha$, $\mu^{e,n} = \mu^{n,e} = 0$ and $\mu^{n,n} = 1 - \alpha$. This yields a power of $1 + 1/\alpha$, as claimed, for both the in- and out-degrees. Model C can easily be generalized to include the parameters of the initial weights of the new nodes given in Model B but we omit that here.

Finally, we also describe a general undirected model which we denote Model D. It is a natural variant of Model C. At each time step three numbers $(m^{e,e}, m^{n,e}, m^{n,n})$ are drawn according to some probability distribution. We assume that the three random variables are bounded. In this time step $m^{e,e}$ undirected edges are added between existing nodes in the graph. The endpoints of these edges are chosen independently according to the current total degrees. Likewise, $m^{n,e}$ edges are added between the new node and existing nodes chosen independently according to the current total degrees. Finally, $m^{n,n}$ undirected self loops are added to the new node. We show that this undirected graph evolution process also yields a power law degree distribution. We derive a power of $2 + (2\mu^{n,n} + \mu^{n,e})/(\mu^{n,e} + 2\mu^{e,e})$. Note that the model of Barabási and Albert (1999) has $\mu^{n,n} = \mu^{e,e} = 0$ and $\mu^{n,e} = m$. Substituting this into our general result gives a power of 3 which matches their heuristically derived bound. Note that the natural undirected version of model A has $\mu^{n,e} = 0$ and thus a power of $2 + \mu^{n,n}/\mu^{e,e} = 1 + 1/\alpha$. As with model C, initial weights can easily be incorporated into Model D.

We remark that our conditions for Model C and D are much weaker than the previous known models. For example, previous known models assume that the way in which edges are added are identical at each time. In our models, we only need to assume edges are added in an "asymptotically similar" way.

Scale Invariance. The evolution of massive graphs can be viewed as a process of growing graphs by adding nodes and edges at a time. One way is to divide the time into almost equal units and combine all nodes born in the same unit time into one super-node. The bigger time unit one chooses, the smaller size of the result graph has. This procedure is similar to scaling maps in space. The property is called *scale-free*. A model is called *scale-free* if it generates the scale-free power graphs with high probability. In other words, an evolution model is time scale invariant if we change the time scale by any given factor and examine the scaled graph, then the original graph and the scaled graph should satisfy the power law with the same powers for the in-degrees and out-degrees. Suppose that a "unit" of time is scaled by a factor of c. In other words, we combine all nodes born in previous c-units into one super node. This has the same effect as adding edges c-times in a large one unit. A detailed definition will be given below.

Briefly, we scale time in our model and then show that the degree distribution of Model C is invariant with respect to the time scaling. To begin the discussion, consider a Model C evolution process with parameters $\mu^{n,n}$, $\mu^{n,e}$, $\mu^{e,n}$, and $\mu^{e,e}$ and a bound B on the number of edges added per time step. Suppose the evolution process is run for T time steps and let G_T be the graph generated. Label nodes by the time step in which they are added to the graph. To scale this evolution process by a factor of σ, we begin by aggregating time steps into super steps of σ consecutive time steps. That is, super-step 1 consists of time steps 1 through σ, super-step 2 consists of times steps $\sigma+1$ through 2σ, and so on (where we assume for convenience that σ divides T). The scaled graph $H_\sigma(G_T)$ is created from G_T as follows. All nodes of G_T with a time label in super step τ (i.e., $\sigma(\tau-1)+1,\ldots,\sigma\tau$) are identified with as single node with super step label τ. (If there is no node in G_T with time label in super step τ then no node is created in G_T^σ with label τ.) An edge in G_T from node i to node j gets mapped to an edge in $H_\sigma(G_T)$ from node $\lceil i/\sigma \rceil$ to node $\lceil j/\sigma \rceil$. The morphism H_σ on this evolution process of Model C defines a natural evolution process, which, strictly speaking, is not covered by Model C. Nonetheless, we will show that this evolution process has the same power law asymptotically as a Model C evolution process with parameters $\mu'^{n,n} = \sigma\mu^{n,n}$, $\mu'^{n,e} = \sigma\mu^{n,e}$ $\mu'^{e,n} = \sigma\mu^{e,n}$ $\mu'^{e,e} = \sigma\mu^{e,e}$ and size bound σB. Given our general results on Model C, the latter Model C process has the same power law as the first Model C process (e.g., the power for the out-degree is $2 + (\mu^{n,n} + \mu^{n,e})/(\mu^{e,n} + \mu^{e,e})$) and therefore the time scaled process defined by the morphism H_σ has the same power law as the first Model C process. Thus, the power law degree distribution of a Model C evolution process is invariant with respect to the time scaling defined above.

The rest of the paper is organized as follows. In section 4, we will define Models A,B,C,D, and state our Theorems (Theorems 1,2,3,4) on the power law degree distribution of these models. We also state the scale-free property of these models (Theorem 5). In section 5, we prove Theorem 1 while Theorem 3 and 5 are proved in Section 6. The proofs of Theorems 2 and 4 are omitted.

4. A General Graph Evolution Process

4.1. Notations and Definitions

In Aiello et al. (2000), a random graph model was proposed which can be viewed as a special case of sparse random graphs with given degree sequences. This model involves two parameters, called logsize, denoted by α and log-log growth rate, denoted by β. A random graph under consideration has the following degree distribution: Suppose there are y vertices of degree x where x and y satisfy

$$\log y = \alpha - \beta \log x$$

Such graphs are called *power law graphs* with parameters (α, β). As it turns out that the parameters capture some universal characteristics of massive graphs. Furthermore, from these parameters, various properties of the graph can be derived. For example, for certain ranges of the parameters, we can compute the expected distribution of the sizes of the connected components which almost surely occur with high probability (Aiello et al. 2000).

For a directed graph, the in-degree and out-degree sequence may follow power laws with different powers, as shown in massive graphs such as the Web graphs.

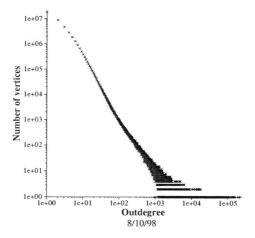

Figure 4.1. The in-degree of call graph

A robust way to generate a power law graph is to consider a random process, which grows the graph by adding one node and some edges at a time. Now we will give the definition of four models.

4.1.1 Model A.
Model A is the basic model which the subsequent models rely upon. It starts with no node and no edge at time 0. At time 1, a node with in-weight 1 and out-weight 1 is added. At time $t+1$, with probability $1-\alpha$ a new node with in-weight 1 and out-weight 1 is added. With probability α a new directed edge uv is added to the

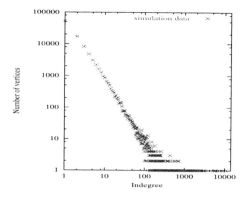

Figure 4.2. Simulation using model A

existing nodes. Here the origin u is chosen with probability proportional to the current out-weight $w_{u,t}^{out} \stackrel{def}{=} 1 + \delta_{u,t}^{out}$ and the destination v is chosen with probability proportional to the current in-weight $w_{v,t}^{in} \stackrel{def}{=} 1 + \delta_{v,t}^{in}$. We note that $\delta_{u,t}^{out}$ and $\delta_{v,t}^{in}$ denote the out-degree of u and the in-degree of v at time t, respectively.

The total in-weight (out-weight) of graph in model A increases by 1 at a time. At time t, both total in-weight and total out-weight are exactly t. So the probability that a new edge is added onto two particular nodes u and v is exactly

$$\alpha \frac{(1 + \delta_{u,t}^{out})(1 + \delta_{v,t}^{in})}{t^2}.$$

The complete analysis will be given completely in next section.

4.1.2 Model B.
Model B is a slight improvement of Model A. Two additional positive constant γ^{in} and γ^{out} are introduced. Different powers can be generated for in-degrees and out-degrees. In addition, the edge density can be independently controlled.

Model B starts with no node and no edge at time 0. At time 1, a node with in-weight γ^{in} and out-weight γ^{out} is added. At time $t + 1$, with probability $1 - \alpha$ a new node with in-weight γ^{in} and out-weight γ^{out} is added. With probability α a new directed edge uv is added to the existing nodes. Here the origin u (destination v) is chosen proportional to the current out-weight $w_{u,t}^{out} \stackrel{def}{=} \gamma^{out} + \delta_{u,t}^{out}$ while the current in-weight is $w_{v,t}^{in} \stackrel{def}{=} \gamma^{in} + \delta_{v,t}^{in}$. Here $\delta_{u,t}^{out}$ is the out-degree of u and $\delta_{v,t}^{in}$ is the in-degree of v at time t, respectively.

In model B, at time t the total in-weight w_t^{in} and the out-weight w_t^{out}) of the graph are random variables. The probability that a new edge is added onto two particular nodes u and v is

$$\alpha \frac{(\gamma^{out} + \delta_{u,t}^{out})(\gamma^{in} + \delta_{v,t}^{in})}{w_t^{in} w_t^{out}}.$$

4.1.3 Model C.
Now we consider Model C, this is a general model with four specified types of edges to be added.

Assume that the random process of model C starts at time t_0. At $t = t_0$, we start with an initial directed graph with some vertices and edges. At step $t > t_0$, a new vertex is added and four numbers $m^{e,e}, m^{n,e}, m^{e,n}, m^{n,n}$ are drawn according to some probability distribution. (Indeed, any bounded distribution is allowed here. It can even be a function of time t as long as the limit distribution exists as t approaches infinity.) We assume that the four random variables are bounded. Then we proceed as follows:

- Add $m^{e,e}$ edges randomly. The origins are chosen with the probability proportional to the current out-degree and the destinations are chosen proportional to the current in-degree.

- Add $m^{e,n}$ edges into the new vertex randomly. The origins are chosen with the probability proportional to the current out-degree and the destinations are the new vertex.

- Add $m^{n,e}$ edges from the new vertex randomly. The destinations are chosen with the probability proportional to the current in-degree and the origins are the new vertex.

- Add $m^{n,n}$ loops to the new vertex.

Each of these random variables has a well-defined expectation which we denote by $\mu^{e,e}, \mu^{n,e}, \mu^{e,n}, \mu^{n,n}$, respectively. We will show that this general process still yields power law degree distributions and the powers are simple rational functions of $\mu^{e,e}, \mu^{n,e}, \mu^{e,n}, \mu^{n,n}$.

4.1.4 Model D.
Model A, B and C are all power law models for directed graphs. Here we describe a general undirected model which we denote by Model D. It is a natural variant of Model C.

We assume that the random process of model C starts at time t_0. At $t = t_0$, we start with an initial undirected graph with some vertices and edges. At step $t > t_0$, a new vertex is added and three numbers $m^{e,e}, m^{n,e}, m^{n,n}$ are drawn according to some probability distribution.

We assume that the three random variables are bounded. Then we proceed as follows:

- Add $m^{e,e}$ edges randomly. The vertices are chosen with the probability proportional to the current degree.
- Add $m^{e,n}$ edges randomly. One vertex of each edge must be the new vertex. The other one is chosen with the probability proportional to the current degree.
- Add $m^{n,n}$ loops to the new vertex.

4.1.5 General notations. For all (directed) graph models A, B, C, D, we denote n_t to be the number of vertices at time t. Let e_t be the number of edges at time t.

For graph models A, B, C, let $d_{i,t}^{in}$ and $d_{j,t}^{out}$ denote the random variables as the number of vertices with in-degree i and out-degree j, respectively. Let $d_{i,j,t}^{joint}$ be the random variable as the number of vertices with in-degree i and out-degree j.

For (undirected) graph model D, let $d_{i,t}$ denote the random variable as the number of vertices with degree i.

4.2. Results and Applications

We first state the theorems that will be proved in latter sections.

Theorem 1 *For model A, the distribution of in-degree and out-degree sequences follow the power law distribution with power $1 + \frac{1}{\alpha}$. The joint distribution of in-degree and out-degree sequence follows the power law distribution with power $2 + \frac{1}{\beta}$. More precisely, we have*

$$Pr(|d_{i,j,t}^{joint} - a_{i,j}t| > \lambda\sqrt{t} + 2) < e^{-\lambda^2/8},$$

$$Pr(|d_{i,t}^{in} - b_i t| > \lambda\sqrt{t} + 2) < e^{-\lambda^2/2},$$

$$Pr(|d_{j,t}^{out} - c_j t| > \lambda\sqrt{t} + 2) < e^{-\lambda^2/2}.$$

where $a_{i,j}, b_i, c_j$ are constants satisfying

$$a_{i,j} = (1-\alpha)\frac{(i+j-2)!\alpha^{i+j-2}}{\prod_{l=2}^{i+j}(1+l\alpha)} = \frac{(\frac{1}{\alpha}-1)\Gamma(\frac{1}{\alpha}+2)}{(i+j)^{\frac{1}{\alpha}+2}} + o_{i+j}(1)$$

$$b_i = (1-\alpha)\frac{(i-1)!\alpha^{i-1}}{\prod_{l=1}^{i}(1+l\alpha)} = \frac{(\frac{1}{\alpha}-1)\Gamma(\frac{1}{\alpha}+1)}{i^{\frac{1}{\alpha}+1}} + o_i(1)$$

$$c_j = (1-\alpha)\frac{(j-1)!\alpha^{j-1}}{\prod_{l=1}^{j}(1+l\alpha)} = \frac{(\frac{1}{\alpha}-1)\Gamma(\frac{1}{\alpha}+1)}{j^{\frac{1}{\alpha}+1}} + o_j(1)$$

Theorem 2 *For model B, the distribution of in-degree sequence follows the power law distribution with power $2 + \frac{\gamma^{in}}{\Delta}$, and the distribution of out-degree sequence follows the power law distribution with power $2 + \frac{\gamma^{out}}{\Delta}$. Here $\Delta = \frac{\alpha}{1-\alpha}$ is the asymptotic edge density. More precisely, we have*

$$Pr(|d_{i,t}^{in} - b_i't| > 2\lambda\sqrt{t}) < e^{-\lambda^2/2},$$

$$Pr(|d_{j,t}^{out} - c_j't| > 2\lambda\sqrt{t}) < e^{-\lambda^2/2}.$$

where b_i', c_j' are constants satisfying

$$\begin{aligned}
b_i' &= (1-\alpha)(\frac{1}{\gamma^{in}} + \frac{1}{\Delta})\prod_{l=1}^{i+1}\frac{l - 2 + \gamma^{in}}{l + \frac{\gamma^{in}}{\alpha}} \\
&= (1-\alpha)(\frac{1}{\gamma^{in}} + \frac{1}{\Delta})\frac{\Gamma(\frac{\gamma^{in}}{\alpha} + 1)}{\Gamma(\gamma^{in} - 1)}\frac{1}{i^{\frac{\gamma^{in}}{\Delta}+2}} + o_i(1) \\
c_j' &= (1-\alpha)(\frac{1}{\gamma^{out}} + \frac{1}{\Delta})\prod_{l=1}^{j+1}\frac{l - 2 + \gamma^{out}}{l + \frac{\gamma^{out}}{\alpha}} \\
&= (1-\alpha)(\frac{1}{\gamma^{out}} + \frac{1}{\Delta})\frac{\Gamma(\frac{\gamma^{out}}{\alpha} + 1)}{\Gamma(\gamma^{out} - 1)}\frac{1}{j^{\frac{\gamma^{out}}{\Delta}+2}} + o_j(1)
\end{aligned}$$

Theorem 3 *For model C, almost surely the out-degree sequence follows the power law distribution with the power $2 + \frac{\mu^{n,n}+\mu^{n,e}}{\mu^{e,n}+\mu^{e,e}}$ where μ's are as defined in 2.1.3.) Almost surely the in-degree sequence follows the power law distribution with the power $2 + \frac{\mu^{n,n}+\mu^{e,n}}{\mu^{n,e}+\mu^{e,e}}$. More precisely, we have*

$$Pr(|d_{i,t}^{in} - b_i''t| > 2M\lambda\sqrt{t}) < e^{-\lambda^2/2},$$

$$Pr(|d_{j,t}^{out} - c_j''t| > 2M\lambda\sqrt{t}) < e^{-\lambda^2/2}.$$

where b_i'', c_j'' satisfy

$$b_i'' = \frac{b''}{i^{2+\frac{\mu^{n,n}+\mu^{e,n}}{\mu^{n,e}+\mu^{e,e}}}} + o_i(1),$$

$$c_j'' = \frac{c''}{j^{2+\frac{\mu^{n,n}+\mu^{e,n}}{\mu^{n,e}+\mu^{e,e}}}} + o_j(1).$$

Here b'', c'', M are constants determined by the joint distribution of $m^{e,e}$, $m^{n,e}, m^{e,n}, m^{n,n}$ of this model, but independent of i and t. (See the proof in section 4 for definitions of b'', c'', M.)

Theorem 4 *For model D, almost surely the degree sequence follows the power law distribution with the power $2 + \frac{2\mu^{n,n} + \mu^{n,e}}{\mu^{n,e} + 2\mu^{e,e}}$. More precisely, we have*

$$Pr(|d_{i,t}^{in} - a_i't| > 2M'\lambda\sqrt{t}) < e^{-\lambda^2/2},$$

where a_i' satisfies

$$a_i' = \frac{a'}{i^{2 + \frac{2\mu^{n,n} + \mu^{n,e}}{\mu^{n,e} + 2\mu^{e,e}}}} + o_i(1).$$

Here a', M' are constants determined by distribution of $(m^{e,e}, m^{n,e}, m^{n,n})$ of this model, but independent of i and t.

Theorem 3 has an important application on "Scale-free" property.

Theorem 5 *Model A, B, C, D are scale-free. Especially almost all previous models (Barabási and Albert 1999, Barabási et al. 1999, Kumar et al. 1999b;a) are scale-free.*

5. Proof of Theorem 1

For models A,B,C,D, we denote $\mathcal{G}_t$ the probability space associated to each graph G_t at time t. As t increases, $\mathcal{G}_t$ can be defined recursively. For each t, let τ_t be a random variable of $\mathcal{G}_t$.

$\{\tau_t\}$ is said to satisfy the *c-Lipschitz condition.* if

$$|\tau_{t+1}(H_{t+1}) - \tau_t(H_t)| \leq c$$

whenever H_{t+1} is obtained from H_t by adding some edges or some vertices at time $t+1$.

This concept is very similar to the vertex or edge Lipschitz condition in classical random graph theory (see Alon and Spencer (1992)). We use the following fact which is from the standard martingale theory.

Lemma 1 *If τ satisfies the c-Lipschitz condition, then we have for every $\lambda > 0$*

$$Pr[|\tau_t - E(\tau_t)| < \lambda\sqrt{t}] < 2e^{-\frac{\lambda^2}{2c^2}}$$

In particular, τ_t is almost surely very close to its expected value $E(\tau_t)$ with an error term $o(t^{\frac{1}{2}+\varepsilon})$ for any $\varepsilon > 0$, as t approaches infinity.

Proof of Theorem 1: Both $\{d_{i,t}^{in}\}$ and $\{d_{j,t}^{out}\}$ satisfy 1-Lipschitz condition. $\{d_{i,j,t}^{joint}\}$ satisfies 2-Lipschitz condition. By Lemma 1, it is enough to compute the corresponding expected values. Here we compute $E(d_{i,j,t}^{joint})$ in detail.

At time 0, there is nothing in graph. At time 1, a node with a loop is added. So we have
$$d^{joint}_{1,1,1} = 1 \text{ and } d^{joint}_{i,j,1} = 0 \text{ for } i > 1 \text{ or } j > 1$$

$i = 1, j = 1$ is special. For $t \geq 1$, we have

$$d^{joint}_{1,1,t+1} = \begin{cases} d^{joint}_{1,1,t} + 1 & \text{w.p. } 1 - \alpha \\ d^{joint}_{1,1,t} - 1 & \text{w.p. } \alpha(2\frac{d^{joint}_{1,1,t}}{t}(1 - \frac{d^{joint}_{1,1,t}}{t}) + \frac{d^{joint}_{1,1,t}}{t^2}) \\ d^{joint}_{1,1,t} - 2 & \text{w.p. } \alpha((\frac{d^{joint}_{1,1,t}}{t})^2 - \frac{d^{joint}_{1,1,t}}{t^2}) \\ d^{joint}_{1,1,t} & \text{otherwise} \end{cases}$$

In general, we have

$$d^{joint}_{i,j,t+1} = \begin{cases} d^{joint}_{i,j,t} + 2 & \text{w.p. } \frac{(i-1)d^{joint}_{i-1,j,t}}{t}\frac{(j-1)d^{joint}_{i,j-1,t}}{t} \\ d^{joint}_{i,j,t} + 1 & \text{w.p. } \alpha\frac{(i-1)d^{join}_{i-1,j,t}}{t}(1 - \frac{(j-1)d^{join}_{i,j-1,t}}{t}) \\ & \quad + \alpha\frac{(j-1)d^{join}_{i,j-1,t}}{t}(1 - \frac{(i-1)d^{join}_{i-1,j,t}}{t}) + \alpha\frac{(i-1)(j-1)d^{join}_{i-1,j-}}{t^2} \\ d^{joint}_{i,j,t} - 1 & \text{w.p. } \alpha\frac{id^{join}_{i,j,t}}{t}(1 - \frac{jd^{join}_{i,j,t}}{t}) + \alpha\frac{jd^{join}_{i,j,t}}{t}(1 - \frac{id^{join}_{i,j,t}}{t}) + \alpha^2 \\ d^{joint}_{i,j,t} - 2 & \text{w.p. } \alpha\frac{ij(d^{join}_{i,j,t})^2}{t^2} - \alpha\frac{ijd^{join}_{i,j,t}}{t^2} \\ d^{joint}_{i,j,t} & \text{otherwise} \end{cases}$$

Let $N_t = (d^{joint}_{i,j,t})_{\text{all } i,j}$ denote the degree distribution at time t. We have

$$E(d^{joint}_{1,1,t+1}|N_t) = d^{joint}_{1,1,t} + 1 - \alpha - \alpha(\frac{2}{t} - \frac{1}{t^2})d^{joint}_{1,1,t}$$

For $(i, j) \neq (1, 1)$, similarly, we have

$$E(d^{joint}_{i,j,t+1}|N_t) = d^{joint}_{i,j,t} + \frac{\alpha}{t}((i-1)(1 - \frac{j}{t})d^{joint}_{i-1,j,t} + (j-1)(1 - \frac{i}{t})d^{joint}_{i,j-1,t} - (i + j - \frac{ij}{t})d^{joint}_{i,j,t})$$

Hence we have the following recurrence formula:

$$E(d^{joint}_{1,1,t+1}) = E(d^{joint}_{1,1,t})(1 - \alpha(\frac{2}{t} - \frac{1}{t^2})) + 1 - \alpha$$

For $(i, j) \neq (1, 1)$, we have

$$E(d^{joint}_{i,j,t+1}) = E(d^{joint}_{i,j,t})(1 - \alpha\frac{(i+j)}{t} + \alpha\frac{ij}{t^2})$$
$$+ \frac{(i-1)\alpha}{t}(1 - \frac{j}{t})E(d^{joint}_{i-1,j,t})$$
$$+ \frac{(j-1)\alpha}{t}(1 - \frac{i}{t})E(d^{joint}_{i,j-1,t})$$

To examine the asymptotic behavior of $E(d_{i,j,t}^{joint})$, we want to express

$$E(d_{i,j,t}^{joint}) = a_{i,j}t + c_{i,j,t},$$

where $c_{i,j,t} = o(t)$ is a lower order term. To choose an appropriate value for $a_{i,j}$, we substitute it into above recurrence formula and let t approach infinity. We obtain

$$a_{1,1} = \frac{1-\alpha}{1+2\alpha}$$

For $(i,j) \neq (1,1)$ we have

$$a_{i,j} = \alpha \frac{(i-1)a_{i-1,j} + (j-1)a_{i,j-1}}{1 + (i+j)\alpha}$$

The solution to the above recurrence is the following:

$$\begin{aligned}
a_{i,j} &= \frac{(1-\alpha)(i+j-2)!\alpha^{i+j-2}}{\prod_{k=2}^{i+j}(1+k\alpha)} \\
&= \frac{(\frac{1}{\alpha}-1)\Gamma(\frac{1}{\alpha}+2)}{(i+j)^{\frac{1}{\alpha}+2}} + o_{i+j}(1)
\end{aligned}$$

for all i, j.

It suffices to establish an upper bound for $c_{i,j,t}$. In fact, we will show that $c_{i,j,t} \leq 2$. This will be proved by induction. When $i = j = 1$, $c_{1,1,t}$ satisfies the following recurrence formula

$$c_{1,1,t+1} = c_{1,1,t}(1 - \alpha(\frac{2}{t} - \frac{1}{t^2})) + \alpha \frac{1-\alpha}{1+2\alpha} \frac{1}{t}$$

Since $c_{1,1,1} = \frac{3\alpha}{1+2\alpha} < 2$, by induction on t, we have

$$c_{1,1,t+1} \leq 2(1 - \alpha(\frac{2}{t} - \frac{1}{t^2})) + \alpha \frac{1-\alpha}{1+2\alpha} \frac{1}{t} \leq 2.$$

For $i \geq 2$ or $j \geq 2$, we can calculate the recurrence formula of $c_{i,j,t}$ and then prove $c_{i,j,t} \leq 2$ for all i, j and t in a similar way.

Other two inequalities can be proved analogously. Actually there is no need to redo the entire proof. We observe that

$$\begin{aligned}
d_{i,t}^{in} &= \sum_{j \geq 0} d_{i,j,t}^{joint} \\
d_{j,t}^{out} &= \sum_{i \geq 0} d_{i,j,t}^{joint}.
\end{aligned}$$

It is easy to see that they are all of the right order of magnitude. □

The proof of Theorem 2 is similar and will be omitted. Next section, we will prove Theorems 3 and 5.

6. The Proofs of Theorems 3 and 5

We first prove the following lemma.

Lemma 2 *If a sequence a_t satisfies the recursive formula*

$$a_{t+1} = (1 - \frac{b_t}{t})a_t + c_t \text{ for } t \geq t_0$$

where $\lim_{t \to \infty} b_t = b > 0$ and $\lim_{t \to \infty} c_t = c$ exists. Then $\lim_{t \to \infty} \frac{a_t}{t}$ exists and

$$\lim_{t \to \infty} \frac{a_t}{t} = \frac{c}{1+b}$$

Proof: Since $\lim_{t \to \infty} b_t = b > 0$, there exists a t_1 satisfying $b_t > 0$ for all $t > t_1$. Notice

$$\frac{c}{1+b}(t+1) = (1 - \frac{b}{t})\frac{c}{1+b}t + c.$$

We have

$$|a_{t+1} - \frac{c}{1+b}(t+1)| = |(1 - \frac{b_t}{t})(a_t - \frac{c}{1+b}t) + (b_t - b)\frac{c}{1+b} + c_t - c|$$

$$\leq |a_t - \frac{c}{1+b}t| + s_t$$

where $s_t = |(b_t - b)\frac{c}{1+b} + c_t - c| \to 0$ as t approaches infinity.
Now we use this inequality recursively. We have

$$|a_t - \frac{c}{1+b}t| \leq |a_{t_1} - \frac{c}{1+b}t_1| + \sum_{k=t_1}^{t-1} s_k = o(t).$$

Hence the limit $\lim_{t \to \infty} \frac{a_t}{t}$ exists and $\lim_{t \to \infty} \frac{a_t}{t} = \frac{c}{1+b}$. □

We now proceed to prove Theorem 3.

Proof of Theorem 3: Only bounded number of edges are added at a time in model C. Let's denote this bound by M. Now both $d_{i,t}^{in}$ and $d_{j,t}^{out}$ satisfy M-Lipschitz condition. By Lemma 1, it is enough to show that the following limits exist.

$$\lim_{t \to \infty} \frac{E(d_{i,t}^{in})}{t} = \frac{b''}{i^{2 + \frac{\mu^{n,n} + \mu^{e,n}}{\mu^{n,e} + \mu^{e,e}}}} + o_i(1). \tag{4.1}$$

$$\lim_{t \to \infty} \frac{E(d_{j,t}^{out})}{t} = \frac{c''}{j^{2 + \frac{\mu^{n,n} + \mu^{n,e}}{\mu^{e,n} + \mu^{e,e}}}} + o_j(1). \tag{4.2}$$

where b'', c'' are some constants independent of i, j.

We prove the equation (4.1). The proof of (4.2) is similar and will be omitted.

We assume that at time t, with probability $p^t_{i'j'k'l'}$, $m^{e,e} = i'$, $m^{n,e} = j'$, $m^{e,n} = k'$, $m^{n,n} = l'$. The probability that a vertex of in-degree $i - s$ becomes a vertex i is exactly

$$\binom{i'+j'}{s}\left(\frac{i-s}{e_t}\right)^s\left(1-\frac{i-s}{e_t}\right)^{i'+j'-s}. \qquad (4.3)$$

The probability that a vertex of in-degree i becomes a vertex of in-degree $i + s$ is exactly

$$\binom{i'+j'}{s}\left(\frac{i}{e_t}\right)^s\left(1-\frac{i}{e_t}\right)^{i'+j'-s}. \qquad (4.4)$$

The new vertex is a vertex of in-degree i is exactly

$$\sum_{k'+l'=i, i', j'} p^t_{i'j'k'l'},$$

which is assume to be well-behaved. So its limit as t approaches ∞ exists. We denote it by p_i.

By linearity of the conditional expectation, we have

$$\begin{aligned}
E(d^{in}_{i,t+1}|G_t) &= \sum_{s\geq 1} d^{in}_{i-s,t} \sum_{i',j',k'l'} \binom{i'+j'}{s}\left(\frac{i-s}{e_t}\right)^s\left(1-\frac{i-s}{e_t}\right)^{i'+j'-s} p^t_{i'j'k'l'} \\
&\quad - d^{in}_{i,t} \sum_{s\geq 1}\sum_{i',j',k'l'} \binom{i'+j'}{s}\left(\frac{i}{e_t}\right)^s\left(1-\frac{i}{e_t}\right)^{i'+j'-s} p^t_{i'j'k'l'} \\
&\quad + d^{in}_{i,t} + \sum_{k'+l'=i,i',j'} p^t_{i'j'k'l'} \\
&= d^{in}_{i,t}\left(1 - \frac{i}{e_t}\sum_{i',j',k'l'}(i'+j')p^t_{i'j'k'l'}(1+o(1))\right) \\
&\quad + d^{in}_{i-1,t}\frac{i-1}{e_t}\sum_{i',j',k'l'}(i'+j')p^t_{i'j'k'l'}(1+o(1)) + p^+_i(1+o(1)) \\
&= d^{in}_{i,t}\left(1 - i\frac{\mu^{n,e}+\mu^{e,e}+o(1)}{(\mu^{n,e}+\mu^{e,e}+\mu^{e,n}+\mu^{n,n})t}\right) \\
&\quad + d^{in}_{i-1,t}(1 - i\frac{\mu^{n,e}+\mu^{e,e}+o(1)}{(\mu^{n,e}+\mu^{e,e}+\mu^{e,n}+\mu^{n,n})t}) + p^+_i(1+o(1))
\end{aligned}$$

Since $m^{e,e}, m^{n,e}, m^{e,n}, m^{n,n}$ are bounded by M, we have $p_i = 0$, for $i > M$. We derive the following recurrence formula.

$$E(d^{in}_{i,t+1}) = E(d^{in}_{i,t})(1 - i\frac{\mu^{n,e}+\mu^{e,e}+o(1)}{(\mu^{n,e}+\mu^{e,e}+\mu^{e,n}+\mu^{n,n})t})$$

$$+E(d_{i-1,t}^{in})\frac{(1+o(1))(i-1)(\mu^{n,e}+\mu^{e,e})}{(\mu^{n,e}+\mu^{e,e}+\mu^{e,n}+\mu^{n,n})t})+o(1) \quad (4.5)$$

for $i > M$.

By induction on i and Lemma 2, equation (4.5) implies $\lim_{t\to\infty}\frac{E(d_{i,t}^{in})}{t}$ exists. Let's denote it by b_i''. b_i'' satisfies

$$b_i'' = \frac{\frac{b_{i-1}''(i-1)(\mu^{n,e}+\mu^{e,e})}{\mu^{n,e}+\mu^{e,e}+\mu^{e,n}+\mu^{n,n}}}{1+\frac{i(\mu^{n,e}+\mu^{e,e})}{(\mu^{n,e}+\mu^{e,e}+\mu^{e,n}+\mu^{n,n})}}$$

$$= \frac{i-1}{i+1+\frac{\mu^{e,n}+\mu^{n,n}}{\mu^{n,e}+\mu^{e,e}}}b_{i-1}''$$

Hence, we have,

$$b_i'' = \frac{i-1}{i+1+\frac{\mu^{e,n}+\mu^{n,n}}{\mu^{n,e}+\mu^{e,e}}}b_{i-1}''$$

$$= b_M'' \prod_{k=M+1}^{i} \frac{k-1}{k+1+\frac{\mu^{e,n}+\mu^{n,n}}{\mu^{n,e}+\mu^{e,e}}}$$

$$= b_M'' \frac{(i-1)!\Gamma(M+2+\frac{\mu^{e,n}+\mu^{n,n}}{\mu^{n,e}+\mu^{e,e}})}{(M-1)!\Gamma(i+2+\frac{\mu^{e,n}+\mu^{n,n}}{\mu^{n,e}+\mu^{e,e}})}$$

$$\approx \frac{b''}{i^{2+\frac{\mu^{e,n}+\mu^{n,n}}{\mu^{n,e}+\mu^{e,e}}}}$$

where
$b'' = b_M'' \frac{\Gamma(2M+2+\frac{\mu^{e,n}+\mu^{n,n}}{\mu^{n,e}+\mu^{e,e}})}{(2M-1)!}$ is a constant.

Equation (4.1) is proved.

By Lemma 1, almost surely the out-degree sequence of G_t satisfies the power law distribution with power $2+\frac{\mu^{e,n}+\mu^{n,n}}{\mu^{n,e}+\mu^{e,e}}$.

Similarly, we can show almost surely the in-degree sequence of G_t satisfies the power law distribution with power $2+\frac{\mu^{n,e}+\mu^{n,n}}{\mu^{e,n}+\mu^{e,e}}$. □

The proof of Theorem 4 is similar to this one and will be omitted.

Next, we will prove Theorem 5.

Proof of Theorem 5: Model A and the previous models of (Barabási and Albert 1999, Barabási et al. 1999, Kumar et al. 1999b;a) are the special cases of Model C. We prove that Model C has the scale-free property. The proofs for Models A, B and D are similar and will be omitted.

We suppose that the evolution G_T is scaled by a factor of σ. (See section 1 for the definition.) The scaled evolution $H_\sigma(G_T)$ is not exactly covered by Model C. But it is naturally approximated by an evolution $G_{T'}$ of Model C with parameters $\mu'^{n,n} = \sigma\mu^{n,n}$, $\mu'^{n,e} = \sigma\mu^{n,e}$ $\mu'^{e,n} = \sigma\mu^{e,n}$ $\mu'^{e,e} = \sigma\mu^{e,e}$ and size bound σM. Given our general results on Model C, the latter Model C process has the same power law as the first Model C process (e.g., the power for the out-degrees is $2 + (\mu^{n,n} + \mu^{n,e})/(\mu^{e,n} + \mu^{e,e})$). Hence, it is enough to show that both the scaled evolution $H_\sigma(G_T)$ and the approximating evolution $G_{T'}$ have the same power for the out-degrees (and in-degrees).

The evolution $H_\sigma(G_T)$ only differs from $G_{T'}$ in the way of adding edges. At each time unit, edges are added simultaneously in $G_{T'}$ while some edges in $H_\sigma(G_T)$ are added simultaneously and some are added sequentially. By examining the proof of Theorem 3, we find that the probabilities given in equations (4.3) and (4.4) are different. However, the main terms (occurred at $s=1$) stay the same. From the proof of Theorem 3, we conclude that both evolutions give the same power for the out-degrees as well as for the in-degrees. □

7. Problems and Remarks

In this paper, we use techniques in random graph theory to analyze power law graphs. The analysis of the evolution of power law graphs is considerably harder than that for the earlier model of random graphs with given degree sequence (Aiello et al. 2000, Molloy and Reed 1995; 1998) since nodes which acquire a relatively large degree early on in the process have an advantage and are substantially different from the nodes that are added later on. Furthermore, the error estimates that are induced by large degrees might dominate some behavior that occur in the large part of graphs with small degrees. Numerous problems remain unsolved several of which we mention here:

- In model A, we obtained the joint distribution of in- and out-degree. In general, it is true that $\lim_{t\to\infty} \frac{d_{i,j}^{joint}}{t}$ exists via the martingale theory. We denote it by $f(i,j)$. What is the asymptotic behavior of $f(i,j)$?

  Question: *Is there a simple asymptotic form of $f(i,j)$ (such as the one given in Theorem 1) for models B, C, D?*

- We only consider cases of adding nodes and edges at a time. This is consistent with some applications like a co-stars graph. However, for many other applications, such as web graphs, there are often some destructions—deleting edges and nodes. Those destructions would affect the power law to some extent. Some cases were considered in Kleinberg et al.'s paper (Kleinberg et al. 1999). Simulations (and heuristic calculations) suggest that the in-degrees also follows the power law (see Kleinberg et al. (1999)). It would be of interest to analyze the more general model when deleting edges are allowed.

Question: *Can results for model C and D be extended allowing node deletion in the random process?*

Bibliography

J. Abello, A. Buchsbaum, and J. Westbrook. A functional approach to external graph algorithms. In *Proc. 6th European Symposium on Algorithms*, pages 332–343, 1998.

W. Aiello, F. Chung, and L. Lu. A random graph model for massive graphs. In *Proceedings of the Thirty-Second Annual ACM Symposium on Theory of Computing*, pages 171–180, 2000.

N. Alon and J.H. Spencer. *The Probabilistic Method*. Wiley and Sons, 1992.

A. Barabási, R. Albert, and H. Jeong. Scale-free characteristics of random networks: The topology of the world wide web. *Physica A*, 272: 173–187, 1999.

A.-L. Barabási and R. Albert. Emergence of scaling in random networks. *Science*, 286:509–512, 1999.

B. Bollobás. *Modern Graph Theory*. Springer-Verlag, 1998.

A. Broder, R. Kumar, F. Maghoul, P. Raghavan, S. Rajagopalan, R. Stata, A. Tomkins, and J. Wiener. Graph structure in the web. *Computer Networks*, 33:309–321, 2000.

L. Egghe and R. Rousseau. *Introduction to Informetrics: Quantitative Methods in Library, Documentation and Information Science*. Elsevier, 1990.

A.A. Fairthore. Empirical hyperbolic distributions (bradford zipf mandelbrot) for bibliometric description and prediction. *Journal of Documentation*, 25:319–343, 1969.

M. Faloutsos, P. Faloutsos, and C. Faloutsos. On power-law relationships of the internet topology. In *Proceedings of the ACM SIGCOM Conference*, pages 251–262, 1999.

N. Gilbert. A simulation of the structure of academic science. *Sociological Research Online*, 2, 1997.

B.M. Hill. Zipf's law and prior distributions for the composition of a polulation. *J. Amer. Statis. Association*, 65:1220–1232, 1970.

J. Kleinberg, S.R. Kumar, P. Raghavan, S. Rajagopalan, and A.S. Tomkins. The web as a graph: Measurements, models and methods. In *Proceedings of the 5th Annual International Conference on Combinatorics and Computing (COCOON)*, 1999.

M. Koenig and T. Harrell. Lotka's law, price's urn and electronic publishing. *Journal of the American Society for Information Science*, June:386–388, 1995.

R. Kumar, P. Raghavan, S. Rajagopalan, D. Sivakumar, A. Tomkins, and E. Upfal. Stochastic models for the web graph. In *Proceedings of the 41st Annual Symposium on Foundations of Computer Science(FOCS 2000)*, pages 57–65, 2000.

S.R. Kumar, P. Raghavan, S. Rajagopalan, and A. Tomkins. Extracting large-scale knowledge bases from the web. In *Proceedings of the 25th VLDB Conference*, pages 639–650, 1999a.

S.R. Kumar, P. Raghavan, S. Rajagopalan, and A. Tomkins. Trawling the web for emerging cyber communities. In *Proceedings of the 8th World Wide Web Conference (WWW8)*, 1999b.

A.J. Lotka. The frequency distribution of scientific productivity. *The Journal of the Washington Academy of the Sciences*, 16:317, 1926.

L. Lu. The diameter of random massive graphs. In *Proceedings of the Twelfth ACM-SIAM Symposium on Discrete Algorithms(SODA 2001)*, pages 912–921, 2001.

H.A. Makse, S. Havlin, and H.E. Stanley. Modelling urban growth patterns. *Nature*, 377:608–612, 1995.

B.B. Mandelbrot. *The Fractal Geometry of Nature*. W. H. Freeman and Company, 1977.

C. Martindale and A.K. Konopka. Oligonucleotide frequencies in DNA follow a Yule distribution. *Computer & Chemistry*, 20(1), 1996.

G.A. Miller, E.B. Newman, and E.A. Friedman. Length-frequency statistics for written english. *Information and Control*, 1:370–389, 1958.

M. Molloy and B. Reed. A critical point for random graphs with a given degree sequence. *Random Structures and Algorithms*, 6(2):161–179, 1995.

M. Molloy and B. Reed. The size of the giant component of a random graph with a given degree sequence. *Combin. Probab. Comput.*, 7: 295–305, 1998.

L.J. Murphy. Lotka's law in the humanities? *Journal of the American Society for Information Science*, Nov.-Dec.:461–462, 1973.

M. Newman, S. Strogatz, and D. Watts. Random graphs with arbitrary degree distribution and their applications. *Physical Review E*, 2001. To appear.

Center for next generation internet, 2000.

R. Rousseau and Q. Zhang. Zipf's data on the frequency of chinese words revisited. *Scientometrics*, 24:201–220, 1992.

M. Schroeder. *Fractals, Caos, Power Laws*. W. H. Freeman and Company, 1991.

Z.K. Silagadze. Citations and the Zipf-Mandelbrot's law. *Complex Syst.*, 11:487–499, 1997.

H.A. Simon. *Models of Man, Social and Rational*. Wiley, 1957.

J. Tuldava. The frequency spectrum of text and vocabulary. *Journal of Quantitative Linguistics*, 3:38–50, 1996.

G.U. Yule. *Statistical Study of Literary Vocabulary*. Cambridge Univ. Press, 1944.

G.K. Zipf. *Human behaviour and the principle of least effort*. Hafner, 1949.

Chapter 5

PROPERTY TESTING IN MASSIVE GRAPHS

Oded Goldreich
Department of Computer Science and Applied Mathematics, Weizmann Institute of Science, Rehovot, Israel
oded@wisdom.weizmann.ac.il

Abstract We consider the task of evaluating properties of graphs that are too big to be even scanned. Thus, the input graph is given in form of an oracle which answers questions of the form *is there an edge between vertices u and v*, or *who is the ith neighbor of v*. Our task is to determine whether a given input graph has a predetermined property or is "relatively far" from any graph having the property. Distance between graphs is measured as the fraction of the possible queries on which the corresponding oracles, representing the two graphs, differ. We show that randomized algorithms of running-time substantially smaller than the size of the input graph may reach (with high probability) a correct verdict regarding whether the graph has some predetermined property (such as being bipartite) or is far from having it.

Keywords: Bipartite graphs, Property testing, Randomized algorithms.

1. Introduction

Suppose we are given a huge graph representing some binary relation over a huge data-set (see below), and we need to determine whether the graph (equivalently, the relation) has some predetermined property. Since the graph is huge, we cannot or do not want to even scan all of it (let alone processing all of it). The question is whether it is possible to make meaningful statements about the entire graph based only on a "small portion" of it. Of course, such statements will at best be approximations. But in many settings approximations are good enough.

As a motivation, let us consider a well-known example in which fast approximations are possible and useful. Suppose that some cost function is defined over a huge set, and that one wants to obtain the average cost of an element in the set. To be more specific, let $\mu : S \to [0,1]$ be a cost function, and suppose we want to estimate $\overline{\mu} \stackrel{\text{def}}{=} \frac{1}{|S|} \sum_{x \in S} \mu(x)$. Then, uniformly (and independently) selecting $m \stackrel{\text{def}}{=} O(\epsilon^{-2} \log(1/\delta))$ sample points, $x_1, ..., x_m$, in S we obtain with probability at least $1 - \delta$ an estimate of $\overline{\mu}$ within $\pm \epsilon$. That is,

$$\mathbf{Pr}_{x_1,...,x_m \in S}\left[\left|\frac{1}{m}\sum_{i=1}^{m}\mu(x_i) - \overline{\mu}\right| > \epsilon\right] < \delta.$$

Graphs capture more complex features of a data-set; that is, relations among pairs of elements (rather then functions of single elements). Specifically, a symmetric binary relation $R \subseteq S \times S$ is represented by a graph $G = (S, R)$, where the elements of S are called **vertices** and the elements in R are called **edges**. In this survey, we focus on two types of graphs:

1 **Dense graphs**: Such graphs have many edges; specifically, $|R| = \Omega(|S|^2)$ (say $|R| > 0.1 \cdot |S|^2$). A natural representation of such graphs is by an oracle which on query a pair $(u, v) \in S \times S$ responds with a bit indicating whether $(u, v) \in R$ or not.

Such an oracle may be either a look-up table to which we have "direct access" (i.e., we can obtain the (u,v)-entry in unit cost) or a fast procedure which allows us to determine whether $(u, v) \in R$.

2 **Bounded-degree graphs**: Such graphs have few edges, and furthermore each vertex in them has few neighbors; that is, for some small d (say $d = 10$), $|\{v : (u,v) \in R\}| \leq d$ holds for every $u \in S$. A natural representation of such graphs (of degree bound d) is by an oracle which on query a pair $(u,i) \in S \times \{1,...,d\}$ responds with the name of the ith neighbor of u (or with a special symbol if u has less than i neighbors). Again, such an oracle may be either a look-up table or a fast procedure.

Each of these two types of graphs gives rise to a natural notion of inspecting a portion of the graph: Asking queries to the corresponding oracle (defined above). Also, each of these types of graphs gives rise to a natural notion of distance (among graphs): Specifically, two dense graphs (over vertex set S) are considered relatively close if they differ on $o(|S|^2)$ edges, whereas two bounded-degree graphs are considered relatively close if they differ on $o(|S|)$ edges. A natural notion of approximation emerges: *A graph approximately has a predetermined property if it is relatively close to a graph having this property.* This leads to the following definition.

Testing graph properties – informal definition. A graph property is a set of graphs closed under graph isomorphism (renaming of vertices). Let $\mathcal{P}$ be such a property. A $\mathcal{P}$-tester is a *randomized* algorithm that is given oracle access to a graph, and has to determine whether the graph is in $\mathcal{P}$ or is far from being in $\mathcal{P}$. The type of oracle and distance-measure depend on the model, and we focus on two such models:

1. The adjacency predicate model: Here the $\mathcal{P}$-tester is given oracle access to a function $f : S \times S \to \{0, 1\}$ which represents the graph $G_f = (S, f^{-1}(1))$, where

$$f^{-1}(1) \stackrel{\text{def}}{=} \{(u,v) \in S \times S : f(u,v) = 1\}.$$

The tester is also given a distance parameter $\epsilon > 0$, and is required to accept f if $G_f \in \mathcal{P}$ and reject f if G_f differs in more than $\epsilon |S|^2$ *edges* from any graph in $\mathcal{P}$. (The tester is allowed to err, in each of these cases, with some small probability.)

2. The incidence function model: Here the $\mathcal{P}$-tester is given oracle access to a function $f : S \times \{1, ..., d\} \to S \cup \{\bot\}$ which represents the graph $G_f = (S, \{(u,v) : u \in S, v \in \Gamma_f(u)\})$, where

$$\Gamma_f(u) \stackrel{\text{def}}{=} \{v : \exists i\, f(u, i) = v\}.$$

The tester is also given a distance parameter $\epsilon > 0$, and is required to accept f if $G_f \in \mathcal{P}$ and reject f if G_f differs in more than $\epsilon d |S|$ *edges* from any graph in $\mathcal{P}$. (Again, a small error probability is allowed.)

Thus, in both cases, rejection is required only in case the corresponding representation is ϵ-far from having the property $\mathcal{P}$. In both cases ϵ-far refers to difference of more than ϵ fraction of the entries in the oracle. The adjacency predicate model is most adequate for testing of dense graphs, whereas the incidence function model is adequate for testing bounded-degree graphs.

The definition of property testing is a relaxation of the standard definition of a decision task (where one is required to decide correctly on ALL graphs): The tester is allowed arbitrary behavior when the graph does not have the property, and yet is ϵ-close to a graph having the property. Thus, a property tester may be far more efficient than a standard decision procedure (for the same property). We shall see that this is indeed the case for a variety of graph properties. But before doing so we wish to further discuss the notion of approximation underlining the above definition.

Firstly, being close to a graph which has the property is a notion of approximation which, in certain applications, may be of great value. Furthermore, in some cases, being close to a graph having the property translates to a standard notion of approximation (see Section 3.2). In other cases, it translates to a notion of "dual approximation" (see, again, Section 3.2).

Secondly, in some cases, we may be forced to take action, without having time to run a decision procedure, while given the option of modifying the graph in the future, at a cost proportional to the number of added/omitted edges. For example, suppose we are given a graph which represents some design problem, where Bipartite graphs correspond to good designs and changes in the design correspond to edge additions/omissions. Using a Bipartiteness tester, we may (with high probability) accept any good design, while rejecting designs which will cost a lot to modify. That is, we may still accept bad designs, but only such which are close to being good and thus will not cost too much to later modify.

Thirdly, we may use the property tester as a preliminary stage before running a slower exact decision procedure. In case the graph is far from having the property, with high probability, we obtain an indication towards this fact, and save the time we might have used running the decision procedure. Furthermore, in case the tester has one-sided error (i.e., it always accepts a graph having the property), we have obtain an absolutely correct answer without running the slower decision procedure at all. The saving provided by using a property tester as a preliminary stage may be very substantial in many natural settings where *typical* graphs either have the property or are very far from having the property. Furthermore, *if* it is *guaranteed* that graphs either have the property or are very far from having it *then* we may not even need to run the decision procedure at all.

Organization. In Section 2, we redefine the two models of testing graph properties discussed informally above. In Section 3 (respectively, Section 4) we survey results in the first (respectively, second) model. We will contrast the two models by considering the complexity of testing bipartiteness in both of them. Additional models for testing graph properties are discussed in Section 5. A wider perspective on property testing is provided in Section 6.

2. Testing Graph Properties – Two Models

We assume some familiarity with basic notions of graph theory and graph algorithms (cf. Even (1979)). We switch to more standard notations for graphs (i.e., denote the vertex and edge sets V and E, respectively). Furthermore, we typically assume that the vertex set V equals $\{1, ..., |V|\}$. Here we discuss undirected graphs, and so both representations presented below are redundant (since each edge appears twice).

Two natural representations of a graph are offered by its adjacency matrix and by its incidence list. Correspondingly, we consider two representations of graphs by functions.

1. An N-vertex graph, $G = (V, E)$, can be represented by the *adjacency predicate*, $f : V \times V \to \{0, 1\}$, so that $(u, v) \in E$ if and only if $f(u, v) = 1$.

2. An N-vertex graph of degree bound d, $G = (V, E)$, can be represented by the *incidence function*, $g : V \times \{1, ..., d\} \to V \cup \{0\}$, so that $g(u, i) = v$ if v is the ith vertex incident at u, and $g(u, i) = 0 \notin V$ if u has less than i neighbors.

As usual, the choice of representation has a fundamental impact on the potential algorithm. Here the impact is even more dramatic since we seek algorithms which only inspect a relatively small fraction of the object (graph represented by a function). Furthermore, there is another fundamental impact of the choice of representation on the task of property testing. This has to do with our definition of distance, which is relative to the size of the domain of the function. In particular, distance ϵ in the first representation means a symmetric difference of $2\epsilon \cdot N^2$ edges, whereas in the second representation this means a symmetric difference of $2\epsilon \cdot dN$ edges. (In both cases, the extra factor 2 is due to the redundant representation which is adopted for sake of simplicity.)

Definition 1 (testing graph properties): *For any $m \in \mathbf{N}$, let $[m] \stackrel{\text{def}}{=} \{1, 2, ..., m\}$. Let $\mathcal{P}$ be a graph property.*

1. *A $\mathcal{P}$-tester for the adjacency predicate model is a randomized algorithm that on input a size parameter $N \in \mathbf{N}$, distance parameter $\epsilon > 0$, and oracle access to a function $f : [N] \times [N] \to \{0, 1\}$, with probability at least $2/3$, accepts if f represents a graph in $\mathcal{P}$ and rejects if f is ϵ-far from any graph in $\mathcal{P}$. That is,*

   - *If $G = ([N], f^{-1}(1))$ is in $\mathcal{P}$ then the algorithm accepts with probability at least $2/3$.*
   - *If for every $h : [N] \times [N] \to \{0, 1\}$ such that $([N], h^{-1}(1)) \in \mathcal{P}$, the functions f and h differ in more than $\epsilon \cdot N^2$ entries then the algorithm rejects with probability at least $2/3$.*

2. *A $\mathcal{P}$-tester for d-bounded incidence function model is a randomized algorithm that on input a size parameter $N \in \mathbf{N}$, distance parameter $\epsilon > 0$, and oracle access to a function $f : [N] \times [d] \to \{0\} \cup [N]$ with probability at least $2/3$ accepts if f represents a graph in $\mathcal{P}$ and rejects if f is ϵ-far from any graph in $\mathcal{P}$. That is, letting $\Gamma_g(u) \stackrel{\text{def}}{=} \{v : \exists i\, g(u,i) = v\}$,*

   - *If $G = ([N], \{(u,v) \in [N] \times [N] : v \in \Gamma_f(u)\})$ is in $\mathcal{P}$ then the algorithm accepts with probability at least $2/3$.*

   - *If for every $h : [N] \times [d] \to \{0\} \cup [N]$ such that $([N], \{(u,v) \in [N] \times [N] : v \in \Gamma_h(u)\}) \in \mathcal{P}$, the functions f and h differ in more than $\epsilon \cdot dN$ entries then the algorithm rejects with probability at least $2/3$.*

As usual, the error probability may be decreased by successive applications of the tester.

The first model (i.e., adjacency predicate) is most appropriate for dense graphs (i.e., $|E| = \Omega(|V|^2)$), whereas the second model (i.e., d-bounded incidence function) is applicable and most appropriate for graphs of degree bound d. Below, we demonstrate the difference between the two representations by considering the task of testing whether a graph is Bipartite.

We will be focus on the query and time complexity of such testers, as a function of N and ϵ. By query complexity we mean the number of oracle queries made by the algorithm. In evaluating the running time, we count each query at unit cost. [1]

We stress that the testing algorithms are allowed to be randomized, and that this is of key importance for achieving query complexity which is significantly lower than the size of the graph. A deterministic tester for any "non-degenerate" property needs to query the oracle on a constant fraction of its domain, and so is of little interest in our context.

3. The First Model (Adjacency Predicates)

In this section we consider the representation of N-vertex graphs by adjacency predicates mapping pairs $\{1, 2, ..., N\} \times \{1, 2, ..., N\}$ to $\{0, 1\}$. Recall that in this model, distance between graphs refers to the fraction of different edges over $N^2/2$.

---

[1] Alternatively, one may consider a RAM model of computation, in which trivial manipulation of vertices (e.g., reading/writing a vertex and comparing vertices) is performed at unit cost.

3.1. Some Known Results

Testers of complexity which depends only on the distance parameter, ϵ, are known for several natural graph properties (Goldreich et al. 1998b, Alon et al. 1999). In particular, the following properties can be tested in query-complexity poly$(1/\epsilon)$ and time complexity exp(poly$(1/\epsilon)$) (cf. Goldreich et al. (1998b)): [2]

- k-Colorability [3], for any fixed $k \geq 2$. The query-complexity is poly(k/ϵ), and for $k = 2$ the running-time is $\widetilde{O}(1/\epsilon^3)$, where by $\widetilde{O}(m)$ we mean $O(m \cdot \text{poly}(\log m))$. In case the graph is k-colorable the tester always accepts, whereas in case the graph is ϵ-far from k-colorable the tester rejects with probability at least $\frac{2}{3}$ and furthermore supplies a small counterexample (in the form of a small subgraph which is not k-colorable).

  The 2-Colorability (equivalently, Bipartite) Tester is presented in Subsection 3.3. An improved analysis has been recently obtained by Alon and Krivelevich (1999).

- ρ-Clique, for any $\rho > 0$. That is, does the N-vertex graph have a clique (i.e., a set of vertices with edges among each pair in it) of size ρN.

- ρ-CUT, for any $\rho > 0$. That is, does the N-vertex graph have a cut of size at least ρN^2. A generalization to k-way cuts works within query-complexity poly$((\log k)/\epsilon)$.

- ρ-Bisection, for any $\rho > 0$. That is, can the vertices of the N-vertex graph be partitioned into two equal parts with at most ρN^2 edges going between them.

All the above property testing problems are special cases of the **General Graph Partition Testing Problem**, parameterized by a set of lower and upper bounds. In this problem one needs to determine whether there exists a k-partition of the vertices so that the number of vertices in each part as well as the number of edges between each pair of parts falls between the corresponding lower and upper bounds (in the set of parameters). For example, ρ-clique is expressible as a 2-partition in which one part has ρN vertices, and the number of edges in this part

---

[2] Except for Bipartite testing, running-time of poly$(1/\epsilon)$ is unlikely, as it will imply $\mathcal{NP} \subseteq \mathcal{BPP}$.

[3] A graph is k-colorable if its vertices can be partitioned into k parts so that there are no edges among vertices residing in the same part.

is $\binom{\rho N}{2}$. A tester for the general problem has also been presented in (Goldreich et al. 1998b): The tester uses $\tilde{O}(k^2/\epsilon)^{2k+O(1)}$ queries, and runs in time exponential in its query-complexity.

Going beyond the general graph partition problem. Although many natural graph properties can be formulated as partition problems, many more cannot. Among these we mention a few classes which certainly do not exhaust all graph properties.

- Many natural graph properties are very easy to test in the adjacency predicate model. A very partial list includes Connectivity, Hamiltonicity, Cycle-freeness and Planarity (cf. (Goldreich et al. 1998b)). The reason being that for these properties either every N-vertex graph is at distance at most $O(1/N)$ from a graph having the desired property (and so for $\epsilon = \Omega(1/N)$ the trivial algorithm which always accepts will do), or the property holds only for sparse graphs (and so for $\epsilon = \Omega(1/N)$ one may reject any non-sparse graph).

- On the other hand, there are ("unnatural") graph properties in $\mathcal{NP}$ which are extremely hard to test; namely, any testing algorithm must inspect at least $\Omega(N^2)$ of the vertex pairs (Goldreich et al. 1998b).

- Alon et al. (1999) have recently suggested to study graph properties through the type (or "logical complexity") of a formula defining the property. Specifically, they considered graph properties expressible by first order formulae. They showed that every graph property expressible by such formula of the form $\exists^*\forall^*$ is testable in complexity depending only on ϵ (alas the dependency many be terrible; e.g., a tower of poly$(1/\epsilon)$ exponentiations). In contrast, they also showed that there exists a (natural) graph property expressible by a first order formula of the form $\forall^+\exists^+$ which cannot be tested within complexity depending only on ϵ.

In view of the above, we believe that providing a characterization of graph properties, according to the complexity of testing them, may be very challenging.

From testing to searching. Most graph properties discussed above are in $\mathcal{NP}$. Furthermore, in these cases the NP-witness for G having property $\mathcal{P}$ is a natural structure in the graph; for example, in case of the General Graph Partition Problem the witness is merely an adequate partition of the vertices. Interestingly, our testers for (all cases of) the General Graph Partition Problem, can be modified into algorithms which provide such approximate NP-witnesses. That is, if the graph has the desired (partitioning) property, then the testing algorithm may actually output auxiliary information that allows to construct, in $\text{poly}(1/\epsilon) \cdot N$-time, a partition which approximately obeys the property. For example, for ρ-CUT, we can construct a partition with at least $(\rho - \epsilon) \cdot N^2$ crossing edges.

One-sided error probability. The k-Colorability tester has one-sided error: It always accepts k-colorable graphs. Furthermore, when rejecting a graph, this tester always supplies a $\text{poly}(1/\epsilon)$-size subgraph which is not k-colorable. The other algorithms for all the other cases of the General Graph Partition Problem discussed above, have two-sided error. This is unavoidable within $o(N)$ query-complexity.

3.2. Testing versus deciding and other forms of approximation

We shortly discuss the relationship of the notion of approximation underlying the definition of testing graph properties to more traditional notions. The latter include exact decision as well as other notions of approximation.

Relation to recognizing graph properties. Our notion of testing a graph property $\mathcal{P}$ is a *relaxation* of the notion of *deciding the graph property* $\mathcal{P}$ which has received much attention in the last three decades (Lovász and Young 1991). In the classical problem there are no margins of error – one is required to accept all graphs having property $\mathcal{P}$ and reject all graphs which lack it. In 1975, Rivest and Vuillemin (1976) resolved the Aanderaa–Rosenberg Conjecture (Rosenberg 1973), showing that any deterministic procedure for deciding any non-trivial monotone N-vertex graph property must examine $\Omega(N^2)$ entries in the adjacency matrix representing the graph. The query complexity of randomized decision procedures was conjectured by Yao to be $\Omega(N^2)$. Progress towards this goal was made by Yao (1987), King (1991) and Hajnal (1991) culminating in an $\Omega(N^{4/3})$ lower bound. This stands in striking contrast to the testing results of (Goldreich et al. 1998b) mentioned above, by which some non-trivial monotone graph properties can be *tested* by examining a constant number of locations in the matrix.

Application to the standard notion of approximation. The relation of testing graph properties to the standard notions of approximation is best illustrated in the case of Max-CUT. Any tester for the class ρ-cut, working in time $T(\epsilon, N)$, yields an algorithm for approximating the maximum cut in an N-vertex graph, up to additive error ϵN^2, in time $\frac{1}{\epsilon} \cdot T(\epsilon, N)$. Thus, for any constant $\epsilon > 0$, using the above tester of (Goldreich et al. 1998b), we can approximate the size of the max-cut to within ϵN^2 in constant time. This yields a **constant time approximation scheme** (i.e., to within any constant relative error) for dense graphs, improving over previous work of Arora et al. (1995) and de la Vega (1994) who solved this problem in polynomial-time (i.e., in $O(N^{1/\epsilon^2})$–time and $\exp(\widetilde{O}(1/\epsilon^2)) \cdot N^2$–time, respectively). In the latter works the problem is solved by actually constructing approximate max-cuts. Finding an approximate max-cut does not seem to follow from the mere existence of a tester for ρ-cut; yet, as mentioned above, the tester in (Goldreich et al. 1998b) can be used to find such a cut in time linear in N.

Relation to "dual approximation" (cf. (Hochbaum and Shmoys 1987; 1988)). To illustrate this relation, we consider the ρ-Clique Tester mentioned above. The traditional notion of approximating Max–Clique corresponds to distinguishing the case in which the max-clique has size at least ρN from, say, the case in which the max-clique has size at most $\rho N/2$. On the other hand, when we talk of testing "ρ-Cliqueness", the task is to distinguish the case in which an N-vertex graph has a clique of size ρN from the case in which it is ϵ-far from the class of N-vertex graphs having a clique of size ρN. This is equivalent to the "dual approximation" task of distinguishing the case in which an N-vertex graph has a clique of size ρN from the case in which any ρN subset of the vertices misses at least ϵN^2 edges. To demonstrate that these two tasks are vastly different we mention that whereas the former task is NP-Hard, for $\rho < 1/4$ (see (Bellare et al. 1998, Håstad 1996b;a)), the latter task can be solved in $\exp(O(1/\epsilon^2))$-time, for any $\rho, \epsilon > 0$. We believe that there is no absolute sense in which one of these approximation tasks is more important than the other: Each of these tasks may be relevant in some applications and irrelevant in others.

3.3. Testing bipartiteness

The bipartite tester is extremely simple: It selects a tiny, random set of vertices and checks whether the induced subgraph is bipartite.

Algorithm 1 (Bipartite Tester in the first model (Goldreich et al. 1998b)): *On input N, d, ϵ and oracle access to an adjacency predicate of an N-vertex graph, $G = (V, E)$:*

  1 Uniformly select a subset of $\widetilde{O}(1/\epsilon^2)$ vertices of V.

2 Accept if and only if the subgraph induced by this subset is Bipartite.

Step (2) amounts to querying the predicate on all pairs of vertices in the subset selected at Step (1), and testing whether the induced graph is bipartite (e.g., by running BFS; see (Even 1979)). As will become clear from the analysis, it actually suffice to query only $\widetilde{O}(1/\epsilon^3)$ of these pairs. We comment that a more complex analysis due to Alon and Krivelevich (1999) implies that the above algorithm is a Bipartite Tester even if one selects only $\widetilde{O}(1/\epsilon)$ vertices (rather than $\widetilde{O}(1/\epsilon^2)$) in Step (1).

Theorem 1 (Goldreich et al. (1998b)): *Algorithm 1 is a Bipartite Tester (in the adjacency predicate model). Furthermore, the algorithm always accepts a Bipartite graph, and in case of rejection it provides a witness of length* poly$(1/\epsilon)$ *(that the graph is not bipartite).*

Proof: Let R be the subset selected in Step (1), and G_R the subgraph induced by it. Clearly, if G is bipartite then so is G_R, for any R. The point is to prove that if G is ϵ-far from bipartite then the probability that G_R is bipartite is at most $1/3$. Thus, from this point on we assume that at least ϵN^2 edges have to be omitted from G to make it bipartite.

We view R as a union of two disjoint sets U and S, where $t \stackrel{\text{def}}{=} |U| = O(\epsilon^{-1} \cdot \log(1/\epsilon))$ and $m \stackrel{\text{def}}{=} |S| = O(t/\epsilon)$. We will consider all possible partitions of U, and associate a partial partition of V with each such partition of U. The idea is that in order to be consistent with a given partition, (U_1, U_2), of U, all neighbors of U_1 (respectively, U_2) must be placed opposite to U_1 (respectively, U_2). We will show that, with high probability, most high-degree vertices in V do neighbor U and so are forced by its partition. Since there are relatively few edges incident to vertices which do not neighbor U, it follows that with very high probability each such partition of U is detected as illegal by G_R. Details follow, but before we proceed let us stress the key observation: It suffices to rule out relatively few (partial) partitions of V (i.e., these induced by partitions of U), rather than all possible partitions of V.

We use the notations $\Gamma(v) \stackrel{\text{def}}{=} \{u : (u,v) \in E\}$ and $\Gamma(X) \stackrel{\text{def}}{=} \cup_{v \in X} \Gamma(v)$. Given a partition (U_1, U_2) of U, we define a (possibly partial) partition, (V_1, V_2), of V so that $V_1 \stackrel{\text{def}}{=} \Gamma(U_2)$ and $V_2 \stackrel{\text{def}}{=} \Gamma(U_1)$ (assume, for simplicity that $V_1 \cap V_1$ is indeed empty). As suggested above, if one claims that G can be "bipartited" with U_1 and U_2 on different sides then $V_1 = \Gamma(U_1)$ must be on the opposite side to U_2 (and $\Gamma(U_1)$ opposite to U_1). Note that the partition of U places no restriction on vertices which have no neighbor in U. Thus, we first ensure that *almost all* "influential" (i.e., "high-degree") vertices in V have a neighbor in U.

Technical Definition 3.1 (high-degree vertices and good sets): *We say that a vertex $v \in V$ is of* high-degree *if it has degree at least $\frac{\epsilon}{3}N$. We call U* good *for V if all but at most $\frac{\epsilon}{3}N$ of the high-degree vertices in V have a neighbor in U.*

We comment that NOT insisting that a good set U neighbors *all* high-degree vertices allows us to show that a random U of size unrelated to the size of the graph is good. (If we were to insist that a good U neighbors *all* high-degree vertices then we would have had to use $|U| = \Omega(\log N)$.)

Claim 3.2 *With probability at least 5/6, a uniformly chosen set U of size t is good.*

Proof: For any high-degree vertex v, the probability that v does not have any neighbor in a uniformly chosen U is at most $(1 - \epsilon/3)^t < \frac{\epsilon}{18}$ (since $t = \Omega(\epsilon^{-1} \log(1/\epsilon))$). Hence the expected number of high-degree vertices which do not have a neighbor in a random set U is less than $\frac{\epsilon}{18} \cdot N$, and the claim follows by Markov's Inequality. □

Technical Definition 3.3 (disturbing a partition of U): *We say that an edge* disturbs *a partition (U_1, U_2) of U if both its end-points are in the same $\Gamma(U_i)$, for some $i \in \{1,2\}$.*

Claim 3.4 *For any good set U and any partition of U, at least $\frac{\epsilon}{3}N^2$ edges disturb the partition.*

Proof: Each partition of V has at least ϵN^2 violating edges (i.e., edges with both end-points on the same side). We upper bound the number of these edges which are not disturbing. Actually, we upper bound the number of edges which have an end-point not in $\Gamma(U)$.

- The number of edges incident to high-degree vertices which do not neighbor U is bounded by $\frac{\epsilon}{3}N \cdot N$ (at most $\frac{\epsilon}{3}N$ such vertices each having at most N incident edges).

- The number of edges incident to vertices which are not of high-degree vertices is bounded by $N \cdot \frac{\epsilon}{3}N$ (at most N such vertices each having at most $\frac{\epsilon}{3}N$ incident edges).

This leaves us with at least $\frac{\epsilon}{3}N^2$ violating edges connecting vertices in $\Gamma(U)$ (i.e., edges disturbing the partition of U). □

The theorem follows by observing that G_R is bipartite only if either (1) the set U is not good; or (2) the set U is good and there exists a partition of U so that none of the disturbing edges occurs in G_R. Using Claim 3.2 the probability of event (1) is bounded by 1/6; and using Claim 3.4 the probability of event (2) is bounded by the probability that there exists a partition of U so that none of the corresponding $\geq \frac{\epsilon}{3}N^2$ disturbing edges has both edge-point in S. Actually, we pair the m vertices of S, and consider the probability that none of these pairs is a disturbing edge for a partition of U. Thus the probability of event (2) is bounded by

$$2^{|U|} \cdot \left(1 - \frac{\epsilon}{3}\right)^{m/2} < \frac{1}{6}$$

where the inequality holds since $m = \Omega(t/\epsilon)$. The theorem follows. ∎

Comment. The procedure employed in the proof yields a poly$(1/\epsilon) \cdot N$-time algorithm for 2-partitioning a bipartite graph so that at most ϵN^2 edges lie within the same side. This is done by running the tester, determining a partition of U (defined as in the proof) which is consistent with the bipartite partition of R, and partitioning V as done in the proof (with vertices which do not neighbor U, or neighbor both U_1, U_2, placed arbitrarily). Thus, the placement of each vertex is determined by inspecting at most $\widetilde{O}(1/\epsilon)$ entries of the adjacency matrix.

4. The Second Model (Incidence Functions)

In this section we consider the representation of N-vertex graphs of degree bound d by incidence functions mapping pairs $\{1, 2, ..., N\} \times \{1, 2, ..., d\}$ to $\{0, 1, 2, ..., N\}$. Recall that in this model, distance between graphs refers to the fraction of different edges over $dN/2$.

4.1. Some known results

Testers of complexity which depends only on the distance parameter, ϵ, are known for several natural graph properties (cf. Goldreich and Ron (1997)). In particular, the following properties can be tested in time (and thus query-complexity) poly(d/ϵ):

- *Connectivity.* The tester runs in time $\widetilde{O}(1/\epsilon)$. In case the graph is connected the tester always accepts, whereas in case the graph is ϵ-far from being connected the tester rejects with probability at least $\frac{2}{3}$ and furthermore supplies a small counter-example to connectivity (in the form of an induced subgraph which is disconnected from the rest of the graph).

- *k-edge-connectivity.* The tester runs in time $\widetilde{O}(k^3 \cdot \epsilon^{-3})$. For $k = 2, 3$ improved testers have running-times $\widetilde{O}(\epsilon^{-1})$ and $\widetilde{O}(\epsilon^{-2})$, respectively. Again, k-edge-connected graphs are always accepted, and rejection is accompanied by a counter-example.

- k-vertex-connectivity, for $k = 2, 3$. The testers run in time $\widetilde{O}(\epsilon^{-k})$.

- *Cycle-Freeness.* The tester runs in time $\widetilde{O}(\epsilon^{-3})$. Unlike all other algorithms, this tester has two-sided error probability, which is unavoidable for testing this property within $o(\sqrt{N})$ queries.

The complexity of Bipartiteness testing is considered in Subsections 4.2 and 4.3: We survey an $\Omega(\sqrt{N})$ lower bound on the query complexity of any tester (Goldreich and Ron 1997), and a matching upper bound (Goldreich and Ron 1999). The lower bound stands in sharp contrast to the situation in the first model (i.e., representation by adjacency predicates), where Bipartite testing is possible in poly($1/\epsilon$)-time. The query complexity upper bound (for the incidence function representation) is obtained by a natural Bipartite-tester of running time (and query complexity) $\widetilde{O}(\text{poly}(1/\epsilon) \cdot \sqrt{N})$. We stress that, for $\epsilon > N^{-\Omega(1)}$, the testers in both models are faster than the decision procedure.

4.2. A lower bound on the complexity of testing bipartiteness

In contrast to Theorem 1, under the incidence function representation there exists no Bipartite tester of complexity independent of the graph size.

Theorem 2 (Goldreich and Ron (1997)): *Testing Bipartiteness* (with constant ϵ and d) *requires* $\Omega(\sqrt{N})$ *queries* (in the incidence function model).

Proof Idea: For any (even) N, we consider the following two families of graphs:

1. The first family, denoted $\mathcal{G}_1^N$, consists of all degree-3 graphs which are composed by the union of a Hamiltonian cycle and a perfect matching. That is, there are N edges connecting the vertices in a cycle, and the other $N/2$ edges are a perfect matching.

2. The second family, denoted $\mathcal{G}_2^N$, is the same as the first *except* that the perfect matchings allowed are restricted as follows: the distance on the cycle between every two vertices which are connected by an perfect matching edge must be odd.

Clearly, all graphs in $\mathcal{G}_2^N$ are bipartite. One first proves that almost all graphs in $\mathcal{G}_1^N$ are far from being bipartite. Afterwards, one proves that a testing algorithm that performs less than $\alpha \sqrt{N}$ queries (for some constant $\alpha < 1$) is not able to distinguish between a graph chosen randomly from $\mathcal{G}_2^N$ (which is always bipartite) and a graph chosen randomly from $\mathcal{G}_1^N$ (which with high probability will be far from bipartite). Loosely speaking, this is done by showing that in both cases the algorithm is unlikely to encounter a cycle (among the vertices it has inspected). ∎

4.3. An algorithm for testing bipartiteness

The lower bound of Theorem 2 is essentially tight. Furthermore, the following natural algorithm constitutes a Bipartite tester of running time $\text{poly}((\log N)/\epsilon) \cdot \sqrt{N}$.

Algorithm 2 (Bipartite Tester in the second model (Goldreich and Ron 1999)):
On input N, d, ϵ and oracle access to an incidence function for an N-vertex graph, $G = (V, E)$, of degree bound d, repeat $T \stackrel{\text{def}}{=} \Theta(\frac{1}{\epsilon})$ times:

1. *Uniformly select s in V.*

2. *(Try to find an odd cycle through vertex s):*

   (a) *Perform $K \stackrel{\text{def}}{=} \text{poly}((\log N)/\epsilon) \cdot \sqrt{N}$ random walks starting from s, each of length $L \stackrel{\text{def}}{=} \text{poly}((\log N)/\epsilon)$.*

   (b) *Let R_0 (respectively, R_1) denote the vertices set reached from s in an even (respectively, odd) number of steps in any of these walks.*

   (c) *If $R_0 \cap R_1$ is not empty then* reject.

If the algorithm did not reject in any one of the above T iterations, then it accepts.

Theorem 3 (Goldreich and Ron (1999)): *Algorithm 2 is a Bipartite Tester (in the incidence function model). Furthermore, the algorithm always accepts a Bipartite graph, and in case of rejection it provides a witness of length* $\text{poly}((\log N)/\epsilon)$ *(that the graph is not bipartite).*

Motivation – the special case of rapid mixing graphs.. The proof of Theorem 3 is quite involved. As a motivation, we consider the special case where the graph has a "rapid mixing" feature. It is convenient to modify the random walks so that at each step each neighbor is selected with probability $1/2d$, and otherwise (with probability at least $1/2$) the walk remains in the present vertex. Furthermore, we will consider a single execution of Step (2) starting from an arbitrary vertex, s, fixed in the rest of the discussion. The rapid mixing feature we assume is that, for every vertex v, a (modified) random walk of length L starting at s reaches v with probability approximately $1/N$ (say, up-to a factor of 2). Note that if the graph is an expander then this is certainly the case (since $L \geq O(\log N)$).

The key quantities in the analysis are the following probabilities, referring to the parity of the length of a path obtained from the random walk by omitting the self-loops (transitions which remain at current vertex). Let $p^0(v)$ (respectively, $p^1(v)$) denote the probability that a (modified) random walk of length L starting at s reaches v while making an even (respectively, odd) number of real (i.e., non-self-loop) steps. By the rapid mixing assumption we have (for every $v \in V$)

$$\frac{1}{2N} < p^0(v) + p^1(v) < \frac{2}{N} \qquad (5.1)$$

We consider two cases regarding the sum $\sum_{v \in V} p^0(v) p^1(v)$ – In case the sum is (relatively) "small", we show that V can be 2-partitioned so that there are relatively few edges between vertices placed in the same side, which implies that G is close to be bipartite. Otherwise (i.e., when the sum is not "small"), we show that with significant probability, when Step (2) is started at vertex s it is completed by rejecting G. The two cases are presented in greater detail in the following (corresponding) two claims.

Claim 4.1 *Suppose $\sum_{v \in V} p^0(v) p^1(v) \leq \epsilon/50N$. Let $V_1 \stackrel{\text{def}}{=} \{v \in V : p^0(v) < p^1(v)\}$ and $V_2 = V \setminus V_1$. Then, the number of edges with both end-points in the same V_σ is bounded above by ϵdN.*

Proof Sketch: Consider an edge (u,v) where, without loss of generality, both u and v are in V_1. Then, both $p^1(v)$ and $p^1(u)$ are greater than $\frac{1}{2} \cdot \frac{1}{2N}$. However, one can show that $p^0(v) > \frac{1}{3d} \cdot p^1(u)$: Observe that a walk of length $L-1$ with path-parity 1 ending at u is almost as likely as such a walk having length L, and that once such a walk reaches u it continues to v in the next step with probability exactly $1/2d$. Thus, such an edge contributes at least $\frac{(1/4N)^2}{3d}$ to the sum $\sum_{v \in V} p^0(v) p^1(v)$. The claim follows. ∎

Claim 4.2 *Suppose $\sum_{v \in V} p^0(v) p^1(v) \geq \epsilon/50N$, and that Step (2) is started with vertex s. Then, with probability at least $2/3$, the set $R_0 \cap R_1$ is not empty* (and rejection follows).

Proof Sketch: Consider the probability space defined by an execution of Step (2) with start vertex s. We define random variables $\zeta_{i,j}$ representing the event that the vertex encountered in the Lth step of the ith walk equals the vertex encountered in the Lth step of the jth walk, and that the ith walk corresponds to an even-path whereas the jth to an odd-path. Then

$$\begin{aligned}
\mathbf{E}(|R_0 \cap R_1|) &> \sum_{i \neq j} \mathbf{E}(\zeta_{i,j}) \\
&= K(K-1) \cdot \sum_{v \in V} p^0(v) p^1(v) \\
&> \frac{500N}{\epsilon} \cdot \sum_{v \in V} p^0(v) p^1(v) \\
&\geq 10
\end{aligned}$$

where the second inequality is due to the setting of K, and the third to the claim's hypothesis. Intuitively, we expect that with high probability $|R_0 \cap R_1| > 0$. This is indeed the case, but proving it is less straightforward than it seems, the problem being that the $\zeta_{i,j}$'s are not pairwise independent. Yet, since the sum of the covariances of the dependent $\zeta_{i,j}$'s is quite small, Chebyshev's Inequality is still very useful (cf. (Alon and Spencer 1992,Sec. 4.3)). Specifically, letting $\mu \stackrel{\text{def}}{=} \sum_{v \in V} p^0(v) p^1(v)$ ($= \mathbf{E}(\zeta_{i,j})$), and $\overline{\zeta}_{i,j} \stackrel{\text{def}}{=} \zeta_{i,j} - \mu$, we get:

$$\begin{aligned}
\mathbf{Pr}\left(\sum_{i \neq j} \zeta_{i,j} = 0\right) &< \frac{\mathbf{Var}(\sum_{i \neq j} \zeta_{i,j})}{(K^2 \mu)^2} \\
&= \frac{1}{K^4 \mu^2} \cdot \left(\sum_{i,j} \mathbf{E}(\overline{\zeta}_{i,j}^2) + 2 \sum_{i,j,k} \mathbf{E}(\overline{\zeta}_{i,j} \overline{\zeta}_{i,k})\right) \\
&< \frac{1}{K^2 \mu} + \frac{2}{K \mu^2} \cdot \mathbf{E}(\zeta_{1,2} \zeta_{1,3})
\end{aligned}$$

For the second term, we observe that $\mathbf{Pr}(\zeta_{1,2} = \zeta_{2,3} = 1)$ is upper bounded by the probability that $\zeta_{1,2} = 1$ times the probability that the Lth vertex of the first walk appears as the Lth vertex of the third path. Using the rapid mixing hypothesis, we upper bound the latter probability by $2/N$, and obtain

$$\begin{aligned}
\mathbf{Pr}(|R_0 \cap R_1| = 0) &< \frac{1}{K^2 \mu} + \frac{2}{K \mu^2} \cdot \mu \cdot \frac{2}{N} \\
&< \frac{1}{3}
\end{aligned}$$

where the last inequality uses $K < N/4$, $\mu \geq \epsilon/50N$ and $K^2 \geq 6 \cdot 50N/\epsilon$. The claim follows. ∎

Beyond rapid mixing graphs.. The proof in (Goldreich and Ron 1999) refers to a more general sum of products; that is,

$$\sum_{v \in V} p_{\text{odd}}(v) p_{\text{even}}(v),$$

where $U \subseteq V$ is an appropriate set of vertices, and $p_{\text{odd}}(v)$ (respectively, $p_{\text{even}}(v)$) is the probability that a random walk (starting at s) passes through v after more than $L/2$ steps and the corresponding path to v has odd (respectively, even) parity. Much of the analysis in (Goldreich and Ron 1999) goes into selecting the appropriate U (and an appropriate starting vertex s), and pasting together many such U's to cover all of V. Loosely speaking, U and s are selected so that there are few edges from U and the rest of the graph, and $p_{\text{odd}}(u) + p_{\text{even}}(u) \approx 1/\sqrt{|V| \cdot |U|}$, for every $u \in U$. The selection is based on the "combinatorial treatment of expansion" of Mihail (1989). Specifically, we use the counterpositive of the standard analysis, which asserts that rapid mixing occurs when all cuts are relatively large, to assert the existence of small cuts which partition the graph so that vertices reached with relatively high probability (in a short random walk) are on one side and the rest of the graph on the other. The first set corresponds to U above and the cut is relatively small with respect to U. A start vertex s for which the corresponding sum is big is shown to cause Step (2) to reject (when started with this s), whereas a small corresponding sum enables to 2-partition U while having few violating edges among the vertices in each part of U.

The actual argument of (Goldreich and Ron 1999) proceeds in iterations. In each iteration a vertex s for which Step (2) accepts with high probability is fixed, and an appropriate set of remaining vertices, U, is found. The set U is then 2-partitioned so that there are few violating edges inside U. Since we want to paste all these partitions together, U may not contain vertices treated in previous iterations. This complicates the analysis, since it must refer to the part of G, denoted H, not treated in previous iterations. We consider walks over an (imaginary) Markov Chain representing the H-part of the walks performed by the algorithm on G. Statements about rapid mixing are made with respect to the Markov Chain, and linked to what happens in random walks performed on G. In particular, a subset U of H is determined so that the vertices in U are reached with probability $\approx 1/\sqrt{|V| \cdot |U|}$ (in the chain) and the cut between U and the rest of H is small. Linking the sum of products defined for the chain with the actual walks performed by the algorithm, we infer that U may be partitioned with few violating edges inside it. Edges to previously treated parts of the graphs are charged to these parts, and edges to the rest of H \ U are accounted for by using the fact that this cut is small (relative to the size of U).

5. Two Other Models

So far our discussion was confined to undirected graphs. Yet, the two models (above) extend naturally to the case of directed graphs. Some of the results for the undirected case extend easily to the directed case (e.g., testing directed connectivity). A basic task in the directed graph models is testing whether a given directed graph is acyclic (i.e., has no directed cycles). Bender and Ron have presented a Acyclicity-tester of poly($1/\epsilon$) complexity in the adjacency predicate model, and showed that no such tester may exist in the incidence list model (Bender and Ron 2000).

In our discussion above (as well as in the next section) we have linked the issue of representation to the distance measure. That is, when representing a graph by an oracle from some domain D to a range R, we have considered the relative distance of graphs as the fraction of different entries in their representation divided by the size of the domain (i.e., $|D|$). This (quite natural) convention is abandoned by Parnas and Ron who developed a more general model by decoupling the representation of the graph from the distance measure (Parnas and Ron 1999): Whatever is the mechanism of accessing the graph, the distance between graphs is defined as the number of edges in their symmetric difference. (The relative distance may be defined as the latter quantity divided by the total number of edges in both graphs.) The new model allows to treat well the case of sparse graphs which are not of bounded-degree – a case that was not treated in a satisfactory manner in either of the previous two models. Many of the testers for bounded-degree graphs can be extended to the case of sparse graphs in the new model. Furthermore, the following problem of estimating the diameter of a graph is shown to be solvable in this model within complexity poly($1/\epsilon$): Given a diameter parameter D and a distance parameter ϵ, determine whether the graph has diameter at most D or is ϵ-far from any graph of diameter at most $2D + 2$ (i.e., one has to add more than $\epsilon \cdot |E|$ edges in order to obtain from the input graph $G = (V, E)$ a graph of diameter at most $2D + 2$). (We comment that in the "bounded-degree model" and for $\epsilon > 1/D$ this task can be easily reduced to testing connectivity.)

6. A Wider Perspective

Our formulation of testing graph properties (in both the adjacency predicate and incident list models) is a special case of property testing of arbitrary functions.

Definition 2 (property tester (Rubinfeld and Sudan 1996)): *Let S be a finite set, and $\mathcal{P}$ a subset of functions mapping S to $\{0,1\}^*$. A (property) tester for $\mathcal{P}$ is a probabilistic oracle machine, M, which given a distance parameter $\epsilon > 0$ and oracle access to an arbitrary function $f : S \to \{0,1\}^*$ satisfies the following two conditions:*

1. (the tester accepts every f in $\mathcal{P}$): *If $f \in \mathcal{P}$ then* $\mathbf{Pr}\left(M^f(\epsilon)=1\right) \geq \frac{2}{3}$.

2. (the tester rejects every f that is far from $\mathcal{P}$): *If $|\{x \in S : f(x) \neq g(x)\}| > \epsilon \cdot |S|$, for every $g \in \mathcal{P}$, then* $\mathbf{Pr}\left(M^f(\epsilon)=1\right) \leq \frac{1}{3}$.

Property testing (as just defined) emerges naturally in the context of program checking (Blum et al. 1993, Lipton 1989, Gemmell et al. 1991, Rubinfeld and Sudan 1996) and probabilistically checkable proofs (PCP) (Babai et al. 1991b;a, Feige et al. 1996, Arora and Safra 1998, Arora et al. 1998, Bellare et al. 1993, Bellare and Sudan 1994, Bellare et al. 1996; 1998, Håstad 1996b; 1997). Specifically, in the context of program checking, one may choose to test that the program satisfies certain properties before checking that it computes a specified function. This paradigm has been followed both in the theory of program checking (Blum et al. 1993, Rubinfeld and Sudan 1996), and in practice where often programmers first test their programs by verifying that the programs satisfy properties that are known to be satisfied by the function they compute. In the context of probabilistically checkable proofs, the property tested is being a codeword with respect to a specific code. This paradigm, explicitly introduced in (Babai et al. 1991a), has shifted from testing codes defined by low-degree polynomials (Babai et al. 1991b;a, Feige et al. 1996, Arora and Safra 1998, Arora et al. 1998) to testing Hadamard codes (Arora et al. 1998, Bellare et al. 1993, Bellare and Sudan 1994, Bellare et al. 1996, Kiwi 1996, Trevisan 1998), and recently to testing the "long code" (Bellare et al. 1998, Håstad 1996b; 1997, Trevisan 1998).

Much of the work cited above deals with the development and analysis of testers for algebraic properties; specifically, linearity, multi-linearity, and low-degree polynomials (Blum et al. 1993, Lipton 1989, Babai et al. 1991b;a, Feige et al. 1996, Gemmell et al. 1991, Rubinfeld and Sudan

1996, Arora and Safra 1998, Arora et al. 1998, Bellare et al. 1993, Bellare and Sudan 1994, Bellare et al. 1996). The study of property testing as applied to combinatorial properties was initiated by Goldreich et al. (1998b). Specifically, they initiated the study of property testing in the adjacency predicate model. The study of property testing in the incidence function model was latter initiated in (Goldreich and Ron 1997). We comment that testing of combinatorial properties other than ones related to graphs has been considered in (Ergun et al. 1998, Goldreich et al. 1998a, Dodis et al. 1999): Specifically, these works consider the task of testing whether a function $f : \Sigma^n \to R$ is monotone (with respect to orderings of both Σ and R). [4]

Further generalization. We mention that an even more general formulation of property testing was suggested in (Goldreich et al. 1998b):

> Let $\mathcal{P}$ be a fixed property of functions, and f be an unknown function. The goal is to determine (possibly probabilistically) if f has property $\mathcal{P}$ or if it is far from any function which has property $\mathcal{P}$, where distance between functions is measured with respect to some distribution D on the domain of f. Towards this end, one is given examples of the form $(x, f(x))$, where x is distributed according to D. One may also be allowed to query f on instances of one's choice.

The above formulation is inspired by the PAC learning model (Valiant 1984). In fact, property testing is related to variants of PAC learning as has been shown in (Goldreich et al. 1998b) and (Kearns and Ron 1998) (the results in (Goldreich et al. 1998b) are generic and (Kearns and Ron 1998) aims at obtaining better results for properties which are related to concept classes extensively investigated in the machine learning literature). The general formulation above allows the consideration of arbitrary distributions (rather than uniform ones), and of testers which utilize only randomly chosen instances (rather than being able to query instances of their own choice).

Bibliography

N. Alon, E. Fischer, M. Krivelevich, and M. Szegedy. Efficient testing of large graphs. In *Proceedings of the 40th IEEE Symp on Foundation of Computer Science*, pages 656–666, 1999.

N. Alon and M. Krivelevich. Testing k-colorability, 1999. Preprint.

---

[4]That is, if $x_i \leq_\Sigma y_i$, for every i, then a monotone f should satisfy $f(x_1 \cdots x_n) \leq_R f(y_1 \cdots y_n)$. Ergun et al. (1998) deal only with the case $n = 1$, Goldreich et al. (1998a) focuses on the case $\Sigma = R = \{0, 1\}$, whereas Dodis et al. (1999) deal with the general case. Specifically, Dodis *et. al.* provide a monotonicity tester with complexity $O(\frac{n}{\epsilon} \cdot (\log |\Sigma|) \cdot (\log |R|))$.

N. Alon and J.H. Spencer. *The Probabilistic Method*. John Wiley & Sons, Inc., 1992.

S. Arora, D. Karger, and M Karpinski. Polynomial time approximation schemes for dense instances of NP-hard problems. In *Proceedings of the Twenty-Seventh Annual ACM Symposium on the Theory of Computing*, pages 284–293, 1995.

S. Arora, C. Lund, R. Motwani, M. Sudan, and M. Szegedy. Proof verification and intractability of approximation problems. *Journal of the ACM*, 45:501–555, 1998.

S. Arora and S. Safra. Probabilistic checkable proofs: A new characterization of NP. *Journal of the ACM*, 45:70–122, 1998.

L. Babai, L. Fortnow, L. Levin, and M. Szegedy. Checking computations in polylogarithmic time. In *Proceedings of the Twenty-Third Annual ACM Symposium on Theory of Computing*, pages 21–31, 1991a.

L. Babai, L. Fortnow, and C. Lund. Non-deterministic exponential time has two-prover interactive protocols. *Computational Complexity*, 1: 3–40, 1991b.

M. Bellare, D. Coppersmith, J. Håstad, M. Kiwi, and M. Sudan. Linearity testing in characteristic two. *IEEE Transactions on Information Theory*, 42:1781–1795, 1996.

M. Bellare, O. Goldreich, and M. Sudan. Free bits, pcps and non-approximability – Towards tight results. *SICOMP*, 27:804–915, 1998.

M. Bellare, S. Goldwasser, C. Lund, and A. Russell. Efficient probabilistically checkable proofs and applications to approximation. In *Proceedings of the Twenty-Fifth Annual ACM Symposium on the Theory of Computing*, pages 294–304, 1993.

M. Bellare and M. Sudan. Improved non-approximability results. In *Proceedings of the 26th Annual ACM Symposium on the Theory of Computing*, pages 184–193, 1994.

M. Bender and D. Ron. Testing acyclicity of directed graphs in sublinear time. In *Proceedings of ICALP*, pages 809–820, 2000.

M. Blum, M. Luby, and R. Rubinfeld. Self-testing/correcting with applications to numerical problems. *Journal of Computer and System Sciences*, 47:549–595, 1993.

W. F. de la Vega. MAX-CUT has a randomized approximation scheme in dense graphs. *Random Structures and Algorithms*, 1994. To appear.

Y. Dodis, O. Goldreich, E. Lehman, S. Raskhodnikova, D. Ron, and A. Samorodnitsky. Improved testing algorithms for monotonocity. In *Proceedings of Random99*, pages 97–108, 1999.

F. Ergun, S. Kannan, S. R. Kumar, R. Rubinfeld, and M. Viswanathan. Spot-checkers. In *Proceedings of the 30th Annual ACM Symposium on Theory of Computing*, pages 259–268, 1998.

S. Even. *Graph Algorithms*. Computer Science Press, 1979.

U. Feige, S. Goldwasser, L. Lovász, S. Safra, and M. Szegedy. Approximating clique is almost NP-complete. *Journal of the ACM*, 43: 268–292, 1996.

P. Gemmell, R. Lipton, R. Rubinfeld, M. Sudan, and A. Wigderson. Self-testing/correcting for polynomials and for approximate functions. In *Proceedings of the Twenty-Third Annual ACM Symposium on Theory of Computing*, pages 32–42, 1991.

O. Goldreich, S. Goldwasser, E. Lehman, and D. Ron. Testing monotinicity. In *Proceedings of the 39th IEEE Symp on Foundation of Computer Science*, pages 426–435, 1998a.

O. Goldreich, S. Goldwasser, and D. Ron. Property testing and its connection to learning and approximation. *Journal of the ACM*, pages 653–750, 1998b.

O. Goldreich and D. Ron. Property testing in bounded degree graphs. In *Proceedings of the 29th ACM Symp. on Theory of Computing*, pages 406–415, 1997.

O. Goldreich and D. Ron. A sublinear bipartite tester for bounded degree graphs. *Combinatorica*, 19:1–39, 1999.

P. Hajnal. An $\Omega(n^{4/3})$ lower bound on the randomized complexity of graph properties. *Combinatorica*, 11:131–144, 1991.

J. Håstad. Clique is hard to approximate within $n^{1-\epsilon}$. In *Proceedings of the 37th IEEE Symp on Foundation of Computer Science*, pages 627–636, 1996a.

J. Håstad. Testing of the long code and hardness for clique. In *Proceedings of the 28th ACM Symp. on Theory of Computing*, pages 11–19, 1996b.

J. Håstad. Getting optimal in-approximability results. In *Proceedings of the 29th ACM Symp. on Theory of Computing*, pages 1–10, 1997.

D. S. Hochbaum and D. B. Shmoys. Using dual approximation algorithms for scheduling problems: Theoretical and practical results. *Journal of the Association for Computing Machinery*, 34:144–162, 1987.

D. S. Hochbaum and D. B. Shmoys. A polynomial approximation scheme for machine scheduling on uniform processors: Using the dual approximation approach. *SIAM Journal on Computing*, 17:539–551, 1988.

M. Kearns and D. Ron. Testing problems with sub-learning sample complexity. In *Proceedings of the 11th COLT*, pages 268–277, 1998.

V. King. An $\Omega(n^{5/4})$ lower bound on the randomized complexity of graph properties. *Combinatorica*, 11:23–32, 1991.

M. Kiwi. *Probabilistically Checkable Proofs and the Testing of Hadamard-like Codes*. PhD thesis, Massachusetts Institute of Technology, 1996.

R. J. Lipton. New directions in testing, 1989. Unpublished manuscript.

L. Lovász and N. Young. Lecture notes on evasiveness of graph properties. Technical Report TR–317–91, Computer Science Department, Princeton University, 1991.

M. Mihail. Conductance and convergence of Markov chains – A combinatorial treatment of expanders. In *Proceedings 30th Annual Symp on Foundations of Computer Science*, pages 526–531, 1989.

M. Parnas and D. Ron. Testing the diameter of graphs. In *Proceedings of Random99*, pages 85–96, 1999.

R. L. Rivest and J. Vuillemin. On recognizing graph properties from adjacency matrices. *Theoretical Computer Science*, 3:371–384, 1976.

A. L. Rosenberg. On the time required to recognize properties of graphs: A problem. *SIGACT News*, 5:15–16, 1973.

R. Rubinfeld and M. Sudan. Robust characterization of polynomials with applications to program testing. *SIAM Journal on Computing*, 25:252–271, 1996.

L. Trevisan. Recycling queries in PCPs and in linearity tests. In *Proceedings of the 30th ACM Symp. on Theory of Computing*, 1998.

L.G. Valiant. A theory of the learnable. *Communications of the ACM*, 27:1134–1142, 1984.

A. C. C. Yao. Lower bounds to randomized algorithms for graph properties. In *Proceedings of the Twenty-Eighth Annual Symposium on Foundations of Computer Science*, pages 393–400, 1987.

III

STRING PROCESSING AND DATA COMPRESSION

Chapter 6

STRING PATTERN MATCHING FOR A DELUGE SURVIVAL KIT

Alberto Apostolico
*Department of Computer Sciences, Purdue University, West Lafayette, IN 47907, USA
and Dipartimento di Elettronica e Informatica, Università di Padova, Padova, Italy*
axa@cs.purdue.edu

Maxime Crochemore
Institut Gaspard Monge, Université de Marne-la-Vallée, F-93160 Noisy-le-Grand, France
Maxime.Crochemore@univ-mlv.fr

Abstract String Pattern Matching concerns itself with algorithmic and combinatorial issues related to matching and searching on linearly arranged sequences of symbols, arguably the simplest possible discrete structures. As unprecedented volumes of sequence data are amassed, disseminated and shared at an increasing pace, effective access to, and manipulation of such data depend crucially on the efficiency with which strings are structured, compressed, transmitted, stored, searched and retrieved. This paper samples from this perspective, and with the authors' own bias, a rich arsenal of ideas and techniques developed in more than three decades of history.

Keywords: String matching, Searching, Indexing, Fingerprinting, Regular expressions, Filtration, Inference, Periodicities, Compression, encoding, Probabilistic automata, Markov chains, Antidictionaries, Association rules.

1. Introduction

This chapter reviews a number of rather ubiquitous primitives related to matching and searching with some elementary discrete structures such as strings, regular expressions, and other aggregates, that are likely to be of relevance, directly or indirectly, in the current and future infrastructures of very large volumes of data. In that context, massive, scattered and diverse information repositories will pose increasing needs for novel approaches to their management by means of compression, inference, comparison and retrieval, mining, and related principles and techniques. Without pretending to be exhaustive, the selection of topics presented in this chapter was inspired by two main principles beside the authors' own bias. The first one, was to recognize that the data flood is forcing a paradigm shift to take place, whereby the previous ambition to organize and funnel to the user as much data as possible is being changed into that of limiting and filtering what the limited ultimate bandwidth, the user himself, may actually intake. The second, and related principle, is that, in computer science jargon, search by value is going to be increasingly replaced by search by contents and, in turn, by search by meaning, in the future. It is believed that, while eminently syntactic in nature, most of the primitives considered here shall still form the core of the semantic capabilities subtending automated association generation and other similar techniques of filtration and inference.

Problems of matching and searching, and the combinatorial properties that support their efficient solutions, may be classified according to a number of paradigms. One way to classify these problems is according to the type of structure (strings, arrays, trees, etc.) in terms of which they are posed. Another is according to the model of computation used, e.g., serial or parallel. Yet another one is according to whether the manipulations that one seeks to optimize need be performed on-line, off-line, in real time, etc. One could distinguish further between matching and searching and, within the latter, between exact and approximate searches, or vice versa. The classification used here privileges certain aspects of exact or approximate searching, combinatorial issues such as the identification of periodicities, symmetries and other regularities, efficient implementations of ancillary functions such as compression and encoding, etc., that are perceived as most relevant in the current context. Due to space limitations we emphasize here problems on strings, but it should be clear that most problems (albeit not their solutions) translate straightforwardly to more complicated structures.

This chapter is organized as follows. In the next section, we review some fundamental facts about regularities that manifest themselves in the form of repetitive substructures. In Section 3, we address issues of searching and indexing: we describe there two central tools for these tasks, suffix and subword automata, and consider their implementation issues in massive data contexts. Section 4 deals with basic problems of counting substring statistics and estimating empirical probabilities in early probabilistic models. In Section 5, we address issues of filtering, fingerprinting and related compaction techniques that variously enter data reduction, certification, watermarking, but also approximate patterns comparison and search. In Section 6, we consider problems of compression, mining for associations and other inference issues in strings.

Some preliminary notational conventions follow. Given an alphabet Σ, we use Σ^+ to denote the free semigroup generated by Σ, and set $\Sigma^* = \Sigma^+ \cup \{\lambda\}$, where λ is the empty word. An element of Σ^* is called a *string* or *sequence* or *word*, and is denoted by one of the letters s, u, v, w, x, y and z. The same letters, upper case, are used to denote *random* strings. We write $x = x_1 x_2 ... x_n$ when giving the symbols of x explicitly. The number n of symbols that form x is called the *length* of x and denoted by $|x|$. If $x = vwy$, then w is a *substring* or *factor* of x and the integer $1 + |v|$ is its *(starting) position* in x. Let $I = [i, j]$ be an interval of *positions* of a string x. We say that a substring w of x *begins* in I if I contains the starting position of w, and that it *ends* in I if I contains the position of the last symbol of w.

2. Basic Regularities and Their Detection

It is customary to distinguish among three types of information: syntactic, semantic, and pragmatic, the last one being an attempt to describe the understanding of meaning as a natural process. As much as we would like to get to this third level, it is likely that we shall only be able to occasionally grasp at the second one using tools and methods of the first. In this section, we see that even restricting to syntactic regularities does not make the job trivial.

Syntactic regularities in strings play a pervasive role in many facets of data analysis. Searching for repeated patterns, periodicities, symmetries, cadences, and other similar forms or unusual patterns in objects is a recurrent task in the compression of data, symbolic dynamics, genome studies, intrusion detection, and countless other activities. In many applications, such regularities represent redundancies and, as such, are sought to be removed. This is the case of Data Compression. In textual substitution methods, for example, strings that appear many times in a subject can be economically replaced by pointers to a single common copy. In many other applications, these kind of regularities are sought as carriers of information. This display of duality for information in this context has been known and debated since early years (Shannon and Weaver 1949, Brillouin 1971).

We concentrate here on a very restricted class of regularities such as cadences, periods, squares, repetitions, palindromes and approximate versions thereof. The first thing to be said is that there are *avoidable* and *unavoidable* such regularities (see, e.g., (Bean et al. 1979, Lothaire 1997)).

2.1. Unavoidable regularities

One remarkable application of the Pigeon Hole Principle leads to establish that if the set of natural numbers $\mathcal{N}$ is partitioned into k classes, then one of the classes contains arbitrarily long arithmetic progressions.

More precisely, we say that the integers $t_1 < t_2 < ... < t_n$ are a *cadence* for word $x_1 x_2 ... x_r$ if $x_{t_1} = x_{t_2} = ... = x_{t_n}$. In this case we also say that n is the *order* of the cadence.

Let now S be a finite subset of $\mathcal{N}$. A cadence of type S is a cadence of the form $\alpha S + \beta$ (i.e., an arithmetic cadence with common difference α when $\alpha, \beta > 0$) For example, ab*babbabbaab* shows an arithmetic cadence of order 4 with $\alpha = 3, \beta = 3, S = \{1, 2, 3\}$. The following theorems hold (see Lothaire (1997)).

Theorem 1 If A is an alphabet with k letters and n is an integer, there is an integer $N = N(k, n)$ such that every word of length $\geq N$ has an arithmetic cadence of order n

Theorem 2 Let S be any finite subset of $\mathcal{N}$ and A an alphabet with k letters. There exists an integer N depending only of S and k such that every word of length $\geq N$ has a cadence of type S.

2.2. Some avoidable regularities: Periods, palindromes and squares

Periods and periodicities are pervasive notions of string algorithmics. A string z has a *period* w if z is a prefix of w^k for some integer $k > 0$ and some non-empty prefix w of z. Alternatively, a string w is a period of a string z if $z = w^l v$ and v is a possibly empty prefix of w. Often when this causes no confusion, we will use the word "period" also to refer to the length or *size* $|w|$ of a period w of z. A string may have several periods. The shortest period (or period length) of a string z is called *the period* of z. Clearly, a string is always a period of itself. This period is called the trivial period.

A germane notion is that of a border. We say that a non-empty string w is a *border* of a string z if z starts and ends with an occurrence of w. That is, $z = uw$ and $z = wv$ for some possibly empty strings u and v. Clearly, a string is always a border of itself. This border is called the trivial border. The implications of these notions on fast string searching are well understood. In fact, it is not difficult to see that two consecutive occurrences of a word may overlap only if their distance equals one of the periods of w.

A string can have many periods, and corresponding borders. The smallest (resp. longer border) period is *the* period (resp., *the* border) of the string. For example *abaabaababaabaababaabaabaab* has borders at *abaabaab*, *abaab* and *ab*.

Once we know how to compute all periods of a string then we also know how to compute all initial palindromes of a string. A *palindrome* is a string that reads the same forward and backward, i.e., $w = w^R$, where w^R is the *reverse* of string w. For this, we run the algorithm on $w!w^R$ where ! is not in the alphabet. Better palindrome detectors are known. In 1976, G. Manacher showed that all initial palindromes –in fact, all palindromes if one just lets the algorithm go on– of a string can be found in linear time (Manacher 1975).

A string can avoid having any nontrivial period but will not take two periods for long. We give here a weak version of an important result known as the "periodicity lemma" (Fine and Wilf 1965, Lyndon and Schutzenberger 1962).

Lemma 1 If w has two periods of length p and q and $|w|$ is at least $p + q$ then w has period $\gcd(p, q)$.

Proof. Assume w.l.o.g. $p > q$ and consider w_i for arbitrary i. We have that either $i - q \geq 1$ or $i + p \leq n$. In the first case, $x_i = x_{i-q} = x_{i-q+p}$, in the second case $x_i = x_{i+p} = x_{i+p-q}$. Thus, $p - q$ is a period. Repeating the treatment on the pair $p, p - q$ leads to the claim. ◇

The computation of the longest borders (and corresponding periods) of all prefixes of a string is afforded in overall linear time and space. We report one such construction in Figure 6.1, for the convenience of the reader, but refer for details and proofs of linearity to discussions of "failure functions" and related constructs such as found in, e.g., (Aho and Corasick 1975, Apostolico and Galil 1997, Crochemore and Rytter 1994).

Once the period structure of the pattern is unveiled, this immediately yields a linear time string searching algorithm. The key element of the algorithm is to maintain, during a text scanning, the notion of the longest prefix of the pattern matched so far, and use the border table to jump over intermediate non-viable candidates. These developments will be discussed some more later, in connection with subword automata.

Let $\pi(w)$ denote the shortest non-zero period length of w. A string w such that $|w| \geq 2\pi(w)$ is said to be *periodic*. By the periodicity lemma, in a periodic string w, all periods lengths that are smaller than $|w|/2$, must be multiples of *the period* length $\pi(w)$. A string w such that setting

```
procedure maxborder ( y )
  begin
    bord[0] ← −1; r ← −1;
    for m = 1 to h do
      while r ≥ 0 and y_{r+1} ≠ y_m do
        r ← bord[r];
      endwhile
      r = r + 1;  bord[m] = r
    endfor
  end
```

Figure 6.1. Computing the longest borders for all prefixes of y

$w = v^k$ implies $k = 1$ is called *primitive*. A *square* is a string w in the form $w = vv$ with v a primitive string. It is natural to wonder whether squares represent avoidable or unavoidable regularities. As is readily seen, on an alphabet of two symbols we can only build a very short string not containing any squares, i.e., a *square-free* string. In fact, in the first three steps we must generate either 010 or 101, at which point adding, say, 0 to 010, introduces the square 00 while adding 1 yields 0101.

At the beginning of the century, A. Thue (Thue 1906; 1912) found that over an alphabet of at least 3 symbols he could build an indefinitely long square-free string. This was achieved by giving a square-free *morphism*, i.e., a rewriting rule that when applied to a square-free string would preserve square-freeness. The morphism considered by Thue is: $rew(a) = abcab$, $rew(b) = acabcb$ and $rew(c) = acbcacb$. Later, S. Istrail (see Berstel (1979)) gave a more compact morphism that is square free if started on the letter a: $rev(a) = abc$, $rew(b) = ac$, and $rew(c) = b$. As for a binary alphabet, it is possible to show that we can build infinite *cube-free* strings, with obvious meaning.

There are, in principle, about $n^2/2$ possible ways to choose indices i and j for the starting and ending positions of a substring in a string of n symbols, and these might all correspond to distinct strings. Is it possible to have as many squares? As it turns out, there can be only $O(n \log n)$ squares. One way to prove this is by giving an algorithm that enumerates all the squares. M. Crochemore showed in 1981 (Crochemore 1981) that this number of squares is also tight: the *Fibonacci* strings, defined by $F_0 = a$, $F_1 = b$, and $F_i = F_{i-1}F_{i-2}$, attain this bound.

There are several efficient or optimal serial (Main and Lorentz 1984, Rabin 1985, Crochemore 1981, Apostolico and Preparata 1983, Gusfield

2.3. Quasiperiods and covers

In the Summer of 1990, A. Ehrenfeucht suggested that some repetitive structures defying the classical characterizations of periods and repetitions could be captured by resort to a germane notion of "quasiperiod". Apostolico and Ehrenfeucht (1993) defined *quasiperiodic* strings as *strings which are entirely covered by occurrences of another (shorter) string*. They also gave an $O(n \log^2 n)$ time algorithm to find all maximal quasiperiodic substrings within a given string. Apostolico et al. (1991) gave an $O(n)$ time algorithm that finds the quasiperiod of a given string, namely the *shortest string* that covers the string in question. This algorithm was subsequently simplified and improved by Breslauer (1992) who gave an $O(n)$ time on-line algorithm, and parallelized by Breslauer (1994) and Iliopoulos and Park (1994), the latter giving an optimal-speedup $O(\log \log n)$ time parallel CRCW-PRAM algorithm. Moore and Smyth (1994) gave an $O(n)$ time algorithm that finds *all strings* that cover a given string. These developments eventually led to the study by Iliopoulos et al. (1993) and by Ben-Amram et al. (1994) of covers which are not necessarily aligned with the ends of the string being covered, but are rather allowed to overflow on either side. The sequential algorithm for this problem takes $O(n \log n)$ time (Iliopoulos et al. 1993) and the parallel counterpart (Ben-Amram et al. 1994) achieves an optimal speedup taking $O(\log n)$ time, but using superlinear space.

To understand these developments, it is convenient to modify slightly the notion of a period. A non-empty string u, $|u| \leq |w|$, will be called *a period* of w if w is a substring of u^k, for some integer $k \geq 1$. Clearly, if u is a period of w, then its length $|u|$ is a period length of w, since $|u|$ is a period length of u^k. Moreover, if $u = xy$, then any *rotation* yx of u is also a period of w since $(yx)^{k+1} = y(xy)^k x = yu^k x$ contains w as a substring. A period u of w that is also a prefix of w is called a *left aligned period*. Clearly, given any period length $\pi > 0$ of w, the prefix $w_{[1...\pi]}$ is a left aligned period of w.

A period u is in fact a *regular cover* of w, where occurrences of u appear in w spaced exactly $|u|$ positions apart (other occurrences are also allowed) and the occurrences on the sides can overflow. Given any period u of w, consider the rotation $\hat{u}$ of u such that $\hat{u}$ is also a prefix of w (in other words, $\hat{u}$ is the rotation of u that is a left aligned period of

w). If $w = \hat{u}^k$ for some integer k, namely if the regular cover of w by u is also right aligned, then w is said to have an *aligned regular cover* u. If w has no proper aligned regular covers (w itself is always a cover) then w is primitive.

2.3.1 General covers. One may generalize the notion of a period u that covers w with regular occurrences that are $|u|$ positions apart in w, to covers where the occurrences of u in w are not required to be uniformly spaced, and are allowed, in addition, to overflow on either side. For example, the string $w =$ 'aabaabab' may be covered by occurrences of $u =$ 'aba', but the positions of these occurrences in w are not regular and in fact *aba* is not a period of w. This type of covers were called *general covers* in Iliopoulos et al. (1993) where a covering string such as our u above is also termed *a seed* of w.

2.3.2 Aligned covers. Some notable families of covers result by considering covering strings u for w that are not necessarily regularly spaced but are aligned on both sides of w and are not allowed to overflow. Such strings u are said to be aligned covers of w. Given the similarity between non-regular covers and regular covers (periods), aligned covers u of w were named *quasiperiods* of w by Apostolico and Ehrenfeucht (1993). In addition, strings that do not have any non-trivial (shorter) aligned covers were called *superprimitive* and strings that have shorter aligned covers were termed *quasiperiodic*. Observe that any periodic string is also quasiperiodic, but not every quasiperiodic string is periodic. Most of our treatment here is confined to aligned covers.

We describe next few easy facts about periods, borders, and aligned covers.

Lemma 2 If a string z is an aligned cover for a string w then z is a border of w.

Proof. Since the first symbol of w must be covered by z, the string w must start with an occurrence of z. Since the last symbol of w must also be covered by z, the string w must also end with an occurrence of z. That is, z is a border of w. ◇

Note that by this last fact any cover of a string w can be represented by a single integer that is the length of the border of w.

Lemma 3 If a string z covers a string w, then z covers also any possible border v of w such that $|v| \geq |z|$.

Proof. Given any prefix of w, it is covered by z except possibly at most the last $|z| - 1$ symbols of the prefix. Similarly, given any suffix of w, it is covered by z except possibly at most the first $|z| - 1$ symbols of the suffix. Since v is a border of w, it is both a prefix and a suffix, and it must be covered by z. ◇

Lemma 4 Every string has a unique quasiperiod.

Proof. Assume that a string w is covered by two strings u and v, and let w.l.o.g. $|u| \leq |v|$. By Lemma 2, v is a border of w. By Lemma 3, u covers w. Since $u \neq v$, then v is quasiperiodic. ◇

Lemma 5 If a string w has a border z such that $2|z| \geq |w|$, then z covers w.

Proof. z covers the first half of w since it is a prefix of w and the last half of w since it is also a suffix. Therefore, all symbols of w are covered by z. ◇

3. Modeling, Counting, Estimating and Scoring

In many applications, repetitions of substrings and other substructures represent redundancies and, as such, may be sought just so as to be removed. This is the case of Data Compression. In textual substitution methods, for example, strings that appear many times in a subject can be economically replaced by pointers to a single common copy. In many other applications, these same kinds of regularities are sought as carriers of information. In applications ranging from Consumer Prediction to Data Mining, Intrusion Detection and Security, Protein and other Biological Sequence Classification, the idea is to infer a consistent behavior from some protocol of past records and then use it to predict future behavior or detect malicious practices. This entails some notion of sequence similarity, whereby having established some set of behavioral sequences as constituting the normal profile, any new sequence can be compared to the dictionary and possibly classified or spotted as abnormal. Learning takes place in general from both positive and negative samples.

As already mentioned, this display of somewhat of a duality for the notion of information has been sensed and debated for decades (Shannon and Weaver 1949, Brillouin 1971). In Shannon's terms, for instance, the self-information of string x relative to a given source P is measured by $-\log P(x)$. This notion is central to coding: the mean codelength of any Uniquely Decipherable Code for strings of the same length is lower bounded by the entropy, the mean of self information. For Brillouin, information is related to redundancy and negentropy, entropy is chaos.

Either way, in our applications we do not know the source probabilities, which are in fact fictitious entities or models. One pervasive problem is therefore to estimate the probabilities from the observed strings, to be used in the design of codes for compression or other purposes. The domains in which this need arises are countless: Prediction, Inference, Modeling, Learning, and Universal Coding, to quote a few. From an information theoretic standpoint, an important question is how to define a notion of information relative to a class of sources. From the standpoint of Pattern Matching, interesting questions revolve around how computationally expensive it is to estimate probabilities and related deviations within a given class. Below, we consider some preliminary counts and statistical computations. Later in this chapter, we will also consider issues of modeling by Markov Chains and related Finite State Automata sources.

3.1. Basic string counts and statistics

The tree T_x is a remarkable compendium of the structure of a string. It can be immediately adapted to solve problems such as finding the longest repeated substring, the longest substring common to many strings, or finding squares or palindromes, etc. To find squares, for instance, it suffices to note the following:

Lemma 6 There is a square in x iff there is a node μ in T_x such that the subtree rooted at μ contains two consecutive leaves i and j such $j - i \leq |w(\mu)|$, where $w(\mu)$ denotes the word on the path from the root to μ.

In fact, if $j - i \leq |w(\mu)|$ as stated, then the two occurrences of $w(\mu)$ at i and j are adjacent or overlap, whence we must have a square. We leave it for the reader to show that in the converse of the proof leaves i and j are indeed consecutive as claimed. The reader might also find it interesting to derive a similar criterion for the detection of palindromes on the tree of $x\$x^R$.

Also the count of occurrences of all substrings of a string x is an easy application of T_x. The number of occurrences (with overlap) of a string w of x is trivially given by the number of leaves reachable from the node closest to the locus of w in T_x, and this is irrespective of whether or not w ends in the middle of an arc. Thus, labeling every internal node α of T_x with the number $c(\alpha)$ of the leaves in the subtree rooted at α yields these statistics for all substrings of x.

The problem becomes more involved if we wanted to build a similar index for the statistics without overlap, in which we count, for each substring, its maximum number of nonoverlapping occurrences. It is seen that this transition induces a twofold change in the structure: on the one hand, the weight in each node does no longer necessarily coincide with the number of leaves; on the other, extra nodes must be introduced to account for changes in the statistics that occur in the middle of arcs. The efficient construction of this augmented index in minimal form (i.e., with the minimum possible number of unary nodes) is quite elaborate (Apostolico and Preparata 1996). For a string x, the resulting structure is denoted $\hat{T}(x)$ and called the *Minimal Augmented Suffix Tree* of x. It is not difficult to build $\hat{T}_x$ in $O(n^2)$ time and space by embedding the necessary weighting as part of the iterated suffix insertion procedure, hence at an expected cost of $O(n \log n)$ (Apostolico and Szpankowski 1992). The time required by the construction given in Apostolico and Preparata (1996) is instead $O(n \log^2 n)$ in the worst case. The number of auxiliary nodes can be bounded by $O(n \log n)$, but it is not clear that such a bound is tight.

Consider for a moment the problem of defining and computing empirical probabilities. One problem here is that the notion of empirical probability is not straightforward. Fortunately, empirical *conditional* probabilities often turn out to be less controversial. One ingredient in the computation of empirical probabilities is the count of occurrences of a string in another string or set of strings. We concentrate on this problem first. Since there can be $O(n^2)$ distinct substrings in a string of n symbols, a table storing the number of occurrences of all substrings of the string might take up in principle $\Theta(n^2)$ space. However, we just saw that linear time and space suffice to build an index suitable to return, for any string w, its χ_w count in x. Here we want to analyze this fact a little more closely. We begin by formulating a "left-context" property, symmetric to one already seen, and conveniently adapted from Blumer et al. (1985).

Given two words x and y, let the *start-set* of y in x be the set of *occurrences* of y in x, i.e., $pos_x(y) = \{i : y = x_i...x_j\}$ for some i and j, $1 \leq i \leq j \leq n$. Two strings y and z are equivalent on x if $pos_x(y) = pos_x(z)$. The equivalence relation instituted in this way is denoted by $\equiv_x$ and partitions the set of all strings over Σ into equivalence classes. Recall that the *index* of an equivalence relation is the number of equivalence classes in it.

Lemma 7 The index k of the equivalence relation $\equiv_x$ obeys $k < 2n$.

Lemma 7 is established in analogy to its right-context counterpart seen in connection with dawgs. In the example of the string *abaababaabaababaababa*, for instance, $\{ab, aba\}$ forms one such C_i-class and so does $\{abaa, abaab, abaaba\}$. Lemma 7 suggests that we might only need to compute empirical probabilities for $O(n)$ substrings in a string with n symbols. The considerations developed earlier make this statement more precise and in fact give one possible proof of it.

We are now ready to consider more carefully the notion of empirical probability. One way to define the empirical probability of w in x is to take the ratio of the count of the number χ_w to $|x| - |w| + 1$, where the latter is interpreted as the maximum number of possible starting positions for w in x. For w and v much shorter than x we have that the difference between $|x| - |w| + 1$ and $|x| - |wv| + 1$ is negligible, which means that the probabilities computed in this way and relative to words that end in the middle of an arc do not change, i.e., we only need to compute those associated with strings that end at a node of the compact T_x.

This notion of empirical probability, however, assumes that every position of x compatible with w length-wise is an equally likely candidate. This is not the case in general, since the maximum number of possible occurrences of one string within another string depends crucially on the compatibility of self-overlaps. For example, the pattern *aba* could occur at most once every two positions in *any* text, *abaab* once every three, etc. Compatible self-overlaps for a string z depend on the structure of the periods of z. An alternative count can be defined as follows.

Definition The maximum possible number of occurrences of a string w into another string x is equal to $(|x| - |w| + 1)/|u|$, where u is the smallest period of w.

According to this definition, in order to compute the empirical probabilities of, say, all prefixes of a string we need to know the borders or periods of all those prefixes. In fact, we know we can manage to carry out *all* the updates relative to the set of prefixes of a same string in *overall* linear time, thus in amortized constant time per update.

The construction of Fig. 6.1 may be applied, in particular, to each suffix suf_i of a string x while that suffix is being inserted as part of the direct tree construction. This would result in an annotated version of T_x in overall quadratic time and space in the worst case. Note that, unlike in the case of empirical probabilities previously considered, the period —and thus also the empirical probabilities according to our definition above— may change along an arc of T_x, so that we may need to compute explicitly all $\Theta(n^2)$ of them. However, if we were interested in such probabilities only at the nodes of the tree, then these could still be computed in overall linear time. The key to this latter fact is to run a suitably adapted version of `maxborder` walking on suffix links "backward", i.e., traversing them in their reverse direction, beginning at the root of T_x and then going deeper and deeper into the tree. One way to visualize this process is as follows. Imagine first the "co-tree" of T_x formed by the reversed suffix links: we can visit such a structure depth first and simultaneously run a procedure much similar to `maxborder` to assign periods to all nodes of T_x. Correctness rests on the fact that for any word w the periods of w and w^R coincide. We shall see shortly that in situations of interest to us we can limit computation to the nodes of T_x.

Lemma 8 *The set of empirical probabilities of all (short) words of x that have a proper locus in T_x can be computed in linear time and space.*

Consider now conditional empirical probabilities, defined as the ratio between the observed occurrences of $s\sigma$ and the occurrences of $s*$, denoting string s followed by any other symbol. The first thing to observe is that the value of this ratio persists along each arc of the tree, i.e.,

$$\tilde{P}(\sigma|s) = \chi_s/\chi_{s\sigma} = 1$$

for any word s ending in the middle of an arc of T_x and followed there by a symbol σ.

Let ν' be the locus of string s'. Recall that $sext[\nu', \sigma]$ is the node ν which is the locus of the shortest extension of as' having a proper locus in T_x. Setting $sext$ links is an easy linear post-processing of T_x. Along these lines, attaching empirical conditional probabilities only to the branching nodes of T_x is doable and suffices. As there are $O(n)$ such nodes, and the alphabet is finite, the collection of all conditional probability vectors for *all* subwords of x takes only linear space.

Lemma 9 *The set of empirical conditional probabilities of all (short) words of a string x over a finite alphabet can be computed in linear time and space.*

An important class of applications, which includes some core tasks of molecular sequence analysis and information retrieval, involves counting, estimating and comparing to expectation not the occurrence number of a word in a text but rather the number of sequences in a given family that contains that word. With some provisos, the constructions just highlighted may be adapted to deal with this notion. The reader is encouraged to develop the details.

3.2. Global detectors of unusual words

As mentioned, the identification of strings that are, by some measure, redundant or rare in the context of larger sequences is variously pursued in order to compress data, unveil structure, infer minimal or compact descriptions, and for purposes of feature extraction and classification. Once a statistical index is built and empirical probabilities are computed, the next step is thus to annotate it with the expected values and variances and measures of discrepancy thereof, under some adopted probabilistic model. This may be still rather bulky in practice. For a given probabilistic model and measure of departure from expected frequency, it is possible to come up with an "observed" string such that all of its $\Theta(n^2)$ substrings are surprisingly over- or under-represented. This means that a table of the "surprising" substrings of a string can contain in principle a number of entries quadratic in the length of that string. As it turns out, it is possible to show that under several accepted measures of frequency deviation, the candidates over- or under-represented words are restricted to the $O(n)$ words that end at internal nodes of a compact suffix tree. as opposed to the $\Theta(n^2)$ possible substrings. Combined with some of the constructions discussed earlier in this section, this leads to the design of global detectors for unusual words that take linear space and linear time to build (Apostolico et al. 1996).

To make our discussion more precise, we need to agree on some measure of "surprise". Perhaps the naivest possible measure is the difference: $\delta_w = f_w - (n-|w|+1)\hat{p}$, where $\hat{p}$ is the product of symbol probabilities for w and $Z|w$ takes the value f_w. Let us say that an over-represented (respectively, under-represented) word w in some class C is δ-significant if no extension (respectively, prefix) of w in C achieves at least the same value of $|\delta|$.

Theorem 3 *The only over-represented δ-significant words in x are the $O(n)$ ones that have a locus in T_x. The only under-represented δ-significant words are the ones that represent one unit-symbol extensions of words that have a locus in T_x.*

Proof. We prove first that no over-represented δ-significant word of x may end in the middle of an arc of T_x. Specifically, any over-represented δ-significant word in x has a proper locus in T_x. Assume for a contradiction that w is a δ-significant over-represented word of x ending in the middle of an arc of T_x. Let $z = wv$ be the shortest extension of w with a defined locus in T_x, and let $\hat{q}$ be the probability associated with v. Then, $\delta_z = f_z - (n - |z| + 1)\hat{p}\hat{q} = f_z - (n - |w| - |v| + 1)\hat{p}\hat{q}$. But we have, by construction, that $f_z = f_w$. Moreover, $\hat{p}\hat{q} < \hat{p}$, and $(n - |w| - |v| + 1) < (n - |w| + 1)$. Thus, $\delta_z > \delta_w$. For this specification of δ, it is easy to prove symmetrically that the only candidates for δ-significant under-represented words are the words ending precisely one symbol past a node of T_x. ◇

It is possible to prove similar properties for more sophisticated measures of surprise characterized by definitions of δ of the more general form: $\delta_w = (f_w - E_w)/N_w$, where: (a) f_w is the frequency or count of the number of times that the word w appears in the text; (b) E_w is the typical or average nonnegative value for f_w (and E is often chosen to be the expected value of the count); (c) N_w is a nonnegative normalizing factor for the difference. (The N is often chosen to be the standard deviation for the count.)

Once one is restricted to the branching nodes of T_x or their one-symbol extensions, it becomes even possible to compute all typical count values E (usually expectation) and their normalizing factors N (usually standard deviation) and other measures discussed earlier in overall linear time and space. For strings emitted by a source with i.i.d. symbols, this is easy to see for expectations but becomes more complicated with variances. To see this, let x be the observed string and $y = y_1 y_2 \ldots y_m$ ($m < (n+1)/2$) be an arbitrary but fixed pattern. For $i \in \{1, 2, \ldots, n - m + 1\}$, define $Z_i | y$ to be 1 if y occurs in X starting at position i and 0 otherwise. We are interested in the expected value and variance of $Z|y$, the total number of occurrences of y in X:

$$Z|y = \sum_{i=1}^{n-m+1} Z_i|y.$$

It is immediate that $E[Z|y] = (n - m + 1)\hat{p}$, where, with p_i denoting the probability for any given k that $X_k = y_i$, $\hat{p} = \Pi_{i=1}^{m} p_i$.

For any symbol a in Σ, computing the expected value $Z|ya$ from $\hat{p}$ and the probability of a is trivially done in constant time. Thus, the expected values associated with all prefixes of a string can be computed in linear time. Attaching these values to the nodes of T_x is easily accomplished in linear time by walking backward on suffix links.

For $m \leq (n+1)/2$, it is possible to express the variance in the following form (the case $m > (n+1)/2$ is quite similar) (Apostolico et al. 1996):

$$Var(Z|y) = (n - m + 1)\hat{p}(1 - \hat{p}) - \hat{p}^2(2n - 3m + 2)(m - 1)$$

$$+ 2\hat{p} \sum_{l=1}^{s_m} (n - m + 1 - d_l) \Pi_{j=m-d_l+1}^{m} p_j$$

where the d_l's are the *periods* of y that satisfy

$$1 \leq d_1 < d_2 < \ldots < d_{s_m} \leq \min(m - 1, n - m).$$

Suppose that we wanted to compute the variance of $Z|y$ for all substrings y of x in accordance to the formula above. Applying the formula from scratch to each substring would require time $\Theta(|x|^3)$, since the number of possible distinct words appearing as substrings of x may be quadratic in $|x|$. In Apostolico et al. (1996), the variance is computed for all prefixes of a string y in overall time $O(|y|)$, by making crucial use of a recurrence that speeds up computation of the term

$$B(m) = \sum_{l=1}^{s_m}(n-m+1-d_l)\Pi_{j=m-d_l+1}^{m} p_j.$$

In this expression, $B(m)$ refers to the prefix $y_1 y_2 ... y_m$ of some string y, $S(m) = \{b_{l,m}\}_{l=1}^{s_m}$ is the set of borders "at m" associated with the periods of $y_1 y_2 ... y_m$ and $bord(m)$ is the longest border of $y_1 y_2 ... y_m$. By a simple adaptation of the maxborder it is possible to derive $B(m)$ quickly from knowledge of $bord(m)$ and of the previously computed values $B(1), B(2), ..., B(m-1)$. Specifically, letting the border associated with period d_l at position m to be

$$b_{l,m} = m - d_l,$$

the following expression of $B(m)$ holds:

$$B(m) = (n - 2m + 1 + bord(m))\Pi_{j=bord(m)+1}^{m} p_j$$

$$+ 2(bord(m) - m) \sum_{l=1}^{s_{bord(m)}} \Pi_{j=b_{l,bord(m)}+1}^{m} p_j$$

$$+ (\Pi_{j=bord(m)+1}^{m} p_j) \, B(bord(m)),$$

where the fact that $B(m) = 0$ for $bord(m) \leq 0$ yields the initial conditions. Note that each product of probabilities can be extracted in constant time from a precomputed table containing the products of the probabilities of all consecutive prefixes of x. From knowledge of $n, m, bord(m)$ and these prefix probability products, the first term of $B(m)$ is computed in constant time. Except for $(bord(m) - m)$, the second term is essentially a sum of probability products taken over all distinct borders of $y_1 y_2 ... y_m$. Thus, given such a sum and $B(bord(m))$ at this point enables one to compute $B(m)$ whence also the variance, in constant time. Maintaining knowledge of the value of such sums during the computation of longest borders is easy, since the value of the sum

$$T(m) = \sum_{l=1}^{s_{bord(m)}} \Pi_{j=b_{l,bord(m)}+1}^{m} p_j$$

obeys the recurrence:

$$T(m) = T(bord(m)) \cdot \Pi_{j=bord(m)+1}^{m} p_j + \Pi_{j=bord(bord(m))+1}^{m} p_j,$$

with $T(m) = 0$ for $bord(bord(m)) \leq 0$. In conclusion, the following holds.

Theorem 4 *Under the independently distributed source model, the mean and variances of all prefixes of a string can be computed in time and space linear in the length of that string.*

Application of this treatment to every suffix of a string yields the mean and variance of all substrings in overall optimal quadratic time. From what we have seen, the quest for surprising words under this model can be limited to those ending at the internal nodes of T_x. Since also the variances can be computed with our recurrence traveling backward on suffix links, this results in a global detector of unusual words in linear time and space.

4. Filtering, Fingerprinting and Approximate Searching

The underlying theme of this section is the derivation of succinct albeit possibly approximate representations of objects. Hashing is one obvious way to do this. In an early approach to fast string searching, Karp, Miller and Rosenberg (cf. (Crochemore and Rytter 1991b)) introduced a strategy based on some notion of a label or signature for the substrings of a string x, also called naming or numbering, as follows. First, generate the list of labels for individual characters, giving as a name to each character the position of its first occurrence in x. Next, perform approximately $\log |x|$ stages, as follows. At the k-th stage, compose all pairs of labels (l_i, l_{i+2^k}), sort them in lexicographic order and relabel each pair (whence also the substring it denotes) by the position of its first occurrence in the sorted list. If this process is performed on the concatenation of a pattern y and the text, then the occurrences of y can be intercepted subsequently by looking for positions of x with appropriate labels. We leave the details to the reader. Among its many virtues (Crochemore and Rytter 1991b), this encoding has recently proved useful in capturing distant relationships among files for compression purposes (Bentley and McIlroy 1999).

Another notable approach to pattern searches based on hash signatures is due to Karp and Rabin (1987). The idea here is to first filter out candidates and then check them individually for exact matching. This philosophy represents a precursor for many strategies dealing with massive data.

In the filtering stage, the pattern y is hashed into a number and then a window of size $|y|$ is slided on the text while the hash values of the

p	0	1	2	3	4	5	6	7	8	9	10	11	12	13	14	15	16	17	18	19
$x[p]$	n	o	␣	d	e	f	e	n	s	e	␣	f	o	r	␣	s	e	n	s	e
$h(x[p...p+4])$	8	8	6	28	9	18	28	26	22	12	17	24	16	0	1	9	—	—	—	—

Figure 6.2. Illustrating Karp-Rabin's algorithm.

corresponding substrings are computed. To be effective in this context, the hash function must be highly discriminating for strings. At the same time, it should be quickly computed and updated in the transition from one text window to the next. This is met by assimilating the symbols of Σ with integers and defining the hash value h for string u by

$$h(u) = \left(\sum_{i=0}^{|u|-1} u[i] \times d^{|u|-1-i} \right) \bmod q,$$

where q and d are two constants. Then, for each string $v \in \Sigma^*$, and symbols $a', a'' \in \Sigma$, $h(va'')$ is computed from $h(a'v)$ by the formula

$$h(va'') = ((h(a'v) - a' \times d^{|v|}) \times d + a'') \bmod q.$$

During the search for pattern x, it suffices to compare the value $h(y)$ with the hash value associated with each substring of length m of text x. If these two values are equal, that is, in case of collision, it is still necessary to check whether the substring is equal to x or not by direct symbol comparisons.

Convenient values for d are the powers of 2; in this case, all products by d are computed as shifts on integers. The value of q is generally a large prime (such that the quantities $(q-1) \times d + |\Sigma| - 1$ and $|\Sigma| \times q - 1$ do not cause overflows), but it can also be the value of the implicit modulus supported by integer operations. The operation of the algorithm is illustrated in Figure 6.2, searching for the pattern $y = sense$ in the text $x = no\ defense\ for\ sense$. Here, symbols are assimilated with their ASCII codes (hence $|\Sigma| = 256$), and the values of q and d are set respectively to 31 and 2. This is a valid choice when the maximal integer is $2^{16} - 1$. The value of $h(y)$ is $(115 \times 16 + 101 \times 8 + 110 \times 4 + 115 \times 2 + 101) \bmod 31 = 9$. Since only $h(y[4...8])$ and $h(x[15...19])$ are equal to $h(y)$, only two substrings of x need to be checked. The worst-case complexity of this string-searching is quadratic, but a prudent choice of the values for q and p leads to $O(m+n)$ expected time.

Signatures may be used to obtain substrings that encapsulate a given text, but also strings that depart significantly from it. This is the general problem of *inverse pattern matching* (Amir et al. 1997a), that refers to the task of inferring from a given text string x a short pattern string y such that y is, by some measure, most typical (or, alternatively, most anomalous) in the context of x. This problem arises in a wide variety of applications and takes up numerous flavors, among which in particular those based on signatures or frequencies of pattern occurrences. When such occurrences need not be exact, alternative measures of typicality can be based on some notion of similarity among strings, such as the Hamming (Hamming 1950) or Levenshtein (Levenshtein 1966) distances. Given a text string x and an integer m, for example, one might ask for a pattern y that scores the smallest (or largest) total number of mismatches when aligned with all substrings of x. Noteworthy variants of the problem arise when the constraint is added that y must be a substring of x, or, symmetrically, that y must not have any occurrence in x. Efficient (occasionally, optimal) sequential algorithms for the problem and its variants were provided in (Amir et al. 1997a, Gąsieniec et al. 1997). Computations of these and similar "distance preserving signatures" (see e.g. Greene et al. (1994)) find use in disparate contexts, including information retrieval, data compression, computer security and molecular biology. In the two latter fields, in particular, highly anomalous patterns are also often sought, e.g., in intrusion (Russel and Gangemi, Sr 1991) or plagiarism detection, in the synthesis of molecular probes in genome sequencing by hybridization (Alberts et al. 1989), in designing control (inactive) antisense oligonucleotides, etc.

As an example, we illustrate the simplest (min) inverse pattern matching problem, which is defined as follows: given a text string $x = x_1 \cdots x_n$ and positive integer $m \leq n$, we want to produce a pattern string $y_{min} = y_1 \cdots y_m$ (of length m) where $ham(y_{min}, x) \leq ham(y, x)$ for all strings $y \in \Sigma^m$. The symmetric *(Max) Inverse Pattern Matching Problem* seeks instead a pattern y_{Max} such that $ham(y_{Max}, x) \geq ham(y, x)$ with respect to all $y \in \Sigma^m$. Both versions of the problem are solved by the same basic strategy. The naive algorithm for the min inverse pattern matching problem is computing the Hamming distance for every possible substring of length m, and choosing the minimum. This algorithm is clearly bad since it takes exponential time. However, an optimal algorithm for solving the problem is readily set up. The idea is to "synthesize" y by choosing its characters one at a time, in such a way that each character will maximize the matches when meeting the positions of the text it will face.

The most difficult variant of the problem is the Max external one, in which y is required not to appear in x. However, also this variant has been shown to have an optimal linear time solution (Gąsieniec et al. 1997).

4.1. Approximate searches

A natural departure from the problem of *exact* string searching, consists in assuming that a symbol can (perhaps only at some definite positions) match a small group of other symbols. At one extreme we may have, in addition to the symbols in the input alphabet Σ, a *don't care* symbol ϕ with the property that ϕ matches any other character in Σ. This gives rise to variants of string searching where, in principle, ϕ appears (i) only in the pattern, (ii) only in the text or (iii) both in pattern and text. Here we briefly address the main variant (i).

One approach to this variant is to try and extend one of the fast string searching algorithms by accommodating don't cares in the pattern. However, the obvious transitivity on character equality, that subtends those and other exact string searching, is lost with don't cares. Some partial recovery is possible when the number and positions of don't cares is fixed. In this case, one may think of adapting some multiple pattern automaton of the kind discussed earlier.

Manber and Baeza-Yates (1991) considered the case where the pattern embeds a string of at most k don't cares, i.e., has the form $y = u\phi^i v$, where $i \leq k$, $u, v \in \Sigma^*$ and $|u| \leq m$ for some given k, m. Their algorithm is off-line in the sense that the text x is preprocessed to build the suffix array associated with it. This operation costs $O(n \log n)$ time in the worst case. Once this is done, the problem reduces to one of efficient implementation of 2-dimensional orthogonal range queries.

A landmark paper by Fischer and Paterson (1974) exposed the similarity of string searching to multiplication, thereby obtaining a number of interesting algorithms for exact string searching and some of its variants. It is not difficult to see that string matching problems can be rendered as special cases of a general linear product. Given two vectors X and Y, their *linear product* with respect to two suitable operations $\otimes$ and $\oplus$, is denoted by $X \overset{\otimes}{\oplus} Y$, and is a vector $Z = Z_0 Z_1 \ldots Z_{m+n}$ where $Z_k = \bigoplus_{i+j=k} X_i \otimes Y_j$ for $k = 0, \ldots, m + n$. If we interpret $\oplus$ as the boolean $\wedge$ and $\otimes$ as the symbol equivalence $\equiv$, then a match of the reverse Y^R of Y, occurs ending at position k in X, where $m \leq k \leq n$, if $[X_{k-m} \ldots X_k] \equiv [Y_m \ldots Y_0]$, that is, with obvious meaning, if $(X \overset{\equiv}{\wedge} Y)_k = \text{TRUE}$. This observation brings string searching into the family of boolean, polynomial and integer multiplications thereby leading quickly to an $O(n \log m \log \log m)$ time solution even in the presence of don't cares, provided that the size of Σ is fixed.

Some central notions of similarity are based on three basic *edit* operations on strings. Given any string w we consider the *deletion* of a symbol from w, the *insertion* of a new symbol in w and the *substitution* of one of the symbols of w with another symbol from Σ. It may be assumed that each edit operation has an associated nonnegative real number representing the *cost* of that operation, so that the cost of deleting from w an occurrence of symbol a is denoted by $D(a)$, the cost of inserting some symbol a between any two consecutive positions of w is denoted by $I(a)$ and the cost of substituting some occurrence of a in w with an occurrence of b is denoted by $S(a,b)$.

Letting now x and y be two strings of respective lengths $|x| = n$ and $|y| = m \leq n$, the *string editing problem* for input strings x and y consists in finding a sequence of edit operations or *edit script* Γ of minimum cost that transforms y into x. The cost of Γ is the *edit distance from y to x*. Edit distances where individual operations are assigned integer or unit costs occupy a special place. Such distances are often called Levenshtein distances, since they were introduced by W. Levenshtein in connection with error correcting codes (Levenshtein 1966). String editing finds applications in a broad variety of contexts, ranging from speech processing to geology, from text processing to molecular biology.

It is not difficult to see that the general (i.e., with unbounded alphabet and unrestricted costs) problem of edit distance computation is solved by a serial algorithm in $\Theta(mn)$ time and space, through dynamic programming. Due to its widespread application of the problem, however, such a solution and a few basic variants were discovered and published in a diverse literature (cf., e.g. Apostolico and Giancarlo (1998)). An $\Omega(mn)$ lower bound was established for the problem by Wong and Chandra for the case where the queries on symbols of the string are restricted to tests of equality. For unrestricted tests, a lower bound $\Omega(n \log n)$ was given by Hirschberg. Algorithms slightly faster than $\Theta(mn)$ were devised by Masek and Paterson, through resort to the so-called "Four Russians Trick". The "Four Russians" are Arlazarov, Dinic, Kronrod, and Faradzev. Along these lines, the total execution time becomes $\Theta(n^2/\log n)$ for bounded alphabets and $O(n^2(\log \log n)/\log n)$ for unbounded alphabets. The method applies only to the classical Levenshtein distance metric, and does not extend to general cost matrices. To this date, the problem of finding either tighter lower bounds or faster algorithms is still open. Details and references can be found in, e.g., (Apostolico and Galil 1997, Apostolico and Giancarlo 1998)).

The computation of edit distances by dynamic programming is readily set up. For this, let $C(i,j)$, $(0 \leq i \leq |y|, 0 \leq j \leq |x|)$ be the minimum cost of transforming the prefix of y of length i into the prefix of x of length j. Let w_k denote the kth symbol of string w. Then $C(0,0) = 0$, $C(i,0) = C(i-1,0) + D(y_i)$ $(i = 1, 2, ..., |y|)$, $C(0,j) = C(0, j-1) +$

$I(x_j)$ $(j = 1, 2, ..., |x|)$, and $C(i, j)$ will be given by

$$\min\{C(i-1, j-1) + S(y_i, x_j),\ C(i-1, j) + D(y_i),\ C(i, j-1) + I(x_j)\}$$

for all i, j, ($1 \leq i \leq |y|, 1 \leq j \leq |x|$). Observe that, of all entries of the C-matrix, only the three entries $C(i-1, j-1)$, $C(i-1, j)$, and $C(i, j-1)$ are involved in the computation of the final value of $C(i, j)$. Hence $C(i, j)$ can be evaluated row-by-row or column-by-column in $\Theta(|y||x|) = \Theta(mn)$ time. An optimal edit script can be retrieved at the end by backtracking through the local decisions that were made by the algorithm.

A few important problems are special cases of string editing, including the *longest common subsequence* problem, *local alignment*, i.e., the detection of local similarities of the kind sought typically in the analysis of molecular sequences such as DNA and proteins, and some important variants of *string searching with errors*, or searching for approximate occurrences of a pattern string in a text string. As highlighted in the following brief discussion, a solution to the general string editing problem implies typically similar bounds for all these special cases.

In many cases of great practical interest, such as e.g., with genomic sequence analysis, the space occupied by the edit distance matrix is unbearable and linear space methods are sought. We refer to (Apostolico 1996, Apostolico and Giancarlo 1998) for details and references.

Sequence similarity is a natural and useful filter for extracting matching information from huge data repositories. some of the fastest and most efficient searches routines work by first detecting regions of strong local resemblance, using conceptual tools of the kind represented by the following lemma.

Lemma 10 If x and y match with at most k differences, then x and y must have at least one identical substring of length $r = \lfloor max\{|x|, |y|\}/(k+1) \rfloor$

Proof. Let w.l.o.g. $|x| = max\{|x|, |y|\}$, and divide x into consecutive intervals of length r. In the alignment, each interval aligns to some part of y, determining $k + 1$ subalignments. If each of these subalignments contained at least one error, then we would have more than k errors. Thus, at least one of the intervals must match exactly a corresponding interval of y. ◇

More about searching with errors is said in the next subsection.

4.2. String searching with errors

Consider the problem of computing, for every position of the text string x, the best edit distance achievable between y and a substring w

of x ending at that position. Under the unit cost criterion, a solution is readily derived from the recurrence for string editing. The first obvious change consists in setting all costs to 1 except that $S(y_i, x_j) = 0$ for $y_i = x_j$. Thus, we have now, for all i, j, $(1 \leq i \leq |y|, 1 \leq j \leq |x|)$,

$$C(i,j) = \min\{C(i-1, j-1) + 1, \ C(i-1, j) + 1, \ C(i, j-1) + 1\}.$$

A second change affects the initial conditions, so that we have now $C(0,0) = 0$, $C(i,0) = i$ $(i = 1, 2, ..., m)$, $C(0,j) = 0$ $(j = 1, 2, ..., n)$. This has the effect of setting to zero the cost of prefixing y by any prefix of x. In other words, any prefix of the text can be skipped free of charge in an optimum edit script.

The computation of C is then performed in much the same way as before, thus taking $\Theta(|y||x|) = \Theta(mn)$ time. This time around, we are interested in the entire last row of matrix C at the outset.

In practice, it is often more interesting to locate only those segments of x that present a high similarity with y under the adopted measure. Formally, given a pattern y, a text x and an integer k, this restricted version of the problem consists in locating all terminal positions of substrings w of x such that the edit distance between w and y is at most k. The recurrence given above will clearly produce this information. However, there are more efficient methods to deal with this restricted case. In fact, a time complexity $O(kn)$ and even sublinear expected time are achievable. We refer to, e.g., (Apostolico and Galil 1997, Crochemore and Rytter 1994) for detailed discussions. In the following, we review some basic principles subtending an $O(kn)$ algorithm for string searching with k differences. Note that when k is a constant the corresponding time complexity is linear.

The crux of the method is to limit computation to $O(k)$ elements in each diagonal of the matrix C. These entries will be called *extremal* and may be defined as follows: a diagonal entry is d-extremal if it is the deepest entry on that diagonal to be given value d ($d = 1, 2, ..., k$). Note that a diagonal might not feature any, say, 1-extremal entry, in which case it would correspond to a perfect match of the pattern. The identification of d-extremal entries proceeds from extension of entries already known to be $(d-1)$-extremal. Specifically, assume we knew that entry $C(i, j)$ is $(d-1)$-extremal. Then, any entry reachable from $C(i, j)$ through a unit vertical, horizontal or diagonal-mismatch step possibly followed by a maximal diagonal stream of matches is d-extremal at worst. In fact, the cost of a diagonal stream of matches is 0, whence the cost of an entry of the type considered cannot exceed d. On the other hand, that cost cannot be smaller than $d - 1$, otherwise this would contradict the assumption $C(i,j) = d-1$. Let entries reachable from a $(d-1)$-extremal entry $C(i,j)$ through a unit vertical, horizontal or diagonal-mismatch step be called *d-adjacent*. Then the following program encapsulates the basic computations.

  Algorithm k-err :
  element array $x[1:n], y[1:m], C[0:m; 0:n]$; integer k
  begin
  ($PHASE$ 1 : initializations)
    set first row of C to 0;
    find the boundary set S_0 of 0-extremal entries
    by exact string searching;
  ($PHASE$ 2 : identify k-extremal entries)
    for $d = 1$ **to** k **do**
      begin
      walk one step horizontally, vertically and
      (on mismatch) diagonally
      from each $(d-1)$-extremal entry in set $S_{(d-1)}$
      to find d-adjacent entries;
      from each d-adjacent entry, compute the farthest
      d-valued entry reachable diagonally from it;
      end
    for $i = 1$ **to** $n - m + 1$ **do**
      begin
      select lowest d-entry on diagonal i
      and put it into the set S_d of d-extremal entries
      end
  end.

It is easy to check that the algorithm performs k iterations in each one of which it does essentially a constant number of manipulations on each of the n diagonals. In turn, each one of these manipulations takes constant time except at the point where we ask to reach the farthest d-valued entry from some other entry on a same diagonal. We would know how to answer quickly that question if we knew how to handle the following query: given two arbitrary positions i and j in the two strings y and x, respectively, find the longest common prefix between the suffix of y that starts at position i and the suffix of x that starts at position j. In particular, our bound would follow if we knew how to process each query in constant time. It is not known how that could be done without preprocessing becoming somewhat heavy. On the other hand, it is possible to have it such that *all* queries have a cumulative amortized cost of $O(kn)$. This possibility rests on efficient algorithms for performing *lowest common ancestor* queries in trees. Space limitations do not allow us to belabor this point any further.

In massive applications, even time $O(kn)$ may be prohibitive. Using filtration methods it is possible to set up sublinear expected time queries. As already highlighted, one possibility is to first look for regions with exact replicas of some pattern segment and then scrutinize those regions. Another, is to look for segments of the text that are within a small distance of some fixed segments of the pattern. Some of the current top performers in molecular database searches are engineered around these ideas (Altschul et al. 1990, Ukkonen 1993, Baeza-Yates and Perleberg 1992, Chang and Lawler 1994). In fact, the whole issue of filtration search may be regarded as a form of pattern discovery (Brazma et al. 1998c;b; 1997; 1998a), probably a fundamental application of future Pattern Matching and one that is discussed more extensively later in this chapter.

The special case where insertions and deletions are forbidden is also solved by an algorithm very similar to the above and within the same time bound. This variant of the problem is often called string searching *with mismatches*. A probabilistic approach to this problem is implicit in Chang and Lawler (1994). When k cannot be considered a constant, an interesting alternative results from Abrahamson's approach to multiple-value string searching (Abrahamson 1987) which results in an algorithm of time $O(nm^{1/2} \log m \log \log^{1/2} m)$.

5. Compressing, Learning, Mining, and Discovering

Data compression brings savings in storage space and transmission time, two commodities in increasingly scarce supply. From the perspective of the data flood ahead, compression also helps in the formation of succinct descriptors and models, thereby helping in overcoming the ultimate limitations imposed by the narrow bandwidth of the final user. Because of this, efficient, innovative compression methods will continue to play an important role.

Of the two main broad classes of compression, standard *lossy* methods such as Mpeg, Jpeg, Wavelets etc. have a definite numerical flavor and derive a limited influence from Pattern Matching. By contrast, nearly every present and future *lossless* method will use more or less sophisticated Pattern Matching techniques. Among the basic methods in this class, we find Run-Length and Huffman Encoding, the latter being further subdivided into static and dynamic codes, Arithmetic Codes, Macro Schemes such as the Ziv-Lempel methods underlying *compress, gzip* and other popular tools, the more recent Burrows-Wheeler transform subtending *bzip*, Predictive Codes, etc. These and others are reviewed in this section.

5.1. Standard compression methods

We outline here some classical yet practical text compression algorithms. Algorithmic efficiency is but one of the parameters against which the efficiency of a method is assessed. The final compression ratio is equally, if not more, important. This latter depends on the nature of the input data. Typically, the final size of compressed textfiles vary from 30% to 50% of the size of the input.

In standard lossless compression, two main strategies are applied. The first strategy is a statistical method that takes into account the frequencies of symbols to build a uniquely decipherable code optimal with respect to the compression. This is considered in Subsection 5.1.1. Subsection 5.1.2 presents a refinement of the coding algorithm of Huffman based on the binary representation of numbers. Huffman codes contain new codewords for the symbols occurring in the text. In this method, fixed-length blocks of bits are encoded by different codewords. In the second strategy, repeated substrings of variable-length from the text are spotted and suitably encoded. This will be seen in Subsection 5.1.3. Due to its ability to capture context dependency, this second strategy often provides better compression ratios.

5.1.1 Huffman coding. The Huffman method is an optimal statistical coding, in which each character or fixed block of characters of the text is replaced by a *codeword* in such a way, that longer and longer codewords are assigned to rarer and rarer characters. The method works for any block length, however, the running time grows exponentially with the length.

The Huffman algorithm uses *prefix* codes, i.e., sets of words in which no word is a *prefix* of another. The advantage with such codes is that decoding is *instantaneous*, in the sense that it can be carried out while the encoded string is being received.

A prefix code on the alphabet $\{0, 1\}$ is represented in a natural way by a binary digital trie in which the leaves are labeled by the original characters, and the path from the root to a character spells out the characters codeword. The specific assignment of codewords depends on the frequencies of the individual characters. The complete compression algorithm consists of three stages: count of character frequencies, construction of the prefix code, encoding of the text. The last two steps use information computed by their preceding step. Decoding is a simple exercise.

The static Huffman method has two main drawbacks: first, if the frequencies of characters in the source text are not known *a priori*, then the input text has to be read twice; second, the coding tree must be included in the compressed file. This is avoided by dynamically updating the coding tree for the consecutive prefixes of the text while consecutive symbols are processed. A combinatorial property of the tree allows its efficient updates. By mimicking the coding process, decoding will expose the tree precisely in the same order.

5.1.2 Arithmetic coding. In arithmetic coding, symbols are treated as digits of a numeration system, and texts as decimal parts of numbers between 0 and 1. The interval $[0,1[$ is first partitioned into $|\Sigma|$ subintervals of size proportional to the probabilities or frequencies of symbols. The same partition is then recursively applied to subintervals as consecutive text symbols are read, thereby mapping the text itself into some subinterval of $[0,1[$. Compression is achieved because highly probable texts end up mapped in wider intervals thus requiring fewer bits in their description.

Formally, let the interval associated with symbol $a_i \in \Sigma$ ($1 \leq i \leq \|\Sigma\|$) be denoted $I(a_i) = [l_i, h_i[$. The intervals satisfy the conditions: $l_1 = 0$, $h_{|\Sigma|} = 1$, and $l_i = h_{i-1}$ for $1 < i \leq |\Sigma|$. Note that $I(a_i) \cap I(a_j) = \emptyset$ if $a_i \neq a_j$.

The encoding consists in computing the interval corresponding to the input text. We begin with the initial interval $[0,1[$. The generic step deals with a symbol a_i of the source text by transforming the current interval $[l, h[$ into $[l', h'[$ where $l' = l + (h-l) * l_i$ and $h' = l + (h-l) * h_i$. From a theoretical standpoint, l alone would suffice to encode the input text.

The decoding phase recapitulates the encoding. Specifically, the first step of decoding consists in identifying the symbol a_i such that $l \in I(a_i)$. At that point, l is replaced by

$$l' \leftarrow \frac{l - l_i}{h_i - l_i},$$

and the process is repeated until $l = 0$. The main problem with arithmetic coding is coping with finite precision while performing arithmetics on real numbers.

5.1.3   LZW coding.   Ziv and Lempel designed a class of compression methods based on the idea of self reference: while the textfile is scanned, substrings or *phrases* are identified and stored in a dictionary, and whenever, later in the process, a phrase or concatenation of phrases is encountered again, this is compactly encoded by suitable pointers (Lempel and Ziv 1976, Ziv and Lempel 1977; 1978). Of the several existing versions of the method, we describe below the one known as Lempel-Ziv-Welsh method, which is incarnated by the `compress` feature under the UNIX operating system.

For the encoding, a dictionary is initialized with all the characters of the alphabet. At the generic iteration, we have just read a segment w of the text. With a the symbol following this occurrence of w, we now proceed as follows: If wa is in the dictionary we read the next symbol, and repeat with segment wa instead of w. If, on the other hand, wa is not in the dictionary, then we append the dictionary index of w to the output file, and add wa to the dictionary; then reset w to a and resume processing from the text symbol following a. Once w is initialized to be the first symbol of the source text, "w belongs to the dictionary" is established as an invariant in the above loop.

Decoding is symmetric, in particular, the dictionary is recovered while the decompression process runs. The basic routine is as follows. We start with a basic dictionary of symbols. Then, when we read the encoding c from the compressed file, we write to the output file the segment w having index c in the dictionary, and add to the dictionary the word wa where a is the first letter of the next segment. Except for a special case, note that we can infer the appropriate dictionary index for wa. A very special case requiring extra care occurs if the symbol a is also the first symbol of w. It is indeed related to squares that occur in the text. We leave the analysis of this case and its (easy) recovery for an exercise.

5.1.4   The Burrows-Wheeler transform.   A recent, imaginative approach due to Burrows and Wheeler (1994) successfully exploits the delicate interplay between locality of reference and pointer size. Assuming an input string $x = dadcbbe$, the encoding performs the following steps. First, we build a table of the cyclic shifts of x, as follows.

S0	*d a d c b b e*
S1	*a d c b b e d*
S2	*d c b b e d a*
S3	*c b b e d a d*
S4	*b b e d a d c*
S5	*b e d a d c b*
S6	*e d a d c b b*

Next, these rotations are lexicographically sorted, resulting in the table:

$$
\begin{array}{ll}
S1 & a\ d\ c\ b\ b\ e\ d \\
S4 & b\ b\ e\ d\ a\ d\ c \\
S5 & b\ e\ d\ a\ d\ c\ b \\
S3 & c\ b\ b\ e\ d\ a\ d \\
S0 & d\ a\ d\ c\ b\ b\ e \\
S2 & d\ c\ b\ b\ e\ d\ a \\
S6 & e\ d\ a\ d\ c\ b\ b
\end{array}
$$

It turns out that strings like the string $y = dcbdeab$ in the last column are highly compressible, e.g., by run-length. In fact, the first column contains sorted symbols that are each immediately adjacent in x to the corresponding symbol in the last column. It is expected then that, in correspondence with a run on the first column, the last one also contains a run. Note that it is possible to go back from the last column y to the first column $y' = abbcdde$ simply by sorting y. More importantly, from knowledge of y, y' and of the rank i of the original string in the sorted list, it is possible to reconstruct the original sequence x. This is achieved by setting up a suitable transformation vector T that tells, for each row j, where in x is row $j + 1$. This vector can be figured out by looking at y and y' as shown in the table below.

$$
\begin{array}{llll}
0 & S1 & d & a \\
1 & S4 & c & b \\
2 & S5 & b & b \\
3 & S3 & d & c \\
4 & S0 & e & d \\
5 & S2 & a & d \\
6 & S6 & b & e
\end{array}
$$

Clearly, we have $T(4) = 0$ since c moves, but what about row 1? The b there could go to either row 0 or 1. The important property is, since y' is sorted then rows beginning with a same character are also sorted. Thus, the first b in row 1 moves to row 0, the second b comes from row 6. The final touch of the method is to perform move-to-front encoding of y. In practice, all 256 codes are kept in a list, and each time a character is to be output, its position is sent to the list, then moved to the front. The result is a string with many of 0's and small integers, which can be compressed using entropy encoders. For example, $y = tttWtwttt$ would be encoded as $[116, 0, 0, 88, 1, 119, 1, 0, 0]$.

The sorting inherent to the Burrows-Wheeler method is suitably implemented with suffix arrays, resulting in a relatively fast process.

5.2. Data compression using antidictionaries

Yet another basic text compression method, called DCA, uses some "negative" information about the text, which is described in terms of antidictionaries (Crochemore et al. 1998b;a; 2000). Contrary to the Ziv and Lempel methods that are centered on dictionaries or sets of words occurring as substrings in the text, this method takes advantage from words that *do not* occur as substrings in the text and are said to be *forbidden*. It is natural to call such sets of words *antidictionaries*.

5.2.1 Encoding and decoding.

Let x be the text on a binary alphabet and let $F(x)$ be the set of substrings of x. For instance, if $x = 01001010$ then $F(x) = \{\varepsilon, 0, 1, 00, 01, 10, 001, 010, 100, 101, \ldots\}$. The antidictionary AD is a *factor code* (no word of the set is a substring of another word of the set) included in $\Sigma^* \setminus F(x)$. For example, $\{000, 10101, 11\}$ is an antidictionary for $x = 01001010$.

The compression algorithm processes the input file on-line. At the generic step, we have read some prefix w of x, and inspect the symbol, say, a, that immediately follows w. If there exists a word $u \in AD$ that is a suffix of wa, then the symbol a is deleted, since it is predictable through resort to the antidictionary. The compression algorithm based on this principle is listed below. In order to be able to decode the output of the encoder, an additional mechanism is necessary. To simplify the exposition, we assume here that the encoder produces also the length of the original text. The decoder works in a fashion which is dual to the encoder, and is presented immediately following it. It uses its knowledge of the length in order to decide when to halt.

The advantage of having a factor code as antidictionary is that the test at Line 3 in the decoder can be satisfied by only one word va. Therefore, no useless word is stored in the antidictionary.

5.2.2 Implementing finite antidictionaries.

The antidictionary queries invoked by the above algorithms are implemented as follows. Starting with the trie of words in the antidictionary, the automaton $\mathcal{A}(AD)$ is built that accepts all strings of which no substring appears in the antidictionary. This is an application of the Aho-Corasick algorithm to the trie, and results in a linear-time algorithm. With this automaton in place, and while reading the text to encode, whenever a transition leads to a state associated with a word of the antidictionary the decoder outputs the dual symbol.

The automaton $\mathcal{A}(AD)$ can be easily transformed into a (finite-state) transducer $\mathcal{T}(AD)$ that realizes the compression algorithm. The decom-

ENCODER (anti-dictionary AD, word $x \in \{0,1\}^*$)
1. $\gamma \leftarrow \varepsilon$;
2. **for** $a \leftarrow$ first to last symbol of x
3.   **if** for any suffix v of the processed text, $v0, v1 \notin AD$
4.   output a;
5. **return** $(|x|, \gamma)$;

DECODER (anti-dictionary AD, integer n, word $\gamma \in \{0,1\}^*$)
1. $w \leftarrow \varepsilon$;
2. **while** $|w| < n$
3.   **if** for some suffix v of w and some $a \in \{0,1\}$, $va \in AD$
4. $w \leftarrow w \cdot \neg a$;
5.   **else**
6. $b \leftarrow$ next symbol of γ;
7. $w \leftarrow w \cdot b$;
8. **return** (w);

Figure 6.3. Antidictionary based compression

pression may be similarly realized by a dual transducer, which is obtained by interchanging input and output labels in the first transducer (with an additional halting instruction to stop the decoding).

The automaton $\mathcal{A}(AD)$ (or the transducer $\mathcal{T}(AD)$) has an interesting synchronization property, which makes it possible to develop algorithms to search compressed texts or to design parallel version of the encoding and decoding algorithms. With k the maximal length of words in AD, this property is as follows: given any two paths $(q_1, a_1, q_2) \cdots (q_k, a_k, q_{k+1})$ and $(q'_1, a_1, q'_2) \cdots (q'_k, a_k, q'_{k+1})$ having the same label $a_1 \cdots a_k$, then the two ending states q_{k+1} and q'_{k+1} coincide. Thus, the encoding of a part of the text certainly depends on its left context, but this is limited to up to a length of k only.

5.2.3   How to build antidictionaries.

In practical applications, the antidictionary is not given *a priori* but it must be derived either from the text to be compressed or from a family of texts produced by the same source as the one producing the text. There exist several criteria to build efficient antidictionaries, that variously depend on different aspects or parameters that one wishes to optimize in the compression process. In turn, each criterion gives rise to a different algorithm and implementation.

The general methods to build antidictionaries are based on data structures that store substrings of words, such as suffix tries, suffix trees, dawgs, and suffix or factor automata. As seen, some notion of a suffix link always accompanies these structures, and such a notion proves essential also in designing efficient algorithms to build representations of sets of minimal forbidden words in term of tries or trees. The antidictionary constructions set up along these lines take time linear in the length of the text to be compressed.

A rough solution to control the size of antidictionaries would be obviously by bounding the length of the words that are admitted in it. A better solution in the static compression scheme is to prune the trie of the antidictionary on the basis of a tradeoff between the space of the trie to be transmitted and the gain in compression. However, the first solution is enough to get compression rates that reach asymptotically the entropy for balanced sources, even if this is not true for general sources. Both solutions can be engineered to run in linear time.

A further sophistication considers self-compressed antidictionaries, and yields best compression ratios for the method.

5.2.4 Variations. The static compression scheme presented above requires to read the text twice. Several variations and improvements can be elaborated upon based on clever combinations of two features suitably injected in the model, namely, statistical filters and dynamic implementations. These are classical features, often included in most data compression methods.

Statistical considerations can be used in the construction of antidictionaries. If a forbidden word is responsible for erasing few bits of the text in the compression algorithm while its description as an element of the antidictionary is "expensive", then the compression rate improves by excluding that word from the antidictionary. On the other hand, one can introduce in the antidictionary a word that is not forbidden but occurs very rarely in the text. In this case, the compression algorithm may produce some errors in predicting the next letter. In order to keep a lossless compression scheme, encoder and decoder must be adapted to manage such errors. Typical errors occur in the case of antidictionaries built for fixed sources as well as in the dynamic approach. Even with errors, assuming that they are rare with respect to the longest word (length) of the antidictionary, the compression scheme may be shown to preserve the synchronization property.

5.3. Searching compressed text

For data stored in compressed form, navigation through compressed databases poses additional questions of string pattern matching. The first question is whether it may be more efficient to decompress the data before processing a search or other standard query or, given the possibility, it might be more expedient to perform the query directly on the compressed data. The answer depends of course on the particular problem instance, as well as on such variables as compression method, algorithmic complexity, memory space available, etc. Among the various methods, the Ziv-Lempel family of compressors have received the largest attention, beginning with studies by Amir et al. (1994) and Farach and Thorup (1994). Along these lines, string search in compressed text was developed for the paradigm by Ziv and Lempel (1977) and its subsequent variant by Welch (1984). The complexities for the searches are respectively of $O(n \log n' + m)$ and $O(n \log m + m)$, where n' is the size of the decompressed text and m the size of the pattern. Thus, compared to linear time string searching in plain texts, an extra log factor emerges. For large patterns, it makes sense to consider instances of the problem where also the pattern is compressed. This case was studied by Gąsieniec and Rytter (1999), who gave algorithms respectively of time $O((n+m)^5)$ and $O((n+m) \log^c(n+m))$ (with c a positive constant) for the LZ and LZW compressors.

Searching files compressed by Huffman encoding is a classical problem treated, e.g., in Moura et al. (1998). Shibata et al. (1999) give a linear-time searching algorithm for files compressed by using antidictionaries.

Mixed techniques have also been developed in which the compression is designed to reduce the searching time. Examples of this approach may be found in Manber (1994) and Navarro and Raffinot (1999). The main drawbacks with the technique is that it often leads to less efficient compression and that of course it will not work with text compressed by standard methods.

5.4. Learning probabilistic automata and modeling by Markov chains

Compression is but one of the domains within which the need arises to develop models of sources. In fact, as already mentioned, the statistical modeling of sequences is a central paradigm of machine learning that finds multiple uses in many domains. The probabilistic automata typically built in these contexts are subtended by uniform, fixed-memory Markov models. In practice, such automata tend to be bulky and computationally imposing both during their synthesis and use. In Ron et al.

(1996), much more compact, tree-shaped variants of probabilistic automata are described which assume an underlying Markov process of variable memory length. These variants, called *PST*s were successfully applied to learning and prediction of protein families in Bejerano and Yona (1999).

In one such automaton, each edge is labeled by a symbol, each node corresponds to a unique string —the one obtained by traveling from that node to the root— and nodes are weighted by a probability vector giving the distribution over the next symbol. The construction starts with a tree consisting of just the root node (i.e., the tree associated with the empty word) and adds paths as follows. It considers the substrings from a family S of strings in order of increasing length. For each substring s considered, it is checked whether there is some symbol σ in the alphabet for which the empirical probability of observing it in S after s is significant and significantly different from the probability of observing it after the longest suffix $suf(s)$ of s. Whenever these conditions hold, the path relative to the substring (and possibly its necessary but currently missing ancestors) is added to the tree.

Given now a string, its weighting by a tree is done by scanning the string one letter after the other while assigning a probability to every symbol, in succession. The probability of a symbol is calculated by walking down the tree in search for the longest suffix that appears in the tree and ends immediately before that symbol, and multiplying the corresponding conditional probability. Since, following each input symbol, the search for the deepest node must be resumed from the root, this process cannot be carried out on-line nor in linear-time in the length of the tested sequence.

As is easy to see, the process of learning the automaton from a given training set S of sequences requires $\Theta(Ln^2)$ worst-case time, where n is the total length of the sequences in S and L is the length of a longest substring of S to be considered for a candidate state in the automaton. Once the automaton is built, predicting the likelihood of a query sequence of m characters may cost time $\Theta(m^2)$ in the worst case. A more efficient computation of empirical probabilities and conditional probabilities, of the kind described in an earlier section of this chapter, leads to equivalent automata that can be learned in time linear in the input size, and will subsequently predict a string of m symbols in $O(m)$ time. We refer to Apostolico and Bejerano (2000) for details.

5.5. Episodes and automatic association generation

Many interesting problems can be cast in the emerging contexts of *data mining* and *information extraction*. As is well known, while traditional data base queries aim at retrieving records based on their isolated contents, these contexts focus on the identification of patterns occurring across records, and aim at the retrieval of information based on the discovery of interesting rules present in large collection of data. Central to these developments is the notion of an *association rule*, which is an expression of the form $S_1 \to S_2$ where S_1 and S_2 are sets of data attributes endowed with sufficient *confidence* and *support*. Sufficient support for a rule is achieved if the number of records whose attributes include $S_1 \cup S_2$ is at least equal to some pre-set minimum value. Confidence is measured instead in terms of the ratio of records having $S_1 \cup S_2$ over those having S_1, and is considered sufficient if this ratio meets or exceeds some pre-set minimum. Clearly, a statistic of the number of records endowed with the given attributes must be computed as a preliminary step, and this is often a bottleneck for the process of information extraction. We refer to Agrawal et al. (1993) and Piatesky-Shapiro and Frawley (1991) for a broader discussion of these concepts.

Some of the considerations developed earlier in this chapter may be regarded from a perspective of automatic generation of association rule. Lemma 7, for instance, can be rephrased by saying that for every word ending in the middle of an arc in T_x, a rule is exposed whereby any occurrence of that word in x *implies* an occurrence also of its extension to the nearest node. From this perspective, the construction of the tree may be regarded as a means for the discovery of this rule.

In a real discovery, though, we do not know *a priori* the rule that will be discovered. Along these lines, looking for squares, palindromes, etc. is only half a discovery, in so far as the "rule" (e.g., ww, ww^R) which we are after is known beforehand. Even so, some mild extensions of this problem may already fit the mining paradigms.

For example, consider the problem of finding, for a given text string x of n symbols and an integer constant d, and for any pair (y, z) of subwords of x, the number of times that y and z occur in tandem (i.e., with no intermediate occurrence of either one in between) within a distance of d positions of x. Although in principle there might be n^4 distinct subword pairs in x, Lemma 7 tells us that it suffices to consider a family of only n^2 such pairs, with the property that for any neglected pair (w', z'), there is a corresponding pair (y, z) contained in our family and such that: (*i*) w' is a prefix of w and z' is a prefix of z, and (*ii*) the tandem index of (w', z') equals that of (w, z). We leave it as an exercise for the reader to find an efficient algorithm for the construction of the table of all such tandem indices. The particularization of the problem to the tandem index of occurrences of the same pattern, which is in fact a relaxed square detection problem, has also been studied recently (Brodal et al. 1999).

Amir et al. (1997b) have used tries to organize and speed up the discovery of association rules in a typical data base, the entries of which are sets of attributes. The first step consists in transforming each record into a string by numbering the different attributes. Next, every set is considered as a string sorted by order of the attribute number. At this point, a trie is built by incremental insertion of all i-elements sorted sets for $i = 1, 2, ... i_{max}$, in succession, where i_{max} is some suitable bound. The nodes of the trie are weighted by the count of the number of records leading to each node (a measure of the support for that node). The data structure at the outset encodes all potential *covers*, a cover in this context being a set of attributes with support exceeding a certain *minsupport* value. To generate associations, one observes that once an association of the form $S \rightarrow \{a\}$ is generated for an attribute, this gives a handle to narrowing down the space of potential attributes of the form $\{a, b\}$, in the sense that only if both associations $S \cup \{a\} \rightarrow \{a\}$ and $S \cup \{b\} \rightarrow \{b\}$ exist, one can hope for association $S \rightarrow \{a, b\}$ to exist. This leads to the following scheme for association generations.

- For each node of the trie, let $s = s_1 s_2 ... s_k$ be the label of the path from the root to that node. Extract, in succession, each s_i and check the resulting string $\bar{s}$ for its support. Whenever the ratio $supp(s)/supp(s_1...s_{i-1}s_{i+1}...s_k) \geq minconf$ then $S - S_i \rightarrow S_i$ is an *association rule*.

- We now have association rules with only one set on the right hand side. These rules are now combined to generate multiple rules. That is, for every pair of rules, generate a new rule with a consequent of size 2, and test its confidence level. Repeat the process to obtain rules with consequents of increasing size.

Other discoveries can be modeled in terms of the detection of special kinds of subsequences. A pattern $y = y_1 \ldots y_m$ occurs as a *subsequence* of a text $x = x_1 \ldots x_n$ iff there exist indices $1 \leq i_1 < i_2 < \cdots < i_m \leq n$ such that $x_{i_1} = y_1, x_{i_2} = y_2, \cdots, x_{i_m} = y_m$; in this case we also say that the substring $w = x_{i_1} x_{i_1+1} \ldots x_{i_m}$ of x is a *realization* of y beginning at position i_1 and ending at position i_m in x. Given two strings $x = x_1 \ldots x_n$ and $y = y_1 \ldots y_m$ over an alphabet Σ, the problem of testing whether y occurs as a subsequence of x is trivially solved in linear time. It is also known that a simple $O(n \log |\Sigma|)$ time preprocessing of x makes it easy to decide subsequently for any x and in at most $|y| \log |\Sigma|$ character comparisons, whether y is a subsequence of x. These problems become more complicated if one asks instead whether y occurs as a subsequence of some substring w of x of bounded length. One way to answer the question is by identifying all distinct minimal realizations w of y. By a realization w being minimal with respect to x, it is meant that y is not a subsequence of any proper substring of w. Variants of this problem arise in numerous applications, ranging from information retrieval and mining recurrent events in telecommunications (see, e.g., Mannila et al. (1995)) to molecular sequence analysis (see, e.g., Waterman (1995)) and intrusion and misuse detection in a computer system. Algorithms for the so-called *episode matching* (Mannila et al. 1995) problem, which consists in finding the *earliest* occurrences of y in all minimal realizations w of y in x have been given in Das et al. (1997). An occurrence $i_1 i_2 \ldots i_m$ of y in a realization w is an earliest occurrence if the string $i_1 i_2 \ldots i_m$ is lexicographically smallest with respect to any other possible occurrence of y in x. The algorithms in Das et al. (1997) perform within roughly $O(nm)$ time, without resorting to any auxiliary structure or index based on the structure of the text.

Many modern pattern or *motif* characterizations and discovery algorithms will come from the flourishing area of Bioinformatics, a microcosmos within which most problems of managing the data and information flood find early and somewhat controlled reflections (see, e.g., (Brazma et al. 1998c;b; 1997; 1998a)). Prominent in this context is the issue of aligning multiple sequences (Apostolico and Giancarlo 1998). This application is explosive in computational demand and is typically approached by way of heuristics. These, in turn, are variously centered around ideas of hinging putative alignments around similar subpatterns of various kinds. One difficulty in this regard is the lack of a unified notion of global comparison, which compounds with the inherent intractability of most exact methods. One way to approach the problem is then to look for "anchor" sets of consecutive columns where a same (short) pattern seems to appear in all sequences. Recursively hinging a global solution around these anchors gives a handle for divide and conquer heuristics. The discovery of anchor patterns fits somewhat into the paradigm of association rule generation. These patterns can be sought among the substrings or subsequences of the sequences, or combinations thereof. For example, one could use the labeling of Karp, Miller, and Rosenberg to label substrings and then look for regions with a concentration of identical labels. A variation on this theme is due to Sagot et al.

(1997) and is based on the notion of a *model* (direct product of subsets of the alphabet) that extends the notion of a consensus sequence. Models capture the similarity between some categories of symbols as is the case with amino acids in the comparison of proteins. For fixed lengths, there is a linear-time algorithm to generate all the models common to a set of strings on the basis of hypotheses on two parameters: a *quorum* for the number of implied sequences, and the maximum acceptable number of errors between the models and their actual occurrences.

Bibliography

K. Abrahamson. Generalized string matching. *SIAM J. Computing*, 16: 1039–1051, 1987.

R. Agrawal, T. Imielinski, and A. Swami. Mining association rules between sets of items in large databases. In *Proc. ACM SIGMOD*, pages 207–216, 1993.

A.V. Aho and M.J. Corasick. Efficient string matching. *C. ACM*, 18: 333–340, 1975.

B. Alberts, D. Bray, J. Lewis, M. Raff, K. Roberts, and J.D. Watson. *Molecular Biology of the Cell*. Garland Publishing, 1989.

S. Altschul, W. Gish, W. Miller, E.W. Myers, and D. Lipman. Basic linear alignment search tool. *J. Mol. Biology*, 215:403–410, 1990.

A. Amir, A. Apostolico, and M. Lewenstein. Inverse pattern matching. *J. of Algorithms*, 24:325–339, 1997a.

A. Amir, G. Benson, and M. Farach. Let sleeping files lie: pattern matching in z-compressed files. In *Proc. of 5th Annual ACM-SIAM Symposium on Discrete Algorihms*, 1994.

A. Amir, R. Feldman, and R. Kashi. A new and versatile method for association generation. *Information Systems*, 1997b. To appear. Preliminary version appeared in PKDD 97.

A. Apostolico. Optimal Parallel Detection of Squares in Strings. *Algorithmica*, 8:285–319, 1992.

A. Apostolico. String editing and longest common subsequences. In G. Rozenberg and A. Salomaa, editors, *Handbook of Formal Languages*, volume II, pages 361–398. Springer-Verlag, 1996.

A. Apostolico and G. Bejerano. Optimal amnesic probabilistic automata or how to learn and classify proteins in linear time and space. In *Proceedings of RECOMB 2000*, pages 25–32, 2000.

A. Apostolico, M.E. Bock, S. Lonardi, and X. Xu. Efficient detection of unusual words. Technical report, Purdue University Computer Science Department, 1996. To appear in Journal of Computational Biology.

A. Apostolico, D. Breslauer, , and Z. Galil. Optimal Parallel Algorithms for Periods, Palindromes and Squares. In *Proc. 19th International Colloquium on Automata, Languages, and Programming*, volume 623 of *Lecture Notes in Computer Science*, pages 296–307. Springer-Verlag, 1992.

A. Apostolico and A. Ehrenfeucht. Efficient Detection of Quasiperiodicities in Strings. *Theoret. Comput. Sci.*, 119:247–265, 1993.

A. Apostolico, M. Farach, and C.S. Iliopoulos. Optimal Superprimitivity Testing for Strings. *Inform. Process. Lett.*, 39:17–20, 1991.

A. Apostolico and Z. Galil, editors. *Pattern Matching Algorithms*. Oxford University Press, 1997.

A. Apostolico and R. Giancarlo. Sequence alignment in molecular biology. *Journal of Computational Biology*, 5:173–196, 1998.

A. Apostolico and F. P. Preparata. Optimal off-line detection of repetitions in a string. *Theoret. Comput. Sci.*, 22:297–315, 1983.

A. Apostolico and F. P. Preparata. Data structures and algorithms for the strings statistics problem. *Algorithmica*, 15:481–494, 1996.

A. Apostolico and W. Szpankowski. Self-alignment in words and their applications. *J. Algorithms*, 13:446–467, 1992.

R. Baeza-Yates and C. Perleberg. Fast and practical approximate string matching. In *Proc. III Symp. on Combinatorial Pattern matching*, Lecture Notes in Computer Science, pages 185–92. Springer-Verlag, 1992.

D. R. Bean, A. Ehrenfeucht, and G.F. McNulty. Avoidable patterns in strings of symbols. *Pacific J. Math.*, 85:261–294, 1979.

G. Bejerano and G. Yona. Modeling protein families using probabilistic suffix trees. In S. Istrail, P. Pevzner, and M. Waterman, editors, *Proceedings of RECOMB99*, pages 15–24. ACM Press, 1999.

A. Ben-Amram, O. Berkman, C. Iliopolous, and K. Park. Computing the Covers of a String in Linear Time. In *Proc. 5th ACM-SIAM Symp. on Discrete Algorithms*, pages 501–510, 1994.

J. Bentley and D. McIlroy. Data compression using long common strings. In *Proceedings of the IEEE Data Compression Conference*, pages 287–295, 1999.

J. Berstel. Sur les mots sans carré définis par un morphism. In *Proc. 6th International Colloquium on Automata, Languages, and Programming*, volume 71 of *Lecture Notes in Computer Science*, pages 16–25. Springer-Verlag, 1979.

A. Blumer, J. Blumer, A. Ehrenfeucht, D. Haussler, M.T. Chen, and J. Seiferas. The Smallest Automaton Recognizing the Subwords of a Text. *Theoretical Computer Science*, 40:31–55, 1985.

A. Brazma, I. Jonassen, I. Eidhammer, and D. Gilbert. Approaches to the automatic discovery of patterns in biosequences. *Journal of Computational Biology*, 5:279–306, 1998a.

A. Brazma, I. Jonassen, J. Vilo, and E. Ukkonen. Pattern discovery in biosequences. In *Proceedings of Fourth International Colloquium on Grammatical Inference (ICGI-98)*, volume 1433 of *Lecture Notes in Computer Science*, pages 255–270. Springer-Verlag, 1998b.

A. Brazma, I. Jonassen, J. Vilo, and E. Ukkonen. Predicting gene regulatory elements in silico on a genomic scale. *Genome Research*, 8: 1202–1215, 1998c.

A. Brazma, J. Vilo, E. Ukkonen, and K. Valtonen. Data mining for regulatory elements in yeast genome. In *Fifth International Conference on Intelligent Systems for Molecular Biology (ISMB-97)*, pages 65–74. AAAI Press, 1997.

D. Breslauer. An On-Line String Superprimitivity Test. *Inform. Process. Lett.*, 44:345–347, 1992.

D. Breslauer. Testing String Superprimitivity in Parallel. *Inform. Process. Lett.*, 49:235–241, 1994.

L. Brillouin. *Science and Information Theory*. Academic Press, 1971.

G.S. Brodal, R. Lyngso, C.N.S. Pedersen, and J. Stoye. Finding maximal pairs with bounded gap. In *Proc. 10th Combinatorial Pattern Matching*, volume 1645 of *Lecture Notes in Computer Science*, pages 342–351. Springer-Verlag, 1999.

M. Burrows and D. J. Wheeler. A block-sorting lossless data compression algorithm. Technical Report 124, Digital Equipments Corporation, 1994.

W.I. Chang and E.L. Lawler. Sublinear expected time approximate string matching and biological applications. *Algorithmica*, 12:327–44, 1994.

M. Crochemore. An optimal algorithm for computing the repetitions in a word. *Inform. Process. Lett.*, 12:244–250, 1981.

M. Crochemore, F. Mignosi, and A. Restivo. Automata and Forbidden Words. *Information Processing Letters*, 67:111–117, 1998a.

M. Crochemore, F. Mignosi, and A. Restivo. Minimal Forbidden Words and Factor Automata. In L. Brim, J. Gruska, and J. Slatuška, editors, *MFCS'98*, volume 1450 of *Lecture Notes in Computer Science*, pages 665–673. Springer-Verlag, 1998b.

M. Crochemore, F. Mignosi, A. Restivo, and S. Salemi. Text Compression Using Antidictionaries, 2000. DCA home page at URL http://www-igm.univ-mlv.fr/~mac/REC/DCA.html.

M. Crochemore and W. Rytter. Efficient parallel algorithms to test square-freeness and factorize strings. *Inform. Process. Lett.*, 38:57–60, 1991a.

M. Crochemore and W. Rytter. Usefulness of the Karp-Miller-Rosenberg algorithm in parallel computations on strings and arrays. *Theoret. Comput. Sci.*, 88:59–82, 1991b.

M. Crochemore and W. Rytter. *Text Algorithms*. Oxford University Press, 1994.

G. Das, R. Fleischer, L. Gąsieniek, D. Gunopulos, and J. Kärkkäinen. Episode matching. In A. Apostolico and J. Hein, editors, *Proceedings of the 8th Annual Symposium on Combinatorial Pattern Matching (CPM'97)*, volume 1264 of *Lecture Notes in Computer Science*, pages 12–27. Springer-Verlag, 1997.

M. Farach and M. Thorup. String matching in Lempel-Ziv compressed strings. In *Proc. of 27th Symposium on Theory of Computing*, 1994.

N.J. Fine and H.S. Wilf. Uniqueness Theorems for Periodic Functions. *Proc. Amer. Math. Soc.*, 16:109–114, 1965.

M.J. Fischer and M.S. Paterson. String matching and other products. In R.M. Karp, editor, *Complexity of Computation*, volume 7, pages 113–125. SIAM-AMS Proceedings, 1974.

L. Gąsieniec, P. Indyk, and P. Krysta. External inverse pattern matching. In *Proceedings of the 8th Annual Symposium on Combinatorial Pattern Matching*, volume 1264 of *Lecture Notes in Computer Science*, pages 90–101. Springer-Verlag, 1997.

L. Gąsieniec and W. Rytter. Almost optimal fully lzw-compressed pattern matching. In J. Storer, editor, *Data Compression Conference*, 1999.

D. Greene, M. Parnas, and F. Yao. Multi-index hashing for information retrieval. In *Proc. 35th Annual Symposium on Foundations of Computer Science*, pages 722–731, 1994.

D. Gusfield and J. Stoye. Linear time algorithms for finding and representing all tandem repeats in a string. Technical Report CSE-98-4, Department of Computer Science, University of California, Davis, 1998a.

D. Gusfield and J. Stoye. Simple and flexible detection of contiguous repeats using a suffix tree. In *9th CPM 98*, volume 1448 of *Lecture Notes in Computer Science*, pages 140–152. Springer-Verlag, 1998b.

R. W. Hamming. Error detecting and error correcting codes. *Bell System Tech. J.*, 29:147–160, 1950.

C.S. Iliopoulos, D.W.G. Moore, and K. Park. Covering a String. In *Proc. 4th Symp. on Combinatorial Pattern Matching*, volume 684 of *Lecture Notes in Computer Science*, pages 54–62. Springer-Verlag, 1993.

C.S. Iliopoulos and K. Park. An Optimal $O(\log \log n)$-time Algorithm for Parallel Superprimitivity Testing. *J. Korea Information Science Society*, 21:1400–1404, 1994.

R. Karp and M.O. Rabin. Efficient randomized pattern matching algorithms. *IBM J. Res. Dev.*, 31:249–260, 1987.

A. Lempel and J. Ziv. On the complexity of finite sequences. *IEEE Trans. on information Theory*, 22:75–81, 1976.

V.I. Levenshtein. Binary codes capable of correcting deletions, insertions and reversals. *Soviet Phys. Dokl.*, 6:707–710, 1966.

M. Lothaire. *Combinatorics on Words*. Cambridge University Press, second edition, 1997.

R. C. Lyndon and M. P. Schutzenberger. The equation $a^m = b^n c^p$ in a free group. *Michigan Math. J.*, 9:289–298, 1962.

M.G. Main and R.J. Lorentz. An $o(n \log n)$ algorithm for finding all repetitions in a string. *J. of Algorithms*, pages 422–432, 1984.

G. Manacher. A new Linear-Time On-Line Algorithm for Finding the Smallest Initial Palindrome of a String. *J. Assoc. Comput. Mach.*, 22:346–351, 1975.

U. Manber. A text compression scheme that allows fast searching directly in the compressed file. In M. Crochemore and D. Gusfield, editors, *Proceedings of the 5th Annual Symposium on Combinatorial Pattern Matching*, volume 807 of *Lecture Notes in Computer Science*, pages 113–124. Springer-Verlag, 1994.

U. Manber and R. Baeza-Yates. An algorithm for string matching with a sequence of don't cares. *Inform. Process. Lett.*, 37:133–136, 1991.

H. Mannila, H. Toivonen, and A.I. Vercamo. Discovering frequent episodes in sequences. In *Proceedings of the 1st International Conference on Knowledge Discovery and Data Mining (KDD'95)*, pages 210–215. AAAI Press, 1995.

D. Moore and W.F. Smyth. Computing the Covers of a String in Linear Time. In *Proc. 5th ACM-SIAM Symp. on Discrete Algorithms*, pages 511–515, 1994.

E. Moura, G. Navarro, N. Ziviani, and R. Beaza-Yates. Direct pattern matching on compressed texts. In *Proc. SPIRE'98*, pages 90–95. IEEE CS Press, 1998.

G. Navarro and M. Raffinot. A general practical approach to pattern matching over ziv-lempel compressed text. In *Proceedings CPM'pp*, pages 14–36, 1999.

G. Piatesky-Shapiro and W.J. Frawley, editors. *Knowledge Discovery in Databases*. AAAI Press/MIT Press, 1991.

M. Rabin. Discovering repetitions in strings. In A. Apostolico and Z. Galil, editors, *Combinatorial Algorithms on Words*, pages 279–288. Springer-Verlag, 1985.

D. Ron, Y. Singer, and N. Tishby. The Power of Amnesia: Learning Probabilistic Automata with Variable Memory Length. *Machine Learning*, 25:117–150, 1996.

D. Russel and G.T. Gangemi, Sr. *Computer Security Basics*. O'Reilly and Associates, Inc., 1991.

M.-F. Sagot, A. Viari, and H. Soldano. Multiple sequence comparison — A peptide matching approach. *Theoret. Comput. Sci.*, 180:115–137, 1997.

C.E. Shannon and W. Weaver. *The Mathematical Theory of Communication*. University of Illinois Press, 1949.

Y. Shibata, M. Takeda, A. Shinohara, and S. Arikawa. Pattern matching in text compressed by using antidictionaries. In M. Crochemore and M. Paterson, editors, *Combinatorial Pattern Matching*, volume 1645 of *Lecture Notes in Computer Science*, pages 37–49. Springer-Verlag, 1999.

A. Thue. Über unendliche zeichenreihen. *Norske Vid. Selsk. Skr. Mat. Nat. Kl. (Cristiania)*, 7:1–22, 1906.

A. Thue. Über die gegenseitige lage gleicher teile gewisser zeichenreihen. *Norske Vid. Selsk. Skr. Mat. Nat. Kl. (Cristiania)*, 1:1–67, 1912.

E. Ukkonen. Approximate string matching and the q-gram distance. In R. Capocelli, A. De Santis, and U. Vaccaro, editors, *SEQUENCES II - Methods in Communication, Security, and Computer Science*, pages 300–312. Springer-Verlag, 1993.

M. Waterman. *Introduction to Computational Biology*. Chapman and Hall, 1995.

T.A. Welch. A technique for high performance data compression. *IEEE Trans. on Computers*, 17:8–19, 1984.

J. Ziv and A. Lempel. A universal algorithm for sequential data compression. *IEEE Trans. on Inform. Theory*, IT-23:337–343, 1977.

J. Ziv and A. Lempel. Compression of individual sequences via variable-rate coding. *IEEE Trans. on Inform. Theory*, 24:530–536, 1978.

Chapter 7

SEARCHING LARGE TEXT COLLECTIONS

Ricardo Baeza-Yates
Dept. of Computer Science, Universidad de Chile, Santiago, Chile
rbaeza@dcc.uchile.cl

Alistair Moffat
Dept. Computer Science and Software Engineering, The University of Melbourne, Victoria 3010, Australia
alistair@cs.mu.oz.au

Gonzalo Navarro
Dept. of Computer Science, Universidad de Chile, Santiago, Chile
gnavarro@dcc.uchile.cl

Abstract In this chapter we present the main data structures and algorithms for searching large text collections. We emphasize inverted files, the most used index, but also review suffix arrays, which are useful in a number of specialized applications. We also cover parallel and distributed implementations of these two structures. As an example, we show how mechanisms based upon inverted files can be used to index and search the Web.

Keywords: Inverted files, Index, Search, Web crawlers, Suffix arrays, Supraindex.

1. Introduction

The amount of textual information available in electronic form is growing at a staggering rate. The best example of this growth is the World-Wide Web, which is estimated to provide access to at least three terabytes of text (that is, three million megabytes). Even in commercial and private hands text collection sizes which were unimaginable a few years ago are common now, and the challenge is to efficiently search those text collections to find the documents or pages that are of interest to the user. Given that brute force scanning of a large document collection is not a viable option except for occasional one-off queries, some kind of index structure is necessary. Indeed, there is nothing novel at all in the notion of an index, and most of us make regular use of them in libraries, books, and telephone directories. In Sections 2 and 3 we describe inverted files and suffix arrays respectively, which are the two standard mechanisms for providing such indexes.

After a query is resolved, we still have the problem of determining which documents are more relevant for the user. That problem depends on the information retrieval (IR) model chosen by the system. Although an introduction to one of the main IR models is given in Section 2, that problem and other problems related to retrieval evaluation, query languages and operations, user interfaces and feedback, and so on, are beyond the scope of this chapter. We refer the reader to (Baeza-Yates and Ribeiro-Neto 1999, Lesk 1997, Witten et al. 1999) for a full exposition of IR and further algorithmic details, as well as related topics such as digital and conventional library systems.

The number of users searching the Web, online bibliographical systems, or digital libraries is also increasing, which adds extra pressure to the efficiency requirements of information servers. In addition, increasingly sophisticated search requirements imply more costly searching algorithms, and even with clever implementation heuristics and fast hardware there is a limit to what can be achieved. One way to extend this limit is the use of parallel and distributed computers. Processors can be added to a system as the requirements increase, resulting in an almost unlimited capacity to cope with more users posing more complex queries on larger text collections. An overview of the issues relevant to parallel and distributed computations for information retrieval is given in Section 4, together with the solutions that are currently proposed.

Finally, using as a case study one of the most topical applications of text retrieval, Section 5 describes the techniques involved in the design of Web search engines, and discusses the particular challenges that arise in web searching.

2. Inverted Files

The standard structure used to provide content-based access to large document collections is the *inverted file index* (Baeza-Yates and Ribeiro-Neto 1999, Frakes and Baeza-Yates 1992, Harman and Candela 1990, Witten et al. 1999). This section describes the components of an inverted file index, shows how they can be economically stored, and then describes the querying process that allows the index to be used to satisfy a user's information need.

2.1. Basic Concepts

An inverted file index consists of two principal components: a *vocabulary* of the index *terms* that appear in the collection, together with, for each term, some associated facts; and in a separate structure, for each term an *inverted list* that records the locations in the collection at which that term appears. For example, if the term *cat* appears in the first, third, and fourth documents of a collection, then the vocabulary for the collection contains a record for *cat* that, amongst other things, records the number of documents that contains it and the location on disk of its inverted list; and an inverted list is stored that, as a minimum, records the document numbers $1, 3, 4$.

Table 7.1. Example of an inverted file index.

Document Collection		Vocabulary		Inverted lists
Doc.	Text	Word	Freq.	
1	fat rat eat cat	bat	3	$\langle 2, 5, 6 \rangle$
2	fat rat eat bat	cat	3	$\langle 1, 3, 4 \rangle$
3	mat cat eat rat	eat	6	$\langle 1, 2, 3, 5, 6, 8 \rangle$
4	mat cat fat cat	fat	4	$\langle 1, 2, 4, 5 \rangle$
5	fat rat eat fat mat bat	mat	4	$\langle 3, 4, 5, 7 \rangle$
6	bat eat rat	rat	7	$\langle 1, 2, 3, 5, 6, 7, 8 \rangle$
7	rat sat mat	sat	1	$\langle 7 \rangle$
8	rat eat rat eat rat eat rat			

Table 7.1 gives a more complete example of a document collection and the vocabulary and inverted lists that result. This structure clearly provides rudimentary query support, since a query *cat* AND *rat* designed to retrieve the documents that contain both of the terms *cat* and *rat* can be resolved by searching the vocabulary for the two terms, fetching their respective inverted lists $\langle 1, 3, 4 \rangle$ and $\langle 1, 2, 3, 5, 6, 7, 8 \rangle$, and then taking the intersection of the sets of document numbers to obtain $\langle 1, 3 \rangle$. Disjunctive queries (using the operator OR) and negations (using a NOT operator) can be similarly handled, taking respectively the union of inverted lists, and the complement. Queries formed using AND, OR, and NOT are known as *Boolean* queries, since each document in the collection either is or is not an answer to the query.

Although they are straightforward to process, Boolean queries are not the preferred querying mechanism for large document collections.

Instead, *ranked* queries are used. In a ranked query a real-valued similarity score is calculated between the query and each of the documents in the collection, and the top r ranked documents are presented as the answers, with r being a parameter set by the user. Provided a suitable heuristic is used, ranked queries provide considerably more flexibility than do Boolean queries, since the formalism of a precise query language is avoided and it becomes much more natural for a user to simply type a list of terms that describe their information need without consideration for the operators that should group them. For example, the intent of the query: *indexing, searching, querying, techniques, algorithms, methods, processes, large, big, gigabyte, megabyte, document, text, collection, database, repository, archive* is abundantly clear, yet it probably would not be if parentheses and Boolean operators were to be inserted. The flexibility offered to users in selecting a value of r is also of considerable benefit – they can set r to be low, and view a short answer list that may have high *precision* but low *recall*, or they can increase r to obtain higher recall at possibly the expense of lower precision.

The desire to use ranked queries does, however, raise the issue of how to calculate the similarity score. One standard mechanism that has been in use in various forms for more than 30 years is the *vector space* model, of which the *cosine rule* is an example (Salton and McGill 1983). If term t appears in document d a total of $f_{d,t}$ times, and at least once in f_t of the N documents in the collection, then one instantiation of the cosine rule calculates the similarity $S_{q,d}$ between query q and document d as:

$$S_{q,d} = \left(\sum_{t \in q} w_{d,t} \cdot w_{q,t}\right) / (W_d \cdot W_q) \qquad w_t = 1 + \log_e(N/f_t)$$
$$w_{d,t} = r_{d,t} \qquad w_{q,t} = r_{q,t} \cdot w_t$$
$$r_{d,t} = 1 + \log_e f_{d,t} \qquad r_{q,t} = 1 + \log_e f_{q,t}$$
$$W_d = \left(\sum_t w_{d,t}^2\right)^{1/2} \qquad W_q = \left(\sum_t w_{q,t}^2\right)^{1/2}.$$

The quantity $S_{q,d}$ lies between zero and one (with one indicating maximal similarity) and corresponds to the cosine of the angle in n-dimensional space between a query vector described by $\langle w_{q,t} \rangle$ and a document vector $\langle w_{d,t} \rangle$, where n is the number of distinct terms in the collection. (Note that there are many alternative formulations that have been proposed (Salton and Buckley 1988, Zobel and Moffat 1998).)

Experimental studies (see, for example, the proceedings of the sequence of TREC Text Retrieval conferences (Harman 1995)) have shown that formulations such as the one shown above yield relatively good retrieval effectiveness, and on queries ranging from 10 to 50 or more terms, and a collection of approximately 2 GB of English text, approximately half of the first $r = 100$ documents can be expected to be relevant to the query. (Of course, it must be noted that performance is highly variable, and for very specific queries with only a handful of answers in the collection, precision at $r = 100$ cannot be this high.)

Further gains in retrieval effectiveness (that is, changes that either boost precision, the fraction of retrieved documents that are relevant, or boost recall, the fraction of relevant documents that are retrieved) can be obtained by a number of additional heuristics. One, known as *pivoting*, biases the document weights W_d so as to obtain what are determined experimentally to be "fairer" normalization values (Singhal et al. 1996). A second technique is *thesaural expansion* – searching for each query term in a thesaurus (and possibly thereby also disambiguating homonyms) and extending the query by including all of the possible alternatives for each term. A third technique that has yielded improved retrieval effectiveness is *relevance feedback*. In automatic feedback systems, the query is evaluated initially with a small pool depth r, and one or more top ranked documents used as the seed for a second query that is taken deeper with a larger r to actually generate answers. The rationale behind this approach is that the highest-scoring document is probably relevant, and inclusion of some or all of its terms in an extended query will hence probably be of benefit. Manual feedback systems allow the user to intervene and distinguish between documents that are relevant, and should be added to the query in a positive sense, and those that are irrelevant, and should be added to the query in a negative sense. After the user views a subset of highly ranked documents the query is reevaluated, using modified term weights w_t that incorporate the positive and negative feedback.

A fourth technique that has been used to enhance effectiveness – with mixed results, it should be noted – is the inclusion of phrases as query terms. The discussion above implicitly assumed that both documents and queries were bags of words. In fact they are not, and word ordering can be critical. Consider, for example, the queries *cat eat rat* and *rat eat cat* applied to the example collection shown in Table 7.1 – one would reasonably expect that the two queries would return different documents in the top-ranked position in the answer list. This intuition has been confirmed by experiments, and regarding word pairs and triples as the index "terms" used to compose documents and queries has led to significant improvements in retrieval effectiveness in simple retrieval models, but only small gains when other sophisticated techniques have already been used (Mitra et al. 1997).

Alternatives to the *document-based* retrieval model described are also possible. One alternative is a *locality-based* model, which scores word locations rather than documents themselves, summing contributions from any nearby query terms, with the contribution decreasing as the number of intervening words increases (de Kretser and Moffat 1999). This mechanism has the desirable side effect of allowing enhanced document presentation, as the document viewer can be initiated at the exact location that triggered the high similarity score, rather than at the beginning of the document. When documents may be many tens of kilobytes long, this is a distinct advantage.

Another approach to the same problems is *passage retrieval*, in which the document source is broken into (possibly overlapping) windows of text of roughly the same size, with indexing and querying performed on the basis of these passages (Kaszkiel et al. 1999).

2.2. Index Representation

Having described the overall process of ranked information retrieval, we now turn to some of the technical detail. In this subsection we examine the inverted index, and show that a document-level index of the type illustrated in Table 7.1 requires a remarkably small overhead as a fraction of the indexed text, just 5–10% for typical collections and typical documents. The next subsection then shows how these compact indexes are used to allow evaluation of ranked queries; and the third subsection considers the problem of actually constructing the inverted index for a large text.

In generating the example of Table 7.1 it was natural to arrange the document numbers in each inverted list into increasing order, and this same ordering can be used to provide economical storage, by recording the difference or *d-gap* between consecutive document numbers. For example, in Table 7.1 the inverted list for the term *cat* would be represented as the sequence of d-gaps $\langle 1, 2, 1 \rangle$. Then, because the sum of the f_t different d-gaps that occur in a list of f_t document identifiers cannot exceed N, the number of documents in the collection, it is possible to code the integer d-gaps in a manner that provides short codewords for the many small d-gaps, and long codewords for the relatively infrequent large d-gap values.

A number of suitable codes are described in detail in Witten et al. (1999), here we briefly describe one of these – the *Golomb* code (Gallager and Van Voorhis 1975, Golomb 1966). Golomb codes are particularly suited to integers distributed according to a geometric probability distribution, in which a gap of length x occurs with probability $Pr(x) = (1-p)^{x-1}p$ for some fixed probability p. This distribution may be assumed if, for example, the f_t documents that contain a given term t are a random subset of the N documents in the collection. The Golomb code for an integer $x \geq 1$ is then given by a unary code for the quotient $q = (x - 1)$ div b (that is, $q - 1$ one-bits followed by a zero-bit) and a $\log_2 b$-bit binary code for the remainder $r = (x - 1) \mod b$. The simplest situation is when b is a power of two, and in this case the codes are also known as *Rice* codes. Table 7.2 shows examples of codewords when $b = 4$ and $b = 8$. Note that the gaps in the codewords are shown only to allow the reader to discern the break between unary and binary parts, and are not present in the actual codes. When b is not a power of two adjustments are made to the binary part so as to make some codewords one bit shorter and not waste any combinations.

Table 7.2. Example Rice/Golomb codewords.

Gap x	Codeword, $b = 4$	Codeword, $b = 8$
1	0 00	0 000
2	0 01	0 001
3	0 10	0 010
4	0 11	0 011
10	110 01	10 001
20	11110 11	110 011
30	11111110 01	1110 101
40	1111111110 11	11110 111

As can be seen in Table 7.2, small values of b give shorter codewords for small gaps x than do large values of b, but achieve this at the expense of generating longer codewords for large values of x than do large-b codes. Indeed, b should be chosen as a function of p, the probability governing the geometric distribution. A good first-order approximation is that

$$b = \frac{\ln 2}{p} \approx 0.69 \cdot \frac{N}{f_t}$$

where, as before, N is the number of documents in the collection and f_t is the number of them that contain (one or more occurrences of) the term t.

Using Golomb codes, and with the value of b set locally for each inverted list (so that common terms with long inverted lists use small values of b), typical document-level indexes occupy about 6 bits per pointer (Witten et al. 1999). And, since each term in a document of a few kilobytes typically appears approximately twice on average, the number of pointers to be stored in the inverted file (given by $\sum_{t=1}^{n} f_t$ for a collection of n distinct terms) is about half the number of words in the collection. Including white space, each word accounts for about 5–6 bytes in a source document, so one pointer is stored for every approximately

10–12 bytes of source text, and the cost of storing that pointer is around 6 bits.

To this cost must be added the bits required to store the $f_{d,t}$ component – the within-document frequency value necessary for ranked queries. For example, the inverted list for *cat* in the collection of Table 7.1 must be extended to

$$\langle ((1,1),(2,1),(1,2)\rangle\;,$$

with, for example, the last pair indicating that *cat* appears twice in the document whose number (d-gap) is one greater then the previous document containing it. The cost of adding the $f_{d,t}$ values is small, and even using a unary code accounts for at most one bit per term appearance, or, using the same estimated values, 2 bits per pointer. Hence, for large collections consisting of documents each a few kilobytes long, one byte of index information is required for each 10–12 bytes of source text, a very modest overhead. Use of more sophisticated compression techniques allows the cost to be reduced below 7 bits per pointer (Witten et al. 1999). It is also worth noting that these estimated costs are all-inclusive, and even frequent words such as *the* and *it*, and numbers such as *28* and *1999* are indexed. Indeed, by employing a suitable compression regime the cost of storing inverted lists for common words is dramatically reduced compared to naive storage methods, and little additional saving is achieved by using a *stop list* to prevent them from being indexed.

The final cost is storage for the vocabulary. For a large collection this is a minor overhead. For example, a 1 GB collection will contain approximately 200 million words, but almost certainly will contain fewer than one million distinct terms, particularly if case-folding and stemming are used to map document words onto their lexicographic roots. Even allowing 12–16 bytes per distinct term (to store the term itself, plus the corresponding f_t value, plus a pointer to its inverted list) the vocabulary contributes only a further approximately 1% of the source text.

While the discussion in this section has been focussed on representing inverted lists as a sequence of d-gaps, other organizations for the inverted lists are also possible. Persin et al. (1996) describe a frequency-sorted index organization that retains the same information, but sorts the pairs in each list into decreasing order of within-document frequency $f_{d,t}$. The advantage of this approach is that if only the most "important" documents for each term are to be processed, and importance can be gauged by the magnitude of the $f_{d,t}$ value in conjunction with the term weight w_t, then only a prefix of each inverted list need be processed.

Another alternative index representation is the blocked index described by Anh and Moffat (1998). Rather than store d-gaps that are relative to the previous document number, this index representation stores gaps relative to document numbers at fixed intervals. It is then possible to identify the document number corresponding to a certain pointer without fully processing all of the previous codewords. Moreover, the codes that are processed are done so in a byte-by-byte sense rather than a bit-by-bit sense, and so considerable reductions in computation times can be achieved.

A further variant of inverted file indexing worthy of note is *block addressing*. Block addressing was first proposed in a system called *Glimpse* (Manber and Wu 1994). The text is logically divided in blocks, and the index does not point to exact word positions, but only to the blocks where the word appears. Space is saved because there are fewer blocks than text positions (and hence the pointers are shorter), and also because all the occurrences of a given word in a single text block are referenced only once. In simplest form, each file in a collection of files is considered to be a block, and in large collections, multiple files may be placed into the same logical block.

Searching in a block addressing index is similar to searching in a full inverted one. The pattern is searched in the vocabulary and a list of blocks where the pattern appears is retrieved. However, to obtain the exact pattern positions in the text, a sequential search over the qualifying blocks becomes necessary, since the stored index is very coarse-grained. In a sense, the system makes use of exhaustive search, but with a filter-index that allows some or most of the blocks to be avoided. That is, the reduction in index space requirements is obtained at the expense of higher search costs.

Block addressing was analyzed by Baeza-Yates and Navarro (1997), who showed both analytically and experimentally that a block addressing index can achieve both sublinear space overhead and sublinear query time, whereas inverted indexes pointing to words or documents achieve only the second goal.

2.3. Query Processing

Given a document-level index augmented by within-document $f_{d,t}$ values, how then should a ranked query over a set of terms be evaluated? And how much more costly is it to evaluate a ranked query than to evaluate a Boolean query? These two issues are considered in this subsection.

As for Boolean queries, ranked queries are evaluated by processing inverted lists. The standard mechanism maintains an *accumulator* variable A_d for each document d, in which a partial similarity score between d and the query q is built up. As each inverted list is processed, a contribution is added to A_d for each document d that appears in that inverted list. This process is summarized in the following pseudo-code.

process_ranked_query(q, r) {
    for each document $d \in \{1, \ldots, N\}$ do
        set $A_d \leftarrow 0$
    for each term $t \in q$ do {
        calculate w_t and $w_{q,t}$
        for each document $d \in$ the inverted list for term t do {
            calculate $w_{d,t}$
            set $A_d \leftarrow A_d + w_{d,t} \cdot w_{q,t}$
        }
    }
    for each document $d \in \{1, \ldots, N\}$ do
        set $A_d \leftarrow A_d/W_d$
    calculate and return the values of d corresponding to
        the r largest values of A_d
}

For small collections this mechanism is straightforward in both implementation and execution. It requires that the document length normalization values W_d be precomputed and available in a file, but this is readily done at index creation time.

For large collections with millions of documents there are a number of potential problems with this approach. One obvious issue is that care needs to be taken with the final extraction of the r largest accumulator values, and a partial sort (rather than a full sort of all N values) should be performed. Suitable techniques are discussed by Witten et al. (1999,Chapter 4).

Another drawback is the need for memory space to support the accumulators. For a collection of N documents either an array A of N floating point values is required, or, if it can be assumed that most of the accumulators remain unused throughout the calculation, a dynamic structure must be manipulated. Unfortunately, in typical ranked queries of 20–30 words, a substantial fraction of the documents in the collection will contain at least one of the query terms, and result in non-zero accumulators. Moreover, the need for accumulators is on a per-query basis, and sharing between users is not possible. For example, if a system is being designed to allow (say) 100 simultaneous users access to (say) a collection of 10 million documents, use of an array data structure (with four bytes per accumulator) results in a memory consumption of 4×10^7 bytes $\times 100 = 4$ gigabytes, a rather large amount by any standards.

The final issue to be considered is not unrelated to the memory problem, and that is the cost of performing floating point arithmetic for every value in what are potentially very long inverted lists. If a user inadvertently uses the word *the* in a query – and, given that we regard the query as simply being another piece of text, there is no reason why the user should penalized if they do use such words – a small similarity contribution might need to be calculated and added to the accumulator of every document in the collection.

To sidestep these latter two problems a number of *pruning* strategies have been devised (Anh and Moffat 1998, Buckley and Lewit 1985, Moffat and Zobel 1996, Persin et al. 1996). Query terms are processed in decreasing order of weight w_t (which is generally in increasing order of term frequency f_t), and heuristic rules are applied to determine whether or not a document should be allowed an accumulator variable, and whether or not any more of this inverted list, or any more inverted lists at all, should be processed.

The use of such strategies – coupled in some cases with frequency-sorted inverted lists (Persin et al. 1996) – has been shown experimentally to allow greatly accelerated query processing speeds, within substantially less memory resources, and usually at negligible (sometimes zero) cost in terms of degraded retrieval effectiveness. Any large-scale system being designed for high throughput rates will need to make use of some kind of pruning mechanism, even if it is as simple as only processing a subset of the query terms. Indeed, the pruning thresholds can be set on a query-by-query basis, and users willing to pay extra for more resource consumption, or users making use of the service late at night at off-peak rates, might be given more generous thresholds than budget-minded and sleep-minded users.

Reorganizing the list representation to facilitate the evaluation of ranked queries does, of course, impact upon Boolean query evaluation. Frequency-sorted indexes lead to more complex Boolean query processing. On the other hand, the mechanisms used by Anh and Moffat (1998) in their blocked inverted lists also accelerate Boolean query handling.

2.4. Index Construction

The implementer of a system based upon inverted files will also face a further vexing question: how to construct the index? Writing a cross-reference-generator program is a standard exercise in second-year data structures and algorithms subjects, but the obvious solutions break down when applied to large volumes of data – they use in-memory data structures of size comparable to or larger than the volume of text being processed, and access those data structures in a non-sequential manner.

Efficient methods have been developed that allow more economical index construction (Moffat and Bell 1995, Witten et al. 1999). Unsurprisingly, they also rely on compression of the various components to reduce the volume of data that must be stored in main memory; and reorganize the operations required so as to make use of sequential disk-based processing. In this section we briefly sketch one mechanism that has been used to index (or *invert*) multi-gigabyte texts within a few tens of megabytes of main memory and only a small amount of extra disk space above and beyond the space required by the final compressed inverted index. This process is sketched in the pseudo-code that follows.

        create_inverted_index(T) {
            while text remains in T do {
                fill the in-memory buffer M with $\langle d, t, f_{d,t} \rangle$ triples
                sort M by the t component of each triple
                write M to disk in compressed form as a run
            }
            read the first block of each of the runs into memory
            while at least one of the runs remains unemptied do {
                append the $\langle d, t, f_{d,t} \rangle$ triple with the smallest t, d to the current output block,
                    recompressing on the fly
                      if the output block is full, write it to a vacant slot on disk
            }
            permute the blocks of the index to bring them into order
            release any free space at the end of the index
        }

The basis of the method is a process that creates a file of $\langle d, t, f_{d,t} \rangle$ triples in document number order (that is, by increasing d) and then reorders that file into term order (increasing t) using a combination of in-memory sorting and in-place multi-way external merging. The sorting phase is then followed by an in-place permutation of the fixed-length blocks that collectively comprise the compressed inverted file (Moffat and Bell 1995). We now examine these steps one by one.

First, the text is read in document order and parsed into terms. A bounded amount of memory is set aside as a buffer to collect $\langle d, t, f_{d,t} \rangle$ triples, and every time that array is full it is sorted using Quicksort, and written to disk in a compressed format. Each partial index – or *run* –

is padded to be a multiple of a fixed block length so that all of the runs stored on disk are block-aligned.

Once all of the text has been processed in this way, the resultant runs are combined in a multiway merge. Just one block of each run is resident in memory at any given time, and so the memory requirement is modest. As the merge proceeds, output blocks are produced and written back to disk to any available slots within the file. No padding is required in the output blocks, meaning that the writing process tends to consume block slots at a slower rate than they are freed by the reading process, and a vacant slot is almost always available somewhere. The compression also tends to improve between input and output stages, as a result of more data for each term being available, further contributing to this effect. And in any case, if a vacant slot is not available at the time one is needed, that block can always be written at the end of the file.

Finally, once all of the runs have been exhausted, the index is complete, but with its blocks permuted throughout the file space allocated to the index rather than stored sequentially, and with unused block slots interspersed. An in-place permutation is then used to reorder the blocks. The permutation requires that each block of the index – typically a kilobyte or so – be read once and written once in a non-sequential order, and so is disk-intensive, but only for a relatively short time. At the conclusion of the block reordering any now-unused space at the end of the file is released.

This discussion has assumed that the collection is static. While this assumption is often valid – for example, when an information collection is to be distributed on a CD-ROM – dynamic collections must also be managed. In a dynamic collection the updates can be of two varieties: existing documents might be edited and perhaps deleted; and new documents might be added as the collection grows.

In the latter of these two cases, while it is tempting to develop a process that inserts single documents as they become available, the resource costs of doing so are very high. Consider, for example, the process of adding a document. As many as several hundred inverted lists must be fetched from disk, extended by a few bits or bytes, and then written back. Even discounting the complexity associated with managing a large set of bit-variable records, the sheer cost of reading and writing inverted lists means that the per-document disk operations required by this process might consume many seconds, severely limiting the rate at which documents can be added. Instead, updates should be batched into a small *stop press* collection that is searched exhaustively during querying. Periodically the stop press is indexed using the mechanisms described above, and then the index for the main collection and the index for the stop press combined using a sequential merge. Only by batching updates in this way can the per-document costs be kept modest.

The more complex update operations are edits and deletes on existing documents, since they involve contraction and expansion in the middle of inverted file entries. One pragmatic solution to this problem is to simply append the revised version of the document as if it were new, and retain the old version with a flag marking it as defunct. This approach automatically creates a record of all alterations to a document, which, with disk space being relatively cheap, may be an attractive additional benefit.

2.5. Other Issues

In the earlier parts of this section we have focussed on the index, and shown how it can be compressed and used to resolve queries. The text of the document collection should also, of course, be stored in compressed form if the collection is to be maintained economically. However, standard utilities such as *gzip* cannot be used, as they code each symbol of the file in a context established by the aggregate of all of the preceding symbols, meaning that it would not be possible to decode individual documents when they are required. Instead, *semi-static* compression mechanisms are used (Witten et al. 1999), in which a preliminary pass is made to accumulate symbol frequencies, and then a second pass made to actually compress the text with respect to those frequencies. During this second pass the bit addresses at which the documents commence are noted in an index file, which is used to facilitate random-access decoding during query processing.

A variety of compression models can be used in this semi-static mode. One attractive option is word-based parsing, in which the text is broken into a sequence of words, and the words are regarded as being emitted from a zero-order Markov model (Moura et al. 1998, Witten et al. 1999, Zobel and Moffat 1995). While the compression effectiveness attained with a zero-order word-based model is not as good as can be achieved with higher order character-based models, the word-based model has two distinct advantages: the loss in effectiveness that results from coupling the model with Huffman coding rather than arithmetic coding is small, and Huffman coding is considerably faster; and the memory space required during decoding operations can also be managed in a useful manner (Moffat et al. 1997).

A further useful side-effect of the word-based model is that if the set of codewords is calculated using radix-256 or radix-128 codeletters, fast compressed-text searching is possible (Moura et al. 1998). Use of a radix-256 code and a spaceless words model allows compression effectiveness only slightly inferior to a radix-2 code, and, because the codewords are all byte-aligned, allows pattern matching on words and phrases by simply compressing them too, and using any standard pattern matching software. This mechanism does, however, allow false matches, since it is impossible to know whether any particular byte in the compressed text is the first byte of a codeword. This alignment problem is solved with the use of a radix-128 code, and storing each codeletter in an 8-bit byte. Setting the top bit in the first byte of each codeword then allows straightforward identification of word alignments, at a cost of using $8/7 = 14\%$ more space than previously.

For typical document collections storing ASCII English text, the word-based model compresses to as little as 25% of the original size can be achieved (Moura et al. 1998, Witten et al. 1999) (better than Ziv-Lempel techniques), with document access times essentially unaffected (Zobel and Moffat 1995). That is, a complete indexed retrieval system can be constructed in 30–35% of the space required by the original source text.

In closing this section on inverted file indexing it is also relevant to comment on one other type of indexing that enjoyed a period of support in the 1980s – *signature files*. In a signature file index a probabilistic hash-filter is used to support conjunctive Boolean queries, with operations performed on *bit slices*. As we have seen, however, ranked queries are the preferred retrieval mechanism, and signature files do not readily support them; and even if the only queries to be performed are conjunctive Boolean, it is not clear that signature files enjoy any space or speed advantage over compressed inverted files (Zobel et al. 1998).

3. Suffix Arrays

Inverted indexes are appropriate for typical text retrieval queries on natural language texts. That is, they assume that the text is composed of words; that it follows some statistical properties that ensure, for example, that the vocabulary is not very large and that the most frequent words are rarely queried; that the queries aim at retrieving whole words or phrases; that if more complex patterns are searched for (say, regular expressions) then they are to be matched against whole words or at least subwords; and that the user is not interested in more complex queries. Under those restrictions, inverted indexes offer a data structure well suited to the problem to be solved, are cheap to build and space-economical, and allow ranking techniques to be used in order to identify the documents most relevant to a given query.

The suffix array is a data structure aimed at a different type of search. It does not require that the text be composed of natural language words, and instead handles any sequences of symbols. It is also able to perform more complex queries at the pattern matching level including simple words and phrases, regular expressions, and in the presence of errors. Suffix arrays can also find the longest repeated string in a text, and even compare the whole text against itself to find interesting auto-repetitions with and without errors.

The suffix array can be applied to classical information retrieval, but it is less adequate than inverted indexes: it is more expensive to build and maintain, powerful in aspects that are marginal in classical scenarios (for example, flexible pattern matching) and less powerful in crucial aspects (such as classical relevance ranking). The real interest for suffix arrays is in other applications, including genetic databases (Baeza-Yates and Gonnet 1990, Gonnet 1992, Manber and Myers 1993) (where the texts are DNA or protein sequences), intrusion detection (where the text is a sequence of events along the time), oriental languages (Nagao and Mori 1994) (where word delimiters are not so clear), agglutinating languages (where words are long and carry a more complex meaning, as in Finnish), and linguistic analysis of the text statistics (Gonnet et al. 1992a) (to detect plagiarism, for instance).

We start by presenting the suffix array structure together with the strategy for searching simple patterns. Then we consider the problem of constructing suffix arrays. Finally, we discuss more complex queries. For simplicity we have used natural language examples, and it should be noted throughout that suffix arrays can be used on any sequence of symbols.

3.1. Structure

Consider the text example shown in Figure 7.1. We have selected a number of *index points* in the text. Index points are the text positions that the suffix array will be able to retrieve. In applications where there are no words it is common to consider that all symbols (in the example, characters) are index points, while if we consider only word beginnings, as in the example, we obtain a functionality similar to inverted indexes. Indeed, with word indexing we can choose to omit stop-words if we wish.

Each index point defines a text *suffix*, that is, a string which starts at the index point and continues until the end of the text. In the figure we have listed the suffixes that are defined by the index points. Notice that the end of the text is signaled by a special character "$", which is smaller than all the other characters.

SEARCHING LARGE TEXT COLLECTION

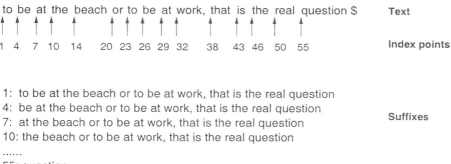

Figure 7.1. A text example, the selected index points and the resulting suffixes.

The suffix array is, in essence, an array recording the positions of all of the index points, ordered *lexicographically* by the text suffix they point to. Figure 7.2 shows the suffix array of our example.

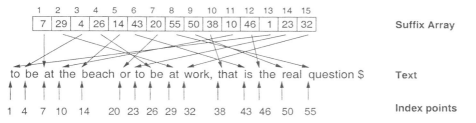

Figure 7.2. Suffix array for the example text.

The goal of this organization is to permit binary searching the query in the suffix array rather than sequentially searching it in the text. The key idea is that we look for text substrings that start at index points, and any such substring is a *prefix* of a text suffix.

Imagine that we are interested in finding the occurrences of the word "be" in the text. Since the index points are lexicographically sorted in the suffix array, all the text suffixes that start with "be" are placed in a continuous range of positions in the suffix array. Both extremes of this range of positions can be binary searched in the suffix array.

The extremes are obtained as follows. First, we search for the pattern "be$". The binary search starts at position 8 in the array. The suffix pointed to is "question$" (index point 55), which is lexicographically greater than our search pattern and therefore we consider array position 4. This time the suffix is "be at..." (index point 26), which is still greater than our pattern. We then consider array position 2, which points to "at work,..." (index point 29), which is lexicographically smaller than our pattern. The binary search continues until we determine that the first suffix larger than or equal to our pattern is at suffix array position 3. To look for the right extreme of the suffix array we perform a similar search for the smallest pattern that is larger than any suffix starting with "be...". This pattern is "bf$", whose search ends at suffix array position 6. Therefore all the occurrences of "be" in the text are in suffix array positions 3 to 5. The corresponding text positions are 4, 26 and 14.

Notice that searching for the occurrences of the pattern "be" is not the same as searching for the word "be". For example, we have retrieved "beach" as well. Suffix arrays do not consider words, and therefore they find any occurrence of our string. Moreover, if we had indexed all text positions it would also have found the string "be" inside words. If we want to retrieve the word "be" we need to search for the pattern "be " (one space at the end), and filter the text to convert all separators (like comma) to spaces. This filtering can be done on the fly, without need to modify the text.

Note also that we need to access the text at search time to compare the query. Moreover, the accesses to the text are essentially at random. A first consequence is that the text must be available at search time, unlike most inverted index scenarios. A second, more serious consequence, is that in practical situations, in which it may impossible to fit either of the suffix array or the text in main memory, the random accesses will be to secondary memory, and so will be very expensive.

Supra-indexes have been proposed to alleviate this problem (Baeza-Yates et al. 1996). A supra-index over a suffix array is obtained by a sampling process: the suffix array is logically divided in blocks of b entries, and the first entry of each block is selected for the supra-index. For each selected entry, the first ℓ characters of the corresponding suffix are stored in the supra-index. The result is a coarser version of the suffix array which, by appropriate selection of b and ℓ, fits in main memory. Figure 7.3 shows a supra-index built for our example.

The binary search now starts in the supra-index in main memory. At the end, two blocks of the suffix array are identified as holding the initial and final positions of the answer range (in fact, since the suffixes of the supra-index are limited at ℓ characters, two blocks may need to be considered for each extreme in the worst case). Only after these bounds have been identified is the disk accessed. The relevant blocks of the suffix array are brought into main memory and the binary searches are completed with the initial algorithm. In the example, the first block (1 to 3) is read to determine the left extreme, and the second block (4 to 6)

SEARCHING LARGE TEXT COLLECTION

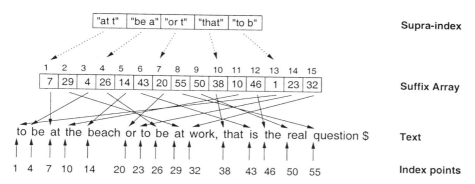

Figure 7.3. Suffix array and a supraindex that takes four characters each three entries. The dotted arrows can be computed as required, and are not stored.

for the right extreme. The search inside the blocks still needs to access the text at random disk positions.

Most of the accesses to the suffix array on disk have been eliminated, as well as some of the random accesses to the text on disk. In practice this reduces search time to 25% to 30% of its original cost. Note that we have assumed that the final block of the suffix array fits in main memory. In huge texts it is possible that the combined goal of having the supra-index of $n\ell/b$ characters and a suffix array block of b entries in memory cannot be achieved for any b. In this case a supra-supra-index can be used, leading to a hierarchical scheme much like Prefix B-trees (Bayer and Unterauer 1977). Another possible choice are the String B-trees (Ferragina and Grossi 1999), which have predictable performance and provide few disk accesses.

The expensive part of the search is still the random accesses to the text in the final search in the suffix array block. In this case, deviating from standard binary search may be better. Binary search is the optimal strategy when accessing all the cells of the array has the same cost. In our case, each cell of the suffix array has a different access cost which is not fixed but depends on the current position of the disk head, that is, on the last element accessed in the suffix array block. For instance, we may prefer to access the suffix array at position $b/3$ instead of $b/2$ because the disk head is currently much closer to the text pointed by position $b/3$. The optimal search strategy can be precomputed at suffix array construction time and stored together with each suffix array block at little space penalty. Reductions of 60% over the original search times are reported (Navarro et al. 2000).

If we are indexing natural language, an interesting choice is that the supra-index does not divide the suffix array at regular intervals and takes

ℓ characters, but that it cuts the blocks where the first word of the suffix pointed changes and stores the word in the supra-index (Barbosa and Ziviani 1995). Figure 7.4 illustrates this arrangement. Since the supra-index becomes similar to the text vocabulary, the result is much like a full inverted index, where the suffix array blocks correspond to the lists of occurrences for each word. The main difference is that the lists of occurrences are not sorted by text positions but lexicographically. In this scheme, simple words can be searched using the supra-index only (the suffix array block is directly the answer) without accessing the text. Searching phrases is simpler with this scheme than with inverted indexes.

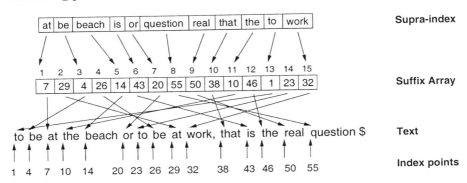

Figure 7.4. Our suffix array and a word supraindex. This time the arrows need to be stored.

With regard to space requirements, the suffix array requires a pointer for each selected index point of the text. Considering that pointers take four bytes, this means that four times the text size is required to index all the text positions, while if indexing the beginnings of non-stopwords the structure needs 30% to 40% of the text size, still rather more than the equivalent document-level compressed inverted file. The extra space required by supra-indexes is negligible.

3.2. Construction

The easiest way to build a suffix array is via an indirect sort of the pointers. That is, first the index points of the text are identified and stored (in increasing text position order) in the suffix array, and then any sorting algorithm is applied on the suffix array. In this sorting process, two suffix array positions are compared by considering the lexicographical relationship between the text suffixes indexed by the two pointers.

Therefore, a suffix array on a text of n index points can be built in $O(n \log n)$ comparisons, where each comparison compares two *strings*. The cost of a random string comparison is $O(1)$ at the beginning of the sorting process (since on average $\sigma/(\sigma-1)$ comparisons are necessary to differentiate two random text positions, where σ is the alphabet size). However, as the sorting process continues, strings that are increasingly similar are compared. On average, two text suffixes that are neighbors in the final suffix array share $O(\log n)$ common letters (Szpankowski 1992), which shows that the cost in terms of number of character comparisons is between $O(n \log n)$ and $O(n \log^2 n)$. Moreover, in some special cases where, for instance, a complete document is repeated in the collection, the performance can be very bad because very long common substrings exist. The worst case is to build the suffix array for a text like "aaaaaa...", which takes $O(n^2 \log n)$ time. A possible solution to this problem is to weaken the structure so that only the first ℓ characters of the suffixes are considered. In this case the worst case cost can be made $O(n\ell \log n)$, and we can search for patterns of length up to ℓ.

A different construction algorithm which builds the complete suffix array (not the weaker ℓ-limited version) in $O(n \log n)$ worst case time and $O(n \log \log n)$ on average is presented by Manber and Myers (1993). The idea is to perform a sequence of bucketsorts, sorting the suffixes according to longer and longer initial prefixes. At the beginning of the i-th iteration the suffixes are already sorted by their first 2^{i-1} letters, and at the end of that iteration are sorted by their first 2^i letters. Hence, in $\lceil \log_2 n \rceil$ iterations the array is totally sorted. Moreover, since on average $O(\log n)$ initial characters are necessary to distinguish between neighboring suffixes of a text, on average only $O(\log \log n)$ iterations are necessary. The complexity follows from the fact that each bucketsort takes $O(n)$ time.

The algorithm starts by sorting the suffixes according to their first letter. Since there are only σ possible different values, the bucketsort is simple. At iteration $i > 0$, we refine the buckets obtained at iteration $i-1$. The sort for the first 2^i letters uses the fact that *all* the suffixes are already sorted by their first 2^{i-1} letters. In particular, if two index points j_1 and j_2 are currently in the same bucket, then they share their first 2^{i-1} symbols, and their first 2^i symbols can be written as xy and xz, where $|x| = |y| = |z| = 2^{i-1}$. Comparing them is the same as comparing y with z. And because these are of length 2^{i-1}, they are already sorted somewhere, since they also start suffixes. Their relative order depends on the buckets in which the index points ($j_1 + 2^{i-1}$ and $j_2 + 2^{i-1}$) are in this iteration.

In particular, if $j_1 + 2^{i-1}$ is currently in the first bucket of the suffix array, then j_1 should, after the next iteration, be in the first sub-bucket of its current bucket. This idea is used to refine all the current buckets in a linear pass over the suffix array. We refer the reader to the original article (Manber and Myers 1993) for the remaining details. The method can be adapted to the case where not all the letters are index points, simply by replacing the first iteration of the algorithm with an ad-hoc sort.

Faster algorithms have been presented recently (Itoh and Tanaka 1999, Larsson and Sadakane 1999, Sadakane 1998). These are hybrids between bucket sorting (as in the Manber and Myers algorithm) and standard sorting algorithms. Despite the fact that they have the same asymptotic time requirements, in practice the new methods are many times faster.

3.2.1 Sorting Large Suffix Arrays. The previous methods work well if the text and the suffix array fit in main memory. However, when this is not the case they suffer from the same problem identified above with respect to the search algorithms: the text is accessed in an essentially random manner on disk. And even if a sorting algorithm designed for secondary memory is used, the problem of random text accesses via the pointers remains.

The following algorithm (Gonnet et al. 1992a) builds a suffix array without random access to the text or suffix array. The text is cut in blocks and the suffix array is incrementally built for the blocks. That is, at iteration i the algorithm obtains the suffix array for the first i blocks of the text. The block size is selected so that the suffix array for one text block can be built in main memory using the previous algorithms.

The construction of the first block is trivial. At iteration $i > 0$, we have already on disk the suffix array of the first $i-1$ text blocks. We bring the i-th text block into memory and build its suffix array. The problem now is to merge the new small suffix array with the large suffix array for the previous text. A classical merge for two sorted sequences cannot be used because it would imply comparing text positions from the first $i-1$ text blocks, which are on disk. However, the i-th text block is in memory and the idea of the merge can be carried out after all. If we are able to determine, for each consecutive pair of positions in the current small suffix array, how many (consecutive) positions of the large suffix array lie between them, then we can perform the merge without accessing any text. The problem is then how to compute those *counters* that tell, in other words, how many suffixes of the first $i-1$ text blocks lie between each pair of positions of the small suffix array.

The key idea is that the $i-1$ text blocks can be used for this purpose rather than their corresponding suffix array. We simply traverse the first $i-1$ text blocks sequentially, and search each text suffix in the small suffix array When the location of that suffix has been determined – that is, that it lies between two given positions – the corresponding counter in incremented. And to search in the small suffix array we just need the i-th text block, which is in memory.

Overall, the process for the i-th step is: (1) build the small suffix array for the i-th text block; (2) initialize the counters; (3) read all the first $i-1$ text blocks and search each of their suffixes in the small suffix array, incrementing for each the corresponding counter; (4) use the counters to merge the small suffix array and the large suffix array representing the first $i-1$ text blocks.

If the text is of size n and we have $O(M)$ main memory, $O((n/M)^2)$ blocks are processed, for a total CPU time of $O(n^2 \log M/M)$. This is greater than the $O(n \log n)$ complexity of the Manber and Myers algorithm, but is much faster in practice because all of the operations are either sequential, or carried out in memory. Other algorithms and a comparison of them, including the algorithm just presented, can be found in (Clauser and Ferragina 2000).

3.3. Complex Searching

Although we have explained only how to search for simple queries, suffix arrays can do much more. Suffix arrays are closely related to suffix trees (Apostolico 1985). A suffix tree is a *trie* data structure built on the suffixes of the text, and the suffix array is obtained by collecting the trie leaves in order. A suffix tree can be built in $O(n)$ time and takes $O(n)$ space. Simple strings of length m can be searched in $O(m)$ time. Regular expressions can be searched in $O(n^\alpha)$ average time, where $0 < \alpha < 1$ is a constant dependent on the regular expression (Baeza-Yates and Gonnet 1996). The algorithm backtracks in the trie looking for text substrings that can match the regular expression. It is possible to adapt the search to allow a limited number of errors (for example, missing characters, extra characters, or altered characters) in the text occurrences at $O(\exp(m))$ or $O(n^\alpha)$ time (Navarro and Baeza-Yates 1999, Ukkonen 1993). The suffix tree has been used in many different applications, including finding the longest common auto-repetition in the text, and matching the whole text against itself with or without errors (Apostolico 1985, Baeza-Yates and Gonnet 1990, Gonnet et al. 1992a;b).

The main drawback of suffix trees is that they take 12 to 20 times the text size (Kurtz 1999), which considerably limits their applicability. Using clever schemes, on average the space can be reduced to 5 times the text (Clark and Munro 1996). On the other hand, suffix arrays can simulate *any* algorithm designed on suffix trees at an $O(\log n)$ extra time penalty, a rather modest overhead for most applications. The key idea of the simulation is that each node in a suffix tree correspond to an interval in the corresponding suffix array, and hence any movement in the suffix tree can simulated by two binary searches in the suffix array. And the benefit that arises from this tradeoff is clear – a suffix array requires just 4 times the memory space of the text being indexed, even if every character of the text is an index point.

4. Parallel and Distributed Indexes

Parallel or distributed processing can be used for two independent tasks: index construction and querying. In some cases, like suffix arrays, it is possible that the bottleneck is in the construction rather than in the querying process. In other cases, the construction is not the problem, but the index has to be distributed to speed up queries, whose performance may be critical to the success of the system.

We start by giving a quick overview of the efficiency measures in parallel computing. Then, we identify three different formats in which an index may be distributed. This is independent on whether the index is generated sequentially or in parallel. Some of the formats can also be queried sequentially or in parallel, although others are specifically designed for a parallel environment. Later, we discuss how to generate and query each type of index, sequentially or in parallel, and analyze the advantages and disadvantages of each choice.

4.1. Efficiency Models

The main difference between "parallel" and "distributed" computing is that in the first the processors can share the data and in the second they cannot. More specifically, there is a moderate to high transmission cost to move data from one processor to another, which ranges from a cost similar to that of accessing secondary storage (Anderson et al. 1995) (in a local ATM switch, for instance) to several seconds (in the Internet, for example). We present our algorithms at a high level trying to address both scenarios, but we are careful in pointing out the key factors that make a parallel or a distributed implementation preferable in each case.

The obvious measure for the efficiency of a parallelized index construction is related to the total elapsed time. This can be expressed with a formula called *speedup*, defined as the time taken by the best sequential algorithm divided by the time taken by the parallel algorithm. The speedup is optimal when it reaches P, the number of processors.

When it comes to measure the efficiency of a parallel querying scheme, two possible alternatives exist. A first one, called *throughput*, is related to the amount of queries answered per time unit and in the ideal case it reaches P times that of a sequential algorithm. A second one, called *response time*, is the amount of time taken to solve a query and ideally reaches $1/P$ of that of a sequential algorithm. For instance, if two processors take queries from different users and solve them sequentially, the throughput doubles but the response time for each user does not change since their query has been solved sequentially. If, on the other hand, both processors cooperate to solve each query, the response time may be halved, but the overhead costs necessary to divide the query and integrate the solutions may reduce the throughput. Which of the two measures is more important largely depends on the application. Increasing the throughput may be more important when there are many inexpensive queries per second and therefore the sequential response time is acceptable. Reducing response time may more important if there are a small number of expensive queries to be processed and the goal is to reduce the time for each.

4.2. Index Distribution

In a distributed environment, an index can be stored in three formats, illustrated in Figure 7.5.

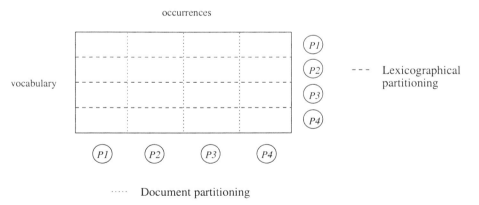

Figure 7.5. Three ways to distribute an index. A centralized index is not partitioned, a document partitioning splits the index by documents, and a lexicographical partitioning splits the index by lexicon.

Centralized index: this is the conventional arrangement in which the index is stored in a single location.

Document partitioned index: the text collection is distributed across a set of machines, and a local index is separately built for the subcollection maintained at each machine. Each index is totally independent from the rest.

Lexicographically partitioned index: the text collection is distributed but the index is global and each machine is responsible for completely handling the queries that lie in some subset of the vocabulary. Each machine knows which of the other machines is responsible for each subset of the query terms. For simplicity we assume that the sets are lexicographical ranges, but other partitions are possible.

Even a centralized index can be searched in parallel, as we see later. The second choice is clearly the simplest way to take advantage of the presence of many processors to answer queries. The most challenging architecture is the third one, in which there is a single index distributed across all the machines.

An important kind of distributed index arises in loosely coupled heterogeneous databases that for efficiency, formatting or administrative reasons cannot cooperate at index construction time. In this case the only scheme that can be applied is document partitioning.

4.3. Parallel Index Generation

The easiest case of parallel index generation is that of document partitioned indexes, since each processor indexes its local text using a sequential algorithm.

A centralized index, on the other hand, can be generated in parallel as well. Normally, the process of producing a centralized index is just one step over that of producing a lexicographically distributed index. We just have to concatenate the lists of occurrences stored in the different processors to obtain the global list of occurrences. In the case of a suffix array this corresponds to concatenating the local suffix arrays obtained.

The most complex algorithms are those to generate lexicographically distributed indexes. We divide them according to the type of index that is to be produced.

4.3.1 Inverted Indexes. We describe an algorithm presented in (Ribeiro-Neto et al. 1998; 1999). We start with the text distributed across the machines. Each processor obtains the vocabulary of its local text. Once this is done the processors engage in a merging process to obtain the complete vocabulary and the global information necessary to distribute it across the processors. The merging of the vocabulary is done by pairwise merging: first each processor numbered $2i+1$ sends its vocabulary to that numbered $2i$, then each processor numbered $4i+2$ sends its vocabulary to that numbered $4i$, and so on until a single processor holds all the vocabulary. This is possible because the vocabulary is small.

Once a processor holds the whole vocabulary it assigns a subset of the words to each other, and broadcasts that information. Now each process knows which lists of occurrences must be stored at each other processor, so they obtain the occurrences from the text and send them to the final owners. There are different choices on when to send those lists, that range from sending each occurrence as it is found to building the local inverted file and then distributing the lists.

4.3.2 Suffix Arrays.
The construction of distributed suffix arrays is essentially a distributed sorting algorithm. From the local texts each processor builds a local suffix array and the problem is then how to obtain a globally sorted suffix array from the locally sorted pieces. Different algorithms based in different sorting algorithms (mergesort and variants of quicksort) are proposed in (Kitajima et al. 1997; 1996, Navarro et al. 1997).

The main complication of the suffix array construction on top of the classical sorting problem is that two suffixes cannot be directly compared since they are pointers to possibly remote text. Using a general sorting scheme and making remote requests each time a comparison is made is impractical. This is generally solved by storing the first characters of each suffix together with the pointer. If the suffixes are long enough this suffices to solve most of the comparisons locally and only a few of them need remote access to the text. Moreover, the suffixes can be kept in the final index to aid in the querying process, as explained in the next section. Of course there is a space-time tradeoff: longer suffixes reduce the number of comparisons but increase the storage required, and they also pose higher communication overheads.

An algorithm to build suffix arrays which is not based in classical sorting is presented in (Kitajima and Navarro 1999). The algorithm is a parallelization of the original sequential Manber and Myers algorithm to build the suffix array (Manber and Myers 1993).

4.4. Parallel Querying

The querying process used is, of course, determined by the index arrangement. The next three subsections describe the mechanisms possible for parallel querying in each of the three index models described above.

4.4.1 Centralized Index. Parallelism can be used even with a centralized index. The simplest way is to have a threaded database server where many processes answer queries independently. Even if there is a single processor to run the threads the throughput can be increased by parallelizing the CPU and I/O tasks. In general, however, handling text indexes is an I/O intensive task, and therefore the number of I/O channels and disks must increase in accordance to the intended gain in throughput. Consideration must also be given to the extent to which multi-threading disrupts the caching and buffering operations on single processor machines. Running too many threads can have a disruptive effect, resulting in a high proportion of cache misses and a reduction in overall throughput.

On the other hand, if response times are to be reduced, parallelizing the processing of each individual query is necessary. A query may consist of a number of different words or patterns that must be searched for and then later combined via Boolean operations or ranking. The set of patterns to search can be partitioned among the processors, so that each one accesses the shared index to retrieve the relevant lists of term appearances (be it from an inverted list or a suffix array).

Once the lists of occurrences have been obtained, the most common combining operations are set unions and intersections (for Boolean operators) and sorting (for ranking or suffix arrays). A considerable literature on basic parallel algorithms for set manipulation and sorting has been developed (see, for example, (Jájá 1992, Quinn 1994)), and many of those techniques are applicable.

In a distributed scenario the communication costs are too high to use a really centralized index. The obvious alternative is replication: there is an exact replica of the index in each machine. Queries can be distributed dynamically according to the workload. Although, again, this increases the throughput of the whole system, the response time does not change. Partitioning the query among processors in a distributed environment is precisely the goal of the two distributed querying schemes that are described next.

4.4.2 Document Partitioned Index. The basic strategy in a document partitioned index is to broadcast every query to every processor. Each processor solves the query for its local text, using its local index. Once the query is solved they send back their results to the originating processor, which collates the results and delivers a set of answers to the user.

This scheme is extremely simple to implement and parallelizes the query very well: not only the throughput increases but also the query time is reduced. There are, however, two reasons why perfect speedup in response time cannot be achieved. The first is that the query time using indexes is normally sublinear. That is, if n is the global text size and $f(n)$ is the time to answer a query in such text, then $f(n) = o(n)$, and this immediately implies that $f(n/P) = \omega(f(n)/P)$. In other words, halving the text does not mean that the query time will halve too. The second reason is that there is an overhead in transmitting and assembling the results.

Assembling the results may be simple or complex depending on the type of query. Since the documents or text positions delivered correspond to disjoint portions of the space, Boolean queries require no more than the concatenation of the individual result sets. The problem is more complex with ranked queries (de Kretser et al. 1998). First, there is a performance problem because only a few documents transmitted by each processor will actually qualify after being compared to the rest, an effect that worsens as the number of processors grows. A possible solution to the problem is a lazy scheme where the querying processor asks progressively fewer and fewer results of the processors, stopping when it has enough to answer the query. Another solution is that the processors locally remove the documents that will probably not qualify after a global comparison, therefore transmitting $O(R/P)$ results, where the final list must have R elements. This is a heuristic and may yield incorrect results, especially if one processor has results much more relevant than the rest.

This is related to a more serious problem, which is the *consistency* of the ranking. Each processor has ranked the query according to its parameters of *local* term frequency and inverse document frequency. The result is not the same if a global ranking is applied. One solution is, at indexing time, to compute and distribute the global frequencies of the terms to all the processors. This scheme, however, is less than satisfactory because the processors are no longer independent in administering their local text collection. In particular, it becomes a hybrid with a lexicographically partitioned index. Another solution is to construct a central vocabulary (rather than a central index) after each of the separate indexes has been constructed; another is to add a further synchronization step at query time in which each processor is first polled to determine local term frequencies and weights for the query terms, and then global query term weights are broadcast back to each of processors so as to ensure that all ranking scores are compatible (de Kretser et al. 1998).

Finally, we consider the best way to partition the documents across the machines, when there is a choice. The simplest alternative and one of the best in practice is a random distribution. This ameliorates the problem of ranking inconsistencies, because the global term frequencies should be similar to the local ones. The workloads also tend to be uniform, since none of the machines will be consistently more loaded than others simply because they hold all of the more popular documents.

The opposite choice is a domain-driven partition: a subject or domain is assigned to each machine and all the documents relevant to that domain are administered by that machine. This makes the relevance ranking very different and does not guarantee a uniform workload. Moreover, even when the workload is uniform the response time is normally not improved because most of the work for a given query is done by a single machine. This distribution, however, is sometimes chosen for administrative reasons or to avoid broadcasting every query to every processor. In fact, this design gets closer to a lexicographically partitioned index, which is covered in the next section.

4.4.3 Lexicographically Partitioned Index.

In this type of index one processor takes full responsibility of the processing of each query term, and the set of processors involved must either later cooperate somehow to perform the ranking or Boolean operations, or transfer the partial results back to the querying processor for collation.

The machine that originates the query knows, for each pattern of the query, which machines are responsible for the involved terms. Therefore, the query is sent only to the responsible machines. In the simplest scenario, those machines send back their result to the originating processor, which is responsible for assembling the results. The assembly is more complex now, as it can involve performing Boolean operations or ranking among the lists.

Simple queries, such as single words and patterns posed to inverted indexes or suffix arrays, may require no further processing. This gives maximum throughput gain, but the response time is unaltered. On the other hand, in complex queries with large intermediate results the most costly part of the process may be the assembly of the occurrences. It is possible that the processors cooperate in this part too, by merging, intersecting or sorting the lists pairwise in parallel instead of giving all the work to the originating processor. For example, if a conjunctive query with four terms was split among processors 1 to 4, then after obtaining the lists processors 1 and 2 can intersect their lists while processors 3 and 4 do the same, and only then the final processor intersects the two remaining lists.

Globally consistent ranking is not a problem in this type of index, because the processor that solves a query has global information about it. To obtain the global ranked results, the processors send back the lists of documents ranked for the term(s) they have solved. A global merging process is carried out, quite similar to set union but taking care of producing the highest ranked documents first. This is more difficult than for a document partitioned index because each document may appear in many lists (if it was selected by many query terms), very high in some lists and very low in others. In a document partitioned index each document appears in one list and producing the ranked result is trivial. The technique of lazy merging is more difficult to apply and less efficient in a lexicographically partitioned index.

A specific problem of suffix arrays with this scheme is that they need to access the text in order to answer queries. Since the text pointed by the array is not necessarily local, accessing remote text becomes necessary. This problem can be alleviated by storing supra-indexes or the first characters of all the suffixes, but in general some remote text has to be accessed to complete the answer. Other indexes that have problems with this distribution strategy are variations on the inverted index that need direct access to the text, for instance to solve phrase queries.

Finally, notice that the querying scheme for the lexicographically partitioned index is similar to the design shown to partition a query in a centralized index, as each processor takes care of some patterns of the query.

4.5. Comparison

It is clear that if there is only one machine, be it with one or many processors, the only choice is a centralized index, where the sole design decision is how to distribute the index and text across the disk array to improve performance. Queries are solved atomically or are split into components depending on whether maximizing throughput or minimizing response time is important. The only situation in which a document-partitioned index might be used is when the single collection becomes physically too large to be managed, either because of storage constraints on specific devices (memory or disk), or because of the cost of constructing a monolithic index.

In a distributed environment, however, we have presented three choices of partitioning the index: replicated, partitioned by documents and partitioned by lexicon. Replication is the best alternative in terms of performance, as it can behave like any of the other two. Of course, replication is only applicable if the collection is small enough to fit in a single machine. However, some replication can be included in the other two designs to improve the efficiency and make the system fault tolerant.

Of the remaining two alternatives, it is not immediately clear which is the best choice, and studies have reached different conclusions under alternative cost scenarios – Tomasic and Garcia-Molina (1993) show that document partitioning performs better on large conjunctive queries where the terms do not distribute uniformly in the query, and Ribeiro-Neto and Barbosa (1998) show that lexicographical partitioning is better when the terms distribute uniformly in the queries. This is clear, since conjunctive queries can be solved locally under a document partitioning scheme and this reduces communication costs. Moreover, if the terms are not uniformly distributed, then under document partitioning only a few processors will work, freeing the others to process new queries.

First consider single term queries. A document partitioning index does not achieve optimum throughput, because the local times are not simply the global times divided by P and because of the effort to assemble the final list of results. However, it will improve the response time because the work to answer the query is effectively divided among many processors. A lexicographically partitioned index, on the other hand, achieves optimum throughput if the queries distribute uniformly in the vocabulary, because there is virtually no overhead (unless direct access to remote text is necessary, as explained): the machine that takes care of the query does all the job. Response times, however, are not improved because a single machine is solving each query. Therefore, the question about which partitioning scheme is better boils down to a question of which efficiency measure is important for the application. For many cheap queries the lexicographical partition is probably better, while for few costly queries the choice should be document partition.

Extra considerations are necessary for more complex queries, however. Consider the case of a conjunction of many simple terms, whose final result is much smaller than the list of occurrences of each individual term. In this case a document partitioned index performs much better. The reason is that all the intersections are carried out locally, and it only transmits the final results corresponding to its subset of the text collection. The lexicographically partitioned index, on the other hand, has to intersect long lists which are not in the same machine, so no matter how the intersections are done, long occurrence lists must be transmitted. A similar, although less critical problem, arises with disjunctive queries, since the document partitioned index does not transmit redundant information across the network and the lexicographical partitioned index may transmit many times the same documents because different query terms appear on them.

Finally, we consider ranking. A first problem of document partitioned indexes is the possibility of inconsistent ranking. As explained, one solution is to have a global vocabulary, which is indeed a hybrid solution between a document and a lexicographical partition. Apart from this issue, ranking is simpler and more efficient in a document partitioned index if there are many terms involved, otherwise the situation is the same as for a single term query.

To summarize, a document partitioned index seems to be the best choice in many different scenarios. Enriching it with global inverse document frequency information is a good choice to ensure correct ranking, although it becomes more complex to administer. On the other hand, a lexicographically partitioned index is a better solution when the queries are simple (for example, single words) and hence cheap to solve, and the critical issue is to solve many queries per second. This case is very important, since it comprises a large fraction of most Web querying applications.

5. Case Study: Searching the Web

The World-Wide Web (WWW, or simply Web) was created at the end of the 1980s (Berners-Lee et al. 1994), and no one who was present at its conception could possibly have imagined its impact just ten years later. The amount of textual data alone is estimated in the order of at least three terabytes, and other media, such as images, audio, and video, are also available, using even more space than the text portion. Thus, the Web can be seen as a very large, unstructured but ubiquitous database. This triggers the need for efficient tools to manage, retrieve, and filter information. Similar problems are also becoming important in large Intranets or in text mining applications.

In this chapter we focus on text, because although there are techniques to search for images and other non-textual data, they cannot be applied (yet) in large scale. While searching Web pages, one implicit problem is that logically a document could be just part of a Web page (for example, a summary of an article in the set of summaries for a magazine issue) or split across many Web pages (for example, a section of an article). However, this cannot be easily known and we have to use the physical division as document space. We also emphasize syntactic search. That is, we search for Web documents that have user-specified words or patterns in their text. Words or patterns may or may not reflect the intrinsic semantics of the text. An alternative approach to syntactic search is to do a natural language analysis of the text. Although the techniques to preprocess natural language and extract the text semantics are not new, they are not yet very effective and are costly for large amounts of data.

In addition, in most cases they are only effective with well structured text, a thesaurus, and other contextual information.

There are basically three different forms of searching the Web. Two of them are well known and are used frequently by most users. The first is to use search engines that index a portion of the Web documents as a full-text database. The second is to use Web directories, which classify selected Web documents by subject. The third, not yet widely available, is to exploit the hyperlink structure present in Web documents. The scope of this chapter is related to the first case. We refer the reader to Baeza-Yates and Ribeiro-Neto (1999) for information on other techniques.

We now mention the main problems posed by the Web. We can divide them in two classes: problems with the data itself and problems regarding the user and his interaction with the retrieval system. The problems related to the data are:

- Distributed data: due to the intrinsic nature of the Web, data spans over many computers and platforms. These computers are interconnected with no predefined topology and the available bandwidth and reliability on the network interconnections vary widely.

- High percentage of volatile data: due to Internet dynamics, new computers and data can be added or removed easily (it is estimated that 40% of the Web changes every month (Kahle 1997)). We also have dangling links and relocation problems when domain or file names change or disappear. In addition, many pages are generated on demand, and might be missed by automatic software.

- Large volume: the exponential growth of the Web poses scaling issues that are difficult to cope with. Current estimations of the number of Web pages is at least 800 million (Lawrence and Giles 1999).

- Unstructured and redundant data: most people say that the Web is a distributed hypertext. However, this is not exactly so. Any hypertext has a conceptual model behind it, which organizes and adds consistency to the data and the hyperlinks. That is hardly true in the Web, even for individual documents. In addition, each HTML page is not well structured and some people use the term *semi-structured data*. Moreover, much Web data is repeated (mirrors or copies) or very similar. Approximately 20% of Web pages are (near) duplicates (Broder et al. 1997, Shivakumar and García-Molina 1998). Semantic redundancy can be even larger.

- Quality of data: the Web can be considered as a new publishing media. However, there is, in most cases, no editorial process. So, data can be even false, invalid (for example, because it is too old), poorly written or, typically, with many errors from different sources (typos, grammatical mistakes, OCR errors, and so on). In particular, around 1 out of every 200 common words and 1 out of every 3 foreign surnames, have at least one typo.

- Heterogeneous data: in addition of having to deal with multiple media types and hence with multiple formats, we also have different languages and, what is worse, different alphabets, some of them very large (for example, Chinese or Japanese Kanji).

The second class of problems are those faced by the user during the interaction with the retrieval system. There are basically two problems: (1) how to specify a query and (2) how to interpret the answer provided by the system. Without taking into account the semantic content of a document, it is not easy to precisely specify a query, unless it is very simple. Further, even if the user is able to pose the query, the answer might be a thousand Web pages. How do we handle a large answer? How do we rank the documents? How do we select the documents that really are of interest to the user? In addition, a single document could be large. How do we browse efficiently in large documents? These problems are still not solved completely (Baeza-Yates and Ribeiro-Neto 1999).

Next, we cover different architectures of retrieval systems that model the Web as a full-text database. One main difference between standard IR systems and the Web is that, in the Web, all queries must be answered without accessing the text (that is, only the indexes are available). To do otherwise would require either storing locally a copy of all of the indexed Web pages (too expensive) or accessing remote pages through the network at query time (too slow). This difference has an impact in the indexing and searching algorithms, as well as in the query languages made available.

Most indexes use variants of the inverted file already mentioned before. Some search engines eliminate stopwords to reduce the size of the index, but if the index is stored using the techniques summarized earlier, the actual saving is relatively small. Also, it is important to remember that a logical view of the text is indexed. Normalization operations may include removal of punctuation and multiple spaces and folding of uppercase to lowercase letters. To give the user some idea about each document retrieved, the index is complemented with a short description of each Web page (creation date, size, the title and the first lines or a few headings are typical). Assuming that 500 bytes are required to store the URL and the description of each Web page, we need 50GB to store the description for 100 million pages. As the user initially receives only a subset of the complete answer of each query, the search engine usually keeps the whole answer set in memory for a short time, to avoid having to recompute it if the user asks for more documents.

By using the compression techniques already discussed, the index size can be reduced to under 10% of the text (Witten et al. 1999). A query is answered by doing a binary search on the vocabulary of the inverted file. If we are searching multiple words, the results have to be combined to generate the final answer. This step is efficient if each word is relatively rare. Another possibility is to compute the complete answer while the user requests more Web pages, using a lazy evaluation scheme.

5.1. Centralized Architecture

Most search engines use a centralized crawler-indexer architecture. Crawlers are programs (software agents) that traverse the Web sending new or updated pages to a main server where they are indexed. Crawlers are also called robots, spiders, wanderers, walkers, and knowbots. In spite of their name, a crawler does not actually move to and run on remote machines. Rather, the crawler runs on a local system and sends requests to remote Web servers. The index is used in a centralized fashion to answer queries submitted from different places in the Web. Figure 7.6 shows the software architecture of a search engine based on the AltaVista architecture (AltaVista 1996). It has two parts: one that deals with the users, consisting of the user interface and the query engine and another that consists of the crawler and indexer modules.

The main problem faced by this architecture is the gathering of the data, because of the highly dynamic nature of the Web, the saturated communication links, and the high load at Web servers. Another important problem is the volume of the data. In fact, the crawler-indexer architecture may not be able to cope with Web growth in the near future. Particularly important is good load balancing between the different activities of a search engine, internally (answering queries and indexing) and externally (crawling).

The largest search engines, considering Web coverage in January of 2000, were Fast (Fast 1998), AltaVista (AltaVista 1996), Northern Light (Northern 1997), Google (Google 1998), and Snap (Snap 1998), in that order. According to recent studies, these engines cover less than 30% of all Web pages (Lawrence and Giles 1999).

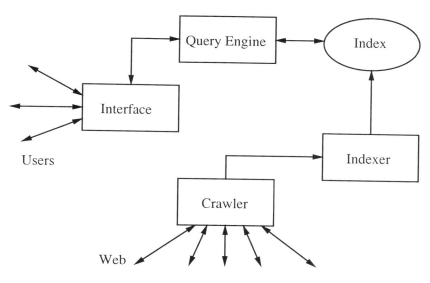

Figure 7.6. Typical crawler-indexer architecture.

5.2. Distributed Architecture

There are several variants of the crawler-indexer architecture. Among them, the most important is Harvest (Bowman et al. 1994). Harvest uses a distributed architecture to gather and distribute data, which is more efficient than the crawler architecture. The main drawback is that Harvest requires the coordination of several Web servers.

The Harvest distributed approach addresses several of the problems of the crawler-indexer architecture, such as: (1) Web servers receive request from different crawlers, increasing their load; (2) Web traffic increases because crawlers retrieve entire objects, but most of their content is discarded; and (3) information is gathered independently by each crawler, without coordination between all the search engines.

To solve these problems, Harvest introduces two main elements: gatherers and brokers. A gatherer collects and extracts indexing information from one or more Web servers. Gathering times are defined by the system and are periodic (that is, there are harvesting times as the name of the system suggests). A broker provides the indexing mechanism and the query interface to the data gathered. Brokers retrieve information from one or more gatherers or other brokers, updating their indexes incrementally. Depending on the configuration of gatherers and brokers, different improvements on server load and network traffic can be achieved. For example, a gatherer can run on a Web server, generating no external traffic for that server. Also, a gatherer can send information to several

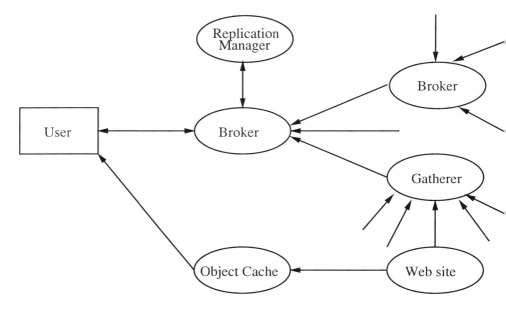

Figure 7.7. Harvest architecture.

brokers, avoiding work repetition. Brokers can also filter information and send it to other brokers. This design allows the sharing of work and information in a very flexible and generic manner. An example of the Harvest architecture is shown in figure (Bowman et al. 1994).

One of the goals of Harvest is to build topic-specific brokers, focusing the index contents and avoiding many of the vocabulary and scaling problems of generic indexes. Harvest includes a distinguished broker that allows other brokers to register information about gatherers and brokers. This is useful to search for an appropriate broker or gatherer when building a new system. The Harvest architecture also provides replicators and object caches. A replicator can be used to replicate servers, enhancing user-base scalability. For example, the registration broker can be replicated in different geographic regions to allow faster access. Replication can also be used to divide the gathering process among many Web servers. Finally, the object cache reduces network and server load, as well as response latency when accessing Web pages. More details on the system can be found in (Bowman et al. 1994).

5.3. Crawling the Web

In this section we discuss how to crawl the Web, as there are several techniques. The simplest is to start with a set of URLs and from there extract other URLs which are followed recursively in a breadth-first or depth-first fashion. For that reason, search engines allow users to submit top Web sites that will be added to the URL set. A variation is to start with a set of populars URLs, because we can expect that they have information frequently requested. Both cases work well for one crawler, but it is difficult to coordinate several crawlers to avoid visiting the same page more than once. Another technique is to partition the Web using country codes or Internet names, and assign one or more robots to each partition, and explore each partition exhaustively.

Considering how the Web is traversed, the index of a search engine is analogous to the stars in the sky. What we see never did exist, as the light has traveled different distances to reach our eye. Web pages referenced in the index were also explored at different dates and they may not exist any more. Nevertheless, when we retrieve a page, we obtain the actual content of it. How fresh are the Web pages referenced in an index? The pages will be from one day to two months old. For that reason, most search engines show in the answer the date when the page was indexed. The percentage of invalid links stored in search engines vary from 2 to 9%. User submitted pages are usually crawled after a few days or weeks. Starting there, some engines traverse the whole Web site, while others select just a sample of pages or pages up to a certain depth. Non-submitted pages will wait from weeks up to a couple of months to be detected. There are some engines that learn the change frequency of a page and visit it accordingly (Coffman et al. 1997). They may also crawl more frequently popular pages (for example, pages having many links pointing to them). Overall, the current fastest crawlers are able to traverse up to several million Web pages per day using many processes.

The order in which the URLs are traversed is important. As already mentioned, the links in a Web page can be traversed breadth first or depth first. Using a breadth first policy, we first look at all the pages linked by the current page, and so on. This matches well Web sites that are structured by related topics. On the other hand, the coverage will be wide but shallow and a Web server can be bombarded with many rapid requests. In the depth first case, we follow the first link of a page and we do the same on that page until we cannot go deeper, returning recursively. This provides a narrow but deep traversal. Only recently, some research on this problem has appeared (Cho et al. 1998), showing that good ordering schemes can make a difference if crawling better pages first.

6. Concluding Remarks

There are many research problems in searching large text collections. Many of them are related to the Web, which is certainly the application which currently enjoys the highest profile. Some important trends are:

- **Distributed architectures:** New distributed schemes to traverse and search the Web must be devised to cope with its growth. This will have an impact on current crawling and indexing techniques, as well as caching techniques for the Web. Currently it remains unclear whether server capacity or network bandwidth will be the bottleneck limiting performance.

- **Indexing:** Which is the best logical view for the text? What should be indexed? How can good compression schemes be exploited to achieve fast searching and reduced network traffic? And how can word lists, URL tables, and so on be efficiently and effectively compressed and updated?

- **Searching:** How to cope with syntactical uncertainty in the user query and with typing errors in databases with low quality control such as the Web? How can we allow for more flexible pattern searching while keeping at the same time search efficiency and user friendliness in the query language? How can we index to search for flexible patterns? How does this affect the precision and recall of the search?

- **Duplicated data:** Good mechanisms to detect and eliminate repeated Web pages (or pages that are syntactically very similar) are needed. Initial approaches are based on resemblance measures using document fingerprints (Broder et al. 1997, Broder 1997). This is related to an important problem in databases: finding similar objects.

An important issue to be settled in the future is a standard protocol to query search engines. One proposal for such a protocol is STARTS (Gravano et al. 1997), which allows users to choose the best sources for querying, evaluate the query at those sources, and merge the query results. This protocol would make easier to build metasearchers, but at the same time that outcome could also be given as a reason for not having a standard, to prevent metasearchers from profiting from the work done by search engines and Web directories. This is a particular case of the federated searching problem from heterogeneous sources, as it is called in the database community (Quass et al. 1995). Federated searching is a problem already studied in the case of the Web, including discovery and ranking of sources (Chawathe et al. 1994, Gravano et al. 2000, Yuwono and Lee 1997). These issues are also very important for digital libraries (Powell and Fox 1998) and visualization (Alonso and Baeza-Yates 1998). A related topic is metadata standards for the Web. XML helps (Deutsch et al. 1998, Goldman et al. 1998, Khare and Rifkin 1977), but semantic integration is still needed.

Hyperlinks can also be used to infer information about the Web. Although this is not exactly searching the Web, it is an important trend called Web mining. Traditionally, Web mining had been focused on text mining, that is, extracting information from Web pages. However, the hyperlink structure can be exploited to obtain useful information. For example, the ParaSite system (Spertus 1997) uses hyperlink information to find pages that have moved, related pages, and personal Web pages. Hyperlinks can also be used for ranking, and has also been used to find communities and similar pages (Gibson et al. 1998, Kleinberg 1998). Other results on exploiting hyperlink structure can be found in (Chakrabarti et al. 1998, Marchiori 1997, Pirolli et al. 1996). Further improvements in this problem include Web document clustering (Broder et al. 1997, Chen 1997, Weiss et al. 1996) (already mentioned), connectivity services (for example, asking which Web pages point to a given page (Bharat et al. 1998)), automatic link generation (Green 1998), extracting information (Bonnet and Tomasic 1998, Brin 1998), and so on.

More information on searching and crawling the Web can also be found in the chapters in this handbook by Marc Najork and Allan Heydon, titled *High-Performance Web Crawling* and by Andrei Broder and Monika Henzinger, titled *Algorithmic Aspects of Information Retrieval on the Web*. These chapters also cover the main challenges: ranking, duplicate filtering, and search-by-example.

Bibliography

O. Alonso and R. Baeza-Yates. A model for visualizing large answers in WWW. In *XVIII Int. Conf. of the Chilean CS Society*, pages 2–7, Antofagasta, Chile, 1998. IEEE CS Press.

AltaVista. AltaVista: Main page. http://www.altavista.com, 1996.

T. Anderson, D. Culler, and D. Patterson. A case for NOW (network of workstations). *IEEE Micro*, 15(1):54–64, 1995.

V. N. Anh and A. Moffat. Compressed inverted files with reduced decoding overheads. In Croft et al. (1998), pages 290–297.

A. Apostolico. The myriad virtues of subword trees. In *Combinatorial Algorithms on Words*, NATO ISI Series, pages 85–96. Springer-Verlag, 1985.

R. Baeza-Yates, E. Barbosa, and N. Ziviani. Hierarchies of indices for text searching. *Information Systems*, 21(6):497–514, 1996.

R. Baeza-Yates and G. Gonnet. All-against-all sequence matching. Dept. of Computer Science, University of Chile, 1990.

R. Baeza-Yates and G. Gonnet. Fast text searching for regular expressions or automaton searching on a trie. *J. of the ACM*, 43(6):915–936, 1996.

R. Baeza-Yates and G. Navarro. Block-addressing indices for approximate text retrieval. In *Proc. ACM CIKM'97*, pages 1–8, 1997. Extended version to appear in *JASIS*.

R. Baeza-Yates and B. Ribeiro-Neto. *Modern Information Retrieval*. Addison-Wesley & ACM Press, Harlow, UK, 1999.

E. Barbosa and N. Ziviani. From partial to full inverted lists for text searching. In R. Baeza-Yates and U. Manber, editors, *Proc. 2nd South American Workshop on String Processing, WSP'95*, pages 1–10, 1995.

R. Bayer and K. Unterauer. Prefix B-trees. *ACM TODS*, 2(1):11–26, Mar 1977.

Tim Berners-Lee, Robert Cailliau, Ari Luotonen, Henrik Frystyk Nielsen, and Arthur Secret. The World-Wide Web. *Comm. of the ACM*, 37(8):76–82, 1994.

Krisha Bharat, Andrei Broder, Monika Henzinger, Puneet Kumar, and Suresh Venkatasubramanian. The connectivity server: fast access to linkage information on the Web. In *7th WWW Conf.*, Brisbane, Australia, April 1998.

Philippe Bonnet and Anthony Tomasic. Partial answers for unavailable data sources. In *Workshop on Flexible Query-Answering Systems*, pages 43–54, 1998.

C. Mic Bowman, Peter B. Danzig, Darren R. Hardy, Udi Manber, and Michael F. Schwartz. The Harvest information discovery and access system. In *Proc. 2nd Inter. World Wide Web Conf.*, pages 763–771, October 1994.

S. Brin. Extracting patterns and relations from the World Wide Web. In *Workshop on Web Databases*, Valencia, Spain, March 1998.

A. Broder, S. Glassman, M. Manasse, and G. Zweig. Syntactic clustering of the Web. In *6th Int'l WWW Conference*, pages 391–404, Santa Clara, CA, USA, April 1997.

A.Z. Broder. On the resemblance and containment of documents. In *Conf. on Compression and Complexity of Sequences*, pages 21–29, Salerno, Italy, 1997. IEEE Computer Society.

C. Buckley and A. F. Lewit. Optimization of inverted vector searches. In *Proc. 8th Annual International ACM SIGIR Conference on Research and Development in Information Retrieval*, pages 97–110, Montreal, Canada, June 1985. ACM Press, New York.

S. Chakrabarti, B. Dom, P. Raghavan, S. Rajagopalan, D. Gibson, and J. Kleinberg. Automatic resource compilation by analyzing hyperlink structure and associated text. In *7th WWW Conference*, pages 65–74, Brisbane, Australia, April 1998.

S. Chawathe, H. García-Molina, J. Hammer, K. Ireland, Y. Papakonstantinou, J. Ullman, and J. Widom. The TSIMMIS Project: Integration of Heterogeneous Information Sources. In *Proc. of IPSJ Conference*, pages 7–18, October 1994.

C. Chen. Structuring and visualizing the WWW by generalized similarity analysis. In *8th ACM Conference on Hypertext and Hypermedia*, pages 177–186, Southampton, England, 1997.

J. Cho, H. García-Molina, and L. Page. Efficient crawling through URL ordering. In *7th WWW Conference*, Brisbane, Australia, April 1998.

D. R. Clark and J. I. Munro. Efficient suffix trees on secondary storage. In *Proceedings of the 7th ACM-SIAM Annual Symposium on Discrete Algorithms*, pages 383–391, Atlanta, Georgia, 1996.

A. Clauser and P. Ferragina. On constructing suffix arrays in external memory. *Algorithmica*, 2000. (to appear). Part of this work appeared in European Symposium on Algorithms, LNCS 1643, 1999.

E. G. Coffman, Z. Liu, and R. R. Weber. Optimal robot scheduling for Web search engines. Technical Report 3317, INRIA, France, December 1997.

W. B. Croft, A. Moffat, C. J. van Rijsbergen, R. Wilkinson, and J. Zobel, editors. *Proc. 21st Annual International ACM SIGIR Conference* on *Research and Development in Information Retrieval*, Melbourne, Australia, August 1998. ACM Press, New York.

O. de Kretser and A. Moffat. Effective document presentation with a locality-based similarity heuristic. In M. Hearst, F. Gey, and R. Tong, editors, *Proc. 22nd Annual International ACM SIGIR Conference on Research and Development in Information Retrieval*, pages 113–120, New York, August 1999. ACM Press.

O. de Kretser, A. Moffat, T. Shimmin, and J. Zobel. Methodologies for distributed information retrieval. In M. P. Papazoglou, M. Takizawa, B. Krämer, and S. Chanson, editors, *Proc. 18th International Conference on Distributed Computing Systems*, pages 66–73, Los Alamitos, CA, May 1998. IEEE Computer Society.

Alin Deutsch, Mary Fernández, Daniela Florescu, Alon Levy, and Dan Suciu. A query language for XML. http://www.research.att.com/~mff/xml/w3cnote.html, 1998.

Fast. Fast: Main page. http://alltheweb.com, 1998.

P. Ferragina and R. Grossi. The string B-tree: A new data structure for string search in external memory and its applications. *J. Assoc. Comput. Mach.*, 46:236–280, 1999.

W. B. Frakes and R. Baeza-Yates, editors. *Information Retrieval: Data Structures and Algorithms*. Prentice-Hall, Englewood Cliffs, New Jersey, 1992.

R. G. Gallager and D. C. Van Voorhis. Optimal source codes for geometrically distributed integer alphabets. *IEEE Transactions on Information Theory*, IT–21(2):228–230, March 1975.

D. Gibson, J. Kleinberg, and P. Raghavan. Inferring Web communities from link topologies. In *9th ACM Conference on Hypertext and Hypermedia*, Pittsburgh, USA, 1998.

R. Goldman, J. McHugh, and J. Widom. Lore: A database management system for XML. Technical report, Stanford University Database Group, 1998.

S. W. Golomb. Run-length encodings. *IEEE Transactions on Information Theory*, IT–12(3):399–401, July 1966.

G. Gonnet. A tutorial introduction to Computational Biochemistry using Darwin. Technical report, Informatik E.T.H., Zuerich, Switzerland, 1992.

G. Gonnet, R. Baeza-Yates, and T. Snider. *New indices for text: Pat trees and Pat arrays*, chapter 3, pages 66–82. In , Frakes and Baeza-Yates (1992), 1992a.

G. Gonnet, M. Cohen, and S. Benner. Exhaustive matching of the entire protein sequence database. *Science*, 256:1443–45, 1992b.

Google. Google: Main page. http://www.google.com, 1998.

L. Gravano, K. Chang, H. García-Molina, C. Lagoze, and A. Paepcke. STARTS: Stanford protocol proposal for Internet retrieval and search. Technical report, Stanford University, Digital Library Project, 1997. http://www-db.stanford.edu/~gravano/starts.html.

Luis Gravano, Hector García-Molina, and Ant hony Tomasic. GlOSS: Text-source discovery over the Internet. *ACM Transactions on Database Systems*, 2000. To appear.

S.J. Green. Automated link generation: can we do better than term repetition. In *7th WWW Conference*, Brisbane, Australia, 1998.

D. K. Harman. Overview of the second text retrieval conference (TREC-2). *Information Prcessing & Management*, 31(3):271–289, May 1995.

D. K. Harman and G. Candela. Retrieving records from a gigabyte of text on a minicomputer using statistical ranking. *Journal of the American Society for Information Science*, 41(8):581–589, August 1990.

H. Itoh and H. Tanaka. An efficient method for in memory construction of suffix arrays. In IEEE CS Press, editor, *Proc. 6th Symp. on String Processing and Information Retrieval, SPIRE'99*, pages 81–89, Cancun, Mexico, 1999.

J. Jájá. *An Introduction to Parallel Algorithms*. Addison-Wesley, 1992.

B. Kahle. Archiving the Internet. http://www.alexa.com/~brewster/essays/sciam_article.html, 1997.

M. Kaszkiel, J. Zobel, and R. Sacks-Davis. Efficient passage ranking for document databases. *ACM Transactions on Information Systems*, 17(4):406–439, October 1999.

R. Khare and A. Rifkin. XML: A door to automated Web applications. *IEEE Internet Computing*, 1(4):78–86, 1977.

J. Kitajima and G. Navarro. A fast distributed suffix array generation algorithm. In *Proc. 6th Symp. on String Processing and Information*

Retrieval, SPIRE'99, pages 97–104, Cancun, Mexico, 1999. IEEE CS Press.

J. Kitajima, G. Navarro, B. Ribeiro-Neto, and N. Ziviani. Distributed generation of suffix arrays: a quicksort-based approach. In *Proc. 4th South American Workshop on String Processing, WSP'97*, pages 53–69. Carleton University Press, 1997.

J. Kitajima, B. Ribeiro-Neto, and N. Ziviani. Network and memory analysis in distributed parallel generation of PAT arrays. In *Proc. 8th Brazilian Symposium on Computer Architectures - High-Performance Processing*, pages 193–202. Brazilian Computer Society - SBC, 1996.

Jon Kleinberg. Authoritative sources in a hyperlinked environment. In *Proc. of the 9th ACM-SIAM Symposium on Discrete Algorithms*, pages 668–677, San Francisco, USA, Jan 1998.

Stefan Kurtz. Reducing the space requirement of suffix trees. *Software – Practice and Experience*, 29(13):1149–1171, 1999.

N. J. Larsson and K. Sadakane. Faster suffix sorting. Technical Report LU-CS-TR:99-214, LUNDFD6/NFCS3140/1–20/(1999), Department of Computer Science, Lund University, Sweden, May 1999.

S. Lawrence and L. Giles. Accessability and distribution of information on the Web. *Nature*, July 1999.

M. Lesk. *Practical Digital Libraries: Books, Bytes, and Bucks*. Morgan Kaufmann, San Francisco, 1997.

U. Manber and E. Myers. Suffix arrays: a new method for on-line string searches. *SIAM Journal on Computing*, pages 935–948, 1993.

Udi Manber and S. Wu. GLIMPSE: A tool to search through entire file systems. In *Usenix Winter 1994 Technical Conference*, pages 23–32, San Francisco, CA, January 1994.

M. Marchiori. The quest for correct information on the Web: Hyper search engi nes. In *6th WWW Conf.*, pages 265–274, Santa Clara, CA, USA, 1997.

M. Mitra, C. Buckley, A. Singhal, and C. Cardie. An analysis of statistical and syntactic phrases. In *Proc. RIAO'97 Conf. on Computer-Assisted Information Searching on the Internet*, pages 200–214, June 1997.

A. Moffat and T. A. H. Bell. In-situ generation of compressed inverted files. *Journal of the American Society for Information Science*, 46(7): 537–550, August 1995.

A. Moffat and J. Zobel. Self-indexing inverted files for fast text retrieval. *ACM Transactions on Information Systems*, 14(4):349–379, October 1996.

A. Moffat, J. Zobel, and N. Sharman. Text compression for dynamic document databases. *IEEE Transactions on Knowledge and Data Engineering*, 9(2):302–313, March 1997.

E. S. de Moura, G. Navarro, N. Ziviani, and R. Baeza-Yates. Fast searching on compressed text allowing errors. In Croft et al. (1998), pages 298–306.

M. Nagao and S. Mori. A new method of n-gram statistics for large number of n and automatic extraction of words and phrases from large text data of japanese. In *Proc. COLING'94*, pages 611–615, 1994.

G. Navarro and R. Baeza-Yates. A new indexing method for approximate string matching. In Maxime Crochemore and Mike Paterson, editors, *Combinatorial Pattern Matching (CPM'99)*, number 1645 in LNCS, pages 163–185, Warwick, UK, July 1999. Springer Verlag.

G. Navarro, E. Barbosa, R. Baeza-Yates, W. Cunto, and N. Ziviani. Binary searching with non-uniform costs and its application to text retrieval. *Algorithmica*, 2000. To appear. Preliminary version in *Proc. ESA'95*, LNCS, Springer Verlag.

G. Navarro, J. Kitajima, B. Ribeiro-Neto, and N. Ziviani. Distributed generation of suffix arrays. In *Proc. CPM'97*, LNCS 1264, pages 102–115, 1997.

Northern. Northern Light: Main page. http://www.northernlight.com, 1997.

M. Persin, J. Zobel, and R. Sacks-Davis. Filtered document retrieval with frequency-sorted indexes. *Journal of the American Society for Information Science*, 47(10):749–764, October 1996.

Peter Pirolli, James Pitkow, and Ramana Rao. Silk from a sow's ear: Extracting usable structures from the Web. In *Proc. of the ACM SIGCHI Conference on Human Factors in Computing Systems*, pages 118–125, Zurich, Switzerland, May 1996. ACM Press.

J. Powell and E. Fox. Multilingual federated searching across heterogeneous collections. *D-Lib Magazine*, September 1998.

D. Quass, A. Rajaraman, Y. Sagiv, J. Ullman, and J. Widom. Querying semistructured heterogeneous information. In *Deductive and Object-Oriented Databases, Proc. of the DOOD '95 Conference*, pages 319–344, Singapore, December 1995. Springer-Verlag.

M. J. Quinn. *Parallel Computing: Theory and Practice*. McGraw-Hill, second edition, 1994.

B. Ribeiro-Neto and R. Barbosa. Query performance for tightly coupled distributed digital libraries. In *Proc. ACM Int. Conference on Digital Libraries (DL'98)*, pages 182–190, 1998.

B. Ribeiro-Neto, J. Kitajima, G. Navarro, C. Sant'Ana, and N. Ziviani. Parallel generation of inverted lists for distributed text collections. In Y. Eterovic, editor, *Proc. XVIII Chilean Computer Science Society, SCCC'98*, pages 149–157. IEEE CS Press, 1998.

B. Ribeiro-Neto, E. Moura, M. Neubert, and N. Ziviani. Efficient distributed algorithms to build inverted files. In *Proc. ACM SIGIR'99*, pages 105–112, 1999.

K. Sadakane. A fast algorithm for making suffix arrays and for Burrows-Wheeler transformation. In *Proc. IEEE Data Compression Conference (DCC'98)*, pages 129–138, 1998.

G. Salton and C. Buckley. Term-weighting approaches in automatic text retrieval. *Information Processing & Management*, 24(5):513–523, 1988.

G. Salton and M. J. McGill. *Introduction to Modern Information Retrieval*. McGraw-Hill, New York, 1983.

N. Shivakumar and H. García-Molina. Finding near-replicas of documents on the Web. In *Workshop on Web Databases*, Valencia, Spain, March 1998.

A. Singhal, C. Buckley, and M. Mitra. Pivoted document length normalization. In H.-P. Frei, D. Harman, P. Schäuble, and R. Wilkinson, editors, *Proc. 19th Annual International ACM SIGIR Conference on Research and Development in Information Retrieval*, pages 21–29, New York, August 1996. ACM Press.

Snap. Snap: Main page. http://www.snap.com, 1998.

Ellen Spertus. ParaSite: Mining structural information on the Web. In *6th Int'l WWW Conference*, Santa Clara, CA, USA, April 1997.

W. Szpankowski. Probabilistic analysis of generalized suffix trees. In *Proc. CPM'92*, pages 1–14, 1992. LNCS 644.

A. Tomasic and H. Garcia-Molina. Performance of inverted indices in shared-nothing distributed text document information retrieval systems. In *Proc. Int. Conf. on Parallel and Distributed Information Systems, PDIS'93*, 1993.

E. Ukkonen. Approximate string matching over suffix trees. In *Proc. CPM'93*, pages 228–242, 1993.

R. Weiss, B. Vélez, M. Sheldon, C. Nemprempre, P. Szilagyi, and D.K. Gifford. HyPursuit: A hierarchical network engine that exploits content-link hypertext clustering. In *7th ACM Conference on Hypertext and Hypermedia*, pages 180–193, Washington, D.C., USA, 1996.

I. H. Witten, A. Moffat, and T. C. Bell. *Managing Gigabytes: Compressing and Indexing Documents and Images*. Morgan Kaufmann, San Francisco, second edition, 1999.

B. Yuwono and D.L. Lee. Server ranking for distributed text retrieval systems on the Internet. In *5th Int. Conf. on Databases Systems for Advanced Applications (DASFAA)*, pages 41–50, Melbourne, Australia, 1997.

J. Zobel and A. Moffat. Adding compression to a full-text retrieval system. *Software—Practice and Experience*, 25(8):891–903, August 1995.

J. Zobel and A. Moffat. Exploring the similarity space. *SIGIR Forum*, 32(1):18–34, Spring 1998.

J. Zobel, A. Moffat, and K. Ramamohanarao. Inverted files versus signature files for text indexing. *ACM Transactions on Database Systems*, 23(4):453–490, December 1998.

Chapter 8

DATA COMPRESSION

David Salomon
Computer Science Department, California State University, Northridge, CA 91330, USA
david.salomon@csun.edu

Abstract The exponential growth of computer applications in the last three decades of the 20th century has resulted in an explosive growth in the amounts of data moved between computers, collected, and stored by computer users. This, in turn, has created the field of *data compression*. Practically unknown in the 1960s, this discipline has now come of age. It is based on information theory, and has proved its value by providing us with fast, sophisticated methods capable of high compression ratios.

This chapter tries to achieve two purposes. Its main aim is to present the principles of compressing different types of data, such as text, images, and sound. Its secondary goal is to outline the principles of the most important compression algorithms. The main sections discuss statistical compression methods, dictionary-based methods, methods for the compression of still images, of video, and of audio data. In addition, there is a short section devoted to wavelet methods, since these seem to hold much promise for the future.

Keywords: Data compression, Information theory, Entropy, Digital images.

1. Introduction

Those who use modems for data communications (and there are many) recognize the importance of speed. Downloading a 1 Mbyte file at 9,600 baud (bits/s) takes at least 15 minutes, while at 56K baud it takes only 3 minutes; a significant difference. The increase in modem speeds over the years has mostly been due to hardware improvements, but data compression made a significant contribution. Modern modems compress the contents of a file before it is sent, and decompress it while it is being received. If the file is compressed to half its original size, then compression effectively doubles the transmission speed. Data communications is thus one field where compression plays an important role, and where fast, sophisticated compression methods can noticeably raise the quality of life (at least the lives of modem users).

People collect data, digital and nondigital, all the time, and compression can help in this area too. A large lawyer's office may accumulate several file cabinets' worth of documents each year. The employees may decide to save space by digitizing old documents, storing them on a large disk or on CD-ROMs, and destroying the originals. After a few years, however, they may notice that even the disks or CD-ROMs occupy a significant amount of space. Instead of getting rid of old disks, such an organization may decide to keep the digitized data in compressed form, where it normally occupies a fraction of its original size.

An extreme example of data accumulation is the FBI. At its Washington, D. C. headquarters, the FBI has more than an acre of file cabinets containing its fingerprint collection. The original fingerprints are in the form of inked impressions on cards. There currently are about 200 million cards in this collection (although "only" 29 million of them are distinct) and new cards are added at the rate of 30,000–50,000 a day! The print of one finger is a complex image, and a single card contains several prints of fingers and palms. Digitizing such a card therefore results in 10 Mb of data, so efficient compression is extremely important. Since the prints are images, not text, it is acceptable to lose some data (data to which the eye is not sensitive) in the compression process. Such *lossy* compression typically results in compression factors of 20 to 1.

These examples, of data communications and archiving, are just two out of many applications of *data compression*, a relatively new and fast-growing field. This chapter describes several techniques and several specific methods for compressing data, but this introduction tries to answer the basic question: How is it possible to represent the same data with fewer bits?

The general answer is: It is possible to compress data because the original representation of the data is inefficient and contains redundant information. Compressing can be done in many different ways, but they are all based on reducing or even completely eliminating, the redundancy in the original data. Data that do not have any redundancy cannot be compressed. Any attempt to compress it, however efficient the method, would result in new data with about the same size. This fact also explains why recursive compression is impossible. Imagine a data file A of size 1 Mb, getting compressed to another file B of half the size. What prevents us from feeding B to the same compression program, to obtain a file C that's half the size of B. The answer is, of course, that B is already compressed, any redundancy in the original file A has already been removed, and therefore B uses an efficient representation and cannot be further compressed. File C would have about the same size as B, and any attempt to compress it further would result in files of about the same size.

It is easy to understand intuitively why conventional data representations are redundant. We start with text, which is commonly used in computer applications. In a text file, each character of text is represented by the seven bits of its ASCII code. (There is an eighth bit, for increased reliability, but this is irrelevant to us.) A fixed-size code makes it easy to process text but is inefficient. In typical English text, the letters "e", "t", and "o" are common whereas "z" and "q" are rare. A more efficient representation of text might use variable-size codes for the characters such that common characters would have short codes and rare characters would be assigned long codes. A compression method based on this principle can be simple, fast and efficient. The results, however, depend on the data being compressed. Data with relatively few "e" and "t" and with many instances of "q" and "z" will not compress well and may even cause expansion. Understanding this fact is a key to a full understanding of data compression.

The preceding paragraph suggests a more general approach to compressing text. Instead of assigning codes to characters based on their frequencies in typical English text, why not assign them codes based on their frequencies in the particular file being compressed. A two-pass approach suggests itself immediately. The input file is first read and the number of occurrences of each character is counted. This is used to calculate character frequencies and to assign variable-size codes to the characters based on these frequencies. The second pass reads the same file again and performs the actual compression using the codes determined by the first pass. Such a method is simple to implement and is efficient. However, it is slow, since the input file typically resides on a disk, and reading a disk is much slower than performing calculations. Another, small, disadvantage of this approach is that the table of codes has to be included at the start of the compressed file, for the use of the decoder. This table, however, is normally small and does not degrade the overall compression by much.

An *adaptive* approach to compression is generally complex but can produce better results. The idea is to read the next character from the input file and write its current code on the compressed file. The encoder then increments the character's frequency by one, which changes the probabilities of all the characters. The last step is to recompute, if necessary, the codes assigned to the characters according to the new probabilities. An adaptive algorithm must be designed such that the decoder will be able to mimic the encoder's operations and compute the new codes in the same way.

Other types of data also exhibit redundancy and images provide a good example. Digital images are very common in computer applications. Such an image is a rectangular array with r rows and c columns of dots called *pixels* (for "picture elements"). The product $r \times c$ is called the *resolution* of the image. Even a low-resolution image contains hundreds of rows and columns, so the total number of pixels is at least in the hundreds of thousands and is normally in the millions. This is why images tend to be big and can benefit from compression. A pixel is represented by a number indicating its color or shade of gray, and the redundancy of this representation is a result of the fact that adjacent pixels tend to have similar colors. This fact lies behind all image compression methods and is considered the principle of image compression. We say that a typical image has *spatial redundancy*.

Here is a simple approach to image compression based on pixel differences. The image is scanned row by row and each row is scanned from left to right (this is called *raster order* scan). Each pixel p is subtracted from its predecessor q and the difference $d = q - p$ is assigned a variable-size code. The principle of image compression tells us that most differences will be small numbers, so we can assign short codes to small differences and long codes to large differences. The more redundancy there is in the image, the better will the compression be. This simple method can be extended to a two-pass method and also to an adaptive method.

Not all adjacent pixels in an image are similar. Some images may exhibit much redundancy while others may have next to none. An important fact about image compression (and about data compression in general) is that images that people are interested in tend to have redundancy. An image with no redundancy is generally not interesting, so no one tries to compress it. An example of such an image is a random image. In a random image, adjacent pixels may have any colors. Some adjacent pixels may accidentally be similar while others may be very different. Such an image does not satisfy the principle of image compression and therefore will not compress under any compression algorithm. In general, random data—be it text, images, or anything else—cannot be compressed, since it has no redundancy.

The next type of data that we consider is a movie. Computer movies have become very popular recently and can, of course, benefit from compression. A movie is a set of still images (called *frames*) with audio. If we ignore the audio part, we can say that a movie has three dimensions.

Each frame has two dimensions and the third dimension is the time. Thus, a movie has two types of redundancies, a spatial (or *intraframe*) redundancy, since adjacent pixels in each frame tend to be similar, and a temporal (or *interframe*) redundancy, since adjacent frames also tend to be similar. A compression method for a movie can start by using an image compression method to compress the first frame (intraframe coding), then compress the frames that follow by encoding the differences between each frame and its predecessor (interframe coding). In principle, interframe coding can continue until a completely different frame is encountered (the start of a new scene). In practice, small errors and loss of information in interframe coding may accumulate, so interframe coding is only used for a few frames. Section 6 discusses the video aspects of the compression of movies.

Audio data is rapidly becoming popular. Sound is captured by a microphone, which converts it into a voltage that varies with time. To digitize the sound, this voltage is measured periodically and each measurement converted into a number (a sound sample). The set of samples becomes the audio file for the sound. Similar to images, adjacent sound samples tend to be similar, and this redundancy is used for audio compression. Better compression can be achieved by eliminating sound samples to which the ear is not sensitive. Section 7 discusses the properties of the human audio system and how they can be used to determine what sounds are inaudible.

The concept of redundancy and, consequently the entire discipline of data compression, are based on information theory, the mathematical theory created in 1948 by Claude Shannon. We tend to consider information an abstract notion, one of those concepts that everyone is familiar with but no one can precisely define or quantify. The approach taken by Shannon was to define the information contents of a message as the amount of surprise in the message. A long message whose contents is already known contains no surprises and thus has little or no information content. In principle, such a message can be expressed in a small number of bits, i.e., it can be highly compressed. As an example, consider a large handbook on gardening. Such a document is bound to have the word "plant" many times. The first time the word is found, it may be written on the compressed file as is, i.e., with no compression. Each subsequent occurrence of the word can be compressed by substituting a pointer to the previous occurrence. If the word is very common, the distance between consecutive occurrences would be short, leading to small pointers (expressed by few bits) and thus to better compression. This is the principle of many *dictionary-based* compression algorithms (Section 3).

Another aspect of information theory is its use of the *logarithm* as the function of information. In order to understand the relation between information content and the logarithm function, let's imagine a two-player game where player A picks up an integer x in the range $[1, 64]$ and player B has to guess its value by asking A questions. The fewer questions needed, the more points B receives. It is possible to use linear guessing, where B asks: Is x 1? Is x 2? and so on. However, much faster guessing is achieved by dividing the range into two equal parts with each question. Thus, B can start by asking: Is $x > 32$. If the answer is Yes, then x must be in the range $[33, 64]$ and B divides this range by 2 and asks: Is $x > 48$? It is easy to see that a maximum of six questions is needed in this case, because splitting the interval $[1, 64]$ in two six times leaves an interval of one number. Since $64 = 2^6$ we get $6 = \log_2 64$.

An important result of information theory (by no means its only important result) is the concept of *entropy* and the conclusion that data can be compressed down to its entropy but not more. We assume that the data to be compressed consists of the symbols $a_1, a_2, \ldots, a_n$, where symbol a_i appears in the data with probability P_i (the symbols can be codes of text characters, values of pixels, sound samples, or anything else). The entropy of a_i is defined as $-P_i \log_2 P_i$, and information theory tells us that this is the smallest number of bits needed, on average, to represent the symbol. The words "on average" are important, since certain data files can be compressed more than their entropy. A file consisting of 5000 occurrences of the letter "x", for example, can be compressed to one "x" followed by the number 5000. A compression program that can compress data to its entropy limit is called an *entropy coder*.

We can gain better insight into the meaning of entropy and what it tells us, by considering a two-symbol alphabet. Let a_1 and a_2 be the only two symbols (with probabilities P_1 and P_2) in a hypothetical alphabet. Without any compression, we can simply assign them the 1-bit codes 0 and 1. Since P_1 and P_2 are probabilities, they must add up to unity, which implies that the entropy of this alphabet is $-P_1 \log_2 P_1 - (1 - P_1) \log_2(1 - P_1)$. We first look at the extreme case where the probabilities differ widely. If P_1 is, say, 99% (and P_2 is, therefore, just 1%), then the total entropy is 0.08. It is possible, in principle, to compress such data to 0.08 bits per symbol (a compression factor is $1/0.08 = 12.5$), although in practice we may not know how to do that. Examining the other extreme case, where the probabilities are equal, we get an entropy of $-0.5 \log_2 0.5 - (1 - 0.5) \log_2(1 - 0.5) = -\log_2 0.5 = 1$, indicating that at least one bit is needed to represent a symbol, and therefore the data cannot be compressed at all.

This result is true in general, not just for two symbols. Imagine a set of data files where symbols appear with equal probabilities. Files in this set cannot, in general, be compressed. Regardless of the compression method used, some of these files may compress, others may expand, but most will stay the same size. An example of such a file is random data. In a truly random data file, all symbols have the same probability, and we already know that random data cannot be compressed. On the other hand, data where symbols have skewed probabilities can, with the right algorithm, compress well.

Before we turn to the details of the various techniques and methods used in compressing data, here are some commonly used terms. The *compression factor* has already been mentioned. It is the size of the original file divided by the size of the compressed file. The better the compression, the larger this number. The *compression ratio* is the inverse of the compression factor. When compression performance is expressed by means of the compression ratio, we expect a small number, close to zero. In image compression, the term "bpp" (bits per pixel) is also used to indicate compression performance.

The *compressor* (or *encoder*) is the program that does the compression. Decompression is done by a *decompressor* or *decoder*. The original data to be compressed is called the *input file*, and the result of a compression is a *compressed file*. The terms *adaptive compression* and *lossy compression* have already been mentioned. The opposite of lossy compression is, of course, *lossless compression*.

2. Statistical Methods

Any data file consists of elementary symbols, and the statistical distribution of those symbols can be used to compress the data. It has been mentioned in the introduction that a uniform distribution results in no compression, whereas a skewed distribution is the key to achieving efficient compression. This section describes three methods for statistical compression, namely Huffman coding, facsimile compression, and arithmetic coding. The principles of each are described, as well as important variants. The less important details, however, are omitted, and in the case of facsimile compression, the complete code tables are not listed. Both Huffman coding and the fax compression standards (as well as many other compression methods) use variable-size, prefix codes, so we start with a short survey of these codes.

2.1. Prefix codes

Imagine an alphabet with just four symbols A, B, C, and D. Messages (texts) in this alphabet are strings consisting of just these four symbols. Without compression, we can simply assign these symbols the four 2-bit codes 00, 01, 10, and 11. In order to compress such texts, we should assign the symbols variable-size codes based on the symbols' frequencies (or probabilities) of occurrence. To focus our discussion, we assume that the probabilities of A, B, C, and D are 40%, 30%, 20%, and 10%, respectively. In order to assign codes to the symbols, we follow the obvious rule that commonly-occurring symbols should get short codes. Based on this rule we can assign, for example, the codes 1, 01, 10, and 010 to the four symbols.

The average length of these codes is easy to calculate in the case of a large file, where the four symbols appear with the probabilities above. In 40% of the cases the symbol is A, so the code is one bit long. In 10% of the cases the symbol is D, so the code is three bits long. The average code length is thus

$$0.4 \times 1 + 0.3 \times 2 + 0.2 \times 2 + 0.1 \times 3 = 1.7 \, \text{bits/symbol}.$$

Since the average code length is less than two bits/symbol, compression is achieved. We can even tell that the compression factor will be $2/1.7 \approx 1.18$. There remain, however, two problems. The first one is obvious; our variable-size codes have been chosen tentatively and may not be the best (i.e., the shortest) ones. There may even be more than one set of the best codes, since the codes themselves are not important, just their lengths. The Huffman method of this section shows how to select a set of the best codes.

The other problem is more subtle. We have to make sure that our codes can be decoded unambiguously. In order to understand this problem, consider the following 10-symbol string $AABABCDCAB$. This string is easy to encode and results in the 17-bit string

$$1\,1\,01\,1\,01\,10\,010\,10\,1\,01$$

(without the spaces). When we try to decode it, however, we run into a problem. The first 1 can be either the code of A or the start of the code of C. We therefore read the next bit and, since it is also a 1, we conclude that the first symbol is A and the second 1 belongs to the second symbol. To decode the second 1, we have to read the next (third) bit. It is a 0, so the second symbol may be C, but it may also be A (and the 0 is the start of the next code). Our simple, tentative code is therefore ambiguous.

The reason this code is ambiguous is that the code of A is 1 and the code of C starts with 1. This also points the way to removing the ambiguity. Once we decide to assign 1 as the code of A, no other code should start with 1; they should all start with 0. In general, we end up with the following *prefix rule*: Once a certain bit pattern has been chosen as the code of a symbol, no other code should start with that pattern (the pattern should not be the *prefix* of any other code). Codes that satisfy this rule are called *prefix codes* and can be decoded unambiguously.

Our next try is to assign the code 1 to A and assign codes that start with 0 to the other three symbols. If we decide to assign B the code 00, the remaining two symbols should have codes that start with 01. We can simply assign them the two codes 010 and 011. This is a set of prefix codes, so any compressed data can be decompressed unambiguously, but the average code length is now

$$0.4 \times 1 + 0.3 \times 2 + 0.2 \times 3 + 0.1 \times 3 = 1.9 \text{ bits/symbol}.$$

We definitely need an algorithm that, given the symbols' probabilities, will produce a set of the shortest prefix codes. Such an algorithm was developed by David Huffman in 1952.

For related material on the subject of this section, see Elias (1975).

2.2. Huffman coding

The only input to this algorithm is a set S of n symbols. For each symbol, its name and probability (or, equivalently, frequency of occurrence) are included in the set. The algorithm writes the symbols' names in the code-tree that it generates, but otherwise the names are not used. The probabilities are used by the algorithm to construct a binary tree that, once completed, makes it easy to compute a set of prefix codes for the n symbols. The key to the algorithm is the observation that small probabilities should be assigned long codes and should therefore be located low in the tree. The algorithm therefore starts by selecting any two of the smallest probabilities A and B (there may be more than one such pair). The two are placed at the lowest level of the (so far empty) tree, and are combined into a binary subtree whose root is named AB. The pair of symbols corresponding to probabilities A and B is then deleted from S and is replaced by a new, temporary symbol with probability $A + B$. The size of S is decremented by 1. The algorithm executes this step a total of $n - 1$ times, until only one symbol, with probability 1, remains in S. The resulting binary tree is called a *Huffman tree*. It is used, in the second part of the algorithm, to assign codes to the n symbols. n nodes of this tree correspond to the original n symbols, and the remaining nodes correspond to temporary symbols.

This algorithm is easy to visualize with an example. Given the set of six probabilities $A = 0.25$, $B = 0.25$, $C = 0.2$, $D = 0.1$, $E = 0.1$, and $F = 0.1$, Figure 8.1 shows the five iterations in the construction of two different Huffman trees. Many different trees can be constructed, since the steps are not unique; different choices can be made in each step. In the first step, for example, symbols E and F were arbitrarily selected. This step could have selected symbols D and E, or D and F, resulting in different trees. The second step of Figure 8.1a has selected symbols D and EF, in contrast to the second step of Figure 8.1b, where symbols C and D were selected. Another choice was to place C to the left of D instead of to its right. The opposite choice would have resulted in a different Huffman tree. Each choice results in a different tree and therefore in a different set of codes. However, all these sets of codes have the same average size, and all are codes with the minimum average size for the given symbol probabilities.

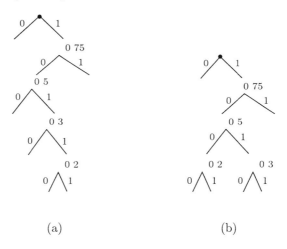

(a)   (b)

Figure 8.1. Two Huffman Code Trees.

The second part of the algorithm assigns the actual codes to the symbols. The code for symbol s is assigned by sliding down the tree from the root to node s and appending one bit to the code for each branch encountered. A zero bit is appended for a left branch, and a one-bit for a right branch (the opposite choice, of course, is also valid and results in the inverse codes). The tree of Figure 8.1a results in the six codes $A = 0$, $B = 11$, $C = 100$, $D = 1010$, $E = 10110$, and $F = 10111$. Notice that they are prefix codes. Their average size is

$$0.25 \times 1 + 0.25 \times 2 + 0.2 \times 3 + 0.1 \times 4 + 0.1 \times 5 + 0.1 \times 5 = 2.75 \text{ bits/symbol}.$$

The reader should do the same calculation for the tree of Figure 8.1b. Notice that these codes have different sizes (the code of B is 1-bit long and there are no 5-bit codes) but the average code size is the same.

Once the codes have been assigned, the Huffman encoder uses them to compress the input file. The encoder inputs the next symbol, uses it to search the code table, finds the code for the symbol, and appends it to the compressed file. As has been mentioned in the Introduction, there are three approaches to determining the Huffman codes of a set of symbols. (1) It is possible to select a set of "training" input files, count symbol frequencies, and use them to compute a set of codes. (2) A two-pass job reads the input once to count symbol frequencies and determine the Huffman codes for the specific input, then reads the input again to do the actual compression. (3) An adaptive encoder assigns and reassigns codes to the symbols as it goes along. The more symbols have been read and compressed, the better the codes get. This type of encoder is discussed in Section 2.3.

The Huffman decoder is also simple, but it needs to know the codes. When approach 1 above is used, the codes are fixed, and both encoder and decoder have the permanent code table built-in. When a two-pass job is used, the encoder has to write the code table at the start of the compressed file, for the decoder's use. An adaptive decoder mimics the operations of the encoder and always assigns the same codes as the encoder.

Once the decoder knows the codes, it constructs the Huffman tree, and reads the input file bit by bit. It starts at the root and slides down one edge for each bit input. If the bit is zero (one), the decoder selects the left (right) edge. When the decoder arrives at a node for a symbol, it writes the symbol on the decompressed file, goes back to the root, and inputs the next bit.

For related material on the subject of this section, see Huffman (1952).

2.3. Adaptive Huffman coding

Using a permanent code table is fast but inefficient, whereas performing a two-pass compression job is efficient but slow. The adaptive Huffman algorithm is a compromise. It inputs the source file once, so it is fast, and it uses the symbols input so far to assign Huffman codes that vary during the compression process and adapt themselves to the symbol probabilities of the particular data being compressed.

When compression starts, the Huffman tree is empty; no probabilities are known, so no codes are assigned to symbols. When a symbol is seen for the first time, its raw code (i.e., its name) is written on the compressed file, and the symbol is added to the tree with a frequency count of 1. Since its frequency is low, the symbol should be assigned a long code. This is achieved by appending the symbol at the bottom of the tree. Each time the symbol is encountered again, its code is written on the compressed file and its count is incremented by 1. If the count is high enough, the tree is rearranged and the symbol is moved higher in the tree, where its new code is shorter. Here is the algorithm in more detail.

The first symbol is read and its raw code (name) is written on the compressed file. The symbol is added to the (so far empty) Huffman tree, and is assigned a code. Since the symbol is the only one in the tree, its code is just one bit. The second symbol is now read. If it is identical to the first, the 1-bit code is written and the symbol's frequency is incremented by 1. Otherwise, the raw code of the second symbol is written on the compressed file, and the symbol itself is appended to the Huffman tree and is assigned a code.

In general, when a symbol is input, the Huffman tree is searched. If the symbol is not found, its raw code is written, and the symbol is appended to the tree with a frequency count of 1. This assigns a Huffman code to the symbol, and also necessitates a rearrangement of the tree.

If the symbol is found in the tree, its current code is written on the compressed file and its frequency is incremented. Since the Huffman tree is based on frequencies (or, equivalently, on probabilities), incrementing a frequency may cause a Huffman tree to become a non-Huffman tree. In such a case, the tree has to be rearranged, thereby changing symbols' positions and therefore their codes.

Before looking at the details of Huffman tree rearranging, we have to mention the concept of an *escape symbol*. The discussion above shows that the compressed file contains raw symbol codes and Huffman codes. The decoder should be able to distinguish between the two, so the encoder precedes each raw code with a special code; the escape symbol. The problem is that the Huffman codes of the symbols vary all the time, which makes it impossible to assign a fixed code to the escape symbol. Its code has to change all the time, and one way to handle this situation is to place the escape symbol in the Huffman code tree, at a fixed position, say, at the bottom-left corner of the tree. Since the escape symbol is included in the tree, it has a code. Since its position is fixed, it can easily be located and its code found. In order to keep the escape symbol at this position, it should be assigned a probability of zero.

In order to understand tree rearranging, we first have to be clear about what it means for a tree to be a Huffman tree. The principle of the Huffman algorithm is to start with the least-probable symbols and place them at the bottom of the code tree. When the tree is complete, the most-probable symbols are found at its top, close to the root. However, a look at the two trees of Figure 8.1 shows that equi-probable symbols can be located on different levels, while symbols with different probabilities can be found on the same level and are not sorted horizontally by probability. These observations lead us to the following, rigorous definition of a Huffman code tree:

Definition: A Huffman code tree is a binary tree such that when it is traversed from bottom to top, and on each level from left to right, the symbols encountered are found to be sorted by increasing probabilities.

This definition is restrictive, since it requires the symbols on each level to be sorted from left to right by increasing probabilities. This restriction, however, makes it easy to check a given tree, find out whether it is still a valid Huffman tree, and if not, rearrange it. This also guarantees that the escape symbol, whose probability is zero, will always stay at the bottom-left corner of the tree.

For related material on the subject of this section, see Knuth (1985) and Vitter (1987).

2.4. Facsimile compression

Facsimile (fax) machines have become a way of life for many people. They make it possible to send documents over existing telephone lines at speeds of about a minute per page. Since fax machines are made by many manufacturers, there must be international standards that specify all aspects of their operation. This section describes the compression standard used by fax machines, but before getting to the details of this standard, it is important to understand how fax machines work and why compression is necessary.

A fax machine must be able to handle any documents, not just text. The machine therefore does not check the contents of the document. It scans the document row by row, converting each row to a large number of small, black and white dots called *pels*. Each dot becomes a bit (zero for white and one for black), and those bits are transmitted on the telephone line. The receiving fax machine assembles the bits into rows of pels and prints the received document row by row. The only problem with this process is the large number of pels. The scanning standard specifies 1664 pels for an 8.2-inch-wide scan line. The number of scan lines varies, since there are three scanning modes. In the medium-resolution mode there are 1956 lines for a 10-inch-high page. The total number of pels per page is therefore about 3.255 million! Notice that this number depends on the scanning mode and on the page size, but does not depend on the contents of the document. Sending 3.255 million bits at 14,400 baud (bits/s) takes 226 seconds, or close to four minutes. It seems too much to ask a person to wait four minutes for one page, which is why compression is essential.

Facsimile compression standards have been developed by the International Telecommunications Union (ITU). This agency of the United Nations is responsible for developing standards for all aspects of telecommunications. Its standards are only recommendations; they are not mandatory, but they are normally adopted by all manufacturers. The official name of the current facsimile compression standard is recommendation T.4, but it is commonly referred to as the *Group 3* method. (Another fax compression standard, recommendation T.6 or Group 4, is briefly mentioned below.)

The Group 3 method is based on the fact that pels are rarely independent. If we select a pel at random in a document, chances are that its immediate neighbors would be identical (i.e., would have the same color). Thus, a typical scan line tends to have *runs* of identical pels. The Group 3 method counts the lengths of runs of black and white pels on a scan line, and writes Huffman codes of those runs on the compressed file. Thus, the method is a modification of the original Huffman algorithm, a modification where codes are assigned to runs of pels, rather than to individual pels. This modification is called *Modified Huffman* or MH. Notice that each scan line is coded separately.

The Huffman codes for the runs were determined by the ITU using a set of training documents that include (1) a typed business letter (in English), (2) a hand-drawn circuit diagram, (3) a French invoice (printed and typed), (4) a report densely typed in French, (5) a technical article with figures and equations (printed in French), (6) a graph with printed captions in French, (7) a dense document in Kanji, and (8) a handwritten memo (in English) with very large white-on-black letters. These documents were scanned, all the runs of black and white pels were counted, and Huffman codes were assigned to those runs as well as to runs that did not happen to appear in these particular documents. These codes are permanent. Every fax machine has the entire table of codes built in.

Since runs can be long, the number of Huffman codes was limited by using two code tables. The first table is used for short runs, from 0 to 63 pels (these are known as *termination codes*), and the second table assigns *makeup codes* for runs lengths that are multiples of 64. Some of the codes are listed in Table 2.4. Part (a) of the table lists 15 termination codes. Runs of 2 and 3 black pels turned out to be the most common in the training documents, so they were assigned the shortest codes. Part

DATA COMPRESSION

Table 8.1. Some Huffman Codes Used in Group 3 Compression.

Run length	White pels	Black pels	Run length	White pels	Black pels
0	00110101	0000110111	64	11011	0000001111
1	000111	010	128	10010	000011001000
2	0111	11	192	010111	000011001001
3	1000	10	256	0110111	000001011011
4	1011	011	320	00110110	000000110011
5	1100	0011	...	...	...
6	1110	0010			
7	1111	00011			
8	10011	000101			
9	10100	000100			
10	00111	0000100			
11	01000	0000101			
12	001000	0000111			
...	...	...			
62	00110011	000001100110			
63	00110100	000001100111			
	(a)			(b)	

(b) of the table lists five of the makeup codes used for long runs. The convention used by Group 3 is that each makeup code must be followed by one termination code. Thus, for example, a run of five white pels is coded as 1100, but a run of 65 white pels is coded as 11011|000111 (the makeup code for 64 white pels followed by the termination code of one white pel) and a run of 64 white pels is coded as 11011|00110101 (the makeup code for 64 white pels followed by the termination code of zero white pels).

It has already been mentioned that each scan line is coded separately. One reason for this is that the first pel of the next line and the last pel of the current line are not correlated. A more important reason, however, is reliability. The individual code bits are sent on a telephone line and may get corrupted during transmission. It is not hard to see that even a single bad bit may cause many errors in the receiver. This is why Group 3 defines a special end-of-line (EOL) code that should appear at the end of each scan line. The EOL code is the 12 bits 000000000001 (this code is usually followed by a 1-bit tag, described below). This special bit pattern is none of the Huffman codes in the fax tables, and it cannot occur inside a validly-encoded scan line. The EOL code enables the decoder to recognize the end of a scan line and also to synchronize itself after an error has occurred. The standard also requires that the entire transmission start with an EOL, and every page must terminate with a *return to control* (RTC) code that consists of six consecutive EOLs (without the tag bits).

The Group 3 method has a two-dimensional option that is briefly mentioned here. The idea is that consecutive scan lines tend to be similar, so a line can be efficiently encoded by encoding the differences between it and its predecessor. In principle, the first line should be encoded as described here (one-dimensional encoding), and all the remaining lines can be encoded in two-dimensions. Two-dimensional encoding can thus be very efficient but is also very sensitive to errors. Any errors occurring when line i is decoded (due to bits corrupted during transmission) will affect all the lines following i. This is why in practice, two-dimensional encoding can only be used for a few lines, and the encoder should then encode the next line in one dimension. To implement this option, the Group 3 standard requires an extra *tag* bit to follow each EOL. A tag of 1 indicates that the next line is encoded in one dimension. A tag of 0 indicates a two-dimensional encoding of the next line. The encoder encodes a line in one dimension, and follows with up to three lines encoded in two dimensions. If the transmitting fax machine decides not to use this option (because of a noisy phone line), it can use one-dimensional encoding for all the scan lines, and it sets all the tags to 1.

The ITU has also developed the T.6 recommendation, commonly known as Group 4. This method uses two-dimensional encoding exclusively, and is used on certain fax machines.

For related material on the subject of this section, see Anderson and et al. (1987), Hunter and Robinson (1980), Marking (1990), and McConnell (1992).

2.5. Arithmetic coding

The length of the Huffman code of a symbol depends on the symbol's probability of occurrence. The smaller the probability, the longer the code. The codes are assigned by sliding down the Huffman code tree from the root to each of the leaves. As a result, each code consists of an integer number of bits, so its length is an integer. The Huffman method is simple but not ideal, since the ideal length of a code is, in general, not an integer.

An example is the set of three symbols, A, B, and C, with probabilities 0.5, 0.3, and 0.2, respectively. The ideal lengths of their codes should be $-\log_2 0.5 = 1$, $-\log_2 0.3 = 1.74$, and $-\log_2 0.2 = 2.32$, respectively, since these lengths produce an average code size of $0.5 \times 1 + 0.3 \times 1.74 + 0.2 \times 2.32 = 1.485$ bits/symbol, equal to the entropy of the set. The Huffman method, however, assigns A a 1-bit code and B and C 2-bit codes. The average length of the codes is therefore $0.5 \times 1 + 0.3 \times 2 + 0.2 \times 2 = 1.5$ bits/symbol, slightly higher.

Arithmetic coding is a statistical compression method that can code a symbol in a noninteger number of bits. Given a set of symbols whose probabilities are known, this method compresses a string of symbols from the set such that the number of bits per symbol equals the entropy of the set (except for small computational errors caused by the limited precision of computer arithmetic). Thus, arithmetic coding is an *entropy encoding* method.

Given a string of symbols, arithmetic coding compresses it into a string of bits. It is helpful to think of the compressed file as a single real number, a fraction in the interval $[0, 1)$. This interval is "open" at 1 and "closed" at 0. Arithmetic coding shows how to associate any given input string of symbols with a subinterval of $[0, 1)$, and the output of the algorithm is a number (any number) in this subinterval. Before looking into the details of the method, it is important to understand subintervals. Perhaps the simplest way to identify a subinterval is by specifying its lower and upper limits (a natural alternative is the lower limit and the width). The point is that these numbers (the limits and the width) are fractions and thus get longer as the subinterval shrinks in size. A large subinterval may be specified by the short numbers $[0.5, 0.7)$, a smaller subinterval may be specified by the longer limits $[0.48, 0.51)$, and a very small subinterval may be specified by the long fractions $[0.49065124, 0.490651245)$. The same is true of the width of the subinterval.

Given that the output of arithmetic coding is a real number (a fraction) in some subinterval of the interval $[0, 1)$, we can now describe the principle of the method. The input is read symbol by symbol and bits are computed and appended to the output for each symbol. The new bits, however, are not the code of the current symbol, since symbols do not have individual codes. Instead, the new bits increase the precision of the output, and thus specify a smaller subinterval. Since arithmetic coding is a statistical method, the sizes of the subintervals involved should depend on the probabilities of the input symbols. If the current input symbol has high probability, few bits should be generated and appended to the output; the current subinterval should therefore shrink just a bit. The lower the probability of the current symbol, the more bits should be computed and appended to the output, reducing the size of the current subinterval.

Thus, the principle of arithmetic coding is: Initially, the output is any number in the interval $[0, 1)$. As input symbols are read and processed, the output becomes a number in a smaller subinterval. If the current symbol has large probability, the current subinterval should shrink a little. If the probability of the current symbol is small, the current subinterval should shrink more. This principle can be rephrased for the ideal case where the input string is infinite. Given an infinite input string, arithmetic coding encodes it into an infinite output string, such that for any finite string S at the beginning of the input, a subinterval is determined whose width equals the probability of S. The final output is the real number at the intersection of all those subintervals. The binary expansion of this number is normally infinite.

Once this principle is grasped, the details of the method are easy to follow. We assume that the input alphabet consists of n symbols with known probabilities of occurrence. The original interval $[0, 1)$ is divided into n subintervals whose widths are proportional to the probabilities of the input symbols (the order of the subintervals is irrelevant). We define the original interval $[0, 1)$ as our initial subinterval. Each time a symbol is read, the current subinterval is divided into n subsubintervals whose lengths are also proportional to the probabilities of the input symbols. One of those subsubintervals, the one corresponding to the current symbol, is selected and it becomes the current subinterval. This way, the current subinterval shrinks all the time and the amount of shrink depends on the probability of the current symbol. The more probable the symbol, the less the shrinkage of the current subinterval.

When the entire input file has been read and processed this way, the encoder ends up with a small subinterval. The final output can be any number in this subinterval, but it is easier for the encoder to output the lower limit of the interval. This limit is the smallest, but not always the shortest, number in the subinterval. As an example, consider the interval $[0.599, 0.611)$. The lower limit, 0.599, is the smallest number in this interval, but the shortest number is 0.6.

Figure 8.2 illustrates subinterval division and shrinkage. We again assume an alphabet of three symbols A, B, and C with probabilities 0.5, 0.3, and 0.2, respectively. The entire interval $[0, 1)$ is selected as the initial subinterval and is divided into three subsubintervals whose widths are proportional to the probabilities of the three symbols. If the first symbol read is A, then subsubinterval $[0.5, 1)$ is selected and it becomes the current subinterval. It is divided into the three subsubintervals $[0.5, 0.6)$, $[0.6, 0.75)$, and $[0.75, 1)$. If the second input symbol is B, then subsubinterval $[0.6, 0.75)$ is selected and it becomes the current subinterval. It is divided into three new subsubintervals $[0.6, 0.63)$, $[0.63, 0.675)$, and $[0.675, 0.75)$. If the third symbol that's read is C, then subsubinterval $[0.6, 0.63)$ is selected and it becomes the current subin-

DATA COMPRESSION

terval. If there is more input, this subinterval is divided into the three smaller subsubintervals [0.6, 0.606), [0.606, 0.615), and [0.615, 0.63). The final output can be any number in the current subinterval [0.6, 0.63), but is normally the lower bound 0.6.

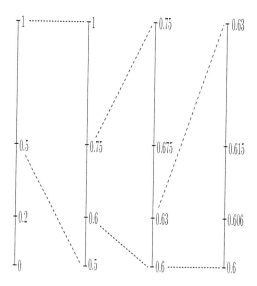

Figure 8.2. Shrinking Subintervals.

The output file also has to include the symbols and their probabilities, for the use of the decoder. The decoder starts by reading the symbols and probabilities, and dividing the interval [0, 1) into n subintervals (in our example, $n = 3$) according to the probabilities. The decoder then reads the first digit of the output. It is 6, so the decoder knows that the output is 0.6.... The output is thus a number in the subinterval [0.5, 1), implying that the first symbol is A. The decoder then eliminates A from the output by subtracting from the output the lower limit, 0.5, of the subinterval of A, and dividing the result by the width of that subinterval (also 0.5). The result is $(0.6 - 0.5)/0.5 = 0.2$, which is inside the subinterval of B. Thus, B becomes the second symbol to be decoded. The decoder subtracts the lower limit of B (0.2) from the current output (also 0.2), and divides by the width of the subinterval of B (0.3). The result is zero, which is inside the subinterval of C. This is why C becomes the third symbol to be decoded.

Notice that the zero does not necessarily indicate the end of the decoding process. The input string $ABCCC$, for example, is also compressed into the number 6. When the decoder decodes that 6, it somehow has to know that when the output has been reduced to zero, there are still three Cs remaining to be decoded. The solution is to add a special symbol, an end-of-input (eoi), to the alphabet. This symbol is assigned a very small probability. When the encoder senses the end of the input data, it encodes the eoi symbol. The decoder knows this, and it stops when it decodes the eoi, not when the compressed file is reduced to zero.

This algorithm is simple in principle, but impractical. If the input string is thousands of symbols long, the output can easily be thousands of bits long. It is impractical to perform arithmetic operations on very long numbers (integers or real), which is why the basic method described here has to be modified in order to be practical. The practical implementation of arithmetic coding is outside the scope of this text, but one point can be mentioned. In the example above, once C has been input and encoded, the interval $[0.6, 0.63)$ becomes the current subinterval. At that point it is clear that the first digit of the output (6) is not going to change. Thus, the encoder does not have to keep the 6, and it can write this digit on the compressed file. The result is that the encoder has to keep only a limited number of digits at any time.

The arithmetic coding algorithm as described so far assumes that the probabilities of occurrence of the symbols are known. In general, each data file to be compressed has different probabilities, so an *adaptive* algorithm, where the probabilities are computed and revised all the time, is useful. Such an algorithm has two parts, a *model* and an *encoder*. The model reads the current symbol from the input file and sends it to the encoder. The model then uses the symbol to update the table of symbol probabilities. The updated table is used by the encoder to encode the next symbol.

The fact that the current symbol is encoded with the old probability table makes it possible for the decoder to mimic the operations of the encoder. When the decoder is ready to decode the current symbol, it uses the existing probability table, decodes the symbol, then invokes the model to update the table. We say that the encoder and decoder work in *lockstep*.

The simplest model updates symbol probabilities by counting the number of times each symbol has been read so far from the input file. If 500 symbols have been read so far, and 23 of them were As, then the probability of A is $23/500 = 0.046$. If the next symbol is an A, then the new probability of A is increased by the model to $24/501 = 0.0479$ and the probabilities of the other symbols are decreased slightly. When this model starts, it knows nothing about the real probabilities of the symbols in the particular file being input. As more and more symbols are input and processed, the probabilities computed by the model improve steadily.

The problem of *zero probabilities* should be mentioned. Suppose, for example, that 100 symbols have been read from the input file before the first "q" is encountered. If the model uses symbol counts to compute probabilities, the probability of "q" would be zero for the first 100 symbols. The problem is that the arithmetic encoder divides the current subinterval into n subsubintervals whose widths are proportional to the probabilities of the symbols, and a symbol with zero probability corresponds to a subsubinterval of zero width; a nonexisting subsubinterval. Also, the decoder has to divide by the width of a subinterval, so it cannot be zero. The reader should also recall that the entropy of a symbol determines the smallest number of bits into which it can be encoded. The entropy depends on $-\log_2 P$ and this quantity is undefined for $P = 0$ and becomes very large for small values of P. The zero probabilities problem can be solved by initializing all the symbol counts to 1 instead of to 0.

More sophisticated models use the context of a symbol to estimate its probability (we use the term *predict* instead of "estimate the probability of"). The simplest context modeler uses *digrams* to predict symbols. A digram is a pair of consecutive symbols.

A good example is the letter "u", which is rare in English. Its probability of occurrence is about 2%. However, if the current symbol is "q", there is a 99% probability that the symbol that follows will be "u." If the next symbol is, in fact, a "u". the context modeler assigns it a probability according to how many times the digram "qu" has been seen in the past. If, however, the next symbol is different, say, an "a", the modeler assigns it a probability according to how many times the digram "qa" has been seen. This number is normally zero or very small, but in certain texts it may be large (for example, it may be part of the word "qaaba" that's common in Moslem religious texts).

For related material on the subject of this section, see Moffat et al. (1998) and Witten et al. (1987).

3. Dictionary Methods

Statistical compression methods use symbol probabilities to reduce the redundancy of data and thereby compress it. Dictionary-based methods, on the other hand, reduce data redundancy by discovering identical fragments of data. In typical English text, for example, the 4-symbol string "the␣" is common. A dictionary-based compression method saves such a string in a table (the dictionary) and replaces future occurrences of the string with a pointer to the location of the string in the table. The longer the string, the better compression is achieved by replacing it with a pointer. The more common a string is, the more compression is achieved by the use of the dictionary.

The entire field of dictionary compression is based on the pioneering work of J. Ziv and A. Lempel in the 1970s. They developed the first two dictionary compression methods, known today as LZ77 and LZ78. Many variations and extensions of these methods have since been developed by researchers and have, as a result, names that start with the initials LZ. This section illustrates the dictionary approach to compression by describing the LZ77 and LZW methods. The former is the first dictionary method and is simple. The latter is the most-commonly used dictionary method.

3.1. LZ77

The main data structure used by LZ77 is a large array divided into two parts, a (short) look-ahead buffer and a (long) search buffer.

⟵coded text··· | search buffer | look-ahead buffer | ··· ⟵ text to be read

Symbols from the input file are read and shifted into this array from right to left. Each symbol moves through the look-ahead buffer, then through the search buffer, and out the left end. Alternatively, we can think of the input file as a long, static string, and imagine the array sliding along this string from left to right. This is why LZ77 is sometimes called a *sliding buffer* algorithm.

The general compression step is illustrated with the Basque tongue twister

        `Akerrak adarrak okerrak ditu.  Okerrak adarrak akerrak ditu.`

(Rough translation from Basque: Billy-goat's horns are twisted. Twisted horns are owned by billy-goat.) Suppose that at a certain point, the array contains

coded text···Ak | errak␣adarrak␣okerrak␣ditu␣O | kerrak␣adarrak␣ake | rrak␣ditu···

The text in the search buffer has already been compressed (the last symbol that has been processed is the "O" of "Okerrak") and the look-ahead buffer contains text ("kerrak...") to be compressed shortly. The encoder tries to match as many symbols as possible from the look-ahead buffer with a string of symbols from the search buffer. The best match (in fact, the only match) is with the 7-symbol string `kerrak␣` that starts 13 positions from the dividing line. Once the encoder finds the best matching string, it prepares the 3-field *token* (13, 7, a). The first element

DATA COMPRESSION

of the token is an offset pointing to the start of the matching string in the search buffer. The second element is the number of symbols matched. The third element is the symbol that follows the matched string in the look-ahead buffer. This symbol is the leftmost a of adarrak. This token is written on the compressed file instead of the eight (not seven) symbols kerrak⊔a. The data in the array is then shifted eight positions to the left (or, alternatively, the array is shifted to the right), to produce

coded text···Akerrak⊔ad | arrak⊔okerrak⊔ditu⊔0kerrak⊔a | darrak⊔akerrak⊔dit | u···

The next symbol to be encoded is the leftmost symbol in the look-ahead buffer. This is the d of darrak. Unfortunately, there is no d in the search buffer (we wish our search buffer would be a bit longer, because we have just lost the only d in it), so the d cannot be matched with anything, and the encoder outputs the token (0, 0, d) as the compressed code of d. It is now clear why the third field of a token is necessary. It serves a useful purpose in cases where no match is found. The data is shifted one position, resulting in

coded text···Akerrak⊔ada | rrak⊔okerrak⊔ditu⊔0kerrak⊔ad | arrak⊔akerrak⊔ditu | ···

The next string to be matched in the look-ahead buffer is arrak⊔ake···. Again we have just lost an excellent match because our search buffer is too short. The search buffer has four a's but none is followed by an r, so the next step results in the token (2, 1, r). Any of the four a's in the search buffer can be selected as the matching string, but our offset (2) indicates that the nearest one was selected. The advantage of selecting the nearest match is that it results in the smallest offset. Sophisticated versions of LZ77 may assign variable-size codes to the offsets (where small offsets are assigned short codes), and write the code on the compressed file, instead of the offset itself. In such a case, small offsets improve the overall compression somewhat.

It is clear from this example that a long search buffer provides a better chance of finding a long match. However, the longer the search buffer, the bigger the offset has to be. Large offsets result in long tokens, so the length of the search buffer is a compromise between better matches and shorter tokens. The search buffer is normally thousands of bytes long, but the length of the look-ahead buffer should only be tens of bytes. If the search buffer is $2^{11} = 2048$ bytes long and the length of the look-ahead buffer is 31 bytes, then the size of an offset is 11 bits and the match length (the second field of the token) is a 5-bit number (between 0 and 31). If each symbol is coded in eight bits, then the total length of a token is $11 + 5 + 8 = 24$ bits; equivalent to three ASCII codes. If a token matches more than three symbols, compression results, but it is possible to end up with expansion.

There is another reason why a very long search buffer is inefficient. Experience shows that most matches are short range. It is rare to have a string of symbols that will not match anything seen recently, but will match a string seen long ago.

The courageous reader should work out the LZ77 tokens resulting from the input string

kakaktua kakak kakekku kenal kakaktua kakak kakekmu.

(An Indonesian tongue twister roughly translated "my granpa's older brother's parrot knows your granpa's older brother's parrot.")

For related material on the subject of this section, see Ziv and Lempel (1977).

3.2. LZW

The downside of LZ77 is the length of its tokens. A token has three fields, one of which is a symbol, and is therefore longer than one symbol. Thus, a token that matches just one or two symbols causes expansion instead of compression. The LZ78 method (not described here) uses two-field tokens and is therefore more efficient. The LZW method improves on LZ78 and uses one-field tokens. LZW is described here, since it is a popular compression method, used in many applications.

The main idea behind LZW is to make sure that each token matches at least one symbol. As a result, this method does not use a look-ahead buffer, a search buffer, or a sliding window. Instead, it maintains a dictionary where each entry is a string of symbols. If the next n input symbols match a dictionary entry but the next $n+1$ symbols do not, the n-symbol string is compressed by writing on the compressed file a pointer to that entry. The pointer can be considered a one-field token.

One problem is to make sure that there will always be a match. The next n input symbols must match a dictionary entry for some positive integer n. This is solved by initializing the dictionary to all the symbols of the alphabet. If the alphabet consists of bytes, the dictionary should be initialized to all 256 possible bytes. They are stored in dictionary entries 0 through 255. If the alphabet is ASCII characters, the dictionary should be initialized to all 128 ASCII characters. This way, at least one symbol will match some dictionary entry.

The dictionary starts with a certain number of symbols, but it should grow during compression. More and more strings should be appended to it, but not at random. The strings added to the dictionary should be taken from the input file, since such strings have a good chance of matching future pieces of the input. Another desirable feature is to add longer and longer strings, since they produce better compression. Also, the strings should be added in a way that the decoder would be able to mimic. The only way for the decoder to decompress the data is to initialize and maintain the dictionary in the same way as the encoder.

The method used by LZW to append strings to the dictionary is simple. Once an n-symbol string has been matched with a dictionary entry, a new, $n+1$-symbol string is appended to the dictionary. This string consists of the n matched symbols plus the symbol that follows them in the input file. Here is a more formal description of this method.

The encoder looks for the largest n such that the next n input symbols match a dictionary entry, but the next $n+1$ symbols do not. We denote the n-symbol input string by N and the symbol following it by s. The encoder outputs a pointer to the dictionary entry containing N, and appends string Ns to the dictionary. It then skips the next n symbols in the input, so that symbol s is the next one to be compressed.

A simple example may serve to illustrate the operation of the encoder. We imagine an alphabet with the five symbols "A", "B", "D", "O", and "Y", stored in entries 0 through 4, respectively. The input data to be compressed is the short string "YABBADABBADO." The first symbol input is "Y." It is found in the dictionary (in entry 4), but string "YA" is not found. The encoder therefore outputs 4 and appends string "YA" to the dictionary (at entry 5). The next symbol is "A." it is found in the dictionary entry 0, but string "AB" is not found. The encoder therefore outputs 0 and appends string "AB" to the dictionary (at entry 6). The complete output is 4, 0, 1, 1, 0, 2, 6, 8, 2, 3. The final dictionary is shown in Table 8.2.

Table 8.2. An LZW Dictionary.

0	A	4	Y	8	BA	11	ABB
1	B	5	YA	9	AD	12	BAD
2	D	6	AB	10	DA	13	DO
3	O	7	BB				

Even this short example shows the main shortcoming of LZW; the slow growth of its dictionary. Each new string added to the dictionary is at most one symbol longer than existing strings. In our example, the first 3-symbol string appended to the dictionary (at entry 11) is the "ABB" of "DABBA." Other dictionary-based compression methods, such as LZMW and LZAP, don't suffer from this limitation.

The LZW decoder starts by initializing its dictionary to all the symbols of the alphabet. If different alphabets may be used, the encoder has to write all the alphabet symbols at the start of the compressed file. The decoder then reads the compressed file pointer by pointer, and uses each pointer to retrieve a string from the dictionary. If the current string retrieved is C, then string Cs should be appended by the decoder to its dictionary. However, symbol s is currently unknown, since it is the first symbol of the *next* string to be decoded. The decoder therefore (1) saves string C, (2) reads the next pointer, (3) uses it to retrieve the next string N from the dictionary, (4) isolates the first symbol s of N, and (5) appends string Cs to the dictionary.

The discussion above shows that the LZW dictionary is a list of strings that can have any size. Such a list is not simple to implement in practice and slow to operate on. However, the LZW dictionary has a property that suggests the use of a different data structure. Any string appended to the dictionary has the form Cs where C is already in the dictionary and s is a symbol. Because of this property, the dictionary is implemented in practice as a data structure called a *trie*.

The last point to discuss is dictionary overflow. Regardless of the amount of memory reserved for the dictionary, it may overflow if the input file is large. This problem is common to many dictionary methods and is not LZW specific. Two common solutions are: (1) Delete the dictionary and start afresh with a new dictionary. (2) Stop appending strings to the dictionary, and in the future use only existing dictionary entries.

The first solution makes sense, since we know that input strings tend to be matched by strings input in the recent past. A large, full dictionary has many strings that were input in the distant past, and they tend to be poor matches for newly input strings. The second solution is easier to implement.

For related material on the subject of this section, see Miller and Wegman (1985), Phillips (1992), Welch (1984).

4. Image Compression

A digital image is a rectangular array of dots, or picture elements, called *pixels*. Such an image can be generated in the computer, using software for drawing or painting, or it can be generated outside the computer (with a digital camera, or by scanning a document), then input as a file. Fast, efficient compression of images is important because (1) images are so popular, (2) image files tend to be large, and (3) images, unlike text, lend themselves to *lossy compression*.

DATA COMPRESSION

The size of an image file is easy to estimate. An image of size $1K \times 1K = 2^{10} \times 2^{10}$ has $2^{20} = 1M$ pixels. That's more than a million! Most current digital cameras generate even bigger images. The number of bits required to represent a pixel depends on the number of colors. A 256-color image has 8-bit pixels, but it is common to have images where each pixel is represented by three bytes. The uncompressed size of such an image is 3 Mb, so its efficient compression can save more than 2 Mb.

The lossy compression of images is based on the nature of human vision. Given a large image with many colors, the brain cannot memorize the precise colors of all the pixels, and therefore cannot tell if the color of a pixel (or several pixels) has been changed.

This section starts by looking at different types of images. Following this, various approaches to image compression are outlined, as well as the concept of *progressive compression*. The section ends with a description of JPEG, a popular image compression method commonly used in web pages.

4.1. Image types

There are two main types of images, *continuous-tone* and *discrete-tone*. There are important differences between them, and as a result, compression methods may be developed specifically for each of these types.

Imagine a picture of a natural scene, showing terrain, mountains, trees, sky, and clouds. When looking at the sky part of this image, we see that the colors of adjacent pixels do not differ much. The same is true for the pixels of a cloud, of a mountain, and of most other natural objects. This type of image has areas where the pixels in an area have the same general color, and where the boundaries between areas are not always well defined. The shape of an area is also generally ill defined; it is not a simple geometric shape such as a rectangle or a circle. Such an image is classified as continuous tone.

An image of an artificial scene, on the other hand, is substantially different. Such an image may consist of a room, pieces of furniture, and other man-made objects. This type of image is also divided into areas where the pixels of an area have similar colors. However, the boundaries between the areas are sharp and well defined, and the areas themselves may have simple geometric shapes, such as a square or an ellipse. This is an example of a discrete-tone image.

Depending on the number of colors, images can also be classified into *bi-level*, *grayscale*, and *color*. A bi-level image has two colors. They are normally black and white, but they can also be referred to as foreground (black) and background (white), or as 1 (black) and 0 (white). A pixel in a grayscale image is represented by a single number that specifies the shade of gray (or of another color) of the pixel. In a color image, each pixel is represented by three numbers, specifying the three components of its color. It turns out that a complete specification of a color requires three numbers. Many times, these are the intensities of the red, green, and blue components of the color (the RGB color system), but they can have different meanings, such as hue, lightness, and saturation (the HLS color system), or luminance and chrominance (the YCbCr color system).

Regardless of its type, an image can be compressed because its original representation is redundant; it is easy to work with, but it is not the most efficient. The discussion of continuous-tone and discrete-tone images illustrates the redundancy in digital images. As we move along an image, examining pixels, we observe that, in general, the color of a pixel is identical or similar to those of its near neighbors. We say that the pixels of an image are not completely independent. There are certain relations between them, and as a result they are *correlated*.

Pixel correlation is also called *spatial redundancy*, to distinguish it from *temporal redundancy*, which is correlation between frames of a movie.

The correlation between pixels is the basis of the *general principle of image compression*, a principle that states:

A pixel in an image tends to be similar to its immediate neighbors.

This principle lies at the bottom of all image compression methods, even methods based on completely different approaches, or methods that employ different mathematical tools. The next section is devoted to a description of the main approaches to image compression.

4.2. Approaches to image compression

Run Length Encoding. This approach can be used to compress a bi-level image. In such an image, there is no such thing as similar pixels. Two pixels can either be identical or different. Thus, the principle of image compression tells us that the immediate neighbors of a pixel P in a bi-level image tend to be identical to P. This suggests the use of *run lengths* to compress such an image. The image is scanned row by row (*raster order* scan) and the lengths of runs of identical pixels counted. If a row consists, for example, of 17 black pixels, followed by 56 white pixels, followed by 8 black ones, etc., then the numbers $17, 56, 8, \ldots$ become a compressed representation of the row. These numbers should be assigned prefix codes, and it is these codes that are actually written on

DATA COMPRESSION

the compressed file. The international standard for facsimile compression (Section 2.4) uses this method.

Context Counting. It is possible to use arithmetic coding to compress an image. Arithmetic coding, however, requires knowledge of the probabilities of the individual symbols being compressed. In a bi-level image, there are two such symbols, a black pixel and a white pixel. A two-pass method can count their numbers in the first pass, calculate their probabilities between the passes, and use the second pass to actually compress the image. However, such a method is normally too slow to be useful in practice. Faster results are obtained with adaptive arithmetic coding. Pixels are read and sent one by one to the arithmetic encoder, each with its probability. The pixels are also counted and the counts are used to predict pixels (i.e., to estimate their probabilities) "on the fly." Imagine a situation where 1128 black pixels and 972 white pixels have been read and processed so far. The next pixel is now read. If it is black, it is sent to the arithmetic encoder with a probability of $1128/(1128+972) = 0.537$. If it is white, its probability is estimated as $972/(1128+972) = 0.463$.

The principle of image compression tells us that *all* the immediate neighbors of a pixel P, not just those located on the same image row, tend to be correlated with P. Better predictions can therefore be obtained if we compare the current pixel to some of its previously-seen neighbors, and estimate its probability from these neighbors.

A context counting compression method uses this approach to compress bi-level images. Such a method scans the image pixel by pixel and looks at the same n neighbors of each pixel. These n neighbors become the *context* of the current pixel. Since a pixel in a bi-level image is one bit, the context is an n-bit integer. The assumption is that if the current pixel is black and has a context C, then it is likely that several (perhaps even many) of the black pixels that have already been seen, have the same context. The method therefore counts how many times each possible context has appeared with a black pixel and how many times it has appeared with a white pixel.

Four different contexts are shown in Figure 8.3, where X indicates the current pixel. The context of 8.3a is three pixels, so there can be eight such contexts. The contexts of 8.3b,c have five and six pixels, respectively. There are more of these contexts, but they also use neighbors that are not immediate and therefore feature weaker correlation. The context of Figure 8.3d is especially interesting. It is symmetric around the current pixel, and therefore seems ideal. It cannot be used in practice, however, since it uses neighbors that have not been processed yet, and therefore are unknown to the decoder.

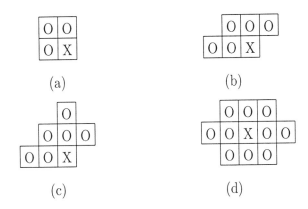

Figure 8.3. Four Pixel Contexts.

The JBIG standard for compressing bi-level images uses this approach.

Context prediction. This approach is for grayscale images. The context of a pixel is used to *predict* the value of the pixel. This can be done by, for example, averaging the values of the pixels of the context. The predicted value of a pixel P is then subtracted from P. The principle of image compression implies that the difference will normally be a small number, although it can sometimes be large. Prefix codes are assigned to all the possible values of the difference such that short codes are assigned to small values.

Experiments with many images suggest that the values of the difference tend to be distributed according to the *Laplace distribution* (see Salomon (2000)) for a discussion of this distribution). This fact can be used to assign a probability to each value of the difference, and encode the differences with adaptive arithmetic coding. This is the principle of the MLP image compression method.

Quantization. This approach to image compression is based on losing data. It results in lossy compression. The simplest implementation of this approach is called *scalar quantization*. A compression method based on scalar quantization examines individual pixels one by one and deletes some bits from each pixel. If the image has 8-bit pixels, it is possible to remove the least-significant four bits of each pixel, and end up with 4-bit pixels. The compression factor obtained is 2 (this is a rare case where the compression performance is known in advance), not very impressive, but the loss of data is significant, and may be prohibitive. Trimming the value of a pixel from eight to four bits reduces the number of colors (or grayscales) from 256 to just 16. This is why *vector quantization*, not scalar quantization, is used in practice.

Given an image, a simple, intuitive vector quantization method partitions it into small blocks, typically 2×2 or 4×4 pixels each. Each block is considered a vector, hence the name "vector quantization." The encoder keeps a list of vectors (called a *codebook*) and compresses each block by writing on the compressed file a pointer to the block in the codebook. The decoder is different and is much simpler. It reads pointers from the compressed file, following each pointer to a block in the codebook, and appending the block to the image-so-far.

In the case of 2×2 blocks, each block (vector) consists of four pixels. If each pixel is one bit, then a block is four bits long and there are only $2^4 = 16$ different blocks. It is easy to store such a small, permanent codebook in both encoder and decoder. However, a pointer to a block in such a codebook is, of course, four bits long, so nothing is gained by replacing blocks with pointers. If each pixel is k bits, then each block is $4k$ bits long and there are 2^{4k} different blocks. The codebook's size increases very fast with k (for $k=8$, for example, it is $256^4 = 2^{32} = 4$ Tera entries), but the point is that we again replace a block of $4k$ bits with a $4k$-bit pointer, thereby achieving no compression. This is true for blocks of any size.

To get compression, we have to go back to the principle of image compression. This principle tells us that image pixels are correlated, which leads to the conclusion that a given image may not contain every possible block. Given an image with 8-bit pixels, the number of 2×2 blocks is $2^{2\cdot 2\cdot 8} = 2^{32} \approx 4.3$ billion, but any particular image may contain just a few thousand different blocks. Based on this, it is easy to propose a better version of vector quantization. This algorithm starts with an empty codebook and scans the image block by block. The codebook is searched for each block. If the block is already in the codebook, the encoder outputs a pointer to the block in the (dynamically growing) codebook. If the block is not in the codebook, it is appended to the codebook and a pointer is output.

This method is simple but inefficient because each block added to the codebook has to be written on the compressed file, so the decoder can append it to *its* codebook. In extreme cases, this may even cause expansion. Another problem is that the codebook's size increases during compression, so the pointers have to get longer.

As a result of these problems, any practical method for vector quantization of images must be lossy. If lossy compression is acceptable, the codebook's size can be greatly reduced. Here is an intuitive, lossy, vector quantization method for images. Start by selecting a large number of "training" images, and analyze them to find the M most-common blocks in them (where M is selected by the implementor). Use these M blocks to construct a "standard" codebook and build it into both encoder and decoder. To compress an image, scan it block by block, and for each block find the codebook entry that best matches it. A pointer to that entry should then be written on the compressed file. The size of the pointer is, of course, $\lceil \log_2 M \rceil$, so the compression ratio (which is known in advance) is

$$\frac{\lceil \log_2 M \rceil}{\text{block size}}.$$

This intuitive method suffers from two problems. One is how to match image blocks to codebook entries. There is a need for a simple, fast process to measure the "distance" between an image block and a codebook entry. Once such a process is developed, a given image block is matched with the codebook entry closest to it. We denote by $B = (b_1, b_2, \ldots, b_n)$ a block of image pixels, and by $C = (c_1, c_2, \ldots, c_n)$ a codebook entry (each is a vector of n bits). We denote the "distance" between them $d(B, C)$. Three common ways to measure it are

$$\begin{align} d_1(B, C) &= \sum_{i=0}^{n} |b_i - c_i| \\ d_2(B, C) &= \sum_{i=0}^{n} (b_i - c_i)^2, \\ d_3(B, C) &= \max_{i=0}^{n} |b_i - c_i| \end{align} \quad (8.1)$$

The first two measures are easy to visualize in the case $n = 3$. Measure $d_1(B, C)$ is the distance between the two three-dimensional vectors B and C when moving along the coordinate axes. Measure $d_2(B, C)$ is the square of the Euclidean (straight line) distance between the two vectors. The third measure $d_3(B, C)$ is easy to interpret for any n. It finds the component where B and C differ most, and it returns this difference. The distances $d_i(B, C)$ can also be considered measures of *distortion*.

The other problem with this method is the quality of the codebook. In the case of 2×2 blocks with 8-bit pixels the total number of blocks is $2^{32} \approx$ 4.3 billion. If we decide to limit the size of our codebook to, say, a million entries, it will contain only 0.023% of the total number of blocks (and still be 32 million bits, or about 4 Mb long). Using this codebook to compress an "atypical" image may result in a large distortion regardless of the distortion measure used. When the compressed image is decompressed, it may look so different from the original as to render this intuitive quantization method useless.

A natural way to solve this problem is to modify the original codebook entries in order to *adapt* them to the particular image being compressed. The final codebook will have to be included in the compressed file but, since it has been adapted to the image, it may be small enough to produce good compression, yet close enough to the image blocks to end up with a decompressed image that's acceptable to the user. Such an adaptive algorithm has been developed by Linde, Buzo, and Gray. It is known as the LBG algorithm and it is the basis of many vector quantization methods for the compression of images and sound

Image Transforms. Mathematical entities may have more than one representation. A simple example is real numbers. We normally use decimal numbers, numbers represented in base 10, but we know that real numbers can be represented in base 2 (binary) and in other bases. Transforming an entity from one representation to another may reveal properties that were hidden in the original representation, and may thus help to solve a problem. Perhaps the simplest example is transforming the decimal number 1024 to binary, where its value is $10,000,000,000_2 = 2^{10}$. In decimal, this number is not round and does not seem special. In binary, however, it has a special value, which explains its common use in computers (this number is denoted by "K", and is used to specify the size of data files and memory).

Another simple, illustrative example is operations on Roman numerals. The ancient Romans had a hard time trying to operate on such numbers, but when *we* have to multiply two Roman numerals, we can easily do it by (1) transforming them into modern (Arabic) notation, (2) multiplying them in the usual way, and (3) transforming the product back into a Roman numeral. Here's a simple example.

$$\text{CXXV} \times \text{XI} \to 125 \times 11 = 1375 \to \text{MCCCLXXV}.$$

An image can be compressed because its pixels are not independent; they are correlated. A random image cannot be compressed because its pixels are independent of each other; they are *decorrelated*. This points the way to image compression by means of a transform. If we can transform the pixels of an image to a representation where they are decorrelated, we have effectively compressed the image.

We start with a simple transform that's easy to visualize geometrically. We take a grayscale image, where each pixel is a single number denoted by P_{ij}, and scan it in raster order, row by row, constructing pairs of adjacent pixels. The first few pairs are (P_{11}, P_{12}), (P_{13}, P_{14}), $(P_{15}, P_{16}), \ldots$. If the image has N pixels, there will be $N/2$ pairs. We consider each pair a point in two-dimensional space. Thus the pair of pixels $(96, 88)$ becomes the point with x coordinate 96 and y coordinate

88. This seems meaningless, since 96 and 88 are grayscale values, not coordinates, but it makes sense when we consider that they are the values of *adjacent* pixels. The principle of image compression tells us that such values tend to be similar. When a pair of identical or similar values becomes the (x, y) coordinates of a point, the point lies on or near the 45° line $y = x$. When we plot all $N/2$ pairs of the image, we end up with a cloud of points, most of which are located on or near this line (Figure 8.4a).

The points are now transformed by rotating them 45° clockwise. Figure 8.4b shows that this brings most points close to the x axis, where their y coordinates are zero or very small numbers. In addition, it is easy to see that the x coordinates of the points do not change much. The rotation has shrunk the values of half the pixels without much affecting the other half.

We denote a general point by (x, y) and its image after a 45° rotation by (x^*, y^*). Equation 8.2 is the mathematical expression of this rotation

$$\begin{aligned}
(x^*, y^*) &= (x, y) \begin{pmatrix} \cos 45° & -\sin 45° \\ \sin 45° & \cos 45° \end{pmatrix} \\
&= (x, y) \frac{1}{\sqrt{2}} \begin{pmatrix} 1 & -1 \\ 1 & 1 \end{pmatrix} \\
&= (x, y) \mathbf{R},
\end{aligned} \tag{8.2}$$

where the rotation matrix $\mathbf{R}$ is orthonormal (i.e., the dot product of a row with itself is one, the dot product of different rows is zero, and the same is true for columns).

If we now write the N new values on a file, the file will be smaller than the original image file; the image has been compressed. Since each of the N new values may require a different number of bits, they should be assigned prefix codes and the codes, not the values, should be written on the file. Better results can be obtained if all the x coordinates are written, encoded, on the file first, followed by all the y coordinates (also encoded). It is clear that the two sets of transformed values are substantially different and should be assigned different sets of prefix codes. Also, the set of y coordinates consists mostly of small numbers and can be further compressed by quantizing its values.

Decompressing is done by reading the compressed file and rotating each pair of input values 45° counterclockwise. The inverse transformation is

$$(x, y) = (x^*, y^*) \mathbf{R}^{-1} = (x^*, y^*) \mathbf{R}^T = (x^*, y^*) \frac{1}{\sqrt{2}} \begin{pmatrix} 1 & 1 \\ -1 & 1 \end{pmatrix} \tag{8.3}$$

(The inverse of an orthonormal matrix is its transpose.)

DATA COMPRESSION

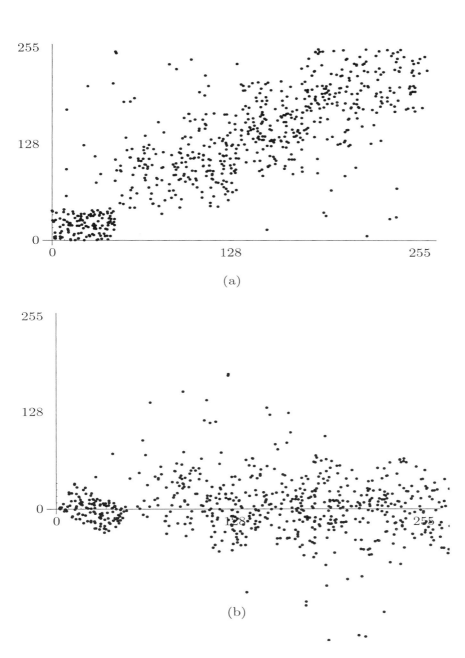

Figure 8.4. Rotating a Cloud of Points.

This process can easily be extended to three, or even more, dimensions. It is possible to scan an image and construct $N/3$ triplets of adjacent pixels. If each triplet is interpreted as a point in three-dimensional space, the points will be concentrated about the line $x = y = z$, a line that makes 45° angles with all three coordinate axes. The transform should rotate this line so that it coincides with the x axis. The new, rotated points have small y and z coordinates, while their x coordinates remain about the same. This process transforms two-thirds of the original N pixels to smaller values. The discrete cosine transform (DCT, Section 4.5) is an example of a transform based on rotation.

Not every transform uses rotations. Transforming image pixels should be a simple, fast operation, so practical image compression methods use *linear transforms*. Each transformed value t_i in a linear transform is a weighted sum of pixel values p_j, where each pixel is multiplied by a weight (or a transform coefficient) w_{ij}. Thus, $t_i = \sum_j p_j w_{ij}$, for $i, j = 1, 2, \ldots, n$. For $n = 4$, this is expressed in matrix notation

$$\begin{pmatrix} t_1 \\ t_2 \\ t_3 \\ t_4 \end{pmatrix} = \begin{pmatrix} w_{11} & w_{12} & w_{13} & w_{14} \\ w_{21} & w_{22} & w_{23} & w_{24} \\ w_{31} & w_{32} & w_{33} & w_{34} \\ w_{41} & w_{42} & w_{43} & w_{44} \end{pmatrix} \begin{pmatrix} p_1 \\ p_2 \\ p_3 \\ p_4 \end{pmatrix}.$$

For a general n, we can express a linear transform by $\mathbf{T} = \mathbf{W} \cdot \mathbf{P}$. Each row of the transform matrix $\mathbf{W}$ is called a "basis vector."

The main task in selecting a transform is the determination of the values of the weights w_{ij}. The basic principle is that the first transformed value t_1 should be large, while the remaining values $t_2, t_3, \ldots$ should be small. The basic transform $t_i = \sum_j p_j w_{ij}$ shows that t_i will be large when each weight w_{ij} reinforces the corresponding pixel p_j. This, of course, happens when the two vectors w_{ij} and p_j have similar values and signs. Conversely, t_i will be small if each weight w_{ij} is small and half of them have the opposite sign of p_j. As a result, if we end up with a large t_i, we conclude that the basis vector w_{ij} resembles the vector of pixels p_j. A small t_i, on the other hand, implies that w_{ij} and p_j have different shapes. We therefore consider the basis vectors w_{ij} tools that extract features from the pixel vector.

It is obvious that, in practice, the weights should be independent of the data items. Otherwise, the weights would have to be written on the compressed file, for the use of the decoder. This, plus the fact that pixels are normally nonnegative, suggests a natural way to select the basis vectors. The first vector, the one that produces t_1, should have positive, perhaps even identical, values. This will reinforce the nonnegative values of the pixels being transformed. In each of the other

DATA COMPRESSION

vectors, half the elements should be positive and the other half should be negative. When such a vector is multiplied by nonnegative pixels, it tends to produce a small transformed value.

We consider the basis vectors tools for extracting features from the pixel vector. A good choice is therefore basis vectors that are very different from each other, and thus extract different features. This naturally suggests the choice of basis vectors that are mutually orthogonal. If the transform matrix $\mathbf{W}$ is orthogonal, the transform itself is called *orthogonal*. All the transforms described in this section are orthogonal. Another idea that helps to select the basis vectors is that they should feature higher and higher frequencies, thus extracting higher-frequency features from the pixels as they are being applied.

The observations above are satisfied by the orthogonal matrix

$$\mathbf{W} = \begin{pmatrix} 1 & 1 & 1 & 1 \\ 1 & 1 & -1 & -1 \\ 1 & -1 & -1 & 1 \\ 1 & -1 & 1 & -1 \end{pmatrix}. \tag{8.4}$$

The first basis vector (the top row of $\mathbf{W}$) consists of all 1's, so it has zero frequency. Each of the vectors that follow has two $+1$s and two -1s, so they produce small transformed values, and their frequencies (the number of sign changes along the basis vector) get higher. This matrix is similar to the *Walsh-Hadamard transform*, a simple, well-known linear transform.

As an example, we transform the vector of four pixels $(4, 6, 5, 2)$

$$\begin{pmatrix} 1 & 1 & 1 & 1 \\ 1 & 1 & -1 & -1 \\ 1 & -1 & -1 & 1 \\ 1 & -1 & 1 & -1 \end{pmatrix} \begin{pmatrix} 4 \\ 6 \\ 5 \\ 2 \end{pmatrix} = \begin{pmatrix} 17 \\ 3 \\ -5 \\ 1 \end{pmatrix}.$$

The results are encouraging, since t_1 is large (compared to the original pixels), and two of the remaining t_i's are small. Other, more sophisticated transforms exist and are used in practice.

Fractal methods. The image is partitioned into parts (overlapping or not) that are processed one by one. Suppose, for example, that the next unprocessed image part is part number 15. The algorithm tries to match it with parts 1–14 that have already been processed. If part 15 can be expressed, for example, as a combination of parts 5 (scaled) and 11 (rotated), then only the few numbers needed to specify the combination should be written on the compressed file. If part 15 cannot be expressed as a combination of already-processed parts, it is written on the compressed file in raw format. In each case, part 15 is declared processed and the algorithm moves to the next part.

This approach is the basis of the various *fractal* methods for image compression. It applies the principle of image compression to image parts instead of to individual pixels. When applied in this way, the principle tells us that "interesting" images (i.e., those that we want to compress and save) have a certain amount of *self similarity*. Various parts of the image are identical or similar to the entire image or to other parts.

Separating the bitplanes. The approaches discussed so far are for grayscale images, but the great majority of digital images in use today are color. Recall that a pixel in such an image is represented by three numbers, specifying the three color components. The simplest approach for compressing a color image is to separate the three color components and compress each separately, as a grayscale image. Thus, if the RGB color space is used, a color image can be separated into three images, with shades of red, green, and blue, respectively, and each compressed separately with a grayscale compression method. A more sophisticated approach is to convert the original color components of an image to YCbCr (luminance and chrominance). It is known that the eye is sensitive to small variations in luminance but not in chrominance, so a compression algorithm can remove more data from the chrominance parts, and produce better (lossy) compression this way, without significant degradation of the quality of the reconstructed image.

For related material on the subject of this section, see Gonzalez and Woods (1992), Howard and Vitter (1992b), Howard and Vitter (1992a), Linde et al. (1980), and Salomon (2000).

4.3. Progressive compression

This is the case where the compressed file is created by the compressor in layers. The first layer contains the most important image information, and each subsequent layer adds details. This kind of compression is especially attractive when compressed images are transmitted over a communications line and are decompressed and viewed in real time. When such an image is received and decompressed, the decoder can quickly display a rough version of the entire image, then improve the display as more and more of the image is being received and decompressed. A user watching the image develop on the screen can normally recognize most of the image features after only 5–10% of it has been decompressed.

DATA COMPRESSION

This should be compared to raster-scan image compression. When an image is raster scanned and compressed, a user normally cannot tell much about the image when only 5–10% of it has been decompressed and displayed. Since images exist to be viewed by humans, progressive compression is attractive even in cases where it is slower or less efficient than non-progressive compression.

It is beneficial to think of progressive decoding as the process of improving image features over time, and this can be done in three ways:

1. Encode low-frequency image data first, then follow with higher-frequency details. An observer watching such an image as it is being decoded sees the image progressing from blurred to sharp. Methods that work this way typically feature medium speed encoding and slow decoding.

2. Start with a gray image and add colors or shades of gray to it. An observer watching such an image being decoded will see all the image details from the start, and will see them improve as more color is continuously added to them. Vector quantization methods use this kind of progressive compression. Such a method normally features slow encoding and fast decoding.

3. Encode the image in layers, where early layers consist of a small number of large, low-resolution pixels, and later layers have smaller, higher-resolution pixels. When a user watches such an image as it is being decoded, they see more details added to the image over time. Thus, such a method adds details (or resolution) to the image as it is being decompressed. This way of progressively encoding an image is called *pyramid coding* or *hierarchical coding*. Most progressive image compression methods employ this principle, so the rest of this section discusses approaches to pyramid coding.

Assuming that the image size is $2^n \times 2^n = 4^n$ pixels, the simplest progressive compression method is to calculate each pixel of layer $i-1$ as the average of a group of 2×2 pixels of layer i. Thus, layer n is the entire image, layer $n-1$ contains $2^{n-1} \times 2^{n-1} = 4^{n-1}$ large pixels of size 2×2, and so on, down to layer 1, with $4^{n-n} = 1$ large pixel, representing the entire image. If the image isn't too large, all the layers can be saved in memory. They are then written on the compressed file in reverse order, starting with layer 1. The single pixel of layer 1 is the "parent" of the four pixels of layer 2, each of which is the parent of four pixels in layer 3, and so on. This method is simple and intuitive, but very inefficient, since the average of integers is not always an integer and since the total number of pixels in the pyramid is

$$4^0 + 4^1 + \cdots + 4^{n-1} + 4^n = \frac{(4^{n+1}-1)}{3} \approx \frac{4^{n+1}}{3} \approx 1.33 \times 4^n = 1.33(2^n \times 2^n).$$

33% more than the original number!

A simple way to bring the total number of pixels in the pyramid down to 4^n is to include only three of the four pixels of a group in layer i, and to compute the value of the 4th pixel using the parent of the group (from the preceding layer, $i-1$) and its three siblings.

A better method is to let the parent of a group help in calculating the values of its four children. This can be done by calculating the differences between the parent in layer $i-1$ and its children in layer i, and writing the differences (suitably coded) in layer i of the compressed file. The decoder decodes the differences, then uses the parent from layer $i-1$ to compute the values of the four pixels.

Some improvement can be achieved if the parent is used to help calculate the values of three child pixels, and then these three plus the parent are used to calculate the value of the fourth pixel of the group. If the four pixels of a group are a, b, c, and d, then their average is $v = (a+b+c+d)/4$. The average becomes part of layer $i-1$, and layer i need only contain the three differences $k = a-b$, $l = b-c$, and $m = c-d$. Once the decoder has read and decoded the three differences, it can use their values, together with the value of v from the previous layer to compute the values of the four pixels of the group. Calculating v by a division by 4 still causes the loss of two bits, but this 2-bit quantity can be isolated before the division, and retained by encoding it separately, following the three differences.

The parent pixel of a group does not have to be its average. One alternative is to select the maximum (or the minimum) pixel of a group as the parent. This has the advantage that the parent is identical to one of the pixels in the group. The encoder has to encode just three pixels in each group and the decoder decodes three pixels (or differences) and uses the parent as the fourth pixel, to complete the group. When encoding consecutive groups in a layer, the encoder should alternate between selecting the maximum and the minimum as parents, since always selecting the same creates progressive layers that are either too dark or too bright.

4.4. JPEG

The acronym JPEG stands for "Joint Photographic Experts Group." Its development was a joint effort of several international standards organizations, a project that started in mid 1987 and took four years to complete. The JPEG standard is intended for the compression of grayscale and color continuous-tone still images (not bi-level), but because of its popularity it is used for the compression of other types of images (not just continuous-tone), and especially in web pages.

JPEG is mostly a lossy compression method; it has a lossless mode but that mode is inefficient and has never been popular. An important feature of JPEG is the use of many parameters. They make it possible to adjust the amount of the data lost (and thus the compression performance) over a very wide range. JPEG also supports progressive and hierarchical modes (not discussed here), and these are also controlled by parameters. The following sentence gives an idea of the power of JPEG lossy compression. When an image is lossily compressed by JPEG with a typical compression factor of 10–20 and then decompressed, the eye normally cannot detect any degradation in either the colors or the overall image quality.

The main compression steps of JPEG are listed here. Following this, the important steps are described in some detail.

1. If the image is in color, it is (optionally) transformed from RGB into a luminance/chrominance color space such as YCbCr (this step is skipped for grayscale images). It has already been mentioned that the eye is sensitive to small variations in luminance but not in chrominance. As a result, the chrominance component can "afford" to lose much data, and thus be highly compressed, without significantly impairing the overall image quality. Even though this step is optional, it is important, since the rest of JPEG processes each color component separately. Without this color transformation, none of the three color components will tolerate much loss, resulting in less compression.

2. If the image is in color and if the user selects the hierarchical compression mode, the two chrominance components are downsampled. One low-resolution pixel is created in this step from either two or four original image pixels. This reduces the image to 2/3 or 1/2 its original size, respectively. The luminance component is not downsampled.

3. The pixels in each of the three color components are partitioned into blocks (called *data units*) of 8×8 pixels each. If the number of image rows or columns is not a multiple of 8, the bottom row and the rightmost column are duplicated as many times as necessary. The encoder processes the image unit by unit in one of two modes. In the non-interleaved mode, the encoder processes all the data units of the first image component, then the data units of the second component, and finally those of the third component. In the interleaved mode the encoder scans the image in raster order and processes the three data units of the three components in each step in the scan.

4. The processing of a data unit starts by applying the *discrete cosine transform* (DCT, Section 4.5) to it. The result is an 8×8 data unit of transform coefficients, of which the one at the top-left corner (the DC component of the transform) is normally large and the remaining 63 (the AC coefficients) are small. This step is the "heart" of JPEG (and of many other image compression methods), but more steps follow.

5. Each of the 64 transform coefficients in a data unit is divided by a number called its *quantization coefficient* (QC), and the result is then rounded to an integer. This is the lossy step of JPEG, where information is irretrievably lost. Each of the 64 QCs is a JPEG parameter that can be input and revised by the user. In practice, most JPEG implementations use the QC tables provided by the JPEG standard for the luminance and chrominance color components.

6. The 64 quantized frequency coefficients (now integers) of each data unit are encoded using a combination of RLE and Huffman coding. This step is described in some detail below. The JPEG standard also has an option (rarely used in practice) where an arithmetic coding variant known as the *QM coder* can be used instead of Huffman coding.

7. The last step adds headers and all the JPEG parameters used, and outputs the result to the compressed file. This file may be in one of three formats (1) the *interchange* format, in which the file contains the compressed image and all the tables needed by the decoder (mostly quantization tables and tables of Huffman codes), (2) the *abbreviated* format for compressed image data, where the file contains the compressed image and may contain no tables (or just a few tables), and (3) the *abbreviated* format for table-specification data, where the file contains just tables, and no compressed image. The second format is used in cases where the user knows that the encoder and decoder have the same tables built-in. The third format is used in cases where many images have been compressed by the same encoder, using the same tables. When those images need to be decompressed, they are sent to a decoder preceded by one file with table-specification data.

The JPEG decoder performs the reverse steps. It decodes the 64 quantized frequency coefficients of a data unit, dequantizes them, and performs the inverse DCT. The result is different from the original data unit, because of the quantization step, but the eye cannot normally detect the difference.

4.5. The DCT

The discrete cosine transform (DCT) is used by many compression algorithms for still images (and also movies), because it is simple and

DATA COMPRESSION

efficient. The DCT is described here both as a rotation and as a linear transform where each basis vector extracts higher-frequency features from the transformed data.

We start with a description of the one-dimensional DCT. Consider the eight cosine waves $w(f) = \cos(f\theta)$, for $0 \leq \theta \leq \pi$, with frequencies $f = 0, 1, \ldots, 7$. These are waves with increasing frequencies. The first one, $w(0) = \cos(0) = 1$ is a horizontal line, so its frequency is zero. The second one, $w(1) = \cos(\theta)$, is a cosine function, so its frequency is 2π. The third one, $w(2) = \cos(2\theta)$, has frequency 4π. Each of the eight waves $w(f)$ is sampled at the eight points

$$\theta = \frac{\pi}{16}, \frac{3\pi}{16}, \frac{5\pi}{16}, \frac{7\pi}{16}, \frac{9\pi}{16}, \frac{11\pi}{16}, \frac{13\pi}{16}, \frac{15\pi}{16},$$

and the eight values are used to construct one basis vector $\mathbf{w}_f$. The resulting eight vectors $\mathbf{w}_f$, $f = 0, 1, \ldots, 7$ are shown in Table 4.5. They constitute the transformation matrix $\mathbf{W}$ of the one-dimensional DCT.

Table 8.3. The One-Dimensional DCT.

θ	0.196	0.589	0.982	1.374	1.767	2.160	2.553	2.945
$\cos 0\theta$	1.	1.	1.	1.	1.	1.	1.	1.
$\cos 1\theta$	0.981	0.831	0.556	0.195	-0.195	-0.556	-0.831	-0.981
$\cos 2\theta$	0.924	0.383	-0.383	-0.924	-0.924	-0.383	0.383	0.924
$\cos 3\theta$	0.831	-0.195	-0.981	-0.556	0.556	0.981	0.195	-0.831
$\cos 4\theta$	0.707	-0.707	-0.707	0.707	0.707	-0.707	-0.707	0.707
$\cos 5\theta$	0.556	-0.981	0.195	0.831	-0.831	-0.195	0.981	-0.556
$\cos 6\theta$	0.383	-0.924	0.924	-0.383	-0.383	0.924	-0.924	0.383
$\cos 7\theta$	0.195	-0.556	0.831	-0.981	0.981	-0.831	0.556	-0.195

It can be shown that the $\mathbf{w}_i$ vectors are orthonormal because of the particular choice of the eight sample points. Since this transformation matrix is orthonormal, it is a rotation matrix. Thus, we can interpret the one-dimensional DCT as a rotation in eight dimensions.

The one-dimensional DCT has another interpretation. We can consider the eight orthonormal vectors $\mathbf{w}_i$ tools to extract features of various frequencies from the data being transformed. The first vector is constant and positive. Each of the remaining seven vectors has four positive and four negative elements, so it transforms the data vector into a small number. Also, successive vectors have increasing frequencies (measured as the number of sign changes along a vector)

As an example, we transform the eight correlated data values

$$\mathbf{P} = (0.6, 0.5, 0.4, 0.5, 0.6, 0.5, 0.4, 0.55).$$

The result of the transform $\mathbf{T} = \mathbf{W} \cdot \mathbf{P}$ is the coefficient vector

$$\mathbf{T} = (0.506, 0.0143, 0.0115, 0.0439, 0.0795, -0.0432, 0.00478, -0.0077).$$

The first coefficient is not much different from the elements of $\mathbf{P}$, but the other seven are much smaller. We can obtain compression by simply writing the eight values on the compressed file, where they will occupy less space than the eight components of $\mathbf{P}$. However, much better compression can be obtained if the eight coefficients of $\mathbf{T}$ are quantized before they are output. The power of the DCT is demonstrated by the fact that the inverse DCT can reconstruct the eight values of $\mathbf{P}$ to a high degree of accuracy even if the eight transform coefficients are coarsely quantized.

In practice, the one-dimensional DCT is calculated by

$$G_f = \frac{1}{2} C_f \sum_{t=0}^{7} p_t \cos\left(\frac{(2t+1)f\pi}{16}\right), \tag{8.5}$$

where

$$C_f = \begin{cases} \frac{1}{\sqrt{2}}, & f = 0, \\ 1, & f > 0, \end{cases} \quad \text{for } f = 0, 1, \cdots, 7.$$

This starts with a set of eight data values p_t (pixels, sound samples, or other data) and produces a set of eight DCT coefficients G_f. The calculation is straightforward and there are ways to speed it up. The *inverse* DCT (IDCT) is calculated by

$$p_t = \frac{1}{2} \sum_{j=0}^{7} C_j G_j \cos\left(\frac{(2t+1)j\pi}{16}\right), \quad \text{for } t = 0, 1, \cdots, 7. \tag{8.6}$$

The following experiment illustrates the power of the DCT. We start with the set of eight data items $\mathbf{P} = (12, 10, 8, 10, 12, 10, 8, 11)$, apply the one-dimensional DCT to it, and end up with the eight coefficients

$$28.6375, 0.571202, 0.46194, 1.757, 3.18198, -1.72956, 0.191342, -0.308709.$$

These can be used to precisely reconstruct the original data (except for small roundoff errors caused by limited machine precision). Our goal, however, is to compress the data even more by quantizing the coefficients. We first quantize them to $28.6, 0.6, 0.5, 1.8, 3.2, -1.8, 0.2, -0.3$, and apply the IDCT to get back

$$12.0254, 10.0233, 7.96054, 9.93097, 12.0164, 9.99321, 7.94354, 10.9989.$$

DATA COMPRESSION

We then quantize the coefficients even more, to 28, 1, 1, 2, 3, −2, 0, 0, and apply the IDCT to get back

12.1883, 10.2315, 7.74931, 9.20863, 11.7876, 9.54549, 7.82865, 10.6557.

Finally we quantize the coefficients to 28, 0, 0, 2, 3, −2, 0, 0, and still get back from the IDCT the sequence

11.236, 9.62443, 7.66286, 9.57302, 12.3471, 10.0146, 8.05304, 10.6842,

where the largest difference between an original value (12) and a reconstructed one (11.236) is 0.764 (or 6.4% of 12).

The principle of image compression tells us that the pixels of an image are correlated in two dimensions, not just in one dimension (a pixel is correlated with its four immediate neighbors, not with just two). This is why image compression methods use the two-dimensional DCT. For an $n \times n$ data unit, this transform is expressed

$$G_{ij} = \frac{1}{\sqrt{2n}} C_i C_j \sum_{x=0}^{n-1} \sum_{y=0}^{n-1} p_{xy} \cos\left(\frac{(2y+1)j\pi}{2n}\right) \cos\left(\frac{(2x+1)i\pi}{2n}\right), \quad (8.7)$$

for $0 \le i, j \le n-1$. JPEG uses $n = 8$. The decoder reconstructs a block of 8×8 (approximate or precise) data values by computing the inverse DCT (IDCT), whose expression is

$$p_{xy} = \frac{1}{4} \sum_{i=0}^{7} \sum_{j=0}^{7} C_i C_j G_{ij} \cos\left(\frac{(2x+1)i\pi}{16}\right) \cos\left(\frac{(2y+1)j\pi}{16}\right), \quad (8.8)$$

where

$$C_f = \begin{cases} \frac{1}{\sqrt{2}}, & f = 0, \\ 1, & f > 0. \end{cases}$$

The two-dimensional DCT can be interpreted as two separate rotations, each in eight dimensions. The first rotation considers each row of the data unit a point in 8-dimensional space, and it rotates the point. The result is a row where the first element is dominant and the remaining seven elements are small. The second rotation considers each of the eight columns a point, and rotates them. The result is one large element at the top-left corner, with 63 smaller elements elsewhere.

4.6. Encoding a data unit

After an 8×8 data unit of DCT coefficients G_{ij} is computed, it is quantized. This is the step where data are lost. Each number in the

DCT coefficients matrix is divided by the corresponding quantization coefficient (QC) from the particular quantization table used, and the result is rounded to the nearest integer. Each color component has its own quantization table and, in principle, all 64×3 quantization coefficients can be specified and fine-tuned by the user for maximum compression. In practice, few users have the time or expertise to experiment with so many parameters, so JPEG software normally uses the default quantization tables. Two recommended tables, for the luminance and the chrominance components, are part of the JPEG standard. They are the result of many experiments performed by the JPEG committee.

If the quantization is done right, very few nonzero numbers will be left in the DCT coefficients data unit after quantization, and they will typically be concentrated at the upper left corner. These numbers are the output of JPEG but they are further compressed before being written on the output file. The JPEG standard specifies three techniques to compress the 8×8 data unit of integers:

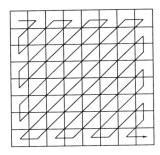

Figure 8.5. A Zigzag Path.

1. The 64 numbers are collected by scanning the data unit in zigzag order (Figure 8.5). This results in a string of 64 numbers that starts with some nonzeros, may have several runs of zeros, and typically ends with a long run of zeros. This last run of zeros is compressed by replacing it with a special end-of block (EOB) code. Anything preceding this last run is compressed further before being written on the compressed file.

2. The nonzero numbers are compressed using Huffman coding. For each nonzero number Y, the JPEG encoder finds the number Z of consecutive zeros preceding Y, and uses the pair (Y,Z) to locate a Huffman code in a special table. That code is written on the compressed file. (The actual coding is slightly more complex.)

3. The DC coefficient is treated differently from the AC coefficients. The DC coefficient is a multiple of the average value of the 64 original

pixels constituting the data unit. Analysis of many images indicates that in a continuous-tone image, adjacent data units are normally correlated in the sense that the average values of the pixels in such units are close. This implies that the DC coefficients of adjacent data units don't differ much. JPEG outputs the first one (encoded), followed by *differences* (also encoded) of the DC coefficients of consecutive data units.

For related material on the subject of this section, see Ahmed et al. (1974), Pennebaker and Mitchell (1992), Rao and Yip (1990), Wallace (1991), and Zhang (1990).

5.   Wavelet Methods

The wavelet transform is, in a general sense, an extension of the Fourier transform. Fourier analysis is useful in many areas of science and engineering, and wavelets are finding more and more applications. Wavelet methods are used in digital signal processing, computer graphics, data compression, and various other fields. This short section introduces the Haar transform, the simplest wavelet. We first show how the Haar transform is used to compress grayscale images, then show how it can be applied to color images.

We start with a single row of n pixels. For simplicity we assume that n is a power of 2. (If n has a different value, the data can be extended by appending zeros. After decompression, the extra zeros are removed.) Recall that any image transform uses a reversible process to convert the original n values into one large value, followed by $n-1$ small values. The Haar transform achieves this by repeatedly computing averages and differences. The process is illustrated here for $n = 8$.

Consider the array of eight values $(1, 2, 3, 4, 5, 6, 7, 8)$. We first compute the four averages $(1+2)/2 = 3/2$, $(3+4)/2 = 7/2$, $(5+6)/2 = 11/2$, and $(7+8)/2 = 15/2$. It is impossible to reconstruct the original eight values from these four averages, so we also compute the four differences $(1-2)/2 = -1/2$, $(3-4)/2 = -1/2$, $(5-6)/2 = -1/2$, and $(7-8)/2 = -1/2$. These differences are called *detail coefficients*. We can think of the averages as a coarse resolution representation of the original image, and of the details as the data needed to reconstruct the original image from this coarse resolution. If the pixels of the image are correlated, the coarse representation will resemble the original pixels while the details will be small. This explains why the Haar transform is useful for compression.

It is easy to see that the array

$$(3/2, 7/2, 11/2, 15/2, -1/2, -1/2, -1/2, -1/2)$$

of the four averages and four differences contains enough information to reconstruct the original eight values. Since the last four components of this array tend to be small, we repeat the process on the four averages, the large components of our array. They are transformed into two averages and two differences, yielding the array

$$(10/4, 26/4, -4/4, -4/4, -1/2, -1/2, -1/2, -1/2).$$

The next, and last, iteration of this process transforms the first two components of the new array into one average (the average of all eight components of the original array) and one difference

$$(36/8, -16/8, -4/4, -4/4, -1/2, -1/2, -1/2, -1/2).$$

The last array is the *Haar wavelet transform* of the original data items.

In the case of eight data values, this process iterates three times. In the general case of n values, the number of iterations is $\log_2 n$. It is also easy to show that the total number of operations is $2(n-1)$. For comparison, the fast Fourier transform requires $n \log n$ operations.

Because of the differences, the Haar transform tends to generate numbers smaller than the original pixel values, so the result is easy to compress with run length encoding, perhaps in combination with Huffman coding. Lossy compression can be obtained if some of the smaller differences are quantized or even converted to zero.

It is useful to associate with each iteration a quantity called *resolution*, that is defined as the number of remaining averages at the end of the iteration. The resolutions after each of the three iterations above are $4(= 2^2)$, $2(= 2^1)$, and $1(= 2^0)$. Each component of the wavelet transform should be normalized by dividing it by the square root of the resolution. Thus, our example Haar transform becomes

$$\left(\frac{36/8}{\sqrt{2^0}}, \frac{-16/8}{\sqrt{2^0}}, \frac{-4/4}{\sqrt{2^1}}, \frac{-4/4}{\sqrt{2^1}}, \frac{-1/2}{\sqrt{2^2}}, \frac{-1/2}{\sqrt{2^2}}, \frac{-1/2}{\sqrt{2^2}}, \frac{-1/2}{\sqrt{2^2}}\right).$$

If this version of the Haar transform is used, then ignoring the smallest differences is the best choice for lossy wavelet compression, since it causes the smallest loss of image information.

The Haar transform is easy to generalize from a single row to an entire image. This can be done in several ways, two of which, the *Standard decomposition* and the *Pyramid decomposition*, are briefly mentioned here.

Standard decomposition starts by computing the Haar transform of every row of the image. This results in a transformed image where the first column contains averages and all the other columns contain

DATA COMPRESSION

differences (Figure 8.6a). The same transform is then applied to all the columns. This results in one average value at the top-left corner, with the rest of the top row containing averages of differences, and with all other pixel values transformed into differences (Figure 8.6b).

Pyramid decomposition computes the wavelet transform of the image by alternating between rows and columns. The first step is to calculate averages and differences for all the rows (just one iteration, not the entire Haar transform). This creates averages in the left half of the image and differences in the right half (Figure 8.6c). The second step is to calculate averages and differences for all the columns, which results in averages in the top-left quadrant of the image and differences elsewhere (Figure 8.6d). Steps 3 and 4 operate on the rows and columns of that quadrant, resulting in averages concentrated in the top-left subquadrant. Pairs of steps are repeatedly executed on smaller and smaller subsquares, until only one average is left, at the top-left corner of the image, and all other pixel values have been reduced to differences (Figure 8.6e,f).

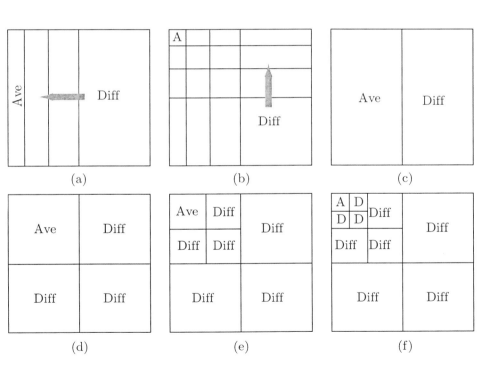

Figure 8.6. (a–b) Standard Decomposition. (c–f) Pyramid Decomposition.

The transforms described in Section 4.2 are orthogonal. They transform the original pixels into a few large numbers and many small numbers. In contrast, wavelet transforms, such as the Haar transform, are *subband transforms*. They partition the image into regions such that one region contains large numbers (averages in the case of the Haar transform) and the other regions contain small numbers (differences). However, these regions, which are called subbands, are more than just sets of large and small numbers. They reflect different geometrical features of the image. To illustrate what this means, we examine a small, mostly-uniform image with one vertical line and one horizontal line. Figure 8.7a shows an 8×8 image with pixel values of 12, except for a vertical line with pixel values of 14 and a horizontal line with pixel values of 16.

```
12 12 12 12 14 12 12 12       12 12 13 12 0 0 2 0       12 12 13 12 0 0 2 0
12 12 12 12 14 12 12 12       12 12 13 12 0 0 2 0       12 12 13 12 0 0 2 0
12 12 12 12 14 12 12 12       12 12 13 12 0 0 2 0       14 14 14 14 0 0 0 0
12 12 12 12 14 12 12 12       12 12 13 12 0 0 2 0       12 12 13 12 0 0 2 0
12 12 12 12 14 12 12 12       12 12 13 12 0 0 2 0       0  0  0  0  0 0 0 0
16 16 16 16 14 16 16 16       16 16 15 16 0 0 2̲ 0       0  0  0  0  0 0 0 0
12 12 12 12 14 12 12 12       12 12 13 12 0 0 2 0       4̲ 4̲ 2̲ 4̲  0 0 4 0
12 12 12 12 14 12 12 12       12 12 13 12 0 0 2 0       0  0  0  0  0 0 0 0
           (a)                           (b)                       (c)
```

Figure 8.7. An 8×8 Image and its Subband Decomposition.

Figure 8.7b shows the results of applying the Haar transform once to the columns of the image. The right half of this figure (the differences) is mostly zeros, reflecting the uniform nature of the image. However, traces of the vertical line can easily be seen (the notation 2̲ indicates a negative difference). Figure 8.7c shows the results of applying the Haar transform once to the rows of Figure 8.7b. The upper-right subband now contains traces of the vertical line, whereas the lower-left subband shows traces of the horizontal line. These subbands are denoted by HL and LH, respectively (Figure 8.6f). The lower-right subband, denoted by HH, reflects diagonal image features (which our example image lacks). Most interesting is the upper-left subband, denoted by LL, that consists entirely of averages. This subband is a one-quarter version of the entire image, containing traces of both the vertical and the horizontal lines.

Figure 8.8 shows four levels of subbands, where level 1 contains the detailed features of the image (also referred to as the high-frequency or fine-resolution wavelet coefficients) and the top level, level 4, contains the

DATA COMPRESSION

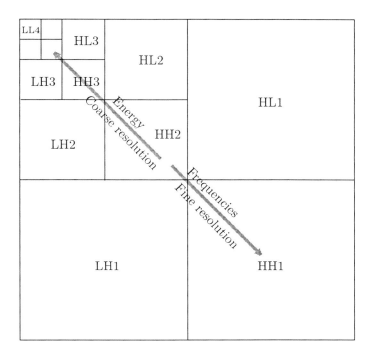

Figure 8.8. Subbands and Levels in Wavelet Decomposition.

coarse image features (low-frequency or coarse-resolution coefficients). It is clear that the lower levels can be quantized coarsely without much loss of important image information, while the higher levels should be quantized finely. The subband structure is the basis of all the image compression methods that use the wavelet transform.

The Haar transform, like any compression method for grayscale images, can be extended to color (three-component) images by separating the three components, then transforming and compressing each individually. If the compression is lossy, it makes sense to convert the three image components from their original color representation (normally RGB), to the YIQ color representation. The Y component of this representation is called *luminance*, and the I and Q (the chrominance) components are responsible for the color information. The advantage of this color representation is that the human eye is most sensitive to Y and least sensitive to Q. A lossy method should therefore leave the Y component alone, delete some data from the I, and delete more data from the Q components, resulting in good compression and in a loss to which the eye is not that sensitive.

For related material on the subject of this section, see Stollnitz et al. (1996).

6. Video Compression

Image compression is important because image files tend to be large. A video file consists of a large number of still images (called *frames*) plus an audio part, which is why efficient, lossy video compression is as important as image compression. Today, with so many popular multimedia applications, video files are common. It is easy to shoot a movie with a video camera, save it on a disk, edit it with the help of video editing software, send the result to other computers, and store it permanently, in compressed format, in an archive such as a CD-ROM or DVD. This is why video compression, and especially an international standard for it, is so important.

Image compression is based on the spatial redundancy of an image, caused by the correlation between pixels. Video compression is based on the spatial redundancy in each frame, as well as on *temporal redundancy* between frames. The latter redundancy stems from the fact that a video frame tends to be similar to its immediate neighbors. A typical technique for video compression should therefore start by encoding the first frame using a compression method for still images. It should then encode each successive frame by locating and encoding the differences between the frame and its predecessor. If a frame is very different from its predecessor (as, for example, the first frame of a shot), it should be coded independently of any other frames. In video compression terminology, a frame that's coded using its predecessor is called *inter frame* (or just *inter*), while a frame that's coded independently is called *intra frame* (or just *intra*).

Video compression is normally lossy. Encoding a frame F_i in terms of its predecessor F_{i-1} introduces some distortions. As a result, encoding frame F_{i+1} in terms of F_i increases the distortion. A frame may lose some bits even if the compression method is lossless. This may happen during transmission or after a long shelf stay. If a frame F_i has lost some bits, then all the frames following it, up to the next intra frame, will be decoded improperly, possibly even with accumulated errors. This is why intra frames should be used from time to time, not just at the start of a shot. In video compression terminology, an intra frame is labeled I and an inter frame is labeled P (for *predictive*).

In practice, an inter frame can be encoded based on one of its predecessors and also on one of its *successors*. We know that an encoder should not use any information that's not available to the decoder, but video compression is special because of the large quantities of data involved. We usually don't mind if the encoder is slow, but the decoder has to be fast. A typical example is video recorded on a hard disk or on a DVD, to be played back. The encoder can take minutes or hours to encode the data. The decoder, however, has to play it back at the correct frame rate (so many frames per second) so it has to be fast. This is why a typical video decoder works in parallel. It has several decoding circuits working simultaneously on several frames.

With this in mind we can imagine a situation where the encoder encodes frame 2 based on both frames 1 and 3, and writes the frames on the compressed file in the order 1, 3, 2. The decoder reads them in this order, decodes frames 1 and 3 in parallel, outputs frame 1, then decodes frame 2 based on frames 1 and 3. The frames should, of course, be clearly tagged (or time stamped). A frame that's encoded based on both past and future frames is labeled B (for *bidirectional*).

Encoding a frame based on its successor makes sense in cases where the movement of an object in the movie gradually uncovers a background area. Such an area may be only partly known in the current frame but may be better known in the next frame. Thus, the next frame is a natural candidate for predicting this area in the current frame.

The concept of a B frame is so useful, that most frames in a compressed video presentation may be of this type. We therefore end up with a sequence of compressed frames of the three types I, P, and B. An I frame is decoded independently of any other frame. A P frame is decoded using the preceding I or P frame. A B frame is decoded using the preceding *and* following I or P frames.

We start with a few intuitive video compression methods.

Subsampling: The encoder selects every other frame and writes it on the compressed stream. This yields a compression factor of 2. The decoder inputs a frame and duplicates it to create two frames. This, of course, may result in the loss of too much data (for example, the first frame of a shot may be lost), while producing mediocre compression.

Differencing: A frame is compared to its predecessor. If the difference between them is small (just a few pixels), the encoder encodes the pixels that are different by writing two items on the compressed file for each pixel: Its image coordinates and the difference between the values of the pixel in the two frames. If the difference between the frames is large, the current frame is written on the output in raw format.

A lossy version of differencing looks at the amount of change in a pixel. If the difference between the intensities of a pixel in the preceding frame and in the current frame is smaller than a certain threshold, the pixel is considered unchanged.

Motion Compensation: Anyone who has watched movies knows that the difference between consecutive frames is small because it is the result of moving the scene, the camera, or both between frames. This feature can therefore be exploited to get better compression. If the encoder discovers that a part M of the preceding frame has been rigidly moved to a different location in the current frame, then M can be compressed by writing the following three items on the compressed file: Its previous location, its current location, and information identifying the boundaries of M.

For related material on the subject of this section, see Manning (1998).

6.1. MPEG

The MPEG standard for video compression is the result of a 10-year effort by an international group of experts under the auspices of the ISO (International Standardization Organization) and the IEC (International Electrotechnical Committee). The name MPEG is an acronym for Moving Pictures Experts Group. The MPEG standard specifies methods for the compression of digital images and sound, and for the synchronization of the two. Currently there are several MPEG standards. MPEG-1 is intended for intermediate data rates, on the order of 1.5 Mbit/s. MPEG-2 is intended for high data rates of at least 10 Mbit/s. MPEG-3 was intended for HDTV compression but was found to be redundant and was merged with MPEG-2. MPEG-4 is intended for very low data rates of less than 64 Kbit/s. A third international body, the ITU-T, has been involved in the design of both MPEG-2 and MPEG-4. This section describes the main features of image compression (not audio compression) used by MPEG-1.

Before we look at the details of MPEG, it helps to understand the meaning of the words "intermediate data rate." We consider a typical example of video with a resolution of 360×288, a depth of 24 bits per pixel, and a refresh rate of 24 frames per second. The image part of this video requires $360 \times 288 \times 24 \times 24 = 59{,}719{,}680$ bits/s. To get an estimate of the size of the audio part, we assume two sound tracks (stereo sound), each sampled at 44KHz with 16-bit samples. The data rate is $2 \times 44{,}000 \times 16 = 1{,}408{,}000$ bits/s. The total is about 61.1 Mbit/s, and this is supposed to be compressed by MPEG-1 to an intermediate data rate of about 1.5 Mbit/s (the size of the sound track alone), a compression factor of more than 40! Another aspect is the decoding speed. An MPEG-compressed movie may end up being stored on a CD-ROM or a DVD and has to be decoded and played in real time.

MPEG uses its own vocabulary. An entire movie is considered a *video sequence*. It consists of *pictures*, each having three *components*, one luminance (Y) and two chrominance (Cb and Cr). The luminance component contains the black-and-white picture and the chrominance components provide the color hue and saturation. Each component is a rectangular array of *samples* and each row of the array is called a *raster line*. A *pel* is the set of three samples. The eye is sensitive to small spatial variations of luminance, but is less sensitive to similar changes in chrominance. As a result, MPEG-1 samples the chrominance components at half the resolution of the luminance component. The term *intra* is used, but the terms *inter* and *nonintra* are used interchangeably.

6.2. MPEG-1 main components

MPEG uses the concepts of I, P, and B pictures, discussed earlier. The pictures are arranged in groups, where a group can be open or closed. In a closed group, P and B pictures are decoded only from other pictures in the group. In an open group, they can be decoded from pictures outside the group. Different regions of a B picture may use different pictures for their decoding. A region may be decoded from some preceding pictures, from some following pictures, from both types, or from none. Similarly, a region in a P picture may use several preceding pictures for its decoding, or use none at all, in which case it is decoded using MPEG's intra methods.

The basic building block of an MPEG picture is the *macroblock*. It consists of a 16×16 block of luminance (grayscale) samples (divided into four 8×8 blocks) and two 8×8 blocks of the matching chrominance samples; a total of six blocks. The MPEG compression of a macroblock consists mainly of passing each of the six blocks through a discrete cosine transform, which creates decorrelated values, then quantizing and encoding the results. It is very similar to JPEG compression, the main differences being that different quantization tables and different code tables are used in MPEG for intra and nonintra, and the rounding is done differently.

When a picture is encoded in nonintra mode (i.e., it is encoded by means of another picture, normally its predecessor), the MPEG encoder generates the differences between the pictures, then applies the DCT to the differences. In such a case, the DCT does not contribute much to the compression because the differences are already decorrelated. Nevertheless, the DCT is useful even in this case, since it is followed by quantization, and the quantization in nonintra coding can be quite coarse.

MPEG performs the important lossy step of quantization by dividing each of the DCT coefficients (an integer) by a quantization coefficient taken from a quantization table. The result is, in general, a noninteger and has to be rounded. MPEG specifies default quantization tables, but custom tables can also be used.

The integers resulting from the quantization process are further encoded, using the nonadaptive Huffman method and Huffman code tables that were calculated by gathering statistics from many training image sequences. The particular code table being used depends on the type of the picture being encoded.

For related material on the subject of this section, see Mitchell et al. (1997) and MPEG (1998).

7. Audio Compression

Image compression is important because images require a lot of storage space. An average-size book, consisting of about a million characters, requires a million bytes for storage. However, adding one image to such a book can easily double its size. Audio files can be as big as image files. Even more important, multimedia applications combine images and audio, thereby resulting in huge video files, making efficient compression a must. This section starts with a short introduction to sound and digitized audio. It then describes some properties of the human auditory system (ear and brain) that make it possible to lossily compress audio files without losing audio quality. Next, two simple audio compression methods, μ-law and ADPCM, are described.

7.1. Digital audio

People who are used to images painted in the traditional way on paper or canvas, may have a hard time getting used to digital images and their pixels. Similarly, people tend to think of sound as a continuous entity, and may have trouble getting used to *digitized sound*. However, sound can be digitized and broken up into individual numbers, called *samples*, without loss of information. Here is how it's done.

When sound is played into a microphone, it is converted into a voltage that varies continuously with time. Such voltage is the *analog* representation of the sound. Digitizing sound is done by measuring the voltage at many points in time, converting each measurement into a number, and writing the numbers on a file. This process is called *sampling*. The sound wave is sampled, and the samples become the digitized sound. The device used for sampling is called an analog-to-digital converter (ADC).

DATA COMPRESSION

Since the sound samples are numbers, they are easy to edit. However, the main use of a sound file is to play it back. This is done by converting the numeric samples back into voltages that are fed continuously into a speaker. The device that does that is called a digital-to-analog converter (DAC). It is intuitively clear that a high sampling rate results in better sound reproduction, but also in many more samples and therefore in bigger files. Thus, the main problem in sound sampling is how often to sample a given sound.

The solution is to identify the highest frequency contained in the sound, and to sample the sound at a little over twice this frequency. This sampling rate is called the Nyquist rate, and it guarantees true reproduction of the sound.

The frequency range of human hearing is normally from 20 Hz to 22,000 Hz, depending on the person and on age. (The range of human voice, by the way, is much more limited. It ranges from about 500 Hz to about 2 KHz.) When sound is digitized at high fidelity, it should therefore be sampled at a little over the Nyquist rate of 2×22,000 = 44,000 Hz. This is why high-quality digital sound is sampled at 44,100 Hz. A lower sampling rate results in distortions, while higher sampling rates do not improve the quality of the played-back sound.

The second problem in sound sampling is the sample size. Each sample becomes a number, but how large should this number be? It is clear that the bigger the sample size, the more tones can be distinguished. In most audio systems used in practice, sound samples are either 8 or 16 bits. With 8-bit samples, there can be 256 sound intensities. Doubling the sample size to 16 bits, doubles the size of the audio file but produces 64K sound levels (that's more than 65,000). With 8-bit samples, feeble sounds may be converted to zero and played back as silence. With 16-bit samples, such sounds may be converted to small but nonzero samples and thus played back at their correct level.

Sound sampled at 44,100 Hz with 8-bit samples results in 44,100 bytes per second. One minute of sound generates 2.646 million bytes. With 16-bit samples, the size doubles. Stereo sound again doubles the storage requirements. Thus, stereo sound with 16-bit samples generates 44,100× 2×2 = 176,400 bytes per second or 10.584 million bytes every minute!

Audio sampling is also called *pulse code modulation* (PCM). We have all heard of AM and FM radio. These terms stand for *amplitude modulation* and *frequency modulation*, respectively. They indicate methods to modulate (i.e., to include binary information in) continuous waves. The term *pulse modulation* refers to techniques for converting a continuous wave to a stream of binary numbers. PCM is one of several possible pulse modulation methods.

For related material on the subject of this section, see Pohlmann (1985).

7.2. The human auditory system

The ear responds to frequencies in the range 20–20,000 Hz, but its sensitivity to the amplitude of the sound in this range is not uniform. It depends on the frequency, and experiments indicate that in a quiet environment the ear's sensitivity is maximal for frequencies in the range 2 KHz to 4 KHz. For any given frequency, there is a *hearing threshold*, below which the ear does not respond to sound.

The existence of such a threshold suggests a natural approach to lossy audio compression. Simply delete all audio samples that are below the threshold. Since the threshold depends on the frequency, the encoder needs to know, at all times, the frequency spectrum of the sound being compressed. The encoder therefore has to save several of the previously-input audio samples at any time ($n - 1$ samples, where n is either a constant or a user-controlled parameter). When the current sample is input, the first step is to transform the n samples to the frequency domain by means of a Fourier transform. The result is m values (called *signals*) that indicate the strength of the sound at m different frequencies. If a signal for frequency f is smaller than the hearing threshold at f, the signal can be deleted.

In addition to the threshold, two more properties of the human auditory system are exploited in audio compression. They are *frequency masking* and *temporal masking*.

Frequency masking (also known as *auditory masking*) occurs when a soft sound x that we can normally hear (because it is loud enough) is masked by a louder sound y at a nearby frequency. A sophisticated audio compression algorithm should be able to identify such a case and delete x, since it is inaudible.

Temporal masking may occur when a loud sound x of frequency f is preceded or followed in time by a soft sound y at a nearby (or the same) frequency. If the time interval between the sounds is short, sound y may be inaudible and can be deleted.

7.3. μ-Law companding

This is an international standard that employs a simple, logarithm-based function to lossily compress digitized audio samples. The compression is done by quantizing each sample, but the quantization is nonlinear. μ-Law is the compression method used by ISDN (integrated services digital network) digital telephony services. The process starts when a voice signal is sent on the telephone line. The ISDN hardware samples the signal 8000 times per second, and generates 14-bit samples.

The μ-Law standard uses nonlinear quantization because experiments indicate that the low amplitudes of speech signals contain more information than the high amplitudes. Imagine an audio signal sent on a telephone line and digitized to 14-bit samples. The louder the speech, the higher the amplitude, and the bigger the value of the sample. Since high amplitudes are less important, they can be coarsely quantized. The largest 14-bit sample is $2^{14} - 1 = 16{,}384$. If it is quantized to 255 (the largest 8-bit number), then the compression factor is $14/8 = 1.75$. When the samples are decoded, a code of 255 is decoded into a value very different from the original 16,384. We say that large samples end up with high quantization noise because of the coarse quantization. Smaller samples should be finely quantized, so they end up with low noise.

The μ-law encoder inputs 14-bit samples and outputs 8-bit codewords. The telephone signals are sampled at 8 KHz (8000 times per second), so the μ-law encoder receives $8000 \times 14 = 112{,}000$ bits/s. At a compression factor of 1.75, the encoder outputs 64,000 bits/s. The μ-law algorithm is part of the more general G.711 standard, that also specifies output rates of 48 Kbps and 56 Kbps.

The μ-law encoder receives a 14-bit *signed* input sample x. This is in the range $[-8192, +8191]$. The sample is normalized to the interval $[-1, +1]$ and the encoder uses the logarithmic expression

$$\operatorname{sgn}(x)\frac{\ln(1+\mu|x|)}{\ln(1+\mu)}, \quad \text{where } \operatorname{sgn}(x) = \begin{cases} +1, & x > 0, \\ 0, & x = 0, \\ -1, & x < 0, \end{cases}$$

(and μ is a positive integer) to compute and output an 8-bit code in the same interval $[-1, +1]$. The output is then scaled to the range $[-256, +255]$. Figure 8.9 shows this output as a function of the input for the three μ values 25, 255, and 2555. It is clear that large values of μ cause coarser quantization for larger samples (those that correspond to high sound amplitudes). Such values allocate more bits to the smaller, more important samples. The G.711 standard recommends the use of $\mu = 255$. The diagram shows only the nonnegative values of the input (i.e., from 0 to 8191). The negative side of the diagram has the same shape but with negative inputs and outputs.

The following illustrates the nonlinear nature of the μ-law. The two (normalized) input samples 0.15 and 0.16 are transformed to outputs 0.6618 and 0.6732. The difference between the outputs is 0.0114. On the other hand, the two input samples 0.95 and 0.96 (bigger inputs but

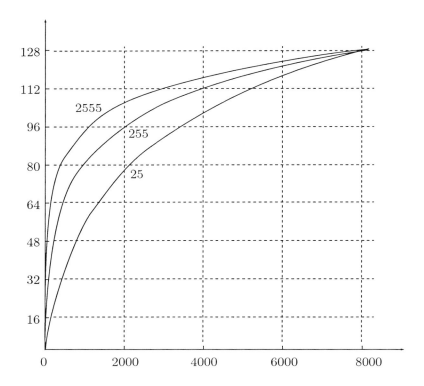

Figure 8.9. The μ-Law For μ Values of 25, 255, and 2555.

with the same difference) are transformed to 0.9908 and 0.9927. The difference between these two outputs is 0.0019; much smaller.

Logarithms are slow to calculate, so the μ-law encoder performs much simpler calculations that result in an approximation.

For related material on the subject of this section, see ITU-T (1989) and Shenoi (1995).

7.4. ADPCM audio compression

Sound can be compressed because audio samples, much like pixels in an image, are correlated. The simplest way to exploit this correlation is to subtract adjacent samples and encode the differences, which tend to be small numbers. Any audio compression method based on this principle is called DPCM (differential pulse code modulation). Such methods, however, are inefficient, since they do not adapt themselves to the varying magnitudes of the audio stream. Better results are achieved by an adaptive version, and any such version is called ADPCM.

DATA COMPRESSION

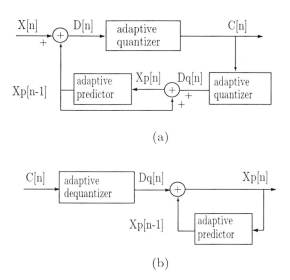

Figure 8.10. ADPCM Encoder (a) and Decoder (b).

ADPCM uses the previous sample (or several previous samples) to predict the current sample (this is similar to adaptive image compression). It then computes the difference between the current sample and its prediction, and quantizes the difference. For each input sample X[n], the output C[n] of the encoder is a certain number of quantization levels. The decoder multiplies this number by the quantization step (and may add half the quantization step, for better precision), to obtain the reconstructed audio sample. The method is efficient because the quantization step is modified all the time, by both encoder and decoder, in response to the varying magnitudes of the input samples. It is also possible to adaptively modify the prediction method.

Various ADPCM methods differ in the way they predict the current sound sample and in the way they adapt to the input (by changing the quantization step size and/or the prediction method).

Figure 8.10a,b shows the general organization of an ADPCM encoder and decoder. Notice that they share two functional units, a feature that helps in both software and hardware implementations. The adaptive quantizer receives the difference D[n] between the current input sample X[n] and the prediction Xp[n − 1]. The quantizer computes and outputs the quantized code C[n] of X[n]. The same code is sent to the adaptive dequantizer (the same dequantizer used by the decoder), which produces the next dequantized difference value Dq[n]. This value is added to the previous predictor output Xp[n − 1], and the sum Xp[n] is sent to the predictor to be used in the next step.

Better prediction would be obtained by feeding the actual input X[n] to the predictor. However, the decoder wouldn't be able to mimic that, since it does not have X[n]. We see that the basic ADPCM encoder is simple, and the decoder is even simpler. It inputs a code C[n], dequantizes it to a difference Dq[n], which is added to the preceding predictor output Xp[n − 1] to form the next output Xp[n]. The next output is also fed into the predictor, to be used in the next step.

8. Epilogue

The main aim of this chapter has been to show that data compression is not done by squeezing bits out of the data, or by applying pressure on it. Data can be compressed only because its original representation is inefficient and contains redundancies. Whatever method is used to compress a given data file, it does it by reducing or completely removing the redundancy from the data. The original representation of data is generally inefficient because this makes it easy to operate on the data. Thus, it is easy to deal with text if characters have fixed-size codes, but variable-size codes are shorter on average. Image processing becomes easy if each pixel is represented by the same number of bits, but a representation based on correlations between pixels is shorter.

The secondary aim of the chapter is to illustrate some of the important methods for compressing different types of digital data such as text, images, video, and sound. Today, after more than 30 years of research in this field, many more compression methods are known, both lossy (for image, video, and audio data), and lossless (mostly for text but also for still images). Certain methods are developed with speed in mind, while others are intended for maximum compression, regardless of the execution time. Most compression algorithms are in the public domain and anyone is free to implement and use them, while some methods have been patented and their use requires a license.

Claude Shannon based his definition of information on the amount of surprise contained in a piece of data. Surprise is also what we feel, when looking at the history of data compression. In addition to the "traditional" statistical and dictionary-based compression methods, many interesting, original, and efficient methods have been developed, based on concepts and ideas that do not seem directly related to compression. Examples of such ideas are finite-state automata, fractals, and wavelets, to mention just a few. Based on past experience, we can confidently predict that this trend will continue, and that in the future we will continue to be surprised by unexpected approaches to, and methods for, data compression

Bibliography

Ahmed, N., Natarajan, T., and Rao, R. K. (1974). Discrete cosine transform. *IEEE Transactions on Computers*, C-23:90–93.

Anderson, K. L. and et al. (1987). Binary-image-manipulation algorithm in the image view facility. *IBM Journal of Research and Development*, 31(1):16–31.

Elias, P. (1975). Universal Codeword Sets and Representations of the Integers. *IEEE Transactions on Information Theory*, IT-21(2):194–203.

Gonzalez, R. C. and Woods, R. E. (1992). *Digital Image Processing*. Reading, MA, Addison-Wesley.

Howard, P. G. and Vitter, J. S. (1992a). Error modeling for hierarchical lossless image compression. In Storer, J., editor, *Proceedings of the 1992 Data Compression Conference*, pages 269–278. Los Alamitos, CA, IEEE Computer Society Press.

Howard, P. G. and Vitter, J. S. (1992b). New methods for lossless image compression using arithmetic coding. *Information Processing and Management*, 28(6):765–779.

Huffman, D. (1952). A Method for the Construction of Minimum Redundancy Codes. *Proceedings of the IRE*, 40(9):1098–1101.

Hunter, R. and Robinson, A. H. (1980). International digital facsimile coding standards. *Proceedings of the IEEE*, 68(7):854–867.

ITU-T (1989). *CCITT Recommendation G.711: Pulse Code Modulation (PCM) of Voice Frequencies*.

Knuth, D. E. (1985). Dynamic huffman coding. *Journal of Algorithms*, 6:163–180.

Linde, Y., Buzo, A., and Gray, R. M. (1980). An algorithm for vector quantization design. *IEEE Transactions on Communications*, COM-28:84–95.

Manning (1998). URL http://lemontree.web2010.com/dvideo/.

Marking, M. P. (1990). Decoding group 3 images. *The C Users Journal*, pages 45–54, June.

McConnell, K. R. (1992). *FAX: Digital Facsimile Technology and Applications*. Norwood, MA, Artech House.

Miller, V. S. and Wegman, M. N. (1985). Variations on a theme by ziv and lempel. In Apostolico, A. and Galil, Z., editors, NATO ASI series, F12, *Combinatorial Algorithms on Words*, pages 131–140. Springer, Berlin.

Mitchell, J. L., Pennebaker, W. B., Fogg, C. E., and LeGall, D. J., editors (1997). *MPEG Video Compression Standard*. New York, Chapman and Hall and International Thomson Publishing.

Moffat, A., Neal, R. M., and Witten, I. H. (1998). Arithmetic coding revisited. *ACM Transactions on Information Systems*, 16(3):256–294.

MPEG (1998). URL http://www.mpeg.org/.

Pennebaker, W. B. and Mitchell, J. L., editors (1992). *JPEG Still Image Data Compression Standard*. New York, Van Nostrand Reinhold.

Phillips, D. (1992). Lzw data compression. *The Computer Application Journal, Circuit Cellar Inc.*, 27:36–48.

Pohlmann, K. (1985). *Principles of Digital Audio*. Indianapolis, IN, Howard Sams and Co.

Rao, K. and Yip, P. (1990). *Discrete Cosine Transform—Algorithms, Advantages, Applications*. London, Academic Press.

Salomon, D. (2000). *Data Compression: The Complete Reference*. New York, NY, Springer-Verlag.

Shenoi, K. (1995). *Digital Signal Processing in Telecommunications*. Upper Saddle River, NJ, Prentice Hall.

Stollnitz, E. J., DeRose, T. D., and Salesin, D. H. (1996). *Wavelets for Computer Graphics*. San Francisco, CA, Morgan Kaufmann.

Vitter, J. S. (1987). Design and analysis of dynamic huffman codes. *Journal of the ACM*, 34(4):825–845.

Wallace, G. K. (1991). The jpeg still image compression standard. *Communications of the ACM*, 34(4):30–44.

Welch, T. A. (1984). A technique for high-performance data compression. *IEEE Computer*, 17(6):8–19.

Witten, I. H., Neal, R. M., and Cleary, J. G. (1987). Arithmetic coding for data compression. *Communications of the ACM*, 30(6):520–540.

Zhang, M. (1990). *The JPEG and Image Data Compression Algorithms.* PhD thesis.

Ziv, J. and Lempel, A. (1977). A universal algorithm for sequential data compression. *IEEE Transactions on Information Theory*, IT-23(3):337–343.

IV

EXTERNAL MEMORY ALGORITHMS AND DATA STRUCTURES

Chapter 9

EXTERNAL MEMORY DATA STRUCTURES

Lars Arge
Department of Computer Science, Duke University, Durham, NC 27708-0129 USA
large@cs.duke.edu

Abstract In many massive dataset applications the data must be stored in space and query efficient data structures on external storage devices. Often the data needs to be changed dynamically. In this chapter we discuss recent advances in the development of provably worst-case efficient external memory dynamic data structures. We also briefly discuss some of the most popular external data structures used in practice.

Keywords: I/O model, B-trees, Interval trees, Planar point location, Kinetic data structures, Proximity queries.

1. Introduction

Massive datasets often need to be stored in space efficient data structures on external storage devices. These structures are used to store a dynamically changing dataset such that queries can be answered efficiently. Many massive dataset applications involve geometric data (for example, points, lines, and polygons) or data which can be interpreted geometrically. Such applications often perform queries which correspond to searching in massive multidimensional geometric databases for objects that satisfy certain spatial constraints. Typical queries include reporting the objects intersecting a query region, reporting the objects containing a query point, and reporting objects near a query point.

While development of practically efficient (and ideally also multi-purpose) external memory data structures (or *indexes*) has always been a main concern in the database community, most data structure research in the algorithms community has focused on worst-case efficient internal memory data structures. Recently, however, there has been some cross-fertilization between the two areas. In this chapter we discuss recent advances in the development of worst-case efficient external memory data structures. We will concentrate on data structures for geometric objects but mention other structures when appropriate. We also briefly discuss some of the most popular external data structures used in practice.

Model of computation. Accurately modeling memory and disk systems is a complex task (Ruemmler and Wilkes 1994). The primary feature of disks we want to model is their extremely long access time relative to that of internal memory. In order to amortize the access time over a large amount of data, typical disks read or write large blocks of contiguous data at once and therefore the standard two-level disk model has the following parameters (Aggarwal and Vitter 1988, Vitter and Shriver 1994, Knuth 1998):

N = number of objects in the problem instance;
T = number of objects in the problem solution;
M = number of objects that can fit into internal memory;
B = number of objects per disk block;

where $B < M < N$. An *I/O operation* (or simply *I/O*) is the operation of reading (or writing) a block from (or into) disk. Refer to Figure 9.1. Computation can only be performed on objects in internal memory. The measures of performance in this model are the number of I/Os used to solve a problem, as well as the amount of space (disk blocks) used and the internal memory computation time.

Several authors have considered more accurate and complex multi-level memory models than the two-level model. An increasingly popular approach to increase the performance of I/O systems is to use several disks in parallel so work has especially been done in multi disk models. See e.g. the recent survey by Vitter (1999a). We will concentrate on the two-level one-disk model, since the data structures and data structure design techniques developed in this model often work well in more

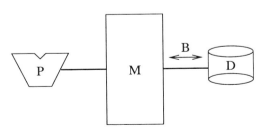

Figure 9.1. Disk model; An I/O moves B contiguous elements between disk and main memory (of size M).

complex models. For brevity we will also ignore internal computation time.

Outline of chapter. The rest of this chapter is organized as follows. In Section 2 we discuss the B-tree, the most fundamental (one-dimensional) external data structure, as well as recent variants and extensions of the structure. In Section 3 we illustrate some of the important techniques and ideas used in the development of provably I/O-efficient data structures for higher-dimensional problems. We do so through a discussion of a data structure for the stabbing query problem. In Section 4 we discuss external point location and a general method for obtaining a dynamic data structure from a static one. In Section 5 and Section 6 we discuss data structures for 3-sided and general (4-sided) two-dimensional range searching, respectively, and in Section 7 we survey various extensions of these structures. Section 8 contains a survey of external data structures for proximity queries, and in Section 10 we discuss the so-called buffer trees, which can often be used in I/O-efficient algorithms.

Several of the worst-case efficient structures we consider are simple enough to be of practical interest. Still, there are many good reasons for developing simpler (heuristic) and general purpose structures without worst-case performance guarantees, and a large number of such structures have been developed in the database community. Even though the focus of this chapter is on provably worst-case efficient data structures, in Section 9 we give a short survey of some of the major classes of such heuristic-based structures. The reader is referred to recent surveys for a more complete discussion (Agarwal and Erickson 1999, Gaede and Günther 1998, Nievergelt and Widmayer 1997).

Throughout the chapter we assume that the reader is familiar with basic internal memory data structures and design and analysis methods, such as balanced search trees and amortized analysis—see e.g. Cormen et al. (1990).

2. B-Trees

The B-tree is the most fundamental external memory data structure (Bayer and McCreight 1972, Comer 1979, Knuth 1998, Huddleston and Mehlhorn 1982). The B-tree corresponds to an internal memory balanced search tree. It uses linear space—$O(N/B)$ disk blocks—and supports insertions and deletions in $O(\log_B N)$ I/Os. One-dimensional range queries, asking for all elements in the tree in a query interval $[q_1, q_2]$, can be answered in $O(\log_B N + T/B)$ I/Os, where T is the number of reported elements.

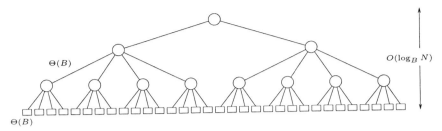

Figure 9.2. B-tree; All internal nodes (except possibly the root) have fan-out $\Theta(B)$ and there are $\Theta(N/B)$ leaves. The tree has height $O(\log_B N)$.

The space, update, and query bounds obtained by the B-tree are the bounds we would like to obtain in general for more complicated problems. The bounds are significantly better than the bounds we would obtain if we just used an internal memory data structure and virtual memory. The $O(N/B)$ space bound is obviously optimal and the $O(\log_B N + T/B)$ query bound is optimal in a comparison model of computation. Note that the query bound consists of an $O(\log_B N)$ search-term corresponding to the familiar $O(\log N)$ internal memory search-term,[1] and an $O(T/B)$ reporting-term accounting for the $O(T/B)$ I/Os needed to report T elements. Recently, the above bounds have been obtained for a number of problems (e.g. Arge and Vitter 1996, Arge et al. 1999b, Vengroff and Vitter 1996, Agarwal et al. 2000b, Callahan et al. 1995, Govindarajan et al. 2000) but higher lower bounds have also been established for some problems (Subramanian and Ramaswamy 1995, Arge et al. 1999b, Hellerstein et al. 1997, Kanellakis et al. 1996, Koutsoupias and Taylor 1998, Samoladas and Miranker 1998, Kanth and Singh 1999). We discuss these results in later sections.

B-trees come in several variants, like B^+ and B^* trees (see e.g. Bayer and McCreight 1972, Comer 1979, Huddleston and Mehlhorn 1982, Arge and Vitter 1996, Knuth 1998, Agarwal et al. 1999, and their references). A basic B-tree is a $\Theta(B)$-ary tree (with the root possibly having smaller degree) built on top of $\Theta(N/B)$ leaves. The degree of internal nodes, as well as the number of elements in a leaf, is typically kept in the range $[B/2\ldots B]$ such that a node or leaf can be stored in one disk block. All leaves are on the same level and the tree has height $O(\log_B N)$—refer to Figure 9.2. In the most popular B-tree variants, the N data elements are stored in the leaves (in sorted order) and each internal node holds $\Theta(B)$ "routing" (or "splitting") elements used to guide searches. As we will see in later sections, it can sometimes be useful to use a B-

---

[1] We use $\log N$ to denote $\log_2 N$.

tree with fan-out $\Theta(B^c)$ for some constant $0 < c \leq 1$. If we keep $\Theta(N/B)$ leaves, every such tree will use $O(N/B)$ space and have height $O(\log_{B^c} N) = O(\log_B N)$.

To answer a range query $[q_1, q_2]$ on a B-tree we first search down the tree for q_1 and q_2 using $O(\log_B N)$ I/Os, and then we report the elements in the $O(T/B)$ leaves between the leaves containing q_1 and q_2. We perform an insertion in $O(\log_B N)$ I/Os by first searching down the tree for the relevant leaf l. If there is room for the new element in l we simply store it there. If not, we *split* l into two leaves l' and l'' of approximately the same size and insert the new element in the relevant leaf. The split of l results in the insertion of a new routing element in the parent of l, and thus the need for a split may propagate up the tree. Propagation of splits can often be avoided by *sharing* some of the (routing) elements of the full node with a non-full sibling. A new (degree 2) root is produced when the root splits and the height of the tree grows by one. Similarly, we can perform a deletion in $O(\log_B N)$ I/Os by first searching for the relevant leaf l and then removing the deleted element. If this results in l containing too few elements we either *fuse* it with one of its siblings (corresponding to deleting l and inserting its elements in the sibling), or we perform a *share* operation by moving elements from a sibling to l. As splits, fuse operations may propagate up the tree and eventually result in the height of the tree decreasing by one.

In internal memory, an N element search tree can be built in optimal $O(N \log N)$ time simply by inserting the elements one by one. In external memory we would use $O(N \log_B N)$ I/Os to build a B-tree using the same method. Interestingly, this is not optimal since Aggarwal and Vitter (1988) showed that sorting N elements in external memory takes $\Theta(\frac{N}{B} \log_{M/B} \frac{N}{B})$ I/Os. We can build a B-tree in the same bound by first sorting the elements and then build the tree level-by-level bottom-up.

2.1. B-Tree Variants and Extensions

Recently, several important variants and extensions of B-trees have been considered. In the following we further discuss weight- and level-balanced B-trees, persistent B-trees, as well as string B-trees.

Weight-balanced B-trees. The *weight-balanced B-tree* developed by Arge and Vitter (1996) are similar to normal B-trees in that all leaves are on the same level and rebalancing is done by splitting and fusing nodes. However, instead of requiring the degree of a node to be $\Theta(B^c)$, we require the weight (or size) of a node v to be $\Theta(B^{ch})$ if v is the root of a subtree of height h. The weight of v is defined as the number of elements in the leaves of the subtree rooted in v. The constraint actually means that v has degree $\Theta(B^c)$ and thus the tree has height $O(\log_B N)$. It also means that the children of v are of approximately the same size $\Theta(B^{c(h-1)})$. In normal B-trees their sizes can differ by a factor exponential in h. Weight-balanced B-trees can be viewed as an external version of BB[α]-trees (Nievergelt and Reingold 1973)—however, weight-balanced B-trees have also been used as a simple alternative to BB[α]-trees in internal memory structures (Arge and Vitter 1996).

After performing an insertion or deletion in a leaf l of a weight-balanced B-tree the weight constraint may be violated in nodes on the path from the root to l. In order to rebalance the tree we perform a split or fuse operation on each of these $O(\log_B N)$ nodes. A key property of a weight-balanced B-tree is that after performing a rebalance operation (split or fuse) on a weight $\Theta(B^{ch})$ node v, $\Theta(B^{ch})$ updates have to be performed below v before another rebalance operation needs to be performed on v. This means that even if the cost of a rebalance operation is $O(B^{ch})$ I/Os, the amortized complexity of an update remains $O(\log_B N)$. The cost of a rebalance operation could for example be $O(B^{ch})$ if v stores a size $\Theta(B^{ch})$ secondary structure that needs to be rebuilt when v splits (for example, a structure on the $\Theta(B^{ch})$ elements below v). The property also suggests a simple rebalancing strategy based on *partial-rebuilding* (see e.g. Overmars 1983); Instead of splitting or fusing nodes on the path from the root to l, we can simply rebuild the tree rooted in the highest unbalanced node on this path. Since the (sub-)tree can be rebuilt in a linear number of I/Os we obtain an $O(\log_B N)$ amortized update bound. Weight-balanced B-trees have been used in numerous efficient data structures, most recently in an elegant so-called *cache-oblivious* B-tree structure by Bender et al. (2000). This structure obtains B-tree-like update and query bounds without explicitly using the (possibly unknown) block size B (see also Frigo et al. 1999). We will discuss other applications in later sections.

Level-balanced B-trees. Apart from the operations discussed above, we sometimes need to be able to perform *divide* and *merge* operations on B-trees. A divide operation at element x constructs two trees containing all elements less than and greater than x, respectively. A merge operation performs the inverse operation. A divide operation can be performed in $O(\log_B N)$ I/Os by first splitting all nodes on the path from the root to the leaf containing x, constructing two trees, and then performing fuse/share operations on the relevant subset of the same nodes in order to reestablish the B-tree invariant for the two trees. Similarly, a merge operation can also be performed in $O(\log_B N)$ I/Os using $O(\log_B N)$ split/share operations (Mehlhorn 1984).

In some applications we need to be able to traverse a path in a B-tree from a leaf to the root. To do so we need a *parent-pointer* from each node to its parent. Maintaining such pointers during a rebalance operation (split, fuse or merge) on a node v requires $\Theta(B)$ I/Os since we need to update parent pointers of $\Theta(B)$ of v's children. This results in a B-tree update, divide, or merge operation taking $O(B \log_B N)$ I/Os. However, using simple modifications of standard B-trees or weight-balanced B-trees, update operations can still be performed in $O(\log_B N)$ I/Os since it can be guaranteed that $\Theta(B)$ updates have to be performed below a node v between rebalance operations on v.

Recently, Agarwal et al. (1999) developed a variant of B-trees in which divide and merge operations can also be supported I/O-efficiently while maintaining parent pointers. The main idea in the so-called *level-balanced B-trees* is to use a global balance condition instead of the local degree or weight conditions used in B-trees or weight-balanced B-trees. More precisely, a constraint is imposed on the number of nodes on each level of the tree. When the constraint is violated the whole subtree at that level and above is rebuilt. The structure uses $O(N/B)$ space, supports query in $O(\log_B N)$ I/Os, and update, divide, and merge operations in $O(\log_B^2 N)$ I/Os amortized.[2] Level-balanced B-trees e.g. have applications in dynamic maintenance of planar st-graphs (Agarwal et al. 1999).

Persistent B-trees. In some database applications we need to be able to update the current database while querying both the current and earlier versions of the database (data structure). One simple but very inefficient way of supporting this functionality is to copy the whole data structure every time an update is performed. Another and much more efficient way is through the (partially) *persistent* technique (Sarnak and Tarjan 1986, Driscoll et al. 1989), also sometimes referred to as the *multiversion* method (Becker et al. 1996, Varman and Verma 1997). Instead of making copies of the structure, the idea in this technique is to maintain one structure at all times but for each element keep track of the time interval at which it is really present in the structure. A B-tree can be made persistent as follows: Each data element is augmented with an *existence interval* consisting of the time at which the element was inserted and (possibly) the time at which it was deleted. We say that an element is *alive* in its existence interval. All elements are stored

---

[2]The precise bounds are actually slightly better and more complicated (Agarwal et al. 1999).

in a slightly modified B-tree where we also associate a *node existence interval* with each node. Apart from the normal B-tree constraint on the number of elements in a node, we also maintain that a node contains $\Theta(B)$ alive elements in its existence interval. This means that for a given time t, the nodes with existence intervals containing t make up a B-tree on the elements alive at that time. Thus we can perform range queries in $O(\log_B N + T/B)$ on any version (at any time) of the tree as usual (remembering to disregard dead elements in the visited nodes). Here N is the number of updates performed.

An insertion in a persistent B-tree is performed almost like a normal insertion. We first find the relevant leaf l and if there is room for it we insert the new elements. Otherwise we have an *overflow* and to handle this we first copy all alive elements in l and make the current time the endpoint of the existence interval of l (corresponding to deleting l at the current time). Depending on how many elements we copied, we either construct one new leaf on them, split them into two equal size groups and construct two new leaves on them, or we copy the alive elements from one of l's siblings and construct one or two leaves out of all the copied elements—this corresponds to performing split or fuse operations on the alive elements in l and its sibling. In all cases we make sure that there is room for $\Theta(B)$ future updates in each of the new leaves. We then insert the new element in the relevant leaf and set the start time of the existence interval of all new leaves to the current time. Finally, we insert references to the new leaves in l's parent and (persistently) delete the reference to l. This may result in similar overflow operations cascading up one path to the root of the structure.

In order to perform a deletion we first update the existence interval of the relevant element in leaf l. As the element is not deleted, we do not need to perform a fuse operation as in a normal B-tree. However, the deletion may result in l containing less than the minimum allowed number of alive elements. If this is the case we copy the alive elements from l and one of its siblings and construct one or two new leaves as during an insertion. We also update the references in l's parent as previously, possibly resulting in similar updates up one path of the tree.

Both insertions and deletions can be handled in $O(\log_B N)$ I/Os since in both cases we touch a constant number of nodes on the $O(\log_B N)$ level of the structure. In total we construct $O(N/B)$ leaves since we construct $O(1)$ new leaves only when $\Theta(B)$ updates have been performed on an existing leaf. A similar argument can be applied to the nodes on each level of the tree and thus we can prove that the structure uses $O(N/B)$ space in total.

Several times in later sections we will construct a data structure by performing N insertion and deletions on an initially empty persistent B-tree, and then use the resulting (static) structure to answer queries. Using the above update algorithms, the construction takes $O(N \log_B N)$ I/Os. Utilizing the *distribution-sweeping* technique, Goodrich et al. (1993) showed how to construct the structure (perform the N updates without doing queries) more efficiently in $O(\frac{N}{B} \log_{M/B} \frac{N}{B})$ I/Os. Their method requires that every pair of elements in the structure can be compared—even a pair of elements not present in the structure at the same time. Unfortunately, as we will see in later sections, when working with geometric objects (such as line segments) we will not always be able to compare any two elements. It should be noted that the $O(N \log_B N)$ construction algorithm—that is, the update algorithm described above— also requires every pair of elements to be comparable, since elements can be used as routing elements in the internal nodes of the structure long after they have been deleted. Thus when performing an update or query with element e at time t, we might have to compare e with elements not alive at time t. However, by storing data elements in all nodes of the tree (not just the leaves) and using slightly different update algorithms, we can eliminate this problem such that the $O(N \log_B N)$ algorithm only compares elements present in the structure at the same time (Arge and Teh 2000).

String B-trees. In the B-tree variants discussed so far, the elements—and thus the routing elements in internal nodes—have been of unit size. In string applications a data element (string of characters) can often be arbitrarily long or different elements can be of different length. This means that we cannot use the strings as routing elements and at the same time maintain a large fan-out of internal nodes. We could store pointers to strings in the internal nodes and obtain fan-out $\Theta(B)$ but searching would then be inefficient since we could be forced to perform a lot of I/Os to route a query through a node. Ferragina and Grossi (1995) (see also Ferragina and Grossi 1996) recently presented an elegant solution to this problem called the *string B-tree*. From a high-level point of view, a string B-tree on K strings of total length N is just a B-tree built on N pointers to the N suffixes of the K strings in lexicographical order. To route a query string q through the $\Theta(B)$ string pointers in an internal node, each such node contains a *blind trie* data structure. A blind trie is a variant of the compacted trie (Knuth 1998, Morrison 1968), which fits in one disk block. Routing q through a node v requires one I/O to load the blind trie, as well as some extra I/Os to scan parts of q and the strings corresponding to the pointers stored in

v. However, since the scanned parts of q correspond to parts which will not be scanned again further down the tree, we can charge the I/Os to those parts of q and obtain an optimal $O(\log_B N + |q|/B)$ search bound. Ferragina and Grossi also showed how to insert or delete a string q in $O(|q|\log_B N)$ I/Os amortized. Other results on string B-trees and external string processing have been obtained by Crauser and Ferragina (1999), Ferragina and Luccio (1998), Farach et al. (1998) and Arge et al. (1997).

3. Interval Management

After considering the one-dimensional B-trees, we now turn to data structures for more complicated and higher-dimensional problems like range searching. In internal memory many elegant data structures have been developed for such problems—see e.g. the recent survey by Agarwal and Erickson (1999). Unfortunately, most of these structures are not efficient when mapped to external memory—mainly because they are normally based on binary trees. The main challenge when developing efficient external structures is to use B-trees as base structures, that is, to use multiway trees instead of binary trees. Recently, some progress has been made in the development of provably I/O-efficient data structures based on multi-way trees. In this section we illustrate some of the techniques and ideas used in the development of these structures through the *stabbing query problem*. The stabbing query problem is the problem of maintaining a dynamically changing set of (one-dimensional) intervals such that given a query point q all intervals containing q can be reported efficiently.

The static version of the stabbing query problem (the set of intervals is fixed) can easily be solved I/O-efficiently using a sweeping idea and a persistent B-tree (Arge et al. 1999b, Chazelle 1986, Ramaswamy 1997). To illustrate this, consider sweeping N intervals along the x-axis starting at $-\infty$, inserting each interval in a B-tree when its left endpoint is reached and deleting it again when its right endpoint is reached. To answer a stabbing query with q we simply have to report all intervals in the B-tree at "time" q—refer to Figure 9.3. Thus following the discussion in Section 2, the structure uses $O(N/B)$ space and can be constructed in $O(\frac{N}{B} \log_{M/B} \frac{N}{B})$ I/Os. Queries can be answered in $O(\log_B N + T/B)$ I/Os.

Following earlier attempts of Kanellakis et al. (1996) (see also Subramanian and Ramaswamy 1995, Ramaswamy and Subramanian 1994, Blankenagel and Güting 1990, Icking et al. 1987), a dynamic structure for the problem was developed by Arge and Vitter (1996). This structure

EXTERNAL MEMORY DATA STRUCTURES

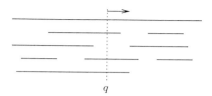

Figure 9.3. Static solution to stabbing query problem using persistence.

can be viewed as an external version of the interval tree (Edelsbrunner 1983a;b). It consists of a fan-out $\Theta(\sqrt{B})$ weight-balanced B-tree T on the endpoints of the intervals (the base tree), with the intervals stored in secondary structures associated with the internal nodes of T as described below. A range X_v (containing all points below v) can be associated with each node v in a natural way. This range is subdivided into $\Theta(\sqrt{B})$ subranges associated with the children of v. For illustrative purposes we call the subranges *slabs* and the left (right) endpoint of such a slab a *slab boundary*. Refer to Figure 9.4. The $\Theta(\sqrt{B}^2) = \Theta(B)$ contiguous sets of slabs are called *multislabs*. An example of a multislab is $X_{v_2} X_{v_3}$ in Figure 9.4. We assign an interval I to the node v where I contains one or more of the slab boundaries of v but not any of the slab boundaries associated with v's parent. Each node v of T contains $\Theta(B)$ secondary structures used to store the set of intervals I_v assigned to v; a *left slab list* and a *right slab list* for each of the $\Theta(\sqrt{B})$ slabs, a *multislab list* for each of the $\Theta(B)$ multislabs, as well as an *underflow structure*. A right slab list contains intervals from I_v with the right endpoint in the corresponding slab, sorted according to the right endpoint. Similarly, a left slab list contains intervals with the left endpoint in a slab, sorted according to the left endpoint. A multislab list stores intervals which span the corresponding multislab but not any wider multislab. If the number of intervals stored in a multislab list is less than B we instead store them in the underflow structure. This means that the underflow

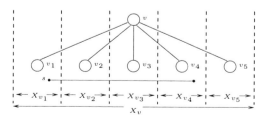

Figure 9.4. Node v in base tree of external interval tree. The range X_v associated with v is divided into 5 slabs.

Figure 9.5. Querying a node with q.

structure contains $O(B^2)$ intervals. An interval is thus stored in at most three structures; two slab lists and possibly in a multislab list or the underflow structure. For example, interval s in Figure 9.4 is stored in the left slab list of the first slab and in the right slab list of the fourth slab, as well as in the multislab list corresponding to the second and third slab. Thus the structure uses $O(N/B)$ space.

In order to answer a stabbing query q we search down T for the leaf containing q, reporting all the relevant intervals among the intervals stored in secondary structures of the node we pass. In node v we report all intervals in multislab lists containing q, as well as all intervals in the underflow structure containing q. We also traverse and report the intervals in the right (left) slab list of the slab containing q from the largest toward the smallest—according to right (left) endpoint—until we meet an interval that does not contain q. No other intervals in the list can contain q—refer to Figure 9.5. If T_v is the number of intervals reported in v we use $O(T_v/B)$ I/Os to report intervals from the slab and multislab lists. There is no $O(\log_B N)$-term since we do not search in any of the lists. If we implement the underflow structure using the static structure based on a persistent B-tree described above, we can also find the relevant intervals in this structure in $O(\log_B B^2 + T_v/B) = O(1 + T_v/B)$ I/Os. Since there are $O(\log_B N)$ nodes on the search path, we in total use $O(\sum_v (1 + T_v/B)) = O(\log_B N + T/B)$ I/Os to answer a query.

To insert a new interval we first use $O(\log_B N)$ I/Os to search down T for the node where the interval needs to be inserted in secondary structures. In this node we insert the interval in a left and right slab list and possibly in a multislab list. If these lists are implemented using B-trees we can do so in $O(\log_B N)$ I/Os. We may also need to insert the interval in the underflow structure. The structure is static but since it has size $O(B^2)$ we can use a global rebuilding idea to make it dynamic (Overmars 1983); we simply store the update in a special "update block" and once B updates have been collected we rebuild the structure using $O(\frac{B^2}{B} \log_{M/B} \frac{B^2}{B})$ I/Os. Assuming $M > B^2$, that is, that the internal memory is capable of holding B blocks, this is $O(B)$ and we obtain an $O(1)$ amortized update bound. Arge and Vitter (1996) have shown how to make this worst-case, even without the assumption on the main memory size. To complete the insertion, we also need to insert the new endpoints in the base tree T and rebalance the tree using split and share operations. Performing split or share operations may be costly since they result in the need for restructuring of the secondary structures. However, since this restructuring can be performed in a linear number of I/Os in the size of the secondary structures and as T is implemented as a weight-balanced B-tree (Section 2), we can obtain an $O(1)$ amortized I/O bound for a rebalance operation. Thus in total we can perform an insertion in $O(\log_B N)$ I/Os amortized. The bound can even be made worst-case using standard lazy rebuilding techniques. Deletions can be handled in $O(\log_B N)$ I/Os in a similar way. Variants of the external interval tree structure—as well as experimental results on applications of it in isosurface extraction[3]—have been considered by Chiang and Silva (Chiang and Silva 1997, Chiang et al. 1998, Chiang and Silva 1999).

The above solution to the stabbing query problem illustrates many of the problems encountered when developing I/O-efficient dynamic data structures, as well as the techniques commonly used to overcome these problems. As already discussed, the main problem is that in order to be efficient, external tree data structures need to have large fan-out. In the above example this resulted in the need for what we called multislabs. To handle multislabs efficiently we used the notion of underflow structure, as well as the fact that we could decrease the fan-out of T to $\Theta(\sqrt{B})$ while maintaining the $O(\log_B N)$ tree height. The underflow structure—implemented using sweeping and a persistent B-tree—solved a static version of the problem on $O(B^2)$ interval in $O(1 + T_v/B)$ I/Os. The structure was necessary since if we had just stored the intervals in multislab lists we might have ended up spending $\Theta(B)$ I/Os to visit the $\Theta(B)$ multislab lists of a node without reporting more than $O(B)$ intervals in total. This would have resulted in an $\Omega(B \log_B N + T/B)$ query bound. We did not store intervals in multislab lists containing $\Omega(B)$ intervals in the underflow structure, since the I/Os spent on visiting such lists during a query can always be charged to the $O(T_v/B)$-term in the query bound. The idea of charging some of the query cost to the output size is often called *filtering* (Chazelle 1986), and the idea of using a static structure on $O(B^2)$ elements in each node has been called the *bootstrapping* paradigm (Vitter 1999a;b). Finally, the ideas of weight-balancing and global rebuilding were used to obtain worst-case efficient update bounds. In Section 5 we will discuss another example of the use of all the above ideas.

4. Planar Point Location

The *planar point location* problem is defined as follows: Given a planar subdivision with N vertices (i.e., a decomposition of the plane into polygonal regions induced by a straight-line planar graph), construct a

---

[3]Based on a sweeping idea and a persistent list, Agarwal et al. (1998) described an efficient static structure for terrain contour line extraction.

data structure so that the face containing a query point $p = (x, y)$ can be reported efficiently. We will concentrate on the problem of finding the first segment of the subdivision hit by a vertical ray emanating at p (a *vertical ray shooting query*)—refer to Figure 9.6. After answering this query, the face containing the query point can easily be found (Overmars 1985).

In internal memory, a lot of work has been performed on the point location problem—see e.g. the survey by Snoeyink (1997). Sarnak and Tarjan (1986) presented a very simple solution to the static problem based on persistence. Their solution is similar to the static solution to the interval management problem discussed in the previous section. It is based on the fact that a vertical line l imposes a natural order on the segments in the subdivision intersected by l. This means that if we sweep the subdivision from left the right ($-\infty$ to ∞) with a vertical line, inserting a segment in a persistent search tree when its left endpoint is encountered and deleting it again when its right endpoint is encountered, we can answer a point location query $p = (x, y)$ by searching for the position of y in the tree at "time" x. Note that in this method the elements (segments) present in the persistent structure at different times cannot necessarily be compared. As discussed in Section 2, this means that we cannot use the $O(\frac{N}{B} \log_{M/B} \frac{N}{B})$ algorithm of Goodrich et al. (1993) to construct the same structure in external memory but have to use the less efficient $O(N \log_B N)$ I/O algorithm. However, we do obtain a linear space external point location data structure that answers queries in $O(\log_B N)$ I/Os. Goodrich et al. (1993) discussed another $O(\log_B N)$ query data structure based on a parallel fractional cascading technique by Tamassia and Vitter (1996). They did not analyze how many I/Os are needed to construct the structure. Several structures which can answer a batch of queries I/O-efficiently have also been proposed (Goodrich

Figure 9.6. Vertical ray shooting query with p.

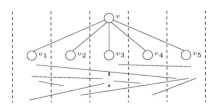

Figure 9.7. Answering a query on segments in I_v. Answer can be in two slab lists and $O(B)$ multislab lists.

et al. 1993, Arge et al. 1995; 1998, Crauser et al. 1998, Vahrenhold and Hinrichs 2000)

Recently, progress has been made in the development of I/O-efficient dynamic point location structures. In the dynamic problem we can change the subdivision dynamically (insert and delete edges/segments and vertices). Agarwal et al. (1999) developed a dynamic structure for *monotone* subdivisions and Arge and Vahrenhold (2000) developed a structure for the general problem. Both structures are based on the external interval tree structure described in the previous section. The main idea is to store the segments of the subdivision (or rather their projection onto the x-axis) in a structure very similar to an interval tree. Doing so a query with $p = (x, y)$ can be answered similarly to a stabbing query with x, except that in each node v visited by the query procedure a ray shooting query is answered on the segments in I_v. The global ray shooting query can then be answered by choosing the lowest segment among the $O(\log_B N)$ segments found this way. We can answer a query on the segments in I_v by answering the query on the segments in two slab lists and $O(B)$ multislab lists (refer to Figure 9.7). Using ideas also utilized in several internal memory structures (Cheng and Janardan 1992, Baumgarten et al. 1994), we can answer queries on the slab lists in $O(\log_B N)$ I/Os with a slightly modified B-tree (Agarwal et al. 1999). It is also easy to answer a ray shooting query on a multislab list in $O(\log_B N)$ I/Os using a B-tree storing the segments in y-order. However, if we query each of the $\Theta(B)$ multislab lists individually we will end up using $O(B \log_B N)$ I/Os to answer the query in v. Agarwal et al. (1999) improved this to $O(\log_B N)$, obtaining an overall query bound of $O(\log_B^2 N)$, by storing the segments in all multislab lists in one combined structure as described below.

Given two segments in the same multislab list we can easily determine which segment is above the other (formally a segment s is above a segment t if there exists a vertical line l intersecting both s and t such that the intersection between l and s is above the intersection between l and t). On the other hand, two segments in different multislab lists might not be comparable (if they cannot be intersected by the same vertical line) and therefore we cannot just build a B-tree on the segments in all multislab lists of a node v and use that to answer a query. Agarwal et al. (1999) used the fact that the segments only have endpoints on $\Theta(\sqrt{B})$ different lines (we imagine cutting the segments in the multislabs at slab boundaries) to construct an efficient structure. They also used that, as shown by Arge et al. (1995), N segments in the plane can be sorted in $O(\frac{N}{B} \log_{M/B} \frac{N}{B})$ I/Os—a set of segments is sorted if for any two comparable segments s and t, if s is above t then s appears after t in the sorted order. More precisely, the *multislab list structure* is constructed as follows: Let $\mathcal{R}$ denote the sorted set of multislab segments in a node. We first construct a fan-out $\sqrt{B}$ B-tree on $\mathcal{R}$. For a node w in the tree, let $\mathcal{R}_w$ denote the subsequence of $\mathcal{R}$ stored in the subtree rooted at w. To guide the processing of queries, we store certain segments of $\mathcal{R}_w$ in each internal node w; let $w_1, \ldots, w_{\sqrt{B}}$ denote the children of an internal node w. For $1 \leq i,j \leq \sqrt{B}$, we define μ_{ij} to be the maximal segment of $\mathcal{R}_{w_i}$ that intersects the jth vertical slab. We store all $\Theta(B)$ segments μ_{ij} at w in $O(1)$ blocks. In this way the structure requires $O(N/B)$ space and can be constructed in $O(\frac{N}{B} \log_{M/B} \frac{N}{B})$ I/Os. To answer a query with $p = (x,y)$, we follow a path from the root to a leaf z of the B-tree so that $\mathcal{R}_z$ contains the result of the query. At each node w visited by the procedure we do the following: If p lies in the interior of the rth slab, we define $E_w = \{\mu_{ir} \mid 1 \leq i \leq \sqrt{B}\}$. The definition of μ_{ij} ensures that if μ_{ir} is the first (lowest) segment of E_w intersected by an upward ray emanating in p, then the tree rooted at w_i contains the first segment of $\mathcal{R}$ hit by an upward ray emanating in p. We therefore visit w_i next. In this way a query can be answered in $O(\log_B N)$ I/Os. One way of thinking of the multislab list structure is as a fan-out $\sqrt{B}$ B-tree for each of the $\sqrt{B}$ slabs, all stored in the same structure; When answering a query in the rth slab, E_w of all nodes make up a fan-out $\sqrt{B}$ B-tree on the segments intersecting the slab.

The main problem in making the above point location structure dynamic is making the multislab list structure dynamic. The problem is that inserting a new segment may change the total order $\mathcal{R}$ considerably; refer to Figure 9.8. Agarwal et al. (1999) used special features of monotone subdivisions to limit such changes and obtained an $O(\log_B^2 N)$ multislab list structure update bound. This is also the global update bound since only one multislab list structure needs to be updated when performing an insertion or deletion and since the rest of the structure can be easily updated in $O(\log_B N)$ I/Os using standard B-tree and weight-balanced B-tree techniques. Arge and Vahrenhold (2000) extended the structure to work for general subdivisions. To do so they used a new general dynamization technique discussed in the next subsection. Using

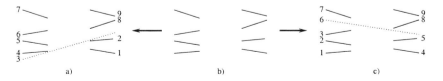

Figure 9.8. Inserting a new segment may change total order significantly. b) Original segments. a) and c) illustrate different insertions and resulting total orders.

this method on the multislab list structure they first developed a dynamic version of this structure, which supports updates in $O(\log_B N)$ I/Os and answers queries in $O(\log_B^2 N)$ I/Os. Using a technique similar to fractional cascading (Chazelle and Guibas 1986, Mehlhorn and Näher 1990) they improved the query performance to $O(\log_B N)$, obtaining a linear space point location structure supporting updates and queries in $O(\log_B^2 N)$ I/Os.

4.1. The logarithmic method

The general method for transforming a static external memory data structure into an efficient dynamic structure is an external version of the *logarithmic method* (Bentley 1979) (see also Overmars 1983). In internal memory, the main idea in this method is to partition the set of N elements into $\log N$ subsets of exponentially increasing size 2^i, $i = 0, 1, 2, \ldots$, and build a static structure $\mathcal{D}_i$ for each of these subsets. Queries are then performed by querying each $\mathcal{D}_i$ and combining the answers, while insertions are performed by finding the first empty $\mathcal{D}_i$, discarding all structures $\mathcal{D}_j$, $j < i$, and building $\mathcal{D}_i$ from the new element and the $\sum_{l=0}^{i-1} 2^l = 2^i - 1$ elements in the discarded structures.

To make the logarithmic method I/O-efficient we need to decrease the number of subsets to $\log_B N$, which in turn means increasing the size of $\mathcal{D}_i$ to B^i. When doing so $\mathcal{D}_j$, $j < i$, does not contain enough objects to build $\mathcal{D}_i$ (since $1 + \sum_{l=0}^{i-1} B^l < B^i$). However, it turns out that if we can build a static structure I/O-efficiently enough, this problem can be resolved and we can make a modified version of the method work in external memory. Consider a static structure $\mathcal{D}$ that can be constructed in $O(\frac{N}{B} \log_B N)$ I/Os and that answers queries in $O(\log_B N)$ I/Os (note that $O(\frac{N}{B} \log_{M/B} \frac{N}{B}) = O(\frac{N}{B} \log_B N)$ if $M > B^2$). We partition the N elements into $\log_B N$ sets such that the ith set has size *less than* $B^i + 1$ and construct an external memory static data structure $\mathcal{D}_i$ for

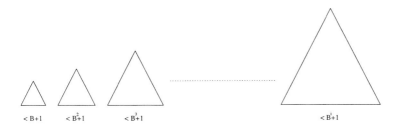

Figure 9.9. Logarithmic method; $\log_B N$ structures—$\mathcal{D}_i$ contains less than $B^i + 1$ elements. $\mathcal{D}_1, \mathcal{D}_2, \ldots, \mathcal{D}_j$ do not contain enough elements to build $\mathcal{D}_{j+1}$ of size B^{j+1}.

each set—refer to Figure 9.9. To answer a query, we simply query each $\mathcal{D}_i$ and combine the results using $O(\sum_{j=1}^{\log_B N} \log_B |\mathcal{D}_j|) = O(\log_B^2 N)$ I/Os. We perform an insertion by finding the first structure $\mathcal{D}_i$ such that $\sum_{j=1}^{i} |\mathcal{D}_j| \leq B^i$, discarding all structures $\mathcal{D}_j$, $j \leq i$, and building a new D_i from the elements in these structures using $O((B^i/B) \log_B B^i) = O(B^{i-1} \log_B N)$ I/Os. Now because of the way $\mathcal{D}_i$ was chosen, we know that $\sum_{j=1}^{i-1} |\mathcal{D}_j| > B^{i-1}$, which means that at least B^{i-1} objects are moved from lower indexed structures to $\mathcal{D}_i$. If we divide the $\mathcal{D}_i$ construction cost between these objects, each object is charged $O(\log_B N)$ I/Os. Since an object never moves to a lower indexed structure we can at most charge it $O(\log_B N)$ times during N insertions. Thus the amortized cost of an insertion is $O(\log_B^2 N)$ I/Os. Note that the key to making the method work is that the factor of B we lose when charging the construction of a structure of size B^i to only B^{i-1} objects is offset by the $1/B$ factor in the construction bound. Arge and Vahrenhold (2000) show how deletions can also be handled I/O-efficiently using a global rebuilding idea.

5. 3-Sided Planar Range Searching

In Section 3 we discussed the stabbing query problem. This problem is equivalent to performing *diagonal corner queries*—a special case of *2-sided range queries*—on a set of points in the plane. Consider mapping an interval $[x, y]$ to the point (x, y) in the plane. Finding all intervals containing a query point q then corresponds to finding all points (x, y) such that $x \leq q$ and $y \geq q$. Refer to Figure 9.10. In this section we consider the more general *3-sided planar range searching* problem: Given a set of points in the plane the solution to a 3-sided query (q_1, q_2, q_3) consists of all points (x, y) with $q_1 \leq x \leq q_2$ and $y \geq q_3$. Refer to Figure 9.11.

Following several earlier attempts (Ramaswamy and Subramanian 1994, Subramanian and Ramaswamy 1995, Blankenagel and Güting 1990,

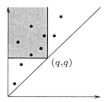

Figure 9.10. Diagonal corner query.

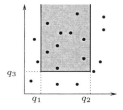

Figure 9.11. 3-sided query.

Icking et al. 1987), Arge et al. (1999b) developed an optimal dynamic structure for the 3-sided planar range searching problem. The structure uses many of the ideas already discussed for the interval tree structure in Section 3: *Bootstrapping* using a static structure, *filtering*, and a *weight-balanced* B-tree. In fact, the I/O-optimal static solution to the problem can be obtained using the same persistence idea as the one used in the interval case. This time we imagine sweeping the plane with a horizontal line from $y = \infty$ to $y = -\infty$ and inserting the x-coordinate of points in a persistent B-tree as they are met. To answer a query (q_1, q_2, q_3) we perform a one-dimensional range query $[q_1, q_2]$ on the B-tree at "time" q_3. Following the discussion in Section 2, the structure obtained this way uses linear space and queries can be answered in $O(\log_B N + T/B)$ I/Os. It can be constructed in $O(\frac{N}{B} \log_{M/B} \frac{N}{B})$ I/Os.

Using the static solution and the general dynamization method discussed in the previous section, we can immediately obtain a dynamic solution to the problem with $O(\log_B^2 N)$ query and update bounds. The optimal dynamic structure however is an external version of the internal memory *priority search tree* structure (McCreight 1985). Like the external interval tree, the external priority search tree consists of a base B-tree on the x-coordinates of the points. As previously, each internal node corresponds naturally to an x-range, which is divided into $\Theta(B)$ slabs by the x-ranges of its children. In each node v we store $O(B)$ points for each of v's $\Theta(B)$ children v_i, namely the B points with the highest y-coordinates in the x-range of v_i (if existing) that have not been stored in ancestors of v. We store the $O(B^2)$ points in the linear space static structure discussed above (the "$O(B^2)$-structure") such that a 3-sided query on them can be answered in $O(T/B)$ I/Os. As in Section 3, we can update the $O(B^2)$-structure in $O(1)$ I/Os using an "update block" and a global rebuilding technique. Since every point is stored in precisely one $O(B^2)$-structure, the structure uses $O(N/B)$ space in total.

To answer a 3-sided query (q_1, q_2, q_3) we start at the root of the external priority search tree and proceed recursively to the appropriate subtrees; when visiting a node v we query the $O(B^2)$-structure and report the relevant points, and then we advance the search to some of the children of v. The search is advanced to child v_i if v_i is either along the leftmost search path for q_1 or the rightmost search path for q_2, or if the entire set of points corresponding to v_i in the $O(B^2)$-structure were reported—refer to Figure 9.12. The query procedure reports all points in the query range since if we do not visit child v_i corresponding to a slab completely spanned by the interval $[q_1, q_2]$, it means that at least one of the points in the $O(B^2)$-structure corresponding to v_i does not satisfy the query. This in turn means that none of the points in the sub-

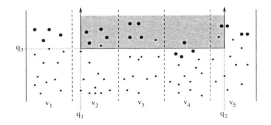

Figure 9.12. Internal node v with children v_1, v_2, ..., v_5. The points in bold are stored in the $O(B^2)$–structure. To answer a 3-sided query we report the relevant of the $O(B^2)$ points and answer the query recursively in v_2, v_3, and v_5. The query is not extended to v_4 because not all of the points from v_4 in the $O(B^2)$–structure satisfy the query.

tree rooted at v_i can satisfy the query. That we use $O(\log_B N + T/B)$ I/Os to answer a query can be seen as follows. In every internal node v visited by the query procedure we spend $O(T_v/B)$ I/Os, where T_v is the number of points reported. There are $O(\log_B N)$ nodes visited on the search paths in the tree to the leaf containing q_1 and the leaf containing q_2 and thus the number of I/Os used in these nodes adds up to $O(\log_B N + T/B)$. Each remaining visited internal node v is not on the search path but it is visited because $\Theta(B)$ points corresponding to it were reported when we visited its parent. Thus the cost of visiting these nodes adds up to $O(T/B)$, even if we spend a constant number of I/Os in some nodes without finding $\Theta(B)$ points to report. Note how we again are using the filtering idea, that is, we are are charging some of our search cost to the points we output as a result of the query.

To insert a point $p = (x,y)$ in the external priority search tree we search down the tree for the leaf containing x, until we reach the node v where p needs to be inserted in the $O(B^2)$–structure. Insertion of p in v (may) result in the $O(B^2)$–structure containing one too many points from the slab corresponding to the child v_j containing x. Therefore, apart from inserting p in the $O(B^2)$–structure, we also remove the point p' with the lowest y-coordinate among the points corresponding to v_j. We insert p' recursively in the tree rooted in v_j. Since we use $O(1)$ I/Os in each of the nodes on the search path, the insertion takes $O(\log_B N)$ I/Os. We also need to insert x in the base B-tree. This may result in split and/or fuse operations and each such operation may require rebuilding an $O(B^2)$–structure. Using weight-balanced B-tress, Arge et al. (1999b) showed how the rebalancing after an insertion can be performed in $O(\log_B N)$ I/Os worst case. Deletions can be handled in $O(\log_B N)$ I/Os in a similar way.

6. General Planar Range Searching

After discussing 2- and 3-sided planar range queries we are now ready to consider general (4-sided) range queries. Given a set of points in the plane we want to be able to find all points contained in a query rectangle. While linear space and $O(\log_B N + T/B)$ query structures exist for the two special cases, Subramanian and Ramaswamy (1995) proved that one cannot obtain an $O(\log_B N + T/B)$ query bound using less than $\Theta(\frac{N/B \log(N/B)}{\log \log_B N})$ disk blocks.[4] This lower bound holds in a natural external memory version of the *pointer machine model* (Chazelle 1990). A similar bound in a slightly different model where the search component of the query is ignored was proved by Arge et al. (1999b). This *indexability model* was defined by Hellerstein et al. (1997) and considered by several authors (Kanellakis et al. 1996, Koutsoupias and Taylor 1998, Samoladas and Miranker 1998). Note that linear space and logarithmic query structures for the *range counting problem* (where only the number of points in the query rectangle, and not the points themselves, need to be reported) can be developed in a slightly different model of computation (see Agarwal et al. 2001a, Zhang et al. 2001, and references therein).

Based on a sub-optimal linear space structure for answering 3-sided queries, Subramanian and Ramaswamy (1995) developed the *P-range tree* that uses optimal $O(\frac{N/B \log(N/B)}{\log \log_B N})$ space but uses more than the optimal $O(\log_B N + T/B)$ I/Os to answer a query. Using their optimal structure for 3-sided queries, Arge et al. (1999b) obtained an optimal structure. We discuss the structure below. In practical applications involving massive datasets it is often crucial that external data structures use linear space. We discuss this further in Section 9. Grossi and Italiano (1999a;b) developed the elegant linear space *cross-tree* data structure which answers queries in $O(\sqrt{N/B} + T/B)$ I/Os. This is optimal for linear space data structures—as e.g. proven by Kanth and Singh (1999). The *O-tree* of Kanth and Singh (1999) obtains the same bounds using ideas similar to the ones used by van Kreveld and Overmars (1991) in *divided k-d trees*. Below, after discussing the $O(\log_B N + T/B)$ query structure, we also discuss the cross-tree further.

Logarithmic query structure. The $O(\log_B N + T/B)$ query data structure is based on ideas from the corresponding internal memory data structure due to Chazelle (1986). It uses both the external interval tree discussed in Section 3 and the external priority search tree discussed

---

[4] In fact, this bound even holds for a query bound of $O(\log_B^c N + T/B)$ for any constant c.

in Section 5 The structure consists of a fan-out $\log_B N$ base tree over the x-coordinates of the N points. As usual an x-range is associated with each node v and it is subdivided into $\log_B N$ slabs by v's children $v_1, v_2, \ldots, v_{\log_B N}$. We store *all* the points in the x-range of v in four secondary data structures associated with v. Two of the structures are priority search trees for answering 3-sided queries—one for answering queries with the opening to the left and one for queries with the opening to the right. We also store the points in a linear list sorted by y-coordinate. For the fourth structure, we imagine linking together for each child v_i the points in the x-range of v_i in y-order, producing a polygonal line monotone with respect to the y-axis. We project all the segments produced in this way onto the y-axis and store them in an external interval tree. With each segment endpoint we also store a pointer to the corresponding point in a child node. Since we use linear space on each of the $O(\log_{\log_B N}(N/B)) = O(\log(N/B)/\log\log_B N)$ levels of the tree, the structure uses $O(\frac{N/B \log(N/B)}{\log\log_B N})$ disk blocks in total.

To answer a 4-sided query $q = (q_1, q_2, q_3, q_4)$ we first find the topmost node v in the base tree where the x-range $[q_1, q_2]$ of the query contains a slab boundary. Consider the case where q_1 lies in the x-range of v_i and q_2 lies in the x-range of v_j—refer to Figure 9.13. The query q is naturally decomposed into three parts, consisting of a part in v_i, a part in v_j, and a part completely spanning nodes v_k, for $i < k < j$. The points contained in the first two parts can be found in $O(\log_B N + T/B)$ I/Os using the 3-sided structures corresponding to v_i and v_j. To find the points in the third part we query the interval tree associated with v with the y-value q_2. This way we obtain the $O(\log_B N)$ segments in the structure containing q_2, and thus (a pointer to) the bottommost point contained in the query for each of the nodes $v_{i+1}, v_{i+2}, \ldots, v_{j-1}$. We then traverse the $j - i - 1 = O(\log_B N)$ relevant sorted lists and output the remaining points using $O(\log_B N + T/B)$ I/Os.

To insert or delete a point, we need to perform $O(1)$ updates on each of the $O(\log(N/B)/\log\log_B N)$ levels of the base tree. Each of these updates takes $O(\log_B N)$ I/Os. We also need to update the base tree. Using a weight-balanced B-tree, Arge et al. (1999b) showed how this can be done in $O((\log_B N)(\log \frac{N}{B})/\log\log_B N)$ I/Os.

Linear space structure. The linear space cross-tree structure of Grossi and Italiano (Grossi and Italiano 1999a;b) consists of two levels. The lower level partitions the plane into $\Theta(\sqrt{N/B})$ vertical slabs and $\Theta(\sqrt{N/B})$ horizontal slabs containing $\Theta(\sqrt{NB})$ points each, forming an irregular grid of $\Theta(N/B)$ *basic squares*—refer to Figure 9.14. Each basic square can contain between 0 and $\sqrt{N/B}$ points. The points are grouped

EXTERNAL MEMORY DATA STRUCTURES

and stored according to the vertical slabs—points in vertically adjacent basic squares containing less than B points are grouped together to form groups of $\Theta(B)$ points and stored in blocks together. The points in a basic square containing more than B points are stored in a B-tree. Thus the lower level uses $O(N/B)$ space. The upper level consists of a linear space search structure which can be used to determine the basic square containing a given point—for now we can think of the structure as consisting of a fan-out $\sqrt{B}$ B-tree $\mathcal{T}_V$ on the $\sqrt{N/B}$ vertical slabs and a separate fan-out $\sqrt{B}$ B-tree $\mathcal{T}_H$ on the $\sqrt{N/B}$ horizontal slabs.

In order to answer a query (q_1, q_2, q_3, q_4) we use the upper level search tree to find the vertical slabs containing q_1 and q_3 and the horizontal slabs containing q_2 and q_4 using $O(\log_B N)$ I/Os. We then explicitly check all points in these slabs and report all the relevant points. In doing so we use $O(\sqrt{NB}/B) = O(\sqrt{N/B})$ I/Os to traverse the vertical slabs and $O(\sqrt{NB}/B + \sqrt{N/B}) = O(\sqrt{N/B})$ I/Os to traverse the horizontal slabs (the $\sqrt{N/B}$-term in the latter bound is a result of the slabs being blocked vertically—a horizontal slab contains $\sqrt{N/B}$ basic squares). Finally, we report all points corresponding to basic squares fully covered by the query. To do so we use $O(\sqrt{N/B} + T/B)$ I/Os since the slabs are blocked vertically. In total we answer a query in $O(\sqrt{N/B} + T/B)$ I/Os.

In order to perform an update we need to find and update the relevant basic square. We may also need to split slabs (insertion) or merge slabs

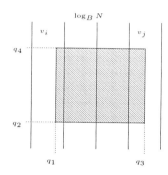

Figure 9.13. The slabs corresponding to a node v in the base tree. To answer a query (q_1, q_2, q_3, q_4) we need to answer 3-sided queries on the points in slab v_i and slab v_j, and a range query on the points in slabs between v_i and v_j.

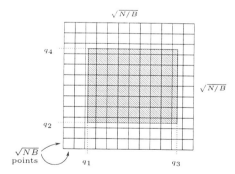

Figure 9.14. Basic squares. To answer a query (q_1, q_2, q_3, q_4) we check points in two vertical and two horizontal slabs, and report points in basic squares completely covered by the query.

with neighbor slabs (deletions). In order to do so efficiently while still being able to answer a range query I/O-efficiently, the upper level is actually implemented using a *cross-tree* $\mathcal{T}_{HV}$. $\mathcal{T}_{HV}$ can be viewed as a cross product of $\mathcal{T}_V$ and $\mathcal{T}_H$: For each pair of nodes $u \in \mathcal{T}_H$ and $v \in \mathcal{T}_V$ on the same level we have a node (u, v) in $\mathcal{T}_{HV}$, and for each pair of edges $(u, u') \in \mathcal{T}_H$ and $(v, v') \in \mathcal{T}_V$ we have an edge $((u, v), (u', v'))$ in $\mathcal{T}_{HV}$. Thus the tree has fan-out $O(B)$ and uses $O((\sqrt{N/B})^2) = O(N/B)$ space. Grossi and Italiano (Grossi and Italiano 1999a;b) showed how we can use the cross-tree to search for a basic square in $O(\log_B N)$ I/Os and how the full structure can be used to answer a range query in $O(\sqrt{N/B} + T/B)$ I/Os. They also showed that if $\mathcal{T}_H$ and $\mathcal{T}_V$ are implemented using weight-balanced B-trees, the structure can be maintained in $O(\log_B N)$ I/Os during an update.

7. Special and Higher-Dimensional Range Searching

As we have seen in the preceding sections, two-dimensional external range searching is theoretically relatively well understood. In contrast, little theoretical work has been done on higher-dimensional range searching and on special cases of higher-dimensional range searching. In this section we survey such results.

Range searching. Vengroff and Vitter (1996) presented a data structure for 3-dimensional range searching with logarithmic query bound. With recent modifications (Vitter 1999a) their structure answers queries in $O(\log_B N + T/B)$ I/Os and uses $O(\frac{N}{B} \log^3 \frac{N}{B} / \log \log_B^3 N)$ space. More generally, they presented structures for answering $(3 + k)$-sided queries (k of the dimensions, $0 \leq k \leq 3$, have finite ranges) in $O(\log_B N + T/B)$ I/Os using $O(\frac{N}{B} \log^k \frac{N}{B} / \log \log_B^k N)$ space.

As mentioned, space use is often as crucial as query time when manipulating massive datasets. The linear space cross-tree of Grossi and Italiano (Grossi and Italiano 1999a;b), as well as the O-tree of Kanth and Singh (1999), can be extended to support d-dimensional range queries in $O((N/B)^{1-1/d} + T/B)$ I/Os. Updates can be performed in $O(\log_B N)$ I/Os. Divide and merge operations can also be performed on the cross-tree in $O((N/B)^{1-1/d} + T/B)$ I/Os and the structure can be used in the design of dynamic data structures for several other problems (Grossi and Italiano 1999a;b).

Halfspace range searching. Given a set of points in d-dimensional space, a halfspace range query asks for all points on one side of a query hyperplane. Halfspace range searching is the simplest form of non-isothetic (non-orthogonal) range searching. The problem was first considered in external memory by Franciosa and Talamo (1994; 1997). Based on an internal memory structure due to Chazelle et al. (1985), Agarwal et al. (2000b) described a simple optimal $O(\log_B N + T/B)$ query, $O(N/B)$ space structure for the 2-dimensional case. The number of I/Os used to construct the structure is $O(N(\log N) \log_B N)$. They also described a structure for the 3-dimensional case, answering queries in $O(\log_B N + T/B)$ expected I/Os but requiring $O((N/B) \log(N/B))$ space. This structure is based on an internal memory result of Chan (2000) and the expected number of I/Os needed to construct the structure is $O((N/B)(\log(N/B)) \log_B N)$.

Based on the internal memory *partition trees* of Matoušek (1992), Agarwal et al. (2000b) also gave a linear space data structure for answering d-dimensional halfspace range queries in $O((N/B)^{1-1/d+\epsilon} + T/B)$ I/Os for any constant $\epsilon > 0$. The structure can be constructed in $O(N \log N)$ I/Os and using partial rebuilding it can support updates in $O((\log(N/B)) \log_B N)$ expected I/Os amortized. Using an improved $O(N \log_B N)$ construction algorithm, Agarwal et al. (2000a) obtained an $O(\log_B^2 N)$ amortized and expected update I/O-bound for the planar case. Agarwal et al. (2000b) also showed how the query bound of the structure can be improved at the expense of extra space. They also discussed how their linear space structure can be used to answer very general queries—more precisely, how all points within a query polyhedron with m faces can be found in $O(m(N/B)^{1-1/d+\epsilon} + T/B)$ I/Os.

Range searching on moving points. Recently there has been an increasing interest in external memory data structures storing continuously moving objects. A key goal is to develop structures that only need to be changed when the velocity or direction of an object changes (as opposed to continuously).

Kollios et al. (1999b) presented initial work on storing moving points in the plane such that all points inside a query range at query time t can be reported in a provably efficient number of I/Os. Their results were improved and extended by Agarwal et al. (2000a) who developed a linear space structure that answers a query in $O((N/B)^{1/2+\epsilon} + T/B)$ I/Os for any constant $\epsilon > 0$. A point can be updated using $O(\log_B^2 N)$ I/Os. The structure is based on partition trees and can also be used to answer queries where two time values t_1 and t_2 are given and we want to find all points that lie in the query range at any time between t_1 and t_2. Using the notion of *kinetic data structures* introduced by Basch et al. (1999), as well as a persistent version of the range searching structure by Arge et al.

(1999b) discussed in Section 6, Agarwal et al. (2000a) also developed a number of other structures with improved query performance. One of these structures has the property that queries in the near future are answered faster than queries further away in time. Further structures with this property were developed by Agarwal et al. (2001b).

8. Proximity Queries

Proximity queries such as nearest neighbor and closest pair queries have become increasingly important in recent years, for example because of their applications in similarity search and data mining.

Callahan et al. (1995) developed the first worst-case efficient external proximity query data structures. Their structures are based on an external version of the *topology trees* of Frederickson (1993) called *topology B-trees*, which can be used to dynamically maintain arbitrary binary trees I/O-efficiently. Using topology B-trees and ideas from an internal structure of Bespamyatnikh (1998), Callahan et al. (1995) designed a linear space data structure for dynamically maintaining the *closest pair* of a set of points in d-dimensional space. The structure supports updates in $O(\log_B N)$ I/Os. The same result was obtained by Govindarajan et al. (2000) using the *well-separated pair decomposition* of Callahan and Kosaraju (Callahan and Kosaraju 1995b;a). Govindarajan et al. (2000) also show how to dynamically maintain a well-separated pair decomposition of a set of d-dimensional points using $O(\log_B N)$ I/Os per update.

Using topology B-trees and ideas from an internal structure due to Arya et al. (1994), Callahan et al. (1995) developed a linear space data structure for the dynamic *approximate nearest neighbor* problem. Given a set of points in d-dimensional space, a query point p, and a parameter ϵ, the approximate nearest neighbor problem consists of finding a point q with distance at most $(1 + \epsilon)$ times the distance of the actual nearest neighbor of p. The structure answers queries and supports updates in $O(\log_B N)$ I/Os. Agarwal et al. (2000a) designed I/O-efficient data structures for answering approximate nearest neighbor queries on a set of moving points.

In some applications we are interested in finding not only the nearest but all the k nearest neighbors of a query point. Based on their 3-dimensional halfspace range searching structure, Agarwal et al. (2000b) described a structure that uses $O((N/B)\log(N/B))$ space to store N points in the plane such that a k nearest neighbors query can be answered in $(\log_B N + k/B)$ I/Os.

9. Practical General-Purpose Structures

Although several of the worst-case efficient (and often optimal) data structures discussed in the previous sections are simple enough to be of practical interest, they are often not the obvious choices when deciding which data structures to use in a real-world application. There are several reasons for this, one of the most important being that in real applications involving massive datasets it is practically feasible to use data structures of size cN/B only for a very small constant c. Since fundamental lower bounds often prevent logarithmic worst-case search cost for even relatively simple problems when restricting the space use to linear, we need to develop heuristic structures which perform well in most practical cases. Space restrictions also motivate us not to use structures for single specialized queries but instead design general structures that can be used to answer several different types of queries. Finally, implementation considerations often motivate us to sacrifice worst-case efficiency for simplicity. All of these considerations have led to the development of a large number of general-purpose data structures that often work well in practice, but which do not come with worst-case performance guarantees. Below we quickly survey the major classes of such structures. The reader is referred to more complete surveys for details (Agarwal and Erickson 1999, Gaede and Günther 1998, Nievergelt and Widmayer 1997, Greene 1989, Orenstein 1990, Samet 1990b).

Range searching in d-dimensions is the most extensively researched problem. A large number of structures have been developed for this problem, including space filling curves (see e.g. Orenstein 1986, Abel and Mark 1990, Asano et al. 1997), grid-files (Nievergelt et al. 1984, Hinrichs 1985), various quad-trees (Samet 1990a;b), kd-B tress (Robinson 1981)—and variants like Buddy-trees (Seeger and Kriegel 1990), hB-trees (Lomet and Salzberg 1990, Evangelidis et al. 1997) and cell-trees (Günther 1989)—and various R-trees (Guttman 1984, Greene 1989, Sellis et al. 1987, Beckmann et al. 1990, Kamel and Faloutsos 1994). Often these structures are broadly classified into two types, namely *space driven* structures (like quad-trees and grid-files), which partition the embedded space containing the data points and *data driven* structures (like kd-B trees and R-trees), which partition the data points themselves.

As mentioned above, we often want to be able to answer a very diverse set of queries, like halfspace range queries, general polygon range queries, and point location queries, on a single data structure. Many of the above data structures can easily be used to answer many such different queries and that is one main reason for their practical success. Recently, there has also been a lot of work on extensions—or even new structures—which also support moving objects (see e.g. Wolfson et al. 1998; 1999, Salzberg and Tsotras 1999, Kollios et al. 1999a, Šaltenis et al. 2000, Pfoser et al. 2000, Tayeb et al. 1998, and references therein) or proximity queries (see e.g. Berchtold et al. 1997; 1998b; 1996, Ciacca et al. 1997, Korn et al. 1996, Papadopoulos and Manolopoulos 1997, Roussopoulos et al. 1995, Seidl and Kriegel 1997, Sproull 1991, White and Jain 1996, Katayama and Satoh 1997, Hjaltason and Samet 1995, Gaede and Günther 1998, Agarwal and Erickson 1999, Nievergelt and Widmayer 1997, and references therein). However, as discussed, most often no guarantee on the worst-case query performance is provided for these structures.

So far we have mostly discussed point data structures. In general, we are interested in storing objects such as lines and polyhedra with a spatial extent. Like in the point case, a large number of heuristic structures, many of which are variations of the ones mentioned above, have been proposed for such objects. However, almost no worst-case efficient structures are known. In practice a *filtering/refinement* method is often used when managing objects with spatial extent. Instead of directly storing the objects in the data structure we store the *minimal bounding (axis-parallel) rectangle* containing each object together with a pointer to the object itself. When answering a query we first find all the minimal bounding rectangles fulfilling the query (the *filtering* step) and then we retrieve the objects corresponding to these rectangles and check each of them to see if they fulfill the query (the *refinement* step). One way of designing data structures for rectangles (or even more general objects) is to transform them into points in higher-dimensional space and store these points in one of the point data structures discussed above (see e.g. Gaede and Günther 1998, Nievergelt and Widmayer 1997, for a survey). However, a structure based on another idea has emerged as especially efficient for storing and querying minimal bounding rectangles. Below we further discuss this so-called R-tree and its many variants.

R-trees. The R-tree, originally proposed by Guttman (1984), is a multiway tree very similar to a B-tree; all leaf nodes are on the same level of the tree and a leaf contains $\Theta(B)$ data rectangles. Each internal node v (except maybe for the root) has $\Theta(B)$ children. For each of its children v_i, v contains the minimal bounding rectangle of all the rectangles in the tree rooted in v_i. An R-tree has height $O(\log_B N)$ and uses $O(N/B)$ space. An example of an R-tree is shown in Figure 9.15. Note that there is no unique R-tree for a given set of data rectangles and that minimal bounding rectangles stored within an R-tree node can overlap.

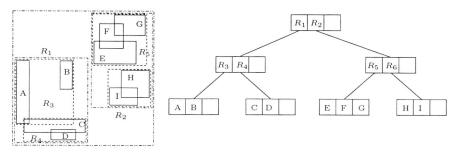

Figure 9.15. R-tree constructed on rectangles A, B, C, ..., I ($B = 3$).

In order to query an R-tree to find, say, all rectangles containing a query point p, we start at the root and recursively visit all children whose minimal bounding rectangle contains p. This way we visit all internal nodes whose minimal bounding rectangle contains p. There can be many more such nodes than actual data rectangles containing p and intuitively we want the minimal bounding rectangles stored in an internal node to overlap as little as possible in order to obtain a query efficient structure.

An insertion can be performed in $O(\log_B N)$ I/Os like in a B-tree. We first traverse the path from the root to the leaf we choose to insert the new rectangle into. The insertion might result in the need for node splittings on the same root-leaf path. As insertion of a new rectangle can increase the overlap in a node, several heuristics for choosing which leaf to insert a new rectangle into, as well as for splitting nodes during rebalancing, have been proposed (Greene 1989, Sellis et al. 1987, Beckmann et al. 1990, Kamel and Faloutsos 1994). The R*-tree variant of Beckmann et al. (1990) seems to result in the best performance in many cases. Deletions are also performed similarly to deletions in a B-tree but we cannot guarantee an $O(\log_B N)$ bound since finding the data rectangle to delete may require many more I/Os. Rebalancing after a deletion can be performed by merging nodes like in a B-tree but some R-tree variants instead delete a node when it underflows and reinsert its children into the tree (often referred to as "forced reinsertion"). The idea is to try to obtain a better structure by forcing a global reorganization of the structure instead of the local reorganization a node merge constitutes.

Constructing an R-tree using repeated insertion takes $O(N \log_B N)$ I/Os and does not necessarily result in a good tree in terms of query performance. Therefore several sorting based $O(\frac{N}{B} \log_{M/B} \frac{N}{B})$ I/O construction algorithms have been proposed (Roussopoulos and Leifker 1985, Kamel and Faloutsos 1993, DeWitt et al. 1994, Leutenegger et al. 1996,

Berchtold et al. 1998a). These algorithms are more than a factor of B faster than the repeated insertion algorithm and several of them produce an R-tree with practically better query performance than an R-tree built by repeated insertion. Still, no better than a linear worst-case query I/O-bound has been proven for any of them. Very recently, however, de Berg et al. (2000) presented an R-tree construction algorithm resulting in an R-tree with provably efficient worst-case query performance measured in terms of certain parameters describing the input data. They also discussed how the structure can be efficiently maintained dynamically.

10. Buffer Trees

In internal memory we can sort N elements in optimal $O(N \log N)$ time using $\Theta(N)$ operations on a dynamic balanced search tree. Using the same algorithm and a B-tree in external memory results in an algorithm using $O(N \log_B N)$ I/Os. This is a factor of $\frac{B \log_B N}{\log_{M/B}(N/B)}$ away from optimal. In order to obtain an optimal sorting algorithm we need a structure that supports updates in $O(\frac{1}{B} \log_{M/B} \frac{N}{B})$ I/Os. The inefficiency of the B-tree sorting algorithm is a consequence of the B-tree being designed to be used in an "on-line" setting where queries should be answered immediately—updates and queries are handled on an individual basis. This way we are not able to take full advantage of the large internal memory. It turns out that in an "off-line" environment where we are only interested in the overall I/O use of a series of operations and where we are willing to relax the demands on the query operations, we can develop data structures on which a series of N operations can be performed in $O(\frac{N}{B} \log_{M/B} \frac{N}{B})$ I/Os in total. To do so we use the *buffer tree* technique developed by Arge (1995a).

Basically the buffer tree is just a fan-out $\Theta(M/B)$ B-tree where each internal node has a buffer of size $\Theta(M)$. The tree has height $O(\log_{M/B} \frac{N}{B})$

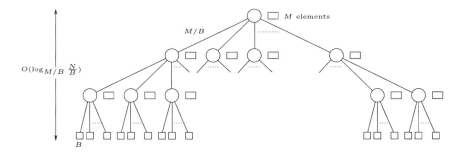

Figure 9.16. Buffer tree; Fan-out M/B tree where each node has a buffer of size M. Operations are performed in a lazy way using the buffers.

refer to Figure 9.16. Operations are performed in a "lazy" manner: In order to perform an insertion we do not (like in a normal B-tree) search all the way down the tree for the relevant leaf. Instead, we wait until we have collected a block of insertions and then we insert this block in the buffer of the root (which is stored on disk). When a buffer "runs full" its elements are "pushed" one level down to buffers on the next level. We can do so in $O(M/B)$ I/Os since the elements in the buffer fit in main memory and the fan-out of the tree is $O(M/B)$. If the buffer of any of the nodes on the next level becomes full by this process, the buffer-emptying process is applied recursively. Since we push $\Theta(M)$ elements one level down the tree using $O(M/B)$ I/Os (that is, we use $O(1)$ I/Os to push one block one level down), we can argue that every block of elements is touched a constant number of times on each of the levels of the tree. Thus, not counting rebalancing, inserting N elements requires $O(\frac{N}{B} \log_{M/B} \frac{N}{B})$ I/Os in total, or $O(\frac{1}{B} \log_{M/B} \frac{N}{B})$ amortized. Arge (1995a) showed that rebalancing can be handled in the same bound.

The basic buffer tree supporting insertions only can be used in an I/O-efficient sorting algorithm. Arge (1995a) showed how deletions and (one-dimensional) range queries can also be supported I/O-efficiently using buffers. The range queries are *batched* in the sense that we do not obtain the result of a query immediately. Instead parts of the result will be reported at different times as the query is pushed down the tree. This means that the data structure can only be used in algorithms where future updates and queries do not depend on the result of the queries. Luckily this is the case in many plane-sweep algorithms (Edelsbrunner and Overmars 1985, Arge 1995a). In general, problems where the entire sequence of updates and queries is known in advance, and the only requirement on the queries is that they must all eventually be answered, are known as *batched dynamic problems* (Edelsbrunner and Overmars 1985). Using the idea of multislabs discussed in Section 3, Arge (1995a) also showed how to implement a buffered segment tree, and Arge et al. (1998) showed how to use this data structure in a technique for solving a general class of high-dimensional problems.

The buffer tree technique has been used to develop several data structures which in turn have been used to develop algorithms in many different areas (Arge et al. 1995; 1997; 1999a, Kumar and Schwabe 1996, Arge 1995b, Fadel et al. 1999, Buchsbaum et al. 2000, van den Bercken et al. 1997; 1998, Hutchinson et al. 1997, Brengel et al. 1999, Sanders 1999). External buffered *priority queues* have been extensively researched because of their applications in graph algorithms. Arge (1995a) showed how to perform deletemin operations on a basic buffer tree in amortized $O(\frac{1}{B} \log_{M/B} \frac{N}{B})$ I/Os. Note that in this case the deletemin occurs right away, that is, it is not batched. This is accomplished by periodically computing the $O(M)$ smallest elements in the structure and storing them in internal memory. Fadel et al. (1999) developed a similar buffered heap. Using a partial rebuilding idea, Brodal and Katajainen (1998) developed a worst-case efficient external priority queue. A sequence of B operations on this structure requires $O(\log_{M/B} \frac{N}{B})$ I/Os. Using the buffer tree technique on a tournament tree, Kumar and Schwabe (1996) developed a priority queue supporting update operations in $O(\frac{1}{B} \log \frac{N}{B})$ I/Os. They also showed how to use their structure in several efficient external graph algorithms (see e.g Abello et al. 1998, Agarwal et al. 1998, Arge et al. 2000, Buchsbaum et al. 2000, Chiang et al. 1995, Hutchinson et al. 1999, Kumar and Schwabe 1996, Maheshwari and Zeh 1999, Munagala and Ranade 1999, Nodine et al. 1996, Ullman and Yannakakis 1991, Maheshwari and Zeh 2001, Feuerstein and Marchetti-Spaccamela 1993, for other results on external graph algorithms and data structures). Note that if the priority of an element is known, an update operation can be performed in $O(\frac{1}{B} \log_{M/B} \frac{N}{B})$ I/Os on a buffer tree using a delete and an insert operation.

11. Conclusions

In this chapter we have discussed recent advances in the development of provably efficient external memory dynamic data structures, mainly for geometric objects. Such structures are often crucial in massive dataset applications. We have discussed some of the most important techniques utilized to obtain efficient structures.

Even though a lot of progress has been made, many problems still remain open. For example, $O(\log_B N)$-query and space efficient structures still need to be found for many higher-dimensional problems. The practical performance of many of the worst-case efficient structures also needs to be researched.

Acknowledgments

The author thanks the National Science Foundation for partially supporting this work through ESS grant EIA–9870734, RI grant EIA–9972879 and CAREER grant EIA–9984099, and Tammy Bailey, Tavi Procopiuc, Jan Vahrenhold, as well as an anonymous reviewer, for comments on earlier drafts of this chapter.

Bibliography

D. J. Abel and D. M. Mark. A comparative analysis of some two-dimensional orderings. *Intl. J. Geographic Informations Systems*, 4

(1):21–31, 1990.

J. Abello, A. L. Buchsbaum, and J. R. Westbrook. A functional approach to external graph algorithms. In *Proc. Annual European Symposium on Algorithms, LNCS 1461*, pages 332–343, 1998.

P. K. Agarwal, L. Arge, G. S. Brodal, and J. S. Vitter. I/O-efficient dynamic point location in monotone planar subdivisions. In *Proc. ACM-SIAM Symp. on Discrete Algorithms*, pages 1116–1127, 1999.

P. K. Agarwal, L. Arge, and J. Erickson. Indexing moving points. In *Proc. ACM Symp. Principles of Database Systems*, pages 175–186, 2000a.

P. K. Agarwal, L. Arge, J. Erickson, P. Franciosa, and J. Vitter. Efficient searching with linear constraints. *Journal of Computer and System Sciences*, 61(2):194–216, 2000b.

P. K. Agarwal, L. Arge, and S. Govindarajan. External range counting. Manuscript, 2001a.

P. K. Agarwal, L. Arge, T. M. Murali, K. Varadarajan, and J. S. Vitter. I/O-efficient algorithms for contour line extraction and planar graph blocking. In *Proc. ACM-SIAM Symp. on Discrete Algorithms*, pages 117–126, 1998.

P. K. Agarwal, L. Arge, and J. Vahrenhold. A time responsive indexing scheme for moving points. Submitted, 2001b.

Pankaj K. Agarwal and Jeff Erickson. Geometric range searching and its relatives. In B. Chazelle, J. E. Goodman, and R. Pollack, editors, *Advances in Discrete and Computational Geometry*, volume 223 of *Contemporary Mathematics*, pages 1–56. American Mathematical Society, Providence, RI, 1999.

A. Aggarwal and J. S. Vitter. The Input/Output complexity of sorting and related problems. *Communications of the ACM*, 31(9):1116–1127, 1988.

L. Arge. The buffer tree: A new technique for optimal I/O-algorithms. In *Proc. Workshop on Algorithms and Data Structures, LNCS 955*, pages 334–345, 1995a. A complete version appears as BRICS technical report RS-96-28, University of Aarhus.

L. Arge. The I/O-complexity of ordered binary-decision diagram manipulation. In *Proc. Int. Symp. on Algorithms and Computation, LNCS*

1004, pages 82–91, 1995b. A complete version appears as BRICS technical report RS-96-29, University of Aarhus.

L. Arge, P. Ferragina, R. Grossi, and J. Vitter. On sorting strings in external memory. In *Proc. ACM Symp. on Theory of Computation*, pages 540–548, 1997.

L. Arge, K. H. Hinrichs, J. Vahrenhold, and J. S. Vitter. Efficient bulk operations on dynamic R-trees. In *Proc. Workshop on Algorithm Engineering, LNCS 1619*, pages 328–347, 1999a.

L. Arge, O. Procopiuc, S. Ramaswamy, T. Suel, and J. S. Vitter. Theory and practice of I/O-efficient algorithms for multidimensional batched searching problems. In *Proc. ACM-SIAM Symp. on Discrete Algorithms*, pages 685–694, 1998.

L. Arge, V. Samoladas, and J. S. Vitter. On two-dimensional indexability and optimal range search indexing. In *Proc. ACM Symp. Principles of Database Systems*, pages 346–357, 1999b.

L. Arge and S.-M. Teh, 2000. Unpublished results.

L. Arge, L. Toma, and J. S. Vitter. I/O-efficient algorithms for problems on grid-based terrains. In *Proc. Workshop on Algorithm Engineering and Experimentation*, 2000.

L. Arge and J. Vahrenhold. I/O-efficient dynamic planar point location. In *Proc. ACM Symp. on Computational Geometry*, pages 191–200, 2000.

L. Arge, D. E. Vengroff, and J. S. Vitter. External-memory algorithms for processing line segments in geographic information systems. In *Proc. Annual European Symposium on Algorithms, LNCS 979*, pages 295–310, 1995. To appear in special issues of *Algorithmica* on Geographical Information Systems.

L. Arge and J. S. Vitter. Optimal dynamic interval management in external memory. In *Proc. IEEE Symp. on Foundations of Comp. Sci.*, pages 560–569, 1996.

S. Arya, D. M. Mount, N. S. Netanyahu, R. Silverman, and A. Wu. An optimal algorithm for approximate nearest neighbor searching. In *Proc. 5th ACM-SIAM Sympos. Discrete Algorithms*, pages 573–582, 1994.

Tetsuo Asano, Desh Ranjan, Thomas Roos, Emo Welzl, and Peter Widmayer. Space-filling curves and their use in the design of geometric data structures. *Theoret. Comput. Sci.*, 181(1):3–15, July 1997.

J. Basch, L. J. Guibas, and J. Hershberger. Data structures for mobile data. *Journal of Algorithms*, 31(1):1–28, 1999.

H. Baumgarten, H. Jung, and K. Mehlhorn. Dynamic point location in general subdivisions. *Journal of Algorithms*, 17:342–380, 1994.

R. Bayer and E. McCreight. Organization and maintenance of large ordered indexes. *Acta Informatica*, 1:173–189, 1972.

B. Becker, S. Gschwind, T. Ohler, B. Seeger, and P. Widmayer. An asymptotically optimal multiversion B-tree. *VLDB Journal*, 5(4):264–275, 1996.

N. Beckmann, H.-P. Kriegel, R. Schneider, and B. Seeger. The R*-tree: An efficient and robust access method for points and rectangles. In *Proc. SIGMOD Intl. Conf. on Management of Data*, pages 322–331, 1990.

M. A. Bender, E. D. Demaine, and M Farach-Colton. Cache-oblivious B-trees. In *Proc. IEEE Symp. on Foundations of Comp. Sci.*, pages 339–409, 2000.

J. L. Bentley. Decomposable searching problems. *Information Processing Letters*, 8(5):244–251, 1979.

S. Berchtold, C. Böhm, D. A. Keim, and H.-P. Kriegel. A cost model for nearest neighbor search in high-dimensional data spaces. In *Proc. ACM Symp. Principles of Database Systems*, pages 78–86, 1997.

S. Berchtold, C. Böhm, and H-P. Kriegel. Improving the query performance of high-dimensional index structures by bulk load operations. In *Proc. Conference on Extending Database Technology, LNCS 1377*, pages 216–230, 1998a.

S. Berchtold, B. Ertl, D. A. Keim, H.-P. Kriegel, and T. Seidl. Fast nearest neighbor search in high-dimensional spaces. In *Proc. Annual IEEE Conference on Data Engineering*, pages 209–218, 1998b.

S. Berchtold, D. A. Keim, and H.-P. Kriegel. The X-tree: An index structure for high-dimensional data. In *Proc. International Conf. on Very Large Databases*, pages 28–39, 1996.

Sergei N. Bespamyatnikh. An optimal algorithm for closets pair maintenance. *Discrete and Computational Geometry*, 19:175–195, 1998.

G. Blankenagel and R. H. Güting. XP-trees—External priority search trees. Technical report, FernUniversität Hagen, Informatik-Bericht Nr. 92, 1990.

K. Brengel, A. Crauser, P. Ferragina, and U. Meyer. An experimental study of priority queues in external memory. In *Proc. Workshop on Algorithm Engineering, LNCS 1668*, pages 345–358, 1999.

G. S. Brodal and J. Katajainen. Worst-case efficient external-memory priority queues. In *Proc. Scandinavian Workshop on Algorithms Theory, LNCS 1432*, pages 107–118, 1998.

Adam L. Buchsbaum, Michael Goldwasser, Suresh Venkatasubramanian, and Jeffery R. Westbrook. On external memory graph traversal. In *Proc. ACM-SIAM Symp. on Discrete Algorithms*, pages 859–860, 2000.

P. Callahan, M. T. Goodrich, and K. Ramaiyer. Topology B-trees and their applications. In *Proc. Workshop on Algorithms and Data Structures, LNCS 955*, pages 381–392, 1995.

P. B. Callahan and S. R. Kosaraju. A decomposition of multidimensional point sets with applications to k-nearest-neighbors and n-body potential fields. *Journal of the ACM*, 42(1):67–90, 1995a.

Paul B. Callahan and S. Rao Kosaraju. Algorithms for dynamic closest-pair and n-body potential fields. In *Proc. 6th ACM-SIAM Sympos. Discrete Algorithms*, pages 263–272, 1995b.

T. M. Chan. Random sampling, halfspace range reporting, and construction of ($\leq k$)-levels in three dimensions. *SIAM Journal of Computing*, 30(2):561–575, 2000.

B. Chazelle and L. J. Guibas. Fractional cascading: I. A data structuring technique. *Algorithmica*, 1:133–162, 1986.

Bernard Chazelle. Filtering search: a new approach to query-answering. *SIAM J. Comput.*, 15(3):703–724, 1986.

Bernard Chazelle. Lower bounds for orthogonal range searching: I. the reporting case. *Journal of the ACM*, 37(2):200–212, April 1990.

Bernard Chazelle, Leonidas J. Guibas, and D. T. Lee. The power of geometric duality. *BIT*, 25(1):76–90, 1985.

S. W. Cheng and R. Janardan. New results on dynamic planar point location. *SIAM J. Comput.*, 21(5):972–999, 1992.

Y.-J. Chiang, M. T. Goodrich, E. F. Grove, R. Tamassia, D. E. Vengroff, and J. S. Vitter. External-memory graph algorithms. In *Proc. ACM-SIAM Symp. on Discrete Algorithms*, pages 139–149, 1995.

Y.-J. Chiang and C. T. Silva. I/O optimal isosurface extraction. In *Proc. IEEE Visualization*, pages 293–300, 1997.

Y.-J. Chiang and C. T. Silva. External memory techniques for isosurface extraction in scientific visualization. In J. Abello and J. S. Vitter, editors, *External memory algorithms*, volume 50 of *DIMACS series in Discrete Mathematics and Theoretical Computer Science*, pages 247–277. American Mathematical Society, 1999.

Y.-J. Chiang, C. T. Silva, and W. J. Schroeder. Interactive out-of-core isosurface extraction. In *Proc. IEEE Visualization*, pages 167–174, 1998.

P. Ciacca, M. Patella, and P. Zezula. M-tree: An efficient access method for similarity search in metric spaces. In *Proc. International Conf. on Very Large Databases*, pages 426–435, 1997.

D. Comer. The ubiquitous B-tree. *ACM Computing Surveys*, 11(2): 121–137, 1979.

T. H. Cormen, C. E. Leiserson, and R. L. Rivest. *Introduction to Algorithms*. The MIT Press, Cambridge, Mass., 1990.

A. Crauser and P. Ferragina. On constructing suffix arrays in external memory. In *Proc. Annual European Symposium on Algorithms, LNCS, 1643*, pages 224–235, 1999.

A. Crauser, P. Ferragina, K. Mehlhorn, U. Meyer, and E. Ramos. Randomized external-memory algorithms for some geometric problems. In *Proc. ACM Symp. on Computational Geometry*, pages 259–268, 1998.

M. de Berg, J. Gudmundsson, M. Hammar, and M. Overmars. On R-trees with low stabbing number. In *Proc. Annual European Symposium on Algorithms*, pages 167–178, 2000.

D. J. DeWitt, N. Kabra, J. Luo, J. M. Patel, and J.-B. Yu. Client-server paradise. In *Proceedings of VLDB Conference*, pages 558–569, 1994.

J. R. Driscoll, N. Sarnak, D. D. Sleator, and R. Tarjan. Making data structures persistent. *Journal of Computer and System Sciences*, 38: 86–124, 1989.

H. Edelsbrunner. A new approach to rectangle intersections, part I. *Int. J. Computer Mathematics*, 13:209–219, 1983a.

H. Edelsbrunner. A new approach to rectangle intersections, part II. *Int. J. Computer Mathematics*, 13:221–229, 1983b.

H. Edelsbrunner and M. Overmars. Batched dynamic solutions to decomposable searching problems. *Journal of Algorithms*, 6:515–542, 1985.

G. Evangelidis, D. Lomet, and B. Salzberg. The hb^π-tree: A multi-attribute index supporting concurrency, recovery and node consolidation. *The VLDB Journal*, 6(1):1–25, 1997.

R. Fadel, K. V. Jakobsen, J. Katajainen, and J. Teuhola. Heaps and heapsort on secondary storage. *Theoretical Computer Science*, 220 (2):345–362, 1999.

M. Farach, P. Ferragina, and S. Muthukrishnan. Overcoming the memory bottleneck in suffix tree construction. In *Proc. IEEE Symp. on Foundations of Comp. Sci.*, pages 174–183, 1998.

P. Ferragina and R. Grossi. A fully-dynamic data structure for external substring search. In *Proc. ACM Symp. on Theory of Computation*, pages 693–702, 1995.

P. Ferragina and R. Grossi. Fast string searching in secondary storage: Theoretical developments and experimental results. In *Proc. ACM-SIAM Symp. on Discrete Algorithms*, pages 373–382, 1996.

P. Ferragina and F. Luccio. Dynamic dictionary matching in external memory. *Information and Computation*, 146(2):85–99, 1998.

E. Feuerstein and A. Marchetti-Spaccamela. Memory paging for connectivity and path problems in graphs. In *Proc. Int. Symp. on Algorithms and Computation, LNCS 762*, pages 416–425, 1993.

P. Franciosa and M. Talamo. Time optimal halfplane search on external memory. Unpublished manuscript, 1997.

P. G. Franciosa and M. Talamo. Orders, k-sets and fast halfplane search on paged memory. In *Proc. Workshop on Orders, Algorithms and Applications (ORDAL'94), LNCS 831*, pages 117–127, 1994.

G. N. Frederickson. A structure for dynamically maintaining rooted trees. In *Proc. ACM-SIAM Symp. on Discrete Algorithms*, pages 175–184, 1993.

M. Frigo, C. E. Leiserson, H. Prokop, and S. Ramachandran. Cache-oblivious algorithms. In *Proc. IEEE Symp. on Foundations of Comp. Sci.*, pages 285–298, 1999.

V. Gaede and O. Günther. Multidimensional access methods. *ACM Computing Surveys*, 30(2):170–231, 1998.

M. T. Goodrich, J.-J. Tsay, D. E. Vengroff, and J. S. Vitter. External-memory computational geometry. In *Proc. IEEE Symp. on Foundations of Comp. Sci.*, pages 714–723, 1993.

S. Govindarajan, T. Lukovszki, A. Maheshwari, and N. Zeh. I/O-efficient well-separated pair decomposition and its applications. In *Proc. Annual European Symposium on Algorithms*, pages 220–231, 2000.

D. Greene. An implementation and performance analysis of spatial data access methods. In *Proc. IEEE International Conference on Data Engineering*, pages 606–615, 1989.

R. Grossi and G. F. Italiano. Efficient cross-tree for external memory. In J. Abello and J. S. Vitter, editors, *External Memory Algorithms*, volume 50 of *DIMACS series in Discrete Mathematics and Theoretical Computer Science*, pages 87–106. American Mathematical Society, 1999a. Revised version available at ftp://ftp.di.unipi.it/pub/techreports/TR-00-16.ps.Z.

R. Grossi and G. F. Italiano. Efficient splitting and merging algorithms for order decomposable problems. *Information and Computation*, 154 (1):1–33, 1999b.

O. Günther. The design of the cell tree: An object-oriented index structure for geometric databases. In *Proc. Annual IEEE Conference on Data Engineering*, pages 598–605, 1989.

A. Guttman. R-trees: A dynamic index structure for spatial searching. In *Proc. SIGMOD Intl. Conf. on Management of Data*, pages 47–57, 1984.

J. M. Hellerstein, E. Koutsoupias, and C. H. Papadimitriou. On the analysis of indexing schemes. In *Proc. ACM Symp. Principles of Database Systems*, pages 249–256, 1997.

K. H. Hinrichs. *The grid file system: Implementation and case studies of applications*. PhD thesis, Dept. Information Science, ETH, Zürich, 1985.

G. R. Hjaltason and H. Samet. Ranking in spatial databases. In *Proc. of Advances in Spatial Databases, LNCS 951*, pages 83–95, 1995.

S. Huddleston and K. Mehlhorn. A new data structure for representing sorted lists. *Acta Informatica*, 17:157–184, 1982.

D. Hutchinson, A. Maheshwari, J-R. Sack, and R. Velicescu. Early experiences in implementing the buffer tree. In *Proc. Workshop on Algorithm Engineering*, pages 92–103, 1997.

D. Hutchinson, A. Maheshwari, and N. Zeh. An external-memory data structure for shortest path queries. In *Proc. Annual Combinatorics and Computing Conference, LNCS 1627*, pages 51–60, 1999.

Ch. Icking, R. Klein, and Th. Ottmann. Priority search trees in secondary memory. In *Proc. Graph-Theoretic Concepts in Computer Science, LNCS 314*, pages 84–93, 1987.

I. Kamel and C. Faloutsos. On packing R-trees. In *Proc. International Conference on Information and Knowledge Management*, pages 490–499, 1993.

I. Kamel and C. Faloutsos. Hilbert R-tree: An improved R-tree using fractals. In *Proc. International Conf. on Very Large Databases*, pages 500–509, 1994.

P. C. Kanellakis, S. Ramaswamy, D. E. Vengroff, and J. S. Vitter. Indexing for data models with constraints and classes. *Journal of Computer and System Sciences*, 52(3):589–612, 1996.

K. V. R. Kanth and A. K. Singh. Optimal dynamic range searching in non-replicating index structures. In *Proc. International Conference on Database Theory, LNCS 1540*, pages 257–276, 1999.

N. Katayama and S. Satoh. The SR-tree: An index structure for high-dimensional nearest-neighbor queries. In *Proc. SIGMOD Intl. Conf. on Management of Data*, pages 369–380, 1997.

D. E. Knuth. *Sorting and Searching*, volume 3 of *The Art of Computer Programming*. Addison-Wesley, Reading MA, second edition, 1998.

G. Kollios, D. Gunopulos, and V. J. Tsotras. Nearest neighbor queries in a mobile environment. In *Proc. International Workshop on Spatio-Temporal Database Management, LNCS 1678*, pages 119–134, 1999a.

G. Kollios, D. Gunopulos, and V. J. Tsotras. On indexing mobile objects. In *Proc. Annu. ACM Sympos. Principles Database Syst.*, pages 261–272, 1999b.

F. Korn, N. Sidiropoulos, C. Faloutsos, E. Siegel, and Z. Protopapas. Fast nearest neighbor search in medical image databases. In *Proc. International Conf. on Very Large Databases*, pages 215–226, 1996.

E. Koutsoupias and D. S. Taylor. Tight bounds for 2-dimensional indexing schemes. In *Proc. ACM Symp. Principles of Database Systems*, pages 52–58, 1998.

V. Kumar and E. Schwabe. Improved algorithms and data structures for solving graph problems in external memory. In *Proc. IEEE Symp. on Parallel and Distributed Processing*, pages 169–177, 1996.

S. T. Leutenegger, M. A. López, and J. Edgington. STR: A simple and efficient algorithm for R-tree packing. In *Proc. Annual IEEE Conference on Data Engineering*, pages 497–506, 1996.

D.B. Lomet and B. Salzberg. The hB-tree: A multiattribute indexing method with good guaranteed performance. *ACM Transactions on Database Systems*, 15(4):625–658, 1990.

A. Maheshwari and N. Zeh. External memory algorithms for outerplanar graphs. In *Proc. Int. Symp. on Algorithms and Computation, LNCS 1741*, pages 307–316, 1999.

A. Maheshwari and N. Zeh. I/O-efficient algorithms for bounded treewidth graphs. In *Proc. ACM-SIAM Symp. on Discrete Algorithms*, pages 89–90, 2001.

J. Matoušek. Efficient partition trees. *Discrete Comput. Geom.*, 8:315–334, 1992.

E.M. McCreight. Priority search trees. *SIAM Journal of Computing*, 14(2):257–276, 1985.

K. Mehlhorn. *Data Structures and Algorithms 1: Sorting and Searching*. Springer-Verlag, EATCS Monographs on Theoretical Computer Science, 1984.

Kurt Mehlhorn and Stefan Näher. Dynamic fractional cascading. *Algorithmica*, 5:215–241, 1990.

D. R. Morrison. PATRICIA: Practical algorithm to retrieve information coded in alphanumeric. *Journal of the ACM*, 15:514–534, 1968.

K. Munagala and A. Ranade. I/O-complexity of graph algorithm. In *Proc. ACM-SIAM Symp. on Discrete Algorithms*, pages 687–694, 1999.

J. Nievergelt, H. Hinterberger, and K.C. Sevcik. The grid file: An adaptable, symmetric multikey file structure. *ACM Transactions on Database Systems*, 9(1):38–71, 1984.

J. Nievergelt and E. M. Reingold. Binary search tree of bounded balance. *SIAM Journal of Computing*, 2(1):33–43, 1973.

J. Nievergelt and P. Widmayer. Spatial data structures: Concepts and design choices. In M. van Kreveld, J. Nievergelt, T. Roos, and P. Widmayer, editors, *Algorithmic Foundations of GIS*, pages 153–197. Springer-Verlag, LNCS 1340, 1997.

M. H. Nodine, M. T. Goodrich, and J. S. Vitter. Blocking for external graph searching. *Algorithmica*, 16(2):181–214, 1996.

J. Orenstein. A comparison of spatial query processing techniques for native and parameter spaces. In *Proc. SIGMOD Intl. Conf. on Management of Data*, pages 343–352, 1990.

J.A. Orenstein. Spatial query processing in an object-oriented database system. In *Proc. ACM SIGMOD Conf. on Management of Data*, pages 326–336, 1986.

M. H. Overmars. Range searching in a set of line segments. In *Proc. 1st Annu. ACM Sympos. Comput. Geom.*, pages 177–185, 1985.

Mark H. Overmars. *The Design of Dynamic Data Structures*. Springer-Verlag, LNCS 156, 1983.

A. Papadopoulos and Y. Manolopoulos. Performance of nearest neighbor queries in R-trees. In *Intl. Conference on Database Theory, LNCS 1186*, pages 394–408, 1997.

D. Pfoser, C. S. Jensen, and Y. Theodoridis. Novel approaches to the indexing of moving objects trajectories. In *Proc. International Conf. on Very Large Databases*, pages 395–406, 2000.

S. Ramaswamy and S. Subramanian. Path caching: A technique for optimal external searching. In *Proc. ACM Symp. Principles of Database Systems*, pages 25–35, 1994.

Sridhar Ramaswamy. Efficient indexing for constraint and temporal databases. In *Proc. International Conference on Database Theory, LNCS 1186*, pages 419–431, 1997.

J.T. Robinson. The K-D-B tree: A search structure for large multidimensional dynamic indexes. In *Proc. SIGMOD Intl. Conf. on Management of Data*, pages 10–18, 1981.

N. Roussopoulos, S. Kelley, and F. Vincent. Nearest neighbor queries. In *Proc. SIGMOD Intl. Conf. on Management of Data*, pages 71–79, 1995.

N. Roussopoulos and D. Leifker. Direct spatial search on pictorial databases using packed R-trees. In *Proc. SIGMOD Intl. Conf. on Management of Data*, pages 17–31, 1985.

Chris Ruemmler and John Wilkes. An introduction to disk drive modeling. *IEEE Computer*, 27(3):17–28, 1994.

B. Salzberg and V. J. Tsotras. A comparison of access methods for time evolving data. *ACM Computing Surveys*, 31(2):158–221, 1999.

H. Samet. *Applications of Spatial Data Structures: Computer Graphics, Image Processing, and GIS*. Addison Wesley, MA, 1990a.

H. Samet. *The Design and Analyses of Spatial Data Structures*. Addison Wesley, MA, 1990b.

V. Samoladas and D. Miranker. A lower bound theorem for indexing schemes and its application to multidimensional range queries. In *Proc. ACM Symp. Principles of Database Systems*, pages 44–51, 1998.

P. Sanders. Fast priority queues for cached memory. In *Proc. Workshop on Algorithm Engineering and Experimentation, LNCS 1619*, pages 312–327, 1999.

N. Sarnak and R. E. Tarjan. Planar point location using persistent search trees. *Communications of the ACM*, 29:669–679, 1986.

B. Seeger and H.-P. Kriegel. The buddy-tree: An efficient and robust access method for spatial data base systems. In *Proc. International Conf. on Very Large Databases*, pages 590–601, 1990.

T. Seidl and H.-P. Kriegel. Efficient user-adaptable similarity search in large multimedia databases. In *Proc. International Conf. on Very Large Databases*, pages 506–515, 1997.

T. Sellis, N. Roussopoulos, and C. Faloutsos. The R^+-tree: A dynamic index for multi-dimensional objects. In *Proc. International Conf. on Very Large Databases*, pages 507–518, 1987.

J. Snoeyink. Point location. In Jacob E. Goodman and Joseph O'Rourke, editors, *Handbook of Discrete and Computational Geometry*, chapter 30, pages 559–574. CRC Press LLC, Boca Raton, FL, 1997.

R. F. Sproull. Refinements to nearest neighbor searching in k-dimensional trees. *Algorithmica*, 6(4):579–589, 1991.

S. Subramanian and S. Ramaswamy. The P-range tree: A new data structure for range searching in secondary memory. In *Proc. ACM-SIAM Symp. on Discrete Algorithms*, pages 378–387, 1995.

R. Tamassia and J. S. Vitter. Optimal cooperative search in fractional cascaded data structures. *Algorithmica*, 15(2):154–171, 1996.

J. Tayeb, O. Ulusoy, and O. Wolfson. A quadtree-based dynamic attribute indexing method. *The Computer Journal*, 41(3):185–200, 1998.

J. D. Ullman and M. Yannakakis. The input/output complexity of transitive closure. *Annals of Mathematics and Artificial Intellegence*, 3: 331–360, 1991.

J. Vahrenhold and K. H. Hinrichs. Planar point-location for large data sets: To seek or not to seek. In *Proc. Workshop on Algorithm Engineering*, 2000.

J. van den Bercken, B. Seeger, and P. Widmayer. A generic approach to bulk loading multidimensional index structures. In *Proc. International Conf. on Very Large Databases*, pages 406–415, 1997.

J. van den Bercken, B. Seeger, and P. Widmayer. A generic approach to processing non-equijoins. Technical Report 14, Philipps-Universität Marburg, Fachbereich Matematik und Informatik, 1998.

M. J. van Kreveld and M. H. Overmars. Divided k-d trees. *Algorithmica*, 6:840–858, 1991.

P. J. Varman and R. M. Verma. An efficient multiversion access structure. *IEEE Transactions on Knowledge and Data Engineering*, 9(3): 391–409, 1997.

D. E. Vengroff and J. S. Vitter. Efficient 3-D range searching in external memory. In *Proc. ACM Symp. on Theory of Computation*, pages 192–201, 1996.

J. S. Vitter. External memory algorithms and data structures. In J. Abello and J. S. Vitter, editors, *External Memory Algorithms*, volume 50 of *DIMACS series in Discrete Mathematics and Theoretical Computer Science*, pages 1–38. American Mathematical Society, 1999a.

J. S. Vitter. Online data structures in external memory. In *Proc. Annual International Colloquium on Automata, Languages, and Programming, LNCS 1644*, pages 119–133, 1999b.

J. S. Vitter and E. A. M. Shriver. Algorithms for parallel memory, I: Two-level memories. *Algorithmica*, 12(2-3):110–147, 1994.

S. Šaltenis, C. S. Jensen, S. T. Leutenegger, and M. A. López. Indexing the positions of continuously moving objects. In *Proc. SIGMOD Intl. Conf. on Management of Data*, pages 331–342, 2000.

D. A. White and R. Jain. Similarity indexing with the SS-tree. In *Proc. Annual IEEE Conference on Data Engineering*, pages 516–523, 1996.

O. Wolfson, A. P. Sistla, S. Chamberlain, and Y. Yesha. Updating and querying databases that track mobile units. *Distributed and Parallel Databases*, 7(3):257–287, 1999.

O. Wolfson, B. Xu, S. Chamberlain, and L. Jiang. Moving objects databases: Issues and solutions. In *Intl. Conf. on Scientific and Statistical Database Management*, pages 111–122, 1998.

D. Zhang, A. Markowetz, V. Tsotras, D. Gunopulos, and B. Seeger. Efficient computation of temporal aggregates with range predicates. In *Proc. ACM Symp. Principles of Database Systems*, pages 237–245, 2001.

Chapter 10

EXTERNAL MEMORY ALGORITHMS

Jeffrey Scott Vitter *
Department of Computer Science, Duke University, Durham, NC 27708-0129, USA
jsv@cs.duke.edu

Abstract   Data sets in large applications are often too massive to fit completely inside the computer's internal memory. The resulting input/output communication (or I/O) between fast internal memory and slower external memory (such as disks) can be a major performance bottleneck. In this paper we survey the state of the art in the design and analysis of *external memory* (or *EM*) *algorithms*, where the goal is to exploit locality in order to reduce the I/O costs.

For sorting and related problems like permuting and fast Fourier transform, the key paradigms include distribution and merging. The paradigm of disk striping offers an elegant way to use multiple disks in parallel. For sorting, however, disk striping can be nonoptimal with respect to I/O, so to gain further improvements we discuss distribution and merging techniques for using the disks independently. We consider EM paradigms for computations involving matrices, geometric data, and graphs, and we look at problems caused by dynamic memory allocation. We report on some experiments in the domain of spatial databases using the TPIE system (Transparent Parallel I/O programming Environment). The newly developed EM algorithms and data structures that incorporate the paradigms we discuss in this chapter are significantly faster than methods currently used in practice.

Keywords: External memory, Disk block, I/O complexity, Out-of-core, Batched sorting, Fourier transform, Parallel disk model, Locality and load balancing, Disk striping, RAID, Batched computational geometry, Batched graph problems, Indivisibility assumption.

---

*Supported in part by Army Research Office MURI grant DAAH04–96–1–0013 and by National Science Foundation research grants CCR–9522047, EIA–9870734, and CCR–9877133. Part of this work was done at INRIA, Sophia Antipolis, France.
*Earlier versions of some of the material in this chapter appeared in (Vitter 1999a; 1998; 1999b;c; 2001).

1. Introduction

In large applications, data sets are often too massive to fit completely inside the computer's internal memory, and they must be stored on larger but slower external memory devices, such as magnetic disks. The *Input/Output* communication (or simply *I/O*) between the fast internal memory and the slow external memory can be a major bottleneck.

Modern programming languages and operating systems support a virtual memory abstraction, in which all of memory is presented as one large uniform address space. In order to cope with the I/O bottleneck, they employ general-purpose caching and prefetching mechanisms, in which the more frequently used data are kept locally in internal memory. If the user or program requests access to data that are not cached, a page fault is generated, and I/O is done to retrieve the data and bring them into internal memory. By their general-purpose nature, caching and prefetching methods cannot be expected to take full advantage of the locality present in every computation. Some computations themselves are inherently non-local, and even with omniscient cache management decisions they are doomed to perform large amounts of I/O and suffer poor performance. Substantial gains in performance may be possible by bypassing the virtual memory system and incorporating locality *directly* into the algorithm design. We refer to such algorithms as *external memory* (or *EM*) *algorithms*. Some authors use the equivalent terms *I/O algorithms* or *out-of-core algorithms*.

In this chapter we discuss several techniques and challenges for how to exploit locality and reduce I/O costs when solving problems in external memory. We concentrate on generic *batched* problems, in which no preprocessing is done and the entire file of data items must be processed, often by streaming the data through the internal memory in one or more passes. For discussion of online problems we refer the reader to the survey by Arge (2001) in this volume and by Vitter (2001). Other chapters in this volume discuss batched and online problems in specific application domains.

We base our approach upon the *parallel disk model* (PDM) described in the next section. PDM provides an elegant and reasonably accurate model for analyzing the relative performance of EM algorithms and data structures. The three main performance measures of PDM are *the number of I/O operations, the disk space usage,* and *the CPU time*. For reasons of brevity, we focus on the first two measures. Most of the algorithms we consider are also efficient in terms of CPU time. We also mention other memory models and practical considerations.

In Section 5 we look at the canonical batched EM problem of external sorting and some related problems. The two important paradigms of distribution and merging account for all well known external sorting algorithms. Sorting with a single disk is now well understood, so we concentrate on the more challenging task of using multiple (or parallel) disks, for which disk striping is nonoptimal. The challenge is to guarantee that the data in each I/O is spread evenly across the disks so that the disks can be used simultaneously. We also cover the fundamental lower bounds on the number of I/Os needed to perform sorting and related batched problems. In Section 6, we discuss grid and linear algebra batched computations.

For most problems, parallel disks can be utilized effectively by means of disk striping or the parallel disk techniques of Section 5, and hence we restrict ourselves starting in Section 7 to the conceptually simpler single-disk case. In Section 7 we mention several effective paradigms for batched EM problems in computational geometry. The paradigms include distribution sweep (for spatial join and finding all nearest neighbors), persistent B-trees (for batched point location and graph drawing), batched filtering (for 3-D convex hulls and batched point location), external fractional cascading (for red-blue line segment intersection), external marriage-before-conquest (for output-sensitive convex hulls), and randomized incremental construction with gradations (for line segment intersections and other geometric problems). In Section 8 we look at EM algorithms for combinatorial problems on graphs, such as list ranking, connected components, topological sorting, and finding shortest paths. One technique for constructing I/O-efficient EM algorithms is to simulate parallel algorithms; sorting is used between parallel steps in order to reblock the data for the simulation of the next parallel step. In Section 9 we discuss EM algorithms that adapt optimally to dynamically changing internal memory allocations. We make concluding remarks in Section 11. Table 10.1 lists several of the EM paradigms discussed in this paper.

In Section 10 we discuss programming environments and tools that facilitate high-level development of efficient EM algorithms. We focus on the TPIE system (Transparent Parallel I/O Environment), which we used in the experiments reported in Section 7.

2. Parallel Disk Model (PDM)

EM algorithms explicitly control data placement and movement, and thus it is important for algorithm designers to have a simple but reasonably accurate model of the memory system's characteristics. Magnetic

Table 10.1. Paradigms for I/O efficiency discussed in this paper.

Paradigm	Reference
Batched dynamic processing	§7
Batched filtering	§7
Batched incremental construction	§7
Disk striping	§4.2
Distribution	§5.1
Distribution Sweeping	§7
Externalization	§7
Fractional cascading	§7
Load balancing	§4
Locality	§4
Marriage before conquest	§7
Merging	§5.2
Parallel simulation	§8
Persistence	§7
Random sampling	§5.1
Scanning (or streaming)	§2.2
Sparsification	§8

disks consist of one or more rotating platters and one read/write head per platter surface. The data are stored on the platters in concentric circles called *tracks*, as shown in Figure 10.1. To read or write a data item at a certain address on disk, the read/write head must mechanically *seek* to the correct track and then wait for the desired address to pass by. The seek time to move from one random track to another is often on the order of 5–10 milliseconds, and the average rotational latency, which is the time for half a revolution, has the same order of magnitude. In contrast, access to data stored in internal memory is about one million times faster! In order to amortize this delay due to seek time and rotational latency, it pays to transfer a large contiguous group of data items, called a *block*. Similar considerations apply to all levels of the memory hierarchy.

Even if an application can structure its pattern of memory accesses to exploit locality and take full advantage of disk block transfer, there is still a substantial *access gap* between internal memory performance and external memory performance. In fact the access gap is growing, since the latency and bandwidth of memory chips are improving more quickly than those of disks. Use of parallel processors further widens the gap. Storage systems such as RAID deploy multiple disks in order to get additional bandwidth (Chen et al. 1994, Hellerstein et al. 1994).

EXTERNAL MEMORY ALGORITHMS

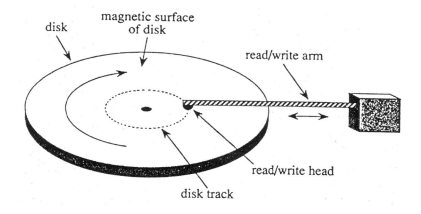

Figure 10.1. Platter of a magnetic disk drive.

In the next section we describe the high-level parallel disk model (PDM), which we use throughout this paper for the design and analysis of EM algorithms and data structures. In Section 2.2 we consider some practical modeling issues dealing with the sizes of blocks and tracks and the corresponding parameter values in PDM. In Section 2.3 we review the historical development of models of I/O and hierarchical memory.

2.1.   PDM and problem parameters

We can capture the main properties of magnetic disks and multiple disk systems by the commonly used *parallel disk model* (PDM) introduced by Vitter and Shriver (1994a):

N = problem size (in units of data items);
M = internal memory size (in units of data items);
B = block transfer size (in units of data items);
D = number of independent disk drives;
P = number of CPUs,

where $M < N$, and $1 \le DB \le M/2$. The data items are assumed to be of fixed length. In a single I/O, each of the D disks can simultaneously transfer a block of B contiguous data items.

If $P \le D$, each of the P processors can drive about D/P disks; if $D < P$, each disk is shared by about P/D processors. The internal memory size is M/P per processor, and the P processors are connected by an interconnection network. For routing considerations, one desired property for the network is the capability to sort the M data items in the collective main memories of the processors in parallel in optimal $O((M/P)\log M)$ time[1]. The special cases of PDM for the case of a single processor ($P = 1$) and multiprocessors with one disk per processor ($P = D$) are pictured in Figure 10.2.

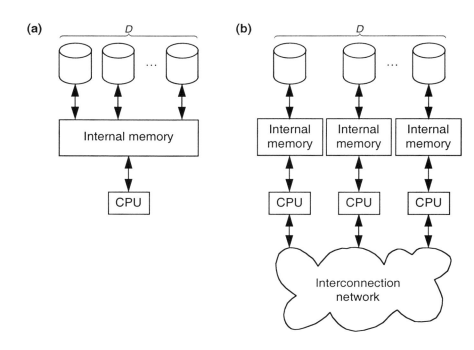

Figure 10.2. Parallel disk model: (a) $P = 1$, in which the D disks are connected to a common CPU; (b) $P = D$, in which each of the D disks is connected to a separate processor.

Queries are naturally associated with online computations, which are the subject of the chapter on external memory data structures, but they can also be done in batched mode. For example, in the batched orthogonal 2-D range searching problem discussed in Section 7, we are given a set of N points in the plane and a set of Q queries in the form of rectangles, and the problem is to report the points lying in each of the Q query rectangles. The number of items reported in response to each query may

---

[1] We use the notation $\log n$ to denote the binary (base 2) logarithm $\log_2 n$. For bases other than 2, the base will be specified explicitly.

vary. We thus need to define two more performance parameters:

Q = number of input queries (for a batched problem);
Z = query output size (in units of data items).

It is convenient to refer to some of the above PDM parameters in units of disk blocks rather than in units of data items; the resulting formulas are often simplified. We define the lower-case notation

$$n = \frac{N}{B}, \quad m = \frac{M}{B}, \quad q = \frac{Q}{B}, \quad z = \frac{Z}{B} \qquad (10.1)$$

to be the problem input size, internal memory size, query specification size, and query output size, respectively, in units of disk blocks.

We assume that the input data are initially "striped" across the D disks, in units of blocks, as illustrated in Figure 10.3, and we require the output data to be similarly striped. Striped format allows a file of N data items to be read or written in $O(N/DB) = O(n/D)$ I/Os, which is optimal.

	$\mathcal{D}_0$	$\mathcal{D}_1$	$\mathcal{D}_2$	$\mathcal{D}_3$	$\mathcal{D}_4$
stripe 0	0 1	2 3	4 5	6 7	8 9
stripe 1	10 11	12 13	14 15	16 17	18 19
stripe 2	20 21	22 23	24 25	26 27	28 29
stripe 3	30 31	32 33	34 35	36 37	38 39

Figure 10.3. Initial data layout on the disks, for $D = 5$ disks and block size $B = 2$. The input data items are initially striped block-by-block across the disks. For example, data items 16 and 17 are stored in the second block (i.e., in stripe 1) of disk $\mathcal{D}_3$.

The three primary measures of performance in PDM are

1 the number of I/O operations performed,

2 the amount of disk space used, and

3 the internal (sequential or parallel) computation time.

For reasons of brevity we shall focus in this paper on only the first two measures. Most of the algorithms we mention run in optimal CPU time, at least for the single-processor case. Ideally algorithms and data structures should use linear space, which means $O(N/B) = O(n)$ disk blocks of storage.

2.2. Practical considerations

Track size is a fixed parameter of the disk hardware; for most disks it is in the range 50–200 KB. In reality, the track size for any given disk depends upon the radius of the track (cf. Figure 10.1). Sets of adjacent tracks are usually formatted to have the same track size, so there are typically only a small number of different track sizes for a given disk. A single disk can have a 3 : 2 variation in track size (and therefore bandwidth) between its outer tracks and the inner tracks.

The minimum block transfer size imposed by hardware is often 512 bytes, but operating systems generally use a larger block size, such as 8 KB. It is possible (and preferable in batched applications) to use logical blocks of larger size (sometimes called clusters) and further reduce the relative significance of seek and rotational latency, but the wall clock time per I/O will increase accordingly. For example, if we set PDM parameter B to be five times larger than the track size, so that each logical block corresponds to five contiguous tracks, the time per I/O will correspond to five revolutions of the disk plus the (now relatively less significant) seek time and rotational latency. If the disk is smart enough, rotational latency can even be avoided altogether, since the block spans entire tracks and reading can begin as soon as the read head reaches the desired track. Once the block transfer size becomes larger than the track size, the wall clock time per I/O grows linearly with the block size.

For best results in batched applications, especially when the data are streamed sequentially through internal memory, the block transfer size B in PDM should be considered to be a fixed hardware parameter a little larger than the track size (say, on the order of 100 KB for most disks), and the time per I/O should be adjusted accordingly. For online applications that use pointer-based indexes, a smaller B value such as 8 KB is appropriate The particular block size that optimizes performance may vary somewhat from application to application.

PDM is a good generic programming model that facilitates elegant design of I/O-efficient algorithms, especially when used in conjunction with the programming tools discussed in Section 10. More complex and precise disk models, such as the ones by Ruemmler and Wilkes (1994), Ganger (1995), Shriver et al. (1998), Barve et al. (1999), and Farach et al. (1998), distinguish between sequential reads and random reads and consider the effects of features such as disk buffer caches and shared buses, which can reduce the time per I/O by eliminating or hiding the seek time. For example, algorithms for spatial join that access preexisting index structures (and thus do random I/O) can often be slower in practice than algorithms that access substantially more data but in a sequential order (as in scanning) (Arge et al. 2000b). It is thus helpful not only to consider the number of block transfers, but also to distinguish between the I/Os that are random versus those that are sequential. In some applications, automated dynamic block placement can improve disk locality and help reduce I/O time (Seltzer et al. 1995).

Another simplification of PDM is that the D block transfers in each I/O are *synchronous*; they are assumed to take the same amount of time. This assumption makes it easier to design and analyze algorithms for multiple disks. In practice, however, if the disks are used independently, some block transfers will complete more quickly than others. We can often improve overall elapsed time if the I/O is done *asynchronously*, so that disks get utilized as soon as they become available. Buffer space in internal memory can be used to queue the read and write requests for each disk.

2.3. Related memory models

The study of problem complexity and algorithm analysis when using EM devices began more than 40 years ago with Demuth's Ph.D. thesis on sorting (Demuth 1956, Knuth 1998). In the early 1970s, Knuth (1998) did an extensive study of sorting using magnetic tapes and (to a lesser extent) magnetic disks. At about the same time, Floyd (1972), Knuth (1998) considered a disk model akin to PDM for $D = 1$, $P = 1$, $B = M/2 = \Theta(N^c)$, for constant $c > 0$, and developed optimal upper and lower I/O bounds for sorting and matrix transposition. Hong and Kung (1981) developed a pebbling model of I/O for straightline computations, and Savage and Vitter (1987) extended the model to deal with block transfer. Aggarwal and Vitter (1988) generalized Floyd's I/O model to allow D simultaneous block transfers, but the model was unrealistic in that the D simultaneous transfers were allowed to take place on a single disk. They developed matching upper and lower I/O bounds for all parameter values for a host of problems. Since the PDM model can be thought of as a more restrictive (and more realistic) version of Aggarwal and Vitter's model, their lower bounds apply as well to PDM. In Section 5.3 we discuss a recent simulation technique due to Sanders et al. (2000); the Aggarwal-Vitter model can be simulated probabilistically by PDM with only a constant factor more I/Os, thus making the two models theoretically equivalent in the randomized sense. Deterministic simulations on the other hand require a factor of $\log(N/D)/\log\log(N/D)$ more I/Os (Armen 1996).

Surveys of I/O models, algorithms, and challenges appear in (Arge 1997; 2001, Gibson et al. 1996, Shriver and Nodine 1996). Several versions of PDM have been developed for parallel computation (Dehne et al. 1999, Li et al. 1996, Sibeyn and Kaufmann 1997). Models of "active disks" augmented with processing capabilities to reduce data traffic to the host, especially during scanning applications, are given in (Acharya et al. 1998, Riedel et al. 1998). Models of microelectromechanical systems (MEMS) for mass storage appear in Griffin et al. (2000).

Some authors have studied problems that can be solved efficiently by making only one pass (or a small number of passes) over the data (Feigenbaum et al. 1999, Henzinger et al. 1999). One approach to reduce the internal memory requirements is to require only an approximate answer to the problem; the more memory available, the better the approximation. A related approach to reducing I/O costs for a given problem is to use random sampling or data compression in order to construct a smaller version of the problem whose solution approximates the original. We refer the reader to other chapters in this volume for related topics; those approaches are orthogonal to the focus of this chapter.

The same type of bottleneck that occurs between internal memory (DRAM) and external disk storage can also occur at other levels of the memory hierarchy, such as between registers and level 1 cache, between level 1 cache and level 2 cache, between level 2 cache and DRAM, and between disk storage and tertiary devices. The PDM model can be generalized to model the hierarchy of memories ranging from registers at the small end to tertiary storage at the large end. Optimal algorithms for PDM often generalize in a recursive fashion to yield optimal algorithms in the hierarchical memory models (Aggarwal et al. 1987b;a, Vitter and Shriver 1994b, Vitter and Nodine 1993). Conversely, the algorithms for hierarchical models can be run in the PDM setting, and in that setting many have the interesting property that they use no explicit knowledge of the PDM parameters like M and B. Frigo et al. (1999) and Bender et al. (2000) develop cache-oblivious algorithms and data structures that require no knowledge of the storage parameters.

However, the match between theory and practice is harder to establish for hierarchical models and caches than for disks. The simpler hierarchical models are less accurate, and the more practical models are architecture-specific. The relative memory sizes and block sizes of the levels vary from computer to computer. Another issue is how blocks from one memory level are stored in the caches at a lower level. When a disk block is read into internal memory, it can be stored in any specified DRAM location. However, in level 1 and level 2 caches, each item can only be stored in certain cache locations, often determined by a hardware modulus computation on the item's memory address. The number of possible storage locations in cache for a given item is called the level of associativity. Some caches are direct-mapped (i.e., with associativity 1), and most caches have fairly low associativity (typically at most 4).

Another reason why the hierarchical models tend to be more architecture specific is that the relative difference in speed between level 1 cache and level 2 cache or between level 2 cache and DRAM is orders of magnitude smaller than the relative difference in latencies between DRAM and the disks. Yet, it is apparent that good EM design principles are useful in developing cache-efficient algorithms. For example, sequential internal memory access is much faster than random access, by about a factor of 10, and the more we can build locality into an algorithm, the faster it will run in practice. By properly engineering the "inner loops", a programmer can often significantly speed up the overall running time. Tools such as simulation environments and system monitoring utilities (Knuth 1999, Rosenblum et al. 1997, Srivastava and Eustace 1994) can provide sophisticated help in the optimization process.

For reasons of focus, we do not consider such hierarchical models and caching issues in this paper. We refer the reader to the following references: Aggarwal et al. (1987a) define an elegant hierarchical memory model, and Aggarwal et al. (1987b) augment it with block transfer capability. Alpern et al. (1994) model levels of memory in which the memory size, block size, and bandwidth grow at uniform rates. Vitter and Shriver (1994b) and Vitter and Nodine (1993) discuss parallel versions and variants of the hierarchical models. The parallel model of Li et al. (1996) also applies to hierarchical memory. Savage (1995) gives a hierarchical pebbling version of (Savage and Vitter 1987). Carter and Gatlin (1998) define pebbling models of nonassociative direct-mapped caches. Rahman and Raman (2000) and Sen and Chatterjee (2000) apply EM techniques to models of caches and translation lookaside buffers. Rao and Ross (1999; 2000) use B-tree techniques to exploit locality for the design of cache-conscious search trees.

3. Fundamental I/O Bounds

The I/O performance of many batched algorithms can be expressed in terms of the bounds for the following three fundamental operations:

1 *Scanning* (a.k.a. *streaming* or *touching*) a file of N data items, which involves the sequential reading or writing of the items in the file.

2 *Sorting* a file of N data items, which puts the items into sorted order.

3 *Outputting* the Z answers to a query in a blocked "output-sensitive" fashion.

We give the I/O bounds for these four operations in Table 10.2. We separately list the special case of a single disk ($D = 1$), since the formulas are simpler and many of the discussions in this paper will be restricted to the single-disk case.

Table 10.2. I/O bounds for the four fundamental operations. The PDM parameters are defined in Section 2.1.

Operation	I/O bound, $D = 1$	I/O bound, general $D \geq 1$
$Scan(N)$	$\Theta\left(\frac{N}{B}\right) = \Theta(n)$	$\Theta\left(\frac{N}{DB}\right) = \Theta\left(\frac{n}{D}\right)$
$Sort(N)$	$\Theta\left(\frac{N}{B} \log_{M/B} \frac{N}{B}\right)$ $= \Theta(n \log_m n)$	$\Theta\left(\frac{N}{DB} \log_{M/B} \frac{N}{B}\right)$ $= \Theta\left(\frac{n}{D} \log_m n\right)$
$Output(Z)$	$\Theta\left(\max\left\{1, \frac{Z}{B}\right\}\right)$ $= \Theta(\max\{1, z\})$	$\Theta\left(\max\left\{1, \frac{Z}{DB}\right\}\right)$ $= \Theta\left(\max\left\{1, \frac{z}{D}\right\}\right)$

The I/O bound $Scan(N) = O(n/D)$, which is clearly required to read or write a file of N items, represents a *linear number of I/Os* in the PDM model. An interesting feature of the PDM model is that almost all nontrivial batched problems require a nonlinear number of I/Os, even those that can be solved easily in linear CPU time in the (internal memory) RAM model. Examples we shall discuss later include permuting, transposing a matrix, list ranking, and several combinatorial graph problems. Many of these problems are equivalent in I/O complexity to permuting or sorting.

The linear I/O bounds for $Scan(N)$ and $Output(Z)$ are trivial. The algorithms and lower bounds for $Sort(N)$ are relatively new and are discussed in later sections. As Table 10.2 indicates, the multiple-disk I/O bounds for $Scan(N)$, $Sort(N)$, and $Output(Z)$ are D times smaller than the corresponding single-disk I/O bounds; such a speedup is clearly the best improvement possible with D disks.

In practice, the logarithmic term $\log_m n$ in the $Sort(N)$ bound is a small constant. For example, in units of items, we could have $N = 10^{10}$, $M = 10^7$, and $B = 10^4$, and thus we get $n = 10^6$, $m = 10^3$, and $\log_m n = 2$, in which case sorting can be done in a linear number of

I/Os. If memory is shared with other processes, the $\log_m n$ term will be somewhat larger, but still bounded by a constant.

It still makes sense to explicitly identify the $\log_m n$ term in the I/O bounds and not hide them within the big-oh or big-theta factors, since the terms can have a significant effect in practice. (Of course, it is equally important to consider any other constants hidden in big-oh and big-theta notations!) The nonlinear I/O bound $\Theta(n \log_m n)$ usually indicates that multiple or extra passes over the data are required. In truly massive problems, the data will reside on tertiary storage. As we suggested in Section 2.3, PDM algorithms can often be generalized in a recursive framework to handle multiple levels of memory. A multilevel algorithm developed from a PDM algorithm that does n I/Os will likely run at least an order of magnitude faster in hierarchical memory than would a multilevel algorithm generated from a PDM algorithm that does $n \log_m n$ I/Os (Vitter and Shriver 1994b).

4. Exploiting Locality and Load Balancing

The key to achieving efficient I/O performance in EM applications is to design the application to access its data with a high degree of locality. Since each read I/O operation transfers a block of B items, we make optimal use of that read operation when all B items are needed by the application. A similar remark applies to write operations. An orthogonal form of locality more akin to load balancing arises when we use multiple disks, since we can transfer D blocks in a single I/O only if the D blocks reside on distinct disks.

An algorithm that does not exploit locality can be reasonably efficient when run on data sets that fit in internal memory, but it will perform miserably when deployed naively in an EM setting, in which virtual memory is used to handle page management. Examining such performance degradation is a good way to put the I/O bounds of Table 10.2 into perspective. In Section 4.1 we examine this phenomenon for the single-disk case, when $D = 1$.

In Section 4.2, we look at the multiple-disk case and discuss the important paradigm of *disk striping* (Kim 1986, Salem and Garcia-Molina 1986), for automatically converting a single-disk algorithm into an algorithm for multiple disks. Disk striping can be used to get optimal multiple-disk I/O algorithms for three of the four fundamental operations in Table 10.2. The only exception is sorting. The optimal multiple-disk algorithms for sorting require more sophisticated load balancing techniques, which we cover in Section 5.

4.1. Locality issues with a single disk

A good way to appreciate the fundamental I/O bounds in Table 10.2 is to consider what happens when an algorithm does not exploit locality. For simplicity, we restrict ourselves in this section to the single-disk case $D = 1$. For many of the batched problems we look at in this paper, such as sorting, FFT, triangulation, and computing convex hulls, it is well known how to write programs to solve the corresponding internal memory versions of the problems in $O(N \log N)$ CPU time. But if we execute such a program on a data set that does not fit in internal memory, relying upon virtual memory to handle page management, the resulting number of I/Os may be $\Omega(N \log n)$, which represents a severe bottleneck.

We would like instead to incorporate locality *directly* into the algorithm design and achieve the desired I/O bound of $O(n \log_m n)$. At the risk of oversimplifying, we can paraphrase the goal of EM algorithm design for batched problems in the following syntactic way: to derive efficient algorithms so that the N and Z terms in the I/O bounds of the naive algorithms are replaced by n and z, and so that the base of the logarithm terms is not 2 but instead m. The relative speedup in I/O performance can be very significant, both theoretically and in practice. For example, the I/O performance improvement can be a factor of $(N \log n)/n \log_m n = B \log m$, which is extremely large.

4.2. Disk striping for multiple disks

It is conceptually much simpler to program for the single-disk case ($D = 1$) than for the multiple-disk case ($D \geq 1$). *Disk striping* (Kim 1986, Salem and Garcia-Molina 1986) is a practical paradigm that can ease the programming task with multiple disks: I/Os are permitted only on entire stripes, one stripe at a time. For example, in the data layout in Figure 10.3, data items 20–29 can be accessed in a single I/O step because their blocks are grouped into the same stripe. The net effect of striping is that the D disks behave as a single logical disk, but with a larger logical block size DB.

We can thus apply the paradigm of disk striping to automatically convert an algorithm designed to use a single disk with block size DB into an algorithm for use on D disks each with block size B: In the single-disk algorithm, each I/O step transmits one block of size DB; in the D-disk algorithm, each I/O step consists of D simultaneous block transfers of size B each. The number of I/O steps in both algorithms is the same; in each I/O step, the DB items transferred by the two algorithms are identical. Of course, in terms of wall clock time, the I/O step in the multiple-disk algorithm will be $\Theta(D)$ times faster than in the single-disk algorithm because of parallelism.

Disk striping can be used to get optimal multiple-disk algorithms for three of the four fundamental operations of Section 3—scanning, online search, and output reporting—but it is nonoptimal for sorting. To see why, consider what happens if we use the technique of disk striping in conjunction with an optimal sorting algorithm for one disk, such as merge sort (Knuth 1998). The optimal number of I/Os to sort using one disk with block size B is

$$\Theta\left(n \log_m n\right) = \Theta\left(n \frac{\log n}{\log m}\right) = \Theta\left(\frac{N}{B} \frac{\log(N/B)}{\log(M/B)}\right). \quad (10.2)$$

With disk striping, the number of I/O steps is the same as if we use a block size of DB in the single-disk algorithm, which corresponds to replacing each B in (10.2) by DB, which gives the I/O bound

$$\Theta\left(\frac{N}{DB} \frac{\log(N/DB)}{\log(M/DB)}\right) = \Theta\left(\frac{n}{D} \frac{\log(n/D)}{\log(m/D)}\right). \quad (10.3)$$

On the other hand, the optimal bound for sorting is

$$\Theta\left(\frac{n}{D} \log_m n\right) = \Theta\left(\frac{n}{D} \frac{\log n}{\log m}\right). \quad (10.4)$$

The striping I/O bound (10.3) is larger than the optimal sorting bound (10.4) by a multiplicative factor of

$$\frac{\log(n/D)}{\log n} \frac{\log m}{\log(m/D)} \approx \frac{\log m}{\log(m/D)}. \quad (10.5)$$

When D is on the order of m, the $\log(m/D)$ term in the denominator is small, and the resulting value of (10.5) is on the order of $\log m$, which can be significant in practice.

It follows that the only way theoretically to attain the optimal sorting bound (10.4) is to forsake disk striping and to allow the disks to be controlled *independently*, so that each disk can access a different stripe in the same I/O step. Actually, the only requirement for attaining the optimal bound is that either reading or writing is done independently. It suffices, for example, to do only read operations independently and to use disk striping for write operations. An advantage of using striping for write operations is that it facilitates the writing of parity information for error correction and recovery, which is a big concern in RAID systems. (We refer the reader to (Chen et al. 1994, Hellerstein et al. 1994) for a discussion of RAID and error correction issues.)

In practice, sorting via disk striping can be more efficient than complicated techniques that utilize independent disks, especially when D is small, since the extra factor $(\log m)/\log(m/D)$ of I/Os due to disk striping may be less than the algorithmic and system overhead of using the disks independently (Vengroff and Vitter 1996). In the next section we discuss algorithms for sorting with multiple independent disks. The techniques that arise can be applied to many of the batched problems addressed later in the paper. Two such sorting algorithms—distribution sort with randomized cycling and simple randomized mergesort—have relatively low overhead and will outperform disk-striped approaches.

5. Sorting and Related Problems

The problem of *external sorting* (or sorting in external memory) is a central problem in the field of EM algorithms, partly because sorting and sorting-like operations account for a significant percentage of computer use (Knuth 1998), and also because sorting is an important paradigm in the design of efficient EM algorithms, as we shall see in Section 8. With some technical qualifications, many problems that can be solved easily in linear time in internal memory, such as permuting, list ranking, expression tree evaluation, and finding connected components in a sparse graph, require the same number of I/Os in PDM as does sorting.

Theorem 1 *(Aggarwal and Vitter 1988, Nodine and Vitter 1995) The average-case and worst-case number of I/Os required for sorting $N = nB$ data items using D disks is*

$$Sort(N) = \Theta\left(\frac{n}{D}\log_m n\right). \qquad (10.6)$$

We saw in Section 4.2 how to construct efficient sorting algorithms for multiple disks by applying the disk striping paradigm to an efficient single-disk algorithm. But in the case of sorting, the resulting multiple-disk algorithm does not meet the optimal $Sort(N)$ bound of Theorem 1. In Sections 5.1 and 5.2 we discuss some recently developed external sorting algorithms that use disks independently. The algorithms are based upon the important *distribution* and *merge* paradigms, which are two generic approaches to sorting. The SRM method and its variants (Barve et al. 1997, Barve and Vitter 1999a, Sanders 2000), which are based upon a randomized merge technique, outperform disk striping in practice for reasonable values of D. All the algorithms use online load balancing strategies so that the data items accessed in an I/O operation are evenly distributed on the D disks. The same techniques can be applied to many of the batched problems we discuss later in this paper.

All the methods we cover for parallel disks, with the exception of Greed Sort in Section 5.2, provide efficient support for writing redundant parity information onto the disks for purposes of error correction and recovery. For example, some of the methods access the D disks independently during parallel read operations, but in a striped manner during parallel writes. As a result, if we write $D-1$ blocks at a time, the exclusive-or of the $D-1$ blocks can be written onto the Dth disk during the same write operation.

In Section 5.3, we will see that if we allow independent reads and writes, we can probabilistically simulate any algorithm written for the Aggarwal–Vitter model discussed in Section 2.3 by use of PDM with the same number of I/Os, up to a constant factor.

In Section 5.4 we consider the situation in which the items in the input file do not have unique keys. In Sections 5.6 and 5.7 we consider problems related to sorting, such as permuting, permutation networks, transposition, and fast Fourier transform. In Section 5.8, we give lower bounds for sorting and related problems.

5.1. Sorting by distribution

Distribution sort (Knuth 1998) is a recursive process in which the data items to be sorted are partitioned by a set of $S-1$ partitioning elements into S buckets. All the items in one bucket precede all the items in the next bucket. We can complete the sort by recursively sorting the individual buckets and concatenating them together to form a single fully sorted list.

One requirement is that we choose the $S-1$ partitioning elements so that the buckets are of roughly equal size. When that is the case, the bucket sizes decrease from one level of recursion to the next by a relative factor of $\Theta(S)$, and thus there are $O(\log_S n)$ levels of recursion. During each level of recursion, we scan the data. As the items stream through internal memory, they are partitioned into S buckets in an online manner. When a buffer of size B fills for one of the buckets, its block is written to the disks in the next I/O, and another buffer is used to store the next set of incoming items for the bucket. Therefore, the maximum number of buckets (and partitioning elements) is $S = \Theta(M/B) = \Theta(m)$, and the resulting number of levels of recursion is $\Theta(\log_m n)$.

It seems difficult to find $S = \Theta(m)$ partitioning elements using $\Theta(n/D)$ I/Os and guarantee that the bucket sizes are within a constant factor of one another. Efficient deterministic methods exist for choosing $S = \sqrt{m}$

partitioning elements (Aggarwal and Vitter 1988, Nodine and Vitter 1993, Vitter and Shriver 1994a), which has the effect of doubling the number of levels of recursion. Probabilistic methods based upon random sampling can be found in Feller (1968). A deterministic algorithm for the related problem of (exact) selection (i.e., given k, find the kth item in the file in sorted order) appears in Sibeyn (1999).

In order to meet the sorting bound (10.6), the formation of the buckets at each level of recursion must be done in $O(n/D)$ I/Os, which is easy to do for the single-disk case. In the more general multiple-disk case, each read step and each write step during the bucket formation must involve on the average $\Theta(D)$ blocks. The file of items being partitioned was itself one of the buckets formed in the previous level of recursion. In order to read that file efficiently, its blocks must be spread uniformly among the disks, so that no one disk is a bottleneck. The challenge in distribution sort is to write the blocks of the buckets to the disks in an online manner and achieve a global load balance by the end of the partitioning, so that the bucket can be read efficiently during the next level of the recursion.

Partial striping is an effective technique for reducing the amount of information that must be stored in internal memory in order to manage the disks. The disks are grouped into clusters of size C and data are written in "logical blocks" of size CB, one per cluster. Choosing $C = \sqrt{D}$ won't change the optimal sorting time by more than a constant factor, but as pointed out in Section 4.2, full striping (in which $C = D$) can be nonoptimal.

Vitter and Shriver (1994a) develop two randomized online techniques for the partitioning so that with high probability each bucket will be well balanced across the D disks. In addition, they use partial striping in order to fit in internal memory the pointers needed to keep track of the layouts of the buckets on the disks. Their first partitioning technique applies when the size N of the file to partition is sufficiently large or when $M/DB = \Omega(\log D)$, so that the number $\Theta(n/S)$ of blocks in each bucket is $\Omega(D \log D)$. Each parallel write operation writes its D blocks in independent random order to a disk stripe, with all $D!$ orders equally likely. At the end of the partitioning, with high probability each bucket is evenly distributed among the disks. This situation is intuitively analogous to the *classical occupancy problem*, in which b balls are inserted independently and uniformly at random into d bins. It is well known that if the load factor b/d grows asymptotically faster than $\log d$, the most densely populated bin contains b/d balls asymptotically on the average, which corresponds to an even distribution. However if the load factor b/d is 1, the largest bin contains $(\ln d)/\ln \ln d$ balls, whereas an average bin contains only one ball (Vitter and Flajolet 1990). Intuitively, the blocks in a bucket act as balls and the disks act as bins. In our case, the parameters correspond to $b = \Omega(d \log d)$, which suggests that the blocks in the bucket should be evenly distributed among the disks.

By further analogy to the occupancy problem, if the number of blocks per bucket is not $\Omega(D \log D)$, then the technique breaks down and the distribution of each bucket among the disks tends to be uneven, causing a bottleneck for I/O operations. For these smaller values of N, Vitter and Shriver use their second partitioning technique: The file is read in one pass, one memoryload at a time. Each memoryload is independently and randomly permuted and written back to the disks in the new order. In a second pass, the file is accessed one memoryload at a time in a "diagonally striped" manner. Vitter and Shriver show that with very high probability each individual "diagonal stripe" contributes about the same number of items to each bucket, so the blocks of the buckets in each memoryload can be assigned to the disks in a balanced round robin manner using an optimal number of I/Os.

DeWitt et al. (1991) present a randomized distribution sort algorithm in a similar model to handle the case when sorting can be done in two passes. They use a sampling technique to find the partitioning elements and route the items in each bucket to a particular processor. The buckets are sorted individually in the second pass.

An even better way to do distribution sort, and deterministically at that, is the BalanceSort method developed by Nodine and Nodine and Vitter (1993). During the partitioning process, the algorithm keeps track of how evenly each bucket has been distributed so far among the disks. It maintains an invariant that guarantees good distribution across the disks for each bucket. For each bucket $1 \leq b \leq S$ and disk $1 \leq d \leq D$, let num_b be the total number of items in bucket b processed so far during the partitioning and let $num_b(d)$ be the number of those items written to disk d; that is, $num_b = \sum_{1 \leq d \leq D} num_b(d)$. By application of matching techniques from graph theory, the BalanceSort algorithm is guaranteed to write at least half of any given memoryload to the disks in a blocked manner and still maintain the invariant for each bucket b that the $\lfloor D/2 \rfloor$ largest values among $num_b(1)$, $num_b(2)$, ..., $num_b(D)$ differ by at most 1. As a result, each $num_b(d)$ is at most about twice the ideal value num_b/D, which implies that the number of I/Os needed to read a bucket into memory during the next level of recursion will be within a small constant factor of optimal.

The distribution sort methods that we mentioned above for parallel disks perform write operations in complete stripes, which makes it easy to write parity information for use in error correction and recovery. But since the blocks written in each stripe typically belong to multiple buckets, the buckets themselves will not be striped on the disks, and we must use the disks independently during read operations. In the write phase, each bucket must therefore keep track of the last block written to each disk so that the blocks for the bucket can be linked together.

An orthogonal approach is to stripe the contents of each bucket across the disks so that read operations can be done in a striped manner. As a result, the write operations must use disks independently, since during each write, multiple buckets will be writing to multiple stripes. Error correction and recovery can still be handled efficiently by devoting to each bucket one block-sized buffer in internal memory. The buffer is continuously updated to contain the exclusive-or (parity) of the blocks written to the current stripe, and after $D-1$ blocks have been written, the parity information in the buffer can be written to the final (Dth) block in the stripe.

Under this new scenario, the basic loop of the distribution sort algorithm is, as before, to read one memoryload at a time and partition the items into S buckets. However, unlike before, the blocks for each individual bucket will reside on the disks in contiguous stripes. Each block therefore has a predefined place where it must be written. If we choose the normal round-robin ordering for the stripes (namely, $\ldots 123 \ldots D 123 \ldots D \ldots$), the blocks of different buckets may "collide", meaning that they need to be written to the same disk, and subsequent blocks in those same buckets will also tend to collide. Vitter and Vitter and Hutchinson (2001) solve this problem by the technique of *randomized cycling*. For each of the S buckets, they determine the ordering of the disks in the stripe for that bucket via a random permutation of $\{1, 2, \ldots, D\}$. The S random permutations are chosen independently. If two blocks (from different buckets) happen to collide during a write to the same disk, one block is written to the disk and the other is kept on a write queue. With high probability, subsequent blocks in those two buckets will be written to different disks and thus will not collide. As long as there is a small pool of available buffer space to temporarily cache the blocks in the write queues, Vitter and Hutchinson show that with high probability the writing proceeds optimally.

We expect that the randomized cycling method or the related merge sort methods discussed at the end of Section 5.2 will be the methods of choice for sorting with parallel disks. Experiments are underway to evaluate their relative performance. Distribution sort algorithms may have an advantage over the merge approaches presented in Section 5.2 in that they typically make better use of lower levels of cache in the memory hierarchy of real systems, based upon analysis of distribution

5.2. Sorting by merging

The *merge* paradigm is somewhat orthogonal to the distribution paradigm of the previous section. A typical merge sort algorithm works as follows (Knuth 1998): In the "run formation" phase, we scan the n blocks of data, one memoryload at a time; we sort each memoryload into a single "run", which we then output onto a series of stripes on the disks. At the end of the run formation phase, there are $N/M = n/m$ (sorted) runs, each striped across the disks. (In actual implementations, we can use the "replacement-selection" technique to get runs of $2M$ data items, on the average, when $M \gg B$ (Knuth 1998).) After the initial runs are formed, the merging phase begins. In each pass of the merging phase, we merge together groups of R runs. For each merge, we scan the R runs and merge the items in an online manner as they stream through internal memory. Double buffering is used to overlap I/O and computation. At most $R = \Theta(m)$ runs can be merged at a time, and resulting number of passes is $O(\log_m n)$.

To achieve the optimal sorting bound (10.6), we must perform each merging pass in $O(n/D)$ I/Os, which is easy to do for the single-disk case. In the more general multiple-disk case, each parallel read operation during the merging must on the average bring in the next $\Theta(D)$ blocks needed for the merging. The challenge is to ensure that those blocks reside on different disks so that they can be read in a single I/O (or a small constant number of I/Os). The difficulty lies in the fact that the runs being merged were themselves formed during the previous merge pass. Their blocks were written to the disks in the previous pass without knowledge of how they would interact with other runs in later merges.

A perfect solution, in which the next D blocks needed for the merge are guaranteed to be on distinct disks, can be devised for the binary merging case $R = 2$ based upon the Gilbreath principle (Gardner 1977, Knuth 1998): The first run is striped in ascending order by disk number, and the other run is striped in descending order. Regardless of how the items in the two runs interleave during the merge, it is always the case that the next D blocks needed for the output can be accessed via a single I/O operation, and thus the amount of internal memory buffer space needed for binary merging can be kept to a minimum. Unfortunately there is no analog to the Gilbreath principle for $R > 2$, and as we have seen above, we need the value of R to be large in order to get an optimal sorting algorithm.

The Greed Sort method of Nodine and Vitter (1995) was the first optimal deterministic EM algorithm for sorting with multiple disks. It handles the case $R > 2$ by relaxing the condition on the merging process. In each step, two blocks from each disk are brought into internal memory: the block b_1 with the smallest data item value and the block b_2 whose largest item value is smallest. If $b_1 = b_2$, only one block is read into memory, and it is added to the next output stripe. Otherwise, the two blocks b_1 and b_2 are merged in memory; the smaller B items are written to the output stripe, and the remaining B items are written back to the disk. The resulting run that is produced is only an "approximately" merged run, but its saving grace is that no two inverted items are too far apart. A final application of Columnsort (Leighton 1985) suffices to restore total order; partial striping is employed to meet the memory constraints. One disadvantage of Greed Sort is that the block writes and block reads involve independent disks and are not done in a striped manner, thus making it difficult to write parity information for error correction and recovery.

Aggarwal and Plaxton (1994) had developed an optimal deterministic merge sort based upon the Sharesort hypercube parallel sorting algorithm (Cypher and Plaxton 1993). To guarantee even distribution during the merging, it employs two high-level merging schemes in which the scheduling is almost oblivious. Like Greed Sort, the Sharesort algorithm is theoretically optimal (i.e., within a constant factor of optimal), but the constant factor is larger than the distribution sort methods.

The most practical method for sorting is based upon the *simple randomized merge sort* (SRM) algorithm of Barve et al. (1997), Barve and Vitter (1999a), referred to as "randomized striping" by Knuth (1998). Each run is striped across the disks, but with a random starting point (the only place in the algorithm where randomness is utilized). During the merging process, the next block needed from each disk is read into memory, and if there is not enough room, the least needed blocks are "flushed" (without any I/Os required) to free up space. Barve et al. (1997) derive an asymptotic upper bound on the expected I/O performance, with no assumptions on the input distribution. A more precise analysis, which is related to the so-called *cyclic occupancy problem*, is an interesting open problem. The cyclic occupancy problem is similar to the classical occupancy problem we discussed in Section 5.2 in that there are b balls distributed into d bins. However, in the cyclical occupancy problem, the b balls are grouped into c chains of length b_1, b_2, ... b_c, where $\sum_{1 \leq i \leq c} b_i = b$. Only the head of each chain is randomly inserted into a bin; the remaining balls of the chain are inserted into the successive bins in a cyclic manner (hence the name "cyclic occupancy"). It is conjectured that the expected maximum bin size in the cyclic occupancy problem is at most that of the classical occupancy problem (Barve et al. 1997, Knuth 1998,problem 5.4.9–28). The bound has been established so far only in an asymptotic sense.

The expected performance of SRM is not optimal for some parameter values, but it significantly outperforms the use of disk striping for reasonable values of the parameters, as shown in Table 10.3. Experimental confirmation of the speedup was obtained on a 500 megahertz CPU with six fast disk drives, as reported by Barve and Vitter (1999a).

Table 10.3. The ratio of the number of I/Os used by simple randomized merge sort (SRM) to the number of I/Os used by merge sort with disk striping, during a merge of kD runs, where $kD \approx M/2B$. The figures were obtained by simulation.

	$D = 5$	$D = 10$	$D = 50$
$k = 5$	0.56	0.47	0.37
$k = 10$	0.61	0.52	0.40
$k = 50$	0.71	0.63	0.51

Barve et al. (2000) and Kallahalla and Kallahalla and Varman (1999) have recently developed competitive and optimal methods for prefetching blocks in parallel I/O systems. Sanders (2000) has shown how to speed up the processing and apply it to external merge sort to get optimal I/O sorting bounds, on the average, for all parameter values. Rather than use random starting points and round-robin stripes as in SRM, Sanders orders the stripes for each run independently, based upon the randomized cycling strategy developed by Vitter and Vitter and Hutchinson (2001) for distribution sort.

5.3. A general simulation

Sanders et al. (2000) and Sanders (2001) give an elegant randomized technique to simulate the Aggarwal-Vitter model of Section 2.3, in which D simultaneous block transfers are allowed regardless of where the blocks are located on the disks. On the average, the simulation realizes each I/O in the Aggarwal-Vitter model by only a constant number of I/Os in PDM. One property of the technique is that the read and write steps use the disks independently. Armen (1996) had earlier shown that deterministic simulations resulted in an increase in the number of I/Os by a multiplicative factor of $\log(N/D)/\log\log(N/D)$.

The technique of Sanders et al. consists of duplicating each disk block and storing the two copies on two independently and uniformly chosen disks (chosen by a hash function). In terms of the occupancy model, each ball (block) is duplicated and stored in two random bins (disks). Let us consider the problem of retrieving a specific set of D blocks from the disks. For each block, there is a choice of two disks from which it can be read. Regardless of which D blocks are requested, Sanders et al. show how to choose the copies that permit the D blocks to be retrieved with high probability in only two parallel I/Os. A natural application of this technique is to the layout of data on multimedia servers in order to support multiple stream requests, as in video-on-demand.

When writing blocks of data to the disks, each block must be written to both the disks where a copy is stored. Sanders et al. prove that with high probability D blocks can be written in $O(1)$ I/O steps, assuming that there are $O(D)$ blocks of internal buffer space to serve as write queues. The read and write bounds can be improved with a corresponding tradeoff in redundancy and internal memory space.

5.4. Handling duplicates

Arge et al. (1993) describe a single-disk merge sort algorithm for the problem of *duplicate removal*, in which there are a total of K distinct items among the N items. It runs in $O(n \max\{1, \log_m(K/B)\})$ I/Os, which is optimal in the comparison model. The algorithm can be used to sort the file, assuming that a group of equal items can be represented by a single item and a count.

A harder instance of sorting called *bundle sorting* arises when we have K distinct key values among the N items, but all the items have different secondary information. Abello et al. (1998) and Matias et al. (2000) develop optimal distribution sort algorithms for bundle sorting using $BundleSort(N, K) = O(n \max\{1, \log_m \min\{K, n\}\})$ I/Os, and Matias et al. (2000) prove the matching lower bound. Matias et al. (2000) also show how to do bundle sorting (and sorting in general) *in place* (i.e., without extra disk space). In distribution sort, for example, the blocks for the subfiles can be allocated from the blocks freed up from the file being partitioned; the disadvantage is that the blocks in the individual subfiles are no longer consecutive on the disk. The algorithms can be adapted to run on D disks with a speedup of $O(D)$ using the techniques described in Sections 5.1 and 5.2.

5.5. Sorting strings

Arge et al. (1997) consider several models for the problem of sorting K strings of total length N in external memory. They develop efficient EM sorting algorithms in these models, making use of the *String B-tree* data structure of Ferragina and Grossi (1995) and the *buffer tree* of Arge (1995a). They also develop a simplified version of the String B-tree for merging called the *lazy trie*. The string sorting problem can be solved in the (internal memory) RAM model in $O(K \log K + N)$ time. By analogy to the problem of sorting integers, it would be natural to expect that the I/O complexity would be $O(k \log_m k + n)$, where $k = \max\{1, K/B\}$. Arge et al. show somewhat counterintuitively that for sorting short strings (i.e., strings of length at most B) the I/O complexity depends upon the total *number of characters*, whereas for long strings the complexity depends upon the total *number of strings*.

Theorem 2 *(Arge et al. 1997) The number of I/Os needed to sort K strings of total length N, where there are K_1 short strings of total length N_1 and K_2 long strings of total length N_2 (i.e., $N = N_1 + N_2$ and $K = K_1 + K_2$), is*

$$O\left(\min\left\{\frac{N_1}{B} \log_m \left(\frac{N_1}{B} + 1\right), K_1 \log_M(K_1 + 1)\right\} \right.$$
$$\left. + K_2 \log_M(K_2 + 1) + \frac{N}{B}\right). \quad (10.7)$$

Lower bounds for various models of how strings can be manipulated are given in Arge et al. (1997). There are gaps in some cases between the upper and lower bounds for sorting.

For further applications involving text strings and text databases, we refer the reader to other chapters in this volume.

5.6. Permuting and transposition

Permuting is the special case of sorting in which the key values of the N data items form a permutation of $\{1, 2, \ldots, N\}$.

Theorem 3 *(Aggarwal and Vitter 1988) The average-case and worst-case number of I/Os required for permuting N data items using D disks is*

$$\Theta\left(\min\left\{\frac{N}{D}, \mathit{Sort}(N)\right\}\right). \quad (10.8)$$

The I/O bound (10.8) for permuting can be realized by using one of the sorting algorithms from Section 5 except in the extreme case

$B \log m = o(\log n)$, in which case it is faster to move the data items one by one in a non-blocked way. The one-by-one method is trivial if $D = 1$, but with multiple disks there may be bottlenecks on individual disks; one solution for doing the permuting in $O(N/D)$ I/Os is to apply the randomized balancing strategies of Vitter and Shriver (1994a).

Matrix transposition is the special case of permuting in which the permutation can be represented as a transposition of a matrix from row-major order into column-major order.

Theorem 4 *(Aggarwal and Vitter 1988) With D disks, the number of I/Os required to transpose a $p \times q$ matrix from row-major order to column-major order is*

$$\Theta\left(\frac{n}{D} \log_m \min\{M, p, q, n\}\right), \qquad (10.9)$$

where $N = pq$ and $n = N/B$.

When B is relatively large (say, $\frac{1}{2}M$) and N is $O(M^2)$, matrix transposition can be as hard as general sorting, but for smaller B, the special structure of the transposition permutation makes transposition easier. In particular, the matrix can be broken up into square submatrices of B^2 elements such that each submatrix contains B blocks of the matrix in row-major order and also B blocks of the matrix in column-major order. Thus, if $B^2 < M$, the transpositions can be done in a simple one-pass operation by transposing the submatrices one-at-a-time in internal memory.

Matrix transposition is a special case of a more general class of permutations called *bit-permute/complement* (BPC) permutations, which in turn is a subset of the class of *bit-matrix-multiply/complement* (BMMC) permutations. BMMC permutations are defined by a $\log N \times \log N$ non-singular 0-1 matrix A and a $(\log N)$-length 0-1 vector c. An item with binary address x is mapped by the permutation to the binary address given by $Ax \oplus c$, where $\oplus$ denotes bitwise exclusive-or. BPC permutations are the special case of BMMC permutations in which A is a permutation matrix, that is, each row and each column of A contain a single 1. BPC permutations include matrix transposition, bit-reversal permutations (which arise in the FFT), vector-reversal permutations, hypercube permutations, and matrix reblocking. Cormen et al. (1999) characterize the optimal number of I/Os needed to perform any given BMMC permutation solely as a function of the associated matrix A, and they give an optimal algorithm for implementing it.

Theorem 5 *(Cormen et al. 1999) With D disks, the number of I/Os required to perform the BMMC permutation defined by matrix A and*

vector c is
$$\Theta\left(\frac{n}{D}\left(1+\frac{\operatorname{rank}(\gamma)}{\log m}\right)\right), \qquad (10.10)$$
where γ is the lower-left $\log n \times \log B$ submatrix of A.

An interesting theoretical question is to determine the I/O cost for each individual permutation, as a function of some simple characterization of the permutation, like number of inversions.

5.7. FFT and permutation networks

Computing the fast Fourier transform (FFT) in external memory consists of a series of I/Os that permit each computation implied by the FFT directed graph (or butterfly) to be done while its arguments are in internal memory. A permutation network computation consists of an oblivious (fixed) pattern of I/Os such that any of the $N!$ possible permutations can be realized; data items can only be reordered when they are in internal memory. A permutation network can be realized by a series of three FFTs (Wu and Feng 1981).

Theorem 6 *With D disks, the number of I/Os required for computing the N-input FFT digraph or an N-input permutation network is $\operatorname{Sort}(N)$.*

Cormen and Nicol (1998) give some practical implementations for one-dimensional FFTs based upon the optimal PDM algorithm of Vitter and Shriver (1994a). The algorithms for FFT are faster and simpler than for sorting because the computation is nonadaptive in nature, and thus the communication pattern is fixed in advance.

5.8. Lower bounds on I/O

In this section we prove the lower bounds from Theorems 1–6 and mention some related I/O lower bounds for the batched problems in computational geometry and graphs that we cover later in Sections 7 and 8.

The most trivial batched problem is that of scanning (a.k.a. streaming or touching) a file of N data items, which can be done in a linear number $O(N/DB) = O(n/D)$ of I/Os. Permuting is one of several simple problems that can be done in linear CPU time in the (internal memory) RAM model, but require a nonlinear number of I/Os in PDM because of the locality constraints imposed by the block parameter B.

The following proof of the permutation lower bound (10.8) of Theorem 3 is due to Aggarwal and Vitter (1988). The idea of the proof is to calculate, for each $t \geq 0$, the number of distinct orderings that are realizable by sequences of t I/Os. The value of t for which the number of distinct orderings first exceeds $N!/2$ is a lower bound on the average number of I/Os (and hence the worst-case number of I/Os) needed for permuting.

We assume for the moment that there is only one disk, $D = 1$. Let us consider how the number of realizable orderings can change as a result of an I/O. In terms of increasing the number of realizable orderings, the effect of reading a disk block is considerably more than that of writing a disk block, so it suffices to consider only the effect of read operations. During a read operation, there are at most B data items in the read block, and they can be interspersed among the M items in internal memory in at most $\binom{M}{B}$ ways, so the number of realizable orderings increases by a factor of $\binom{M}{B}$. If the block has never before resided in internal memory, the number of realizable orderings increases by an extra $B!$ factor, since the items in the block can be permuted among themselves. (This extra contribution of $B!$ can only happen once for each of the N/B original blocks.) There are at most $n + t \leq N \log N$ ways to choose which disk block is involved in the tth I/O. (We allow the algorithm to use an arbitrary amount of disk space.) Hence, the number of distinct orderings that can be realized by all possible sequences of t I/Os is at most

$$(B!)^{N/B} \left(N (\log N) \binom{M}{B} \right)^t. \tag{10.11}$$

Setting the expression in (10.11) to be at least $N!/2$, and simplifying by taking the logarithm, we get

$$N \log B + t \left(\log N + B \log \frac{M}{B} \right) = \Omega(N \log N). \tag{10.12}$$

Solving for t, we get the matching lower bound $\Omega(n \log_m n)$ for permuting for the case $D = 1$. The general lower bound (10.8) of Theorem 3 follows by dividing by D.

Permuting is a special case of sorting, and hence, the permuting lower bound applies also to sorting. In the unlikely case that $B \log m = o(\log n)$, the permuting bound is only $\Omega(N/D)$, and we must resort to the comparison model to get the full lower bound (10.6) of Theorem 1 (Aggarwal and Vitter 1988). In the typical case in which $B \log m = \Omega(\log n)$, the comparison model is not needed to prove the sorting lower bound; the difficulty of sorting in that case arises not from determining the order of the data but from permuting (or routing) the data.

The proof used above for permuting also works for permutation networks, in which the communication pattern is oblivious (fixed). Since the choice of disk block is fixed for each t, there is no $N \log N$ term as there is in (10.11), and correspondingly there is no additive $\log N$ term in the inner expression as there is in (10.12). Hence, when we solve for t, we get the lower bound (10.6) rather than (10.8). The lower bound follows directly from the counting argument; unlike the sorting derivation, it does not require the comparison model for the case $B \log m = o(\log n)$. The lower bound also applies directly to FFT, since permutation networks can be formed from three FFTs in sequence. The transposition lower bound involves a potential argument based upon a togetherness relation (Aggarwal and Vitter 1988).

Arge et al. (1993) show for the comparison model that any problem with an $\Omega(N \log N)$ lower bound in the (internal memory) RAM model requires $\Omega(n \log_m n)$ I/Os in PDM for a single disk. Their argument leads to a matching lower bound of $\Omega(n \max\{1, \log_m(K/B)\})$ I/Os in the comparison model for duplicate removal with one disk.

For the problem of bundle sorting, in which the N items have a total of K distinct key values (but the secondary information of each item is different), (Matias et al. 2000) derive the matching lower bound $BundleSort(N, K) = \Omega(n \max\{1, \log_m \min\{K, n\}\})$. The proof consists of two parts. The first part is a simple proof of the same lower bound as for duplicate removal, but without resorting to the comparison model (except for the pathological case $B \log m = o(\log n)$). It suffices to set (10.11) to be at least $N!/((N/K)!)^K$, which is the maximum number of permutations of N numbers having K distinct values. Solving for t gives the lower bound $\Omega(n \max\{1, \log_m(K/B)\})$, which is equal to the desired lower bound for $BundleSort(N, K)$ when $K = B^{1+\Omega(1)}$ or $M = B^{1+\Omega(1)}$. Matias et al. (2000) derive the remaining case of the lower bound for $BundleSort(N, K)$ by a potential argument based upon the transposition lower bound. Dividing by D gives the lower bound for D disks.

Chiang et al. (1995), Arge (1995b), Arge and Miltersen (1999), and Munagala and Ranade (1999) give models and lower bound reductions for several computational geometry and graph problems. The geometry problems discussed in Section 7 are equivalent to sorting in both the internal memory and PDM models. Problems like list ranking and expression tree evaluation have the same nonlinear I/O lower bound as permuting. Other problems like connected components, biconnected components, and minimum spanning forest of sparse graphs with E edges and V vertices require as many I/Os as E/V instances of permuting V items. This situation is in contrast with the (internal memory) RAM model, in which the same problems can all be done in linear CPU time. (The known linear-time RAM algorithm for minimum spanning tree is randomized.) In some cases there is a gap between the best known upper and lower bounds, which we examine further in Section 8.

The lower bounds mentioned above assume that the data items are in some sense "indivisible", in that they are not split up and reassembled in some magic way to get the desired output. It is conjectured that the sorting lower bound (10.6) remains valid even if the indivisibility assumption is lifted. However, for an artificial problem related to transposition, Adler (1996) showed that removing the indivisibility assumption can lead to faster algorithms. A similar result is shown by Arge and Miltersen (1999) for the decision problem of determining if N data item values are distinct. Whether or not the conjecture is true is a challenging theoretical open problem.

6. Matrix and Grid Computations

Dense matrices are generally represented in memory in row-major or column-major order. Matrix transposition, which is the special case of sorting that involves conversion of a matrix from one representation to the other, was discussed in Section 5.6. For certain operations such as matrix addition, both representations work well. However, for standard matrix multiplication (using only semiring operations) and LU decomposition, a better representation is to block the matrix into square $\sqrt{B} \times \sqrt{B}$ submatrices, which gives the upper bound of the following theorem:

Theorem 7 *(Hong and Kung 1981, Savage and Vitter 1987, Vitter and Shriver 1994a, Womble et al. 1993) The number of I/Os required for standard matrix multiplication of two $K \times K$ matrices or to compute the LU factorization of a $K \times K$ matrix is $\Theta(K^3 / \min\{K, \sqrt{M}\}DB)$.*

Hong and Kung (1981) and Nodine et al. (1991) give optimal EM algorithms for iterative grid computations, and Leiserson et al. (1993) reduce the number of I/Os of naive multigrid implementations by a $\Theta(M^{1/5})$ factor. Gupta et al. (1995) show how to derive efficient EM algorithms automatically for computations expressed in tensor form.

If a $K \times K$ matrix A is sparse, that is, if the number N_z of nonzero elements in A is much smaller than K^2, then it may be more efficient to store only the nonzero elements. Each nonzero element $A_{i,j}$ is represented by the triple $(i, j, A_{i,j})$. Unlike the dense case, in which transposition can be easier than sorting (e.g., see Theorem 4 when $B^2 \leq M$), transposition of sparse matrices is as hard as sorting:

Theorem 8 *For a matrix stored in sparse format and containing N_z nonzero elements, the number of I/Os required to convert the matrix from row-major order to column-major order, and vice-versa, is $\Theta(Sort(N_z))$.*

The lower bound follows by reduction from sorting. If the ith item in the input of the sorting instance has key value $x \neq 0$, there is a nonzero element in matrix position (i, x).

For further discussion of numerical EM algorithms we refer the reader to the survey by Toledo (1999). Some issues regarding programming environments are covered in (Corbett et al. 1996) and Section 10.

7. Batched Problems in Computational Geometry

Problems involving massive amounts of geometric data are ubiquitous in spatial databases (Laurini and Thompson 1992, Samet 1989a;b), geographic information systems (GIS) (Laurini and Thompson 1992, Samet 1989a, van Kreveld et al. 1997), constraint logic programming (Kanellakis et al. 1990; 1996), statistics, virtual reality systems, computer graphics (Funkhouser et al. 1992), and object-oriented databases (Zdonik and Maier 1990). NASA's Earth Observing System project, the core part of the Earth Science Enterprise (formerly Mission to Planet Earth), produces petabytes (10^{15} bytes) of raster data per year (NAS 2001)! Microsoft's TerraServer online database of satellite images is over one terabyte in size (Ter 2001). A major challenge is to develop mechanisms for processing the data, or else much of it will be useless.

For systems of this size to be efficient, we need fast EM algorithms and data structures for basic problems in computational geometry. Luckily, many problems on geometric objects can be reduced to a small core of problems, such as computing intersections, convex hulls, or nearest neighbors. Useful paradigms have been developed for solving these problems in external memory.

For brevity, in the remainder of this paper we deal only with the single-disk case $D = 1$. The single-disk I/O bounds for the batched problems can often be cut by a factor of $\Theta(D)$ for the case $D \geq 1$ by using the load balancing techniques of Section 5. In practice, disk striping (cf. Section 4.2) may be sufficient.

Theorem 9 *The following batched problems involving $N = nB$ input items, $Q = qB$ queries, and $Z = zB$ output items can be solved using*

$$O((n + q) \log_m n + z) \qquad (10.13)$$

I/Os with a single disk:

1 *Computing the pairwise intersections of N segments in the plane and their trapezoidal decomposition,*

2 Finding all intersections between N nonintersecting red line segments and N nonintersecting blue line segments in the plane.

3 Answering Q orthogonal 2-D range queries on N points in the plane (i.e., finding all the points within the Q query rectangles),

4 Constructing the 2-D and 3-D convex hull of N points,

5 Voronoi diagram and triangulation of N points in the plane,

6 Performing Q point location queries in a planar subdivision of size N,

7 Finding all nearest neighbors for a set of N points in the plane,

8 Finding the pairwise intersections of N orthogonal rectangles in the plane,

9 Computing the measure of the union of N orthogonal rectangles in the plane,

10 Computing the visibility of N segments in the plane from a point,

11 Performing Q ray-shooting queries in 2-D Constructive Solid Geometry (CSG) models of size N,

The parameters Q and Z are set to 0 if they are not relevant for the particular problem.

Goodrich et al. (1993), Zhu (1994), Arge et al. (1995), Arge et al. (1998b), and Crauser et al. (1998; 1999) develop EM algorithms for those problems using the following EM paradigms for batched problems:

Distribution sweeping a generalization of the distribution paradigm of Section 5 for "externalizing" plane sweep algorithms;

Persistent B-trees an offline method for constructing an optimal-space persistent version of the B-tree data structure, yielding a factor of B improvement over the generic persistence techniques of Driscoll et al. (1989).

Batched filtering a general method for performing certain simultaneous EM searches in data structures that can be modeled as planar layered directed acyclic graphs; it is useful for 3-D convex hulls and batched point location. Multisearch on parallel computers is considered in (Dittrich et al. 1998).

External fractional cascading an EM analog to fractional cascading on a segment tree, in which the degree of the segment tree is $O(m^\alpha)$

for some constant $0 < \alpha \leq 1$. Batched queries can be performed efficiently using batched filtering; online queries can be supported efficiently by adapting the parallel algorithms of work of Tamassia and Vitter (1996) to the I/O setting.

External marriage-before-conquest an EM analog to the technique of Kirkpatrick and Seidel (1986) for performing output-sensitive convex hull constructions.

Batched incremental construction a localized version of the randomized incremental construction paradigm of Clarkson and Shor (1989), in which the updates to a simple dynamic data structure are done in a random order, with the goal of fast overall performance on the average. The data structure itself may have bad worst-case performance, but the randomization of the update order makes worst-case behavior unlikely. The key for the EM version so as to gain the factor of B I/O speedup is to batch together the incremental modifications.

We focus in the remainder of this section primarily on the distribution sweep paradigm (Goodrich et al. 1993), which is a combination of the distribution paradigm of Section 5.1 and the well known sweeping paradigm from computational geometry (Preparata and Shamos 1985, de Berg et al. 1997). As an example, let us consider computing the pairwise intersections of N orthogonal segments in the plane by the following recursive distribution sweep: At each level of recursion, the region under consideration is partitioned into $\Theta(m)$ vertical *slabs*, each containing $\Theta(N/m)$ of the segments' endpoints. We sweep a horizontal line from top to bottom to process the N segments. When the sweep line encounters a vertical segment, we insert the segment into the appropriate slab. When the sweep line encounters a horizontal segment h, as pictured in Figure 10.4, we report h's intersections with all the "active" vertical segments in the slabs that are spanned *completely* by h. (A vertical segment is "active" if it intersects the current sweep line; vertical segments that are found to be no longer active are deleted from the slabs.) The remaining two end portions of h (which "stick out" past a slab boundary) are passed recursively to the next level, along with the vertical segments. The downward sweep then proceeds. After the initial sorting (to get the segments with respect to the y-dimension), the sweep at each of the $O(\log_m n)$ levels of recursion requires $O(n)$ I/Os, yielding the desired bound (10.13). Some timing experiments on distribution sweeping appear in Chiang (1998). Arge et al. (1998b) develop a unified approach to distribution sweep in higher dimensions.

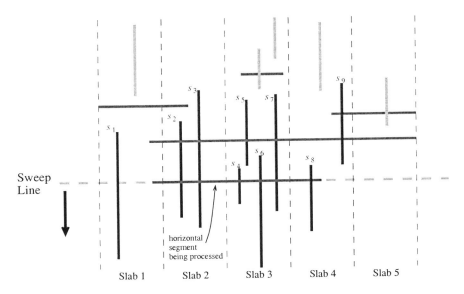

Figure 10.4. Distribution sweep used for finding intersections among N orthogonal segments. The vertical segments currently stored in the slabs are indicated in bold (namely, $s_1, s_2, \ldots, s_9$). Segments s_5 and s_8 are not active, but have not yet been deleted from the slabs. The sweep line has just advanced to a new horizontal segment that completely spans slabs 2 and 3, so slabs 2 and 3 are scanned and all the active vertical segments in slabs 2 and 3 (namely, s_2, s_3, s_4, s_6, s_7) are reported as intersecting the horizontal segment. In the process of scanning slab 3, segment s_5 is discovered to be no longer active and can be deleted from slab 3. The end portions of the horizontal segment that "stick out" into slabs 1 and 4 are handled by the lower levels of recursion, where the intersection with s_8 is eventually discovered.

A central operation in spatial databases is spatial join. A common preprocessing step is to find the pairwise intersections of the bounding boxes of the objects involved in the spatial join. The problem of intersecting orthogonal rectangles can be solved by combining the previous sweep line algorithm for orthogonal segments with one for range searching. Arge et al. (1998b) take a more unified approach using distribution sweep, which is extendible to higher dimensions: The active objects that are stored in the data structure in this case are rectangles, not vertical segments. The authors choose the branching factor to be $\Theta(\sqrt{m})$. Each rectangle is associated with the largest contiguous range of vertical slabs that it spans. Each of the possible $\Theta(\binom{\sqrt{m}}{2}) = \Theta(m)$ contiguous ranges of slabs is called a *multislab*. The reason why the authors choose the branching factor to be $\Theta(\sqrt{m})$ rather than $\Theta(m)$ is so that the number of multislabs is $\Theta(m)$, and thus there is room in internal memory for a buffer for each multislab. The height of the tree remains $O(\log_m n)$.

The algorithm proceeds by sweeping a horizontal line from top to bottom to process the N rectangles. When the sweep line first encounters a rectangle R, we consider the multislab lists for all the multislabs that R intersects. We report all the active rectangles in those multislab lists, since they are guaranteed to intersect R. (Rectangles no longer active are discarded from the lists.) We then extract the left and right end portions of R that partially "stick out" past slab boundaries, and we pass them down to process in the next lower level of recursion. We insert the remaining portion of R, which spans complete slabs, into the list for the appropriate multislab. The downward sweep then continues. After the initial sorting preprocessing, each of the $O(\log_m n)$ sweeps (one per level of recursion) takes $O(n)$ I/Os, yielding the desired bound (10.13).

The resulting algorithm, called Scalable Sweeping-Based Spatial Join (SSSJ) (Arge et al. 1998a;b), outperforms other techniques for rectangle intersection. It was tested against two other sweep line algorithms: the Partition-Based Spatial-Merge (QPBSM) used in Paradise (Patel and DeWitt 1996) and a faster version called MPBSM that uses an improved dynamic data structure for intervals (Arge et al. 1998a). The TPIE system described in Section 10 served as the common implementation platform. The algorithms were tested on several data sets. The timing results for the two data sets in Figures 10.5(a) and 10.5(b) are given in Figures 10.5(c) and 10.5(d), respectively. The first data set is the worst case for sweep line algorithms; a large fraction of the line segments in the file are active (i.e., they intersect the current sweep line). The second data set is a best case for sweep line algorithms, but the two PBSM algorithms have the disadvantage of making extra copies of the rectangles. In both cases, SSSJ shows considerable improvement over the PBSM-based methods. In other experiments done on more typical data, such as TIGER/line road data sets (tig 1992), SSSJ and MPBSM perform about 30% faster than does QPBSM. The conclusion we draw is that SSSJ is as fast as other known methods on typical data, but unlike other methods, it scales well even for worst-case data. If the rectangles are already stored in an index structure, such as the R-tree index structure, hybrid methods that combine distribution sweep with inorder traversal often perform best (Arge et al. 2000b).

For the problem of finding all intersections among N line segments, Arge et al. (1995) give an efficient algorithm based upon distribution sort, but the output component of the I/O bound is slightly nonoptimal: $z \log_m n$ rather than z. Crauser et al. (1998; 1999) attain the optimal I/O bound (10.13) by constructing the trapezoidal decomposition for the intersecting segments using an incremental randomized construction.

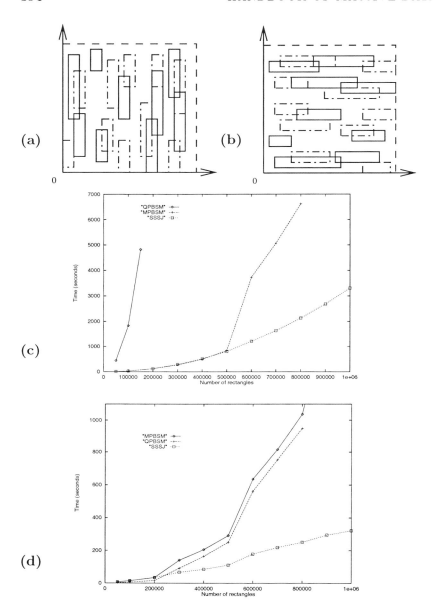

Figure 10.5. Comparison of Scalable Sweeping-Based Spatial Join (SSSJ) with the original PBSM (QPBSM) and a new variant (MPBSM) (a) Data set 1 consists of tall and skinny (vertically aligned) rectangles. (b) Data set 2 consists of short and wide (horizontally aligned) rectangles. (c) Running times on data set 1. (d) Running times on data set 2.

For I/O efficiency, they do the incremental updates in a series of batches, in which the batch size is geometrically increasing by a factor of m.

8. Batched Problems on Graphs

The first work on EM graph algorithms was by Ullman and Yannakakis (1991) for the problem of transitive closure. Chiang et al. (1995) consider a variety of graph problems, several of which have upper and lower I/O bounds related to sorting and permuting. Abello et al. (1998) formalize a functional approach to EM graph problems, in which computation proceeds in a series of scan operations over the data; the scanning avoids side effects and thus permits checkpointing to increase reliability. Kumar and Schwabe (1996), followed by Buchsbaum et al. (2000), develop graph algorithms based upon amortized data structures for binary heaps and tournament trees. Munagala and Ranade (1999) give improved graph algorithms for connectivity and undirected breadth-first search, and Arge et al. (2000a) extend the approach to compute the minimum spanning forest (MSF). Meyer (2001) provides some improvements for graphs of bounded degree. Arge (1995b) gives efficient algorithms for constructing ordered binary decision diagrams. Grossi and Italiano (1999) apply their multidimensional data structure to get dynamic EM algorithms for MSF and two-dimensional priority queues (in which the *delete_min* operation is replaced by $delete\_min_x$ and $delete\_min_y$). Techniques for storing graphs on disks for efficient traversal and shortest path queries are discussed in (Agarwal et al. 1998, Goldman et al. 1998, Hutchinson et al. 1999, Nodine et al. 1996). Computing wavelet decompositions and histograms (Vitter and Wang 1999, Vitter et al. 1998, Wang et al. 2000) is an EM graph problem related to transposition that arises in On-Line Analytical Processing (OLAP). Wang et al. (1998) give an I/O-efficient algorithm for constructing classification trees for data mining.

Table 10.4 gives the best known I/O bounds for
several graph problems, as a function of the number $V = vB$ of vertices and the number $E = eB$ of edges. The best known I/O lower bound for these problems is $\Omega((E/V)Sort(V) = e\log_m v)$, as mentioned in Section 5.8. A sparsification technique (Eppstein et al. 1997) can often be applied to convert I/O bounds of the form $O(Sort(E))$ to the improved form $O((E/V)Sort(V))$. For example, the actual I/O bounds for connectivity and MSF derived by Munagala and Ranade (1999) and Arge et al. (2000a) are $O(\max\{1, \log\log(V/e)\}Sort(E))$. For the MSF problem, we can partition the edges of the graph into E/V sparse subgraphs on V vertices, and then apply the algorithm of Arge et al.

Table 10.4. Best known I/O bounds for batched graph problems for the single-disk case $D = 1$. The number of vertices is denoted by $V = vB$ and the number of edges by $E = eB$. The terms $Sort(N)$ and $BundleSort(N, K)$ are defined in Sections 3 and 5.4. Lower bounds are discussed in Section 5.8.

Graph Problem	I/O Bound, $D = 1$
List ranking, Euler tour of a tree, Centroid decomposition, Expression tree evaluation	$\Theta(Sort(V))$ (Chiang et al. 1995)
Connected components, Minimum spanning forest (MSF)	$O\left(\max\left\{1,\ \log\log\dfrac{V}{e}\right\}\dfrac{E}{V}Sort(V)\right)$ (Arge et al. 2000a, Eppstein et al. 1997) (Munagala and Ranade 1999) (deterministic) $\Theta\left(\dfrac{E}{V}Sort(V)\right)$ (Chiang et al. 1995) (randomized)
Bottleneck MSF, Biconnected components	$O\left(\min\left\{V^2,\ \max\left\{1,\ \log\dfrac{V}{M}\right\}\dfrac{E}{V}Sort(V),\right.\right.$ $\left.\left.(\log B)\dfrac{E}{V}Sort(V) + e\log V\right\}\right)$ (Abello et al. 1998, Chiang et al. 1995) (Eppstein et al. 1997) (Kumar and Schwabe 1996) (deterministic) $\Theta\left(\dfrac{E}{V}Sort(V)\right)$ (Chiang et al. 1995) (Eppstein et al. 1997) (randomized)
Ear decomposition, Maximal matching	$O\left(\min\left\{V^2,\ \max\left\{1,\ \log\dfrac{V}{M}\right\}Sort(E),\right.\right.$ $\left.\left.(\log B)Sort(E) + e\log V\right\}\right)$ (Abello et al. 1998, Chiang et al. 1995) (Kumar and Schwabe 1996) (deterministic) $O(Sort(E))$ (Chiang et al. 1995) (randomized)
Undirected breadth-first search	$O(BundleSort(E, V) + V)$ (Munagala and Ranade 1999)

(*To be continued*)

(2000a) to each subproblem to create E/V spanning forests in a total of $O(\max\{1, \log\log(V/e)\}(E/V)Sort(V))$ I/Os. We can then merge the

(*Table 10.4: continued*)

Graph Problem	I/O Bound, $D = 1$
Undirected single-source shortest paths	$O(e \log e + V)$ (Kumar and Schwabe 1996)
Directed and undirected depth-first search, Topological sorting, Directed breadth-first search, Directed single-source shortest paths	$O\left(\min\left\{\dfrac{ve}{m} + V,\ (V+e)\log v\right\}\right)$ (Buchsbaum et al. 2000, Chiang et al. 1995) (Kumar and Schwabe 1996)
Transitive closure	$O\left(Vv\sqrt{\dfrac{e}{m}}\right)$ (Chiang et al. 1995)

E/V spanning forests, two at a time, in a balanced binary merging procedure by repeatedly applying the algorithm of Arge et al. (2000a). After the first level of binary merging, the spanning forests collectively have at most $E/2$ edges; after two levels, they have at most $E/4$ edges, and so on in a geometrically decreasing manner. The total cost for the final spanning forest is thus $O(\max\{1, \log\log(V/e)\}(E/V)Sort(V))$ I/Os. The reason why sparsification works is that the spanning forest output by each binary merge is only $\Theta(V)$ in size, yet it preserves the necessary information needed for the next merge step. The same approach works for connectivity.

In the case of *semi-external graph problems* (Abello et al. 1998), in which the vertices fit in internal memory but not the edges (i.e., $V \leq M < E$), several of the problems in Table 10.4 can be solved optimally in external memory. For example, finding connected components, biconnected components, and minimum spanning forests can be done in $O(e)$ I/Os when $V \leq M$. The I/O complexities of several problems in the general case remain open, including connected components, biconnected components, and minimum spanning forests in the deterministic case, as well as breadth-first search, topological sorting, shortest paths, depth-first search, and transitive closure. It may be that the I/O complexity for several of these problems is $\Theta((E/V)Sort(V)+V)$. For special cases, such as trees, planar graphs, outerplanar graphs, and graphs of bounded tree width, several of these problems can be solved substantially faster in $O(Sort(E))$ I/Os (Agarwal et al. 1998, Chiang et al. 1995, Maheshwari and Zeh 1999; 2001).

Chiang et al. (1995) exploit the key idea that efficient EM algorithms can often be developed by a sequential simulation of a parallel algorithm for the same problem. The intuition is that each step of a parallel algorithm specifies several operations and the data they act upon. If we bring together the data arguments for each operation, which we can do by two applications of sorting, then the operations can be performed by a single linear scan through the data. After each simulation step, we sort again in order to reblock the data into the linear order required for the next simulation step. In list ranking, which is used as a subroutine in the solution of several other graph problems, the number of working processors in the parallel algorithm decreases geometrically with time, so the number of I/Os for the entire simulation is proportional to the number of I/Os used in the first phase of the simulation, which is $Sort(N) = \Theta(n \log_m n)$. The optimality of the EM algorithm given in Chiang et al. (1995) for list ranking assumes that $\sqrt{m} \log m = \Omega(\log n)$, which is usually true in practice. That assumption can be removed by use of the buffer tree data structure (Arge 1995a). A practical, randomized implementation of list ranking appears in Sibeyn (1997).

Dehne et al. (1997; 1999) and Sibeyn and Kaufmann (1997) use a related approach and get efficient I/O bounds by simulating coarse-grained parallel algorithms in the BSP parallel model. Coarse-grained parallel algorithms may exhibit more locality than the fine-grained algorithms considered in Chiang et al. (1995), and as a result the simulation may require fewer sorting steps. Dehne et al. make certain assumptions, most notably that $\log_m n \leq c$ for some small constant c (or equivalently that $M^c < NB$), so that the periodic sortings can each be done in a linear number of I/Os. Since the BSP literature is well developed, their simulation technique provides efficient single-processor and multiprocessor EM algorithms for a wide variety of problems.

In order for the simulation techniques to be reasonably efficient, the parallel algorithm being simulated must run in $O((\log N)^c)$ time using N processors. Unfortunately, the best known polylog-time algorithms for problems like depth-first search and shortest paths use a polynomial number of processors, not a linear number. P-complete problems like lexicographically-first depth-first search are unlikely to have polylogarithmic time algorithms even with a polynomial number of processors. The interesting connection between the parallel domain and the EM domain suggests that there may be relationships between computational complexity classes related to parallel computing (such as P-complete problems) and those related to I/O efficiency. It may thus be possible to show by reduction that certain groups of problems are "equally hard" to solve efficiently in terms of I/O and are thus unlikely to have solutions as fast as sorting.

9.   Dynamic Memory Allocation

The amount of internal memory allocated to a program may fluctuate during the course of execution because of demands placed on the system by other users and processes. EM algorithms must be able to adapt dynamically to whatever resources are available so as to preserve good performance (Pang et al. 1993a). The algorithms in the previous sections assume a fixed memory allocation; they must resort to virtual memory if the memory allocation is reduced, often causing a severe degradation in performance.

Barve and Vitter (1999b) discuss the design and analysis of EM algorithms that adapt gracefully to changing memory allocations. In their model, without loss of generality, an algorithm (or program) $\mathcal{P}$ is allocated internal memory in phases: During the ith phase, $\mathcal{P}$ is allocated m_i blocks of internal memory, and this memory remains allocated to $\mathcal{P}$ until $\mathcal{P}$ completes $2m_i$ I/O operations, at which point the next phase begins. The process continues until $\mathcal{P}$ finishes execution. The model makes the reasonable assumption that the duration for each memory allocation phase is long enough to allow all the memory in that phase to be used by the algorithm.

For sorting, the lower bound approach of (10.11) implies that

$$\sum_i 2m_i \log m_i = \Omega(n \log n).$$

We say that $\mathcal{P}$ is *dynamically optimal* for sorting if

$$\sum_i 2m_i \log m_i = O(n \log n)$$

for all possible sequences $m_1, m_2, \ldots$ of memory allocation. Intuitively, if $\mathcal{P}$ is dynamically optimal, no other algorithm can perform more than a constant number of sorts in the worst-case for the same sequence of memory allocations.

Barve and Vitter (1999b) define the model in generality and give dynamically optimal strategies for sorting, matrix multiplication, and buffer tree operations. Their work represents the first theoretical model of dynamic allocation and the first algorithms that can be considered dynamically optimal. Previous work was done on memory-adaptive algorithms for merge sort (Pang et al. 1993a, Zhang and Larson 1997) and hash join (Pang et al. 1993b), but the algorithms handle only special cases and can be made to perform nonoptimally for certain patterns of memory allocation.

10. The TPIE External Memory Programming Environment

In this section we describe TPIE (Transparent Parallel I/O Environment)[2] (Arge et al. 1999, TPI 1999, Vengroff and Vitter 1996), which serves as the implementation platform for the experiments described in Section 7 as well as in several of the referenced papers. TPIE is a comprehensive set of C++ templates for EM paradigms and a run-time library. Its goal is to help programmers develop high-level, portable, and efficient implementations of EM algorithms.

There are three basic approaches to supporting development of I/O-efficient code, which we call *access-oriented*, *array-oriented*, and *framework oriented*. TPIE falls primarily into the third category with some elements of the first category. Access-oriented systems preserve the programmer abstraction of explicitly requesting data transfer. They often extend the read-write interface to include data type specifications and collective specification of multiple transfers, sometimes involving the memories of multiple processing nodes. Examples of access-oriented systems include the UNIX file system at the lowest level, higher-level parallel file systems such as Whiptail (Shriver and Wisniewski 1995), Vesta (Corbett and Feitelson 1996), PIOUS (Moyer and Sunderam 1996), and the High Performance Storage System (Watson and Coyne 1995), and I/O libraries MPI-IO (Corbett et al. 1996) and LEDA-SM (Crauser and Mehlhorn 1999).

Array-oriented systems access data stored in external memory primarily by means of compiler-recognized data types (typically arrays) and operations on those data types. The external computation is directly specified via iterative loops or explicitly data-parallel operations, and the system manages the explicit I/O transfers. Array-oriented systems are effective for scientific computations that make regular strides through arrays of data and can deliver high-performance parallel I/O in applications such as computational fluid dynamics, molecular dynamics, and weapon system design and simulation. Array-oriented systems are generally ill-suited to irregular or combinatorial computations. Examples of array-oriented systems include PASSION (Thakur et al. 1996), Panda (Seamons and Winslett 1996) (which also has aspects of access orientation), PI/OT (Parsons et al. 1997), and ViC* (Colvin and Cormen 1998)

---

[2]The TPIE software distribution is available free of charge at http://www.cs.duke.edu/TPIE/ on the World Wide Web.

TPIE (Arge et al. 1999, TPI 1999, Vengroff and Vitter 1996) provides a framework-oriented interface for batched computation as well as an access-oriented interface for online computation. Instead of viewing batched computation as an enterprise in which code reads data, operates on it, and writes results, a framework-oriented system views computation as a continuous process during which a program is fed streams of data from an outside source and leaves trails of results behind. TPIE programmers do not need to worry about making explicit calls to I/O routines. Instead, they merely specify the functional details of the desired computation, and TPIE automatically choreographs a sequence of data movements to feed the computation.

TPIE is written in C++ as a set of templated classes and functions. It consists of three main components: a block transfer engine (BTE), a memory manager (MM), and an access method interface (AMI). The BTE is responsible for moving blocks of data to and from the disk. It is also responsible for scheduling asynchronous read-ahead and write-behind when necessary to allow computation and I/O to overlap. The MM is responsible for managing main memory in coordination with the AMI and BTE. The AMI provides the high-level uniform interface for application programs. The AMI is the only component that programmers normally need to interact with directly. Applications that use the AMI are portable across hardware platforms, since they do not have to deal with the underlying details of how I/O is performed on a particular machine.

We have seen in the previous sections that many batched problems in spatial databases, GIS, scientific computing, graphs, and string processing can be solved optimally using a relatively small number of basic paradigms like scanning (or scanning), multiway distribution, and merging, which TPIE supports as access mechanisms. Batched programs in TPIE thus consist primarily of a call to one or more of these standard access mechanisms. For example, a distribution sort can be programmed by using the access mechanism for multiway distribution. The programmer has to specify the details as to how the partitioning elements are formed and how the buckets are defined. Then the multiway distribution is invoked, during which TPIE automatically forms the buckets and writes them to disk using double buffering.

For online data structures such as hashing, B-trees, and R-trees, TPIE supports more traditional block access like the access-oriented systems.

11. Conclusions

In this paper we have described several useful paradigms for the design and implementation of efficient external memory (EM) algorithms. The problem domains we have considered include sorting, permuting, FFT, scientific computing, computational geometry and spatial databases, graphs, and text. Interesting challenges remain in virtually all these problem domains. One difficult problem is to prove lower bounds for permuting and sorting without the indivisibility assumption. Another promising area is the design and analysis of EM algorithms for efficient use of multiple disks. Optimal bounds have not yet been determined for several basic EM graph problems like topological sorting, shortest paths, breadth-first and depth-first search, and connected components. There is an intriguing connection between problems that have good I/O speedups and problems that have fast and work-efficient parallel algorithms.

A continuing goal is to develop optimal EM algorithms and to translate theoretical gains into observable improvements in practice. For some of the problems that can be solved optimally up to a constant factor, the constant overhead is too large for the algorithm to be of practical use, and simpler approaches are needed. In practice, algorithms cannot assume a static internal memory allocation; they must adapt in a robust way when the memory allocation changes.

Many interesting challenges and opportunities in algorithm design and analysis arise from new architectures being developed, such as networks of workstations, hierarchical storage devices, disk drives with processing capabilities, and storage devices based upon microelectromechanical systems (MEMS). Active (or intelligent) disks, in which disk drives have some processing capability and can filter information sent to the host, have recently been proposed to further reduce the I/O bottleneck, especially in large database applications (Acharya et al. 1998, Riedel et al. 1998). MEMS-based nonvolatile storage has the potential to serve as an intermediate level in the memory hierarchy between DRAM and disks. It could ultimately provide better latency and bandwidth than disks, at less cost per bit than DRAM (Schlosser et al. 2000, Vettiger et al. 2000).

Acknowledgments

The author wishes to thank Lars Arge, Adam Buchsbaum, Jeff Chase, David Hutchinson, Amin Vahdat, and the members of the Center for Geometric Computing at Duke University for several helpful comments and suggestions. Figure 10.1 comes from Cormen et al. (1990). Figure 10.5 is a modified version of a figure in Arge et al. (1998a).

Bibliography

J. Abello, A. Buchsbaum, and J. Westbrook. A functional approach to external graph algorithms. In *Proceedings of the 6th Annual European Symposium on Algorithms*, volume 1461 of *Lecture Notes in Computer Science*, pages 332–343. Springer-Verlag, 1998.

A. Acharya, M. Uysal, and J. Saltz. Active disks: Programming model, algorithms and evaluation. *ACM SIGPLAN Notices*, 33:81–91, 1998.

M. Adler. New coding techniques for improved bandwidth utilization. In *37th IEEE Symposium on Foundations of Computer Science*, pages 173–182, 1996.

P. K. Agarwal, L. Arge, T. M. Murali, K. Varadarajan, and J. S. Vitter. I/O-efficient algorithms for contour line extraction and planar graph blocking. In *Proceedings of the ACM-SIAM Symposium on Discrete Algorithms*, pages 117–126, 1998.

A. Aggarwal, B. Alpern, A. K. Chandra, and M. Snir. A model for hierarchical memory. In *Proceedings of the 19th ACM Symposium on Theory of Computation*, pages 305–314, 1987a.

A. Aggarwal, A. Chandra, and M. Snir. Hierarchical memory with block transfer. In *Proceedings of the 28th Annual IEEE Symposium on Foundations of Computer Science*, pages 204–216, 1987b.

A. Aggarwal and C. G. Plaxton. Optimal parallel sorting in multi-level storage. In *Proceedings of the Fifth Annual ACM-SIAM Symposium on Discrete Algorithms*, pages 659–668, 1994.

A. Aggarwal and J. S. Vitter. The Input/Output complexity of sorting and related problems. *Communications of the ACM*, 31:1116–1127, 1988.

B. Alpern, L. Carter, E. Feig, and T. Selker. The uniform memory hierarchy model of computation. *Algorithmica*, 12:72–109, 1994.

L. Arge. The buffer tree: A new technique for optimal I/O-algorithms. In *Proceedings of the Workshop on Algorithms and Data Structures*, volume 955 of *Lecture Notes in Computer Science*, pages 334–345. Springer-Verlag, 1995a. A complete version appears as BRICS technical report RS-96-28, University of Aarhus.

L. Arge. The I/O-complexity of ordered binary-decision diagram manipulation. In *Proceedings of the International Symposium on Algorithms and Computation*, volume 1004 of *Lecture Notes in Computer Science*, pages 82–91. Springer-Verlag, 1995b.

L. Arge. External-memory algorithms with applications in geographic information systems. In M. van Kreveld, J. Nievergelt, T. Roos, and P. Widmayer, editors, *Algorithmic Foundations of GIS*, volume 1340 of *Lecture Notes in Computer Science*, pages 213–254. Springer-Verlag, 1997.

L. Arge. External memory data structures. In J. Abello, P. M. Pardalos, and M. G. C. Resende, editors, *Handbook of Massive Data Sets*. Kluwer Academic Publisher, 2001. This volume.

L. Arge, G. S. Brodal, and L. Toma. On external memory MST, SSSP and multi-way planar graph separation. In *Proceedings of the Scandinavian Workshop on Algorithms Theory*, 2000a.

L. Arge, P. Ferragina, R. Grossi, and J. Vitter. On sorting strings in external memory. In *Proceedings of the ACM Symposium on Theory of Computation*, pages 540–548, 1997.

L. Arge, K. H. Hinrichs, J. Vahrenhold, and J. S. Vitter. Efficient bulk operations on dynamic R-trees. In *Proceedings of the 1st Workshop on Algorithm Engineering and Experimentation*, pages 328–348, 1999.

L. Arge, M. Knudsen, and K. Larsen. A general lower bound on the I/O-complexity of comparison-based algorithms. In *Proceedings of the 3rd Workshop on Algorithms and Data Structures*, volume 709 of *Lecture Notes in Computer Science*, pages 83–94. Springer-Verlag, 1993.

L. Arge and P. Miltersen. On showing lower bounds for external-memory computational geometry problems. In J. Abello and J. S. Vitter, editors, *External Memory Algorithms*, volume 50, pages 139–159. American Mathematical Society, 1999.

L. Arge, O. Procopiuc, S. Ramaswamy, T. Suel, J. Vahrenhold, and J. S. Vitter. A unified approach for indexed and non-indexed spatial joins. In *Proceedings of the 7th International Conference on Extending Database Technology*, 2000b.

L. Arge, O. Procopiuc, S. Ramaswamy, T. Suel, and J. S. Vitter. Scalable sweeping-based spatial join. In *Proceedings of the 24th International Conference on Very Large Databases*, pages 570–581, 1998a.

L. Arge, O. Procopiuc, S. Ramaswamy, T. Suel, and J. S. Vitter. Theory and practice of I/O-efficient algorithms for multidimensional batched searching problems. In *Proceedings of the ACM-SIAM Symposium on Discrete Algorithms*, pages 685–694, 1998b.

L. Arge, D. E. Vengroff, and J. S. Vitter. External-memory algorithms for processing line segments in geographic information systems. In *Proceedings of the Third European Symposium on Algorithms*, volume 979 of *Lecture Notes in Computer Science*, pages 295–310. Springer-Verlag, 1995. To appear in *Algorithmica*.

C. Armen. Bounds on the separation of two parallel disk models. In *Proceedings of the 4th Workshop on Input/Output in Parallel and Distributed Systems*, pages 122–127, 1996.

R. D. Barve, E. F. Grove, and J. S. Vitter. Simple randomized mergesort on parallel disks. *Parallel Computing*, 23:601–631, 1997.

R. D. Barve, M. Kallahalla, P. J. Varman, and J. S. Vitter. Competitive parallel disk prefetching and buffer management. *Journal of Algorithms*, 36, 2000.

R. D. Barve, E. A. M. Shriver, P. B. Gibbons, B. K. Hillyer, Y. Matias, and J. S. Vitter. Modeling and optimizing I/O throughput of multiple disks on a bus. In *SIGMETRICS Joint International Conference on Measurement and Modeling of Computer Systems*, pages 83–92, 1999.

R. D. Barve and J. S. Vitter. A simple and efficient parallel disk mergesort. In *Proceedings of the 11th Annual ACM Symposium on Parallel Algorithms and Architectures*, pages 232–241, 1999a.

R. D. Barve and J. S. Vitter. A theoretical framework for memory-adaptive algorithms. In *Proceedings of the 40th Annual IEEE Symposium on Foundations of Computer Science*, pages 273–284, 1999b.

M. A. Bender, E. D. Demaine, and M. Farach-Colton. Cache-oblivious B-trees. In *Proceedings of the 41st Annual Symposium on Foundations of Computer Science (FOCS 2000)*, 2000.

A. L. Buchsbaum, M. Goldwasser, S. Venkatasubramanian, and J. R. Westbrook. On external memory graph traversal. In *Proceedings 11th ACM-SIAM Symposium Discrete Algorithms*, 2000.

L. Carter and K. S. Gatlin. Towards an optimal bit-reversal permutation program. In *Proceedings of the IEEE Symposium on Foundations of Computer Science*, pages 544–553, 1998.

P. M. Chen, E. K. Lee, G. A. Gibson, R. H. Katz, and D. A. Patterson. RAID: high-performance, reliable secondary storage. *ACM Computing Surveys*, 26:145–185, 1994.

Y.-J. Chiang. Experiments on the practical I/O efficiency of geometric algorithms: Distribution sweep vs. plane sweep. *Computational Geometry: Theory and Applications*, 8:211–236, 1998.

Y.-J. Chiang, M. T. Goodrich, E. F. Grove, R. Tamassia, D. E. Vengroff, and J. S. Vitter. External-memory graph algorithms. In *Proceedings of the ACM-SIAM Symposium on Discrete Algorithms*, pages 139–149, 1995.

K. L. Clarkson and P. W. Shor. Applications of random sampling in computational geometry, II. *Discrete and Computational Geometry*, 4:387–421, 1989.

A. Colvin and T. H. Cormen. ViC*: A compiler for virtual-memory C*. In *Proceedings of the Third International Workshop on High-Level Programming Models and Supportive Environments*, 1998.

P. Corbett, D. Feitelson, S. Fineberg, Y. Hsu, B. Nitzberg, J.-P. Prost, M. Snir, B. Traversat, and P. Wong. Overview of the MPI-IO parallel I/O interface. In R. Jain, J. Werth, and J. C. Browne, editors, *Input/Output in Parallel and Distributed Computer Systems*, volume 362 of *The Kluwer International Series in Engineering and Computer Science*, chapter 5, pages 127–146. Kluwer Academic Publishers, 1996.

P. F. Corbett and D. G. Feitelson. The Vesta parallel file system. *ACM Transactions on Computer Systems*, 14:225–264, 1996.

T. H. Cormen, C. E. Leiserson, and R. L. Rivest. *Introduction to Algorithms*. MIT Press, 1990.

T. H. Cormen and D. M. Nicol. Performing out-of-core FFTs on parallel disk systems. *Parallel Computing*, 24:5–20, 1998.

T. H. Cormen, T. Sundquist, and L. F. Wisniewski. Asymptotically tight bounds for performing BMMC permutations on parallel disk systems. *SIAM Journal on Computing*, 28:105–136, 1999.

A. Crauser, P. Ferragina, K. Mehlhorn, U. Meyer, and E. A. Ramos. Randomized external-memory algorithms for geometric problems. In *Proceedings of the 14th ACM Symposium on Computational Geometry*, pages 259–268, 1998.

A. Crauser, P. Ferragina, K. Mehlhorn, U. Meyer, and E. A. Ramos. I/O-optimal computation of segment intersections. In *External Memory Algorithms*, DIMACS Series on Discrete Mathematics and Theoretical Computer Science, pages 131–138. American Mathematical Society, 1999.

A. Crauser and K. Mehlhorn. LEDA-SM: Extending LEDA to secondary memory. In J. S. Vitter and C. Zaroliagis, editors, *Proceedings of the Workshop on Algorithm Engineering*, Lecture Notes in Computer Science, pages 228–242. Springer-Verlag, 1999.

R. Cypher and G. Plaxton. Deterministic sorting in nearly logarithmic time on the hypercube and related computers. *Journal of Computer and System Sciences*, 47:501–548, 1993.

M. de Berg, M. van Kreveld, M. Overmars, and O. Schwarzkopf. *Computational Geometry Algorithms and Applications*. Springer-Verlag, 1997.

F. Dehne, W. Dittrich, and D. Hutchinson. Efficient external memory algorithms by simulating coarse-grained parallel algorithms. In *Proceedings of the 9th ACM Symposium on Parallel Algorithms and Architectures*, pages 106–115, 1997.

F. Dehne, W. Dittrich, D. Hutchinson, and A. Maheshwari. Reducing I/O complexity by simulating coarse grained parallel algorithms. In *Proceedings of the International Parallel Processing Symposium*, pages 14–20, 1999.

H. B. Demuth. *Electronic Data Sorting*. PhD thesis, Stanford University, 1956. A shortened version appears in *IEEE Transactions on Computing*, C-34(4), 296–310, April 1985, Special issue on Sorting, E.E. Lindstrom, C.K. Wong, and J.S. Vitter, editors.

D. J. DeWitt, J. F. Naughton, and D. A. Schneider. Parallel sorting on a shared-nothing architecture using probabilistic splitting. In *Proceedings of the First International Conference on Parallel and Distributed Information Systems*, pages 280–291, 1991.

W. Dittrich, D. Hutchinson, and A. Maheshwari. Blocking in parallel multisearch problems. In *Proceedings of the ACM Symposium on Parallel Algorithms and Architectures*, pages 98–107, 1998.

J. R. Driscoll, N. Sarnak, D. D. Sleator, and R. E. Tarjan. Making data structures persistent. *Journal of Computer and System Sciences*, 38: 86–124, 1989.

D. Eppstein, Z. Galil, G. F. Italiano, and A. Nissenzweig. Sparsification—A technique for speeding up dynamic graph algorithms. *Journal of the ACM*, 44:669–696, 1997.

M. Farach, P. Ferragina, and S. Muthukrishnan. Overcoming the memory bottleneck in suffix tree construction. In *Proceedings of the IEEE Symposium on Foundations of Computer Science*, pages 174–183, 1998.

J. Feigenbaum, S. Kannan, M. Strauss, and M. Viswanathan. An approximate l1-difference algorithm for massive data streams. In *Proceedings of the 40th Annual IEEE Symposium on Foundations of Computer Science*, pages 501–511, 1999.

W. Feller. *An Introduction to Probability Theory and its Applications*, volume 1. John Wiley & Sons, third edition, 1968.

P. Ferragina and R. Grossi. The String B-tree: A new data structure for string search in external memory and its applications. In *Proceedings of the 27th Annual ACM Symposium on Theory of Computing*, pages 693–702, 1995. To appear in *Journal of the ACM*.

R. W. Floyd. Permuting information in idealized two-level storage. In R. Miller and J. Thatcher, editors, *Complexity of Computer Computations*, pages 105–109. Plenum, 1972.

M. Frigo, C. E. Leiserson, H. Prokop, and S. Ramachandran. Cache-oblivious algorithms. In *IEEE Symposium on Foundations of Computer Science*, 1999.

T. A. Funkhouser, C. H. Sequin, and S. J. Teller. Management of large amounts of data in interactive building walkthroughs. In *Proceedings of the 1992 ACM SIGGRAPH Symposium on Interactive 3D Graphics*, pages 11–20, 1992.

G. R. Ganger. Generating representative synthetic workloads: An unsolved problem. In *Proceedings of the Computer Measurement Group Conference*, pages 1263–1269, 1995.

M. Gardner. *Magic Show*. Knopf, 1977.

G. A. Gibson, J. S. Vitter, and J. Wilkes. Report of the working group on storage I/O issues in large-scale computing. *ACM Computing Surveys*, 28:779–793, 1996.

R. Goldman, N. Shivakumar, S. Venkatasubramanian, and H. Garcia-Molina. Proximity search in databases. In *Proceedings 24th Int. Conf. Very Large Data Bases*, pages 26–37, 1998.

M. T. Goodrich, J.-J. Tsay, D. E. Vengroff, and J. S. Vitter. External-memory computational geometry. In *IEEE Foundations of Computer Science*, pages 714–723, 1993.

J. L. Griffin, S. W. Schlosser, G. R. Ganger, and D. F. Nagle. Modeling and performance of MEMS-based storage devices. In *Proceedings of the Joint International ACM Conference on Measurement and Modeling of Computer Systems*, 2000.

R. Grossi and G. F. Italiano. Efficient cross-trees for external memory. In J. Abello and J. S. Vitter, editors, *External Memory Algorithms*, volume 50, pages 87–106. American Mathematical Society, 1999.

S. K. S. Gupta, Z. Li, and J. H. Reif. Generating efficient programs for two-level memories from tensor-products. In *Proceedings of the Seventh IASTED/ISMM International Conference on Parallel and Distributed Computing and Systems*, pages 510–513, 1995.

L. Hellerstein, G. Gibson, R. M. Karp, R. H. Katz, and D. A. Patterson. Coding techniques for handling failures in large disk arrays. *Algorithmica*, 12:182–208, 1994.

M. R. Henzinger, P. Raghavan, and S. Rajagopalan. Computing on data streams. In *External Memory Algorithms*, DIMACS Series on Discrete Mathematics and Theoretical Computer Science, pages 107–118. American Mathematical Society, 1999.

J. W. Hong and H. T. Kung. I/O complexity: The red-blue pebble game. In *Proceedings of the 13th Annual ACM Symposium on Theory of Computation*, pages 326–333, 1981.

D. Hutchinson, A. Maheshwari, and N. Zeh. An external memory data structure for shortest path queries. In *Proceedings of the 5th Annual International Conference on Computing and Combinatorics*, volume 1627 of *Lecture Notes in Computer Science*, pages 51–60. Springer-Verlag, 1999.

M. Kallahalla and P. J. Varman. Optimal read-once parallel disk scheduling. In *Proceedings of the Sixth Workshop on Input/Output in Parallel and Distributed Systems*, pages 68–77. ACM Press, 1999.

P. C. Kanellakis, G. M. Kuper, and P. Z. Revesz. Constraint query languages. In *Proceedings of the 9th ACM Conference on Principles of Database Systems*, pages 299–313, 1990.

P. C. Kanellakis, S. Ramaswamy, D. E. Vengroff, and J. S. Vitter. Indexing for data models with constraints and classes. *Journal of Computer and System Science*, 52:589–612, 1996.

M. Y. Kim. Synchronized disk interleaving. *IEEE Transactions on Computers*, 35:978–988, 1986.

D. G. Kirkpatrick and R. Seidel. The ultimate planar convex hull algorithm? *SIAM Journal on Computing*, 15:287–299, 1986.

D. E. Knuth. *Sorting and Searching*, volume 3 of *The Art of Computer Programming*. Addison-Wesley, second edition, 1998.

D. E. Knuth. *MMIXware*. Springer-Verlag, 1999.

V. Kumar and E. Schwabe. Improved algorithms and data structures for solving graph problems in external memory. In *Proceedings of the 8th IEEE Symposium on Parallel and Distributed Processing*, pages 169–176, 1996.

R. Laurini and D. Thompson. *Fundamentals of Spatial Information Systems*. Academic Press, 1992.

F. T. Leighton. Tight bounds on the complexity of parallel sorting. *IEEE Transactions on Computers*, C-34:344–354, 1985.

C. E. Leiserson, S. Rao, and S. Toledo. Efficient out-of-core algorithms for linear relaxation using blocking covers. In *Proceedings of the IEEE Symposium on Foundations of Computer Science*, pages 704–713, 1993.

Z. Li, P. H. Mills, and J. H. Reif. Models and resource metrics for parallel and distributed computation. *Parallel Algorithms and Applications*, 8:35–59, 1996.

A. Maheshwari and N. Zeh. External memory algorithms for outerplanar graphs. In *Proceedings of the 5th Annual International Conference on Computing and Combinatorics*, volume 1627 of *Lecture Notes in Computer Science*, pages 51–60. Springer-Verlag, 1999.

A. Maheshwari and N. Zeh. I/O-efficient algorithms for bounded treewidth graphs. In *Proceedings of the 12th Symposium on Discrete Algorithms*, 2001.

Y. Matias, E. Segal, and J. S. Vitter. Efficient bundle sorting. In *Proceedings of the 11th ACM-SIAM Symposium on Discrete Algorithms*, pages 839–848, 2000.

U. Meyer. External memory BFS on undirected graphs with bounded degree. In *Proceedings of the 12th Symposium on Discrete Algorithms*, 2001.

S. A. Moyer and V. Sunderam. Characterizing concurrency control performance for the PIOUS parallel file system. *Journal of Parallel and Distributed Computing*, 38:81–91, 1996.

K. Munagala and A. Ranade. I/O-complexity of graph algorithms. In *Proceedings of the ACM-SIAM Symposium on Discrete Algorithms*, pages 687–694, 1999.

NASA's Earth Observing System (EOS) web page and NASA Goddard Space Flight Center, 2001.
Online document: http://eospso.gsfc.nasa.gov/.

M. H. Nodine, M. T. Goodrich, and J. S. Vitter. Blocking for external graph searching. *Algorithmica*, 16:181–214, 1996.

M. H. Nodine, D. P. Lopresti, and J. S. Vitter. I/O overhead and parallel VLSI architectures for lattice computations. *IEEE Transactions on Computers*, 40:843–852, 1991.

M. H. Nodine and J. S. Vitter. Deterministic distribution sort in shared and distributed memory multiprocessors. In *Proceedings of the 5th Annual ACM Symposium on Parallel Algorithms and Architectures*, pages 120–129, 1993.

M. H. Nodine and J. S. Vitter. Greed Sort: An optimal sorting algorithm for multiple disks. *Journal of the ACM*, 42:919–933, 1995.

H. Pang, M. Carey, and M. Livny. Memory-adaptive external sorts. In *Proceedings of the 19th Conference on Very Large Data Bases*, pages 618–629, 1993a.

H. Pang, M. J. Carey, and M. Livny. Partially preemptive hash joins. In P. Buneman and S. Jajodia, editors, *Proceedings of the 1993 ACM SIGMOD International Conference on Management of Data*, pages 59–68, 1993b.

I. Parsons, R. Unrau, J. Schaeffer, and D. Szafron. PI/OT: Parallel I/O templates. *Parallel Computing*, 23:543–570, 1997.

J. M. Patel and D. J. DeWitt. Partition based spatial-merge join. In *Proceedings of the ACM SIGMOD International Conference on Management of Data*, pages 259–270, 1996.

F. P. Preparata and M. I. Shamos. *Computational Geometry*. Springer-Verlag, 1985.

N. Rahman and R. Raman. Adapting radix sort to the memory hierarchy. In *ALENEX '00, Workshop on Algorithm Engineering and Experimentation*, Lecture Notes in Computer Science. Springer-Verlag, 2000.

J. Rao and K. Ross. Cache conscious indexing for decision-support in main memory. In *Proceedings of the 25th International Conference on Very Large Databases*, pages 78–89. Morgan Kaufmann Publishers, 1999.

J. Rao and K. A. Ross. Making B$^+$-trees cache conscious in main memory. In *Proceedings of the 2000 ACM SIGMOD International Conference on Management of Data*, SIGMOD Record (ACM Special Interest Group on Management of Data), pages 475–486, 2000.

E. Riedel, G. A. Gibson, and C. Faloutsos. Active storage for large-scale data mining and multimedia. In *Proceedings of the IEEE International Conference on Very Large Databases*, pages 62–73, 1998.

M. Rosenblum, E. Bugnion, S. Devine, and S. A. Herrod. Using the SimOS machine simulator to study complex computer systems. *ACM Transactions on Modeling and Computer Simulation*, 7:78–103, 1997.

C. Ruemmler and J. Wilkes. An introduction to disk drive modeling. *IEEE Computer*, pages 17–28, 1994.

K. Salem and H. Garcia-Molina. Disk striping. In *Proceedings of the 2nd IEEE Data Engineering Conference*, pages 336–342, 1986.

H. Samet. *Applications of Spatial Data Structures: Computer Graphics, Image Processing, and GIS*. Addison-Wesley, 1989a.

H. Samet. *The Design and Analysis of Spatial Data Structures*. Addison-Wesley, 1989b.

P. Sanders. Personal communication, 2000.

P. Sanders. Reconciling simplicity and realism in parallel disk models. In *Proceedings of the 12th Annual ACM-SIAM Symposium on Discrete Algorithms*, 2001.

P. Sanders, S. Egner, and J. Korst. Fast concurrent access to parallel disks. In *Proceedings of the 11th Annual ACM-SIAM Symposium on Discrete Algorithms*, pages 849–858, 2000.

J. E. Savage. Extending the Hong-Kung model to memory hierarchies. In *Proceedings of the 1st Annual International Conference on Computing and Combinatorics*, volume 959 of *Lecture Notes in Computer Science*, pages 270–281. Springer-Verlag, 1995.

J. E. Savage and J. S. Vitter. Parallelism in space-time tradeoffs. *Advances in Computing Research*, 4:117–146, 1987.

S. W. Schlosser, J. L. Griffin, D. F. Nagle, and G. R. Ganger. Designing computer systems with MEMS-based storage. In *Proceedings of the 9th International Conference on Architectural Support for Programming Languages and Operating Systems*, 2000.

K. E. Seamons and M. Winslett. Multidimensional array I/O in Panda 1.0. *Journal of Supercomputing*, 10:191–211, 1996.

M. Seltzer, K. A. Smith, H. Balakrishnan, J. Chang, S. McMains, and V. Padmanabhan. File system logging versus clustering: A performance comparison. In *Usenix Annual Technical Conference*, pages 249–264, 1995.

S. Sen and S. Chatterjee. Towards a theory of cache-efficient algorithms. In *Proceedings of the 11th Annual ACM-SIAM Symposium on Discrete Algorithms*, 2000.

E. Shriver, A. Merchant, and J. Wilkes. An analytic behavior model for disk drives with readahead caches and request reordering. In *Joint International Conference on Measurement and Modeling of Computer Systems*, pages 182–191, 1998.

E. A. M. Shriver and M. H. Nodine. An introduction to parallel I/O models and algorithms. In R. Jain, J. Werth, and J. C. Browne, editors, *Input/Output in Parallel and Distributed Computer Systems*, chapter 2, pages 31–68. Kluwer Academic Publishers, 1996.

E. A. M. Shriver and L. F. Wisniewski. An API for choreographing data accesses. Technical Report PCS-TR95-267, Dept. of Computer Science, Dartmouth College, 1995.

J. F. Sibeyn. From parallel to external list ranking. Technical Report MPI–I–97–1–021, Max-Planck-Institut, 1997.

J. F. Sibeyn. External selection. In *Proceedings of the 1999 Symposium on Theoretical Aspects of Computer Science*, volume 1563 of *Lecture Notes in Computer Science*, pages 291–301, 1999.

J. F. Sibeyn and M. Kaufmann. BSP-like external-memory computation. In *Proceedings of the 3rd Italian Conference on Algorithms and Complexity*, pages 229–240, 1997.

A. Srivastava and A. Eustace. ATOM: A system for building customized program analysis tools. *ACM SIGPLAN Notices*, 29:196–205, 1994.

R. Tamassia and J. S. Vitter. Optimal cooperative search in fractional cascaded data structures. *Algorithmica*, 15:154–171, 1996.

Microsoft's TerraServer online database of satellite images, 2001. Online at http://terraserver.microsoft.com/.

R. Thakur, A. Choudhary, R. Bordawekar, S. More, and S. Kuditipudi. Passion: Optimized I/O for parallel applications. *IEEE Computer*, 29:70–78, 1996.

TIGER/Line (tm). Technical Report 1992, U S. Bureau of the Census, 1992.

S. Toledo. A survey of out-of-core algorithms in numerical linear algebra. In J. Abello and J. S. Vitter, editors, *External Memory Algorithms*, volume 50, pages 161–179. American Mathematical Society, 1999.

TPIE user manual and reference, 1999. The manual and software distribution are available on the web at http://www.cs.duke.edu/TPIE/.

J. D. Ullman and M. Yannakakis. The input/output complexity of transitive closure. *Annals of Mathematics and Artificial Intellegence*, 3: 331–360, 1991.

M. van Kreveld, J. Nievergelt, T. Roos, and P. W., editors. *Algorithmic Foundations of GIS*, volume 1340 of *Lecture Notes in Computer Science*. Springer-Verlag, 1997.

D. E. Vengroff and J. S. Vitter. I/O-efficient scientific computation using TPIE. In *Proceedings of the Fifth NASA Goddard conference on Mass Storage Systems*, pages 553–570, 1996.

P. Vettiger, M. Despont, U. Drechsler, U. Dürig, W. Häberle, M. I Lutwyche, E. Rothuizen, R. Stutz, R. Widmer, and G. K. Binnig. The "Millipede"—more than one thousand tips for future AFM data storage. *IBM Journal of Research and Development*, 44:323–340, 2000.

J. S. Vitter. External memory algorithms. In *Proceedings of the 6th Annual European Symposium on Algorithms*, Lecture Notes in Computer Science. Springer-Verlag, 1998.

J. S. Vitter. External memory algorithms and data structures. In J. Abello and J. S. Vitter, editors, *External Memory Algorithms*, volume 50 of *DIMACS Series on Discrete Mathematics and Theoretical Computer Science*. American Mathematical Society, 1999a.

J. S. Vitter. Online data structures in external memory. In *Proceedings of the 26th Annual International Colloquium on Automata, Languages, and Programming*, pages 119–133, 1999b.

J. S. Vitter. Online data structures in external memory. In *Proceedings of the 6th Biannual Workshop on Algorithms and Data Structures*, 1999c.

J. S. Vitter. External memory algorithms and data structures: Dealing with MASSIVE DATA. *ACM Computing Surveys*, 2001. To appear. Available via the author's web page http://www.cs.duke.edu/~jsv/.

J. S. Vitter and P. Flajolet. Average-case analysis of algorithms and data structures. In J. van Leeuwen, editor, *Handbook of Theoretical Computer Science, Volume A: Algorithms and Complexity*, chapter 9, pages 431–524. North-Holland, 1990.

J. S. Vitter and D. A. Hutchinson. Distribution sort with randomized cycling. In *Proceedings of the 12th ACM-SIAM Symposium on Discrete Algorithms*, 2001.

J. S. Vitter and M. H. Nodine. Large-scale sorting in uniform memory hierarchies. *Journal of Parallel and Distributed Computing*, 17:107–114, 1993.

J. S. Vitter and E. A. M. Shriver. Algorithms for parallel memory I: Two-level memories. *Algorithmica*, 12):110–147, 1994a.

J. S. Vitter and E. A. M. Shriver. Algorithms for parallel memory II: Hierarchical multilevel memories. *Algorithmica*, 12:148–169, 1994b.

J. S. Vitter and M. Wang. Approximate computation of multidimensional aggregates of sparse data using wavelets. In *Proceedings of the ACM SIGMOD International Conference on Management of Data*, pages 193–204, 1999.

J. S. Vitter, M. Wang, and B. Iyer. Data cube approximation and histograms via wavelets. In *Proceedings of the Seventh International Conference on Information and Knowledge Management*, pages 96–104, 1998.

M. Wang, B. Iyer, and J. S. Vitter. Scalable mining for classification rules in relational databases. In *Proceedings of the International Database Engineering & Application Symposium*, pages 58–67, 1998.

M. Wang, J. S. Vitter, L. Lim, and S. Padmanabhan. Wavelet-based cost estimation for spatial queries, 2000. Manuscript.

R. W. Watson and R. A. Coyne. The parallel I/O architecture of the high-performance storage system (HPSS). In *Proceedings of the Fourteenth IEEE Symposium on Mass Storage Systems*, pages 27–44, 1995.

D. Womble, D. Greenberg, S. Wheat, and R. Riesen. Beyond core: Making parallel computer I/O practical. In *Proceedings of the 1993 DAGS/PC Symposium*, pages 56–63, 1993.

C. Wu and T. Feng. The universality of the shuffle-exchange network. *IEEE Transactions on Computers*, C-30:324–332, 1981.

S. B. Zdonik and D. Maier, editors. *Readings in Object-Oriented Database Systems*. Morgan Kauffman, 1990.

W. Zhang and P.-A. Larson. Dynamic memory adjustment for external mergesort. In *Proceedings of the 23rd International Conference on Very Large Data Bases*, pages 376–385, 1997.

B. Zhu. Further computational geometry in secondary memory. In *Proceedings of the International Symposium on Algorithms and Computation*, volume 834 of *Lecture Notes in Computer Science*, page 514. Springer-Verlag, 1994.

V

OPTIMIZATION

Chapter 11

DATA ENVELOPMENT ANALYSIS (DEA) IN MASSIVE DATA SETS

José H. Dulá
School of Business, University of Mississippi, University, MS 38677, USA
jdula@olemiss.edu

Francisco J. López
School of Business, University of Mississippi, University, MS 38677, USA
fjlopez@olemiss.edu

Abstract   Data Envelopment Analysis (DEA) is a clustering methodology for records in data sets corresponding to entities sharing a common list of attributes. Broadly defined, DEA partitions the records into two subsets; those 'efficient' and those 'inefficient.' An efficient record is one which lies on a specific portion of the boundary of a finitely generated polyhedral set in the dimension of the attribute space known as the 'frontier'; inefficient points are those located elsewhere. In traditional applications, DEA frontiers are nonparametric surrogates for unknown theoretical efficiency limits. More generally, however, frontiers are subsets of the boundary defined by extreme elements of the data set. This chapter deals with data envelopment analysis under this broader, more general, definition as it applies to large scale problems.

Keywords: Clustering, Data envelopment analysis, Non-parametric estimation, Linear programming, Convex polyhedral set, Envelopment forms, Decomposition algorithms.

1.   Introduction

Data Envelopment Analysis (DEA), as originally proposed by Charnes *et al.* in 1978 Charnes et al. (1978) , is a non-parametric frontier estimation methodology based on linear programming for measuring relative efficiencies and performance of a collection of related comparable entities (called Decision Making Units or DMUs) in transforming inputs into outputs. DEA is a powerful quantitative tool that provides a means to obtain useful information about efficiency and performance of firms, organizations, and all sorts of functionally similar, relatively autonomous operating units. DEA's domain can be any group of many entities characterized by the same set of multiple attributes. The objective of a DEA study is to assess the efficiency of each entity in relation to its peers. The result of a DEA study is a classification of all entities as either "efficient" or "inefficient."

DEA has been successfully applied in various fields. Particularly noteworthy examples of applications where DEA has shown its effectiveness appear in education, health care, and finance. Education was one of the earliest applications of DEA (see e.g., Bessent et al. (1982)); it served to test and validate it as a tool for analysis and decision support. Education was an ideal introduction for DEA because it typically deals with comparisons of many similar and autonomous 'not for profit' entities described by incommensurate inputs and outputs. The complex nature of the issues involved in health care make it a natural element in DEA's domain. Efficiency and performance assessments of physicians, hospitals, etc., necessarily involve much more than just costs; the prescribed situation for a DEA study. In finance, DEA is ideally suited to compare and assess efficiency of banks and branches since these are well structured, standard, operations that essentially deal with the same types of inputs and outputs somewhat independently of their size and location. It is in finance where DEA finds applications the scales of which test the limits of the performance of the equipment and algorithms. The reader is directed to the many excellent comprehensive expositions that introduce DEA, its history, and its applications; e.g., Dulá (2000), Seiford (1996).

It is not difficult to imagine applications for DEA which vastly exceed what have been until recently the limits of its operational range. Operations involving large numbers of individual records with a common list of multiple, ordinal-valued, attributes per record are usual –and growing larger – and are potential applications for DEA processing. Examples include, personnel records for large agencies; e.g., armed forces; companies and agencies with large "customer" data bases; e.g., credit card companies, credit reporting agencies, etc. The number of records range from several hundreds of thousands to many millions. If a record is an employee in the armed forces, the relevant attributes for a DEA study may include level of education, numbers of years in the service, scores in entrance examinations ("intellectual and skill endowment") and scores on recent performance evaluations which are frequently administered. If a record is a customer with a credit card company, relevant attributes may be: years as a customer; average balance; income; a measure of payment discipline; etc. The technology to apply DEA at these scales has not been available and is only now becoming feasible to entertain studies for such applications.

DEA is about extremes. A DEA study reveals the limits of the levels displayed by the values of the attributes of the entities in the data set. Essentially, a DEA study provides knowledge about the geometrical location of entities' data points with respect to a boundary region of a polyhedral hull, in the space of the attributes. This region is the "performance" or "efficiency" *frontier* of the attributes. Data points with coordinates on the frontier correspond to efficient entities; otherwise they are inefficient. The polyhedral hull, also called the *production possibility set*, is generated using constrained linear combinations of the attribute vectors of all the entities in the study.

The data for a DEA study consists of n entities each characterized by an m-dimensional vector the components of which are the data values for the attributes in the model. An efficiency study involving DEA is computationally intensive. Detecting the location of a point in a polyhedral set can be achieved with linear programming. Since each data point needs to be checked, DEA requires the repeated application of linear programming. As many linear programs as there are entities may need to be solved to perform one DEA study and several analyses may be involved in an elaborate study. Compare this with, say, least-squares regression where, for similar data structures, only a single system of linear equations needs to be solved. In DEA this would be multiplied by the number of iterations to solve an LP and the total number of LPs in the complete study.

The same set of entities can be used to generate many different production possibility sets. The production possibility set which will define the model for a study depends on the type of extreme behavior we seek to identify. This is a modeling decision affected by judgments about each attribute value's merit or desirability; and which limit of its magnitudes are relevant; i.e., upper limits, lower limits, or both. Production possibility sets generated by a given set of entities may be polyhedral cones, polyhedrons with multiple extreme points and directions of recessions and, for the sake of this development, even bounded polytopes.

The evolution of DEA as an analytical tool has been and will remain closely linked to computational and algorithmic advances. Computational issues become increasingly relevant as studies involve larger numbers of entities, as analysts ask more questions about different economic assumptions governing the transformation process, and as the studies become more dynamic requiring more frequent evaluations.

2. The Mathematical Model

The data set, $\mathcal{A}$, for a DEA study consists of the n data points, $a^1, \ldots, a^n$; one for each entity. We organize the data in the following manner:

$$A = \begin{bmatrix} a^1, \cdots, a^n \end{bmatrix}, \quad \text{where,} \quad a^j = \left(a_1^j, \ldots, a_m^j\right)^T; \quad j = 1, \ldots, n,$$

and A is the m by n matrix the columns of which are the data points. In traditional DEA, each data point is composed of two types of components, those pertaining to the m_1 "inputs," $0 \neq x^j \geq 0$, and those corresponding to the m_2 "outputs," $0 \neq y^j \geq 0$. In a more general model we admit any partitioning of the components of the attributes as "desirable" and "undesirable," and permit a category of attributes for which both upper and lower magnitudes are interesting.

Each data point, a^j, collects a list of deterministic values that captures the essence of the process. All n entities will be evaluated using the same m attributes. The choice of these attributes is a modeling decision. The analyst needs to interact closely with the decision maker to arrive at a final list of attributes which will adequately reflect the activity of the entities and provide meaningful measurements of their relative performance.

We adopt here the standard assumption that the data is *reduced*. This is an assumption about the way the data is presented for analysis. In a reduced data set there is no duplication of data points and, for some production possibility sets (e.g., cones), this also means that no two points are multiples of each other.

A major modeling decisions regards the shape of the production possibility set which will be the basis for the study. The production possibility set is a finitely generated polyhedral set defined by the data as follows:

$$\mathcal{P} = \left\{ y \in \Re^m | A\lambda \overset{1}{\leftrightarrow} y, \mathbf{e}^T \lambda \overset{2}{\leftrightarrow} 1, \lambda \geq 0 \right\}$$

for $A \in \Re^{m \times n}$, $\lambda \in \Re^n$, $\mathbf{e} = (1, \ldots, 1)$. We say that the data points, $a^j; j = 1, \ldots, n$, *generate* the set $\mathcal{P}$; so, there is yet another way to refer to the data points; namely as 'generators.' The polyhedral set $\mathcal{P}$ is defined by a linear system where the first m relations are given by the rows of the matrix A. Each of these relations corresponds to one of the attributes in the model. The $m+1$st relation is the *convexity* row. The symbols '$\overset{1}{\leftrightarrow}$', '$\overset{2}{\leftrightarrow}$' are devices to represent the possible relations between the left-hand and right-hand sides of each row of the system. The first symbol applies to the first m rows of the system and the second to the convexity row. Thus:

$$\stackrel{1}{\leftrightarrow} \equiv \begin{cases} \geq \\ \leq \\ = \end{cases} \quad \text{and} \quad \stackrel{2}{\leftrightarrow} \equiv \begin{cases} \cdot \\ = \\ \geq \\ \leq \end{cases}$$

The first group of three relations, '$\stackrel{1}{\leftrightarrow}$,' is assigned depending on the assessment about the attribute to which the row corresponds. Desirable attributes ('outputs,' in traditional DEA) will be related with a '$\geq$' to their right-hand side. If attribute i has this inequality, the production possibility set recedes in the direction of the negative of the ith unit vector implying that the frontier is in the opposite direction. These are attributes where more is better. Symmetrically, undesirable attributes ('inputs') relate with a '$\leq$' to their right hand side; here, "less is more." In traditional DEA there are only these two choices and they can be interpreted in terms of the "free-disposability" assumptions about the attribute Banker et al. (1984). The third possibility, '=,' is interesting because it is involved in the description of the convex hull of the data points, a bounded polyhedron. The frontier of a bounded polyhedron is its entire boundary. This means that the analysis seeks the entities with an attribute mix which is at limiting levels on the two opposing extremes. This is the search for the familiar "outliers" of, say, a multivariate sample. Finding the extreme elements of such a finitely generated polyhedral set is an important problem within the scope of this presentation with applications in statistics, e.g., Gastwirth estimators Gastwirth (1966).

The second group of relations, '$\stackrel{2}{\leftrightarrow}$', involves the convexity row. The choice in DEA for this relation is connected with the "returns to scale" assumption about the model. The four standard models in DEA are:

i) "$\stackrel{2}{\leftrightarrow} \equiv \cdot$". No convexity row: This is the constant returns model also known as the CCR model. This was the original DEA model and it was introduced by Charnes, Cooper, and Rhodes Charnes et al. (1978). Under this assumption the nonnegative scaling of a data point does not affect its relative position to the frontier.

ii) "$\stackrel{2}{\leftrightarrow} \equiv =$". The convexity row is an equality: This is the variable returns, "BCC," model of Banker, Charnes, and Cooper Banker et al. (1984). Under this assumption, uniform scaling beyond extreme observed values for the attributes cannot be assumed to be possible and these extreme values actually help define the frontier.

iii) & iv) " $\stackrel{2}{\leftrightarrow} \equiv \geq$ " or " $\stackrel{2}{\leftrightarrow} \equiv \leq$ ". The convexity row is an inequality: these two cases are the increasing and decreasing returns models of Banker *et al.* Banker et al. (1984) and Seiford and Thrall Seiford and Thrall (1990). These models are used in the case the assumption that specifically scaling down or specifically scaling up is invalid.

In terms of its shape, $\mathcal{P}$ is always a convex polyhedral set. Without the convexity row it is a cone; otherwise, it is a polyhedron with, possibly, many extreme points. The set is unbounded whenever at least one of the first m relations is an inequality since the "slack/surplus" provides a direction of recession. This means that all production possibility sets in DEA are unbounded. The only cases when $\mathcal{P}$ is bounded can occur when there are no inequalities in the first m relations and the convexity row is either an equality or less-than-or-equal to its right-hand side. In the first case $\mathcal{P}$ is simply the convex hull of the data points. With regards to dimension, the production possibility set in DEA always has full dimension, m, since it will contain a full basis of unit vectors, or its negatives, defining its recession cone. Whenever one of the attribute rows is an equality, this assurance of full dimensionality is lost. In this case the dimension of the production possibility set may depend on the dimension of the recession cone or the number of affinely independent columns in the matrix A.

The frontier of a production possibility set is defined by the efficiency postulates for the model (see, e.g., Banker *et al.* Banker et al. (1984)). For example, in the constant returns model in DEA, inefficient points are those that can be expressed as a nonnegative linear combination of the points in the data set such that all input components are less than or equal to its corresponding inputs and all output components are greater than or equal – with at least one relation in either case holding as a strict inequality.

Geometrically, the frontier of a production possibility set is defined by the extreme elements of the data set: extreme rays if $\mathcal{P}$ is a cone and extreme points in all other cases[1] . The frontier is the union of facets (faces, if not a full dimensional object) of the polyhedral set. A fundamental insight is that these extreme elements correspond to actual generators; that is, a subset of the data set will be the extreme elements that define the facets of the frontier. Another fundamental insight is

---

[1] Recall that our data is 'reduced' which means that, besides no duplicates, there is also no proportionality between points when the production possibility set is a cone or when the convexity row is greater than or equal to unity.

that any generator which is extreme to the production possibility set is necessarily on the frontier. Other generators may belong to the frontier, and we will want to identify them, but experience shows that these tend to be rare.

The frontier is part of the boundary of the production possibility set but not all boundary points necessarily belong to the frontier. This absence of an assurance of the equivalence between boundary and frontier represents a complication in DEA that has been recognized since its beginnings. Data points which are on the boundary but which are not on the frontier are called *weakly efficient*. Although also rare, they must be identified so as not to be misclassified as efficient entities.

An important consequence of a DEA study regards what is referred to as "benchmarking." A benchmark is a point in the frontier that is associated with an inefficient entity. A benchmark is used to compare the data point being tested with a point, actual or virtual, which is, in some sense, near and which is on the frontier. Such a benchmark serves as a point of reference for recommendations as to how to attain efficiency status. Ideally, a benchmark would be the point on the frontier closest to the coordinates of the inefficient entity being tested. Unfortunately, finding the "closest" in Euclidean distance is impractical since it involves the solution of a quadratic program. That is why a benchmark in DEA is usually a point in the frontier obtained by extending the point being tested along some fixed direction until it hits the frontier.

A DEA study requires that we locate each of the data points in the production possibility set relative to the frontier. It is required that we learn whether a point is interior or on the boundary, and, if on the boundary, whether it is on the frontier or not. If the point is not on the frontier, the analysis ought to proceed to make benchmarking recommendations. This analysis can be made a number of ways. There are opportunistic schemes to locate some of these points; especially if they are extreme. We describe these in the following section. An inescapable realization about DEA, however, is that the only way to conclusively resolve the efficiency status of all entities an to provide benchmarks for inefficient ones requires the solutions to linear programs.

3. Linear Programming and DEA

A deterministic algorithm to classify all n entities as either efficient or inefficient in a DEA study requires the solution to linear programs. The original formulation proposed by Charnes *et al.* Charnes et al. (1978) was derived from the definition of efficiency as the ratio of the weighted sum of outputs to the weighted sum of inputs; the weights being the variables in the problem. The constraints normalized the efficiency "score" and provided for the problem of weak efficiency by setting the weights to be strictly positive. The original fractional program was recast as an LP applying the familiar transformation of Charnes and Cooper Charnes and Cooper (1962) and appears as follows:

$$\max_{\pi \geq \epsilon e, \sigma \geq \epsilon e} \quad \langle \sigma, y^{j^*} \rangle$$
$$\text{s.t.} \quad -\langle \pi, x^j \rangle + \langle \sigma, y^j \rangle \leq 0, \quad j = 1, \ldots, n;$$
$$\langle \pi, x^{j^*} \rangle = 1.$$

where y^j, x^j are the vectors of outputs and inputs, respectively, for the jth entity and j^* is the index of the entity being tested or "scored." The ϵ, known as the *non-Archimedean constant* appearing in the restriction of the weights, π, σ, is how the restriction of strict positivity of these weights is resolved. This LP is the "constant returns/multiplier form/input oriented," formulation. Immediately we see that a number of LP formulations are possible when we consider the other DEA models; e.g. variable returns, etc.; the opposite orientation ("output oriented"); unoriented "additive" versions Charnes et al. (1985); and for every one of these, there is a dual (known as "envelopment" forms).

The solution to a multiplier LP formulation is less than unity if and only if the entity being tested, j^*, is not on the frontier. The optimal solution provides a supporting hyperplane for the corresponding production possibility set. The support set of such a hyperplane is defined by extreme elements which are, in turn, generators. If the entity being tested is inefficient, these generators compose the *reference set* for the benchmark. A combination of these reference set elements using as coefficients their marginal costs defines the "input-oriented" benchmark. This benchmark is a "virtual" entity (not necessarily part of the data set) on the boundary of the production possibility set. A similar "output-oriented" benchmark can be obtained using the solution to the output-oriented LP. Unoriented forms such as the *additive* model Charnes et al. (1985) are not useful for benchmarking and are used only to resolve efficiency status issues.

The basic DEA procedure is to solve one LP for each of the n entities in the model. This simple prescription for DEA can mask an onerous computational task. Every iteration in the procedure requires the complete solution of a dense LP. Multiplier forms require that the objective function and the last constraint be updated with the attribute values for the entity being evaluated.

Compounding the clear computational burden of solving n LPs for one DEA analysis is the particularly serious problem of degeneracy and cycling. Severe problems with degeneracy, including cycling, have been reported in the literature (Barr and Durchholz (1997)). This concern has lead to works such as Charnes et al. (1993) where an antidegeneracy/cycling linear programming method especially designed for data envelopment analysis is introduced. A factor inducing degeneracy in DEA codes is the duplication of the column of the entity being scored in the constraint matrix of the dual envelopment forms. Another is that the production possibility set can have multiple supporting hyperplanes at its extreme elements.

Linear programs and their roles have evolved since the first formulations by Charnes *et al.* Charnes et al. (1978). There are many formulations with different roles in DEA. The issue of the non-Archimedean constant is now resolved using a two phase method where this condition is relaxed in the first LP and, if the result is inconclusive, a second, additive, LP is solved. Other modifications include removing the data of the entity being tested from the coefficient matrix Dulá and Hickman (1997). One advantage of this is that it may prevent degeneracy. There is much interest in exploring variations especially of the unoriented additive formulationsCooper et al. (1998) Tone (1998)).

It is clear from the description of the basic DEA procedure that there are limits to the scale at which it may be applied. To break out of these limits and include massive applications to DEA's domain, we must resort, not just to faster computers, but to better algorithms and implementations.

4. Data Set Characteristics and Computational Environments

Several factors affect the time required to perform a DEA study. These factors can be grouped in two categories. The first category includes elements that are intrinsic characteristics of the data set. These are the cardinality of the data set, the dimension of each entity, and the "density" of extreme elements explained below. The second category refers to the resources used to solve the problem, which include hardware and LP solvers.

The first of the three characteristics of a data set is its "cardinality." The cardinality of a problem is the number of entities (records) in the data set. Typically this number easily exceeds several hundred in applications. The second is its "dimension." This is the number of attributes common to all entities. Normally this number ranges between five and twenty. The "density" of a data set is the proportion of entities that are classified as efficient. Typically, as the cardinality of the data set increases, densities tend to decrease. Our experience shows that beyond cardinalities of 10,000 entities, we can expect densities below five percent and in some cases much less.

There are three basic computer environments where DEA problems are solved: (1) desktop personal computers (PCs), (2) scientific workstations, and (3) high performance mainframes. These environments are used to run different kinds of software applications that range from macros within spreadsheet involving resident optimizers (e.g., a Visual Basic program using the "Solver" in MS Excel), to highly specialized programs written in a scientific programming language (e.g., Fortran or C) using high performance LP solvers (e.g., IMSL or CPlex) with different degrees of sophistication depending on the algorithm selected.

The time to perform a DEA study is affected by the intrinsic characteristics of the data set, by the hardware, and by the LP solvers used. The impact on times of cardinality and dimension are predictable as we will see in the next section. We will also see that density plays a role in the solution times. The other items have the expected impact: computers have a major impact on solution times; LP implementations vary considerably, and, not surprisingly, 'lower end' solvers (e.g. MS Excel's "Solver") are much slower than, say Cplex. Certainly, the most interesting factor affecting times is the algorithm used to execute the DEA analysis about which there is considerable interest in DEA.

5. Procedures and Algorithms for DEA

There is a variety of algorithms for DEA and they all have one common feature: they involve linear programs. Depending on the particular approach used, algorithms can be classified as follows:

1 <u>Traditional or "naive" procedures</u>. These procedures are based on the direct application of the original definitions and formulations; namely, an entity is on the frontier if its score to the appropriate LP satisfies the necessary and sufficient conditions. We will refer to any algorithm based on this direct application of the definition as "naive."

2 <u>Enhanced naive algorithms</u>. These are naive algorithms enhanced to incorporate and take advantage of clever or opportunistic preprocessors to determine the status of many entities without having to pay the full computational price (i.e., LP solutions). Some ideas that have been proposed and tested include:

   (a) <u>Sorting</u>. Maximum and minimum attribute values over the entire data set can correspond to extreme elements depending on the model Dulá et al. (1992). Therefore, simple sortings based on each attribute can detect the entities that are extreme to their production possibility set. Other preprocessors based on other kind of sortings include Dulá et al. (1992) Rosen et al. (1992).

(b) <u>Translating hyperplanes</u>. Extreme points necessarily belong to the support set of supporting hyperplanes. Therefore, unique maximum and minimum level values of arbitrary hyperplanes correspond to extreme elements of the production possibility set. These are, of course, efficient entities. Procedures based on 'translating' arbitrary hyperplanes are relatively inexpensive since they require only inner products Dulá et al. (1992).

(c) <u>Rotating hyperplanes</u> Dulá and López (2000). A supporting hyperplane "anchored" at an extreme point can be rotated to obtain new supporting hyperplanes that reveal the status of previously unknown extreme points. A systematic approach based on this idea allows the discovery of potentially many extreme points. The computational price amounts to inner products and minimum ratio tests.

(d) <u>"Restricted Basis Entry" (RBE)</u> Ali (1993). When an inefficient entity is identified it can be omitted on any subsequent LP to be solved. This is the idea of RBE which is easy to implement as LPs are iteratively formulated and solved. The systematic application of this approach progressively reduces the size of the LPs that need to be tested.

3 <u>Decomposition Algorithms</u> Barr and Durchholz (1997). An idea to speed up DEA computations is based on combining RBE and data partitioning. The idea applies the principle that if an entity is inefficient with respect to a subset of the entities, it will be inefficient with respect to any superset. An implementation consists of partitioning the data set into uniformly sized "blocks" and applying traditional DEA procedures to them independently identifying inefficient entities within blocks. The procedure proceeds to join processed blocks with unknown status and running these through the analysis. This process is repeated until a final block is created. By then it is expected that the bulk of inefficient entities have been culled. All entities are scored in a second phase using LPs composed of the entities which survived the culling; hopefully, a much smaller LP than would otherwise be used in a naive approach. Several implementational issues must be resolved. Principal among these is the choice of the block size. It turns out

that the performance of decomposition techniques is sensitive to this decision.

4 Frame algorithms Dulá and Helgason (1996) Dulá et al. (1992) Dulá et al. (1997) Dulá and López (2000) Lopez (1999) Venugopal (1995) . A "frame" is a minimal cardinality subset of entities that generate the same polyhedral set as the full data set. The frame is composed of the generators which are extreme elements of the production possibility set. Frame algorithms build the frame sequentially, one element at a time. The process of "erecting" the frame begins with the first extreme element. A linear program and inner products reveal a new extreme element which is added to the frame. The process is repeated until all extreme elements are identified. These algorithms begin with small linear programs the dimension of which grow one column (or row, in a dual) at a time. The final size of the LPs will be determined by the number of efficient elements which are extreme; that is, the size depends directly on the density of the data set. In a second phase all the entities are scored using LPs that omit all non-extreme elements. There are important advantages to frame-based algorithms including avoiding problems with degeneracy, fewer problems with weak efficient entities, and synergies when performing analyses with multiple models taking advantage of the interrelation among frames Dulá and Thrall (1999).

Practitioners have long suspected that the time required to solve DEA problems using a naive approach can be approximated with a quadratic function of the cardinality of the data set. This is confirmed by our testings. Figure 1 depicts this relation for the actual case of five dimensions and 4% density on an SGI Origin supercomputer. The results correspond to a naive implementation enhanced with RBE.

The actual relation between cardinality, n, and time, t, is approximated by the function
$$t = cn^2,$$
where the factor c depends on cardinality and dimension. The factor would be unaffected by density in a pure (unenhanced) naive implementation[2] . Clearly overheads and the hardware and software used will also affect this factor. Table 1.1 shows actual values for c for the case of a DEA problem in five dimensions with a density of around 39% solved

---

[2]Density affects naive implementations enhanced with RBE. The impact is to favor lower densities since this means that more entities are excluded from LPs as the procedure iterates.

on an SGI "Origin" 2000 with eight R10000 processors at 195 Mhz. Our experience indicates that this value reduces and eventually stabilizes as the cardinality increases.

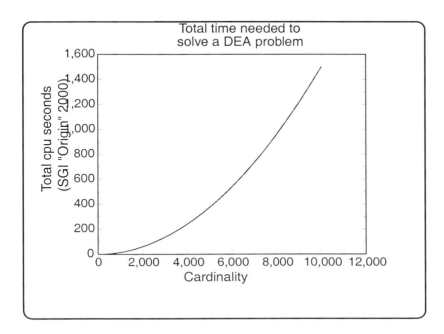

Figure 1.

Table 11.1. Values of factor c for a DEA problem solved on a SGI "Origin" 2000.

Density (Approx)	Number of points	Dimension 5	Dimension 10	Dimension 20
	250	1.07E-5	2.63E-5	8.57E-5
	500	7.32E-6	2.15E-5	8.08E-5
	1000	6.74E-6	1.73E-5	7.11E-5
39%	2000	7.63E-6	1.58E-5	6.48E-5
	5000	6.93E-6	1.79E-5	6.38E-5
	10000	6.99E-6	1.76E-5	6.16E-5
	20000	6.69E-6	1.58E-5	5.79E-5

Experiments comparing the performance of the three basic DEA procedures reveal that frame algorithms are the fastest followed by decomposition schemes and naive based procedures; with the differences becoming more dramatic as the cardinality of the problem increases. This is consistently the case. A typical relation among the three approaches is depicted in Figure 2. The computer used to generate the values in this figure was a SGI "Power Challenge L" with four R8000 processors at 75 Mhz. The problems generating the data for this figure had 10 dimensions and a "low" density category since the number of efficient entities is less than 10%. The information in the figure suggests that frame-based algorithms substantially dominate at large cardinalities and low densities; the typical characteristics of large scale DEA applications. It is important to note that, unlike unenhanced naive procedures, both the decomposition schemes and the frame based algorithms are greatly favored by low densities.

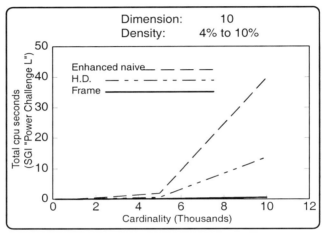

Figure 2.

When an analyst is faced with the prospect of a DEA study, the decisions following the construction of the model, will be about the hardware, algorithm and implementation. Solution times will play a major role in these decisions. It will serve the analyst to have a classification scheme about the size of the problem with standards, benchmarks, and recommendations to plan the DEA study.

6. Classification of DEA Problems

We propose a classification scheme for DEA. To make this classification useful, and stable in a changing technological environment, we will fix the hardware and algorithm using standards that remain contemporary and meaningful to the typical analyst. The classification scheme is based on the time that it takes to process all the entities in the study using a standard, state-of-the-art, desktop computer and the basic naive algorithm implemented using commercially available spreadsheets and bundled optimizers (e.g., VBA in MS Excel) to automate the process.

The proposed classification scheme for DEA is presented next.

Small. These problems can be solved in up to a few minutes (less than ten minutes). Current technology places problems with up to 500 entities in this category. Due to their short solution times no recommendations are necessary.

Medium. Problems in this category can be solved in a few hours; up to 8 hours. The limits for this category are roughly beteen 5000 and 10000 entities. This estimate is affected by several factors, including the dimension of the problem. At this point it pays to implement some of the preprocessors such as sortings and RBE. For problems closer to the upper limit of this category with many dimensions (20 or more) the analyst might consider commercially available packages OnFront (2001). Depending on the frequency with which these problems have to be solved other options should be considered. In dynamic applications where the analysis requires frequent solutions or in studies involving multiple models, it is recommended that more advanced algorithms be employed. Hierarchical decomposition Barr and Durchholz (1997) can reduce times substantially provided the problem is amenable and the block size parameters are optimized. The frame algorithm is faster than decomposition approaches with times between 1/5 and 1/3 those of enhanced naive algorithms especially when densities are below 20%.

Large. Problems will be classified as large if it is possible to obtain only one run per day; that is, times range between 8 and 16 hours. At this scale what-if or dynamic analyses are impractical unless faster computers and better algorithms are used. With current technology, DEA problems start to fall in this category at between 10,000 and 20,000 entities. Our results indicate that frame algorithms can reduce times by one order of magnitude especially when the density is low (less than 5%). For example, a problem with 20,000 entities, dimension ten, and density of 4% on an SGI Origin takes approximately 418 seconds using our implementation of a frame algorithm vs 3,185 seconds using an (RBE enhanced) implementation of the naive algorithm.

Massive. These are problems that are impractical to solve even in the fastest desktop PC (more than 16 hours). Problems in this category have many more than 20,000 entities and, as discussed earlier could include situations involving several hundred thousand entities or more. At this scale it is necessary to use frame-based algorithms. The implementation should incorporate sophisticated linear program solvers, such as Cplex, properly sequenced available preprocessing schemes, and high-end scientific workstations and, for larger problems, supercomputers. Only this way will times be such that what-if and dynamic analyses will be feasible. Although no data is available, it is clear that problems with millions of entities will require load distribution. We will close this section with a short discussion on the impact of parallelization on DEA.

A note about parallelization. There is no reason why LPs have to be solved sequentially in any of the three basic algorithms to solve DEA problems. Therefore, one way to accelerate the performance of procedures such as the naive approaches, decomposition methods, and the frame algorithm is to distribute the load of solving individual linear programs among several processors. In all three algorithms, the LPs can be solved separately and the solutions independently advance the progress of the procedure. There is limited experience on parallelization schemes but indications are that substantial gains in time will be achieved through them. In a nearly asynchronous MIMD implementation, the frame algorithm achieved a speedup factor of 7 to 8 using 14 to 16 processors Dulá et al. (1995). In a more directed study Barr and Durchholz (1997), a decomposition implementation yielded nearly linear speedups. Today's network architectures suggest obvious master/slave distribution scheme with the master assigning LPs to the subordinates and managing the information as it arrives. The coarse grained nature of such a parallelization is a good indication that substantial speed-ups are possible. Since all algorithms for DEA rely in the same way on the solution of linear programs, we might expect that the impact will be comparable across all the procedures. Certainly more work is needed to learn exactly how much can be gained.

7. Conclusion

Until recently the two terms "Data Envelopment Analysis" and "Massive Data Sets" were incongruous. Now, applications over hundreds of thousands of entities are feasible and soon applications with millions of entities will fall within its domain. Being able to identify interesting entities in a data base that are in some sense distinguished because their attribute mix attains interesting magnitude limits, to identify those that do not – and for these to be able to propose changes that will position them in the frontier – to be able to track the progress and evolution of each individual entity; all this knowledge has real value to a firm or agency that makes decisions involving these entities. DEA helps to effectively utilize data and gain advantages from the knowledge about distinguished entities. The value of this kind of decision support grows with the size of the application to which it applies.

Bibliography

A.I. Ali. Streamlined computation for data envelopment analysis. *European Journal of Operational Research*, 64:61–67, 1993.

R.D. Banker, A. Charnes, and W.W. Cooper. Some models for estimating technological and scale inefficiencies in data envelopment analysis. *Management Science*, 30:1078–1092, 1984.

R.S. Barr and M.L. Durchholz. Parallel and hierarchical decomposition approaches for solving large-scale data envelopment analysis models. *Annals of Operations Research*, 73:339–372, 1997.

A. Bessent, W. Bessent, J. Kennington, and B. Reagan. An application of mathematical programming to assess productivity in the houston independent school district. *Management Science*, 28:1355–1367, 1982.

A. Charnes and W.W. Cooper. Programming with linear fractional functionals. *Naval Research Logistics Quaterly*, 9:181–185, 1962.

A. Charnes, W.W. Cooper, B. Golany, L. Seiford, and J. Stutz. Foundations of data envelopment analysis for pareto-koopmans efficient empirical production functions. *Journal of Econometrics*, 30:91–107, 1985.

A. Charnes, W.W. Cooper, and E. Rhodes. Measuring the efficiency of decision making units. *European Journal of Operational Research*, 2: 429–444, 1978.

A. Charnes, J. Rousseau, and J. Semple. An effective non-archimedean anti-degeneracy/ cycling linear programming method especially for data envelopment analysis and like methods. *Annals of Operations Research*, 47:271–278, 1993.

W.W. Cooper, K.S. Park, and J.T. Pastor. Ram: A range adjusted measure of inefficiency for use with additive models and relations to other models and measures in dea. Technical report, IC^2 Institute,

The University of Texas, 1998. To appear in *Journal of Productivity Analysis*.

J.H. Dulá. Data envelopment analysis (dea). In P.M. Pardalos and M.G.C. Resende, editors, *Handbook of Applied Optimization*. Oxford University Press, 2000.

J.H. Dulá and R.V. Helgason. A new procedure for identifying the frame of the convex hull of a finite collection of points in multidimensional space. *European Journal of Operational Research*, 92:352–367, 1996.

J.H. Dulá, R.V. Helgason, and B.L. Hickman. Preprocessing schemes and a solution method for the convex hull problem in multidimensional space. In O. Balci, editor, *Computer Science and Operations Research: New Developments in their Interfaces*, pages 59–70. Pergamon Press, 1992.

J.H. Dulá, R.V. Helgason, and N. Venugopal. A nearly asynchronous parallel lp-based algorithm for the convex hull problem in multidimensional space. In S. Nash and A. Sopher, editors, *The Impact of Emerging Technology on Computer Science and Operations Research*, pages 89–102. Kluwer Academic Publishers, 1995.

J.H. Dulá, R.V. Helgason, and N. Venugopal. An algorithm for identifying the frame of a pointed finite conical hull. *INFORMS Journal of Computing*, 10:323–330, 1997.

J.H. Dulá and B. L. Hickman. Effects of excluding the column being scored from the dea envelopment lp technology matrix. *Journal of the Operational Research Society*, 48:1001–1012, 1997.

J.H. Dulá and F.J. López. Detecting the impact of including and omitting an attribute in dea. Technical report, School of Business Administration, University of Mississippi, 2000.

J.H. Dulá and R.M. Thrall. Accelerating dea over multiple models and forms with frames. Technical report, School of Business Administration, University of Mississippi, 1999.

J. Gastwirth. On robust procedures. *Journal of the American Statistical Association*, 61:929–948, 1966.

F.J. Lopez. *Algorithms to Obtain the Frame of a Finitely Generated Unbounded Polyhedron*. PhD thesis, University of Mississippi, 1999.

OnFront. Emq ab, 2001.

J.B. Rosen, G.L. Xue, and A.T. Phillips. Efficient computation of extreme points of convex hulls in ∇^d. In P.M. Pardalos, editor, *Advances in Optimization and Parallel Computing*, pages 267–292. North Holland, 1992.

L.M. Seiford. Data envelopment analysis: The evolution of the state of the art (1978–1995). *The Journal of Productivity Analysis*, 7:99–137, 1996.

L.M. Seiford and R.M. Thrall. Recent developments in dea: The mathematical programming approach to frontier analysis. *Journal of Econometrics*, 46:7–38, 1990.

K. Tone. A slacks-based measure of efficiency in data envelopment analysis. Technical Report I-98-0001, National Graduate Institute for Policy Studies, 1998.

N. Venugopal. *Determining the Frame of a Pointed Polyhedral Cone*. PhD thesis, Southern Methodist University, 1995.

Chapter 12

OPTIMIZATION METHODS IN MASSIVE DATA SETS

P. S. Bradley
Microsoft Research, Redmond, WA 98052, USA
bradley@microsoft.com

O. L. Mangasarian
Computer Sciences Dept., University of Wisconsin, Madison, WI 53706, USA
olvi@cs.wisc.edu

D. R. Musicant
Computer Sciences Dept., University of Wisconsin, Madison, WI 53706, USA
musicant@cs.wisc.edu

Abstract We describe the role of generalized support vector machines in separating massive and complex data using arbitrary nonlinear kernels. Feature selection that improves generalization is implemented via an effective procedure that utilizes a polyhedral norm or a concave function minimization. Massive data is separated using a linear programming chunking algorithm as well as a successive overrelaxation algorithm, each of which is capable of processing data with millions of points.

Keywords: Vector machines, Nonlinear kernels, Polyhedral norm, Concave function minimization, Linear and quadratic programming, Linear separability, Lagrange multipliers.

1. Introduction

We address here the problem of classifying data in n-dimensional real (Euclidean) space R^n into one of two disjoint finite point sets (i.e. classes).

The support vector machine (SVM) approach to classification (Vapnik 1995, Bennett and Blue 1997, Girosi 1998, Wahba 1999, Cherkassky and Mulier 1998, Schölkopf 1997, Smola 1998) attempts to separate points belonging to two given sets in R^n by a nonlinear surface, often only implicitly defined by a kernel function. Since the nonlinear surface in R^n is typically linear in its parameters, it can be represented as a linear function (plane) in a higher, often much higher, dimensional space, say R^k. Also, the original points of the two given sets can also be mapped into this higher dimensional space. If the two sets are linearly separable in R^k, then it is intuitively plausible to generate a plane mid-way between the furthest parallel planes apart that bound the two sets. Using a distance induced by the kernel generating the nonlinear surface in R^n, it can be shown (Vapnik 1995) that such a plane optimizes the generalization ability of the separating plane. If the two sets are not linearly separable, a similar approach can be used (Cortes and Vapnik 1995, Vapnik 1995) to maximize the distance between planes that bound each set with certain minimal error.

The feature selection problem addressed here is that of discriminating between the two point sets in R^n by a separating plane utilizing as few of the n original problem features as possible. We focus on two approaches. The first approach is motivated by formulating the feature selection problem as the minimization of a concave function over a polyhedral region. A stationary point to this problem is efficiently computed by solving a sequence of linear programs in a successive linearization algorithm. The second approach results from an investigation of the SVM problem formulated as maximizing the margin of separation measured in the 1-norm and ∞-norm.

We also consider a linear programming approach (Bradley and Mangasarian 1998a, Bennett 1998) to SVMs (Vapnik 1995, Wahba 1999, Cortes and Vapnik 1995) for the discrimination between two possibly massive data sets. A proposed approach consists of a novel method for solving linear programs with an extremely large number of constraints that is proven to be monotonic and finite. In the standard SVM formulation (Osuna et al. 1997b;a, Platt 1999) very large quadratic programs are solved. In contrast, the formulation here consists of solving a linear program which is considerably less difficult. For simplicity, our results are given here for a linear discriminating surface, i.e. a separating plane. However, extension to nonlinear surfaces such as quadratic (Mangasarian 1965) or more complex surfaces (Burges 1998) is straightforward.

We next consider the successive overrelaxation (SOR) method for solving massive quadratic programming SVMs. A conventional SVM in its dual formulation contains bound constraints, as well as an equality constraint that requires special treatment in iterative procedures. A very simple convex quadratic program with bound constraints *only* can be obtained by taking the dual of the quadratic program associated with an SVM that maximizes the margin (distance between bounding separating planes) with respect to *all* the parameters determining the bounding planes (Mangasarian and Musicant 1998). This quadratic program can be solved for massive data sets by successive overrelaxation to obtain a linear or nonlinear separating surface (Mangasarian and Musicant 1999).

1.1. Notation and background

- All vectors will be column vectors unless transposed to a row vector by a prime superscript $'$. The scalar (inner) product of two vectors x and y in R^n will be denoted as $x'y$.

- For a vector x in the n-dimensional real space R^n, $|x|$ will denote a vector of absolute values of the components x_i, $i = 1, \ldots, n$ of x.

- For a vector x in R^n, x_+ denotes the vector in R^n with components $\max\{0, x_i\}$.

- For a vector x in R^n, x_* denotes the vector in R^n with components $(x_*)_i$ equal 1 if $x_i > 0$ and 0 otherwise (i.e. x_* is the result of applying the step function to the components of x).

- The base of the natural logarithm will be denoted by ε, and for a vector $y \in R^n$, ε^{-y} will denote a vector in R^n with components ε^{-y_i}, $i = 1, \ldots, n$.

- For $x \in R^n$ and $1 \leq p < \infty$, the norm $\|x\|_p$ will denote the p-norm:

$$\|x\|_p = \left(\sum_{i=1}^n |x_i|^p\right)^{\frac{1}{p}},$$

and

$$\|x\|_\infty = \max_{1 \leq i \leq n} |x_i|.$$

- For a general norm $\|\cdot\|$ on R^n, the *dual norm* $\|\cdot\|'$ on R^n is defined as

$$\|x\|' = \max_{\|y\|=1} x'y.$$

The 1-norm and ∞-norm are dual norms, and so are a p-norm and a q-norm for which $1 \leq p, q \leq \infty$ and $\frac{1}{p} + \frac{1}{q} = 1$.

- The notation $A \in R^{m \times n}$ will signify a real $m \times n$ matrix. For such a matrix A' will denote the transpose of A, A_i will denote the i-th row of A, and $A_{\cdot j}$ will denote the jth column of A.

- A vector of ones in a real space of arbitrary dimension will be denoted by e. A vector of zeros in a real space of arbitrary dimension will be denoted by 0. The identity matrix in a real space of arbitrary dimension will be denoted by I.

- The notation $\arg\min_{x \in S} f(x)$ will denote the set of minimizers of $f(x)$ on the set S.

- We shall employ the MATLAB "dot" notation (MATLAB 1992) to signify application of a function to all components of a matrix or a vector. For example if $A \in R^{m \times n}$, then $A_\bullet^2 \in R^{m \times n}$ will denote the matrix of elements of A squared.

2. Generalized Support Vector Machines for Massive Data Discrimination

We start with a nonlinear separating surface (12.1), implicitly defined by some chosen kernel and by some linear parameters $u \in R^m$ to be determined. These parameters turn out to be closely related to some dual variables. Based on this surface we derive a general convex mathematical program (12.5) that attempts separation via the nonlinear surface (12.1) while minimizing some function f of the parameters u. The function f which attempts to suppress u can be interpreted as minimizing the number of support vectors, or under more conventional assumptions as maximizing the distance between the separating planes in R^k. The choice of f leads to various SVMs. We consider two classes of such machines based on whether f is quadratic or piecewise linear. If we choose f to be a quadratic function generated by the kernel defining the nonlinear surface (12.1), then we are led to the conventional dual quadratic program (12.9) associated with an SVM which requires positive definiteness of this kernel. However the quadratic function choice for f can be divorced from the kernel defining the separating surface and this leads to other convex quadratic programs such as (12.10) *without* making any assumptions on the kernel. Another class of SVMs is generated by choosing a piecewise linear convex function for f and this leads to linear programs such as (12.11) and (12.12), both of which make no assumptions on the kernel.

We begin by defining a general kernel function as follows.

Definition 1 *Let $A \in R^{m \times n}$ and $B \in R^{n \times \ell}$. The **kernel** $K(A, B)$ maps $R^{m \times n} \times R^{n \times \ell}$ into $R^{m \times \ell}$.*

In particular if x and y are column vectors in R^n, then $K(x', A')$ is a row vector in R^m, $K(x', y)$ is a real number and $K(A, A')$ is an $m \times m$ matrix. Note that for our purposes here $K(A, A')$ need not be symmetric in general. Examples of kernels follow, where $a \in R^m$, $b \in R^\ell$, $\mu \in R$ and d is an integer.

Example: **Polynomial Kernel (degree d)** $(AB + \mu ab')_\bullet^d$

Example: **Neural Network Kernel** $(AB + \mu ab')_{\bullet *}$

Example: **Radial Basis Kernel** $\varepsilon^{-\mu \|A'_i - B_{\cdot j}\|^2}$, $i, j = 1, \ldots, m$, $\ell = m$.

Note that our approach allows discontinuous kernels such as the neural network kernel with a discontinuous step function without the need for a smoothing approximation such as the sigmoid or hyperbolic tangent approximation as is usually done (Vapnik 1995, Cherkassky and Mulier 1998).

2.1. GSVM: The general support vector machine

We consider a given set $\mathcal{A}$ of m points in the n-dimensional real feature space R^n represented by the matrix $A \in R^{m \times n}$. Each point A_i, $i = 1, \ldots, m$, belongs to class 1 or class -1 depending on whether D_{ii} is 1 or -1, where $D \in R^{m \times m}$ is a given diagonal matrix of plus or minus ones. We shall attempt to discriminate between the classes 1 and -1 by a nonlinear *separating surface*, induced by some kernel $K(A, A')$, as follows:

$$K(x', A')Du = \gamma, \qquad (12.1)$$

where $K(x', A') \in R^m$, according to Definition 1. The parameters $u \in R^m$ and $\gamma \in R$ are determined by solving a mathematical program, typically quadratic or linear. A point $x \in R^n$ is classified in class 1 or -1 according to whether the *decision function*

$$(K(x', A')Du - \gamma)_*, \qquad (12.2)$$

yields 1 or 0 respectively. The kernel function $K(x', A')$ implicitly defines a nonlinear map from $x \in R^n$ to some other space $z \in R^k$ where k may be much larger than n. In particular if the kernel K is an inner product kernel under Mercer's condition (Courant and Hilbert 1953,pp

138-140),(Vapnik 1995, Cherkassky and Mulier 1998, Burges 1998) (an assumption that we will not make) then for x and y in R^n:

$$K(x,y) = h(x)'h(y), \qquad (12.3)$$

and the separating surface (12.1) becomes:

$$h(x)'h(A')Du = \gamma, \qquad (12.4)$$

where h is a function, not easily computable, from R^n to R^k, and $h(A') \in R^{k \times m}$ results from applying h to the m columns of A'. The difficulty in computing h and the possible high dimensionality of R^k have been important factors in using a kernel K as a generator of an implicit nonlinear separating surface in the original feature space R^n but which is linear in the high dimensional space R^k. Our separating surface (12.1) written in terms of a kernel function retains this advantage and is linear in its parameters, u, γ. We now state a mathematical program that generates such a surface for a general kernel K as follows:

$$\begin{aligned}
\min_{u,\gamma,y} \quad & \nu e'y + f(u) \\
\text{s.t.} \quad & D(K(A, A')Du - e\gamma) + y \geq e \\
& y \geq 0.
\end{aligned} \qquad (12.5)$$

Here f is some convex function on R^m, typically some norm or seminorm, and ν is some positive parameter that weights the separation error $e'y$ versus suppression of the separating surface parameter u. Suppression of u can be interpreted in one of two ways. We interpret it here as minimizing the number of support vectors, i.e. constraints of (12.5) with positive Lagrange multipliers. A more conventional interpretation is that of maximizing some measure of the distance or margin between the bounding parallel planes in R^k, under appropriate assumptions, such as f being a quadratic function induced by a positive definite kernel K as in (12.9) below. As is well known, this leads to improved generalization by minimizing an upper bound on the VC dimension (Vapnik 1995, Schölkopf 1997).

We term a solution of the mathematical program (12.5) and the resulting decision function (12.2) a *generalized support vector machine* (GSVM). In the following section we derive a number of special cases, including the standard SVM. First, however, we note that the mathematical program (12.5) has a solution whenever f is a piecewise-linear or quadratic function bounded below on R^m (Mangasarian 1999b,Proposition 2.1). Note that no convexity of f is required for this existence result. However in applications where duality theory will be invoked, f will need to be convex.

2.2. Quadratic programming support vector machines

We consider specific formulations of the SVM problem, including the standard ones (Vapnik 1995, Cherkassky and Mulier 1998, Burges 1998) and those obtained by setting f of (12.5) to be a convex quadratic function $f(u) = \frac{1}{2}u'Hu$, where $H \in R^{m \times m}$ is some symmetric positive definite matrix. The mathematical program (12.5) becomes the following convex quadratic program:

$$\min_{u,\gamma,y} \nu e'y + \tfrac{1}{2}u'Hu$$
$$\text{s.t.} \quad D(K(A,A')Du - e\gamma) + y \geq e \qquad (12.6)$$
$$y \geq 0.$$

The Wolfe dual (Wolfe 1961, Mangasarian 1969) of this convex quadratic program is:

$$\min_{r \in R^m} \tfrac{1}{2}r'DK(A,A')DH^{-1}DK(A,A')'Dr - e'r$$
$$\text{s.t.} \quad e'Dr = 0 \qquad (12.7)$$
$$0 \leq r \leq \nu e.$$

Furthermore, the primal variable u is related to the dual variable r by:

$$u = H^{-1}DK(A,A')'Dr. \qquad (12.8)$$

If we assume that the kernel $K(A,A')$ is symmetric positive definite and let $H = DK(A,A')D$, then our dual problem (12.7) degenerates to the dual problem of the standard SVM (Vapnik 1995, Cherkassky and Mulier 1998, Burges 1998) with $u = r$:

$$\min_{u \in R^m} \tfrac{1}{2}u'DK(A,A')Du - e'u$$
$$\text{s.t.} \quad e'Du = 0 \qquad (12.9)$$
$$0 \leq u \leq \nu e.$$

The positive definiteness assumption on $K(A,A')$ in (12.9) can be relaxed to positive *semi*definiteness while maintaining the convex quadratic program (12.6), with $H = DK(A,A')D$, as the direct dual of (12.9) without utilizing (12.7) and (12.8). The symmetry and positive semidefiniteness of the kernel $K(A,A')$ for this version of an SVM is consistent with the SVM literature. The fact that $r = u$ in the dual formulation (12.9), shows that the variable u appearing in the original formulation (12.6) is also the dual multiplier vector for the first set of constraints of (12.6). Hence the quadratic term in the objective function of (12.6) can be thought of as suppressing as many multipliers of support vectors as possible. This is another interpretation of the standard SVM that is usually interpreted as maximizing the margin or distance between parallel separating planes.

This leads to the idea of using other values for the matrix H other than $DK(A, A')D$ that will also suppress u. One particular choice is interesting because it puts no restrictions on K: no symmetry, no positive definiteness or semidefiniteness and not even continuity. This is the choice $H = I$ in (12.6) which leads to a dual problem (12.7) with $H = I$ and $u = DK(A, A')'Dr$ as follows:

$$\min_{r \in R^m} \tfrac{1}{2}r'DK(A, A')K(A, A')'Dr - e'r$$
$$\text{s.t.} \quad e'Dr = 0 \quad (12.10)$$
$$0 \leq r \leq \nu e.$$

We note immediately that $K(A, A')K(A, A')'$ is positive semidefinite with no assumptions on $K(A, A')$, and hence the above problem is an always solvable convex quadratic program for any kernel $K(A, A')$. In fact by (Mangasarian 1999b, Proposition 2.1) the quadratic program (12.6) is solvable for *any* symmetric positive definite matrix H, and by quadratic programming duality so is its dual problem (12.7), the solution r of which can be immediately used to generate a decision function (12.2). Thus we are free to choose any symmetric positive definite matrix H to generate an SVM. Experimentation determines the most appropriate choices for H.

We turn our attention to linear programming SVMs.

2.3. Linear programming support vector machines

In this section we consider problems generated from the mathematical program (12.5) by using a piecewise linear function f in the objective function thus leading to linear programs.

An obvious choice for f is the 1-norm of u, which leads to the following linear programming formulation:

$$\min_{u,\gamma,y,s} \nu e'y + e's$$
$$\text{s.t.} \quad D(K(A, A')Du - e\gamma) + y \geq e \quad (12.11)$$
$$-s \leq u \leq s$$
$$y \geq 0.$$

A solution (u, γ, y, s) to this linear program for a chosen kernel $K(A, A')$ will provide a decision function as given by (12.2). This linear program parallels the quadratic programming formulation (12.10) that was obtained as the dual of (12.5) by setting $f(u)$ therein to half the 2-norm squared of u whereas $f(u)$ is set to the 1-norm of u in (12.11). Another linear programming formulation that somewhat parallels the quadratic programming formulation (12.9), which was obtained as the dual of (12.5) by setting $f(u)$ therein to half the 2-norm squared of $K(A, A')^{\frac{1}{2}} Du$, is obtained by setting f to be the 1-norm of $K(A, A')Du$. This leads to the following linear program:

$$\begin{aligned} \min_{u,\gamma,y,s} \quad & \nu e'y + e's \\ \text{s.t.} \quad & D(K(A, A')Du - e\gamma) + y \geq e \\ & -s \leq K(A, A')Du \leq s \\ & y \geq 0. \end{aligned} \quad (12.12)$$

No assumptions of symmetry or positive definiteness on $K(A, A')$ are needed in either of the above linear programming formulations as was the case in the quadratic program (12.9).

It is interesting to note that if the linear kernel $K(A, A') = AA'$ is used in the linear program (12.11) we obtain the high-performing 1-norm linear SVM proposed in (Bredensteiner and Bennett 1998) and utilized successfully in (Bredensteiner 1997, Bennett et al. 1998, Bradley and Mangasarian 1998a). Hence, if we set $w = A'Du$ in (12.11) we obtain (12.26) and (12.28).

This linear programming SVM was implemented in (Mangasarian and Musicant 1999) with nonlinear kernels. A quadratic kernel resulted in better testing set results than a linear kernel did on the liver-disorders data set from the UCI repository (Murphy and Aha 1992). Figure 12.1 depicts a rather complex "checkerboard" example for which no linear classifier can give good separation. Figure 12.2 shows the sharp nonlinear separation obtained by a nonlinear polynomial kernel of degree 6 ($d = 6$):

Polynomial Kernel: $K(A, A') = ((\frac{A}{\lambda} - \rho)(\frac{A}{\lambda} - \rho)' - \mu)_\bullet^d$

The parameter values for λ, ρ, and μ are shown in the caption for Figure 12.2.

We next focus on the feature selection problem in classification.

3. Feature Selection via Concave Minimization and Support Vector Machines

We consider in this section the problem of discriminating between two finite point sets in R^n by a separating plane that utilizes as few of the n features as possible. Two approaches are described.

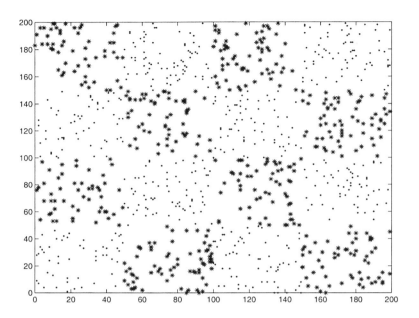

Figure 12.1. Checkerboard training data set

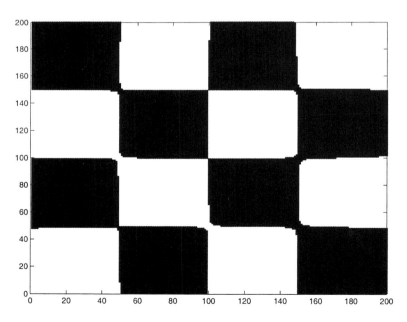

Figure 12.2. Indefinite sixth degree polynomial kernel separation of the checkerboard data set ($\nu = 10,000$, $\lambda = 100$, $\rho = 1$, $d = 6$, $\mu = 0.5$)

The first approach (Mangasarian 1996, Bradley et al. 1998b), described in Section 3.1, involves the minimization of a concave function on a polyhedral set and is based on the following considerations. A plane is constructed such that a weighted sum of distances of misclassified points to the plane is minimized, utilizing as few dimensions of the original feature space R^n as possible. This is achieved by constructing two parallel bounding planes, utilizing a minimal number of dimensions. The two planes determine two opposite halfspaces. Each halfspace mostly contains points of one set. These planes are determined so as to minimize the sum of weighted distances to the bounding plane from points lying in the wrong halfspace. This leads to the minimization of a concave function on a polyhedral set (problems (12.19) and (12.21) below) for which a stationary point can be obtained by solving a few (typically 5 to 7) linear programs in a successive linearization algorithm (Algorithm 2 below). The final separating plane is taken midway between the two bounding parallel planes.

The second approach, that of an SVM (Vapnik 1995, Bennett and Blue 1997, Girosi 1998, Wahba 1999) (described in Section 3.2), constructs the two parallel bounding planes in the full n-dimensional space R^n as in the first approach outlined above, but in addition attempts to push these planes as far apart as possible. The justification for this is to improve the VC dimension (Vapnik 1995), which in turn improves generalization. The typical formulation of the SVM problem uses the 2-norm to measure the distance between the two bounding planes and leads to a quadratic programming problem (Vapnik 1995). We use here instead the 1-norm and the ∞-norm that lead to the single linear programs (12.25) and (12.26) below, respectively. Although improved generalization is the primary purpose of the SVM formulation, it turns out that the linear program (12.26) resulting from employing the ∞-norm to measure the distance between the two bounding planes leads also to a feature selection method, whereas the linear program (12.25) resulting from the use of the 1-norm does not lead to a feature selection method. Note that the norms on w used in (12.25) and (12.26) are dual to those used to measure the distance between the separating planes (Mangasarian 1999a).

3.1. FSV: Feature selection via concave minimization

We consider a feature selection procedure that has been effective in medical and other applications (Bradley et al. 1998b, Mangasarian 1996).

Given m points in R^n represented by the $m \times n$ matrix A and the $m \times m$ diagonal matrix D with $D_{ii} = 1$ if the i-th data point A_i belongs to class 1 and $D_{ii} = -1$ if the i-th data point belongs to class -1, we wish to discriminate between them by a separating plane:

$$P = \{x \mid x \in R^n, x'w = \gamma\}, \qquad (12.13)$$

with normal $w \in R^n$ and 1-norm distance to the origin of $\dfrac{|\gamma|}{\|w\|_\infty}$ (Mangasarian 1999a). We shall attempt to determine w and γ so that the separating plane P defines two open halfspaces $\{x \mid x \in R^n, x'w > \gamma\}$ containing mostly points A_i from class 1, and $\{x \mid x \in R^n, x'w < \gamma\}$ containing mostly points A_i from class -1. Hence we wish to satisfy

$$\begin{array}{ll} A_i'w > \gamma & \text{if } D_{ii} = 1, \\ A_i'w < \gamma & \text{if } D_{ii} = -1, \end{array} \qquad (12.14)$$

to the extent possible. Upon normalization, these inequalities can be equivalently written as:

$$D(Aw - e\gamma) \geq e. \qquad (12.15)$$

Conditions (12.14) or equivalently (12.15) can be satisfied if and only if the convex hulls of the data points belonging to class 1 and the data points belonging to class -1 are disjoint. This is not the case in many real-world applications. Hence, we attempt to satisfy (12.15) in some "best" sense, for example, by minimizing some norm of the violations of (12.15) such as

$$\min_{w,\gamma} f(w,\gamma) = \min_{w,\gamma} \| (D(Aw - e\gamma) - e)_+ \|_1. \qquad (12.16)$$

Recall that for a vector x, x_+ denotes the vector with components $\max\{0, x_i\}$. Two principal reasons for choosing the 1-norm in (12.16) are:

(i) Problem (12.16) is then reducible to a linear program (12.17) with many important theoretical properties making it an effective computational tool (Bennett and Mangasarian 1992).

(ii) The 1-norm is less sensitive to outliers such as those occurring when the underlying data distributions have pronounced tails, hence (12.16) has a similar effect to that of robust regression (Huber 1981),(Hassoun 1995,pp 82-87).

The formulation (12.16) is equivalent to the following robust linear programming formulation (RLP) proposed in (Bennett and Mangasarian

1992) and effectively used to solve problems from real-world domains (Mangasarian et al. 1995):

$$\min_{w,\gamma,y} \quad e'y$$
$$\text{s.t.} \quad D(Aw - e\gamma) + y \geq e \quad (12.17)$$
$$y \geq 0.$$

The linear program (12.17) or, equivalently, the formulation (12.16) define a separating plane P that approximately satisfies the conditions (12.15) in the following sense. For each positive value of y_i, the corresponding data point A_i is in error. Thus, if $D_{ii} = 1$, then the data point A_i lies on the wrong side of the bounding plane $x'w = \gamma + 1$ for class 1; if $D_{ii} = -1$, the data point A_i lies on the wrong side of the bounding plane $x'w = \gamma - 1$ for class -1. Hence the objective function of the linear program (12.17) minimizes the sum of distances, weighted by $\|w\|'$, of misclassified points to their respective bounding planes. The separating plane P (12.13) is midway between the two bounding planes and parallel to them.

We now introduce a feature selection idea (Mangasarian 1996, Bradley et al. 1998b) by attempting to suppress as many components as possible of the normal vector w to the separating plane P that is consistent with obtaining an acceptable separation between the points with class 1 and class -1. We achieve this by introducing an extra term into the objective of (12.17) while weighting the original objective by $\nu > 0$ as follows:

$$\min_{w,\gamma,y} \quad \nu e'y + e'|w|_*$$
$$\text{s.t.} \quad D(Aw - e\gamma) + y \geq e \quad (12.18)$$
$$y \geq 0.$$

Note that the vector $|w|_* \in R^n$ has components which are equal to 1 if the corresponding components of w are nonzero and components equal to zero if the corresponding components of w are zero. Recall that e is a vector of ones and $e'|w|_*$ is simply a count of the nonzero elements in the vector w. Problem (12.18) balances the error in separating the points from class 1 and -1 (the term $e'y$), and the number of nonzero elements of w (the term, $e'|w|_*$). Further, note that if an element of w is zero, the corresponding feature is removed from the problem.

By introducing the variable v we are able to eliminate the absolute value from problem (12.18) which leads to the following equivalent parametric program for $\nu > 0$:

$$\begin{aligned}
\min_{w,\gamma,y,v} \quad & \nu e'y + e'v_* \\
\text{s.t.} \quad & D(Aw - e\gamma) + y \geq e \\
& -v \leq w \leq v \\
& y \geq 0.
\end{aligned} \quad (12.19)$$

Since v appears positively weighted in the objective and is constrained by $-v \leq w \leq v$, it effectively models the vector $|w|$. This feature selection problem will be solved for a value of $\nu > 0$ for which the resulting classification obtained by the midway separating plane (12.13) generalizes best, estimated by a cross-validation tuning procedure. Typically this will be achieved in a feature space of reduced dimensionality, that is $e'v_* < n$ (i.e. the number of features used is less than n).

Because of the discontinuity of the step function term $e'v_*$, we approximate it by a concave exponential on the nonnegative real line (Mangasarian 1996). The approximation of the step vector v_* of (12.19) by the concave exponential:

$$v_* \approx t(v, \alpha) = e - \varepsilon^{-\alpha v}, \alpha > 0, \quad (12.20)$$

leads to the smooth problem (**FSV**:Feature Selection Concave), for $\nu > 0$:

$$\begin{aligned}
\min_{w,\gamma,y,v} \quad & \nu e'y + e'\left(e - \varepsilon^{\alpha v}\right) \\
\text{s.t.} \quad & D(Aw - e\gamma) + y \geq e \\
& -v \leq w \leq v \\
& y \geq 0.
\end{aligned} \quad (12.21)$$

It can be shown (Bradley et al. 1998a,Theorem 2.1) that for a finite value of α (appearing in the concave exponential) the smooth problem (12.21) generates an exact solution of the nonsmooth problem (12.19). We note that this problem is the minimization of a concave objective function over a polyhedral set. Even though it is difficult to find a global solution to this problem, a fast successive linear approximation (SLA) algorithm (Bradley et al. 1998b,Algorithm 2.1) terminates finitely (usually in 5 to 7 steps) at a stationary point which satisfies the minimum principle necessary optimality condition for problem (12.21) (Bradley et al. 1998b,Theorem 2.2). This solution also leads to a sparse w with good generalization properties. We state the SLA algorithm below.

Algorithm 2 Successive Linearization Algorithm (SLA) for FSV (12.21). *Choose $\nu > 0$. Start with a random $(w^0, \gamma^0, y^0, v^0)$. Having*

$(w^i, \gamma^i, y^i, v^i)$ determine the next iterate by solving the linear program:
$(w^{i+1}, \gamma^{i+1}, y^{i+1}, v^{i+1}) \in$

$$\arg \text{vertex} \min_{w,\gamma,y,v} \nu e'y + \alpha(\varepsilon^{-\alpha v^i})'(v - v^i)$$
$$\text{s.t.} \quad D(Aw - e\gamma) + y \geq e \qquad (12.22)$$
$$-v \leq w \leq v$$
$$y \geq 0.$$

Stop when

$$\nu\left(e'(y^{i+1} - y^i)\right) + \alpha(\varepsilon^{-\alpha v^i})'(v^{i+1} - v^i) = 0. \qquad (12.23)$$

Comment: It was empirically determined that a value of $\alpha = 5$ produced satisfactory solutions. The parameter ν is usually chosen so as to maximize the generalization ability of the resulting classifier (as measured by cross-validation (Stone 1974), for instance).

3.2. Feature selection via support vector machines

The SVM idea (Vapnik 1995, Bennett and Blue 1997, Girosi 1998, Wahba 1999), although not originally intended as a feature selection tool, does in fact indirectly suppress components of the normal vector w to the separating plane P (12.13) when an appropriate norm is used for measuring the distance between the two parallel bounding planes for the sets being separated. The SVM approach consists of adding another term, $\|w\|'$, to the objective function of the RLP (12.17) in a similar manner to the appended term $e'|w|_*$ of problem (12.18). Here, $\|\cdot\|'$ is the dual of some norm on R^n used to measure the distance between the two bounding planes. The justification for this term is as follows. The separating plane P (12.13) generated by the RLP linear program (12.17) lies midway between the two parallel planes $w'x = \gamma + 1$ and $w'x = \gamma - 1$. The distance, measured by some norm $\|\cdot\|$ on R^n, between these planes is precisely $\frac{2}{\|w\|'}$ (Mangasarian 1999a, Theorem 2.1). The appended term to the objective function of the RLP (12.17), $\|w\|'$, is twice the reciprocal of this distance, which has the effect of driving the distance between these two planes up to obtain better separation. This results then in the following mathematical programming formulation for the SVM problem:

$$\min_{w,\gamma,y} \nu e'y + \|w\|'$$
$$\text{s.t.} \quad D(Aw - e\gamma) + y \geq e \qquad (12.24)$$
$$y \geq 0.$$

Points A_i appearing in active constraints of the linear program (12.24) with positive dual variables constitute the *support vectors* of the problem. These points are the *only* data points that are relevant for determining the optimal separating plane. Their number is usually small and it is proportional to the generalization error of the classifier (Osuna et al. 1997b).

If we use the 1-norm to measure the distance between the planes, then the dual to this norm is the ∞-norm and accordingly $\|w\|' = \|w\|_\infty$ in (12.24) which leads to the following linear programming formulation:

$$\begin{aligned} \min_{w,\gamma,y,\sigma} \quad & \nu e'y + \sigma \\ \text{s.t.} \quad & D(Aw - e\gamma) + y \geq e \\ & -e\sigma \leq w \leq e\sigma \\ & y \geq 0. \end{aligned} \quad (12.25)$$

Similarly if we use the ∞-norm to measure the distance between the planes, then the dual to this norm is the 1-norm and accordingly $\|w\|' = \|w\|_1$ in (12.24) which leads to the following linear programming formulation:

$$\begin{aligned} \min_{w,\gamma,y,s} \quad & \nu e'y + e's \\ \text{s.t.} \quad & D(Aw - e\gamma) + y \geq e \\ & -s \leq w \leq s \\ & y \geq 0. \end{aligned} \quad (12.26)$$

We note that the first paper on the multisurface method on pattern separation (Mangasarian 1968) also proposed and implemented the idea of separating the bounding planes, just as the SVM approach does.

Usually the SVM problem is formulated using the 2-norm in the objective (Vapnik 1995, Bennett and Blue 1997). Since the 2-norm is dual to itself, it follows that the distance between the parallel planes defining the separating surface is also measured in the 2-norm when this formulation is used.

An experimental comparison between the linear classifiers obtained by solving the FSV problem (12.21), the SVM formulation with a 1-norm minimization of w (12.26), the SVM formulation with an ∞-norm minimization of w (12.25) and the RLP (12.17) is given in (Bradley and Mangasarian 1998a). The FSV (12.21) and the SVM 1-norm (12.26) were the only methods tested that computed classifiers utilizing fewer than the original number of problem features. There was no significant difference in the generalization performance of these classifiers on 6 publicly available data sets (Murphy and Aha 1992).

Classifiers obtained by solving the SVM ∞-norm (12.25) suppressed none of the original problem features for all but the smallest values of ν

(near 0), which in general is of little use because it is often accompanied by poor set separation.

4. Massive Data Discrimination via Linear Support Vector Machines

We consider in this section the problem of discriminating between two massive finite point sets in R^n by a linear programming approach to SVMs. In the standard SVM formulation (Osuna et al. 1997b;a, Platt 1999) large quadratic programs are solved. In contrast, the formulation here consists of solving a linear program which is considerably less difficult. The algorithm creates a succession of sufficiently small linear programs that separate chunks of the data at a time. The key idea is that a small number of support vectors, corresponding to linear programming constraints with positive dual variables, are carried over between the successive small linear programs, each of which contains a chunk of the data. We prove that this procedure is monotonic and terminates in a finite number of steps at an exact solution that leads to an optimal separating plane for the entire data set.

4.1. LSVM: The linear support vector machine

Given m points in R^n represented by the matrix $A \in R^{m \times n}$ with the class label for data point A_i given by the corresponding diagonal element D_{ii} of the diagonal matrix $D \in R^{m \times m}$, we wish to discriminate between points A_i with $D_{ii} = 1$ from those A_i with $D_{ii} = -1$. A separating plane may be efficiently computed by solving the robust linear program (RLP) (12.17) (Bennett and Mangasarian 1992):

$$\min_{w,\gamma,y} \quad e'y$$
$$\text{s.t.} \quad D(Aw - e\gamma) + y \geq e \quad (12.27)$$
$$y \geq 0.$$

The objective function of (12.27) minimizes the sum of distances, weighted by $\|w\|'$, of misclassified points to the bounding planes: $x'w = \gamma + 1$ for points A_i with $D_{ii} = 1$ and $x'w = \gamma - 1$ for points A_i with $D_{ii} = -1$. The linear support vector machine (LSVM) consists of parametrically adding another term, $\|w\|'$, to the objective function of (12.27). Sections 1 and 2 provide detailed justification for this additional term.

Thus, appending $\|w\|_1$ to ν times the objective function of (12.27) with $\nu > 0$ leads to the following LSVM where $e's = \|w\|_1$ at optimality:

$$\begin{aligned} \min_{w,\gamma,y,s} \quad & \nu e'y + e's \\ \text{s.t.} \quad & D(Aw - e\gamma) + y \geq e \\ & -s \leq w \leq s \\ & y \geq 0. \end{aligned} \qquad (12.28)$$

4.2. LPC: Linear Program Chunking

We consider a general linear program

$$\min_x \{c'x \,|\, Hx \geq b\}, \qquad (12.29)$$

where $c \in R^n$, $H \in R^{m \times n}$ and $b \in R^m$. We state now our chunking algorithm and establish its finite termination for the linear program (12.29) where m may be orders of magnitude larger than n. In its dual form our algorithm can be interpreted as a block-column generation method related to column generation methods of Gilmore-Gomory (Gilmore and Gomory 1961), Dantzig-Wolfe (Dantzig and Wolfe 1960), (Chvátal 1983,pp 198-200,428-429) and others (Murty 1983,pp 243-248), but it differs from active set methods (Luenberger 1984,pp 326-330) in that it does not require the satisfaction of a working set of constraints as equalities.

Algorithm 3 LPC: Linear Programming Chunking Algorithm for (12.29) *Let $[H \ b]$ be partitioned into ℓ blocks, possibly of different sizes, as follows:*

$$[H \ b] = \begin{bmatrix} H^1 & b^1 \\ \vdots & \vdots \\ H^\ell & b^\ell \end{bmatrix}.$$

Assume that (12.29) and all subproblems (12.30) below, have vertex solutions. At iteration $j = 1, \ldots$ compute x^j by solving the following linear program:

$$x^j \in \arg \text{vertex} \min \left\{ c'x \;\middle|\; \begin{aligned} H^{(j \bmod \ell)} x &\geq b^{(j \bmod \ell)} \\ \bar{H}^{(j \bmod \ell)-1} x &\geq \bar{b}^{(j \bmod \ell)-1} \end{aligned} \right\}, \qquad (12.30)$$

where $[\bar{H}^0 \ \bar{b}^0]$ is empty and $[\bar{H}^j \ \bar{b}^j]$ is the set of active constraints (that is all inequalities of (12.29) satisfied as equalities by x^j) with positive optimal Lagrange multipliers at iteration j. Stop when $c'x^j = c'x^{j+\tau}$ for some input integer τ. Typically $\tau = 4$.

Theorem 4 Finite Termination of LPC Algorithm *The sequence $\{x^j\}$ generated by the LPC Algorithm 3 has the following properties:*

(i) *The sequence $\{c'x^j\}$ of objective function values is nondecreasing and is bounded above by the minimum of $\min_{x} \{c'x \mid Hx \geq b\}$.*

(ii) *The sequence of objective function values $\{c'x^j\}$ becomes constant, that is: $c'x^{j+1} = c'x^j$ for all $j \geq \bar{j}$ for some $\bar{j} \geq 1$.*

(iii) *For $j \geq \bar{j}$, active constraints of (12.30) at x^j with positive multipliers remain active for iteration $j+1$.*

(iv) *For all $j \geq \tilde{j}$, for some $\tilde{j} \geq \bar{j}$, x^j is a solution of the linear program (12.29) provided all active constraints at x^j have positive multipliers for $j \geq \bar{j}$.*

This theorem is significant not only for the support vector application presented here, but also as a fundamental computational approach for handling linear programs with massive constraints for which the subproblems (12.30) of the LPC Algorithm 3 have vertex solutions. To establish its validity we first prove a lemma.

Lemma 5 *If $\bar{x}$ solves the linear program (12.29) and $(\bar{x}, \bar{u}) \in R^{n+m}$ is a Karush-Kuhn-Tucker (KKT) (Mangasarian 1969) point (i.e. a primal-dual optimal pair) such that $\bar{u}_I > 0$ where $I \subset \{1, \ldots, m\}$ and $\bar{u}_J = 0$, $J \subset \{1, \ldots, m\}$, $I \cup J = \{1, \ldots, m\}$, then*

$$\bar{x} \in \arg\min\{c'x \mid H_I x \geq b_I\} \tag{12.31}$$

where H_I consists of rows H_i, $i \in I$ of H, and b_I consists of elements b_i, $i \in I$.

Proof The KKT conditions (Mangasarian 1969) for (12.29) satisfied by $(\bar{x}, \bar{u})$ are:

$$c = H'\bar{u}, \ \bar{u} \geq 0, \ \bar{u}'(H\bar{x} - b) = 0, \ H\bar{x} - b \geq 0,$$

which under the condition $\bar{u}_I > 0$ imply that

$$H_I \bar{x} = b_I, \ \bar{u}_J = 0, \ H_J \bar{x} \geq b_J.$$

We claim now that $\bar{x}$ is also a solution of (12.31) because $(\bar{x}, \bar{u}_I)$ satisfy the KKT sufficient optimality conditions for (12.31):

$$c = H_I'\bar{u}_I, \ \bar{u}_I > 0, \ H_I \bar{x} = b_I. \diamond$$

Proof of Theorem 4

(i) By Lemma 5, $c'x^j$ is a lower bound for $c'x^{j+1}$. Hence the sequence $\{c'x^j\}$ is nondecreasing. Since the the constraints of (12.30) form a subset of the constraints of (12.29), it follows that $c'x^j \leq \min_x \{c'x \,|\, Hx \geq b\}$.

(ii) Since there are a finite number of (feasible and infeasible) vertices to the original linear program (12.29) as well as of the subproblems (12.30), it follows that from a certain $\bar{j}$ onward, a finite subset of these vertices will repeat infinitely often. Since a repeated vertex gives the same value for $c'x$, it follows, by the nondecreasing property of $\{c'x^j\}$ just established, that all vertices between repeated vertices also have the same objective value $c'x$ and hence: $c'x^j = c'x^{j+1} \leq \min_x \{c'x \,|\, Hx \geq b\}, \ \forall j \geq \bar{j}$.

(iii) Let $\bar{j}$ be as defined in the theorem. Let the index $t \in \{1, \ldots, m\}$ be that of some active constraint at iteration $\bar{j}$ with positive multiplier ($H_t x^{\bar{j}} = b_t$, $u_t^{\bar{j}} > 0$) which has become inactive at the next step, that is: $H_t x^{\bar{j}+1} > b_t$. We then obtain the following contradiction by part (ii) above and the KKT saddlepoint condition:

$$0 = c'x^{\bar{j}+1} - c'x^{\bar{j}} \geq u^{\bar{j}'}(\bar{H}^{\bar{j}} x^{\bar{j}+1} - \bar{b}^{\bar{j}}) \geq u_t^{\bar{j}}(H_t x^{\bar{j}+1} - b_t) > 0.$$

(iv) By (ii) a finite number of vertices repeat infinitely for $j \geq \bar{j}$ all with constant $c'x^j$. Since active constraints with positive multipliers at iteration j remain active at iteration $j+1$ by (iii) and hence have a positive multiplier by assumption of part (iv), the set of active constraints with positive multipliers will remain constant for $j \geq \hat{j}$, for some $\hat{j} \geq \bar{j}$ (because there are a finite number of constraints) and hence x^j will remain a fixed vertex $\bar{x}$ for $j \geq \hat{j}$. The point $\bar{x}$ will satisfy all the constraints of problem (12.29) because all constraints are eventually imposed on the infinitely repeated vertex $\bar{x}$. Hence $c'\bar{x}$ which lower-bounds the minimum of (12.29) is also a minimum of (12.29) because $\bar{x}$ is feasible. Hence the algorithm can be terminated at $j = \bar{j}$. ◇

Remark 6 *We have not excluded the possibility (never observed computationally) that the objective function remains constant over a finite number of iterations then increases. Theorem 4 asserts that eventually the objective function will be constant and equal to the minimum for all iterates $j \geq \bar{j}$. Of course, one can check the satisfaction of the KKT conditions for optimality, although this does not seem to be needed in practice.*

Remark 7 *In order to handle degenerate linear programs, i.e. those with active constraints with zero multipliers at a solution, we have modified the computational implementation of the LPC Algorithm 3 slightly by admitting into $[\bar{H}^j \ \ \bar{b}^j]$ all active constraints at iteration j even if they have zero multipliers. We note that the assumption of part (iv) of Theorem 4 is required to prove that x^j is a solution of (12.29) for all $j > \bar{j}$.*

The LPC Algorithm 3 was experimentally evaluated in (Bradley and Mangasarian 1998b) and some results are summarized here. Applied to a separation task of 200,000 points in R^{32}, the LPC algorithm computed a solution in 1.75 hours while 6.94 hours were required to solve the full data set linear program. The chunk size corresponding to this running time was 25,000 rows (1/8 total data set size). Separation of 500,000 data points via LPC required 25.91 hours and separation of 1 million data points in R^{32} required 231.32 hours.

We note that there exists a chunk size which is empirically best in the sense that it balances the computational overhead of constructing the LPC subproblems (12.30) with the computational overhead of solving the LPC subproblems. For small chunk sizes, the computational burden is dominated by subproblem construction since there are many and solving them is not expensive. For large chunk sizes, the computational burden is dominated by solving large subproblem LPs.

5. Successive Overrelaxation for Support Vector Machines

Successive overrelaxation, originally developed for the solution of large systems of linear equations (Ortega and Rheinboldt 1970, Ortega 1972) has been successfully applied to mathematical programming problems (Cryer 1971, Mangasarian 1977; 1991, Pang 1986, Mangasarian and Leone 1988, Li 1993), some with as many 9.4 million variables (Leone and Tork Roth 1993). By taking the dual of the quadratic program associated with an SVM (Vapnik 1995, Cherkassky and Mulier 1998) for which the margin (distance between bounding separating planes) has been maximized with respect to *both* the normal to the planes as well as their location, we obtain a very simple convex quadratic program with bound constraints *only*. This problem is equivalent to a symmetric mixed linear complementarity problem (i.e. with upper and lower bounds on its variables (Dirkse and Ferris 1995)) to which SOR can be directly applied. This corresponds to solving the SVM dual convex quadratic program for one variable at a time, that is computing one multiplier of a potential support vector at a time.

We again consider the problem of discriminating between m points in the n dimensional real space R^n, represented by the $m \times n$ matrix A, according to membership of each point A_i in the classes 1 or -1 as specified by a given $m \times m$ diagonal matrix D with ones or minus ones along its diagonal. For this problem the standard linear SVM with a linear kernel AA' (Vapnik 1995, Cherkassky and Mulier 1998) is given by the following for some $\nu > 0$:

$$\begin{aligned}\min_{w,\gamma,y} \quad & \nu e'y + \tfrac{1}{2} w'w \\ \text{s.t.} \quad & D(Aw - e\gamma) + y \geq e \\ & y \geq 0.\end{aligned} \qquad (12.32)$$

Note that this is a special case of problem (12.6), with $w = A'Du$, $H = DK(A, A')D$, and $K(A, A') = AA'$. Here w is the normal to the bounding planes:

$$\begin{aligned} x'w - \gamma &= +1 \\ x'w - \gamma &= -1. \end{aligned} \qquad (12.33)$$

The first plane above bounds the class 1 points and the second plane bounds the class -1 points, if the two classes are linearly separable and $y = 0$. If the classes are linearly inseparable then the two planes bound the two classes with a "soft margin" determined by a slack variable $y \geq 0$, that is:

$$\begin{aligned} x'w - \gamma + y_i &\geq +1, \text{ for } x = A_i \text{ and } D_{ii} = +1, \\ x'w - \gamma - y_i &\leq -1, \text{ for } x = A_i \text{ and } D_{ii} = -1. \end{aligned} \qquad (12.34)$$

The one-norm of the slack variable y is minimized with weight ν in (12.32). The quadratic term in (12.32), which is twice the reciprocal of the square of the 2-norm distance $\frac{2}{\|w\|_2}$ between the two planes of (12.33) in the n-dimensional space of $w \in R^n$ for a *fixed* γ, maximizes that distance. In our approach here, which is similar to that of (Boser et al. 1992, Friess et al. 1998, Friess 1998), we measure the distance between the planes in the $(n+1)$-dimensional space of $[w;\gamma] \in R^{n+1}$ which is $\frac{2}{\|[w;\gamma]\|_2}$, instead of $\frac{2}{\|w\|_2}$. Using this measure of distance instead results in our modification of the SVM problem:

$$\begin{aligned} \min_{w,\gamma,y} \quad & \nu e'y + \tfrac{1}{2}(w'w + \gamma^2) \\ \text{s.t.} \quad & D(Aw - e\gamma) + y \geq e \\ & y \geq 0. \end{aligned} \qquad (12.35)$$

The Wolfe duals (Mangasarian 1969,Section 8.2) to the quadratic programs (12.32) and (12.35) are as follows.

$$\max_{u} \; -\frac{1}{2}u'DAA'Du + e'u, \text{ s.t. } e'Du = 0, \; 0 \le u \le \nu e$$
$$(w = A'Du). \tag{12.36}$$

$$\max_{u} \; -\frac{1}{2}u'DAA'Du - \frac{1}{2}u'Dee'Du + e'u, \text{ s.t. } 0 \le u \le \nu e$$
$$(w = A'Du, \; \gamma = -e'Du, \; y = (e - D(Aw - e\gamma))_+). \tag{12.37}$$

The principal difference between these duals is that (12.37) has simple bound constraints only, whereas (12.36) has in addition a computationally complicating equality constraint. We also note that the variables (w, γ, y) of the primal problem (12.35) can be directly computed from the solution u of its dual (12.37) as indicated. However, only the variable w of the primal problem (12.32) can be directly computed from the solution u of its dual (12.36) as shown. The remaining variables (γ, y) of (12.32) can be computed by setting $w = A'Du$ in (12.32), where u is a solution of its dual (12.36), and solving the resulting linear program for (γ, y). Alternatively, γ can be determined by minimizing the expression for $e'y = e'(e - D(Aw - e\gamma))_+$ as a function of the single variable γ after w has been expressed as a function of the dual solution u as indicated in (12.36), that is:

$$\min_{\gamma \in R} e'(e - D(AA'Du - e\gamma))_+. \tag{12.38}$$

We note that formulations (12.36) and (12.37) can be extended to a general nonlinear kernel $K(A, A') : R^{m \times n} \times R^{n \times m} \to R^{m \times m}$ by replacing AA' by the kernel $K(A, A')$. (For more details about kernel function choices, see Section 2) Thus the conventional SVM with a positive semidefinite kernel $K(A, A')$ (Vapnik 1995, Cherkassky and Mulier 1998) is given by the following dual convex quadratic program for some $\nu > 0$ which is obtained from (12.36) by replacing AA' by $K(A, A')$:

$$\min_{u \in R^m} \; \tfrac{1}{2}u'DK(A, A')Du - e'u$$
$$\text{s.t. } \; e'Du = 0 \tag{12.39}$$
$$0 \le u \le \nu e.$$

A solution u of this quadratic program leads to the nonlinear separating surface

$$K(x', A')Du = \gamma, \tag{12.40}$$

where γ is defined in a similar manner to (12.38) above by a solution of:
$$\min_{\gamma \in R} e'(e - D(K(A, A')Du - e\gamma))_+. \tag{12.41}$$

Note that the explicit definition of the nonlinear surface (12.40) in R^n in terms of u and γ obviates the need for computing w, since w is defined only in the higher dimensional space in which the nonlinear surface (12.40) is mapped into a plane.

By a similar approach we obtain the following quadratic dual with bound constraints only and a nonlinear kernel $K(A, A')$ by replacing AA' in (12.37) by $K(A, A')$:

$$\min_u \frac{1}{2} u'D[K(A, A') + ee']Du - e'u, \text{ s.t. } 0 \leq u \leq \nu e$$
$$(\gamma = -e'Du), \tag{12.42}$$

with an explicit expression for γ and the same separating surface (12.40). The formulation (12.42) allows a direct use of the SOR algorithm to solve very large problems.

We mention in passing that another possible change in (12.39) allows the use of possibly indefinite kernels. One particular formulation motivated by (12.10) is the following one which, as in (12.42), incorporates the equality $e'Du = 0$ into the objective function:

$$\min_u \frac{1}{2} u'D[K(A, A')K(A, A')' + ee']Du - e'u, \text{ s.t. } 0 \leq u \leq \nu e$$

$$(\gamma = -e'Du),$$
$$\tag{12.43}$$

with a separating surface different from (12.40):
$$K(x', A')K(A, A')'Du = \gamma. \tag{12.44}$$

Note that the kernel $K(A, A')$ in the formulation (12.43) is completely arbitrary and need not satisfy any positive semidefiniteness condition in order for the objective function of (12.43) to be convex. This makes the separating surface (12.44) quite general.

It is interesting to note that for a linear kernel, the standard SVM problem (12.36) and our modification (12.37) frequently give the same w. For 1,000 randomly generated problems with $A \in R^{40 \times 5}$ and the same ν, only 34 cases had solution vectors w that differed by more than 0.001 in their 2-norm.

The main reason for introducing our modification (12.35) of the SVM is that its dual (12.37) does not contain an equality constraint as does the dual (12.36) of (12.32). This enables us to apply in a straightforward manner the effective matrix splitting methods such as those of (Mangasarian 1977; 1991, Luo and Tseng 1993) that process one constraint of (12.35) at a time through its dual variable, without the complication of having to enforce an equality constraint at each step on the dual variable u. This permits us to process massive data without bringing it all into fast memory. We thus apply the SOR algorithm to the nonlinear SVMs (12.42) and (12.43). Note that the linear SVM is simply the special case of (12.42) where $K(A, A') = AA'$. These problems can be stated as:

$$\min_{u} \frac{1}{2} u'Mu - e'u, \text{ s.t. } u \in S = \{u \mid 0 \leq u \leq \nu e\}, \qquad (12.45)$$

with the symmetric matrix M defined as $D(K(A, A') + ee')D$ and $D(K(A, A')K(A, A')' + ee')D$ respectively for these two problems. Thus M will be positive semidefinite if we assume that $K(A, A')$ is positive semidefinite in the former case and under no assumptions in the latter case. If we decompose M as follows:

$$M = L + E + L', \qquad (12.46)$$

where $L \in R^{m \times m}$ is the strictly lower triangular part of the symmetric matrix M, and $E \in R^{m \times m}$ is the positive diagonal of M, then a necessary and sufficient optimality condition for (12.45) for positive semidefinite M is the following gradient projection optimality condition (Polyak 1987, Luo and Tseng 1993):

$$u = (u - \omega E^{-1}(Mu - e))_\#, \ \omega > 0, \qquad (12.47)$$

where $(\cdot)_\#$ denotes the 2-norm projection on the feasible region S of (12.45), that is:

$$((u)_\#)_i = \begin{cases} 0 & \text{if } u_i \leq 0 \\ u_i & \text{if } 0 < u_i < \nu \\ \nu & \text{if } u_i \geq \nu \end{cases}, \ i = 1, \ldots, m. \qquad (12.48)$$

Our SOR method, which is a matrix splitting method that converges linearly to a point $\bar{u}$ satisfying (12.47), consists of splitting the matrix M into the sum of two matrices as follows:

$$M = \omega^{-1} E(B + C), \text{ s.t. } B - C \text{ is positive definite.} \qquad (12.49)$$

For our specific problem we take:

$$B = (I + \omega E^{-1} L), \ C = ((\omega - 1)I + \omega E^{-1} L'), \ 0 < \omega < 2. \qquad (12.50)$$

This leads to the following linearly convergent (Luo and Tseng 1993,Equation (3.14)) matrix splitting algorithm:

$$u^{i+1} = (u^{i+1} - Bu^{i+1} - Cu^i + \omega E^{-1}e)_{\#}, \qquad (12.51)$$

for which

$$B + C = \omega E^{-1}M, \; B - C = (2-\omega)I + \omega E^{-1}(L - L'). \qquad (12.52)$$

Note that for $0 < \omega < 2$, the matrix $B + C$ is positive semidefinite and matrix $B - C$ is positive definite. The matrix splitting algorithm (12.51) results in the following easily implementable SOR algorithm once the values of B and C given in (12.50) are substituted in (12.51).

Algorithm 8 SOR Algorithm *Choose* $\omega \in (0, 2)$. *Start with any* $u^0 \in R^m$. *Having* u^i *compute* u^{i+1} *as follows:*

$$u^{i+1} = (u^i - \omega E^{-1}(Mu^i - e + L(u^{i+1} - u^i)))_{\#}, \qquad (12.53)$$

until $\|u^{i+1} - u^i\|$ *is less than some prescribed tolerance.*

Remark 9 *The components of* u^{i+1} *are computed in order of increasing component index. Thus the SOR iteration (12.53) consists of computing* u_j^{i+1} *using* $(u_1^{i+1}, \ldots, u_{j-1}^{i+1}, u_j^i, \ldots, u_m^i)$. *That is, the latest computed components of u are used in the computation of* u_j^{i+1}. *The strictly lower triangular matrix L in (12.53) can be thought of as a substitution operator, substituting* $(u_1^{i+1}, \ldots, u_{j-1}^{i+1})$ *for* $(u_1^i, \ldots, u_{j-1}^i)$.

We have immediately from (Luo and Tseng 1993,Proposition 4.21) the following linear convergence result.

Theorem 10 SOR Linear Convergence *The iterates* $\{u^i\}$ *of the SOR Algorithm 8 converge R-linearly to a solution of* $\bar{u}$ *of the dual problem (12.42), and the objective function values* $\{f(u^i)\}$ *of (12.42) converge Q-linearly to* $f(\bar{u})$. *That is for* $i \geq \bar{i}$ *for some* $\bar{i}$:

$$\begin{aligned} \|u^i - \bar{u}\| &\leq \mu \delta^i, \text{ for some } \mu > 0, \; \delta \in (0,1), \\ f(u^{i+1}) - f(\bar{u}) &\leq \tau(f(u^i) - f(\bar{u})), \text{ for some } \tau \in (0,1). \end{aligned} \qquad (12.54)$$

Remark 11 *A significant simplification can be made when a linear kernel is used in (12.45) with* $M = D(AA' + ee')D$. *The matrix M can be rewritten as* HH', *where* $H = D[A \; -e]$. *Even though our SOR iteration (12.53) is written in terms of the full $m \times m$ matrix HH', it can easily be implemented one row at a time without bringing all of the data into memory as follows for* $j = 1, \ldots, m$:

$$u_j^{i+1} = (u_j^i - \omega E_{jj}^{-1}(H_j(\sum_{\ell=1, j>1}^{j-1} H_\ell u_\ell^{i+1} + \sum_{\ell=j}^{m} H_l u_\ell^i - 1)))_{\#}. \qquad (12.55)$$

A simple interpretation of this step is that one component of the multiplier u_j is updated at a time by bringing one constraint of (12.35) at a time.

We benchmarked the SOR algorithm with a linear kernel against the SMO (Platt 1999) and SVM^{light} (Joachims 1999) algorithms using subsets of the UCI Adult data set (Murphy and Aha 1992). For larger versions of the data set, SOR ran almost twice as fast as SMO, and more than an order of magnitude faster than SVM^{light}. SOR produced similar test set accuracies to the other two algorithms. See (Mangasarian and Musicant 1998) for more details.

SOR was also used with a linear kernel on synthetic highly separable massive data sets (Mangasarian and Musicant 1998). The algorithm terminated in 9.7 hours on a one million point data set, and reached 95% of true separability on a 10 million point data set in only 14.3 hours. Finally, SOR with a *nonlinear* kernel was used on a synthetic Gaussian data set (Mangasarian and Musicant 1999). Both a linear and quadratic kernel were used, with the quadratic kernel showing improved test set accuracy over the linear kernel. The quadratic kernel yielded a test set accuracy of 93.4%, whereas the linear kernel produced a test set accuracy of only 81.7% under tenfold cross-validation.

6. Conclusion

We have proposed a direct mathematical programming framework for general SVMs that makes essentially no or few assumptions on the kernel employed. We have derived new kernel-based linear programming formulations (12.11) and (12.12), and a new quadratic programming formulation (12.10) that require no assumptions on the kernel K. These formulations can lead to different but equally satisfactory decision functions as that obtained by the quadratic programming formulation (12.9) for a conventional SVM that requires symmetry and positive definiteness of the kernel. Even for negative definite kernels these new formulations can generate decision functions that separate the given points whereas the conventional SVM does not.

Feature selection that improves generalization was achieved by parametrically minimizing a polyhedral norm or a concave function that suppresses as many components of the features as possible, while generating a separating surface that discriminates effectively between the points of a given set containing two categories of points.

We have described a linear programming chunking algorithm for discriminating between massive data sets. The algorithm, significant in its own right as a linear programming decomposition algorithm, is very effective for discrimination problems with large data sets that may not fit in machine memory and for problems taking excessive time to process. The algorithm uses support vector ideas by keeping only essential data points needed for determining a separating plane. The algorithm can handle extremely large data sets because it deals with small chunks of the data at a time and is guaranteed to terminate in a finite number of steps. Although we have not discussed parallelization here, this can be easily implemented by splitting the data among processors and sharing only support vectors among them. This would allow one to handle problems with extremely large data sets on a network of workstations or PCs.

We have described a powerful iterative method, successive overrelaxation, for the solution of extremely large discrimination problems using SVMs. The method converges considerably faster than other methods that require the presence of a substantial amount of the data in memory. We have solved problems that cannot be directly handled by conventional methods of mathematical programming. The proposed method scales up with no changes and can be parallelized by using techniques already implemented (Leone et al. 1990, Leone and Tork Roth 1993). We have also described how use successive overrelaxation in conjunction with nonlinear kernels to generate nonlinear separating surfaces.

Acknowledgements

This research is supported by National Science Foundation Grants CCR-9322479, CCR-9729842 and CDA-9623632, and by Air Force Office of Scientific Research Grant F49620-97-1-0326.

Bibliography

K. P. Bennett. Combining support vector and mathematical programming methods for induction. Technical report, Department of Mathematical Sciences, Rensselaer Polytechnic Institute, 1998.

K. P. Bennett and J. A. Blue. A support vector machine approach to decision trees. In *Proceedings of IJCNN'98*, pages 2396–2401, 1997.

K. P. Bennett, D. Hui, and L. Auslender. On support vector decision trees for database marketing. Technical Report 98-100, Department of Mathematical Sciences, Rensselaer Polytechnic Institute, 1998.

K. P. Bennett and O. L. Mangasarian. Robust linear programming discrimination of two linearly inseparable sets. *Optimization Methods and Software*, 1:23–34, 1992.

B. E. Boser, I. M. Guyon, and V. N. Vapnik. A training algorithm for optimal margin classifiers. In D. Haussler, editor, *Proceedings of the 5th Annual ACM Workshop on Computational Learning Theory*, pages 144–152. ACM Press, 1992.

P. S. Bradley and O. L. Mangasarian. Feature selection via concave minimization and support vector machines. In J. Shavlik, editor, *Machine Learning Proceedings of the Fifteenth International Conference (ICML '98)*, pages 82–90. Morgan Kaufmann, 1998a.

P. S. Bradley and O. L. Mangasarian. Massive data discrimination via linear support vector machines. Technical Report 98-05, Computer Sciences Department, University of Wisconsin, 1998b. To appear in Optimization Methods and Software.

P. S. Bradley, O. L. Mangasarian, and J. B. Rosen. Parsimonious least norm approximation. *Computational Optimization and Applications*, 11:5–21, 1998a.

P. S. Bradley, O. L. Mangasarian, and W. N. Street. Feature selection via mathematical programming. *INFORMS Journal on Computing*, 10:209–217, 1998b.

E. J. Bredensteiner. *Optimization Methods in Data Mining and Machine Learning*. PhD thesis, Department of Mathematical Sciences, Rensselaer Polytechnic Institute, 1997.

E. J. Bredensteiner and K. P. Bennett. Feature minimization within decision trees. *Computational Optimizations and Applications*, 10: 111–126, 1998.

C. J. C. Burges. A tutorial on support vector machines for pattern recognition. *Data Mining and Knowledge Discovery*, 2:121–167, 1998.

V. Cherkassky and F. Mulier. *Learning from Data - Concepts, Theory and Methods*. John Wiley & Sons, 1998.

V. Chvátal. *Linear Programming*. W. H. Freeman and Company, 1983.

C. Cortes and V. Vapnik. Support vector networks. *Machine Learning*, 20:273–279, 1995.

R. Courant and D. Hilbert. *Methods of Mathematical Physics*. Interscience Publishers, 1953.

C. W. Cryer. The solution of a quadratic programming problem using systematic overrelaxation. *SIAM Journal on Control and Optimization*, 9:385–392, 1971.

G. B. Dantzig and P. Wolfe. Decomposition principle for linear programs. *Operations Research*, 8:101–111, 1960.

S. P. Dirkse and M. C. Ferris. MCPLIB: A collection of nonlinear mixed complementarity problems. *Optimization Methods and Software*, 5:319–345, 1995.

T.-T. Friess. Support vector neural networks: The kernel adatron with bias and soft margin. Technical report, Department of Automatic Control and Systems Engineering, University of Sheffield, 1998.

T.-T. Friess, N. Cristianini, and C. Campbell. The kernel-adatron algorithm: A fast and simple learning procedure for support vector machines. In J. Shavlik, editor, *Machine Learning Proceedings of the Fifteenth International Conference (ICML'98)*, pages 188–196. Morgan Kaufmann, 1998.

P. C. Gilmore and R. E. Gomory. A linear programming approach to the cutting stock problem. *Operations Research*, 9:849–859, 1961.

F. Girosi. An equivalence between sparse approximation and support vector machines. *Neural Computation*, 10:1455–1480, 1998.

M. H. Hassoun. *Fundamentals of Artificial Neural Networks*. MIT Press, 1995.

P. J. Huber. *Robust Statistics*. John Wiley & Sons, 1981.

T. Joachims. Making large-scale support vector machine learning practical. In B. Schölkopf, C. J. C. Burges, and A. J. Smola, editors, *Advances in Kernel Methods - Support Vector Learning*, pages 169–184. MIT Press, 1999.

R. De Leone, O. L. Mangasarian, and T.-H. Shiau. Multi-sweep asynchronous parallel successive overrelaxation for the nonsymmetric linear complementarity problem. *Annals of Operations Research*, 22:43–54, 1990.

R. De Leone and M. A. Tork Roth. Massively parallel solution of quadratic programs via successive overrelaxation. *Concurrency: Practice and Experience*, 5:623–634, 1993.

W. Li. Remarks on convergence of matrix splitting algorithm for the symmetric linear complementarity problem. *SIAM Journal on Optimization*, 3:155–163, 1993.

D. G. Luenberger. *Linear and Nonlinear Programming*. Addison–Wesley, 1984. Second edition.

Z.-Q. Luo and P. Tseng. Error bounds and convergence analysis of feasible descent methods: A general approach. *Annals of Operations Research*, 46:157–178, 1993.

O. L. Mangasarian. Linear and nonlinear separation of patterns by linear programming. *Operations Research*, 13:444–452, 1965.

O. L. Mangasarian. Multi-surface method of pattern separation. *IEEE Transactions on Information Theory*, IT-14:801–807, 1968.

O. L. Mangasarian. *Nonlinear Programming*. McGraw–Hill, 1969. Reprint: SIAM Classic in Applied Mathematics 10, 1994, Philadelphia.

O. L. Mangasarian. Solution of symmetric linear complementarity problems by iterative methods. *Journal of Optimization Theory and Applications*, 22:465–485, 1977.

O. L. Mangasarian. On the convergence of iterates of an inexact matrix splitting algorithm for the symmetric monotone linear complementarity problem. *SIAM Journal on Optimization*, 1:114–122, 1991.

O. L. Mangasarian. Machine learning via polyhedral concave minimization. In H. Fischer, B. Riedmueller, and S. Schaeffler, editors, *Applied Mathematics and Parallel Computing – Festschrift for Klaus Ritter*, pages 175–188. Physica-Verlag, A Springer-Verlag Company, 1996.

O. L. Mangasarian. Arbitrary-norm separating plane. *Operations Research Letters*, 24, 1999a.

O. L. Mangasarian. Generalized support vector machines. In A. Smola, P. Bartlett, B. Schölkopf, and D. Schuurmans, editors, *Advances in Large Margin Classifiers*. MIT Press, 1999b. To appear.

O. L. Mangasarian and R. De Leone. Parallel gradient projection successive overrelaxation for symmetric linear complementarity problems. *Annals of Operations Research*, 14:41–59, 1988.

O. L. Mangasarian and David R. Musicant. Successive overrelaxation for support vector machines. Technical Report 98-18, Computer Sciences Department, University of Wisconsin, 1998. To appear in IEEE Transactions on Neural Networks.

O. L. Mangasarian and David R. Musicant. Data discrimination via nonlinear generalized support vector machines. Technical Report 99-03, Computer Sciences Department, University of Wisconsin, 1999.

O. L. Mangasarian, W. N. Street, and W. H. Wolberg. Breast cancer diagnosis and prognosis via linear programming. *Operations Research*, 43:570–577, 1995.

MATLAB. User's guide, 1992.

P. M. Murphy and D. W. Aha. UCI repository of machine learning databases. Technical report, Department of Information and Computer Science, University of California, 1992.

K. G. Murty. *Linear Programming*. John Wiley & Sons, 1983.

J. M. Ortega. *Numerical Analysis, A Second Course*. Academic Press, 1972.

J. M. Ortega and W. C. Rheinboldt. *Iterative Solution of Nonlinear Equations in Several Variables*. Academic Press, 1970.

E. Osuna, R. Freund, and F. Girosi. Improved training algorithm for support vector machines. In *Proceedings of IEEE NNSP'97*, pages 276–285. IEEE Press, 1997a.

E. Osuna, R. Freund, and F. Girosi. Training support vector machines: An application to face detection. In *IEEE Conference on Computer Vision and Pattern Recognition*, pages 130–136, 1997b.

J.-S. Pang. More results on the convergence of iterative methods for the symmetric linear complementarity problem. *Journal of Optimization Theory and Applications*, 49:107–134, 1986.

J. Platt. Sequential minimal optimization: A fast algorithm for training support vector machines. In B. Schölkopf, C. J. C. Burges, and A. J. Smola, editors, *Advances in Kernel Methods - Support Vector Learning*, pages 185–208. MIT Press, 1999.

B. T. Polyak. *Introduction to Optimization*. Optimization Software, Inc., Publications Division, 1987.

B. Schölkopf. *Support Vector Learning*. R. Oldenbourg Verlag, 1997.

A. J. Smola. *Learning with Kernels*. PhD thesis, Technische Universität Berlin, 1998.

M. Stone. Cross-validatory choice and assessment of statistical predictions. *Journal of the Royal Statistical Society*, 36:111–147, 1974.

V. N. Vapnik. *The Nature of Statistical Learning Theory*. Springer Verlag, 1995.

G. Wahba. Support vector machines, reproducing kernel Hilbert spaces and the randomized GACV. In B. Schölkopf, C. J. C. Burges, and A. J. Smola, editors, *Advances in Kernel Methods - Support Vector Learning*, pages 69–88. MIT Press, 1999.

P. Wolfe. A duality theorem for nonlinear programming. *Quarterly of Applied Mathematics*, 19:239–244, 1961.

Chapter 13

WAVELETS AND MULTISCALE TRANSFORM IN ASTRONOMICAL IMAGE PROCESSING

Fionn Murtagh
School of Computer Science, The Queen's University of Belfast, Belfast BT7 1NN, Northern Ireland
f.murtagh@qub.ac.uk

Jean-Luc Starck
DAPNIA/SEI-SAP, CEA/Saclay, 91191 Gif sur Yvette, France
jstarck@cea.fr

Abstract With the requirements of scientific and medical image database support in mind, we describe a range of useful technologies for storage, transmission and display. These new technologies are all based on discrete wavelet or related multiscale transforms. Other important issues include noise modeling, and the innovative use of entropy for information characterization.

Keywords: Wavelets, Multiscale transforms, Compression entropy, Astronomical data, Pyramidal transform, B-spline, Visualization.

1. Introduction

We take astronomical image processing and handling as the main application area addressed in this article. Generalization of much of what we describe to other scientific or medical imaging is immediate.

For compression and related technologies, quality measures defined globally (mean square error, peak signal-to-noise ratio) are often used, together with visual (subjective) quality. Astronomical imaging offers some interesting advantages, in that quite different, but easily defined, criteria are of great importance. These include positional (astrometric) accuracy, accumulated intensity (photometric) accuracy, and the presence or absence of faint but well-defined image features. An astronomical testbed, therefore, is one which facilitates testing and is generalizable to broad classes of scientific and medical imaging.

From year to year, the quantity of astronomical data increases at an ever growing rate. In part this is due to very large digitized sky surveys in the optical and near infrared, which in turn owes its origin to the development of digital imaging arrays such as CCDs (charge coupled devices). The size of digital arrays continually increases following the demands of astronomical research for obtaining larger quantities of data in shorter time periods.

Images of dimensions 8000 × 8000 are to be produced by the European Southern Observatory's (ESO's) Very Large Telescope (VLT), the construction of which is well underway at Paranal, Chile, with first light in mid-1998. Astronomical data are usually stored as 32 bits per pixel. The MegaCam camera being built for the Canada-France-Hawaii Telescope (CFHT), Hawaii, will use 16000 × 16000 images. In medical imaging, large images are also used. Digitized mammogram images of typical dimensions 4500 × 4500, with dynamic range of 16 bits, imply approximately 50 MB per image.

Astronomy projects such as the European DENIS (DEep Near Infrared Survey of the Southern Sky, 1995–1999, http://www-denis.iap.fr/denis.html) and American 2MASS (Two Micron All Sky Survey, http://pegasus.phast.umass.edu/2mass.html, from 1997) infrared sky surveys, or the Franco-Canadian MegaCam camera (http://www-terapix.iap.fr, from 2001) and the American Sloan Digital Sky Survey (http://www.sdss.org, from 1998), will soon be collectively producing many tens of terabytes of image data. These projects are based on CCD detectors. Independently, the routine and massive digitization of photographic plates has been made possible by the advent of automatic plate scanning machines (MAMA, APM, COSMOS, SuperCOSMOS, APS, PMM, PDSs). These machines allow for the quantification of the truly enormous amount of useful astronomical data represented in a photograph of the sky, and they have aided in realizing the potential of large area photographic sky surveys.

Clearly the storage and manipulation of astronomical data always requires the latest innovation in archiving techniques (12in or $5\frac{1}{4}$in WORMs in the past, CD WORMs or even magnetic disks with RAID technology now, hopefully DVD in the near future). In addition, the simple transfer of such amounts of data over computer networks becomes cumbersome and in some cases practically impossible. Consider, for example, a Schmidt plate image as used for Hubble Space Telescope pointing to and locking on targets: the plate when digitized at 15 micron sampling steps yields a 23,050 square raster image occupying approximately 1.1 Gbytes. On an early Monday morning US eastern time, which was lunchtime in western Europe, in mid-July 2000, we found an Internet data transmission rate of about 40 kbytes/second. One digitized image of the size referred to would take about 8 hours to be transmitted at such a rate.

Facing this extraordinary increase in pixel volume, and taking into account the fact that catalogs produced by extraction of information from pixels can always be locally wrong or incomplete, the needs of the scientist follow two very different directions:

- On the one hand, the development of web technology creates a need for fast access to informative pixel maps, which are more intuitively understandable than the catalogs of reduced data alone.

- On the other hand, quantitative work often requires accurate refinement of astrometry and photometry, or effective redetection of missed objects.

Thus the astronomical community, in common with other communities, is confronted with a critical need for improved data compression, transmission and display techniques. Several techniques have in fact been used or even developed in the field of astronomy, in particular for compression, and these will be especially dealt with in this article.

In Section 2, we first review wavelet and other multiscale transforms, motivating their important role for the handling and interpretation of large images and signals. Section 3 deals with compression, from the application perspective. We address the issue of generally-applicable methods for large, noise-ridden scientific and medical images. The inherent compressibility of an image is discussed. In Section 4, we turn attention to distributed display environments.

2. Wavelet and Related Multiscale Transforms

A wavelet transform means decomposing data – an image or a signal – into fundamental building blocks. In the Fourier transform case, the building block functions are sine and cosine functions. In the wavelet transform case, wavelet functions are used, which oscillate about zero, but also quickly damp down to zero. The wavelet function is therefore localized in time or space, unlike the building block functions, also called basis functions, used in the Fourier transform case.

Wavelets serve to decompose the signal into fine to coarse resolution components, which we hope will reflect well the fine to coarse resolution features in our data.

We find in practice that the decomposition into the new wavelet basis has a clear "energy packing" property. That is, a lot of wavelet coefficients (which, when used with the wavelet basis functions, or fundamental building blocks, allow the original data to be exactly reconstructed) are zero or near-zero in value, while others are correspondingly high valued. It proves effective, under such circumstances, to apply a noise filtering strategy by setting near-zero wavelet coefficients to zero.

The last aspect of the wavelet transform leads to its usefulness for compression.

2.1. Continuous wavelet transform

The discrete wavelet transforms which are, for the most part, used in practice, can all be linked to the continuous wavelet transform. The continuous wavelet transform is useful in its own right. Firstly, it constitutes a framework for defining the wavelet transform, and discrete implementations can subsequently be derived from it. Secondly, it provides a time-frequency analysis method in its own right. In this section we will briefly define the continuous transform and then proceed to discuss widely used discrete transforms.

The Morlet-Grossmann definition (Grossmann et al., 1989) of the continuous wavelet transform for a 1-dimensional signal $f(x) \in L^2(R)$, the space of all square integrable functions, is:

$$W(a,b) = \frac{1}{\sqrt{a}} \int_{-\infty}^{+\infty} f(x)\psi^* \left(\frac{x-b}{a}\right) dx \qquad (13.1)$$

where:

- $W(a,b)$ is the wavelet coefficient of the function $f(x)$
- $\psi(x)$ is the analyzing wavelet (and ψ^* is the conjugate)
- a (> 0) is the scale parameter
- b is the position parameter

For exact reproducibility of the input data from the wavelet coefficients, it can be shown that the wavelet function must be of zero mean. Therefore, a property of the wavelet transform is to furnish zero mean wavelet coefficients. Some practitioners have pursued work with the continuous wavelet transform, but most wavelet and other multiscale transforms have been discrete. We will look at discrete transforms next. In the definition above, and in most definitions in the following sections, we will assume a 1-dimensional signal for notational convenience. Generalization to 2-dimensional (2D) image data is reasonably straightforward.

2.2. Orthogonal discrete wavelet transform

We have already indicated how sinusoidal functions are at the heart of Fourier transforms, whereas many different functions are used in wavelet transforms. To begin with, one set of functions is used to represent smooth or low-frequency parts of a signal, and another set of functions is used to represent detail or high-frequency parts of the signal. Let us denote one of the former by ϕ, which is called a scaling function or a father wavelet. It integrates to 1: $\int \phi(t)dt = 1$. Representing the detail signal is the wavelet function, proper, or mother wavelet, ψ, which is of zero mean: $\int \psi(t)dt = 0$.

The orthogonal wavelet series approximation of a continuous signal $f(x)$ is given by:

$$f(x) = \sum_k s_k \phi_{p,k}(x) + \sum_k d_{p,k} \psi_{p,k}(x) + \sum_k d_{p-1,k} \psi_{p-1,k}(x)$$
$$+ \ldots + \sum_k d_{1,k} \psi_{1,k}(x) \qquad (13.2)$$

where p is the number of resolution scales, and all of our functions are parametrized by k. The coefficients d are the wavelet coefficients. The first term is a very smooth approximation to our original data. The sum of the first two terms is an approximation which is a little less smooth, and we can continue in this way to refine our approximation. We see here a family of multiresolution approximations to our original data. The decomposition is referred to as multiresolution analysis, a term coined by Mallat (1989).

Now, usually, the functions ϕ and ψ corresponding to different resolution levels are not defined analytically. Instead they are defined by a dilation equation, i.e. the function at a resolution level keeps the same overall shape as at the previous level but is expanded or dilated in scale. The following scaling and translation relations hold:

$$\phi_{q,k}(x) = 2^{-q/2}\phi(2^{-q}x - k) = 2^{-q/2}\phi\left(\frac{x - 2^q k}{2^q}\right)$$
$$\psi_{q,k}(x) = 2^{-q/2}\psi(2^{-q}x - k) = 2^{-q/2}\psi\left(\frac{x - 2^q k}{2^q}\right) \qquad (13.3)$$

Let us look at the terms in the right-hand expressions. The $2^{-q/2}$ term is a normalizing one. For both ϕ and ψ, we see that we have expressions of the form $\phi(\frac{x-b}{a})$ and $\psi(\frac{x-b}{a})$. The numerator in the fraction represents a dyadic scale. This means that resolution scales increase in smoothness, or decrease in detail, by a two-fold factor at each level. The b term is a translation term. We are sampling with an increasing sampling gap in the function at each level.

What we have used here is termed the discrete wavelet function. The continuous wavelet transform, CWT, is based on continuous values of the scaling or dilation parameter, a, and of the translation parameter, b.

The discrete wavelet transform is usually used on discrete data, so instead of $f(x)$ we analyse $f_1, f_2, \ldots, f_n$. In equation 13.2, let us put all of the coefficients into a vector, which we will call w. Hence we define this vector as a concatenation of the following terms: $s_k, d_{p,k}, d_{p-1,k}, \ldots, d_{1,k}$ for all k. But there is a mistake of notation here, in that the k terms can be quite different in number at each level. In equation 13.3, you will remember that we had a translation term at each level, which results in halving the number of terms which we are interested in at each level. The level closest to our original data has half the number of terms which we started out with ($q = 1$ in equation 13.3, and the last term in equation 13.2). The next level has half as many terms again, and so on. Therefore, working backwards, we find that the size of vector w above is $n/2 + n/4 + \ldots +$ last smoothed term $= n$. The orthonormal wavelet transform thus is most appropriate when we make use of all possible resolution levels, i.e. $\log_2 n$ resolution levels, with n taken as an integer power of 2. Furthermore we can define a matrix, W, such that $w = Wf$. For examples of W, see Press et al. (1992), Chapter 13.

For calculating the orthogonal wavelet transform, rather than using a matrix product, it is faster to use an embedded loop algorithm, which leads to $O(n)$ computational cost. This is better than the Fast Fourier Transform, which is of $O(n \log n)$ computational cost. Such an algorithm is called a pyramid algorithm, although "triangle" would be a more appropriate characterization for a unidimensional or 1D signal.

2.3. Pyramidal transforms

Mallat's presentation of the wavelet transform, following in the tradition of the pyramid algorithm of Burt and Adelson (1983), was to consider two analysis filters, a low-pass filter L and a high-pass filter H. An operator which decimates by 2 is also considered: delete every other value of the signal. The low-pass filter implements the scaling function. The high-pass filter is applied after each application of the low-pass filter.

A backward pass allows for synthesis of the original data. Conjugate filters H^* and L^* are used, together with an upsampling operator which inserts a zero between every other value of the signal. The relations between H and L, and H^* and L^*, are very particular in the case of the orthogonal (more correctly, orthonormal) wavelet transform. A quadrature mirror filter relationships connects the two filters. Corresponding filters can also be defined for biorthogonal wavelets (described below).

The smooth and detail coefficients, at each scale q, are given by the projection of our function onto the new basis functions.

$$s_{q,k} = \int \phi_{q,k}(x) f(x) dx$$

$$d_{q,k} = \int \psi_{q,k}(x) f(x) dx$$

In the DWT, the corresponding discrete versions of these are used. A list of coefficient values can be seen at the Bath Wavelet Warehouse, http://dmsun4.bath.ac.uk/wavelets/warehouse.html.

The term subband is also used to refer to a given scale. Keeping certain scales gives rise to subband filtering.

The basis functions, by construction, are orthonormal:

$$\int \phi_{q,k}(x) \phi_{q,k'}(x) dx = \delta_{k,k'}$$

$$\int \phi_{q,k}(x) \psi_{q,k'}(x) dx = 0$$

$$\int \psi_{q,k}(x) \psi_{q',k'} = \delta_{q,q'} \delta_{k,k'}$$

where the Dirac delta function is given by $\delta_{i,j} = 1$ if $i = j$ and $= 0$ if $i \neq j$. Again, corresponding discrete values are widely used.

Major examples of orthonormal wavelet functions are the following. The Haar wavelet, developed by Haar in 1910, has compact support, is symmetric and is not continuous. Daublets, named after Ingrid Daubechies, have compact support, and are continuous. Symmlets are of compact support, continuous and are nearly symmetric. Coiflets are nearly symmetric also and have vanishing moments for both ϕ and ψ.

2.4. Biorthogonal wavelets

The wavelet functions used thus far are asymmetric and as a consequence they introduce a phase shift in the coefficients at successive levels. Biorthogonal wavelets avoid these problems, but at the cost of dropping the orthogonality property. Biorthogonal wavelets were introduced by Cohen et al. (1992).

Biorthogonal wavelets use a closely associated set of four functions, ϕ, $\tilde{\phi}$, ψ, and $\tilde{\psi}$. The functions ϕ and ψ we have seen before, and the functions $\tilde{\phi}$ and $\tilde{\psi}$ are termed the dual wavelets. The role of mother and father wavelets, or their duals, can be reversed. The following biorthogonality relationships are satisfied:

$$\int \phi_{q,k}(x)\tilde{\phi}_{q,k'}(x)dx = \delta_{k,k'}$$

$$\int \phi_{q,k}(x)\tilde{\psi}_{q,k'}(x)dx = 0$$

$$\int \psi_{q,k}(x)\tilde{\psi}_{q',k'} = \delta_{q,q'}\delta_{k,k'}$$

The biorthogonal wavelet decomposition is expressed in terms of the dual functions:

$$f(x) = \sum_k s_k \tilde{\phi}_{p,k}(x) + \sum_k d_{p,k}\tilde{\psi}_{p,k}(x) + \sum_k d_{p-1,k}\tilde{\psi}_{p-1,k}(x) + \ldots + \sum_k d_{1,k}\tilde{\psi}_{1,k}(x)$$

An example of biorthogonal wavelets is provided by the B-spline function. A B-spline of degree 0 corresponds to the box function used as the Haar father wavelet; a B-spline of degree 1 (given by the convolution of a B_0-spline with itself) is a triangle function; a B-spline of degree 2 is a quadratic bump function. Different dual functions can be chosen, parameterized by the size of their support. Such wavelets therefore have a nomenclature consisting of two numbers, the first being the degree of the polynomial for the wavelet, and the second being the length of the support of the dual wavelet.

2.5. 2D wavelet transforms

The tensor product of a horizontal 1D wavelet and a vertical 1D wavelet leads to four different types of 2D wavelets:

$\Phi(x,y) = \phi_h(x) \times \phi_v y$ = horizontal father × vertical father
$\Psi_v(x,y) = \psi_h(x) \times \phi_v y$ = horizontal mother × vertical father
$\Psi_h(x,y) = \phi_h(x) \times \psi_v y$ = horizontal father × vertical mother
$\Psi_d(x,y) = \psi_h(x) \times \psi_v y$ = horizontal mother × vertical mother

This family therefore consists of one father wavelet and three mother wavelets, corresponding to detail information in vertical, horizontal and diagonal directions.

Wavelet coefficients, multiplied by each of these four different 2D wavelets, provide respectively

- the smooth image
- vertical detail images

- horizontal detail images
- diagonal detail images

for each level. An example is shown in Figure 13.2.

2.6. Redundant wavelet transforms

Without the orthogonality or biorthogonality constraints discussed above, there is a lot more scope for choosing a wavelet function. In particular if we do not use decimation we potentially have translation invariance, i.e. a shift in the signal does not result in a different pattern of wavelet coefficients. This is a key property, critical in certain problems like time series analysis, and helpful for visual display. With noise filtering, better prediction ensues, and few artifacts such as Gibbs phenomena near discontinuities.

In the à trous ("with holes" in French; Holschneider et al. (1989)) transform, the wavelet coefficients are calculated as the pixelwise differences $d_j = s_j - s_{j-1}$ etc. where j is the resolution level and s_0 is the input data. The B_3 spline is often used as the father wavelet or scaling function. Rather than dilating the wavelet, integer power of 2 increasing gaps ("holes") are used in the convolutions at each stage. A very simple additive decomposition of the data results from this, which is discussed a little further under "cube transform" below.

One of the most common and general approaches for performing a multiscale decomposition is similar to this procedure: blur the image, perhaps follow with subsampling, and then carry out pixel-by-pixel subtraction of one image from another. Such subtraction produces detail signal because we are specifying the signal which existed before the blurring, but which did not survive the blurring.

Figure 13.1 illustrates this, using a B_3 spline function as this "blurring function" – the scaling function. The upper right is the original image. The upper left is the smoothed residual (or "background" or DC – "direct current" – component). The lower images provide the detail coefficients. The original data (upper right) can be seen to be a pixelwise additive decomposition, where the components of this decomposition are delineated as shown.

The original image can be recreated from such a transform. In the case of a transform with subsampling (decimation), the reconstruction of the image uses interpolation. An iterative refinement of the transform data (see Starck et al. (1998a)) can be used to ensure exact reproducibility of the input data. Non-wavelet transforms can be developed from this principle. These include transforms based on median or mathemat-

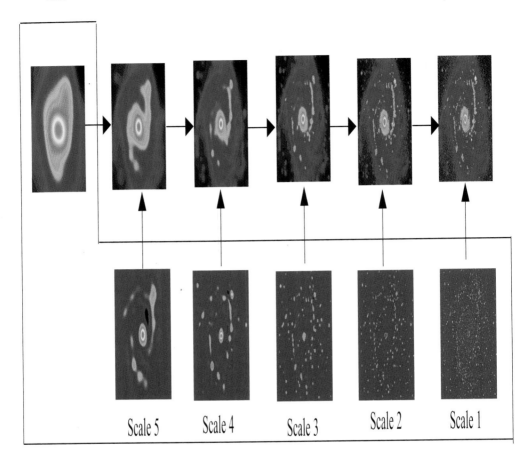

Figure 13.1. The à trous redundant wavelet transform, showing 5 wavelet scales. The B_3 spline is used as scaling function. Upper right: original input image, a spiral galaxy and stars. Delineated area: the wavelet transform.

ical morphology (opening, closing) operators. Again, see Starck et al. (1998a) for examples.

2.7. Non-wavelet multiscale transforms

The pyramidal median transform is a good example of a multiscale transform which is not a wavelet transform. The principle used can be viewed in similar terms to the à trous wavelet transform: take a median smooth, producing s_1, and subtract from the input image to provide scale 1 coefficients. Again median smooth s_1, producing s_2, and scale 2 coefficients are formed from $s_2 - s_1$. We continue in this way for a set number of resolution scales.

Now, the median kernel is doubled for each resolution scale, which would make this whole procedure quite inefficient. This is avoided by keeping the median kernel the same, but by simply making use of a pyramidal transform. Thus, at each resolution scale, decimation is carried out. To ensure exact reconstruction of the original data, we can make use of (i) interpolation when undoing the effects of decimation, and since this will lead to a good approximation of our original data, but an approximation nonetheless, (ii) refining our multiscale coefficients iteratively to ensure that with interpolation a perfect reconstruction is possible. A few iterations suffice, so that there is no great additional computational burden. This multiscale transform method is termed the Pyramidal median transform (PMT) and is further discussed below.

2.8. Output data structures

Output data structures are an interesting way – and very meaningful to the user – to categorize wavelet and other multiscale transforms. In this section, we look exclusively at the 2D case but the same principles hold, mutatis mutandis, for the 1D or 3D cases.

Many discrete wavelet transform algorithms have been described (Starck et al. 1998a). The most widely-known one is probably the orthogonal one, proposed by Mallat (Mallat 1989) and its biorthogonal version. Using this algorithm, the transformation of an image produces an image. Some of the other algorithms do not produce an image, but a pyramid, or cube. So we separate the different wavelet transform algorithms into classes, depending on the output data type. Other multiscale methods, which are nonlinear can also be categorized in this way. We distinguish between the following classes of multiscale transform in the case of 2D images:

1. Transforms which produce *cubes*.

2. Transforms which produce *pyramids*.

3. Transforms which produce *images* (non-redundant transforms).

4. Other classes of output – *half pyramids, directional cubes*.

The first two, and the directional cubes, are redundant (i.e. there are more pixels in the transformation than in the input data).

2.8.1 Cube transform.
For the first class, the input image I can be expressed as (see Bijaoui et al. (1994), Murtagh et al. (1995)):

$$I(x,y) = c_p(x,y) + \sum_{j=1}^{p} w_j(x,y) \qquad (13.4)$$

w_j (for $j = 1 \ldots p$) and c_p represent the transformation of I. c_p is a very smoothed version of the image I, and w_j is the image which contains information at scale j. The transformation is defined by $n = p + 1$ images (n = number of scales). The p images have zero mean (or approximately zero, for nonlinear transforms). Each of them corresponds to the information at a given scale, i.e. structure of a given size in the input image. Compact structures (with size of one or two pixels) will be found at the first scale ($j = 1$). For this class of transformation, which is very redundant, the amount of data is multiplied by the number of scales n. Therefore this takes considerable memory space, but the multiresolution coefficients ($w_j(x,y)$) are easy to interpret.

2.8.2 Pyramidal transform.
The second class of transformation is less redundant. The first scale has the same size as the image, but for the other scales, the number of pixels is reduced by four at each resolution. Thus, if N^2 is the number of pixels of I, the number of pixels of the transformation is $4/3 N^2$.

2.8.3 Non-redundant transform.
The last class is completely non-redundant, and the number of pixels is the same as in the input image. This means that it is an image. Figure 13.2 shows the representation of an image using the Mallat transform (Mallat 1989, Antonini et al. 1992, Daubechies 1988). At a given resolution, the image is shared between four parts. Three subimages correspond to details of the image in the horizontal, vertical, and diagonal directions, and the last part corresponds to the image at a lower resolution. The process can then be repeated on the image at the lower resolution. The Haar wavelet transform, lifting scheme transform (Sweldens 1996), and the G transform (which is a nonlinear transform based on mathematical morphology erosion and dilation operations) produce the same kind of representation.

The Feauveau transform (Feauveau (1990); see Figure 13.3) is not redundant, but the representation is different. There is no prioritized direction, and we have an intermediate resolution (half resolution). Strømme (1999) finds in favor of the Feauveau transform due to the total number of arithmetic operations needed.

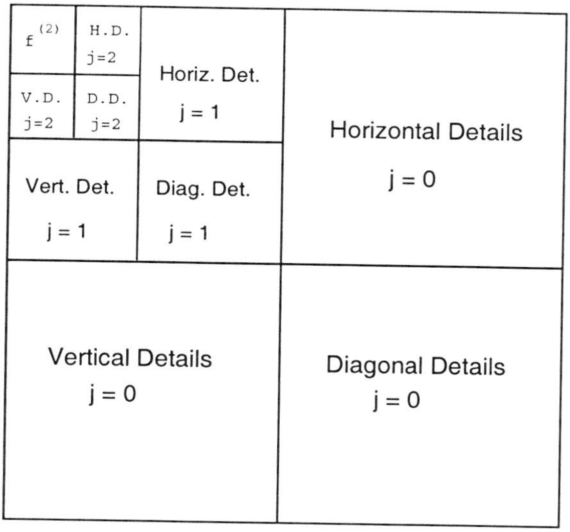

Figure 13.2. Mallat wavelet transform representation of an image.

2.8.4 Other classes of transform output.

The half-pyramidal transform was proposed by Bijaoui et al. (1997). This is relatively close to the pyramidal transform, but the two first scales are not decimated (i.e. they have the same size as the input image). The rationale is to have a compromise between a decimated transform – maybe causing difficulty for feature analysis – and a non-decimated one – giving rise to data redundancy.

In the directional cube transform, just as for the orthogonal Mallat transform, the analysis is directional, but there is no decimation. This means that the number of output bands is equal to the number of scales multiplied by the number of directions, and each band has the same number of pixels as the input image. In practice, two directions are used (vertical and horizontal).

3. Compression

Compression is a critical issue for the storage and transfer of large images. Multiscale transforms may allow a further direct use of compression for image interpretation and understanding, in that support is available for fast generation of low-resolution versions of an image.

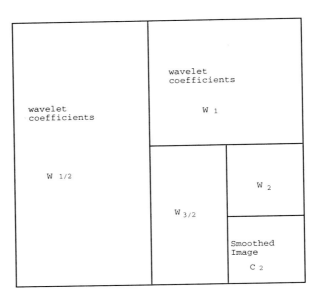

Figure 13.3. Feauveau wavelet transform representation of an image.

3.1. Lossy compression

Compression methods which are in operational use for scientific data include the following (Murtagh et al. 1998, Louys et al. 1999).

1. JPEG: a widely-used single image lossy or lossless compressor, based on blockwise decorrelation of the image data, followed by use of the discrete cosine transform.

2. Wavelet transform: for compression it is clear that a pyramidal transform is necessary. Our experiments were based on a biorthogonal Antonini-Daubechies transform.

3. Fractal coding: the image is decomposed into blocks, and each block is fractal coded.

4. Mathematical morphology: shape information is separated out in such a way that reconstruction of the input image data is possible. This is followed by the usual quadtree and Huffman coding.

5 HCOMPRESS: based on the Haar wavelet transform, this was developed at Space Telescope Science Institute by R. White, following earlier work by G. Richter.

6 Pyramidal median transform, PMT: a multiscale (pyramidal) median transform is the basis of this method (Starck et al. 1996; 1998a). The median provides for robustness. While there is a small price to be paid in the compressor's time requirements, we avoid any serious obstacle by making good use of the decrease of each resolution scale's dimensionality in the pyramid.

References for these approaches can be found in Louys et al. (1999). Other variants are possible. For example, the blockiness observable at very faint intensities following use of the Haar wavelet transform may be remedied through iterative refinement (Bobichon and Bijaoui 1997). This allows the very efficient Haar wavelet transform to be kept in the compressor.

Table 13.1 summarizes properties of these approaches for scientific (astronomy) imagery, with indicative timings. It includes whether there is artifact creation (e.g. blockiness) for the given compression rations. Some of these implementations have implementations available for progressive transmission (in C or in Java), including JPEG, HCOMPRESS and PMT.

A by now very commonly used wavelet compression scheme uses the Mallat approach (Figure 13.2), using the biorthogonal wavelet transform, with an economical storage scheme. The latter is based on a widely used principle in compression of any kind: if similar values are repeated, then we only need to indicate the extent of this similarity. Runlength encoding, for example, exploits runs of similar values. We could do this at each resolution scale or band. The embedded zerotree wavelet (EZW, Shapiro (1993)) scheme uses such a prediction scheme, based on the lowest resolution (top left in Figure 13.2), then predicting all corresponding values at the next higher resolution level, and so on. If a value is zero at a low resolution, then there is a good chance that coefficients at higher resolution in the same spatial position will also be insignificant. This is the basis for fast and effective wavelet compressors.

However, if one seeks to define and remove noise related to data capture, then the Mallat wavelet scheme can lead to aliasing-related damage being done to real signal. To avoid that, a redundant wavelet or other multiscale transform can be used. The cost is somewhat greater computation, especially for the compressor.

Compression can benefit enormously from noise filtering. After all, noise by definition is not compressible, and usually real signal is very compressible. In various case-studies using the PMT, it is clear that (i) the computational requirements are somewhat greater than for other wavelet approaches, especially for compression, but (ii) the performance is superior on scientific and medical images. Quadtree and Huffman coding are used to store the coefficients created. See Murtagh et al. (1998).

Table 13.1. Compression of a 1024 × 1024 integer-2 (16 bit) image. Platform: Sun Ultra-Enterprise, 250 MHz and 1 processor.

	Comp. time (sec)	Decomp. time (sec)	Artifact	Comp. ratio
JPEG	1.17	4.7	Y	<40
Wavelet	45	7.1	Y	270
Fractal	18.3	9	Y	< 30
Math. Morpho.	13	7.86	N	< 210
Hcompress	3.29	2.82	Y	270
Hcompress + iter rec	3.29	77	N	270
PMT	7.8	3.1	N	270

From a web server point of view we would like to compare the performances of the different packages considering two scenarios:

- Archive original and compressed images and distribute both on demand.
- Compress the data before transferring them and let the end-user decompress them at the client side.

This second situation has been studied and is illustrated in Figure 13.4. Considering a network rate of 10 kbits/second and an image of 2

Mbytes, we measured the time necessary to compress, transmit and decompress the image. Methods are ordered from top to bottom according to increasing visual quality of the decompressed image. If we consider 20 seconds to be the maximum delay the end-user can wait for an image to be delivered, only HCOMPRESS and PMT succeed, with fewer artifacts for PMT.

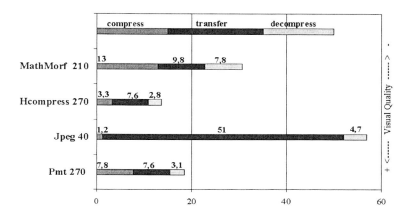

Figure 13.4. Comparison of the overall time for compression, transmission and decompression for distribution of astronomical images using the web: the network rate is taken to be 10 kbits/second, and the image size is 2Mbytes (1024 × 1024 × 2 bytes). The best performing codecs with respect to visual quality are shown towards the bottom of the figure.

3.2. Lossless compression

The compression methods looked at above involve filtering of information which is not considered to be of use. This includes what can demonstrably (or graphical) be shown to be noise. Noise is the part of information which is non-compressible, so that the residual signal is usually very highly compressible. It may be feasible to store such noise which has been removed from images on backing data store, but clearly the access to such information is less than straightforward.

If instead we seek a compression method which is guaranteed not to destroy information, what compression ratios can be expected?

We note firstly that quantization of floating (real) values necessarily involves some loss, which is avoided if we work in integer arithmetic only. This is not a restrictive assumption. Appropriate rescaling of image values may be availed of.

Next we note that Sweldens lifting scheme (Sweldens 1996) provides a convenient algorithmic framework for many wavelet transforms. The low-pass and band-pass operations are replaced by predictor and update operators at each resolution level, in the construction of the wavelet transform. When the input data consist of integer values, the wavelet transform no longer consists of integer values, and so we redefine the wavelet transform algorithm to face this problem. The predictor and update operators use, where necessary, a floor truncation function. The lifting scheme formulas for prediction and updating allow this to be carried out with no loss of information. If this had been done in the usual algorithmic framework, truncation would entail some small loss of information.

For comparison purposes, we use JPEG in lossless mode. We also use the standard Unix gzip command, which implements the Lempel-Ziv run-length encoding scheme, especially widely-used for text data.

As in Louys et al. (1999), we base our experiments on image data from a Schmidt photographic plate in the region of globular cluster M5 (numbered ESO 7992v), scanned with the CAI-MAMA facility. We use subimages of dimensions 1024 × 1024. Sampling is 0.666 arcseconds (i.e. seconds of arc, where 60 seconds equals one minute, and 60 minutes equals one degree) per pixel. Ancillary catalog information was available for our tests. The images are representative of those used in Aladin (Bonnarel et al. 1999), the reference image service of the CDS (Strasbourg Data Centre, Strasbourg Observatory).

Table 13.2 shows our findings, where we see that our integer and lifting scheme wavelet transform approach is a little better than the other methods in terms of compression ratio. Not surprisingly, the compression rate is not large for such lossless compression. Routine mr_lcomp (MR/1 1999) was used for this work. Lossless JPEG suffers additionally from rounding errors.

We now turn attention to usage of a lossless wavelet-based compressor, above and beyond the issues of economy of storage space and of transfer time. The lifting scheme implementation of the Haar wavelet transform presents the particularly appealing property that lower resolution versions of an image are *exactly* two-fold rebinned versions of the next higher resolution level. For aperture photometry (i.e. accumulated intensities, or flux, within set aperture radii) and other tasks, lower level resolution can be used to provide a partial analysis. A low resolution

Table 13.2. Compression of a 1024 × 1024 2-byte integer (16 bit) image. Platform: Sun Ultra-Sparc 250 MHz and 2 processors.

Software	CPU time (sec) Compression time	Decompression time	Compression ratio
JPEG lossless	2.0	0.7	1.6
Lifting scheme with Haar	4.3	4.4	1.7
Gzip (Unix)	13.0	1.4	1.4

level image can be used scientifically since its "big" pixels contain the integrated average of flux covered by them.

The availability of efficiently delivered low resolution images can thus be used for certain scientific objectives. This opens up the possibility for an innovative way to analyze distributed image holdings.

3.3. Compressibility and entropy

We use a new definition of entropy, which incorporates resolution scale and a noise model, to characterize "informative structure" in data. Why this is necessary will be explained in the following.

The term "entropy" is due to Clausius (1865), and the concept of entropy was introduced by Boltzmann into statistical mechanics, in order to measure the number of microscopic ways that a given macroscopic state can be realized. Shannon (1948) founded the mathematical theory of communication when he suggested that the information gained in a measurement depends on the number of possible outcomes out of which one is realized. Shannon also suggested that the entropy can be used for maximization of the bits transferred under a quality constraint. Jaynes (1957) proposed to use the entropy measure for radio interferometric image deconvolution, in order to select between a set of possible solutions that which contains the minimum of information, or following his entropy definition, that which has maximum entropy. In principle, the solution verifying such a condition should be the most reliable. Much work has been carried out in the last 30 years on the use of entropy for the general problem of data filtering and deconvolution.

Traditionally information and entropy are determined from events and the probability of their occurrence. Signal and noise are basic building-blocks of signal and data analysis in the physical and communication sciences. Instead of the probability of an event, we are led to consider the probabilities of our data being either signal or noise.

Consider any data signal with interpretative value. Now consider a uniform "scrambling" of the same data signal. (Starck et al. (1998b), illustrate this with the widely-used Lena test image.) Any traditional definition of entropy, the main idea of which is to establish a relation between the received information and the probability of the observed event, would give the same entropy for these two cases. A good definition of entropy should instead satisfy the following criteria:

1 The information in a flat signal is zero.

2 The amount of information in a signal is independent of the background.

3 The amount of information is dependent on the noise. A given signal Y ($Y = X+$ Noise) doesn't furnish the same information if the noise is high or small.

4 The entropy must work in the same way for a signal value which has a value $B + \epsilon$ (B being the background), and for a signal value which has a value $B - \epsilon$.

5 The amount of information is dependent on the correlation in the signal. If a signal S presents large features above the noise, it contains a lot of information. By generating a new set of data from S, by randomly taking the values in S, the large features will evidently disappear, and this new signal will contain less information. But the data values will be the same as in S.

To cater for background, we introduce the concept of multiresolution into our entropy. We will consider that the information contained in some dataset is the sum of the information at different resolution levels, j. A wavelet transform is one choice for such a multiscale decomposition of our data. We define the information of a wavelet coefficient $w_j(k)$ at position k and at scale j as $I = -\ln(p(w_j(k)))$, where p is the probability of the wavelet coefficient being signal and not noise. Note that this is not the same as probability of occurrence. It also presupposes what we mean by noise, i.e. our noise model. Entropy, commonly denoted as H, is then defined as the sum over all positions, k, and over all scales, j, of all I.

For Gaussian noise we continue in this direction, using Gaussian probability distributions, and find that the entropy, H, is the sum over all positions, k, and over all scales, j, of $(w_j(k)^2)/(2\sigma^2 j)$ (i.e. the coefficient squared, divided by twice the standard deviation squared of a given scale). Sigma, σ, or the standard deviation, is the (Gaussian) measure of the noise. We see that the information is proportional to the energy of the wavelet coefficients. The larger the value of a wavelet coefficient, then the higher will be the contribution to the information furnished by this wavelet coefficient. A range of applications and examples can be found in Starck et al. (1998b) and Starck and Murtagh (1999).

Our entropy definition is completely dependent on the noise modeling. If we consider a signal S, and we assume that the noise is Gaussian, with a standard deviation equal to sigma, we won't measure the same information compared to the case when we consider that the noise has another standard deviation value, or if the noise follows another distribution.

Returning to our example of a signal of substantive value, and a scrambled version of this, we can plot an information versus scale curve (e.g. log(entropy) at each scale using the above definition, versus the multiresolution scale). For the scrambled signal, the curve is flat. For the original signal, it increases with scale.

We can use such an entropy versus scale plot to investigate differences between encrypted and unencrypted signals, to study typical versus atypical cases, and to differentiate between atypical or interesting signals. Experimental assessment of how this works in practice is reported on in Starck et al. (2000).

Among the application studies presented in Starck et al. (2000) are the following. Firstly, the case is assessed of faint astronomical sources in a noisy field. The sky background is given by uniformly distributed pixel values. A number of faint sources – below the noise level – are added. The multiscale entropy is calculated for both cases. The number of faint sources is varied. The multiscale entropy is shown to pick up the presence of such "real" information. The second study takes as its point of departure the compression work reported on in Louys et al. (1999). The "inherent compressibility" of a large number of astronomical images was discussed in that work, based on astronomical criteria of astrometry (positional information), photometry (integrated or accumulated flux intensity information), and faint features (no visual flaws). The "inherent compression ratios" were determined from various compression methods and quantified according to acceptable levels of the astronomical criteria referred to. These inherent or best compression ratios were found to be very well correlated with multiscale entropy. Again, this is indicative support for the usefulness of our approach to quantifying multiscale entropy.

4. Distributed Large Image Display Environments

4.1. Requirements

In order to visualize very large images in a reasonable amount of time, the transmission has to be based on two concepts:

- data compression
- progressive transmission

The latter consists of visualizing quickly a low resolution image, and then increasing the quality with the arrival of new bits. Wavelet based methods become very attractive for this, because they integrate in a natural way this multiresolution concept. A coarse resolution image can be directly obtained without having to decompress the whole file.

But with very large images, a third concept is necessary, which is the region of interest. Indeed, images are becoming so large that it is impossible to display them in a normal display window, and we need to have the ability to focus on a given area of the image at a given resolution. To move from one area to another, or to increase the resolution of a part of the area is a user task, and is a new active element of the decompression.

One possible approach would be to let the server extract the selected area, compress it, and send it to the user. This solution is simple and does not need any development, but presents several drawbacks:

- Server load: the image at full resolution is transmitted. the server must decompress the full size image and re-compress the selected area on each user request.
- Transfer speed: the full image is transmitted.

The principle of a "Large Image Visualization Environment" (LIVE, a prototype of which is available in MR/1 (1999)) is to add the following functionalities:

- Full image display at a very low resolution.
- Image navigation: the user can go up (the quality of an area of the image is improved) or down (return to the previous image). Going up or down in resolution implies a four-fold increase or decrease in the size of what is viewed.

To support this, blockwise and scalewise storage is required (Figure 13.5). A large image (say 4000×4000), which is compressed by blocks (8×8, each block having a size of 500×500), is represented at five resolution levels. The visualization window (of size 256 × 256 in our example) covers the whole image at the lowest resolution level (image size 250 × 250), but only one block at the full resolution (in fact between one and four, depending on the position in the image). The LIVE concept consists of moving the visualization window into this pyramidal structure, without having to load into memory the large image. The image is first visualized at low resolution, and the user can indicate (using the mouse) which part of the visualized subimage he wishes to zoom on. At each step, only wavelet coefficients of the corresponding blocks and of the new resolution level are decompressed.

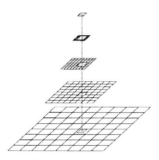

Figure 13.5. An example of the pre-structuring of a large image which is necessary for supporting progressive transmission. The large image is compressed by block, and represented at five resolution levels. At each resolution level, the display window is superimposed at a given position.

Experiments using a similar approach are reported on in Poulakidas et al. (1996).

4.2. Alternative image display paradigms

The system for multiscale storage and transmission of image data, which we have at our disposal, lends itself also to the display paradigm known as foveation, a special case of display with differential detail, which we will briefly discuss in this section.

The center of the human retina has resolution of about 1 arcminute (minutes of arc, i.e. angular distance, where 60 minutes equals one degree) under ideal conditions, and this falls off rapidly outside the central 2 degree region. At 10 degrees eccentricity, the resolution is about 10 arcminutes. A high-resolution display with 1280 pixels across a 60 degree field of view achieves about 3 arcminutes per pixel. This means that outside the foveal region, much of the screen resolution is wasted.

An image with variable spatial resolution is referred to as a *foveated image*, or a *multifoveated image* if there is more than one single foveal region with high resolution. The human visual system has a particular logarithmic resolution-decrease function from the focus of interest (Chang and Yap 1997, Chang 1998).

A foveated image has full resolution in a region of interest, around the point of interest, and downgraded resolution (perhaps decreasing from the region of interest) elsewhere. Chang and Yap (1997) characterize their system which delivers foveated images from an image server as *thinwire visualization*, which refers to the limited bandwidth between server and client, and therefore the interest in economizing on this bandwidth by availing of spatially-variable resolution. In Chang and Yap (1997), a wavelet-based foveation method is described: variable-sized output display pixels (super-pixel sampling) are used, based on a Haar wavelet transform. Chang et al. (1997) describe this approach for display of images from a (geographic) map server.

Differential detail can also be based on a fish-eye view of an image or more generally of data (Furnas 1986, Sakar and Brown 1992). An alternative variable space trade-off between scale and resolution is described in Stone et al. (1994). It is more computationally tractable in a web-based setting. This is a "magic lens", a high-resolution image display window which is passed over the less-enhanced display window. Such selective detail, based on cursor control by the user, is reviewed in Hirtle (1998) using map data.

4.3. Short review of software products

ImagePump (http://www.island.com) is a web-based image database server, written in Java and C++, which supports color, and a range of operations including zooming. It targets prepress and graphics arts users, with typical image sizes of 100-350 MB. This recently introduced (mid-1998) software package has useful interfaces to image and graphic packages from Adobe and others. Pre-calculated pyramidal and tiled subimage structures are availed of, and can be created at runtime.

Live Picture (http://www.livepicture.com) focuses on the market for online catalogs and image materials, in the area of web-based electronic commerce (and can of course cater for scientific or other imagery). It includes an image server with support for multiresolution zooming, 3D images and streaming video delivery technology. At the user end, Java and ActiveX are used.

5. Conclusion

The area we have described is an exciting one. It includes the following principal aspects: theoretical understanding of information and data, statistical modeling of data, information characterization, distributed computing models, and image and data display environments. We have sseen how needs and requirements are continuing to grow in these fields. We have also seen how wavelet and related multiscale signal and image transforms are the basis for innovative solutions.

Bibliography

M. Antonini, M. Barlaud, P. Mathieu, and I. Daubechies. Image coding using wavelet transform. *IEEE Transactions on Image Processing*, 1: 205–220, 1992.

A. Bijaoui, F. Rué, and B. Vandame. Multiscale vision and its application to astronomy. In V. Di Gesù, M.J.B. Duff, A. Heck, M. C. Maccarone, L. Scarsi, and H.U. Zimmermann, editors, *Data Analysis in Astronomy*, pages 337–343. World Scientific, 1997.

A. Bijaoui, J.L. Starck, and F. Murtagh. Restauration des images multi-échelles par l'algorithme à trous. *Traitement du Signal*, 11:229–243, 1994.

Y. Bobichon and A. Bijaoui. A regularized image restoration algorithm for lossy compression in astronomy. *Experimental Astronomy*, 7:239–255, 1997.

F. Bonnarel, P. Fernique, F. Genova, J.G. Bartlett, O. Bienaymé, J. Florsch, and H. Ziaeepour. Aladin: a reference tool for identification of astronomical sources. In D.M. Mehringer, R.L. Plante, and D.A. Roberts, editors, *Astronomical Data Analysis Software and System VIII*, pages 229–232. Astronomical Society of the Pacific, 1999.

P. J. Burt and A. E. Adelson. The laplacian pyramid as a compact image code. *IEEE Transactions on Communications*, 31:532–540, 1983.

E. C. Chang. *Foveation Techniques and Scheduling Issues in Thinwire Visualization*. PhD thesis, Department of Computer Science, New York University, 1998.

E. C. Chang and C. K. Yap. A wavelet approach to foveating images. In *Proc. 13th ACM Symp. Computational Geometry*, pages 397–399. ACM, 1997. Extended version at:
ftp://cs.nyu.edu/pub/local/yap/visual/foveated.ps.gz.

E. C. Chang, C. K. Yap, and T. J. Yen. Realtime visualization of large images over a thinwire. In *Proc. IEEE Visualization*, 1997. Available at: http://www.cz3.nus.edu.sg/~changec/pub.html.

R. Clausius. Annalen der physik, serie 2, 1865.

A. Cohen, I. Daubechies, and J.C. Feauveau. Biorthogonal bases of compactly supported wavelets. *Communications in Pure and Applied Mathematics*, 45:485–560, 1992.

I. Daubechies. Orthogonal bases of compactly supported wavelets. *Communications in Pure and Applied Mathematics*, 41:909–996, 1988.

J. C. Feauveau. *Analyse multirésolution par ondelettes non-orthogonales et bancs de filtres numériques*. PhD thesis, Université Paris Sud, 1990.

G. W. Furnas. Generalized fisheye views. In *Proc. ACM CHI'86 Conf. on Human Factors in Computing Systems*, pages 16–32, 1986.

S. C. Hirtle. The cognitive atlas: using gis as a metaphor for memory. In M. J. Egenhofer and R. G. Golledge, editors, *Spatial and Temporal Reasoning in Geographic Information Systems*, pages 263–271. Oxford University Press, 1998.

M. Holschneider, R. K. Martinet, J. Morlet, and P. Tchamitchian. A real-time algorithm for signal analysis with the help of the wavelet transform. In J. M. Combes, A. Grossmann, and Ph. Tcha-mitchian, editors, *Wavelets: Time-Frequency Methods and Phase-Space*, pages 286–297. Springer-Verlag, Berlin, 1989.

E. T. Jaynes. Information theory and statistical mechanics. *Phys. Rev.*, 106:620–630, 1957.

M. Louys, J. L. Starck, S. Mei, F. Bonnarel, and F. Murtagh. Astronomical image compression. *Astronomy and Astrophysics Supplement*, 136:579–590, 1999.

S. Mallat. A theory for multiresolution signal decomposition: the wavelet representation. *IEEE Transactions on Pattern Analysis and Machine Intelligence*, 11:674–693, 1989.

MR/1. The multiresolution analysis software, version 2.0, and mr/2: Multiresolution entropy, version 1.0, 1999. http://www.multiresolution.com.

F. Murtagh, J. L. Starck, and M. Louys. Very high quality image compression based on noise modeling. *International Journal of Imaging Systems and Technology*, 9:38–45, 1998.

F. Murtagh, J.L. Starck, and A. Bijaoui. Multiresolution in astronomical image processing: a general framework. *The International Journal of Image Systems and Technology*, 6:332–338, 1995.

A. Poulakidas, A. Srinivasan, O. Egecioglu, O. Ibarra, and T. Yang. Experimental studies on a compact storage scheme for wavelet-based multiresolution subregion retrieval. In *Proc. NASA 1996 Combined Industry, Space and Earth Science Data Compression Workshop*, Utah, April 1996. Linked from: http://pw1.netcom.com/~tmhenry/thePaper/Sect3.html.

W. H. Press, S. A. Teukolsky, W. T. Vetterling, and B. P. Flannery. *Numerical Recipes: The Art of Scientific Computing*. Cambridge University Press, 2 edition, 1992.

M. Sakar and M.H. Brown. Graphical fisheye views of graphs. In *Proc. CHI'92*, pages 83–91, Monterey, CA, 1992.

C. E. Shannon. A mathematical theory for communication. *Bell Systems Technical Journal*, 27:379–423, 1948.

J. M. Shapiro. Embedded image coding using zerotrees of wavelet coefficients. *IEEE Transactions on Signal Processing*, 41:3445–3462, 1993.

J. L. Starck and F. Murtagh. Multiscale entropy filtering. *Signal Processing*, 76:147–165, 1999.

J. L. Starck, F. Murtagh, and A. Bijaoui. *Image and Data Analysis: The Multiscale Approach*. Cambridge University Press, 1998a.

J. L. Starck, F. Murtagh, and F. Bonnarel. Multiscale entropy for semantic description of images and signals. submitted, 2000.

J. L. Starck, F. Murtagh, B. Pirenne, and M. Albrecht. Astronomical image compression based on noise suppression. *Publications of the Astronomical Society of the Pacific*, 108:446–455, 1996.

J.L. Starck, F. Murtagh, and R. Gastaud. A new entropy measure based on the wavelet transform and noise modeling. *IEEE Transactions on Circuits and Systems II: Analog and Digital Signal Processing*, 45: 1118–1124, 1998b.

M.C. Stone, K. Fishkin, and E.A. Bier. The movable filter as a user interface tool. In *Proc. CHI'94 Conf. Human Factors in Comput. Syst.*, pages 306–312, 1994.

Ø. Strømme. *On the applicability of wavelet transforms to image and video compression.* PhD thesis, University of Strathclyde, February 1999.

W. Sweldens. The lifting scheme: a custom-design construction of biorthogonal wavelets. *Appl. Comput. Harmon. Anal.*, 3:186–200, 1996.

Chapter 14

CLUSTERING IN MASSIVE DATA SETS

Fionn Murtagh
School of Computer Science, The Queen's University of Belfast, Belfast BT7 1NN, Northern Ireland
f.murtagh@qub.ac.uk

Abstract We review the time and storage costs of search and clustering algorithms. We exemplify these, based on case-studies in astronomy, information retrieval, visual user interfaces, chemical databases, and other areas. Theoretical results developed as far back as the 1960s still very often remain topical. More recent work is also covered in this article. This includes a solution for the statistical question of how many clusters there are in a dataset. We also look at one line of inquiry in the use of clustering for human-computer user interfaces. Finally, the visualization of data leads to the consideration of data arrays as images, and we speculate on future results to be expected here.

Keywords: Clustering, Massive data sets.

> 'Now', said Rabbit, 'this is a Search, and I've Organised it –'
> 'Done what to it?' said Pooh.
> 'Organised it. Which means – well, it's what you do to a Search, when you don't all look in the same place at once.'
> A.A. Milne, *The House at Pooh Corner* (1928) – M.S. Zakaria

1. Introduction

Nearest neighbor searching is considered in sections 2 to 6 for one main reason: its utility for the clustering algorithms reviewed in sections 7 to 10 and, partially, 11. They are the building blocks for the most efficient implementations of hierarchical clustering algorithms, and they can be used to speed up other families of clustering algorithms.

The best match or nearest neighbor problem is important in many disciplines. In statistics, k-nearest neighbors, where k can be 1, 2, etc., is a method of non-parametric discriminant analysis. In pattern recognition, this is a widely-used method for unsupervised classification (see Dasarathy (1991)). Nearest neighbor algorithms are the building block of clustering algorithms based on nearest neighbor chains; or of effective heuristic solutions for combinatorial optimization algorithms such as the traveling salesman problem, which is a paradigmatic problem in many areas. In the database and more particularly data mining field, NN searching is called similarity query, or similarity join (Bennett et al. 1999).

In section 2, we begin with data structures where the objective is to break the $O(n)$ barrier for determining the nearest neighbor (NN) of a point. A database record or tuple may be taken as a point in a space of dimensionality m, the latter being the associated number of fields or attributes. These approaches have been very successful, but they are restricted to low dimensional NN-searching. For higher dimensional data, a wide range of bounding approaches have been proposed, which remain $O(n)$ algorithms but with a low constant of proportionality.

We assume familiarity with basic notions of similarity and distance, the triangular inequality, ultrametric spaces, Jaccard and other coefficients, normalization and standardization. For an implicit treatment of data theory and data coding, see Murtagh and Heck (1987). Useful background reading can be found in Arabie et al. (1996). In particular output representational models include discrete structures, e.g. rooted labeled trees or dendrograms, and spatial structures (Arabie and Hubert 1996), with many hybrids.

Sections 2 to 6 relate to nearest neighbor searching, an elemental form of clustering, and a basis for clustering algorithms to follow. Sections 7 to 11 review a number of families of clustering algorithm. Sections 12 to 14 relate to visual or image representations of data sets, from which a number of interesting algorithmic developments arise.

2. Binning or Bucketing

In this approach to NN-searching, a preprocessing stage precedes the searching stage. All points are mapped onto indexed cellular regions of space, so that NNs are found in the same or in closely adjacent cells. Taking the plane as as example, and considering points (x_i, y_i), the maximum and minimum values on all coordinates are obtained (e.g. $(x_j^{\min}, y_j^{\min})$). Consider the mapping (Fig. 14.1)

$$x_i \longrightarrow \{\lfloor (x_{ij} - x_j^{\min})/r \rfloor\}$$

CLUSTERING IN MASSIVE DATA SETS

where constant r is chosen in terms of the number of equally spaced categories into which the interval $[x_j^{\min}, x_j^{\max}]$ is to be divided. This gives to x_i an integer value between 0 and $\lfloor (x_{ij}^{\max} - x_{ij}^{\min})/r \rfloor$ for each attribute j. $O(nm)$ time is required to obtain the transformation of all n points, and the result may be stored as a linked list with a pointer from each cell identifier to the set of points mapped onto that cell.

$$x^{\min}, y^{\min} = 0, 0, \ x^{\max}, y^{\max} = 50, 40, \ r = 10$$

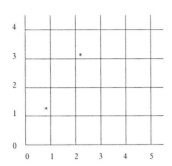

Point (22,32) is mapped onto cell (2,3);
point (8,13) is mapped onto cell (0,1).

Figure 14.1. Example of simple binning in the plane.

NN-searching begins by finding the closest point among those which have been mapped onto the same grid cell as the target point. This gives a current NN point. A closer point may be mapped onto some other grid cell if the distance between target point and current NN point is greater than the distance between the target point and any of the boundaries of the cell containing it. Some further implementation details can be found in Murtagh (1993b).

A powerful theoretical result regarding this approach is as follows. For uniformly distributed points, the NN of a point is found in $O(1)$, or constant, expected time (see Delannoy (1980), or Bentley et al. (1980), for proof). Therefore this approach will work well if approximate uniformity can be assumed or if the data can be broken down into regions of approximately uniformly distributed points.

Simple Fortran code for this approach is listed, and discussed, in Schreiber (1993). The search through adjacent cells requires time which increases exponentially with dimensionality (if it is assumed that the number of points assigned to each cell is approximately equal). As a result, this approach is suitable for low dimensions only. Rohlf (1978) reports on work in dimensions 2, 3, and 4; and Murtagh (1983) in the plane. Rohlf also mentions the use of the first 3 principal components to approximate a set of points in 15-dimensional space.

From the constant expected time NN search result, particular hierarchical agglomerative clustering methods can be shown to be of linear expected time, $O(n)$ (Murtagh 1983). The expected time complexity for Ward's minimum variance method is given as $O(n \log n)$. Results on the hierarchical clustering of up to 12,000 points are discussed.

The limitation on these very appealing computational complexity results is that they are only really feasible for data in the plane. Bellman's curse of dimensionality manifests itself here as always. For dimensions greater than 2 or 3 we proceed to the situation where a binary search tree can provide us with a good preprocessing of our data.

3. Multidimensional Binary Search or kD Tree

A multidimensional binary search tree (MDBST) preprocesses the data to be searched through by two-way subdivision, and subdivisions continue until some prespecified number of data points is arrived at. See example in Fig. 14.2. We associate with each node of the decision tree the definition of a subdivision of the data only, and we associate with each terminal node a pointer to the stored coordinates of the points. Using the approximate median of projections keeps the tree balanced, and consequently $O(\log n)$ levels, at each of which $O(n)$ processing is required. Hence the construction of the tree takes $O(n \log n)$ time.

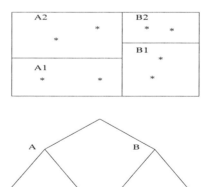

Figure 14.2. A MDBST using planar data.

The search for a NN then proceeds by a top-down traversal of the tree. The target point is transmitted through successive levels of the tree using the defined separation of the two child nodes at each node. On arrival at a terminal node, all associated points are examined and a current NN selected. The tree is then backtracked: if the points associated with any node could furnish a closer point, then subnodes must be checked out.

The approximately constant number of points associated with terminal nodes (hyper-rectangular cells in the space of points) should be greater than 1 in order that some NNs may be obtained without requiring a search of adjacent cells (other terminal nodes). Friedman et al. (1977) suggest a value of the number of points per bin between 4 and 32 based on empirical study.

The MDBST approach only works well with small dimensions. To see this, consider each coordinate being used once and once only for the subdivision of points, i.e. each attribute is considered equally useful. Let there be p levels in the tree, i.e. 2^p terminal nodes. Each terminal node contains approximately c points by construction and so $c2^p = n$. Therefore $p = \log_2 n/c$. As sample values, if $n = 32768, c = 32$, then $p = 10$. I.e. in 10-dimensional space, using a large number of points associated with terminal nodes, more than 30000 points will need to be considered. For high dimensional spaces, two alternative MDBST specifications are as follows.

All attributes need not be considered for splitting the data if it is known that some are of greater interest than others. Linearity present in the data may manifest itself via the variance of projections of points on the coordinates; choosing the coordinate with greatest variance as the discriminator coordinate at each node may therefore allow repeated use of certain attributes. This has the added effect that the hyper-rectangular cells into which the terminal nodes divide the space will be approximately cubical in shape. In this case, Friedman et al. (1977) show that search time is $O(\log n)$ on average for the finding of a NN. Results obtained for dimensionalities of between 2 and 8 are reported on in Friedman et al. (1977), and in the application of this approach to minimal spanning tree construction in Bentley and Friedman (1978). Lisp code for the MDBST is discussed in Broder (1990).

The MDBST has also been proposed for very high dimensionality spaces, i.e. where the dimensionality may be greater than the number of points, as could be the case in a keyword-based system. Keywords (coordinates) are batched, and the following decision rule is used: if some *one* of a given batch of node-defining discriminating attributes is present, then take the left subtree, else take the right subtree. Large n, well in excess of 1400, was stated as necessary for good results (Weiss 1981, Eastman and Weiss 1982). General guidelines for the attributes which define the direction of search at each level are that they be related, and the number chosen should keep the tree balanced. On intuitive grounds, our opinion is that this approach will work well if the clusters of attributes, defining the tree nodes, are mutually well separated.

An MDBST approach is used by Moore (1998) in the case of Gaussian mixture clustering. Over and above the search for nearest neighbors based on Euclidean distance, Moore allows for the Mahalanobis metric, i.e. distance to cluster centers which are "corrected" for the (Gaussian) spread or morphology of clusters. The information stored at each node of the tree includes covariances. Moore (1998) reports results on numbers of objects of around 160,000, dimensionalities of between 2 and 6, and speedups of 8-fold to 1000-fold. Pelleg and Moore (1999) discuss results on some 430,000 two-dimensional objects from the Sloan Digital Sky Survey (see section 10 below).

4. Projections and Other Bounds

4.1. Bounding using Projection or Properties of Metrics

Making use of bounds is a versatile approach, which may be less restricted by dimensionality. Some lower bound on the dissimilarity is efficiently calculated in order to dispense with the full calculation in many instances.

Using projections on a coordinate axis allows the exclusion of points in the search for the NN of point x_i. Points x_k, only, are considered such that $(x_{ij} - x_{kj})^2 \leq c^2$ where x_{ij} is the jth coordinate of x_i, and where c is some prespecified distance (see Fig. 14.3).

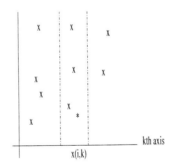

Figure 14.3. Two-dimensional example of projection-based bound. Points with projections within distance c of given point's (*) projection, alone, are searched. Distance c is defined with reference to a candidate or current nearest neighbor.

Alternatively more than one coordinate may be used. The prior sorting of coordinate values on the chosen axis or axes expedites the finding of points whose full distance calculation is necessitated. The preprocessing required with this approach involves the sorting of up to m sets of coordinates, i.e. $O(mn \log n)$ time.

Using one axis, it is evident that many points may be excluded if the dimensionality is very small, but that the approach will disimprove as the latter grows. Friedman et al. (1975) give the expected NN search time, under the assumption that the points are uniformly distributed, as $O(mn^{1-1/n})$. This approaches the brute force $O(nm)$ as n gets large. Reported empirical results are for dimensions 2 to 8.

Marimont and Shapiro (1979) extend this approach by the use of projections in subspaces of dimension greater than 1 (usually about $m/2$ is suggested). This can be further improved if the subspace of the principal components is used. Dimensions up to 40 are examined.

The Euclidean distance is very widely used. Two other members of a family of Minkowski metric measures require less computation time to calculate, and they can be used to provide bounds on the Euclidean distance. We have:

$$d_1(x, x') \geq d_2(x, x') \geq d_\infty(x, x')$$

where d_1 is the Hamming distance defined as $\sum_j | x_j - x'_j |$; the Euclidean distance is given by the square root of $\sum_j (x_j - x'_j)^2$; and the Chebyshev distance is defined as $\max_j | x_j - x'_j |$.

Kittler (1978) makes use of the following bounding strategy: *reject all points y such that $d_1(x,y) \geq \sqrt(m)\delta$* where δ is the current NN d_2-distance. The more efficiently calculated d_1-distance may thus allow the rejection of many points (90% in 10-dimensional space is reported on by Kittler). Kittler's rule is obtained by noting that the greatest d_1-distance between x and x' is attained when

$$| x_j - x'_j |^2 = d_2^2(x, x')/m$$

for all coordinates, j. Hence $d_1(x, x') = d_2(x, x')/\sqrt{m}$ is the greatest d_1-distance between x and x'. In the case of the rejection of point y, we then have:

$$d_1(x, y) \leq d_2(x, y)/\sqrt{m}$$

and since, by virtue of the rejection,

$$d_1(x, y) \geq \sqrt{m}\delta$$

it follows that $\delta \leq d_2(x, y)$.

Yunck (1976) presents a theoretical analysis for the similar use of the Chebyshev metric. Richetin et al. (1980) propose the use of both bounds. Using uniformly distributed points in dimensions 2 to 5, the latter reference reports the best outcome when the rule: *reject all y such that* $d_\infty(x,y) \geq \delta$ precedes the rule based on the d_1-distance. Up to 80% reduction in CPU time is reported.

4.2. Bounding using the Triangular Inequality

The triangular inequality is satisfied by distances: $d(x,y) \leq d(x,z) + d(z,y)$, where x, y and z are any three points. The use of a *reference point*, z, allows a full distance calculation between point x, whose NN is sought, and y to be avoided if

$$\mid d(y,z) - d(x,z) \mid \geq \delta$$

where δ is the current NN distance. The set of all distances to the reference point are calculated and stored in a preprocessing step requiring $O(n)$ time and $O(n)$ space. The above cut-off rule is obtained by noting that if

$$d(x,y) \geq \mid d(y,z) - d(x,z) \mid$$

then, necessarily, $d(x,y) \geq \delta$. The former inequality above reduces to the triangular inequality irrespective of which of $d(y,z)$ or $d(x,z)$ is the greater.

The set of distances to the reference point, $\{d(x,z) \mid x\}$, may be sorted in the preprocessing stage. Since $d(x,z)$ is fixed during the search for the NN of x, it follows that the cut-off rule will not then need to be applied in all cases.

The single reference point approach, due to Burkhard and Keller (1973), was generalized to multiple reference points by Shapiro (1977). The sorted list of distances to the first reference point, $\{d(x,z_1) \mid x\}$, is used as described above as a preliminary bound. Then the subsequent bounds are similarly employed to further reduce the points requiring a full distance calculation. The number and the choice of reference points to be used is dependent on the distributional characteristics of the data. Shapiro (1977) finds that reference points ought to be located away from groups of points. In 10-dimensional simulations, it was found that at best only 20% of full distance calculations were required (although this was very dependent on the choice of reference points).

Hodgson (1988) proposes the following bound, related to the training set of points, y, among which the NN of point x is sought. Determine in advance the NNs and their distances, $d(y, NN(y))$ for all points in the training set. For point y, then consider $\delta_y = \frac{1}{2}d(y, NN(y))$. In seeking $NN(x)$, and having at some time in the processing a candidate NN, y', we can exclude all y from consideration if we find that $d(x, y') \leq \delta_{y'}$.

In this case, we know that we are sufficiently close to y' that we cannot improve on it.

We return now to the choice of reference points: Ruiz (1986) proposes the storing of inter-point distances between the members of the training set. Given x, whose NN we require, some member of the training set is used as a reference point. Using the bounding approach based on the triangular inequality, described above, allows other training set members to be excluded from any possibility of being NN(x). Micó et al. (1992) and Ramasubramanian and Paliwal (1992) discuss further enhancements to this approach, focused especially on the storage requirements.

Fukunaga and Narendra (1975) make use of both a hierarchical decomposition of the data set (they employ repeatedly the k-means partitioning technique), and bounds based on the triangular inequality. For each node in the decomposition tree, the center and maximum distance to the center of associated points (the "radius") are determined. For 1000 points, 3 levels were used, with a division into 3 classes at each node.

All points associated with a non-terminal node can be rejected in the search for the NN of point x if the following rule (Rule 1) is not verified:

$$d(x,g) - r_g < \delta$$

where δ is the current NN distance, g is the center of the cluster of points associated with the node, and r_g is the radius of this cluster. For a terminal node, which cannot be rejected on the basis of this rule, each associated point, y, can be tested for rejection using the following rule (Rule 2):

$$\mid d(x,g) - d(y,g) \mid \geq \delta.$$

These two rules are direct consequences of the triangular inequality.

A branch and bound algorithm can be implemented using these two rules. This involves determining some current NN (the bound) and subsequently branching out of a traversal path whenever the current NN cannot be bettered. Not being inherently limited by dimensionality, this approach appears particularly attractive for general purpose applications.

Other rejection rules are considered by Parsi and Kanal (1985). A simpler form of clustering is used in the variant of this algorithm proposed by Niemann and Goppert (1988). A shallow MDBST is used, followed by a variant on the branching and bounding described above.

Bennett et al. (1999) use the nearest neighbor problem as a means towards solving the Gaussian distribution mixture problem. They consider a preprocessing approach similar to Fukunaga and Narendra (1975) but with an important difference: to take better account of cluster structure in the data, the clusters are multivariate normal but not necessarily of diagonal covariance structure. Therefore very elliptical clusters are allowed. This in turn implies that a cluster radius is not of great benefit for establishing a bound on whether or not distances need to be calculated. Bennett et al. (1999) address this problem by seeking a stochastic guarantee on whether or not calculations can be excluded. Technically, however, such stochastic bounds are not easy to determine in a high dimensional space.

An interesting issue raised in Beyer et al. (1999) is discussed also by Bennett et al. (1999): if the ratio of the nearest and furthest neighbor distances converges in probability to 1 as the dimensionality increases, then is it meaningful to search for nearest neighbors? This issue is not all that different from saying that neighbors in an increasingly high dimensional space tend towards being equidistant. In section 5, we will look at approaches for handling particular classes of data of this type.

4.3. Fast Approximate Nearest Neighbor Finding

Kushilevitz et al. (1998), working in Euclidean and L_1 spaces, propose fast approximate nearest neighbor searching, on the grounds that in systems for content-based image retrieval, approximate results are adequate. Projections are used to bound the search. Probability of successfully finding the nearest neighbor is traded off against time and space requirements.

5. The Special Case of Sparse Binary Data

"High-dimensional", "sparse" and "binary" are the characteristics of keyword-based bibliographic data, with maybe values in excess of 10000 for both n and m. Such data is usually stored as list data structures, representing the mapping of documents onto index terms, or vice versa. Commercial document collections are usually searched using a Boolean search environment. Documents associated with particular terms are retrieved, and the intersection (AND), union (OR) or other operations on such sets of documents are obtained. For efficiency, an *inverted file* which maps terms onto documents must be available for Boolean retrieval. The efficient NN algorithms, to be discussed, make use of both the document-term and the term-document files.

The usual algorithm for NN-searching considers each document in turn, calculates the distance with the given document, and updates the NN if appropriate. This algorithm is shown schematically in Fig. 14.4

CLUSTERING IN MASSIVE DATA SETS

Usual algorithm:

```
Initialize current NN
For all documents in turn do:
... For all terms associated with the document do:
... ... Determine (dis)similarity
... Endfor
... Test against current NN
Endfor
```

Croft's algorithm:

```
Initialize current NN
For all terms associated with the given document do:
... For all documents associated with each term do:
... ... For all terms associated with a document do:
... ... ... Determine (dis)similarity
... ... Endfor
... ... Test against current NN
... Endfor
Endfor
```

Perry-Willett algorithm:

```
Initialize current NN
For all terms associated with the given document, i, do:
... For all documents, i', associated with each term, do:
... ... Increment location i' of counter vector
... Endfor
Endfor
```

Figure 14.4. Algorithms for NN-searching using high-dimensional sparse binary data.

(top). The inner loop is simply an expression of the fact that the distance or similarity will, in general, require $O(m)$ calculation: examples of commonly used coefficients are the Jaccard similarity, and the Hamming (L_1 Minkowski) distance.

If $\bar{m}$ and $\bar{n}$ are, respectively, the average numbers of terms associated with a document, and the average number of documents associated with a term, then an average complexity measure, over n searches, of this usual algorithm is $O(n\bar{m})$. It is assumed that advantage is taken of some packed form of storage in the inner loop (e.g. using linked lists).

Croft's algorithm (see Croft (1977), and Fig. 14.4) is of worst case complexity $O(nm^2)$. However the number of terms associated with the document whose NN is required will often be quite small. The National Physical Laboratory test collection, for example, which was used in Murtagh (1982) has the following characteristics: $n = 11429$, $m = 7491$, $\bar{m} = 19.9$, and $\bar{n} = 30.4$. The outermost and innermost loops in Croft's algorithm use the document-term file. The center loop uses the term-document inverted file. An average complexity measure (more strictly, the time taken for best match search based on an average document with associated average terms) is seen to be $O(\bar{n}\bar{m}^2)$.

In the outermost loop of Croft's algorithm, there will eventually come about a situation where – if a document has not been thus far examined – the number of terms remaining for the given document do not permit the current NN document to be bettered. In this case, we can cut short the iterations of the outermost loop. The calculation of a bound, using the greatest possible number of terms which could be shared with a so-far unexamined document has been exploited by Smeaton and van Rijsbergen (1981) and by Murtagh (1982) in successive improvements on Croft's algorithm.

The complexity of all the above algorithms has been measured in terms of operations to be performed. In practice, however, the actual accessing of term or document information may be of far greater cost. The document-term and term-document files are ordinarily stored on direct access file storage because of their large sizes. The strategy used in Croft's algorithm, and in improvements on it, does not allow any viable approaches to batching together the records which are to be read successively, in order to improve accessing-related performance.

The Perry-Willett algorithm (see Perry and Willett (1983)) presents a simple but effective solution to the problem of costly I/O. It focuses on the calculation of the number of terms common to the given document x and each other document, y, in the document collection. This set of values is built up in a computationally efficient fashion. $O(n)$ operations are subsequently required to determine the (dis)similarity, using another vector comprising the total numbers of terms associated with each document. Computation time (the same "average" measure as that used above) is $O(\bar{n}\bar{m} + n)$. We now turn attention to numbers of direct-access reads required.

In Croft's algorithm, all terms associated with the document whose NN is desired may be read in one read operation. Subsequently, we require $\bar{n}\bar{m}$ reads, giving in all $1+\bar{n}\bar{m}$. In the Perry-Willett algorithm, the outer loop again pertains to the one (given) document, and so all terms associated with this document can be read and stored. Subsequently, $\bar{m}$ reads, i.e. the average number of terms, each of which demands a read of a set of documents, are required. This gives, in all, $1+\bar{m}$. Since these reads are very much the costliest operation in practice, the Perry-Willett algorithm can be recommended for large values of n and m. Its general characteristics are that (i) it requires, as do all the algorithms discussed in this section, the availability of the inverted term-document file; and (ii) it requires in-memory storage of two vectors containing n integer values.

6. Hierarchical Agglomerative Clustering

The algorithms discussed in this section can be characterized as *greedy* (Horowitz and Sahni 1979). A sequence of irreversible algorithm steps is used to construct the desired data structure.

We will not review hierarchical agglomerative clustering here. For essential background, the reader is referred to Murtagh and Heck (1987), Gordon (1999), or Jain and Dubes (1988). This section borrows on Murtagh (1992).

One could practically say that Sibson (1973) and Defays (1977) are part of the prehistory of clustering. Their $O(n^2)$ implementations of the single link method and of a (non-unique) complete link method, respectively, have been widely cited.

In the early 1980s a range of significant improvements were made to the Lance-Williams, or related, dissimilarity update schema (de Rham 1980, Juan 1982), which had been in wide use since the mid-1960s. Murtagh (1985) presents a survey of these algorithmic improvements. We will briefly describe them here. The new algorithms, which have the potential for *exactly* replicating results found in the classical but more computationally expensive way, are based on the construction of *nearest neighbor chains* and *reciprocal* or mutual NNs (NN-chains and RNNs).

A NN-chain consists of an arbitrary point (a in Fig. 14.5); followed by its NN (b in Fig. 14.5); followed by the NN from among the remaining points (c, d, and e in Fig. 14.5) of this second point; and so on until we necessarily have some pair of points which can be termed reciprocal or mutual NNs. (Such a pair of RNNs may be the first two points in the chain; and we have assumed that no two dissimilarities are equal.)

a   b   c   d   e

Figure 14.5. Five points, showing *NN*s and *RNN*s.

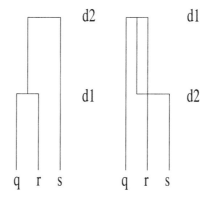

Figure 14.6. Alternative representations of a hierarchy with an inversion. Assuming dissimilarities, as we go vertically up, criterion values (d_1, d_2) decrease. But here, undesirably, $d_2 > d_1$.

In constructing a NN-chain, irrespective of the starting point, we may agglomerate a pair of RNNs as soon as they are found. What guarantees that we can arrive at the same hierarchy as if we used traditional "stored dissimilarities" or "stored data" algorithms? Essentially this is the same condition as that under which no inversions or reversals are produced by the clustering method. Fig. 6 gives an example of this, where s is agglomerated at a lower criterion value (i.e. dissimilarity) than was the case at the previous agglomeration between q and r. Our ambient space has thus contracted because of the agglomeration. This is due to the algorithm used – in particular the agglomeration criterion – and it is something we would normally wish to avoid.

This is formulated as:

Inversion impossible if: $d(i,j) < d(i,k)$ or $d(j,k) \Rightarrow d(i,j) < d(i \cup j, k)$

This is essentially Bruynooghe's *reducibility property* (Bruynooghe (1977); see also Murtagh (1984)). Using the Lance–Williams dissimilarity update formula, it can be shown that the minimum variance method does not give rise to inversions; neither do the linkage methods; but the median and centroid methods cannot be guaranteed not to have inversions.

To return to Fig. 14.5, if we are dealing with a clustering criterion which precludes inversions, then c and d can justifiably be agglomerated, since no other point (for example, b or e) could have been agglomerated to either of these.

The processing required, following an agglomeration, is to update the NNs of points such as b in Fig.14.5 (and on account of such points, this algorithm was dubbed *algorithme des célibataires* in de Rham (1980)). The following is a summary of the algorithm:

NN-chain algorithm

Step 1 Select a point arbitrarily.

Step 2 Grow the NN-chain from this point until a pair of RNNs are obtained.

Step 3 Agglomerate these points (replacing with a cluster point, or updating the dissimilarity matrix).

Step 4 From the point which preceded the RNNs (or from any other arbitrary point if the first two points chosen in Steps 1 and 2 constituted a pair of RNNs), return to Step 2 until only one point remains.

In Murtagh (1983; 1984; 1985) and Day and Edelsbrunner (1984), one finds discussions of $O(n^2)$ time and $O(n)$ space implementations of Ward's minimum variance (or error sum of squares) method and of the centroid and median methods. The latter two methods are termed the UPGMC and WPGMC criteria by Sneath and Sokal (1973). Now, a problem with the cluster criteria used by these latter two methods is that the reducibility property is not satisfied by them. This means that the hierarchy constructed may not be unique as a result of inversions or reversals (non-monotonic variation) in the clustering criterion value determined in the sequence of agglomerations.

Murtagh (1984) describes $O(n^2)$ time and $O(n^2)$ space implementations for the single link method, the complete link method and for the weighted and unweighted group average methods (WPGMA and UPGMA). This approach is quite general vis à vis the dissimilarity used and can also be used for hierarchical clustering methods other than those mentioned.

Day and Edelsbrunner (1984) prove the exact $O(n^2)$ time complexity of the centroid and median methods using an argument related to the combinatorial problem of optimally packing hyperspheres into an m-dimensional volume. They also address the question of metrics: results are valid in a wide class of distances including those associated with the Minkowski metrics.

The construction and maintenance of the nearest neighbor chain as well as the carrying out of agglomerations whenever reciprocal nearest neighbors meet, both offer possibilities for parallelization. Implementations on a SIMD machine were described by Willett (1989).

Evidently both coordinate data and graph (e.g., dissimilarity) data can be input to these agglomerative methods. Gillet et al. (1998) in the context of clustering chemical structure databases refer to the common use of the Ward method, based on the reciprocal nearest neighbors algorithm, on data sets of a few hundred thousand molecules.

Applications of hierarchical clustering to bibliographic information retrieval are assessed in Griffiths et al. (1984). Ward's minimum variance criterion is favored.

From details in White and McCain (1997), the Institute of Scientific Information (ISI) clusters citations (science, and social science) by first clustering highly cited documents based on a single linkage criterion, and then four more passes are made through the data to create a subset of a single linkage hierarchical clustering.

7. Graph Clustering

Hierarchical clustering methods are closely related to graph-based clustering. For a start, a dendrogram is a rooted labeled tree. Secondly, and more importantly, some methods like the sinlge and complete link methods can be displayed as graphs, and are very closely related to mainstream graph data structures.

An example of the increasing prevalence of graph clustering in the context of data mining on the web is presented in Fig. 14.7: Amazon.com provides information on what other books were purchased by like-minded individuals.

The single link method was referred to in the previous section, as a widely-used agglomerative, hence hierarchical, clustering method. Rohlf (1982) reviews algorithms for the single link method with complexities ranging from $O(n \log n)$ to $O(n^5)$. The criterion used by the single link

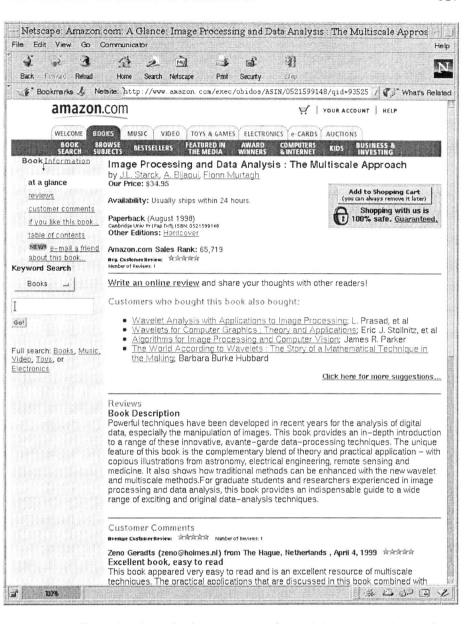

Figure 14.7. Example of graph clustering in a data mining perspective at Amazon.com: "Customers who bought this book also bought..."

method for cluster formation is weak, meaning that noisy data in particular give rise to results which are not robust.

The minimal spanning tree (MST) and the single link agglomerative clustering method are closely related: the MST can be transformed irreversibly into the single link hierarchy Rohlf (1973). The MST is defined as of minimal total weight, it spans all nodes (vertices) and is an unrooted tree. The MST has been a method of choice for at least four decades now either in its own right for data analysis (Zahn 1971), as a data structure to be approximated (e.g. using shortest spanning paths, see Murtagh (1985), p. 96), or as a basis for clustering. We will look at some fast algorithms for the MST in the remainder of this section.

Perhaps the most basic MST algorithm, due to Prim and Dijkstra, grows a single fragment through $n-1$ steps. We find the closest vertex to an arbitrary vertex, calling these a fragment of the MST. We determine the closest vertex, not in the fragment, to any vertex in the fragment, and add this new vertex into the fragment. While there are fewer than n vertices in the fragment, we continue to grow it.

This algorithm leads to a unique solution. A default $O(n^3)$ implementation is clear, and $O(n^2)$ computational cost is possible (Murtagh (1985), p. 98).

Sollin's algorithm constructs the fragments in parallel. For each fragment in turn, at any stage of the construction of the MST, determine its closest fragment. Merge these fragments, and update the list of fragments. A tree can be guaranteed in this algorithm (although care must be taken in cases of equal similarity) and our other requirements (all vertices included, minimal total edge weight) are very straightforward. Given the potential for roughly halving the data remaining to be processed at each step, not surprisingly the computational cost reduces from $O(n^3)$ to $O(n^2 \log n)$.

The real interest of Sollin's algorithm arises when we are clustering on a graph and do not have all $n(n-1)/2$ edges present. Sollin's algorithm can be shown to have computational cost $m \log n$ where m is the number of edges. When $m \ll n(n-1)/2$ then we have the potential for appreciable gains.

The MST in feature spaces can of course make use of the fast nearest neighbor finding methods studied earlier in this article. See section 4.4 of Murtagh (1985) for various examples.

Other graph data structures which have been proposed for data analysis are related to the MST. We know, for example, that the following subset relationship holds:

$$\text{MST} \subseteq \text{RNG} \subseteq \text{GG} \subseteq \text{DT}$$

where RNG is the relative neighborhood graph, GG is the Gabriel graph, and DT is the Delaunay triangulation. The latter, in the form of it's dual, the Voronoi diagram, has been used for analyzing the clustering of galaxy locations. References to these and related methods can be found in Murtagh (1993a).

8. Nearest Neighbor Finding on Graphs

Clustering on graphs may be required because we are working with (perhaps complex non-Euclidean) dissimilarities. In such cases where we must take into account an edge between each and every pair of vertices, we will generally have an $O(m)$ computational cost where m is the number of edges. In a metric space we have seen that we can look for various possible ways to expedite the nearest neighbor search. An approach based on visualization – turning our data into an image – will be looked at below. However there is another aspect of our similarity (or other) graph which we may be able to turn to our advantage. Efficient algorithms for sparse graphs are available. Sparsity can be arranged – we can threshold our edges if the sparsity does not suggest itself more naturally. A special type of sparse graph is a planar graph, i.e. a graph capable of being represented in the plane without any crossovers of edges.

For sparse graphs, algorithms with $O(m \log \log n)$ computational cost were described by Yao (1975) and Cheriton and Tarjan (1976). A short algorithmic description can be found in Murtagh (1985) (pp. 107–108) and we refer in particular to the latter.

The basic idea is to preprocess the graph, in order to expedite the sorting of edge weights (why sorting? – simply because we must repeatedly find smallest links, and maintaining a sorted list of edges is a good basis for doing this). If we were to sort all edges, the computational requirement would be $O(m \log m)$. Instead of doing that, we take the edge set associated with each and every vertex. We divide each such edge set into groups of size k. (The fact that the last such group will usually be of size $< k$ is taken into account when programming.)

Let n_v be the number of incident edges at vertex v, such that $\sum_v n_v = 2m$.

The sorting operation for each vertex now takes $O(k \log k)$ operations for each group, and we have n_v/k groups. For all vertices the sorting requires a number of operations which is of the order of $\sum_v n_v \log k = 2m \log k$. This looks like a questionable – or small – improvement over $O(m \log m)$.

Determining the lightest edge incident on a vertex requires $O(n_v/k)$ comparisons since we have to check all groups. Therefore the lightest edges incident on all vertices are found with $O(m/k)$ operations.

When two vertices, and later fragments, are merged, their associated groups of edges are simply collected together, therefore keeping the total number of groups of edges which we started out with. We will bypass the issue of edges which, over time, are to be avoided because they connect vertices in the same fragment: given the fact that we are building an MST, the total number of such edges-to-be-avoided cannot surpass $2m$.

To find what to merge next, again $O(m/k)$ processing is required. Using Sollin's algorithm, the total processing required in finding what to merge next is $O(m/k \ \log n)$. The total processing required for grouping the edges, and sorting within the edge-groups, is $O(m \log k)$, i.e. it is one-off and accomplished at the start of the MST-building process.

The total time is $O(m/k \ \log n) + O(m \log k)$. Let's fix $k = \log n$ (aha!). Then the second term dominates and gives overall computational complexity as $O(m \log \log n)$.

This result has been further improved to near linearity in m by Gabow et al. (1986), who develop an algorithm with complexity $O(m \log \log \log \ldots n)$ where the number of iterated log terms is bounded by m/n.

Motwani and Raghavan (1995) (chapter 10) base a stochastic $O(m)$ algorithm for the MST on random sampling to identify and eliminate edges that are guaranteed not to belong to the MST.

Let's turn our attention now to the case of a planar graph. For a planar graph we know that $m \leq 3n - 6$ for $m > 1$. (For proof, see for example Tucker, 1980, or any book on graph theory).

Referring to Sollin's algorithm, described above, $O(n)$ operations are needed to establish a least cost edge from each vertex, since there are only $O(n)$ edges present. On the next round, following fragment-creation, there will be at most ceil($n/2$) new vertices, implying of the order of $n/2$ processing to find the least cost edge (where ceil is the ceiling function, or smallest integer greater than the argument). The total computational cost is seen to be proportional to: $n + n/2 + n/4 + \ldots = O(n)$.

So determining the MST of a planar graph is linear in numbers of either vertices or edges.

Before ending this review of very efficient clustering algorithms for graphs, we note that algorithms discussed so far have assumed that the similarity graph was undirected. For modeling transport flows, or economic transfers, the graph could well be directed. Components can be defined, generalizing the clusters of the single link method, or the complete link method. Tarjan (1983) provides an algorithm for the latter agglomerative criterion which is of computational cost $O(m \log n)$.

9. K-Means and Family

The non-technical person more often than not understands clustering as a partition. K-means looked at in this section, or the distribution mixture approach looked at section 10, provide solutions.

A mathematical definition of a partition implies no multiple assignments of observations to clusters, i.e. no overlapping clusters. Overlapping clusters may be faster to determine in practice, and a case in point is the one-pass algorithm described in Salton and McGill (1983). The general principle followed is: make one pass through the data, assigning each object to the first cluster which is close enough, and making a new cluster for objects that are not close enough to any existing cluster.

Broder et al. (1997) use this algorithm for clustering the web. A feature vector is determined for each HTML document considered, based on sequences of words. Similarity between documents is based on an inverted list, using an approach like those described in section 5. The similarity graph is thresholded, and components sought.

Broder (1998) solves the same clustering objective using a thresholding and overlapping clustering method similar to the Salton and McGill one. The application described is that of clustering the Altavista repository in April 1996, consisting of 30 million HTML and text documents, comprising 150 GBytes of data. The number of serviceable clusters found was 1.5 million, containing 7 million documents. Processing time was about 10.5 days. An analysis of the clustering algorithm used by Broder can be found in Borodin et al. (1999), who also considers the use of approximate minimal spanning trees.

The threshold-based pass of the data, in its basic state, is susceptible to lack of robustness. A bad choice of threshold leads to too many clusters or two few. To remedy this, we can work on a well-defined data structure such as the minimal spanning tree. Or, alternatively, we can iteratively refine the clustering. Partitioning methods, such as k-means, use iterative improvement of an initial estimation of a targeted clustering.

A very widely used family of methods for inducing a partition on a data set is called k-means, c-means (in the fuzzy case), Isodata, competitive learning, vector quantization and other more general names (non-overlapping non-hierarchical clustering) or more specific names (minimal distance or exchange algorithms).

The usual criterion to be optimized is:

$$\frac{1}{|I|} \sum_{q \in Q} \sum_{i \in q} \|\vec{i} - \vec{q}\|^2$$

where I is the object set, $|\,.\,|$ denotes cardinality, q is some cluster, Q is the partition, and q denotes a set in the summation, whereas $\vec{q}$ denotes some associated vector in the error term, or metric norm. This criterion ensures that clusters found are compact, and therefore assumed homogeneous. The optimization criterion, by a small abuse of terminology, is ofter referred to as a minimum variance one.

A necessary condition that this criterion be optimized is that vector $\vec{q}$ be a cluster mean, which for the Euclidean metric case is:

$$\vec{q} = \frac{1}{|q|} \sum_{i \in q} \vec{\imath}$$

A batch update algorithm, due to Lloyd (1982), Forgy (1965), and others, makes assignments to a set of initially randomly-chosen vectors, $\vec{q}$, as step 1. Step 2 updates the cluster vectors, $\vec{q}$. This is iterated. The distortion error, equation 1, is non-increasing, and a local minimum is achieved in a finite number of iterations.

An online update algorithm is due to MacQueen (1976). After each presentation of an observation vector, $\vec{\imath}$, the closest cluster vector, $\vec{q}$, is updated to take account of it. Such an approach is well-suited for a continuous input data stream (implying "online" learning of cluster vectors).

Both algorithms are gradient descent ones. In the online case, much attention has been devoted to best learning rate schedules in the neural network (competitive learning) literature: Darken and Moody (1991; 1992), Darken et al. (1992), Fritzke (1997).

A difficulty, less controllable in the case of the batch algorithm, is that clusters may become (and stay) empty. This may be acceptable, but also may be in breach of our original problem formulation. An alternative to the batch update algorithm is Späth (1985) exchange algorithm. Each observation is considered for possible assignment into any of the *other* clusters. Updating and "downdating" formulas are given by Späth. This exchange algorithm is stated to be faster to converge and to produce better (smaller) values of the objective function. Over decades of use, we have also verified that it is a superior algorithm to the minimal distance one.

K-means is very closely related to Voronoi (Dirichlet) tesselations, to Kohonen self-organizing feature maps, and various other methods.

The batch learning algorithm above may be viewed as

1. An assignment step which we will term the E (estimation) step: estimate the posteriors,

$$P(\text{observations} \mid \text{cluster centres})$$

2 A cluster update step, the M (maximization) step, which maximizes a cluster center likelihood.

Neal and Hinton (1998) cast the k-means optimization problem in such a way that both E- and M-steps monotonically increase the maximand's values. The EM algorithm may, too, be enhanced to allow for online as well as batch learning (Sato and Ishii 1999).

In Thiesson et al. (1999), k-means is implemented (i) by traversing blocks of data, cyclically, and incrementally updating the sufficient statistics and parameters, and (ii) instead of cyclic traversal, sampling from subsets of the data is used. Such an approach is admirably suited for very large data sets, where in-memory storage is not feasible. Examples used by Thiesson et al. (1999) include the clustering of a half million 300-dimensional records.

10. Fast Model-Based Clustering

It is traditional to note that models and (computational) speed don't mix. We review recent progress in this section.

10.1. Modeling of Signal and Noise

A simple and applicable model is a distribution mixture, with the signal modeled by Gaussians, in the presence of Poisson background noise.

Consider data which are generated by a mixture of $(G-1)$ bivariate Gaussian densities, $f_k(x;\theta) \sim \mathcal{N}(\mu_k, \Sigma_k)$, for clusters $k = 2, \ldots, G$, and with Poisson background noise corresponding to $k = 1$. The overall population thus has the mixture density

$$f(x;\theta) = \sum_{k=1}^{G} \pi_k f_k(x;\theta)$$

where the mixing or prior probabilities, π_k, sum to 1, and $f_1(x;\theta) = \mathcal{A}^{-1}$, where $\mathcal{A}$ is the area of the data region. This is the basis for *model-based clustering* (Banfield and Raftery 1993, Dasgupta and Raftery 1998, Murtagh and Raftery 1984, Banerjee and Rosenfeld 1993).

The parameters, θ and π, can be estimated efficiently by maximizing the mixture likelihood

$$L(\theta, \pi) = \prod_{i=1}^{n} f(x_i; \theta),$$

with respect to θ and π, where x_i is the i-th observation.

Now let us assume the presence of two clusters, one of which is Poisson noise, the other Gaussian. This yields the mixture likelihood

$$L(\theta,\pi) = \prod_{i=1}^{n}\left[\pi_1 \mathcal{A}^{-1} + \pi_2 \frac{1}{2\pi\sqrt{|\Sigma|}}\exp\left\{-\frac{1}{2}(x_i-\mu)^T\Sigma^{-1}(x_i-\mu)\right\}\right],$$

where $\pi_1 + \pi_2 = 1$.

An iterative solution is provided by the expectation-maximization (EM) algorithm of Dempster et al. (1977). We have already noted this algorithm in informal terms in the last section, dealing with k-means. Let the "complete" (or "clean" or "output") data be $y_i = (x_i, z_i)$ with indicator set $z_i = (z_{i1}, z_{i2})$ given by $(1,0)$ or $(0,1)$. Vector z_i has a multinominal distribution with parameters $(1; \pi_1, \pi_2)$. This leads to the *complete data log-likelihood*:

$$l(y,z;\theta,\pi) = \Sigma_{i=1}^{n}\Sigma_{k=1}^{2} z_{ik}[\log \pi_k + \log f_k(x_k;\theta)]$$

The E-step then computes $\hat{z}_{ik} = E(z_{ik} \mid x_1, \ldots, x_n, \theta)$, i.e. the posterior probability that x_i is in cluster k. The M-step involves maximization of the *expected complete data log-likelihood*:

$$l^*(y;\theta,\pi) = \Sigma_{i=1}^{n}\Sigma_{k=1}^{2} \hat{z}_{ik}[\log \pi_k + \log f_k(x_i;\theta)].$$

The E- and M-steps are iterated until convergence.

For the 2-class case (Poisson noise and a Gaussian cluster), the complete data likelihood is

$$L(y,z;\theta,\pi) = \prod_{i=1}^{n}\left[\frac{\pi_1}{\mathcal{A}}\right]^{z_{i1}}\left[\frac{\pi_2}{2\pi\sqrt{|\Sigma|}}\exp\left\{-\frac{1}{2}(x_i-\mu)^T\Sigma^{-1}(x_i-\mu)\right\}\right]^{z_{i2}}$$

The corresponding expected log-likelihood is then used in the EM algorithm. This formulation of the problem generalizes to the case of G clusters, of arbitrary distributions and dimensions.

Fraley (1999) discusses implementation of model-based clustering, including publicly available software.

In order to assess the evidence for the presence of a signal-cluster, we use the *Bayes factor* for the mixture model, M_2, that includes a Gaussian density as well as background noise, against the "null" model, M_1, that contains only background noise. The Bayes factor is the posterior odds for the mixture model against the pure noise model, when neither is favored a priori. It is defined as $B = p(x|M_2)/p(x|M_1)$, where $p(x|M_2)$ is the *integrated likelihood* of the mixture model M_2, obtained by integrating over the parameter space. For a general review of Bayes factors, their use in applied statistics, and how to approximate and compute them, see Kass and Raftery (1995).

We approximate the Bayes factor using the *Bayesian Information Criterion* (BIC) (Schwarz 1978). For a Gaussian cluster and Poisson noise, this takes the form:

$$2\log B \approx BIC = 2\log L(\hat{\theta}, \hat{\pi}) + 2n \log \mathcal{A} - 6\log n,$$

where $\hat{\theta}$ and $\hat{\pi}$ are the maximum likelihood estimators of θ and π, and $L(\hat{\theta}, \hat{\pi})$ is the maximized mixture likelihood.

A review of the use of the BIC criterion for model selection – and more specifically for choosing the number of clusters in a data set – can be found in Fraley and Raftery (1999).

An application of mixture modeling and the BIC criterion to gamma-ray burst data can be found in Mukherjee et al. (1998). So far around 800 observations have been assessed, but as greater numbers become available we will find the inherent number of clusters in a similar way, in order to try to understand more about the complex phenomenon of gamma-ray bursts.

10.2. Application to Thresholding

Consider an image or a planar or 3-dimensional set of object positions. For simplicity we consider the case of setting a single threshold in the image intensities, or the point set's spatial density.

We deal with a combined mixture density of two *univariate* Gaussian distributions $f_k(x;\theta) \sim \mathcal{N}(\mu_k, \sigma_k)$. The overall population thus has the mixture density

$$f(x;\theta) = \sum_{k=1}^{2} \pi_k f_k(x;\theta)$$

where the mixing or prior probabilities, π_k, sum to 1.

When the mixing proportions are assumed equal, the log-likelihood takes the form

$$l(\theta) = \sum_{i=1}^{n} \ln \left[\sum_{k=1}^{2} \frac{1}{2\pi\sqrt{|\sigma_k|}} \exp\left\{ -\frac{1}{2\sigma_k}(x_i - \mu_k)^2 \right\} \right]$$

The EM algorithm is then used to iteratively solve this (Celeux and Govaert 1995). This method is used for appraisals of textile (jeans and other fabrics) fault detection in Campbell et al. (1999). Industrial vision inspection systems potentially produce large data streams, and fault detection can be a good application for fast clustering methods. We are currently using a mixture model of this sort on SEM (scanning electron microscope) images of cross-sections of concrete to allow for subsequent characterization of physical properties.

Image segmentation, per se, is a relatively straightforward application, but there are novel and interesting aspects to the two studies mentioned. In the textile case, the faults are very often perceptual and relative, rather than "absolute" or capable of being analyzed in isolation. In the SEM imaging case, a first phase of processing is applied to de-speckle the images, using multiple resolution noise filtering.

Turning from concrete to cosmology, the Sloan Digital Sky Survey (SDSS 1999) is producing a sky map of more than 100 million objects, together with 3-dimensional information (redshifts) for a million galaxies. Pelleg and Moore (1999) describe mixture modeling, using a k-D tree preprocessing to expedite the finding of the class (mixture) parameters, e.g. means, covariances.

11. Noise Modeling

In Starck et al. (1998) and in a wide range of papers, we have pursued an approach for the noise modeling of observed data. A multiple resolution scale vision model or data generation process is used, to allow for the phenomenon being observed on different scales. In addition, a wide range of options are permitted for the data generation transfer path, including additive and multiplicative, stationary and non-stationary, Gaussian ("read out" noise), Poisson (random shot noise), and so on.

Given point pattern clustering in two- or three-dimensional spaces, we will limit our overview here to the Poisson noise case.

11.1. Poisson noise with few events using the à trous transform

If a wavelet coefficient $w_j(x,y)$ is due to noise, it can be considered as a realization of the sum $\sum_{k \in K} n_k$ of independent random variables with the same distribution as that of the wavelet function (n_k being the number of events used for the calculation of $w_j(x,y)$). This allows comparison of the wavelet coefficients of the data with the values which can be taken by the sum of n independent variables.

The distribution of one event in wavelet space is then directly given by the histogram H_1 of the wavelet ψ. As we consider independent events, the distribution of a coefficient w_n (note the changed subscripting for w, for convenience) related to n events is given by n autoconvolutions of H_1:

$$H_n = H_1 \otimes H_1 \otimes ... \otimes H_1$$

For a large number of events, H_n converges to a Gaussian.

Figure 14.8. Data in the plane. The 256 × 256 image shows 550 "signal" points – two Gaussian-shaped clusters in the lower left and in the upper right – with in addition 40,000 Poisson noise points added. Details of recovery of the clusters is discussed in Murtagh and Starck (1998).

Fig. 14.8 shows an example of where point pattern clusters – density bumps in this case – are sought, with a great amount of background clutter. Murtagh and Starck (1998) refer to the fact that there is no computational dependence on the number of points (signal or noise) in such a problem, when using a wavelet transform with noise modeling.

Some other alternative approaches will be briefly noted. The Haar transform presents the advantage of its simplicity for modeling Poisson noise. Analytic formulas for wavelet coefficient distributions have been derived by Kolaczyk (1997), and Jammal and Bijaoui (1999). Using a new wavelet transform, the Haar à trous transform, Zheng et al. (1999) appraise a denoising approach for financial data streams, – an important preliminary step for subsequent clustering, forecasting, or other processing.

11.2. Poisson noise with nearest neighbor clutter removal

The wavelet approach is certainly appropriate when the wavelet function reflects the type of object sought (e.g. isotropic), and when superimposed point patterns are to be analyzed. However, non-superimposed point patterns of complex shape are very well treated by the approach described in Byers and Raftery (1998). Using a homogeneous Poisson noise model, they derive the distribution of the distance of a point to its kth nearest neighbor.

Next, Byers and Raftery (1998) consider the case of a Poisson process which is signal, superimposed on a Poisson process which is clutter. The kth nearest neighbor distances are modeled as a mixture distribution: a histogram of these, for given k, will yield a bimodal distribution if our assumption is correct. This mixture distribution problem is solved using the EM algorithm. Generalization to higher dimensions, e.g. 10, is also discussed.

Similar data was analyzed by noise modeling and a Voronoi tesselation preprocessing of the data in Allard and Fraley (1997). It is pointed out there how this can be a very useful approach where the Voronoi tiles have meaning in relation to the morphology of the point patterns. However, it does not scale well to higher dimensions, and the statistical noise modeling is approximate. Ebeling and Wiedenmann (1993), reproduced in Dobrzycki et al. (1999), propose the use of a Voronoi tesselation for astronomical X-ray object detection and characterization.

12. Cluster-Based User Interfaces

Information retrieval by means of "semantic road maps" was first detailed by Doyle (1961). The spatial metaphor is a powerful one in human information processing. The spatial metaphor also lends itself well to modern distributed computing environments such as the web. The Kohonen self-organizing feature map (SOM) method is an effective means towards this end of a visual information retrieval user interface. We will also provide an illustration of web-based semantic maps based on hyperlink clustering.

The Kohonen map is, at heart, k-means clustering with the additional constraint that cluster centers be located on a regular grid (or some other topographic structure) and furthermore their location on the grid be monotonically related to pairwise proximity (Murtagh and Pajares

1995). The nice thing about a regular grid output representation space is that it lends itself well as a visual user interface.

Fig. 14.9 shows a visual and interactive user interface map, using a Kohonen self-organizing feature map (SOM). Color is related to density of document clusters located at regularly-spaced nodes of the map, and some of these nodes/clusters are annotated. The map is installed as a clickable imagemap, with CGI programs accessing lists of documents and – through further links – in many cases, the full documents. In the example shown, the user has queried a node and results are seen in the right-hand panel. Such maps are maintained for (currently) 12000 articles from the *Astrophysical Journal*, 7000 from *Astronomy and Astrophysics*, over 2000 astronomical catalogs, and other data holdings. More information on the design of this visual interface and user assessment can be found in Poinçot et al. (1998; 1999).

Figure 14.9. Visual interactive user interface to the journal *Astronomy and Astrophysics*. Original in color.

Guillaume (Guillaume and Murtagh 1999) developed a Java-based visualization tool for hyperlink-based data, consisting of astronomers, astronomical object names, article titles, and with the possibility of other objects (images, tables, etc.). Through weighting, the various types of links could be prioritized. An iterative refinement algorithm was developed to map the nodes (objects) to a regular grid of cells, which as for the Kohonen SOM map, are clickable and provide access to the data represented by the cluster. Fig. 10 shows an example for an astronomer (Prof. Jean Heyvaerts, Strasbourg Astronomical Observatory).

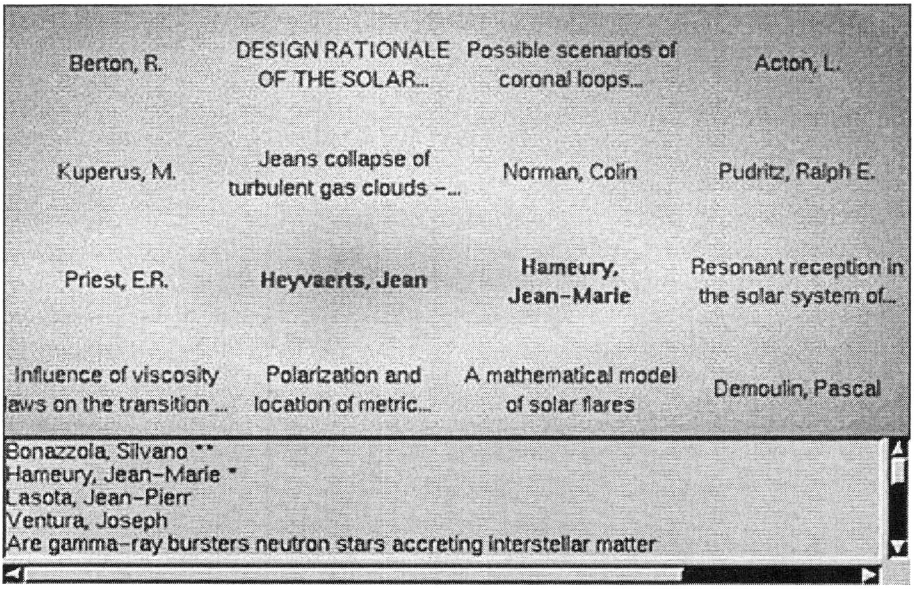

Berton, R.	DESIGN RATIONALE OF THE SOLAR...	Possible scenarios of coronal loops...		Acton, L.
Kuperus, M.	Jeans collapse of turbulent gas clouds –...	Norman, Colin		Pudritz, Ralph E.
Priest, E.R.	Heyvaerts, Jean	Hameury, Jean-Marie		Resonant reception in the solar system of...
Influence of viscosity laws on the transition ...	Polarization and location of metric...	A mathematical model of solar flares		Demoulin, Pascal

Bonazzola, Silvano **
Hameury, Jean-Marie *
Lasota, Jean-Pierr
Ventura, Joseph
Are gamma-ray bursters neutron stars accreting interstellar matter

Figure 14.10. Visual interactive user interfaces, based on graph edges. Map for astronomer Jean Heyvaerts. Original in color.

These new cluster-based visual user interfaces are not computationally demanding. They are not however scalable in their current implementation. Document management (see e.g. Cartia (1999)) is less the motivation as is instead the interactive user interface.

13. Images from Data

It is quite impressive how 2D (or 3D) image signals can handle with ease the scalability limitations of clustering and many other data processing operations. The contiguity imposed on adjacent pixels bypasses the need for nearest neighbor finding. It is very interesting therefore to consider the feasibility of taking problems of clustering massive data sets into the 2D image domain. We will look at a few recent examples of work in this direction.

Church and Helfman (1993) address the problem of visualizing possibly millions of lines of computer program code, or text. They consider an approach borrowed from DNA sequence analysis. The data sequence is tokenized by splitting it into its atoms (line, word, character, etc.) and then placing a dot at position i, j if the ith input token is the same as the jth. The resulting dotplot, it is argued, is not limited by the available display screen space, and can lead to discovery of large-scale structure in the data.

When data do not have a sequence we have an invariance problem which can be resolved by finding some row and column permutation which pulls large array values together, and perhaps furthermore into proximity to an array diagonal. Berry et al. (1996) have studied the case of large sparse arrays. Gathering larger (or nonzero) array elements to the diagonal can be viewed in terms of minimizing the envelope of nonzero values relative to the diagonal. This can be formulated and solved in purely symbolic terms by reordering vertices in a suitable graph representation of the matrix. A widely-used method for symmetric sparse matrices is the Reverse Cuthill-McKee (RCM) method.

The complexity of the RCM method for ordering rows or columns is proportional to the product of the maximum degree of any vertex in the graph representing the array values and the total number of edges (nonzeroes in the matrix). For hypertext matrices with small maximum degree, the method would be extremely fast. The strength of the method is its low time complexity but it does suffer from certain drawbacks. The heuristic for finding the starting vertex is influenced by the initial numbering of vertices and so the quality of the reordering can vary slightly for the same problem for different initial numberings. Next, the overall method does not accommodate dense rows (e.g., a common link used in every document), and if a row has a significantly large number of nonzeroes it might be best to process it separately; i.e., extract the dense rows, reorder the remaining matrix and augment it by the dense rows (or common links) numbered last. Elapsed CPU times for a range of arrays and permuting methods are given in Berry et al. (1996), and as an indication show performances between 0.025 to 3.18 seconds for permuting a 4000×400 array. A review of public domain software for carrying out SVD and other linear algebra operations on large sparse data sets can be found in section 8.3 of Berry et al. (1999).

Once we have a sequence-respecting array, we can immediately apply efficient visualization techniques from image analysis. Murtagh et al. (2000) investigate the use of noise filtering (i.e. to remove less useful array entries) using a multiscale wavelet transform approach.

An example follows. From the Concise Columbia Encyclopedia (1989 2nd ed., online version) a set of data relating to 12025 encyclopedia entries and to 9778 cross-references or links was used. Fig. 14.11 shows a 500 × 450 subarray, based on a correspondence analysis (i.e. ordering of projections on the first factor).

This part of the encyclopedia data was filtered using the wavelet and noise-modeling methodology described in Murtagh et al. (2000) and the outcome is shown in Fig. 14.12. Overall the recovery of the more apparent alignments, and hence visually stronger clusters, is excellent. The first relatively long "horizontal bar" was selected – it corresponds to column index (link) 1733 = geological era. The corresponding row indices (articles) are, in sequence:

```
SILURIAN PERIOD
PLEISTOCENE EPOCH
HOLOCENE EPOCH
PRECAMBRIAN TIME
CARBONIFEROUS PERIOD
OLIGOCENE EPOCH
ORDOVICIAN PERIOD
TRIASSIC PERIOD
CENOZOIC ERA
PALEOCENE EPOCH
MIOCENE EPOCH
DEVONIAN PERIOD
PALEOZOIC ERA
JURASSIC PERIOD
MESOZOIC ERA
CAMBRIAN PERIOD
PLIOCENE EPOCH
CRETACEOUS PERIOD
```

The work described here is based on a number of technologies: (i) data visualization techniques; (ii) the wavelet transform for data analysis; and (iii) data matrix permuting techniques. The wavelet transform has linear computational cost in terms of image row and column dimensions, and is independent of the pixel values.

CLUSTERING IN MASSIVE DATA SETS 533

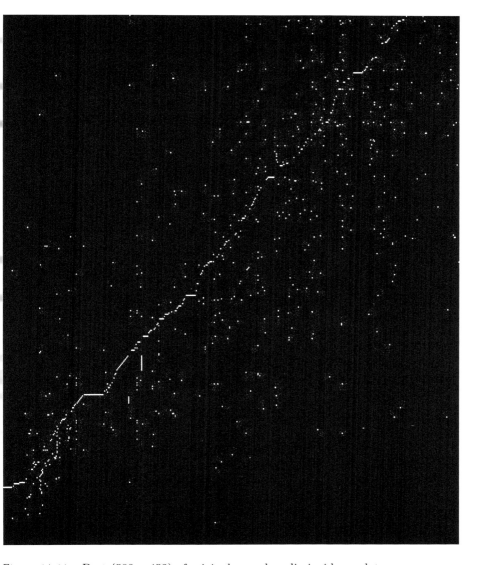

Figure 14.11. Part (500 × 450) of original encyclopedia incidence data array.

14. Conclusion

Viewed from a commercial or managerial perspective, one could justifiably ask where we are now in our understanding of problems in this area relative to where we were back in the 1960s? Depending on our answer to this, we may well proceed to a second question: Why have

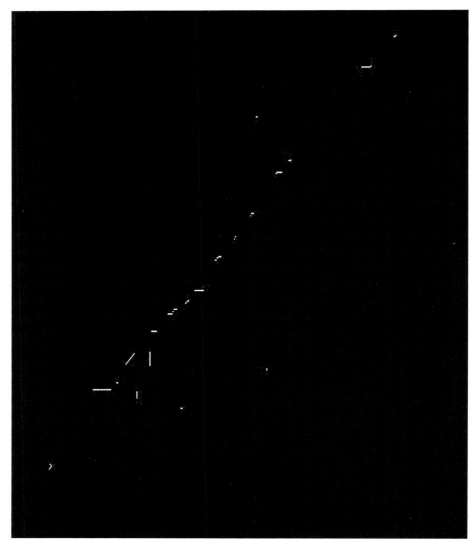

Figure 14.12. End-product of the filtering of the array shown in the previous Figure.

all important problems not been solved by now in this area – are there major outstanding problems to be solved?

As described in this chapter, a solid body of experimental and theoretical results have been built up over the last few decades. Clustering remains a requirement which is a central infrastructural element of very many application fields.

There is continual renewal of the essential questions and problems of clustering, relating to new data, new information, and new environments. There is no logjam in clustering research and development simply because the rivers of problems continue to broaden and deepen. Clustering and classification remain quintessential issues in our computing and information technology environments (Murtagh 1998).

Acknowledgements

Some of this work, in particular in sections 9 to 12, represents various collaborations with the following: J.L. Starck, CEA; A. Raftery, University of Washington; C. Fraley, University of Washington and MathSoft Inc.; D. Washington, MathSoft, Inc.; Ph. Poinçot and S. Lesteven, Strasbourg Observatory; D. Guillaume, Strasbourg Observatory, University of Illinois and NCSA.

Bibliography

D. Allard and C. Fraley. Non-parametric maximum likelihood estimation of features in spatial point processes using Voronoi tessellation. *Journal of the American Statistical Association*, 92:1485–1493, 1997.

P. Arabie and L. J. Hubert. *An overview of combinatorial data analysis*, pages 5–63. In , Arabie et al. (1996), 1996.

P. Arabie, L. J. Hubert, and G. De Soete, editors. *Clustering and Classification*. Singapore: World Scientific, 1996.

S. Banerjee and A. Rosenfeld. Model-based cluster analysis. *Pattern Recognition*, 26:963–974, 1993.

J. D. Banfield and A. E. Raftery. Model-based Gaussian and non-Gaussian clustering. *Biometrics*, 49:803–821, 1993.

K. P. Bennett, U. Fayyad, and D. Geiger. Density-based indexing for approximate nearest neighbor queries. Technical report, Microsoft, 1999. Microsoft Research Technical Report MSR-TR-98-58.

J. L. Bentley and J. H. Friedman. Fast algorithms for constructing minimal spanning trees in coordinate spaces. *IEEE Transactions on Computers*, C-27:97–105, 1978.

J. L. Bentley, B. W. Weide, and A. C. Yao. Optimal expected time algorithms for closest point problems. *ACM Transactions on Mathematical Software*, 6:563–580, 1980.

M. W. Berry, Z. Drmač, and E. R. Jessup. Matrices, vector spaces, and information retrieval. *SIAM Review*, 41:335–362, 1999.

M. W. Berry, B. Hendrickson, and P. Raghavan. Sparse matrix reordering schemes for browsing hypertext. In J. Renegar, M. Shub, and S. Smale, editors, *Lectures in Applied Mathematics (LAM) Vol. 32: The Mathematics of Numerical Analysis*, pages 99–123. American Mathematical Society, 1996.

K. Beyer, J. Goldstein, R. Ramakrishnan, and U. Shaft. When is nearest neighbor meaningful? In *Proceedings of the 7th International Conference on Database Theory (ICDT)*, Jerusalem, Israel, 1999. in press.

A. Borodin, R. Ostrovsky, and Y. Rabani. Subquadratic approximation algorithms for clustering problems in high dimensional spaces. In *Proc. 31st ACM Symposium on Theory of Computing (STOC-99)*, 1999.

A. J. Broder. Strategies for efficient incremental nearest neighbor search. *Pattern Recognition*, 23:171–178, 1990.

A. Z. Broder. *On the resemblance and containment of documents*, pages 21–29. IEEE Computer Society, 1998.

A. Z. Broder, S. C. Glassman, M. S. Manasse, and G. Zweig. Syntactic clustering of the web. In *Proc. Sixth International World Wide Web Conference*, pages 391–404, 1997.

M. Bruynooghe. Méthodes nouvelles en classification automatique des données taxinomiques nombreuses. *Statistique et Analyse des Données*, (3):24–42, 1977.

W. A. Burkhard and R. M. Keller. Some approaches to best-match file searching. *Communications of the ACM*, 16:230–236, 1973.

S. D. Byers and A. E. Raftery. Nearest neighbor clutter removal for estimating features in spatial point processes. *Journal of the American Statistical Association*, 93:577–584, 1998.

J. G. Campbell, C. Fraley, D. Stanford, F. Murtagh, and A. E. Raftery. Model-based methods for textile fault detection. *International Journal of Imaging Science and Technology*, 10:339–346, 1999.

Cartia. Mapping the information landscape, client-server software system, 1999. Caria, Inc.

G. Celeux and G. Govaert. Gaussian parsimonious clustering models. *Pattern Recognition*, 28:781–793, 1995.

D. Cheriton and D. E. Tarjan. Finding minimum spanning trees. *SIAM Journal on Computing*, 5:724–742, 1976.

K. W. Church and J. I. Helfman. Dotplot: a program for exploring self-similarity in millions of lines of text and code. *Journal of Computational and Graphical Statistics*, 2:153–174, 1993.

W. B. Croft. Clustering large files of documents using the single-link method. *Journal of the American Society for Information Science*, 28:341–344, 1977.

C. Darken, J. Chang, and J. Moody. Learning rate schedules for faster stochastic gradient search. In *Neural Networks for Signal Processing 2, Proceedings of the 1992 IEEE Workshop*, Piscataway, 1992. IEEE Press.

C. Darken and J. Moody. Note on learning rate schedules for stochastic optimization. In Lippmann, Moody, and Touretzky, editors, *Advances in Neural Information Processing Systems 3*. Morgan Kaufmann, Palo Alto, 1991.

C. Darken and J. Moody. Towards faster stochastic gradient search. In Hanson Moody and Lippmann, editors, *Advances in Neural Information Processing Systems 4*, San Mateo, 1992. Morgan Kaufman.

B. V. Dasarathy. *Nearest Neighbor (NN) Norms: NN Pattern Classification Techniques*. IEEE Computer Society Press, New York, 1991.

A. Dasgupta and A. E. Raftery. Detecting features in spatial point processes with clutter via model-based clustering. *Journal of the American Statistical Association*, 93:294–302, 1998.

W. H. E. Day and H. Edelsbrunner. Efficient algorithms for agglomerative hierarchical clustering methods. *Journal of Classification*, 1:7–24, 1984.

C. de Rham. La classification hiérarchique ascendante selon la méthode des voisins réciproques. *Les Cahiers de l'Analyse des Données*, V:135–144, 1980.

D. Defays. An efficient algorithm for a complete link method. *Computer Journal*, 20:364–366, 1977.

C. Delannoy. Un algorithme rapide de recherche de plus proches voisins. *RAIRO Informatique/Computer Science*, 14:275–286, 1980.

A. R. Dempster, N. M. Laird, and D. B. Rubin. Maximum likelihood from incomplete data via the EM algorithm. *Journal of the Royal Statistical Society, Series B*, 39:1–22, 1977.

A. Dobrzycki, H. Ebeling, K. Glotfelty, P. Freeman, F. Damiani, M. Elvis, and T. Calderwood. *Chandra Detect 1.0 User Guide*. Chandra X-Ray Center, 1999. Smithsonian Astrophysical Observatory, Version 0.9.

L. B. Doyle. Semantic road maps for literature searchers. *Journal of the ACM*, 8:553–578, 1961.

C. M. Eastman and S. F. Weiss. Tree structures for high dimensionality nearest neighbor searching. *Information Systems*, 7:115–122, 1982.

H. Ebeling and G. Wiedenmann. Detecting structure in two dimensions combining voronoi tessellation and percolation. *Physical Review E*, 47:704–714, 1993.

E. Forgy. Cluster analysis of multivariate data: efficiency vs. interpretability of classifications. *Biometrics*, 21:768, 1965.

C. Fraley. Algorithms for model-based Gaussian hierarchical clustering. *SIAM Journal of Scientific Computing*, 20:270–281, 1999.

C. Fraley and A. E. Raftery. How many clusters? which clustering method? answers via model-based cluster analysis. *The Computer Journal*, 41:578–588, 1999.

J. H. Friedman, F. Baskett, and L. J. Shustek. An algorithm for finding nearest neighbors. *IEEE Transactions on Computers*, C-24:1000–1006, 1975.

J. H. Friedman, J. L. Bentley, and R. A. Finkel. An algorithm for finding best matches in logarithmic expected time. *ACM Transactions on Mathematical Software*, 3:209–226, 1977.

B. Fritzke. Some competitive learning methods, 1997.

K. Fukunaga and P. M. Narendra. A branch and bound algorithm for computing k-nearest neighbors. *IEEE Transactions on Computers*, C-24:750–753, 1975.

V. J. Gillet, D. J. Wild, P. Willett, and J. Bradshaw. Similarity and dissimilarity methods for processing chemical structure databases. *The Computer Journal*, 41:547–558, 1998.

A. D. Gordon. *Classification*. Champman and Hall, 2 edition, 1999.

A. Griffiths, L. A. Robinson, and P. Willett. Hierarchic agglomerative clustering methods for automatic document classification. *Journal of Documentation*, 40:175–205, 1984.

D. Guillaume and F. Murtagh. Clustering of XML documents. *Computer Physics Communications*, 1999. submitted.

M. E. Hodgson. Reducing the computational requirements of the minimum-distance classifier. *Remote Sensing of Environment*, 25:117–128, 1988.

E. Horowitz and S. Sahni. *Fundamentals of Computer Algorithms*. London: Pitman, 1979.

A. K. Jain and R. C. Dubes. *Algorithms for Clustering Data*. Englewood Cliffs: Prentice-Hall, 1988.

G. Jammal and A. Bijaoui. Multiscale image restoration for photon imaging systems. In *SPIE Conference on Signal and Image Processing: Wavelet Applications in Signal and Image Processing, VII*, July 1999.

J. Juan. Programme de classification hiérarchique par l'algorithme de la recherche en chaîne des voisins réciproques. *Les Cahiers de l'Analyse des Données*, VII:219–225, 1982.

R. E. Kass and A. E. Raftery. Bayes factors. *Journal of the American Statistical Association*, 90:773–795, 1995.

J. Kittler. A method for determining k-nearest neighbors. *Kybernetes*, 7:313–315, 1978.

E. D. Kolaczyk. Nonparametric estimation of gamma-ray burst intensities using haar wavelets. *Astrophysical Journal*, 483:340–349, 1997.

E. Kushilevitz, R. Ostrovsky, and Y. Rabani. Efficient search for approximate nearest neighbors in high-dimensional spaces. In *Proc. of 30th ACM Symposium on Theory of Computing (STOC-30)*, 1998.

P. Lloyd. Least squares quantization in pcm. *IEEE Transactions on Information Theory*, 1982. Technical note, Bell Laboratories, 1957.

J. MacQueen. Some methods for classification and analysis of multivariate observations. In *Proceedings of the Fifth Berkeley Symposium on Mathematical Statistics and Probability*, volume 1, pages 281–297, Berkeley, 1976. University of California Press.

R. B. Marimont and M. B. Shapiro. Nearest neighbor searches and the curse of dimensionality. *Journal of the Institute of Mathematics and Its Applications*, 24:59–70, 1979.

L. Micó, J. Oncina, and E. Vidal. An algorithm for finding nearest neighbors in constant average time with a linear space complexity. In *The 11th International Conference on Pattern Recognition*, volume II, pages 557–560, New York, 1992. IEEE Computer Science Press.

A. Moore. Very fast EM-based mixture model clustering using multiresolution kd-trees. *Neural Information Processing Systems*, December 1998.

R. Motwani and P. Raghavan. *Randomized Algorithms*. Cambridge University Press, 1995.

S. Mukherjee, E. D. Feigelson, G. J. Babu, F. Murtagh, C. Fraley, and A. Raftery. Three types of gamma-ray bursts. *The Astrophysical Journal*, 508:314–327, 1998.

F. Murtagh. A very fast, exact nearest neighbor algorithm for use in information retrieval. *Information Technology*, 1:275–283, 1982.

F. Murtagh. Expected time complexity results for hierarchic clustering algorithms which use cluster centers. *Information Processing Letters*, 16:237–241, 1983.

F. Murtagh. Complexities of hierarchic clustering algorithms: state of the art. *Computational Statistics Quarterly*, 1:101–113, 1984.

F. Murtagh. *Multidimensional Clustering Algorithms*. Würzburg: Physica-Verlag, 1985.

F. Murtagh. Comments on 'parallel algorithms for hierarchical clustering and cluster validity'. *IEEE Transactions on Pattern Analysis and Machine Intelligence*, 14:1056–1057, 1992.

F. Murtagh. Multivariate methods for data analysis. In A. Sandqvist and T.P. Ray, editors, *Central Activity in Galaxies: From Observational Data to Astrophysical Diagnostics*, pages 209–235, Berlin, 1993a. Springer-Verlag.

F. Murtagh. Search algorithms for numeric and quantitative data. In A. Heck and F. Murtagh, editors, *Intelligent Information Retrieval: The Case of Astronomy and Related Space Sciences*, pages 49–80, Dordrecht, 1993b. Kluwer Academic.

F. Murtagh. Foreword to the special issue on clustering and classification. *The Computer Journal*, 41:517, 1998.

F. Murtagh and A. Heck. *Multivariate Data Analysis*. Dordrecht: Kluwer Academic, 1987.

F. Murtagh and M. H. Pajares. The Kohonen self-organizing map method: an assessment. *Journal of Classification*, 12:165–190, 1995.

F. Murtagh and A.E. Raftery. Fitting straight lines to point patterns. *Pattern Recognition*, 17:479–483, 1984.

F. Murtagh and J. L. Starck. Pattern clustering based on noise modeling in wavelet space. *Pattern Recognition*, 31:847–855, 1998.

F. Murtagh, J. L. Starck, and M. Berry. Overcoming the curse of dimensionality in clustering by means of the wavelet transform. *The Computer Journal*, 43:107–120, 2000.

R. Neal and G. Hinton. A view of the EM algorithm that justifies incremental, sparse, and other variants. In M. Jordan, editor, *Learning in Graphical Models*, pages 355–371, Dordrecht, 1998. Kluwer Academic Publisher.

H. Niemann and R. Goppert. An efficient branch-and-bound nearest neighbor classifier. *Pattern Recognition Letters*, 7:67–72, 1988.

B. K. Parsi and L. N. Kanal. An improved branch and bound algorithm for computing k-nearest neighbors. *Pattern Recognition Letters*, 3: 7–12, 1985.

D. Pelleg and A. Moore. Accelerating exact k-means algorithms with geometric reasoning. In *Proceedings KDD-99, Fifth ACM SIGKDD International Conference on Knowledge Discovery and Data Mining*, San Diego, August 1999.

S. A. Perry and P. Willett. A review of the use of inverted files for best match searching in information retrieval systems. *Journal of Information Science*, 6:59–66, 1983.

P. Poinçot, S. Lesteven, and F. Murtagh. A spatial user interface to the astronomical literature. *Astronomy and Astrophysics Supplement*, 130:183–191, 1998.

P. Poinçot, S. Lesteven, and F. Murtagh. Maps of information spaces: assessments from astronomy. *Journal of the American Society for Information Science*, 1999. submitted.

V. Ramasubramanian and K. K. Paliwal. An efficient approximation-algorithm for fast nearest-neighbor search based on a spherical distance coordinate formulation. *Pattern Recognition Letters*, 13:471–480, 1992.

M. Richetin, G. Rives, and M. Naranjo. Algorithme rapide pour la détérmination des k plus proches voisins. *RAIRO Informatique/Computer Science*, 14:369–378, 1980.

F. J. Rohlf. Algorithm 76: Hierarchical clustering using the minimum spanning tree. *The Computer Journal*, 16:93–95, 1973.

F. J. Rohlf. A probabilistic minimum spanning tree algorithm. *Information Processing Letters*, 7:44–48, 1978.

F. J. Rohlf. Single link clustering algorithms. In P. R. Krishnaiah and L.N. Kanal, editors, *Handbook of Statistics*, volume 2, pages 267–284, Amsterdam, 1982. North-Holland.

E. V. Ruiz. An algorithm for finding nearest neighbors in (approximately) constant average time. *Pattern Recognition Letters*, 4:145–157, 1986.

G. Salton and M. J. McGill. *Introduction to Modern Information Retrieval*. New York: McGraw-Hill, 1983.

M. Sato and S. Ishii. Reinforcement learning based on on-line EM algorithm. In M.S. Kearns, S.A. Solla, and D.A. Cohn, editors, *Advances in Neural Information Processing Systems 11*, pages 1052–1058, Cambridge, 1999. MIT Press.

T. Schreiber. Efficient search for nearest neighbors. In A.S. Weigend and N.A Gershenfeld, editors, *Predicting the Future and Understanding the Past: A Comparison of Approaches*, New York, 1993. Addison-Wesley.

G. Schwarz. Estimating the dimension of a model. *The Annals of Statistics*, 6:461–464, 1978.

SDSS. Sloan digital sky survey, 1999.

M. Shapiro. The choice of reference points in best-match file searching. *Communications of the ACM*, 20:339–343, 1977.

R. Sibson. Slink: an optimally efficient algorithm for the single link cluster method. *The Computer Journal*, 16:30–34, 1973.

A. F. Smeaton and C. J. van Rijsbergen. The nearest neighbor problem in information retrieval: an algorithm using upperbounds. *ACM SIGIR Forum*, 16:83–87, 1981.

P. H. A. Sneath and R. R. Sokal. *Numerical Taxonomy*. San Francisco: W. H. Freeman, 1973.

H. Späth. *Cluster Dissection and Analysis: Theory, Fortran Programs, Examples*. Chichester: Ellis Horwood, 1985.

J. L. Starck, F. Murtagh, and A. Bijaoui. *Image and Data Analysis: The Multiscale Approach*. New York: Cambridge University Press, 1998.

R. E. Tarjan. An improved algorithm for hierarchical clustering using strong components. *Information Processing Letters*, 17:37–41, 1983.

B. Thiesson, C. Meek, and D. Heckerman. Accelerating EM for large databases. Technical report, Microsoft, 1999. Microsoft Research Technical Report MST-TR-99-31.

S. F. Weiss. A probabilistic algorithm for nearest neighbor searching. In R. N. Oddy and et al., editors, *Information Retrieval Research*, pages 325–333, London, 1981. Butterworths.

H. D. White and K. W. McCain. Visualization of literatures. *Annual Review of Information Science and Technology (ARIST)*, 32:99–168, 1997.

P. Willett. Efficiency of hierarchic agglomerative clustering using the icl distributed array processor. *Journal of Documentation*, 45:1–45, 1989.

A. C. Yao. An $o(|e|\log\log|v|)$ algorithm for finding minimum spanning trees. *Information Processing Letters*, 4:21–23, 1975.

T. P. Yunck. A technique to identify nearest neighbors. *IEEE Transactions on Systems, Man, and Cybernetics*, SMC-6:678–683, 1976.

C. T. Zahn. Graph-theoretical methods for detecting and describing gestalt clusters. *IEEE Transactions on Computers*, C-20:68–86, 1971.

G. Zheng, J. L. Starck, J.G. Campbell, and F. Murtagh. Multiscale transforms for filtering financial data streams. *Journal of Computational Intelligence in Finance*, 7, 1999.

VI

DATA MANAGEMENT

Chapter 15

MANAGING AND ANALYZING MASSIVE DATA SETS WITH DATA CUBES

Mirek Riedewald
Computer Science Department, University of California, Santa Barbara CA 93106 USA
mirek@cs.ucsb.edu

Divyakant Agrawal
Computer Science Department, University of California, Santa Barbara CA 93106 USA
agrawal@cs.ucsb.edu

Amr El Abbadi
Computer Science Department, University of California, Santa Barbara CA 93106 USA
amr@cs.ucsb.edu

Abstract Data cubes combine an easy-to-understand conceptual model with an implementation that enables the fast summarization of large data sets. This makes them a powerful tool for supporting the interactive analysis of massive data collections like data warehouses and digital libraries. This article surveys some of the recent developments in data cube research. We mainly focus on techniques for fast aggregation in data warehousing environments. This includes work on group-by and range queries, approximate query responses, and compression. Since sparse high-dimensional data cubes are of increasing interest, issues related to them are explicitly discussed.

Keywords: Data cube, OLAP, Data warehousing, Range queries, Compression

1. Introduction

Statistical databases, data warehouses, and digital libraries all have to deal with massive data sets. Advancing information technology increases the demand for a growing amount of information to be maintained for interactive analysis. While this article mainly focuses on data warehousing, the discussion of the problems and techniques is generally transferable to any application domain where interactive analysis of massive data sets is a major concern.

Data warehouses collect large amounts of data from diverse sources and maintain them, usually in a centralized location. They can easily reach a size in the order of Terabytes and keep on growing. This is due to the fact that modern information technology enables collecting and storing data from virtually any source. Human analysts can not "digest" this huge amount of information at a detailed level. Consequently the system needs to provide a summarized view of the data that is of interest for a certain task. For instance, in a company's data warehouse in order to find out in which region a certain product does not sell well, it is not necessary to look at each single sales transaction. The analyst rather needs the aggregate (sum) of all sales for certain regions and time periods. While such a summarized high-level view allows the discovery of phenomena in the data collection, it does not facilitate discovering their *causes*. In the example, for finding out *why* the product does not sell well, it might be necessary to increase the level of detail until the cause is found. Consequently, data warehouses have to provide high-level aggregate views of the data as well as the possibility to *drill down* to the detailed base information.

This article's emphasis is on aggregation of data and interactive analysis. In data warehousing environments the term Online Analytical Processing (OLAP) is used for this type of applications. To facilitate the analysis process, OLAP tools provide a multidimensional view of the data. More precisely, the data is conceptually modeled as a multidimensional hyper-rectangle, or *data cube* for short. The d functional attributes that describe a data item are the *dimensions* of the data cube. Some of them are hierarchical, e.g., year-quarter-month-day for the time dimension (sometimes multiple hierarchies for an attribute exist, e.g., week-day is another hierarchy for the time dimension). A d-tuple of dimension values defines a *cell* of the data cube, similar to a multidimensional array. Cells store the value of an additional attribute, the *measure* attribute. A simple data warehouse example could contain a data cube for sales. This cube has the dimensions Product, Time (date of sale), and Location (place of sale) and the measure attribute Amount (value of sales transaction). Hence for each product, time and location the corresponding amount of sales is stored. Figure 15.1 shows the data cube. If there is no sales transaction for a certain combination of dimension values, the corresponding cell is *empty*. Here empty cells contain the value 0. Another data cube could record the cost of marketing campaigns for the time and product dimension. An analyst can use the data in the two cubes for finding out if a marketing campaign was successful by comparing the sales figures before and after the campaign. He/she can also find out if there is a correlation between the amount of money invested and the success. This can be done for all products together, or for groups of products or even single products. By combining hypothetical with historical data in the cube, an analyst can predict future events, e.g., the success of a future marketing campaign.

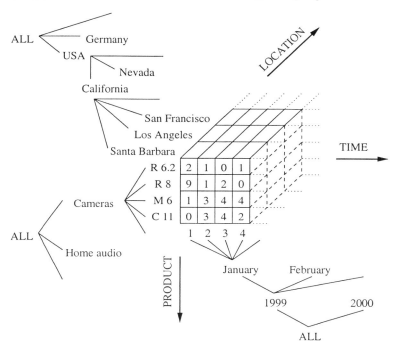

Figure 15.1. Data cube example

This simple example already gives a flavor of possible queries. In general the following query types are typical in OLAP applications (cf. (Chaudhuri and Dayal 1997) for the first three items).

- Roll-up, drill-down: increases, respective decreases, level of aggregation along one or more attribute hierarchies (e.g., monthly instead of daily sales figures).

- Slice-and-dice: selects attributes (dimensions) and data items (cells) of interest.

- Pivot: re-orients multidimensional view of the data.

- "What-if": incorporates hypothetical data (i.e., updates the values of designated cells) into the data cube in order to make predictions, etc.

The data cube model simplifies the analysis process by providing more intuition than collections of "flat" tables. Queries can be described as regions in the data cube. By augmenting the data cube with additional cells, group-bys, cross-tabs and sub-totals are easily included into the model (CUBE operator as proposed in (Gray et al. 1997)). Also, whenever data cubes can be implemented as multidimensional arrays, the above queries can be efficiently supported. For instance (Sarawagi and Stonebraker 1994) analyzes different schemes for storing multidimensional arrays in secondary (hard disk) and tertiary memory (e.g., tapes). Note, however, that data cubes are not necessarily implemented as multidimensional arrays, especially when the data cube is sparse, i.e., contains a high percentage of empty cells. "What-if" queries require special support for fast updates.

This article surveys techniques for interactive analysis with data cubes. In Section 2 relevant properties of OLAP applications and general solutions for supporting interactive analysis are discussed. Since the term "data cube" is used to denote different entities in the literature, it is formalized for this article in Section 3. Sections 4, 5, and 6 survey data cube techniques for interactive analysis based on this unified terminology. The main focus is on support for aggregate range queries, aggregation over different dimensions, and compression of the data cube. Section 7 concludes this article.

2. Interactive Analysis of Massive Data Sets

When designing algorithms for interactive analysis of massive data sets, the right balance between query, update, and storage cost (measured in response time or storage units, respectively) has to be found. Since queries clearly dominate in the analysis process, supporting fast queries is the top priority for any OLAP tool. Response times should be in the order of seconds, minutes at most, even for complex queries. What about updates? First of all, to enable "what-if" analysis and applications where the use of the latest data is crucial (e.g., stock trading) efficient updates are mandatory. Even systems that process updates offline in batches benefit from faster updates. They shrink the update window and thus increase the availability of the data and/or enable the processing of more updates. Also, as a future trend it can be expected that globalization will blur the clear distinction between day and night access patterns to the data warehouse. The space consumption of a technique seems to be of low importance. This is basically true since hard disks and optical storage devices offer huge amounts of space at low cost. Still there is reason for concern. Typically, the cheaper the unit of space, the slower the access to the storage medium. Also, techniques that try to speed up queries without taking space aspects into account could lead to *database explosion* (see (Pendse and Creeth 2000)), a case when the generated amount of data is too large to fit on any storage medium. This effect is discussed in more detail in Section 4.

The above concern can be summarized as an optimization problem between query, update and storage costs. *In general the goal of interactive data analysis techniques is to minimize the query cost subject to constraints for update and storage costs.* In the following general techniques for reducing the query cost are discussed.

If a query has to collect the required information from multiple sources, the response time will be affected by network traffic and the load on the sources (the slowest source determines the response time). Concentrating all data in a single location avoids this performance penalty. For instance, data warehouses store data from different operational systems, e.g., marketing, finance, and production department, who view the same product from different perspectives. By combining the departmental data in a single location, cross-referencing becomes possible. In Figure 15.2 the components of a data warehouse that are relevant for our discussion are shown. Data of interest from the sources is extracted, cleaned and transformed to be integrated into the warehouse. Whenever the source data changes, the warehouse has to be refreshed, either periodically (batch updates) or instantly. The analysis can then be based solely on the warehouse data. The operational systems, which are optimized toward heavy loads of short transactions do not suffer a slowdown from handling complex analysis queries. Also, while maintaining historical data of committed transactions causes (typically unnecessary) overhead for operational systems, it is mandatory for analysis purposes in an enterprise.

The cost of complex queries can be dramatically reduced by storing *pre-computed* aggregate values. For instance, pre-computed views that combine frequently referenced information from multiple tables are a common technique to avoid expensive joins in relational database systems. Other useful values to be pre-computed are aggregates for contiguous regions of the data cube. Whenever pre-computation is considered, the problem of finding the "optimal" set of pre-computed values has to be solved. The more pre-computed values are stored, the higher the storage costs. What is perhaps more serious is that pre-computed values depend on the base values they aggregate. Hence updates to a single

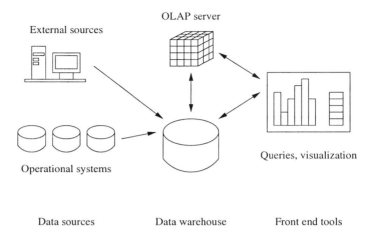

Figure 15.2. Typical data warehouse (details omitted)

base value have to be propagated to all pre-computed aggregates that depend on them, resulting in increased update costs.

In the case of ad hoc queries, the optimization problem becomes even harder. Typically the next query issued to the system depends on the result of the previous one(s). For instance, an analyst might first look at the sales figures aggregated over all products. When some interesting phenomenon is detected, e.g., low sales in a certain month, he/she drills down to product categories and even single products to find out what caused the phenomenon. After finishing this process, the analyst might start all over by looking at a completely different region of the data cube. Recent work models this special type of behavior with *hierarchical range queries* (Koudas et al. 2000). A set of ranges is hierarchical, if any two members are either disjoint or one is contained in the other. However, the problem of predicting OLAP queries in general is far from being solved.

Systems that support interactive data analysis are increasingly confronted with high dimensionality of the data space. In order to examine complex dependencies between different attributes, a corresponding number of dimensions is required to define the data cube. For instance the TPC-H benchmark (Transaction Processing Performance Council 2001) that models a typical commercial setting uses 16 attributes for its Lineitem table. Census data sets, like for instance the Current Population Survey March Questionnaire Supplement (Person Data Files for 1995) (U.S. Census Bureau 2001), have 300 or more functional attributes. Despite of being massive, the high dimensionality causes the data set to be sparse, i.e., a high percentage of the cells in the data cube will be empty. In Table 15.1 eight possible census attributes and their domains, most of them as defined by the US Bureau of Census, are listed. A corresponding data cube that measures the number of people for these eight dimensions has about $2*10^{14}$ cells. Still, even storing data about the whole world's population (less than 10 billion people) will not fill more than 0.005% of the cells with values. The non-empty cells are typically not uniformly distributed over the data cube. Thus, simply implementing sparse data cubes as multidimensional arrays is not feasible. Techniques that work well for dense data cubes are often not applicable in the case of sparsity. Also, in general sparsity results in less efficient algorithms. In Sections 4 and 5 this is discussed in more detail.

Table 15.1. Attributes and their domains for a hypothetical census data set

Attribute	Domain	Attribute	Domain
Age	0 - 150	Income	0 - 99999
Weight	1 - 500	Race	1 - 5
Family type	1 - 5	Marital status	1 - 7
Class of worker	0 - 8	Education	1 - 17

A straightforward idea for dealing with sparsity is to compress the data cube. Compression not only reduces the space consumption, it can also result in considerable savings in query response time when a compact representation of the requested data is accessed and transferred from secondary or tertiary storage (e.g., from disk or tape). Note, that compression typically implies approximate query responses. Lossy compression techniques inherently provide approximations. To make approximate answers useful, they should be accompanied with some kind of quality indication. This could be absolute or probabilistic *error bounds* which enable a user or application to judge the accuracy of the result. In general *absolute* error bounds are preferable, because they guarantee that the exact result is within the bounds. *Probabilistic* bounds like confidence intervals only provide the information that the exact result is within the bounds with a certain probability less than 100% (typically 95%). Unfortunately there is no "best" approximation technique, i.e., one has to carefully select a suitable technique and approximation granularity depending on the application. Also, the definition of what "sufficient accuracy" means can be very specific for an application or even query. For instance the (fairly inaccurate) answer $100,000 < x < 10,000,000$ might be sufficiently accurate when one is looking for a major vendor of a product and a major vendor is defined as one who sells more than 100,000 units. However, the (seemingly very accurate) answer $99.9 < y < 100$ and $99.8 < z < 100$ is not accurate enough to answer the question if y is greater than z.

3. What is a Data Cube?

Data cubes are used by OLAP tools to simplify the analysis of multidimensional data sets, i.e., where data items are described by a large number of attributes. Materialized data cubes are maintained at the server site or even on the clients (depending on the size of the cube and the client's power). They are built based on the information in the data warehouse (see Figure 15.2). Updates to the warehouse have to be reflected in the data cubes. User queries can be efficiently answered by OLAP tools using data cubes. Before discussing how data cubes support interactive analysis, we have to formalize the term itself. We first give an overview of terminology used in the literature.

In (Pendse and Creeth 2000) the term *hypercube* refers to data structures where all functional attributes of interest constitute the dimensions of a single hyper-rectangle and where the single cells store the values of measure attributes depending on the dimension values. However, hypercubes, especially for sparse data sets, might be implemented using a different underlying model. *Multicube* structures (Pendse and Creeth 2000) consist of a collection of smaller hyper-rectangles, each of which is a hypercube for a subset of the dimensions of interest. In the research community the term *data cube* either refers to the hypercube as described above, or to such a hypercube which is augmented by additional cells that store pre-computed aggregates.

We define a **data cube** for d functional attributes of interest as follows. The d functional attributes constitute the dimensions of a d-dimensional hyper-rectangle, the *data cube*. The side-length of the data cube in a dimension is identical to the domain size of the corresponding attribute. Each d-tuple of dimension values defines a unique cell in the data cube. Consequently, a data cube with domain sizes $N_1, N_2, \ldots, N_d$ has $N_1 \cdot N_2 \cdots N_d$ cells. A cell stores the value of a measure attribute. The letter n denotes the number of *non-empty* cells. Note, that this model corresponds to the hypercube in (Pendse and Creeth 2000). It does not imply how the data cube is *implemented*, e.g., as a multidimensional array or a relational table.

The term "data cube" is often used in the literature to refer to the result of the CUBE operator as proposed by Gray et al. (Gray et al. 1997). We believe that the term *extended data cube* provides more intuition and therefore use this term for the result of the CUBE operator. The extended data cube is discussed in more detail in the next section.

4. Extended Data Cubes for OLAP

The extended data cube is the result of a d-dimensional generalization of the GROUP BY operator. It contains the results of grouping by all subsets of the dimensions. In the example in Table 15.2 the extended data cube for a data cube with the three dimensions Time (T), Product (P), and Location (L) and the measure attribute Sale is shown. Note that the extended data cube is a collection of 2^d related data cubes, called *cuboids*. The single tables list all non-empty cells of the cuboids. The core cuboid in Table 15.2(a) corresponds to the data cube as defined in this article. The other 2^d-1 cuboids (Tables 15.2(b)–15.2(h)) are the results of grouping the original data set in the data cube by subsets of the set of dimensions. They therefore contain pre-computed aggregate information for regions of the data cube. In relational database terminology the core cuboid is the base relation and the other cuboids are *views* generated by the corresponding GROUP BY queries. For faster access specialized index structures were developed for the extended data cube (Johnson and Shasha 1997; 1999) (using a notion of cube forests), and for materialized views (Kotidis and Roussopoulos 1998) (taking advantage of commonalities between different cuboids/views).

The multidimensional data cube model provides a more compact description of the extended data cube by combining all 2^d cuboids into a single hyper-rectangle. Each dimension is augmented by a single value, "ALL", which represents the aggregation over the complete dimension. The additional surface layers are "padded" to the data cube and store the corresponding cuboids. For instance the tuples in Table 15.2(d) are stored in the cells (T,P,L,Sale)=(T=ALL,P,L,Sale). Figure 15.3 illustrates this for the example from Table 15.2.

The extended data cube obviously requires additional storage. Assuming for simplicity that the data cube has N cells per dimension, the extended data cube will have $(N+1)^d$ cells. Let n again denote the number of non-empty cells in the data cube. In the worst case all cells of the extended data cube are non-empty. The factor by which the space consumption increases is therefore bounded by $(N+1)^d/n$. For dense data cubes $n \approx N^d$ and therefore the bound is approximately $(1+1/N)^d$, i.e., typically close to 1. However, for sparse data cubes the space factor can explode. The reason is that the GROUP BY on sparse data cubes does not necessarily lead to a considerable amount of aggregation, i.e., many of the cuboids have a number of non-empty cells that is in the order of n. And since there is an exponential number of cuboids, the space overhead grows exponentially with the dimensionality. This effect, termed *database explosion*, is explained in more detail in (Pendse and

Table 15.2. Extended data cube as relational tables

T	P	L	Sale
98	R8	SB	2
98	R8	SF	3
98	M6	LA	1
99	R8	SB	2
99	R8	SF	3
99	M6	LA	4
00	R8	SF	3
00	M6	LA	1

(a) Given data set.

T	P	Sale
98	R8	5
98	M6	1
99	R8	5
99	M6	4
00	R8	3
00	M6	1

(b) GROUP BY T,P.

T	L	Sale
98	SB	2
98	SF	3
98	LA	1
99	SB	2
99	SF	3
99	LA	4
00	SF	3
00	LA	1

(c) GROUP BY T,L.

P	L	Sale
R8	SB	4
R8	SF	9
M6	LA	6

(d) GROUP BY P,L.

T	Sale
98	6
99	9
00	4

(e) GROUP BY T.

P	Sale
R8	13
M6	6

(f) GROUP BY P.

L	Sale
SB	4
SF	9
LA	6

(g) GROUP BY L.

Total Sale
19

(h) GROUP BY {}.

Creeth 2000). It can already be observed in our (fairly dense) example. Tables 15.2(a) and 15.2(c) contain the same number of non-empty cells, i.e., the aggregation along the product dimension does not result in any aggregation. Higher dimensionality and greater sparsity tend to amplify the space explosion effect. Beyer and Ramakrishnan (Beyer and Ramakrishnan 1999) observe this for a real data set. Database explosion is an example that despite low storage costs, pre-computation has not become an easy task. From the above discussion, it follows that computing the extended data cube for sparse data sets is a major algorithmic challenge. Here we only provide an overview of various proposed techniques. For details the reader is advised to refer to the cited papers.

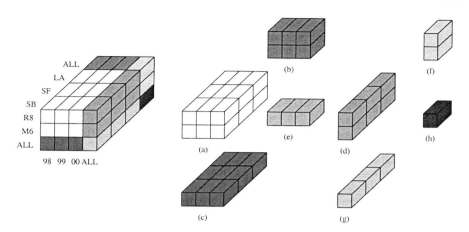

Figure 15.3. Extended data cube as multidimensional cube, cuboids of Table 15.2 indicated

In (Beyer and Ramakrishnan 1999) a technique is proposed that allows to only generate those cuboids or partitions of cuboids that perform "enough" aggregation. This directly addresses the reason for database explosion. In the paper related algorithms for computing the extended data cube are discussed as well.

Another way to avoid database explosion is to select and pre-compute only those cuboids that promise the most benefit. Assuming a certain query workload one can try to find a set of cuboids that minimizes the query costs subject to memory (storage) or maintenance (update) cost constraints. Harinarayan et al. (Harinarayan et al. 1996) develop a lattice framework for expressing dependencies between the cuboids. It captures for instance, that the cuboid in Table 15.2(f) could be computed from the cuboid in Table 15.2(b), but not from the one in Table 15.2(c). The framework is used to develop a greedy heuristic that determines sets of cuboids that result in "almost optimal" tradeoffs between query cost and storage overhead (e.g., minimize query cost subject to storage constraint). Gupta et al. (Gupta et al. 1997) extend this technique to include the selection of index structures for the cuboids. Later Shukla et al. (Shukla et al. 2000) extend the approach of (Harinarayan et al. 1996) to multicube systems.

Smith et al. (Smith et al. 1998) generalize the cuboid selection problem of (Harinarayan et al. 1996) to a view element selection framework. Based on that framework an algorithm is developed which for a given population of queries selects the optimal non-redundant set of view elements that minimizes the query costs. Also, a greedy heuristic is proposed that selects redundant view elements in order to minimize query costs subject to storage constraints.

In (Gupta 1997) a theoretical framework for selection of views in a data warehouse is developed. The goal is to minimize the sum of query and update cost, given a certain amount of storage. Another solution that especially addresses the selection of cuboids for high dimensionality, taking hierarchies into account, is presented in (Baralis et al. 1997). Note that the problem of which cuboids to pre-compute and to store is a special case of the general *view selection problem*. The goal is to find an "optimal" set of views such that some cost function is minimized subject to (resource) constraints, e.g., minimize the query cost given constraints on update and/or storage cost. A recent work in this field that contains references to other related work is (Gupta and Mumick 1999). In general, in order to be effective, cuboid/view selection algorithms need to have knowledge about the expected query workload. Note also, that the optimization problem is too complex to be solved exactly (e.g., $\mathcal{NP}$ complete) and therefore requires heuristic solutions.

5. Supporting Efficient Range Queries on Data Cubes

The aggregate range query is an important tool in OLAP applications. It selects ranges on the dimensions and aggregates the measure attribute's values for all selected cells of the data cube. Finding the "total sales volume for last month in California" and finding the "average income of people between 30 and 35 years" are examples of aggregate range queries. In the data cube model a range query corresponds to a hyper-rectangular region, which will be referred to as *query-region*. Note, that the query does not have to specify a range for each dimension. Unspecified ranges imply the selection of the whole domain of the dimension. In the following sections we survey techniques for efficient support of aggregate range queries. First dense data cubes are discussed, then sparse ones. If not stated otherwise, the techniques are illustrated and explained for the operator SUM.

5.1. Techniques for Dense Data Cubes

Dense data cubes can be stored and accessed like multidimensional arrays. Thus each cell can be located in memory by using a (simple) function for computing its position depending on the dimension values, i.e., an index structure and additional index accesses are not necessary. Also, while for instance in relational tables dimension values are stored repeatedly for tuples they occur in, this is not necessary for an array where the position of a cell implicitly defines its dimension values. Note further, that an update will never change the structure of the array, only a value in a cell. This is different for sparse data cubes where empty and non-empty cells have to be treated differently (e.g., empty cells are not stored) and updates can lead to expensive restructuring operations as known from index structures. Consequently, algorithms that exploit advantageous properties of dense data cubes are more efficient on such cubes than general algorithms.

But, are dense data cubes practically relevant? With increasing dimensionality the number of data cube cells grows exponentially. Hence high-dimensional data cubes will be sparse. However, data distributions in practice are typically skewed and contain clusters with a higher density of non-empty cells. When the data cube is partitioned appropriately, some of the partitions might be dense enough to be handled like dense (sub)cubes. Corresponding approaches are presented in Section 6. Also, not all queries select a range in each dimension. Typically an analyst will specify selection conditions only for a subset of the dimensions. One could therefore maintain cubes for such subsets. This results in smaller and denser data cubes, the motivation for designing systems that use multicubes (cf. Section 3). The extended data cube (cf. Section 4) is another example for where dense data cubes occur. The cuboids which are the result of grouping by small numbers of dimensions (the cuboids with the highest levels of aggregation) typically are dense. By replacing or complementing dense cuboids with the corresponding cuboids that support aggregate range queries, the response time for this type of query on the extended data cube is reduced. Aggregate range queries can also be included into the framework for the optimal selection of cuboids to materialize by adjusting the cost formulas of the optimization problem.

In the following, techniques that speed up range queries on dense data cubes are discussed. The common idea is to replace values in the data cube by pre-computed aggregate values. We refer to such a data cube with pre-computed values as the *pre-aggregated* data cube. Whenever required for clarity, the term *original* data cube is used to denote the data cube without pre-computed values. Note, that a single cell in a pre-aggregated data cube can contain the aggregate over a set of cells in the original data cube. This set is referred to as the *aggregation region* of the cell. While pre-aggregation speeds up queries, the update cost increases. An update that only affects a single cell in the original data cube affects all cells in the pre-aggregated cube whose aggregation region contains the updated cell.

The following notation will be used. For simplicity and without loss of generality assume that each of the d dimensions of the original data cube has the domain $\{0, 1, \ldots, N-1\}$, i.e., the data cube has N^d cells. Out of these cells n are non-empty, where $n \approx N^d$ because the cube is dense. The query and update cost are expressed in terms of the number of accessed and affected cells, respectively. This cost model is independent of implementation details and quite realistic since dense data cubes are handled like multidimensional arrays. Let q, u, and s denote the costs of query, update and space, respectively. If not otherwise stated, those costs are worst case costs. Clearly, using the original data cube results in costs $(q, u, s) = (N^d, 1, N^d)$. Since the data cube is dense, it holds that $(q, u, s) \approx (n, 1, n)$.

Let A denote the original data cube. Its cells $c = [c_1, \ldots, c_d]$, where the c_i are elements of the domain of the corresponding dimension, contain the measure attribute's value $A[c]$. For two cells e and f, with $e : f$ a hyper-rectangular region is denoted, i.e., the set of all cells c that satisfy $e_i \leq c_i \leq f_i$ for all $1 \leq i \leq d$. Cell e is the *anchor*, f is the *endpoint* of the region. Cells in region $e : f$ that have at least one dimension value in common with the anchor cell, i.e., lie in the same surface, are *border cells* of that region. All others are *inner cells*. Consequently anchor and endpoint of the data cube are $[0, \ldots, 0]$ and $[N-1, \ldots, N-1]$, respectively. Its inner cells c are all those with $c_i > 0$ for all $1 \leq i \leq d$. All others are border cells (including the anchor). If $e = [0, \ldots, 0]$, region $e : f$ is a *prefix region*. The set of values in region $e : f$ is denoted $A[e] : A[f]$, and op($A[e] : A[f]$) is the result of applying aggregate operator op to those values. Consequently, SUM($A[e] : A[f]$) denotes a range sum query. We refer to SUM($A[0, \ldots, 0] : A[f]$) as a *prefix sum*.

The techniques below are explained for the aggregate operator SUM but are applicable to any invertible operator (e.g., COUNT) and any operator that can be expressed with invertible operators (e.g., AVG (average)). The techniques provide a spectrum of tradeoffs of query and update costs and most of them do not require more space than the original data cube.

In (Ho et al. 1997a) Ho et al. propose a technique, which we term the Prefix Sum (PS) technique, to speed up range sum queries at the cost of more expensive updates. The idea is to use a pre-aggregated data cube PS which stores the prefix sums of the original data cube. More formally, for all cells c, $PS[c] = $ SUM($A[0, \ldots, 0] : A[c]$). By combining up to 2^d of those prefix sums any range sum query can be answered. Figure 15.4 shows a 2-dimensional example. While the query accesses all selected cells on the original data cube, on the pre-aggregated cube at most $2^2 = 4$ accesses are necessary. The approach is mainly hampered by its high update costs of N^d in the worst case (update to upper left cell). The overall worst case costs of the PS technique are $(q, u, s) = (2^d, N^d, N^d) \approx (2^d, n, n)$ (since $n \approx N^d$ for dense data cubes). The

MANAGING AND ANALYZING MASSIVE DATA SETS

authors propose an efficient batch updating technique. However, for dynamic environments the update cost is still too high.

A	0	1	2	3	4	5	6	7	8
0	3	5	1	2	2	4	6	3	3
1	7	3	2	6	8	7	1	2	4
2	2	4	2	3	3	3	4	5	7
3	3	2	1	5	3	5	2	8	2
4	4	2	1	3	3	4	7	1	3
5	2	3	3	6	1	8	5	1	1
6	4	5	2	7	1	9	3	3	4
7	2	4	2	2	3	1	9	1	3
8	5	4	3	1	3	2	1	9	6

Query: SUM(A[4,2]:A[6,5])
Result: 1+3+3+4+3+6+1+8+2+7+1+9 = 48

PS	0	1	2	3	4	5	6	7	8
0	3	8	9	11	13	17	23	26	29
1	10	18	21	29	39	50	57	62	69
2	12	24	29	40	53	67	78	88	102
3	15	29	35	51	67	86	99	117	133
4	19	35	42	61	80	103	123	142	161
5	21	40	50	75	95	126	151	171	191
6	25	49	61	93	114	154	182	205	229
7	27	55	69	103	127	168	205	229	256
8	32	64	81	116	143	186	224	257	290

SUM(A[4,2]:A[6,5]) = PS[6,5]-PS[3,5]-PS[6,1]+PS[3,1]
Result: 154-86-49+29 = 48

Figure 15.4. Original data cube A and corresponding prefix cube PS with range sum query

Geffner et al. address the problem of the high update costs in (Geffner et al. 1999a;b). Their technique retains the constant query time but considerably reduces the update costs at a cost of additional space. In (Riedewald et al. 2000b) Riedewald et al. improve this technique by guaranteeing at least as efficient queries and updates while removing the storage overhead. We describe their Space-Efficient Relative Prefix Sum (SRPS) technique which is based on similar ideas. The reason for the high update costs of PS are the dependencies of the pre-aggregated values. Some of these dependencies are removed in SRPS by partitioning the data cube into smaller hyper-rectangular chunks, called *boxes*. Essentially inner cells of a box store prefix sums local to the box, while the border cells include outside values into their summation. Figure 15.5 shows examples for aggregation regions of box cells. Note that any *prefix* range query can be answered by accessing up to 2^d cells ($2^d - 1$ border cells, 1 inner cell) in the box the endpoint of the query-region falls into. Figure 15.6 shows an example for an SRPS cube and how a prefix sum can be partitioned such that the corresponding box values provide the value. Compared to the PS technique more values are accessed to obtain a prefix sum, but it is still much less costly than adding the values of the original data cube on-the-fly. Arbitrary range sum queries are answered by combining up to 2^d prefix sums (like for the PS technique). The cascading updates are restricted to the box that contains the updated cell. Outside the box only some of the border cells of other boxes are affected as shown for an example in Figure 15.6. The authors show analytically that the optimal tradeoff of query and update costs occurs for choosing the side-length of a box to be $\sqrt{N}$. Then the worst case costs are $(q, u, s) = (4^d, 2^d N^{d/2}, N^d) \approx (4^d, 2^d \sqrt{n}, n)$. Note, that the PS

562   HANDBOOK OF MASSIVE DATA SETS

technique is a special case of SRPS, namely when choosing a box size of 1.

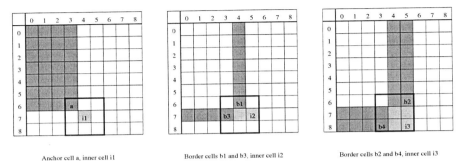

Figure 15.5.   Aggregation regions of cells for SRPS

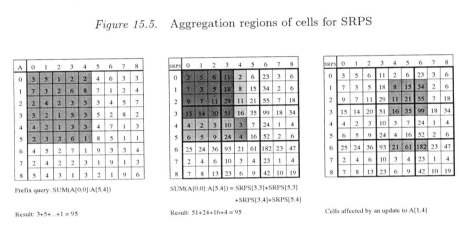

Figure 15.6.   Original data cube A and corresponding SRPS cube with range sum query (middle) and update (right)

What-if analysis and very dynamic environments demand even faster updates. For those applications Geffner et al. (Geffner et al. 2000) propose a technique that balances query and update costs. Both costs are polylogarithmic in the data size, however, at least twice the space of the original data cube is required. In (Riedewald et al. 2000b) an improved technique is proposed that eliminates the storage overhead while at the same time not increasing the cost for queries and updates. An outline of this technique, the Space-Efficient Dynamic Data Cube (SDDC), is given here. To facilitate the description, we first introduce a simplified version of SDDC, the *Basic* SDDC technique. It illustrates how SDDC works, but has high update costs. The Basic SDDC technique partitions the data cube into boxes as done by SRPS. Also, the aggregation regions of the border cells of a box are defined as for SRPS. The number of boxes, in contrast, is bounded by a constant (e.g., divide each dimension in half).

Consequently, the side-length of a box is linear in the domain size. For inner cells of a box, instead of storing local prefix sum values, a recursive approach is taken. To be more specific, the same partitioning into boxes is recursively applied to the region of inner box cells. The recursive partitioning defines a tree. Queries and updates are processed by descending this tree. Figure 15.7 shows a Basic SDDC cube; in Figure 15.8 the tree structure is illustrated and how a query is processed. Note, that the update cost is high, since already at the root level $O(N^{d-1})$ border cells might be affected by an update (e.g., update to cell [1, 1]). This is due to the fact that border cell values are cumulative. SDDC reduces the update costs by encoding the values in the surfaces of border cells. For 2-dimensional data cubes the two 1-dimensional border cell surfaces are stored in balanced binary trees that store intermediate sums and are embedded into the space of the border cells. The surfaces of data cubes with $d > 2$ dimensions are recursively encoded as $(d-1)$-dimensional data cubes. All values "fit" into the space of those border cells whose values they encode. That way the SDDC technique guarantees worst case costs of $(q, u, s) = (2^d \log^d N, \log^d N, N^d) \approx (\frac{2^d}{d^d} \log^d n, \frac{1}{d^d} \log^d n, n)$. The main disadvantage is that due to its recursive nature the technique is not easy to analyze (e.g., for average costs) and to implement.

A	0	1	2	3	4	5	6	7	8	9
0	3	5	1	2	2	4	6	3	3	1
1	7	3	2	6	8	7	1	2	4	2
2	2	4	2	3	3	3	4	5	7	4
3	3	2	1	5	3	5	2	8	2	1
4	4	2	1	3	3	4	7	1	3	2
5	2	3	3	6	1	8	5	1	1	2
6	4	5	2	7	1	9	3	3	4	1
7	2	4	2	2	3	1	9	1	3	3
8	5	4	3	1	3	2	1	9	6	5
9	6	1	2	4	2	1	3	1	5	2

Prefix query: SUM(A[0,0]:A[7,8])

Result: 3+5+...+3 = 256

SDDC	0	1	2	3	4	5	6	7	8	9
0	3	5	6	8	10	17	6	9	12	13
1	7	3	2	11	8	33	1	2	7	2
2	9	4	2	3	3	50	4	5	7	4
3	12	9	5	28	14	69	7	15	35	7
4	16	2	1	6	3	86	7	1	11	2
5	18	19	29	54	74	126	25	45	65	77
6	4	5	2	14	1	28	3	3	10	1
7	6	4	2	8	3	42	9	1	13	3
8	11	13	7	30	7	60	13	9	39	9
9	17	1	2	7	2	76	3	1	9	2

SUM(A[0,0]:A[7,8]) = SDDC[5,5]+SDDC[7,5]+SDDC[4,8]+SDDC[6,8]+SDDC[7,8]

Result: 126+42+65+10+13 = 256

Figure 15.7. Original data cube A and corresponding Basic SDDC cube with range sum query

Chan and Ioannidis (Chan and Ioannidis 1999b) apply ideas from bitmap index design to the range query problem and define an interesting class of so called Hierarchical Cubes (HC). The data cube is hierarchically partitioned into smaller boxes of equal size. According to the partitioning, the cells are assigned to classes. Two techniques, Hierarchical Rectangle Cubes (HRC) and Hierarchical Band Cubes (HBC) are proposed that define aggregation regions for the cells depending on the classes. Thus, by having different hierarchical partitionings and differ-

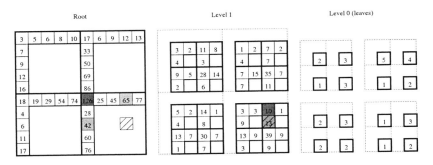

Figure 15.8. Tree structure of the Basic SDDC cube (query endpoint hatched, accessed cells shaded)

ent ways of assigning aggregation regions, a variety of tradeoffs between query and update cost can be generated. Figure 15.9 shows an example for HBC when each dimension is recursively split in half. The shading in the middle cube indicates the level of the cell (darkest at root level). In the right cube it indicates how the query result is obtained as the sum of two pre-computed values that are the range sum for the corresponding region. Like the other techniques, Hierarchical Band Cubes compute the result for arbitrary range sum queries by combining up to 2^d prefix sums. All Hierarchical Cubes have a storage cost of $N^d \approx n$. The formulas for query and update costs are complex (see (Chan and Ioannidis 1999b)). Finding optimal configurations therefore requires an experimental evaluation.

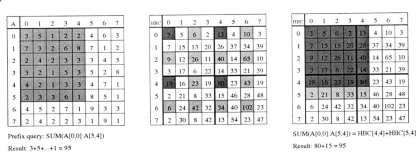

Figure 15.9. Original data cube A and a possible HBC cube

All of the techniques presented in this section so far handle data cubes of any dimensionality by dealing with all dimensions "at the same time" and treating the dimensions uniformly. The algorithms are typically complex (e.g., (Chan and Ioannidis 1999b, Geffner et al. 2000)) and it is difficult to prove their correctness and to analyze their performance. Riedewald et al. (Riedewald et al. 2001) address this problem with the

Iterative Data Cubes (IDC) technique. The proposed approach provides a modular framework for combining one-dimensional pre-aggregation techniques to create space-optimal high-dimensional pre-aggregated data cubes. By selecting the appropriate one-dimensional techniques, specific properties of the dimensions can be taken into account, e.g., hierarchies. The iterative pre-aggregation process essentially reduces the high-dimensional pre-aggregation problem to the one-dimensional case. Thus the space of possible tradeoffs between query and update costs is restricted compared to more general approaches. However, it is shown that the tradeoffs of previously proposed techniques are matched.

While the above techniques address the problem of finding good cost tradeoffs between queries and updates, Ho et al. (Ho et al. 1997b) examine tradeoffs for query and storage cost. They analyze and develop covering codes to support a generalization of range queries, called *partial sum queries*. A partial sum query can select dimension "ranges" which are not contiguous.

In (Ho et al. 1997a) a technique for the efficient computation of range max queries is proposed. The data cube is recursively partitioned into smaller hyper-rectangular chunks (similar to a multidimensional quad tree). For each chunk the position of the maximum value in the chunk is stored. The corresponding tree is traversed top-down, using pruning techniques to reduce the number of accesses. Lee et al. (Lee et al. 2000) improve on this result. By using a different pruning strategy they obtain constant query cost on the average and logarithmic update cost in the size of the data cube (for fixed dimensionality).

5.2. Techniques for Sparse Data Cubes

For sparse data cubes the number of non-empty cells n is much smaller than the size of the data cube N^d. Applying one of the techniques from Section 5.1 fills a large number of the non-empty cells and the pre-aggregated cube will be dense. Computing, storing and maintaining a dense pre-aggregated data cube of size N^d is infeasible already for medium dimensionality and small dimension domains. For instance, a 6-dimensional data cube with only 50 values per dimension has already more than 10^{10} cells. Consequently, sparse high-dimensional data cubes require different techniques.

The theory of processing range queries on sparse data sets has been extensively studied by computational geometry and database researchers. Here only the most relevant results are surveyed. Lower bounds for the problem were proven by Fredman (Fredman 1981), Willard (Willard 1986), and Chazelle (Chazelle 1990). Their results are shown in Table 15.3. The cost of an algorithm is measured as the number of arithmetic operations for obtaining the query result (e.g., number of necessary additions). To make the bounds as general as possible, access costs are not taken into account. The lower bounds for semigroups are valid for aggregate operators that are not invertible (e.g., MAX and MIN for any numerical attribute). The results for groups are valid for invertible aggregate operators like SUM and COUNT when the inverse elements exist in the attribute's domain (and therefore also for AVG which can be computed from SUM and COUNT). In (Chazelle 1988) Chazelle presents algorithms with the best known asymptotic upper bounds for operators that are not invertible, i.e., for semigroups (Tables 15.4 and 15.5). It is distinguished between the static (no updates) and the dynamic (updates allowed) case. Note, that by taking advantage of an operator's properties better algorithms are possible, e.g., the cost for MAX is lower than for SUM. The bounds for the dynamic case are obtained for a weak machine model. For models closer to a real computer it might be possible to lower the costs. For further details the reader is advised to refer to the original papers. While the cited results provide valuable insights to the range query problem, they are typically not of high practical relevance for data cubes. For instance a space consumption of $n \log^{d-2} n$ can easily be infeasible for high-dimensional data cubes. Also, the theoretical results often assume unbounded dimension sizes. However, by taking advantage of the fact that in practice data types are of bounded size (e.g., real numbers are mapped to floating point numbers of 8 bytes size) more efficient algorithms can be designed.

Table 15.3. Lower bounds for orthogonal range searching (costs in terms of arithmetic operations over the (semi)group).

Setting	Lower bound
Commutative semigroup[a] (query, insert, delete)	$q + u \in \Omega(\log^d n)$
Abelian group[b] (query, insert, delete)	$q + u \in \Omega(\log^d n)$
Faithful semigroup[c] (query, m units of storage)	$q(m) \in \Omega\left(\left(\frac{\log n}{\log(\frac{2m}{n})}\right)^{d-1}\right)$
Faithful semigroup (query, insert)	$q + u \in \Omega\left(\left(\frac{\log n}{\log \log n}\right)^d\right)$

[a]This bound is only valid for algorithms that are correct for any commutative semigroup.
[b]The pre-computed values have to be orthogonal range sums.
[c]The bound is tight for $m \in \Omega(n \log^{d-1+\epsilon} n)$ (fixed $\epsilon > 0$), and for $m \in O(n)$ for some semigroups like (N, MAX).

Table 15.4. Best known tradeoffs for orthogonal range searching (semigroup, static case, i.e., no updates)[a].

Aggregate	Query	Space
COUNT	$O(\log^{d-1} n)$	$O(n \log^{d-2} n)$
MAX/MIN	$O(\log^{d-1+\epsilon} n)$	$O(n \log^{d-2} n)$
	$O(\log^{d-1} n \log \log n)$	$O(n \log^{d-2} n \log \log n)$
	$O(\log^{d-1} n)$	$O(n \log^{d-2+\epsilon} n)$
SUM	$O(\log^{d+\epsilon} n)$	$O(n \log^{d-2} n)$
	$O(\log^{d} n \log \log n)$	$O(n \log^{d-2} n \log \log n)$
	$O(\log^{d} n)$	$O(n \log^{d-2+\epsilon} n)$

[a]$\epsilon > 0$.

Table 15.5. Best known tradeoffs for orthogonal range searching (semigroup, dynamic case, i.e., including updates).

Aggregate	Query	Update	Space
COUNT	$O(\log^{d} n)$	$O(\log^{d} n)$	$O(n \log^{d-2} n)$
MAX/MIN	$O(\log^{d+1} n)$	$O(\log^{d+1} n)$	$O(n \log^{d-2} n)$
SUM	$O(\log^{d+2} n)$	$O(\log^{d+2} n)$	$O(n \log^{d-2} n)$

A straightforward practical solution for sparse data cubes is to only store the non-empty cells (all other cells are implicitly known to be empty). Index structures can be used to speed up queries and updates. Possible index structures are surveyed in (Gaede and Günther 1998). More recent results are proposed in (Böhm and Kriegel 2000, Chan and Ioannidis 1999a, Ester et al. 2000, Markl et al. 1999, O'Neil and Quass 1997). This approach is followed by Relational OLAP (ROLAP) systems. Interestingly, for high dimensionality even the most sophisticated index structures can be outperformed by a simple sequential scan of the complete data set. (Böhm and Kriegel 2000, Weber et al. 1998) are only two of the many examples of techniques that try to tackle this so called "curse of dimensionality". Alternatively one can try to retain as many of the advantages of implementing multidimensional OLAP systems similar to arrays. To make this feasible, the multidimensional arrays use techniques that compress sparse regions. This is done by Multidimensional OLAP (MOLAP) systems. However, special support for range queries is typically not provided. (Berson and Smith 1997, Pendse and Creeth 2000) discuss the differences between ROLAP and MOLAP in more detail. In Section 6 the aspect of compression will be discussed. In the following we present a specialized technique for support of range queries on sparse data cubes.

In (Riedewald et al. 2000a) Riedewald et al. propose a technique, pCube, that combines features of an index structure with special support for interactive analysis. The idea is to recursively partition the data cube into smaller hyper-rectangular chunks. This defines a tree where the children of a node cover parts of their parents chunk. The region of the data cube that corresponds to a certain node is termed its *node-region*. Each node stores simple statistics about the non-empty cells in its node-region, e.g., SUM, COUNT, MAX, MIN of the values. This information is referred to as the *node-information*. A query processes the tree top-down, starting at the root and then descending to all children whose node-region intersects the query-region. Based on the information in the nodes visited so far, an approximate answer is returned. The node-information is chosen in such a way, that the approximate answer is accompanied with *absolute* error bounds. The further the query descends the tree, the more accurate the result. Figure 15.10 shows an example for a simple quad-tree-like pCube whose nodes only store the sum of the values in the node-regions. The query-region is shaded, the output is given in the figure. Note, that the approximate result is obtained by assuming uniform distribution of the values. To compute the error bounds, the extreme cases are evaluated (all/no non-empty cells fall into the query-region). A user or application can select an individual tradeoff between query cost and accuracy. When the query reaches the leaf nodes, the exact result is obtained smoothly from the same structure. The continuous feedback is especially helpful for interactive analysis. However, good worst case costs as for dense data cubes are not guaranteed. The space is linear in the number of non-empty cells. Typically less than twice the space of the set of non-empty cells is consumed. The query cost is proportional to the number of nodes whose node-region is intersected by the query-region. Savings can be expected for practical applications when the query-region completely contains a node-region. Then this node's information already provides the exact query result for its node-region without further descending its subtree (e.g., upper left of the internal nodes in Figure 15.10). The update cost is linear in the height of the tree for sets of statistical values that are *self-maintainable* with respect to insertions and deletions. A set of aggregate functions is self-maintainable if the new values of the functions can be computed solely from the old values and from the changes caused by the update operation (see (Mumick et al. 1997)). Consequently COUNT and SUM are self-maintainable. AVG is not self-maintainable, but can be made so by computing it from SUM and COUNT. MAX, MIN, and median are not self-maintainable (except for the trivial case that all base values are stored). However, pCube can efficiently handle MAX and MIN due to its hierarchical structure. Note that pCube is an example for compression on different levels which frees the system or developer from choosing the "best" granularity of the approximation. Other compression techniques are discussed in the next section.

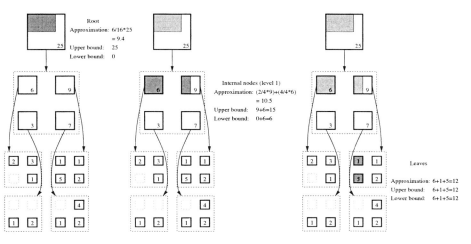

Figure 15.10. Processing a query on a simple pCube for aggregate operator SUM

6. Compressed Data Cubes and Approximate Answers

The goal of compression, also called data reduction, is to find a smaller representation of the data set that leads to less space consumption and also reduces access costs by avoiding access to and transfer of "unimportant" data. In the New Jersey Data Reduction Report (Barbará et al. 1997) various compression techniques are surveyed. Those techniques are naturally related to *approximate* responses to queries. The main issue for all compression techniques is to find one that fits the data (e.g., dimensionality, skew, sparsity) and the application (e.g., ratio of queries versus updates, type of queries). In addition some techniques require the selection of an appropriate approximation granularity (e.g., number of coefficients that describe the data set). In the following compression techniques which were proposed for data cubes are surveyed.

A general technique to compress a sparse data cube is to partition it into smaller chunks. The chunks are either obtained by partitioning according to the data distribution or according to the geometry of the data cube, independent of where the non-empty cells are (often regular partitioning). Dense and sparse chunks are each encoded by specialized techniques. For instance Goil et al. (Goil and Choudhary 1997) propose a bit-encoded sparse structure for sparse chunks and compare it to other sparse approaches. Their technique takes advantage of bit operations which can be processed very efficiently. In (Goil and Choudhary 1999) the same authors extend the technique for use on parallel machines. Zhao et al. (Zhao et al. 1997) propose storing dense chunks as multidimensional arrays, while sparse chunks are encoded using chunk-offset compression. The DROLAP technique (Cheung et al. 1999) identifies dense regions in the data cube. Those are stored as multidimensional arrays (MOLAP) and indexed with a multidimensional index structure (e.g., an R-tree (Guttman 1984)). Non-empty cells in sparse regions are stored in relational tables with supporting index structures (ROLAP). As long as most non-empty cells occur in clusters and those clusters are dense enough, the approach can combine the advantages of both worlds – ROLAP and MOLAP. Chunking in general will be most efficient when regions in the data cube exist that have properties that can be exploited by specialized algorithms which are more efficient than general ones.

The techniques proposed in (Barbará and Sullivan 1997, Barbará and Wu 1999b) also chunk the data cube. Dense chunks are approximated by statistical models, e.g., regression or loglinear models. Outliers from dense chunks, i.e., values that are not well enough approximated by the model, and values from sparse chunks are retained separately. Queries are answered by approximating each single value in the query-region based on the models. By reading outliers and retained values from sparse chunks, the initial approximate result can be refined. When a certain per-cell relative error is maintained, error bounds can be returned. However, maintaining the per-cell error under updates might require frequent recomputation of the approximation model. The approach is most efficient for low to medium dimensionality and when the non-empty cells occur in dense clusters. An advantage of using parametric approximation models is that the information provided by the parameters can be exploited for data mining (Barbará and Wu 1999a).

An established data reduction technique that works well for one-dimensional data sets is the histogram. (Poosala 1997) provides an overview of the field, and (Aboulnaga and Chaudhuri 1999, Jagadish et al. 1998, Koudas et al. 2000) some of the recent results. For high dimensionality there are two general approaches. The first is to maintain a one-dimensional histogram for each of the dimensions independently. The multidimensional distribution is approximated by combining the histograms using the attribute value independence assumption. Unfortunately correlations and dependencies between the dimensions are lost, i.e., one can expect high approximation errors. As an alternative high-dimensional histograms (Poosala and Ioannidis 1997) that retain the dependencies between the dimensions are proposed. The disadvantage of this approach is, that the histogram buckets have to be small in order to provide a good approximation quality. Consequently the number of buckets must grow exponentially with increasing dimensionality (up to almost the size of the set of non-empty cells). Lee et al. (Lee et al. 1999) show that previous multidimensional histogram techniques are only useful up to three or four dimensions. By compressing the set of histogram buckets with the discrete cosine transform more buckets can be maintained, resulting in good approximations for up to ten dimensions. A technique for selecting histograms for the extended data cube is proposed in (Poosala and Ganti 1999). In general, present histogram techniques are best suited for low to medium dimensionality, i.e., up to about ten dimensions. Histograms are typically not efficiently incrementally maintainable and updates have to be performed in batches. Also, histograms only offer a more or less fixed accuracy. If a user requires more accuracy for a certain query, the query has to start all over on the uncompressed data cube.

Vitter et al. (Vitter and Wang 1999, Vitter et al. 1998) propose approximating data cubes using the wavelet transform. The technique in (Vitter et al. 1998) is intended for dense data cubes, while (Vitter and Wang 1999) explicitly deals with the aspect of sparsity. Wavelets offer a compact representation with good approximation on multiple levels of resolution. This theoretically offers different tradeoffs between query cost and accuracy. However, since the proposed techniques do not return error bounds, the feature can not be taken advantage of in practice. Also, to efficiently handle sparse cubes, the algorithm in (Vitter and Wang 1999) drops less important coefficients during the computation. Thus the accuracy is restricted by the method and depends on the data set. For higher accuracy the query must access the uncompressed data cube.

A technique by Shanmugasundaram et al. (Shanmugasundaram et al. 1999) achieves very high compression ratios of 1000 and more. The statistical structure of the data is modeled by estimating the probability density. This is done by finding clusters in the data set and approximating them by a multivariate Gaussian probability density function. The answer to a query is obtained by integrating over this approximated density function. Consequently the answer is approximate. Probabilistic error bounds can be returned. However, these bounds are not specific to the query, but rather of the form "90% of all queries that contain at least 1000 points are at most 10% off the exact result". The technique is best suited for low-dimensional clustered data sets. Updates that trigger the recomputation of the clusters and the approximation model are expensive. Like for histograms and wavelet transform, the accuracy is fixed by the technique and the data set. An interesting feature is that all attributes are treated equally. There is no distinction between dimensions and measure attribute. Thus one can also select on the "measure attribute" (e.g., include only sales higher than 100).

As pointed out before, an important requirement for any technique that returns approximate results is that an indication of the accuracy be returned as well. In the following two examples are discussed that are not specifically designed for data cubes, but which prove useful for the field as well.

The main goal of the CONTROL project at the University of California at Berkeley (Hellerstein et al. 1999) is to develop algorithms that provide support for interactive analysis. Queries continuously return results which are approximate answers with probabilistic error bounds, i.e., confidence intervals. Traditional algorithms which are optimized to minimize the time to completion of the query are typically not suitable for returning useful intermediate results. Consequently new algorithms for interactive analysis are developed, e.g., ripple joins (Haas and Hellerstein 1999). Techniques for the AQUA system (Acharya et al. 1999) have similar features in the sense that fast approximate results with probabilistic error bounds are returned to the user. The approach, however, is different from CONTROL. AQUA uses synopsis data structures, more precisely histograms and samples. Thus, while CONTROL algorithms can smoothly run till completion and return the exact results, in AQUA the accuracy of the result and the possible refinement steps are determined by the maintained synopsis structures. In both cases the user benefits from getting an early indication about the expected result. The error bounds, however, are only *probabilistic* (confidence intervals) and hence it is not guaranteed that the final result will be within the bounds, especially for skewed distributions and small result sets. In that case *absolute* error bounds as returned by techniques proposed in (Riedewald et al. 2000a, Rowe 1985) are preferable. On the other hand, whenever probabilistic bounds are appropriate, they can be expected to be tighter than their absolute counterparts, since the latter have to take the extremes into account.

7. Conclusion

This article surveyed techniques for interactively analyzing massive data sets using data cubes. We mainly focussed on support for general aggregation (extended data cube), range queries and compression. While data cubes in general are useful for all types of OLAP queries like roll-up, drill-down, slice-and-dice, pivot, and "what-if", we did not discuss specific techniques for all these queries in detail. For instance, for roll-up and drill-down queries OLAP systems must provide special support for attribute hierarchies. This by itself is a field of active research (Chen et al. 2000, Jagadish et al. 1999). One of the major problems, not only in OLAP applications, but for database systems in general, is how to efficiently deal with aggregation on high-dimensional sparse data sets. While some aspects of pre-computation, e.g., which of the cuboids of an extended data cube to materialize, are relatively well understood, the general problem of how much and what to pre-compute still remains an open problem. It is necessary to find the right balance between query, update, and storage cost. This is of particular importance for sparse data cubes.

Acknowledgments

This work was partially supported by NSF grants EIA-98-18320, IIS-98-17432, and IIS-99-70700.

Bibliography

A. Aboulnaga and S. Chaudhuri. Self-tuning histograms: Building histograms without looking at data. In *Proc. Int. Conf. on Management of Data (SIGMOD)*, pages 181–192, 1999.

S. Acharya, P. B. Gibbons, V. Poosala, and S. Ramaswamy. Join synopses for approximate query answering. In *Proc. Int. Conf. on Management of Data (SIGMOD)*, pages 275–286, 1999.

E. Baralis, S. Paraboschi, and E. Teniente. Materialized view selection in a multidimensional database. In *Proc. Int. Conf. on Very Large Databases (VLDB)*, pages 156–165, 1997.

D. Barbará et al. The new jersey data reduction report. *Data Engineering Bulletin*, 20(4), 1997.

D. Barbará and M. Sullivan. Quasi-cubes: Exploiting approximations in multidimensional databases. *SIGMOD Record*, 26(3):12–17, 1997.

D. Barbará and X. Wu. Using approximations to scale exploratory data analysis in datacubes. In *Proc. Int. Conf. on Knowledge Discovery and Data Mining (SIGKDD)*, pages 382–386, 1999a.

D. Barbará and X. Wu. Using loglinear models to compress datacubes. Technical report, George Mason University, 1999b.

A. Berson and S. J. Smith. *Data Warehousing, Data Mining, and OLAP*. McGraw-Hill, 1997.

K. Beyer and R. Ramakrishnan. Bottom-up computation of sparse and iceberg CUBEs. In *Proc. Int. Conf. on Management of Data (SIGMOD)*, pages 359–370, 1999.

C. Böhm and H.-P. Kriegel. Dynamically optimizing high-dimensional index structures. In *Proc. Int. Conf. on Extending Database Technology (EDBT)*, pages 36–50, 2000.

C. Y. Chan and Y. E. Ioannidis. An efficient bitmap encoding scheme for selection queries. In *Proc. Int. Conf. on Management of Data (SIGMOD)*, pages 215–226, 1999a.

C.-Y. Chan and Y. E. Ioannidis. Hierarchical cubes for range-sum queries. In *Proc. Int. Conf. on Very Large Databases (VLDB)*, pages 675–686, 1999b. Extended version published as Technical Report, Univ. of Wisconsin, 1999.

S. Chaudhuri and U. Dayal. An overview of data warehousing and olap technology. *SIGMOD Record*, 26(1):65–74, 1997.

B. Chazelle. A functional approach to data structures and its use in multidimensional searching. *SIAM Journal on Computing*, 17(3):427–462, 1988.

B. Chazelle. Lower bounds for orthogonal range searching: II. The arithmetic model. *Journal of the ACM*, 37(3):439–463, 1990.

Q. Chen, U. Dayal, and M. Hsu. An OLAP-based scalable web access analysis engine. In *Proc. Int. Conf. on Data Warehousing and Knowledge Discovery (DaWaK)*, pages 210–223, 2000.

D. W. Cheung, B. Zhou, B. Kao, K. Hu, and S. D. Lee. DROLAP — a dense-region based approach to on-line analytical processing. In *Proc. Int. Conf. on Database and Expert Systems Applications (DEXA)*, pages 761–770, 1999.

M. Ester, J. Kohlhammer, and H.-P. Kriegel. The dc-tree: A fully dynamic index structure for data warehouses. In *Proc. Int. Conf. on Data Engineering (ICDE)*, pages 379–388, 2000.

M. L. Fredman. A lower bound on the complexity of orthogonal range queries. *Journal of the ACM*, 28(4):696–705, 1981.

V. Gaede and O. Günther. Multidimensional access methods. *ACM Computing Surveys*, 30(2):170–231, 1998.

S. Geffner, D. Agrawal, and A. El Abbadi. The dynamic data cube. In *Proc. Int. Conf. on Extending Database Technology (EDBT)*, pages 237–253, 2000.

S. Geffner, D. Agrawal, A. El Abbadi, and T. Smith. Relative prefix sums: An efficient approach for querying dynamic OLAP data cubes. In *Proc. Int. Conf. on Data Engineering (ICDE)*, pages 328–335, 1999a.

S. Geffner, M. Riedewald, D. Agrawal, and A. El Abbadi. Data cubes in dynamic environments. *Data Engineering Bulletin*, 22(4):31–40, 1999b.

S. Goil and A. Choudhary. BESS: Sparse data storage of multidimensional data for OLAP and data mining. Technical report, Northwestern University, 1997.

S. Goil and A. Choudhary. An infrastructure for scalable parallel multidimensional analysis. In *Proc. Int. Conf. on Scientific and Statistical Database Management (SSDBM)*, pages 102–111, 1999.

J. Gray, S. Chaudhuri, A. Bosworth, A. Layman, D. Reichart, M. Venkatrao, F. Pellow, and H. Pirahesh. Data cube: A relational aggregation operator generalizing group-by, cross-tab, and sub-totals. *Data Mining and Knowledge Discovery*, pages 29–53, 1997.

H. Gupta. Selection of views to materialize in a data warehouse. In *Proc. Int. Conf. on Database Theory (ICDT)*, pages 98–112, 1997.

H. Gupta, V. Harinarayan, A. Rajaraman, and J. D. Ullman. Index selection for OLAP. In *Proc. Int. Conf. on Data Engineering (ICDE)*, pages 208–219, 1997.

H. Gupta and I. S. Mumick. Selection of views to materialize under a maintenance cost constraint. In *Proc. Int. Conf. on Database Theory (ICDT)*, pages 453–470, 1999.

A. Guttman. R-trees: A dynamic index structure for spatial searching. In *Proc. Int. Conf. on Management of Data (SIGMOD)*, pages 47–57, 1984.

P. J. Haas and J. M. Hellerstein. Ripple joins for online aggregation. In *Proc. Int. Conf. on Management of Data (SIGMOD)*, pages 287–298, 1999.

V. Harinarayan, A. Rajaraman, and J. D. Ullman. Implementing data cubes efficiently. In *Proc. Int. Conf. on Management of Data (SIGMOD)*, pages 205–216, 1996.

J. M. Hellerstein et al. Interactive data analysis: The CONTROL project. *IEEE Computer*, pages 51–59, August 1999.

C. Ho, R. Agrawal, N. Megiddo, and R. Srikant. Range queries in OLAP data cubes. In *Proc. Int. Conf. on Management of Data (SIGMOD)*, pages 73–88, 1997a.

C. Ho, J. Bruck, and R. Agrawal. Partial-sum queries in OLAP data cubes using covering codes. In *Proc. Symp. on Principles of Database Systems (PODS)*, pages 228–237, 1997b.

H. V. Jagadish, N. Koudas, S. Muthukrishnan, V. Poosala, K. Sevcik, and T. Suel. Optimal histograms with quality guarantees. In *Proc. Int. Conf. on Very Large Databases (VLDB)*, pages 275–286, 1998.

H. V. Jagadish, L. V. S. Lakshmanan, and D. Srivastava. What can hierarchies do for data warehouses? In *Proc. Int. Conf. on Very Large Databases (VLDB)*, pages 530–541, 1999.

T. Johnson and D. Shasha. Some approaches to index design for cube forests. *IEEE Data Engineering Bulletin*, 20(1):27–35, 1997.

T. Johnson and D. Shasha. Some approaches to index design for cube forests. *Data Engineering Bulletin*, 22(4):31–40, 1999.

Y. Kotidis and N. Roussopoulos. An alternative storage organization for ROLAP aggregate views based on cubetrees. In *Proc. Int. Conf. on Management of Data (SIGMOD)*, pages 249–258, 1998.

N. Koudas, S. Muthukrishnan, and D. Srivastava. Optimal histograms for hierarchical range queries. In *Proc. Symp. on Principles of Database Systems (PODS)*, pages 196–204, 2000.

J.-H. Lee, D.-H. Kim, and C.-W. Chung. Multi-dimensional selectivity estimation using compressed histogram information. In *Proc. Int. Conf. on Management of Data (SIGMOD)*, pages 205–214, 1999.

S. Y. Lee, T. W. Ling, and H. G. Li. Hierarchical compact cube for range-max queries. In *Proc. Int. Conf. on Very Large Databases (VLDB)*, pages 232–241, 2000.

V. Markl, F. Ramsak, and R. Bayer. Improving OLAP performance by multidimensional hierarchical clustering. In *Proc. Int. Database Engineering and Applications Symp. (IDEAS)*, pages 165–177, 1999.

I. S. Mumick, D. Quass, and B. S. Mumick. Maintenance of data cubes and summary tables in a warehouse. In *Proc. Int. Conf. on Management of Data (SIGMOD)*, pages 100–111, 1997.

P. E. O'Neil and D. Quass. Improved query performance with variant indexes. In *Proc. Int. Conf. on Management of Data (SIGMOD)*, pages 38–49, 1997.

N. Pendse and R. Creeth. The OLAP report. Parts available online in the current edition. http://www.olapreport.com/Analyses.htm, 2000.

V. Poosala. *Histogram-Based Estimation Techniques in Databases*. PhD thesis, Univ. of Wisconsin-Madison, 1997.

V. Poosala and V. Ganti. Fast approximate answers to aggregate queries on a data cube. In *Proc. Int. Conf. on Scientific and Statistical Database Management (SSDBM)*, pages 24–33, 1999.

V. Poosala and Y. E. Ioannidis. Selectivity estimation without the attribute value independence assumption. In *Proc. Int. Conf. on Very Large Databases (VLDB)*, pages 486–495, 1997.

M. Riedewald, D. Agrawal, and A. El Abbadi. pCube: Update-efficient online aggregation with progressive feedback and error bounds. In *Proc. Int. Conf. on Scientific and Statistical Database Management (SSDBM)*, pages 95–108, 2000a.

M. Riedewald, D. Agrawal, and A. El Abbadi. Flexible data cubes for online aggregation. In *Proc. Int. Conf. on Database Theory (ICDT)*, 2001. To appear.

M. Riedewald, D. Agrawal, A. El Abbadi, and R. Pajarola. Space-efficient data cubes for dynamic environments. In *Proc. Int. Conf. on Data Warehousing and Knowledge Discovery (DaWaK)*, pages 24–33, 2000b.

N. C. Rowe. Antisampling for estimation: An overview. *IEEE Transactions on Software Engineering*, 11(10):1081–1091, 1985.

S. Sarawagi and M. Stonebraker. Efficient organization of large multidimensional arrays. In *Proc. Int. Conf. on Data Engineering (ICDE)*, pages 328–336, 1994.

J. Shanmugasundaram, U. Fayyad, and P. S. Bradley. Compressed data cubes for OLAP aggregate query approximation on continuous dimensions. In *Proc. Int. Conf. on Knowledge Discovery and Data Mining (SIGKDD)*, pages 223–232, 1999.

A. Shukla, P. Deshpande, and J. F. Naughton. Materialized view selection for multi-cube data models. In *Proc. Int. Conf. on Extending Database Technology (EDBT)*, pages 269–284, 2000.

J. R. Smith, V. Castelli, A. Jhingran, and C.-S. Li. Dynamic assembly of views in data cubes. In *Proc. Symp. on Principles of Database Systems (PODS)*, pages 274–283, 1998.

Transaction Processing Performance Council. TPC-H benchmark (1.1.0), 2001. Available at http://www.tpc.org.

U.S. Census Bureau. http://www.census.gov, 2001.

J. S. Vitter and M. Wang. Approximate computation of multidimensional aggregates of sparse data using wavelets. In *Proc. Int. Conf. on Management of Data (SIGMOD)*, pages 193–204, 1999.

J. S. Vitter, M. Wang, and B. Iyer. Data cube approximation and histograms via wavelets. In *Proc. Intl. Conf. on Information and Knowledge Management (CIKM)*, pages 96–104, 1998.

R. Weber, H.-J. Schek, and S. Blott. A quantitative analysis and performance study for similarity-search methods in high-dimensional spaces. In *Proc. Int. Conf. on Very Large Databases (VLDB)*, pages 194–205, 1998.

D. E. Willard. Lower bounds for dynamic range query problems that permit subtraction. In *Proc. Int. Coll. on Automata, Languages and Programming*, number 226 in LNCS, pages 444–453, 1986.

Y. Zhao, P. M. Deshpande, and J. F. Naughton. An array-based algorithm for simultaneous multidimensional aggregates. In *Proc. Int. Conf. on Management of Data (SIGMOD)*, pages 159–170, 1997.

Chapter 16

DATA SQUASHING: CONSTRUCTING SUMMARY DATA SETS

William DuMouchel
AT&T Labs Research
Florham Park, NJ 07932, USA
dumouchel@research.att.com

Abstract   A "large dataset" is here defined as one that cannot be analyzed using some particular desired combination of hardware and software because of computer memory constraints. DuMouchel et al. (1999) defined "data squashing" as the construction of a substitute smaller dataset that leads to approximately the same analysis results as the large dataset. Formally, data squashing is a type of lossy compression that attempts to preserve statistical information. To be efficient, squashing must improve upon the common strategy of taking a random sample from the large dataset. Three recent papers on data squashing are summarized and their results are compared.

Keywords: Sampling, substitute dataset, moment matching

1.   Introduction

One of the chief obstacles to effective data mining is the clumsiness of managing and analyzing data in very large files. The process of model search and model fitting often require many passes over a large dataset, or random access to the elements of a large dataset. Many statistical fitting algorithms assume that the entire dataset being analyzed fits into computer memory, restricting the number of feasible analyses. Here we define "large dataset" as one that cannot be analyzed using some particular desired combination of hardware and software because of computer memory constraints. There are two basic approaches to this problem: either switch to a different hardware/software/analysis strategy or else substitute a smaller dataset for the large one. Here we assume that the former strategy is unavailable or undesirable and consider ways of constructing a smaller substitute dataset. This latter approach was named data squashing by DuMouchel et al. (1999). Formally, data squashing is a form of lossy compression that attempts to preserve statistical information. Suppose that the original or "mother" dataset is a matrix Y having N rows or entities and n columns or variables. The squashed dataset is a matrix X having M rows and $n + 1$ columns, where $M << N$. The extra column in X is a column of weights, $w_i, i = 1, \ldots, M$, where $w_i > 0$ and $\sum_i w_i = N$. It is assumed that M is small enough so that X can be processed by the desired hardware/software, and that the software can make appropriate use of the weight variable. The n-dimensional distribution of the rows of X weighted by the w_i is intended to approximate the distribution of the rows of Y well enough that statistical analysis of X is an acceptable substitute for the desired analysis of Y.

There are two trivial forms of data squashing that can often be used as comparison or baseline methods. The first is simple random sampling, in which X consists of a random sample of M rows of Y, each given weight $w_i = N/M$. The biggest disadvantage of this strategy is the inaccuracy introduced by sampling variance. Dividing a sample size by 100 multiplies most variances of estimates by 100 as well. With very large initial sample sizes, this may not be a problem for simple estimates such as overall means or proportions or sample correlations. However, for many business purposes, the detection of small differences, or the detection of trends in a small subset of the overall population, is crucial to the success of the data mining project. In such cases the equivalent of throwing away 99% of the data will be unacceptable. The second trivially easy data squashing method might be called unique row extraction, in which X consists of the set of unique rows of Y, and w_i is the multiplicity of the i-th row of X in Y. If the resulting X is small enough to fit in memory, then we have what might be called perfect or lossless squashing. (One might round each quantitative element of Y slightly before extracting the unique rows, so that the rounded values are still considered fully informative for the purposes of statistical analyses, thereby reducing M, the number of unique rows, and thus X.) For a nontrivial application of squashing, we must have a situation where the X from unique row extraction is too large to analyze with the desired hardware and software and where also the analysis results from simple random samples of size small enough to be so analyzed are deemed to be too variable for the purposes of the analyses.

This chapter summarizes the results of three recent papers on data squashing. In addition to DuMouchel et al. (1999), we consider Madigan et al. (1999) and Owen (1999). All these papers are experimental in nature, and none of them make a conclusive case that the data squashing concept is a significant contribution to the practice of data mining. But they do show that for certain analysis goals data squashing can be at least two orders of magnitude more efficient than random sampling. The three papers describe very different methodologies and theoretical rationales for data squashing. Comparing and contrasting them helps to define the limits of data squashing and to suggest future directions for data squashing research.

2. Three Data Squashing Methods

DuMouchel et al. (1999) presents a theoretical framework and justification for data squashing involving a Taylor series representation for the likelihood function from an arbitrary modeling problem. The Taylor series describes the local behavior of the log likelihood function for each fixed parameter value as the continuous variables in X vary. As such, the theory assumes that each column of X must have a restricted range to achieve an accurate approximation, leading to a strategy of defining multivariate bins in every dimension and repeating the data squashing independently within each bin. The conclusion is that if a low-order Taylor series can approximate the contribution to the log likelihood within each bin, then a strategy of choosing X and w_i to match low-order moments within each bin will allow the squashed dataset to approximately duplicate the corresponding analysis of Y. See DuMouchel et al. (1999) for details of this theoretical justification. An implementation of this method involves separate construction of a weighted set of points for each region of Y defined by fixed values of the categorical variables and fixed ranges of all continuous variables. These points might be constructed to match, for example, the means and covariance matrix of the points in the corresponding rows of Y. The construction involves a search in the constrained $m(Q+1)$-dimensional space defined by the m weights and mQ variable values if there are m points to construct and Q continuous variables. The constraints come about because the weights should be positive and the continuous variable values should lie within their respective fixed ranges.

The model likelihood-based method of Madigan et al. (1999) attempts to build a squashed dataset that approximates a specific likelihood function directly, rather than rely on the general Taylor series argument for approximating all possible likelihood functions. In (Madigan et al. 1999) the authors choose logistic regression for a fixed response variable as the specific model around which the squashing is structured. The resulting squashed dataset may not be as useful as a generically squashed dataset for all possible analyses, but may be more accurate for analysis models similar to logistic regression, for example other classification methods involving the same response variable. The method of approximation involves matching a likelihood profile, that is a vector of values of the log likelihood function at K values in the p-dimensional parameter space of the logistic regression coefficients. The K parameter vectors must be chosen so that a smooth function of p variables can be approximately identified by the corresponding K function values. This is achieved by using the techniques of quadratic response surface estimation from the theory of statistical design of experiments. Whereas DuMouchel et al. (1999) explicitly partitions the data space into bins based on compact regions of the original variables, Madigan et al. (1999) partitions the same data space into regions defined by clustering the likelihood profile vectors contributed by each data point in the mother dataset. Data points having very similar likelihood profiles are deemed equivalent and are merged into a single squashed data point by taking their mean. The corresponding w_i is the number of points so merged. The computations require two one-pass algorithms involving the mother dataset, one to get an approximate estimate of the all-data logistic regression coefficients, and one to perform an approximate clustering of the N likelihood profiles.

In spite of the use of the word "likelihood", the empirical likelihood method of Owen (1999) has more in common with the moment matching method of DuMouchel et al. (1999) than with the model likelihood-based method of Madigan et al. (1999). The methodology of Owen (1999) also directly matches moments of the original variables in the Y dataset to moments in the X dataset. However, Owen (1999) avoids the computationally intensive constrained nonlinear optimizations that Madigan et al. (1999) uses to find solutions. Instead, Owen (1999) starts out with a simple random sample to get the X-values and reweights these sampled points to fit the required moments. The estimation of the w_i involves the maximization of the product of the wi among all weight vectors that satisfy the moment equalities, an algorithm called empirical likelihood estimation and described in Owen (1990). The result is a greatly reduced computational effort. The corresponding downside is primarily that a simple random sample is not an efficient choice of rows of X, so that M must necessarily be large to match a given number of moments.

Table 16.1. Characteristics of the three squashing investigations

Method	Moment Matching within Bins DuMouchel et al. (1999)	Model Likelihood Based Madigan et al. (1999)	Emperical Likelihood Moment Match. (Owen 1999)
Response-variable specific?	No	Yes	Yes[1]
Type of functions matched	Moments of raw data of	Likelihood function profile	Moments of raw data
No. of matched functions	$\approx Mn$	149 M (10 coefs)	$\approx 2n$
Achieve exact match?	No	No	Yes
Generate pseudovalues and weights for rows?	Generate both	Generate both	Sample pseudovalues, generate weights
Computational techniques	Moment computations, constrained nonlinear least squares	One-pass Log. Regr., Likelihood profiles, one-pass clustering	Moment computations, empirical likelihood maximization
Comp. Effort	High	Medium	Low
Largest Y-matrix used in examples	745,000 rows 8 columns	745,000 rows 8 columns	92,000 rows 39 columns
Analyses investigated	Logistic Regression (10 coefs and 48 coefs)	Log.Regr.(5 & 10 coefs) Log. Regr. variable sel. Neural Network classif.	Log. Regr. (39 coefs) Boosted decision trees
Reduction factor (range)	43 – 341	100	11.5 – 92
Efficiency vs. SRS for Log. Regr. coefs.	Up to 656 (10 coefs) Up to 86 (48 coefs)	$\approx 10{,}000$ (5 coefs) Up to 8,100 (10 coefs)	About 4 (39 coefs)
Eff. for alternative analyses	NA	> 1000? (variable sel.) 16 (neural network)	≈ 1 (boosted trees) ≈ 1
Investigate/discuss classification accuracy?	No	Not much	Yes
Simulation-based choice of tuning constants?	No	Yes	No
Factors investigated	Binning strategies, reduction factor, no. of moments fitted	Likelihood profile settings, initial coefficient estimator	Reduction factor

[1] As described, (Owen 1999) specifies the response, but method is easily generalized to match arbritary moments.

For example, Owen (1999) reports an example in which it was not possible to reweight a sample of 500 points to match just 78 moments using the empirical likelihood algorithm. Perhaps because of this problem, Owen (1999) chooses to estimate only very low-order moments and also only estimates M points globally, rather than repeat the estimation and moment matching separately within many disjoint regions of the n-dimensional space.

3. Detailed Comparison of the Three Squashing Investigations

Table 16.1 provides a comparison of the three squashing methodologies for each of 15 characteristics. This section discusses each row of Table 16.1 in turn.

1. Response-variable specific? As discussed above, the original algorithm (DuMouchel et al. 1999) makes no assumption as to which variable is a response, although the example evaluations of squashing's accuracy all use the same response variable. It is assumed but not proven that similar accuracy would be attained for models using other response variables with the same squashed dataset. On the other hand, the algorithm in (Madigan et al. 1999) is very clearly tuned to the particular response variable defined by the likelihood function used to create likelihood profiles. It is assumed that analyses of the squashed dataset involving other response variables would match those of the mother dataset much more poorly. Owen's discussion in (Owen 1999) focuses on a particular response variable, in particular the way preliminary data transformations and imputation of missing values are carried out, and in his choice of moments to match. However, the empirical likelihood squashing methodology seems perfectly general. If the moments and cross-moments being matched are symmetrically defined as to all the variables, there is no reason to think that the squashing would preferentially work better for any one response variable.

2. Type of functions matched. As discussed above, DuMouchel et al. (1999) and Owen (1999) match moments of the raw data, while Madigan et al. (1999) operates in a sort of dual parameter space, matching points with similar likelihood profiles.

3. No. of matched functions. As discussed in more detail in the paper, the algorithm of DuMouchel et al. (1999) estimates weighted pseudopoints separately within each of many bins. The number of pseudopoints constructed within each bin rises proportional to the log of the number of mother data points in the bin. For more populated bins, the number of moments that the construction attempts to match rises to "use up" the degrees of freedom available within the bin. The result is that a bin having only one pseudopoint will match the means of each variable only, while bins with many pseudopoints may involve higher order moments even including all 4th-order moments and cross moments. As a rough approximation, the number of moments is about Mn, if there are M pseudopoints and n variables. Since the implementation described by Owen (1999) does not involve separate estimations within bins, and further focuses on estimating the moments that naïve Bayes classifiers would use, there are only $2n$ moments being fit, irrespective of M. In the Madigan et al. (1999) method, there are K values per likelihood profile and the clustering attempts to match them separately for each of M pseudopoints. The product MK is often about $Mp2$, since it takes about $p2$ design points to estimate a quadratic response surface in p dimensions. In their 10-parameter example, Madigan et al. (1999) uses $K = 149$.

4. *Achieve exact match?* While the DuMouchel et al. (1999) and Madigan et al. (1999) methods involve a great many functions of the data in the matching process, there is no attempt to achieve an exact or globally optimal solution to the matching equations. On the other hand, Owen (1999) requires an exact solution to its specified optimization problem, and therefore requires a much larger M to match a given number of functions. The results in rows 11 and 12 of Table 16.1 seem to indicate that the former strategy is more effective.

5. *Generate pseudovalues and weights for rows?* As discussed above, DuMouchel et al. (1999) and Madigan et al. (1999) estimate both the values of X and the weights, while Owen (1999), obtains the values of X from a random sample and estimates the weights only. Although this may be inefficient in requiring a larger M, it saves much computation and also ensures that only actual mother data values enter into the squashed dataset. (No families with 2.2 children!)

6. *Computational techniques.* As discussed above, the DuMouchel et al. (1999) method must collect very many moments and cross moments from the binned mother data and then solve constrained nonlinear least squares problems to match the moments. The Owen (1999) algorithm collects fewer moments and uses the empirical likelihood method to match these moments by reweighting a random sample. The likelihood profile method of Madigan et al. (1999) avoids iterative computations entirely with two one-pass algorithms, the more laborious second one requiring the computation of a likelihood profile for each point and then immediately assigning it to one of M clusters.

7. *Computational effort.* For the same size M and n, we estimate that DuMouchel et al. (1999) requires the most computational effort to produce the squashed dataset, while Owen (1999) requires the least.

8. *Largest Y-matrix used in the examples.* DuMouchel et al. (1999) and Madigan et al. (1999) each use the same dataset having about 745,000 rows and 8 columns. The dataset in (Owen 1999) has 92,000 rows and 39 columns. All three squashing methods scale up linearly with respect to the number of rows. Scaling up with the number of columns is more problematical. In unreported preliminary calculations, the efficiency of the DuMouchel et al. (1999) and the Madigan et al. (1999) methods drops off as n increases up to 8, so that, for example, one cannot say how well a 39-column implementation of either of these two squashing methods would work.

9. *Analyses investigated.* All three papers use logistic regression as the primary analysis for examples. DuMouchel et al. (1999) provides results for both a main-effects model having 10 coefficients and a second-order model having 48 coefficients. Madigan et al. (1999) uses two main-effects models having 5 and 10 coefficients, respectively. In addition, Madigan et al. (1999) investigates the behavior of all-subsets logistic regression variable selection for simulated datasets having 100,000 rows and 5 coefficients, as well as the behavior of a neural network (Venables and Ripley 1997), having two input units, one hidden layer with three units, and a single dichotomous output unit. In their neural network example, the squashed dataset is constructed using a logistic regression profile likelihood, after which its ability to duplicate the neural network of the mother dataset is evaluated. Owen (1999) presents a main-effects logistic regression with 39 coefficients. It also uses the same squashed datasets to estimate boosted decision trees (Friedman 1999a;b) and compare them to those estimated from the mother dataset.

10. *Reduction factor (range).* The reduction factor is the ratio N/M of the number of rows in Y to the number of rows in X. Examples in DuMouchel et al. (1999) range from 43 to 341, and in Owen (1999) the reduction factor ranges from 11.5 to 92. All the examples in Madigan et al. (1999) have reduction factor equal 100, except for the neural network example reduction factor, which is 10.

11. *Efficiency vs. SRS for Log. Regr. coefs.* The statistical efficiency achieved for the purpose of estimating p regression coefficients, compared to that expected from random sampling, is defined as

$$\text{eff} = Np/(M \sum_{j=1}^{p} (b_j - \beta_j)^2 / \sigma_j^2)$$

where b_j is the estimate of the j-th regression coefficient based on the squashed dataset, β_j is the estimated coefficient from the mother dataset, and σ_j is the standard error of the coefficient based on estimation from the mother dataset. If the squashed dataset is a random sample from the mother dataset, the expected value of *eff* is 1. Each of the three papers presents various results with various values of *eff*, depending on different reduction factors, binning algorithms, likelihood profile settings, and so forth. The values in row 11 of Table 16.1 are the maximum reported efficiencies for the indicated value of p. Here we see dramatic differences in the accuracy of the three methods of creating squashed datasets. The model likelihood-based squashing of Madigan et al. (1999) is far more accurate for logistic regression coefficients than the other two methods, and the moment matching within bins of DuMouchel et al. (1999) is more accurate than the empirical likelihood method of Owen (1999). However, in general accuracy decreases with increasing number of coefficients p, and dataset dimension n, so caution in interpreting these numbers is warranted. It is probable that the likelihood profile method benefits "unfairly" from being based on the very model that it is being evaluated on.

12. *Efficiency for alternate analyses.* For the two papers that presented analyses other than logistic regression, the efficiency of squashing drops dramatically. Owen was unable to detect any consistent estimation advantage of empirical likelihood squashing over simple random sampling for the boosted decision tree analyses. The efficiency of boosting in a three-dimensional neural network simulated dataset with a reduction factor of 10 was about 16 using the likelihood profile method. The definition of efficiency for the ability to replicate a neural network was not based on coefficients but was based on the average squared difference in predictions between the neural network based on the mother dataset and that from the squashed dataset.

13. *Investigate/discuss classification accuracy?* DuMouchel et al. (1999) and Madigan et al. (1999) focus almost entirely on evaluating data squashing based on accuracy of parameter estimation or comparison of predictions based on Y versus those based on X. This corresponds with the stated goal of data squashing of being able to duplicate the results of an analysis of the large dataset, as opposed to how well one can predict or classify new data. In a high noise environment, prediction or classification error may be quite large even if there is no error in estimating model parameters. Owen (1999) emphasizes this point and the related one that in such cases there may be diminishing returns in any improvement over a simple random sample, if the only goal is to predict or classify new data. The paper shows that ROC curves, for example,

may hardly change even though squashing provides a 4 to 1 improvement in the efficiency of parameter estimation compared to a simple random sample.

14. Simulation-based choice of tuning constants? Madigan et al. (1999) includes the results of preliminary experiments comparing ways of choosing the configuration of parameter values at which the likelihood profile is evaluated. This may have enabled that method to be better tuned than the other two methods.

15. Factors investigated. DuMouchel et al. (1999) reports results from several choices of bin definitions, number of moments fitted, and reduction factor, for both the main effects and second-order logistic regression model. However, these factors were not set up in a factorial design as were the likelihood profile settings comparisons reported in Madigan et al. (1999). Only the reduction factor was varied in Owen (1999).

4. Discussion

The good performance of the model likelihood-based squashing is impressive, making us eager to better understand its limits. Can it scale up to larger numbers of parameters and variables? The length of a likelihood profile vector would need to be greatly increased to handle either the 48-parameter model of DuMouchel et al. (1999) or the 39-parameter model of Owen (1999). Would the one-pass clustering of the profile vectors still work so well? It is remarkable that replacing each cluster by its average data vector leads to very accurate squashing. DuMouchel et al. (1999) assessed the seemingly similar strategy of replacing each Y-matrix bin by the mean of all the data falling in the bin. The results were horrible (far worse than for simple random samples) for both partitions used, one having 394 bins and the other having 3,710 bins in the eight-dimensional data space. This failure is presumably because the set of bin means has much reduced variance for each variable compared to Y, resulting in some severely biased regression coefficients. Yet even crude one-pass clusters in the 149-dimensional likelihood profile space allow the use-the-mean rule to retain almost full accuracy for coefficient estimation. An important question for future research is whether the likelihood profile method can be extended to models for the joint distribution of all variables, avoiding the need to specify a response variable.

Although the method of Owen (1999) seemed to be inefficient compared to the other squashing methods, perhaps it can be modified to improve performance and still retain the benefits of quick computation and having the elements of X be legitimate elements of Y. Reweighting a stratified sample from Y may greatly improve the performance of the empirical likelihood method. Some as yet unpublished experimentation of our own indicates that taking a random point from each Y-data bin works much better than taking the bin means. The initial weights are the bin sizes, and these weights can be iteratively improved to allow matching of moments or other functions, either by empirical likelihood estimation or by quadratic programming methods. Along these same lines, a method similar to stratified sampling and having about the same goals as data squashing is called delegate sampling, proposed by Breiman and Friedman (1984).

DuMouchel et al. (1999) and Owen (1999) did not address the problem of missing data, while Owen (1999) addressed it in an ad-hoc manner, dropping some variables having lots of missing data, and devising an imputation scheme for filling in missing values in the other variables. For some purposes, it might be desired to have the same distribution of missing data in the squashed data set as in the mother dataset, for example if one wanted to be able to build models for missingness. The original concept of data squashing in DuMouchel et al. (1999) was as a generic data description module, independent of whatever subsequent analyses are planned. It is assumed that an analyst exploring the squashed dataset wants to see everything, warts and all, including the prevalence and patterns of missing data. This highlights the need for squashing techniques that smoothly handle both categorical and continuous variables. The likelihood profile method has a potential weakness here, in that Madigan et al. (1999) does not state what should be done if the clustering of likelihood profiles leads to a pooling of data having different categorical values. (It never happened in their examples.)

Concerning the general research methodology of all three papers, they all used logistic regression as their primary analysis example. This is not because the authors felt a great need for new ways to fit logistic regression to huge datasets. To the contrary, widely available programs like SAS proc logistic (SAS Institute 1998) can handle very large datasets because they do not keep the entire dataset in memory and quickly compute the coefficients in just a few passes over the data. This is convenient for data squashing research, since it allows easy computation of the results on the large dataset to evaluate each squashing technique. Logistic regression, especially a second-order logistic regression that estimates quadratic and interaction terms such the 48-coefficient model in DuMouchel et al. (1999) is viewed as an easy-to-work-with proxy for other highly nonlinear methods for which there may be no available software that avoids the need to keep all the data in memory. To evaluate the performance of data squashing on such latter methods, we must necessarily restrict the mother dataset to be of manageable size. An interesting research question is how to extrapolate the efficiency measurements in a squashing evaluation. If, for a certain squashing method, $eff = 100$ when $N = 105$ and $M = 103$, what does that say about eff when $N = 107$ and $M = 104$? Of course, there can be many alternative definitions of efficiency besides eff as defined in the discussion of row 11 of Table 16.1. Regression coefficients from multiparameter models are very sensitive to near collinearities in the data, making the value of eff somewhat unstable in such datasets. Such instability could be viewed as a proper challenge for a good squashing technique to meet, since many of the nonlinear techniques such as neural networks or decision trees that one might want to apply to the squashed dataset also have similar instabilities due to their dependence on local properties of the data. A more stable and also more generally applicable measure of squashing efficiency is one that focuses on stability of predictions:

$$pred.eff = \frac{\sum_{i=1}^{N}(p_i^{SRS} - p_i^Y)^2}{\sum_{i=1}^{N}(p_i^X - p_i^Y)^2},$$

where each p_i denotes a prediction of the mean response for row i of the mother dataset, and the subscripts X, Y, and SRS denote predictions based on parameters estimated from the squashed dataset of size M, the mother dataset, and a simple random sample of size M, respectively. For further accuracy in estimating $pred.eff$, the numerator and denominator could each be estimated from the mean of several samples. Note that this measure focuses on the ability of the squashed data to replicate analyses of the mother data, not the ability to predict new data. There will be a different value of $pred.eff$ for each combination of predictive method and response variable that is evaluated.

In summary, these papers show that data squashing can be a great improvement over random sampling when the number of variables is not too great or the response variable is fixed. Unfortunately, the need for squashed datasets is greatest when there are tens, hundreds or even thousands of variables. In such diverse datasets usually there are many potential modeling projects involving many choices of response variables. It would be extremely valuable to have a single dataset of manageable size, produced by an enterprise-wide data warehouse resource but available for easy analysis throughout the organization, with the assurance that even reasonably complex relationships among the variables are quite likely to be replicable in the original dataset. We can only hope that the papers reviewed here stimulate more research into extending the reach of data squashing methods.

Bibliography

W. DuMouchel, C. Volinsky, T. Johnson, C. Cortes, and D. Pregibon. Squashing flat files flatter. In *Proceedings of the Fifth ACM Conference on Knowledge Discovery and Data Mining*, pages 6–15, 1999.

J. Friedman. Greedy function approximation: A stochastic boosting machine. Technical report, Department of Statistics, Stanford University, 1999a.

J. Friedman. Stochastic gradient boosting. Technical report, Department of Statistics, Stanford University, 1999b.

D. Madigan, N. Raghavan, W. DuMouchel, M. Nason, C. Posse, and G. Ridgeway. Likelihood-based data squashing: A modeling approach to instance construction. Technical report, AT&T Labs Research, 1999.

A. Owen. Empirical likelihood ratio confidence regions. *The Annals of Statistics*, 18:90–120, 1990.

A. Owen. Data squashing by empirical likelihood. Technical report, Department of Statistics, Stanford University, 1999.

SAS Institute. SAS Users Manual, 1998.

W.N. Venables and B.D. Ripley. *Modern Applied Statistics with S-Plus*. Springer-Verlag, 1997.

Chapter 17

MINING AND MONITORING EVOLVING DATA

Venkatesh Ganti
Department of Computer Sciences, University of Wisconsin, Madison, WI 53706 USA
vganti@cs.wisc.edu

Raghu Ramakrishnan
Department of Computer Sciences, University of Wisconsin, Madison, WI 53706 USA
raghu@cs.wisc.edu

Abstract   Data mining algorithms have been the focus of much recent research. The initial spurt of research on data mining algorithms typically considered static datasets. In practice, the input data to a data mining process resides in a large data warehouse whose data is kept up-to-date through periodic or occasional insertion and deletion of sets of tuples. Consequently, several issues that arise in a dynamically evolving database have recently begun to receive widespread attention. In this article, we survey research on two important issues: (1) exploiting the systematic data evolution for efficiently maintaining data mining models, and (2) monitoring changes in data characteristics. We classify research addressing these two problems based on a few distinguishing characteristics, and then briefly discuss all techniques captured by this classification.

Keywords: Data mining, Dynamic databases, Evolving data, Monitoring data characteristics.

1. Knowledge Discovery and Data Mining

Advances in database technology over the last decade have led to ubiquitous deployment of database systems for processing customer transactions, and for managing large amounts of data collected from several types of business-to-consumer interactions. Over the years, large businesses have realized that the data they accumulate is only as valuable as the business intelligence, or understanding of their enterprise, that they can extract from it. Consequently, they have invested significant time and money to build large *data warehouses* to consolidate historical data collected from customer interactions. The resulting spotlight on the analysis of these huge data repositories for decision support has led to the emergence of *knowledge discovery in databases (KDD)*. With roots in machine learning, pattern recognition, and statistics, KDD focuses on extracting interesting actionable patterns from large datasets.

Consider the following data analysis problems that managers face while making business decisions. In a bid to improve cross-selling, a recommendation agent at Amazon.com wants to know what other products to recommend to a customer whose browsing behavior and purchasing history are known. In a cost-cutting example, an analyst at AT&T may want to model historical call data records to detect patterns of fraud. In a promotional campaign, a marketing analyst at Walmart may want to know what groups of (related) products are purchased together so that promoting sales on one product increases sales on other products in the group.

In all three cases, Amazon.com, AT&T, and Walmart periodically update their data repositories with large amounts of information: browsing behavior, call data records, customer transactions, etc. Over a period of time they accumulate huge data repositories. The goal of the KDD process is to extract actionable nuggets of information from data that may be used in gaining business intelligence.

The knowledge discovery process consists of four broad steps (Fayyad et al. 1996). In the first *data selection* step, we identify the relevant attributes and the target dataset from the raw data. In the *data transformation and cleaning* step, we remove noise and outliers, transform field values to common units, generate new fields by combining existing fields, and bring data into a (relational) schema that is used as input to the third *data mining* step. The data transformation and cleaning step might also involve a denormalization of the underlying relations. In the *data mining* step, we usually fit models to data and discover interesting characteristics or patterns. The computed models constitute potential knowledge to be gleaned from the data. The class of models being considered determines the type of data characteristics that are highlighted by the data mining process. For instance, the class of *frequent itemset* models capture correlations between attribute values, the class of *cluster* models capture natural groups of records in the data, and the *decision tree* models capture characteristics that predict the behavior of future records. In the fourth *evaluation* step, we present the extracted patterns in an understandable form to the end user, say, through visualization. The results of any step in the KDD process might lead us back to an earlier step in order to redo the process with the new knowledge gained. In this article, we concentrate on the data mining step. The data mining techniques we discuss in this article reinforce techniques developed for other steps in the overall knowledge discovery process.

Significant progress has been made in the machine learning, the statistics, and the KDD communities on building sophisticated models to highlight interesting patterns in the data. Whereas machine learning and statistics communities focused mostly on improving the quality of models, the KDD community has also emphasized the scalability of model construction algorithms to analyze large datasets.

1.1. Data Mining Models

In this section, we first explain what we mean by data mining models. We then describe three classical data mining models—*frequent itemsets, decision trees, and clusters.*

Given a dataset, the goal of a typical data mining algorithm is to efficiently construct a "good" model that extracts interesting characteristics in the dataset (Fayyad et al. 1996). The goal of the data mining activity determines the purpose of a data mining model, *prediction* or *data exploration*, which in turn affects the data characteristics encapsulated by the model. Consider the history of customer purchases, browsing behavior, and their demographic information in the Amazon.com database. A predictive decision tree model induced from this database is able to *predict* the behavior of a future customer, say, whether the customer is expected to buy a book or not. In such a model, characteristics that are predictive of the behavior (e.g., the attribute that describes whether a customer subscribed to National Geographic or not) are emphasized. On the other hand, a clustering model induced on the same database to assist in exploring the browsing behavior of groups of customers characterizes the overall data distribution and the interactions among *all* attributes. Thus, functionalities of data mining models and, consequently, the data characteristics that they encapsulate are affected by the goals of the knowledge discovery process.

Several data mining models with a variety of functionalities and encapsulation characteristics have been proposed. For example, frequent itemsets, decision trees, clusters, neural networks, support vector machines, linear discriminant analysis, logistic regression, linear regression, regression trees have been used for a broad range of knowledge discovery tasks. Among these models, frequent itemsets, decision trees, and clusters have received significant attention from the database and the data mining communities because of their effectiveness for various KDD tasks and the feasibility to scalably induce them from large datasets. In this article, we study these three classes of models, and describe them in detail.

2. Preliminaries

We now introduce a few standard terms that we require later, and then introduce a hypothetical running example. We use this example to illustrate various concepts and ideas.

Definition 2.1 Given two sets S_1 and S_2, the *multiset union* $S_1 \uplus S_2$ of S_1 and S_2 consists of an element once for each of its occurrences in S_1 or S_2. The *multiset difference* $S_1 \ominus S_2$ deletes from S_1 an element once for each of its occurrences in S_2.

A *partially ordered* set $\langle Q; \preceq \rangle$ consists of a non-empty set Q and a reflexive, antisymmetric, transitive binary relation $\preceq$ on Q. Let $\langle Q; \preceq \rangle$ be a partially ordered set and let $H \subseteq Q$. An element $a \in Q$ is called a *lower bound* of H if $a \preceq h$ for all $h \in H$. A lower bound a of H is the *greatest lower bound* of H if, for any lower bound b of H, we have $b \preceq a$. We denote the greatest lower bound of H by $\bigwedge H$. Similarly, an element $a \in Q$ is called an *upper bound* of H if $h \preceq a$ for all $h \in H$. An upper bound a of H is the *least upper bound* of H if, for any upper bound b of H, we have $a \preceq b$. We denote the least upper bound of H by $\bigvee H$.

A partially ordered set $\langle Q; \preceq \rangle$ is called a *meet-semilattice* if for all $a, b \in Q$, $\bigwedge \{a, b\}$ exists. $\langle Q; \preceq \rangle$ is a called *lattice* if $\bigvee \{a, b\}$ and $\bigwedge \{a, b\}$ exist for all $a, b \in Q$. ⊙

For the remainder of this article, we use the term *tuple* generically to stand for the basic unit of information in the data, e.g., a customer transaction consisting of a set of products purchased, a database record, or an n-dimensional point. The context usually disambiguates the type of information unit being referred to. We use the term *attribute space* to refer to the domain over which tuples are defined.

Example 2.1 E-businesses and retail businesses typically gather the demographic information of all their registered customers and maintain their day-to-day customer interactions. Suppose Amazon.com accumulates the following hypothetical database through its day-to-day business activity. The first relation R_1 shown in Table 17.1 consists of the demographic information of all its registered customers. In addition, suppose the business has collected information as to whether each customer ever subscribed to the monthly *National Geographic* magazine. We show a hypothetical example R_2 in Table 17.2. Observe that the subscription relation has a foreign key relationship with the demographic relation. Subscription behavior with respect to the demographic characteristics is obtained by the foreign key join (R_1.CID = R_2.CID) of the two relations R_1 and R_2 on the CID column. The third relation captures the purchasing history of customers. A customer's visit to the virtual store generates a basket consisting of all the books purchased by the customer. The third example relation R_3 shown in Table 17.3 consists of a hypothetical market basket relation.

Table 17.1. R_1: Demographic Relation

CID	Ethnicity	Age	Children	Salary
1	American	33	0	$50K$
2	Indian	31	1	$55K$
3	American	36	1	$45K$
4	Chinese	25	2	$50K$
5	Indian	30	0	$52K$
6	American	36	0	$65K$
7	American	25	0	$110K$
8	Chinese	36	1	$70K$
9	American	40	2	$38K$
10	American	31	1	$48K$
11	Indian	25	0	$65K$
12	American	45	1	$80K$
13	Indian	23	2	$70K$
14	Chinese	45	0	$50K$

2.1. Frequent Itemset Models

A *market basket* is a collection of items purchased by a customer in one *transaction*, which is a well-defined business activity. As mentioned above, retailers accumulate massive market basket datasets. Given these market basket databases, one data mining problem is to mine for relationships between sets of items in terms of customer purchases. These relationships may then be used for recommending books to customers, or for launching more effective marketing campaigns. As an example, a significant number of customers may purchase J.R.R.Tolkien's fantasy novels the *Hobbit* and the *Lord of the Rings* together. Therefore, a customer who bought the Hobbit may be recommended the Lord of the Rings. A popular way of exploring market basket datasets for interesting item correlations is to find sets of items, or *itemsets*, that appear together in significantly many transactions (Agrawal et al. 1993). Such

Table 17.2. R_2: Magazine Subscribers Relation

CID	Subscription	
1	no	
2	no	
3	no	
4	no	
5	no	
6	no	
7	yes	DEMOGRAPHIC
8	no	INFORMATION FROM R_1
9	no	
10	no	
11	no	
12	yes	
13	no	
14	no	

Table 17.3. R_3: A Market-basket relation: D^T.

TID	CID	Book	Price	Date
101	1	Crime and punishment (CP)	25	1/10/99
101	1	War and peace (WP)	20	1/10/99
101	1	The Idiot (I)	30	1/10/99
102	1	The Mother (M)	50	2/5/99
102	1	The Idiot	25	2/5/99
103	2	Crime and Punishment	25	3/1/99
103	2	The Mother	500	3/1/99
103	2	The Idiot	100	3/1/99

frequently occurring itemsets are called *frequent itemsets*, and the set of all frequent itemsets constitutes the *frequent itemsets model*. We now formally describe this model and then illustrate these definitions through our running hypothetical example.

Definition 2.2 Let $\mathcal{I} = \{i_1, \ldots, i_n\}$ be a set of literals called *items*. A *transaction* and an *itemset* are subsets of $\mathcal{I}$. A transaction P is said to *contain* an itemset X if $X \subseteq P$. Let D be a set of transactions. The *frequency* $\sigma_D(X)$ of an itemset X in D is the fraction of the total number of transactions in D that contain X: $\sigma_D(X) \stackrel{\text{def}}{=} \frac{|\{P : P \in D : X \subseteq P\}|}{|D|}$.[1]

An itemset X is said to be a k-itemset if $|X| = k$. The *itemset lattice* $\langle 2^{\mathcal{I}}, \subset \rangle$ is the lattice formed by the set of all possible itemsets and the

---
[1] Note that this is sometimes called the *relative frequency* in earlier literature.

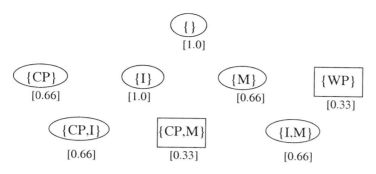

Figure 17.1. An example lattice of itemsets

subset relation. An itemset X is said to be an *ancestor* of another itemset Y if $X \subset Y$, and Y is said to be a *descendant* of X.

Let m_s $(0 < m_s < 1)$ be a constant called the *minimum frequency threshold*. An itemset X is said to be *frequent* in D if $\sigma_D(X) \geq m_s$. The *frequent itemset model* is the set $L(D, m_s)$ of all itemsets that are frequent on D; formally, $L(D, m_s) = \{X : X \subseteq \mathcal{I}, \sigma_D(X) \geq m_s\}$. The *negative border* $NB^-(D, m_s)$ of D at minimum frequency threshold m_s is the set of all infrequent itemsets whose proper subsets are all frequent. Formally, $NB^-(D, m_s) = \{X : X \subseteq \mathcal{I}, \sigma_D(X) < m_s \wedge \forall Y \subset X, \sigma_D(Y) \geq m_s\}$. ⊙

The frequency of itemsets satisfies the following, often called *anti-monotonicity*, property: $X \subseteq Y \Rightarrow \sigma_D(X) \geq \sigma_D(Y)$. All efficient scalable algorithms for mining frequent itemsets exploit the anti-monotonicity property (e.g., (Agrawal et al. 1996, Brin et al. 1997)).

The frequent itemset model may be used to deduce additional information like *association rules*. Informally, an association rule captures the relationship between two sets of items X and Y where the likelihood of a customer purchasing the set Y of items given that she purchased the set X of items is high (greater than a threshold called the *minimum confidence* threshold).

Definition 2.3 Association Rule: Let $\mathcal{I}$ be a set of items, D be a set of transactions, and let m_s and m_c $(0 < m_s, m_c < 1)$ be two constants. A rule of the form $X \Rightarrow Y$ where X and Y are itemsets such that the $\sigma_D(X \cup Y) \geq m_s$ and $\frac{\sigma_D(Y)}{\sigma_D(X \cup Y)} \geq m_c$ is called an *association rule*. ⊙

We now illustrate the above concepts through the hypothetical market-basket relation shown in Table 17.3. The interesting part (frequent itemsets and negative border itemsets) of the lattice is shown in Figure 17.1. For the sake of clarity, we do not show the descendant relationships (that

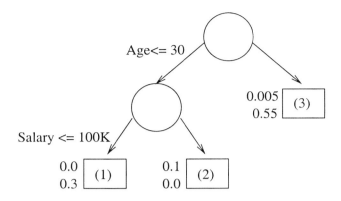

Figure 17.2. A Hypothetical Decision Tree

is, edges) between itemsets and also denote the novels by the following abbreviations: WP for "War and Peace," CP for "Crime and Punishment," I for "The Idiot," and M for "The Mother." If we set the minimum frequency threshold to 50%, then itemsets {CP}, {I}, {M},{M,I} and {CP,I} are frequent. And, if the minimum confidence threshold is set to 66% then {I $\Rightarrow$ M} is an association rule because the frequency of {I,M} is 0.66 and $\frac{\sigma(\{I,M\})}{\sigma(\{I\})} = 0.66$. The set {{WP}, {CP,M}} constitutes the negative border at 50% minimum frequency threshold. In Figure 17.1, each frequent itemset is depicted within an oval boundary and each negative border itemset within a rectangular boundary. Frequencies of these itemsets are shown below them.

2.2. Decision Tree Models

Suppose that Amazon.com wants to use a new mailing list, it purchased, with addresses and demographic information of potential customers to promote travel-related books. We would like to label each person in the new mailing list based on the likelihood that she will buy travel-related books being promoted by Amazon.com. Our goal now is to use information about current customers to identify potential customers in the new mailing list. The relation shown in Table 17.2 describes whether each current customer ever subscribed to the National Geographic magazine. Assuming that subscription to National Geographic indicates a high level of interest in travel-related books, we build a model that predicts the level of interest, on travel-related books, of a new potential customer.

The problem can be formulated as follows. We are given a *database*. The records have one designated categorical[2] attribute called the *dependent attribute*; the other attributes are called *predictor attributes*. We also call the dependent attribute the *class label attribute*. The goal is to build a model that takes as input the values of predictor attributes and computes a value for the dependent attribute. The problem of building a good predictive model is called the *classification problem*.[3]

Many classification models have been proposed in the literature: neural networks, genetic algorithms, bayesian methods, log-linear and other statistical methods, decision tables and tree-structured models called *decision tree models*. (For excellent reviews of classification methods, see Weiss and Kulikowski (1991), Michie et al. (1994), Hand (1997).) Decision tree models are especially attractive in a data mining environment for several reasons. First, due to their intuitive representation, the resulting classification model is easy to understand by humans. Second, decision trees are non-parametric and thus especially suited for exploratory knowledge discovery. Third, the accuracy of decision trees is comparable or superior to other classification models (Murthy 1995, Lim et al. 1997). Last, there exist fast scalable algorithms to construct decision trees from very large training databases. We now formally describe the classification problem and the class of decision tree models.

Let $X_1, \ldots, X_m$ be random variables where X_t has domain $\mathcal{D}(X_t)$. Let C be a categorical random variable with domain $\mathcal{D}(C)$. The X_i correspond to the *predictor attributes* and C is the dependent attribute, the *class label*. Table 17.2 shows an example training database of magazine subscribers with 14 records ($N = 14$). There are four predictor attributes[4] ($m = 4$): *Ethnicity*, *Age*, *Number of Children*, and *Salary*. *Age*, *Number of Children*, and *Salary* are numerical attributes, whereas *Ethnicity* is a categorical attribute with three categories: American, Indian, and Chinese. The class label has two categories indicating whether the customer subscribes to the National Geographic magazine.

A *classifier d* is a function from the domain of the predictor attributes to the domain of the class label. Formally:

$$d : \mathcal{D}(X_1) \times \cdots \times \mathcal{D}(X_m) \mapsto \mathcal{D}(C).$$

A *decision tree* is a special type of classifier. It is a directed, acyclic graph T in the form of a tree. The root of the tree does not have any

---

[2] Attributes whose domain is totally ordered are called *numeric*, whereas attributes whose domain is not ordered are called *categorical* (sometimes, also called *nominal*).
[3] If the dependent attribute is numerical, it is called a *regression problem*.
[4] Since the customer identifier (CID) is a unique arbitrarily assigned identifier for each tuple, we do not use it as a predictive attribute.

Cluster 1	$\langle 2, 1700 \rangle$ $\langle 3, 2000 \rangle$ $\langle 4, 2300 \rangle$
Cluster 2	$\langle 10, 1800 \rangle$ $\langle 12, 2100 \rangle$ $\langle 11, 2040 \rangle$
Cluster 3	$\langle 2, 100 \rangle$ $\langle 3, 200 \rangle$ $\langle 3, 150 \rangle$

Figure 17.3. Clusters Table

incoming edges. Every other node has exactly one incoming edge and may have outgoing edges. If a node has no outgoing edges it is called a *leaf node*, otherwise it is called an *internal node*. Each leaf node is labeled with one class label and each internal node n is labeled with one predictor attribute X_n called the *splitting attribute*.[5] Each edge (n, n') from an internal node n to one of its children n' has a predicate $q_{(n,n')}$ associated with it where $q_{(n,n')}$ involves only the splitting attribute X_n of node n. The set of predicates Q_n on the outgoing edges of an internal node n must be *non-overlapping* and *exhaustive*. A set of predicates Q is *non-overlapping* if the conjunction of any two predicates in Q evaluates to false. A set of predicates Q is *exhaustive* if the disjunction of all predicates in Q evaluates to true. We will call the set of predicates Q_n on the outgoing edges of an internal node n the *splitting predicates of n*. An example decision tree model induced on the relation obtained from the equi-join on CID of R_1 (Table 17.1) and R_2 (Table 17.2) is depicted in Figure 17.2.

2.3. Cluster Models

An analyst may want to identify natural groups of customers in the Amazon.com database based on their purchasing behavior which, let us suppose, is captured by the following two-column table: the *number of books* in a transaction and the *total price* of all books in it. This table is computed using an SQL query on the market basket relation (Table 17.3) and the pricing catalog in the Amazon.com database. The materialization of relevant information is typically done in the first data selection step of the knowledge discovery process. Tuples in this table are separable into three clusters as shown in Figure 17.3 where tuples

---

[5]There are split selection methods that generate linear combination splits that involve more than one predictor attribute. In this dissertation, we concentrate on decision trees whose splitting criteria involve a single predictor attribute as defined in this section.

in each cluster are more similar to tuples in the same cluster than to tuples in other clusters. (Data in Figure 17.3 does not match that in Table 17.3 because the earlier figure accommodated only a few transactions.) Informally, Cluster 1 is characterized by a high cost for a small number of books; Cluster 2 by a high cost for a large number of books; Cluster 3 by a low cost for a small number of books.

The goal in *clustering* is to automatically partition the data into natural groups, called *clusters*, where tuples in a cluster are more "similar" to each other than to tuples in other clusters. The notion of similarity between tuples is usually captured by a *distance* function, and the quality of clustering is usually measured by a distance-based criterion function (e.g., the weighted total or average distance between pairs of points in clusters). And, the goal of a clustering algorithm is to determine a good—as determined by the criterion function—partition of the dataset into clusters. A *cluster model* consists of all the clusters identified in the data. Since the clustering problem has been considered in several domains, many definitions exist which sometimes influence the algorithms as well. Since we discuss several incremental clustering algorithms in this article, we do not attempt to constrain the problem definition to that of a specific approach.

3. Mining Environments and Opportunities

Revisiting the three examples discussed earlier, Amazon.com, AT&T, and Walmart periodically update their data repositories with large amounts of information: browsing behavior, call data records, customer transactions, etc. Thus these data repositories evolve over time. We now consider two data mining applications that are typical in such scenarios. Suppose an analyst is modelling the AT&T database for detecting fraudulent call patterns. Processes causing fraud change dynamically because people continuously find newer ways of breaking the system. Therefore, the analyst may only be interested in modelling the most recently collected set of call data records, say, those collected in the past three months. In another application, the analyst may be interested in analyzing the impact of a marketing campaign targeted at increasing the sales of cell phones and related services. The goal here is to study the increase in the targeted service and the impact (decrease or increase) on other related services, say, wireless internet connectivity and long distance telephone activity. Therefore, the analyst wants to "compare" the customer behavior in call data records collected before the campaign with the behavior in data collected after the campaign. Going one step further, the analyst wants to know the features in customer data that account for the "most difference" between the data collected before and after the campaign.

Thus, data evolution gives rise to several mining and monitoring challenges: how do we maintain data mining models on interesting time-varying subsets of an evolving database? How do we monitor changes in "characteristics" of data collected during different periods? How do we detect trends in data characteristics over time? In this article, we survey techniques that have been developed for addressing the model maintenance and change monitoring problems.

Even though analysis of time series or similar type of data is also very important (see, for example, Chatfield (1984), Box et al. (1994), Weigend and Gerschenfeld (1994)), we do not discuss them. A very large body of work in Statistics and related literature addresses the analysis of time series (or similar type of) data, and hence has to be the focus of a separate article by itself for any reasonable coverage. Even for the closely related problem of detecting "sequential patterns" arising in the context of market basket analysis, we merely point the interested reader to the original papers (Agrawal and Srikant 1995, Srikant and Agrawal 1996).

We now formally define the model maintenance and change detection problems. A *model* $M(D)$ is a data mining model (belonging to one of several classes, such as decision tree models or cluster models) induced from the dataset D. By *data characteristics* of D, we mean the "knowledge" or "patterns" encapsulated in a model $M(D)$.

Definition 3.1 Let D_1 be a dataset and let D_2 be a new dataset resulting from an update to D_1. We say that D_1 and D_2 are two consecutive datasets being modelled.

Model Maintenance: The goal of model maintenance is to update a model $M(D_1)$ on D_1 to a model $M(D_2)$ that reflects the insertion and deletion of tuples.[6]

Change Detection: The goal of change detection is to quantify the difference, in terms of their data characteristics, between D_1 and D_2. ⊙

The relationship between the two consecutive datasets D_1 and D_2 considered in the literature is tied very closely to the type of data evo-

---
[6] Note that a straight forward model maintenance algorithm is to rebuild the model $M(D_2)$ from scratch. However, the goal here is to be much more efficient than rebuilding the model from scratch.

lution. The type of evolution and the relationship between consecutive datasets is the topic of the next section.

3.1. Types of Data Evolution

In this section, we discuss three types of data evolution that have been considered in research on evolving data. All of them deal with scenarios where tuples are either added to the database or existing tuples are deleted, but do not allow in-place modification of tuples.

3.1.1 Insertion-only Evolution.
In this model, the database is a single entity which evolves through insertions of new sets of tuples. Once a new set of tuples is inserted into the database, they are no longer distinguished in any way from the current set of tuples in the database.

Formally, let D be the current database snapshot. We assume that the database D has evolved through a sequence of additions of sets $d_1^+, \ldots, d_k^+, \ldots$ of tuples. Therefore, $D = d_1^+ \uplus \cdots \uplus d_k^+ \cdots$. Let d^+ be the set of tuples being inserted into D. Then, $D' = D \uplus d^+$ is the new database snapshot.

Once a current database snapshot D_1 is updated to D_2, the logical order in the arrival of tuples is ignored. Consequently, only one type of relationship between the two consecutive datasets D_1 and D_2 (see Definition 3.1) has been considered in the literature: $D_2 = D_1 \uplus d^+$ where d^+ is the newly added set of tuples.

3.1.2 Adhoc Evolution.
This model differs from the insertion-only type of evolution in that it also allows tuples to be deleted from the database. Tuples in the database are not logically ordered according to their arrival time.

Formally, let D be the current database snapshot. We assume that the database D has evolved through a sequence of additions and deletions of sets $d_1^+, d_1^-, \ldots, d_k^+, d_k^-, \ldots$ of tuples. Therefore, $D = d_1^+ \ominus d_1^- \uplus \cdots \uplus d_k^+ \ominus d_k^- \cdots$. Let d^+ be the set of tuples being inserted into D, and d^- be the set being deleted from D. Then, $D' = D \uplus d^+ \ominus d^-$ is the new database snapshot.

As in the case of insertion-only evolution, only one type of relationship between the two consecutive datasets D_1 and D_2 of Definition 3.1 has been considered in the related literature: $D_2 = D_1 \uplus d^+ \ominus d^-$.

3.1.3 Systematic Block Evolution.
In this model, the database is treated as a sequence of *blocks* where each block is a set of tuples that have been simultaneously added to the database. The database snapshot D consists of a (conceptually infinite) sequence of

blocks $D_1, \ldots, D_k, \ldots$ where each block D_k is associated with an *identifier k*. We assume without loss of generality that all identifiers are natural numbers and that they increase in the order of their arrival. Unless otherwise mentioned, we use t to denote the identifier of the "latest" block D_t. We call the sequence of all blocks $D_1, \ldots, D_t$ currently in the database the *current database snapshot*.

A variety of possibilities for the two consecutive datasets D_1 and D_2 of Definition 3.1 have been discussed in the context of systematic evolution. In recent work, we explored through the DEMON framework the selection (via the class of regular expression constraints) of interesting time-varying subsets of blocks in the dataset for mining (Ganti et al. 2000). At any instant, D_1 is the subset of blocks selected (by the constraint) from the current database snapshot and D_2 is the subset selected from the next database snapshot resulting from the addition of a new block. For instance, if blocks of data are appended every day to the database, an analyst may want to maintain a data mining model on blocks selected by the following constraint: blocks collected on alternate days over a sliding window consisting of the most recent 30 days.

Even though the specification of time-varying subsets has originally been proposed in the context of the model maintenance problem, it is applicable even for the change detection problem since the techniques developed for measuring changes in data characteristics just operate on two datasets without relying on their relationship with each other. We now discuss the specification of a broad class of such interesting time-varying subsets.

Specification of Time-varying Subsets. We take a two-step approach for selecting time-varying subsets. First, we identify the span of the data of interest, which may vary as new blocks are inserted. Second, we select a subset of blocks within this span. Following this two-step approach, we introduce the *data span dimension* to allow temporal restrictions on the data being mined and then further refine these restrictions through the notion of a *block selection sequence*.

Data Span Dimension: When mining systematically evolving data, some applications require mining all data accumulated thus far, whereas some other applications are interested in mining only the most recently collected portion of the data.

As an example, consider an application that extracts from a large database of documents a set of document clusters, each consisting of a set of documents related to a common concept (Willett 1988). Occasionally, a new block of documents is added to the database, necessitating an update of the document clusters. Typical applications in this domain are interested in clustering the entire collection of documents.

In a different application consider the database of a hypothetical toy store which is updated daily. Suppose the set of frequent itemsets discovered from the database is used by an analyst to devise marketing strategies for new toys. The model obtained from all the data may not interest the analyst for the following reasons. (1) Popularity of most toys is short-lived. Part of the data is "too old" to represent the current customer patterns, and hence the information obtained from this part is stale and does not buy any competitive edge. (2) Mining for patterns over the entire database may dilute some patterns that may be visible if only the most recent window of data, say, the latest 28 days, is analyzed. The marketing analyst may be interested in precisely these patterns to capitalize on the latest customer trends.

To capture these two different requirements, we introduce the *data span dimension*, which offers two options. In the *unrestricted window (UW)* option, the relevant data consists of all the data collected so far. In the *most recent window (MRW)* option, a specified number w of the most recently collected blocks of data are selected as input to the data mining activity. We call the parameter w the *window size*. Formally, let $D_1, \ldots, D_t$ be the current database snapshot. Then the *unrestricted window* (denoted $D[1,t]$) consists of all the blocks in the snapshot. If $t \geq w$ the *most recent window* (denoted $D[t-w+1, t]$) of size w consists of the blocks $D_{t-w+1}, \ldots, D_t$; otherwise, it consists of the blocks $D_1, \ldots, D_t$. In the remainder of the article, we assume without loss of generality that $t \geq w$. The special case $t < w$ can be handled easily.

Block Selection Sequence: We now introduce an additional selection constraint called the *block selection predicate* that can be applied in conjunction with the options on the data span dimension to achieve a fine-grained block selection. The following hypothetical applications (of interest to a marketing analyst) defined on the toy store database motivate the finer-level block selection.

1 The analyst wants to model data collected on all Mondays to analyze sales immediately after the weekend. The required blocks are selected from the unrestricted window by a predicate that marks all blocks added to the database on Mondays.

2 The analyst is interested in modelling data collected on all Mondays in the past 28 days (corresponding to the last 4 weeks). In this case, a predicate that marks all blocks collected on Mondays in the most recent window of size 28 selects the required blocks.

3 The analyst wants to model data collected on the same day of the week as today within the past 28 days. The required blocks are selected from the most recent window of size 28 by a predicate that, starting from the beginning of the window, marks all blocks added every seventh day.

Note that the block selection predicate is independent of the starting position of the window in the first and second applications whereas in the third application, it is defined relative to the beginning of the window and thus moves with the window. We now define the *block selection sequence (BSS)* to capture the intuition behind predicates that marks relevant blocks in the above examples. Informally, the BSS is a bit sequence of 0's and 1's; a 1 in the position corresponding to a block indicates that the block is selected for mining, and a 0 indicates that the block is left out.[7]

Definition 3.2 Let $D[1,t] = \{D_1, \ldots, D_t\}$ be the current database snapshot and let $D[t-w+1, t]$ be the most recent window of size w. A *window-independent BSS* is a sequence $\langle b_1, \ldots, b_t, \ldots \rangle$ of 0/1 bits. A *window-relative BSS* is a sequence $\langle b_1, \ldots, b_w \rangle$ of bits ($b_i \in \{0,1\}$), one per block in the most recent window. ⊙

Note: It is straight-forward to introduce a logical notion of blocks over an insertion-only evolutionary model. For instance, we may consider all tuples inserted during a continuous period of time (e.g., a day or a week) to constitute a block. Also, when tuples contain the timestamp of their arrival, each block may be defined by a selection predicate on the timestamp attribute.

4. Overview of the Article

We now enumerate the problem space and outline the body of work that we are going to discuss in this article. As shown in Figure 17.4, the enumeration is based on identifying three dimensions: the problem being addressed, the type of data evolution, and the class of data mining models. We then enumerate the scenarios considered for each problem.

For each of the two problems, we show the interesting scenarios that have been considered in the literature. As shown in Figure 17.4, the combinations resulting from the types of data evolution have not been

---

[7]Further refining the data selection to allow a user-defined predicate p on top of a block selection sequence is done as follows. Each block d is split into two blocks d_p and $d_{p'}$; d_p contains all tuples in d that satisfy the predicate p and $d_{p'}$ contains all remaining tuples in d. $d_{p'}$ is now ignored.

Problem	Class of Models		Type of Data Evolution
Model Maintenance	Frequent Itemsets, Decision Trees, Clusters	×	Insertion-only Evolution, Adhoc Evolution, Systematic Evolution
Change Detection	Frequent Itemsets, Decision Trees, Clusters, Partitioning Grid		

Figure 17.4. Enumeration of the Problem Space

considered for the change detection problem for the following reason. The types of evolution differ in the flexibility they provide for selecting subsets of the database. However, unlike the problem of model maintenance, which requires different algorithms and ideas to exploit this flexibility, the problem of change detection does not. The methodologies developed for measuring differences between two datasets can be applied in any case once the relevant subsets of a database are identified.

Note that the *partition grid*[8], which is not a classical data mining model, has been employed to capture the characteristics of datasets for the change detection problem. Several methods have shown that comparing distributions of two datasets within each region in the grid yields a fairly good idea of the difference between two datasets.

4.1. Model Maintenance

We consider recent scalable algorithms on incrementally maintaining data mining models, with a particular emphasis on frequent itemsets, decision trees, and clusters. In Figure 17.5, we list all papers that studied the problem of scalable incremental maintenance of frequent itemsets, decision trees, and clusters. The figure shows that most work has been focused on maintaining models under insertion-only and adhoc types of evolution. Note that algorithms that have been designed for these types of evolution may be used for maintaining models under systematic evolution. However, as discussed later, they may not be efficient for maintaining models with respect to a broad class of block selection sequences. We discuss all algorithms listed here in Section 5

Besides this body of work emphasizing scalability, incremental model maintenance under adhoc evolution of a broad variety of model classes

---

[8] A partitioning grid is a partition of the entire attribute space into disjoint hyper-rectangles and each hyper-rectangle summarizes the distribution of data that belong to it.

	Insertion-only	Ad hoc	Systematic
Frequent Itemsets	DIC: (Brin et al. 1997) Carma: (Hidber 1999)	FUP: (Cheung et al. 1996a), (Cheung et al. 1997), (Cheung et al. 1996b), TBAR: (Thomas et al. 1997), BORDERS: (Feldman et al. 1997), Delta: (Pudi and Haritsa 2000)	GEMM (Ganti et al. 2000)
Decision Trees		Boat: (Gehrke et al. 1999)	GEMM
Clusters	BIRCH: (Zhang et al. 1996) Clustering Framework: (Bradley et al. 1998)	Incremental DBScan: (Ester et al. 1998)	GEMM

Figure 17.5. Scalable Algorithms for Incremental Model Maintenance

has been addressed in the Machine Learning literature. However, scalability with respect to data size has not been an issue for these algorithms because they typically worked with small datasets. Below, we briefly discuss this work.

Utgoff developed ID5, the first algorithm for incrementally maintaining decision trees in a dynamic environment (Utgoff 1989). ID5 restructures an existing tree to reflect the characteristics of the newly added block of data. However, it cannot handle numeric attributes and assumes that the complete database fits in main memory. Utgoff et al. extended this work to allow numeric attributes as well (Utgoff et al. 1997); they presented a series of restructuring operations that can be used to update a decision tree construction algorithm for a dynamically changing database. This technique also assumes that the entire database fits in main memory.

For the class of cluster models, several incremental algorithms within the context of conceptual clustering have been proposed in 1980's (e.g., Lebowitz (1982), Schlimmer and Fisher (1986), Lebowitz (1987), Fisher (1987)). The goal of these algorithms was to develop a classification scheme that assigns objects to a concept or a cluster based on the values that the object takes on a set of attributes. Whenever new objects are added to the database, this clustering scheme is updated. How-

Data Characteristics	Method
Frequent itemsets	FOCUS: (Ganti et al. 1999)
Decision tree	Misclassification error: (Breiman et al. 1984), (Loh and Vanichsetakul 1988), (Loh and Shih 1997) Chi-squared metric: (D'Agostino and Stephens 1986), FOCUS: (Ganti et al. 1999)
Clusters	FOCUS: (Ganti et al. 1999)
Partitioning Grid	Chi-squared metric: (D'Agostino and Stephens 1986), Data Spheres: (Johnson and Dasu 1998) FOCUS: (Ganti et al. 1999)

Figure 17.6. Change Detection Methods

ever, all these algorithms assume that the dataset fits in main memory. More recently, Charikar et al. proposed a model for evaluating an incremental algorithm for clustering a dynamic set of points in a metric space (Charikar et al. 1997). They propose new incremental clustering algorithms, which optimize one among a few specialized criterion functions, with (provably) good quality guarantees. Besides assuming that the dataset fits in main memory so that they can randomly access sets of tuples in a cluster, they have neither evaluated these algorithms nor demonstrated the practical value of their guarantees through even a prototypical implementation.

4.2. Change Detection

In Figure 17.6, we outline the work that addressed the problem of detecting or monitoring changes in data characteristics. As mentioned earlier, data characteristics are encapsulated by a model induced from a dataset. The following classes of models have typically been employed for this purpose in the context of the change detection problem: *frequent itemsets, decision trees, clusters, and a partitioning grid.*

Besides the body of work listed in Figure 17.6, there has been a significant amount of work in the market basket domain on monitoring individual itemsets over a period of time (Agrawal and Psaila 1995,

Arning et al. 1996, Chakrabarti et al. 1998), and on characterizing each itemset with a measure of informativeness (Webb 1995, Silbershatz and Tuzhilin 1996, Fukuda et al. 1996, Rastogi and Shim 1998, Subramonian 1998). In either case, they compare the frequency of an itemset in the current snapshot with either its frequency in an earlier snapshot or its expected frequency to quantify the deviation. The magnitude of deviation determines the degree of change or the degree of suprise, whichever is the case.

Finally, we note that some of the methodologies (FOCUS, data spheres, chi-squared metric) listed in Figure 17.6, may be used to measure differences between datasets arising from a context other than an evolutionary scenario. For instance, we are able to measure the difference between two datasets, both collected at Amazon.com, consisting of customer purchases from Wisconsin and California, respectively.

5. Model Maintenance Algorithms

In this section, we discuss incremental algorithms for maintaining data mining models under all three types of evolution.

5.1. Insertion-only Evolution

All algorithms that we discuss here maintain the classes of frequent itemset models or cluster models.[9] Even though they resulted from the drive for developing scalable algorithms for computing the set of frequent itemsets from a very large (static) dataset, they are based on an incremental assimilation paradigm. Therefore, they can be extended easily to maintain models under the insertion-only type of evolution. The high level structure of these algorithms is as follows. While scanning a dataset, they maintain a model that is "close" to the exact model on the data. The maintained model is continuously updated as new tuples are scanned. Additional information may have to be maintained to facilitate these updates. At the end of the scan, additional processing, sometimes involving more scans, may be required to determine the exact model. We describe the static versions and then the (usually straight-forward) extensions for incremental maintenance.

5.1.1 Frequent Itemset Models. All algorithms we discuss for maintaining frequent itemset models under insertion-only (and ad-hoc) evolution take the following approach. Using the set of frequent itemsets from the old database, they generate potentially frequent can-

---

[9]Similar algorithms do not exist for the class of decision tree models.

didate itemsets and count their frequencies in the updated database to check if they are frequent. If some of these candidate itemsets are found to be frequent, the process of candidate generation and counting is iterated. Each iteration typically requires a scan of the entire database. Algorithms differ in the number of times they iterate and the number of candidate itemsets they generate. The smaller the number of iterations and the number of candidate itemsets, the more efficient the algorithm.

The candidate generation almost always involves the "pruned" *prefix join* of two input sets of itemsets. The prefix join of two sets of itemsets L_1 and L_2 extends pairs of itemsets X and Y one from each input operand (say, $X \in L_1$ and $Y \in L_2$) by one more item if the following two conditions are satisfied. First, the sizes of X and Y must be equal, i.e., $|X| = |Y|$. Second, if their sizes are equal to k then their $k-1$ prefixes must match. That is, X and Y should be of the form $\{a_1, \ldots, a_{k-1}, a_k\}$ and $\{a_1, \ldots, a_{k-1}, a_{k+1}\}$. If these two conditions are met the itemset $\{a_1, \ldots, a_k, a_{k+1}\}$ is added to result of the prefix join of L_1 and L_2. An itemset Z in the prefix join of L_1 and L_2 is *pruned* out if all proper subsets of Z are not frequent. All itemsets in the prefix join that are not pruned constitute the pruned prefix join of L_1 and L_2 (and also the set of candidate itemsets).

While counting frequencies of a set L of itemsets, we need to determine for each transaction P all itemsets in L that are contained in P. Since this operation is performed, repeatedly in some cases, for each transaction in the database it has to be fast. For this purpose, Agrawal et al. (1996) have proposed the hash tree, and Mueller (1995) proposed the prefix tree. Both data structures maintain a set of itemsets and provide an efficient way of determining all itemsets that are contained in a given transaction. All market basket analysis algorithms we discuss in this article assume that itemsets are maintained in one of these two data structures while counting their frequencies in a database.

Note: Maintenance algorithms for insertion-only environments maintain the index of the transaction when we begin counting an itemset. Therefore, there is an implicit assumption that transactions in a dataset are logically ordered. However, when we delete a transaction we typically do not have its position in this logically ordered multiset of transactions. Therefore, these algorithms cannot maintain the set of frequent itemsets under arbitrary deletions of transactions.

Notation: We use the following notation throughout this section. Let D be the old dataset and $L(D, m_s)$ be the set of frequent itemsets computed from D at a minimum frequency threshold m_s. Let d^+ be the set of new transactions being added to D. Let $L_k(D, m_s)$ represent the set of k-itemsets that are frequent over D at the minimum support threshold m_s. We first describe the algorithms in the static context and then discuss how insertions may be handled.

DIC. Brin et al. proposed an "optimistic" dynamic itemset counting algorithm DIC (Brin et al. 1997). The intuition behind their algorithm is as follows. Typically, itemsets can be judged to be frequent or not by counting their frequency on a part of the dataset. Based on this intuition, they optimistically start counting itemsets if all their subsets are "suspected" to be frequent instead of being surely frequent. Therefore, even before the first scan is complete, new candidate itemsets are generated using the set of itemsets that are either suspected to be frequent or surely frequent, and begin to be counted.[10]

DIC designates *stop points* at intervals of every M transactions. For each itemset X being counted, the stop point from where X began to be counted and its current support are maintained. DIC starts counting the frequencies of all 1-itemsets. During a data scan, DIC pauses at each stop point and checks if new itemsets need to be added. At any designated stop point, a candidate itemset X is *suspected* to be frequent if X's support in the subset of D enclosed between the stop point from where X was counted and the current stop point (wrap around from the beginning of the file if the end of file lies between the two) is greater than m_s. New candidates are generated from the pruned prefix join of the set of newly suspected-to-be-frequent itemsets with the union of the sets of frequent and old suspected-to-be-frequent itemsets. These new candidates are then counted from that stop point. Thus, itemsets can be counted starting at any stop point and not just from the beginning of the dataset. DIC maintains and updates the count of an itemset X until its support in the entire dataset is counted, i.e., until the scan comes back to the stop point where X was first added to the set of candidates. At this point, we know for sure whether X is frequent or not.

When the set of frequent itemsets has to be updated to reflect the addition of a set d^+ of new tuples to D, we can treat the end of D as another stop point in the dataset $D \uplus d^+$. We generate a set C of new candidate itemsets by joining $L(D, m_s)$ with itself.[11] We merely resume the static version of DIC from this new stop point. After d^+ is scanned,

---
[10]Most previous algorithms count frequencies of itemsets over the entire database before generating and counting new candidate itemsets.
[11]Note that C is the negative border $NB^-(D, m_s)$.

if new itemsets in C are frequent, we generate candidate itemsets again and scan $D \uplus d^+$ pausing at all stop points.

Let the set of frequent itemsets shown in Figure 17.1 be computed from a dataset D. Suppose a new set d^+ of transactions is added to D. The frequencies in d^+ of all negative border itemsets (because they are candidate itemsets) are counted. Suppose {WP} is frequent in d^+. Then, we begin to count frequencies of the new set of candidates {{CP,WP}, {I,WP}, {M,WP}} while scanning $D \uplus d^+$. If none of these itemsets is frequent in the updated database, then the update procedure stops.

Carma. Hidber proposed an online algorithm Carma for mining frequent itemsets (Hidber 1999). The primary objective of Carma is to provide continuous feedback to the user while a dataset is being munged. Ultimately, if the user is willing to wait long enough, the exact answer is produced. Along the way, however, the user is always provided with an estimate of the set of frequent itemsets. Carma is similar to DIC in that new itemsets may start being counted in the middle of a scan. During the first scan, Carma identifies a set S of potentially frequent itemsets. For each itemset v, it maintains the following information to compute the deterministic lower and upper bounds of its frequency.

count(v) : Frequency of v since v was inserted into S
firstTrans(v) : Index of the transaction when we v was inserted into S
maxMissed(v) : Upper bound on the number of occurrences of v before v was inserted

Suppose the ith transaction in the dataset is being read, then for any itemset $v \in S$, we have the following deterministic bounds on the frequency of v.

$$\frac{count(v)}{i} \leq \sigma(v) \leq \frac{maxMissed(v) + count(v)}{i}$$

As the scan continues, the information associated with itemsets is updated, and new itemsets may be added to or deleted from S. $count(v)$ of an itemset $v \in S$ is incremented if a new transaction contains v. $maxMissed(v)$ is updated such that it is equal to $\lfloor (i-1) * m_s \rfloor + |v| - 1$. Note that $count(v) + maxMissed(v)$ is an upper bound for the frequency of v in D. We begin to count the frequency of a new itemset w if the upper bounds of all its proper subsets are above the minimum threshold. As more of the dataset is scanned, the interval enclosing the frequency of an itemset typically shrinks. At any point, the current set of itemsets along with their frequency intervals provides the user an estimate of the final answer. A second scan of the dataset is required to determine the frequencies of itemsets exactly. For a discussion on generalizations to cases where the frequency threshold may be varied, we refer the reader to the original paper.

Carma can easily be modified to incrementally maintain frequent itemsets when new transactions are being added to the dataset. We merely pretend that the first scan of the dataset is not yet complete and scan the set of newly inserted transactions. We then scan the entire dataset, if necessary, again to determine the exact set of frequent itemsets and their frequencies. Except for the lower and upper bounds associated with itemsets being counted, the example discussed for DIC also illustrates Carma.

5.1.2   Cluster Models.

We now discuss two algorithms (BIRCH (Zhang et al. 1996) and the clustering framework (Bradley et al. 1998)). Even though these algorithms were proposed in the static context they can maintain cluster models under the insertion-only type of evolution. They take a two-phase approach. In the first phase, the dataset is summarized into "sub-clusters." The second phase merges these sub-clusters into the required number of clusters using one of several traditional clustering algorithms. (See Duda and Hart (1973), Fukunaga (1990) for an overview of several algorithms.) The intuition behind the pre-clustering approach taken by these algorithms is explained by the following analogy. Suppose each tuple describes the location of a marble. Given a large number of marbles distributed on the floor, these algorithms replace dense regions of marbles with tennis balls where each tennis ball is a *sub-cluster* of a cluster of marbles. The number of tennis balls is a controllable parameter and the space required for representing a tennis ball is much smaller than that required for representing the collection of marbles. Therefore, it is possible to cluster these tennis balls using one's own favorite clustering algorithm, e.g., K-Means (Duda and Hart 1973, Fukunaga 1990). The fact that the set of sub-clusters is maintainable incrementally can be exploited to derive efficient model maintenance algorithms.

BIRCH. The intuition behind the BIRCH algorithm is that since a cluster corresponds to a dense region of objects, it can be treated collectively through a summarized representation called its *cluster feature (CF)*. A CF of a cluster is the triple consisting of the number of points in the cluster, the centroid of the cluster, and the radius of the cluster. CFs are efficient for two reasons. First, they occupy much less space than the naive representation which maintains all objects in a cluster. Second, CFs are sufficient for calculating all inter-cluster and intra-cluster measurements involved in making clustering decisions. Moreover, these calculations can be performed much faster than using all objects in clusters. For instance, distances between clusters, radii of clusters, CFs—and hence other properties—of merged clusters can all be computed very quickly from the CFs of individual clusters.

In Figure 17.3, the CFs for the two clusters 1 and 2 consisting of 2-dimensional points are $(3, \langle 3, 2000 \rangle, 734.85)$ and $(3, \langle 11, 1980 \rangle, 129.62)$. The CF of the cluster obtained by merging clusters 1 and 2 can be obtained from their respective CFs.

Since the entire dataset usually does not fit into main memory, BIRCH incrementally evolves clusters by sequentially scanning the dataset. At any stage, the next tuple in the dataset is inserted into the cluster closest to it as long as the insertion of the new tuple into the cluster does not deteriorate its "quality." Otherwise, the new tuple forms its own cluster. Typically, the radius of a cluster is used to indicate its quality. The cluster is then updated to reflect the insertion of the new tuple. To expedite the search for the closest cluster BIRCH organizes all clusters at the leaf nodes of a height-balanced tree, called a *CF-tree*, which functions as an index on the clusters. (We skip the details of the CF-tree here.) Overall, the concepts of CF and CF-tree enable BIRCH to achieve scalability and speed while ensuring good clustering quality.

The incremental maintenance of cluster models under the insertion-only evolution relies on maintaining the CF-tree. When a new set d^+ of tuples is added to the database D, we pretend as if we are resuming the clustering process which has been suspended after scanning the initial portion D of $D \uplus d^+$. The second phase of BIRCH is then invoked on the updated CF-tree to obtain the updated cluster model. Since the second phase works on the in-memory set of cluster features, it is very fast. Therefore, the first phase dominates the overall resource requirements. Because the incremental version of BIRCH significantly reduces the time to obtain the updated set of sub-clusters, it is much more efficient than re-clustering the entire dataset.

However, BIRCH suffers from the following drawback. If a tuple is re-inserted into the CF-tree, it may not be inserted into the sub-cluster that it was inserted into the first time. The implication is that it is not possible to maintain sub-clusters under deletions of tuples because the CF-vector that would be modified to reflect the deletion may not be the one into which the tuple was inserted into in the first place.

The Clustering Framework. Bradley et al. (1998) proposed a framework for scaling up a broad class of iterative clustering algorithms (e.g., K-Means (Forgy 1965, MacQueen 1967)). Like BIRCH, the clustering framework takes a summarization-based approach and dense regions of tuples are represented by their cluster features. However, they further generalize by classifying sub-clusters into three different classes. The generalization allows them to integrate prior information about the location and the structure of clusters with the summarization process.

The basic idea behind the framework is to identify the *discardable set of tuples*, the *compressible set of tuples*, and the *main-memory set of tuples*. A tuple is discardable if its membership in a cluster can be ascertained; only the CF of all discardable tuples in a cluster is retained and the actual tuples are discarded. A tuple is compressible if it is not discardable but belongs to a very tight sub-cluster consisting of a set of tuples that always share cluster membership, i.e., if they move from one cluster to another they all move together. Such a sub-cluster of compressible tuples is also summarized using their CF. A tuple is a main-memory tuple if it is neither discardable nor compressible. Main-memory tuples are either noise tuples or border tuples of a cluster, and are retained in memory. The iterative clustering algorithm then moves only the main-memory tuples and the CFs of compressible tuples between clusters till the criterion function is optimized.

The discardable, compressible, and main-memory sets of tuples can be maintained incrementally whenever new sets of tuples are added to the database. As in the case of BIRCH, the second clustering phase using the compressed representation is much faster than the pre-clustering phase. Thus, incremental maintenance is much better than re-clustering the entire dataset from scratch.

5.2.   Adhoc Evolution

We now discuss algorithms for maintaining frequent itemset models, decision trees, and cluster models. Wherever possible, we illustrate them through examples.

5.2.1   Frequent Itemset Models.   All incremental algorithms for maintaining frequent itemsets are based on one or more of the following three properties. All three properties have first been observed and exploited in the context of computing the set of frequent itemsets from a static dataset of transactions.

1 **Anti-monotonicity**: If X and Y are itemsets such that $X \subseteq Y$, then frequencies of X and Y over any dataset D satisfy the following relationship: $\sigma_X(D) \geq \sigma_Y(D)$. Agrawal et al. have first

exploited this property for computing frequent itemsets (Agrawal et al. 1996).

2. **Frequent Partitions**: Let D_1 and D_2 be two datasets. If an itemset is frequent (at a certain minimum frequency threshold m_s) over $D_1 \uplus D_2$ then its frequency in at least one of D_1 or D_2 is greater than or equal to m_s. Savasere et al. have exploited this property to design an efficient two-pass algorithm for computing frequent itemsets (Savasere et al. 1995).

3. **Negative Border Sufficiency**: To compute the set $L(D, m_s)$ of frequent itemsets of D at a minimum frequency threshold m_s, it is necessary and sufficient to determine the negative border $NB^-(D, m_s)$. Toivonen relied on this property to develop a two pass sampling-based algorithm for computing frequent itemsets (Toivonen 1996).

Mostly, we stick to the notation introduced while describing algorithms for maintaining frequent itemsets models under insertion-only evolution. In addition, let d^- denote the set of transactions being deleted from the current database snapshot D.

The maintenance algorithms that we discuss below do not assume a logical order among transactions in the dataset. Therefore, maintenance under deletion of transactions is similar to that under insertions. For each algorithm, we first discuss maintenance under insertions and then extend it to deletions.

FUP and its Derivatives. The FUP algorithm and its derivatives (Cheung et al. 1996a; 1997; 1996b) are the first to specifically address the problem of incrementally maintaining frequent itemsets. These algorithms exploit properties (1) and (2). The basic idea behind the FUP algorithm is that if an itemset is frequent in the updated database, then it has to be frequent in either the old database or the new set of transactions. Therefore, all itemsets that become frequent due to the addition of the new set have to be frequent in the new set. This observation is used by FUP to prune the number of itemsets whose frequencies in the old database are counted. Since the old database is typically expected to be larger than the new set, reduction in the number of itemsets whose frequencies are counted against the old database reduces the overall maintenance time.

FUP makes several iterations over the database. In each iteration, it scans the entire database (including the new set of transactions and the old database) to count frequencies of a few candidate itemsets. In the kth iteration, FUP scans the increment d^+ to count the frequencies of all candidate k-itemsets. From property (2), we know that if an itemset X becomes frequent in $D \uplus d^+$ due to the addition of d^+, then it has to be frequent in d^+. If new k-itemsets (not in $L_k(D,m_s)$) become frequent, frequencies in D of these new k-itemsets are counted by scanning D. At the end of the scan, we know $L_k(D \uplus d^+, m_s)$. New candidate $(k+1)$-itemsets are generated through the pruned prefix join of $L_k(D \uplus d^+, m_s)$ with itself. This procedure is iterated until no more new frequent itemsets are found.

As an example, let the set of frequent itemsets computed from D be as shown in Figure 17.1. Suppose a new set of transactions d^+ is added to a dataset D. FUP first scans d^+ to count frequencies of all 1-itemsets. If {WP} along with other 1-itemsets is frequent in d^+, then D is scanned to count its frequency in D. After this scan, we know the frequencies of all 1-itemsets in $D \uplus d^+$. We now generate new candidates {CP,WP}, {I,WP}, {M,WP} and count their frequencies in d^+. If none of them is frequent in d^+, we do not scan D at all. The update is complete.

Maintaining the set of frequent itemsets when a set of transactions is deleted is very similar. Only difference is that we decrement the frequencies of all itemsets that are contained in a transaction being deleted. Cheung et al. further optimized FUP to maintain models under deletions (Cheung et al. 1997; 1996b), and to maintain association rules in the presence of hierarchies over the set of items (Cheung et al. 1996b).

BORDERS. BORDERS is a negative-border based algorithm for maintaining frequent itemsets (Aumann et al. 1999).[12] This algorithm exploits properties (1) and (3). The basic observation is that the negative border $NB^-(D, m_s)$ can be used to detect whether new itemsets become frequent due to the addition and deletion of transactions, and to identify such new itemsets that cause changes to the frequent itemset model. Specifically, if a new itemset $X \notin L(D, m_s)$ becomes frequent due to the database update, then some itemset $Y \subseteq X$ is in $NB^-(D, m_s)$. From property (1), we know that Y should also be frequent in the updated database. Consequently, only descendants of such *promoted* negative border itemsets are potential candidates for becoming frequent. Therefore, the promoted negative border sets focus our attention on the appropriate set of itemsets that need to checked.

BORDERS maintains the set of negative border itemsets $NB^-(D, m_s)$ in addition to the set of frequent itemsets $L(D, m_s)$. When a set of trans-

---

[12] A preliminary version of this algorithm appeared in (Feldman et al. 1997).

actions d^+ is added to D, BORDERS updates the frequencies of all current frequent and negative border itemsets $L(D, m_s) \cup NB^-(D, m_s)$ by scanning d^+ once. Due to the database update, the following cases occur: current frequent itemsets may no longer be frequent in $D \uplus d^+$ or itemsets in the negative border may become frequent in $D \uplus d^+$. If the frequency of a currently frequent itemset X falls below m_s, then X is deleted from $L(D, m_s)$. Also, if all subsets of X are still frequent in $D \uplus d^+$, then X is added to the negative border. If the frequency of an itemset $X \in NB^-(D, m_s)$ increases above m_s (that is, X is a promoted negative border set), then it is possible that some descendants of X may also be frequent. Therefore, BORDERS first identifies all promoted negative border itemsets and then generates potential descendant candidate itemsets from the promoted set. Frequencies of these candidates are then counted to identify the actual frequent itemsets. The descendant candidate generation and frequency counting steps are iterated until there are no more new frequent itemsets. Those candidate itemsets that are not frequent but all of whose proper subsets are frequent in the updated database are added to the negative border. We now describe the iterative process.

In the first iteration, S_1 is the set of promoted negative border itemsets. In the ith iteration, the set C_{i+1} of potentially frequent candidate itemsets is the pruned prefix join of S_i with $L(D, m_s) \cup S_1 \cup \cdots \cup S_i$. Frequencies of all itemsets in C_{i+1} are counted by scanning the updated database. All itemsets in C_{i+1} whose frequencies exceed m_s constitute the set S_{i+1}.

As an example, consider the addition of d^+ to the dataset D that yields the set of frequent itemsets in Figure 17.1. The negative border is the set {{CP,M},{WP}}. The frequencies of all negative border and frequent itemsets are updated by scanning d^+. Suppose the negative border itemset {WP} is promoted. Then, the new set of candidates {{WP,CP}, {WP,I},{WP,M}} is computed; their frequencies are counted by scanning $D \uplus d^+$. Suppose none of these candidates is frequent. Because all proper subsets of each of these candidates are frequent, they are added to the new negative border $NB^-(D \uplus d^+, m_s)$ completing the update process.

Maintaining a set of frequent itemsets when a set of transactions is deleted is very similar to maintenance under insertion. The procedure differs only in the update to frequencies of current frequent and negative border itemsets. We decrement the frequencies of all itemsets that are contained in a transaction being deleted.

TBAR. TBAR is also a negative-border based algorithm and is similar to BORDERS.[13] This algorithm exploits properties (1), (2), and (3). TBAR updates the frequencies of all current frequent itemsets $L(D, m_s)$ and negative border itemsets $NB^-(D, m_s)$ by scanning d^+ to identify all itemsets in $L(D, m_s)$ that are no longer frequent and the set N of promoted itemsets in $NB^-(D, m_s)$. Then, they compute the set of frequent itemsets $L(d^+, m_s)$ and the negative border $NB^-(d^+, m_s)$ from the increment d^+. TBAR computes candidate itemsets that are potentially frequent in $D \uplus d^+$ and then counts their frequencies through a single scan of $D \uplus d^+$. Frequencies of all candidate itemsets are now counted through a single scan of $D \uplus d^+$.

The set of candidate itemsets is a "pruned closure" of the set N of promoted negative border itemsets with respect to the set of frequent itemsets $L(D, m_s)$. The *pruned closure* of N with respect to $L(D, m_s)$ is obtained by the following iterative process similar to that in BORDERS. Let $S = \phi$. In the first iteration, S_1 is the set N of promoted negative border itemsets. In the ith iteration, if any itemset $X \in S_i$ also belongs to $NB^-(d^+, m_s)$, then X is removed from S_i and added to S. (This pruning step is an improvement over BORDERS.) The set S_{i+1} of potentially frequent candidate itemsets is the pruned prefix join of S_i with $L(D, m_s) \cup S_1 \cup \cdots \cup S_i$. The iterative process stops when $S_{k+1} = \phi$ for some k. The union $S \cup S_1 \cup \cdots \cup S_k$ of all candidate itemsets is the pruned closure. The frequencies of all itemsets in S is also counted because they may belong to the updated negative border.

Considering the example in Figure 17.1, TBAR updates the frequencies of all itemsets shown in the figure by scanning d^+. Suppose the itemset {WP} is promoted. Then, the set of frequent and negative border itemsets in d^+ are computed. Suppose $L(d^+, 0.33)$ is {{CP}, {I}, {M}, {WP}, {CP,I}, {I,M}} and $NB^-(d^+, 0.33)$ is {{CP,M}, {WP,CP}, {WP,I}, {WP,M}}. Now, the pruned closure of {{WP}} with respect to the old frequent itemsets is {{WP}, {WP,CP}, {WP,I}, {WP,M}}. The updated database is scanned once to count frequencies of these itemsets. The set {{WP,CP}, {WP,I}, {WP,M}} of itemsets is added to the updated negative border.

Maintenance under deletion of transactions is very similar to that under insertion except that frequencies of itemsets that are contained in a deleted transaction are decremented. Since frequent itemsets and the negative border are both computed for the increment d^+, if the size of d^+ is very large (say, comparable to D) then TBAR could be

---

[13]TBAR and BORDERS have been developed concurrently.

almost as expensive as computing frequent itemsets, from scratch, over the updated database.

DELTA. Pudi and Haritsa (2000) put all the above ideas together in another algorithm DELTA, which further optimizes BORDERS and TBAR algorithms by reducing the number of candidate itemsets generated and the number of scans of the increment d^+. They achieve this by exploiting properties (1), (2), and (3).

DELTA updates frequencies of all current frequent itemsets $L(D, m_s)$ and negative border itemsets $NB^-(D, m_s)$ to identify all itemsets in $L(D, m_s)$ that are no longer frequent and the set N of promoted itemsets in $NB^-(D, m_s)$. A set N' of new candidate itemsets are generated through the prefix join of N and $L(D, m_s) \cup N$. The frequencies of these candidates in the increment d^+ are then counted by scanning d^+. Because frequency in D of any itemset $X \in N'$ is less than m_s, from property (2) we know that if X has to be frequent in $D \uplus d^+$ then it has to be frequent in d^+. Therefore, any itemset in N' that is not frequent is deleted to obtain another set of itemsets, say N''. Now, DELTA computes the closure of N'' with respect to $L(D, m_s)$. The computation of the closure is similar to that of the computation of the pruned closure in TBAR except that candidate itemsets in the negative border of d^+ are not pruned (because we have not computed $NB^-(d^+, m_s)$). The frequencies of all itemsets in N' and the closure of N'' are counted by scanning the updated database $D \uplus d^+$. Since DELTA prunes the closure of N without computing the negative border of d^+, it is efficient even for very large size of d^+. Maintaining the set of frequent itemsets when transactions are deleted is very similar except that frequencies of itemsets contained in a deleted transaction are decremented.

Revisiting our example in Figure 17.1, DELTA updates the frequencies of all itemsets shown in the figure by scanning d^+. Suppose the set N of promoted sets be {{WP}}. The set N' is {{WP,CP},{WP,I},{WP,M}}. We count the frequencies in d^+ of all itemsets in N'. If all of them are less than the minimum frequency threshold, N'' is empty and so is its closure. D is scanned to count frequencies of itemsets in N', and the negative border is updated.

5.2.2 Decision Tree Models.

We now describe a scalable algorithm for incrementally maintaining decision tree models under adhoc evolution. Recall that each internal node in a decision tree is associated with a splitting attribute and a splitting predicate. A splitting predicate for a numerical splitting attribute is of the form $X \leq v$; v is called the *split point*. For the example tree in

Figure 17.2, the splitting predicate for the root is $Age \leq 30$. A splitting predicate for a categorical attribute divides the domain of the attribute into two sets.

BOAT. BOAT is a scalable decision tree construction algorithm that can also incrementally maintain decision trees. In this article, we only discuss the incremental aspect of the algorithm. For other details, we refer the reader to Gehrke et al. (1999).

The incremental maintenance algorithm is based on the belief that in most cases, the structure of a decision tree at least at the upper levels does not change drastically when sets of tuples are inserted or deleted. That is, splitting predicates on categorical attributes do not change, and the split points on numerical splitting predicates changes only slightly. BOAT significantly reduces the amount of work to be done when updates to the database cause minor changes in the structure of the tree. In particular, most effort is concentrated where the decision tree changes significantly.

As with most incremental maintenance algorithms, BOAT also annotates each internal node in the tree with additional information. First, if the splitting attribute is numerical it maintains a confidence interval around the split point. The significance of the interval is that if the data distribution remains similar then the probability with which the split point on that attribute remains within that interval is high (adjustable while the interval is being constructed). Second, BOAT maintains either in main memory or on disk all tuples that fall within this interval. Third, it maintains a histogram on the range of the splitting predicate where each bucket of the histogram contains the class distribution of tuples that fall within that bucket. Fourth, it also maintains the best splitting criteria at that node for all other attributes.

When a set of tuples inserted into or deleted from the database is available, we pour them down the tree and recursively adjust each internal node. The adjustment involves two steps. First, using all the annotated information and the set of tuples that reach the node, we classify the amount of change at the node into one of two categories: *drastic* or *moderate*. We consider the amount of change at a node to be *drastic* if the splitting attribute itself changes or the split point moves outside the confidence interval; otherwise, it is *moderate*. (We skip the details of the categorization.) Second, depending on the amount of change, we adjust the splitting criterion at the node. The amount of work required for adjusting the criteria depends on the amount of change.

If the change is drastic, we reconstruct the entire subtree underneath that node using the static version of the BOAT algorithm. Under a moderate change, if the splitting attribute is categorical then we do not need to do anything. If the splitting attribute is numerical, a new split point within the confidence interval is computed. Because of this adjustment, tuples that might have gone to the left child have to effectively be deleted from the left subtree and inserted into the right subtree. Since all tuples that fall within the confidence interval are maintained at the node, this task can easily be managed. In addition, we also update the information that annotates the node.

5.2.3 Cluster Models.

We now discuss an incremental algorithm for maintaining cluster models under the adhoc evolutionary model, which addresses the maintenance of a specific density-based notion of clusters proposed by Ester et al. (1995) in the context of spatial data. Therefore, this incremental algorithm is very tightly coupled with the density-based notion of clusters and the DBScan algorithm, which has been developed for identifying such clusters in a dataset.

Incremental DBScan. We first introduce the density-based notion of clusters proposed by Ester et al. (1995), and then discuss the incremental maintenance of such clusters.

Intuitively, a cluster is a maximal set of tuples within a dense region with no holes. Given a distance function d, a constant ϵ (a positive real number) and a minimum density threshold m (a positive integer), a tuple t_1 is said to be in the ϵ-neighborhood of another tuple t_2 if the distance $d(t_1, t_2)$ between them is less than ϵ; a tuple in a set S of tuples is called a *core* tuple if the number of tuples in S within its ϵ-neighborhood is greater than m. A tuple t_{n+1} is *density-connected* to a core tuple t_1 if there exists a sequence of core tuples $t_2, \ldots, t_n$ such that t_{i+1} is in the ϵ-neighborhood of t_i, for $1 \leq i \leq n$. A density-based notion of a cluster is a *maximal set S* of tuples that satisfies the following two properties. First, any pair of tuples in S are density-connected to a core tuple in S. Second, any tuple t that is within the ϵ-neighborhood of a core tuple in S is also in S. All tuples in a cluster that are not core tuples are called *border tuples*. Also note that a cluster is uniquely identified by any one of its core tuples.

Given a dataset D of tuples, DBScan discovers the set of clusters in D and the set of *noise* tuples that do not belong to any cluster (Ester et al. 1995). The incremental DBScan algorithm is based on the observation that when tuples are inserted into or deleted from D, the update only affects the neighborhoods of the inserted or deleted tuples. That is, only clusters that overlap the ϵ-neighborhood of an inserted or deleted tuple will be affected.

Suppose a new tuple t is inserted into a dataset D. Due to the insertion, tuples in the ϵ-neighborhood of t may become core tuples because the numbers of tuples in their ϵ-neighborhoods have increased by one. The set U_{ins} of core tuples in $D \uplus \{t\}$ within the ϵ-neighborhoods of these *new* core tuples may potentially change their cluster membership due to the insertion. There are only four types of changes: *new noise tuples*, *creation* of a new cluster, *absorption* into an existing cluster, and *merger* of two clusters. Note that a subset of a cluster can never move from one cluster to another because any core tuple uniquely identifies the cluster, and hence either all or none of the core tuples move. As described below, the properties of the set U_{ins} determines the type of change.

1. **Noise**: U_{ins} is empty. That is, the insertion of t did not create new core tuples. t is a noise tuple and does not belong to any cluster.

2. **Creation**: U_{ins} contains only core tuples that did not belong to any cluster before t was inserted. A new cluster containing these tuples along with t is created.

3. **Absorption**: U_{ins} contains core tuples that are members of exactly one cluster. Then, t and possibly some noise tuples are absorbed into the cluster.

4. **Merger**: U_{ins} contains core objects which are members of more than one cluster. All these clusters and the new tuple t are merged.

Suppose a tuple t is deleted from a dataset D. Due to the deletion, tuples in the ϵ-neighborhood of t may no longer be core tuples because their ϵ-neighborhood densities have decreased by one. If these ex-core tuples are binding tuples in their ϵ-neighborhoods to the cluster, then the cluster may either be reduced due to their removal or be split into two or more smaller clusters. The set U_{del} of core tuples in $D \ominus \{t\}$ within the ϵ-neighborhoods of these ex-core tuples are the tuples to be checked first. As in the case of insertion, U_{del} indicates the type of change to the set of clusters.

1. **Removal**: U_{del} is empty. Then t is deleted from D.

2. **Reduction**: All tuples in U_{del} are within the ϵ-neighborhoods of each other. Then, t is deleted from D and a few tuples in the ϵ-neighborhood of t may have to be removed from their cluster.

3. **Potential Split**: U_{del} contains tuples that are not within ϵ-neighborhoods of each other. Since all tuples in U_{del} previously belonged

to the same cluster, say, C before t was deleted, we need to check if pairs of tuples in U_{del} are density-connected to a core tuple in C. Otherwise, C has to be split. The new set of clusters is found by running DBScan, starting with U_{del}, on tuples in the cluster C.

Incremental DBScan relies on an index that can efficiently retrieve all tuples within a certain region. Insertion or deletion of a tuple triggers a series of region queries because update sets (U_{ins} and U_{del}) and the immediate neighbors of tuples in these sets are computed. Updating the cluster model when tuples are deleted is on the average more expensive than updating the model when new tuples are inserted; a deletion requires more region queries because many tuples in the cluster may have to be retrieved to ascertain whether a cluster has to be split or not.

5.3. Systematic Evolution

We now describe an algorithm GEMM for maintaining models over time-varying subsets of a systematically evolving database (Ganti et al. 2000). GEMM is a generic algorithm in that it can maintain any class of models. Besides the data span option and the block selection sequence, GEMM also takes as input another algorithm that can maintain models under tuple insertions. It then builds upon the input algorithm to maintain models under systematic insertions and deletions of blocks of tuples. By encapsulating peculiarities in maintaining each class of models in the input parameter, GEMM is able to maintain several classes of data mining models.

The input algorithm for maintaining models under tuple insertions achieves the following purpose. Let $\mathcal{M}$ be the class of data mining models and $A_{\mathcal{M}}$ be an incremental model maintenance algorithm for maintaining models under tuple insertions. That is, given a dataset D, a model $M(D)$ on D, and a new set d of tuples, $A_{\mathcal{M}}(D, M(D), d)$ returns the updated model $M(D \uplus d)$. GEMM can be instantiated with $A_{\mathcal{M}}$ to derive a model maintenance algorithm with respect to any block selection sequence and either option on the data span dimension. Note that $A_{\mathcal{M}}$ does not need to maintain models under deletions of tuples. However, GEMM instantiates algorithms that can maintain models under systematic deletions. We now discuss GEMM for both options on the data span dimension.

5.3.1 Unrestricted Window. It is straight-forward to extend a maintenance algorithm for adhoc evolution to maintain models with respect to a BSS over the unrestricted window option. A simple strategy is as follows.

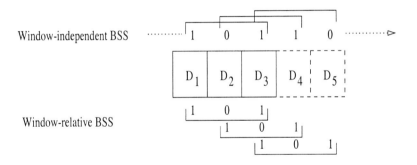

Figure 17.7. Most Recent Window

Let $M(D[1,t], b)$ be the current model on $D[1,t]$ with respect to the block selection sequence b. When a new block D_{t+1} is added to $D[1,t]$ and $b_{t+1} = 1$, the current model is updated using the incremental algorithm $A_\mathcal{M}$ to derive the new model $M(D[1, t+1], b)$. If $b_{t+1} = 0$, the current model carries over to the new snapshot.

5.3.2 Most Recent Window. Let $D[1,t]$ be the current database snapshot and w be the window size. For both window independent and window relative block selection sequences, the central idea in GEMM is as follows. Starting with the block D_{t-w+1}, the window $D[t-w+1, t]$ of size w evolves in w steps as each block D_{t-w+i}, $1 \leq i \leq w$, is added to the database. Therefore, the required model for the window $D[t-w+1, t]$ can be incrementally evolved using $A_\mathcal{M}$ in w steps. For example, the window D_3, D_4, D_5 in Figure 17.7 evolves in three steps starting with D_3, and consequently the model on D_3, D_4, D_5 can be built in three steps. The implication is that at any point, we have to maintain models for all future windows—windows which become current at a later instant $t' > t$—that overlap with the current window.

Suppose the current window c_w is $D[t-w+1, t]$. There are $(w-1)$ future windows that overlap with $D[t-w+1, t]$. We incrementally evolve models (using $A_\mathcal{M}$) for all such future windows. For each future window $f_i = D[i+t-w+1, i+t]$, $0 < i < w$, we maintain the model with respect to an "appropriate" BSS for the prefix $D[i+t-w+1, t]$ of f_i that overlaps with c_w. (The choice of the appropriate BSS for each prefix depends on the type of BSS, and is explained later.) Since there are $w-1$ future windows overlapping with the current window, we maintain $w-1$ models in addition to the required model on the current window. Whenever a new block is added to the database shifting the window to $D[t-w+2, t+1]$, the model corresponding to the suffix

$D[t-w+2,t]$ of c_w is updated "appropriately" using $\mathcal{A}_\mathcal{M}$ to derive the required model on the new window $D[t-w+2,t+1]$.

The remaining models in the collection are also updated but note that they are not required immediately. Hence the time by when the required model is available is determined by just one update. Moreover, updates to all models other may be piggybacked. Note that we also need additional (disk) space for maintaining at most w models. Since the amount of space occupied by a model is several orders of magnitude less than that required for the data in each block, the overhead is insignificant. We now illustrate through examples the derivation of block selection sequences with respect to which we maintain models.

Window-independent BSS. Consider the snapshot $D[1,3]$ shown in Figure 17.7 with $w=3$ and the window-independent BSS $\langle b_1, b_2, \ldots \rangle = \langle 10110 \cdots \rangle$ (shown above the window). The current model on $D[1,3]$ is extracted from the blocks D_1 and D_3. After D_4 is added, the window shifts right and the new model on $D[2,4]$ is extracted from the blocks D_3 and D_4. We observe that the new model can be obtained by updating (using $\mathcal{A}_\mathcal{M}$) the model extracted from D_2 and D_3 (the prefix of $D[2,4]$ that overlaps with $D[1,3]$). The observation here is that the relevant set of blocks (for the model extracted from $D[2,3]$) is selected from $D[1,3]$ by projecting the two bits b_2 and b_3 from the original BSS $\langle 10110 \cdots \rangle$, and by padding the projection b_2, b_3 with a zero bit in the leftmost place to derive $\langle 0, b_2, b_3 \rangle$.

Window-relative BSS. Consider the database snapshot $D[1,3]$ shown in Figure 17.7 with $w=3$ and the window-relative BSS=$\langle 101 \rangle$. The current model on $D[1,3]$ is extracted from the blocks D_1 and D_3. When D_4 is added, the window shifts right and the new model on $D[2,4]$ is extracted from the blocks D_2 and D_4. Observe that the new model can be obtained by updating (using $\mathcal{A}_\mathcal{M}$) the model extracted from the block D_2. The observation is that the relevant set of blocks for extracting the model from the overlap between $D[1,3]$ and $D[2,4]$) is selected from $D[1,3]$ by the BSS $\langle 0,1,0 \rangle$—obtained by right-shifting the original BSS $\langle 101 \rangle$ once and padding the leftmost bit with a zero.

Note: Even though algorithms for maintaining models under insertion-only and adhoc types of evolution may be applicable even in a systematic evolutionary scenario, they will not be as efficient. As an example, consider the maintenance of a model on data collected on every alternate day within the last four weeks. The relevant subsets of any consecutive snapshots are totally disjoint. In such cases, maintenance algorithms designed for insertion-only or adhoc evolutionary scenarios have to touch the entire dataset at least once before the required model is produced.

6. Change Detection

In this section, we discuss several methods for measuring changes in data characteristics, called *deviation*. The intuition behind all these approaches is that interesting characteristics in a dataset are captured by a model induced from the dataset. Therefore, we can measure deviation between datasets in terms of the models they induce. Since the data characteristics encapsulated in a model depend on the class of models to which it belongs, the deviation between two datasets also depends on the class of models considered. We first describe the FOCUS framework that generalizes many of these methods, and then discuss other methods.

6.1. The FOCUS framework

In recent work, we developed the FOCUS framework for measuring differences in data characteristics, as captured by a wide variety of models including the three popular classes of models: *frequent itemsets, decision tree models, clusters*. In addition to computing the deviation between two datasets, the FOCUS framework may be used to compute its statistical significance, and to automatically determine the regions where two datasets differ the most. We now illustrate the concepts and ideas behind the deviation computation using the class of decision tree models. For a complete formal description on computing deviations through several classes of models, we refer the readers to the original paper (Ganti et al. 1999).

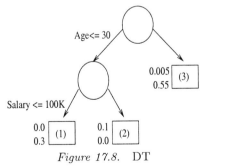

Figure 17.8. DT

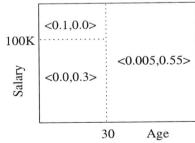

Figure 17.9. DT:two-components

Let a decision tree constructed from the join of the hypothetical magazine subscription relation R_2 shown in Table 17.2 and the demographic relation R_1 shown in Table 17.1 be as shown in Figure 17.8. Recall that the class label attribute in the subscription relation described whether a customer purchased a magazine or not, and hence has a binary domain ("YES" or "NO"). For the sake of clarity, we use C_1 for "YES" and C_2 for "NO." The decision tree consists of three leaf nodes. The class distribution at each leaf node is shown beside it (on the left side) with the top (bottom) number denoting the fraction of database tuples that belong to class C_1 (C_2, respectively). For instance, the fractions of database tuples that belong to the classes C_1 and C_2 in the leaf node (1) are 0.0 and 0.3, respectively. Each leaf node in the decision tree corresponds to two regions (one region for class C_1 and one region for class C_2), and each region is associated with the fraction of tuples in the dataset that map into it; this fraction is called the *measure* of the region. Generalizing from this example, each leaf node of a decision tree for k classes is associated with k regions in the attribute space each of which is associated with its measure. These k regions differ only in the class label attribute. In fact, the set of regions associated with all the leaf nodes partition the attribute space.

We call the set of regions associated with all the leaf nodes in the decision tree model the *structural component* of the model. We call the set of measures associated with each region in the structural component the *measure component* of the model. The property that a model consists of structural and measure components, called the *two-component property*, is exhibited by several classes, such as frequent itemsets, decision trees, and clusters. Figure 17.9 shows the set of regions in the structural component of the decision tree in Figure 17.8 where the two regions corresponding to a leaf node are collapsed together for clarity in presentation. The two measures of a leaf node are shown as an ordered pair, e.g., the ordered pair $\langle 0.0, 0.3 \rangle$ consists of the measures for the two collapsed regions of the leaf node (1) in Figure 17.8.

We now illustrate the idea behind the computation of deviation between two datasets over a set of regions. Let D_1 and D_2 be two datasets. Given a region and the measures of that region from the two datasets, the *deviation* between D_1 and D_2 with respect to the region is a function (e.g., absolute difference) of the two measures; we call this function the *difference function*. A generalization to the deviation over a set of regions is a "combination" of all their deviations at each region; we represent this combination of deviations by a function called the *aggregate function*, e.g., sum.

If two datasets D_1 and D_2 induce decision tree models with identical structural components, we can combine the two ideas—the two-component property and the deviation with respect to a set of regions—to compute their deviation as follows: the deviation between D_1 and D_2

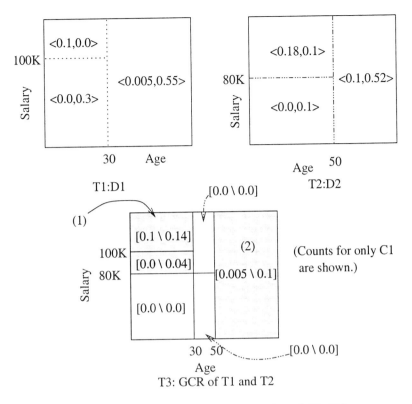

Figure 17.10. Decision Tree Model: $T_3 = \bigwedge(T_1, T_2)$

is the deviation between them with respect to the set of regions in their (identical) structural components.

However, the decision tree models induced by two distinct datasets typically have different structural components, and hence the simple strategy described above for computing deviations may not apply. Therefore, we first make their structural components identical by "extending" them. The extension operation relies on the structural relationships between models, and involves refining the two structural components by splitting regions until the two sets become identical. Intuitively, the refined set of regions is the finer partition obtained by overlaying the two partitions of the attribute space induced by the structural components of both decision trees. We call the refined set of regions the *greatest common refinement* (GCR) of the two structural components. For instance, in Figure 17.10, T_3 is the GCR of the two trees T_1 induced by D_1 and T_2 induced by D_2. In each region of the GCR T_3, we show a hypothetical set of measures (only for class C_1) from the datasets D_1 and D_2. For

instance, the measures for the region "salary $\geq 100K$ and age < 30"
for the class C_1 from D_1 and D_2 are 0.0 and 0.04, respectively. For
the classes of frequent itemsets, decision trees, clusters, and partitioning
grids, the GCR of two models can be shown to exist always.

To summarize, the deviation between two datasets D_1 and D_2 is computed as follows. The structural components of the two decision tree models are extended to their GCR. Then, the deviation between D_1 and D_2 is the deviation between them over the set of all regions in the GCR. In Figure 17.10, if the difference function is the absolute difference and the aggregate function is the sum then (part of) the deviation between D_1 and D_2 over the set of all C_1 regions is given by the sum of deviations at each region in T_3: $|0.0 - 0.0| + |0.0 - 0.04| + |0.1 - 0.14| + |0.0 - 0.0| + |0.0 - 0.0| + |0.005 - 0.1| = 0.175$.

6.2. Misclassification Error

A traditional approach to judge whether characteristics in a new dataset differ from the old dataset is to quantify how well a model induced from an old dataset fits a new dataset. Essentially, the question is cast as "By how much does the old model misrepresent the new data?" The *misclassification error* has been widely used in the context of decision trees (Breiman et al. 1984, Loh and Vanichsetakul 1988, Loh and Shih 1997). We show that the misclassification error can be captured as a special case of the FOCUS framework by carrying the structural component of the old model over to the new dataset and appropriately choosing f and g.

Informally, the misclassification error of a decision tree T on a dataset D is the fraction of tuples in D that T misclassifies. Formally, let T be a decision tree induced from a dataset D_1. Let D_2 be a second dataset. For each tuple $t \in D_2$, let $C' = T(t)$ be the class label predicted by T for t. If the true class C of t is different from C' then t is said to be *misclassified* by T. The misclassification error $ME^T(D_2)$ of T with respect to D_2 is the fraction of the number of tuples in D_2 misclassified by T.

$$ME^T(D_2) \stackrel{\text{def}}{=} \frac{|\{t \in D_2 \text{ and } T \text{ misclassifies } t\}|}{|D_2|}$$

The misclassification error of T on a dataset D_2 can be computed from the FOCUS framework as follows. Let T' be another decision tree whose structural component is identical to that of T but whose measure component is defined with respect to D_2. Let the difference function be the absolute difference and the aggregate function be the sum function. Then, $ME^T(D_2) = \frac{1}{2} \cdot deviation_{\{D_1, D_2\}}(T, T')$.

Consider the decision tree T shown in Figure 17.11. Suppose the numbers shown in each region correspond to the exact and predicted fractions of tuples in a dataset D. The misclassification error is $\frac{1}{2} \cdot (|0.1 - 0.09| + |0.0 - 0.01| + |0.005 - 0.01| + |0.0 - 0.01| + |0.3 - 0.29| + |0.55 - 0.545|)$.

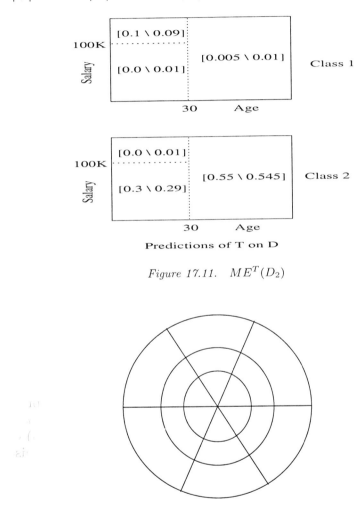

Figure 17.11. $ME^T(D_2)$

Figure 17.12. A hypothetical data spheres grid

6.3. Grid-based Methods

Several grid-based methods have been proposed to measure differences between the characteristics of two datasets. The general methodology is as follows. The attribute space is partitioned into a grid consisting

MINING AND MONITORING EVOLVING DATA

of disjoint (usually hyper-rectangular) regions. In the language of the FOCUS framework, this grid is the structural component of the models induced from both datasets. Given a dataset, each region is associated with a "summary" of all tuples in the dataset that belong to the region. The set of summaries of all regions in the grid is the measure component. The deviation between the two datasets is an aggregate of the difference between their summaries over each region. The methods differ in the types of partitions they use, the summaries they construct, and the procedure for computing the difference over a region.

6.3.1 Chi-squared Metric.

Let D_1 and D_2 be two datasets, and let G be a grid with the set $\{r_1, \ldots, r_k\}$ of disjoint regions. Any arbitrary partition of the attribute space may be used. However, a partition of the space into hyper-rectangular regions is typically employed. All tuples in a dataset D that belong to a region r are "summarized" by the measure $m_r(D)$ of r with respect to D. Let the measure components of G with respect to D_1 and D_2 be $\{m_{r_1}(D_1), \ldots, m_{r_k}(D_1)\}$ and $\{m_{r_1}(D_2), \ldots, m_{r_k}(D_2)\}$, respectively. Then, the chi-squared metric $X^2(D_1, D_2)$ is defined as follows.

$$X^2(D_1, D_2) = \sum_{i=1}^{k} |D_2| \cdot \frac{(m_{r_i}(D_1) - m_{r_i}(D_2))^2}{m_{r_i}(D_1)}$$

The chi-squared metric is typically employed to test the hypothesis that a dataset follows a certain expected distribution. The value assumed by the metric is then statistically tested against standard chi-squared distributions before the hypothesis is accepted or rejected. In such a hypothesis-testing setting, $m_{r_i}(D_1)$ is the expected measure of r_i for any dataset that has the same characteristics as D_1, and $m_{r_i}(D_2)$ is the observed measure. $X^2(D_1, D_2)$ is indicative of whether D_2 has the same characteristics as those of D_1.

We note that the chi-squared test can also be applied in the context of decision tree models because a decision tree partitions the attribute space into hyper-rectangular regions. Recall that each leaf node in the decision tree is associated with a set of regions, one per class label. The sets of expected and observed measures of these regions are the measure components of T with respect to D_1 and D_2, respectively. Using these measures, the chi-squared metric can be computed in a straightforward way.

We now illustrate the computation using the grid shown in Figure 17.11. Assuming that D_2 has one thousand tuples, the chi-squared metric is: $1000 \cdot (\frac{(0.1-0.09)^2}{0.09} + \frac{(0.0-0.01)^2}{0.01} + \frac{(0.005-0.01)^2}{0.01} + \frac{(0.0-0.01)^2}{0.01} + \frac{(0.3-0.29)^2}{0.29} + \frac{(0.55-0.545)^2}{0.545})$.

Note that the chi-squared metric is undefined if some of the expected measures for regions in the grid are zeroes. One standard strategy employed in such cases is to increase the measure m such that $|D_2| \cdot m = c$, for some non-zero positive number c (D'Agostino and Stephens 1986).

6.3.2 Data Spheres.

Johnson et al. proposed another grid-based method to measure differences between the characteristics of two datasets (Johnson and Dasu 1998). Unlike the chi-squared metric, they devise their own efficient data dependent grid construction procedure. They handle categorical attributes by dividing the dataset into "sub-populations." All tuples that consist of the same combination of categorical attribute values are treated together as a sub-population, and differences between the corresponding sub-populations of two datasets are measured independently. For instance, if *ethnicity* and *age group* are two categorical attributes taking 3 values each, then each dataset is divided into 9 sub-populations. The actual difference meaurement procedure itself assumes numeric attributes only. In the discussion below, we assume that data is described only by numeric attributes.

The gridding of the attribute space is done as follows. They compute the "trimmed centroid" of all tuples in the dataset by removing, for each attribute, a small percentage (one to two percent) of outlying tuples that assume extreme values on that attribute. The centroid of all remaining tuples is the *trimmed centroid*. With the trimmed centroid as the center, several concentric spherical layers are constructed such that each layer has approximately the same number of tuples. Note that these equiquantile layers form a data dependent partition. That is, the partition may be different if we use the second dataset instead of the first to construct the partition. This dependence has not been discussed in the original paper, and we assume that the partition is always based on the first dataset. The set of concentric spheres is further divided using a set of radial hyper-planes (that contain the center of the sphere). A hypothetical data spheres partition of a 2-dimensional attribute space into 18 regions in three different layers is shown in Figure 17.12.

Given a dataset D, they compute, for each region in the grid, the number, the centroid, and the covariance matrix of all tuples in D that belong to the region. The set of these three statistics is called the *profile* of the region with respect to D. The set of all profiles is the equivalent of the measure component.

Given two datasets, they use two complementary statistical tests on each region to compare the profiles from the two datasets. Each of these tests by itself may conclude that two datasets are similar even if they do not have the same distribution. Therefore, relying on multiple tests decreases the chance of concluding that two datasets are similar when they actually are not. The first is the *multi-nominal test for proportions*, which compares the proportions of tuples belonging to a region within a sub-population (Rao 1965). The second test is the *Mahalanobis D^2 test*, which is used to establish the closeness of multivariate means in the two profiles corresponding to a region (Rao 1965). We skip the details of these statistical tests.

7. Conclusions

In this article, we presented a comprehensive survey of research on mining and monitoring evolving data. We classified existing research based on the type of data evolution they consider, the class of data mining models they study, and the knowledge discovery problem they address. Specifically, all research on mining dynamically evolving data can be classified into three types of evolution: insertion-only, ad hoc, and systematic. The research that we discussed mostly addressed three classes of data mining models: frequent itemsets, decision trees, and clusters; and, on two problems: maintaining models and detecting changes in data characteristics. We also provided pointers to related research that is not covered by this classification.

Bibliography

R. Agrawal, T. Imielinski, and A. Swami. Mining association rules between sets of items in large databases. In *Proc. of the ACM SIGMOD Conference on Management of Data*, pages 207–216, Washington, D.C., May 1993.

R. Agrawal, H. Mannila, R. Srikant, H. Toivonen, and A. I. Verkamo. Fast Discovery of Association Rules. In Usama M. Fayyad, Gregory Piatetsky-Shapiro, Padhraic Smyth, and Ramasamy Uthurusamy, editors, *Advances in Knowledge Discovery and Data Mining*, chapter 12, pages 307–328. AAAI/MIT Press, 1996.

R. Agrawal and G. Psaila. Acive data mining. *Proceedings of the first international conference on knowledge discovery and data mining*, 1995.

R. Agrawal and R. Srikant. Mining Sequential Patterns. In *Proc. of the 11th Int'l Conference on Data Engineering*, Taipei, Taiwan, March 1995.

A. Arning, R. Agrawal, and P. Raghavan. A linear method for deviation detection in large databases. In *Proc. of the 2nd Int'l Conference* on *Knowledge Discovery in Databases and Data Mining*, Portland, Oregon, August 1996.

Y. Aumann, R. Feldman, O. Lipsttat, and H. Manilla. Borders: An efficient algorithm for association generation in dynamic databases. *Journal of Intelligent Information Systems*, April 1999.

G. Box, G. Jenkins, and G. Reinsel. *Time Series Analysis: Forecasting and Control.* Prentice Hall, 3rd edition edition, 1994.

P. Bradley, U. Fayyad, and C. Reina. Scaling clustering algorithms to large databases. In *The Fourth International Conference on Knowledge Discovery and Data Mining*, August 27-31 1998.

L. Breiman, J. H. Friedman, R. A. Olshen, and C. J. Stone. *Classification and Regression Trees.* Wadsworth, Belmont, 1984.

S. Brin, R. Motwani, J. D. Ullman, and S. Tsur. Dynamic itemset counting and implication rules for market basket data. In *Proc. of the ACM SIGMOD Conference on Management of Data*, May 1997.

S. Chakrabarti, S. Sarawagi, and B. Dom. Mining surprising patterns using temporal description length. In *Proceedings of the 24th International Conference on Very Large Databases*, pages 606–617, August 1998.

M. Charikar, C. Chekuri, T. Feder, and R. Motwani. Incremental clustering and dynamic information retrieval. In *Proceedings of the 29th Annual ACM Symposium on Theory of Computing*, 1997.

C. Chatfield. *The Analysis of Time Series: an Introduction.* Chapman and Hall, 1984.

D. Cheung, J. Han, V. Ng, and C. Wong. Maintenance of discovered association rules in large databases: An incremental updating technique. In *Proceedings of the twelfth international conference on data engineering (ICDE)*, February 1996a.

D. Cheung, S. Lee, and B. Kao. A general incremental technique for maintaining discovered association rules. In *Proceedings of the fifth DASFAA Conference*, April 1997.

D. Cheung, T. Vincent, and W. Benjamin. Maintenance of discovered knowledge: A case in multi-level association rules. In *Proceedings of the second international conference on knowledge discovery in databases*, August 1996b.

R. B. D'Agostino and M. A. Stephens. *Goodness-of-fit techniques.* New York: M.Dekker, 1986.

Richard Duda and Peter Hart. *Pattern Classification and Scene analysis.* Wiley, 1973.

M. Ester, H. P. Kriegel, J. Sander, M. Wimmer, and X. Xu. Incremental clustering for mining in a data warehousing environment. In *Proceedings of the 24th International Conference on Very Large Databases*, pages 323–333, August 1998.

M. Ester, H. P. Kriegel, J. Sander, and X. Xu. A density-based algorithm for discovering clusters in large spatial databases with noise. In *Proc. of the 2nd Int'l Conference on Knowledge Discovery in Databases and Data Mining*, Portland, Oregon, August 1995.

U. M. Fayyad, G. P. Shapiro, P. Smyth, and R. Uthurusamy, editors. *Advances in Knowledge Discovery and Data Mining.* AAAI/MIT Press, 1996.

R. Feldman, Y. Aumann, A. Amir, and H. Mannila. Efficient algorithms for discovering frequent sets in incremental databases. *Workshop on Research issues on Data Mining and Knowledge Discovery*, 1997.

D. H. Fisher. Knowledge acquisition via incremental conceptual clustering. *Machine Learning*, pages 139–172, 1987.

E. Forgy. Cluster analysis of multivariate data: Efficiency vs. interpretability of classifications. *Biometrics*, 21, 1965.

T. Fukuda, Y. Morimoto, S. Morishita, and T. Tokuyama. Data mining using two-dimensional optimized association rules: Scheme, algorithms, and visualization. In *Proceedings of the 1996 international conference on the management of data (SIGMOD)*, pages 13–23, Montreal, 1996.

K. Fukunaga. *Introduction to Statistical Pattern Recognition.* Academic Press, San Diego, CA, 1990.

V. Ganti, J. Gehrke, and R. Ramakrishnan. Demon–mining and monitoring evolving data. In *Proceedings of the sixteenth international conference on data engineering (ICDE)*, pages 439–448, San Diego, CA, March 2000.

V. Ganti, J. Gehrke, R. Ramakrishnan, and W.-Y. Loh. A framework for measuring changes in data characteristics. In *Proceedings of the 18th Symposium on Principles of Database Systems*, 1999.

J. Gehrke, V. Ganti, R. Ramakrishnan, and W.-Y. Loh. BOAT–optimistic decision tree construction. In *Proceedings of the ACM SIGMOD International Conference on Managment of Data*, June 1999.

D. J. Hand. *Construction and Assessment of Classification Rules*. John Wiley & Sons, Chichester, England, 1997.

C. Hidber. Online association rule mining. In *Proceedings ACM SIGMOD International Conference on Management of Data, Philadephia, Pennsylvania*, pages 145–156, June 1-3 1999.

T. Johnson and T. Dasu. Comparing massive high-dimensional datasets. In *Proceedings, Fourth International Conference on Knowledge Discovery and Data Mining*, pages 229–233, August 1998.

M. Lebowitz. Correcting erroneous generalizations. *Cognition and Brain Theory*, 5:367–381, 1982.

M. Lebowitz. Experiments with incremental concept formation. *Machine Learning*, 2:103–138, 1987.

T.-S. Lim, W.-Y. Loh, and Y.-S. Shih. An empirical comparison of decision trees and other classification methods. Technical Report 979, Department of Statistics, University of Wisconsin, Madison, June 1997.

W.-Y. Loh and Y.-S. Shih. Split selection methods for classification trees. *Statistica Sinica*, 7(4), October 1997.

W.-Y. Loh and N. Vanichsetakul. Tree-structured classification via generalized disriminant analysis (with discussion). *Journal of the American Statistical Association*, 83:715–728, 1988.

J. MacQueen. Some methods for classification and analysis of multivariate observations. In L.M. Le Cam and J. Neyman, editors, *Proceedings of the fifth berkeley symposium on mathematical statistics and probability*, volume I. University of California Press, 1967.

D. Michie, D. J. Spiegelhalter, and C. C. Taylor. *Machine Learning, Neural and Statistical Classification*. Ellis Horwood, 1994.

A. Mueller. Fast sequential and parallel algorithms for association rule mining: A comparison. Technical report, University of Maryland, August 1995.

S. K. Murthy. *On growing better decision trees from data*. PhD thesis, Department of Computer Science, Johns Hopkins University, Baltimore, Maryland, 1995.

V. Pudi and J. Haritsa. Incremental mining of association rules. Technical report, DSL, Indian Institute of Science, Bangalore, 2000.

C. Rao. *Statistical Inference and its Applications*. John Wiley, 1965.

R. Rastogi and K. Shim. Mining optimized association rules with categorical and numeric attributes. In *Proceedings of the 14th international conference on data engineering*, pages 503–512, 1998.

A. Savasere, E. Omiecinski, and S. Navathe. An efficient algorithm for mining association rules in large databases. In *Proc. of the VLDB Conference*, Zurich, Switzerland, September 1995.

J. C. Schlimmer and D. Fisher. A case study of incremental concept induction. In *Proceedings of the Fifth International Machine Learning Workshop*, pages 496–501, Philadelphia, PA, 1986.

A. Silbershatz and A. Tuzhilin. What makes patterns interesting in knowledge discovery systems. *IEEE Transactions on Knowledge and Data Engineering*, 8(6), 1996.

R. Srikant and R. Agrawal. Mining Sequential Patterns: Generalizations and Performance Improvements. In *Proc. of the Fifth Int'l Conference on Extending Database Technology (EDBT)*, Avignon, France, March 1996.

R. Subramonian. Defining diff as a data mining primitive. In *Proceedings, Fourth International Conference on Knowledge Discovery and Data Mining*, pages 334–338, New York City, New York, August 27-31 1998.

S. Thomas, S. Bodagala, K. Alsabti, and S. Ranka. An efficient algorithm for the incremental updation of association rules in large databases. In *Proceedings of 3rd International Conference on Knowledge Discovery in Databases*, 1997.

H. Toivonen. Sampling large databases for association rules. In *Proc. of the 22nd Int'l Conference on Very Large Databases*, pages 134–145, Mumbai (Bombay), India, September 1996.

P. E. Utgoff. Incremental induction of decision trees. *Machine Learning*, 4:161–186, 1989.

P. E. Utgoff, N. C. Berkman, and J. A. Clouse. Decision tree induction based on efficient tree restructuring. *Machine Learning*, 29:5–44, 1997.

G. I. Webb. Opus: An efficient admissible algorithm for unordered search. *Journal of Artificial Intelligence Research*, 3:431–465, 1995.

A. Weigend and N. Gerschenfeld. *Time Series PRediction: Forecasting the Future and Understanding the Past.* Addison Wesley, 1994.

S. M. Weiss and C. A. Kulikowski. *Computer Systems that Learn: Classification and Prediction Methods from Statistics, Neural Nets, Machine Learning, and Expert Systems.* Morgan Kaufman, 1991.

P. Willett. Recent trends in hierarchical document clustering: A critical review. *Information Processing and management*, 24(5):577–597, 1988.

T. Zhang, R. Ramakrishnan, and M. Livny. BIRCH: An efficient data clustering method for very large databases. In *Proc. of the ACM SIGMOD Conference on Management of Data*, Montreal, Canada, June 1996.

Chapter 18

DATA QUALITY IN MASSIVE DATA SETS

Michael F. Goodchild
National Center for Geographic Information and Analysis, and Department of Geography, University of California, Santa Barbara, CA 93106-4060, USA
good@ncgia.ucsb.edu

Keith C. Clarke
National Center for Geographic Information and Analysis, and Department of Geography, University of California, Santa Barbara, CA 93106-4060, USA
kclarke@geog.ucsb.edu

Abstract   All data contain errors, and large spatial data sets are especially prone because they contain data from multiple sources, and use different assumptions about structure and semantics. The general issue is one of data quality assurance, defined in terms of lineage, completeness, logical consistency, attribute accuracy, and positional accuracy. We review a series of quality metrics suitable for empirical description of data quality, and consider some of the special issues of quality related to spatial data, especially the need to incorporate visualizations of data quality into graphics and maps. We conclude that data quality is an essential component of software for spatial data handling, including geographic information systems.

Keywords: Data quality, Metrics, Spatial data, Geographic information systems.

1.   The Quality Problem

Data sets capture facts and enable their management and retrieval. Yet in almost all instances there exists some external reference to which the data set's version of the facts can be compared. A data set created by digitizing the contents of a published book can be checked against the original; a measurement of temperature recorded in a data set can be checked against an independent measurement; and a real-estate agent's record of the price of a house purchase can be compared against the local tax assessor's data set. Data *quality* is quantifiable from the results of such checks, and data set contents that fail to match their own or independent reference sources are said to contain *errors*, or to be of poor quality. Measures of data quality can be devised, based on the frequency of errors, or on their magnitudes. Where there is no reference source of equal or superior quality, then the fact being recorded is based on inadequate definitions and is inherently vague and error-prone. For example, there is no way of checking the quality of the statement "it is cold here" against measurements of temperature, or location, although the statement may be a correct representation of what was said, and captured into the data set. Since a perfect match between data and the real world is generally impossible, we conclude that error as defined here must be endemic in data sets.

Many different terms are associated with quality, or the lack of it, and there is little consensus about their precise meanings. Imprecision, inaccuracy, inexactness, vagueness, uncertainty, unreliability, and incompleteness all imply lack of quality in some sense. Nevertheless, the terms capture divergent forms of variation between data set and reference, or different sources of difference.

Data quality is an important issue for massive data sets, because poor quality implies that decisions based on data set content will also be poor, and because massive data sets may have been assembled quickly, from multiple sources, at multiple scales, from sources with inherent vagueness, or with little concern for quality. Massive data sets once gloriously isolated by their size or complexity now find themselves open to searching and use by millions over the World Wide Web, regardless of their quality. High quality can be expensive, particularly if it involves human intervention in verification and if many or all data set records have to be checked.

Poor-quality can itself result in high costs, which may exceed the costs of correction. Data sets may be used for regulation, where poor-data quality may be the cause of costly litigation, particularly if it can be shown that the developers and users of data sets failed to take adequate actions to maximize quality or to deal with the known consequences of poor quality. Cartographic examples of missing map features or mislocated buildings abound, as in the case of the 1998 ski-lift accident in Italy, or the 1999 accidental bombing of the Chinese Embassy in Belgrade. Poor quality data sets used for scientific research cast doubt on the quality of the resulting scientific conclusions. Users of poor quality data sets quickly become frustrated once products are found to be unreliable. Errors and uncertainties *propagate* from the data set to products and decisions derived from it, including answers to queries, results of analysis, and transformations. Users of data sets need to know something of the inherent quality of a data set's contents in order to assess the fitness of the data set for specific purposes, and to determine the quality of products derived from the data set. Such information can be communicated in the form of text, but visualization also provides an important tool for informing users about quality.

This chapter is structured as follows. The next section deals with the description and representation of quality in data sets, and the techniques that have been devised for communicating knowledge of quality through visualization. This is followed by a section on the implications of quality, with discussion of the state of knowledge in propagation. The chapter uses the example of spatial data sets frequently, in part because research on them has advanced to a significant degree, and many results have been incorporated into standards and practice; and in part because quality has added dimensions and significance for spatial data.

2. Elements of Quality

One of the most comprehensive analyses of data set quality is found in Federal Information Processing Standard 173, otherwise known as the Spatial Data Transfer Standard (www.fgdc.gov; for a more extensive discussion of the elements of spatial data quality see Guptill and Morrison (1995). Devised in the mid 1980s, it identifies five components of quality for spatial data, as follows:

- *Lineage*, defined as information about the process of creation of the data set, such as the instruments used to make measurements, the identities of individuals and agencies responsible for creation, and the standards used to define the data set's contents. By knowing such details, it is possible in many cases to make inferences about quality. For example, knowing the identity of the instrument used to acquire measurements often allows the user to make meaningful estimates of their accuracy. Lineage also serves another useful purpose by providing feedback – for example, if serious errors are found in data it might be possible to link them to specific faults in the production process. It is the data set lineage that answers science's call for documentation permitting repeatability of experimental results, and therefore the independent confirmation of findings necessary for the scientific method.

- *Completeness*, or the degree to which the data set captures all of the expected data. Completeness is often linked to the currency of the data, or the degree to which the data represent current conditions, or conditions that existed at some point in the past and for which the data set is intended to form a complete representation. Currency is a significant problem for digital data sets, especially when the date for which the data are intended to be valid differs from the date of construction of the data set, or if either of these dates are not precisely defined, or if different versions of the data set are not clearly identified. Completeness can also refer to spatial extent, the number of available attributes actually included, and to known missing data. Many data sets for the United States, for example, actually exclude Alaska, Hawaii, and the United States Territories, and variable numbers of attributes for each state.

- *Logical consistency.* This refers to the internal consistency of the data, and the data set's adherence to its own defined rules. For example, logical consistency is violated if an object has two unique identifiers, or if the value of an attribute falls outside its defined domain. In spatial data sets, there can be logical inconsistency between the geometric content of a data set (a point lies inside the boundary defining California) and the topological content (the point has an attribute indicating it is in Nevada). If the rules are well-defined, then it is in principle possible to detect errors of logical consistency without human intervention, and it may also be possible to correct them. Such corrections require their own rules (does geometry over-ride topology, forcing the attribute to be changed to California, or does topology over-ride geometry, forcing the point to be moved to the geometric center of Nevada?), and it is difficult to avoid rules that create their own conflicts (moving the point may be problematic if it is connected to another object – for example, if the point is part of a lake shoreline).

- *Attribute accuracy.* This refers to the accuracy of the recorded attributes associated with each object. In a spatial data set, each object – a road, a mountain, a lake, a city, a house – will have certain defined attributes. These might include a unique identification number, a name, or in the case of a city the current population. Attributes can be differentiated in various ways by type. They may be *qualitative* (e.g., name) or *quantitative* (e.g., population count), and more elaborate schemes exist (see, for example, Chrisman (1997)). From the perspective of quality, it is important to distinguish between cases where an attribute can be only *right* or *wrong*, and cases where it is possible to define degrees of correctness. In the former instance, quality is best measured by the proportion of errors, but in the latter case many methods are available for measuring quality, and many of these are discussed in the next section. For example, a misspelled name of a city is more right (and possibly open to automatic correction) than a name that is completely wrong (e.g., in the case of Pittsburgh, *Pittsburg* is less erroneous than *Pittston*). Finally, correctness may be defined with reference not to reality but to the measurement process. The debate over the use of sampling in the Year 2000 U.S. Census, for example, has led to legislative prohibition of the methods that could have provided the most accurate results given the available budget. Yet the census itself assumes that the population's street addresses on April 15th, 2000 are their actual "places" as far as the federal government is concerned.

- *Positional accuracy.* The position recorded for an object on the Earth's surface can never be perfectly accurate, since the instruments available for measuring position (surveying instruments, or the Global Positioning System) have limited accuracy, and the positions of the reference objects (the Poles, Equator, and Greenwich Meridian) are also not perfectly defined. Even well-known positional reference systems, such as the latitude and longitude of geographic coordinates, require, at the minimum, knowledge of the Earth model, its size and shape, and the vertical datum in use. Standard coordinate systems such as the Universal Transverse Mercator have inherent positional accuracies of about 1 part in 2000, with systematic error depending on position. In some cases it may be impossible to separate positional accuracy from attribute accuracy. For example, it may be impossible to determine in the case of a measurement of the elevation of a point above sea level whether the correct elevation has been recorded at the wrong point, or whether the wrong elevation has been recorded at the correct point. Nevertheless, spatial data with only limited positional accuracy or precision, such as digital versions of coarse-scale maps, can still have immense scientific value and may need to be used in combination with data of different levels of quality.

This five-component scheme is recognized by being written into a major U.S. standard, but many other terms have been proposed, often with conflicting definitions, to capture the elements of data quality. Many forms of data are inherently *vague*, because it is impossible to decide with certainty what the correct value should be. For example, it is impossible to determine when something should be described as cold. Such evaluations are often termed *subjective*, because there is no reason to expect any two people to agree on the correct value – they are not *replicable*. Many scientists would argue that such data have no value, but others would argue that vagueness of communication is an indispensable part of human existence.

Empirical scientists often distinguish between *accuracy*, or the degree of agreement between a recorded observation and its true value, and *precision*, or the degree of detail with which the measurement is recorded. A widely recognized principle holds that precision should never exceed accuracy. For example, if a thermometer is capable of measuring to the nearest Celsius degree, then recorded measurements should never include decimal places (e.g., 21 is acceptable but 20.986 is not). But precision is also used to refer to the variation among repeated measurements of the same phenomenon with the same instrument.

3. Description of Quality

3.1. Numeric values

Consider a measuring instrument such as a thermometer, and suppose that it is being used to measure a temperature whose correct value is 21.0 Celsius. The thermometer is inherently inaccurate, and returns a value of 23. By repeatedly comparing true and measured values it is observed that the thermometer's measurements are in error by amounts ranging from -2 to $+2$ Celsius. So a straightforward way to record quality would be by specifying the *range*. In a data set, this could be recorded in the form of additional attributes – for example, as $\langle 23, +2, -2 \rangle$. The query "Is the temperature greater than 26?" would return "no", but the query "Is the temperature less than 22?" would return "maybe".

Range provides an easy means of responding to simple queries, but it is problematic because it provides no information on the relative frequency of large and small errors. In reality, it is almost always true that the thermometer will produce small errors more frequently than large ones. Moreover, if large errors are rare, it will be difficult to provide an accurate estimate of range without making a very large number of tests. Fortunately, it is known that under a wide range of circumstances the relative frequencies of large and small errors are consistent with a simple model, known as the *Gaussian* or *normal* distribution, the *error function*, or the *bell-curve*, and shown in Figure 18.1. As a *probability density function*, the probability of an observation lying between any two values of the x-axis is defined by the area under the curve between those limits. The width of the curve is best defined by the distance between the center and the points of inflection, and is known as the *standard deviation*. The instrument is said to be biased if the mean error is not zero. Finally, the *standard error* or *root mean squared error* (RMSE) is

DATA QUALITY IN MASSIVE DATA SETS

defined as the square root of the mean squared difference from the true value:

$$\text{RMSE} = \left[\frac{1}{n}\sum_i (x_i - X)^2\right]^{1/2},$$

where X is the true value, n is the number of observations, and x_i denotes an observation, when the number of such observations is very large.

The points of inflection shown in Figure 18.1 represent one standard deviation on either side of the mean. Approximately 68% of errors will be smaller than one standard deviation, and 32% will be larger. More useful perhaps is the fact that 95% of errors will lie within 1.96 standard deviations of the mean, or approximately 2 standard deviations. This is the basis for the *confidence limits* commonly heard in association with opinion polls – for example, that the true value "will lie within 2 percentage points 19 times out of 20".

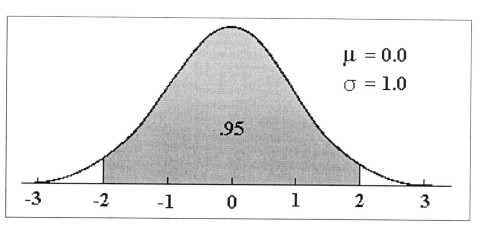

Figure 18.1. The Gaussian distribution with a mean of 0 and a standard deviation of 1, showing the probability of a value lying between 2 standard deviations on either side of the mean. Note the points of inflection (change of curvature from upward-facing to downward-facing) at 1 standard deviation on either side of the mean.

While the Gaussian distribution is often a very accurate model of errors, it is not as easy to apply to the resolution of queries. But with a little effort, it is possible to replace the "maybe" response of the earlier example with a precise estimate of the probability that the temperature is less than 22, given knowledge of the recorded observation, and of the mean and standard deviation of the error distribution.

3.2. Qualitative values

The previous section discussed attributes that have numeric qualities. Suppose now that a data set contains a qualitative attribute, such as the names of streets. In most such cases a simple approach is to estimate the proportion of such attributes that are correct, and the proportion that are in error, and to attach these proportions to the results of queries. A common instance of errors in qualitative data occurs in the accuracy assessment of certain types of maps. For example, a map of land use might be prepared by classifying a remotely sensed scene from a satellite. A scene from the Enhanced Thematic Mapper Plus instrument on the Landsat 7 satellite consists of an array of picture elements (*pixels*) that are approximately square and $15m$ on a side on the Earth's surface. To test the accuracy of the automated classification, a random sample of locations is selected on the ground, and visited to determine the actual land use. Table 18.1 shows a hypothetical result of checking 100 such points.

Table 18.1. Results of accuracy assessment of a map of land use (rows indicate recorded values, columns indicate ground truth).

	Residential	Open space	Agriculture	Woodland	Water	Totals
Residential	33	3	0	1	0	37
Open space	2	24	4	0	0	30
Agriculture	0	1	17	2	0	20
Woodland	0	0	2	5	0	7
Water	0	0	0	0	6	6
Totals	**35**	**28**	**23**	**8**	**6**	**100**

From Table 18.1, it is apparent that the recorded value (row) agreed with ground truth (column) in a total of 85 cases out of 100, since 85 cases lie on the main diagonal of the table. Thus a convenient way of summarizing accuracy is to say that the probability of a randomly chosen point having the correct recorded value is 0.85. But the table clearly contains much more useful information. Water is never confused with any other class, since it is easy to identify correctly in satellite images. Of the 35 points that are truly residential, 33 are correctly classified, but 2 are confused with open space. Agriculture is also sometimes confused with open space, and sometimes with woodland.

Suppose that a user queries the data set to determine the class of land use at a point. In general, we can say that the probability of a correct response is 0.85. But if the land use recorded in the data set is residential, we know from the table that a better estimate of correctness is 33/37, or 0.89, with probabilities of 0.08 that the truth is open space, and 0.03 that the truth is woodland.

3.3. Fuzziness

The previous section was based on an implicit assumption – that the class at a point must be one of the five recognized options. In reality, the area covered by a single pixel may be a mixture, for example at the edge of a lake, so that the truth is not 100% water or 100% woodland but some mixture of the two. Recently, there has been much research on so-called mixture methods, which attempt to identify the percentages of various pure classes present in a mixed pixel (see, for example, Gillespie (1992)).

At a more fundamental level, however, it may be impossible to define such categories as residential precisely, because the term itself implies a mixture of different surfaces: roof, concrete, asphalt, grass, water (pool). Rather, the set of pixels labeled residential is fuzzy, with poorly defined properties. Fuzzy set theory has become popular in recent years as a way of dealing with situations in which assignment to classes is overly restrictive (see, for example, Zhu et al. (1996)).

In fuzzy set theory, membership in a class is measured on a continuous scale that is often constrained to the range [0,1]. A pixel that is most like the pure concept of residential is assigned the highest membership value, while one that has nothing in common with residential is assigned 0. The memberships for a pixel can be conceived as a vector $\{m_1, m_2, \ldots, m_n\}$, where n is the number of classes, and m_i denotes the membership of the pixel in the i-th class.

Fuzzy set theory is attractive in dealing with uncertainty in categorical data because it admits degrees of belonging, and thus approximates the way humans think about classes of land use, or any categories defined by complex or subjective measurements such as soil type, flood risk, or land suitability. An observer might well be able to distinguish between areas that are more residential and areas that are less so, or to agree that the degree of "residentialness" declines as one moves away from a city's center. Reasoning is also possible based on fuzzy sets, using certain axiomatic propositions to manipulate degrees of fuzziness. On the other hand, it seems dubious to claim that a degree of membership assigned by one observer has any meaning to another observer, when neither the class itself nor the scale of measurement of membership are well-defined.

3.4. Metadata

Metadata are defined as data about data; they include descriptions of the general properties of a data set, including its structure, format, language, and definitions; and also information about quality, ownership, and other properties that are useful to potential users. Metadata are analogous to the information in a library's card catalog, or to the information printed at the front of a book, or on the outside of a package.

If a data set is passed without explanation or documentation from one person to another, it can amount to little more than a confusing mass of binary digits. Metadata are "what make data useful" in the words of Francis Bretherton. They allow a user to assess the fitness of data for a particular application, particularly with respect to quality. Lack of metadata can also contribute to lack of quality, if a user makes the wrong assumptions about the data's meaning. For example, a user might see a set of numbers labeled "temperature," and not knowing the scale of measurement might wrongly assume that the scale was Celsius rather than the intended Fahrenheit. In effect, this would introduce an error in every value other than -40.

Quality description is an important component of metadata, especially for spatial data. The Content Standards for Digital Geospatial Metadata, created by the U.S. Federal Geographic Data Committee (www.fgdc.gov), include extensive and precise descriptions of quality, using the five components discussed earlier. The approach has been described as *truth in labeling*, since it attempts to elicit from the creator of the data as much useful information as possible about quality, but sets no absolute standards or thresholds of quality that must be met. Thus a data set with a quality statement that reads "This data set has no quality statement" is fully compliant with the standard, but also has information of value to the user in making decisions about the quality of the data.

Unfortunately, the metadata approach falls far short of a complete solution to the problem of describing quality, for several reasons. First, it favors descriptions that apply uniformly to the entire data set, such as a single measure of positional accuracy. In reality, however, it is common for elements of a data set to have different levels of quality, and quality may need to be defined at the level of the class of object, the individual object, or even the individual attribute. For geospatial data, it is common for quality to vary geographically, and many topographic maps include a much smaller map inset indicating how the quality of the main map varies.

Second, the metadata approach implies that quality can be described adequately without substantial restructuring of the data, by adding appropriate *slots* to the existing data model. Consider the case of a geographic region, such as the Atlantic Ocean, represented in a data set as a *polygon*, a series of points delimiting the ocean's boundary in clockwise or counter-clockwise order. In reality the Atlantic Ocean is not well-defined, and we might wish to describe its quality by adding suitable descriptors to the data set. One way to do this is to create a fuzzy region, by conceiving of a continuous variable p such that the value of p at some point is the degree of membership of that point in the concept *Atlantic Ocean*. To represent the spatial variation of p, however, we would have to abandon the polygon representation, and adopt a raster or some other way of describing what is now a continuous surface. In other words, description of quality has forced a change of data model (Burrough and Frank 1996).

Third, the metadata approach implies that it is possible to achieve a complete description of quality that is intelligible and useful to the user. In practice, description of quality through appropriate models can be exceedingly complex. The Gaussian error model described earlier is among the simplest of statistical models of error, yet even it is a sophisticated statistical concept. In geospatial data, it is common for the error affecting one object to be similar to the error affecting other objects, especially if the two objects are close to each other and if they have been measured by the same process. For example, suppose a map is created from an aerial photograph. One of the sources of positional error is misregistration of the photograph; and this form of error will affect all objects mapped from the same aerial photograph to varying degrees. Positional errors of objects are frequently *correlated*, and the degree of correlation is found to vary inversely with distance.

Because of positive correlations, the *relative* accuracy of the positions of nearby objects can be much higher than the *absolute* accuracy, and much higher than is implied by general descriptive measures such as the RMSE that are contained in metadata. Relative properties such as ground slope can be estimated accurately from digital elevation data even though absolute elevations in the data set are of poor quality, provided errors show strong positive correlations over short distances (Hunter and Goodchild 1997).

Many models of correlated errors exist, but they are not widely known outside the research community, and their use in metadata to describe quality is therefore highly problematic, since most users are not equipped to understand or deal with them. To address this issue, Goodchild et al. (1999) have argued that the concept of metadata should be broadened to include *methods*. Instead of the parameters of a complex error model, a producer should provide code that simulates the error model, producing a sample of versions of the data set that represent the range of variation due to uncertainty. Suppose, for example, that one wished to describe the uncertainty associated with a forecasted high temperature of 25 Celsius. Someone familiar with the Gaussian error model would understand the statement that uncertainty was characterized by a standard error of 2. But the same information is contained in the simulated set $\{26, 24, 23, 28, 21, \ldots\}$ if these are generated using an appropriate code. Goodchild et al. (1999) apply the same concept to the much more complicated case of geospatial data sets, arguing that the concept is no more difficult in the latter case. Although the models are far more complex, they need only be understood by the creators of the data and the simulation code, not by the users.

3.5. Visualization

Visualization provides an attractive medium for communication of information about quality. Visualization has already proven its effectiveness as a way of searching massive data sets for pattern and structure. The existence of uncertainty can be conveyed by removing, blurring, or greying, or by changing the visual depth of objects, bringing certain objects to the front and pushing uncertain objects to the back. Visualization of large spatial data sets as a method of communicating information about geospatial data quality has been the subject of intensive research (Beard et al. 1991, Davis and Keller 1997) and was reviewed more recently by Clarke and Teague (1998).

MacEachren (1992) investigated the use of existing map methods for uncertainty depiction, and introduced the variable of visual focus, shown by crispness, fill clarity, fog, and resolution variation used to adjust the boundaries between map features. More certain objects were depicted as "a sharp, narrow line" and less certain features as "a broad, fuzzy line that fades" toward the periphery. McGranaghan (1993) examined realism and time as potential variables for symbolization. Objects of lower data quality appear more "cartoonish" if data quality is low and more realistic if data quality is high. Time-based methods necessarily involve animation and several methods utilizing time as a cartographic variable were considered, including blinking, fading, and moving. The amount of time a blinking object is present or absent reflects quality information. Fading can be employed by having an object on the map oscillate to reflect quality; McGranaghan showed a stream segment oscillating between green and blue (high confidence) or green and red (low confidence).

Animations showing multiple realizations of a data set, and associated with the range of uncertainty described above, have been employed by Fisher (1993) and Ehlschlaeger et al. (1997). Fisher used animation techniques based on his earlier research to depict positional uncertainty in soil maps. Soil inclusion information, provided by the data producer, is conveyed to the user through an animated soil map that uses randomization to show these inclusions within the predominant soil types. Cells are continuously and randomly selected based on given inclusion rate, producing a stochastic realization of soil type distribution at any point in time. Ehlschlaeger et al. (1997) utilized animation to display multiple stochastic realizations of output from least – cost path analyses based on coarse resolution terrain data. Using multiple possible elevation surfaces, a series of cost surfaces for a least-cost path algorithm were produced showing the resulting shortest path. Each realization was used as a frame to create a smooth animation.

The integration of uncertainty information and data into a single display without graphic overloading was explored by Wittenbrink et al. (1996) through an approach called verity visualization. This method includes uncertainty visualization using uncertainty glyphs, fat surfaces, perturbations, and oscillations. Uncertainty glyphs, using various graphic variables such as size and shape of an icon to depict data attributes, are placed on the visualization or map itself to indicate uncertainty at different locations. Fat surfaces indicate uncertainty in information by presenting a range of data values at each location on the surface. Finally, Clarke et al. (1999) have advocated using visual depth in virtual-reality-based representations of data, so that the "nearness" of the data to the viewer portrays uncertainty using some of the variables already discussed, such as color and animation. So, for example, as the data user zooms in on a feature, it wobbles more or less depending on its uncertainty.

In spite of this promising research, in the case of geospatial data, it is clear that modern cartographic practice has traditionally left little room for uncertainty, and the practices of the past – leaving areas blank, inserting mythical beasts – have now largely disappeared. Research shows that users need to be cued to expect uncertainty, but that once appropriate instructions have been given, have no difficulty associating grayness, blurring, or even shaking with uncertainty (MacEachren 1992). As the research in this area yields results of use in everyday practice, two types of user interfaces between the data and the uncertainty seem possible. In the first, the treatment of uncertainty is as another layer of the map, subject to viewing, and use in analytical operations. In this method, the use of multiple realizations, all of equal possibility given the error bounds, is one way to show uncertainty and estimate its propagation into results (Journel 1996). Alternatively, uncertainty can be integrated into the visual display of the information, and activated by the data user when it becomes of concern during the analysis of information. Either way, the revised role for uncertainty in the use of data from massive data sets is significantly enhanced. Visualization offers a promising suite of methods for informing the data user about uncertainty.

4. Working with Poor Data

References have been made to queries based on uncertain data. More generally, the term propagation refers to the impact of uncertain or erroneous data on the results of query, analysis, and modeling. For example, consider a square parcel of land $100m$ on a side, with a true area of 1 hectare. Suppose that the corner points are inaccurately surveyed, with a mean error of 0 and a standard error of $2m$ in both coordinates. If the errors are uncorrelated, it is possible to compute the standard error in the estimate of area (Chrisman and Yandell 1989); in this case, the result is $200m^2$. If the errors have a perfect positive correlation (in other words, are identical), then the error in area is 0, since the square moves under error as a rigid body without rotation or warping. Thus the manner in which error propagates into the result – the estimate of area – depends on the nature of the error.

The classical theory of measurement provides a basis for analysis of propagation in numeric data. Suppose that some scalar measurement, such as a measurement of temperature using a thermometer, is distorted by an error generated by the measuring instrument. The apparent value of temperature x' can be represented as the sum of a true value x and a distortion δx. If some manipulation of x is required, the theory of measurement error provides a simple basis for estimating how error in x will propagate through the manipulation, and thus for estimating error in the products of manipulation (Taylor 1982) (see Heuvelink (1998), and Heuvelink et al. (1989), for discussions of this in the context of geospatial data). Suppose that the manipulation is a simple squaring,

$y = x^2$, and write δy as the distortion that results. Then:

$$y + \delta y = (x + \delta x)^2$$
$$= x^2 + 2x\delta x + \text{terms of order } \delta x^2.$$

Ignoring higher-order terms, we have:

$$\delta y = 2x \delta x.$$

More generally, given a measure of uncertainty in x such as its standard error σ_x, the uncertainty in some $y = f(x)$, denoted by σ_y, is given by:

$$\sigma_y = \frac{df}{dx} \sigma_x.$$

The analysis can be readily extended to the multivariate case and the associated partial derivatives.

In most cases, however, the analysis that results in the product y will be much too complex to represent as a single function f, and in cases where a function exists it may be non-differentiable. Simulation provides an alternative that is more general, more straightforward conceptually, and also more suited to non-numeric data. A series of inputs is generated, representing the variation in the data due to uncertainty, error, or poor quality. Each input is then analyzed, to create a series of outputs. Uncertainty in the output can be represented through some measure, such as RMSE, or by visualization.

5.   Final Comments

Quality remains a major issue for users of massive data sets, especially when the data were created by people or processes remote from the user. Humans are often faced with having to take information at face value, and have developed complex arrangements and conventions as the basis for trust. For example, we trust information we read in certain newspapers because we trust the newspaper's staff and news-gathering processes.

Many of these conventions fail in the case of digital data. Electronic networks make it easy for data sets to be copied and transferred without identification of the creator, and make it easy for data from different sources to be merged, creating products with heterogeneous quality. Metadata are expensive to create, and owners of data often lack the motivation to create them in advance of use. Finally, few software products offer the ability to handle information on quality, or to propagate it appropriately to new data or results of analysis. Nevertheless, much progress has been made in recent years, and new products now becoming available are much more likely to provide metadata services, and to support handling, visualizing, and propagating information about quality.

Bibliography

M.K. Beard, B.P. Buttenfield, and S.B. Clapham. NCGIA Research Initiative 7: Visualization of spatial data quality. Technical Report 91-26, National Center for Geographic Information and Analysis, 1991.

P.A. Burrough and A.U. Frank. *Geographic objects with indeterminate boundaries*. Taylor and Francis, 1996.

N.R. Chrisman. *Exploring geographic information systems*. Wiley, 1997.

N.R. Chrisman and B. Yandell. Effects of point error on area calculations. *Surveying and Mapping*, 48:241–246, 1989.

K. Clarke, P.D. Teague, and H.G. Smith. Virtual depth-based representation of cartographic uncertainty. In W. Shi, M.F. Goodchild, and P.F. Fisher, editors, *Proceedings of the International Symposium on Spatial Data Quality '99*, pages 253–259, 1999.

K.C. Clarke and P.D. Teague. Cartographic symbolization of uncertainty. In *Proceedings, ACSM Annual Conference*, 1998. CD-ROM.

T.J. Davis and C.P. Keller. Modelling and visualizing multiple spatial uncertainties. *Computers and Geosciences*, 23:397–408, 1997.

C.R. Ehlschlaeger, A.M. Shortridge, and M.F. Goodchild. Visualizing spatial data uncertainty using animation. *Computers and Geosciences*, 23:387–395, 1997.

P.F. Fisher. Visualizing uncertainty in soil maps by animation. *Cartographica*, 30:20–27, 1993.

A.R. Gillespie. Spectral mixture analysis of multispectral thermal infrared images. *Remote Sensing of Environment*, 42:137–145, 1992.

M.F. Goodchild, A.M. Shortridge, and P. Fohl. Encapsulating simulation models with geospatial data sets. In K. Lowell and A. Jaton, editors, *Spatial accurary assessment: Land information uncertainty in natural resources*, pages 131–138. Ann Arbor Press, 1999.

S.C. Guptill and J.L. Morrison. *Elements of spatial data quality*. Elsevier, 1995.

G.B.M. Heuvelink. *Error propagation in environmental modelling with GIS*. Taylor and Francis, 1998.

G.B.M. Heuvelink, P.A. Burrough, and A. Stein. Propagation of errors in spatial modelling with GIS. *International Journal of Geographical Information Systems*, 3:303–322, 1989.

G.J. Hunter and M.F. Goodchild. Modeling the uncertainty of slope and aspect estimates obtained from spatial databases. *Geographical Analysis*, 29:35–47, 1997.

A.G. Journel. Modelling uncertainty and spatial dependence: Stochastic imaging. *International Journal of Geographical Information Systems*, 10:517–522, 1996.

A.M. MacEachren. Visualizing uncertain information. *Cartographic Perspectives*, 13:10–19, 1992.

M. McGranaghan. A cartographic view of spatial data quality. *Cartographica*, 30:8–19, 1993.

J.R. Taylor. *An introduction to error analysis: The study of uncertainties in physical measurements*. University Science Books, 1982.

C.M. Wittenbrink, A.T. Pang, and S. Lodha. Glyphs for visualizing uncertainty in vector fields. *IEEE Transactions on Visualization and Computer Graphics*, 2:266–279, 1996.

A.X. Zhu, L.E. Band, B. Dutton, and T.J. Nimlos. Automated soil inference under fuzzy logic. *Ecological Modelling*, 90:123–145, 1996.

Chapter 19

DATA WAREHOUSING

Theodore Johnson
AT&T Labs Research, Florham Park, NJ 07932, USA
johnsont@research.att.com

Abstract A *data warehouse* is a repository for information that is collected, cleaned, and made available for analysis. A well run data warehouse makes many analyses easy to run, because many complex details have been taken care of already. In this chapter, we survey issues related to building, maintaining, and querying a data warehouse, and recent research that has been performed to address these issues.

Keywords: Data warehousing, Querying, Online transaction processing, Data quality.

1. Introduction

A *data warehouse* generally refers to any collection of data which is made available for analysis. However, one expects the data warehouse to have certain properties and to provide certain services, for example see (Chaudhuri and Dayal 1997, Hammer et al. 1995). The data warehouse services are geared towards making perhaps widely disparate data readily available for complex analysis. We can categorize data warehousing activities as:

- **Data gathering:** The data in a data warehouse is usually derived from a wide variety of sources. Often, just collecting data from multiple sources into a single location is a valuable service. However, the different sources may have widely varying data schemas, may use inconsistent terminology, may use inconsistent keys, and may store the data in a variety of ways (relational DB/object DB/flat file, ascii/unicode, etc.). A further service is to transform the data to simplify data integration. Finally, data quality problems are common, and incoming data needs to be carefully monitored and cleaned.

- **Data organization:** The attributes of a data table often consist of *dimensional* attributes which classify the tuple, and *measure* attributes which describe the event that the tuple refers to. Normalizing the data table creates a *star schema*, in which *dimension* tables radiate from the central *fact table*. In many cases, the data warehouse administrator knows that certain types of queries are common. To speed up these queries, one can create *materialized views* that fully or partially answer the query. One particular type of view is a *data cube*, in which data is aggregated at all granularities of the dimensions. Special databases (*multidimensional databases*) have been developed that store and query data cubes efficiently.

  The data in the warehouse is frequently temporal, and represents a history of past behavior. Often, queries should run in the past rather than in the present.

- **Data maintenance:** The data warehouse must incorporate new data as it is received. The temporal nature of much of the data in the warehouse makes the update simple. For example, to maintain the most recent 2 years worth of data, one needs only to load the new data and discard (or move to tape) the oldest data. However, complex views might be computed over this 2-year window, and the view must be updated to be consistent with the change in the warehouse.

- **Querying:** The users of a data warehouse often wish to make complex analyses. A large class of these queries ask for simple aggregates (counts, sums, and so on) at different levels of granularity (the data cube queries). Other queries are similar, but require more complex aggregates (top ten, moving average, etc.). More complex queries might require complex grouping, or data transformations.

An alternative to building an explicitly materialized data warehouse is to use *mediators* to provide a (non-materialized) view of one or more data sets (Wiederhold 1992). However, as Hammer et al. (1995) and others argue, a cleaned, organized, and readily available data source is a valuable resource. Another reason for building a data warehouse is to support the unusual and data intensive queries typical of data analysts (French 1995). For example, a data warehouse usually needs only the coarsest level of concurrency control (if any is needed at all), but typically needs extensive indexing for accelerate analysis queries. An On-Line Transaction Processing (OLTP) database needs fine-grained concurrency control and a few carefully chosen indices (because index maintenance substantially increases update costs). Data warehouse queries are often very long running and access a large portion of the database. These long-lived transactions conflict with the many short-lived update transactions of an OLTP system, and therefore must be executed on a separate database.

In the remainder of this chapter, we survey the current technology in data gathering, organization, maintenance, and querying in a data warehouse setting, and present prominent research in detail.

2. Data Gathering

The first step towards building a data warehouse is to gather the data. If all goes well, the data is fetched from fully cooperative data sources that use the latest database technology and transmit their data over very high speed data lines. All of the source databases use consistent schemas, keys, and semantics. When the source databases receive new relevant data, an update event is *triggered* (see Chakravarthy and Lomet (1992), Dayal (1998), and Hanson and Widom (1993)), and the relevant new data is sent to the data warehouse.

However, usually things are not so simple. The data sources will typically use a variety of database systems (relational/object-oriented/IMS) if they use databases at all (e.g., web log file dumps). These systems will present their data in a wide variety of formats. Many of these systems do no support triggers, and those that do might not be willing to use them (and suffer their overhead) to support an external activity. Many of these data sources do not support queries, and many that do will not allow the data warehouse maintainer to submit queries (for performance and security reasons). Usually the data sources use inconsistent schemas and have inconsistent semantics. Often the data is dirty.

To gather data, the data warehouse maintainer must first find a way to extract data from the sources. This might be accomplished by collecting periodic data dumps (transmitted by tape or FTP), or by arranging for the data warehouse to query the source database. After the data is extracted, one usually needs to transform the data so that it becomes consistent with (and thus can be joined with) the other data in the warehouse.

2.1. Mediators

The data sources are usually maintained by more or less autonomous entities. Therefore, one is likely to encounter a wide variety of *heterogeneity* in the data (Wiederhold 1993):

- **Hardware and operating system:** Different computing platforms use different representations of the same atomic objects. One common example is the big-endian versus little-endian representation of integers. String data may be encoded using a variety of character sets (ascii, extended ascii, or unicode).

- **Organizational Models:** The data source may use a relational database, an object-oriented database, a legacy database, or no database at all.

- **Heterogeneity in representation:** The same object may be represented in a variety of ways. Examples include 5-digit versus 9-digit zip codes, month/day/year versus day/month/year date representations, English versus metric measure, and so on.

- **Heterogeneity of meaning:** Often, an attribute refers to a local classification or judgement. For example, the color of a car, the creditworthyness of a customer, and so on. These attributes cannot be cleanly mapped into another representation.

- **Heterogeneity of scope:** Databases are defined over particular collections of objects. A contract employee might not be in a payroll database even though she effectively works full time at a company. A customer who subscribes to a wireless service might not subscribe to a cable television service.

- **Level of abstraction:** Attributes or data may be reported at different levels of aggregation. For example, location may be reported by street address, zip code, or city of residence. Sales may be reported at the detail level, or by a daily aggregate.

- **Temporal validity:** The data that is reported by a source is often temporal, and is a *snapshot* of the history of the database. For example, the payroll database contains the current salaries of the current employees. A data warehouse needs to add temporal information.

The data warehouse maintainer needs to devise a system that will perform the data transformations necessary to remove or ameliorate the heterogeneity of the data. Often this is done by a complex programming effort using procedural programs written in e.g. Perl or C. Fortunately, extensive research has been devoted to methods for automating these translation tasks. A *mediator* is a program that performs data translations using declarative rules to define the transformations. Mediators often translate queries as well as data, giving a user transparent access to disparate data sources. Current mediator research includes TSIM-MIS (Y. Papakonstantinou 1996; 1995), Garlic (M. Carey et al. 1995), Pegasus (R. Ahmed et al. 1991), and Hermes (Subrahmanian 1995), and the works of (Hammer and McLeod 1993, Litwin et al. 1990, W. Kim et al. 1993).

For a specific example of a mediator, lets consider the work of Abiteboul, Cluet, and Milo in developing a mediator for text file resident data (Abiteboul et al. 1993). Many text files contain useful data, but the data does not have a schema and is not organized into a simple record format. However, many text files can be parsed by using lexical analysis tools such as Yacc or Bison. The observation in (Abiteboul et al. 1993) is that by incorporating relation forming rules into the lexer actions, text file resident data can be transformed into a relation.

To illustrate, we'll use the example presented in Abiteboul et al. (1993). A declaration of (Unix shell) aliases has the format of a list of alias declarations separated by semicolons, with the alias separated from its value by a comma. For example, mail, m; ll, ls -l. The Yacc lexical specification is

```
Aliases: Aliases Alias |
         Alias
Alias:   String ',' String ';'
```

The structure returned by the YACC specification is a list of Aliases nodes, each of which points to an Alias node, each of which contains a pair or strings. However, our database needs a relation of the pairs as input. By using the right collection of actions, we can accomplish this:

```
Aliases:   Aliases Alias { $$ = Union($1,Set($2) ) } |
           Alias { $$ =Set($1) }
Alias:     String ',' String ';' {
           $$ = new(Alias, Tuple(Name,$1,Definition,$2) ) }
```

The mediator defines a **view**, i.e., the Aliases relation is defined as a computation over another data set. The Yacc specification defines a way to create a **materialized view**, or a view whose value is computed. However, it is possible to use the Aliases relation without materializing it. Furthermore, it is possible to perform query optimizations on this type of view definition. For example if we ask for the Definitions of all aliases whose Name is "lm", we can rewrite the action of the first Aliases rule to be,

```
Aliases: Aliases Alias {
        $$ = Union($1,Project(Definition,
            ( Select(Name,'lm', Set($2))) ) } |
```

2.2. Data Quality

A difficult problem in data warehouses is the problem of data quality. A very common data glitch is that data is simply entered wrong (e.g., a help desk records an address incorrectly). Similarly, some fields or records may be missing. Partially resolved heterogeneities as discussed in Section 2.1 can cause unpleasant surprises. Other problems include (Dasu and Johnson 1999) unreported changes in the schema, unreported changes in measurement scale or format, temporary reversion to defaults, switched fields, extra or missing fields, missing and default values, gaps in time series, inconsistent recency in subpopulations, or duplicate or missing data on subpopulations. Many of these problems relate to the fact that the data sources provide information that is highly processed, and subtle changes in the processing often are not reported to the data warehouse manager.

Although maintaining data quality is by far the most time consuming activity in most data warehouse installations, surprisingly little research has been done on the topic. In this section, we describe three methods aimed at improving data quality. See also the chapter in this book, *Data Quality* by Michael F. Goodchild and Keith C. Clark.

2.2.1 Merge/Purge. The *merge/purge* problem (Hernandez and Stolfo 1995) is to join two collections of records when the join key is dirty. A typical example is when the key defines a customer, who is represented by a name and address. Frequently names and address are inconsistently reported and incorrectly transcribed. The *merge* aspect of the problem refers to performing a join, for example to collect a history of customer interaction reports. The *purge* aspect is to remove duplicates, for example from a mailing list. Related research includes work on the *semantic integration* problem (Kent 1991), and on the *instance identification* (Wang and Madnick 1989).

Because we cannot hope for an exact match between keys, we can instead try to find the best match (or best k matches) for every record in a relation as determined by the degree of similarity between two keys. There are many choices for a distance function, and the best one to use depends on the application. For example, the string editing distance between two attributes is a natural default choice. However, one might want to allow letter transposition as an editing operation. Similarly, first names and middle names are often abbreviated, so the abbreviation should be a close match. After rating the similarity between all attributes that compose a key, the result is summed to compute the overall similarity. Again, this summation might be complex and application dependent.

The default algorithm for performing a merge/purge join is compare every record from the first table to every record from the second. However, in very large databases this algorithm will be far too slow. Instead, we need a way to reduce the size of the search space.

Hernandez and Stolfo (1995) propose that a *hash key* be computed from the key used to perform the merge/purge join. This key can be a soundex of a subset of the key attributes, the first few letters of one or more attributes, and so on. After this key is computed, there are two complementary approaches for performing the merge/purge join.

- **Sorting:** If similar records are likely to produce hash keys that are close in their sort order, then one approach is to compute the hash keys, sort them, and compare records whose hash keys are within distance w in the sorted list.

- **Hashing:** Records are thrown into buckets based on the value of their hash key, and a default merge/purge join is performed in each bucket.

These two techniques can be combined, e.g. the merge/purge join in each hash bucket can be performed by the sorting method. For improved robustness, multiple passes can be made, each with a different hash key. Hernandez and Stolfo report that using multiple sort keys but with a small window gives the best results.

2.2.2 Consistency constraints.

Most DMBS systems allow the user to specify *consistency constraints* on the relations in the database. These constraints might be simple (e.g., age is an integer, and $0 \leq \text{age} \leq 150$), or more complex (e.g., an employee earns less than his manager). During data ingest, records that fail these constraints are not inserted into the database but rather are written to a log record for closer examination. Many data quality problems can be caught with these constraints. However, many data quality problems are more subtle.

One can frequently characterize the expected value of an attribute. For example, one would be surprised to find a sales clerk earning a six figure salary, or to find the president of the company earning minimum wage. Through a statistical analysis, one might find that an employee's salary fits reasonable well to a model that depends on the other values of the attribute. For example, we might find that

$$\log(\text{Salary}) = \beta_1 \text{Experience} + \beta_2 \text{Rank} + \beta_3 \text{Department}$$

Of course, salaries are not fully determined by experience, rank, and department so we expect some amount of variation around the mean. The typical variation is expressed as a confidence interval. If an employee's salary is outside of its confidence interval, then we might be suspicious of the accuracy of the tuple.

Hou and Zhang (1995) describe an implementation in which they integrate this kind of statistical consistency constraint into an experimental database, *CASE-DB*. They describe a collection of models that incorporate both numerical and categorical data. Because one should expect outliers in any data set, tuples that fail the statistical constraints are not ejected from the database, but are marked as suspicious.

2.2.3 Population constraints. One problem with using statistical constraints is that robust constraints may be very hard to find in actual data sets. They are parametric, and depend on carefully constructed models. The actual data may not fit the models (e.g. not have a Normal error distribution), and the correct models may evolve over time. A non-parametric model would be more robust and perhaps easier to specify.

Dasu and Johnson (1999) apply a data mining technique to identify data that may be corrupted. Their technique uses a method that determines whether two data sets have the same multivariate distribution of data. A scalable space partitioning method is used to partition the data sets. If two data sets have the same distribution, then the counts of the data points from the two data sets in each partition should be similar, and the points in each partition should have similar distributions. The authors provide statistical tests that determine whether the data sets have the same distribution. One aspect of this technique is that the test can be performed on subregions of the entire data set.

As in (Hou and Zhang 1995), the authors assume that data sets change their aggregate statistical behavior slowly. Their data set comparison technique compares a previous collection of data with a new one. The portions of the new data that are significantly different than the old can be isolated for further analysis.

3. Data Organization

A data warehouse can contain a wide variety of data, ranging from the simple and highly structured (e.g., relational tables containing customer interactions) to the complex and/or highly unstructured (images, web pages, ad-hoc data). A large body of work has been developed for data warehouses of relational data and as most data warehouses contain this type of data, we generally assume that the warehouse stores relational data.

3.1. Multidimensional Data

The data in a warehouse is often viewed as *multidimensional* data (Chaudhuri and Dayal 1997). The attributes of a relation are classified as either *dimension* or *measure* attributes. The dimension attributes classify the object that the tuple represents, while the measure attributes represent the unique properties of the object. Often, several attributes represent successively refined classifications of the object. These attributes are considered to be a single *hierarchical dimension*.

A common relational representation of multidimensional data is the *star schema*. Each dimension is represented by a separate *dimension table*, and contains the unique values of the dimension. The warehoused data is represented by a central *fact table*, which contains foreign keys that link the fact table to the dimension tables, and also the measure attributes. Figure 19.1 shows an example of a fact table Orders (derived from Chaudhuri and Dayal (1997)) about product orders, with the dimensions Product, Location, and Date.

The advantages of the star schema include the usual advantages of relation normalization: reduced space use (by removing redundancies), eliminating update anomalies, and so on. However, the star schema is a special type of normalization. Conventional third normal form schemas usually create too many tables and require too many joins to process queries (Colliat 1996). In addition, the user might wish to browse and query the dimension tables themselves.

The dimension tables can also be normalized. For example, the type of product determines the category, which in turn determines the industry. Breaking the product table into three tables (one for the product, category, and industry, respectively) we can save space (especially if there are multiple category related attributes, e.g. a manager associated with each location) and avoid update anomalies. This type of schema is called a *starflake* schema because the dimension tables branch away from the central fact table. In a *fact constellation*, multiple fact tables share the same dimension tables. For example, the Shipments fact table

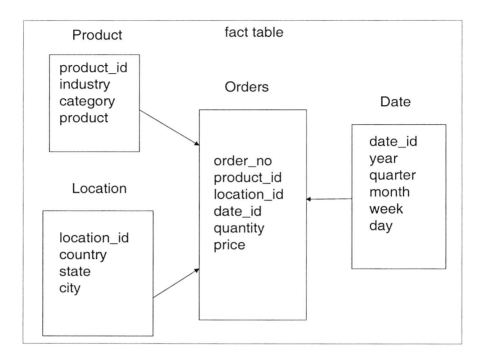

Figure 19.1. A star schema on the orders table.

would share the Product, Location, and Date dimension tables with the Orders fact table.

Although star and snowflake schemas are designed to support dimensional data, dimensions and their hierarchies are not first class objects in SQL databases. As a result, it can be difficult to write queries that range over the hierarchies, and it can be difficult to represent heterogeneous hierarchies.

Jagadish et al. (1999) propose an SQL extension that makes hierarchies a first class entity. A dimension is a collection of tables, organized as a directed acyclic graph. Each of these tables must have as one of its attributes a *hierarchical attribute*. The hierarchical attribute has some special properties (which we describe shortly) which tie the dimension tables together and express the hierarchy.

Two examples of dimensions (derived from Jagadish et al. (1999)) are shown in Figure 19.2. The Location dimension is comprised of four tables representing four classes of regions, where each region has a manager (this type of ancillary information can also be represented using conventional star and snowflake schemas). Notice that although the address determines the city, the loc1 and loc2 tables are not associated by a foreign key, but rather through the hierarchical attribute locID. The Product dimension is more complex, and represents a heterogeneous dimension. The product (e.g. clothing) can be either formal or casual. Formal clothes can be either ties or suits (we do not show the hierarchy under casual clothes). Ties have one type of description, while suits have another. All of these dimension tables are associated using the hierarchical attribute pID. The ability to represent heterogeneous dimensions is very useful because real life objects often refuse to fit into simple taxonomies.

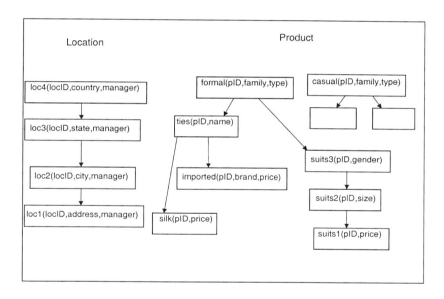

Figure 19.2.   Heterogenous dimension tables.

The set of values of the hierarchical attribute for the Location dimension is the collection of locID values in the tables loc1 through loc4. This collection can be accessed and queried. A hierarchical attribute can be represented by its path through the dimension DAG. For example, a value of locID in loc4 might be country=USA, while a descendent value in loc2 might be country=USA::state=NJ::city=Florham Park. Hierarchical attributes have a set of special predicates defined on them, ($<$, $>$, $<<$, $>>$), which correspond to child-of, parent-of, descendent-of, and ancestor-of. For example, both country=USA $>>$ country=USA::state=NJ::city=Florham Park and country=USA $>$ country=USA::state=NJ::city=Florham Park are true.

Jagadish et al. (1999) propose a small extension to SQL that allows the expression of complex queries over dimensional data. One extension is the predicate on hierarchical attributes. The other extension, DIMENSION, declares that a tuple variable ranges over the values of the hierarchical attribute of a dimension.

One example query is, "find all locations managed by Hamid Ahmadi". This query is expressed by

```
Select      L.locID
Dimensions  Location L
Where       L.manager = "Hamid Ahmadi"
```

Another example query is, "Find all products that grossed over $100,000 in sales". This can be expressed by

```
Select      P.pID
Dimensions  Product P
From        Sales
Where       Sales.pID <<= P.pID
Group By    P.pID
Having      sum(dollarAmt) > 100000
```

3.2. Data Cubes

Star and snowflake schemas are useful for extracting individual tuples from a data warehouse, but often users prefer to browse and analyze summaries of the date. A convenient method for representing these summaries is by a *data cube*, i.e., a collection of aggregates at all levels of granularity (Gray et al. 1996). The name "data cube" is derived from a way to visualize the multidimensional aggregate. An example data cube is shown in Figure 19.3, over the fact table in Figure 19.1. There is one cell for each value of the dimension tables, each containing a summary of the tuples of the fact table (for example, count) with the same values in their dimension attributes. A data cube also contains all lower dimensional summary tables (or *subcubes*), for example on product and date, on product alone, etc. Other common aggregates are sum, average, min, and max.

A data cube on d dimensions is defined by 2^d group-by queries. A special value, ALL, is used to represent the full cube and all subcubes in

DATA WAREHOUSING

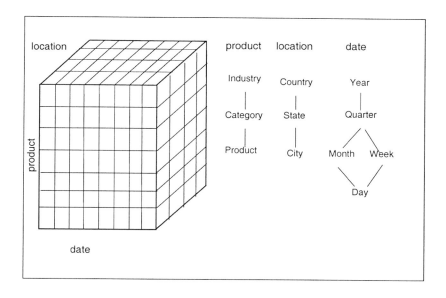

Figure 19.3. A data cube on the Orders fact table.

a single table. An example of a query that would generate a data cube on the Orders fact table is

```
Select    ALL, ALL, ALL, count(*)
From      Orders
              Union
Select    product_id, ALL, ALL, count(*)
From      Orders
Group By  product_id
              Union
...
              Union
Select    ALL, location_id, date_id, count(*)
From      Orders
Group By  location_id, date_id
              Union
Select    product_id, location_id, date_id, count(*)
```

```
From      Orders
Group By  product_id, location_id, date_id
```

The first group-by creates the *grand total* or *ALL* subcube, while the last one creates the *base subcube*. Gray et al. (1996) propose that SQL be extended with the Cube By keyword, which would create a data cube. For example, creating the data cube on the Orders fact table is

```
Select    product_id, location_id, date_id, count(*)
From      Orders
Cube By   product_id, location_id, date_id
```

The dimensions in a data cube often contain a hierarchy. For example, the product dimension contains the hierarchy (industry, category, product). Some dimensions have a *lattice* structure, for example the date dimension as is illustrated in Figure 19.3. Weeks do not fall neatly into months, so weeks and months are disjoint ways of categorizing days (week boundaries do not match quarter boundaries, but in this example we have forced a fit). If the dimensions have a hierarchical or lattice structure, the user usually wants all aggregation at all levels of the granularity of all of the dimensions. If dimension i has l_i levels, then there are $\prod_{i=1}^{d}(l_i + 1)$ different group-bys in the data cube. For example, the Orders data cube has $(3 + 1)(3 + 1)(5 + 1) = 96$ subcubes. An example of a subcube created using hierarchy and lattice dimensions is:

```
Select    industry, category, ALL,
          country, ALL, ALL,
          year, quarter, month, ALL, day, count(*)
From      Orders
Group By  industry, category, country, year, quarter,
          month, day
```

A typical query on a data cube specifies a subcube to query, and a restriction on the range on the dimensional attributes. For example, one might ask for weekly sales in New Jersey of clothing during 1998. The structure of the data cube encourages the user to explore interesting features of the data. For example, the user might *roll up* the query by asking for sales in the United States, or might *drill down* by asking for daily sales.

The types of aggregates stored in a data cube are classified as *distributive*, *algebraic*, or *holistic*. Distributive and algebraic aggregates can be summed from smaller parts (algebraic aggregates depend on other aggregates). The common aggregate functions (count, sum, min, max, average) are all distributive or algebraic. Holistic aggregates in general need to process all fact table tuples within the group. Examples of holistic aggregates are most-frequent, top-ten, and median. Distributive and algebraic aggregates can be efficiently computed and maintained, while the computation and update of holistic aggregates is difficult to do efficiently.

3.2.1 Multidimensional data models.

While multidimensional data-bases are a useful concept, they are not as solidly defined as, say, relational databases. It can be difficult to define complex operations over multidimensional data, and many multidimensional database tools use ad-hoc methods for defining queries.

To address this problem, several researchers have developed formal models of multidimensional data (Pederson and Jensen 1999, Kimball 1996, Agrawal et al. 1997b, Rafanelli and Shoshani 1990, Gyssens and Lakshmannan 1997, Gyssens et al. 1996, Li and Wang 1996, Lehner 1998, Datta and Thomas 1997). These works try to address several issues, including what it means to restructure multidimensional data, and making measure and dimension attributes interchangeable. Restructuring means that we might want to use more or fewer dimensions to classify the data. The issue of whether an attribute is a measure or a dimensional attribute can be illustrated by considering the Orders fact table (see Figure 19.1). It might be the case that the user might want to know the average price grouped by quantity. In this case, quantity changes from being a measure to being a dimension.

Gyssens and Lakshmannan (1997) propose a relatively simple but yet powerful model for multidimensional databases, so we choose it for discussion. Given a fact table with attributes $R = \{A_1, \ldots, A_m\}$, we choose a collections of dimensions $D_1, \ldots, D_k$, where each $D_i \subseteq R$ and $D_i \cap D_k = \emptyset$. That is, the dimensions are non-overlapping subsets of the attributes of the fact table. The remaining attributes are the measure attributes M. This defines a multidimensional schema S.

Next, we distribute the fact table tuples among the cells defined by the dimensions. Given a fact table R and multidimensional schema S, there will be a collection of values R_{D_i} for each dimension D_i. The cells of the multidimensional table is the cross product $R_{D_1} \times \cdots \times R_{d_k}$. We extend R to R' by creating a dummy tuple containing NULL values in the measure attributes whenever there is a cell in the multidimensional table without a corresponding tuple with the same values of the dimensional attributes in R. Finally, we create the multidimensional table $tab_S(R)$ by creating a cell for every unique value of $R_{D_1} \times \cdots \times R_{d_k}$, and put in each cell the set of corresponding tuples from R' projected to the measure attributes.

That is, each cell contains something (perhaps a NULL tuple), and in general a cell contains a set of tuples.

The advantage of this scheme is that many operations on the multidimensional table can be defined largely in terms of operations on the fact tables. For example, let $tab_{S_1}(R_1)$ and $tab_{S_2}(R_2)$ be two multidimensional tables. Then $tab_{S_1}(R_1) \times tab_{S_2}(R_2) = tab_{S_1 \cup S_2}(R_1 \times R_2)$.

Gyssens and Lakshmannan (1997) show how the *fold* and *unfold* operators can be defined. The fold operator removes a dimension from a multidimensional table, while the unfold operator adds one. These can be defined as modifications to the multidimensional schema. The authors also define a mechanism for aggregating the measure tuples in each cell, and for *classification* which allows tuples from R to match with more than one cell in the multidimensional table.

A more complex multidimensional data model is described in Pederson and Jensen (1999). The authors try to address the problem of modeling real-world data with a dimensional table (their motivating example is a medical diagnosis database). They found standard multidimensional databases need more flexibility in their semantics about dimensions. Their extensions include uncertain diagnoses, fact tables whose dimensional attributes might exist at various levels in the dimensional hierarchy (i.e., not just at a leaf), and temporal dimension tables.

The issue of temporal dimension tables deserves some attention. The fact table in a data warehouse often represents historical records. For example, in the Orders fact table, one of the dimensions is the date on which the order was placed. The dimension tables usually change their state over time, so we need to know the date of the fact table tuple to join it with a dimension table tuple.

Conventional relations in a DBMS are *snapshot* relations, because without temporal information, they represent a collection of objects at an instant in time. A *temporal database* (C. S. Jensen et al. 1994) associates a time interval with each tuple. Our concern is with *valid time*, or the lifetime of the object represented by the tuple.

Some dimensions can be treated as snapshot relations. For example, in the orders database products might change over time. However, new products can be given new product IDs. In other cases, the object remains the same but changes its properties over time. One example is a relation that describes properties of customers (or suppliers, etc.). Customers might move, change their telephone numbers, change their relationship with a company, change their demographic information, and so on. However, it is the same customer after the change. These records must be marked with an interval of validity. Joining a fact table with a temporal dimension table requires that the timestamp of the fact table tuple lie within the valid time of the dimension table tuple as well as a matching foreign key.

A data warehouse model that accounts for the temporal nature of the dimensions is the *chronicle* data model (Jagadish et al. 1995). A chronicle is an ordered set of timestamped records. The records in a chronicle are joined with temporal dimension tables and aggregated. The emphasis of Jagadish et al. (1995) is on efficient maintenance of the aggregates.

3.3. Data Cube Storage

Data cubes have become popular in part because they promise fast interactive results. As we have defined the data cube, all results have been computed, so evaluating a query is often just a matter of retrieving the data from the cube. However, the data cube might become very large, increasing storage costs and also data retrieval times. Several alternative data cube storage architectures have been proposed to speed up access times and/or reduce storage costs.

3.3.1 ROLAP. One approach to storing a data cube is to use a conventional relational DBMS. This type of database is termed a *ROLAP* database (Relational On-Line Analytical Processing). The multidimensional view of the data may be on the fact table (which is often stored in order to answer ad-hoc queries), or on the most refined collection of aggregates in the data cube (i.e., grouped by every dimension), the *base subcube*. These tables usually have many indices built on them, e.g., one for each dimension.

One problem with a ROLAP approach is that the base table can be very large, and user queries are often at a relatively high level of aggregation. Colliat (1996) gives an example of a six dimensional database whose base table contains 122 million rows and occupies 17 Gbytes. Many aggregation queries would require accessing the entire table (because the indices are not clustered), and take a very long time to complete.

To speed up query evaluation, we can create tables that represent the entire data cube (i.e., every granularity of aggregation). However, the entire data cube is often very large, especially when the number of dimensions is large. Ross and Srivastava (1997) cite examples where the full data cube is 11 times larger than the fact table.

One option is to materialize a subset of the subcubes of the full data cube. Harinarayan et al. (1996) describe a method for determining an optimal such subset.

The key idea is to generate the object to be optimized. Let C_1 and C_2 be two subcubes, where C_1 is defined by the group-by attributes in G_1, and C_2 is similarly defined by G_2. If $G_1 \subset G_2$, then we can compute C_1 using the data in C_2. Given a collection of group-by attributes G, we can define a lattice $L(G)$ whose nodes are subsets of G. Given nodes G_1 and G_2 in $L(G)$, we draw an edge from G_1 to G_2 if $G_1 \subset G_2$ and $|G_1| + 1 = |G_2|$. An example lattice on the Orders data cube is shown in Figure 19.4. By convention, we put the ALL subcube at the top, and the base subcube at the bottom (in keeping with the terms roll-up and drill-down).

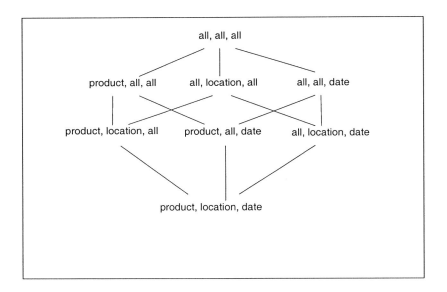

Figure 19.4. Lattice of subcubes of the Orders data cube.

Given a total space budget of S, we would like to compute a set of subcubes to materialize that minimizes the query execution time. Therefore, we need to know three things: the size of each subcube, the likelyhood that each subcube is queried, and the time to answer a query for a subcube from one of the subcube's descendents. The sizes of the subcubes can be estimates using the techniques described in Shukla et al. (1996). Harinarayan et al. (1996) use very simple models of query cost: every subcubes is accessed equally often, and the cost of materializing a subcube is its size divided by the size of its largest materialized descendent in $L(G)$ (more complex models are clearly possible).

The problem of picking an optimal set of views to materialize is NP-complete. However, Harinarayan et al. (1996) found that the following greedy algorithm works well: At each step, choose to materialize the subcube that has the maximum benefit (reduction in materialization cost for other subcubes divided by the subcube's size)

Many extensions to this model are possible. The flat dimensions can be replaced by hierarchical or lattice dimensions, and the problem definition remins the same. In Gupta et al. (1997), the results of Harinarayan et al. (1996) are extended to account for index selection on the materialized subcubes. Baralis et al. (1997) find an optimal set of subcubes for a representative set of queries. The sublattice defined by these queries is much smaller that the full lattice $L(G)$, so the optimization is much faster. In Shukla et al. (1998), a faster heuristic for subcube selection is proposed (the heuristic chooses the smallest subcubes first). Gupta (1997) and Theodoratos and Sellis (1997) give a general framework for this type of view selection algorithm.

3.3.2 Molap.

The highly structured nature of a multidimensional database suggests that a specialized storage structure might have better performance than a ROLAP database. Gray et al. (1996) suggest storing base subcube data in a multidimensional array. If the values of the dimension attributes do not change, they can be translated to a contiguous set of integers. The aggregated values of the measure attributes can then be packed into a multidimensional array. This representation can be space efficient because the dimension attributes are not stored in the data cube (but a mapping table needs to be stored for each dimension). In addition, data cube entry lookup is very fast as the address of the entry can be computed by the integer representation of the dimension values. Additional subcubes can also be materialized.

However, the multidimensional array representation of a data cube has two weaknesses. First, good performance generally depends on the array being main-memory resident. Sarawagi and Stonebraker (1994) evaluate a collection of techniques for efficient secondary storage of large multidimensional arrays. The basic problem is to chop the array into *chunks*, which are small subarrays that can fit into a block of memory. For an example, let us assume that the array has three dimensions, of sizes 100, 2000, and 8000 respectively. Suppose that a block can hold 8000 array entries. One possible chunk size is $20 \times 20 \times 20$. If the user access a rectangle of size $100 \times 100 \times 100$, then the number of chunks to be fetched from disk is minimal (probably $(5+1)^3 = 196$ blocks which is close the the $5^3 = 125$ minimum). However, if the user requests a $5 \times 5 \times 4000$ rectangle, then $(1+1)*(1+1)*(200+1) = 804$ blocks will be fetched, which is much larger than the minimum 13. Clearly, the optimal chunking depends on the access pattern. Sarawagi and Stonebraker (1994) present heuristics for finding a good chunking given a list of likely rectangle shapes that the users will request. The authors also examine issues in chunk layout on disk to minimize seek time, and in storing multiple copies fo the array. However, multidimensional array storage is a difficult problem to solve satisfactorily.

Another weakness of the multidimensional array representation of a data cube is that the base subcube is likely to be *sparse*. Colliat (1996) reports that a 20% storage utilization in the base subcube is common. Other researchers (Shukla et al. 1996, Ross and Srivastava 1997, Beyer and Ramakrishnan 1999) report similar or more extreme sparsity (e.g., 10^{-7}% sparsity in one example based on actual data). Because the multidimensional array represents all empty as well as filled cells, it can be enormously space inefficient.

One approach is to create a hybrid storage structure that uses multidimensional arrays wherever the data is dense. Colliat (1996) described the storage structure of the Essbase multidimensional database. Essbase engineers observed that in many data sets, some collections of dimensions are dense (every cell is filled) while other collections of dimensions are sparse (only a few combinations of the sparse dimensions exist in the fact table). For example, a marketing database might have dimensions Channels, Products, Markets, Date, and Scenario. The collection of dimensions Channels, Products, and Markets are sparse (i.e., a product is usually distributed through only a few channels into only a few markets), while Date and Scenario are dense (sales and sale estimates are recorded for every day). Essbase builds an index on dimensions Channels, Products, and Markets. Each entry in the index points to a multidimensional array with dimensions Date and Scenario. The index on the sparse dimensions is usually small enough to be memory resident, and the multidimensional arrays are usually compact enough to be fetched with one disk read. Colliat (1996) reports storage space reduction as well as greatly decreased query times as compared to a ROLAP database.

Another hybrid array based storage structure divides the multidimensional array into chunks, but stores the chunks differently depending on whether they are dense (e.g., 40% or more of the cells are filled) or sparse (Goil and Choudhary 1999, Zhao et al. 1997). The chunking of the multidimensional array can be thought of as an index. Each entry

DATA WAREHOUSING

in the top-level points to the storage location of a chunk, and indicates the storage type (sparse or dense). Dense chunks are stored is the usual way, but sparse chunks are stored as a list of filled cells. Because of the chunking, each of the dimensions has a small range. Therefore the dimension attributes of the data cube cells can be represented compactly, e.g. through bit packing.

Aggregation is performed by marching along the rolled-up dimensions and summing chunks. For example, suppose that the Orders data cube is stored as a chunked multidimensional array, and we want to compute the subcube grouped by Product and Date. For each set of chunks with the same attribute range on the Product and Date dimensions, we sum these chunks to create a chunk of the Product, Date subcube. In (Zhao et al. 1997), this technique is developed into a full cube computation algorithm for MOLAP databases.

3.3.3 Hierarchical storage.

In addition to ROLAP and MOLAP multidimensional data base storage, several alternative structures have been proposed, which we summarize here. Each of these techniques makes use of hierarchical decompositions of the data in some way.

Ho et al. (1997) propose structures for computing range aggregates over multidimensional data sets. They assume that the dimensional attributes are (or can be mapped to) contiguous integers, and that the base subcube is dense. At every cell in the multidimensional array, they compute the prefix sum over the base subcube. For example, suppose that we have a three dimensional array whose values in the base subcube are $s(x, y, z)$. Then the prefix sum at (x_0, y_0, z_0) is

$$p(x_0, y_0, z_0) = \sum_{x=0}^{x_0} \sum_{y=0}^{y_0} \sum_{z=0}^{z_0} s(x, y, z)$$

Once the prefix sum array has been computed, range sums can be computed quickly by using the principle of inclusion-exclusion. Figure 19.5 illustrates the process. To compute the sum over the shaded region on the left, we take the prefix sum of the lower right hand side region and subtract the prefix sums of the upper region and of the left hand side region. The prefix sum of the region defined by the upper right hand point of the desired range sum has been subtracted twice, so we add it back in. In d dimensions, 2^d prefix sums must be accessed. Ho et al. (1997) also give an efficient algorithm for computing the prefix sum array.

This multidimensional prefix sum array can be chunked (for more efficient updates, or for secondary storage). Each chunk contains a summary prefix sum, and the computation of a range sum uses sums from

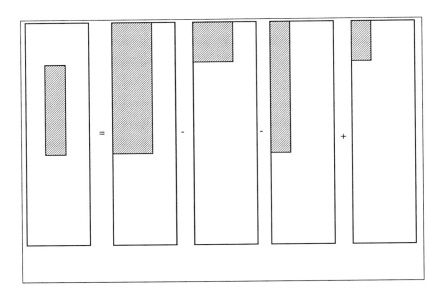

Figure 19.5. Computing range sums from prefix sums using inclusion-exclusion.

chunk summaries and from prefix-summed chunks. Ho et al. (1997) also propose a structure for computing a max or min aggregate. The idea is to build a multidimensional index over the dimension space. The index must partition the data space (so one cannot use R-trees). Each interior node stores the max (or min) over the region it covers. A range query executes by taking the min (or max) over all maximal interior nodes that cover the query region.

Johnson and Shasha (1996; 1997) describe an index structure that stores the full data cube directly in a search structure. A *cube forest*, illustrated in Figure 19.6, is a template for building indices. The tree rooted at Z represents an index whose first attribute is Z. For every unique value of Z, there are subindices on A, B, and C respectively. In each of the subindices, we have an index to the aggregates of the subcubes on two attributes, ZA, ZB, and ZC. In general, each node represents a new subindex, and the set of attribute names along a path from a root represents the indexed subcube. A cube forest is generated recursively. For example, the cube forest on attributes A, B, C are the

DATA WAREHOUSING

three rightmost trees. The cube forest on Z, A, B, C is generated by creating node Z and making it a parent of the trees in the cube forest on A, B, C, and then also adding the trees in the cube forest on A, B, C. Every subcube is represented by a node in the cube forest.

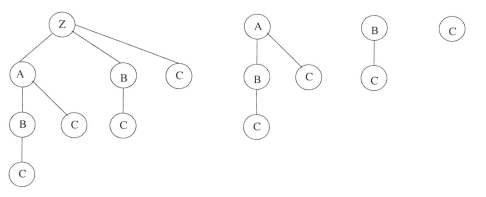

Figure 19.6. A cube forest over dimensions Z,A,B,C.

Each tree in a cube forest represents an index structure. The index stores aggregates as well as pointers to children. An example is shown in Figure 19.7. This forest has two index templates (not generated by the cube forest algorithm), which generate two indices when instantiated with fact table data. The quantity inside the node is the aggregated value, while the quantity outside the node is the value used for indexing. The index on the left creates two subindices for each unique value of A, as is indicated by the tree template.

Johnson and Shasha (1996) describe several extensions to the cube forest, including the handling of hierarchical dimensions, tree pruning (which is similar to materialized view selection (Harinarayan et al. 1996), see Section 3.3.1), a B-tree based index structure, and query and update algorithms.

Roussoupolos et al. propose an R-tree based method for data cube storage, which they term the *cubetree* (Roussopoulos et al. 1997, Kotidis and Roussopoulos 1998). They extend the attribute range of each dimension with an ALL attribute, and store the aggregates in an R-tree. The method of packing an R-tree can have a significant effect on performance. As is the case with chunked multidimensional arrays, it is difficult to optimize for all access patterns simultaneously. For example, suppose that the data cube is over dimensions A, B, C, and the sort order is ABC. Then, the entries for the A, B, and AB subcubes tend

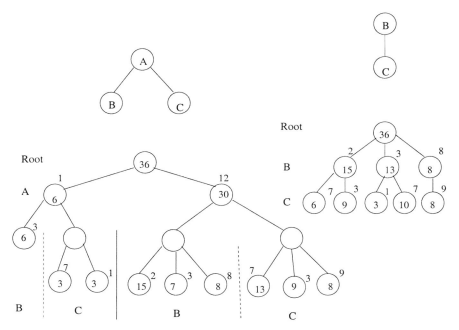

Figure 19.7. Indices instantiated from cube forest templates and fact table data.

to be spread throughout the R-tree. The authors propose a variety of techniques for reducing this problem. For example, one can change the sort order to be A, AB, ABC. That is, an entry (a, ALL, ALL) is ordered before any tuple (a', b', ALL), and so on.

3.4. Computing Data Cubes

Materializing a full data cube can be a computationally intensive task, as the preceding discussion should indicate. As the number of dimensions increases, the number of different subcubes increases exponentially. In general, not all of these subcubes can fit into memory simultaneously, and would benefit from different sort orders. In this section, we survey the work done to compute ROLAP data cubes. The researchers who have proposed MOLAP and other data cube representations have proposed specialized algorithms for their structures (see Zhao et al. (1997), Ho et al. (1997), Johnson and Shasha (1996), and Roussopoulos et al. (1997)).

The data cube computation algorithms take advantage of the cube lattice structure (see Figure 19.4), and of four techniques: making use of compatible sort orders, using hash tables, partitioning the input, and computing a subcube from a descendent subcube.

The *pipesort* algorithm (S. Agrawal et al. 1996) attempts to reuse sort orders. For example, suppose that we are building a data cube on dimensions A,B,C,D. If we sort the fact table on A,B,C,D, then we can compute the subcube on A,B,C,D by summing the measure aggregates while the dimension attributes have the same value, and outputting the subcube tuple when the dimension attributes change. Next, we can observe that we can compute the subcubes on A,B,C, A,B, and A at the same time and in the same way. For the next run, we can observe that we do not need to sort the fact table, we can work from any of the subcubes that we have created (as long as the aggregates are commutative or algebraic). Given these tools, we need to develop a plan to materialize the data cube.

S. Agrawal et al. (1996) give the following algorithm for computing a pipesort plan. Each edge in the data cube lattice (e.g, see Figure 19.4) is labeled with two costs: the cost of computing the parent from the child if the parent can use the sort order used to compute the child, and the cost of computing the parent from the child if the sort order cannot be reused (although the child might be partially sorted for the parent, allowing the use of a fast sorting algorithm). The problem reduces to covering each node with an in-edge in a way that minimizes the total cost (with obvious restrictions on the applicable sort orders). The authors give algorithms for solving this problem.

An extension to the pipesort algorithm is the *overlap* algorithm (S. Agrawal et al. 1996). The authors make more explicit use of the observation that a sort order for a child might be partially sorted for the parent. Suppose that the data is sorted for the child in the order ABCD. If the parent is a subcube over ABD, then the parent cannot be computed directly from the child's sort order. However, the child's sort order is roughly sorted for the parent, i.e. on AB. The overlap technique buffers the tuples from the child that have the same AB values, sorts them on D, and then uses the pipesort technique to compute the parent. Again, a computation plan must be generated. For each edge, we need an estimate of the amount of data to be buffered to compute the parent from the child if the sort orders are not compatible (this can be estimated using the techniques described in Shukla et al. (1996)). The optimization now is to compute as many subcubes as possible subject to memory constraints.

The *pipehash* algorithm (S. Agrawal et al. 1996) does not sort the data, and instead relies on hash tables to perform aggregation. As in the overlap algorithm, the authors try to compute as many subcubes from the same data stream as possible, by applying a tuple from the data stream to hash tables for several subcubes. The authors make a further optimization, that after a child cube is computed, it is in memory and can be used to compute parent subcubes. The authors give an algorithm that minimizes the amount of data to be scanned, subject to memory constraints.

Ross and Srivastava (1997) observe that the pipesort, pipehash, and overlap algorithms might not work well for sparse data cube. One problem is that a parent subcube might be nearly the same size as the child. In this case, buffering strategies do not work well. Another problem is that as the number of dimensions increases, the number of sorts required can become very large (exponential in the number of dimensions). These authors propose the *partitioned-cube* algorithm, which uses a partitioning strategy to reduce the number of I/Os.

The partitioned-cube algorithm takes as input a collection of tuples to be cubed, R, and a set of dimensions to compute subcubes over, $D_1, \ldots, D_k$ (there might be restrictions on the subcubes to compute). If R can fit into memory, then the data cube is computed in memory. Otherwise, partitioned-cube takes the following steps:

1. Partition R into n sets, using dimension D_1 as the partitioning attribute.

2. For each $i = 1, \ldots, n$,

    (a) Compute the subcubes that include D_1 using partitioned-cube with partition R_i and dimensions $(D_1, \ldots, D_k)$.

    (b) Append the output of the subcubes to their respective relations.

3. Use partitioned-cube to compute the subcubes on dimensions $(D_2, \ldots, D_k)$ using the subcube on $(D_1, \ldots, D_k)$ as input.

The dimension for partitioning was chosen to be D_1 for clarity, in general it can be any of the dimensions. After a few partitioning steps, R will fit into memory and the results can be processed in-memory.

Beyer and Ramakrishnan (1999) observe that in a very sparse data cube, most of the groups in the base subcube have only one fact table tuple in them. Furthermore, many rollups are required before reaching a parent group that has more than one fact table tuple in the group. They propose a top-down[1] evaluation strategy for computing the data cube. The first step is to compute the grand total (all aggregate). For each dimension, the fact table is partitioned on each value of the attribute,

DATA WAREHOUSING

and the top-down cube algorithm is recursively applied (avoiding subcubes that have already been computed). Whenever the top-down cube algorithm reaches a group that contains a single fact table, it outputs all child groups and returns.

4. Data Maintenance

The data warehouse periodically (or continually) receives new data, which must be added to the warehouse. Often, adding data to the data warehouse is not a complex problem. The data in a warehouse is often temporal, with the newest data arriving in each batch. The data warehouse administrator can *horizontally partition* the fact table by the time dimension. The new data is loaded into the partition for the temporally newest data (there might be some tuples to load into older partitions). Relation partitions often have their own indices, so this partitioning strategy minimizes the amount of index updating that is required A data warehouse often maintains a moving window on the most recent data, for example a two year window on customer transactions. The oldest partition can be dropped from the relation in order to save space.

A data warehouse often contains many *materialized views* on the fact tables. These views often have a temporal component, and can be partitioned in the same way that the fact table is. For example, the Orders fact table has Date as one dimension. Any subcube of the Orders data cube can be temporally partitioned, greatly simplifying maintenance.

However, some views are more complex and do not lend themselves to such simple maintenance strategies. For example, one view might be a count of the number of each product sold during the last 10 days. In addition, updates to the fact table often contain stragglers, which must be inserted into an older partition. Instead of recomputing the view from the partition, it is desirable to have an *incremental view maintenance* algorithm which will make small modifications to the materialized view.

Blakeley et al. (1986) propose a method for maintaining views defined by select-project-join queries. Suppose that view V is defined over relations S and R by the formula

$$V = \pi_Y(\sigma_X(S \times R)) \qquad (19.1)$$

The σ_X predicate contains the selections that define the join. Suppose that the tuples in A_R are added to S, and A_R are added to R. Then we can use algebraic transformations to deduce that the value of the updated view, V' is

$$V' = \pi_Y(\sigma_X((S \cup A_S) \times (R \cup A_R)))$$
$$= V \cup \pi_Y(\sigma_X(A_S \times R)) \cup \pi_Y(\sigma_X(S \times A_R)) \cup$$
$$\pi_Y(\sigma_X(A_S \times A_R)) \quad (19.2)$$

If A_R and A_S are small, it is much more efficient to update V by using Formula 19.2 than by recomputing the view from its definition. This formula can be extended to handle deleted tuples also. Let D_S and D_R be the tuples deleted from S and R, respectively. Let $S^- = S - D_S$ and $R^- = R - D_R$. Then we can again use algebraic simplification to deduce that (Hanson 1986):

$$V' = \pi_Y(\sigma_X((S - D_S \cup A_S) \times (R - D_R \cup A_R)))$$
$$= V - \pi_Y(\sigma_X(S^- \times D_R)) - \pi_Y(\sigma_X(D_S \times R^-))$$
$$-\pi_Y(\sigma_X(D_S \times D_R)) \cup \pi_Y(\sigma_X(S^- \times A_R))$$
$$\cup \pi_Y(\sigma_X(A_S \times R^-)) \cup \pi_Y(\sigma_X(A_S \times A_R)) \quad (19.3)$$

The update to V is phrased in terms of S^- and R^-, which are computed first. The update expressions are computed and applies to V. There is one detail that requires further attention when tuples are deleted from R or S. A tuple might occur in V because of multiple derivations from R and S. Suppose that there are two tuples r_1 and r_2 in R that join with tuple s in S to create tuple v in V. If r_1 is deleted from R, then Formula 19.3 will delete v from V even though v can still be derived from r_2 and s. The problem is that relational algebra operates over sets instead of multisets (i.e., no duplicates allowed). The solution is to use a counting algorithm: Each tuple v in V has a count associated with it, of the number of ways that v can be derived from R and S. Each of the update terms compute the number of ways each tuple in the term can be derived. These counts are added to or subtracted from the counts in V. If the count of a tuple in V becomes zero, the tuple is deleted.

There are many variations of the view maintenance problem (Gupta and Mumick 1995). Views can be updated by changes to the fact tables, or by changes to the view definition. The view might be expressed in a variety of languages. Special view maintenance algorithms might make use of constraints or keys. Some views can be *self-maintainable* (Gupta et al. 1996), meaning that they can be updated using only the view and the updates, not the fact table.

A collection of research has studied view maintenance specialized for data warehouse maintenance. One problem with the view maintenance is that updates may be occurring to the fact tables as the view updates are being computed. In (Agrawal et al. 1997a, Zhuge et al. 1995), algorithms for handling concurrent source updates are presented. Quass and Widom (1997) present algorithms to permit concurrent update and querying of a data warehouse.

When a fact table is updated, its indices need to be updated also. However, updating the indices can cause a very large amount of random I/O. Jagadish et al. (1997) present a technique that batches the index updates, reducing the I/O cost. Mumick et al. (1997) present algorithms for incremental update of ROLAP data cubes and subcubes (most MOLAP storage structures have specialized update algorithms, see Section 3.3.2). Garcia-Molina et al. (1998) examine the problem of expiring data in a warehouse. The issue is that expired data might be needed to maintain views, so all dependent data must be expired before a tuple can be deleted. Pfoser and Jensen (1999) adapt view maintenance algorithms to maintaining temporal data that is partitioned into current and historical tuples.

5. Queries

Users of data warehouses often want to ask complex analysis queries on the data. We have already discussed one type of complex query, the data cube queries. In this section, we examine techniques developed to support complex ad-hoc queries, which are not well supported by data cubes.

5.1. Indices and Joins

As with any other type of database, data warehouses can benefit from the use of indices, e.g. B-trees, R-trees, etc. OLAP queries tend to have special requirements, which has led to some types of indices specialized for data warehouses.

Sarawagi (1997) discusses the relative merits of indexing techniques for OLAP data. Some desirable properties are

- The ability to select arbitrary subsets from several of many dimensions. For example, a selection might be for sales of shoes on Tuesdays from the Orders fact table. Several, but not all, of the dimensions are restricted, and the Date dimension is restricted in a complex way (one that does not follow the hierarchy).

- The ability to index the dimension hierarchy. For example, we might want to change the selection to sales of clothing during any August.

- The ability to handle batch updates.

- The ability to supply a good traversal order.

- No dependence on data sparsity or density.

Conventional B-tree indices can index multiple dimensions, but only by a lexicographic sort order over the multiple dimensions. The B-trees are therefore not symmetric. Performing selections on multiple dimensions can be expensive because the list of records returned by the multiple indices must be merged, but the lists do not have a common sort order.

The special MOLAP storage structures described in Section 3.3.2 implicitly make use of indices. The Essbase storage engine (Colliat 1996) builds a B-tree type index to point to dense arrays. This indexing scheme is rigid, and can result in poor performance if some assumptions are not met. First, one must guess which are the sparse and which are the dense dimensions, and hope that the assumed sparsity and density actually occurs in the data set. Second, if a query does not restrict one of the sparse dimensions, every dense array must be read into main memory. Third, if the index on the sparse dimensions is large, it will not fit into memory causing many I/Os during a search.

The chunked multidimensional array structures (Goil and Choudhary 1999, Zhao et al. 1997) use a grid file to index the stored chunks. While this structure is symmetric, it depends on a minimal level of density (at least some chunks should be dense, or ROLAP storage is better). In addition, it suffers from the *curse of dimensionality* as the number of partitions increases exponentially with the number of dimensions. Given a fixed budget of chunks to store, the number of partitions per dimension decreases exponentially with the number of dimensions.

The multidimensional nature of OLAP data suggests using a multidimensional search structure, such as an R-tree (Guting 1994), as the index. In Section 3.3.3, we discuss the work of Roussopoulos et al. (1997) and Kotidis and Roussopoulos (1998) to index a data cube. Sarawagi (1997) also discusses issues in using an R-tree to index OLAP data. However R-trees are also known to suffer the curse of dimensionality, and have poor performance for more than about four dimensions. While considerable work has been done to index high dimensional data (see for example Berchtold et al. (1998)), these methods usually perform poorly if only a few of many dimensions are restricted.

O'Neil and Quass (1997) present and analyze a collection of indices that are specially designed for data warehouses, which they call *variant indices*. The common feature of variant indices is that they are inherently sequential – each entry in the variant index has a positional mapping to a tuple in the fact table. While this property might seem to inhibit random access search, it allows the simple combination of multiple index results (because they share a common sort order), and it provides an plan to access the fact table tuples once all the indexing predicates have been evaluated. The positional reference to a record in a table is usually called a *Record ID (RID)*. Variant indices provide sorted lists of RIDS that satisfy the indexing predicate.

A *bitmap* is a string of bits (zero or one). When used for indexing, there is a one-to-one mapping between the bit position in the bitmap and the record position in a relation. For example, the 33rd bit in the bitmap represents the 33rd record in the relation (whose RID is 33). Each bitmap has a predicate associated with it. A bit in the bitmap is set to one if the corresponding record satisfies the predicate, else the bit is reset (to zero). A *bitmap index* is a collection of bitmaps that index a relation. Typically a bitmap index will index the value of an attribute, with one bitmap per attribute value.

The main advantage of bitmap indices is the fast evaluation of Boolean expressions over a collection of predicates. To compute the RIDs that satisfy P or Q, we fetch the bitmap for P, the bitmap for Q, and perform word-wise Boolean or operations. The result is the desired RID list (the *foundset*). Arbitrarily complex Boolean expressions involving predicates on many different attributes can be used. Another advantage of bitmap indices is that count aggregates can be evaluated quickly – we just need to count the number of set bits in the foundset. We can do the count in a byte-wise manner with an array that contains the number of set bits in the array index.

Bitmap indices have the disadvantage that they do not scale well when the indexed attribute has many unique values. This disadvantage manifests itself by large storage costs and high costs to evaluate large range predicates. However, these costs can be reduced by the use of bitmap compression.

Other variant indices can be useful if the cardinality of the attribute is large. A *projection* index contains a list of values of the indexed attribute, with a positional correspondence between the position in the projection index list and the position of the record in the relation. A projection index is similar to a vertical partition of the relation, but the projection index is used for indexing; the attribute still exists in the relation. A projection index can support large range queries efficiently, producing a bitmap of the selected records. Projection indices can also be used for sum aggregates.

If the indexed attribute is an integer (or can be mapped to one), then we can define a *bitslice* index on the attribute. A bitslice index is a collection of bitmaps in which each bitmap represents a bit in the binary representation of the attribute. For example, suppose that the indexed attribute is an integer in the range 0 through 7. Then the bitslice index has three bitmaps, B_0, B_1, and B_2. If a record has the value 5 for the indexed attribute, then its entry in B_0 and B_2 are set to 1, while its entry in B_1 is reset to 0.

Bitslice indices can support efficient range queries, i.e. one can write a small predicate that expresses the range (O'Neil and Quass 1997, Chan and Ioannidis 1998). Bitslice queries can also support efficient sum aggregates. The foundset is and-ed with each bitmap in the bitslice index. Then, we count the set bits in each bitmap and take a weighted sum.

Projection indices can be used as *join indices*, which index a table R based on its join with a table S. When a fact table is stored using a star schema, the variant indices can be used as join indices in a natural way: dimension table entries have variant indices which indicate the matching tuples in the fact table.

Although bitslice indices can support range queries, computing a range query predicate can require a complex Boolean expression that accesses the bitslice index bitmaps multiple times. Several researchers have investigated ways to obtain faster range predicates from bitslice indices.

Wu and Buchmann (1998) observe that the range query predicate can often be simplified by Boolean expression simplification. For example, the range predicate "$X \geq 128$" reduces to testing if B_7 is set. The authors next observe that bitslice indices can represent hierarchical dimensional attributes that are not integers, by mapping the attribute values to integers. They give an algorithm for performing this mapping that minimizes the Boolean expressions for common hierarchical ranges.

Chan and Ioannidis (1998) propose a generalization of the bitslice index. First, they propose *attribute decomposed* bitmaps. The idea is to represent an integer in a variable-radix numbering system, and to use a bitmap index for each digit of the representation. For example, the range on numbers 0 ... 59 can be represented in a $(3, 4, 5)$ number system. For example, The number 48 would be represented by $(2, 1, 3)$ because $2 * (4 * 5) + 1 * (5) + 3 = 48$. For each digit, they propose a *range* representation. That is, the predicate that defines the bitmap is $x \geq n$ instead of $x = n$. By using these representations, they develop an algorithm that creates an efficient predicate for evaluating range queries.

Another method for the efficient storage of bitmap indices over an attribute with a high cardinality is to use bitmap compression. Johnson (1999) makes a performance study of bitmap compression algorithms. The author finds that compressed bitmaps can be as space efficient as bitslice or projection indices. This study also examines the performance of algorithms for performing Boolean operations using compressed bitmaps, and finds that performing or operations with sparse bitmaps can be very fast.

Some data warehouses are very large, suggesting that *summary* indices can be effective. Johnson (1998) proposes an index on relation partitions, rather than tuples. A two phase search is used: first the partitions that contain relevant data are identified, then the partitions are opened and searched. The author uses a bitmap index for the summary, and provides algorithms for efficient search and update. Moerkotte (1998) summarizes table partitions with aggregate values. For example, the summary can be the minimum and maximum timestamp of the tuples in the partition. These summaries can determine which partitions contain tuples whose timestamp lies within a time range. If the partition lies completely within the range, other summaries, such as count, can be used directly without accessing the partition. The author extends this idea by attaching predicates to the summaries, and discusses their use in query optimization.

5.2. Query Processing for OLAP Data Warehouses

An extensive literature has been developed on techniques for optimizing OLAP queries. Many ad-hoc data analysis queries have a complex nested structure, make frequent use of aggregation with a group-by, and access materialized views. In this section we briefly review the techniques proposed in the literature.

5.2.1 Queries with group-by.
A common feature of OLAP queries is that a quantity is aggregated, grouped by a key value. While the grouping implies that the relation needs to be sorted (or otherwise grouped), the output of a group-by is usually much smaller than the input relation. This suggests that group-by queries might be able to take advantage of some special query optimization and evaluation strategies.

Suppose that we are given the following query:

```
Select    SG₁, ..., SGₙ, SA₁, ..., SAₘ
From      R₁, ..., Rₖ
Where     C₀ AND ··· AND Cₗ
Group By  G₁, ... Gₙ
```

The SG_i are attributes selected from group-by attributes G_i, SA_i are aggregated values, R_i are the source relations, and C_i are the selection conditions. A standard way to evaluate this type of query (Graefe 1993) is to perform the join, project to the relevant attributes, sort the relation on the group-by attributes, and compute the aggregates values (in a manner similar to the pipesort algorithm, see section 3.4). Alternatively, we can build a hash table on the group-by attributes and collect the aggregate values on the fly. A range of hybrid strategies are possible.

Shatdal and Naughton (1995) evaluate algorithms for evaluating aggregation with group-by queries on a parallel processor. The authors assume that the relation to be aggregated is available and distributed among a number of processors. There are two basic strategies: 1) redistribute the tuples of the relation among the processors so that each has a partition, then aggregate locally or 2) aggregate locally, redistribute the result, and aggregate again. The authors find that the first technique works well if the number of groups is large, while the second technique works well if the number of groups is small. They also propose some hybrid and adaptive strategies.

Yan and Larson (1994) show the conditions in which a group-by may be pushed past a join (the idea being that the group-by result is smaller than the relation). That is, they replace some of the relations R_i in the query by the result of another group-by query. Generally, the later joins must be on foreign keys. Chaudhuri and Shim (1994) show how to make a cost-based optimization of this type of query transformation. For example, pushing group-by past a join can affect the optimal join order. Yan and Larson (1995) observe that if a join is highly selective, it might be better to pull the group-by above the join (e.g., the group-by might be below the join because the query is a nested query where one of the relations to join is the result of group-by with aggregation). They discuss query transformations which push group-by in both directions. Gupta et al. (1995) generalize group-by to the generalized projection (GP) operator, and provide rules moving the GP operators in the query tree.

5.2.2 Nested queries. A *nested* query is one that contains a subquery. An example is Muralikrishna (1992):

```
Select     Dept.name
From       Dept
Where      Dept.workstations <
               (Select count(EMP.*) from EMP
                Where Dept.name = Emp.Deptname)
```

As written, this type of query requires individualized processing (the nested query) for every tuple in the Dept relation. The query processing is likely to be slow because relational databases are optimized for set-oriented processing. The query is rewritten by unrolling it into a sequence of single-block queries (Muralikrishna 1992, Dayal 1987) (we note that the join must be an outer join):

```
TEMP(dname,dcount) =
Select     Deptname, count(EMP.*)
From       EMP
Group By   Deptname

Select     Dept.name
From       Dept, TEMP
Where      Dept.name = TEMP.dname (*) and
           Dept.workstations < TEMP.dcount
```

5.2.3 Using materialized views. A data warehouse often contains several materialized views. Many user queries will directly reference the materialized views. However, a natural question is whether materialized views not directly referenced in the query can help to answer the query.

Many OLAP queries request data from a data cube, e.g. aggregations grouped by a subset of the dimensions. As is discussed in Section 3.3.1, many of these subcubes might not be materialized. Instead, the subcube is materialized using data from one of its child subcubes. The data cube computation algorithms in Section 3.4 include view choice heuristics. Generally, the smallest child is chosen.

Chaudhuri et al. (1995) examine the problem of using materialized views to optimize queries. They point out that even if a query references a materialized view, it might be better to use the table that the view is derived from (for example, the base table might have a useful index that the materialized view does not). They treat materialized views as predicates that the query might make use of, and develop a cost-based optimization. Gupta et al. (1995) use a query transformation algorithm to determine if a query involving aggregation can be answered using a view (i.e. it can if the query can be transformed to incorporate the view definition). Srivastava et al. (1996) use a semantic approach to determine if a query can be answered using a view. Their conditions reduce to ensuring that the view has not lost any essential information: no necessary columns have been projected out, the selection conditions did not throw out any needed tuples, and the granularity of aggregation is at least as fine as the query result.

5.3. User-defined Functions and Aggregates

Many of the aggregates that users need are *holistic* aggregates, such as top-ten or median. Alternatively, the aggregate might be a sum, but each summand involves a complex logic to compute. In these cases, the user might wish to perform the aggregation using a *user defined function* (UDF) or a *user defined aggregate* (UDAF). Database extensions for temporal financial data are described in Chandra and Segev (1993).

A user-defined function is a routine written in a procedural language, for example C++ or Java. It is often the case that a function that is easy to specify in a procedural language is extremely difficult to express in a declarative one, especially a special purpose language such as SQL. The function might be part of an existing code base, so that reusing the function is safer than writing a new version in an internal database language.

Unfortunately, the function might perform arbitrary actions. It might issue system calls, issue subqueries, perform lengthy computations, or allocate large blocks of memory (and all of these are reasonable actions depending on the context). In addition, the UDF might need to use *scratchpad* memory that persists between calls to the UDF (for example to cache expensive computations). Writing a UDF is a complex process, and is usually done by specialty software vendors, or by an experienced Data Base Administrator (DBA).

A UDF must be *registered* with the DBMS before it can be used. The registration includes linkage and parameter passing information, as well as information about the UDF's behavior: does it make SQL queries, does it open files, does it need scratchpad memory, and so on. See (Jaedicke and Mitschang 1998, Chamberlin 1996) for examples of defining UDFs. In (Godfrey et al. 1998) different architectures for adding UDFs and UDAFs to an extensible database are evaluated.

A user defined aggregate function is registered with the DBMS in a manner similar to that of a UDF. However, a UDAF has three procedures associated with it: An initialization procedure, a procedure for adding another value to the aggregate, and a function that extracts a final value from the aggregate.

A significant problem with using UDFs and UDAFs is that it can be difficult to optimize queries involving them. Query optimizers need an estimate of the execution time and space usage of the UDFs and UDAFs. These estimates can be included in the registration information. If the UDF is a predicate, then an estimate of the selectivity is needed also. Aoki (1999) describes a method for estimating predicate selectivity for a user defined type by using user-defined index structures.

Several authors have researched the problem of optimizing queries with expensive predicates (Chaudhuri and Shim 1996, Hellerstein 1994, Hellerstein and Stonebraker 1993). The predicates can be pushed above or below joins, and the order in which predicates are evaluated can be interchanged. The idea is to optimize a tradeoff of predicate cost versus predicate selectivity. Seshadri et al. (1997) describe Predator, an extensible database systems which performs query optimizations over user defined types.

5.4. Complex Queries

If the DBMS does not provide a mechanism for efficiently computing the aggregate needs for a user's analysis, then the only recourse is to user defined functions, or possible to exporting the data from the database and analyzing it with procedural programs written in Perl, Java, C++, and so on. However, it would be better if the DBMS would compute the desired output directly. In this section, we survey research into query languages and evaluation techniques for complex queries.

5.5. Built-in Complex Aggregation Functions

As we discussed in Section 3.2.1, many data sets in a data warehouse are temporal. A user often needs to compute a *temporal aggregate*, that is, one which changes over time. Put another way, there is an additional group-by attribute, namely the time span during which the aggregate value was constant.

Lets consider an example (Kline and Snodgrass 1995) of a temporal salary database Employed, with attributes Name, Salary, Dept, Start-Time, and EndTime. The pair of attributes StartTime and EndTime define the duration of time for which this attribute was valid (the *valid time*). The currently valid tuples have an EndTime value of infinity.

Suppose that we would like to know the average salary by department. The query would be:

```
Select     Dept, Avg(Salary)
From       Employed
Group By   Dept
```

The equivalent temporal aggregation query is:

```
Select     Dept, TimeRange, Avg(Salary)
From       Employed
Group By   Dept, TimeRange
```

Where TimeRange is a maximal time range during which the department's average salary is constant.

Computing a temporal aggregate can be difficult. For each group of tuples, the time line needs to be partitioned by the start times and the stop times of the tuples in the group. Long lived tuples can affect many time ranges, leading to an $O(N^2)$ algorithm,

Kline and Snodgrass (1995) propose an aggregation tree for efficiently computing a temporal aggregate. The index is a binary (or m-ary) tree that indexes constant regions on the time line. The leaves of the tree correspond to the constant regions. Interior nodes represent the regions of their (contiguous) children. For storing aggregates, the idea is to take advantage of the way that ancestor nodes summarize child nodes.

Suppose that we want to store a count aggregate. To insert a range (T_{hi}, T_{lo}), we descend the tree to insert the break in the constant regions at T_{lo} if the break does not already exist. We do the same for T_{hi}. Then for every maximal (highest) node in the tree that is covered by the range (T_{lo}, T_{hi}), we increment the count at the node. To obtain a count at a time point, descent the tree and sum the counts. Figure 19.8 shows an example of an aggregation tree. We note the close resemblance between this temporal aggregation tree and the range-max index proposed in Ho et al. (1997) (see Section 3.3.3).

Fang et al. (1998) have defined **iceberg** queries to be group-by queries with a monotone constraint, i.e. of the form

```
Select      SG₁, ..., SGₙ, SA₁, ..., SAₘ
From        R₁, ..., Rₖ
Where       C₀ AND ··· AND Cₗ
Group By    G₁, ... Gₙ
Having      C
```

The constraint C must be **monotone**, meaning that C is true when N tuples are in the group, then C is true when $N + M$ tuples are in the group, $M \geq 0$. Examples of monotone constraints are count(*) > 5, and sum(Salary) > 50000 (assuming that there are no negative salaries). Examples of iceberg queries include finding all products with more than $1,000,000 in sales during a month.

Conventional group-by query processing can evaluate iceberg queries, but they might have performance problems. In many cases the number of groups that satisfy the Having constraint is small compared to the number of groups. In addition, the number of groups might be very large (in Fang et al. (1998), one example of an iceberg query is a step in comparing large collections of documents for duplicates). The idea

DATA WAREHOUSING

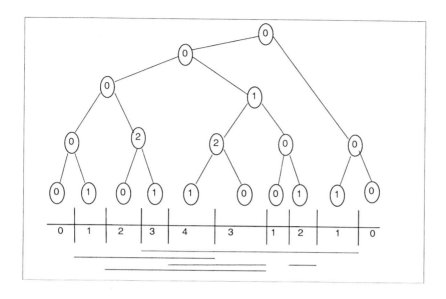

Figure 19.8. A tree structure for efficient time interval aggregation.

behind iceberg queries is to perform processing for the groups that satisfy the Having clause only (if possible).

Two basic techniques are used. The first is to sample the relation and select the groups that satisfy a scaled Having constraint. If the constraint is on the count of the tuples in the group, the sampled relation is likely to find the groups that constraint on the full relation. However, further processing is needed to eliminate groups that do not meet the constraint (*false positives*) and find groups that meet the constraint on the full but not the sampled relation (*false negatives*). The second technique is to hash the groups into buckets (i.e., coarser groups) and select the buckets that meet the constraint for further processing (and eliminate the false positives). We note the similarity to the technique used for the merge/purge problem (Hernandez and Stolfo 1995) discussed in Section 2.2.1.

A variety of hybrid strategies are proposed that combine the basic strategies. For example, sampling can be used to identify a set of groups (the *heavy* groups) that are likely to meet the constraint. For a second scan over the relation, the hashing technique is used, except that the heavy groups are aggregated separately. The idea behind this approach is to reduce the number of false positives that the hash technique returns.

A popular holistic aggregate is a *top ten* aggregate, for example "For each state, what are the top ten products in terms of sales?". More generally, one can ask for the top-K, where K might be very large. Carey and Kossmann (1998) have examined the problem of efficiently processing queries that have a Stop After K clause, meaning to stop output after K tuples have been produced. This problem expresses the top-K aggregate when there is also an Order By clause in the query.

A top-K query can be computed by producing the entire output relation, sorting it, and return the first K tuples. However, the authors propose efficient operators that can be put into the query plan. If the output relation is already sorted, or has an index on the sort-by attribute, then the query plan returns the first K tuples that are generated. If K is small, a main memory priority queue is used. If K is large and there is no sort order or index to take advantage of, then the output relation is partitioned on the sort-by attributes. The output is generated by sorting the smallest partition, the second smallest, and so on until K tuples have been produced.

Another useful holistic is a *median*, which is the value such that half the tuples have a larger value and half have a smaller. More generally, *quantiles* are the values at arbitrary cutoff points, such as 10% or 75% (medians are the 50% quantile). Computing quantiles exactly generally takes memory proportional to the number of unique values. However, *approximate* quantiles can be computed in linear time and in a small amount of memory. The quantiles are approximate in that the value that you get might not be ranked, e.g. 50%, but it will be ranked in the interval $[50\% - \epsilon, 50\% + \epsilon]$.

Manku et al. (1998) present a linear time approximate quantile algorithm that executed in small memory. The main idea is to partition the relation and compute quantiles on each partition. Two quantiles can be summed in an approximate manner: the result is an unevenly distributed set of quantiles, where there is some error in the approximation (about 1 quantile worth). The algorithm keep a running sum of the quantiles as they are computed.

5.5.1 Query languages for complex aggregation. Some authors have examined the problem of extending SQL to better express complex data analysis queries. One example of this type of extension is the Cube By declaration (see Section 3.2). In this section, we survey some other extensions.

Several researchers (Dayal 1983, Hsu and Parker 1995, Rao et al. 1996) have proposed extending SQL to include quantified predicates. An example of a query with a quantified predicate is Rao et al. (1996), "Find all patient-disease pairs (p,d) such that patient p has all symptoms associated with disease d". The quantification is that patient p has *all* of the symptoms. Other quantifiers are none, some, not all, and X%.

This query can be expressed using, for example, a nested aggregation query (Rao et al. 1996):

```
Select     P.name, D.name
From       PatientSymptoms P, DiseaseSymptoms D
Where      P.symptom = D.symptom
Group By   P.name, D.name
Having     count(P.symptom) =
              (Select count(D1.symptom)
               From DiseaseSymptoms D1 where D1.name =
               D.name)
```

Using quantified predicates, a much simpler and more direct expression is possible:

```
Select     P.name, D.name
From       PatientSymptoms P, DiseaseSymptoms D
Where      all
              (Select P1.symptom from PatientSymptoms P1
              where P.name = P1.name)
                 IS A
              (Select D1.symptom from DiseaseSymptoms D1
              where D.name = D1.name)
```

Rao et al. (1996) present an algorithm for evaluating queries with quantified predicates. They report an implementation that is orders of magnitude faster for some queries than commercial DBMSs that evaluated equivalent queries.

Chatziantoniou and Ross (1996), Chatziantoniou (1999b;a), Johnson and Chatziantoniou (1999a) presents an SQL extension, called *EMF-SQL* that expresses multiple aggregation features in a simple way. The main idea is to define *grouping variables*, or sets of tuples associated with each group. Membership in a grouping variable is determined by constraints, where the constraints can use the group-by attributes as constants. An example query is, "for each customer and for each month, report the sales to that customer in New York, in New Jersey, and also report the three month moving average of the sales to the customer":

```
select Customer, Month, sum(X.Price), sum(Y.Price),
avg(Z.Price) from Sales
Where Year = 1998
Group By Product, Month ; X, Y, Z
Such That   (X.Customer=Customer and X.Month=Month and
             X.State = 'NY'),
            (Y.Customer=Customer and Y.Month=Month and
             X.State = 'NJ'),
            (Z.Customer=Customer and Z.Month ≤ Month and
             Z.Month ≥ Z.Month-2 )
```

The grouping variable X represents sales to a particular customer during a particular month and in New York, as its constraint in the Such That clause indicates. The grouping variable Z is the set of all sales to the customer within the last three months. Aggregate values of other grouping variables can also be used in constraints. For example, we can make Y represent sales in NJ with a larger price than any in NY by adding the following constraint to Y: Y.Price > max(X.Price). Chatziantoniou has developed an efficient hash-based query evaluation algorithm that is orders of magnitude faster than commercial DBMSs on equivalent queries (Chatziantoniou 1999b;a).

A powerful feature of EMF-SQL is its ability to aggregate over tuples "outside the group" (see also Gyssens and Lakshmannan (1997) in Section 3.2.1). Grouping variable Z associates tuples from several months with a particular group, allowing a simple expression of a running average. In (Johnson and Chatziantoniou 1999b), EMF-SQL is extended to permit the expression of most holistic aggregates, including median and most-frequent. In (Chatziantoniou et al. 1999), EMF-SQL is expressed as an algebra. Grouping variables are represented with a special operator which joins a relation with a collection of groups. The groups can be derived by arbitrary relational operators, permitting great flexibility in the expression of OLAP queries.

6.   Conclusion

A well run data warehouse makes many difficult analyses easy to specify and execute, by taking care of many complex details related to data gathering, cleaning, integration, modeling, maintenance, and querying. However, these issues become the problems for the data warehouse. Fortunately, data warehousing has become an active research area with many results and tools for addressing each of these issues.

Bibliography

S. Abiteboul, S. Cluet, and T. Milo. Querying and updating the file. In *Proc. 19th VLDB Conf.*, pages 73–84, 1993.

D. Agrawal, A. El Abbadi, A. Singh, and T. Yurek. Efficient view maintenance at data warehouses. In *Proc. ACM SIGMOD Conf.*, pages 417–427, 1997a.

R. Agrawal, A. Gupta, and S. Sarawagi. Modeling multidimensional databases. In *Proc. IEEE Intl. Conf. on Data Engineering*, pages 232–243, 1997b.

P. Aoki. How to avoid building DataBlades that know the value of everything and the cost of nothing. In *Proc. Intl. Conf. Scientific and Statistical Database Management*, pages 122–135, 1999.

E. Baralis, S. Paraboschi, and E. Teniente. Materialized views selection in a multidimensional database. In *Proc. 23rd VLDB Conf.*, pages 156–165, 1997.

S. Berchtold, C. Bohm, and H.-P. Kriegel. The pyramid-tree: Breaking the curse of dimensionality. In *ACM SIGMOD Conf.*, pages 142–153, 1998.

K. Beyer and R. Ramakrishnan. Bottom-up computation of sparse and iceberg CUBEs. In *Proc. ACM SIGMOD Conf.*, pages 359–370, 1999.

J.A. Blakeley, P.-A. Larson, and F.W. Tompa. Efficiently updating materialized views. In *Proc. ACM SIGMOD Conf.*, pages 61–71, 1986.

C. S. Jensen et al. A glossary of temporal database concepts. *ACM SIGMOD Record*, 23(1):52–64, 1994.

M.J. Carey and D. Kossmann. Reducing the braking distance of an SQL engine. In *Proc. 24th VLDB Conf.*, pages 158–169, 1998.

S. Chakravarthy and D. Lomet, editors. *Special Issue in Active Databases*, volume 4 of *Data Engineering Bulletin*. IEEE, 1992.

D. Chamberlin. *Using the New DB2*. Morgan Kaufman, 1996.

C.Y. Chan and Y.E. Ioannidis. Bitmap design and evaluation. In *Proc. ACM SIGMOD Conf.*, pages 355–366, 1998.

R. Chandra and A. Segev. Managing temporal financial data in an extensible database. In *Proc. VLDB Conf.*, pages 302–313, 1993.

D. Chatziantoniou. Ad Hoc OLAP : Expression and Evaluation. In *IEEE International Conference on Data Engineering*, page 250, 1999a.

D. Chatziantoniou. Evaluation of Ad Hoc OLAP : In-Place Computation. In *ACM/IEEE International Conference on Scientific and Statistical Database Management*, pages 34–43, 1999b.

D. Chatziantoniou, T. Johnson, and Samuel Kim. On Modeling and Processing Decision Support Queries. Submitted for publication, 1999.

Damianos Chatziantoniou and Kenneth Ross. Querying Multiple Features of Groups in Relational Databases. In *22nd VLDB Conference*, pages 295–306, 1996.

S. Chaudhuri and U. Dayal. On overview of data warehousing and OLAP technology. *ACM SIGMOD Record*, 26(1):65–74, 1997.

S. Chaudhuri, R. Krishnamurthy, S. Potamianos, and K. Shim. Optimizing queries with materialized views. In *Proc. IEEE Intl. Conf. Data Engineering*, pages 190–200, 1995.

S. Chaudhuri and K. Shim. Including group-by in query optimization. In *Proc. 20th VLDB Conf.*, pages 354–366, 1994.

S. Chaudhuri and K. Shim. Optimization of queries with user-defined predicates. In *Proc. 22nd VLDB Conf.*, pages 87–98, 1996.

G. Colliat. OLAP, relational, and multidimensional database systems. *ACM SIGMOD Record*, 25(3):64–69, 1996.

T. Dasu and T. Johnson. Hunting of the Snark: Finding data glitches using data mining techniques. In *1999 MIT Conf. on Information Quality*, pages 52–61, 1999.

A. Datta and H. Thomas. A conceptual model for on-line analytical processing in decision support databases. In *Proc. of WITS*, pages 91–100, 1997.

U. Dayal. Processing queries with quantifiers. In *Proc. ACM Symp. Principles of Database Systems*, pages 125–136, 1983.

U. Dayal. Of nests and trees: A unified approach to processing queries that contain nested subqueries, aggregates, and quantifiers. In *Proc. VLDB Conf.*, pages 197–208, 1987.

U. Dayal. Active database management systems. In *Proc. 3rd Intl. Conf. on Data and Knowledge Bases*, pages 150–169, 1998.

M. Fang, N. Shivakumar, H. Garcia-Molina, R. Motwani, and J.D. Ullman. Computing iceberg queries efficiently. In *Proc. 24th VLDB Conf.*, pages 299–310, 1998.

C.D. French. "one size fits all" database architectures do not work for DSS. In *Proc. ACM SIGMOD Conf.*, pages 449–450, 1995.

H. Garcia-Molina, W. Labio, and J. Yang. Expiring data in a warehouse. In *Proc. 24th VLDB Conf.*, pages 500–511, 1998.

Michael Godfrey, Tobias Mayr, Praveen Seshadri, and Thorsten von Eicken. Secure and portable database extensibility. In *Proc ACM SIGMOD Conf.*, pages 390–401, 1998.

S. Goil and A. Choudhary. An infrastructure for scalable parallel multidimensional analysis. In *Proc. 11th Intl. Conf. Scientific and Statistical Database Management*, pages 102–111, 1999.

G. Graefe. Query evaluation techniques for large databases. *ACM Computing Surveys*, 25(2), 1993.

J. Gray, A. Bosworth, A Layman, and H. Pirahesh. Data cube: A relational aggregation operator generalizing group-by, cross-tab, and sub-totals. In *Proc. IEEE Intl. Conf. on Data Engineering*, pages 152–159, 1996.

A. Gupta, V. Harinarayan, and D. Quass. Aggregate-query processing in data warehousing environments. In *Proc. 21st VLDB Conf.*, pages 358–369, 1995.

A. Gupta, H.V. Jagadish, and I.S. Mumick. Data integration using self-maintainable views. In *Proc. 5th Intl. Conf. Extending Database Technology*, pages 140–144, 1996.

A. Gupta and I.S. Mumick. Maintenance of materialized views: Problems, techniques, and applications. *Data Engineering Bulletin*, 18(2): 3–18, 1995.

H. Gupta. Selections of views to materialise in a data warehouse. In *Proc. 6th. ICDT*, pages 98–112, 1997.

H. Gupta, V. Harinarayan, A. Rajaraman, and J. Ullman. Index selection for OLAP. In *Proc. 13th IEEE Intl. Conf. on Data Engineering*, pages 208–219, 1997.

R.H. Guting. An introduction to spatial database systems. *VLDB Journal*, 3(4):357–399, 1994.

M. Gyssens, L.V.S. Lakshmanan, and I.N. Subramanian. Tables as a paradigm for querying and restructuring. In *Proc. ACM Symp. on Principles of Database Systems*, pages 93–103, 1996.

M. Gyssens and L.V.S. Lakshmannan. A foundation for multidimensional databases. In *Proc. 23rd VLDB Conf.*, pages 106–115, 1997.

J. Hammer, H. Garcia-Molina, J. Widom, and W. Labio. The Stanford data warehousing project. *IEEE Data Engineering Bulletin*, 18(2): 41–48, 1995.

J. Hammer and D. McLeod. An approach to resolving semantic heterogeneity in a federation of autonomous, hetergeneous, database systems. *Int'l. Journal of Intelligent and Cooperative Information Systems*, 2:51–83, 1993.

E. Hanson. A performance analysis of view materialization strategies. In *Proc. ACM SIGMOD Conf.*, pages 440–453, 1986.

E. Hanson and J. Widom. An overview of production rules in database systems. *The Knowledge Engineering Review*, 8(2):121–143, 1993.

V. Harinarayan, A. Rajaraman, and J. Ullman. Implementing data cubes efficiently. In *Proc. ACM SIGMOD Conf.*, pages 205–216, 1996.

J.M. Hellerstein. Predicate migration placement. In *Proc. ACM SIGMOD Conf.*, pages 9–18, 1994.

J.M. Hellerstein and M. Stonebraker. Predicate migration: Optimization queries with expensive predicates. In *Proc. ACM SIGMOD Conf.*, pages 267–276, 1993.

M.A. Hernandez and S.J. Stolfo. The merge/purge problem for large databases. In *Proc ACM SIGMOD Conf.*, pages 127–138, 1995.

C.-T. Ho, R. Agrawal, N. Megiddo, and R. Srikant. Range queries in OLAP data cubes. In *Proc. ACM SIGMOD Conf.*, pages 73–88, 1997.

W.-C. Hou and Z. Zhang. Enhancing database correctness. In *Proc. ACM Sigmod Conf.*, pages 223–232, 1995.

P.-Y. Hsu and D.S. Parker. Improving SQL with generalized quantifiers. In *Proc. IEEE Intl. Conf. Data Engineering*, pages 298–305, 1995.

M. Jaedicke and B. Mitschang. On parallel processing of aggregate and scalar functions in object-relational DBMS. In *Proc. ACM SIGMOD Conf.*, pages 379–389, 1998.

H.V. Jagadish, L.V.S. Lakshmanan, and D. Srivastava. What can hierarchies do for data warehouses? In *Proc. 25th Intl. VLDB Conf.*, pages 530–541, 1999.

H.V. Jagadish, I.S. Mumick, and A. Silberschatz. View maintenance issues for the chronicle data model. In *Proc. 14th ACM Symp. Principles of Database Systems*, pages 113–124, 1995.

H.V. Jagadish, P.P.S Narayan, S. Seshardi, S. Sudarshan, and R. Kanneganti. Incremental organization for dara recording and warehousing. In *Proc. 23rd VLDB Conf.*, pages 16–25, 1997.

T. Johnson. Coarse indices for a tape-based data warehouse. In *Proc. IEEE Intl. Conf. Data Engineering*, pages 231–240, 1998.

T. Johnson. Performance measurements of compressed bitmap indices. In *Proc. Intl. VLDB Conf.*, pages 278–289, 1999.

T. Johnson and D. Chatziantoniou. *Databases in Telecommunications*, chapter Joining Very Large Data Sets, pages 118–132. Lecture Notes in Computer Science, Vol. 1819. Springer, 1999a.

T. Johnson and D. Chatziantoniou. Extending Complex Ad Hoc OLAP. In *Conf. Information and Knowledge Management*, pages 170–179, 1999b.

T. Johnson and D. Shasha. Hierarchically split cube forests for decision support: description and tuned design. Dept. of Computer Science tr727, New York University, www.cs.nyu.edu, November 1996.

T. Johnson and D. Shasha. Some approaches to index design for cube forest. *IEEE Data Engineering Bulletin*, 20(1):27–35, 1997.

W. Kent. The breakdown of the information model in multi-database systems. *ACM SIGMOD Record*, 20(2):10–15, 1991.

R. Kimball. *The Data Warehouse Toolkit*. Wiley, 1996.

N. Kline and R.T. Snodgrass. Computing temporal aggregates. In *Proc. IEEE Intl. Conf. on Data Engineering*, pages 222–231, 1995.

Y. Kotidis and N. Roussopoulos. An alternative storage organization for ROLAP aggregate views based on cubetrees. In *Proc. ACM SIGMOD Conf.*, pages 249–258, 1998.

W. Lehner. Modeling large scale OLAP scenarios. In *Proc. EDBT Conf.*, pages 154–167, 1998.

C. Li and X.S. Wang. A data model for supporting on-line analytical processing. In *Proc. Conference on Information and Knowledge Management*, pages 81–88, 1996.

W. Litwin, L. Mark, and N. Roussopoulos. Interoperability of multiple autonomous databases. *Computing Surveys*, 22:267–293, 1990.

M. Carey et al. Towards hetergenous multimedia information systems: The Garlic approach. In *Proc. Research Issues in Data Engineering - Distributed Object Management Workshop*, pages 124–131, 1995.

G.S. Manku, S. Rajagopalan, and B.G. Linsday. Approximate medians and other quantiles in one pass and with limited memory. In *Proc. ACM SIGMOD Conf.*, pages 426–435, 1998.

G. Moerkotte. Small materialized aggregates: A light weight index structure for data warehousing. In *Proc. 24th VLDB Conf.*, pages 476–487, 1998.

I.S. Mumick, D. Quass, and B.S. Mumick. Maintenance of data cubes and summary tables in a warehouse. In *Proc. ACM SIGMOD Conf.*, pages 100–111, 1997.

M. Muralikrishna. Improved unnesting algorithms for join aggregate SQL queries. In *Proc. 18th VLDB Conf.*, pages 79–90, 1992.

P.E. O'Neil and D. Quass. Improved query performance with variant indexes. In *Proc. SIGMOD Conf.*, pages 38–49, 1997.

T.B. Pederson and C.S. Jensen. Multidimensional data modeling for complex data. In *Proc. IEEE Intl. Conf. on Data Engineering*, pages 336–345, 1999.

D. Pfoser and C. Jensen. Incremental join of time-oriented data. In *Proc. 11th Intl. Conf. Scientific and Statistical Database Management*, pages 232–243, 1999.

D. Quass and J. Widom. On-line warehouse view maintenance. In *Proc. ACM SIGMOD Conf.*, pages 393–404, 1997.

R. Ahmed et al. The Pegasus heterogenous multidatabase system. *IEEE Computer*, 24:19–27, 1991.

M. Rafanelli and A. Shoshani. STORM: A statistical object representation model. In *Proc. of the Scientific and Statistical Database Management Conf.*, pages 14–29, 1990.

S.G. Rao, A. Badia, and D. Van Gucht. Providing better support for a class of decision support queries. In *Proc. ACM SIGMOD Conf.*, pages 217–227, 1996.

K.A. Ross and D. Srivastava. Fast computation of sparse datacubes. In *Proc. 23rd VLDB Conf.*, pages 116–125, 1997.

N. Roussopoulos, Y. Kotidis, and M. Roussopoulos. Cubetree: Organization of and bulk incremental updates on the data cube. In *Proc. ACM SIGMOD conf.*, pages 89–99, 1997.

S. Agrawal et al. On the computation of multidimensional aggregates. In *Proc. 22nd VLDB Conf.*, pages 506–521, 1996.

S. Sarawagi. Indexing OLAP data. *Data Engineering Bulletin*, 20(1): 36–34, 1997.

S. Sarawagi and M. Stonebraker. Efficient organization of large multidimensional arrays. In *Proc. IEEE Intl. Conf. Data Engineering*, pages 328–336, 1994.

P. Seshadri, M. Livny, and R. Ramakrishnan. The case for enhanced abstract data types. In *Proc. VLDB Conf.*, pages 66–75, 1997.

A. Shatdal and J.F. Naughton. Adaptive parallel aggregation algorithms. In *Proc. ACM SIGMOD Conf.*, pages 104–114, 1995.

A. Shukla, P. Deshpande, and J.F. Naughton. Materialized view selection for multidimensional datasets. In *Proc. 24th VLDB*, pages 488–499, 1998.

A. Shukla, P. Deshpande, J.F. Naughton, and K. Ramasamy. Storage estimation for multidimensional aggregates in the presence of hierarchies. In *Proc. 22nd VLDB Conf.*, pages 522–531, 1996.

D. Srivastava, S. Dar, H.V. Jagadish, and A.Y. Levy. Answering queries with aggregation using views. In *Proc. 22nd VLDB Conf.*, pages 318–329, 1996.

V.S. Subrahmanian. HERMES: A heterogenous reasoning and mediator system. http://www.cs.umd.edu/projects/hermes/overview/-paper, 1995.

D. Theodoratos and T.K. Sellis. Data warehouse configuration. In *Proc. 23rd VLDB Conf.*, pages 126–135, 1997.

W. Kim et al. On resolving schematic heterogeneity in multidatabase systems. *Distributed and Parallel Databases*, 2:251–279, 1993.

Y.R. Wang and S.E. Madnick. The inter-database instance identification problem in integrating autonomous systems. In *Proc. 6th Intl. Conf. on Data Engineering*, 1989.

G. Wiederhold. Mediators in the architecture of future information systems. *IEEE Computer*, 25(3):38–49, 1992.

G. Wiederhold. Intelligent integration of information. In *Proc. ACM SIGMOD*, pages 434–437, 1993.

M.-C. Wu and A.P. Buchmann. Encoded bitmap indexing for data warehouses. In *Proc. IEEE Intl. Conf. Data Engineering*, pages 220–230, 1998.

H. Garcia-Molina Y. Papakonstantinou, S. Abiteboul. Object fusion in mediator systems. In *Proc. 22nd VLDB Conf.*, pages 413–424, 1996.

J. Widom Y. Papakonstantinou, H. Garcia-Molina. Object exchange across heterogenous information sources. In *Proc. ICDE Conf.*, pages 251–260, 1995.

W.P. Yan and P.-A. Larson. Performing group-by before join. In *Proc. IEEE Intl. Conf. Data Engineering*, pages 89–100, 1994.

W.P. Yan and P.-A. Larson. Eager aggregation and lazy aggregation. In *Proc. 21st VLDB Conf.*, pages 345–357, 1995.

Y. Zhao, P.M. Deshpande, and J.F. Naughton. An array-based algorithm for simultaneous multidimensional aggregates. In *Proc. ACM SIGMOD Conf.*, pages 159–170, 1997.

Y. Zhuge, H. Garcia-Molina, J. Hammer, and J. Widom. View maintenance in a warehousing environment. In *Proc. ACM SIGMOD Conf.*, pages 316–327, 1995.

Chapter 20

AGGREGATE VIEW MANAGEMENT IN DATA WAREHOUSES

Yannis Kotidis
AT&T Labs – Research, Florham Park, NJ 07932, USA
kotidis@research.att.com

Abstract Materialized views and their potential have been recently rediscovered for the content of OLAP and data warehousing. A flurry of papers has been generated on how views can be used to accelerate ad-hoc computations over massive datasets. In this chapter we introduce and comment on some main-stream approaches for defining, computing, using and maintaining materialized views with aggregations in a large data warehouse.

Keywords: Data warehousing, Aggregate views, Dynamic caching.

1. Introduction

Data warehousing is a collection of decision support technologies that aim at enabling an enterprise to make better and faster decisions (Chaudhuri and Dayal 1997). Finding the right information at the right time is a necessity for survival in today's competitive marketplace and this area has enjoyed an explosive growth in the past few years. Data warehousing products have been successfully deployed in data rich industries ranging from retail to financial services and telecommunications.

From a historical perspective, the data warehousing area is a byproduct of tremendous advances in information technology. During the last few decades there has been an increasing movement to computerize every possible business process. However, most of the applications developed over the years were mostly stand-alone efforts. As a result there was virtually no integration among different applications, even within the domain of a single enterprise. The situation was even more dramatic from the data point of view. The same data item could have inconsistent definitions, meanings and representations along different platforms.

Although relational databases were at first believed to provide direct access to a single unambiguous copy of the data, it was very soon realized that they couldn't support both operational and decision-support users. Operational workload consists of many concurrent short transactions that access and modify a few records at a time. In addition depending on the application it might have strict real-time requirements. The obvious examples are ATM and airline reservation systems. Such applications automate structured and repetitive tasks that require detailed up to date information. Decision support applications on the other hand use tactical information that answers "who" and "what" questions about past events (OLAP Council 2001a). This requires a stable view of the underlying data that can only be obtained using costly relational locking mechanisms, unless a second copy of the information is used. Furthermore, decision support often requires historical data that is usually unavailable in operational databases that only deal with recent data. Finally, operational databases are designed to reflect the operational semantics of their applications and to maximize transaction throughput. Decision support analysis on the other hand is typically ad-hoc, based on multidimensional business models and operations that require different data models and new access methods tailored for query intensive applications.

In the broadest sense, a data warehouse is a single, integrated informational store that provides stable, point-in-time data for decision support applications (Barquin and Edelstein 1997). Unlike traditional database systems that automate day-to-day operations, a data warehouse provides an environment in which an organization can evaluate and analyze its enterprise data over time. Since data might be coming from different legacy systems within the organization, significant cleansing, transformation and reconciliation may be necessary before it is loaded in the repository. Data cleaning involves operations varying from data integrity tests and simple transformation tasks (e.g. inconsistent field lengths or descriptions) to more advanced tools that look for suspicious statistical patterns in the data.

2. Materialized Aggregate Views for Better Performance

Most data warehouses adopt a multidimensional approach for representing the data. The origins of this practice go back to PC spreadsheet programs that were extensively used by business analysts. More advanced multidimensional access is now achieved through interfaces that provide "On Line Analytical Processing" (OLAP), which involves interactive access to a wide variety of possible views of the information. OLAP software allows the data to be rendered across any dimension and at any level of aggregation. For example in a sales data warehouse customer, product, location and time of sale may be the dimensions of interest. In order to extract this information on the fly, the navigation tool interacts with the data warehouse, which provides the core data that is being analyzed. Examples of possible computations include among others:

- *multidimensional ratios* like "show me the contribution to weekly profit made by all items sold in NJ between May 1 and May 7".

- *quantiles* e.g. "show my top-5 selling products across the state of NJ".

- *statistical profiles* e.g. "show sales by store for all stores in the bottom 5% of sales".

- *comparisons* e.g. "sales in this fiscal period versus last period".

The main cost in terms of the time consumed of executing this type of queries is not only doing the actual arithmetic, but also of retrieving the data items (or "cells' in the multidimensional dialect) that affect the calculated functions. For a large data warehouse, executing queries with aggregations against the detailed records takes hours, simply because of the volume of records that are being accessed. As a result most implementations facilitate some form of pre-aggregation in order to support complex data-intensive queries in a interactive fashion. In relational databases, materialized derived relations (views) have long been proposed to speed up query processing. In the data warehouse, these views store redundant, aggregated information and are commonly referred to as *summary tables* (Chaudhuri and Dayal 1997). A materialized view that contains highly consolidated information is typically much smaller than the base relations used to store all detailed records. As a result, querying the view instead of the base records offers several orders of magnitude faster query speeds.

Since materialized views promise high performance improvements over accessing detailed records they are a valuable component in the design of a data warehouse. They might, for example, include high level consolidations, which are bound to be needed for reports or ad-hoc analyzes, and which involve too much data to be calculated on the fly. For example in a sales data warehouse, a large majority of the queries may involve transactions over the last few weeks or months. Having these sales summarized in a view significantly speeds up many queries.

If query response time is the only concern, an eager policy of materializing all possible aggregations that might be requested will yield an excellent effect on performance since each query will require a minimum amount of data movement and on the fly calculations. Unfortunately this plan is not viable, as the number of possible aggregate views is exponential in the number of dimensions that the dataset is analyzed on. Thus, materializing all possible views demands an enormous amount of pre-processing cost and disk volume to compute and store the aggregations. Furthermore, much like a cache, the views get dirty whenever the data warehouse tables are modified. For most organizations, the maintenance process is happening in a daily fashion, usually overnight. In order to correctly reflect the underlying data, the views have to be updated in a process that is known as view maintenance. Having many views materialized in the system lengthens the duration of the daily update process and reduces the warehouse "on-line" period, i.e. the portion of time the system is available for analysis.

Materialized views with aggregates introduce new challenges in the design and use of the data warehouse. These challenges include:

- identifying the views that we want to materialize.
- efficiently computing these views from the raw data
- exploiting the views to answer queries
- efficiently maintaining the views when new data is shipped to the data warehouse.

Before getting to see some of the approaches for dealing with these problems we will make a brief introduction on the star schema organization that is frequently used to model a relational data warehouse and the data cube operator that provides broadly accepted framework for describing materialized aggregate views.

3. A Relational Data Warehouse Architecture

Information within the data warehouse is modeled in a multidimensional fashion. Formally, a dimension is a structural attribute that lists members of similar type (OLAP Council 2001a). The time dimension is such an example, common among most data warehouse applications. Dimension members allow us to pin-point the raw data for analysis. We can think of them as indices to a virtual multi-dimensional array of numeric attributes that are the target of the analysis. These attributes are often called *measures* and are used as input to the aggregation functions. Such measures include sales, revenue, inventory e.t.c.

Often dimension's members are organized on parent-child relationships, like for example all days, weeks, months, quarters and years along the time dimension. The resulting relations form a *hierarchy*. Dimension members along the hierarchy consolidate data of all the members which are their children. Hierarchies provide the means to encode our domain knowledge and address the data at different levels of aggregation. This is achieved through the *drill-down* and *roll-up* operators. By drilling-down on the aggregate data we get a more detailed view of the information. For example starting from the total sales per year, we drill-down and ask for a monthly breakdown of sales for the last year and then examine the actual transactions for a month with irregular volume of sales. Roll-up is the opposite operation where information is examined at progressively coarser granularity.

In a relational implementation, the data warehouse is usually organized in a specialized lay-out that is known as the *star schema* (Kimball 1996). In its simplest form there is a central table F called the fact table that contains the facts, i.e the transactional information that is of interest. Each record in F consists of two parts. The first part is a list of attributes that are foreign keys to the satellite dimension tables discussed bellow and identify a unique position for the record in the multidimensional space in which we analyze the data. The second part contains a list of measures that provide the numeric information on which analysis is being performed. For each dimension in the dataset there is a single table that contains information for the specific dimension. This table holds a primary key that is used to link dimension-specific attributes with the fact table.

Figure 20.1 shows a star-schema organization for a data warehouse that analyzes sales data. The data contains sales information for parts that are being purchased from different suppliers and sold to customers. In this simplified example each transaction records a sale to a customer. A record contains the part identifier *partkey*, the identifier of the supplier *suppkey* from which the part was ordered and also the identifier of the customer *custkey* who bought the part. There is also a numeric attribute *amount* that stores how much the customer paid for the product. The star schema provides a clean lay-out in which dimension specific attributes are hidden from the facts and are encapsulated in the dimension tables. For example the customer table contains all customer attributes, like name and contact info. Furthermore, it allows for easy schema modifications, fast loading of data and incremental updates.

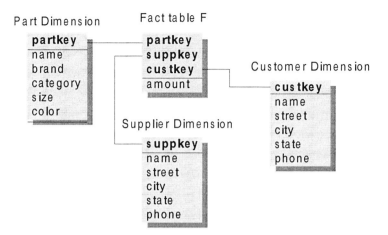

Figure 20.1. A Simple Star Schema Organization

The simplest form of analysis against the data is to aggregate the measure(s) on one or more selected dimensions. For example query "find the total sales per customer" is written in SQL:

q_1: ```
select custkey, sum(amount) as total_sales
from F
group by custkey
```

while query "find the total sales per customer and product" is translated to:

$q_2$: ```
select partkey,custkey, sum(amount) as total_sales
from F
group by partkey,custkey
```

More expressiveness can be achieved by combining facts with information stored in the dimension tables. For example query "find our customers in NJ along with their aggregated sales" is stated as:

q_3: ```
select customer.custkey, customer.name, sum(amount) as
total_sales from F, customer
where F.custkey = customer.custkey
and customer.state = 'NJ'
group by customer.custkey, customer.name
```

This type of queries consist of joins of the fact table with a number of dimension tables, with possible selections on some dimension attributes and the aggregation of one or more of the measures grouped by a subset
```

of attributes from the dimension tables. These queries are known as Select-Project-Join (SPJ) queries.

The group-by operator in SQL is frequently used to compute such aggregations among ad-hoc groups. Gray et al. (1996) introduced the data cube CUBE operator as a generalization of the traditional group by syntax. Their motivation was that certain types of data analysis are difficult to be expressed in SQL. Examples that are common in report writing are the histogram, cross-tabulation, roll-up, drill-down and sub-total constructs. In technical terms, the data cube is a redundant multidimensional projection of a relation. It computes all possible group by SQL operators among the dimensions. The data cube of our three dimensional example will be the union of all 7 aggregate queries that we can construct by grouping on different combinations of attributes *partkey, suppkey, custkey*, plus the value that is derived when we aggregate over all data:

```
select sum(amount) as total_sales
from F
```

In general for n dimensions, the data cube computes 2^n group bys, corresponding to all possible combinations of the selected dimensions. Each one of these group bys can be realized as a materialized view. In the following section we address the problem of efficiently computing these views.

4. Efficient Computation of Views with Aggregates

For the following discussion we assume that only aggregates based on the dimensions keys are of interest, like in queries q_1, q_2. However the results can be extended when groupings are performed on dimension members that are part of a hierarchy.

The straightforward way to compute all these views is to rewrite the CUBE query as a collection of 2^n SQL queries and execute them separately, like for instance queries q_1 and q_2 of the previous example. However, even for small number of dimensions, this naive computation of the views is unrealistic, because of the mere size of the raw data that needs to be accessed over and over again to compute the aggregates. Apart from the I/O overhead an independent computation of the aggregates does not realize possible overlap among the calculations that are needed for the views. For example after query q_2 is executed and the result is materialized as view u, we can rewrite the view that corresponds to query q_1 as:

```
create view v as
select custkey, sum(u.total_sales) as total_sales
from u
group by custkey
```

What allows this rewriting is a special property of the *sum()* function that allows aggregates to be further aggregated. Such functions allow the input set to be partitioned into disjoint sets that can be aggregated separately and later combined. Aggregate functions with this property are called *distributive* and are characterized from the fact that no state-information is needed for the function to summarize a sub-aggregation. Other distributive functions include *count()*, *min()* and *max()*.

Often functions of interest can be expressed by combining two or more distributive functions. The avg() function for example can be computed using sum() and count(). Functions that require a fixed size state information to describe a sub-aggregate are called algebraic and can be also optimized when computing the aggregates for multiple views in parallel. Other algebraic functions include center of mass, standard deviation, maxN() (N largest values) and minN(). However for functions like *median()* there is no bound on the size of state information that is needed to describe a sub-aggregate. Such functions are often referred to as *holistic* and void all optimizations that we describe bellow.

For non-holistic aggregate functions we can share I/O and computations among multiple views of the data cube. These gains are realized by exploiting well-defined dependencies among the views that are best depicted in the lattice (Harinarayan et al. 1996) of Figure 20.2, for our three dimensional example. Each node in the lattice represents a view that aggregates data over the attributes present in that node; for example node (*partkey*) is view v. The node labeled as *none* represents a view that computes a single super-aggregate over all input cells. For our sales dataset this will be the overall volume of sales for all records in the fact table. The lattice representation is based on a derived-from relationship, which defines a partial ordering $\preceq$ among the views. For two views v and u, $v \preceq u$ if and only if v can be computed from the tuples of u for any instance of the dataset. For example $(partkey) \preceq (partkey, custkey)$ but not vise-versa. In the lattice notation there is a downward path from u to v if and only if $v \preceq u$.

Typically group-bys are being computed either by sort-based or by hash-based methods as discussed in (Graefe 1993). Agrawal et al. (1996) describes how these methods can be extended for multiple group-by computations, like the case of the data cube. Using the authors' notation

AGGREGATE VIEW MANAGEMENT IN DATA WAREHOUSES

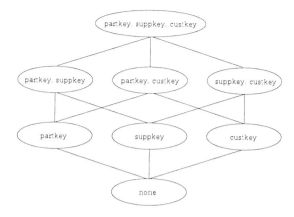

Figure 20.2. The Data Cube Lattice for three Dimensions

there are four possible optimizations that can be incorporated in the computation of the aggregate views:

- **smallest-parent:** this optimization implies that a view should be computed from the smallest previously computed view. For example view (*partkey*) can be computed from any of (*partkey, custkey*), (*partkey, suppkey*) and (*partkey, custkey, suppkey*). If we have an estimate for the sizes of these views, we can pick the smallest one for computing (*partkey*).

- **cache-results:** this optimization favors in-memory results of a previous computation to derive other views and thus reduce disk I/O.

- **amortize-scans:** when data has to be read from disk, the cost of I/O should be amortized by computing as many views as possible. For instance if view (*partkey, suppkey*) was previously stored on disk, we may compute all of (*partkey*), (*suppkey*), (*none*) in one scan over (*partkey, suppkey*).

- **share sorts and partitions:** this optimizations are specific to the algorithm used for computing the views. When sort-based algorithms are employed we can share the cost of sorting among multiple views, or in some cases exploit partially matching sort orders. For example if the records of view (*partkey, suppkey*) are sorted on low-*partkey*, low-*suppkey*, i.e. first on *partkey* and then on *suppkey* for tuples with the same part identifier, we can then compute view (*partkey*) with no additional sorts in the following manner: we first project out the *suppkey* column. This will give as a sorted list (with duplicates) of *partkey* values along with partially aggregated measures. We then simply merge these sub-aggregates for every distinct *partkey* value. Partially matching sort orders are also useful. If for example view (*partkey*, *suppkey*, *custkey*) is sorted on attribute order, we can compute (*partkey*, *custkey*) by partitioning on the first attribute and independently sorting on *custkey*. Similar optimizations can be performed when the aggregates are computed using hash-based algorithms by sharing the partitioning cost among multiple views.

For datasets that are much larger than the available main memory these optimizations are often contradictory. For instance we might decide to sort a memory resident view instead of using one that has a desirable sort order but is stored on disk. Comparing the sort based techniques against hash-based computations of the aggregates, Agrawal et al. (1996) conclude that sorting works better for sparse datasets, while hashing methods gain from decreasing levels of sparseness. In practice performance strongly depends on the amount of memory that is available to hold intermediate results and auxiliary data structures (e.g hash tables) and the distribution of values of the dataset. Real-world data however, for many application domains, are often very large and sparse. Ross and Srivastava (1997) argue that none of the previous algorithms is very efficient for sparse datasets, especially when the base relation is much larger than main memory. For these cases, they propose an algorithm that follows a divide-and-conquer strategy to split the problem into several simpler computations of sub-cubes that are then computed by multiple in-memory sorts. Other techniques for computing multiple aggregate views in parallel can be found at (Beyer and Ramakrishnan 1999, Li et al. 1999, Muto and Kitsuregawa 1999).

These algorithms work well for Relational OLAP (ROLAP) systems that store their data in conventional relational tables. For Multidimensional OLAP (MOLAP) systems, that store their data in sparse multidimensional arrays rather than in tables, Zhao et al. (1997) proposed a different array-based algorithm. For the synthetic and rather dense datasets that they used for their experiments the authors showed that the MOLAP approach can be significantly faster that the ROLAP table-based algorithms. The benefit comes from the fact that the array representation allows direct access to individual cells. However, multidimensional arrays have limitations when dealing with sparse high dimensional datasets.

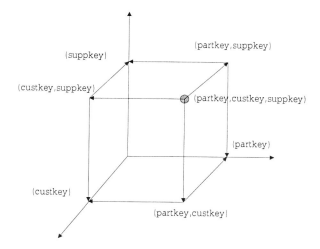

Figure 20.3. Multi-dimensional Projections of an Input Cell

4.1. Estimating the Size of the Views

An unexpected behavior of multidimensional aggregations is that the amount of possible results is many times larger than the detailed data. Figure 20.3 shows an input record of the fact table of Figure 20.1 in the three dimensional space defined by the keys: *partkey, suppkey, custkey*. Each possible aggregation across these dimensions can be seen as a projection of this point in the sub-space of the participating dimensions. For example all group by (*partkey, custkey*) aggregates will be projected in the corresponding 2-dimensional plane in this space. Some of these projections will be overlapped with projections of other input points, however each input record may introduce as many as seven new aggregates in the worst case. Subsequently, data in each of these sub-spaces will be much denser and voluminous than the input data, which is expected to be fairly sparse for many real datasets (Colliat 1996).

Before getting to compute all or some of the views of the data cube, we need an estimate of the disk space required accommodating the derived aggregates. This information comes handy for tuning the algorithms used to compute the aggregates to decide on hashing values, sizes of disk runs for sorting e.t.c. The size of the views depends on the number of dimensions involved and the distribution of values along their domains. For simplicity we make the assumption that all dimensions have the same domain size D. For a view with i dimensions the maximum number of ways that values from these dimensions can be combined and return an aggregate is D^i, which is the volume of an i-dimensional array of

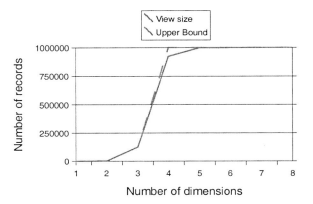

Figure 20.4. View Sizes for Uniform Distribution

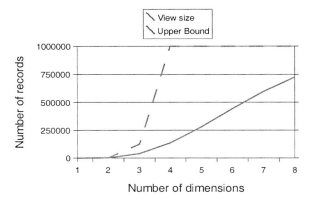

Figure 20.5. View Sizes for 80-20 Self-similar Distribution

length D per side. Since each input record generates at most one new entry in this array, the size of the view is also bounded by the size of the fact table F, denoted as $|F|$. Therefore, a crude upper-bound for the size of the view is $min(D^i, |F|)$. Figure 20.4 plots the size of the views for an eight dimensional uniform dataset where D is 50 and the fact table contains 1 million records. The dotted line represents the previous upper bound. In real-datasets the result is strongly affected by any skewness observed. In Figure 20.5 we depict another dataset, in which the values along each dimension are following an independent 80-20 self-similar distribution (Gray et al. 1994): 20% of each dimension's domain was present in 80% of the input records. The resulting views are now much smaller that the previous uniform case, as aggregates fall into existing computed cells.

For estimating the size of the views we can take a random sample of the dataset, compute the views on that and linearly scale up the estimate to the whole dataset. Unfortunately this estimate is very crude as scaling produces a biased estimator for the size of the views. Another problem with sampling is that even if a small fraction of the base data is accessed, it is still relatively expensive, as every sampled tuple may require one page access, depending on the strategy used. Shukla et al. (1996) propose another strategy for estimating the size of the views using a probabilistic counting algorithm (Flajolet and Martin 1985) that counts the number of distinct elements observed in a multi-set. The algorithm makes a single pass over the fact table F and produces an estimate using a fixed amount of memory. It maintains a bit vector $B[0 \ldots L-1]$, where L is a parameter, depending on the available memory. It also requires a hashing function $h()$ that uniformly distributes the input values from F into range $[0 \ldots 2^L - 1]$. For each tuple x read from F, we set bit $B[i]$, where i is the position of the least significant bit in the binary representation of $h(x)$. For a perfectly uniform hashing function $h()$, $B[i]$ is set with probability $\frac{n}{2^{i+1}}$. Therefore, we can use the position of the leftmost zero bit in B to estimate the number of distinct elements in the input. This estimate is typically within a factor of 2 from the actual value, however precision can be improved using multiple hashing functions and bitmaps.

5. View Selection for Data Warehouses

For n dimensions there are 2^n different views that we can materialize and use for future queries. Even for a small number of dimensions the possible number of views will typically allow just a small fraction of them to be materialized. Picking the right subset of them is a non-trivial task because of the dependencies among the views that are depicted in the lattice of Figure 20.2. Based on the $\preceq$ partial order, a materialized view may be used to answer queries on other views too. Thus, that we may decide to materialize an infrequently used view if it allows us to answer queries on many other views with acceptable overhead. Furthermore, we may also include a view that is of no interest at query time but allows fast updates on other views as discussed in section 8.

View selection is usually modeled as the problem of finding those views that minimize query response time under some resource constraint. The resource can be the available disk space, or the time that we can spend on maintaining the views when the underlying data gets refreshed, or both. Extensive materialization will lead to diminishing query response gains and unacceptable disk-space and maintenance-time overhead. Fig-

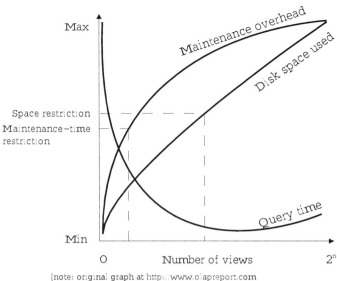

Figure 20.6. View Selection Tradeoffs

ure 20.6 shows query response, disk space and maintenance cost as more views get materialized. The graphs imply that there is an optimum amount of views that deliver query response that is fairly close to what we could have achieved with a full materialization, but with much smaller resource consumption. It is interesting that adding extra views, will in some cases deliver worst query performance than a more conservative partial materialization (OLA 2001b). The reason is that as database size increases a smaller subset of the views will remain in memory buffers. In such cases it might be prudent to maintain fewer views and do aggregations on the fly, when required.

Karloff and Mihail (1999) show that no polynomial time, in the number of views approximation (with respect to query response time), algorithm exists for the view selection problem unless $P \neq NP$. This means that every polynomial-time algorithm will output solutions with query response arbitrary worst compared to the optimal selection. Thus, most studies focus on special cases of practical significance based on heuristics that guide the selection process.

Roussopoulos (1982) first explored the problem of selecting a set of materialized views (with no aggregations) for answering queries under the presence of updates and a global space constraint. Harinarayan et al. (1996) model the view selection problem using the lattice framework. They use the benefit that we gain by materializing a view as a heuristic for choosing among the candidate views. Intuitively the benefit of a view depicts how materializing the view improves the cost of evaluating other non-materialized views including itself. More formally if $\mathcal{V}$ is a set of views that are already materialized, then the benefit $B(v, \mathcal{V})$ of adding a new view v in $\mathcal{V}$ is defined as follows: for each view $u \preceq v$ let $C_\mathcal{V}(u)$ be the cost of computing u from $\mathcal{V}$ and $C_v(u)$ the cost of computing u from v, after v is materialized. Then the benefit of v with respect to $\mathcal{V}$ is given from the formula:

$$B(v, \mathcal{V}) = \sum_{u: u \preceq v \wedge C_v(u) < C_\mathcal{V}(u)} (C_\mathcal{V}(u) - C_v(u))$$

Set $\mathcal{V}$ initially contains the top view from the lattice as there is no other view that can be used to answer queries to that view. Then, a greedy algorithm adds more views to $\mathcal{V}$ until a disk space constraint is fulfilled. The benefit metric can be easily modified to take into account possible knowledge of the query pattern on the views: we simply multiply the costs $C_{v/\mathcal{V}}(u)$ with the expected frequency rate of queries on that view.

The greedy algorithm picks a set of views $\mathcal{V}$ with at least 63% the benefit of an optimal solution. More formally if B_{greedy} is the benefit of k views chosen by the greedy algorithm and B_{opt} is the benefit of an optimal set of k views then $B_{greedy}/B_{opt} \geq 1 - \frac{1}{e}$. Notice that only the benefit of the solution is bound to be relatively close to that of the optimal selection and there is no guarantee on the query response times for the chosen views. A drawback of this algorithm is that it is, in most cases, impractically slow. Each greedy step is quadratic to the number of views and this yields an $O(2^{3d})$ worst case running cost. Shukla, Deshpande and Naughton recently showed (Shukla et al. 1998) that a simple algorithm that picks views according to their size achieves the same benefit bound, for many practical cases.

A follow-up research (Gupta et al. 1997) investigates the combined view and index selection problem under a given space constraint. For materialized views that are very large, having them pre-computed and stored in summary tables is wasteful, unless we index them in order to support fast access to individual records. The authors present a family of algorithms of increasing time complexity that consider also indices (B-trees) for the selected views.

Smith et al. (1998) propose a method of decomposing the views into a hierarchy of view elements that correspond to partial and residual aggregations. Their algorithm picks a non-redundant set of such elements and minimizes query response time. If extra space is available a second algorithm releases the non-redundancy requirement and focuses on selecting a set of elements that minimizes query processing for a target storage bound. The main drawback of this approach is the extremely high complexity of the decomposition step, which makes the algorithms impractical even for a limited number of dimensions/views.

Selecting a view set to materialize and possibly some indices on them, is just the tip of the iceberg. Clearly, query performance is improved as more views are materialized. With the cost of disks constantly dropping, disk storage constraint is no longer the limiting factor in the view selection but the time to refresh the materialized views during updates. More materialization implies a larger maintenance window. This update window is the major data warehouse parameter, constraining over-materialization, as seen in Figure 20.6. The work we discussed so far ignores the maintenance cost of the views. Gupta (1997) provides a theoretical-framework for the general view selection problem and presents polynomial-time algorithms for some special cases, which lower-bound the benefit of the optimal solution. In particular, he considers the case where both query response time and the maintenance cost is to be minimized for a bounded space. This framework is extended in Gupta (1999) to address the problem of selecting views to materialize under the constraint of a given amount of total maintenance time. Baralis et al. (1997) and Yang et al. (1997) present various algorithms for minimizing the response time and the maintenance overhead without any resource constraint. Labio et al. (1997) use an A^* search, similar to that of Roussopoulos (1982) to pick the best set of views when only the maintenance cost is to be minimized. Finally, Theodoratos and Sellis (1997) define the *Data Warehouse configuration problem* as a state-space optimization problem where the maintenance cost of the views needs to be minimized, while all the queries can be answered by the selected views. They propose a genetic algorithm that gradually refines a sub-optimal selection by making local configuration changes.

6. Dynamic Caching of Views

The idea behind view selection is that a fairly small number of views may provide substantial performance boost for many complex analytical queries. In many cases however users submit their queries interactively, i.e they do not have a predetermined set of queries in mind, but rather they are making up their queries on the way based on the feedback they get from the system. This type of analysis often results in querying the data in surprising ways that are not best supported from the materialized views selected by the previous algorithms. In addition decision support queries typically return relatively small results containing few interesting aggregates. Query: "find the total sales for the last 5 years in all stores in NJ" is a fine example of that. Processing this query requires scanning and aggregating lots of detailed records, while the result is just a single value.

Furthermore, as users query patterns and data trends change overtime and as the data warehouse is evolving with respect to new business requirements that continuously emerge, even the most fine-tuned selection of views that we might have obtained at some point, will very quickly become outdated. This means that the selected set of views should be monitored and re-calibrated if query performance is not satisfactory. This task for a complex data warehouse where many users with different profiles submit their queries is rather complicated and time consuming. In addition, the maintenance window, the disk space restrictions and other important operational parameters of the system may also change. For example, an unexpected large volume of daily updates will throw the selected set of views as not update-able unless some of these views are discarded.

An observation from Figure 20.6 is that the selection process is guided with respect to two constrains: the available disk space to store the aggregates and the required maintenance window when the views get refreshed. When optimizing for both space and maintenance-time it is likely that the selection will fully utilize only one of them. In Figure 20.6 the maintenance-time constrain is stricter and does not allow us to add extra views, even-though we can afford the disk space. At query time this extra space is waisted even-though it could be used to temporary stage other aggregates.

To summarize our discussion the following postulates are made:

- ad-hoc analysis is in many cases unpredictable. It might be hard to find a set of views that fits all users.

- query results are often relatively small, as they contain aggregated data

- data is relatively static with only infrequent updates that are happening in predetermined intervals

- a static selection of views can not fully utilize the disk-space and maintenance-time restrictions of the system

These observations suggest that a query result caching architecture is particularly well suited for a data warehouse environment. The cache manager utilizes a dedicated disk space for storing computed aggregates that are further engaged for answering new queries. The problem differs from traditional caching in the following aspects:

- cached aggregates have different sizes and computation costs. An aggregate query may yield a result as big as the top view of the lattice and as small as a single aggregate. Furthermore, it is far more expensive to recompute results of a high level of aggregation since they require scanning and processing lots of detailed records. This implies that LRU or LFU replacement policies are probably not well suited for managing the cache.

- cached results are often not independent. We have already discussed dependencies of materialized aggregates in the lattice framework of Figure 20.2. Drill-down and roll-up queries that are common in OLAP analysis tend to fill the cache with results of different aggregation levels. An effective caching architecture should understand and exploit the dependencies among the aggregates. For instance a more detailed cached result may be used to answer a coarser aggregate query in the future.

- cached results get dirty when the underlying data is modified. Traditional caching typically invalidates dirty objects when data changes at the sources. However, this practice is not efficient for disk resident materialized aggregates with potentially large re-computation overhead. Assuming updates are happening in a periodic fashion we need algorithms that will efficiently maintain the cache with respect to the changes and the dependencies among the cached results. If the whole cache can not be updated within the allowed maintenance period we would have to choose among the cached results and discard some of the aggregates.

- testing whether the cache content can be used to answer a query can be hard as discussed in the next section. Thus, we need practical implementations that will restrict the form of queries admitted in the cache to allow for fast look-ups at query time.

A framework for caching and reusing results in relational database systems has been presented in Sellis (1988). The WATCHMAN cache manager, introduced in Scheuermann et al. (1996) uses replacement and admission techniques tuned for analytical workload and is used in the dynamic caching system of Shim et al. (1999). The cache manager dynamically maintains the content of the cache by deciding whether a newly computed query result should be admitted and if so which already cached result should be evicted to free some space. The admission and replacement algorithms are based on a profit metric, which considers for each result the average rate of reference, its size and its re-computation cost. Kotidis and Roussopoulos (1999) extends this metric to take into account the content of the cache and the maintenance cost of the result when the base data is updated. The intuition is that if two results are cached as views v and u and both have fairly large re-computation costs but $v \preceq u$ then v should probably get a smaller profit score, since it can always be re-computed from the cache using u without accessing the detailed records. Similarly, we might want to credit u with higher profit value based on the number of cached results that can be recomputed from u or be maintained from u in the hybrid maintenance scheme discussed in section 8.

Testing whether a cached result of an arbitrary aggregation query can be used to process a new query is NP-hard (Yan and Larson 1995) in general. To overcome this difficulty dynamic caching systems restrict the form of queries that are maintained in the cache. Kotidis and Roussopoulos (1999) brakes an incoming query into a set of *multidimensional query fragments*. These are SPJ-queries where each selection predicate is of the form *attribute = literal*. The cache manager uses a network of multi-dimensional indices organized in a lattice topology for locating cached results that can answer a new query. Shim et al. (1999) uses a broader class of *canonical* queries. These are aggregation queries of the form:

```
select selection_list, aggr_list
from table_list
where join_condition
and select_condition
group by groupby_list
```

The *join_condition* is a list of equality predicates among attributes of the fact and the dimension tables connected by AND. Predicates in *select_condition* involve only single attributes and literals connected with one of the $<, >, \leq, \geq, =, \neq, between$ operators. Such a query is transform into queries q_1 and q_2 where q_1 contains no selection clause and q_2 returns the same result as q when evaluated over the result set of q_1. q_1 is called the base query and if no joins are present it corresponds to one of the nodes of the data cube lattice.[1].

Deshpande et al. (1998) introduces a different caching architecture where data are organized in the form of *chunks*. These chunks corresponds to partitions created when a uniform multidimensional grid is imposed on-top of the dataset. Answering a query translates into finding the appropriate chunks that contain the requires aggregates. Two potential drawbacks of this approach is that chunking might not work well on skewed datasets and also that it requires the data warehouse tables to be stored using a specialized *chunked file organization* (Sarawagi and Stonebraker 1994).

7. Answering Queries Using Views

In many cases, queries on the data warehouse can be answered using the materialized aggregate views without accessing the detailed records. Given a query q and a view v, checking if all records of q are stored in v (along with possible uninteresting tuples) is reduced to the query containment problem and is well known to be NP-hard (Abiteboul and Duschka 1998, Kolaitis and Vardi 1998, Kolaitis et al. 1998). For the special case of SPJ-views there are algorithms (Larson and Yang 1985, Yang and Larson 1987, Chaudhuri et al. 1995, Levy et al. 1995, Srivastava et al. 1996) that can be used to optimize the execution of user queries against the views.

The data cube views have a very restricted form that makes the problem somehow easier, since they contain no joins and selections but only groupings over dimension's keys. The most common way to use such a view is to roll up by grouping on additional columns and add possibly some selection filters. Even in this case we should make sure that the aggregate functions used in the view can be safely rolled up for answering the query. For example a view that computes both *count()* and *avg()* can be rolled up for computing *sum()* for a query but for other statistical functions rolling up the aggregates may not be possible.

In the general case the views that have been materialized in the data warehouse may contain selections and joins between the fact and some of the dimension tables. For example assume that our star schema of Figure 20.1 has been extended to include the time dimension and a materialized view v stores the total sales by quarter for each part:

```
v: create view v as
   select part.partkey, time.quarter, time.year, sum(sales)
   as total_sales
   from F, part, time
   where F.partkey = part.partkey and
   F.timekey = time.timekey
   group by part.partkey, time.quarter, time.year
```

Assume now a query q that requests the total sales per part for a specific category of parts (e.g. "auto-parts") for the year 2000:

q: ```
select part.partkey, sum(sales) as total_sales
from F, part, time
where F.partkey = part.partkey and
F.timekey = time.timekey and
part.category = ''auto-parts'' and
time.year=2000 group by part.partkey
```

in the presence of $v$ the query can be rewritten as following:

$q(v)$: ```
select v.partkey,sum(v.total_sales) as total_sales
from v, part,
where v.partkey = part.partkey
and v.year = 2000
and part.category = ''auto-parts''
group by v.partkey
```

In this rewriting we avoid joining with the fact and the time tables since the view contains the necessary information to roll up from quarters to years, however we still have to do a join with the dimension table for parts to get the category attribute. Such a join is called a joinback (Bello et al. 1998) and is made possible because we know that the *partkey* attribute in the view is a primary key on that table and thus no information is lost when using the view. We would like to point out here that any rewriting functionality should be integrated within the query optimizer that knows about integrity constrains, value distributions and possible indexes on the views and the tables. For example it might be faster to query the fact table through an appropriate index than rewriting a query to use a large unindexed materialized view.

The hierarchical and functional relationships of attributes stored in the dimension tables should be taken into account when doing roll ups or joinbacks. For the previous example we used the fact that *partkey* is the primary key for the part dimension to infer the missing category attribute from the part table. Functional dependency information is also necessary when rolling up aggregates. For example if city uniquely identifies a state in the customer table it is valid to roll up sum of sales by city to sum of sales by state. Sometimes domain knowledge is necessary to correctly interpret the results. For example if we do not sell to all cities in NJ, then the previous aggregates will refer to a subset of cities in the NJ. Similarly if a part belongs to multiple categories (this requires some modifications in our schema) then rolling up the sum of sales per category to compute the overall sales gives an incorrect result.

8. Updating the Views

As changes are made to the base tables of the data warehouse, the materialized aggregate views must also be updated to reflect the new state of the detailed records. The data warehouse tables are themselves views of multiple external data sources that periodically ship their updates to the repository. Changes are not made immediately but are deferred and applied in large batches during a down-time period. This not only allows the data warehouse to provide a consistent snapshot during analysis but makes maintenance more efficient using bulk load and update techniques. In the database literature there is an abundance of work related to view maintenance[2]. Frequently, only a small part of the view changes in response to changes in the base relations, or similarly few of the changes affect the view. In such cases it might be faster to compute only the changes in the view to update its materialization. This is called incremental view maintenance and uses a delta paradigm to represent changes of the base relations that are then used to update the view(s). This implies that we have access to the set of changes that have happened to the base data. Unfortunately this is not always the case, especially for views over distributed data sources that often don't use relational databases. Furthermore, even if the data sources are willing to report changes, querying them during maintenance may be prohibitively expensive (Quass et al. 1996).

For our discussion, we assume that all aggregate views are updated with respect to changes in the fact table, after the fact table itself has been updated. Different policies are implemented, depending on the types of updates and the properties of the aggregate functions that are computed by the views. One can always recompute the aggregate views from scratch, using techniques described in section 4, every time the fact and/or the dimension tables are modified. This approach ultimately leads to unacceptable performance as the size of the data warehouse increases over time.

Table 20.1. Materialized View v on *partkey*

partkey	total_sales	max_sale
100	28	4
101	11	5
102	5	5
103	17	8
104	22	8

Table 20.2. F^+: insertion in the fact table

partkey	suppkey	custkey	amount
102	...	...	7
100	...	...	2
103	...	...	1
102	...	...	3
106	...	...	9
102	...	...	5

Recently we have seen incremental update algorithms (Gupta et al. 1993, Griffin and Libkin 1995, Jagadish et al. 1995, Quass 1996, Mumick et al. 1997, Roussopoulos et al. 1997, Kotidis and Roussopoulos 1998) that handle views with aggregations. These algorithms avoid full recomputation of the views by using appropriate maintenance expressions in response to changes in the dataset. Often such changes involve only insertions, however the update process should be able to handle both insertions and deletions. The delta paradigm divides these changes into two sets F^+ and F^-. Set F^+ contains all new tuples inserted in the fact table, while F^- all deleted records. Updates (modifications) are expressed in this framework as a deletion followed by an insertion. For a materialized view v our goal is to create an expression that will correctly reflect changes F^+ and F^- to v. This may or may not be possible depending on the aggregate functions that are computed by the view and the type of changes that we want to apply. All distributive functions like *count()*, *sum()*, *max()* can be refreshed incrementally when only insertions are allowed. For example assume that the following view is materialized in the data warehouse:

```
v: create view v as
   select partkey,sum(amount) as total_sales, max(amount)
   as max_sale from F
   group by partkey
```

Table 20.3. Summary delta table $delta_v^+$

partkey	total_sales	max_sale
100	2	2
102	15	7
103	1	1
106	9	9

Table 20.4. Updated View

partkey	total_sales	max_sale
100	30	4
101	11	5
102	20	7
103	18	8
104	22	8
106	9	9

The view computes the maximum and total sales per part from the transactions stored in the fact table F. Tables 20.1, 20.2 provide a snapshot for this view along with a set of insertions F^+ that we want to apply. Both $sum()$ and $max()$ allow further aggregation based on the new data. Based on this property we can create a summary delta table (Mumick et al. 1997, Roussopoulos et al. 1997) for this view as follows:

$delta_v^+$: create view $delta_v^+$ as
    select partkey,sum(amount) as total_sales, max(amount)
    as max_sale from F^+
    group by partkey

$delta_v^+$ is shown in Table 20.3 and applies the definition of the view on the new records only. We can now *merge* the old snapshot of the view with the newly computed aggregates in the following way:

1. if for a record in $delta_v^+$ there is no a record in v with the same *partkey* value we copy the record to the view. This is the case for the new entry with *partkey*=106.

2. if there is already an entry in v, for the $sum()$ function we add the two aggregates and update the value stored in v, while for the $max()$ function we keep the maximum of the two.

Table 20.4 shows the refreshed view after changes in F^+ have been applied. In order to implement the merge procedure we can open a cursor in $delta_v^+$ and check each record against the view. For this process to work efficiently there should be an index on the primary key *partkey* of the view. However, for large batches of insertions checking the index over and over again will yield a substantial overhead. If on the other hand group bys are computed using a sort-based algorithm we could take advantage of matching sort orders (like the case of Tables 20.1, 20.3) and simply merge the aggregates by sequentially scanning the records of both tables (Roussopoulos et al. 1997).

Table 20.5. Update policies for aggregate views

policy	description
re-computation	recompute from F
incremental	incrementally from $delta_v^+$
hybrid	update $u : v \preceq u$, recompute v from u

In the presence of multiple views we can use optimizations similar to those employed for computing the views at the fist place. For example we can compute $delta_v^+$ from the summary-delta table $delta_u^+$ of another view u if $v \preceq u$ and the delta table of u is smaller that F^+. Alternatively, if u has already been updated, we can re-compute view v from u and avoid the overhead of materializing its summary-delta table. This implies a hybrid maintenance scheme, in which some of the views are updated incrementally from the deltas, others are recomputed from F and others are re-computed from other views. The decision is based on a cost-based optimization that takes into account the time that we can spend in maintaining the views, the disk space available for temporal results and the data structures that we use for storing and indexing the views and the fact table. Kotidis and Roussopoulos (1999) describes such a hybrid update algorithm.

Table 20.5 summarizes the alternative ways that view v can be maintained with respect to changes in F^+. We would like to stretch out that incremental updates are not a panacea for the view maintenance problem. Complex maintenance expressions over unindexed data/deltas may result to thrashing and void any benefits from incremental computations of the aggregate functions as shown in Kotidis and Roussopoulos (1998).

Incremental and hybrid update algorithms can be used for algebraic functions too, however their efficiency is questionable if the function requires significant state information to describe a sub-aggregate. For example $avg()$ can be maintained using partial $sum()$ and $count()$ information, but for $maxN()$ the implementation will be very inefficient.

If we try to extend the delta-paradigm for deletions we will find that the results do not hold for all distributive functions. For instance if F^- contains just a single record with $partkey=104$ and $amount=8$ this tells us that the entry with the maximum sales for this part is deleted from the fact table. However, we have no way of knowing what the new maximum sale for this part is without looking back at the fact table. For the $sum()$ function the situation is slightly better. Each deleted record can be expressed as an insertion where the value of the measure is negated. This yields correct output from the merging step if we remove from v

records whose *sum()* aggregate is zero, as they correspond to entries that contained some value but their individual records are now deleted from F. Thus, we can treat zero as a special value that indicates when all tuples in a group have been deleted. One however should restrain from using such deductions because they are highly application depended and can lead to unpredictable results. In order to consistently maintain the *sum()* aggregate in the presence of deletions we can include a *count()* function in the view definition. When *count()* becomes zero, we can safely deduct that the record can be removed from the view.

Mumick et al. (1997) defines a function as self-maintenable if its new value can be computed solely from the old value and the changes to the base records with respect to new updates. Functions can be self-maintenable with respect to insertions F^+, deletions F^- or both. All distributive functions are by definition self-maintenable with respect to insertions and all self-maintenable functions must be distributive. Adding extra information can make a distributive function like *sum()* self-maintenable with respect to deletions but this is not always the case. For instance if *count()*> 0 we still need to look in F to find the new value for *max()* and *min()* after the maximum or the minimum value is deleted.

9. Final Comments

Materialized views and their implications have been recently rediscovered for the content of OLAP and data warehousing. A flurry of papers has been generated on how views can be used to accelerate ad-hoc computations over massive datasets. Picking and materializing the right set of views, with respect to the workload and the disk space and maintenance time constraints has also received considerable attention. Recent approaches seem to agree that their use is best exemplified in a dynamic environment, in which a materialized set of views is reconciled with each new query. Substantial effort has also been given for optimizing their computation and maintenance tasks. From the commercial side materialized views are eventually getting the attention they deserve in products like Oracle 8.1, IBM DB2 (Zaharioudakis et al. 2000) and HP Intelligent Warehouse. Materialized views with their versatility and potential introduce new challenges with interesting research and engineering questions. Taking a leap from the centralized data warehouse model and moving in a distributed and possibly mobile world, we are faced with new challenges in view management. Thus, the excitement about materialized views and their applications is expected to continue for the foreseeable future.

Bibliography

S. Abiteboul and O. M. Duschka. Complexity of Answering Queries Using Materialized Views. In *Proceedings of the Seventeenth ACM SIGACT-SIGMOD-SIGART Symposium on Principles of Database Systems*, pages 254–263, Seattle, Washington, June 1998.

S. Agrawal, R. Agrawal, P. Deshpande, A. Gupta, J. Naughton, R. Ramakrishnan, and S. Sarawagi. On the Computation of Multidimensional Aggregates. In *Proc. of 22nd VLDB conference*, pages 506–521, Bombay, India, August 1996.

E. Baralis, S. Paraboschi, and E. Teniente. Materialized View Selection in a Multidimensional Database. In *Proc. of the 23th International Conference on VLDB*, pages 156–165, Athens, Greece, August 1997.

R. Barquin and H. Edelstein, editors. *Building, Using and Managing the Data Warehouse*. The Data Warehousing Institute Series. Prentice Hall PTR, 1997.

Randall G. Bello, Karl Dias, Alan Downing, James Feenan Jr., William D. Norcott, Harry Sun, Andrew Witkowski, and Mohamed Ziauddin. Materialized Views in Oracle. In *Proceedings of 24rd International Conference on Very Large Data Bases*, pages 659–664, New York City, New York, August 1998.

Kevin S. Beyer and Raghu Ramakrishnan. Bottom-Up Computation of Sparse and Iceberg CUBEs. In *Proceedings ACM SIGMOD International Conference on Management of Data*, pages 359–370, Philadephia, Pennsylvania, June 1999.

S. Chaudhuri and U. Dayal. An Overview of Data Warehousing and OLAP Technology. *SIGMOD Record*, 26(1), September 1997.

S. Chaudhuri, R.i Krishnamurthy, S. Potamianos, and K. Shim. Optimizing Queries with Materialized Views. In *Proceedings of the Eleventh International Conference on Data Engineering*, pages 190–200, Taipei, Taiwan, March 1995.

G. Colliat. OLAP, Relational and Multidimensional Database Systems. *SIGMOD Record*, 25(4):64–69, Sept 1996.

P. M. Deshpande, K. Ramasamy, A. Shukla, and J.F. Naughton. Caching Multidimensional Queries Using Chunks. In *Proceedings of the ACM SIGMOD*, pages 259–270, Seattle, Washington, June 1998.

P. Flajolet and G. N. Martin. Probabilistic Counting Algorithms for Database Applications. *Journal of Computer and System Sciences*, 31(2):182–209, 1985.

G. Graefe. Query Evaluation Techniques for Large Databases. *ACM Computing Surveys*, 25(2):73–170, 1993.

J. Gray, A. Bosworth, A. Layman, and H. Piramish. Data Cube: A Relational Aggregation Operator Generalizing Group-By, Cross-Tab, and Sub-Totals. In *Proc. of the 12th ICDE*, pages 152–159, New Orleans, February 1996. IEEE.

J. Gray, P. Sundaresan, S. Englert, K. Baclawski, and P. Weiberger. Quickly Generating Billion-Record Synthetic Databases. In *Proc. of the ACM SIGMOD*, pages 243–252, Minneapolis, May 1994.

T. Griffin and L. Libkin. Incremental Maintenance of Views with Duplicates. In *Proceedings of the ACM SIGMOD*, pages 328–339, San Jose, CA, May 1995.

A. Gupta, I.S. Mumick, and V.S. Subrahmanian. Maintaining Views Incrementally. In *Proceedings of the ACM SIGMOD Conference*, pages 157–166, Washington, D.C., May 1993.

H. Gupta. Selections of Views to Materialize in a Data Warehouse. In *Proceedings of ICDT*, pages 98–112, Delphi, January 1997.

H. Gupta. Selection of Views to Materialize Under a Maintenance Cost Constraint. In *Proceedings of ICDT*, Jerusalam, Israel, January 1999.

H. Gupta, V. Harinarayan, A. Rajaraman, and J. Ullman. Index Selection for OLAP. In *Proceedings of ICDE*, pages 208–219, Burmingham, UK, April 1997.

V. Harinarayan, A. Rajaraman, and J. Ullman. Implementing Data Cubes Efficiently. In *Proc. of ACM SIGMOD*, pages 205–216, Montreal, Canada, June 1996.

H. Jagadish, I. Mumick, and A. Silberschatz. View Maintenance Issues in the Chronicle Data Model. In *Proceedings of PODS*, pages 113–124, San Jose, CA, 1995.

H. J. Karloff and M. Mihail. On the Complexity of the View-Selection Problem. In *Proceedings of the 18th ACM SIGACT-SIGMOD-SIGART Symposium on Principles of Database Systems*, pages 167–173, Philadelphia, Pennsylvania, May 1999.

R. Kimball. *The Data Warehouse Toolkit*. John Wiley & Sons, 1996.

P. G. Kolaitis, D. L. Martin, and M. N. Thakur. On the Complexity of the Containment Problem for Conjunctive Queries with Built-in Predicates. In *Proceedings of the Seventeenth ACM SIGACT-SIGMOD-SIGART Symposium on Principles of Database Systems*, pages 197–204, Seattle, Washington, June 1998.

P. G. Kolaitis and M. Y. Vardi. Conjunctive-Query Containment and Constraint Satisfaction. In *Proceedings of the Seventeenth ACM SIGACT-SIGMOD-SIGART Symposium on Principles of Database Systems*, pages 205–213, Seattle, Washington, June 1998. ACM Press.

Y. Kotidis and N. Roussopoulos. An Alternative Storage Organization for ROLAP Aggregate Views Based on Cubetrees. In *Proceedings of the ACM SIGMOD International Conference on Management of Data*, pages 249–258, Seattle, Washington, June 1998.

Yannis Kotidis and Nick Roussopoulos. DynaMat: A Dynamic View Management System for Data Warehouses. In *Proceedings of the ACM SIGMOD International Conference on Management of Data*, pages 371–382, Philadelphia, Pennsylvania, June 1999.

W. Labio, D. Quass, and B. Adelberg. Physical Database Design for Data Warehouses. In *Proceedings of ICDE*, pages 277–288, Birmingham, U.K., April 1997.

P.-Å. Larson and H. Z. Yang. Computing Queries from Derived Relations. In *Proceedings of the 11th VLDB Conference*, pages 259–269, Stockholm, Sweden, 1985.

A. Y. Levy, A. O. Mendelzon, Y. Sagiv, and D. Srivastava. Answering Queries Using Views. In *Proceedings of the Fourteenth ACM SIGACT-SIGMOD-SIGART Symposium on Principles of Database Systems*, pages 95–104, San Jose, California, May 1995.

Jianzhong Li, Doron Rotem, and Jaideep Srivastava. Aggregation Algorithms for Very Large Compressed Data Warehouses. In *Proceedings of 25th International Conference on Very Large Data Bases*, pages 651–662, Edinburgh, Scotland, September 1999.

I. S. Mumick, D. Quass, and B. S. Mumick. Maintenance of Data Cubes and Summary Tables in a Warehouse. In *Proceedings of the ACM SIGMOD Conference*, pages 100–111, Tucson, Arizona, May 1997.

Seigo Muto and Masaru Kitsuregawa. A Dynamic Load Balancing Strategy for Parallel Datacube Computation. In *DOLAP '99*, pages 67–72, Kansas City, Missouri, November 1999.

The OLAP Council.
http://www.olapcouncil.org, 2001a.

The OLAP Report.
http://www.olapreport.com, 2001b.

D. Quass. Maintenance Expressions for Views with Aggregation. In *Proceedings of VIEWS 96*, pages 110–118, Montral, Canada, June 1996.

D. Quass, A. Gupta, I.S. Mumick, and J. Widom. Making Views Self-Maintainable for Data Warehousing. In *Proceedings of the Fourth International Conference on Parallel and Distributed Information Systems*, pages 158–169, Miami Beach, Florida, December 1996.

K.A. Ross and D. Srivastava. Fast Computation of Sparse Datacubes. In *Proceedings of the 23th VLDB Conference*, pages 116–125, Athens, Greece, Auguost 1997.

N. Roussopoulos. View Indexing in Relational Databases. *ACM Transactions on Database Systems*, 7(2):258–290, June 1982.

N. Roussopoulos, Y. Kotidis, and M. Roussopoulos. Cubetree: Organization of and Bulk Incremental Updates on the Data Cube. In *Proceedings of the ACM SIGMOD International Conference on Management of Data*, pages 89–99, Tucson, Arizona, May 1997.

S. Sarawagi and M. Stonebraker. Efficient Organization of Large Multidimensional Arr ays. In *Proceedings of ICDE*, pages 328–336, Houston, Texas, 1994.

P. Scheuermann, J. Shim, and R. Vingralek. WATCHMAN: A Data Warehouse Intelligent Cache Manager. In *Proceedings of the 22th VLDB Conference*, pages 51–62, Bombay, India, September 1996.

T. K. Sellis. Intelligent Caching and Indexing Techniques for Relational Database Systems. *Information Systems*, 13(2):175–185, 1988.

Junho Shim, Peter Scheuermann, and Radek Vingralek. Dynamic Caching of Query Results for Decision Support Systems. In *SSDBM*, pages 254–263, Cleaveland, Ohio, July 1999.

A. Shukla, P. Deshpande, J. Naughton, and K. Ramasamy. Storage Estimation for Multidimensional Aggregates in the Presense of Hierarchies. In *Proc. of VLDB*, pages 522–531, Bombay, India, August 1996.

A. Shukla, P.M. Deshpande, and J.F. Naughton. Materialized View Selection for Multidimensional Datasets. In *Proceedings of the 24th VLDB Conference*, pages 488–499, New York City, New York, August 1998.

J. R. Smith, C. Li, V. Castelli, and A. Jhingran. Dynamic Assembly of Views in Data Cubes. In *PODS*, pages 274–283, Seattle, Washington, June 1998.

D. Srivastava, S. Dar, H.V. Jagadish, and A. Y. Levy. Answering Queries with Aggregation Using Views. In *Proceedings of 22th International Conference on Very Large Data Bases*, pages 318–329, Mumbai (Bombay), India, September 1996.

D. Theodoratos and T. Sellis. Data Warehouse Configuration. In *Proc. of the 23th International Conference on VLDB*, pages 126–135, Athens, Greece, August 1997.

Weipeng P. Yan and Per-Åke Larson. Eager Aggregation and Lazy Aggregation. In *Proceedings of 21th International Conference on Very Large Data Bases*, pages 345–357, Zurich, Switzerland, September 1995.

H. Z. Yang and Per-Åke Larson. Query Transformation for PSJ-Queries. In *Proceedings of 13th International Conference on Very Large Data Bases*, pages 245–254, Brighton, England, September 1987.

J. Yang, K. Karlapalem, and Q. Li. Algorithms for Materialized View Design in Data Warehousing Environment. In *Proceedings of the 23th VLDB Conference*, pages 136–145, Athens, Greece, Augoust 1997.

Markos Zaharioudakis, Roberta Cochrane, George Lapis, Hamid Pirahesh, and Monica Urata. Answering Complex SQL Queries Using Automatic Summary Tables. In *Proceedings of ACM SIGMOD International Conference on Management of Data*, pages 105–116, Dallas, Texas, May 2000.

Y. Zhao, P.M. Deshpande, and J.F. Naughton. An Array-Based Algorithm for Simultaneous Multidimensional Aggregates. In *Proceedings of the ACM SIGMOD Conference*, pages 159–170, Tucson, Arizona, May 1997.

Chapter 21

SEMISTRUCTURED DATA AND XML

Dan Suciu
Computer Science Department, University of Washington, Seattle, WA 98195, USA
suciu@cs.washington.edu

Abstract The distinguishing feature of semistructured data is that the schema is embedded with the data. The main challenge is to cope with the additional flexibility without sacrificing efficiency. We introduce semistructured data by presenting a syntax and describing the datamodel. We discuss some query languages designed for semistructured data and address some systems issues, such as storage and XML compression.

Keywords: Massive data sets, Telecommunications, Telephone billing systems, Datastore, I/O

1. The Semistructured Data Model

There exists today an increasing amount of data that cannot be easily managed in a relational or object-oriented database. Such data includes biological data (ACeDB (Thierry-Mieg and Durbin 1992), EMBL(Higgins et al. 1992)), data stored as structured documents (SGML (Goldfarb 1990)), data resulting from the integration of heterogeneous sources, or just data fetched from the Web. Today, such data is processed directly by applications, not by general purpose data management systems. The semistructured data model is the database community's attempt to apply traditional database management techniques to such data. While initially considered a niche application, after the release in 1998 of the XML 1.0 Recommendation by the World Wide Web Consortium (Consortium 1998) the demand for semistructured data management techniques and systems has increased dramatically. Today all major database vendors are providing XML tools and interfaces, and new startups have emerged offering a large array of XML management products.

The distinguishing feature of semistructured data is that the schema is embedded with the data. We say that the data is "schema-less" or "self-describing". By contrast, in relational data the schema is kept separately from the data, in the catalog. In semistructured data every data item carries, at least conceptually, the schema with it. Even when a separate schema is given, such as an XML-Schema for an XML document, the data remains self-describing and the schema is only used for validation purposes. As a consequence the data could become arbitrarily irregular, since there is no discipline one has to obey when creating the data. Such data is easiest represented in an ASCII format, e.g. as XML files or in some other syntax, but, as we shall see, can be stored reasonably efficiently in other formats too.

The main challenge in semistructured data is to cope with the additional flexibility without sacrificing efficiency. In an extreme case, a semistructured data may have any structure, no matter how irregular: a system must be able to process it without making any a priori assumption on its structure, and one needs to be willing to pay a price in performance. In practice, however, the data is often more regular, and users expect to see better performance in such cases. At the other extreme the data may have the regular structure of a, say, relational database: ideally a system's performance should match in this case that of relational databases. Coping with the flexibility in structure while at the same time providing decent runtime performance is one of the main challenges in semistructured data.

We introduce semistructured data by presenting a syntax and describing the datamodel. We discuss next some query languages designed for semistructured data. Finally we address some systems issues: storage of semistructured data and XML compression.

1.1. Syntax

While XML is by far the most popular concrete syntax for semistructured data, we introduce semistructured data using an alternative, abstract syntax. We do so because the abstract syntax corresponds to a simple and elegant data model; by contrast, as we shall see, XML has some features that do not map easily to the model.

In any strongly type system we have to describe the type of a data instance first before we can create instances of that data. For example, one defines the following type in C:

```
struct { char name[20];   int tel;   char email[40]; }
```

then create objects of that type, like the following record:

```
{"Alan", 5557786, "agb@abc.com"}
```

SEMISTRUCTURED DATA AND XML

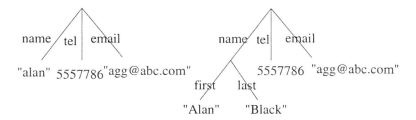

Figure 21.1. Tree representations of simple semistructured data instances.

In semistructured data the type is described together with the data. In the syntax we are using throughout this chapter, the record above becomes:

    {name: "Alan", tel: 5557786, email: "agb@abc.com"}

Data is described as a set of label-value pairs, such as name: "Alan". Quite complex data instances can be expressed that way, as we illustrate in the following.

The values may be strings, like "Alan", integers, like 5557786, or other structures as in:

    {name: {first: "Alan", last: "Black"},
     tel: 5557786,
     email: "agb@abc.com"
    }

Notice that the same data can also be represented as a tree with labeled edges, see Figure 21.1.

Duplicates labels are allowed in this syntax, as in:

    {name: "Alan", tel: 5557786, tel: 5558762 }

Representing Relational Databases. Semistructured data can easily represent regular (i.e. strongly typed) data. We illustrate here with the case of relational data. Every relational database has a schema: for example r1(a,b,c) r2(c,d). In these expressions, r1 and r2 are the names of two relations, and a,b,c and c,d are the column names of the two relations. In practice we also have to specify the types of those columns. An instance of such a schema consists of two tables whose data conforms to this specification, for example:

r1:	a	b	c
	a1	b1	c1
	a2	b2	c2

r2:	c	d
	c2	d2
	c3	d3
	c4	d4

```
{people:
   {person:
        {name: "Alan", phone: 5557786, email: "agg@abc.com"},
    person:
        {name: {first: "Sarah", last: "Green"},
         phone: 5556877,
         email: "sara@math.xyz.edu"
        },
    person:
        {name: "Fred", phone: 5556312, phone: 5554321,
      email: "fred@dot.com", height: 183}
    person:
        {name: "John", phone :5551414, email: "jhh@abc.com", height: 188}
    person:
        {name: {first: "Susan", last: "Greer"},
         email: "susan@cs.xyz.edu"
        }
   }
}
```

(a)

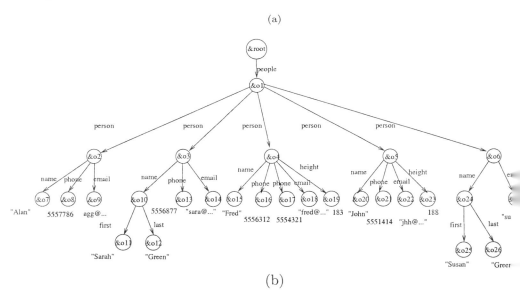

(b)

Figure 21.2. An example of semistructured data. Textual representation (a) and tree representation (b).

The same data can be represented as semistructured data:

```
{db:
   { r1: {row: { a: a1, b: b1,  c: c1},
          row: { a: a2, b: b2,  c: c2}
          },
     r2: {row: { c: c2,  d: d2},
          row: { c: c3,  d: d3},
          row: { c: c4,  d: d4}
          }
   }
}
```

Notice how some schema components, such as the attribute names a, b, c, d, are now replicated at each data item.

Characteristics of semistructured data. The above illustrates that one can easily describe sets of tuples. But one can easily describe sets of heterogeneous elements as well. The novelty in semistructured data is precisely its ability to accommodate variations in structure. Fig. 21.2 illustrates an example of a semistructured data instance consisting of a heterogeneous collection of person objects, both written in our syntax and represented as a tree.

Examining this example, we find several characteristics that distinguishes semistructured data from other data models:

- missing fields; e.g. all persons in Fig. 21.2 have a phone attribute except for the last one, where this field is missing.

- additional fields; e.g. some persons have the additional height attribute[1]

- multiple fields; e.g. the third person has two phone attributes.

- fields with different types in different objects; e.g. the name is a string in some objects and a complex object in other objects.

- heterogeneous collections; e.g. people is a heterogeneous collection of persons.

1.2. The object exchange model, OEM

The semistructured data model in use is the *Object Exchange Model* (OEM), originally defined for Tsimmis, a system for integrating heterogeneous data sources (Garcia-Molina et al. 1997). Several variants of the model exists: we follow here the variant defined for Lore (Abiteboul et al. 1997).

An OEM database consists of a collection of objects, each carrying a unique object identifier (*oid*). There are two kinds of objects: *atomic* objects and *complex* objects. Each object has a *value*, as follows. The value of an atomic object is of a basic atomic type, e.g. `integer`, `real`, `string`, `image`. The value of a complex object is a set of *(label, oid)* pairs, where *label* is a string and *oid* is an object identifier. One or more objects are distinguished entry points in the database and have externally known names. In this chapter we assume a unique distinguished entry point, and refer to it as the *root*.

Equivalently an OEM database is a graph, in which objects form the nodes and there is an edge from $o1$ to $o2$ if $o1$ is a complex object whose value includes a pair $(l, o2)$. The graph is labeled, with labels both on edges and some nodes: l is the label associated to the edge $(o1, o2)$, while every atomic object o is labeled with its atomic value. The graph is also rooted, i.e. it has a distinguished node called the root.

We extend the simple syntax in Sec. 1.1 to represent object identifiers arbitrary graphs, by allowing an object identifier to optionally precede an object's value. Object identifiers are written with a leading & sign. For example the data in Fig. 21.2 can be written as:

```
&root
  {people:
     &o1 {person:
          &o2 {name: &o7 "Alan", phone: &o8 5557786, email: &o9 "agg@abc.com"},
          person:
          &o3 {name: &o10 {first: &o11 "Sarah", last: &o12 "Green"},
               phone: &o13 5556877,
               email: &o14 "sara@math.xyz.edu"
              },
          person:
          &o4 {name: &o15 "Fred", phone: &o16 5556312, phone: &o17 5554321,
               email: &o18 "fred@dot.com", height: &o19 183}
          person:
          &o5 {name: &o20 "John", phone: &o21 5551414,
               email: &o22 "jhh@abc.com", height: &o23 188}
          person:
          &o6 {name: &o4 {first: &o25 "Susan", last: &o26 "Greer"},
               email: &o27 "susan@cs.xyz.edu"
              }
         }
  }
```

Notice that even atomic objects receive oids.

Graphs with cycles are easily represented by referring to other oids. For example Fig. 21.3 illustrates an OEM graph and its textual representation. The data represents a collection of three persons, in which `Mary` has two children, `John` and `Jane`. . We can see that oids are used

SEMISTRUCTURED DATA AND XML 749

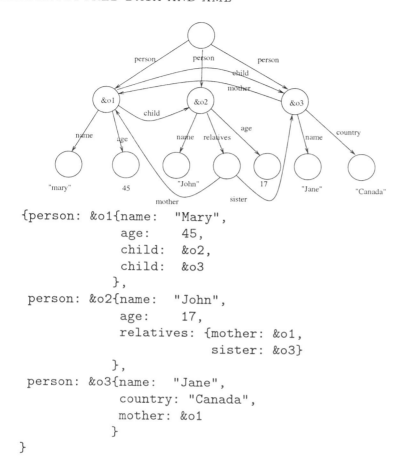

```
{person: &o1{name:    "Mary",
             age:     45,
             child:   &o2,
             child:   &o3
            },
 person: &o2{name:    "John",
             age:     17,
             relatives: {mother: &o1,
                         sister: &o3}
            },
 person: &o3{name:    "Jane",
             country: "Canada",
             mother: &o1
            }
}
```

Figure 21.3. An OEM graph and its textual representation.

in two ways. When they are followed by a value, like person: &o1{...} then they are considered to be a *definition*. When they are not followed by a value, like in mother: &o1 they are considered a reference. Oid definitions must be unique, and every reference must be to some defined oid.

Notice that defining an oid is optional: when it is missing the system will assign one automatically. For example {a:&o1{b:&o2 5}} and {a:{b:5}} denote isomorphic graphs, as does {a:&o1{b:5}}.

1.2.1   Object equality.   OEM defines a shallow equality operator on objects: two objects are equal if and only if their object identifiers are equal. As a consequence OEM collections behave like bags, not sets. For example the object:

{a: {b: 3}, a: {b: 3}}

cannot be simplified to

{a: {b: 3}}

because the system assigns a distinct oid to each {b: 3}, making duplicate elimination impossible.

This behavior is suitable from many applications, but some require however a deep equality operator and duplicate elimination. For semistructured data, deep equality cannot be defined to be graph isomorphism: we want the object $\{a:\{b:3\}, a:\{b:3\}\}$ to be equal to $\{a:\{b:3\}\}$, although the two tree are not isomorphic. We give the definition below, from Buneman et al. (2000).

Definition 1 Value-equality *for trees,* $t \equiv t'$, *is defined as follows. When t and t' are atomic values,* $t \equiv t'$ *if those values are the same. If* $t = \{l_1 : t_1, \ldots, l_m : t_m\}$ *and* $t' = \{l'_1 : t'_1, \ldots, l'_n : t'_n\}$, *then* $t \equiv t'$ *if the following two conditions hold:*

- *for each* $i \in \{1, \ldots, m\}$, *there exists* $j \in \{1, \ldots, n\}$ *s.t.* $l_i = l'_j$ *and* $t_i \equiv t'_j$, *and*

- *for each* $j \in \{1, \ldots, n\}$, *there exists* $i \in \{1, \ldots, m\}$ *s.t.* $l_i = l'_j$ *and* $t_i \equiv t'_j$.

The definition only works for tree objects. For arbitrary graphs one can use as deep equality *graph bisimulation* (Buneman et al. 2000).

2. XML

XML 1.0 is the recommendation of the World Wide Web Consortium (W3C) to complement HTML for data exchange on the Web. While XML has many uses, we discuss here only its application as a *data exchange format*. For such applications XML serves as a concrete syntax for semistructured data and, in this role, XML becomes a universal data exchange format for the Web. For example in a business to business transaction the data being exchanged is most likely to be in XML format.

XML is defined as a certain subset of SGML (Goldfarb 1990) and was designed by choosing a subset of SGML constructs that were simple and sufficient for document markup on the Web. In this section we will describe only those constructs that are useful in its role as a data exchange format, and omit several features that are only relevant to

```
<people>
  <person>
    <name> Alan </name>
    <phone> 5557786 </phone>
    <email> "agg@abc.com" </email>
  </person>
  <person>
    <name> <first> Sara </first> <last> Green </last> </name>
    <phone> 5556877 </phone>
    <email> sara@math.xyz.edu </email>
  </person>
  <person>
    <name> Fred </name>
    <phone> 5556312 </phone>
    <phone> 5554321 </phone>
    <email> fred@dot.com </email>
    <height> 183 </height>
  </person>
  <person>
    <name> John </name>
    <phone> 5551414 </phone>
    <email> jhl@abc.com </email>
    <height> 188 </height>
  </person>
  <person>
    <name> <first> Susan </first>
           <last> Greer </last>
    </name>
    <email> susan@cs.xyz.edu </email>
  </person>
</people>
```

Figure 21.4. Example of XML Data.

document markup. We refer the interested reader to the original XML reference (Consortium 1998) or to one of the numerous XML textbooks for the additional text oriented features.

2.1. XML syntax

XML Basics. Consider the semistructured data instance in Fig. 21.2. Its equivalent representation in XML is given in Fig. 21.4. Here `<person>`, `<phone>`, etc are called *tags*. Tags must appear in pairs: a *begin tag* (e.g. `<person>`) and an *end tag* (e.g. `</person>`). Tags are enclosed in angle brackets, while end tags start with a `/`. The text enclosed between a begin tag and an end tag, including the two tags, is called *an element*, while the text without the two enclosing tags is called the *element content*. We call the *element type* the name of the tag; our example has elements of types `person`, `name`, `phone`, `email`, etc. Elements may be nested: for example the first element of type `person` contains three subelements. Finally, the free text inside elements is called PCDATA, from *parsed character data*.

As an abbreviation, an empty element of the form `<a> </a>` can be written as `<a/>`.

XML Attributes. XML allows us to associate *attributes* with elements. Here we have to be a little careful with terminology. Attributes, in the sense used in relational databases, have so far been expressed in XML by tags. XML uses the term *attribute* for what is sometimes called a *property* in data models. In XML, attributes are defined as *(name,value)* pairs. In the example below attributes are used to specify the language or the currency:

```
<product>
   <name language="French">trompette a six trous</name>
   <price currency="Euro"> 420.12 </price>
   <address format="XLB56" language="French">
       <street>31 rue Croix-Bosset</street>
       <zip>92310</zip> <city>Svres</city>
       <country>France</country>
   </address>
</product>
```

Here the `name` element has an attribute `language` whose value is the string `"French"`. As with tags, users may define arbitrary attributes, like `language`, `currency`, and `format` above. The value of an attribute is always a string, and must be enclosed in quotation marks.

Attributes are a consequence of XML's origin as a document markup language and introduce ambiguity in representing information, when XML is used as a data exchange format. For example we could represent the information about Alan as:

```
<person> <name> Alan </name>
   <age> 24 </age>
   <email> agb@abc.com </email>
</person>
```

or as:

```
<person name="Alan" age="42" email="agb@abc.com"/>
```

or as:

```
<person age="42">
    <name> Alan </name>
    <email> agb@abc.com </email>
</person>
```

Attributes are more restrictive than elements however. A given attribute may only occur once within a tag, while sub-elements of the same type may be repeated. Also the value associated with an attribute is restricted to be a string (i.e. an atomic value), while the content of an element may either PCDATA or other subelements (i.e. either an atomic value or a complex object). A last distinction is that attributes are unordered while elements are ordered. That is the two elements below are distinct:

```
<person> <name> Alan </name> <age> 24 </age> </person>
<person> <age> 24 </age> <name> Alan </name> </person>
```

while the two elements below are structurally equal:

```
<person name="Alan" age="42"/>
<person age="42" name="Alan"/>
```

Well-Formed XML Documents. So far we have presented the basic XML syntax. There are only two constraints that have to be satisfied: tags have to nest properly, and attributes have to be unique. An XML document satisfying these two constraints is said to be *well-formed*. Being well-formed is a very weak constraint; it does little more than ensure that XML data will parse into a labeled tree.

XML References. The implicit XML data model is a tree. Often however one needs to describe a graph instead, and for that one can use references. We illustrate with the graph in Fig 21.3: the corresponding XML file is in Fig. 21.5.

The assumption here is that the `id` attribute defines an oid while the attributes `idref, mother, sister` define references to oids. In database terminology `id` is a key, while `idref, mother, sister` are foreign keys. The user specifies in the DTD (Sec. 2.2) which attributes are keys, which are references, and which are neither.

This technique for representing references is sometimes called *value-based*. Following reference edges amounts to performing a join. For example if we have a set of `person` elements and a set of `relatives` elements, in order to find everyone's mother we have to join the two sets on the condition `id=mother`. An implementation might use an index on the `id` attribute (e.g. a hash table) to facilitate such joins.

```
<person id="o1" haircolor="brown">
    <name> Mary </name>
    <age>  45 </age>
    <child idref="o2">
    <child idref="o3">
</person>
<person id="o2">
    <name> John </name>
    <age>  17 </age>
    <relatives mother="o1",  sister="o3"/>
</person>
<person id="o3">
    <name> Jane </name>
    <country> Canada </country>
    <mother idref="o1"/>
</person>
```

Figure 21.5. An XML fragment with references.

The XML language offers only minimal support for references: it only checks if all keys are unique and if reference attributes point to existing keys in the document. More powerful constraints will be included in XML-Schema (Beech et al. 1999, Biron and Malhotra 1999), a standard that will supersede DTDs.

Finally we note that the way references are defined in XML makes the analogy to a graph less clean than in the OEM data model. For example consider the semistructured data instance in Figure 21.6. We could encode this in XML either as:

```
<a> <b id="o123"> some string </b> </a>
<a c="o123"/>
```

and assume that the attribute c is a reference attribute, or as:

```
<a b="o123"/>
<a> <c id="o123"> some string </b> </a>
```

assuming that b is now a reference attribute.

2.2.   Document type declarations (DTDs)

A Document Type Declaration (DTD) serves as grammar for the underlying XML document, and it is part of the XML language definition. To some extent a DTD can also serve as schema for the data represented

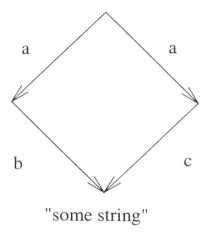

Figure 21.6. Illustration for XML references.

by the XML document; hence our interest here. As schema formalisms DTDs are somehow unsatisfactory and currently a new proposal is being developed, called XML-Schema (Beech et al. 1999, Biron and Malhotra 1999), that will fix some of the DTDs' limitations in representing schemas.

A Simple DTD. The DTD below describes the structure of the XML document in Fig. 21.4.

```
<!DOCTYPE people [
   <!ELEMENT people (person*)>
   <!ELEMENT person (name,age,email)>
   <!ELEMENT name (#PCDATA)>
   <!ELEMENT age (#PCDATA)>
   <!ELEMENT email (#PCDATA)>
]>
```

The first line says that the root element has type `people`. The next five lines are markup declarations stating that an element of type `people` contains an arbitrary number of elements of type `person`, each containing `name`, `age`, and `email` elements. The latter contain only character data (no elements). Here `person*` is a regular expression, meaning any number of person elements. Other regular expressions are `e+` (one or more occurrences), `e?` (zero or one), `e | e'` (alternation) and `e,e'` (concatenation).

DTDs as Grammars. A DTD is precisely a context-free grammar for the document. In particular the example above imposes that <name>, <age>, and <email> appear *in that order* in a <person> element. Grammars can be recursive, like in the following DTD describing binary trees:

```
<!ELEMENT node (leaf | (node,node))>
<!ELEMENT leaf (#PCDATA)>
```

An example of such an XML document is:

```
<node>
    <node>
        <node> <leaf> 1 </leaf> </node>
        <node> <leaf> 2 </leaf> </node>
    </node>
    <node>
        <leaf> 3 </leaf>
    </node>
</node>
```

DTDs as Schemas. DTDs can be used to a certain extent as schemas. The simple DTD above requires that person elements have a name field. In this sense they resemble types, though less expressive than one may wish. For example consider the relational schema r1(a,b,c), r2(c,d). We would like to define a DTD describing all valid XML representations of relational data of that schema (as described in Section 1.1). A first problem we encounter is that we need two structures for the element row: one for that occurring under r1, another for that under r2. This is not possible in DTDs, since each element has a unique definition, regardless of its context[2]. For that purpose we change slightly the representation by renaming rows in r1 to r1Row and rows in r2 to r2Row. The corresponding DTD is:

```
<!DOCTYPE db [
  <!ELEMENT db (r1, r2)>
  <!ELEMENT r1 (r1Row*)>
  <!ELEMENT r2 (r2Row*)>
  <!ELEMENT r1Row (a,b,c)>
  <!ELEMENT r2Row (c,d)>
  <!ELEMENT a (#PCDATA)>
  <!ELEMENT b (#PCDATA)>
  <!ELEMENT c (#PCDATA)>
  <!ELEMENT d (#PCDATA)>
```

]>

The DTD correctly constrains r1Row elements to contain three components a, b, c and r2Row elements to contain the components c, d. But it also forces a certain order on these elements, and there is no easy way to remove this order constraint in a DTD[3].

It is easy to describe optional, or repeated components in a DTD. For example one could modify r1Row to:

<!ELEMENT r1Row (a,b?,c+)>

stating that a is required and unique, that b is optional, and that c is required, but may have several occurrences.

Declaring Attributes in DTDs. DTDs also allow us to assert the type of attributes. For example the following DTD fragment describes the attributes of the XML fragment in Fig. 21.5.

```
<!ATTLIS person     id        ID    #REQUIRED
                    haircolor CDATA $IMPLIED>
<!ATTLIS child      idref     IDREF #IMPLIED>
<!ATTLIS relatives  mother    IDREF #REQUIRED
                    sister    IDREF #IMPLIED
                    brother   IDREF #IMPLIED>
```

An element of type person may have two attributes, id and haircolor. The first is required and its type is ID (i.e. it is a key). The second one is optional and its type is CDATA (i.e. any string). Similarly, an element of type relatives may have up to three attribuges, mother, sister, and brother, all of type IDREF (i.e. a reference, also called a foreign key) and all optional.

Valid XML Documents. Recall that a well-formed XML document is one which has matching markups and unique attributes in each element. A *valid* XML document is one which, in addition, has a DTD, and which conforms to that DTD. That is, elements may be nested only in the way described by the DTD, and may have only the attributes allowed by the DTD. In addition, the values of attributes of type ID must define unique keys across the entire XML document, and all values of attributes of type IDREF must be existing keys.

3. Query Languages

One could use a general purpose programming language for processing semistructured data and XML. However, there are well known advantages from using a special purpose, declarative query language instead, such as: conciseness, guaranteed termination, low complexity, and, especially, potential for optimization.

In relational databases the standard query language is SQL (Contributor), while in object-oriented databases the standard query language is OQL (Cattell 1996). None is directly applicable to semistructured data or XML for two reasons. First, semistructured data is self-describing, hence there is often no a priori knowledge about its structure. In particular, it can be arbitrarily deep, hence a query language must be capable to retrieve data at any depth. Second, a query language for semistructured data must be able to *construct* its answer as an instance of semistructured data, and this is different from constructing an answer that is an instance of a relational data, or an object-oriented data.

Most query languages proposed for semistructured data or XML are declarative and inspired by SQL or OQL. They differ mostly in the way in which they deal with the two extensions. We illustrate here Lorel, UnQL, and XML-QL. A fourth language, XSL, is not declarative but uses structural recursion to traverse the XML document. We will illustrate throughout this section with the semistructured data instance in Fig. 21.7 (using XML syntax).

Lorel. Abiteboul et al. (1997) was obtained by extending OQL to semistructured data. Consider first:

```
select  X.author
from    biblio.book X, X.year Y
where   Y = "1999"
```

Like in SQL and OQL, a query has three clauses: `select`, `from`, and `where`. In the `from` clause the variables X and Y are bound, as follows. First we start from the root of the database and follow all paths labeled `biblio.book`: X is bound successively to each object at the end of such a path. For each such X, the variable Y is bound to the objects rechable from X by a `year` link respectively. In the `where` clause we filter out only those bindings from which the year is `1999`. Finally, in the `select` clause we return the `author` value of X.

Readers familiar with OQL will immediately notice the similarity between this query and OQL. However there are two important differences. First, in the way Lorel handles types. OQL is strongly typed; for example we would be guaranteed that X has a `year` component, otherwise the expression `X.year Z` would raise a type error. By contrast, Lorel handles semistructured data, and X may be bound to an object that doesn't have a `year`. No type error is reported, but such an X will not be returned in the answer. Similarly, it may happen that X has multiple

```
<biblio>  <book>    <author> Smith </author>
                   <author> Young </author>
                   <title> Query Processing </title>
                   <year> 1999 </year>
          </book>
          <paper>  <author> Mark </author>
                   <author> Smith </author>
                   <title> Indexes in Databases </title>
                   <year> 1982 </year>
                   <comments> An excellent overview </comments>
          </paper>
          <paper>  <author> Smith </author>
                   <title> Queries for Dummies </title>
                   <year> 1972 </year>
                   <year> 1988 </year>
                   <comments> This paper is really shallow </comments>
          </paper>
          <paper>  <author ID="o162"> <name> Alan </name>
                                      <email> agg@abc.com </email>
                   </author>
                   <title> Query Debugging </title>
                   <year> 1988 </year>
          </paper>
          <book>   <author> Alan </author>
                   <title> The Query Handbook </title>
                   <publisher> DB Publishing Co </publisher>
                   <pages> 592 </pages>
          </book>
</biblio>
```

Figure 21.7. Semistructured data instance.

year attributes. In that case the where condition checks if any of these values is equal to 1999. In other words Lorel's semantic is *existential*, since it checks whether there *exists* some Y s.t. Y = ''1999''. Second, Lorel differs from OQL in the way it does coercions on atomic values. Its data repository is in binary format, and atomic values include string, integer, etc. Often the same data type has different representations. For example the year 1999 can be represented as an integer in some part of the data, or as the string "1999" in some other part. Lorel has a set of rules for doing coercions between datatypes, which applies both to the values in the database and the constants in the query, in order to deal with such differences in representation. For example the query could as well been written as:

```
select  X.author
from    biblio.book X, X.year Y
where   Y = 1999
```

Here 1999 is an integer, rather than a string. The answer to this query will be the same as the answer to the previous one.

Lorel has a few useful abbreviations that allow us to avoid introducing variables, similar to those in SQL. Our query can be written as:

```
select biblio.book.author
where  biblio.book.year = "1999"
```

The general rule is that a variable is introduced for all longest common prefixes in different paths in the query: in this case a variable X is introduced for biblio.book. When no variables are introduced by the from clause, it can be dropped from the query. We refer the reader to Abiteboul et al. (1997) for a complete discussion.

In order to navigate to arbitrary depth in the data, Lorel introduced *regular path expressions*. For example, the following query returns all comments in books, regardless of their depth:

```
select X
from biblio.book.%.comment X
```

Here % is a wildcard, matching an arbitrarily long path in the data. Thus, all comments will be returned, regardless of how deep they are.

UnQL. The language UnQL (Buneman et al. 1996; 2000) adapted the comprehension syntax (Buneman et al. 1994) to query semistructured data. Its central features are patterns and templates. Patterns are used to navigate through the data, while templates are used to construct results. For example the previous query would be written in UnQL as:

```
select Y
where {biblio: {book: {author: Y, year: 1999}}}
```

Ther expression {biblio: {book: {author: Y, year: 1999}}} is called a *pattern*. The syntax of a pattern is similar to that of semistructured data, but in addition a pattern may use variables, like Y in the pattern above. The query returns all authors that published books in 1999.

In UnQL we use templates to control how the answer will be constructed. Consider the following query:

```
Q1 = select {result: {author: X, year: Y}}
     where  {biblio: {paper: {author: X, year: Y}}}
```

This returns a set of paper authors together with the years they published, and the result is:

```
{result: {author: "Mark",  year: 1982},
 result: {author: "Smith", year: 1982},
 result: {author: "Smith", year: 1972},
 result: {author: "Smith", year: 1988},
 result: {author: {name: "Alan", email: "agg@abc.com"}, year: 1988}
}
```

Note that one author may occur several times, once for each year in which he published. To group results by author one uses a nested query:

```
select {result: {author: X, select {year: Y}
                 where {biblio: {paper: {author: X, year: Y}}}
       }}
       where  {biblio: {paper: {author: X}}}
```

The answer is:

```
{result: {author: "Mark",  year: 1982},
 result: {author: "Smith", year: 1982, year: 1972, year:1988},
 result: {author: {name: "Alan", email: "agg@.abc.com"}, year: 1988}
}
```

Notice that the query performs a join on the **author** value by using the same variable X twice. UnQL uses the value equality predicate from Definition 1. The way the query is formulated it also includes all paper authors whose paper do not have a **year:** in that case the list of years for that authors will be empty. Readers familiar with relational databases will recognize that the semantics is that of a left outer join.

XML-QL. The query language XML-QL (Deutsch et al. 1999b) was specifically designed for XML, and combines features from UnQL and StruQL (another query language for semistructured data (Fernandez et al. 1998)). It uses the OEM datamodel instead of XML's model: in particular it assumes an unordered tree.[4] As a simple illustration, consider the UnQL query Q1 above. It is expressed in XML-QL as:

```
Q2 = where <biblio> <paper> <author> $X </author>
                           <year> $Y </year>
                   </paper>
             </biblio>  in "www.abc/biblio.xml"
      construct <result> <author> $X </author> <year> $Y </year> </result>
```

Like UnQL, XML-QL has patterns in the `where` clause, and templates in the `construct` clause.

Joins are used in XML-QL for data integration. The following XML-QL query integrates books from two sources, based on their `isbn`:

```
where <biblio> <book> <author> $X </> <isbn> $N </> </> </> in "www.abc/biblio.xml"
      <catalog> <book> <isbn> $N </> <price> $P </> </> </> in "www.abc/catalog.x
      $P > 100.00
construct <result> <author> $X </> </result>
```

The query returns all authors of books over 100.00 USD. Notice that XML-QL allows us to abbreviate `</>` any end-tag.

Like UnQL, XML-QL allows nested queries to control grouping. The following query uses a subquery to reverse the hierarchy in `biblio`, from a group by publications to a group by authors.

```
where <biblio> <$T> <author> $X </> </> </> in "www.abc/biblio.xml"
construct <result> <author> $X </>
                   where <biblio>
                           <$T> <author> $X </>
                                <title> $Y </>
                           </>
                         </> in "www.abc/biblio.xml"
                   construct <$T> $Y </>
           </>
```

The answer will look like:

```
<result> <author> Smith </author>
         <book> ... </book>
         <paper> ... < /paper>
         <paper> ... </paper>
         <book> ... </book>
         ...
```

```
</result>
<result> <author> Mark </author>
         ...
</result>
```

Note the use of the *tag variable* $T. It is bound to a tag, and copied to the answer.

Declarative Semantics. All query languages described so far have a simple, declarative semantics. We describe that semantics in detail for XML-QL, then comment how it can be adapted to Lorel and UnQL.

Consider a simple XML-QL query Q, consisting of a single block:

```
Q =  where W construct C
```

The where clause, W, consists of several patterns to be matched against several data sources, and of several conditions on the variables. The input to the query consists of an XML tree for each data source. Assume first that the query does not contain other nested subqueries, i.e. C is a template without subqueries.

Let $\bar{V} = (\$X_1, \$X_2, \ldots, \$X_k)$ be all variables occurring in the patterns in W. In XML-QL all variables in C are required to appear in $\bar{V}$. The query's semantics is described in two parts.

The first part constructs the bindings of all variables. It results is a relation T with one column for each variable in $\bar{V}$:

	X_1	X_2	...	X_k
r_1:	v_{11}	v_{12}	...	v_{1k}
	...	...	...	...
r_n:	v_{n1}	v_{n2}	...	v_{nk}

Each row in T corresponds to a binding of all variables in $\bar{V}$. Since we assume an unordered data model, this table is unordered. Let $r_1, r_2, \ldots r_n$ be all rows in T, in some arbitrary order.

In the second part we consider each row r_i in T, $i = 1, n$, and compute $C[r_i]$, i.e. the XML fragment obtained by substituting the variables in C with their values in r_i. Then the query's answer is obtained by concatenating these individual results:

$$\text{Answer} = C[r_1]C[t_2]\ldots C[r_n] \quad (21.1)$$

Note that the order here is arbitrary: the result is an unordered tree.

For example consider the XML-QL query Q2 above. It's where clause has two variables, $X, $Y, and the first part of the semantics results in the following table:

$X	$Y
"Mark"	1982
"Smith"	1982
"Smith"	1972
"Smith"	1988
&o162	1988

Note that the "value" of a variable is an atomic value, when that variable is bound to an atomic object, or an object identifier, when that variable is bound to a complex object (e.g. last row).

The template in the `construct` clause is:

```
<result> <author> $X </author> <year> $Y </year>
```

hence the answer to the query is:

```
<result> <author> Mark  </author> <year> 1982 </year> </result>
<result> <author> Smith </author> <year> 1982 </year> </result>
<result> <author> Smith </author> <year> 1972 </year> </result>
<result> <author> Smith </author> <year> 1988 </year> </result>
<result> <author> <name>Alan</name>
                 <email>agg@abc.com</email>
        </author>
        <year>1988 </year>
</result>
```

This simple semantics can be easily extended. First, to deal with nested subqueries in C we simply apply the semantics recursively to $C[r_i]$. To deal with an ordered data model, it suffices to define a certain order for the table T, based on the order of the input XML documents, then preserve that order in Eq.(21.1).

UnQL has an unordered data model with deep equality given by Definition 1. Here Eq.(21.1) changes to:

$$\text{Answer} \;=\; C[r_1] \cup C[t_2] \cup \ldots \cup C[r_n]$$

where $\cup$ indicates that duplicates are to be eliminated. Lorel is also based on the OEM model, and has in addition to the type `oid`, a set type. The answer of a query is defined to be a set, hence Eq.(21.1) changes to:

$$\text{Answer} \;=\; \{C[r_1], C[t_2], \ldots, C[r_n]\}$$

XSL. A W3C recommendation, XSL is a stylesheet language for XML (Clark 1999). Its primary use is that of transforming XML to HTML, in order to be displayed. XSL can express however general XML to XML transformations, and thus can serve as an XML transformation and query language. Unlike the previous languages XSL does not have a declarative semantics, but relies on a recursive definition using patterns and templates. We start by illustrating with a simple XSL program, which returns all authors in the input document:

```
<xsl:template>
    <xsl;apply-templates/>
</xsl:template>
<xsl:template match="comments">
   <result> <xsl:value-of/> </result>
</xsl:template>
<xsl:template match="biblio">
   <answer> <xsl:apply-templates/> </answer>
</xsl:template>
```

An XSL program consists of a collection of *templates*. Each template has the form:

```
<xsl:template match=XPath-Expr>  Template-Body  </xsl:template>
```

Here `XPath-Expr` is a pattern (and may be missing), while `Template-Body` is an XML fragment interspersed with XSL directives. An XSL program views an XML document as a tree, and acts like a recursive function on that tree, starting at the root node. At each node the function tries to match that node with all `XPath-Expr` patterns of all templates, in reverse order of the templates. At the first match it finds, it evaluates the `Template-Body`. The `Template-Body` usually contains some XML fragments, which the XSL processor simply returns as the result of the recursive function; but `Template-Body` may also contain some `<xsl:apply-templates/>` directive(s), which causes the recursive function to be applied recursively on all children of the current node. Another directive, `<xsl:value-of/>` determines the XSL processor to return the content of the current node.

In our example, the root node is an element of type `biblio`, hence the last template is triggered: this returns

```
<answer> ... </answer>
```

with the content obtained by processing all `biblio`'s children recursively. All these children are of type `book` or `paper`: in each case, the only template that matches is the first template (since a missing match=XPath-Expr means that any node matches), This determines the children to be processed recursively, etc. The recursive process stops at comments nodes, where the value of the comment is returned. On the XML data in Fig. 21.7, the XSL program returns:

```
<answer> <result> An excellent overview </result>
         <result> This paper is really shallow </result>
</answer>
```

XSL uses recursion to traverse the data to an arbitrary depth: the XSL example above returns all comments, regardless of the depth at which they occur. Recursion always terminates when it is applied exclusively to children. In XSL however it is possible to apply the recursion to an arbitrary node, and, hence, write non-terminating programs. Such programs are, of course, impossible to "optimize" in the way database systems optimize queries.

4. Storage

All modern database systems store the data according to their types. For example a record with three integer fields will be stored as 12 consecutive bytes, while a set can be stored as an array together with an integer representing its length. What is important is that the type has to be known in advance, before the data can be stored. By contrast, semistructured data and XML are self-describing, and do not have an a priori type. Even XML documents with a DTD cannot be considered "typed" for this purpose, because a DTD does not prescribe any particular storage method.

We describe here a few techniques for storing XML data. Depending on the back-end storage, we can classify the techniques as follows.

- storing in an ASCII text file
- storing in relations
- storing in an object-oriented database (or object repository)

We discuss here the first two storage methods and refer the reader to Christophides et al. (1994) for a presentation of the third.

4.1. ASCII text files

The XML document is stored directly, in an ASCII file. An application needs to parse the file in order to accesses the underlying data: this results in sequential access only. Additional indexes may be provided for direct access to certain subelements.

This naive method can be quite effective for applications accessing the XML data in read-only mode. Although XML is verbose, the space requirement turns out to be small compared to the other methods described below, because relational and object-oriented systems introduce a lot of space overhead. Moreover, this method also has the advantage that it preserves the order of the XML elements, and, overall, is a faithful representation of the XML standard. By contrast the methods below abstract the XML document as a tree, thus loosing some information (e.g. white-spaces, entities, CDATA sections), which may be important for some applications.

The main disadvantages of this method is that it doesn't support updates, and that the XML data has to be parsed by the application each time it needs to access it: this can be prohibitively expensive. Also, one has to implement a query processor from scratch. By contrast the other storage methods described below can use the query engine of an existing system.

4.2. Relational store

We discuss here three alternatives for mapping XML data to relations.

Ternary Relation. Since every semistructured data is a labeled graph, its edges can be represented in a ternary relation. This method has been described in Florescu and Kossmann (1999). For example the data in Fig 21.2 will be represented by:

Ref

soid	label	doid
&root	"people"	&o1
&o1	"person"	&o2
&o1	"person"	&o3
&o1	"person"	&o4
&o1	"person"	&o5
&o1	"person"	&o6
&o2	"name"	&o7
&o2	"phone"	&o8
&o2	"email"	&o9
&o3	"name"	&o10
...	...	...

Val

oid	value
&o7	"Alan"
&o8	5557786
&o9	"agg@abc.com"
...	...

The table Ref stores the edges, while the table Val stores the values.

Queries on the original semistructured data can be directly translated into SQL queries[5] over Ref and Val. For example the following UnQL query

```
Q3 = select X
```

```
where    {people.person: {name: "Alan", phone: X}}
```

will be translated into:

```
select v_x.value
from Ref e_people, Ref e_person, Ref e_name, Ref e_phone, Val v_alan, val v_x
where e_people.soid = "o1" and e_people.label = "people" and
      e_people.doid = e_person.soid and e_person.label = "person" and
      e_person.doid = e_name.soid and e_name.label = "name" and
      e_person.doid = e_phone.soid and e_phone.label = "phone" and
      e_name.doid = v_alan.oid and v_alan.value = "Alan" and
      e_phone.doid = v_x.oid
```

This example highlights the main disadvantage of this method: simple patterns may get translated into large number of joins (in this example a five-way join). Yet, with appropriate indexes, experiments reported in Florescu and Kossmann (1999) show that queries like the above can be answered with reasonable performance by commercial RDBMS.

In practice, several improvements are possible, as described in Florescu and Kossmann (1999). First, for applications where the order among siblings is important one can add another column to Ref storing an integer representing that edge's order number. Second, one needs to split the Val table according to the type of the atomic value: that is all string values will be stored in a table ValString, all integer values will be stored in a table ValInteger, etc. As an optimization Florescu and Kossmann (1999) report significant speedups if the Val tables are inlined in the Ref relation. Thus, Ref has now multiple additional columns, one for each atomic data type, storing the value of doid. The example above becomes:

Ref

soid	label	doid	order	type	valString	valInteger
&root	"people"	&o1	0	0	null	null
&o1	"person"	&o2	1	0	null	null
...	...	...	...	...	...	...
&o2	"name"	&o7	0	1	"Alan"	null
&o2	"phone"	&o8	1	2	null	5557786
&o2	"email"	&o9	2	1	"agg@abc.com"	null
...	...	...	...	...	...	...

The first three columns are as before. The order column is illustrated in the last three rows showing the children of &o2: their order value is 0, 1, 2 respectively. The type indicates the type of the doid object: 0 means that it is a complex object, 1 that it is a string, 2, that it is an integer. Referring again to the last three rows, &o7 is an atomic object of type string, hence its type is 1, while &o8 is of type integer, hence its type is 2.

The main advantage of this technique is that it uses a generic relational schema, independent of the semistructured data. Also, queries are easily translated into SQL, and updates are supported. The main disadvantage is that it does not cluster objects well. One element may be split into many tuples in the `Ref` relation. As a consequence one often requires many joins to answer queries. The performance degrades significantly for queries that need to reconstruct the entire XML document Florescu and Kossmann (1999), precisely due to the large number of joins needed.

Using DTDs.   When the XML data to be stored has a DTD, then it is possible to derive a relational schema from that DTD. Referring to the data in Fig. 21.2, assume the following DTD[6]:

```
<!ELEMENT person (name, phone*, email, height?)>
<!ELEMENT name   (first,last)>
<!ELEMENT phone  (#PCDATA)>
<!ELEMENT email  (#PCDATA)>
<!ELEMENT height (#PCDATA)>
```

The reader may have notice that the data in Fig. 21.2 is actually not valid w.r.t. this DTD, because the DTD only allows structured `names`. We will correct this below, but for the time being we will show how XML data that is valid wrt. this DTD can be stored in tables. Following Shanmugasundaram et al. (1999), the first step in deriving a schema is to construct a DTD graph: this is illustrated in Fig. 21.8. The nodes are the elements in the DTD and the edges represent the containment relationship and a information on cardinality: a "*" symbol means that the subelement may occur zero or more times, "?" means that it may occur zero or one time, while no symbol means that the element occurs exactly once. The DTD graph looses any other information, such as more refined information on cardinality (e.g. two or more occurrences) or ordering information among subelements. The DTD-graph in our example is a tree, but this is not true in general, when elements are recursive.

Next a relational schema is derived from the DTD graph as follows. For each of the following elements we create a relation: for the root element, for every element with an incoming edge labeled *, and for every element with more than one incoming edge (this happens for example in a DTD graph with cycles). The other relations are "in-lined". The relation's attributes are then derived from the DTD-graph straightforwardly. Our running example will make this clear. We create relations for `person` and `phone`, and their schemas are:

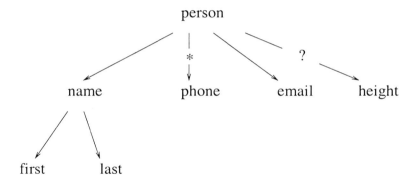

Figure 21.8. DTD Graph.

person(<u>persId</u>, name.first, name.last, email, height)
phone(<u>phoneId</u>, persId, value)

The persId attribute in phone is a foreign key in person.

Query translation is obvious. The query Q3 above becomes the following SQL statement:

```
select phone.value
from person, phone
where person.name = "Alan" and person.persId = phone.persId
```

This method has a clear advantage over the ternary-relation storage, since it tends to map an entire XML elements to one record. This results in better clustering and in fewer joins for the translated queries. This works well when a DTD is available which expresses faithfully the structure of the XML data.

As we have seen however DTDs sometimes cannot describe the intended structure well. To see such an example let us examine what happens if we use the correct DTD for the data in Fig. 21.2. We want to allow name to be either an element containing two subelements of type first and last, or a #PCDATA. We attempt to describe this as:

```
<!ELEMENT name ((#PCDATA | (first,last)))>    --- incorrect !
```

but unfortunately this is an incorrect DTD because DTDs impose restrictions on how one can combine #PCDATA with other subelements, and we can only describe name as:

```
<!ELEMENT name ((#PCDATA | first | last)*)>
```

Many DTDs are designed like this in practice. The new DTD leads us to the following relational schema:

```
person(persId, email, height)
name.PCDATA(id, persId, value)
name.first(id, persId, first)
name.last(id, persId, last)
phone(phoneId, persId, value)
```

The translated SQL query becomes now more complex, since we have to join person with name.PCDATA, name.first, name.last, and phone. In general, a relational schema derived from a DTD often results in more tables than necessary, because DTDs were designed as a validation tool, not as a storage description method.

Direct Mapping. When a DTD is not available or when it does not describe the structure well, it may be more reasonably to infer the relational schema directly from the data. This method is described in Deutsch et al. (1999c) and often results in quite natural mappings. For example the data in Fig. 21.2 can be represented as:

People1

oid	name	phone1	phone2	email	height
&o2	"Alan"	5557786	null	"agg@abc.com"	null
&o4	"Fred"	5556312	5554321	"fred@dot.com"	183
&o5	"John"	5551414	null	"jhh@abc.com"	188

People2

oid	name.first	name.last	phone	email
&o3	"Sarah"	"Green"	5556877	"sara@math.xyz.edu"
&o6	"Susan"	"Greer"	null	"susan@cs.xyz.edu"

We have inferred here a "good" relational schema for the given semistructured data. The person objects needed to be split into two tables because the structure of their name field differs: the first table contains those persons whose name is a #PCDATA, the second those whose name consists of a first and last field. Notice also the way we have treated phone elements: since all persons in the data had 1 or at most 2 phones, we introduced two relational attributes phone1 and phone2, rather than storing phones in a separate relation.

Query translation into SQL is straightforward. The query above becomes:

```
(select People1.phone1 as phone
 from   People1
 where  People1.name = "Alan") union
(select People1.phone2 as phone
 from   People1
 where  People1.name = "Alan")
```

Only `People1` is queried, because objects in `People2` have a structured `name` subobject and would not match the original query.

Before discussing how one discovers such a schema, let us see how updates can be handled. The easiest case is when the update matches the existing relational schema. For example consider the insertion of the following person:

{person: {name: "Bill", email: "bill@billco.com"}}

This is translated directly into an insertion into `People1`, which becomes:

People1

oid	name	phone1	phone2	email	height
&o2	"Alan"	5557786	null	"agg@abc.com"	null
&o4	"Fred"	5556312	5554321	"fred@dot.com"	183
&o5	"John"	5551414	null	"jhh@abc.com"	188
&o20	"Bill"	null	null	"bill@billco.com"	null

Now suppose that we insert the person:

{person: {name: "Bill", email: "bill@billco.com"}, salary: 55000}

We can only store in `People1` the `name` and `email`. A solution for storing `salary`, as well as any other additional labels is to add two generic tables to the relational schema, `Ref` and `Val`, similar to the ternary-table storage technique. These will store the *overflow* part of the data, that does not fit the regular structure imposed by `Person1` and `Person2`. We could then store the overflow part as:

Ref

soid	label	doid
&o20	salary	&o30

Val

oid	value
&o30	55000

Thus any update can be handled: if it does not fit the fixed schema, it is redirected to the overflow relations. The presence of the overflow relations however complicates the query translation, and the resulting SQL queries will be efficient only if `Ref` and `Val` are small (ideally empty). As more updates are done, and `Ref`, `Val` increase in size, a difficult reorganization of the relational storage may be needed.

Finally we discuss how to derive the relational schema in the first place. This can be done with data mining, as described in Deutsch et al. (1999c). In semistructured data, data mining searches for tree patterns Wang and Liu (1998). We change perspectives a bit and assume the given semistructured data consists of a large collection of root objects:

in the example in Fig. 21.2 the root objects would be the five `person` objects. The data mining algorithm looks for tree patterns with high support. A *tree pattern* here is simply a tree, p, and we say that p occurs in some object o if p is a subtree of o. The *support* of p is then simply the number of root objects in which p occurs. Illustrating with the data in Fig. 21.2, the following two patterns:

```
p1 = {person: {name, phone, phone, email, height}}
p2 = {person: {name: {first, last}, phone, email}}
```

have supports 1 and 1 respectively. Patterns with high support can be found using a standard a priori algorithm Agrawal et al. (1993), Agrawal and Srikant (1994), Wang and Liu (1998).

Patterns with high support are natural candidates for relation schemas. Continuing our example, patterns `p1`, `p2` produce precisely the relations `Person1` and `Person2` above. In general we cannot create a new relation for every pattern with high support, since this would result in too many relations with overlapping data. Instead we have to select a subset of patters that cover the semistructured data objects best. For details we refer the reader to Deutsch et al. (1999c).

5. XML Mediators

The main use of XML will be as a syntax for describing *views* of data residing in other formats. In data exchange, for instance, the data provider translates its relational or legacy data into XML, while the data consumer translates XML into its own internal representation.

A system module that translates between data models is called a *mediator*. We illustrate here a mediator system, SilkRoute Fernandez et al. (2000b), for exporting relational data into XML.

Consider a relational database with the following schema:

StoreUnit(<u>sId</u>, name, phone)
BookEntries(<u>bId</u>, title, authors)
StoreBookRelation(<u>sId, bId</u>, price, stock)

`StoreUnit` holds information about bookstores and `BookEntries` about books. The relation `StoreBookRelation` tells us which store sells what book, together with the selling price and current stock.

We want to export this data in XML and decide to group books by stores. That is, the exported XML view will have elements of the form:

```
<store> <name>  ... </name>
        <phone> ... </phone>
        <book>  ... </book>
```

```
            <book> ... </book>
            <book> ... </book>
            ...
   </store>
```

Each book element will have the form:

```
<book> <title> ... </title>
       <authors> ... </authors>
       <price> ... </price>
</book>
```

In each store we decide to include only books with `stock > 0`.

What is important here is that the schema of the relational data differs substantially from that of the XML view. The relational schema is flat, normalized, and often proprietary. There are well understood design methodologies that corporations follow in order to arrive to their relational schema. Business partners wish to exchange their relational data without revealing the underlying schema. For example the fact that `price` is an attribute of `StoreBookRelation` rather than `BookEntries` could be an indication that the company plans to charge in the future different prices at different stores – a piece of information it may not want to make public. By contrast the XML structure (e.g. described by a DTD) is nested, unnormalized, and public. Usually such a DTD is reached by agreement between two or more business partners.

One needs to specify the translation from the relational data to the XML view. In SilkRoute this is done through a query that combines the languages SQL and XML-QL. In our example the query is:

```
M = from StoreUnit $S
    construct
      <store>
        <name>  $S.name   </name>
        <phone> $S.phone  </phone>
        (from   BookEntries $B, StoreBookRelation $SB
         where  $S.sId = $SB.sId and
                $SB.bId = $B.bId and
                $SB.stock > 0
         construct <book>
                     <title>   $B.title   </title>
                     <authors> $B.authors </authors>
                     <price>   $SB.price  </price>
                   </book>
        )
```

```
            </store>
```

The from and where clauses have a syntax resembling SQL, while the construct clause contains XML patterns, like XML-QL.

One possibility is to execute this query and obtain the XML representation of the entire database, but this is probably a bad idea. Instead, a mediator allows the XML view to be *virtual*. When users pose queries on the XML view, SilkRoute rewrites them into queries over the relational schema. To see such a rewriting, consider the following XML-QL query asking for all bookstores which sell "The Java Programming Language" under $25:

```
Q = where <store>
            <name> $N </name>
            <book>
              <title> The Java Programming Language </title>
              <price> $P </price>
            </book>
          </store> in M,
          $P < 25
    construct <result> $N </result>
```

When presented with this query the mediator will compose it with M. After some algebraic manipulations and simplifications the resulting query is:

```
Q' = from StoreUnit $S, BookEntries $B, StoreBookRelation $SB
     where $S.sid = $SB.sid and $SB.bid = $B.bid
       and $B.title = "The Java Programming Language"
       and $SB.stock > 0 and $SB.price < 25
     construct <result> $S.name </result>
```

This query now can be executed directly by a SQL engine (except for the construct part which requires some XML formatting, called *tagging* in Shanmugasundaram et al. (2000)). Moreover only a few bookstore names need to be actually transmitted from the server to the client.

Note that a mediator system like SilkRoute allows us to process XML queries entirely with a relational database system. The only additional work one has to do is tagging.

The composition algorithm in SilkRoute Fernandez et al. (2000b) is designed specifically for XML queries. A composition algorithm for a more general query language for the Tsimmis project is described in Papakonstantinou et al. (1996).

6. Data Compression

Application specific data exchange formats have been around for many years and have processed it with proprietary tools. It may sound attractive to migrate such legacy data to XML, because then one could use general-purpose XML tools that are now becoming available. In addition to bringing extra features that are often useless in such applications, the may disadvantage of XML is that it is verbose, because tags are written explicitly and twice (begin tag and end tag), and because all data values are stored as strings. There exists the concern that processing the XML data will be substantially less efficient than processing the application specific formats, especially in data exchange and data archiving, where size matters. An obvious thing to try is to compress the data, using some general purpose compressor like `gzip`. But even then the compressed XML file is larger than the compressed application specific format. There is really no technical incentive to migrate application specific formats to XML.

The solution is to design a specialized XML compressor. We describe here `XMill` Liefke and Suciu (1999), a compressor that reduces the size of the XML files below that of the application specific formats. It does so by exploiting the structure of the XML file.

Let us illustrate with a simple example: Web Log files. Virtually every Web server logs its traffic, for security purposes, and this data can be (and often is) analyzed. Each line in the log file represents an HTTP request. A typical entry in such a log file is:

```
202.239.238.16|GET / HTTP/1.0|text/html|200|1997/10/01-00:00:02|-|4478|-|-|http:
//www.net.jp/|Mozilla/3.1[ja](I)
```

This is an example of an application specific data format. Different formats are currently in use: here we use a variation on Apache's *Custom Log Format*[7]. Each line is a record with eleven fields delimited by |: host, request line, content type, etc. Hence, the file's structure is very simple, with records with a fixed number of variable-length fields[8].

Collected over long periods of time, Web logs can take huge amounts of space. In our example we only considered a file with 100000 entries as the one above. Its size is almost 16MB, and `gzip` shrinks it to 1.6MB:

```
weblog.dat:   15.9MB     weblog.dat.gz:  1.6MB
```

Applications processing such Web logs are brittle, and in general not portable, since different vendors use different formats. To gain flexibility, we may consider converting the Web log into XML with the following format:

```
<apache:entry>
 <apache:host>202.239.238.16</apache:host>
 <apache:requestLine>GET / HTTP/1.0</apache:requestLine>
 <apache:contentType>text/html</apache:contentType>
 <apache:statusCode>200</apache:statusCode>
 <apache:date>1997/10/01-00:00:02</apache:date>
 <apache:byteCount>4478</apache:byteCount>
 <apache:referer>http://www.net.jp/</apache:referer>
 <apache:userAgent>Mozilla/3.1[ja](I)</apache:userAgent>
</apache:entry>
```

Applications are now easier to write. However the size increases substantially:

```
weblog.xml:   24.2MB        weblog.xml.gz:   2.1MB
```

We want to gain XML's flexibility without using more space. An obvious idea is to assign integer codes (1, 2, 3, ...) to the XML tags, and use a single character for closing tags. A more interesting idea is to separate the XML *structure*, i.e. its tags encoded now as numbers, from its *content*, i.e. the data values, and compress the structure and the content independently with `gzip`. The structure contains many repeated substrings, because most entries will have the same, or almost the same subelements: `gzip`, which is based on Ziv-Lempel LZ77 Ziv and Lempel (1977), compresses such repeated strings very well. The `XMill` command

```
xmill -p // weblog.xml
```

performs precisely these steps and results in a smaller compressed file:

```
        weblog1.xmi:        1.75MB
```

The next idea is to compress data values separately, based on their tags: that is, all host values are compressed together, all request lines are compressed together, etc. We save more space, because each different type of data value is compressed by `gzip` better separately than together with values of a different type. This behavior is the default in `XMill` and can be achieved by:

```
    xmill   weblog.xml
```

This reduces the size even further:

```
        weblog2.xmi:        1.33MB
```

We now use less space than the original gzipped file (`weblog.dat.gz`).

```
-p//apache:host=>seqcomb(u8 "." u8 "." u8 "." u8)
-p//apache:userAgent=>seq(e "/" e)
-p//apache:byteCount=>u
-p//apache:statusCode=>e
-p//apache:contentType=>e
-p//apache:requestLine=>seq("GET " rep("/" e) " HTTP/1." e)
-p//apache:date=>seq(u "/" u8 "/" u8 "-" u8 ":" di ":" di)
-p//apache:referer=>or(seq("file:" t)
                      seq("http://" or(seq(rep("." e) "/" rep("/" e))
rep("." e))) t)
```

Figure 21.9. XMill settings `settings.pz` for efficient compression of Web log data.

We can do quite a lot better than that. The idea is to inspect carefully each field and use a specialized compressor for it. For example the `<apache:host>` is usually (or always) an IP address, hence can be stored as four unsigned bytes; most entries in `<apache:requestLine>` start with `GET` and end in `HTTP/1.0` (some in `HTTP/1.1`): these substrings can be factored out. Other improvements are also possible. The corresponding settings for XMill are shown in Fig. 21.9, which describes specialized compressors for eight of the eleven fields. The XMill command line is:

```
xmill -f settings.pz weblog.xml
```

This reduces the compressed size to:

```
weblog3.xmi:      0.82MB
```

Note that this is about half the original gzipped file. This achieves our goal: the compressed XML-ized data can be stored in less space than the compressed original data, while applications gain in flexibility[9].

The Web log is a simple example illustrating column-wise compression applied to XML. The second example is much more complex. SwissProt is a well-maintained database for representing protein structure [10]. It uses a specific data format, called EMBL Higgins et al. (1992), for representing information about genes and proteins, illustrated in Fig. 21.10. Consider a conversion of the original EMBL data into XML as shown in Fig. 21.11.

The original file has 98MB and the XML-ized version had 165MB; `gzip` reduces the files to 16MB and 19MB, respectively:

```
sprot.dat:        98MB       sprot.xml:        165MB
sprot.dat.gz:     16MB       sprot.xml.gz:     19MB
```

```
ID    108_LYCES     STANDARD;      PRT;    102 AA.
AC    Q43495;
DT    15-JUL-1999 (Rel. 38, Created)
DT    15-JUL-1999 (Rel. 38, Last sequence update)
DT    15-JUL-1999 (Rel. 38, Last annotation update)
DE    PROTEIN 108 PRECURSOR.
OS    Lycopersicon esculentum (Tomato).
OC    Eukaryota; Viridiplantae; Streptophyta; Embryophyta; Tracheophyta;
OC    euphyllophytes; Spermatophyta; Magnoliophyta; eudicotyledons;
OC    core eudicots; Asteridae; euasterids I; Solanales; Solanaceae;
OC    Solanum.
RN    [1]
RP    SEQUENCE FROM N.A.
RC    STRAIN=CV. VF36; TISSUE=ANTHER;
RX    MEDLINE; 94143497.
RA    CHEN R., SMITH A.G.;
RL    Plant Physiol. 101:1413-1413(1993).
DR    EMBL; Z14088; CAA78466.1; -.
DR    MENDEL; 8853; LYCes;1133;1.
KW    Signal.
FT    SIGNAL        1      30         POTENTIAL.
FT    CHAIN        31     102         PROTEIN 108.
FT    DISULFID     41      77         BY SIMILARITY.
FT    DISULFID     51      66         BY SIMILARITY.
FT    DISULFID     67      92         BY SIMILARITY.
FT    DISULFID     79      99         BY SIMILARITY.
SQ    SEQUENCE   102 AA;  10576 MW;  AFA4875A CRC32;
      MASVKSSSSS SSSSFISLLL LILLVIVLQS QVIECQPQQS CTASLTGLNV CAPFLVPGSP
      TASTECCNAV QSINHDCMCN TMRIAAQIPA QCNLPPLSCS AN
//
```

Figure 21.10. Example of SwissProt data.

```
<Entry id="108_LYCES" class="STANDARD" mtype="PRT" seqlen="102">
  <AC>Q43495</AC>
  <Mod date="15-JUL-1999" Rel="38" type="Created"></Mod>
  <Mod date="15-JUL-1999" Rel="38" type="Last sequence update"></Mod>
  <Mod date="15-JUL-1999" Rel="38" type="Last annotation update"></Mod>
  <Descr>PROTEIN 108 PRECURSOR</Descr>
  <Species>Lycopersicon esculentum (Tomato)</Species>
  <Org>Eukaryota</Org> <Org>Viridiplantae</Org>   ...   <Org>Solanum</Org>
  <Ref num="1" pos="SEQUENCE FROM N.A">
     <Comment>STRAIN=CV. VF36</Comment>
     <MedlineID>94143497</MedlineID>
     <Author>CHEN R</Author> <Author>SMITH A.G</Author>
     <Cite>Plant Physiol. 101:1413-1413(1993)</Cite>
  </Ref>
  <EMBL prim_id="Z14088" sec_id="CAA78466"></EMBL>
  <MENDEL prim_id="8853" sec_id="LYCes" status="1133"></MENDEL>
  <Keyword>Signal</Keyword>
  <Features>  <SIGNAL from="1" to="30">   <Descr>POTENTIAL</Descr>   </SIGNAL>
              <CHAIN from="31" to="102">  <Descr>PROTEIN 108</Descr>  </CHAIN>
              ...
  </Features>
</Entry>
```

Figure 21.11. XML Representation of SwissProt entry

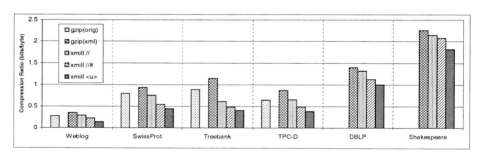

Figure 21.12. Compression Results

The three steps above result in the following improvements in size:

```
        sprot1.xmi:       15MB
        sprot2.xmi:       11MB
        sprot3.xmi:       8.6MB
```

Note that the last file is obtained after fine-tuning XMill on the SwissProt data.

Fig. 21.12 shows how XMill performs on a variety of XML data sets: Weblog and SwissProt were described before, Treebank Marcus et al. (1993) is a large collection of parsed English sentences from the Wall Street Journal stored in a Lisp like notation, converted to XML, TPC-

D(XML) is an XML representation of the TPC-D benchmark database, DBLP is the popular database bibliography database[11], and Shakespeare is a corpus of marked-up Shakespeare plays. The bar heights indicate the compression ratio, measured in bits/bytes: the lower the bars, the better the compression. For example 2 bits/bytes means that the size of the compressed file is 1/4 of the uncompressed file.

For each data set the four connected bars in Fig. 21.12 represent the compression ratio for gzip, and for XMill under the three settings described earlier. The right-most (smallest) bar requires user intervention, the previous one is XMill's default setting. Whenever a non-XML source was available the side bar on the left represents the relative size of the source file compressed with gzip (i.e. the height of the bar is size(gzip(orig))/(8∗size(XML))). For the first four data sets (which had more data and less text), XMill's compresses under the default setting to 45%-60% the size of gzip. Using semantic compressors, XMill reduced the size to 35%-47% of gzip's. For the more text-like data sets, XMill performs only slightly better than gzip. setting, XMill compressed the XML file better than gzip compressed the original file.

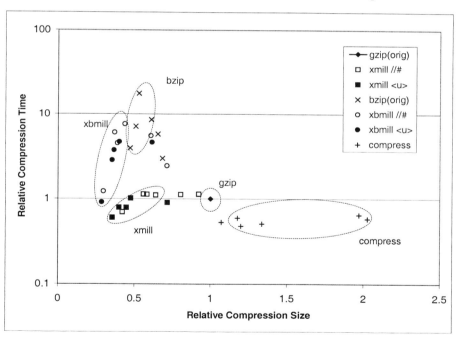

Figure 21.13. Compression rate vs. time.

Time/Space Tradeoff. Different general-purpose compressors offer a variety of time/space tradeoffs. Fig. 21.13 shows this tradeoff for the following compressors, on the same datasets described above: gzip, compress, bzip, XMill, and Xbmill. Here compress is the popular Unix tool that is faster than gzip but achieves worse compression rates; bzip is a general-purpose compressor that achieves better compression rates but is excessively slow; XMill is the XML specialized compressor using gzip as its internal compressor; finally Xbmill is like XMill but uses Xbmill as its internal compressor. The blobs highlight the "data-like" XML data sets (Weblog, SwissProt, Treebank, and TPC-D). The diagram shows that XMill offers the best overall time/space tradeoff for XML data.

7. Further Reading

Much of the material covered in this chapter is described in greater detail in Abiteboul et al. (1999). Several tutorials on semistructured data and XML are available Abiteboul (1997), Buneman (1997), Suciu (1999b;a), Florescu and Simeon (2000). A survey about database techniques for the World-Wide Web can be found in Florescu et al. (1998a).

Some general-purpose semistructured data management system are discussed in the literature: Lore Abiteboul et al. (1997), McHugh and Widom (1999), Tsimmis Papakonstantinou et al. (1995), Garcia-Molina et al. (1997), Yat Cluet et al. (1998), Christophides et al. (2000). Query processing and optimization is described in Lore. Two research prototypes addressed specifically the problem of Web site management: Strudel Fernandez et al. (1998; 2000a) and Araneus Atzeni et al. (1997).

Several query languages have been discussed for XML. For a survey and comparison see Deutsch et al. (1999a), Bonifati and Ceri (2000).

From a theoretical perspective a key feature distinguishing query languages for semistructured data is the presence of regular path expressions. It has been shown in Fernandez et al. (2000a) that such languages correspond to first order logic extended with transitive closure: such a logic was studied by Immerman Immerman (1987), and is known to be in NLOGSPACE. Thus, query languages with regular path expressions are probably less expressive than Datalog, which can express PTIME-complete queries. Some properties that were undecidable for datalog become decidable for certain such languages: for example containment of conjunctive queries with regular expressions is decidable Florescu et al. (1998b), D.Calvanese et al. (1998).

Schemas for semistructured data have received a lot of attention. Graph Schemas are introduced in Buneman et al. (1997) to express some known structure on the data. Data Guides Goldman and Widom (1997), Nestorov et al. (1997b) are extracted automatically from the data and represent in some sense the "tightest" graph schema for a given data. More complex formalisms, generalizing both graph schemas and DTDs, are described in Beeri and Milo (1999). Some forms of semantic optimizations, using schemas or DTDs are described in Fernandez and Suciu (1998), Liefke (1999).

A semistructured data may have more than one schema. This creates the intriguing possibility of extracting the schema automatically from the data. The *schema extraction* problem is defined and studied in Nestorov et al. (1997a), and for DTDs in Garofalakis et al. (2000).

Indexes for semistructured data are discussed in Goldman and Widom (1997), Nestorov et al. (1997b) (dataguides are a form of indexes) and Milo and Suciu (1999a).

A schema for semistructured data can be used to analyze the query. For example one can *infer* the possible type(s) of each variable in the query: this problem is studied in Milo and Suciu (1999b). Queries expressing XML transformations, like XSL or XML-QL, often need to construct results that are valid wrt. a given output DTD. The *typechecking* problem is studied in Milo et al. (2000), Papakonstantinou and Vianu (2000).

Bibliography

S. Abiteboul, P. Buneman, and D. Suciu. *Data on the Web : From Relations to Semistructured Data and Xml*. Morgan Kaufmann, 1999.

S. Abiteboul, D. Quass, J. McHugh, J. Widom, and J. Wiener. The Lorel query language for semistructured data. *International Journal on Digital Libraries*, 1(1):68–88, April 1997.

Serge Abiteboul. Querying semi-structured data. In *Proceedings of the International Conference on Database Theory*, pages 1–18, Deplhi, Greece, 1997. Springer-Verlag.

R. Agrawal, T. Imielinski, and A. Swami. Mining association rules between sets of items in large databases. In *Proceedings of ACM SIGMOD Conference on Management of Data*, pages 207–216, Washington, DC, 1993.

Rakesh Agrawal and Ramakrishnan Srikant. Fast algorithms for mining association rules in large databases. In *Proceedings of VLDB*, pages 487–499, Santiago de Chile, Chile, September 1994. Morgan Kaufmann.

P. Atzeni, G. Mecca, and P. Merialdo. To weave the Web. In *VLDB*, pages 206–215, 1997.

D. Beech, S. Lawrence, M. Maloney, N. Mendelsohn, and H. Thompson. Xml schema part 1: Structures, May 1999. http://www.w3.org/TR/xmlschema-1/.

Catriel Beeri and Tova Milo. Schemas for integration and translation of structured and semi-structured data. In *Proceedings of the International Conference on Database Theory*, pages 296–313, 1999.

P. Biron and A. Malhotra. Xml schema part 2: Datatypes, May 1999. http://www.w3.org/TR/xmlschema-2/.

A. Bonifati and S. Ceri. Comparative analysis of five XML query languages. *SIGMOD Record*, 29(1):68–79, March 2000.

P. Buneman, M. Fernandez, and D. Suciu. Unql: A query language and algebra for semistructured data based on structural recursion. *VLDB Journal*, 9(1):76–110, 2000.

P. Buneman, L. Libkin, D. Suciu, V. Tannen, and L. Wong. Comprehension syntax. *SIGMOD Record*, 23(1):87–96, March 1994.

Peter Buneman. Tutorial: Semistructured data. In *Proceedings of ACM Symposium on Principles of Database Systems*, pages 117–121, 1997.

Peter Buneman, Susan Davidson, Mary Fernandez, and Dan Suciu. Adding structure to unstructured data. In *Proceedings of the International Conference on Database Theory*, pages 336–350, Deplhi, Greece, 1997. Springer Verlag.

Peter Buneman, Susan Davidson, Gerd Hillebrand, and Dan Suciu. A query language and optimization techniques for unstructured data. In *Proceedings of ACM-SIGMOD International Conference on Management of Data*, pages 505–516, 1996.

R. G. G. Cattell, editor. *The Object Database Standard: ODMG-93*. Morgan Kaufmann, San Mateo, California, 1996.

V. Christophides, S. Abiteboul, S. Cluet, and M. Scholl. From structured documents to novel query facilities. In Richard Snodgrass and Marianne Winslett, editors, *Proceedings of 1994 ACM SIGMOD International Conference on Management of Data*, pages 313–324, Minneapolis, Minnesota, May 1994.

V. Christophides, S. Cluet, and J. Simeon. On wrapping query languages and efficient xml integration. In *Proceedings of SIGMOD*, Dallas, TX, May 2000.

James Clark. XSL transformations (XSLT) specification, 1999. available from the W3C, http://www.w3.org/TR/WD-xslt.

S. Cluet, C. Delobel, J. Simeon, and K. Smaga. Your mediators need data conversion ! In *Proceedings ACM-SIGMOD International Conference on Management of Data*, pages 177–188, 1998.

World Wide Web Consortium. Extensible markup language (xml) 1.0, 1998. http://www.w3.org/TR/REC-xml.

Hugh Darwen (Contributor) and Chris J. Date. *Guide to the Sql Standard : A User's Guide to the Standard Database Language Sql*. Addison-Wesley, 1997.

D.Calvanese, G.Giacomo, and M.Lenzerini. On the decidability of query containment under constraints. In *PODS*, pages 149–158, 1998.

A. Deutsch, M. Fernandez, D. Florescu, A. Levy, D. Maier, and D. Suciu. Querying XML data. *IEEE Data Engineering Bulletin*, 22(3):10–18, 1999a.

A. Deutsch, M. Fernandez, D. Florescu, A. Levy, and D. Suciu. A query language for XML. In *Proceedings of the Eights International World Wide Web Conference (WWW8)*, pages 77–91, Toronto, 1999b.

A. Deutsch, M. Fernandez, and D. Suciu. Storing semistructured data with STORED. In *Proceedings of the ACM SIGMOD International Conference on Management of Data*, pages 431–442, 1999c.

M. Fernandez, D. Florescu, A. Levy, and D. Suciu. Declarative specification of web sites with strudel. *VLDB Journal*, 9(1):38–55, 2000a.

M. Fernandez, D. Suciu, and W. Tan. SilkRoute: trading between relations and XML. In *Proceedings of the WWW9*, pages 723–746, Amsterdam, 2000b.

Mary Fernandez, Daniela Florescu, Jaewoo Kang, Alon Levy, and Dan Suciu. Catching the boat with Strudel: experience with a Web-site management system. In *Proceedings of ACM-SIGMOD International Conference on Management of Data*, pages 414–425, 1998.

Mary Fernandez and Dan Suciu. Optimizing regular path expressions using graph schemas. In *Proceedings of the International Conference on Data Engineering*, pages 14–23, 1998.

D. Florescu, A. Levy, and A. Mendelzon. Database techniques for the World-Wide Web: a survey. *SIGMOD RECORD*, 27:59–75, 1998a.

D. Florescu and J. Simeon. XML data: from research to standards, 2000. Tutorial presented at VLDB'2000.

Daniela Florescu and Donald Kossmann. Storing and querying xml data using an rdbms. *IEEE Data Engineering Bulletin*, 22(3), 1999.

Daniela Florescu, Alon Levy, and Dan Suciu. Query containment for conjunctive queries with regular expressions. In *Proceedings of the ACM SIGACT-SIGMOD Symposium on Principles of Database Systems*, pages 139–148, 1998b.

H. Garcia-Molina, Y. Papakonstantinou, D. Quass, A. Rajaraman, Y. Sagiv, J. Ullman, and J. Widom. The TSIMMIS project: Integration of heterogeneous information sources. *Journal of Intelligent Information Systems*, 8(2):117–132, March 1997.

M. Garofalakis, A. Gionis, R. Rastogi, S. Seshadri, and K. Shim. XTRACT: A system for extracting document type descriptors from XML documents. In *Proceedings of SIGMOD*, pages 165–176, Dallas, TX, 2000.

C. F. Goldfarb. *The SGML Handbook*. Clarendon Press, Oxford, 1990.

Roy Goldman and Jennifer Widom. DataGuides: enabling query formulation and optimization in semistructured databases. In *Proceedings of Very Large Data Bases*, pages 436–445, September 1997.

D. G. Higgins, R. Fuchs, P. J. Stoehr, and G. N. Cameron. The EMBL data library. *Nucleic Acids Research*, 20:2071–2074, 1992.

Neil Immerman. Languages that capture complexity classes. *SIAM Journal of Computing*, 16:760–778, 1987.

H. Liefke and D. Suciu. XMill: an efficent compressor for XML data. In *Proceedings of SIGMOD*, pages 153–164, Dallas, TX, 1999.

Hartmut Liefke. Horizontal query optimization on ordered semistructured data. In *Proceedings of WebDB*, pages 61–68, 1999.

M. Marcus, B. Santorini, and M.A.Marcinkiewicz. Building a large annotated corpus of English: the Penn Treenbak. *Computational Linguistics*, 19, 1993.

J. McHugh and J. Widom. Query optimization for XML. In *Proceedings of VLDB*, pages 315–326, Edinburgh, UK, September 1999.

T. Milo, D. Suciu, and V. Vianu. Typechecking for xml transformers. In *Proceedings of the ACM Symposium on Principles of Database Systems*, pages 11–22, Dallas, TX, 2000.

Tova Milo and Dan Suciu. Index structures for path expressions. In *Proceedings of the International Conference on Database Theory*, pages 277–295, 1999a.

Tova Milo and Dan Suciu. Type inference for queries on semistructured data. In *Proceedings of the ACM Symposium on Principles of Database Systems*, pages 215–226, 1999b.

S. Nestorov, S. Abiteboul, and R. Motwani. Inferring structure in semistructured data. In *Proceedings of the Workshop on Management of Semi-structured Data*, 1997a. Available from http://www.research.att.com/~suciu/workshop-papers.html.

S. Nestorov, J. Ullman, J. Wiener, and S. Chawathe. Representative objects: concise representation of semistructured, hierarchical data. In *International Conference on Data Engineering*, pages 79–90, 1997b.

Y. Papakonstantinou, S. Abiteboul, and H. Garcia-Molina. Object fusion in mediator systems. In *Proceedings of Very Large Data Bases*, pages 413–424, September 1996.

Y. Papakonstantinou, H. Garcia-Molina, and J. Widom. Object exchange across heterogeneous information sources. In *IEEE International Conference on Data Engineering*, pages 251–260, March 1995.

Y. Papakonstantinou and V. Vianu. DTD inference for views of XML data. In *Proceedings of PODS*, pages 35–46, Dallas, TX, 2000.

J. Shanmugasundaram, E. Shekita, R. Barr, M. Carey, B. Lindsay, H. Pirahesh, and B. Reinwald. Efficiently publishing relational data as xml documents. In *Proceedings of VLDB*, pages 65–76, Cairo, Egipt, September 2000.

J. Shanmugasundaram, K. Tufte, G. He, C. Zhang, D. DeWitt, and J. Naughton. Relational databases for querying XML documents: limitations and opportunities. In *Proceedings of VLDB*, pages 302–314, Edinburgh, UK, September 1999.

D. Suciu. From semistructured data to XML, 1999a. Tutorial presented at VLDB'1999.

D. Suciu. Managing Web Data (tutorial). In *Proceedings SIGMOD*, page 510, 1999b.

J. Thierry-Mieg and R. Durbin. Syntactic Definitions for the ACEDB Data Base Manager. Technical Report MRC-LMB xx.92, MRC Laboratory for Molecular Biology, Cambridge,CB2 2QH, UK, 1992.

Ke Wang and Huiqing Liu. Discovering typical structures of documents: a road map approach. In *ACM SIGIR Conference on Research and Development in Information Retrieval*, pages 146–154, August 1998.

J. Ziv and A. Lempel. A universal algorithm for sequential data compression. *IEEE Transactions on Information Theory*, 23(3):337–343, 1977.

VII

ARCHITECTURE ISSUES

Chapter 22

OVERVIEW OF HIGH PERFORMANCE COMPUTERS

Aad J. van der Steen
Dept. of Computational Physics, Utrecht University, 3508 TD Utrecht, The Netherlands
steen@phys.uu.nl

Jack Dongarra
Computer Science Department, University of Tennessee and Oak Ridge National Laboratory, Oak Ridge, Tennessee, USA
dongarra@cs.utk.edu

Abstract The overview given here concentrates on the computational capabilities of the systems discussed. To do full justice to all assets of present days high-performance computers one should list their I/O performance and their connectivity possibilities as well. However, the possible permutations of configurations even for one model of a certain system often are so large that they would multiply the volume of this report, which we tried to limit for greater clarity. So, not all features of the systems discussed will be present. Still we think (and certainly hope) that the impressions obtained from the entries of the individual machines may be useful to many. We also omitted some systems that may be characterized as "high-performance" in the fields of database management, real-time computing, or visualization. Therefore, as we try to give an overview for the area of general scientific and technical computing, systems that are primarily meant for database retrieval like the AT&T GIS systems or concentrate exclusively on the real-time user community, like Concurrent Computing Systems, are not discussed in this report. Although most terms will be familiar to many readers, we still think it is worthwhile to give some of the definitions in section 2 because some authors tend to give them a meaning that may slightly differ from the idea the reader already has acquired.

Keywords: Flynn taxonomy, Shared and distributed memory machines, High performance fortran, Parallel virtual machine, Message passing interface, RISC processor.

1. Introduction

Before going on to the descriptions of the machines themselves, it is important to consider some mechanisms that are or have been used to increase the performance. The hardware structure or *architecture* determines to a large extent what the possibilities and impossibilities are in speeding up a computer system beyond the performance of a single CPU. Another important factor that is considered in combination with the hardware is the capability of compilers to generate efficient code to be executed on the given hardware platform. In many cases it is hard to distinguish between hardware and software influences and one has to be careful in the interpretation of results when ascribing certain effects to hardware or software peculiarities or both. In this chapter we will give most emphasis to the hardware architecture. For a description of machines that can be considered to be classified as "high-performance" one is referred to (Culler et al. 1998, van der Steen 1995).

2. The Main Architectural Classes

Since many years the taxonomy of Flynn (1972) has proven to be useful for the classification of high-performance computers. This classification is based on the way of manipulating of instruction and data streams and comprises four main architectural classes. We will first briefly sketch these classes and afterwards fill in some details when each of the classes is described.

- **SISD** machines: These are the conventional systems that contain one CPU and hence can accommodate one instruction stream that is executed serially. Nowadays many large mainframes may have more than one CPU but each of these execute instruction streams that are unrelated. Therefore, such systems still should be regarded as (a couple of) SISD machines acting on different data spaces. Examples of SISD machines are for instance most workstations like those of DEC, Hewlett-Packard, and Sun Microsystems. The definition of SISD machines is given here for completeness' sake. We will not discuss this type of machines in this report.

- **SIMD** machines: Such systems often have a large number of processing units, ranging from 1,024 to 16,384 that all may execute the same instruction on different data in lock-step. So, a single instruction manipulates many data items in parallel. Examples of SIMD machines in this class are the CPP DAP Gamma II and the Alenia Quadrics.

Another subclass of the SIMD systems are the vector-processors. Vector-processors act on arrays of similar data rather than on single data items using specially structured CPUs. When data can be manipulated by these vector units, results can be delivered with a rate of one, two and — in special cases — of three per clock cycle (a clock cycle being defined as the basic internal unit of time for the system). So, vector processors execute on their data in an almost parallel way but only when executing in vector mode. In this case they are several times faster than when executing in conventional scalar mode. For practical purposes vector-processors are therefore mostly regarded as SIMD machines. An example of such systems is for instance the Hitachi S3600.

- **MISD** machines: Theoretically in these type of machines multiple instructions should act on a single stream of data. As yet no practical machine in this class has been constructed nor are such systems easily to conceive. We will disregard them in the following discussions.

- **MIMD** machines: These machines execute several instruction streams in parallel on different data. The difference with the multi-processor SISD machines mentioned above lies in the fact that the instructions and data are related because they represent different parts of the same task to be executed. So, MIMD systems may run many sub-tasks in parallel in order to shorten the time-to-solution for the main task to be executed. There is a large variety of MIMD systems and especially in this class the Flynn taxonomy proves to be not fully adequate for the classification of systems. Systems that behave very differently like a four-processor NEC SX-5 vector system and a thousand processor SGI/Cray T3E fall both in this class. In the following we will make another important distinction between classes of systems and treat them accordingly.

- **Shared-memory systems**: Shared-memory systems have multiple CPUs all of which share the same address space. This means that the knowledge of where data is stored is of no concern to the user as there is only one memory accessed by all CPUs on an equal basis. Shared memory systems can be both SIMD or MIMD. Single-CPU vector processors can be regarded as an example of the former, while the multi-CPU models of these machines are examples of the latter. We will sometimes use the abbreviations SM-SIMD and SM-MIMD for the two subclasses.

- **Distributed-memory systems**: In this case each CPU has its own associated memory. The CPUs are connected by some network and may exchange data between their respective memories when required. In contrast to shared-memory machines the user must be aware of the location of the data in the local memories and will have to move or distribute these data explicitly when needed. Again, distributed-memory systems may be either SIMD or MIMD. The first class of SIMD systems mentioned which operate in lock step, all have distributed memories associated to the processors. As we will see, distributed-memory MIMD systems exhibit a large variety in the topology of their connecting network. The details of this topology are largely hidden from the user which is quite helpful with respect to portability of applications. For the distributed-memory systems we will sometimes use DM-SIMD and DM-MIMD to indicate the two subclasses.

As already alluded to, although the difference between shared and distributed-memory machines seems clear cut, this is not always entirely the case from the user's point of view. For instance, the late Kendall Square Research systems employed the idea of "virtual shared-memory" on a hardware level. Virtual shared-memory can also be simulated at the programming level: A specification of High Performance Fortran (HPF) was published in 1993 (Forum 1993) which by means of compiler directives distributes the data over the available processors. Therefore, the system on which HPF is implemented in this case looks like a shared-memory machine to the user. Other vendors of Massively Parallel Processing systems (sometimes called MPP systems), like HP and SGI/Cray, also support proprietary virtual shared-memory programming models due to the fact that these physically distributed memory systems are able to address the whole collective address space. So, for the user such systems have one *global address space* spanning all of the memory in the system. We will say a little more about the structure of such systems in section 7. In addition, packages like TreadMarks (Amza et al. 1996) provide a virtual shared-memory environment for networks of workstations.

Another trend that has came up in the last few years is *distributed processing*. This takes the DM-MIMD concept one step further: instead of many integrated processors in one or several boxes, workstations, mainframes, etc., are connected by (Gigabit) Ethernet, Fiber Channel, ATM, or otherwise and set to work concurrently on tasks in the same program. Conceptually, this is not different from DM-MIMD computing, but the

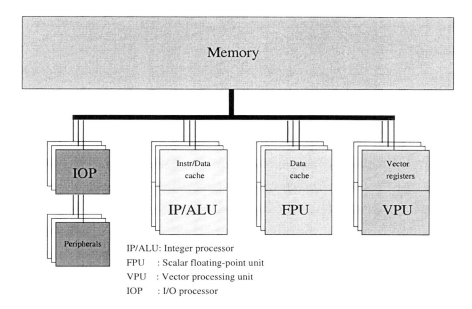

Figure 22.1. Block diagram of a vector processor.

communication between processors is often orders of magnitude slower. Many packages to realize distributed computing are available. Examples of these are PVM (standing for Parallel Virtual Machine) (Geist et al. 1994), and MPI (Message Passing Interface, (Snir et al. 1998, Gropp et al. 1998)). This style of programming, called the "message passing" model has becomes so much accepted that PVM and MPI have been adopted by virtually all major vendors of distributed-memory MIMD systems and even on shared-memory MIMD systems for compatibility reasons. In addition there is a tendency to cluster shared-memory systems, for instance by HiPPI channels, to obtain systems with a very high computational power. E.g., the NEC SX-5, and the SGI/Cray SV1 have this structure. So, within the clustered nodes a shared-memory programming style can be used while between clusters message-passing should be used.

3. Shared-memory SIMD Machines

This subclass of machines is practically equivalent to the single-processor vector-processors, although other interesting machines in this subclass have existed (viz. VLIW machines (van der Steen 1990)). In the block diagram in Figure 22.1 we depict a generic model of a vector architecture. The single-processor vector machine will have only one of

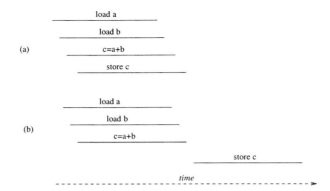

Figure 22.2. Schematic diagram of a vector addition. Case (a) when two load- and one store pipe are available; case (b) when two load/store pipes are available.

the vector-processors depicted and the system may even have its scalar floating-point capability shared with the vector processor (as was the case in some SGI/Cray systems). It may be noted that the VPU does not show a cache. The majority of vector-processors do not employ a cache anymore. In many cases the vector unit cannot take advantage of it and execution speed may even be unfavorably affected because of frequent cache overflow.

Although vector-processors have existed that loaded their operands directly from memory and stored the results again immediately in memory (CDC Cyber 205, ETA-10), all present-day vector-processors use vector registers. This usually does not impair the speed of operations while providing much more flexibility in gathering operands and manipulation with intermediate results.

Because of the generic nature of Figure 22.1 no details of the interconnection between the VPU and the memory are shown. Still, these details are very important for the effective speed of a vector operation: when the bandwidth between memory and the VPU is too small it is not possible to take full advantage of the VPU because it has to wait for operands and/or has to wait before it can store results. When the ratio of arithmetic to load/store operations is not high enough to compensate for such situations, severe performance losses may be incurred. The influence of the number of load/store paths for the dyadic vector operation $c = a + b$ (a, b, and c vectors) is depicted in Figure 22.2. Because of the high costs of implementing these data paths between memory and the VPU, often compromises are sought and the number of systems that have the full required bandwidth (i.e., two load operations and one store operation at the *same* time) is limited. In fact, in the vector systems marketed today this high bandwidth thus not occur any longer. Vendors rather rely on additional caches and other tricks to hide the lack of bandwidth.

The VPUs are shown as a single block in Figure 22.1. Yet, again there is a considerable diversity in the structure of VPUs. Every VPU consists of a number of vector functional units, or "pipes" that fulfill one or several functions in the VPU. Every VPU will have pipes that are designated to perform memory access functions, thus assuring the timely delivery of operands to the arithmetic pipes and of storing the results in memory again. Usually there will be several arithmetic functional units for integer/logical arithmetic, for floating-point addition, for multiplication and sometimes a combination of both, a so-called compound operation. Division is performed by an iterative procedure, table look-up, or a combination of both using the add and multiply pipe. In addition, there will almost always be a mask pipe to enable operation on a selected subset of elements in a vector of operands. Lastly, such sets of vector pipes can be replicated within one VPU (2 up to 16-fold replication occurs). Ideally, this will increase the performance per VPU by the same factor provided the bandwidth to memory is adequate.

4. Distributed-memory SIMD Machines

Machines of this type are sometimes also known as *processor-array* machines (Hockney and Jesshope 1987). Because the processors of these machines operate in lock-step, i.e., all processors execute the same instruction at the same time (but on different data items), no synchronization between processors is required. This greatly simplifies the design of such systems. A *control processor* issues the instructions that are to be executed by the processors in the processor array. All currently available DM-SIMD machines use a front-end processor to which they are connected by a data path to the control processor. Operations that cannot be executed by the processor array or by the control processor are offloaded to the front-end system. For instance, I/O may be through the front-end system, by the processor array machine itself or both. Figure 22.3 shows a generic model of a DM-SIMD machine of which actual models will deviate to some degree. Figure 22.3 might suggest that all processors in such systems are connected in a 2-D grid and indeed, the interconnection topology of this type of machines always includes the 2-D grid. As opposing ends of each grid line are also always connected the topology is rather that of a torus. For several machines this is not the only interconnection scheme: They might also be connected in 3-D, diagonally, or more complex structures.

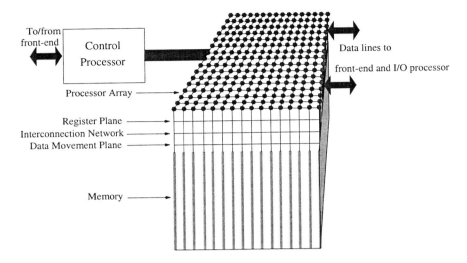

Figure 22.3. A generic block diagram of a distributed-memory SIMD machine.

It is possible to exclude processors in the array from executing an instruction on certain logical conditions, but this means that for the time of this instruction these processors are idle (a direct consequence of the SIMD-type operation) which immediately lowers the performance. Another factor that may adversely affect the speed occurs when data required by processor i resides in the memory of processor j (in fact, as this occurs for all processors at the same time this effectively means that data will have to be permuted across the processors). To access the data in processor j, the data will have to be fetched by this processor and then send through the routing network to processor i. This may be fairly time consuming. For both reasons mentioned DM-SIMD machines are rather specialized in their use when one wants to employ their full parallelism. Generally, they perform excellently on digital signal and image processing and on certain types of Monte Carlo simulations where virtually no data exchange between processors is required and exactly the same type of operations is done on massive datasets with a size that can be made to fit comfortable in these machines.

The control processor as depicted in Figure 22.3 may be more or less intelligent. It issues the instruction sequence that will be executed by the processor array. In the worst case (that means a less autonomous control processor) when an instruction is not fit for execution on the processor array (e.g., a simple print instruction) it might be offloaded to the front-end processor which may be much slower than execution on the control processor. In case of a more autonomous control processor this can be avoided thus saving processing interrupts both on the front-end and the control processor. Most DM-SIMD systems have the possibility to handle I/O independently from the front/end processors. This is not only favorable because the communication between the front-end and back-end systems is avoided. The (specialized) I/O device for the processor-array system is generally much more efficient in providing the necessary data directly to the memory of the processor array. Especially for very data-intensive applications like radar- and image processing such I/O systems are very important.

A feature that is peculiar to this type of machines is that the processors sometimes are of a very simple bit-serial type, i.e., the processors operate on the data items bitwise, irrespective of their type. So, e.g., operations on integers are produced by software routines on these simple bit-serial processors which takes at least as many cycles as the operands are long. So, a 32-bit integer result will be produced two times faster than a 64-bit result. For floating-point operations a similar situation holds, be it that the number of cycles required is a multiple of that needed for an integer operation. As the number of processors in this type of systems is mostly large (1024 or larger, the Alenia Quadrics is a notable exception, however), the slower operation on floating-point numbers can be often compensated for by their number, while the cost per processor is quite low as compared to full floating-point processors. In some cases, however, floating-point coprocessors are added to the processor-array. Their number is 8–16 times lower than that of the bit-serial processors because of the cost argument. An advantage of bit-serial processors is that they may operate on operands of any length. This is particularly advantageous for random number generation (which often boils down to logical manipulation of bits) and for signal processing because in both cases operands of only 1–8 bits are abundant. As the execution time for bit-serial machines is proportional to the length of the operands, this may result in significant speedups.

5. Shared-memory MIMD Machines

In Figure 22.1 already one subclass of this type of machines was shown. In fact, the single-processor vector machine discussed there was a special case of a more general type. The figure shows that more than one FPU and/or VPU may be possible in one system.

The main problem one is confronted with in shared-memory systems is that of the connection of the CPUs to each other and to the memory. As more CPUs are added, the collective bandwidth to the memory ideally should increase linearly with the number of processors, while each pro-

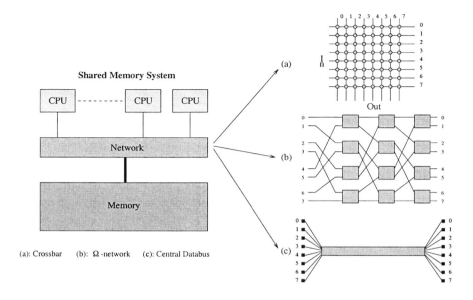

Figure 22.4. Some examples of interconnection structures used in shared-memory MIMD systems.

cessor should preferably communicate directly with all others without the much slower alternative of having to use the memory in an intermediate stage. Unfortunately, full interconnection is quite costly, growing with $\mathcal{O}(n^2)$ while increasing the number of processors with $\mathcal{O}(n)$. So, various alternatives have been tried. Figure 22.4 shows some of the interconnection structures that are (and have been) used.

As can be seen from Figure 22.4, a crossbar uses n^2 connections, an Ω-network uses $n \log_2 n$ connections, while, with the central bus, there is only one connection. This is reflected in the use of each connection path for the different types of interconnections: for a crossbar each data path is direct and does not have to be shared with other elements. In case of the Ω-network there are $\log_2 n$ switching stages and as many data items may have to compete for any path. For the central data bus all data have to share the same bus, so n data items may compete at any time.

The bus connection is the least expensive solution, but it has the obvious drawback that bus contention may occur thus slowing down the computations. Various intricate strategies have been devised using caches associated with the CPUs to minimize the bus traffic. This leads however to a more complicated bus structure which raises the costs. In practice it has proved to be very hard to design buses that are fast enough, especially where the speed of the processors has been increasing very quickly and it imposes an upper bound on the number of processors thus connected that in practice appears not to exceed a number of 10-20. In 1992, a new standard (IEEE P896) for a fast bus to connect either internal system components or to external systems has been defined. This bus, called the Scalable Coherent Interface (SCI) should provide a point-to-point bandwidth of 200-1,000 MB/s. It is in fact used in the HP Exemplar systems, but could also be used within a network of workstations. The SCI is much more than a simple bus and it can act as the hardware network framework for distributed computing, see James et al. (1990).

A multi-stage crossbar is a network with a logarithmic complexity and it has a structure which is situated somewhere in between a bus and a crossbar with respect to potential capacity and costs. The Ω-network as depicted in figure 22.4 is an example. Commercially available machines like the IBM RS/6000 SP, the SGI Origin2000, and the Cenju-4 use such a network structure, but a number of experimental machines also have used this or a similar kind of interconnection. The BBN TC2000 that acted as a virtual shared-memory MIMD system used an analogous type of network (a Butterfly-network) and it is quite conceivable that new machines may use it, especially as the number of processors grows. For a large number of processors the $n \log_2 n$ connections quickly become more attractive than the n^2 used in crossbars. Of course, the switches at the intermediate levels should be sufficiently fast to cope with the bandwidth required. Obviously, not only the *structure* but also the *width* of the links between the processors is important: a network using 16-bit parallel links will have a bandwidth which is 16 times higher than a network with the same topology implemented with serial links.

In all present-day multi-processor vector-processors crossbars are used. This is still feasible because the maximum number of processors in a system is still rather small (32 at most presently). When the number of processors would increase, however, technological problems might arise. Not only it becomes harder to build a crossbar of sufficient speed for the larger numbers of processors, the processors themselves generally also increase in speed individually, doubling the problems of making the speed of the crossbar match that of the bandwidth required by the processors.

Whichever network is used, the type of processors in principle could be arbitrary for any topology. In practice, however, bus structured machines do not have vector processors as the speeds of these would grossly mismatch with any bus that could be constructed with reasonable costs. All available bus-oriented systems use RISC processors. The local caches of the processors can sometimes alleviate the bandwidth problem if the data access can be satisfied by the caches thus avoiding references to the memory.

The systems discussed in this subsection are of the MIMD type and therefore different tasks may run on different processors simultaneously. In many cases synchronization between tasks is required and again the interconnection structure is here very important. Most vector-processors employ special communication registers within the CPUs by which they can communicate directly with the other CPUs they have to synchronize with. A minority of systems synchronize via the shared memory. Generally, this is much slower but may still be acceptable when the synchronization occurs relatively seldom. Of course in bus-based systems communication also has to be done via a bus. This bus is mostly separated from the data bus to assure a maximum speed for the synchronization.

6. Distributed-memory MIMD Machines

The class of DM-MIMD machines is undoubtly the fastest growing part in the family of high-performance computers. Although this type of machines is more difficult to deal with than shared-memory machines and DM-SIMD machines. The latter type of machines are processor-array systems in which the data structures that are candidates for parallelization are vectors and multi-dimensional arrays that are laid out automatically on the processor array by the system software. For shared-memory systems the data distribution is completely transparent to the user. This is quite different for DM-MIMD systems where the user has to distribute the data over the processors and also the data exchange between processors has to be performed explicitly. The initial reluctance to use DM-MIMD machines seems to have been decreased. Partly this is due to the now existing standard for communication software (Geist et al. 1994, Snir et al. 1998, Gropp et al. 1998) and partly because, at least theoretically, this class of systems is able to outperform all other types of machines.

The advantages of DM-MIMD systems are clear: the bandwidth problem that haunts shared-memory systems is avoided because the bandwidth scales up automatically with the number of processors. Furthermore, the speed of the memory which is another critical issue with shared-memory systems (to get a peak performance that is comparable to that of DM-MIMD systems, the processors of the shared-memory machines should be very fast and the speed of the memory should match it) is less important for the DM-MIMD machines, because more processors can be configured without the afore mentioned bandwidth problems.

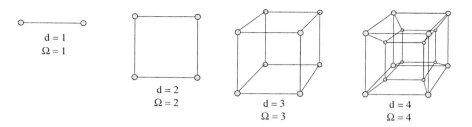

Figure 22.5. 1-, 2-, 3-, and 4-dimensional hypercube connections

Of course, DM-MIMD systems also have their disadvantages: The communication between processors is much slower than in SM-MIMD systems, and so, the synchronization overhead in case of communicating tasks is generally orders of magnitude higher than in shared-memory machines. Moreover, the access to data that are not in the local memory belonging to a particular processor have to be obtained from non-local memory (or memories). This is again on most systems very slow as compared to local data access. When the structure of a problem dictates a frequent exchange of data between processors and/or requires many processor synchronizations, it may well be that only a very small fraction of the theoretical peak speed can be obtained. As already mentioned, the data and task decomposition are factors that mostly have to be dealt with explicitly, which may be far from trivial.

It will be clear from the paragraph above that also for DM-MIMD machines both the topology and the speed of the data paths are of crucial importance for the practical usefulness of a system. Again, as in the section on SM-MIMD systems, the richness of the connection structure has to be balanced against the costs. Of the many conceivable interconnection structures only a few are popular in practice. One of these is the so-called hypercube topology as depicted in Figure 22.5.

A nice feature of the hypercube topology is that for a hypercube with 2^d nodes the number of steps to be taken between any two nodes is at most d. So, the dimension of the network grows only logarithmically with the number of nodes. In addition, theoretically, it is possible to simulate any other topology on a hypercube: trees, rings, 2-D and 3-D meshes, etc. In practice, the exact topology for hypercubes does not matter too much anymore because all systems in the market today employ what is called "wormhole routing". This means that when a message is sent from node i to node j, a header message is sent from i to j, resulting in a direct connection between these nodes. As soon as this connection is established, the data proper is sent through this connection without disturbing the operation of the intermediate nodes.

Except for a small amount of time in setting up the connection between nodes, the communication time has become virtually independent of the distance between the nodes. Of course, when several messages in a busy network have to compete for the same paths, waiting times are incurred as in any network that does not directly connect any processor to all others and often rerouting strategies are employed to circumvent busy links.

A fair amount of massively parallel DM-MIMD systems seem to favor a 2- or 3-D mesh (torus) structure. The rationale for this seems to be that most large-scale physical simulations can be mapped efficiently on this topology and that a richer interconnection structure hardly pays off. However, some systems maintain (an) additional network(s) besides the mesh to handle certain bottlenecks in data distribution and retrieval (Horie et al. 1991).

A large fraction of systems in the DM-MIMD class employ crossbars. For relatively small amounts of processors (in the order of 64) this may be a direct or 1-stage crossbar, while to connect larger numbers of nodes multi-stage crossbars are used, i.e., the connections of a crossbar at level 1 connect to a crossbar at level 2, etc., instead of directly to nodes at more remote distances in the topology. In this way it is possible to connect in the order of a few thousands of nodes through only a few switching stages. In addition to the hypercube structure, other logarithmic complexity networks like Butterfly, Ω, or shuffle-exchange networks are often employed in such systems.

As with SM-MIMD machines, a node may in principle consist of any type of processor (scalar or vector) for computation or transaction processing together with local memory (with or without cache) and, in almost all cases, a separate communication processor with links to connect the node to its neighbors. Nowadays, the node processors are mostly off-the-shelf RISC processors sometimes enhanced by vector processors. A problem that is peculiar to this DM-MIMD systems is the mismatch of communication vs. computation speed that may occur when the node processors are upgraded without also speeding up the intercommunication. In some cases this may result in turning computational-bound problems into communication-bound problems.

7. CC-NUMA Machines

As already mentioned in the introduction, a trend can be observed to build systems that have a rather small (up to 16) number of RISC processors that are tightly integrated in a cluster, a Symmetric Multi-Processing (SMP) node. The processors in such a node are virtually al-

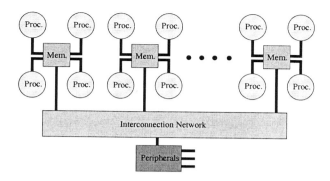

Figure 22.6. Block diagram of a system with a "hybrid" network: clusters of four CPUs are connected by a crossbar. The clusters are connected by a less expensive network, e.g., a Butterfly network

ways connected by a 1-stage crossbar while these clusters are connected by a less costly network. Such a system may look as depicted in Figure 22.6. Note that in Figure 22.6 all CPUs in a cluster are connected to a common part of the memory. This is similar to the policy mentioned for large vector-processor ensembles mentioned above but with the important difference that all of the processors can access all of the address space. Therefore, such systems can be considered as SM-MIMD machines. On the other hand, because the memory is physically distributed, it cannot be guaranteed that a data access operation always will be satisfied within the same time. Therefore such machines are called CC-NUMA systems where CC-NUMA stands for Cache Coherent Non-Uniform Memory Access. The term "Cache Coherent" refers to the fact that for all CPUs any variable that is to be used must have a consistent value. Therefore, it must be assured that the caches that provide these variables are also consistent in this respect. There are various ways to ensure that the caches of the CPUs are coherent. One is the *snoopy bus protocol* in which the caches listen in on transport of variables to any of the CPUs and update their own copies of these variables if they have them. Another way is the *directory memory*, a special part of memory which enables to keep track of the all copies of variables and of their validness.

For all practical purposes we can classify these systems as being SM-MIMD machines also because special assisting hardware/software (such as a directory memory) has been incorporated to establish a single system image although the memory is physically distributed.

8. Recount of (almost) Available Systems

In this section we give a recount of all types of systems as discussed in previous sections. When vendors market more than one type of machine we will discuss them in distinct subsections. So, for instance, we will discuss NEC systems under entries, SX-5 and Cenju-4 because they have a very different structure.

The systems are presented alphabetically. The "Machine type" entry shortly characterizes the type of system as discussed previously: Processor Array, CC-NUMA, etc.

8.1. The Alenia Quadrics

Machine type: Processor array. **Models**: Quadrics Qx, QHx, $x = 1, \ldots, 16$.
Front-end: Almost any workstation.
Operating system: Internal OS transparent to the user, Unix on front-end.
Connection structure: 3-D mesh (see remarks).
Compilers: TAO: a Fortran 77 compiler with some Fortran 90 and some proprietary array extensions.
Vendors information Web page: www.quadrics.com
Year of introduction: 1994.

System parameters

Model	Qx	QHx
Clock cycle	40 ns	40 ns
Theor. peak performance		
Per Proc. (32-bits)	50 Mflop/s	50 Mflop/s
Maximal (32-bits)	6.4 Gflop/s	100 Gflop/s
Main memory	$\leq$2 GB	$\leq$32 GB
No. of processors	8–128	128–2048
Communication bandwidth		
Per Proc.	50 MB/s	50 MB/s
Aggregate local	$\leq$6 GB/s	$\leq$96 GB/s
Aggregate non-local	$\leq$1.5GB/s	$\leq$24 GB/s

Remarks: The Quadrics is a commercial spin-off of the APE-100 project of the Italian National Institute for Nuclear Physics. Systems are available in multiples of 8 processor nodes in the Q-model where up to 16 boards can be fitted into one crate or in multiples of 128 nodes in the QH-model by adding up to 15 crates to the minimal 1-crate system. The interconnection topology of the Quadrics is a 3-D grid with interconnections to the opposite sides (so, in effect a 3-D torus). The 8-node floating-point boards (FPBs) are plugged into the crate backplane which provides point-to-point communication and global control distribution. The FPBs are configured as 2^3 cubes that are connected to the other boards appropriately to arrive at the 3-D grid structure.

The basic floating-point processor, the so-called MAD chip, contains a register file of 128 registers. Of these registers the first two hold permanently the values 0 and 1 to be able to express any addition or multiplication as a "normal operation", i.e., a combined multiply-add operation, where a multiplication is of the form, $a \times b + 0$ and an addition is $a \times 1 + b$. In favorable circumstances the processor can therefore deliver two floating-point operations per cycle. Instructions are centrally issued by the controller at a rate of one instruction every two clock cycles.

Communication is controlled by the Memory Controller and the Communication Controller which are both housed on the backplane of a crate. When the Memory Controller generates an address it is decoded by the Communication Controller. In case non-local access is desired, the Communication Controller will provide the necessary data transmission. The memory bandwidth per processor is 50 MB/s which means that every 2 cycles an operand can be shipped into or out of a processor. The bandwidth for non-local communication turns out to be only four times smaller than local memory access.

The Quadrics communicates with the front-end system via a T805 transputer-based interface system, called the Local Asynchronous Interface (LAI). The interface can write and read the memories of the nodes and the Controller. Presently, the bandwidth of the interface to the front-end processor is not very large (1 MB/s). It is expected that this can be improved by about a factor of 30 in the near future. I/O has to be performed via the front-end system and will therefore be relatively slow.

The TAO language has several extensions to employ the SIMD features of the Quadrics. Firstly, floating-point variables are assumed to be local to the processor that owns them, while integer variables are assumed to be global. Local variables can be promoted to global variables. Other extensions are the ANY, ALL, and WHERE/END WHERE keywords that can be used for global testing and control. Processors that not meet a global condition effectively skip the operation(s) that are associated with it. For easy referencing nearest-neighbor locations special constants LEFT, RIGHT, UP, DOWN, FRONT, and BACK are provided. In addition, new data types and operators on these data types are supported together with overloading of operators. This enables very concise code for certain types of calculations.

Measured performances: No measured performances have been reported for this machine.

8.2. The Avalon A12

Machine type: RISC-based distributed-memory multi-processor.
Models: Avalon A12.
Operating system: AVALON micro kernel based Unix (Image compatible with Digital Unix).
Connection structure: Multistage variable (see remarks).
Compilers: Fortran 77, Fortran 90 extensions, HPF, ANSI C.
Vendors information Web page: www.teraflop.com/
Year of introduction: 1996.

System parameters

Model	A12
Clock cycle	2.5 ns
Theor. peak performance	
Per Proc. (64-bits)	800 Mflop/s
Maximal	1.3 Tflop/s
Memory/node	≤ 1 GB
Memory/maximal	1.7 TB
No. of processors	12–1680
Communication bandwidth	
Point-to-point	128–400 MB/s
Bisectional (maximal)	10 GB/s

Remarks: The A12 is be based on the DEC Alpha 21164 RISC processor. The processor used in the system has a clock cycle of 2.5 ns. However, most of the information given at the vendors Web page still uses the data for a node processor with a 3.3 ns clock to describe the configuration properties. The Web information is therefore internally somewhat inconsistent. Because the Alpha 21164 has dual floating-point arithmetic pipes it will deliver a theoretical peak performance of 800 Mflop/s. The maximum configuration of the system is given as 1680 processors The first and second level cache reside on chip, a 1 MB third level cache is provided on each A12 CPU card. The bandwidth to/from the first level cache is sufficient to transport two operands to the CPU and to ship one result back in one cycle. The second level cache has two-thirds of this bandwidth, while the third level cache has the capability of providing an 64-bit word every two cycles. The bandwidth to/from memory is 400 MB/s or one 64-bit word every 8 cycles. The memory has two-way interleaved banks of a memory that can be up to 1 GB/node.

Each CPU card contains a Alpha 21164 processor, the third level or B cache and the local memory for that node. Twelve CPU cards can be housed in one crate which has a full crossbar backplane. This yields a inter-node bandwidth of slightly under 400 MB/s between the cards within one crate. Apart from the 12 slots for CPU cards, there are two extra dual channel slots that can accommodate communication cards that provide the connections with other crates. For the in-crate crossbar CMOS technology is used. However, for the inter-crate connections ECL logic is employed. The actual connections between crates are made by coaxial cables. This way of connection provides a large flexibility in the overall interconnection topology: one could build trees or toruses or a secondary level crossbar (in the last case one crate should be filled entirely with communication cards to build a 144 processor system). The communication speed between crates is less fast (but still respectable): 128 MB/s. Various configurations are described at the Web-address given above.

I/O can be configured in various ways: It is possible to put 32-bit or 64-bit PCI expansion cards on each CPU card to obtain what Avalon calls "Type 1 I/O nodes". Also, a direct switch connection via a variant of the communication card can be made to the outside world. Depending on the number of cards the bandwidth is 400 or 800 MB/s for this type 3 I/O node. The type 2 I/O node is in fact a dedicated TCP/IP connection as needed for the control workstation required by the system.

Measured Performances: A 140-node A12 was installed by the end of 1996 at Los Alamos National Laboratory of which the processors had a faster clock: 1.88 ns, instead of 2.5 ns. In Dongarra (1999) a speed of 48.6 Gflop/s was reported for this configuration on the solution of a full linear system of order 62720, 33% of the Theoretical Peak Performance.

8.3. The Cambridge Parallel Processing Gamma II

Machine type: Processor array.
Models: Gamma II Plus 1000, Gamma II Plus 4000.
Front-end: DEC, HP, or Sun workstation, stand-alone for dedicated applications.
Operating system: Internal OS transparent to the user, Unix on front-end.
Connection structure: 2-D mesh, row- and column data paths (see remarks).
Compilers: FORTRAN-PLUS (a Fortran 77 compiler with some Fortran 90 and some proprietary array extensions), C++.

Vendors information Web page: www.cppus.com
Year of introduction: 1995.

System parameters

Model	Gamma II Plus 1000	Gamma II Plus 4000
Clock cycle	33 ns	33 ns
Theor. peak performance		
Per Proc. (32-bits)	0.6 Mflop/s	0.6 Mflop/s
Maximal (32-bits)	0.6 Gflop/s	2.4 Gflop/s
1-bit (Gop/s)	30.7	122.8
8-bit (Gop/s)	30.7	122.8
Program memory	$\leq$4 MB	$\leq$4 MB
No. of processors	$\leq$128 MB	$\leq$512 MB
Internal communication speed		
Across row, column	120 MB/s	480 MB/s
Memory to PE	3.84 GB/s	15.4 GB/s

Remarks: In November 1995 the new Gamma II Plus models have been announced by CPP. In essence there is not much difference with its predecessor the DAP Gamma. However, the clock cycle has tripled to 33 ns with an equivalent rise in the peak performance of the systems.

The Gamma II is presented as the fourth generation of this type of machine. Indeed, the macro architecture of the systems has hardly changed since the first ICL DAP (the first generation of this system) was conceived. As in the ICL DAP in the Gamma 1000 models the 1024 processors are ordered in a 32×32 array, while the Gamma 4000 has 4096 processors arranged in a 64×64 square.

The systems are able to operate byte parallel on appropriate operands to speed up floating-point operations, however, for logical operations bit-wise operations are possible, which makes the machines quite fast in this respect. As the byte parallel code consists of separate sequences of microcode instructions, the bit processor plane and the byte processor plane are in fact independent and can work in parallel. This is also the case for I/O operations. Also character-handling can be done very efficiently. This is the reason that Gamma systems are often used for full text searches.

As in all processor-array machines, the control processor (called the Master Control Unit (MCU) in the DAP) has a separate memory to hold program instructions while the data are held in the data memory associated with each Processing Element (PE) in the processor array. So, for a Gamma 1000 with 128 MB of data memory each PE has 128 KB of data memory directly associated to it. To access data in other

PEs memories these must be brought up to the data routing plane and shifted to the appropriate processor.

As already mentioned under the heading of the connection structure, there are two ways of connecting the PEs. One is the 2-D mesh that connects each element to its North-, East-, West-, and South neighbor. In addition there are row- and column data paths that enable the fast broadcast of a row or column to an entire matrix by replication. Conversely, they can be used for row or column-wise reduction of matrix objects into a column or row-vector of results from, e.g., a summing or maximum operation.

Separate I/O processors and disk systems can be attached to the Gamma directly thus not burdening the front-end machine (and the connection between front-end and Gamma) with I/O operations and unnecessary data transport. One of these I/O devices is the GIOC that can transport data to the data memory at a sustained rate of 80 MB/s transposing the data to the vertical storage mode of the data memory on the fly. Also, a direct video interface is available to operate a frame buffer.

A nice (non-standard) feature of the FORTRAN-PLUS compiler is the possibility to use logical matrices as indexing objects for computational matrix objects. This enables a very compact notation for conditional execution on the processor array. Since 1997 also C++ is available.

Measured Performances: In Flanders (1991) the speed of matrix multiplication on various DAP models (precursors of the Gamma systems) is analyzed. The documentation states 32-bit floating-point add speed of 1.68 Gflop/s on 4096 PEs, while a 32-bit 1,024 complex FFT would attain 2.49 Gflop/s. No independent performance figures for the Gamma II Plus systems are available.

8.4. The C-DAC PARAM OpenFrame system

Machine type: RISC-based distributed-memory multi-processor.
Models: PARAM OpenFrame 9000 system.
Operating system: PARAS micro kernel based Unix (compatible with Sun's Solaris).
Connection structure: Multistage variable (see remarks).
Compilers: Fortran 77, Fortran 90, HPF, ANSI C, C++.
Vendors information Web page: www.soft.net/cdac/
Year of introduction: 1996.

System parameters

Model	OpenFrame 9000
Clock cycle	3 ns (see remarks)
Theor. peak performance	
Per Proc. (64-bits)	600 Mflop/s (see remarks)
Maximal	600 Gflop/s (see remarks)
Memory/node	—
Memory/maximal	—
No. of processors	1–1024
Communication bandwidth	
Point-to-point	80 MB/s
Aggregate bandwidth	3.2 GB/s

Remarks: The OpenFrame 9000 system is the fourth generation of CDAC machines that is developed by CDAC, the Centre for Development of Advanced Computing, an institute in India that has as its mission to develop an manufacture "state-of-the-art open architecture supercomputers". This system is the second generation that is marketed abroad. In the predecessor, the PARAM 9000/SS SuperSPARC II processors were used. In the present model Sun UltraSPARCs or DEC/Compaq Alpha chips are employed or even mixtures of these. Also, the maximum possible number of processors has been increased from 200 to 1024 in replicatable units of 32 processors. As the type of processor and the clock cycle are not fixed, no theoretical peak performance can be specified. When the same basic processors are assumed as Sun employs in most of its Enterprise servers, the estimates as given in the parameter list above seem more or less indicative. The documentation also does not reveal any details of the type of the interconnection network except that cut-through wormhole routing is used. A point-to-point communication bandwidth of 80 MB/s (bi-directional) is quoted. The aggregate bandwidth for a maximal configuration is 3.2 GB/s.

The amount of available software shows that the PARAM OpenFrame 9000 is not a first-generation system. Apart from Fortran 77, Fortran 90, HPF, and C++ are available and the CORE, MPI, and PVM message passing interfaces are available. There is a parallel debugger, a proprietary performance evaluation tool called AIDE, while TOTALVIEW can be delivered on request.

In addition, a library of parallel routines, PARUL, is available. This library contains PVM versions of dense linear algebra routines, eigenvalue routines, and FFTs.

Measured Performances: No measured performances of the PARAM OpenFrame 9000 are available at this moment for any configuration.

8.5. The Compaq/DEC GS60/140

Machine type: RISC-based shared-memory multiprocessor.
Models: GS60, GS140, Cluster.
Operating system: Digital Unix (DEC's flavor of Unix).
Compilers: Fortran 77, HPF, C, C++.
Vendors information Web page: www.digital.com/info/hpc
Year of introduction: 1998.

System parameters

Model	GS60	GS140	Cluster
Clock cycle	1.66 ns	1.66 ns	1.66 ns
Theor. peak performance			
Per Proc.	1.2 Gflop/s	1.2 Gflop/s	1.2 Gflop/s
Maximal	7.2 Gflop/s	16.8 Gflop/s	67.2 Gflop/s
Memory	$\leq$ 12 GB	$\leq$ 28 GB	$\leq$ 112 GB
No. of processors	$\leq$ 6	$\leq$ 14	$\leq$ 56
Memory bandwidth			
Processor/memory	1.87 GB/s	1.87 GB/s	1.87 GB/s
Between cluster nodes	—	—	100 MB/s

Remarks: The GS60 and GS140 are almost identical to their predecessors, the AlphaServers 8200 and 8400. The difference lies in the processor that is used: instead of the Alpha 21164 at a clock rate of 1.6 ns in the new systems an Alpha 21264 with a clock rate of 1.66 ns is used. Note that this leads to a *decrease* in the theoretical peak performance from 1.25 to 1.2 Gflop/s per processor. However, Compaq claims that the new processor will generally give a performance *increase* of a factor 2.5 with respect to the Alpha 21164 which is not unlikely with the improvements made in the chip.

The GS60 and GS140 are symmetric multi-processing systems. The GS60 model is a somewhat smaller copy of the GS140 model: in the GS60 a maximum of 6 CPUs can be accommodated while this number is 14 for the GS140 model. Also, there is room for at most 12 GB of memory in the GS60 while the GS140 can house 28 GB. However, the amount of CPUs and memory is not independent. For instance, the GS140 has 9 system slots. One of these is reserved for I/O and one will have to contain at least one CPU module which can contain 1 or 2 CPUs. From the remaining slots 7 can be used either for memory or for a CPU module. So, one has to choose for either higher computational power or for more memory. This can potentially be a problem for large applications that require both.

The GS systems (GS stands for Global Server) can be clustered using PCI bus MemoryChannel link cables that are connected to a hub. The systems need not be of the same model. The bandwidth of this interconnect is slightly over 100 MB/s. Up to four systems can be coupled in this way. To support this kind of cluster computing, HPF and optimized versions of PVM and MPI are available.

Measured Performances: For the the GS60 and GS140 no performance figures are available yet.

8.6. The Fujitsu AP3000

Machine type: RISC-based distributed-memory multi-processor.
Models: AP3000.
Operating system: Cell OS (transparent to the user) and Solaris (Sun's Unix variant) on the front-end system.
Connection structure: 2-D torus.
Compilers: Parallel Fortran/AP, Fortran 90, HPF, C, C++.
Vendors information Web page: www.fujitsu.com
Year of introduction: 1996.

System parameters

Model	AP3000
Clock cycle	3.3 ns
Theor. peak performance	
Per Proc. (64-bits)	600 Mflop/s
Maximal	614 Gflop/s
Memory/node	≤ 2 GB
Memory/maximal	≤ 2 TB
No. of processors	4–1024
Communication bandwidth	
Point-to-point	200 MB/s

Remarks: The AP3000 is the successor of the earlier AP1000 system. Although the name could suggest otherwise, few characteristics of the AP1000 have been retained except that Sun SPARC processors are used in the nodes. No front-end processor is required anymore as in the former system.

Also the communication network has been simplified considerably with respect to that in the earlier model: where three different networks were present in the AP1000 (see Horie et al. (1991)), in the AP3000 the nodes are connected in a 2-D torus structure with a bi-directional bandwidth of 200 MB/s and there is a separate control network. The maximum amount of memory is huge: a full 1024 node system can accommodate 2 TB.

Another difference with the AP1000 system is that the fastest nodes (the U300 nodes described here) can have either 1 or 2 CPUs as opposed to only one CPU in the AP1000. The two CPUs share the on-board memory.

The available software for the AP3000 is extensive: Parallel Fortran/AP is a Fortran 77 with extensions that offers a shared-memory-like programming model for the system. In addition, HPF is available and the machine can also be used with a message passing model as customized MPI/AP and PVM/AP are offered. As sequential languages to be used with the message passing libraries Fortran 90, C and C++ are available.

Measured Performances: The system has been announced in March 1996 and installations have been done in Japan, the University of Singapore and at the Australian National University but as yet no performance figures are published.

8.7. The Fujitsu VPP700 series

Machine type: Distributed-memory vector multi-processor.
Models: VX-E, VPP300-E, VPP700-E.
Operating system: UXP/V (a V5.4 based variant of Unix).
Connection structure: Full distributed crossbar.
Compilers: Fortran 90/VP (Fortran 90 Vector compiler), Fortran 90/VPP (Fortran 90 Vector Parallel compiler), C/VP (C Vector compiler), C, C++.
Vendors information Web page: www.fujitsu.com
Year of introduction: VX, VPP300: 1995, VPP700: 1996.

System parameters

Model	VX-E	VPP300-E	VPP700-E
Clock cycle	6.6 ns	6.6 ns	6.6 ns
Theor. peak performance			
Per Proc. (64-bits)	2.4 Gflop/s	2.4 Gflop/s	2.4 Gflop/s
Maximal	9.6 Gflop/s	38.4 Gflop/s	614.4 Gflop/s
Memory/node	$\leq$ 2 GB	$\leq$ 2 GB	$\leq$ 2 GB
Memory/maximal	$\leq$ 8 GB	$\leq$ 32 GB	$\leq$ 512 GB
No. of processors	1–4	1–16	8–256
Memory bandw./proc.	19.6 GB/s	19.6 GB/s	19.6 GB/s
Communication bandwidth			
Point-to-point	615 MB/s	615 MB/s	615 MB/s

Remarks: The VX-E, VPP300-E, and VPP700-E systems (with E for extended) are "midlife kickers": minor extensions of the VX, VPP300, and VPP700. The only difference is a slightly faster system clock: 6.6 ns in the E models instead of the 7 ns in the former systems. There are no architectural changes. The VPP300 is a successor to the earlier VPP500. It is a much cheaper CMOS implementation of its predecessor with some important differences. First, no VPX200 front-end system is required anymore. Second, the crossbar that is used to connect the vector nodes is distributed. Therefore, the cost of a system is scalable: one does not need to buy a complete enclosure with the full crossbar for only a few nodes. The VX series is in fact a smaller version of the VPP300 with a maximum of 4 processors. Both the VX machines and the VPP300 systems are air-cooled.

The architecture of the VPP300 nodes is almost identical to that of the VPP500: Each node, called a Processing Element (PE) in the system is a powerful (2.4 Gflop/s peak speed with a 6.6 ns clock) vector processor in its own right. The vector processor is complemented by a Large Instruction Word scalar processor with a peak speed of 300 Mflop/s. The scalar instruction format is 64 bits wide and may cause the execution of three operations in parallel. Each PE has a memory of up to 2 GB while a PE communicates with its fellow PEs at a point-to-point speed of 570 MB/s. This communication is cared for by separate Data Transfer Units (DTUs). To enhance the communication efficiency, the DTU has various transfer modes like contiguous, stride, sub array, and indirect access. Also translation of logical to physical PE-ids and from Logical in-PE address to real address are handled by the DTUs. When synchronization is required each PE can set its corresponding bit in the Synchronization Register (SR). The value of the SR is broadcast to all PEs and synchronization has occurred if the SR has all its bits set for the relevant PEs. This method is comparable to the use of synchronization registers in shared-memory vector processors and much faster than synchronizing via memory.

The VPP700 is a logical extension of the VPP300. While the processors in the latter machine are connected by a full crossbar, the maximum configuration of a VPP700 consists of 16 clusters of 16 processors connected by a level-2 crossbar. So, a fully configured VPP700 consists in fact of 16 full VPP300s. Because the diameter of the network is 2 (for the larger configurations) instead of 1 as in the VPP300, the communication time between processors will be slightly larger. At the moment this worst case increase is not exactly known to the author.

The Fortran compiler that comes with the VPP300/700 has extensions that enable data decomposition by compiler directives. This evades in many cases restructuring of the code. The directives are different from those as defined in the High Performance Fortran Proposal but it should be easy to adapt them. Furthermore, it is possible do define parallel regions, barriers, etc., via directives, while there are several intrinsic functions to enquire about the number of processors and to execute POST/WAIT commands. Furthermore, also a message passing programming style is possible by using the PVM or PARMACS communication libraries that are available. Of course the software for the VPP700 and the VPP300 is exactly the same and the systems can run each others executables.

Measured Performances: Of the VX-E, VPP300-E, and VPP700-E no performance figures are known but in Dongarra (1999) results for the VX, the VPP300, and the VPP700 are given. The speed for solving dense linear system of sizes 28,800 59,200, and 111,360 was 8.6, 34.1, and 213 Gflop/s on a 4 proc. VX, a 16 proc. VPP300, and a 116 proc. VPP700, respectively.

8.8. The Hitachi S3600 series

Machine type: Vector-processor
Models: S3600/120, S3600/140, S3600/160, S3600/180
Operating system: VOS3/HAP/ES (IBM MVS compatible) and OSF/1
Compilers: FORT77/HAP vectorizing Fortran 77, C, C++.
Vendors information Web page: none.
Year of introduction: 1994.

System parameters

Model	S3600/120	S3600/140	S3600/160	S3600/180
Clock cycle VPU	4 ns	4 ns	4 ns	4 ns
Clock cycle scal. proc.	8 ns	8 ns	8 ns	8 ns
Theor. peak perform.	0.25 Gflop/s	0.25 Gflop/s	1.0 Gflop/s	2.0 Gflop/s
Main memory	128-256 MB	256-512 MB	256-512 MB	512-1024 MB
Extended memory	$\leq$ 6 GB	$\leq$ 16 GB	$\leq$ 16 GB	$\leq$ 16 GB

Remarks: The S3600 system is the only single-CPU vector-processor that is still marketed and it might be withdrawn soon in favor of the Hitachi SR8000 (see below).
The speed differences between the different models stem from replication of the multiply/add pipe in the models S3600/120-180. The /160 and /180 models have respectively two- and four-fold sets of a separate add- and a multi-functional multiply/add vector pipes. This should lead to a maximum of 3 results per clock cycle per pipe set. So, contrary to the information given by the vendor, the maximum performance of, e.g., the /180 should in some situations be 3 Gflop/s instead of 2.

All configurations of the S3600, as in its direct predecessor the S-820, are air cooled while most machines in this class rely at least on water cooling.

Unlike the S-820 series, the S3600 series is also marketed worldwide, and not only in Japan.

Measured performances: In Dongarra (1999) a speed of 851 Mflop/s for the solution of a full linear system of order 1000 is reported for the S3600/160. The S3600/180 attains a performance of 1672 Mflop/s on the same problem.

8.9. The Hitachi SR8000

Machine type: RISC-based distributed-memory multi-processor.
Models: SR8000.
Operating system: HI-UX/MPP (Micro kernel Mach 3.0).
Connection structure: Multi-dimensional crossbar (see remarks).
Compilers: Fortran 77, Fortran 90, Parallel Fortran, HPF, C, C++.
Vendors information Web page:
www.hitachi.co.jp/Prod/comp/hpc/eng/sr81e.html
Year of introduction: 1998.

System parameters

Model	SR8000
Clock cycle	4.0 ns
Theor. peak performance	
Per Proc. (64-bits)	8 Gflop/s
Maximal	1 Tflop/s
Memory/node	$\leq$ 8 GB
Memory/maximal	$\leq$ 1 TB
No. of processors	4–128
Communication bandwidth	
Point-to-point	1 GB/s

Remarks: The SR8000 is the third generation of distributed-memory parallel systems of Hitachi. It is to replace both its direct predecessor, the SR2201 and the late top-vector-processor, the S-3800 (see 9).

The basic node processor is a 4 ns clock PowerPC node with major enhancements made by Hitachi. E.g., a hardware barrier synchronization is added and the additions required for "Pseudo Vector Processing" (PVP). The latter means that for operations on long vectors one does not incur the detrimental effects of cache misses that often ruin the performance of RISC processors unless code is carefully blocked and unrolled. This facility was already available on the SR2201 and experiments have shown that this idea seems to work well (see Hit (2001)).

The peak performance per basic processor, or IP, can be attained with 2 simultaneous multiply/add instructions resulting in a speed of 1 Gflop/s. However, eight basic processors are coupled to form one processing node all addressing a common part of the memory. For the user this node is the basic computing entity with a peak speed of 8 Gflop/s. Hitachi refers to this node configuration as COMPAS, Co-operative Micro-Processors in single Address Space. In fact this is a kind of SMP clustering as discussed in sections 2 and 7. A difference with most of these systems is that for the user the individual processors in a cluster node are not accessible. Every node also contains an SP, a system processor that performs system tasks, manages communication with other nodes and a range of I/O devices.

The SR8000 has a multi-dimensional crossbar with a bi-directional link speed of 1 GB/s. From 4–8 nodes the cross-section of the network is 1 hop. For configurations 16–64 it is 2 hops and for a 128-node system it is 3 hops.

Like in some other systems as the SGI/Cray T3E (8.17), Meiko CS-2 (8.12), and the NEC Cenju-4 (8.13), one is able to directly access the memories of remote processors. Together with the fast hardware-based barrier synchronization this should allow for writing distributed programs with very low parallelization overhead.

The following software products are supported in addition to those already mentioned above: PVM, MPI, PARMACS, Linda, and FORGE90. In addition several numerical libraries like NAG and IMSL are offered.

Measured Performances: As of January 1999 the first installations were planned. A maximal configuration has just been placed at the University of Tokyo Computing Centre. At this moment no performance figures are available.

8.10. The HP Exemplar V2500

Machine type: RISC-based CC-NUMA system.
Models: Exemplar V2500.
Operating system: HP-UX (HP's usual Unix flavor).
Connection structure: Ring.
Compilers: Fortran 77, Fortran 90, Parallel Fortran, HPF, C, C++.
Vendors information Web page: www.hp.com
Year of introduction: 1998.

System parameters

Model	Exemplar V2500
Clock cycle	2.27 ns
Theor. peak performance	
Per Proc. (64-bits)	1.76 Gflop/s
Maximal	225.3 Gflop/s
Memory/node	≤ 1 GB
Memory/maximal	≤ 16 GB
No. of processors	2–128
Communication bandwidth	
Aggregate (per cabinet)	15.36 GB/s
Aggregate (inter-cabinet)	3.84 GB/s

Remarks: The V2500 is the latest in the series of Exemplar systems that have been offered first by Convex and later by HP since 1995 (see section 9). The architecture, however, has not radically changed: up to 32 PA-RISC 8500 chips are clustered via a crossbar to form an SMP node. The PA-RISC 8500 CPUs run at a clock cycle of 2.27 ns. As a CPU contains 2 floating-point units that are able do execute a combined floating multiply-add instruction, in favorable circumstances four flops/cycle can be achieved and a Theoretical Peak Performance of 1.76 Gflop/s per CPU can be attained. Per SMP node the peak speed is 56.32 Gflop/s.

Up to four SMP nodes can be coupled by a so-called SCA HyperLink, uni-directional SCI rings with an aggregate bandwidth of 3.84 GB/s, while the aggregate bandwidth within an SMP node is 15.36 GB/s. The HyperLinks tolerate multiple outstanding requests and, in addition, there is a "HyperLinkcache" that both help in hiding the communication latency in inter-node communication.

As in the former systems a shared memory parallel model is supported. HP is a partner in the OpenMP organization and will therefore make available this style of shared-memory parallel programming in addition to (and later on instead of) its proprietary parallel model. The shared-memory parallelism is not confined to the SMP nodes: a multi-node system can be addressed globally making the Exemplar a CC-NUMA system. The memory latency within and between nodes differs by about a factor of 3–3.5.

Measured Performances: In Dongarra (1999) a speed of 31.59 Gflop/s is reported for a 1-cabinet, 32 processor system when solving a 41,000-order dense linear system, an efficiency of 56% on this problem.

8.11. The IBM RS/6000 SP

Machine type: RISC-based distributed-memory multi-processor.
Models: IBM RS/6000 SP.
Operating system: AIX (IBM's Unix variant).
Connection structure: Ω-switch.
Compilers: XL Fortran (Fortran 90), HPF, XL C, C++.
Vendors information Web page:
www.rs6000.ibm.com/hardware/largescale/index.html.
Year of introduction: 1998 (POWER3 SMP), 1997 (P2SC).

System parameters

Model	RS/6000 SP POWER3 SMP	RS/6000 SP P2SC
Clock cycle	3 ns	6.25 ns
Theor. peak perform.		
Per Proc. (64-bits)	666 Mflop/s	640 Mflop/s
Maximal	variable (see remarks)	variable (see remarks)
Memory/node	≤ 4 GB	$\leq 1/2$ GB (see remarks)
Memory/maximal	≤ 0.5 TB	≤ 1 TB
No. of processors	8–512	8–512
Comm. bandwidth		
Point-to-point	160 MB/s	160 MB/s

Remarks: The variety in the types of nodes that are available for the RS/6000 SP is short of bewildering (IBM provides a 48 page document for selecting the appropriate nodes in an SP system). We only discuss the subset that is most relevant for scientific and technical computation, i.e., the P2SC thin nodes and the POWER3 SMP thin and wide nodes. The P2SC nodes can deliver 4 floating-point results per clock cycle. As the fastest of the P2SC nodes has a clock cycle of 6.25 ns, it has a peak performance of 640 Mflop/s while a single POWER3 processor can attain 666 Mflop/s at maximum with 2 floating-point units. Another difference is that the P2SC nodes have a primary data cache of 128 KB while it is only 64 KB on the POWER3 chip. On the other hand, the P2SC has no secondary cache while the POWER3 has an up to 16 MB secondary cache.

IBM positions the P2SC-based and POWER3 systems primarily for the technical/scientific market. In the parameter list above we included the presently fastest P2SC and the POWER3 processors. POWER3s can be combined into a 2-way SMP cluster with a peak performance of 1.33 Gflop/s.

The SP configurations are housed in columns, "tall" or "short" frames, of which the tall frames can contain 8–16 processor nodes and short frames half of the tall frames. How many actually are installed depends on the type of node employed: a thin node occupies half of the space of a wide node. Although the processors in these nodes are basically the same there are some differences. At the time of writing no 6.25 ns clock P2SC wide nodes were available yet. The fastest in this class feature a clock cycle of 7.4 ns giving a peak speed of 540 Mflop/s. For the POWER3 nodes there is no difference in speed between thin and wide nodes. Each frame is recommended to be configured with at least one wide node, although a frame completely filled with thin nodes seems possible according to the documentation.

POWER3 wide nodes have 10 PCI slots against only 2 PCI slots in the thin node. The P2SC-based nodes use MicroChannel instead of PCI busses and wide nodes have a double amount of MicroChannel slots (8 instead of 4) as compared to the thin nodes. Furthermore, the maximum memory of a P2SC wide node can be 2 GB whereas the maximum for thin nodes is 1 GB. For POWER3 nodes there is no difference between thin and wide nodes with respect to the maximum amount of memory. IBM envisions the wide node more or less as a server for a frame and recommends configurations of one wide node packaged with 14 thin nodes per column (although this may differ with the needs of the user). The RS/6000 SP is accessed through a front-end control workstation that also monitors system failures. Failing nodes can be taken off line and exchanged without interrupting service. In addition, file servers can be connected to the system while every node can have up to 2 GB. This can greatly speed up applications with significant I/O requirements.

The so-called high-performance switch that connects the nodes is an Ω-switch as described in section 5 and, although we mentioned only the highest speed option for the communication, the high-performance switch, there is a wide range of other options that could be chosen instead: Ethernet, Token Ring, FDDI, etc., are all possible. The best measured speeds of the high-performance switch are about 110 MB/s in point-to-point communication while a bandwidth for the communication ports of 160 MB/s is quoted for P2SC nodes and of 480 MB/s for POWER3 nodes. Unfortunately, the online (semi)technical information of IBM is not very helpful in providing more detailed information with regard to the other properties of the switch so, for instance, a bisection bandwidth cannot be provided at this point. The high-performance switch has some redundancy built into it for greater reliability.

Applications can be run using PVM or MPI. Also High Performance Fortran is supported, both a proprietary version and a compiler from the Portland Group. IBM uses its own PVM version from which the data format converter XDR has been stripped. This results in a lower overhead at the cost of generality. Also the MPI implementation, MPI-F, is optimized for the RS/6000 SP systems.

Commercially, systems up to 512 nodes are marketed, but larger systems are possible. In fact, the POWER3 node is a first commercial spin-off of the ASCI Blue Pacific system with more than 1300 processors (see ASCI (2001)).

Measured Performances: In Dongarra (1999) a performance of 547.0 Gflop/s for a 475 604e based node (1900 processor) system is reported for solving a 244000-order dense linear system, while a POWER3 based system with 1344 processors attained a speed of 468.2 Gflop/s on a similar problem of order 205000. This amounts to efficiencies of 43 and 52%, respectively.

8.12. The Meiko Computing Surface 2

Machine type: Distributed-memory multi-vector-processor.
Models: Computing Surface 2.
Operating system: Internal OS transparent to the user, Solaris (Sun's Unix variant) on the front-end system.
Connection structure: Multistage crossbar.
Compilers: Extended Fortran 77, ANSI C.
Vendors information Web page: www.meiko.com.
Year of introduction: 1994.

System parameters

Model	Computing Surface 2
Clock cycle	20 ns
Theor. peak performance	
Per Proc. (64-bits)	200, 40 Mflop/s
Maximal	204.8 Gflop/s
Memory/node	32–128, 32–512 MB
Memory/maximal	$\leq$128 GB
No. of processors	8–1024
Communication bandwidth	
Point-to-point (bi-directional)	50 MB/s

Remarks: The CS-2 features 8-1,024 processor elements (PEs) which can be either scalar or vector nodes. Apart from a separate communications module, these PEs contain either a SuperSPARC or a SuperSPARC + 2 μVP vector-processors. The speed of a scalar PE is estimated to be 40 Mflop/s (at a 20 ns clock) and 200 Mflop/s for the vector PEs for 64-bit precision. The μVP modules are manufactured by Fujitsu. The speed at 32-bit precision is doubled with respect to 64-bit operation and, unlike the early Fujitsu VP products, use IEEE 754 floating-point format. The memory has 16 banks and to avoid memory bank conflicts the CS-2 has the interesting option to have scrambled allocation of addresses, thus guaranteeing good access at potential problematic strides 2, 4, etc.

The point-to-point communication speed is 100 MB/s (50 MB/s in each direction). Because the communication happens through multi-level crossbars, called "layers" by Meiko, the aggregate bandwidth of the system scales with the number of PEs, with a latency of 200 ns per layer. As the maximum configuration of the machine contains 1,024 PEs, the theoretical peak performance at 64-bit precision is about 200 Gflop/s. It is possible to connect each PE to its own I/O devices to have scalable parallel I/O with the scaling of other resources.

The Portland Group which has won some renown for its excellent i860 compilers has developed the compilers for the CS-2. These include Fortran 77, Fortran 90, and ANSI C. The compiler offers HPF data distribution directives. Furthermore, some optimized standard linear algebra and FFT routines are offered via a proprietary numerical library.

In the USA the machine is marketed by Meiko. In 1996 Meiko has merged with Alenia, the same firm that also markets the Alenia Quadrics. Although the new marketing policy never has been made clear, it may be assumed that Alenia will market the system in Europe and the rest of the world (see www.quadrics.com).

Measured Performances: In Dongarra (1999) a speed of 5.0 Gflop/s on a 64 processor CS-2 is reported for the solution of an order 18688 dense linear system. From the NAS parallel benchmarks (NPB 1997) some results on a 128 processor machine are given for class B problems: EP took 21.16 seconds while 6.52 seconds was measured for the MG problem.

8.13. The NEC Cenju-4

Machine type: RISC-based distributed-memory multi-processor.
Models: Cenju-4.
Operating system: Cenjuiox (Mach micro-kernel based Unix flavor).
Connection structure: Multistage crossbar.
Compilers: Fortran 77, Fortran 90, HPF (subset), ANSI C.
Vendors information Web page:
kiefer.gmd.de:8002/popcorn/services/Overview.html.
Year of introduction: 1998.

System parameters

Model	Cenju-4
Clock cycle	5 ns
Theor. peak performance	
Per Proc. (64-bits)	400 Mflop/s
Maximal	410 Gflop/s
Memory/node	$\leq$512 MB
Memory/maximal	$\leq$512 GB
No. of processors	8–1024
Communication bandwidth	
Point-to-point	200 MB/s

Remarks: The name Cenju-4 suggests that there have been predecessors, Cenju-1, Cenju-2, and Cenju-3. This is indeed the case but the first two systems have only been used internally by NEC for research purposes and were never officially marketed. The Cenju-3 was also placed externally but, again, mostly for evaluation purposes. The same is the case for the present Cenju-4: it is not actively marketed, although NEC will have no objections to selling it. Officially, the Cenju-series is regarded by NEC as systems to gain experience in massively parallel computing and to develop the proper tools for it.

The Cenju-4 is based on the MIPS R10000 RISC processor. All processors have, apart from their on-chip 32 KB primary data and instruction cache, a secondary cache of 1 MB to mitigate the problems that arise in the high data usage of the CPU.

The interconnection type used in the Cenju is a multistage crossbar build from 4×4 modules that are pipelined. So, in a full configuration the maximal number of levels in the crossbar to be traversed is six. The peak transfer rate of the crossbar is quoted as 200 MB/s irrespective of the data placement. Preliminary measurements of the author of this report show that the practical transfer rate for point-to-point communication is at least 175 MB/s with MPI; a quite high efficiency.

The system needs a front-end processor like the NEC EWS4800 (functionally equivalent to Silicon Graphics workstations) or SUN. The I/O requirements have to be fulfilled by the front-end system as the Cenju does not have local (distributed) I/O capabilities.

There is some software support that should make the programmer's life somewhat easier. The library PARALIB/CJ contains proprietary functions for forking processes, barrier synchronization, remote procedure calls, and block transfer of data. Like on the Cray T3E (8.17), the Hitachi SR8000 (8.9), and on the Meiko CS-2 (8.12) the programmer has the possibility to write/read directly to/from non-local memories which avoids much message passing overhead.

Measured Performances: No systematic performance measurement have been done yet on the Cenju-4. However, from comparative studies it seems that the speed on some applications is presently about 2/3 of an equivalent SGI R10000 node due to a different compiler technology (Cassirer and Steckel 1998). Nagel reports a speed of 90-100 Mflop/s for in-cache matrix-matrix multiplication in Fortran 90 per node (Nagel 1998).

8.14. The NEC SX-5 series

Machine type: Shared-memory multi-vector-processor.
Models: SX-5/8 SX-5/16, SX-5M.
Operating system: Super-UX (Unix variant based on BSD V.4.3 Unix).
Compilers: Fortran 90, HPF, ANSI C, C++.
Vendors information Web page: www.ess.nec.de → SX-5 Series.
Year of introduction: 1998.

System parameters

Model	SX-5/8	SX-5/16	SX-5Mx
Clock cycle	4 ns	4 ns	4 ns
Theor. peak performance Per Proc. (64-bits)	8 Gflop/s	8 Gflop/s	8 Gflop/s
Maximal Single frame:	64 Gflop/s	128 Gflop/s	—
Multi frame:	—	—	4 Tflop/s
Main Memory (per frame)	≤64 GB	≤128 GB	≤4 TB
No. of processors	4–8	8–16	8–512

Remarks: The SX-5 series is offered is three models: a single-frame model that can house at most 8 CPUs (SX-5/8), a single-frame model containing up to 16 CPUs (SX-5/16), and multi-frame models (SX-5Mx) where $x = 2, \ldots, 32$ in which 2–32 single-frame systems are coupled into a larger system. There are two ways to couple the SX-5 frames in a multi-frame configuration: NEC provides a full crossbar, the so-called IXS crossbar to connect the various frames together at a speed of 16 GB/s for point-to-point out-of-frame communication (128 GB/s bi-sectional bandwidth for a maximum configuration). In addition, a HiPPI interface is available for inter-frame communication at lower cost and speed.

Every CPU contains 16 functional unit pipe sets. As the clock cycle is 4 ns and each pipe set is able to deliver 2 floating-point results per cycle, the total maximum performance is 8 Gflop/s per CPU. The bandwidth from memory to the CPUs is 32 64-bit words per cycle per CPU. With a replication factor of 16 pipe sets this is enough to provide two operands per pipe set but it is not sufficient to transport the results back to the memory at the same time. So, some trade-offs with the re-use of operands have to be made to attain the peak performance.

In contrast with its predecessor, the SX-4, the SX-5 is not offered anymore with faster, but more expensive and bulkier SRAM memory. The systems are exclusively manufactured with Synchronous DRAM memory.

The technology used is CMOS. This lowers the fabrication costs and the power consumption appreciably (the same approach is being used in the Fujitsu VPP700 and the SGI/Cray SV1, see section 8.7 and 8.16) and all models are air cooled.

For distributed computing there is an HPF compiler and for message passing optimized MPI (MPI/SX) is available.

Measured Performances: The first systems just have been delivered (Dec. 1998). Therefore, no independent performance figures are available at this time.

8.15. The Silicon Graphics Origin series

Machine type: RISC-based CC-NUMA system.
Models: Origin 200, Origin 2000.
Operating system: IRIX (SGI's Unix variant).
Connection structure: Crossbar, hypercube (see remarks).
Compilers: Fortran 77, Fortran 90, HPF, C, C++, Pascal.
Vendors information Web page:
www.sgi.com/Products/hardware/servers/index.html.
Year of introduction: 1996.

System parameters

Model	Origin 200	Origin 2000
Clock cycle	4 ns	4 ns
Theor. peak performance		
Per Proc. (64-bits)	500 Mflop/s	500 Mflop/s
Maximal	2 Gflop/s	64 Gflop/s
Memory	≤4 GB	≤256 GB
No. of processors	1–4	2–128
Communication bandwidth		
Point-to-point	780 MB/s	780 MB/s
Aggregate peak	3.1 GB/s	99.8 GB/s
Bisectional	1.6 GB/s	82 GB/s

Remarks: The Origin 2000 is the latest high-end parallel server marketed by SGI. The basic processor is the MIPS R10000 which is now offered at a clock cycle of 4 ns. A maximum of 128 processors can be configured in the system. The interconnection is somewhat hybrid: 4 CPUs on two node cards can communicate directly with the memory partitions of each other via the hub, a 4-ported non-blocking crossbar. Hubs can be coupled to other hubs in a hypercube fashion. Although the standard maximal configuration of an Origin2000 contains 128 processors, on special request larger systems can be build with the same technology. This was for instance done for the ASCI Blue Mountain project (ASCI 2001).

The machine is a typical representative of the CC-NUMA class of systems. The memory is physically distributed over the node boards but there is one system image. Because of the structure of the system, the bi-sectional bandwidth of the system remains constant from 4 processors on: 82 GB/s. This is a large improvement over the earlier PowerChallenge systems which possessed a 1.2 GB/s bus.

The Origin 200 is a smaller configuration, using the same crossbar as the Origin 2000 but without the need for the hypercube connections used in the latter. Effectively, it is a SMP system because of the uniform access of the memory modules. Therefore, also the bi-sectional bandwidth is identical to the point-to-point bandwidth: 1.6 GB/s.

Parallelization is done either automatically by the (Fortran or C) compiler or explicitly by the user, mainly through the use of directives. All synchronization, etc., has to be done via memory. This may cause potentially a fairly large parallelization overhead. Also a message passing model is allowed on the Origin using the optimized SGI versions of PVM, MPI, and the SGI/Cray-specific `shmem` library. Programs implemented in this way will possibly run very efficiently on the system.

A nice feature of the Origins is that it may migrate processes to nodes that should satisfy the data requests of these processes. So, the overhead involved in transferring data across the machine are minimized in this way. The technique is reminiscent of the late Kendall Square Systems although in these systems the data were moved to the active process. SGI claims that the time for non-local memory references is on average about 3 times longer than for local memory references.

Measured Performances: In Dongarra (1999) a speed of 1.608 Tflop/s out of 2.52 was measured on a system with 5040 processors on the ASCI Blue Mountain machine for the solution of a linear system with a size of 374400, an efficiency of 64%. Furthermore, an extensive benchmark report from the EuroBen Foundation is available for the regular Origin2000 configuration (van der Steen 1998).

8.16. The SGI/Cray Research Inc. SV1

Machine type: Shared-memory multi-vector-processor.
Models: SGI/Cray SV1-A, SV1-1, SV1 Supercluster.
Operating system: UNICOS (Cray Unix variant).
Compilers: Fortran 90, C, C++, Pascal, ADA.
Vendors information Web page: www.sgi.com/sv1/
Year of introduction: 1998.

System parameters

Model	SGI/Cray SV1-A	SGI/Cray SV1-1	SV1 Supercluster
Clock cycle	3.33 ns	3.33 ns	3.33 ns
Theor. peak perform.			
Per Proc.	1.2/4.8 Gflop/s	1.2/4.8 Gflop/s	1.2/4.8 Gflop/s
Maximal	19.2 Gflop/s	38.4 Gflop/s	1.2 Tflop/s
Memory	$\leq$16 GB	$\leq$32 GB	$\leq$1 TB
No. of processors	8–16	8–32	$\leq$1024
Memory bandwidth			
Memory–Cache	7.7 GB/s	7.7 GB/s	7.7 GB/s
Cache–CPU	9.6 GB/s	9.6 GB/s	9.6 GB/s
Aggregate	30.7 GB/s	61.4 GB/s	—
Node–node (bi-direct.)	—	—	1 GB/s

Remarks: The SGI/Cray SV1 is the successor both to the CMOS-based Cray J90 and the Cray T90 which was based on ECL technology. The SV1 systems are CMOS-based and therefore much cheaper to manufacture than the ECL-based systems. In this respect SGI is following the trend set in by Fujitsu and NEC a few years ago with their vector systems (see 8.7 and 8.14).

The single-cabinet configurations come in two sizes, the SV1-A and the SV1-1 that can house 4 and 8 processor boards, respectively. Each processor board contains 4 CPUs that can deliver a peak rate of 4 floating-point operations per cycle, amounting to a theoretical peak performance of 1.2 Gflop/s per CPU. However, 4 CPUs can be coupled *across* CPU boards in a configuration to form a so-called Multi Streaming Processor (MSP) resulting in a processing unit that has effectively a theoretical peak performance of 4.8 Gflop/s. The configuration into MSPs and/or single CPU combinations can be done via software at start-up time. The vector start-up time for the single CPUs is smaller than for MSPs, so for small vectors single CPUs might be preferable while for programs containing long vectors the MSPs should be of advantage. The number of combinations that can be made is large but at least 8 CPUs must be configured as single 1.2 Gflop/s CPUs. So a full SV1-1 cabinet may be configured as 32 single 1.2 Gflop/s CPUs or as 1–6 MSPs with the remaining processors as single CPUs.

Another new feature in the SV1 is a combined scalar and vector cache of 256 KB per CPU. This cache is important because the bandwidth of 7.7 GB/s per CPU board amounts to only 0.8 eight-byte operands per cycle. The cache can ship 1 operand per cycle to a CPU. This relative bandwidth is much smaller than what was offered in the former Cray systems which makes the cache all the more important.

Like in the NEC SX-5 single cabinets can be combined to form a cluster (Supercluster in SGI/Cray terminology) by a so-called GigaRing. The GigaRing, which is also used to coupled I/O sub-systems, is comprised of two counter-rotating rings with a bandwidth of 1 GB/s each. Where the systems in a cabinet are SM-MIMD systems, a multi-cabinet Supercluster is an DM-MIMD system and can be operated in parallel only by some parallel programming model like MPI or HPF.

Measured Performances: The importance of the cache is well illustrated by the performance of a matrix-matrix multiplication as occurs in the EuroBen Benchmark. With a single processor (called Single Stream Processing, SSP, by SGI) and with the cache a peak speed of 999 Mflop/s is observed at a matrix order of $n = 300$ and decreasing to a speed of 666 Mflop/s at an order of $n = 1000$. Without the cache the speed reaches at an order of $n = 300$ a speed of 625 Mflop/s and slowly increases to about 650 Mflop/s at $n = 1000$. In MSP mode this behavior is similar: with cache the speed at $n = 100$ is 2.61 Gflop/s, decreasing to 1.41 Gflop/s at $n = 1000$. Without the cache the observed speed at $n = 100$ is 1.0 Gflop/s and rises to 1.4 Gflop/s at $n = 1000$. This means that for modestly sized problems the cache can boost the performance with a factor 1.5–2. The efficiency in MSP mode is presently not too high: just over 50% in a favorable situation. As the MSP facility is very new, one may expect that the efficiency will increase considerably in the near future.

8.17. The SGI/Cray Research Inc. T3E

Machine type: RISC-based distributed-memory multi-processor.
Models: T3E, T3E-900, T3E-1200.
Operating system: UNICOS/mk (micro kernel-based Unix).
Connection structure: 3-D Torus.
Compilers: Fortran 77, Fortran 90, HPF, ANSI C, C++.
Vendors information Web page:
`www.cray.com/products/systems/crayt3e/`

OVERVIEW OF HIGH PERFORMANCE COMPUTERS

Year of introduction: T3E, T3E-900: 1996, T3E-1200: 1997.

System parameters

Model	T3E-900	T3E-1200
Clock cycle	2.2 ns	1.67 ns
Theor. peak performance		
Per Proc. (64-bits)	900 Mflop/s	1200 Mflop/s
Maximal	1843 Gflop/s	2458 Gflop/s
Memory/node	≤ 2 GB	≤ 2 GB
Memory/maximal	≤ 4 TB	≤ 4 TB
No. of processors	6–2048	6–2048
Communication bandwidth		
Point-to-point	300 MB/s	300 MB/s

Remarks: The T3E is the second generation of DM-MIMD systems from SGI/Cray. Lexically, it follows in name after its predecessor T3D which name referred to its connection structure: a 3-D torus. In this respect it has still the same interconnection structure as the T3D. In many other respects, however, there are quite some differences. A first and important difference is that no front-end system is required anymore (although it is still possible to connect to SGI/Cray vector systems). The systems up to 128 processors are air-cooled. The larger ones, from 256-2048 processors, are liquid cooled.

The T3E uses the DEC Alpha 21164 for its computational tasks just like the Avalon A12 (see 8.2). In 1997, a T3E-1200 was introduced that uses 21164A processors at a clock rate of only 1.67 ns but that is identical in any other aspect to the T3E-900. SGI stresses, that the processors are encapsulated in such a way that they can be exchanged easily for any other (faster) processor as soon as this would be available without affecting the macro-architecture of the system.

Each node in the system contains one processing element (PE) which in turn contains a CPU, memory, and a communication engine that takes care of communication between PEs. The bandwidth between nodes is quite high: 300 MB/s, bi-directional. Like the T3D, the T3E has hardware support for fast synchronization. E.g., barrier synchronization takes only one cycle per check.

In the micro-architecture most changes have taken place with the transition from the T3D to the T3E. Firstly, there is only one CPU per node instead of two, which removes a source of asymmetry between processors. Secondly, the new node processor has a 96 KB 3-way set-associative secondary cache which may relieve some of the problems of data fetching that were present in the T3D where only a primary cache was present. Third, the Block Transfer Engine has been replaced by a set of E-registers and streaming registers that are much more flexible and that remove some odd restrictions on the size of shared arrays and the number processes when using Cray-specific PVM. An interesting additional feature is the availability of 32 contexts per processor which opens the door for multiprocessing.

In the T3D all I/O had to be handled by the front-end, a system at least from the Cray Y-MP/E class. In the T3E distributed I/O is present. For every 8 PEs an I/O channel can be configured in the air-cooled systems and 1 I/O channel per 16 nodes in the liquid-cooled systems. The maximum bandwidth for a channel is about 1 GB/s, the actual speed will be in the order of 700 MB/s.

The T3E supports various programming models. Apart from PVM3 and MPI for message passing and HPF for data distribution, a Cray proprietary one-sided communication library, the so-called shmem library can be employed for message passing. In addition, the BSP library (see Hill et al. (1997)), also a one-sided message passing library is available. The shmem library is implemented close to the hardware and shows a very low latency of only 1.6 μs.

Measured Performances: In Dongarra (1999) a speed of 1.127 Tflop/s is reported for the solution of a dense linear system of order 148800 on a T3E-1200 with 1488 processors. The efficiency for such an exercise is 63%.

8.18. The Sun E10000 Starfire

Machine type: RISC-based shared-memory multiprocessor.
Models: E10000 Starfire.
Operating system: Solaris (Sun's Unix flavor).
Compilers: Fortran 77, Fortran 90, C, C++.
Vendors information Web page:
www.sun.com/servers/ultra_enterprise/10000/
Year of introduction: 1997.

System parameters

Model	E10000
Clock cycle	4 ns
Theor. peak performance	
Per Proc. (64-bits)	500 Mflop/s
Maximal	32 Gflop/s
Main Memory	$\leq$ 64 GB
No. of processors	16–64

Remarks: The Starfire E10000 is the largest of a series of Ex000 servers, where x can be 3, 4, 5, 6, 10. We only discuss this largest model as Sun has clearly positioned this machine themselves as a system for large-scale high performance computing. The basic processor is a 4 ns cycle Ultra-SPARC processor with a theoretical peak performance of 500 Mflop/s. Up to 64 processors are connected by a 64×64 crossbar, the largest crossbar employed commercially. This crossbar, called the Gigaplane-XB, also makes it different from the lower-end models from the Ex000 series as these systems use a bus interconnect between processors. The system is built up from system boards each containing up to 4 processors, 2 level-2 caches ($\leq$ 4 MB) and 4 memory banks that plug into the Gigaplane crossbar which thus acts as a backplane. The caches are kept coherent by a "snoopy bus" protocol, i.e., each cache is aware of the (in)validation of data by continuous monitoring the data on the backplane and updating their copies accordingly.

The Gigaplane crossbar connects to the processors with separate data and address lines which recognizes the fact that most data transfers are essentially point-to-point transfers while addresses often have to be broadcasted to many or all processors. The effective aggregate bandwidth for data is 102.4 GB/s with a point-to-point speed of 1.6 GB/s (theoretical peak).

The Starfire is a typical SMP machine with provisions for shared-memory parallelism in the Fortran and C(++) compilers by directives in the source code. Sun has joined the OpenMP consortium for standardizing the shared-memory programming model. Of course it is possible to cluster E10000s servers and use such a cluster in a DM-MIMD way.

Measured Performances: In Dongarra (1999) a speed of 26.45 Gflop/s is reported for a 64 processor machine in solving an order 19968 linear system. The efficiency for this problem is 83%. In Dongarra (1999) also results for a 4-way cluster with a clock cycle of 3 ns are reported. This 256 processor system reached a speed of 123.9 Gflop/s in solving a linear system of order 80640. This amounts to an efficiency of 72%.

8.19. The Tera MTA

Machine type: Distributed-memory multi-processor.
Models: MTA-xC, $x = 1, \ldots, 256$.
Operating system: Unix BSD4.4 + proprietary micro kernel.
Compilers: Fortran 77 (Fortran 90 extensions), HPF, ANSI C, C++.
Vendors information Web page: www.tera.com/
Year of introduction: 1997.

System parameters

Model	MTA-xC
Clock cycle	3 ns
Theor. peak performance	
Per Proc. (64-bits)	1 Gflop/s
Maximal	256 Gflop/s
Main Memory	$\leq$ 256 GB
No. of processors	16–256

Remarks: Although the memory in the MTA is physically distributed, the system is emphatically presented as a shared-memory machine (with non-uniform access time). The latency incurred in memory references is hidden by *multi-threading*, i.e., usually many concurrent program threads (instruction streams) may be active at any time. Therefore, when for instance a load instruction cannot be satisfied because of memory latency the thread requesting this operation is stalled and another thread of which an operation can be done is switched into execution. This switching between program threads only takes 1 cycle. As there may be up to 128 instruction streams per processor and 8 memory references can be issued without waiting for preceding ones, a latency of 1024 cycles can be tolerated. References that are stalled are retried from a retry pool. A construction that worked out similarly was to be found in the late Stern Computing Systems SSP machines (see in section 9).

The connection network connects a 3-D cube of p processors with sides of $p^{1/3}$ of which alternately the x- or y axes are connected. Therefore, all nodes connect to four out of six neighbors. In a p processor system the worst case latency is $4.5p^{1/3}$ cycles; the average latency is $2.25p^{1/3}$ cycles. Furthermore, there is an I/O port at every node. Each network port is capable of sending and receiving a 64-bit word per cycle which amounts to a bandwidth of 5.33 GB/s per port. In case of detected failures, ports in the network can be bypassed without interrupting operations of the system.

Although the MTA should be able to run "dusty-deck" Fortran programs because parallelism is automatically exploited as soon as an opportunity is detected for multi-threading, it may be (and often is) worthwhile to explicitly control the parallelism in the program and to take advantage of known data locality occurrences. MTA provides handles for this in the form of compiler directives, library routines, including synchronization, barrier, and reduction operations on defined groups of threads. Controlled and uncontrolled parallelism approaches may be freely mixed. Furthermore, each variable has a full/empty bit associated with it which can be used to control parallelism and synchronization with almost zero overhead. HPF will also be supported for SPMD-style programming.

Measured Performances: The company has presently delivered a 4-processor system to the San Diego Supercomputing Center. This system runs at a clock cycle of 4.4 ns instead of the planned 3 ns. Consequently, the peak performance of a processor is 450 Mflop/s. Using the EuroBen Benchmark (van der Steen 1991) a performance of 388 Mflop/s out of 450 Mflop/s was found for an order 800 matrix-vector multiplication, an efficiency of 86%. On 4 processors a speed of 1 Gflop/s out of 1.8 Gflop/s was found, an efficiency of 56% on the same problem. For 1-D FFTs up to 1 million elements a speed of 106 Mflop/s was found on 1 processor and the about the same speed on 4 processors due to an insufficient availability of parallel threads.

9. Systems that Disappeared from the List

As already stated in the introduction the list of systems is not complete. On one hand this is caused by the sheer number of systems that are presented to the market and are often very similar to systems described above (for instance, the Volvox system not listed was very similar but not equivalent to the former C-DAC system and there are numerous other examples). On the other hand, there are many systems that are still in operation around the world, often in considerable quantities that for other reasons are excluded. The most important reasons are:

- The system is not marketed anymore. This is generally for one of two reasons:
   - The manufacturer is out of business.
   - The manufacturer has replaced the system by a newer model of the same type or even of a different type.
- The system has become technologically obsolete in comparison to others of the same type. Therefore, listing them is not sensible anymore.

Below we present a table of systems that fall into one of the categories mentioned above. We think this may have some sense to those who come across machines that are still around but are not the latest in their fields. It may be interesting at least to have an indication how such systems compare to the newest ones and to place them in context.

It is good to realism that although systems have disappeared from the section above they still may exist and are actually sold. However, their removal stems in such cases mainly from the fact that they are not serious candidates for high-performance computing anymore.

The table is, again, not complete and admittedly somewhat arbitrary. The data are in a highly condensed form: the system name, system type, theoretical maximum performance of a fully configured system, and the reason for their disappearance is given. The arbitrariness lies partly in the decision which systems are still sufficiently of interest to include and which are not.

We include also both the year of introduction and the year of exit of the systems when they were readily accessible. These time-spans could give a hint of the dynamics that governs this very dynamical branch of the the computer industry.

- **Machine**: The Alex AVX 2.
    - **Year of introduction**: 1992.
    - **Year of exit**: 1997.
    - **Type**: RISC-based distributed-memory multi-processor.
    - **Theoretical Peak performance** : 3.84 Gflop/s.
    - **Reason for disappearance**: System is obsolete, there is no new system planned.

- **Machine**: Alliant FX/2800.
    - **Year of introduction**: 1989.
    - **Year of exit**: 1992.
    - **Type**: Shared-memory vector-parallel, max. 28 processors.
    - **Theoretical Peak performance**: 1120 Mflop/s.
    - **Reason for disappearance**: Manufacturer out of business.

- **Machine**: The AxilSCC.
    - **Year of introduction**: 1996.
    - **Year of exit**: 1997.
    - **Type**: RISC-based distributed-memory system, max. 512 processors.
    - **Theoretical Peak performance** : 76.8 Gflop/s.
    - **Reason for disappearance**: System is not marketed anymore by Axil.

- **Machine**: BBN TC2000.
    - **Year of introduction**: ?

– **Year of exit**: 1990.
– **Type**: Virtual shared-memory parallel, max. 512 processors.
– **Theoretical Peak performance**: 1 Gflop/s.
– **Reason for disappearance**: Manufacturer has discontinued marketing parallel computer systems.

- **Machine**: Cambridge Parallel Processing DAP Gamma.
    - **Year of introduction**: 1986.
    - **Year of exit**: 1995.
    - **Type**: Distributed-memory processor array system, max. 4096
    - processors. **Theoretical Peak performance**: 1.6 Gflop/s (32-bit).
    - **Reason for disappearance**: replaced by newer Gamma II Plus series (8.3).

- **Machine**: C-DAC PARAM 9000/SS.
    - **Year of introduction**: 1995.
    - **Year of exit**: 1997.
    - **Type**: Distributed-memory RISC based system, max. 200 processors.
    - **Theoretical Peak performance**: 12.0 Gflop/s.
    - **Reason for disappearance**: replaced by newer PARAM OpenFrame series (8.4).

- **Machine**: Convex SPP-1000/1200/1600.
    - **Year of introduction**: 1995 (SPP-1000).
    - **Year of exit**: 1996 (SPP-1600).
    - **Type**: Distributed-memory RISC based system, max. 128 processors.
    - **Theoretical Peak performance**: 25.6 Gflop/s
    - **Reason for disappearance**: replaced by newer HP Exemplar V2500 system (8.10).

- **Machine**: Cray Computer Corporation Cray-2.
    - **Year of introduction**: 1982.
    - **Year of exit**: 1992.

- **Type**: Shared-memory vector-parallel, max. 4 processors.
- **Theoretical Peak performance**: 1.95 Gflop/s.
- **Reason for disappearance**: Manufacturer out of business.

- **Machine**: Cray Computer Corporation Cray-3.
    - **Year of introduction**: 1993.
    - **Year of exit**: 1996.
    - **Type**: Shared-memory vector-parallel, max. 16 processors.
    - **Theoretical Peak performance**: 16 Gflop/s.
    - **Reason for disappearance**: Manufacturer out of business.

- **Machine**: Cray Research Inc. APP.
    - **Year of introduction**: 1993.
    - **Year of exit**: 1995.
    - **Type**: Shared-memory RISC based system, max. 84 processors.
    - **Theoretical Peak performance**: 6.7 Gflop/s.
    - **Reason for disappearance**: Product line discontinued, gap was filled by Cray J90 (see below).

- **Machine**: Cray T3D.
    - **Year of introduction**: 1994.
    - **Year of exit**: 1996.
    - **Type**: Distributed-memory RISC based system, max. 2048 processors.
    - **Theoretical Peak performance**: 307 Gflop/s.
    - **Reason for disappearance**: replaced by newer Cray T3E (8.17).

- **Machine**: Cray T3E Classic.
    - **Year of introduction**: 1996.
    - **Year of exit**: 1997.
    - **Type**: Distributed-memory RISC based system, max. 2048 processors.
    - **Theoretical Peak performance**: 1228 Gflop/s.

- **Reason for disappearance**: replaced Cray T3Es with faster clock. (8.17).

- **Machine**: Cray J90.
  - **Year of introduction**: 1994.
  - **Year of exit**: 1998.
  - **Type**: Shared-memory vector-parallel, max. 32 processors.
  - **Theoretical Peak performance**: 6.4 Gflop/s.
  - **Reason for disappearance**: replaced by newer SGI/Cray SV1 (8.16).

- **Machine**: Cray Research Inc. Cray Y-MP, Cray Y-MP M90.
  - **Year of introduction**: 1989 (Cray Y-MP).
  - **Year of exit**: 1994 (Cray Y-MP M90).
  - **Type**: Shared-memory vector-parallel, max. 8 processors.
  - **Theoretical Peak performance**: 2.6 Gflop/s.
  - **Reason for disappearance**: replaced by newer C90 (see below).

- **Machine**: Cray Y-MP C90.
  - **Year of introduction**: 1994.
  - **Year of exit**: 1996.
  - **Type**: Shared-memory vector-parallel, max. 16 processors.
  - **Theoretical Peak performance**: 16 Gflop/s.
  - **Reason for disappearance**: replaced by newer T90 (see below).

- **Machine**: Cray Y-MP T90.
  - **Year of introduction**: 1995.
  - **Year of exit**: 1998.
  - **Type**: Shared-memory vector-parallel, max. 32 processors.
  - **Theoretical Peak performance**: 58 Gflop/s.
  - **Reason for disappearance**: replaced by newer SGI/Cray SV1 (8.16).

- **Machine**: Digital Equipment Corp. Alpha farm.
  - **Year of introduction**: —.

- **Year of exit**: 1994.
- **Type**: Distributed-memory RISC based system, max. 4 processors.
- **Theoretical Peak performance**: 0.8 Gflop/s.
- **Reason for disappearance**: replaced by newer AlphaServer clusters (see below).

- **Machine**: Digital Equipment Corp. AlphaServer 8200 & 8400.
  - **Year of introduction**: —.
  - **Year of exit**: 1998.
  - **Type**: Distributed-memory RISC based systems, max. 6 processors
  - (AlphaServer 8200) or 14 (AlphaServer 8400). **Theoretical Peak performance**: 7.3 Gflop/s, resp. 17.2 Gflop/s.
  - **Reason for disappearance**: replaced by newer Compaq GS60 and GS140 (8.5).

- **Machine**: Fujitsu AP1000.
  - **Year of introduction**: 1991.
  - **Year of exit**: 1996.
  - **Type**: Distributed memory RISC based system, max. 1024 processors.
  - **Theoretical Peak performance**: 5 Gflop/s.
  - **Reason for disappearance**: replaced by the AP3000 systems (8.6).

- **Machine**: Fujitsu VPP500 series.
  - **Year of introduction**: 1993.
  - **Year of exit**: 1995.
  - **Type**: Distributed-memory multi-processor vector-processors, max.
  - 222 processors. **Theoretical Peak performance**: 355 Gflop/
  - **Reason for disappearance**: replaced by the VPP300/700 series (8.7).

- **Machine**: Fujitsu VPX200 series.
  - **Year of introduction**: —.

- **Year of exit**: 1995.
  - **Type**: Single-processor vector-processors.
  - **Theoretical Peak performance**: 5 Gflop/s.
  - **Reason for disappearance**: replaced by the VPP300/700 series (8.7).

- **Machine**: Hitachi S-3800 series.
  - **Year of introduction**: 1993.
  - **Year of exit**: 1998.
  - **Type**: Shared-memory multi-processor vector-processors, max. 4
  - processors. **Theoretical Peak performance**: 32 Gflop/s.
  - **Reason for disappearance**: Replaced by the SR8000 (8.9).

- **Machine**: Hitachi SR2001 series.
  - **Year of introduction**: 1994.
  - **Year of exit**: 1996.
  - **Type**: Distributed-memory RISC based system, max. 128 processors.
  - **Theoretical Peak performance**: 23 Gflop/s.
  - **Reason for disappearance**: Replaced by successor SR2201 (see below).

- **Machine**: Hitachi SR2201 series.
  - **Year of introduction**: 1996.
  - **Year of exit**: 1998.
  - **Type**: Distributed-memory RISC based system, max. 1024 processors.
  - **Theoretical Peak performance**: 307 Gflop/s.
  - **Reason for disappearance**: Replaced by the newer SR8000 (8.9).

- **Machine**: HP/Convex C4600 series.
  - **Year of introduction**: 1994.
  - **Year of exit**: 1997.
  - **Type**: Shared-memory vector-parallel, max. 4 processors.

- **Theoretical Peak performance**: 3.24 Gflop/s.
- **Reason for disappearance**: HP does not market the vector product line anymore.

- **Machine**: IBM ES/9000 series.
  - **Year of introduction**: 1991.
  - **Year of exit**: 1994.
  - **Type**: Shared-memory vector-parallel system, max. 6 processors.
  - **Theoretical Peak performance**: 2.67 Gflop/s.
  - **Reason for disappearance**: IBM does not pursue high-performance computing by this product line anymore.

- **Machine**: IBM SP1 series.
  - **Year of introduction**: 1992.
  - **Year of exit**: 1994.
  - **Type**: Distributed-memory RISC based system, max. 64 processors.
  - **Theoretical Peak performance**: 8 Gflop/s.
  - **Reason for disappearance**: Replaced by the newer RS/6000 SP (8.11).

- **Machine**: Intel Paragon XP.
  - **Year of introduction**: 1992.
  - **Year of exit**: 1996.
  - **Type**: Distributed-memory RISC based system, max. 4000 processors.
  - **Theoretical Peak performance**: 300 Gflop/s.
  - **Reason for disappearance**: Except for a non-commercial research system (the ASCI Option Red system at Sandia National Labs.) Intel is not in the business of high-performance computing anymore.

- **Machine**: Kendall Square Research KSR2.
  - **Year of introduction**: 1992.
  - **Year of exit**: 1994.

- **Type**: Virtually shared-memory parallel, max. 1088 processors.
- **Theoretical Peak performance**: 400 Gflop/s.
- **Reason for disappearance**: Kendall Square has terminated its business.

- **Machine**: Kongsberg Informasjonskontroll SCALI.
  - **Year of introduction**: 1996.
  - **Year of exit**: 1997.
  - **Type**: Distributed-memory RISC based system, max. 512 processors.
  - **Theoretical Peak performance**: 76.8 Gflop/s.
  - **Reason for disappearance**: Kongsberg does not market the system anymore.

- **Machine**: MasPar MP-1, MP-2.
  - **Year of introduction**: 1991 (MP-1).
  - **Year of exit**: 1996.
  - **Type**: Distributed-memory processor array system, max. 16384
  - processors. **Theoretical Peak performance**: 2.4 Gflop/s (64-bit, MP-2).
  - **Reason for disappearance**: Systems are not marketed anymore.

- **Machine**: Matsushita ADENART.
  - **Year of introduction**: 1991.
  - **Year of exit**: 1997.
  - **Type**: Distributed-memory RISC based system, 256 processors.
  - **Theoretical Peak performance**: 2.56 Gflop/s.
  - **Reason for disappearance**: Machine is obsolete and no new systems are developed in this line.

- **Machine**: Meiko CS-1 series.
  - **Year of introduction**: 1989.
  - **Year of exit**: 1995.
  - **Type**: Distributed-memory RISC based system.

- **Theoretical Peak performance**: 80 Mflop/s per processor.
- **Reason for disappearance**: Replaced by the newer CS-2 (8.12).

■ **Machine**: nCUBE 2S.
- **Year of introduction**: 1993.
- **Year of exit**: 1998.
- **Type**: Distributed-memory system, max. 8192 processors.
- **Theoretical Peak performance**: 19.7 Gflop/s.
- **Reason for disappearance**: NCUBE has withdrawn from the scientific and technical market. The nCUBE 2S is now offered as a parallel multimedia server.

■ **Machine**: nCUBE 3.
- **Year of introduction**: — .
- **Year of exit**: —
- **Type**: Distributed-memory system, max. 10244 processors.
- **Theoretical Peak performance**: 1 Tflop/s.
- **Reason for disappearance**: Was announced several times as the successor of the nCUBE 2S (see above) but was never realized.

■ **Machine**: NEC Cenju-3.
- **Year of introduction**: 1994.
- **Year of exit**: 1996.
- **Type**: Distributed-memory system, max. 256 processors.
- **Theoretical Peak performance**: 12.8 Gflop/s.
- **Reason for disappearance**: replaced by newer Cenju-4 series (8.13).

■ **Machine**: NEC SX-3R.
- **Year of introduction**: 1993.
- **Year of exit**: 1996.
- **Type**: Shared-memory multi-processor vector processors, max. 4
- processors.
Theoretical Peak performance: 25.6 Gflop/s.

- **Reason for disappearance**: replaced by newer SX-4 series (see below).

- **Machine**: NEC SX-4.
  - **Year of introduction**: 1995.
  - **Year of exit**: 1996.
  - **Type**: Distributed-memory cluster of SM-MIMD vector processors,
  - max. 256 processors. **Theoretical Peak performance**: 1 Tflop/s.
  - **Reason for disappearance**: replaced by newer SX-5 series (8.14).

- **Machine**: Parsys SN9000 series.
  - **Year of introduction**: 1993.
  - **Year of exit**: 1995.
  - **Type**: Distributed-memory RISC based system, max. 2048.
  - **Theoretical Peak performance**: 51.2 Gflop/s.
  - **Reason for disappearance**: Replaced by the newer TA9000 (but see below).

- **Machine**: Parsys TA9000 series.
  - **Year of introduction**: 1995.
  - **Year of exit**: 1996.
  - **Type**: Distributed-memory RISC based system, max. 512 processors.
  - **Theoretical Peak performance**: 119.3 Gflop/s.
  - **Reason for disappearance**: Parsys does not offer complete system anymore. Instead it sells node cards based on the TA9000 for embedded systems.

- **Machine**: Parsytec GC/Power Plus.
  - **Year of introduction**: 1993.
  - **Year of exit**: 1996.
  - **Type**: Distributed-memory RISC based system.
  - **Theoretical Peak performance**: 266.6 Mflop/s per processor.

- **Reason for disappearance**: System has been replaced by the Parsytec CC systems (see below).

■ **Machine**: Parsytec CC series.
  - **Year of introduction**: 1996.
  - **Year of exit**: 1998.
  - **Type**: Distributed-memory RISC based system.
  - **Theoretical Peak performance**: unspecified.
  - **Reason for disappearance**: Vendor has withdrawn from the High-Performance computing market.

■ **Machine**: Siemens-Nixdorf VP2600 series.
  - **Year of introduction**: ?
  - **Year of exit**: 1995.
  - **Type**: Single-processor vector-processors.
  - **Theoretical Peak performance**: 5 Gflop/s.
  - **Reason for disappearance**: replaced by the VPP300/700 series (8.7).

■ **Machine**: Silicon Graphics PowerChallenge.
  - **Year of introduction**: 1994.
  - **Year of exit**: 1996.
  - **Type**: Shared-memory multi-processor, max. 36 processors.
  - **Theoretical Peak performance**: 14.4 Gflop/s.
  - **Reason for disappearance**: replaced by the SGI Origin 2000 (8.15).

■ **Machine**: Stern Computing Systems SSP.
  - **Year of introduction**: 1994.
  - **Year of exit**: 1996.
  - **Type**: Shared-memory multi-processor, max. 6 processors.
  - **Theoretical Peak performance**: 2 Gflop/s.
  - **Reason for disappearance**: Vendor terminated its business just before delivering first systems.

■ **Machine**: Thinking Machine Corporation CM-2(00).
  - **Year of introduction**: 1987.

- **Year of exit**: 1991.
  - **Type**: SIMD parallel machine with hypercube structure, max. 64K
  - processors. **Theoretical Peak performance**: 31 Gflop/s.
  - **Reason for disappearance**: was replaced by the newer CM-5 (but see below).

- **Machine**: Thinking Machine Corporation CM-5.
  - **Year of introduction**: 1991.
  - **Year of exit**: 1996.
  - **Type**: Distributed-memory RISC based system, max. 16K processors.
  - **Theoretical Peak performance**: 2 Tflop/s.
  - **Reason for disappearance**: Thinking Machine Corporation has stopped manufacturing hardware and hopes to keep alive as a software vendor.

10. Systems under Development

Although we mainly want to discuss real, marketable systems and not experimental, special purpose, or even speculative machines, we want to include a section on systems that are in a far stage of development and have a fair chance of reaching the market. For inclusion in section 3 we set the rule that the system described there should be on the market within a period of 6 months from announcement. The systems described in this section will in all probability appear within one year from the publication of this report. However, there are vendors who do not want to disclose any specific data on their new machines until they are actually beginning to ship them. We recognize the wishes of such vendors (it is generally wise not to stretch the expectation of potential customers too long) and will not disclose such information.

Below we discuss systems that may lead to commercial systems to be introduced on the market between somewhat more than half a year to a year from now. The commercial systems that result from it will sometimes deviate significantly from the original research models depending on the way the development is done (the approaches in Japan and the USA differ considerably in this respect) and the user group which is targeted.

The year 1998 has been fruitful in terms of new systems: 3 large vendors came out with new systems: Hitachi with the SR8000, NEC with the SX-5, and SGI/Cray with the SV1. However, new ASCI contracts have been made and other vendors are due to come up with new systems.

10.1. Compaq/DEC

The present Compaq/DEC GS60 and GS140 servers (8.5) are in fact relabeled AlphaServers 8200 and 8400. They are still bus-based systems that cannot be scaled up effectively. As Compaq/DEC has won an ASCI contract to build a 30 Tflop/s machine by 2001, it is clear that the structure of the system has to change. Although such plans are not officially released it seems certain that a multi-level crossbar based on the fast DEC Memory Channel will be used. For instance, using an 8×8 crossbar an aggregate bandwidth of 8 GB/s per crossbar level could be reached. Using the next generation EV7 Alpha chip at a 0.9 ns clock cycle an 8-way cluster would provide a 35.2 Gflop/s building block. The system will probably be presented as a CC-NUMA machine.

10.2. Fujitsu

Fujitsu is the only large vector-processor vendor that did not present a new system by the end of 1998. The introduction of the successor to the VPP700 is expected in the spring of this year. As usual Fujitsu has been extremely tight-lipped about the new systems and no details are known. As the VPP700 has been a reasonably successful machine it stands to reason that the structure of its successor will be (almost) the same: vector-processors connected by a multi-level crossbar. The speed of both components will have undoubtly increased. In contrast to NEC and SGI/Cray, Fujitsu did not present a shared-memory image for the first-level clusters in its VPP700, except by its proprietary extended Fortran and HPF. In the next generation system this omission will very probably be mended.

10.3. Hewlett-Packard

The HP Exemplar systems consist of 32-CPU cluster nodes that are connected by 1 GB/s SCI rings to increase the system size (8.10). Although the speed of the PA-RISC processors have increased much over the years, the bandwidth of the rings stayed the same. This is obviously becoming a bottleneck in the scalability of the HP systems. Therefore, in the near future a new system with a different structure may be expected. In keeping with the majority of systems that are built presently one may expect that also HP will decide to connect its SMP clusters by a multi-level crossbar. As the point-to-point bandwidth between clusters certainly not will decrease, e.g., a 16×16 crossbar would need at least an aggregate bandwidth of 16 GB/s. As the present Exemplars already are CC-NUMA systems, the next generation certainly will also be of this class. At the moment HP keeps the options open whether the next system will be based on the next generation of the PA-RISC system or that the new IA-64 chip will be used.

10.4. IBM

IBM was, together with Intel and SGI/Cray one of the first ASCI contractors and has as such built the ASCI Blue Pacific machine with a peak speed of more than 1 Tflop/s. This system is based on the POWER3 chip (see 8.11). In following contracts IBM ventures to make a 100 Tflop/s system based on the successor of the POWER3 chip and with a higher level of clustering. At this moment the ASCI Blue machine has four-way clustered nodes. This might stay the same in the near future. With an increase in the clock speed from 3.3 to 2.2 ns a single CPU with four floating-point units already would have a peak speed of 1.8 Gflop/s which would give an four-way node a performance of 7.2 Gflop/s. It is to be expected that the switching fabric also will be altered. The High-Performance switch, as it is, would clearly form a bottleneck for the scalability. So, as in most other new systems (and in the present RS/6000 SP) a multi-level crossbar is highly probable. A new system could for instance consist of packages of 4 four-way nodes of approximately 30 Gflop/s peak performance which in turn are four-way connected, etc.

10.5. SGI

After the introduction of the SGI/Cray SV1 a new RISC processor based machine may be expected to replace the successful Origin2000 system. As SGI has loosened its ties with MIPS, the current source of Origin CPUs, SGI might choose another chip to base the Origin successor on. As SGI is also developing on Intel platforms now, the Intel IA-64 might be a possible choice when no MIPS processors would be employed. Whatever is chosen, the nodes need a to have speed of 1 Gflop/s or greater to be competitive with similar systems available around the year 2000. In most other respects the system might have largely the structure of the Origin2000. According to the SGI road map, the T3E line will at least for the next generation not be fused with the Origin successor and so will have a distinct architecture. What is sure is that the new system will retain the successful CC-NUMA concept of the present machine.

For SGI the integration of the machine lines is of extreme importance. Only a few vendors have as much as 2 product lines that are marketed (Fujitsu, Hitachi, NEC) where it is not always clear what are the benefits of one product line over the other. SGI still maintains no less than three product lines in high-performance computing, each with its own virtues and customer base. It is a highly complicated balancing act to reduce the product lines to two or even one without disappointing a part of the customers.

Acknowledgments

It is not possible to thank all people that have been contributing to this overview. Many vendors and people interested in this project have been so kind to provide me with the vital information or to correct us when necessary. Therefore, we will have to thank them here collectively but not less heartily for their support.

Special thanks are due to Greg Astfalk of Hewlett-Packard for valuable comments on the architecture section and on the HP Exemplar in particular.

Bibliography

C. Amza, A.L. Cox, S. Dwarkadas, P. Keleher, Honghui Lu, R. Rajamony, Weimin Yu, and W. Zwaenepoel. TreadMarks: Shared Memory Computing on Networks of Workstations. *IEEE Computer*, 29:18–28, 1996.

ASCI. The asci program, 2001. http://www.llnl.gov/asci/.

K. Cassirer and B. Steckel. Block-Structured Multigrid on the Cenju. In *2nd Cenju Workshop*, 1998.

D.E. Culler, J.P. Singh, and A. Gupta. *Parallel Computer Architecture: A Hardware/Software Approach*. Morgan Kaufmann Publishers Inc., 1998.

J.J. Dongarra. Performance of various computers using standard linear equations software. Technical Report CS-89-85, Computer Science Department, Univ. of Tennessee, 1999.

P. Flanders. Matrix Multiplication on 'C' series DAPs, 1991. AMT Document TR40.

M.J. Flynn. Some computer organisations and their effectiveness. *IEEE Trans. on Comp.*, C-21:948–960, 1972.

High Performance Fortran Forum. High Performance Fortran Language Specification. *Scientific Programming*, 2:1–170, 1993.

A. Geist, A. Beguelin, J. Dongarra, R. Manchek, W. Jaing, and V. Sunderam. *PVM: A Users' Guide and Tutorial for Networked Parallel Computing*. MIT Press, 1994.

W. Gropp, S. Huss-Ledermann, A. Lumsdaine, E. Lusk, B. Nitzberg, W. Saphir, and M. Snir. *MPI: The Complete Reference, Vol. 2, The MPI Extensions*. MIT Press, 1998.

J.M.D. Hill, W. McColl, D.C. Stefanescu, M.W. Goudreau, K. Lang, S.B. Rao, T. Suel, T. Tsantilas, and R. Bisseling. BSPlib: The BSP Programming Library. Technical Report PRG-TR-29-9, Oxford University Computing Laboratory, 1997.

2001. www.hitachi.co.jp/Prod/comp/hpc/eng/sr1.html.

R. W. Hockney and C. R. Jesshope. *Parallel Computers II*. Adam Hilger, 1987.

T. Horie, H. Ishihata, T. Shimizu, S. Kato, S. Inano, and M. Ikesaka. AP1000 architecture and performance of LU decomposition. In *Proc. Internat. Symp. on Supercomputing*, pages 46–55, 1991.

D.V. James, A.T. Laundrie, S. Gjessing, and G.S. Sohi. Scalable Coherent Interface. *IEEE Computer*, 23:74–77, 1990.

W.E. Nagel. Applications on the Cenju: First Experience with Effective Performance. In *2nd Cenju Workshop*, 1998.

Web page for the NAS Parallel benchmarks NPB2, 1997. http://science.nas.nasa.gov/Software/NPB/.

M. Snir, S. Otto, S. Huss-Lederman, D. Walker, and J. Dongarra. *MPI: The Complete Reference, Vol. 1, The MPI Core*. MIT Press, 1998.

A.J. van der Steen. Exploring VLIW: Benchmark tests on a Multiflow TRACE 14/300. Technical Report TR-31, Academic Computing Centre Utrecht, 1990.

A.J. van der Steen. The benchmark of the EuroBen Group. *Parallel Computing*, 17:1211–1221, 1991.

A.J. van der Steen, editor. *Aspects of computational science*. 1995.

A.J. van der Steen. Benchmarking the Silicon Graphics Origin2000 System. Technical Report WFI-98-2, Dept. of Computational Physics,

Utrecht University, 1998. The report can be downloaded from: www.phys.uu.nl/~steen/euroben/reports/.

Chapter 23

THE NATIONAL SCALABLE CLUSTER PROJECT: THREE LESSONS ABOUT HIGH PERFORMANCE DATA MINING AND DATA INTENSIVE COMPUTING

Robert Grossman
University of Illinois at Chicago and Magnify Inc., Chicago, IL 60607, USA
grossman@uic.edu

Robert Hollebeek
University of Pennsylvania and Hubs Inc., Philadelphia, PA 19104, USA
bobh@NSCP01.PHYSICS.upenn.edu

Abstract We discuss three principles learned from experience with the National Scalable Cluster Project. Storing, managing and mining massive data requires systems that exploit parallelism. This can be achieved with shared-nothing clusters and careful attention to I/O paths. Also, exploiting data parallelism at the file and record level provides efficient mapping of data-intensive problems onto clusters and is particularly well suited to data mining. Finally, the repetitive nature of data mining demands special attention be given to data layout on the hardware and to software access patterns while maintaining a storage schema easily derived from the legacy form of the data.

Keywords: Data mining, I/O, Distributed computing, Distributed data, Scalable clusters, Parallel computing, Data intensive computing.

1. Introduction

Today a 10 GB data set can easily be produced with a simple mistake, 100 GB to 1 TB data sets are common, and data sets larger than a Terabyte are becoming common. In this paper, we describe three basic lessons we have learned about data mining and data intensive computing on large and massive data sets. First, we give a few definitions. A Terabyte is 1000 Gigabytes and a Petabyte is a 1000 Terabytes. A data set is large if it is larger than most of your friends' data sets. A data set is massive if it is larger than most other data sets you have seen. We assume the data is divided into files, that files are divided into records, and that records contain attributes. By archiving data, we mean storing data in such a way as to provide file-level access to the data. By managing data, we mean storing data in such a way as to provide record-level and attribute-level access to the data. By mining data, we mean looking for patterns, changes, associations, anomalies and other statistically significantly structures in data.

We employ the following three very rough rules of thumb:

- **Rule 1:** You can archive ten to a thousand times more data with a data archival system than you can manage with a data management system, and you can manage $10\times$ to $1000\times$ more data with a data management system than you can mine with a data mining system.

- **Rule 2:** Simple ideas scale the best.

- **Rule 3:** People usually claim that they can mine $10\times$ to $1000\times$ more data than they can by misunderstanding Rule 1.

Rule 1 is a consequence of how systems are designed and how data is moved. A system can index and efficiently access a certain number of elements. For archival systems, these elements are files; for databases systems, these elements are records. Files are typically much larger than records. Most data mining algorithms require scanning or moving the data multiple times, which explains the third part of Rule 1, although some algorithms require only a few scans of the data. Rule 2 is a reflection of how software and systems are typically built. Rule 3 is a consequence of how people are designed.

The National Scalable Cluster Project (NSCP) is an NSF-funded collaboration focused on data mining and data intensive computing of large and distributed data sets. The NSCP has exploited the fact that, beginning approximately in the mid-1990s, workstations were becoming commodities. By forming local clusters of workstations with high performance switches and hubs and by using the appropriate data and compute management software, super-computer scale data intensive computing became practical using clusters. The NSCP-1 Meta-Cluster was completed in 1996 and linked geographically distributed clusters using the commodity Internet. The NSCP-2 Meta-Cluster was completed in 1998 and used OC-3 networks to link the clusters.

In this paper, we describe three lessons about data mining and data intensive computing we have learned while working with the NSCP-1 and NSCP-2 Meta-Cluster. Section 2 contains a brief description of the NSCP Meta-Cluster. Section 3 describes some related work. Sections 4–6 describe the lessons and Section 7 is the conclusion.

2. The NSCP Meta-cluster

The National Scalable Cluster Project (NSCP) collaboration of research groups has pioneered the application of cluster computing and high performance wide area networks to a variety of problems in data mining and data intensive computing, including working with several terabyte size collections. The core research groups from the University of Pennsylvania and the University of Illinois at Chicago work with collaborators at over ten institutions. The project was founded in 1994 by Grossman and Hollebeek.

The NSCP collaborators have assembled local clusters of workstations and connected these local clusters into wide area clusters of clusters or Meta-Clusters. NSCP also developed several software packages for data intensive computing using the Meta-Cluster. The NSCP-1 Meta-Cluster linked geographically distributed clusters using the commodity Internet. The NSCP-2 Meta-Cluster used OC-3 networks to link the clusters. The NSCP-1 and NSCP-2 Meta-Clusters have been used by a variety of scientists and engineers working on applications in high energy physics, computational chemistry, nonlinear simulation, bioinformatics, medical imaging, network traffic analysis, digital libraries of video data, and economic data.

An NSCP-3 Meta-Cluster is currently being designed and tentatively scheduled for deployment in 2001. The NSCP-3 Meta-Cluster is being designed to exploit wave division multiplexing (WDM) technology. WDM is now being used to greatly increase the available bandwidth on links connecting geographically distributed nodes by packing many wavelengths carrying separated data streams onto a single fiber.

Currently, the NSCP consists of approximately 100 nodes and 3 terabytes of disk geographically distributed among the participating sites, and connected by laboratory, campus, and national ATM networks. NSCP developed software and third party software are provided so that applications can transparently access as many nodes and as much disk as required. NSCP supports one large digital library (500 Gigabytes), two moderate size digital libraries (100 Gigabytes each), and several smaller ones. More details can be found at http://nscp.upenn.edu and http://www.uic.uic.edu.

2.1. An Introduction to HP-D2M2: High Performance Distributed Data Mining and Management

The term "data mining" is often used to refer narrowly to the data-intensive and compute-intensive process of winnowing nuggets of useful information from a data source. We think it is helpful to include in that catch phrase the entire required sequence of processes including system design, system operation and optimization, as well as the algorithm design and the software design required for the grand goal of the process – namely knowledge discovery. Our own interest in data mining extends primarily to its application to very large and/or distributed collections of information. Since the goal of data mining is to extract important and – one hopes – previously unknown information from such collections, doing this quickly and efficiently, or doing it on very large distributed and often disorganized collections of data requires high performance techniques. We focus here on three lessons that we have learned in the past several years through our work in the National Scalable Cluster Project, and the Terabyte Challenge, applying new techniques to large distributed data samples.

The first lesson is that we must parallelize the hardware and software I/O in data intensive computing problems and that in doing so, successive layers of hardware and software need to be well matched to eliminate potential choke-points. The second lesson is that we need to exploit data parallelism, and the third and final lesson is that we must pay particular attention to data layout in both hardware and software.

The phenomenon of rapidly increasing volumes of data is familiar to all by this time. Anyone with a home PC, an office machine, a portable, and perhaps a palm device knows something about how rapidly information can become diffuse and disorganized. Just as we do not always have good methods of organizing our personal piles of paper, or the scraps on our desks, we do not always have effective means of organizing our large distributed digital collections either. As the total volume of information increases, and as it becomes more geographically dispersed, the process of data mining can become very difficult.

The first – and perhaps most important – step in the data mining and knowledge-discovery process is to select and locate relevant and interesting data. We then need to accumulate, aggregate, organize and catalog it. The goal of the process of data mining is to extract new knowledge. Given the large number of potential avenues of investigation, the next step is to automate or semi-automate the extraction of features, clusters, aggregates or other interesting correlations among parts of the data. Fortunately, high performance computers can search through incredible numbers of combinations and possible scenarios even for large collections of data.

Data mining exercises are highly iterative and require human intervention between iterations. Often, knowledge obtained in a first iteration must be fed back into the system, and the next iteration steered in a new direction. Anyone who hopes to simply pour wheat chaff (i.e., raw data) in the top funnel of a giant vat, and sip Cognac dripping from the bottom in a hands-off, completely automated fashion, is setting their sights too high. The best data mining process will require thoughtful collaboration between content experts and experts in the data mining process.

Almost any collection of information has the potential of providing new knowledge, and in some sense, the more information that is collected, the more likely new knowledge is. There is however a temptation to shy away from very large collections due to the technical difficulties of managing and processing them. So the major goal is to recognize the value of growing large collections of information, and then to design new ways in which the relationships contained in that information might be extracted and used.

There are many potential choices for what data one might collect and mine, and it is very important to give some careful thought to selecting a set of information which, at least initially, is thought to contain some useful, valuable and actionable information. It makes little sense to start with dull and uninteresting data since there is no magic to the data mining process even if there is a lot of black art! For beginners, there is one exception to this rule. It may be useful to pick a set of information which contains knowledge you already know, and use the data mining software in an exercise to verify this knowledge. If you have purchasing information for example, take a mix of good and bad customers (defined either in terms of purchases or support costs) and see if the data mining process can segment the information into two such classes. Then, see if it can identify some of the other distinguishing demographics of those two classes. If you are successful, you will certainly bolster your confidence in your ability to properly use the tools and interpret the results.

Once a good data sample has been selected, it is also a good idea to simplify the data by eliminating some variables, and grouping others into classes. It is always possible to carry such variables along temporarily and bring them back into the process later, but when starting out, simplicity is key. This is due, at least in part, to the rapidly rising difficulty and diminishing speed of the data mining process with the dimensionality (in terms of number of variables) and complexity of the data set. Thus, it is better to start simply, and iterate.

3. Related Work

In this section, we describe related work in cluster computing, testbeds, distributed data mining systems, and data management software. One of our model applications is mining high energy physics data. We also describe related work in this area. This section is based in part on (Grossman et al. 1999b;a).

3.1. Cluster Computing

There are also a number of active research projects in cluster computing, including the Berkeley Now Project (Anderson et al. 1995), the Beowolf Project (Becker et al. 1995) at CalTech, and the Seamless Parallel and Distributed Computing Project (Ishikawa et al. 1997) of the Real World Computing Partnership. On the whole, these projects have focused on providing a general computing infrastructure for local clusters, while the NSCP has focused on data intensive computing using wide area meta-clusters built by connecting several geographically distributed local clusters.

3.2. Testbeds

For the past five years, the NSCP-1 and the NSCP-2 Meta-Cluster have provided the infrastructure for a national data intensive and data mining testbed called the Terabyte Challenge (TC). Approximately 25 different applications have been run on the Terabyte Challenge Testbed. The Terabyte Challenge has won multiple awards and has been used as an interoperability testbed for the development of the Predictive Model Markup Language (PMML). PMML is an XML-based language for data intensive computing, which is becoming an industry standard (http://www.dmg.org).

There are a number of testbeds (Foster and Kesselman 1998, Grimsaw and Wulf 1996) for high performance computing, high performance networking, and distributed computing. Perhaps the most similar testbed is the Globus Ubiquitous Supercomputing Testbed Organization (GUSTO) and the Centurion Legion Project Testbed. The Globus Project, led by I. Foster and C. Kesselman, is building middleware for grid-based computation. The Legion Project, led by A. Grimsaw of the University of Virginia, is developing a wide-area system for distributed object computing. In contrast, the Terabyte Challenge Testbed has focused on two objectives: a) providing an application testbed for data intensive computing using local and wide area clusters, especially those requiring high performance wide area networks; and b) a testbed for languages and protocols for data intensive computing and data mining.

3.3. Distributed Data Mining Systems

Papyrus is a layered infrastructure for high performance and wide area data mining and predictive modeling developed by the Laboratory for Advanced Computing at the University of Illinois at Chicago. Papyrus is specifically designed to support clusters of workstations, distributed clusters of workstations (meta-clusters) and distributed clusters of workstations connected with high performance networks (super-clusters).

Several other systems have been developed for distributed data mining. Perhaps the most mature are: the JAM system developed by Stolfo et. al. (Stolfo et al. 1997), the Kensington system developed by Guo et al. (1997), and BODHI developed by Kargupta et al. (1997). These systems differ in data, task and model strategies:

- **Data strategy:** Distributed data mining can choose to move data, to move intermediate results, to move predictive models, or to move the final results of a data mining algorithm. Distributed data mining systems which employ local learning build models at each site and move the models to a central location. Systems that employ centralized learning move the data to a central location for model building. Systems can also employ hybrid learning, that is, strategies that combine local and centralized learning. JAM, Kensington and BODHI all employ local learning.

- **Task strategy:** Distributed data mining systems can choose to coordinate a data mining algorithm over several sites or to apply data mining algorithms independently at each site. With independent learning, data mining algorithms are applied to each site independently. With coordinated learning, one (or more) sites coordinate the tasks within a data mining algorithm across several sites.

- **Model strategy:** Several different methods have been employed for combining predictive models built at different sites. The simplest most common method is to use voting and combine the outputs of the various models with a majority vote (Dietterich 1997). Meta-learning combines several models by building a separate meta-model whose inputs are the outputs of the various models and whose output is the desired outcome (Stolfo et al. 1997).

Knowledge probing considers learning from a black box viewpoint and creates an overall model by examining the input and the outputs to the various models, as well as the desired output (Guo et al. 1997). Multiple models, or what are often called ensembles or committees of models, have been used for quite a while in (centralized) data mining. A variety of methods have been studied for combining models in an ensemble, including Bayesian model averaging and model selection (Raftery et al. 1996), stacking (Wolpert 1992), partition learning (Grossman et al. 1996), and other statistical methods, such as mixture of experts (Xu and Jordan 1993). JAM employs meta-learning, while Kensington employs knowledge probing.

Papyrus is designed to support different strategies for data-parallel, task-parallel and model-parallel computation. For example, in contrast to JAM and Kensington, Papyrus can not only move models from node to node, but can also move data from node to node when that strategy is desired. In contrast to BODHI, Papyrus is built over a data warehousing layer that can move data over both commodity and high performance networks. Also, Papyrus is a specialized system that is designed for clusters, meta-clusters, and super-clusters, while JAM, Kennsington and BODHI are designed for mining data distributed over the internet. All four systems make use of Java. JAM employs Java applets to move machine learning algorithms to distributed data. Kensington uses Java JDBC to mine distributed data. Papyrus uses Java aplets.

Moore et al. (1998) stresses the importance of developing an appropriate storage and archival infrastructure for high performance data mining and discusses some of the efforts of the SDSC in this area. The distributed data mining system developed by Subramonian and Parthasarathy (1998) is designed to work with clusters of SMP workstations and like Papyrus is designed to exploit clusters of workstations. Both this system and Papyrus are designed around data clusters and compute clusters. Papyrus also explicitly supports clusters of clusters and clusters connected with different types of networks.

3.4. Data Management Software

Many of the NSCP-1 and NSCP-2 Meta-Cluster applications used PTool, a light weight, high performance persistent object managed designed for local and wide area clusters and meta-clusters. PTool has successfully managed distributed terabyte sized data collections.

3.5. Data Mining Algorithms

A variety of different strategies are possible in distributed and high performance data mining and predictive modeling. Data may be moved from node to node, predictive models produced by data mining algorithms may be moved from node to node, or the results of applying data mining algorithms to specific data sets may be moved from node to node. The optimal strategy depends upon the amount of data, its distribution, and the performance characteristics of the network connecting the nodes. (Grossman et al. 1996).

Classification and regression trees are standard tools in data mining. Data parallel approaches in high performance computing are also standard. A standard reference for ensembles of trees is Breiman et al. (1984).

3.6. Mining High Energy Physics Data

We have worked intensively using the NSCP-1 and NSCP-2 Meta-Clusters to develop next generation systems for mining high energy physics data. Our viewpoint has evolved over time: we began with the idea of using databases to manage HEP data (1991–1992); we then explored using specialized persistent object managers (1993–1996); we then broadened our perspective to using wide area distributed object management systems (1996–1998); more recently, we have developed a web-inspired infrastructure (1998-present). In each case, we did extensive experiments with NSCP developed software.

There is an extensive literature on computing with high energy physics data. Approximately every 18 months, the Conference in Computing in High Energy Physics (CHEP) meets to explore the state of the art. These proceedings contain a large number of related papers. Several projects deserve special mention. The goal of the RD-45 Project at CERN is to develop a scalable persistent object manager for petabytes of high energy physics data. The goal of the Particle Physics Data Grid led by CalTech is to develop an infrastructure for the widely distributed analysis of high energy physics data. The goal of the Nile Project at Cornell is to provide a data analysis infrastructure for the CLEO detector at Cornell based upon distributed computing.

4. Lesson 1: PIOM: Parallelize I/O and Match

In this section, we wish to emphasize the first of three lessons on high performance large-scale data intensive computing that have been learned in the National Scalable Cluster Project. The first lesson is a reminder that with massive data there is no choice but to balance parallel networks with parallel input/output systems to parallel CPUs (parallel networks to parallel I/O to parallel CPUs).

Cluster architectures are in many ways uniquely suited to data intensive tasks. Using load balancing on parallel nodes, data striping across parallel disks and parallel execution on many nodes driven by parallel aware application programming interfaces (MPI for example), an application can be developed to take advantage of the very large internal bandwidth of a many-node cluster. Distributed applications, and in particular wide area distributed, data intensive applications however must deal with the bandwidths available over wide area networks (WAN) which are typically much lower than processor I/O bus and even local interconnect speeds. There are technologies emerging, such as Wave Division Multiplexing (WDM), which support parallel wide area networks so that geographically distributed clusters can be connected together in a balanced fashion. See the diagram in Figure 23.1.

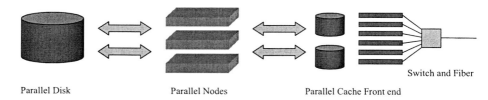

Figure 23.1. Geographically distributed clusters.

To maintain high volume data flow, systems must eliminate bottlenecks between the disk/cpu/network components. The goal is to match the throughput of each stage as the data passes from disk to I/O bus to processor to the network and back again. When the desired speed (as defined by the network) is not achievable with commodity hardware, parallelism is used to multiply the maximum data rate. Using the NSCP/SP2 configurations at Penn as an illustration, a group of 4 processors can be fed by a 100BaseT switch and four Fast Ethernet interfaces or (or, alternatively, by 4 OC3 ATM interfaces and an ATM switch). The OC3 ATM capability (nominally 155 Mbs) is degraded by ATM overhead and system performance, but 100Mbs per processor is achievable with either Ethernet or ATM. For four processors, the resulting 400Mbit/s is roughly 50Mbytes/second. Fifty Mbytes (MB) per second from the four processors can be fed to parallel disks that are connected using Serial Storage Architecture with a transfer capability of 80MB/s. Interprocessor communication on the SP2 switches can sustain 90MB/s. Each SSA loop feeds a single processor, but that loop must contain a sufficiently large number of physical disks (typically ten) to saturate the loop, as the maximum data transfer rate from a single disk is about 8MB/s. The result is a system with several high-speed networks, and where the maximum rate at each stage of the data transfer from disk to network is well matched. This configuration for the parallel data system is scalable by replicating groups of processors (Hollebeek 1998). It is shown schematically in Figure 23.2.

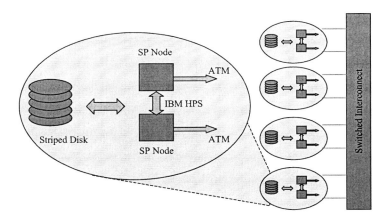

Figure 23.2. Scalable parallel data system.

Terabyte and larger data collections are becoming more common, but we do not have the technology today to casually and quickly explore remote data or to mine data of this scale. Over the next several years, as the amount of WAN dark fiber grows, bandwidth is expected to become a commodity. In this environment, it is possible to imagine a distributed instrument that allows the casual exploration and distributed mining of very large data sets. On the other hand, even with OC-3 links, we found in NSCP that in practice only relatively small amounts of remote data (< 100 GB) could be moved or explored in a casual fashion. At its theoretical limit, an OC-3 can move a Terabyte of data in about 14 hours and a Gigabyte in about a minute. The NSCP-3 Meta-Cluster would be designed to facilitate the casual remote exploration of even larger data sets and the distributed mining of geographically distributed data. We take the viewpoint that the goal of a distributed computation is to complete the computation with sufficient accuracy while moving as little data as possible. Given data sets of the size under discussion, many computations require repetitive movement of 10 GB or more of data. This requires 8.6, 2.1 and 0.5 minutes using an OC-3, OC-12 and OC-48 respectively at its theoretical limit. Much higher bandwidths are available internally within the local clusters, thus after taking advantage of parallel I/O within the cluster, a key requirement is to parallelize the data flow between the distributed components and to carefully match parallel storage to parallel CPU to parallel network paths. Below we explore these requirements. Such a system may require a dedicated hardware system for each local cluster whose function is to matching system internal I/O to external I/O and to maintain parallel streams on the network.

One could argue that a simpler non-parallel high-speed serial network connection would alleviate this problem, and certainly there is continuous progress in designing such serial networks. However, there is some evidence (Bailey and Grossman 1999) that parallel transfer techniques from an application's point of view, are particularly well suited to long haul networks depending on the protocol and it's implementation (e.g. TCP/IP). Therefore, we would argue that there is the potential for interesting new capabilities exploiting parallel links in a WAN environment that we can explore. Further, we expect such parallel links to appear soon in high performance metropolitan area networks and aggregation points and eventually on wide area networks through the use of wave division multiplexing on fiber networks. Effective use of such a network requires careful integration of the wide area network and local networks into the cluster as proposed in this work. Distributed data enabled by parallel network interconnects provides a number of interesting opportunities including much larger scalable bandwidth, the possibility of explicitly using the parallel streams for data striping, or using wavelengths for separated functions (e.g. cluster manager, network management, process migration, and data transfer). All would be useful for data intensive applications. Today we are seeing a rapid growth of optical networks on the wide area and the beginnings of a rapid expansion of local and metropolitan area optical characteristics (Isenberg 1999). Historically, their very high speed and quality has led to their preferred use on long lines for WAN backbones. Wave Division Multiplexing (WDM) is now being used to greatly increase the available bandwidth on these backbones by packing many wavelengths carrying separated data streams onto a single fiber. Until recently, these techniques were too expensive to consider for networking components in our designs, but this has changed with rapid advances in communications markets and hardware, and the reduced cost of this type of equipment.

Video data is likely to play an interesting role in the development of such systems. The real time characteristics of video and the necessity of delivering frames in a uniform stream are related to similar requirements in scalable parallel network designs. Video server applications include video on demand, distance learning, and interactive TV and thus are under rapid development in industry. By adapting some of this software and hardware technology for the transmission of data, we can expect improvements in overall transmission rates, file system and I/O efficiency. To set the scale, a single video stream is 1.5 to 6 Mbit/s, so a large 1000 stream server is comparable to OC48.

Stream starvation in such a server leads to poor file server performance. In video servers, poor performance due to stream starvation is immediately noticeable at the client end as poor video quality. Thus video systems and video server software have often paid particular attention to these issues. Stream starvation can be avoided by buffering data or (inefficiently) designing the server to have much more than the minimum required capacity. Video servers can also incorporate a request control mechanism to prevent overloading and a scheduling mechanism to assure stream delivery. Though the data content is different, many of the ideas used in video servers are also useful in designing data pumps whose goal is to supply a continuous stream of data to a network or remote application.

NSCP has used a component of IBM parallel file systems called Virtual Shared Disk (VSD) (IBM 1994) in implementing parallel access to files for a medical storage project described in Section 6. VSD is high speed and has the ability to support parallel access to files from a single application or from a number of applications running in parallel. We have also used IBM Videocharger software to build a video server and library. The underlying technology in the video server is called Tiger Shark (IBM 1998) and takes advantage of real-time features of the AIX operating system including automatic balancing of loads across disks.

Under development by IBM is integration of the Videocharger and Digital Library system with General Parallel File System (GPFS) which is based on VSD. Thus, we are already beginning to see a merging of parallel high performance file systems with video technologies.

Returning to Lesson 1, the primary message is that rapid manipulation of large data collections requires parallel implementation wherever possible, and that extending and matching the parallelism between all of the components of the system, namely disk, CPU and network is required.

5. Lesson 2: EDT : Exploit Data Parallelism

Consider the problem of building a classifier on a large data set, say a terabyte of data. There are three fundamentally different approaches — sampling, task parallelism and data parallelism:

- Sample, classify, and repeat. With this approach, one samples the data to fit into memory, builds a classifier and validates it on another sample. This process is repeated until a satisfactory classifier is obtained or until 5pm, which ever comes earlier. With this approach megabyte samples are extracted from the terabyte and most of the data is ignored. This is the traditional approach and is usually justified by the observation that memories are getting larger all the time.

- Assign processors to those parts of computation that can be executed in parallel. For example, when building a tree-based classifier, the most expensive part of the computation is the sorting and the computation of the splitting value. This can be easily done in parallel for nodes at the same level. As usual, there can be communication bottlenecks though as the algorithm proceeds. We note that some algorithms for computing tree-based classifiers exploit data structures so that the data can be pre-sorted once, instead of at each node. With is this approach, all the data is examined but much of it must be moved to those processors which are free.

- Partition, classify and build an ensemble. With this approach, one partitions the data, grows a classifier on each partition, and assembles the classifiers to produce an ensemble of classifiers (Grossman et al. 1996, Dietterich 1997). An ensemble of classifiers is collection of classifiers with a rule for combining them. As an example, classifiers can be combined via a majority vote. With this approach, all the data is examined and all of the data is left in place.

Often the simplest approach is the most scalable - generally a data parallel approach is simpler than a task parallel approach. In the NSCP, we have generally used the third approach. In the remainder of this section, we discuss this approach in more detail. One of the reasons that data parallel approaches have not become more popular is that the resulting object is not a classifier itself but rather an ensemble of classifiers, and ensembles are not as well understand as a single classifier. From the appropriate viewpoint however, ensembles are very natural objects. For example, consider a terabyte of data managed by a 100 node cluster, with each node containing 10 Gigabytes of local data, with the problem being to find a tree-based classifier for detecting transaction fraud. It is natural for each node independently and in parallel to construct a classification tree and then to ship the tree to a coordinating node, which assembles the trees into an ensemble of trees. A transaction is evaluated for fraud by evaluating it individually on each of the 100 trees in the ensemble. Each tree individually indicates whether the transaction is fraudulent or not. A simple majority vote is then used to produce the final classification for the ensemble. This is a simple, but effective approach for working with Terabyte size data sets (Grossman et al. 1996). Another advantage of data parallel approaches involving ensembles is that this approach also works naturally for building classifiers on geographically distributed data. For example, in the NSCP, we have used ensembles of tree-based classifiers to analyze health care outcome data that is geographically distributed by the hospitals that produced it.

6. Lesson 3: Pay Attention to PDLDT: Physical Data Layout and Data Transformations

6.1. Legacy Formats and Data Layout

Many projects with large data collections employ some form of database for maintaining that information. Along with the database come legacy data formats in which the data is held, standard reports, and probably considerable numbers of tools and software that accept the particular legacy format as input. Despite the desirability of extracting simplified and optimized forms of the data, a problem occurs when the data collection size reaches the Terabyte level. Implementing Terabyte database systems is difficult, and copying or transforming them is time-consuming. Further, there may not be sufficient facilities to simultaneously maintain both a legacy format and a newer but more efficient format. Thus, there has to be a tradeoff between a new form of database optimized for the data mining process, and using the data in its legacy form as much as possible.

An attractive approach is to design tools that interface the legacy format as much as possible to the data mining process. With the latter approach, the possibility of using the library of software and tools that rely on the legacy form during the mining process remains open. Furthermore, this approach eliminates the need to synchronize the mined data with another perhaps dynamic legacy database and reduces the total volume of stored information since one does not have to have two copies of the data, one in each format. The disadvantage is of course in efficiency of the mining process itself. This is counterbalanced by the efficiency of legacy code reuse. Since high performance means getting accurate information as quickly as possible, sticking with the legacy form is an option that should not be discarded lightly.

A hybrid approach, which we have used in NSCP, is to use index data sets to point back to the legacy format (see `http://nscp.upenn.edu`). In this approach, the legacy format is actually rearranged to optimize it for the mining process, but an index table allows the reconstruction of the original form. This optimizes the performance of the mining process at the slight expense of the other processes. New code for data transforms to legacy formats are required.

6.2. I/O Efficiency through Data Layout

The key point in worrying about the data format is that its layout on the hardware seriously affects the performance of any system which needs to repetitively transfer the data from disk or other storage medium to CPU. As a simple example, consider a system that accumulates transaction records and places them in storage as they are accumulated. The data are time ordered, and this is convenient for answering questions about the time dependence of the data. For example, one can efficiently look up all transactions that occurred last February: Simply extract that portion of the data. Since these records are all adjacent and sorted, this can be done efficiently by the hardware. Suppose, however, that the subject of the data mining query is to know the zipcodes of everyone who had bought a particular product. This would require reading all of the data, making a particular selection (product name) and then accumulating the zipcodes. From a hardware point of view this could be very inefficient and could require reading a complete buffer to extract only a few words. If the database were resorted by product, or if we explicitly produced such a table, we could recover some of this efficiency. This same situation occurs for example in re-sorting astronomical sky images that are collected and stored over time or for high energy particle events that are stored in time sequence as they are collected.

Many database products will handle some of these operations automatically, provided the database does not get too large to fit in memory. Some in fact automatically index frequently used quantities and store hashed index tables in the database or in memory. A more systematic approach is required for very large databases or distributed databases.

We conclude this section with two examples taken from NSCP applications (Hollebeek 1999) of large-scale data using different indexing schemes both scalable and extendible to distributed systems. The first is take from the field of high energy particle physics which accumulates large samples of processed data in several year runs, accumulating 10–100 Terabytes of legacy format data per year. In this case, data transforms were used to access reformatted legacy data. The second comes from storage of digital radiology information. This is an extremely large collection that grows each year at a projected rate five years from now that could exceed 28 Petabytes per year. In this case, index information is extracted as data is collected leading to large meta-data collections.

6.2.1 Particle physics.

Particle Physics detectors have millions of high-speed electronics elements that record the information about collision events occurring with MHz frequency. Therefore, particle detector experiments collect data at rates that can easily result in samples of 10–100 terabytes. Data samples for a CDF experiment at the Fermi National Accelerator Center including raw data, simulated data, and analysis data sets total approximately 100 terabytes. Future accelerators and detectors as well as upgrades to existing detectors will increase this number to the petabyte range and above. Analysis of the data consists of statistical analysis of selected items, extensive analysis of correlations within records often with non-linear combinations of items, and significant amounts of visualization and CPU-intensive Monte Carlo studies.

The traditional data organization in particle physics uses sequential files with fixed-length blocking of variable length records. Data for an entire event is read, and selections of interesting events are made based on typically a small fraction of the total data. It has become common practice for many users to make a single pass through the data to collect summary information, recording the few hundred words per 200,000 bytes that are most useful. While this approach does speed data processing considerably, it requires the user to be very careful in the selection of the initial summary information – it is basically a survival mechanism in the face of a data flood. Other techniques including elimination of some data (data summary tapes or DST's), event selection, and summarization of data, all tend to limit the ability to use any analysis code that relies on full access to the complete data record. Our strategy has been to rewrite the data in a form that makes fixed-length efficient parallel file access possible, while reducing the total amount of data read by a typical analysis program. At the same time, we try to maintain the ability to deal with the data in its original legacy form through data transforms. This has been accomplished by making a new structure for the input data format called "multifile."

The major advantage of the new form was that the data was broken down into commonly used components or objects, making it no longer necessary to read an entire event to begin processing. Instead, only those parts of the event necessary for a particular analysis need be read. The details of the format were arranged to lend themselves to parallel usage either by striping objects across disks or by distributing portions of collections of events across disks. This reorganization of the serial data within a record into a "multifile" structure collects specific types of data across events. Cross-reference tables connect individual data types or objects to companion information from the rest of the event record. Using this scheme, many elements of the event records become fixed length objects thus making parallel I/O easier. Elements that are variable in length within a single event because they contain a variable number of fixed-length elements also become fixed length. The major gain in this technique comes from being able to access only the segments of data that are needed or used in a particular analysis while preserving the remainder of the data for other users and in improving I/O efficiency through fixed blocking. The resulting collection was also implemented in a persistent store making the very large collection appear to be memory-resident using very lightweight database protocols.

6.2.2 Digitized radiology.

Digitized radiology is becoming increasingly common as film methods are replaced by digital X-rays, and increasingly, medical diagnosis relies on intrinsically digital technologies (MRI, CAT, PET). Several archive strategies are being considered, but there is a considerable difference between archival storage and a database that holds the information in a form where it becomes useful as a database. NSCP has prototyped some elements of this problem in order to investigate the issues involved at the hardware and software level, and these designs are now the basis of a Next Generation Internet project being carried out jointly with four hospitals (http://nscp.upenn.edu/ndma). Our goal was to design a database/ archive that could meet the needs of a radiology department, but that would also become a research tool capable of treating the aggregating information as a database. In principle the idea is very simple. What makes this particular task different from the traditional database problem is the size and complexity of the underlying data and that eventually it must be implemented for distributed systems.

A major radiology department can collect more than 6TB of data per year and there are approximately 2000 such units in the US alone, thus one can rapidly accumulate useful data at a scale that exceeds anything we have encountered before. These data have to be stored, manipulated, and retrieved fast enough so that doctors can access a record (image) in a few seconds or minutes at the most. Ideally, the storage system should also be able to scan across the entire contents of a collection for records of interest or even for features embedded in the images.

We have addressed several aspects of this problem in a prototype system called RadAR (radiology archive). The choice of the hardware and the software design was based on the need to have a reliable and fast archive/database consistent with execution of research query functions. Initial tests have shown that the system would be able to easily accumulate 100GB per day per data engine and have remaining capacity for individual queries or for data mining since the internal bandwidth is much higher than required for handling just the input. The system has a very large front-end capacity, and we are confident that the system can scale well and perform as a simultaneous archive and data mining and querying engine. Further tests at the terabyte scale with additional data are the subject of the National Library of Medicine pilot.

In RadAR, data are received from the hospital in a legacy format called DICOM. A multithreaded process running on the input stream extracts the image portion of the record and additional attributes that are likely targets of a later query. From then on, portions of the records are stored in different parts of memory, disk, or tape resident databases, and a meta-database is kept with the location of all attributes of the stored DICOM record. This is similar to the multi-file approach. We have divided the storage space in three major areas. The short-term memory (disk resident) holds only one day's data and is the fastest access medium. The medium term storage is a parallel collection of fast hard disks. This area holds 3 months of data. Finally the long-term memory is a tape library. The long-term storage is however not time ordered, but is sorted. Recently accessed records migrate to short term storage. All processes involved in receiving, migrating, searching, and retrieving are controlled by RadAR. RadAR is a parallel, multi-threaded; object oriented "data server." The front end is web-based. Web techniques will allow an unsophisticated user to execute parallel queries automatically and easily, either through simple forms or through SQL queries. Additional web controlled (Java) processes direct input data to the database and monitor the database activity. Crucial data is mirrored and/or replicated on two systems. The result in this example is a system that simultaneously will store incoming data, construct ongoing meta-data catalogs, and enable use of the sorted collection as a database.

7. Conclusion

We have focused our attention on the lessons to be learned from attempts to manage massive distributed data. The only way to get fast enough performance for very large collections of externally stored information is to implement hardware-level parallelism on as many subsystems as possible. This means that the data should be distributed across parallel disk systems, the database should be parallel-enabled, and the CPUs should be a parallel system with high interprocessor bandwidth. Massive data is intrinsically distributed. It is managed and produced by distributed sensors and systems. Distributed management must play a key role, and when available, distributed systems offer an additional layer of parallelism at the meta-cluster level for parallel storage and analysis. Database systems must assist the data mining process by intelligently storing meta-information and taking advantage of data parallelism. Finally, a key to efficient use of the resulting parallel systems and data parallel analysis is the careful attention to data layout and indexing.

Bibliography

T.E. Anderson, D.E. Culler, D.A. Patterson, and the NOW Team. A case for networks of workstations: NOW. *IEEE Micro*, 15:54–64, 1995.

S. Bailey and R. Grossman. Transport layer multiplexing with psocket. Technical report, NCDM, 1999.

D.J. Becker, T. Sterling, D. Savarese, E. Dorband, U. A. Ranawake, and C.V. Packer. BEOWULF: A parallel workstation for scientific computation. In *Proceedings of the 1995 International Conference on Parallel Processing (ICPP)*, pages 11–14, 1995.

L. Breiman, J.H. Friedman, R.A. Olshen, and C.J. Stone. *Classification and Regression Trees*. Wadsworth, 1984.

T.G. Dietterich. Machine learning research: Four current directions. *AI Magazine*, 18:97–136, 1997.

I. Foster and C. Kesselman. Globus project: A status report. In *Proceedings of the Heterogeneous Computing Workshop*, pages 4–18. IEEE, 1998.

S. Grimsaw and W.A. Wulf. Legion – A view From 50,000 feet. In *Proceedings of the Fifth IEEE International Symposium on High Performance Distributed Computing*. IEEE Computer Society Press, 1996.

R.L. Grossman, S. Bailey, A. Ramu, B. Malhi, H. Sivakumar, and A. Turinsky. Papyrus: A system for data mining over local and wide area clusters and super-clusters. In *Proceedings of Supercomputing*. IEEE, 1999a.

R.L. Grossman, H. Bodek, D. Northcutt, and H.V. Poor. Data mining and tree-based optimization. In E. Simoudis, J. Han, and U. Fayyad, editors, *Proceedings of the Second International Conference on Knowledge Discovery and Data Mining*, pages 323–326. AAAI Press, 1996.

R.L. Grossman, S. Kasif, D. Mon, A. Ramu, and B. Malhi. The preliminary design of Papyrus: A system for high performance distributed data mining over clusters, meta-clusters and super-clusters. In *KDD-98 Workshop on Distributed Data Mining*. AAAI, 1999b.

Y. Guo, S.M. Rueger, J. Sutiwaraphun, and J. Forbes-Millott. Meta-learnig for parallel data mining. In *Proceedings of the Seventh Parallel Computing Workshop*, pages 1–2, 1997.

R. Hollebeek. Data intensive computing, data mining, and economic development. In *Proceedings of Supercomputing*. IEEE, 1998. http://nscp.upenn.edu/hollebeek/talks/sc98.

R. Hollebeek. Data intensive mining. In *Proceedings of Supercomputing*. IEEE, 1999. http://nscp.upenn.edu/hollebeek/talks/sc99.

IBM. *IBM Virtual Shared Disk Users Guide*, 1994.

IBM, 1998. http://www.research.ibm.com/journal/rd/422/haskin.html.

D.S. Isenberg. Dark fiber economics. *America's Network*, 1999.

Y. Ishikawa, M. Sato, T. Kudoh, and J. Shimada. Towards a seamless parallel computing system on distributed environments. In *CPSY. SWOPP'97*, 1997.

H. Kargupta, I. Hamzaoglu, and B. Stafford. Scalable, distributed data mining using an agent based architecture. In D. Heckerman, H. Mannila, D. Pregibon, and R. Uthurusamy, editors, *Proceedings the Third International Conference on Knowledge Discovery and Data Mining*, pages 211–214. AAAI Press, 1997.

R. Moore, T.A. Prince, and M. Ellisman. Data intensive computing and digital libraries. *Communications of the ACM*, 41:56–62, 1998.

A.E. Raftery, D. Madigan, and J.A. Hoeting. Bayesian model averaging for linear regression models. *Journal of the American Statistical Association*, 92:179–191, 1996.

S. Stolfo, A.L. Prodromidis, and P.K. Chan. JAM: Java agents for meta-learning over distributed databases. In *Proceedings of the Third International Conference on Knowledge Discovery and Data Mining*. AAAI Press, 1997.

R. Subramonian and S. Parthasarathy. A framework for distributed data mining. In *International Workshop on Distributed Data Mining (with KDD98)*, 1998.

D. Wolpert. Stacked generalization. *Neural Networks*, 5:241–259, 1992.

L. Xu and M.I. Jordan. EM learning on a generalised finite mixture model for combining multiple classifiers. In *Proceedings of World Congress on Neural Networks*. Erlbaum, 1993.

Chapter 24

SORTING AND SELECTION ON PARALLEL DISK MODELS

Sanguthevar Rajasekaran
Department of CISE, University of Florida, Gainesville, FL 32611, USA
raj@cise.ufl.edu

Abstract Data explosion is an increasingly prevalent problem in every field of science. Traditional out-of-core models that assume a single disk have been found inadequate to handle voluminous data. As a result, models that employ multiple disks have been proposed in the literature. For example, the Parallel Disk Systems (PDS) model assumes D disks and a single computer. It is also assumed that a block of data from each of the D disks can be fetched into the main memory in one parallel I/O operation.

In this article, we survey sorting and selection algorithms that have been devised for out-of-core models assuming multiple disks. We also consider practical implementations of parallel disk models.

Keywords: Parallel disks, Sampling, Sorting and selection algorithms.

1. Introduction

Computing applications have advanced to a stage where the data involved is humongous. The volume of data calls for the use of secondary storage devices such as disks. Even a single disk may not be enough to handle I/O operations efficiently. Thus researchers have proposed models with multiple disks.

One of the models (which refines prior models) that has been studied well is the Parallel Disk Systems (PDS) (Vitter and Shriver 1994). In this model there are D disks and a single computer (sequential or parallel). In one parallel I/O operation, a block of data from each of the D disks can be brought into the main memory of the computer. A block consists of B records. We usually require that $M \geq 2DB$, where M is the internal memory size. At least DB amount of memory is needed to store the data fetched from the disks and the remaining part of the main memory can be used to overlap local computations with I/O operations. Algorithmists have designed algorithms for various fundamental problems on the PDS model. In the analysis of these algorithms, only I/O operations are counted since the local computations are usually very fast.

In this article, we survey sorting and selection algorithms that have been proposed for models with multiple disks. We also investigate the issue of implementing such models in practice. In Section 2 we present details of the PDS model. Sections 3 and 4 are devoted to a survey of sorting and selection algorithms, respectively. In Section 5 we study the problem of implementing parallel disk models in practice. Section 6 concludes the article.

2. Parallel Disk Systems

Here we give more details of the PDS model. A PDS consists of a computer (this could be sequential or parallel) together with D disks. For any given problem, the input will be given in the disks and the output also is expected to be written in the disks. In one I/O operation, a block of B records can be brought into the core memory of the computer from each one of the D disks. In analyzing the time complexity of any algorithm on the PDS model, we consider only the number of parallel I/O operations and neglect the local computation time since the later is usually much smaller.

If M is the internal memory size of the computer, then one usually requires that $M \geq 2DB$. A portion of this memory is used to store operational data whereas the other portion is used for storing prefetched data that enables overlap of local computations with I/O operations. From hereon, M is used to refer to only DB.

The sorting problem on the PDS can be defined as follows. There are a total of N records to begin with so that there are $\frac{N}{D}$ records in each disk. The problem is to rearrange the records such that they are in either ascending order or descending order with $\frac{N}{D}$ records ending up in each disk. For the selection problem also, each disk will have $\frac{N}{D}$ input keys to begin with. The output is through any standard I/O device associated with the computer.

3. Sorting Results on the PDS Model

The problem of external sorting has been widely explored owing to its paramount importance. Given a sequence of n keys, the problem of sorting is to rearrange this sequence in either ascending or descending order. It is easy to show that a lower bound on the number of I/O read steps for parallel disk sorting is[1] $\Omega\left(\frac{N}{DB}\left\lceil\frac{\log(N/B)}{\log(M/B)}\right\rceil\right)$. Here N is the number of records to be sorted and M is the internal memory size of the computer. B is the block size and D is the number of parallel disks used. Numerous asymptotically optimal algorithms that make $O\left(\frac{N}{DB}\left\lceil\frac{\log(N/B)}{\log(M/B)}\right\rceil\right)$ I/O read steps have been proposed (see e.g., Nodine and Vitter (1995), Aggarwal and Plaxton (1994), Arge (1995)).

In the model of Aggarwal and Vitter (1988), each I/O operation results in the transfer of D blocks each block having B records. A refinement of this model was envisioned in Vitter and Shriver (1994). Many asymptotically optimal algorithms have been developed for sorting on this model. Nodine and Vitter's optimal algorithm (Nodine and Vitter 1990) involves the solution of certain matching problems. Aggarwal and Plaxton's optimal algorithm (Aggarwal and Plaxton 1994) is based on the Sharesort algorithm of Cypher and Plaxton (which was originally offered for the hypercube model). An optimal randomized algorithm was given by Vitter and Shriver for disk sorting (Vitter and Shriver 1994). Though these algorithms are highly nontrivial and theoretically interesting, the underlying constants in their time bounds are high.

The algorithm that people use in practice is the simple disk-striped mergesort (DSM) (Barve et al. 1996), even though it is not asymptotically optimal. DSM is simple and the underlying constant is small. In any I/O operation, DSM accesses the same portions of the D disks. This has the effect of having a single disk which can transfer DB records in a single I/O operation. DSM is basically an $\frac{M}{DB}$-way mergesort. To start with, initial runs are formed in one pass through the data. After this, the disks have N/M runs each of length M. Next, $\frac{M}{DB}$ runs are merged at a time. Blocks of any run are uniformly striped across the disks so that in future they can be accessed in parallel utilizing the full bandwidth.

Each phase of merging can be done with one pass through the data. There are $\frac{\log(N/M)}{\log(M/DB)}$ phases and hence the total number of passes made by DSM is $\frac{\log(N/M)}{\log(M/DB)}$, i.e., the total number of I/O read operations called for by DSM is $\frac{N}{DB}\left(1+\frac{\log(N/M)}{\log(M/DB)}\right)$. Note that the constant here is just 1.

If N is a polynomial in M and B is small (which are readily satisfied in practice), the lower bound simply yields $\Omega(1)$ passes. All the optimal algorithms mentioned above make only $O(1)$ passes. Thus the challenge in the design of sorting algorithms lies in reducing this constant. If

$M = 2DB$, the number of passes made by DSM is $1 + \log(N/M)$, which is $\omega(1)$.

Some recent works specifically focus on the design of practical sorting algorithms. For example, Pai et al. (211-239) analyzed the average case performance of a simple merging algorithm, with the help of an approximate model of average case inputs. Barve et al. (1996) have proposed a simple randomized algorithm (SRM) and analyzed its performance. The analysis of SRM involves the solution of certain occupancy problems. The expected number $Reads_{SRM}$ of I/O read operations needed in their algorithm is such that

$$Reads_{SRM} \le \frac{N}{DB} + \frac{N}{DB} \frac{\log(N/M)}{\log kD} \frac{\log D}{k \log \log D} \times \left(1 + \frac{\log \log \log D}{\log \log D} + \frac{1 + \log k}{\log \log D} + O(1)\right). \quad (24.1)$$

SRM merges $R = kD$ runs at a time, for some integer k. The expected performance of SRM is optimal when $R = \Omega(D \log D)$. However, in this case, the internal memory needed is $\Omega(BD \log D)$. They have also compared SRM with DSM through simulations and shown that SRM performs better than DSM.

Recently, Rajasekaran (1998a) has presented an algorithm (called (l, m)-merge sort (LMM)) that is asymptotically optimal when N is a polynomial in M and B is small. The algorithm is as simple as DSM. LMM makes less number of passes through the data than DSM when D is large.

Problems such as FFT computations (see e.g., Cormen (1999)), selection (see e.g., Rajasekaran (1998b)), etc. have also been studied on the PDS model.

Next we present details of the LMM algorithm.

3.1. The (ℓ, m)-Merge Sort (LMM)

Most of the sorting algorithms that have been developed for the PDS are based on merging. To begin with, these algorithms form $\frac{N}{M}$ *runs* each of length M. A run refers to a sorted subsequence. These initial runs can be formed in one pass through the data (or equivalently $\frac{N}{DB}$ parallel I/O operations). Thereafter the algorithms merge R runs at a time. Let a *phase of merges* stand for the task of scanning through the input once and performing R-way merges. Note that each phase of merges will reduce the number of remaining runs by a factor of R. For instance, the DSM algorithm employs $R = \frac{M}{DB}$. The difference among the above sorting algorithms lies in how each phase of merges is done.

LMM of Rajasekaran (1998a) is also based on merging. It fixes $R = \ell$, for some appropriate ℓ. The LMM generalizes such algorithms as the odd-even merge sort, the s^2-way merge sort of Thompson and Kung (1977), and the columnsort algorithm of Leighton (1985).

The well known odd-even mergesort algorithm has $R = 2$. It repeatedly merges two sequences at a time. There are n sorted runs each of length 1 to begin with. Thereafter the number of runs decreases by a factor of 2 with each phase of merges. The odd-even merge algorithm is used to merge any two sequences. A description of odd-even merge follows.

1) Let $U = u_1, u_2, \ldots, u_q$ and $V = v_1, v_2, \ldots, v_q$ be the two sorted sequences to be merged. *Unshuffle* U into two: $U_{odd} = u_1, u_3, \ldots, u_{q-1}$ and $U_{even} = u_2, u_4, \ldots, u_q$. Likewise unshuffle V into V_{odd} and V_{even}.

2) Recursively merge U_{odd} with V_{odd}. Let $X = x_1, x_2, \ldots, x_q$ be the resultant sequence. Also merge U_{even} with V_{even} to get $Y = y_1, y_2, \ldots, y_q$.

3) *Shuffle* X and Y to form the sequence: $Z = x_1, y_1, x_2, y_2, \ldots, x_q, y_q$.

4) Do one step of *compare-exchange operation*, i.e., sort successive subsequences of length two in Z. In particular, sort y_1, x_2; sort y_2, x_3; and so on. The resultant sequence is the merge of U and V.

The zero-one principle can be used to prove the correctness of this algorithm. Thompson and Kung's algorithm (Thompson and Kung 1977) is a generalization of the above algorithm. In this algorithm, R takes on the value s^2 for some appropriate function s of n. At any given time s^2 runs are merged using an algorithm similar to odd-even merge.

LMM generalizes s^2-way merge sort. LMM employs $R = \ell$. The number of runs is reduced by a factor of ℓ by each phase of merges. At any given time, ℓ runs are merged using the (ℓ, m)-merge algorithm. This merging algorithm is similar to the odd-even merge except that in Step 2, the runs are m-way unshuffled (instead of 2-way unshuffling). In Step 3, m sequences are shuffled and also in Step 4, the local sorting is done differently. A description of the merging algorithm follows.

Algorithm (l, m)-merge

1) Let $U_i = u_i^1, u_i^2, \ldots, u_i^r$, for $1 \leq i \leq l$, be the sequences to be merged. When r is *small* use a base case algorithm. Otherwise, unshuffle each

U_i into m parts. I.e., partition U_i into $U_i^1, U_i^2, \ldots, U_i^m$, where $U_i^1 = u_i^1, u_i^{1+m}, \ldots$; $U_i^2 = u_i^2, u_i^{2+m}, \ldots$; and so on.

2) Merge $U_1^j, U_2^j, \ldots, U_l^j$, for $1 \leq j \leq m$, recursively. Let the merged sequences be $X_j = x_j^1, x_j^2, \ldots, x_j^{lr/m}$, for $1 \leq j \leq m$.

3) Shuffle $X_1, X_2, \ldots, X_m$. In particular, form the sequence $Z = x_1^1, x_2^1, \ldots, x_m^1, x_1^2, x_2^2, \ldots, x_m^2, \ldots, x_1^{lr/m}, x_2^{lr/m}, \ldots, x_m^{lr/m}$.

4) At this point the length of the 'dirty sequence' (i.e., unsorted portion) can be shown to be no more than lm. We don't know where the dirty sequence is located. There are many ways to clean up the dirty sequence. One such way is given below.

Let Z_1 denote the sequence of the first lm elements of Z; Let Z_2 denote the next lm elements as Z_2; and so on. Thus Z is partitioned into $Z_1, Z_2, \ldots, Z_{r/m}$. Sort each one of the Z_i's. Then merge Z_1 and Z_2; merge Z_3 and Z_4; etc. Finally merge Z_2 and Z_3; merge Z_4 and Z_5; and so on.

Figure 24.1 illustrates this algorithm.

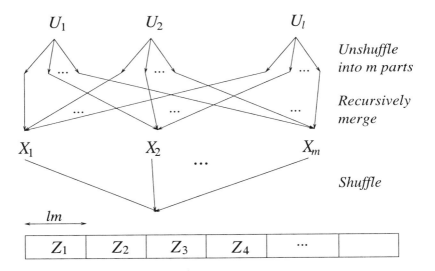

Sort each Z_i ;
Merge Z_1 with Z_2; merge Z_3 with Z_4 ; etc.
Merge Z_2 with Z_3; merge Z_4 with Z_5 ; etc.

Figure 24.1. The (ℓ, m)-merge algorithm

The above algorithm is not specific to any architecture. (The same can be said about any algorithm). Rajasekaran (1998a) gives an implementation of LMM on PDS. The number of I/O operations used in this implementation is $\frac{N}{DB}\left[\frac{\log(N/M)}{\log(\min\{\sqrt{M},M/B\})}+1\right]^2$. This number is a constant when N is a polynomial in M and M is a polynomial in B In this case LMM is optimal. It has been demonstrated that LMM can be faster than the DSM when D is large (Rajasekaran 1998a). Recent implementation results of Cormen and Pearson (1999), Pearson (1999) indicate that LMM is competitive in practice.

3.2. LMM on PDS

In this section we give details of the implementation of LMM on the PDS model. The implementation merges R runs at a time, for some appropriate R. We have to specify how the intermediate runs are stored across the D disks. Number the disks as well as the runs from zero. Each run will be striped across the disks. If $R \geq D$, the starting disk for the ith run is $i \bmod D$, i.e., the zeroth block of the ith run will be in disk $i \bmod D$; its first block will be in disk $(i+1) \bmod D$; and so on. This strategy yields perfect disk parallelism since in one I/O read operation, one block each from D distinct runs can be accessed. If $R < D$, the starting disk for the ith run is $i\frac{D}{R}$. (Assume without loss of generality that D divides R.) Even now, we can obtain $\frac{D}{R}$ blocks from each of the runs in one I/O operation and hence achieve perfect disk parallelism. See Figure 24.2.

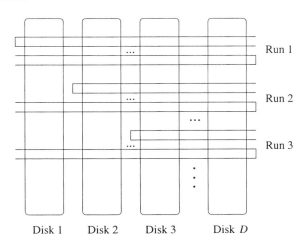

Figure 24.2. Striping of runs

The value of B will be much less than M in practice. For instance, when $\frac{M}{B} > \sqrt{M}$ the number of read passes made by LMM is no more than $\left(2\frac{\log(N/M)}{\log M}+1\right)^2$. However, the case $\frac{M}{B} \leq \sqrt{M}$ is also considered. The number of read passes made by LMM is upper bounded by $\left[\frac{\log(N/M)}{\log(\min\{\sqrt{M},M/B\})}+1\right]^2$ in either case. LMM forms initial runs of length M each in one read pass through the data. After this, the runs will be merged R at a time. Throughout, we use $T(u,v)$ to denote the number of read passes needed to merge u sequences of length v each.

3.2.1 Some special cases. Some special cases will be considered first. The first case is the problem of merging $\sqrt{M}$ runs each of length M, when $\frac{M}{B} \geq \sqrt{M}$. In this case use $R = \sqrt{M}$. This merging can be done using **Algorithm** (l,m)-merge with $l = m = \sqrt{M}$.

Let the sequences to be merged be $U_1, U_2, \ldots, U_{\sqrt{M}}$. In Step 1, each U_i gets unshuffled into $\sqrt{M}$ parts. Each part is of length $\sqrt{M}$. This unshuffling takes one pass. In Step 2, $\sqrt{M}$ merges have to be done. Each merge involves $\sqrt{M}$ sequences of length $\sqrt{M}$ each. There are only M records in each merge and hence all the merges can be done in one pass through the data. Step 3 involves shuffling and Step 4 involves cleaning up. The length of the dirty sequence is $(\sqrt{M})^2 = M$. We can combine these two steps to complete them in one pass through the data. This can be done as follows. Have two successive Z_i's (c.f. **Algorithm** (l,m)-merge) (call these Z_i and Z_{i+1}) at any time in the main memory. Sort Z_i and Z_{i+1} and merge them. Ship Z_i to the disks. Bring in Z_{i+2}, sort it, and merge it with Z_{i+1}. Ship out Z_{i+1}; and so on.

Observe that perfect disk parallelism is maintained throughout. Thus we get:

Lemma 3.1 $T(\sqrt{M}, M) = 3$, if $\frac{M}{B} \geq \sqrt{M}$.

The second special case considered is that of merging $\frac{M}{B}$ runs each of length M, when $\frac{M}{B} < \sqrt{M}$. Employ **Algorithm** (l,m)-merge with $l = m = \frac{M}{B}$. Along similar lines, we can prove the following Lemma.

Lemma 3.2 $T\left(\frac{M}{B}, M\right) = 3$, if $\frac{M}{B} < \sqrt{M}$.

3.2.2 The general algorithm. The general algorithm utilizes the above special cases. The general algorithm is also presented in two cases, one for $\frac{M}{B} \geq \sqrt{M}$ and the other for $\frac{M}{B} < \sqrt{M}$. As usual, initial runs are formed in one pass. After this pass, N/M sorted sequences of length M each remain to be merged.

If $\frac{M}{B} \geq \sqrt{M}$, employ **Algorithm** (l,m)-merge with $l = m = \sqrt{M}$ and $R = \sqrt{M}$. Let K denote $\sqrt{M}$ and let $\frac{N}{M} = K^{2c}$. As a result, $c = \frac{\log(N/M)}{\log M}$. The following relation is easy to see.

$$T(K^{2c}, M) = T(K, M) + T(K, KM) + \cdots + T(K, K^{2c-1}M). \quad (24.2)$$

This relation means that we start with K^{2c} sequences of length M each; we merge K at a time to end up with K^{2c-1} sequences of length KM each; again merge K at a time to end up with K^{2c-2} sequences of length $K^2 M$ each; and so on. Finally there will remain K sequences of length $K^{2c-1}M$ each which are merged. Each phase of merges is done using the **Algorithm** (l, m)-merge with $l = m = \sqrt{M}$.

$T(K, K^i M)$ for any i can be computed as follows. There are K sequences to be merged each of length $K^i M$. Let these sequences be $U_1, U_2, \ldots, U_K$. In Step 1, each U_j is unshuffled into K parts each of size $K^{i-1}M$. This can be done in one pass. Now there are K merging problems, where each merging problem involves K sequences of length $K^{i-1}M$ each. The number of passes needed is $T(K, K^{i-1}M)$. The length of the dirty sequence in Steps 3 and 4 is $\leq K^2 = M$. Thus Steps 3 and 4 can be completed in one pass. Therefore,

$$T(K, K^i M) = T(K, K^{i-1}M) + 2.$$

Expanding this out we see,

$$T(K, K^i M) = 2i + T(K, M) = 2i + 3.$$

We have made use of the fact that $T(K, M) = 3$ (c.f. Lemma 3.1).
Upon substitution of this into Equation 24.2, we get

$$T(K^{2c}, M) = \sum_{i=0}^{2c-1}(2i + 3) = 4c^2 + 4c$$

where $c = \frac{\log(N/M)}{\log M}$. If $N \leq M^3$, the above merging cost is ≤ 24 passes.
We have the following

Theorem 1 *The number of read passes needed to sort N records is $1 + 4\left(\frac{\log(N/M)}{\log M}\right)^2 + 4\frac{\log(N/M)}{\log M}$, if $\frac{M}{B} \geq \sqrt{M}$. This number of passes is no more than $\left[\frac{\log(N/M)}{\log(\min\{\sqrt{M}, M/B\})} + 1\right]^2$.*

Now consider the case $\frac{M}{B} < \sqrt{M}$. **Algorithm** (l,m)-merge can be used with $l = m = \frac{M}{B}$ and $R = \frac{M}{B}$. The following Theorem is proven in a similar fashion.

Theorem 2 *The number of read passes needed to sort N records is upper bounded by $\left[\frac{\log(N/M)}{\log(\min\{\sqrt{M},M/B\})} + 1\right]^2$, if $\frac{M}{B} < \sqrt{M}$.*

Theorems 1 and 2 yield

Theorem 3 *We can sort N records in $\leq \left[\frac{\log(N/M)}{\log(\min\{\sqrt{M},M/B\})} + 1\right]^2$ read passes over the data, maintaining perfect disk parallelism. In other words, the total number of I/O read operations needed is $\leq \frac{N}{DB}\left[\frac{\log(N/M)}{\log(\min\{\sqrt{M},M/B\})} + 1\right]^2$.*

Observation. In **Algorithm** (l,m)-merge, both l and m have to be $\leq \frac{M}{B}$ in order to achieve perfect disk parallelism.

4. Selection Algorithms for the PDS

Given a sequence of n keys and an integer i, $1 \leq i \leq n$, the problem of selection is to identify the ith smallest of the n keys. This important comparison problem has been extensively studied. Numerous asymptotically optimal sequential algorithms have been discovered. Asymptotically optimal algorithms have been presented for varying parallel models as well.

Recently, Rajasekaran (1998b) has given two asymptotically optimal algorithms for selection for the PDS model. The first algorithm is deterministic and the second is randomized. The deterministic algorithm has been implemented and has been shown to perform better in practice than the algorithm of Manku et al. (1998).

In this section a summary of these algorithms is presented.

4.1. A Randomized Algorithm

In this section we describe the randomized selection algorithm of Rajasekaran (1998b). The number of I/O read operations made by the algorithm is $O\left(\frac{N}{DB}\right)$ with high probability.

Numerous selection algorithms proposed in the literature, be they deterministic or randomized, sequential or parallel, are based on sampling.

For instance, Floyd and Rivest's randomized algorithm (Floyd and Rivest 1975) is based on random sampling. The algorithm has the following steps. 1) Select a random sample S of s elements from the input set X; 2) Sort the sample S and find two elements ℓ_1 and ℓ_2 from S such that the ith smallest element of X will have a value in between ℓ_1 and

ℓ_2 and also the number of keys from X that have a value in between ℓ_1 and ℓ_2 is 'small'; 3) Eliminate keys of X that do not have a value in the range $[\ell_1, \ell_2]$; and 4) Perform an appropriate selection in the set of remaining keys.

Sampling techniques have also been used to develop selection algorithms for numerous parallel models of computing. Though sampling based algorithms have sampling as the common theme, they have model-dependent innovations and might employ additional techniques.

A Sampling Lemma. Let Y be a sequence of n numbers from a linear order and let $S = \{k_1, k_2, \ldots, k_s\}$ be a random sample from Y. Also let $k'_1, k'_2, \ldots, k'_s$ be the sorted order of this sample. If r_i is the rank of k'_i in Y, the following lemma provides a high probability confidence interval for r_i. (The rank of any element k in Y is one plus the number of elements $< k$ in Y.)

Lemma 4.1 *For every $\alpha > 0$, Prob.* $\left(|r_i - i\frac{n}{s}| > \sqrt{3\alpha} \frac{n}{\sqrt{s}} \sqrt{\log n} \right) < n^{-\alpha}$.

For a proof of this Lemma see Rajasekaran and Reif (1993).

We say a randomized algorithm uses $\widetilde{O}(g(n))$ amount of any resource (like time, space, etc.) if there exists a constant c such that the amount of resource used is no more than $c\alpha g(n)$ with probability $\geq 1 - n^{-\alpha}$ on any input of length n and for any α (see e.g., Horowitz et al. (1998)). Similar definitions apply to $\widetilde{o}(g(n))$ and other such 'asymptotic' functions. By *high probability* we mean a probability $\geq 1 - n^{-\alpha}$ for any fixed $\alpha \geq 1$ (n being the input size of the problem at hand). Let $B(n,p)$ denote a Binomial distribution with parameters n and p.

Let the input size for the selection problem be N and let $N = M^c$. In practice c can be assumed to be a constant. In today's PC market, M is of the order of megabytes and the disk space is of the order of gigabytes. So it is perhaps safe to assume that c is no more than 3. To begin with each key is *alive* and $n = N$.

Algorithm RSelect

repeat

    Step 1. Let n denote the number of alive keys. If $n \leq M$ then goto Step 6. Each alive key gets included in the sample S with probability $\frac{M}{n}$. Thus the expected number of sample keys is M. The actual number of keys in S can be shown to be $M + \widetilde{o}(M)$. Count the number s of sample keys.

    Step 2. Sort S and pick two keys ℓ_1 and ℓ_2 from S whose ranks in S are $i\frac{s}{n} - \delta$ and $i\frac{s}{n} + \delta$, respectively, for $\delta \geq \sqrt{3\alpha s \log n}$, for any fixed $\alpha \geq 1$.

Step 3. Count the number n_1 of alive keys that are less than ℓ_1 and the number n_2 of alive keys that have a value in the range $[\ell_1, \ell_2]$.

Step 4. If $i < n_1$, or $i > n_1 + n_2$, or $n_2 > \frac{n}{M^{0.4}}$, goto Step 1.

Step 5. Any alive key whose value lies outside the range $[\ell_1, \ell_2]$ dies. Set $i = i - n_1$ and $n = n_2$.

forever

Step 6. Sort the alive keys and output the ith smallest key.

Analysis. The number of sample keys in Step 1 has a binomial distribution $B(n, M/n)$. Applying Chernoff bounds we see that $s = M + \widetilde{o}(M)$. Step 1 takes $O(\frac{n}{DB})$ I/O read operations.

Step 2 does not involve any I/O operations since S is kept in the internal memory.

In Step 3, an application of Lemma 4.1 implies that the number n_2 of keys surviving Step 5 is $\widetilde{O}\left(\frac{n}{\sqrt{s}}\sqrt{\log n}\right) = \widetilde{O}\left(\frac{n}{M^{0.4}}\right)$. It follows that, the number of iterations of the *repeat* loop is $\widetilde{O}(c)$.

In Step 4, we can show that the probability of executing the goto statement is very small.

Step 5 takes $O(\frac{n}{DB})$ I/O read operations.

The number of surviving keys from one iteration of the *repeat* loop to the next decreases by a factor of $M^{0.4}$ with high probability. As a result, the total number of I/O read operations made by the entire algorithm is $\widetilde{O}\left(\frac{N}{DB}\right)$.

Thus the following Theorem follows.

Theorem 4 *Selection from out of N keys can be performed on the PDS using $\widetilde{O}\left(\frac{N}{DB}\right)$ I/O read operations.* □

4.2. A Deterministic Algorithm

In this section we summarize the deterministic selection algorithm of Rajasekaran (1998b). The number of I/O read operations performed by the algorithm is $O\left(\frac{N}{DB}\right)$. The constant in this time bound is small and hence this algorithm has the potential of being practical. Recent implementation results indicate that this algorithm is faster than the algorithm of Manku et al. (1998).

In the development of deterministic selection algorithms also sampling has dominated. For instance, Blum et al.'s algorithm (Blum et al. 1973) partitions the input such that there are 5 elements in each part, finds the median of each part, finds the median M of these medians, splits the input into two groups (those that are $\leq M$ and those that are greater than M), identifies the group that has the key to be selected, and finally performs a selection in the group that contains the key to be selected. We can think of the medians of the 5-element parts as a sample of the input keys and hence the median M of these medians can be expected to be an approximate median of the input keys.

Let X be a collection of n keys whose ith smallest key we are interested in finding. We identify two elements of X such that they will bracket the ith smallest element and also the number of keys of X that have a value in between these two keys is not large.

Partition $X = R_0$ into M-element parts. Sort each part. In each part retain those keys that are at a distance of $\sqrt{M}$ from each other. I.e., keep the keys whose ranks are $\sqrt{M}, 2\sqrt{M}, 3\sqrt{M}, \ldots$. The number of keys in the retained set R_1 is $\frac{n}{\sqrt{M}}$. Now group the elements of R_1 such that there are M elements in each part, sort each part, and retain only every $\sqrt{M}$th element in each part. Call the retained set R_2. Proceed to obtain R_i's in a similar fashion (for $i \geq 3$) until we reach a stage when $|R_j| \leq M$. If $n = M^c$, then clearly, $j = 2c - 2$.

We can use a tree of degree $\sqrt{M}$ to represent this process. There are M input elements in each leaf. R_0 comprises all the elements in the leaves. The root has R_j. Let the root be in level j. Let its children be in level $j - 1$, and so on. The leaves are at level 0. The root has $\sqrt{M}$ children. Each such child has M elements. $\sqrt{M}$ elements are passed on from each child to its parent. In general each node in the tree has M elements. $\sqrt{M}$ elements from out of these will go to its parent. Each node has $\sqrt{M}$ children.

Pick from R_j two elements ℓ_1 and ℓ_2 whose ranks are $i\frac{|R_j|}{n} - \delta$ and $i\frac{|R_j|}{n} + \delta$, respectively.

Assume without loss of generality that $|R_j| = M$. In this case, $|R_i| = M(\sqrt{M})^{j-i}$. Consider an element x whose rank in R_j is q. The rank of q in R_{j-1} will be in the range $[q\sqrt{M}, q\sqrt{M} + \sqrt{M}(\sqrt{M} - 1)]$. This rank is also in the interval $[q\sqrt{M}, q\sqrt{M} + M]$. In other words, there is an uncertainty of M in the rank of x in R_{j-1}. Each child of the root contributes $\sqrt{M} - 1 \approx \sqrt{M}$ to this uncertainty. In general the uncertainty in the rank of x in R_i is contributed to by each node in level i. There are $M^{(j-i)/2}$ nodes at level i. Every such node contributes $\sqrt{M} - 1$ to the uncertainty. As a result, if $U(i)$ is the maximum possible rank of x in R_i, then $U(i)$ satisfies:

$$U(i) \leq \sqrt{M}\, U(i+1) + M^{(j-i+1)/2}$$

which solves to $U(i) \leq M^{(j-i)/2}U(j) + (j-i)M^{(j-i+1)/2}$. When $i = 0$, we get $U(0) \leq M^{j/2}U(j) + jM^{(j+1)/2}$. In other words, $U(0) \leq qM^{c-1} + (2c-2)\frac{n}{\sqrt{M}}$. Note that $c = \frac{\log n}{\log M}$.

Thus if we pick δ to be $(2c-2+\epsilon)\sqrt{M}$, for any $\epsilon > 0$, the rank of ℓ_1 in R_0 will be in the interval

$$\left[i - (2c - 2 + \epsilon)\frac{n}{\sqrt{M}}, i - \epsilon\frac{n}{\sqrt{M}}\right].$$

The rank of ℓ_2 in R_0 will be in the interval

$$\left[i + (2c - 2 + \epsilon)\frac{n}{\sqrt{M}}, i + (4c - 4 + \epsilon)\frac{n}{\sqrt{M}}\right].$$

In summary, we realize that the ith smallest element of R_0 will have a value in the interval $[\ell_1, \ell_2]$ and also the number of keys of R_0 that have a value in the interval $[\ell_1, \ell_2]$ is no more than $(6c - 6 + 2\epsilon)\frac{n}{\sqrt{M}}$.

Let the above process of starting from R_0 and obtaining $R_1, R_2, \ldots, R_j$ be called a *stage of sampling*. The number of I/O read operations needed for a stage is $O\left(\frac{n}{DB}\right)$ where $|R_0| = n$. A stage of sampling corresponds to one iteration of the *repeat* loop of RSelect. Similar sampling techniques have been employed by Munro and Paterson (1980).

Let $K = k_1, k_2, \ldots, k_N$ be the input. Say we are interested in finding the ith smallest key. To begin with each key is alive and $n = N$. A detailed description of the selection algorithm follows.

Algorithm DSelect

repeat

    Step 1. If $n \leq M$ goto Step 3. Perform one stage of sampling in the collection of alive keys. Obtain two keys ℓ_1 and ℓ_2 that will bracket the key to be selected.

    Step 2. Kill the alive keys that have a value outside the range $[\ell_1, \ell_2]$. Count the number n of keys surviving this step.

forever

    Step 3. Sort the alive keys and output the ith smallest element.

Analysis. The number of alive keys reduces by a factor of $\Omega\left(\frac{\sqrt{M}\log M}{\log n}\right)$ from one iteration to the next of the *repeat* loop. As a result, if $N = M^c$, the number of iterations of the *repeat* loop is $O(c)$.

The number of I/O read operations needed in any iteration of the *repeat* loop is $O\left(\frac{n}{DB}\right)$ as has been established before. Since the number of alive keys decreases by a factor of $\Omega\left(\frac{\sqrt{M}\log M}{\log n}\right)$ from one iteration to the next, the total number of I/O read operations is only $O\left(\frac{N}{DB}\right)$. We arrive at the following Theorem.

Theorem 5 *We can perform selection from N given keys on the PDS using $O\left(\frac{N}{DB}\right)$ I/O read operations.* □

5. A Practical Realization of Parallel Disks

Though the existing models with multiple disks address the problem of data explosion, it's not clear how these models can be realized in practice. The assumption of bringing D blocks of data in one I/O operation may not be practical. A new model called a Parallel Machine with Disks (PMD) is proposed in Rajasekaran and Jin (1999) that is a step in the direction of practical realization. A PMD is nothing but a parallel machine where each processor has an associated disk. The parallel machine can be structured or unstructured. The underlying topology of a structured parallel machine could be a mesh, a hypercube, a star graph, etc. Examples of unstructured parallel computers include SMP, NOW, a cluster of workstations (employing PVM or MPI), etc. The PMD is nothing but a parallel machine where we study out-of-core algorithms. In the PMD model we not only count the I/O operations but also the communication steps. One can think of a PMD as a realization of the PDS model.

Every processor in a PMD has an internal memory of size M. In one parallel I/O operation, a block of B records can be brought into the core memory of each processor from its own disk. There are a total of $D = P$ disks in the PMD, where P is the number of processors. Records from one disk can be sent to another via the communication mechanism available for the parallel machine after bringing the records into the main memory of the origin processor. The communication time could potentially be comparable to the time for I/O on the PMD. It is essential therefore to not only account for the I/O operations but also for the communication steps, in analyzing any algorithm's run time on the PMD.

We can state the sorting problem on the PMD as follows. There are a total of N records to begin with. There are $\frac{N}{D}$ records in each disk. The problem is to rearrange the records so that they are in either ascending order or descending order with $\frac{N}{D}$ records ending up in each disk. We assume that the processors themselves are ordered so that the smallest $\frac{N}{D}$ records will be output in the first processor's disk, the next smallest $\frac{N}{D}$ records will be output in the second processor's disk, and so on.

We can apply LMM on a general PMD in which case the number of I/O operations will remain the same, i.e., $\frac{N}{DB}\left[\frac{\log(N/M)}{\log(\min\{\sqrt{M},M/B\})}+1\right]^2$. In particular, the number of passes through the data will be $\left[\frac{\log(N/M)}{\log(\min\{\sqrt{M},M/B\})}+1\right]^2$. Such an application needs mechanisms for (k to k) routing and (k to k) sorting.

The problem of (k to k) routing in the context of a parallel machine is the problem of packet routing where there at most k packets of information originating from each processor and at most k packets are destined for each processor. We are required to send all the packets to their correct destinations as quickly as possible. The problem of (k to k) sorting is one where there are k keys to begin with at each processor and we are required to sort all the keys and send the smallest k keys to processor 1, the next smallest k keys to processor 2, and so on.

Let R_M and S_M denote the time needed for performing one (M to M) routing and one (M to M) sorting on the parallel machine, respectively. Then, in each pass through the data, the total communication time will be $\frac{N}{DM}(R_M + S_M)$, implying that the total communication time for the entire algorithm will be $\leq \frac{N}{DM}(R_M + S_M)\left[\frac{\log(N/MD)}{\log(\min\{\sqrt{MD},MD/B\}}+1\right]^2$.

The following Theorem is proven in Rajasekaran and Jin (1999).

Theorem 6 *The task for sorting on a PMD can be finished in $\frac{N}{DB}\left[\frac{\log(N/MD)}{\log(\min\{\sqrt{MD},MD/B\}}+1\right]^2$ I/O operations. The total communication time is $\leq \frac{N}{DM}(R_M + S_M)\left[\frac{\log(N/MD)}{\log(\min\{\sqrt{MD},MD/B\}}+1\right]^2$.* □

6. Conclusions

In this article we have provided a survey of sorting and selection algorithms that can be found in the literature. In particular, we have summarized the DSM, SRM, and LMM algorithms for sorting. DSM is a simple algorithm that does well in practice but is not asymptotically optimal. SRM is a randomized algorithm. LMM algorithm is asymptotically optimal and is promising to be practical. In addition we have discussed asymptotically optimal (deterministic and randomized) algorithms for selection on the PDS model. We have also studied the problem of realizing parallel disk models in practice.

Bibliography

A. Aggarwal and C. G. Plaxton. Optimal parallel sorting in multi-level storage. In *Proc. Fifth Annual ACM Symposium on Discrete Algorithms*, pages 659–668, 1994.

A. Aggarwal and J. S. Vitter. The input/output complexity of sorting and related problems. *Communications of the ACM*, 31(9):1116–1127, 1988.

L. Arge. The buffer tree: A new technique for optimal I/O-algorithms. In *Proc. 4th International Workshop on Algorithms and Data Structures (WADS)*, pages 334–345, 1995.

R. Barve, E. F. Grove, and J. S. Vitter. Simple randomized mergesort on parallel disks. Technical report, Department of Computer Science, 1996.

M. Blum, R. W. Floyd, V. Pratt, R. L. Rivest, and R. E. Tarjan. Time bounds for selection. *Journal of Computer and System Sciences*, 7: 448–461, 1973.

T. Cormen. Determining an out-of-core fft decomposition strategy for parallel disks by dynamic programming. In Robert S. Schreiber Michael T. Heath, Abhiram Ranade, editor, *Algorithms for Parallel Processing*, pages 307–320. New York: Springer-Verlag, 1999.

T. Cormen and M. D. Pearson, 1999. Personal Communication.

R. W. Floyd and R. L. Rivest. Expected time bounds for selection. *Communications of the ACM*, 18(3):165–172, 1975.

E. Horowitz, S. Sahni, and S. Rajasekaran, editors. *Computer Algorithms*. W. H. Freeman Press, 1998.

T. Leighton. Tight bounds on the complexity of parallel sorting. *IEEE Transactions on Computers*, C34(4):344–354, 1985.

G.S. Manku, S. Rajagopalan, and G. Lindsay. Approximate medians and other quantiles in one pass and with limited memory. In *Proc. of the 1998 ACM SIGMOD International Conference on Management of Data*, pages 426–435, 1998.

J. I. Munro and M. S. Paterson. Selection and sorting with limited storage. *Theoretical Computer Science*, 12:315–323, 1980.

M. H. Nodine and J. S. Vitter. Large scale sorting in parallel memories. In *Proc. Third Annual ACM Symposium on Parallel Algorithms and Architectures*, pages 29–39, 1990.

M. H. Nodine and J. S. Vitter. Greed sort: Optimal deterministic sorting on parallel disks. *Journal of the ACM*, 42(4):919–933, 1995.

V. S. Pai, A. A. Schaffer, and P. J. Varman. Markov analysis of multiple-disk prefetching strategies for external merging. *Theoretical Computer Science*, 128(2):1994, 211-239.

M. D. Pearson. Fast out-of-core sorting on parallel disk systems. Technical report, Dartmouth College, Computer Science, 1999. ftp://ftp.cs.dartmouth.edu/TR/TR99-351.ps.Z.

S. Rajasekaran. A framework for simple sorting algorithms on parallel disk systems. In *Proc. 10th Annual ACM Symposium on Parallel Algorithms and Architectures*, pages 88–97, 1998a.

S. Rajasekaran. Selection algorithms for the parallel disk systems. In *Proc. International Conference on High Performance Computing*, 1998b.

S. Rajasekaran and X. Jin. A practical model for parallel disks. manuscript, 1999.

S. Rajasekaran and J.H. Reif. Derivation of randomized sorting and selection algorithms. In R. Paige, J.H. Reif, and R. Wachter, editors, *Parallel Algorithm Derivation and Program Transformation*, pages 187–205. Kluwer Academic Publishers, 1993.

C. D. Thompson and H. T. Kung. Sorting on a mesh connected parallel computer. *Communications of the ACM*, 20(4):263–271, 1977.

J. S. Vitter and E. A. M. Shriver. Algorithms for parallel memory I: Two-level memories. *Algorithmica*, 12(2-3):110–147, 1994.

VIII
APPLICATIONS

Chapter 25

BILLING IN THE LARGE

Andrew Hume
AT&T Labs Research, Florham Park, NJ 07932, USA
andrew@research.att.com

Abstract There is a growing need for very large databases which are not practical to implement with conventional relational database technology. These databases are characterized by huge size and frequent large updates; they do not require traditional database transactions, instead the atomicity of bulk updates can be guaranteed outside of the database. Given the I/O and CPU resources available on modern computer systems, it is possible to build these huge databases using simple flat files and simply scanning all the data when doing queries. This paper describes Gecko, a system for tracking the state of every call in a very large billing system, which uses sorted flat files to implement a database of about 60G records occupying 2.6TB. We focus on the performance issues, particularly with regard to job management, I/O management and data distribution, and on the tools we built to run the system.

Keywords: Massive data sets, Telecommunications, Telephone billing systems, Datastore, I/O

1. Introduction

Billing is one of the least attractive chores in computer science. For the most part, it is simple, straightforward, full of picky little details, and unchallenging. Generating monthly bills for a few thousand customers each with a few transactions, is fairly easy. And with one exception, it remains straightforward even when the number of customers becomes large, such as for the bigger credit card companies. The main exception is the phone company, where there are several times as many transactions (calls), and each one is for a much lower amount (typically, $0.25-2 versus $20-50).

Indeed, within large telecommunication companies such as AT&T, billing is an extremely hard problem, although some of the reasons might not be obvious. The reasons include

- extremely large number of transactions, say 250-300M, per day.

- large number of customers, say 50-100M.

- complexity and volatility of taxes and tariffs. This applies not only to simply rating each call, which may require access to a database entry for the originator of a call, but to the various calling plans that affect how the bill is calculated from the individual call events. For both of these tasks, new schemes may need to be introduced within days or even hours, and may even need to be applied retroactively, sometimes on a very wide basis.

- supporting customer care. Typically, this means being able to access individual calls made over the last 3-12 months, and being able to adjust bills. There are normally quite stringent performance requirements for accessing the calls.

Surprisingly (to most people), the customer care issue, especially the performance and latency issues, is the dominant driver for system requirements.

The rest of this paper focuses on a system called Gecko that solves, at least from a data point of view, a superset of the billing problem. More exactly, it traces the information flowing through the billing process by maintaining a database of about 60G records, occupying 2.6TB, describing each processing step for each call. Below, we describe how we implemented the database and the architecture of the reporting subsystem. We focus on the performance issues, particularly with regard to process management, I/O management and data distribution. We'll finish with some thoughts on the challenges posed by Gecko to the database community.

2. Gecko

Gecko is a system for monitoring information flowing through much of AT&T's telephone billing systems. The focus for Gecko is residential billing. Billing for business customers follows a process analogous to that for residential customers, but is currently outside of Gecko's purview. The overall flow, as shown in Figure 25.1, is fairly straightforward [1].

As each call is completed, a record is generated by one of about 160 switches distributed geographically around the USA and periodically sent to BILLDATS (located in Arizona). BILLDATS ensures that no

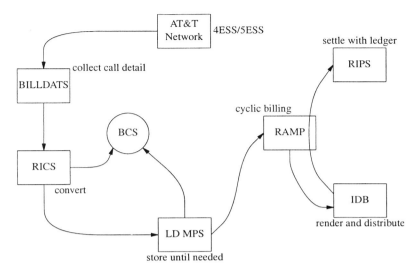

Figure 25.1. High level billing data flow

data from the switches are lost and sends the records to RICS. RICS is primarily concerned with validating the switch records, deciding where (which billing system) to send the records, and converting the records into formats suitable for that biller. (The native switch records have hundreds of different formats.) Records for residential calls are then sent to MPS where they are stored waiting for their billing cycle to occur. Each billing cycle day (22 per month), the residential biller (RAMP) requests all records for about 5% (100/22) of its customers, calculates the bill and sends it to IDB. IDB is responsible for formatting and rendering the bill (either sending the bill electronically to a local telephone company or printing and mailing a paper bill). When bills are paid, RAMP sends the records involved to RIPS. RIPS handles settlement of the calls, which is essentially interfacing with the corporate accounting system (customer X was billed $Y and paid $Z). At various points, records may be sent to the business billers; these are denoted as BCS in Figure 25.1.

Given the complexity of this sort of architecture, and the stunning volumes involved, it is important to address end-to-end completeness and accuracy.

Gecko attempts to answer the question: is every call billed exactly once? It does so in a novel way: it tracks the progress of (or records corresponding to) each call throughout the billing process. Although this seems an obvious solution, the volumes involved have hitherto made this sort of scheme infeasible. Here, we focus on the database underlying

Gecko and ignore the many other difficult aspects of dealing with arcane record formats from unwieldy legacy systems (these have been addressed in Hume et al. (2000)). Therefore, we will assume that the various data feeds have been converted into fixed-length records, or tags, for further processing.

2.1. The Problem

The problem, then, is this: we need to maintain a datastore (database) of tagsets. Each tagset consists of one or more tags sharing the same key (24 bytes including fields like originating number and timestamp). Each day, we add about 3100 files totaling about 1B tags, age off certain tagsets (generally these are tagsets that have been completely processed), and generate various reports about the tagsets in the datastore. The relevant quantitative information is

- each tag is 96 bytes; a tagset is $80 + 24n$ (where the tagset has n tags).

- the datastore will typically contain about 60B tags in 13B tagsets.

- the target for producing reports is about 11 hours; the target for the entire cycle is about 15 hours (allowing some time for user ad hoc queries), with a maximum of 20 hours (allowing some time for system maintenance).

Tagsets which have exited the billing process, that is, they represent calls that have been billed, are eventually aged out of the datastore. Typically, we keep tagsets for 30 days after the exit (in order to facilitate analysis).

Finally, there needs to be a mechanism for examining the tagsets contributing to any particular numeric entry in the reports; for example, if we report that 20,387 tagsets are delayed and were last seen in RICS, the users need to be able examine those tagsets.

3. The Current Architecture

The design we implemented to solve the above problem does not use conventional database technology; as described in Hume and Maclellan (1999), we experimented with an Oracle-based implementation, but it was unsatisfactory. The prototype was built using Oracle 7, organized as a table per call origination date; this facilitated deleting old calls from the database and generating reports by origination date. In several ways, the prototype was unattractive: complete tagsets were not actually stored (just the latest tag), handling cases where the call's origination date changed (because of timezone issues) was difficult, and the whole database schema was tied very closely to the exact reports required. Nevertheless, it worked. The time for an update/report cycle was 16 hours (the prototype ran on a 20 CPU SGI Challenge). By comparison, a version of the flat file scheme described in this paper was able to do the same processing in 3 hours. This was not unexpected; this is not the application relational databases were designed for. The main surprise was how much longer it took, and how fragile the DB architecture/schema was (if we changed the reports to need another field, the flat file code was almost unchanged, but the DB architecture would likely have to be completely redesigned).

Instead, we used sorted flat files and relied on the speed and I/O capacity of modern high-end Unix systems, such as large SGI and Sun systems. The following description describes our current implementation; in some cases, this was rather different than our original design. These differences are described in section 4.

3.1. Data Design

The system supporting the datastore is a Sun E10000, with 32 processors and 6GB of memory, running Solaris 2.6. The datastore disk storage is provided by 16 A3000 (formerly RSM2000) RAID cabinets. Each A3000 has two UltraSCSI controllers mediating access to 5 trays of 7 9.1GB disks. The two controllers (and Solaris) provide seamless access should one controller fail, but otherwise can be considered as independent. Each cabinet is configured as 7 36GB virtual disks (LUNs); each LUN is a 4+1 RAID 5, with a stripe size of 64KB, arranged in columns (that is, each LUN has one disk per tray). The datastore uses 6 of the LUNs in each A3000, 3 on each controller. (The remaining LUN is used for other purposes and is quiescent during the datastore update process.) Each controller has a dedicated UltraSCSI connection to the E10000.

For backup purposes, we have a StorageTek 9310 Powderhorn tape silo with 8 Redwood tape drives. Each drive can read and write at about 10MB/s. The silo has a capacity of 6000 50GB tapes. Because we were a little short on places to plug in SCSI adapters, each pair of drives shares a Fast Wide SCSI bus.

The datastore is organized as 93 filesystems mounted under /gecko/-datastore. Each filesystem is a Veritas filesystem occupying a single LUN. The filesystems each have 52 directories; each directory contains a *partition* of the datastore. Tagsets are allocated to one of these 4836 partitions according to a hash function based on the originating telephone number. This hash works out fairly well: the average partition size is 597MB, the smallest is 504MB, the largest is 2943MB, the 5 percentile is 530MB, the 95 percentile is 740MB. Currently, the datastore is about 2.6TB. Because of Solaris's inability to sustain large amounts of sequential file I/O through the file page cache (which seems fixed as of Solaris 2.8), all the datastore filesystems are mounted as direct I/O; that is, all file I/O on those filesystems bypasses the page cache.

This 'feature' turned out to be a blessing in disguise because it helped us discover an unexpected and deep design paradigm: designing for a scalable cluster of systems networked together is isomorphic to designing for a single system with a scalable number of filesystems. Just as with a cluster of systems, where you try to do nearly all the work locally on each system and minimize the inter-system communications, you arrange that processing of data on each filesystem is independent of processing on any other filesystem. The goal, of course, is to make the design scalable and have predictable performance. In this case, using the system-wide buffer cache would be an unnecessary bottleneck. This isomorphism is so pervasive that when we evaluate design changes, we think of filesystems as having bandwidths and copying files from one filesystem to another is exactly the same as *ftp*ing files over a network.

This isomorphism seems related to the duality discussed in Lauer and Needham (1979), a duality between systems made of a smaller number of larger processes communicating by modest numbers of messages and systems comprising large numbers of small processes communicating via shared memory. By replacing 'shared memory' by 'intrasystem file I/O' and 'message passing' with 'networked file I/O', Lauer and Needham's arguments about the fundamental equivalence of the two approaches seem fresh and persuasive.

Because of our dependence on I/O speed and because datastore I/O is unbuffered, we need to carefully control I/O to the datastore in order to get good I/O performance. This was easy because we use a single tool, *woomera*, to control the 10,000+ tasks needed to update the datastore. The relevant part of *woomera*, which is described in more detail below, is that jobs can be marked as using various resources and it is easy to specify limits on how many tasks sharing a resource may run simultaneously. By marking all the tasks that run on filesystem gb with a resource **LVOLgb**, we can set the limit for the resource *LVOLgb* to one and thus ensure that at most one task will run on that filesystem at any given time.

3.2. Report Architecture

Gecko is required to generate three different reports. Two of the reports describe calls that are still being processed, and the other describes the eventual disposition of calls. This latter report requires a summary of all calls ever processed. The current reporting architecture is a combination of two things. One, the history file, is a summary of all tagsets that have been deleted from the datastore. This summary is a fairly general breakdown that can support a fairly wide range of reports related to the existing reports. Thus, most changes to the reports do not require modifying the history; just its interpretation and tabulation. The history file is currently a few hundred MB in size and grows slowly over time. The second are summaries of tagsets still in the datastore. These latter summaries are never stored for any length of time; they are simply intermediate results for the daily update/report cycle.

Reporting is based on an intermediate report format (IRF), which represents a summary derived from a tagset file. The data for the daily reports comes from two sources: the current datastore, and a history of deleted tagsets. There is a tool to collapse multiple IRF files into a single minimally sized IRF file. The processing steps are:

- in each partition, generate IRF files for the new data file and the deleted tagsets.

- in each filesystem, collapse the deleted tagsets IRF files into a single file.

- collapse all 93 deleted tagsets IRF files and the history file into the new history file.

- generate a report from the history file and the 4836 data IRF files.

- backup the history file to tape.

After the reports are generated, they are shipped to a Web site for access by our customers.

3.3. Process Architecture

The current processing architecture is fairly straightforward. With the exception of the first step, which occurs throughout the day, the remaining steps occur as part of a daily cycle that starts around 0030 hours.

The first step is to distribute the incoming tags out to the datastore. For every source file, files with the same name are created in each filesystem (and not partition) and filled with tags that hash to any partition on that filesystem.

The next step examines all the filesystems, determines which input files will be included in this cycle, and then generates all the tasks to be executed for this update cycle. The 10,000+ tasks are then given to *woomera* for execution.

Within each filesystem, the incoming tags are sorted together in 1GB parcels. The resulting files are merged and then split out into each of the 52 partitions. The end result is the *add file*, a sorted file of tags to be added to each partition's data file. (The 1GB size is an administrative convenience; from this we experimentally tuned the various sorting parameters, most noticeably the amount of memory used.)

The next process, *pu*, updates the partition data file by merging in the new tags, deleting appropriate tagsets, and generating an intermediate report output. The deleted tagsets are put into the *delete file*. Because the underlying filesystem is unbuffered, all tag-related I/O goes through a n-buffering scheme (we currently use triple buffers). We generate a summary description of the delete file.

The next step, performed after all the partitions on a specific filesystem have been processed, rolls up the two different report summaries in all those partitions into two equivalent files for the whole filesystem.

The next step generates the reports for that cycle. First, we combine the summary for the deleted tagsets with the old history file and generate a new history file. Second, we combine that and the 93 filesystem summaries and generate a single set of reports.

Finally, we backup the datastore in two passes. The first pass stores all the add files and delete files. The second pass stores a rotating sixth of the 4836 datastore partitions. (This is exactly analogous to incremental and full backups.)

3.4. Trust but Verify

A significant aspect of our implementation, and one we didn't anticipate, involves performing integrity checks whenever we can. This extends from checking that when we sort several hundred files together, the size of the output equals the sum of the sizes of the input files, to whenever we process tagset files, we verify the format and data consistency. Although this is frankly tedious and modestly expensive, it has been necessary given the number of bugs in the underlying system software and hardware.

4. Current Performance

We can characterize Gecko's performance by two measures. The first is how long it takes to achieve the report and cycle end gates. The second is how fast we can scan the datastore; this is a reasonable estimate for how long it takes to do an ad hoc search/extract.

The report gate is reached when the datastore has been completely updated and we've generated the final reports. Over the last 12 cycles, the report gate ranged between 6.1 and 9.9 wall clock hours, with an average time of 7.6 hours. The cycle end gate is reached after the updated datastore has been backed up and any other housekeeping chores have been completed. Over the last 12 cycles, the cycle end gate ranged between 11.1 and 15.1 wall clock hours, with an average time of 11.5 hours.

We can measure how fast we scan the datastore by running a simple statistical analysis program on each partition and managing it as though we were performing an update cycle. The elapsed wall clock time was 71 minutes, which yields an overall average I/O rate of about 606MB/s.

If we change the analysis program to one performing SQL-like select criteria, then the numbers change to 392MB/s for a null query and 255MB/s for a simple query. (Note that the query program was fairly crude and untuned.)

5. What We Learned along the Way

5.1. Decouple What from How

The natural way to perform the daily update cycle is to have some program take some description of the work and figure out what to do and then do it, much like the Unix tool *make*. We deliberately rejected this scheme in favor of a three part scheme: one program figures out what has to be done (*dsum*), and then gives it to another to schedule and execute (*woomera*), while another program monitors things and tweaks various *woomera* controls (*bludge*). Although superficially more complicated, each component is much simpler to build and maintain and allows reuse by other parts of our process. More importantly, it allows realtime adjustments (e.g. pause all work momentarily) as well as structural constraints (e.g. keep the system load below 60 or no more than 2 tasks running on filesystem **gb**).

The decision to execute everything through *woomera* and manage this by *bludge* has worked out extremely well. We get logging, flexible control, and almost complete independence of the mechanics of job execution and the management of that execution. We can't imagine any real nontrivial production system not using similar schemes. In addition, this has allowed us to experiment with quite sophisticated I/O management schemes; for example, without affecting any other aspect of the daily update cycle, we played with:

- minimizing head contention by allowing only one process per filesystem (by adding a per filesystem resource)

- managing RAID controller load by restricting the number of processes using filesystems associated with specific RAID controllers (by adding a per controller resource)

- managing SCSI bus load (by adding a per SCSI bus resource)

We experimented during production runs, manually adjusting limits and measuring changes in processing rates and various metrics reported by standard performance tools such as *iostat*, *vmstat*, and *mpstat*.

5.2. Cycle Management

Initially, we controlled the job flow by manually setting various resources within *woomera*. After a while, this became mechanical in nature so we automated it as a shell script called *bludge*. Every few minutes, *bludge* analyzes the system activity and determines what state the machine is in, and sets various limits accordingly. For example, the jobs that convert tap files into tags have the resource PRS_LIMIT. When we are in the CPU-intensive part of the update cycle, *bludge* sets the limit for PRS_LIMIT to zero.

Another example is that the datastore update jobs have a resource indicating the filesystem containing the partition, say LVOLgb. *Bludge* tries to make the datastore update finish on each filesystem simultaneously (because this minimizes the total update cycle length). Typically, we run about 50 update jobs simultaneously. So if *bludge* notices that filesystem gb is 70% done and filesystem bf is only 45% done, it will likely set LVOLgb to zero and LVOLbf to 1 or 2 until filesystem bf catches up.

5.3. Recovery

For the first few months of production, we averaged a system crash every 2–3 days. This caused us to quickly develop and exercise effective techniques for restarting our update cycle. The two key concepts were careful logging of program start and end, and arranging that programs like *pu* were transactions that either completed or failed, and if they hadn't done either, then they could be rerun safely.

5.4. Centralizing Tag I/O

All tag I/O flows through one module. While this seems an obvious thing to do, it has meant this module is the most difficult piece of code in the entire project. The most visible benefits have been: performance improvements (such as when we changed from normal synchronous I/O to asynchronous multibuffered I/O) are immediately available to all tools processing tags or tagsets, application ubiquity (files can be transparently interpreted as files of tags or tagsets regardless of what was in the original file), and functional enhancements (such as when we supported internal tagset compression) are immediately available to all tools processing tags or tagsets.

5.5. Weakness of System Hardware/Software

While most people would agree that Gecko is pushing the limits of what systems can deliver, we were surprised by how many system hardware and software problems impacted our production system. Most were a surprise to us, so we'll list a few as a warning to others:

- we originally had fewer, larger filesystems made by striping together 3 36GB LUNs. We expected to get faster throughput, but instead ran into controller throughput bottlenecks and baffling (to both Sun and us) performance results as we varied the stripe width.

- trying to force several hundred MB/s of sequential I/O through the page cache never really worked; it either ran slowly or crashed the system. Apparently, the case of sequentially reading through terabytes of disk was never thought of by the designers of the virtual memory/page cache code. (To be fair, large sequential I/O also seems to confuse system configurers and RAID vendors, who all believe more cache memory will solve this problem.) Tuning various page cache parameters helped a little, but in the end, we just gave up and made the filesystems unbuffered and put double-buffering into our application. (Of course, that didn't help the backup software or any other programs that run on those filesystems, but c'est la vie.)

- we ran into unexplained bottlenecks in the throughput performance of pipes.

- we ran into annoying filesystem bugs (such as reading through a directory not returning all the files in that directory) and features (such as the internal filename lookup cache has a hard coded name length limit; unfortunately all our source filenames, about 60-70 characters long, are longer than that limit!).

- it is fairly easy to make the Solaris virtual memory system go unstable when you have less physical swap space than physical memory. While this seems an easy thing to avoid, it took several months before we found someone at Sun who knew this.

(To be fair, Sun eventually resolved most of these problems, and it is probable that we would have run into similar problems with other vendors' systems.)

5.6. Effective Reporting Architecture

It took three complete redesigns to come up with a good reporting architecture. Although much of the churn was due to changing requirements, we simply did not appreciate the difficulty of this part. The current scheme seems quite good; it runs 10% faster and requires about 7% of the memory of the previous scheme, and also allows efficient investigation of report anomalies.

5.7. Sorting

The initial sorting takes about 25% of the report gate time budget. The original scheme split the source tag files directly into each partition, and then sorted the files within each partition as part of the *pu* process step. This ran into a filename lookup bottleneck. Not only did it require 52 times as many filename lookups (once per partition rather than once per filesystem), these lookups were not cached as the filenames were too long. The current scheme is much better, but we thought of a better scheme, derived from an idea suggested by Ze-wei Chen, but have not implemented it yet. Here, we would split the original source tag files into several buckets (based on ranges of the sorting key) in each filesystem. After we sort each bucket, we can simply split the result out to the partitions appending to the add file. This eliminates the final merge pass and avoids the pipe performance bottleneck.

5.8. Distributed Design

The distributed layout of the datastore has worked out very well. It allows a great degree of parallel processing without imposing a great load on the operating system. Although we have not yet made use of it, it also allows processing distributed across distinct systems as well as filesystems. If we had implemented Gecko on a central server and a number of smaller servers (rather than one big SMP), then the only significant traffic between servers would be the background splitting of tags out to the smaller server throughout the day and copying the rolled up report summaries back to the central server. This latter amounts to only a few hundred MB, which is easily handled by modern LANs.

6. Conclusion

By almost any measure, Gecko solves a very large problem. Using a straightforward architecture and being careful about the details, especially I/O management, we have a system in production today that handles the volumes and meets its deadlines.

The datastore design, a myriad of sorted flat files, has proved to be a good one, even though it isn't a conventional database. It works, and it comfortably beats its processing deadlines.

There are aspects of Gecko, or rather the problem that Gecko solves, that seems to be a challenge to current database technology.

The first is sheer size. Given the number of records and number of updates per day, the cost of maintaining indices is prohibitive. And generating the reports pretty much requires a linear scan anyway. Nevertheless, these requirements aren't so different than those for data warehouses, and thus we expect that they will be resolved eventually.

The second is a bigger problem: the tagset database is not a relational database. Certainly the tags are tuples in the database sense. But nearly all the value and outputs of the Gecko system derives from analysis of tagsets, not individual tags; that is, sequences of time ordered events. Such 'event databases' are becoming increasingly common, appearing in such contexts as analyzing the "customer experience", probing complex processing systems from the outside, and analyzing certain classes of IP (Internet Protocol) packet traces.

The issue is not, of course, whether we can use a relational database to hold this data; we obviously can. The more important issues are

- how do you formulate and express queries? SQL and its brethren deal well with tuples, but event database queries sound more like pattern matching married with temporal constraints. For example, "find tagsets where a certain sequence of tags occurs within a three day period, followed by another sequence spanning at most seven days, and with at least a 30 day gap between the two sequences". The best generalization I have found for formulating these queries is synchronized parallel pattern matching. By this I mean extending the ordinary notion of regular expressions matching text by considering the text to be t n-tuples, rather than t characters. You then take n independent patterns, match them against the t text tuples performing pattern matching for each of the n patterns along each of the corresponding element of the tuples. Finally, you allow inter-pattern synchronization, where you can force alignment between tuples matched by subsets of the n patterns, such as "the last element in the third subexpression in the first pattern must come from the same tuple as the first element in the last subexpression in the third pattern".

- how should you store and index this data? Because most of the analysis is sequence analysis, the traditional techniques of indexing along various fields seems inadequate for all but an initial filtering (if you have cooperative patterns). The Gecko technique of variable length tagsets stored sequentially (and sorted) within a flat file works quite well, but lacks sophistication. The only approach I consider to be an improvement is to automatically generate code for pattern matching; this is an old theme, but still quite effective today (Greer 1999).

Acknowledgments

This work was a team effort; the rest of the Gecko team are Ray Bogle, Scott Daniels, Chuck Francis, Jon Hunt, Angus Maclellan, Pamela Martin, and Connie Smith. There have been many others within the Consumer Billing and Research organizations with AT&T who have helped; in fact, too many to list here. We have had invaluable help from our vendors, but Martin Canoy from Sun has been simply indispensable.

Notes

1. Caution: nearly all organizations are sensitive about about their internal operations, and this is especially true of billing. AT&T is no exception, and in order to preserve some proprietary aspects of the billing process, the process descriptions and quantitative information have deliberately been kept somewhat imprecise, but they are accurate enough for the discussions within this paper. Many of the names that follow are acronyms. Most of the time, the expanded version of the acronym is both obscure and unilluminating; we therefore will treat them simply as names.

Bibliography

R. Greer. Daytona and the Fourth-Generation Language Cymbal. In *ACM SIGMOD Conference*, June 1999.

A. Hume, S. Daniels, and A. Maclellan. Gecko: Tracking a very large billing system. In *Proceedings of the 2000 USENIX Annual Technical Conference*, pages 93–105, 2000.

A. Hume and A. Maclellan. Project Gecko: Pushing the envelope. In *NordU'99 Proceedings*, 1999.

H.C. Lauer and R.M. Needham. On the duality of operating system structures. *Operating Systems Review*, 13(2):3–19, 1979.

Chapter 26

DETECTING FRAUD IN THE REAL WORLD

Michael H. Cahill
Lucent Technologies, New Providence, NJ 07974, USA
mikecahill@lucent.com

Diane Lambert
Bell Labs, Lucent Technologies, Murray Hill, NJ 07974, USA
dl@bell-labs.com

José C. Pinheiro
Bell Labs, Lucent Technologies, Murray Hill, NJ 07974, USA
jcp@bell-labs.com

Don X. Sun
Bell Labs, Lucent Technologies, Murray Hill, NJ 07974, USA
dxsun@bell-labs.com

Abstract   Finding telecommunications fraud in masses of call records is more difficult than finding a needle in a haystack. In the haystack problem, there is only one needle that does not look like hay, the pieces of hay all look similar, and neither the needle nor the hay changes much over time. Fraudulent calls may be rare like needles in haystacks, but they are much more challenging to find. Callers are dissimilar, so calls that look like fraud for one account look like expected behavior for another, while all needles look the same. Moreover, fraud has to be found repeatedly, as fast as fraud calls are placed, the nature of fraud changes over time, the extent of fraud is unknown in advance, and fraud may be spread over more than one type of service. For example, calls placed on a stolen wireless telephone may be charged to a stolen credit card. Finding fraud is like finding a needle in a haystack only in the sense of sifting through masses of data to find something rare. This chapter describes some issues involved in creating tools for building fraud systems that are accurate, able to adapt to changing legitimate and fraudulent behavior, and easy to use.

Keywords: Customer profiles, Customer relationship management, Dynamic databases, Incremental maintenance, Massive data, Sequential updating, Transaction data, Thresholding.

1. Introduction

Fraud is a big business. Calls, credit card numbers, and stolen accounts can be sold on the street for substantial profit. Fraudsters may subscribe to services without intending to pay, perhaps with the intention of re-selling the services, or even the account itself, at a low cost until shut down. Call sell operations may extend their lives by subverting regulatory restrictions that are in place to protect debtors. Gaining access to a telephone or telephone line by physical intrusion still accounts for some fraud. Fraudsters also focus on the people who use and operate the network by applying "social engineering" to instruct an unsuspecting subscriber or operator to unknowingly agree to carry fraudulent traffic. Large profits have justified the growth of a well-organized and well-informed community of fraudsters who are clever and mobile. Fraud is also important to shady organizations that want to communicate without leaving records of their calls that can be traced back to them. Domestically, *Telecom and Network Security Review* (Vol. 4(5), April 1997) estimates that fraud losses in the U.S. telecommunications industry amount to between 4% and 6% of revenue. Internationally, the figures are generally worse, with several new service providers reporting losses over 20%.

Many service providers respond by building fraud control centers. They acquire multimillion dollar network and operations systems, hire and train staff for 24-by-7 operations, educate customers, require the use of Personal Identification Numbers, partner with competitors and law enforcement agencies, perform internal audits, and constantly tune their operations. Automated fraud detection systems may detect calls to certain "hot numbers", simultaneous use of calling cards in distant locations, which is unlikely except in the case of fraud, or other patterns of usage that are known to be associated with fraud. Such efforts have helped to reduce fraud, but the set of fraudsters is continually replenished and fraudsters have been able to continue to operate.

Detecting fraud is hard, so it is not surprising that many fraud systems have serious limitations. Different systems may be needed for different kinds of fraud (calling card fraud, wireless fraud, wireline fraud, subscription fraud), each system having different procedures, different parameters to tune, different database interfaces, different case management tools and different quirks and features. Fraud systems may be awkward to use. If they are not integrated with billing and other databases, then the fraud analyst may waste time on simple tasks, such as pulling relevant data from several disparate databases. Many systems have high false alarm rates, especially when fraud is only a small percentage of all traffic, so the chance of annoying a legitimate customer with a false alarm may be much higher than the chance of detecting fraud. More elaborate systems, such as those based on hidden Markov models, may promise more accuracy, but be useless for realtime detection for all but the smallest service provider. Finally, fraud and legitimate behavior constantly change, so systems that cannot evolve or "learn" soon become outdated. Thus, there is a need for tools for designing accurate and user-friendly systems that can be applied to detecting fraud on different kinds of telecommunications services, that can scale up or down, and that can adapt to the changing behavior of both legitimate customers and fraudsters.

Overwhelming data complicates each step of the design and implementation of a fraud management system, where overwhelming is defined not in terms of an absolute standard but relative to the available computing resources. There can be hundreds of millions of call records available for designing the detection algorithms, but there is no need for real-time performance during the design stage and researchers and developers often have access to powerful computing. In production, there may be only millions of calls per day, but each call has to be screened for signs of fraud quickly, faster than calls are being placed, or else the system may fall behind the traffic flow. Moreover, the computing environment may be designed for processing bills rather than for complicated numerical processing, limiting the kinds of algorithms and models that can be used to detect fraud. Once an account is flagged for fraud, all the calls for the case may need to be re-analyzed to prioritize the cases that require human intervention or analysis. There may not be a huge number of calls in an account with suspicious activity, but specialized algorithms for fitting complex models that take call history and account information into account may be needed to pinpoint fraud accurately. If the case is opened for investigation, then thousands of historical and current call records and other kinds of business and account history may need to be considered to determine the best response. Case management tools are needed to help the analyst sift through that data. All these stages are important, all involve data that can overwhelm the resources available, but the data requirements of each are very different.

This chapter begins by considering the heart of a fraud management system: the fraud detection algorithm. Simply stated, a fraud detection algorithm has two components: (1) a summary of the activity on an account that can be kept current and (2) rules that are applied to account summaries to identify accounts with fraudulent activity. Section 2 describes these components for threshold based fraud detection. Section 3 describes our approach, which is based on tracking each account's behavior in realtime.

Identifying possible cases of fraud automatically is usually not the last step in fraud detection. Often, the fraud cases need to be prioritized to help a supervisor determine which possible case of fraud should be investigated next. The performance of a fraud management system then depends on both the detection step and the prioritization step. The latter step tends to be ignored by fraud system designers, which can lead to unrealistic estimates of performance. Realistic performance analysis is discussed in Section 4. Final thoughts on fraud detection are given in Section 5.

2. Fraud Detection Based on Thresholding

Summarizing account activity is a major step in designing a fraud detection system because it is rarely practical to access or analyze all the call records for an account every time it is evaluated for fraud. A common approach is to reduce the call records for an account to several statistics that are computed each period. For example, average call duration, longest call duration, and numbers of calls to particular countries might be computed over the past hour, several hours, day, or several days. Account summaries can be compared to thresholds each period, and an account whose summary exceeds a threshold can be queued to be analyzed for fraud. Summaries over fixed periods resemble the aggregations of calls used in billing, so software for threshold based fraud detection is not difficult to write or manage. The summaries that are monitored for fraud may be defined by subject matter experts, and thresholds may be chosen by trial and error. Or, decision trees or machine learning algorithms may be applied to a training set of summarized account data to determine good thresholding rules.

Systems based on thresholding account summaries are popular, perhaps because they are easy to program and their logic is easily understood. Thresholding has several serious disadvantages, however. First, thresholds may need to vary with time of day, type of account, and type of call to be sensitive to fraud without setting off too many false alarms for legitimate traffic. Consequently, multivariate rather than univariate statistics must be thresholded. The need for sensitivity and specificity can easily lead to thousands of thresholds that interact with each other and need to be initialized, tuned, and periodically reviewed by an expert to accommodate changing traffic patterns. Accounts with high calling rates or unusual, but legitimate, calling patterns may regularly exceed the thresholds, setting off false alarms. Raising the thresholds reduces the false alarm rate but increases the chances of missing fraud cases. Classifying accounts into segments and applying thresholds to each segment separately may improve performance, but at the expense of multiplying the number of thresholds that have to be managed. Additionally, rules applied to summaries over fixed periods cannot react to fraud until the period is over nor consider calls from previous periods. Such arbitrary discontinuities in account summaries can impair the performance of the system. Perhaps most importantly, sophisticated fraudsters expect thresholds to apply and restrict their activity on any one account to levels that cannot be detected by most thresholding systems. Thus, there are both too many false alarms for legitimate calling and too many missed cases of fraud.

Thresholding can be improved, though. For example, Fawcett and Provost (1997) develop an innovative method for choosing *account-specific thresholds* rather than universal thresholds that apply to all accounts or all accounts in a segment. Their procedure takes daily traffic summaries for a set of accounts that experienced at least 30 days of fraud-free traffic before being hit by fraud and applies a machine learning algorithm to each account separately to develop a set of rules that distinguish fraud from non-fraud for the account. Thus, each account has its own set of rules at this point. The superset of the rules for all accounts is then pruned by keeping only those that apply to or *cover* many accounts, with possibly different thresholds for different accounts. For example, the rule may specify that long international calls indicate fraud, where long might be interpreted as more than three standard deviations above the mean duration for the account during its period of fraud-free traffic. Pruning then proceeds sequentially: a candidate rule is added to the current set of rules if it applies to a minimum number of accounts that have not already been covered by a specified number of rules. The final set of rules, therefore, covers "most" accounts, with the understanding that most of the final rules may be irrelevant for most accounts, but all the final rules are relevant for at least some accounts.

A fraud detection system, which is based on account-specific thresholds, is straight forward to implement for established accounts. The calls for the period of interest are separated by account, account summaries are computed, and then account summaries are compared to account-specific thresholds that were previously computed from training data. This process is similar to billing, which also requires account aggregation and access to account information. The account-specific thresholds can be updated periodically by re-fitting trees and sequentially selecting the account summaries to threshold. Re-training requires more resources than running the detection algorithm does, but re-training may be needed infrequently. Fawcett and Provost (1997) describe an application of their methods to a set of fewer than 1,000 accounts, each of which had at least 30 days of fraud-free activity followed by a period of wireless cloning fraud.

Account-specific thresholding has limitations, though. Perhaps most importantly, a procedure that requires a fixed period, such as 30 days, of uncontaminated traffic for training does not apply to accounts that experience fraud before the training period is over. In subscription fraud, for example, all the calls for an account are fraudulent, so there is no fraud-free period. Moreover, rules that are good for one time period may not be relevant for future time periods because account calling behavior, both fraudulent and legitimate, changes over time. And the basic limitations of thresholding—looking for fraud only at the end of a period and basing fraud detection on calls in only the current period— still apply. Nonetheless, a method that automatically derives account specific thresholds is clearly an important advance in threshold-based fraud detection.

3. Fraud Detection Based on Tracking Account Behavior

3.1. Account Signatures

Like Fawcett and Provost (1997), we believe that fraud detection must be tailored to each account's own activity. Our goals for fraud detection are more ambitious, though. First, fraud detection should be *event-driven*, not time-driven, so that fraud can be detected as it is happening, not at fixed points in time that are unrelated to account activity. Second, fraud detection methods should have *memory* and use all past calls on an account, weighting recent calls more heavily but not ignoring earlier calls. Third, fraud detection must be able to *learn* the calling pattern on an account and *adapt* to legitimate changes in calling behavior. Fourth, and perhaps most importantly, fraud detection must be *self-initializing* so that it can be applied to new accounts that do not have enough data for training.

The basis of our approach to fraud detection is an account summary, which we call an *account signature*, that is designed to track legitimate calling behavior for an account. An account signature might describe which call durations, times-between-calls, days-of-week and times-of-day, terminating numbers, and payment methods are likely for the account and which are unlikely for the account, for example. That is, given a vector of M call variables $\mathbf{X}_n = (X_{n,1}, X_{n,2}, \ldots, X_{n,M})$ for each call n, the likely (and unlikely) values of $\mathbf{X}_n$ are described by a multivariate probability distribution P_n, and an *account signature* is an estimate of P_n for the account. Because fraud typically results in unusual account activity, P_n is the right background to judge fraud against.

Estimating the full multivariate distribution P_n is often impractical, both in terms of statistical efficiency and storage space, so a major task in signature design is to reduce the complexity of P_n. To do that, we rely on the law of iterated probability (Devore 2000), which states that $P_n(\mathbf{X}_n)$ can be expressed as the product

$$P_n(X_{n,1} = x_1) P_n(X_{n,2} = x_2 | X_{n,1} = x_1) \cdots \qquad (26.1)$$
$$P_n(X_{n,M} = x_M | X_{n,1} = x_1, \ldots, X_{n,M-1} = x_{M-1}).$$

The first term in the product represents the marginal distribution of $X_{n,1}$, and each successive term is conditional on all the variables that were entered before it. (The order of the variables is arbitrary.) The last term, for example, implies that the distribution of $X_{n,M}$ depends on the outcomes of the other $M-1$ variables. Thus, there is a different conditional distribution of $X_{n,M}$ for each possible combination of outcomes of all the other variables.

For example, suppose $X_{n,1}$ represents call duration discretized into 10 different intervals, $X_{n,2}$ represents the period of day (peak or off-peak), and $X_{n,3}$ represents direction (incoming or outgoing). Then equation (26.1) requires 31 terms: one for the marginal distribution of duration, 10 for the conditional distributions of time of day given duration, and 20 for the conditional distribution of direction given each possible combination of duration and time of day. Some of these terms might be redundant, however. For example, if the probability that a call is incoming rather than outgoing is different for peak and off-peak hours but independent of call duration, then there are only 2 conditional distributions for call direction, not 20. Then the account signature $\mathcal{A}_n$, which is an estimate of P_n, would be the product of 13 estimated distributions, rather than 31 estimated distributions. Each term in the product $\mathcal{A}_n$ is called a *signature component*. Each signature component has a *signature variable* X_m and, possibly, a set of *conditioning variables*. The set of all signature components summarizes our knowledge about calling behavior, ignoring only those relationships among variables that are unimportant.

3.2. Signature Design

In the applications that we have seen, fast processing has depended on allocating the same, small amount of space to each account signature. Controlling the size of a signature can also contribute to its accuracy and precision. For example, suppose the duration of a call is the same for peak and off-peak hours, but a separate signature component is (needlessly) reserved for each. Then a peak call will not be used to update the signature component for off-peak duration, even though it would be statistically appropriate to do so. Consequently, the signature component for off-peak duration will be estimated from a smaller sample size than it should be, leading to statistically inefficient estimates. Designing a signature amounts to eliminating conditioning variables that do not matter and controlling the amount of space devoted to each remaining term in the product (26.1).

Like most fraud detection systems, we assume that there is a set of *priming data* that consists of all calls for a large number of accounts over a fixed period (say 60 days) that can be used to design signatures. We also assume that we have call records for a second set of accounts that experienced fraud during the period. For subscription fraud, all the calls on the account are fraudulent. For other kinds of fraud, some calls are fraudulent and some are not. Ideally, the fraudulent calls are labelled; otherwise they have to be labelled "by hand". In any case, because one fraudster often affects more than one account and we are interested in characterizing fraud, not a particular fraud user, we collapse all the fraud records for any particular kind of fraud into one set of *target data* and, ultimately, into one *fraud signature*. There may be separate fraud signatures for wireless subscription fraud, calling card fraud, and cloning fraud, but not a separate signature for each case of subscription fraud, for example. Each fraud signature has the same structure as an account signature.

The first design step is to choose the type of representation to be used for each signature variable. A continuous variable, such as call duration or time between calls, might be described by a parametric distribution, reducing each signature component for duration to a vector of estimated parameters, such as a mean and a standard deviation. But often no parametric family is both flexible enough to fit all accounts well and tractable enough to estimate quickly (say, during call setup or teardown). Therefore, we generally take all signature components to be nonparametric. In particular, a continuous variable or ordered categorical variable with many possible values can be discretized, so its signature components are vectors of probabilities over a set of fixed intervals. Alternatively, a continuous variable can be represented as the coefficients of a fixed set of knots for a spline fit to the log density or as a vector of quantiles. A signature variable with many possible unordered categories, such as terminating number or area code, can be represented by the labels and probabilities of the most likely categories. This kind of representation resembles a histogram, but one in which the labels of the bins are not fixed. For illustration, all signature components are assumed to be represented by histograms in this chapter.

There are many criteria for defining histogram bins; see Gibbons et al. (1997) and Ioannidis and Poosala (1999), for example. Many of these criteria are designed to give good performance over all possible values of the signature variable, but in fraud detection only small probability events are important because only small probability events are able to indicate fraud. Since our goal is fraud detection, we choose the bins of the histogram so that, on average, it is as easy as possible to distinguish legitimate calls from fraud calls. For a given signature variable, this can be accomplished by maximizing the average weighted Kullback-Leibler (AWKL) distance from the histogram for an account in the priming data $(p_{i,k})$ to the histogram for the fraud data (f_k), where

$$\text{AWKL} = -\frac{1}{N}\sum_{i=1}^{N}\left(w\sum_{k=1}^{K}p_{i,k}\log\frac{p_{i,k}}{f_k} + (1-w)\sum_{k=1}^{K}f_k\log\frac{f_k}{p_{i,k}}\right)$$

for some w, $0 \leq w \leq 1$, K is the number of bins in the histogram, and N is the number of accounts in the priming data (Chen et al. 2000b). The weight w controls the balance between more informative description of legitimate behavior and better separation of fraud and legitimate behavior. When $w = 0$, the criterion ignores the fraud training data and emphasizes only the ability to represent the behavior of legitimate accounts well. When $w = 1$, the criterion considers only the ability to avoid false alarms. Intermediate values of w balance these two concerns.

The cutpoints $d_1, \ldots, d_{k-1}$ that maximize the AWKL criterion can be found by numerical search, if feasible. If exhaustive search is not feasible, then searching can be limited by requiring minimum widths for the K final bins. For example, call duration might be measured to the nearest second, but each bin might be required to be at least one minute long and endpoints might be restricted to integer minutes.

The AWKL distance can also be used when a signature variable X is represented by something other than a histogram. For example, if the signature represents a continuous distribution with probability density $p_i(x)$ for account i and probability density $f(x)$ for fraud, then the sum over i in the AWKL distance is replaced by an integral over x. An appropriate set of parameters maximizes the integral form of AWKL.

If X is represented by a vector of quantiles, then a change of variables shows that $\int_x p(x)\log(p(x))dx = \int_q \log(p(P^{-1}(u))du$, where P^{-1} is the quantile function defined by $P^{-1}(u) = q$ if $\int_{-\infty}^q p(x)dx = u$, so the AWKL criterion still applies. The quantiles to be used in the signature are again those that maximize AWKL.

Methods for deciding which conditional distributions to keep in a signature are discussed in detail in Chen et al. (2000b). The basic idea involves computing a p-value for a χ^2 test for each account in the training data and keeping only the conditioning variables that are statistically significant for a majority of accounts and highly statistically significant for at least some accounts. The other possible conditioning variables add only noise rather than predictive power to the signature of most accounts. Conditioning variables are added sequentially until the incremental benefit from any of the remaining variables is too small to be statistically significant for a majority of accounts. Loosely stated, a conditioning variable is kept only if it is important for describing many accounts, and very important for at least some accounts.

3.3. Keeping a Signature Current

A key feature of a signature is that it can be updated sequentially, so that it evolves in time (measured in calls) rather than abruptly changing at the end of an arbitrary period. This enables event-driven fraud detection because there is always an up-to-date standard against which fraud can be assessed. Sequential updating is also computationally efficient because it does not require access to a data warehouse, which is often slow, but only access to a data structure that is short enough to store in main memory. In the wireline, calling card and wireless fraud detection systems in which we have applied signatures, each signature can be stored in about the amount of space required to store one call, with careful quantization of probabilities.

Most signature components, such as duration or method of payment, can be considered to be randomly sampled. Thus, they can be updated by exponentially weighted moving averaging. For example, suppose the signature component is a vector of probabilities, call $n+1$ is represented by a vector $\mathbf{X}_{n+1}$ of 0's except for a 1 in the bin that contains the observed value of the call, and A_n is the account's signature component after the previous call n. Then the updated signature component based on call $n+1$ is

$$A_{n+1} = (1-w)A_n + w\mathbf{X}_{n+1},$$

where w determines the rate at which old calls are "aged out" of the signature component and the effect of the current call on the component.

If $w = .05$, the probability assigned to the observed bin increases by the constant amount .05 and the probability of any other bin decreases by a factor of .95. Also, call $n - 10$, which was 10 calls earlier than call n, has about 60% the weight of call n at the time of call n if $w = .05$, and about 82% the weight of call n if $w = .02$. The smaller w, the more stable the signature component. Of course, some care has to be taken to avoid incorporating fraud into the signature; this is discussed in Section 3.4 below.

A variant of exponentially weighted stochastic approximation, which is similar in computational effort to exponentially weighted moving averaging, can be used to update signature components that are vectors of quantiles (Chen et al. 2000a). Signature components that are represented as tail probabilities rather than as complete distributions can be updated by using a move-to-the-front scheme to update the labels of the top categories and a variant of exponentially weighted moving averaging to update their probabilities. (See Gibbons et al. (1997) for the details of one possible approach.) Timing variables, like day-of-week and hour-of-day, are not randomly observed because they are observed "in order"—all the calls for Tuesday of this week are observed before all the calls for Wednesday of this week. Nonetheless, timing distributions can be updated in a way that is similar in spirit and in amount of computation to exponentially weighted moving averaging (Lambert et al. 1999).

Of course, sequential updating requires a starting point or *initial signature* for a new account. One way to do that is to segment the signatures for the accounts in the priming data, basing the segmentation criteria on information in the first one or few calls in an account. Details of one procedure based on statistical testing are given in Chen et al. (2000b); details of another procedure that uses multivariate regression trees are given in Yu and Lambert (1999). These procedures initialize each signature component separately. This gives a huge number of possible initial signatures (products of initial signature components), expressing a huge number of possible calling patterns. That is, a newly active account is assigned to a different segment of customers for each signature component, and each assignment depends only on the first few calls on the account. Finally, note that it is the initialization of signatures from the calls for a set of legitimate accounts, rather than from the calls on the account alone, that enables us to detect subscription fraud, in which every call on an account is fraudulent. Moreover, call-by-call updating ensures that an account with many calls soon evolves to its own "segment", allowing for personalized customer relationship management.

3.4. Using Signatures to Detect Fraud

Scoring a call for fraud is a matter of comparing its probability under the account signature to its probability under a fraud signature. Suppose the account signature after call n is $\mathcal{A}_n$, the fraud signature is $\mathcal{F}$, which is independent of the number of calls on the account, and call $n+1$ with signature variables $\mathbf{x}_{n+1}$ is observed. Then the *call score* for call $n+1$ is defined by

$$C_{n+1} = log(\mathcal{F}(\mathbf{x}_{n+1})/\mathcal{A}_{n+1}(\mathbf{x}_{n+1})).$$

Because $\mathcal{A}_{n+1}$ and $\mathcal{F}$ are products of signature components, the call score is a sum of contributions from the signature components. Note that standard statistical theory implies that the log likelihood ratio C_{n+1} is the best discriminator of fraud ($\mathcal{F}$) from legitimate activity on the account ($\mathcal{A}_n$) when $\mathbf{x}_{n+1}$ is observed (Bickel and Doksum 1976).

The higher the call score, the more suspicious the call. For a call to obtain a high score, it has to be unexpected for the account, so $\mathcal{A}_{n+1}(\mathbf{x}_{n+1})$ must be small. Calls that are not only unexpected under the account signature but also expected under the fraud signature score higher, and are considered more suspicious, than calls that are unexpected for both the account and fraud. Thus, some departures from the signature are more interesting than others from the perspective of fraud management. As constructed, each signature component contributes equally to the fraud score because the multivariate distribution that predicts call $n+1$ is a product of the marginal and conditional distributions represented by the signature components. As a result, some signature components can counteract others. This is not unreasonable. An hour long wireless call may be less suspicious if it originates from a region that is often used by the account. It is, however, possible to weight different signature components differently. Weights might depend on the reliability of the estimated distribution, some subjective information about the value of the signature component for fraud detection, or tuning that optimizes performance on training data. Note that the choice of the fraud distribution also affects the call score. In particular, a uniform fraud distribution implies that all unexpected call characteristics are equally good indicators of fraud.

Call scores serve two purposes. One is to give information that can be used to identify *accounts* that may have fraudulent activity. The other is to identify *calls* that may be suspicious and so should not be used to update the signature. For example, negative scores suggest that the call is not fraudulent so it should be used to update the signature. Calls with high positive scores raise concerns about fraud and, to be safe, should not be used to update the signature. Small positive scores are ambiguous. They might suggest a slight change in calling pattern that resembles fraud, or they might suggest that there is a subtle case of fraud. This ambiguity suggests the following procedure: update the signature if the call score is negative, do not update the signature if the call score is high, and act probabilistically if the call score is positive but small. In the latter case, a signature is updated with a probability that depends on the call score, varying from probability one for a call score that is less than or equal to zero to probability zero for a call score that is sufficiently high.

Egregious cases of fraud may generate calls with scores so high that a service provider may be willing to declare that fraud has occurred with only one call. Usually, however, the evidence from one call alone is not sufficient to identify fraud. If not, it is important to monitor the *score rate*, where score rate is either the average score over the last several calls that have a score above a pre-determined threshold or the average score of calls above a threshold over a specified time period. Some minimal information about previous calls needs to be kept in the signature to calculate the score rates. Accounts with high score rates can then be examined for fraud. The thresholds are not applied to all aggregated calls at the end of a period but rather to aggregated *high scoring* calls at the time of a possibly fraudulent call. We say that an account that exceeds the thresholds on score rates is *flagged*. Flagging filters the set of all accounts, producing a smaller set of accounts that have some evidence of fraud, just as standard account thresholding filters.

Both call scoring and account flagging have parameters that can be thought of as thresholds, but there is a major difference with the kinds of thresholds discussed in Section 2. In standard account thresholding, an account is marked as suspicious if it passes any one of several thresholds. In signature-based thresholding, different thresholds are not developed for different call characteristics or different kinds of accounts, such as business accounts or residential accounts. Instead, thresholds are applied to log-likelihood ratios that are a combination of all call characteristics. In other words, taking log-likelihood ratios converts all the different call characteristics and types of accounts to a standard measurement scale with one common set of thresholds.

Nonetheless, there are thresholds that need to be tuned in a signature-based fraud detection system. These parameters may be chosen to minimize the chance of flagging a legitimate account as fraudulent, subject to a constraint on the probability of missing a legitimate case of fraud in a set of training data. Because there are only a limited number of parameters to tune, the optimal parameters can be found by searching over a grid.

In summary, signatures are the basis of event-driven, adaptive, self-initializing fraud detection. It is event-driven in the sense that it is applied at the time of each call, which is the only time that fraud is active. It is self-learning, in the sense that the basis for comparison, the signature, is updated with each call that is not judged to be suspicious. Moderate changes in behavior are learned more slowly, on average, because the rate at which shifts in behavior are incorporated into a signature depends on the size of the shift. Signature-based fraud detection is also self-initializing in the sense that the first call or two on the account is used to assign signature components to new accounts. Because the signature components are initialized with calling patterns for previous accounts without fraud, it is possible to detect subscription fraud for new accounts. In a sense, the automatic initialization step allows us to start with a procedure for new accounts that is akin to universal thresholding with a huge number of segments. The procedure then naturally evolves to a procedure that is akin to customer specific thresholding for established accounts. Moreover, the threshold limits are placed on likelihood ratios and hence are the same for all segments, thus greatly reducing the number of parameters that have to be tuned in advance.

4. Performance Metrics

The performance of a fraud system is ultimately determined by the losses a service provider is able to prevent, but measuring averted losses, which never occur, is clearly difficult if not impossible. So, instead, service providers use metrics like the detection rate, false alarm rate, average time to detection after fraud starts, and average number of fraud calls or minutes until detection. An ideal fraud detection system would have 0% false alarms and 100% hits with instantaneous detection. But, finding all cases of fraud as soon as fraud starts requires mislabeling many (if not most) legitimate accounts as fraudulent at least once. A realistic, practical fraud system strikes a satisfactory balance of the performance criteria.

First, however, it is important to define the metrics carefully. Traditionally, the false alarm rate has been defined to be the percentage of legitimate accounts mislabelled as fraud. (False alarm rates, like type II errors, are usually quoted as percents, rather than as rates.) If there are 1,000,000 legitimate accounts in the population and 100 of these accounts are falsely labeled as fraud, then the false alarm rate is .01%. False alarms are important at the flagging stage because the goal of that step is to reduce the set of accounts in the population that have to be considered for fraud to just those that had fraud. False alarms are also important if account activity is restricted for flagged accounts because then the goal is to keep the number of legitimate accounts in the population that are needlessly restricted as small as possible.

If the evidence for fraud is ambiguous or attracting customers is costly, the service provider may require that a fraud analyst investigate the case before activity on the account is restricted. In that case, there is at least one queue of flagged accounts and the highest priority case in the queue is investigated whenever a fraud analyst becomes available. A queue may prioritize accounts by the number of fraudulent minutes accumulated to date or by the time of the most recent high scoring call, for example. Performance can then be evaluated after flagging or after prioritization. For example, the *flagging detection rate* is the fraction of compromised accounts in the population that are flagged. The *system detection rate*, which includes the rules used to decide which flagged accounts to open, is the fraction of compromised accounts in the population that are investigated by a fraud analyst. The system and flagging detection rates are equal only when every flagged case of fraud is investigated. Otherwise, the system detection rate is smaller than the flagging detection rate because both detection rates are computed relative to the number of accounts with fraud in the population.

Service providers are also keen to know that their analysts are working on fraud cases, not investigating legitimate accounts. Thus, a different question is what fraction of investigated cases have fraud? The *flagging hit rate* is the fraction of flagged accounts that have fraud, and the *system hit rate* is the fraction of investigated cases that have fraud. One minus the system hit rate is often a good measure of the service provider's perception of the "real false alarm rate," especially since this is the only error rate that the service provider can evaluate easily from experience. That is, only the cases that are acted upon may be of interest to the service provider, not the legitimate cases in the population that were never judged to be suspicious. If 20 cases of fraud are investigated and only 8 turn out to be fraud, then a service provider may feel that the "real false alarm rate" is 60%, even if only .01% of the legitimate accounts in the population are flagged as fraud.

Typically, the flagging false alarm rate, which is computed relative to all legitimate accounts, is much smaller than one minus the system hit rate because so many of the accounts in the population never experience fraud and so few accounts are investigated. Note that the hit rate after the flagging step should be larger than the fraction of accounts in the population that have fraud. Otherwise, flagging is no better than randomly labeling accounts as fraudulent. The difference between the fraction of fraud in the population and the fraction of fraud in flagged accounts is a measure of the efficiency of the fraud detection algorithm. Similarly, the system hit rate should be larger than the flagging hit rate, or else the analyst can find as much fraud by randomly selecting one of the flagged accounts to investigate.

The flagging false alarm rate, detection rate, and hit rate can be estimated by applying the flagging algorithm to a set of training accounts, some of which have fraudulent activity. The only subtlety is that these rates vary over time and should be investigated as a function of time. If there are L_t legitimate accounts on day t and $L_{1,t}$ of these are flagged, then the flagging false alarm rate for day t is $L_{1,t}/L_t$. If there are X_t active cases of fraud on day t and $X_{1,t}$ of these cases are flagged, then the flagging detection rate for day t is $X_{1,t}/X_t$. The flagging hit rate is then $X_{1,t}/(L_{1,t}+X_{1,t})$. Because legitimate and fraudulent behavior is learned over time in a signature-based fraud detection system, each of these performance statistics should improve quickly when the system is still new and then stabilize if the number of new accounts added stabilizes over time.

Some care needs to be taken in counting accounts when estimating system performance metrics. After an account is flagged and put into a priority queue, it can either be opened for investigation by a fraud analyst, it can remain in the queue without exceeding flagging thresholds again, or it can remain in the queue and continue to cross flagging thresholds. If the case is opened, the account is removed from the queue and the analyst decides, perhaps after contacting the account owner, if fraud has been committed. If an account remains in the queue unopened and it is not flagged again, then eventually it is considered uninteresting and reaped from the queue. The simplest rule is to reap an account if it has not been opened and has not been flagged again for a specified a number of days. If an account that is already queued is flagged again, then its priority needs to be re-computed to reflect the continuing suspicious activity.

Simulating system performance, then, requires simulating the prioritization and reaping processes. In practice, re-prioritization may occur whenever another account is flagged, but for simulation purposes it is enough to re-prioritize accounts once a day, for example. Then, for each day, the number of accounts opened, the number of accounts with fraud opened, and the number of active fraud cases can be counted to compute the system hit rate and system detection rate, which are typically the most important parameters to service providers.

Finally, note that realistic performance assessment requires assuming realistic levels of fraud. Performance often appears better for larger fraud rates, but if typically 4% to 6% of all accounts are infected by fraud annually, then assuming that 20% to 30% of all accounts are infected by fraud in three months is not realistic and overstates system performance.

5. Further Thoughts

This chapter describes an approach to fraud detection that is based on tracking calling behavior on an account over time and scoring calls according to the extent that they deviate from that pattern and resemble fraud. Signatures avoid the discontinuities inherent in most threshold-based systems that ignore calls earlier than the current time period. In the applications that we have encountered, a signature can be designed to fit in about the space needed to store one call. Consequently, they can be stored in main memory that is quick to access, rather than in a data warehouse that is slow to access. In one application to wireless fraud detection, we had 33 gigabytes of raw wireless call records for about 1.5 million customers. Each signature required 200 bytes, so the signature database required only about 300 megabytes to store. In an application to domestic wireline calling, there were 90 gigabytes of call records for about 1 million customers. Each signature required only 80 bytes to store, so the signature database required only about 80 megabytes. Because signatures are updated call-by-call, there is no need for offline processing or for using out-of-date customer profiles. Thus, signatures are able to avoid two common complaints about fraud detection systems that profile customers: they cannot scale up and they cannot keep up with the volume of data. Moreover, signatures are initialized from legitimate data from a huge number of possible initializations, so they can detect subscription fraud. This is in contrast to profiling systems, which use only the calls on the account itself to profile customers and, therefore, cannot detect subscription fraud. Finally, the signature evolves with each new call that is not considered fraudulent, so each established customer eventually has its own signature, not a signature designed for a segment of customers. Evolving the customer from an initial segment to a personalized segment is painless—no additional processing, such as re-clustering a database, is necessary.

It is possible to put other kinds of fraud detection algorithms in the signature framework. For example, many service providers keep lists of "hot numbers" that are associated with fraud. These can also be used in a signature-based system, by giving each call a high score when a hot number is called. More importantly, perhaps, it is possible to have scores for warm numbers that are often but not exclusively associated with fraud. These numbers can be assigned a contribution to the log-likelihood that is smaller than that for hot numbers, for example. It is also possible to keep account characteristics that can be derived from call records and that might be useful for fraud detection in the signature. For example, the fact that an account is a coin phone might change the rate at which it is attacked by fraud or the kind of fraud it is likely to be subjected to.

There is much more to fraud detection than the algorithms that score calls and label accounts as suspicious, though. For example, the fraud detection system needs to be able to access calls at the switch, before they are sent to a data warehouse, in order to be real-time or nearly real-time. Putting hooks into a telecommunications network to pull call data from a switch can be extremely difficult. After an account is flagged, investigators need sophisticated tools for case management. These tools must be integrated with billing systems so that the investigator can access payment history. The case management tools must also be integrated with service provisioning systems so that an investigator can enable service restrictions on the account, if appropriate. Putting hooks into these systems is non-trivial at best. Supervisors then need case management tools that allow them to track the performance of the system and to spot new trends in fraud. Tools are needed to ensure that if one account for a customer is closed for fraud, then all accounts for that customer are closed for fraud, for example.

Signatures can be adapted to any kind of fraud in which transactions are made on accounts. This includes credit card fraud, in which the transaction is a purchase, and medical fraud, where the account may be a medical practitioner and the transaction an interaction with a patient or the account may be a patient and the transaction a visit to a medical practitioner. More generally, signatures can be used to predict transaction behavior on accounts. For example, signatures can be used in to track the behavior of visitors at web sites to identify those who are about to make purchases. While further research is needed to work out the details for particular applications, the concept of signature-based predictive tracking is broad and potentially valuable for a wide range of applications.

Acknowledgements

At various times, our thoughts on fraud detection have been influenced by discussions with many people, especially Jerry Baulier (Lucent Technologies), Rick Becker (AT&T Labs), Fei Chen (Bell Labs, Lucent Technologies), Linda Clark (Bell Labs, Lucent Technologies), Virginia Ferrara (Lucent Technologies), Al Malinowski (AT&T), Daryl Pregibon (AT&T Labs) and Allan Wilks (AT&T Labs).

Bibliography

P. J. Bickel and K. A. Doksum. *Mathematical statistics*. Holden-Day, San Francisco, CA, 1976.

Fei Chen, Diane Lambert, and José C. Pinheiro. Sequential percentile estimation. Technical memorandum, Bell Labs, Lucent Technologies, 2000a.

Fei Chen, Diane Lambert, José C. Pinheiro, and Don X. Sun. Reducing transaction databases, without lagging behind the data or losing information. Technical memorandum, Bell Labs, Lucent Technologies, 2000b.

Jay L. Devore. *Probability and Statistics for Engineering and the Sciences*. Wadsworth, Belmont, CA, 5th edition, 2000.

Tom Fawcett and Foster Provost. Adaptive fraud detection. *Data Mining and Knowledge Discovery*, 1:291–316, 1997.

Philip B. Gibbons, Yannis E. Ioannidis, and Viswanath Poosala. Fast incremental maintenance of approximate histograms. In *Proceedings of the 23rd VLDB Conference*, pages 466–475, 1997.

Yannis E. Ioannidis and Viswanath Poosala. Histogram-based approximations to set-valued query answers. In *Proceedings of the 25th VLDB Conference*, pages 174–185, 1999.

Diane Lambert, José C. Pinheiro, and Don X. Sun. Updating timing profiles for millions of customers in real-time. Technical memorandum, Bell Labs, Lucent Technologies, 1999.

Yan Yu and Diane Lambert. Fitting trees to functional data, with an application to time-ofday patterns. *Journal of Computational and Graphical Statistics*, 8(4):749–762, 1999.

Chapter 27

MASSIVE DATASETS IN ASTRONOMY

Robert J. Brunner
California Institute of Technology, Pasadena, CA 91125, USA
rb@astro.caltech.edu

S. George Djorgovski
California Institute of Technology, Pasadena, CA 91125, USA
george@astro.caltech.edu

Thomas A. Prince
California Institute of Technology, Pasadena, CA 91125, USA
prince@srl.caltech.edu

Alex S. Szalay
Johns Hopkins University, Baltimore, MD 21218, USA
szalay@pha.jhu.edu

Abstract Astronomy has a long history of acquiring, systematizing, and interpreting large quantities of data. Starting from the earliest sky atlases through the first major photographic sky surveys of the 20th century, this tradition is continuing today, and at an ever increasing rate.

Like many other fields, astronomy has become a very data-rich science, driven by the advances in telescope, detector, and computer technology. Numerous large digital sky surveys and archives already exist, with information content measured in multiple Terabytes, and even larger, multi-Petabyte data sets are on the horizon. Systematic observations of the sky, over a range of wavelengths, are becoming the primary source of astronomical data. Numerical simulations are also producing comparable volumes of information. Data mining promises to both make the scientific utilization of these data sets more effective and more complete, and to open completely new avenues of astronomical research.

Technological problems range from the issues of database design and federation, to data mining and advanced visualization, leading to a new toolkit for astronomical research. This is similar to challenges encountered in other data-intensive fields today.

These advances are now being organized through a concept of the Virtual Observatories, federations of data archives and services representing a new information infrastructure for astronomy of the 21st century. In this article, we provide an overview of some of the major datasets in astronomy, discuss different techniques used for archiving data, and conclude with a discussion of the future of massive datasets in astronomy.

Keywords: Astronomy, Digital sky survey, Space telescope, Data mining, Knowledge discovery in databases, Clustering analysis, Virtual observatory.

1. Introduction: The New Data-Rich Astronomy

A major paradigm shift is now taking place in astronomy and space science. Astronomy has suddenly become an immensely data-rich field, with numerous digital sky surveys across a range of wavelengths, with many Terabytes of pixels and with billions of detected sources, often with tens of measured parameters for each object. This is a great change from the past, when often a single object or a small sample of objects were used in individual studies. Instead, we can now map the universe systematically, and in a panchromatic manner. This will enable *quantitatively and qualitatively new science,* from statistical studies of our Galaxy and the large-scale structure in the universe, to the discoveries of rare, unusual, or even completely new types of astronomical objects and phenomena. This new digital sky, data-mining astronomy will also enable and empower scientists and students anywhere, without an access to large telescopes, to do first-rate science. This can only invigorate the field, as it opens the access to unprecedented amounts of data to a fresh pool of talent.

Handling and exploring these vast new data volumes, and actually making real scientific discoveries poses a considerable technical challenge. The traditional astronomical data analysis methods are inadequate to cope with this sudden increase in the data volume (by several orders of magnitude). These problems are common to all data-intensive fields today, and indeed we expect that some of the products and experiences from this work would find uses in other areas of science and technology. As a testbed for these software technologies, astronomy provides a number of benefits: the size and complexity of the data sets are nontrivial but manageable, the data generally are in the public-domain, and the knowledge gained by understanding this data is of broad public appeal.

In this chapter, we provide an overview of the state of massive datasets in astronomy as of mid-2001. In Section 2, we briefly discuss the nature of astronomical data, with an emphasis on understanding the inherent complexity of data in the field. In Section 3, we present overviews of many of the largest datasets, including a discussion of how the data are utilized and archived. Section 4 provides a thorough discussion of the virtual observatory initiative, which aims to federate all of the distributed datasets described in Section 3 into a coherent archival framework. We conclude this chapter with a summary of the current state of massive datasets in astronomy.

2. The Nature of Astronomical Data

By its inherent nature, astronomical data are extremely heterogeneous, in both format and content. Astronomers are now exploring all regions of the electromagnetic spectrum, from gamma-rays through radio wavelengths. With the advent of new facilities, previously unexplored domains in the gravitational spectrum will soon be available, and exciting work in the astro-particle domain is beginning to shed light on our Universe. Computational advances have enabled detailed physical simulations which rival the largest observational datasets in terms of complexity. In order to truly understand our cosmos, we need to assimilate all of this data, each presenting its own physical view of the Universe, and requiring its own technology.

Despite all of this heterogeneity, however, astronomical data and its subsequent analysis can be broadly classified into five domains. In order to clarify later discussions, we briefly discuss these domains and define some key astrophysical concepts which will be utilized frequently throughout this chapter.

- Imaging data is the fundamental constituent of astronomical observations, capturing a two-dimensional spatial picture of the Universe within a narrow wavelength region at a particular epoch or instant of time. While this may seem obvious to most people—after all, who hasn't seen a photograph—astrophysical pictures (see, *e.g.*, Figures 27.1 and 27.2) are generally taken through a specific filter, or with an instrument covering a limited range of the electromagnetic spectrum, which defines the wavelength region of the observation. Astronomical images can be acquired directly, *e.g.*, with imaging arrays such as CCDs[1], or synthesized

---

[1] Charge Coupled Device—a digital photon counting device that is superior to photographic images in both the linearity of their response and quantum efficiency

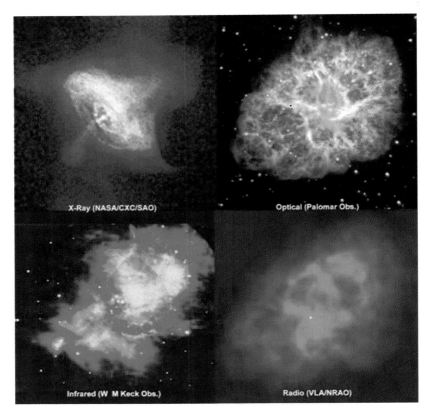

Figure 27.1. A multiwavelength view of the Crab nebula, a supernova remnant that was first sighted by Chinese astronomers in 1054 AD. Clearly demonstrated in this montage of images are the different physical processes that are manifested in the different spectral regions. Image Credit: NASA/CXC/SAO.

from interferometric observations as is customarily done in radio astronomy.

- Catalogs are generated by processing the imaging data. Each detected source can have a large number of measured parameters, including coordinates, various flux quantities, morphological information, and areal extant. In order to be detected, a source must stand out from the background noise (which can be either cosmic or instrumental in origin). The significance of a detection is generally quoted in terms of σ, which is a relative measure of the strength of the source signal relative to the dispersion in the background noise. We note that the source detection process is generally limited both in terms of the flux (total signal over the

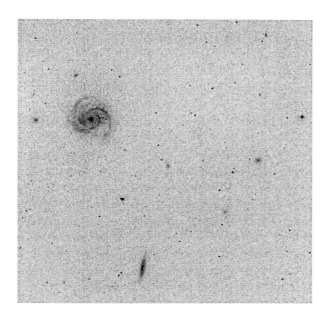

Figure 27.2. An image from the DPOSS survey (see below) of the field of M100, a nearby spiral galaxy. This single image is one millionth of the size of the entire sky.

background) and surface brightness (intensity contrast relative to the background).

Coordinates are used to specify the location of astronomical sources in the sky. While this might seem obvious, the fact that we are sited in a non-stationary reference frame (*e.g.*, the earth rotates, revolves around the sun, and the sun revolves around the center of our Galaxy) complicates the quantification of a coordinate location. In addition, the Earth's polar axis precesses, introducing a further complication. As a result, coordinate systems, like Equatorial coordinates, must be fixed at a particular instant of time (or epoch), to which the actual observations, which are made at different times, can be transformed. One of the most popular coordinate systems is J2000 Equatorial, which is fixed to the initial instant (zero hours universal time) of January 1, 2000. One final caveat is that nearby objects (*e.g.*, solar system bodies or nearby stars) move on measurable timescales. Thus the date or precise time of a given observations must also be recorded.

Flux quantities determine the amount of energy that is being received from a particular source. Since different physical processes emit radiation at different wavelengths, most astronomical images are obtained through specific filters. The specific filter(s) used varies, depending on the primary purpose of the observations and the type of recording device. Historically, photographic surveys used filters which were well matched to the photographic material, and have names like O, E, J, F, and N. More modern digital detectors have different characteristics (including much higher sensitivity), and work primarily with different filter systems, which have names like U, B, V, R, and I, or g, r, i in the optical, and J, H, K, L, M, and N in the near-infrared.

In the optical and infrared regimes, the flux is measured in units of magnitudes (which is essentially a logarithmic rescaling of the measured flux) with one magnitude equivalent to -4 decibels. This is the result of the historical fact that the human eye is essentially a logarithmic detector, and astronomical observations have been made and recorded for many centuries by our ancestors. The zeropoint of the magnitude scale is determined by the star Vega, and thus all flux measurements are relative to the absolute flux measurements of this star. Measured flux values in a particular filter are indicated as $B = 23$ magnitudes, which means the measured B band flux is $10^{0.4 \times 23}$ times fainter than the star Vega in this band. At other wavelengths, like x-ray and radio, the flux is generally quantified in standard physical units such as ergs cm^{-2} s^{-1} Hz^{-1}. In the radio, observations often include not only the total intensity (indicated by the Stokes I parameter), but also the linear polarization parameters (indicated by the Stokes Q, and U parameters).

- Spectroscopy, Polarization, and other follow-up measurements provide detailed physical quantification of the target systems, including distance information (e.g., redshift, denoted by z for extragalactic objects), chemical composition (quantified in terms of abundances of heavier elements relative to hydrogen), and measurements of the physical (e.g., electromagnetic, or gravitational) fields present at the source. An example spectrum is presented in Figure 27.3, which also shows the three optical filters used in the DPOSS survey (see below) superimposed.

- Studying the time domain (see, e.g., Figure 27.4) provides important insights into the nature of the Universe, by identifying moving objects (e.g., near-Earth objects, and comets), variable sources (e.g., pulsating stars), or transient objects (e.g., supernovae, and gamma-ray bursts). Studies in the time domain either require multiple epoch observations of fields (which is possible in the overlap

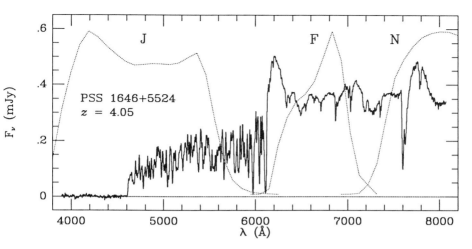

Figure 27.3. A spectrum of a typical $z > 4$ quasar PSS 1646+5524, with the DPOSS photographic filter transmission curves (J, F, and N) overplotted as dotted lines. The prominent break in the spectrum around an observed wavelength of 6000 Angstroms is caused by absorption by intergalactic material (that is material between us and the quasar) that is referred to as the Lyα forest. The redshift of this source can be calculated by knowing that this absorption occurs for photons more energetic than the Lyα line which is emitted at 1216 Angstroms.

regions of surveys), or a dedicated synoptic survey. In either case, the data volume, and thus the difficulty in handling and analyzing the resulting data, increases significantly.

- Numerical Simulations are theoretical tools which can be compared with observational data. Examples include simulations of the formation and evolution of large-scale structure in the Universe, star formation in our Galaxy, supernova explosions, *etc.* Since we only have one Universe and cannot modify the initial conditions, simulations provide a valuable tool in understanding how the Universe and its constituents formed and have evolved. In addition, many of the physical processes that are involved in these studies are inherently complex. Thus direct analytic solutions are often not feasible, and numerical analysis is required.

3. Large Astronomical Datasets

As demonstrated below, there is currently a great deal of archived data in Astronomy at a variety of locations in a variety of different database systems systems. In this section we focus on ground-based surveys, ground-based observatories, and space-based observatories. We do

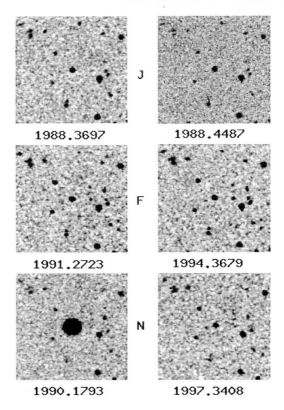

Figure 27.4. Example of a discovery in the time domain. Images of a star, PVO 1558+3725, seen in the DPOSS plate overlaps in J (= green, top) F (= red, middle) and N ($\approx$ near-infrared, bottom). Since the plates for the POSS-II survey were taken at different epochs (*i.e.* they were taken on different days), that can be separated by several years (the actual observational epoch is indicated below each panel), we have a temporal recording of the intensity of the star. Notice how the central star is significantly brighter in the lower right panel. Subsequent analysis has not indicated any unusual features, and, as a result, the cause, amplitude, and duration of the outburst are unknown.

not include any discussion of information repositories such as the Astrophysics Data System[2] (ADS), the Set of Identifications, Measurements, and Bibliography for Astronomical Data[3] (SIMBAD), or the NASA Extragalactic Database[4] (NED), extremely valuable as they are. This review focuses on more homogeneous collections of data from digital sky

---

[2] http://adswww.harvard.edu/
[3] http://cdsweb.u-strasbg.fr/Simbad.html
[4] http://ned.ipac.caltech.edu

surveys and specific missions rather than archives which are more appropriately described as digital libraries for astronomy.

Furthermore, we do not discuss the large number of new initiatives, including the Large-Aperture Synoptic Survey Telescope (LSST), the California Extremely Large Telescope (CELT), the Visible and Infrared Survey Telescope[5] (VISTA), or the Next Generation Space Telescope[6] (NGST), which will provide vast increases in the quality and quantity of astronomical data.

3.1. Ground-Based Sky Surveys

Of all of the different astronomical sources of data, digital sky surveys are the major drivers behind the fundamental changes underway in the field. Primarily this is the result of two factors: first, the sheer quantity of data being generated over multiple wavelengths, and second, as a result of the homogeneity of the data within each survey. The federation of different surveys would further improve the efficacy of future ground- and space-based targeted observations, and also open up entirely new avenues for research.

In this chapter, we describe only some of the currently existing astronomical archives as examples of the types, richness, and quantity of astronomical data which is already available. Due to the space limitations, we cannot cover many other, valuable and useful surveys, experiments and archives, and we apologize for any omissions. This summary is not meant to be complete, but merely illusory.

Photographic plates have long endured as efficient mechanisms for recording surveys (they have useful lifetimes in excess of one hundred years and offer superb information storage capacity, but unfortunately they are not directly computer-accessible and must be digitized before being put to a modern scientific use). Their preeminence in a digital world, however, is being challenged by new technologies. While many photographic surveys have been performed, *e.g.*, from the Palomar Schmidt telescope[7] in California, and the UK Schmidt telescope in New South Wales, Australia, these data become most useful when the plates are digitized and cataloged.

While we describe two specific projects, as examples, several other groups have digitized photographic surveys and generated and archived

---

[5] http://www.vista.ac.uk
[6] http://www.ngst.stsci.edu/
[7] A Schmidt telescope has an optical design which allows it to image a very wide field, typically several degrees on a side. This is in contrast to most large telescopes which have field of views that are measured in arcminutes.

the resulting catalogs, including the Minnesota Automated Plate Scanner[8] (APS; Pennington et al. 1993), the Automated Plate Measuring Machine[9] (APM; McMahon and Irwin 1992) at the Institute of Astronomy, Cambridge, UK, the coordinates, sizes, magnitudes, orientations, and shapes (COSMOS; Yentis et al. 1992) and its successor, SuperCOSMOS[10], plate scanning machines at the Royal·Observatory Edinburgh. Probably the most popular of the digitized sky surveys (DSS) are those produced at the Space Telescope Science Institute[11] (STScI) and its mirror sites in Canada[12], Europe[13], and Japan[14].

DPOSS The Digitized Palomar Observatory Sky Survey[15] (DPOSS) is a digital survey of the entire Northern Sky in three visible-light bands, formally indicated by g, r, and i (blue-green, red, and near-infrared, respectively). It is based on the photographic sky atlas, POSS-II, the second Palomar Observatory Sky Survey, which was completed at the Palomar 48-inch Oschin Schmidt Telescope (Reid et al. 1991). A set of three photographic plates, one in each filter, each covering 36 square degrees, were taken at each of 894 pointings spaced by 5 degrees, covering the Northern sky (many of these were repeated exposures, due to various artifacts such as the aircraft trails, plate defects, *etc.*). The plates were then digitized at the Space Telescope Science Institute (STScI), using a laser microdensitometer. The plates are scanned with 1.0"pixels, in rasters of 23,040 square, with 16 bits per pixel, producing about 1 Gigabyte per plate, or about 3 Terabytes of pixel data in total.

These scans were processed independently at STScI (for the purposes of constructing a new guide star catalog for the HST) and at Caltech (for the DPOSS project). Catalogs of all the detected objects on each plate were generated, down to the flux limit of the plates, which roughly corresponds to the equivalent blue limiting magnitude of approximately 22. A specially developed software package, called SKICAT (Sky Image Cataloging and Analysis Tool; Weir et al. 1995) was used to analyze the images. SKICAT incorporates some machine learning techniques for object classification and measures about 40 parameters for each object in each

---

[8] http://aps.umn.edu/
[9] http://www.ast.cam.ac.uk/~mike/casu/apm/apm.html
[10] http://www.roe.ac.uk/cosmos/scosmos.html
[11] http://archive.stsci.edu/dss/
[12] http://cadcwww.dao.nrc.ca/dss/
[13] http://archive.eso.org/dss/dss/
[14] http://dss.nao.ac.jp
[15] http://dposs.caltech.edu/

band. Star-galaxy classification was done using several methods, including decision trees and neural nets; for brighter galaxies, a more detailed morphological classification may be added in the near future. The DPOSS project also includes an extensive program of CCD calibrations done at the Palomar 60-inch telescope. These CCD data were used both for magnitude calibrations, and as training data sets for object classifiers in SKICAT. The resulting object catalogs were combined and stored in a Sybase relational DBMS system; however, a more powerful system is currently being implemented for more efficient scientific exploration. This new archive will also include the actual pixel data in the form of astrometrically and photometrically calibrated images.

The final result of DPOSS will be the Palomar Norris Sky Catalog (PNSC), which is estimated to contain about 50 to 100 million galaxies, and between 1 and 2 billion stars, with over 100 attributes measured for each object, down to the equivalent blue limiting magnitude of 22, and with star-galaxy classifications accurate to 90% or better down to the equivalent blue magnitude of approximately 21. This represents a considerable advance over other, currently existing optical sky surveys based on large-format photographic plates. Once the technical and scientific verification of the final catalog is complete, the DPOSS data will be released to the astronomical community.

As an indication of the technological evolution in this field, the Palomar Oschin Schmidt telescope is now being retrofitted with a large CCD camera (QUEST2) as a collaborative project between Yale University, Indiana University, Caltech, and JPL. This will lead to pure digital sky surveys from Palomar Observatory in the future.

USNO-A2 The United States Naval Observatory Astrometric (USNO-A2) catalog[16] (Monet et al. 1996) is a full-sky survey containing over five hundred million unresolved sources down to a limiting magnitude of $B \sim 20$ whose positions can be used for astrometric references. These sources were detected by the Precision Measuring Machine (PMM) built and operated by the United States Naval Observatory Flagstaff Station during the scanning and processing of the first Palomar Observatory Sky Survey (POSS-I) O and E plates, the UK Science Research Council SRC-J survey plates, and

---

[16] http://ftp.nofs.navy.mil/projects/pmm/a2.html

the European Southern Observatory ESO-R survey plates. The total amount of data utilized by the survey exceeds 10 Terabytes.

The USNO-A2 catalog is provided as a series of binary files, organized according to the position on the sky. Since the density of sources on the sky varies (primarily due to the fact that our galaxy is a disk dominated system), the number of sources in each file varies tremendously. In order to actually extract source parameters, special software, which is provided along with the data is required. The catalog includes the source position, right ascension and declination (in the J2000 coordinate system, with the actual epoch derived as the mean of the blue and red plate) and the blue and red magnitude for each star. The astrometry is tied to the ACT catalog (Urban et al. 1997). Since the PMM detects and processes at and beyond the nominal limiting magnitude of these surveys, a large number of spurious detections are initially included in the operational catalog. In order to improve the efficacy of the catalog, sources were required to be spatially coincident, within a 2" radius aperture, on the blue and red survey plate.

SDSS The Sloan Digital Sky Survey[17] (SDSS) is a large astronomical collaboration focused on constructing the first CCD photometric survey of the North Galactic hemisphere (10,000 square degrees—one-fourth of the entire sky). The estimated 100 million cataloged sources from this survey will then be used as the foundation for the largest ever spectroscopic survey of galaxies, quasars and stars.

The full survey is expected to take five years, and has recently begun full operations. A dedicated 2.5m telescope is specially designed to take wide field (3 degree x 3 degree) images using a 5 by 6 mosaic of 2048x2048 CCD's, in five wavelength bands, operating in scanning mode. The total raw data will exceed 40 TB. A processed subset, of about 1 TB in size, will consist of 1 million spectra, positions and image parameters for over 100 million objects, plus a mini-image centered on each object in every color. The data will be made available to the public (see, *e.g.*, Figure 27.5 for a public SDSS portal) at specific release milestones, and upon completion of the survey.

During the commissioning phase of the survey data was obtained, in part to test out the hardware and software components. Already, a wealth of new science has emerged from this data.

---

[17] http://www.sdss.org/

The Sloan Digital Sky Survey (SDSS) is a joint project of The University of Chicago, Fermilab, the Institute for Advanced Study, the Japan Participation Group, The Johns Hopkins University, the Max-Planck-Institute for Astronomy (MPIA), the Max-Planck-Institute for Astrophysics (MPA), New Mexico State University, Princeton University, the United States Naval Observatory, and the University of Washington.

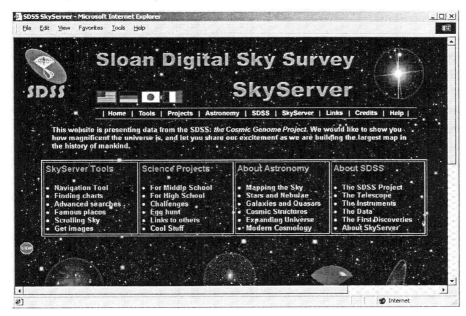

Figure 27.5. A public portal into the Sloan Digital Sky Survey, which is the cosmic equivalent to the Human Genome project.

There is a large number of experiments and surveys at aimed at detecting time-variable sources or phenomena, such as gravitational microlensing, optical flashes from cosmic gamma-ray bursts, near-Earth asteroids and other solar system objects, *etc.* A good list of project websites is maintained by Professor Bohdan Paczynski at Princeton[18]. Here we describe several interesting examples of such projects.

MACHO The MACHO project[19] was one of the pioneering astronomical projects in generating large datasets. This project was designed to look for a particular type of dark matter collectively classified

---
[18]http://astro.princeton.edu/faculty/bp.html
[19]http://wwwmacho.anu.edu.au/

as Massive Compact Halo Objects (*e.g.*, brown dwarfs or planets) from whence the project's name was derived. The signature for sources of this type is the amplification of the light from extragalactic stars by the gravitational lens effect of the intervening MACHO. While the amplitude of the amplification can be large, the frequency of such events is extremely rare. Therefore, in order to obtain a statistically useful sample, it is necessary to photometrically monitor several million stars over a period of several years. The MACHO Project is a collaboration between scientists at the Mt. Stromlo and Siding Spring Observatories, the Center for Particle Astrophysics at the Santa Barbara, San Diego, and Berkeley campuses of the University of California, and the Lawrence Livermore National Laboratory.

The MACHO project built a two channel system that employs eight 2048x2048 CCDs, which was mounted and operated on the 50-inch telescope at Mt. Stromlo. This large CCD instrument presented a high data rate (especially given that the survey commenced in 1992) of approximately several Gigabytes per night. Over the course of the survey, nearly 100,000 images were taken and processed, with a total data volume exceeding 7 Terabytes. While the original research goal of finding microlensing events was realized (essentially by a real-time data-analysis system), the MACHO data provides an enormously useful resource for studying a variety of variable sources. Unfortunately, funding was never secured to build a data archive, limiting the utility of the data primarily to only those members of the MACHO science team. Another similar project is the OGLE-II[20], or the second Optical Gravitational Lensing Experiment, which has a total data volume in excess of one Terabyte.

ROTSE The Robotic Optical Transient Search Experiment[21] is an experimental program to search for astrophysical optical transients on time scales of a fraction of a second to a few hours. While the primary incentive of this experiment has been to find the optical counterparts of gamma-ray bursts (GRBs), additional variability studies have been enabled, including a search for orphan GRB afterglows, and an analysis of a particular type of variable star, known as an RR Lyrae, which provides information on the structure of our Galaxy.

---

[20] http://astro.princeton.edu/~ogle
[21] http://www.umich.edu/~rotse

The ROTSE project initially began operating in 1998, with a fourfold telephoto array, imaging the whole visible sky twice a night to limiting flux limit of approximately 15.5. The total data volume for the original project is approximately four Terabytes. Unlike other imaging programs, however, the large field of view of the telescope results in a large number of sources per field (approximately 40,000). Therefore, reduction of the imaging data to object lists does not compress the data as much as is usual in astronomical data. The data is persisted on a robotic tape library, but insufficient resources have prevented the creation of a public archive.

The next stage of the ROTSE project is a set of four (and eventually six) half meter telescopes to be sited globally. Each telescope has a 2 degree field of view and operations, including the data analysis, and it will be fully automated. The first data is expected to begin to flow during 2001 and the total data volume will be approximately four Terabytes. The limiting flux of the next stage of the ROTSE experiment will be approximately 18 – 19, or more than ten times deeper than the original experiment. The ROTSE (and other variability survey) data will provide important multi-epoch measurements as a complement to the single epoch surveys (*e.g.*, DPOSS, USNOA2, and the SDSS). Other examples of similar programs include the Livermore Optical Transient Imaging System (LOTIS) program[22], which is nearly identical to the original ROTSE experiment, and its successor, Super-LOTIS.

NEAT Near Earth Asteroid Tracking (NEAT) program[23] is one of several programs that are designed to discover and characterize near earth objects (*e.g.*, Asteroids and comets). Fundamentally, these surveys cover thousands of square degrees of the sky every month to a limiting flux depth of approximately 17 – 20, depending on the survey. All together, these programs (which also include Catalina[24], LINEAR[25], LONEOS[26], and Spacewatch[27]) generate nearly 200 Gigabytes of data a night, yet due to funding restrictions, a large part of this data is not archived. NEAT is currently the only program which provides archival access to its data through

---

[22] http://hubcap.clemson.edu/~ggwilli/LOTIS/
[23] http://neat.jpl.nasa.gov/
[24] http://www.lpl.arizona.edu/css/
[25] http://www.ll.mit.edu/LINEAR/
[26] http://asteroid.lowell.edu/asteroid/loneos/loneos_disc.html
[27] http://www.lpl.arizona.edu/spacewatch/

the skymorph project[28] (see Figure 27.6 for the skymorph website). All told, these surveys have around 10 Terabytes of imaging data in hand, and continue to operate. The NEAT program has recently taken over the Palomar Oschin 48-inch telescope, which was used to generate the two POSS photographic surveys, in order to probe wider regions of the sky to an even deeper limiting flux value.

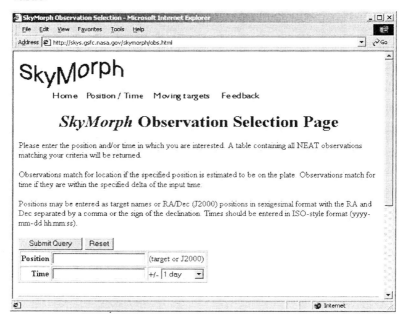

Figure 27.6. The skymorph web-site, which provides access to the data from the NEAT program.

2MASS The Two Micron All-Sky Survey[29] (2MASS) is a near-infrared (J, H, and K_S) all sky survey. The 2MASS project is a collaboration between the The University of Massachusetts, which constructed the observatory facilities and operates the survey, and the Infrared Processing and Analysis Center (Caltech), which is responsible for all data processing and archive issues. The survey began in the spring of 1997, completed survey quality observations in 2000, and is expected to publish the final catalog sometime in 2002.

---

[28] http://skys.gsfc.nasa.gov/skymorph/skymorph.html
[29] http://www.ipac.caltech.edu/2mass/

The 2MASS survey utilized two new, highly automated 1.3m telescopes, one at Mt. Hopkins, AZ and one at CTIO, Chile. Each telescope was equipped with a three channel camera, which uses HgCdTe detectors, and was capable of observing the sky simultaneously at J, H, and K_S. The survey includes over twelve Terabytes of imaging data, and the final catalog is expected to contain more than one million resolved galaxies, and more than three hundred million stars and other unresolved sources to a 10σ limiting magnitude of $K_S < 14.3$.

The 2MASS program is leading the way in demonstrating the power of public archives to the astronomical community. All twelve Terabytes of imaging data are available on near-online tape storage, and the actual catalogs are stored in an Informix backed archive (see Figure 27.9 for the public access to the 2MASS data). When complete, the survey will produce the following data products:

- a digital atlas of the sky comprising more than 1 million 8'x16' images having about 4" spatial resolution in each of the three wavelength bands,

- a point source catalog containing accurate (better than 0.5") positions and fluxes (less than 5% for $K_S > 13$) for approximately 300,000,000 stars.

- an extended source catalog containing positions and total magnitudes for more than 500,000 galaxies and other nebulae.

NVSS The National Radio Astronomical Observatory (NRAO), Very Large Array (VLA), Sky Survey (NVSS) is a publicly available, radio continuum survey[30] covering the sky north of −40 degrees declination. The survey catalog contains over 1.8 millions discrete sources with total intensity and linear polarization image measurements (Stokes I, Q, and U) with a resolution of 45", and a completeness limit of about 2.5 mJy. The NVSS survey is now complete, containing over two hundred thousand snapshot fields. The NVSS survey was performed as a community service, with the principal data products being released to the public by anonymous FTP as soon as they were produced and verified. The primary

---
[30] http://www.cv.nrao.edu/~jcondon/nvss.html

means of accessing the NVSS data products remains the original anonymous FTP service, however, other archive sites have begun to provide limited data browsing and access to this important dataset.

The principal NVSS data products are a set of 2326 continuum map "cubes," each covering an area of 4 degrees by 4 degrees with three planes containing the Stokes I, Q, and U images, and a catalog of discrete sources on these images (over 1.8 million sources are in the entire survey). Every large image was constructed from more than one hundred of the smaller, original snapshot images.

FIRST The Faint Images of the Radio Sky at Twenty-cm (FIRST) survey[31] is an ongoing, publicly available, radio snapshot survey that is scheduled to cover approximately ten thousand square degrees of the North and South Galactic Caps in 1.8" pixels (currently, approximately eight thousand square degrees have been released). The survey catalog, when complete should contain around one million sources with a resolution of better than 1". The FIRST survey is being performed at the NRAO VLA facility in a configuration that provides higher spatial resolution than the NVSS, at the expense of a smaller field of view.

A final atlas of radio image maps with 1.8" pixels is produced by coadding the twelve images adjacent to each pointing center. A source catalog including peak and integrated flux densities and sizes derived from fitting a two-dimensional Gaussian to each source is generated from the atlas. Approximately 15% of the sources have optical counterparts at the limit of the POSS I plates ($E \sim 20$); unambiguous optical identifications ($< 5\%$ false rates) are achievable to $V \sim 24$. The survey area has been chosen to coincide with that of the Sloan Digital Sky Survey (SDSS). It is expected that at the magnitude limit of the SDSS, approximately 50% of the optical counterparts to FIRST sources will be detected. Both the images and the catalogs constructed from the FIRST observations are being made available to the astronomical community via the project web-site (see Figure 27.7 for the public web-site) as soon as sufficient quality-control tests have been completed.

---

[31] http://sundog.stsci.edu/top.html

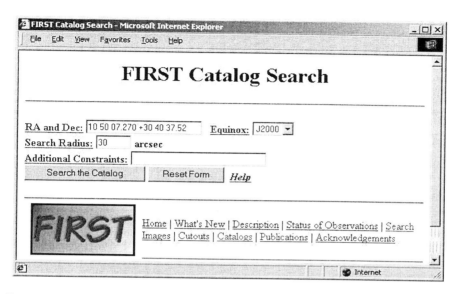

Figure 27.7. The FIRST survey web search engine. Both images and catalog entries can be retrieved.

3.2. Ground-Based Observatory Archives

Traditional ground-based observatories have been saving data, mainly as emergency back-ups for the users, for a significant time, accumulating impressive quantities of highly valuable, but heterogeneous data. Unfortunately, with some notable exceptions, the heterogeneity and a lack of adequate funding have limited the efforts to properly archive this wealth of information and make it easily available to the broad astronomical community. We see the development of good archives for major ground-based observatories as one of the most pressing needs in this field, and a necessary step in the development of the National Virtual Observatory.

In this section, we discuss three specific ground-based observatories: the National Optical Astronomical Observatories (NOAO), the National Radio Astronomical Observatory (NRAO), and the European Southern Observatory (ESO), focusing on their archival efforts. In addition to these three, many other observatories have extensive archives, including the Canada-France-Hawaii telescope[32] (CHFT), the James Clerk Maxwell Telescope[33] (JCMT), the Isaac Newton Group of telescopes

---

[32] http://www.cfht.hawaii.edu/
[33] http://www.jach.hawaii.edu/JACpublic/JCMT/home.html

at La Palma[34] (ING), the Anglo-Australian Observatory[35] (AAT), the United Kingdom Infrared Telescope[36] (UKIRT), and the Australia National Telescope Facility[37] (ATNF).

NOAO The National Optical Astronomy Observatory[38] (NOAO) is a US organization that manages ground-based, national astronomical observatories, including the Kitt Peak National Observatory, Cerro Tololo Inter-American Observatory, and the National Solar Observatory. NOAO also represents the US astronomical community in the International Gemini Project. As a national facility, NOAO telescopes are open to all astronomers regardless of institutional affiliation, and has provided important scientific opportunities to astronomers throughout the world, who would otherwise had little or no opportunity to obtain astronomical observations.

The NOAO has been archiving all data from their telescopes in a program called save-the-bits, which, prior to the introduction of survey-grade instrumentation, generated around half a Terabyte and over 250,000 images a year. With the introduction of survey instruments and related programs, the rate of data accumulation has increased, and NOAO now manages over 10 Terabytes of data.

NRAO The National Radio Astronomy Observatory[39] (NRAO) is a US research facility that provides access to radio telescope facilities for use by the scientific community, in analogy to the primarily optical mission of the NOAO. Founded in 1956, the NRAO has its headquarters in Charlottesville, VA, and operates major radio telescope facilities at Green Bank, WV, Socorro, NM, and Tucson, AZ. The NRAO has been archiving their routine observations and has accumulated over ten Terabytes of data. They also have provided numerous opportunities for surveys, including the previously discussed NVSS and FIRST radio surveys as well as the Green Bank surveys.

ESO The European Southern Observatory[40] (ESO), operates a number of telescopes (including the four 8m class VLT telescopes) at two

---

[34] http://www.ing.iac.es/
[35] http://site.aao.gov.au/AAO/
[36] http://www.jach.hawaii.edu/JACpublic/UKIRT/home.html
[37] http://www.atnf.csiro.au/
[38] http://www.noao.edu/
[39] http://www.nrao.edu/
[40] http://www.eso.org/

observatories in the southern hemisphere: the La Silla Observatory, and the Paranal observatory. ESO is currently supported by a consortium of countries, with Headquarters in Garching, near Munich, Germany.

As with many of the other ground-based observatories, ESO has been archiving data for some time, with two important differences. First, they were one of the earliest observatories to appreciate the importance of community service survey programs (these programs generally probe to fainter flux limits over a significantly smaller area than the previously discussed surveys), which are made accessible to the international astronomical community in a relatively short timescale. Second, appreciating the legacy aspects of the four 8 meter Very Large Telescopes, ESO intentionally decided to break with tradition, and imposed an automatic operation of the telescopes that provides a uniform mechanism for data acquisition and archiving, comparable to what has routinely been done for space-based observatories (see the next section). Currently, the ESO data archive is starting to approach a steady state rate of approximately 20 Terabytes of data per year from all of their telescopes. This number will eventually increase to several hundred Terabytes with the completion of the rest of the planned facilities, including the VST, a dedicated survey telescope similar in nature to the telescope that was built for the SDSS project.

3.3. NASA Space-Based Observatory Archives

With the continual advancement of technology, ground-based observations continue to make important discoveries. Our atmosphere, however, absorbs radiation from the majority of the electromagnetic spectrum, which, while important to the survival of life, is a major hindrance when trying to untangle the mysteries of the cosmos. Thus space-based observations are critical, yet they are extremely expensive. The resulting data is extremely valuable, and all of the generated data is archived. While there have been (and continue to be) a large number of satellite missions, we will focus on three major NASA archival centers: MAST, IRSA, and HEASARC (officially designated as NASA's distributed Space Science Data Services), the Chandra X-ray Observatory archive (CXO), and the National Space Science Data Center (NSSDC).

MAST The Multimission Archive at the Space Telescope Science Institute[41] (MAST) archives a variety of astronomical data, with

---

[41] http://archive.stsci.edu/mast.html

the primary emphasis on the optical, ultraviolet, and near-infrared parts of the spectrum. MAST provides a cross correlation tool that allows users to search all archived data for all observations which contain sources from either archived or user supplied catalog data. In addition, MAST provides individual mission query capabilities. Preview images or spectra can often be obtained, which provides useful feedback to archive users. The dominant holding for MAST is the data archive from the Hubble Space Telescope (see Figure 27.8 for the HST archive web-site). This archive has been replicated at mirror sites in Canada[42], Europe[43] and Japan[44], and has often taken a lead in astronomical archive developments.

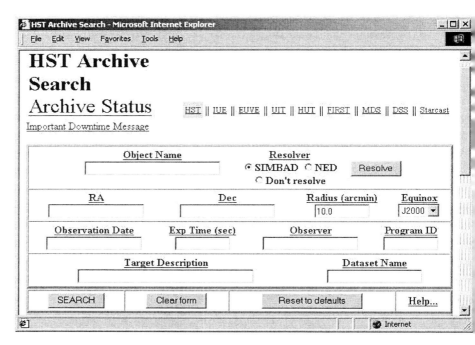

Figure 27.8. The Hubble Space Telescope (HST) archive (part of the MAST holdings) interface.

Based on the archival nature of the requested data, MAST provides data access in a variety of different ways, including intermediate disk staging and FTP retrieval and direct web-based downloads.

---

[42] http://cadcwww.dao.nrc.ca/hst/
[43] http://www.stecf.org/
[44] http://hst.nao.ac.jp/

MAST holdings currently exceed ten Terabytes, including or providing links to archival data for the following missions or projects:

- the Hubble Space Telescope (HST),
- the International Ultraviolet Explorer (IUE),
- Copernicus UV Satellite (OAO-3),
- the Extreme Ultraviolet Explorer (EUVE),
- the two space shuttle ASTRO Missions:
  - the Ultraviolet Imaging Telescope (UIT),
  - the Wisconsin Ultraviolet Photo Polarimetry Experiment (WUPPE),
  - the Hopkins Ultraviolet Telescope (HUT),
- the FIRST (VLA radio data) survey,
- the Digitized Sky Survey,
- the Roentgen Satellite (ROSAT) archive,
- the Far Ultraviolet Spectroscopic Explorer (FUSE),
- Orbiting and Retrieval Far and Extreme Ultraviolet Spectrograph (ORFEUS) missions,
- the Interstellar Medium Absorption Profile Spectrograph(IMAPS) (first flight),
- the Berkeley Extreme and Far-UV Spectrometer (BEFS) (first flight).

The Hubble Space Telescope (HST) is the first of NASA's great observatories, and provides high-resolution imaging and spectrographic observations from the near-ultraviolet to the near-infrared parts of the electro-magnetic spectrum (0.1 – 2.5 microns). It is operated by the Space Telescope Science Institute (STScI) is located on the campus of the Johns Hopkins University and is operated for NASA by the Association of Universities for Research in Astronomy (AURA).

IRSA The NASA Infrared Processing and Analysis Center (IPAC) Infrared Science Archive[45] (IRSA) provides archival access to a variety of data, with a primary focus on data in the infrared portion of the spectrum (see Figure 27.9 for the IRSA archive web-site).

---

[45] http://irsa.ipac.caltech.edu

IRSA has taken a strong leadership position in developing software and Internet services to facilitate access to astronomical data products. IRSA provides a cross-correlation tool allowing users to extract specific data on candidate targets from a variety of sources. IRSA provides primarily browser based query mechanisms, including access to both catalog and image holdings.

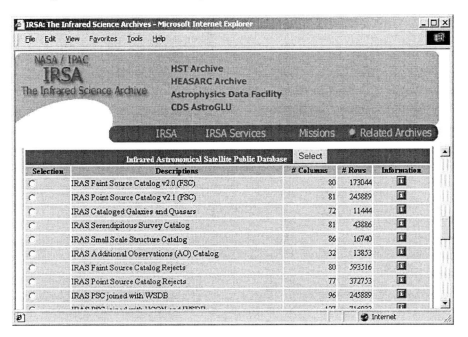

Figure 27.9. The IRSA Public access site. From this web-site, an archive user can query the contents of the publicly available 2MASS data, as well as other datasets that have been ingested into the IRSA Informix database server.

IRSA contains over fifteen Terabytes of data, mostly related to the 2MASS survey, and currently maintains the archives for the following datasets at IPAC:

- the Two Micron All-Sky Survey (2MASS),
- the Space Infrared Telescope Facility (SIRTF),
- the European Infrared Space Observatory (ISO),
- the Midcourse Space Experiment (MSX),
- the Infrared Astronomical Satellite (IRAS).

The Space Infrared Telescope (SIRTF) observatory will be the fourth and final of NASA's great observatories, and will provide imaging and spectrographic observations in the infrared part of the electro-magnetic spectrum (3 – 180 microns). SIRTF is expected to be launched in 2002. The SIRTF science center (SSC) is located on the campus of the California Institute of Technology and is operated for NASA by the Jet propulsion Laboratory.

The Infrared Space Observatory (ISO) operated at wavelengths from 2.5 – 240 microns, obtaining both imaging and spectroscopic data are available over a large area of the sky. The Midcourse Space Experiment (MSX) operated from 4.2 – 26 microns, and mapped the Galactic Plane, the gaps in the IRAS data, the zodiacal background, confused regions away from the Galactic Plane, deep surveys of selected fields at high galactic latitudes, large galaxies, asteroids and comets. The Infrared Astronomical Satellite (IRAS) performed an unbiased all sky survey at 12, 25, 60 and 100 microns, detecting approximately 350,000 high signal-to-noise infrared sources split between the faint and point source catalogs. A significantly larger number of sources ($>$ 500,000) are included in the faint source reject file, which were below the flux threshold required for the faint source catalog.

HEASARC The High Energy Astrophysics Science Archive Research Center[46] (HEASARC) is a multi-mission astronomy archive with primary emphasis on the extreme ultra-violet, X-ray, and Gamma ray spectral regions. HEASARC currently holds data from 20 observatories covering 30 years of X-ray and gamma-ray astronomy. HEASARC provides data access via FTP and the Web (see Figure 27.10), including the Skyview interface, which allows multiple images and catalogs to be compared.

The HEASARC archive currently includes over five Terabytes of data, and will experience significant increases with the large number of high-energy satellite missions that are either currently or soon will be in operation. HEASARC currently provides archival access to the following missions:

- the Roentgen Satellite (ROSAT),
- the Advanced Satellite for Cosmology (ASCA),
- BeppoSAX,
- the Compton Gamma Ray Observatory (CGRO), the second of NASA's great observatories,

---

[46]http://heasarc.gsfc.nasa.gov

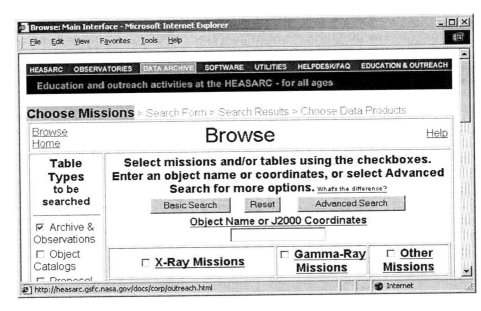

Figure 27.10. The HEASARC public access site, where data from the x-ray and gamma-ray space missions are archived.

- the Extreme Ultraviolet Explorer (EUVE),
- the High Energy Astrophysics Observatory (HEAO 1),
- the Einstein Observatory (HEAO 2),
- the European Space Agency's X-ray Observatory (EXOSAT),
- the Rossi X-ray Timing Explorer (Rossi XTE),
- the Advanced Satellite for X-ray Astronomy (Chandra),
- the X-ray Multiple Mirror Satellite (XMM-Newton),
- the High Energy Transient Explorer (HETE-2),

Several of these missions elicit further discussion. First, the Einstein Observatory[47] was the first fully imaging X-ray telescope put into space, with an angular resolution of a few arcseconds and was sensitive over the energy range 0.2 – 3.5 keV. The ROSAT[48] satellite was an X-ray observatory that performed an all sky survey in the 0.1 – 2.4 keV range as well as numerous pointed observations.

---

[47] http://heasarc.gsfc.nasa.gov/docs/einstein/heao2.html
[48] http://wave.xray.mpe.mpg.de/rosat/

The full sky survey data has been publicly released, and there are catalogs of both the full sky survey (Voges et al. 1999) as well as serendipitous detections from the pointed observations (White et al. 1994). The ASCA[49] satellite operated in the energy range 0.4 – 10 keV, that performed several small area surveys and obtained valuable spectral data on a variety of astrophysical sources. Finally, the Chandra (also see the next section) and XMM-Newton[50] satellites are currently providing revolutionary new views on the cosmos, due to their increased sensitivity, spatial resolution and collecting area.

Skyview[51] is a web-site operated from HEASARC which is billed as a "Virtual Observatory on the Net" (see Figure 27.11). Using public-domain data, Skyview allows astronomers to generate images of any portion of the sky at wavelengths in all regimes from radio to gamma-ray. Perhaps the most powerful feature of the Skyview site is its ability to handle the geometric and coordinate transformations required for presenting the requested data to the user in the specified format.

CDA The Chandra X-ray observatory[52] is third of NASA's great observatories, and provides high-resolution imaging and spectrographic observations in the X-ray part of the electro-magnetic spectrum. Chandra was launched by the Space Shuttle Columbia during July, 1999. Unlike the Hubble data archive, which is part of MAST, and the SIRTF data archive, which will be part of IRSA, the Chandra Data Archive (CDA) is part of the Chandra X-Ray Observatory Science Center (CXC) which is operated for NASA by the Smithsonian Astrophysical Observatory. The data is also archived at HEASARC, which is the relevant wavelength space mission archive.

The Chandra data products can be roughly divided into science-related and engineering data. The engineering data products include all data relating to the spacecraft subsystems: including such quantities as spacecraft temperature and operating voltages. The scientific data products are divided into three categories: primary, secondary, and supporting products. Primary products are generally the most desired, but the secondary products can provide important information required for more sophisticated analysis and,

---

[49] http://www.astro.isas.ac.jp/asca/index-e.html
[50] http://xmm.vilspa.esa.es/
[52] http://chandra.harvard.edu/

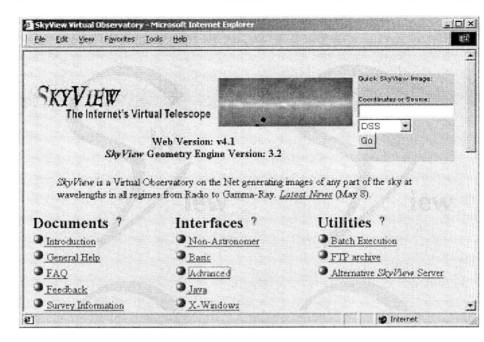

Figure 27.11. The Skyview virtual telescope site, which provides access to a large number of digital imagery and catalog information.

possibly, limited reprocessing and fine-tuning. The actual data can be retrieved via several different mechanisms or media, including web-based downloads, staged anonymous FTP retrieval, or mailed delivery of 8 mm Exabyte, 4 mm DAT, or CDROMs.

NSSDC The National Space Science Data Center[53] (NSSDC) provides network-based and offline access to a wide variety of data from NASA missions, including the Cosmic Background Explorer [54] (COBE), accruing data at the rate of several Terabytes per year. NSSDC was first established at the Goddard Space Flight Center in 1966, and continues to archive mission data, including both independently and not independently (*i.e.* data that requires other data to gain utility) useful data. Currently, the majority of network-based data dissemination is via WWW and FTP, and most offline data dissemination is via CD-ROM. NSSDC is generally regarded as the final repository of all NASA space mission data.

---

[53] http://nssdc.gsfc.nasa.gov/
[54] http://space.gsfc.nasa.gov/astro.cobe

PDS In this chapter, we do not discuss in detail the great wealth of data available on solar system objects. The site for the archival access to scientific data from NASA planetary missions, astronomical observations, and laboratory measurements is the Planetary Data System[55] (PDS). The homepage for the PDS is displayed in Figure 27.12, and provides access to data from the Pioneer, Voyager, Mariner, Magellan, NEAR spacecraft missions, as well as other data on asteroids, comets, and the Planets—including Earth.

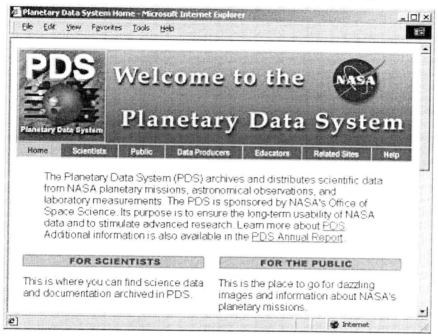

Figure 27.12. The Planetary Data System web-site. From this web-site, users can query and extract archived the enormous quantity of data that has been obtained on the Solar system objects.

4. The Future of Observational Astronomy: Virtual Observatories

Raw data, no matter how expensively obtained, are no good without an effective ability to process them quickly and thoroughly, and to refine the essence of scientific knowledge from them. This problem has suddenly increased by orders of magnitude, and it keeps growing.

---

[55] http://pds.jpl.nasa.gov

A prime example is the efficient scientific exploration of the new multi-Terabyte digital sky surveys and archives. How can one make efficiently discoveries in a database of billions of objects or data vectors? What good are the vast new data sets if we cannot fully exploit them?

In order to cope with this data flood, the astronomical community started a grassroots initiative, the National (and ultimately Global) Virtual Observatory (see, *e.g.*, Brunner et al. 2001,for a virtual observatory conference proceedings). Recognizing the urgent need, the National Academy of Science Astronomy and Astrophysics Survey Committee, in its new decadal survey entitled *Astronomy and Astrophysics in the New Millennium*, recommends, as a first priority, the establishment of a National Virtual Observatory. The NVO will likely grow into a Global Virtual Observatory, serving as the fundamental information infrastructure for astronomy and astrophysics in the next century. We envision productive international cooperation in this rapidly developing new field.

The NVO would federate numerous large digital sky archives, provide the information infrastructure and standards for ingestion of new data and surveys, and develop the computational and analysis tools with which to explore these vast data volumes. It would provide new opportunities for scientific discovery that were unimaginable just a few years ago. Entirely new and unexpected scientific results of major significance will emerge from the combined use of the resulting datasets, science that would not be possible from such sets used singly. The NVO will serve as *an engine of discovery for astronomy* (NVO Informal Steering Committee 2001).

Implementation of the NVO involves significant technical challenges on many fronts: How to manage, combine, analyze and explore these vast amounts of information, and to do it quickly and efficiently? We know how to collect many bits of information, but can we effectively refine the essence of knowledge from this mass of bits? Many individual digital sky survey archives, servers, and digital libraries already exist, and represent essential tools of modern astronomy. However, in order to join or federate these valuable resources, and to enable a smooth inclusion of even greater data sets to come, a more powerful infrastructure and a set of tools are needed.

The rest of this review focuses on the two core challenges that must be tackled to enable the new, virtual astronomy:

1 Effective federation of large, geographically distributed data sets and digital sky archives, their matching, their structuring in new ways so as to optimize the use of data-mining algorithms, and fast data extraction from them.

2 Data mining and "knowledge discovery in databases" (KDD) algorithms and techniques for the exploration and scientific utilization of large digital sky surveys, including combined, multi-wavelength data sets.

These services would carry significant relevance beyond Astronomy as many aspects of society are struggling with information overload. This development can only be done by a wide collaboration, that involves not only astronomers, but computer scientists, statisticians and even participants from the IT industry.

4.1. Architecting the Virtual Observatory

The foundation and structure of the National Virtual Observatory (NVO) are not yet clearly defined, and are currently the subject of many vigorous development efforts. One framework for many of the basic architectural concepts and associated components of a virtual observatory, however, has become popular. First is the requirement that, if at all possible, all data must be maintained and curated by the respective groups who know it best — the survey originators. This requires a fully distributed system, as each survey must provide the storage, documentation, and services that are required to participate in a virtual observatory.

The interconnection of the different archive sites will need to utilize the planned high-speed networks, of which there are several testbed programs already available or in development. A significant fraction of the technology for the future Internet backbone is already available, the problem is finding real-world applications which can provide a sufficient load. A Virtual Observatory, would, of course, provide heavy network traffic and is, therefore, a prime candidate for early testing of any future high-speed networks.

The distributed approach advocated by this framework (see Figure 27.13 for a demonstration) relies heavily on an the ability of different archives to participate in "collaborative querying". This tight integration requires that everything must be built using appropriately developed standards, detailing everything from how archives are "discovered" and "join" the virtual observatory, to how queries are expressed and data is transferred. Once these standards have been developed, implementation (or retrofitting as the case may be) of tools, interfaces, and protocols that operate within the virtual observatory can begin.

The architecture of a virtual observatory is not only dependent on the participating data centers, but also on the users it must support. For example, it is quite likely that the general astronomy public (*e.g.*, amateur

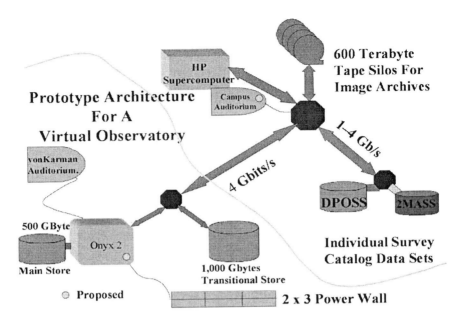

Figure 27.13. One straw-man layout for a virtual observatory prototype, demonstrating the highly distributed nature of the system. Of particular importance are the high speed network interconnections, extremely large storage systems, and high performance compute servers. Image Credit: Dave Curkendall and the ALPHA group, (JPL).

astronomers, K-12 classrooms, *etc.*) would use a virtual observatory in a casual lookup manner (*i.e.* the web model). On the other hand, a typical researcher would require more complex services, such as large data retrieval (*e.g.*, images) or cross-archive joins. Finally, there will also be the "power users" who would require heavy post-processing of query results using super-computing resources (*e.g.*, clustering analysis).

From these user models we can derive "use cases", which detail how a virtual observatory might be utilized. Initially, one would expect a large number of distinct "exploratory" queries as astronomers explore the multi-dimensional nature of the data. Eventually the queries will become more complex and employ a larger scope or more powerful services. This model requires the support of several methods for data interaction: manual browsing, where a researcher explores the properties of an interesting class of objects; cross-identification queries, where a user wants to find all known information for a given set of sources; sweeping queries, where large amounts of data (*e.g.*, large areal extents, rare object searches) are processed with complex relationships; and the creation of new "personal" subsets or "official" data products. This approach leads, by necessity, to allowing the user to perform customizable, complex analysis (*i.e.* data-mining) on the extracted data stream.

4.2. Connecting Distributed Archives

As can be seen from Section 3, a considerable amount of effort has been expended within the astronomical community on archiving and processing astronomical data. On the other hand, very little has been accomplished in enabling collaborative, cross-archive data manipulation (see Figure 27.14). This has been due, in part, to the previous dearth of large, homogeneous, multi-wavelength surveys; in other words, the payoff for federating the disparate datasets has previously been too small to make the effort worthwhile. Here we briefly outline some of the key problem areas (*cf.* Szalay and Brunner 1999) for a more detailed discussion), that must be addressed in order to properly build the foundation for the future virtual observatories.

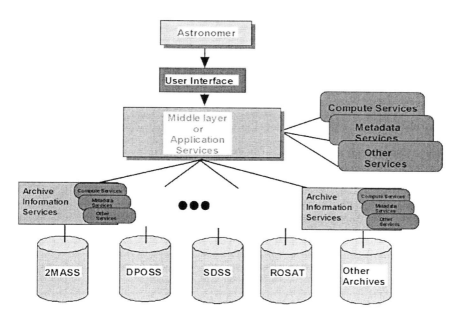

Figure 27.14. A prototype blueprint for the system architecture of a virtual observatory. The key concept throughout our approach is the plug—and—play model where different archives, compute services and user tools all interact seamlessly. This system model is predicated on the universal adoption of standards dictating everything from how archives communicate with each other to how data is transferred between archives, services and users.

4.2.1 Communication Fundamentals. The first requirement for connecting highly distributed datasets is that they must be able to communicate with each other. This communication takes multiple roles, including the initiation of communication, discovering the holdings and capabilities of each archive, the actual process of querying, the streaming of data, and an overall control structure. None of these ideas are entirely new, the general Information Technology field has been confronting similar issues and solutions, such as Grid frameworks (Foster and Kesselman 2001), JINI[56] and the Web services model (see, *e.g.*, the IBM Web Service web-site[57]) for more information) are equally applicable.

Clearly, the language for communicating will be the extensible markup language (XML), using a community defined standard schema. This will allow for control of the inter-archive communication and processing (*e.g.*, the ability to perform basic checkpoint operations on a query: stop, pause, restart, abort, and provide feedback to the end-user). A promising, and simple mechanism for providing the archive communication endpoints is through web services, which would be built using the Simple Object Access Protocol[58] (SOAP), Web Services Description Language[59] (WSDL), and a common Universal Description, Discovery, and Integration[60] (UDDI) registry. An additional benefit of this approach is that pre-existing or legacy archival services can be retrofitted (by mapping a new service onto existing services) in order to participate in collaborative querying.

This model also allows for certain optimizations to be performed depending on the status of the archival connections (*e.g.*, network weather). Eventually, a learning mechanism can be applied to analyze queries, and using the accumulated knowledge gained from past observations (*i.e.* artificial intelligence), queries can be rearranged in order to provide further performance enhancements.

4.2.2 Archival Metadata. In order for the discovery process to be successful, the archives must communicate using shared semantics. Not only must this semantic format allow for the transfer of data contents and formats between archives, but it also should clearly describe the specific services that an archive can support (such as cross-identification or image registration) and the expected format of the input

---

[56] A registered trademark of the SUN Microsystems corporation.
[57] http://www.ibm.com/developerworks/webservices/
[58] http://www.w3.org/TR/2000/NOTE-SOAP-20000508/
[59] http://xml.coverpages.org/wsdl.html
[60] http://www.uddi.org/

and output data. Using the web service model, our services would be registered in a well known UDDI registry, and communicate their capabilities using WSDL. Depending on the need of the consumer, different amounts (or levels) of detailed information might be required, leading to the need for a hierarchical representation. Once again, the combination of XML and a standardized XML Schema language provides an extremely powerful solution, as is easily generated, and can be parsed by machines and read by humans with equal ease. By adopting a standardized schema, metadata can be easily archived and accessed by any conforming application.

4.2.3 High Performance Data Streaming.

Traditionally, astronomers have communicated data either in ASCII text (either straight or compressed), or by using the community standard Flexible Image Transport Standard (FITS). The true efficacy of the FITS format as a streaming format, however, is not clear, due to the difficulty of randomly extracting desired data or shutting off the stream. The ideal solution would pass different types of data (*i.e.* tabular, spectral, or imaging data) in a streaming fashion (similar to MPI—Message Passing Interface), so that analysis of the data does not need to wait for the entire dataset before proceeding. In the web services model, this would allow different services to cooperate in a head-to-tail fashion (*i.e.* the UNIX pipe scenario). This is still a potential concern, as the ability to handle XML encoded binary data is not known.

4.2.4 Astronomical Data Federation.

Separate from the concerns of the physical federation of astronomical data via a virtual observatory paradigm is the issue of actually correlating the catalog information from the diverse array of multiwavelength data (see the skyserver project [61] for more information). While seemingly simple, the problem is complicated by the several factors.

First is the sheer size of the problem, as the cross-identification of billions of sources in both a static and dynamic state over thousands of square degrees in a multi-wavelength domain (Radio to X-Ray) is clearly a computationally challenging problem, even for a consolidated archive. The problem is further complicated by the fact that observational data is always limited by the available technology, which varies greatly in sensitivity and angular resolution as a function of wavelength (*e.g.*, optical-infrared resolution is generally superior to high energy resolution).

---

[61] http://www.skyserver.org/

Furthermore, the quality of the data calibration (either spectral, temporal, or spatial) can also vary greatly, making it extremely difficult to to unambiguously match sources between different wavelength surveys. Finally, the sky looks different at different wavelengths (see, *e.g.*, Figure 27.1), which can produce one-to-one, many-to-one, one-to-one, many-to-many, and even one/many-to-none scenarios when federating multiwavelength datasets. As a result, sometimes the source associations must be made using probabilistic methods (Lonsdale et al. 1998, Rutledge et al. 2000).

4.3. Data Mining and Knowledge Discovery

Key to maximizing the knowledge extracted from the ever-growing quantities of astronomical (or any other type of) data is the successful application of data-mining and knowledge discovery techniques. This effort as a step towards the development of the next generation of science analysis tools that redefine the way scientists interact and extract information from large data sets, here specifically the large new digital sky survey archives, which are driving the need for a virtual observatory (see, *e.g.*, Figure 27.15 for an illustration).

Such techniques are rather general, and should find numerous applications outside astronomy and space science. In fact, these techniques can find application in virtually every data-intensive field. Here we briefly outline some of the applications of these technologies on massive datasets, namely, unsupervised clustering, other Bayesian inference and cluster analysis tools, as well as novel multidimensional image and catalog visualization techniques. Examples of particular studies may include:

1. Various classification techniques, including decision tree ensembles and nearest-neighbor classifiers to categorize objects or clusters of objects of interest. Do the objectively found groupings of data vectors correspond to physically meaningful, distinct types of objects? Are the known types recovered, and are there new ones? Can we refine astronomical classifications of object types (*e.g.*, the Hubble sequence, the stellar spectral types) in an objective manner?

2. Clustering techniques, such as the expectation maximization (EM) algorithm with mixture models to find groups of interest, to come up with descriptive summaries, and to build density estimates for large data sets. How many distinct types of objects are present in the data, in some statistical and objective sense? This would also be an effective way to group the data for specific studies, *e.g.*, some

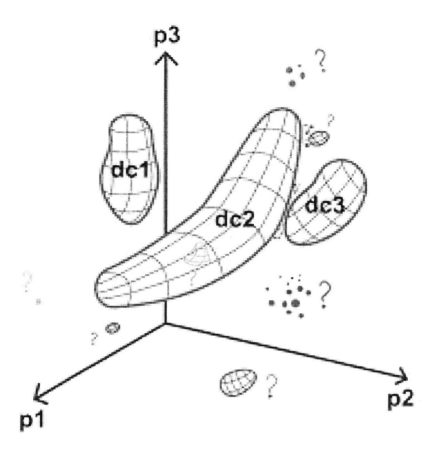

Figure 27.15. A demonstration of a generic machine-assisted discovery problem—data mapping and a search for outliers. This schematic illustration is of the clustering problem in a parameter space given by three object attributes: P1, P2, and P3. In this example, most of the data points are assumed to be contained in three, dominant clusters (DC1, DC2, and DC3). However, one may want to discover less populated clusters (e.g., small groups or even isolated points), some of which may be too sparsely populated, or lie too close to one of the major data clouds. All of them present challenges of establishing statistical significance, as well as establishing membership. In some cases, negative clusters (holes), may exist in one of the major data clusters.

users would want only stars, others only galaxies, or only objects with an IR excess, *etc.*

3 Use of genetic algorithms to devise improved detection and supervised classification methods. This would be especially interesting in the context of interaction between the image (pixel) and catalog (attribute) domains.

4 Clustering techniques to detect rare, anomalous, or somehow unusual objects, *e.g.*, as outliers in the parameter space, to be selected for further investigation. This would include both known but rare classes of objects, *e.g.*, brown dwarfs, high-redshift quasars, and also possibly new and previously unrecognized types of objects and phenomena.

5 Use of semi-autonomous AI or software agents to explore the large data parameter spaces and report on the occurrences of unusual instances or classes of objects. How can the data be structured to allow for an optimal exploration of the parameter spaces in this manner?

6 Effective new data visualization and presentation techniques, which can convey most of the multidimensional information in a way more easily grasped by a human user. We could use three graphical dimensions, plus object shapes and coloring to encode a variety of parameters, and to cross-link the image (or pixel) and catalog domains.

Notice that the above examples are moving beyond merely providing assistance with handling of huge data sets: these software tools may become capable of *independent or cooperative discoveries*, and their application may greatly enhance the productivity of practicing scientists.

It is quite likely that many of the advanced tools needed for these tasks already exist or be under development in the various fields of computer science and statistics. In creating a virtual observatory, one of the most important requirements is to bridge the gap between the disciplines, and introduce modern data management and analysis software technologies into astronomy and astrophysics.

4.3.1 Applied Unsupervised Classification. Some preliminary and illusory experiments using Bayesian clustering algorithms were designed to classify objects present in the DPOSS catalogs (de Carvalho et al. 1995) using the AutoClass software (Cheeseman et al. 1988). The program was able separate the data into four recognizable and astronomically meaningful classes: stars, galaxies with bright central cores, galaxies without bright cores, and stars with a visible "fuzz" around them. Thus, the object classes found by AutoClass are astronomically meaningful—even though the program itself does not know about stars, galaxies and such! Moreover, the two morphologically distinct classes of galaxies populate different regions of the data space, and have systematically different colors and concentration indices, even though AutoClass was not given the color information. Thus, *the program has found astrophysically meaningful distinction between these classes of objects, which is then confirmed by independent data.*

One critical point in constructing scientifically useful object catalogs from sky surveys is the classification of astronomical sources into either stars or galaxies. Various supervised classification schemes can be used for this task, including decision trees (see, *e.g.*, Weir et al. 1995) or neural nets (Odewahn et al. 1992). A more difficult problem is to provide at least rough morphological types for the galaxies detected, in a systematic and objective way, without visual inspection of the images, which is obviously impractical. This actually provides an interesting opportunity—the application of new clustering analysis and unsupervised classification techniques may divide the parent galaxy population into astronomically meaningful morphological types on the basis of the data themselves, rather than some preconceived, human-imposed scheme.

Another demonstration of the utility of these techniques can be seen in Figure 27.16. In this experiment, the Expectation Maximization technique was applied on a star-galaxy training data set of approximately 11,300 objects with 15 parameters each. This is an unsupervised classification method which fits a number of multivariate Gaussians to the data, and decides on the optimal number of clusters needed to describe the data. Monte-Carlo cross validation was used to decide on the optimal number of clusters (see, *e.g.*, Smyth in press). The program found that there are indeed two dominant classes of objects, *viz.*, stars and galaxies, containing about 90% of all objects, but that there are also a half-dozen other significant clusters, most of which correspond to legitimate subclasses such as saturated stars, *etc.* Again, this illustrates the potential of unsupervised classification techniques for objective partitioning of data, identification of artifacts, and possibly even discovery of new classes of objects.

4.3.2   Analyzing Large, Complex Datasets.   Most clustering work in the past has only been applied to small data sets. The main reasons for this are due to memory storage and processing speed. With orders of magnitude improvement in both, we can now begin to contemplate performing clustering on the large scale. However, clustering algorithms have high computational complexity (from high polynomial, order 3 or 4, to exponential search). Hence a rewriting of these algorithms, shifting the focus from performing expensive searches over small data sets, to robust (computationally cheap) estimation over very large data sets is in order.

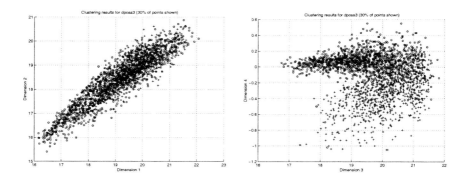

Figure 27.16. An example of unsupervised classification of objects in the star-galaxy training data set from DPOSS, from the experiment using the Expectation Maximization multivariate Gaussians mixture modeling. Two dominant clusters found are shown, encoded as circles (in reality stars) and crosses (galaxies). The plot on the left shows a typical parameter projection in which the two classes are completely blended. The plot on the right shows one of the data projections in which the two classes separate.

The reason we need to use large data sets is motivated by the fact that new classes to be discovered in the data are likely to be rare occurrences (else humans would have surely found them). For example, objects like quasars (extremely luminous, very distant sources that are believed to be powered by supermassive black holes) constitute a tiny proportion of the huge number of objects detectable in our survey, yet they are an extremely important class of objects. Unknown types (classes) of objects that may potentially be present in data are likely to be as rare. Hence, if an automated discovery algorithm is to have any hope of finding them, it must be able to process a huge amount of data, millions of objects or more.

Current clustering algorithms simply cannot run on more than a few thousand cases in less than 10–dimensional space, without requiring weeks of CPU time.

1. Many clustering codes (*e.g.*, AutoClass) are written to demonstrate the method, and are ill-suited for data sets containing millions or billions of data vectors in tens of dimensions. Improving the efficiency of these algorithms as the size and complexity of the datasets is increased is an important issue.

2. With datasets of this size and complexity, multi-resolution clustering is a must. In this regime, expensive parameters to estimate, such as the number of classes and the initial broad clustering are quickly estimated using traditional techniques like K-means

clustering or other simple distance-based method (Duda and Hart 1981). With such a clustering one would proceed to refine the model locally and globally. This involves iterating over the refinements until some objective (like a Bayesian criterion) is satisfied.

3 Intelligent sampling methods where one forms "prototypes" of the case vectors and thus reduces the number of cases to process. Prototypes can be determined based on nearest-neighbor type algorithms or K-means to get a rough estimate, then more sophisticated estimation techniques can refine this. A prototype can represent a large population of examples. A weighting scheme based on number of cases represented by each prototype, as well as variance parameters attached to the feature values assigned to the prototype based on values of the population it represents, are used to describe them. A clustering algorithm can operate in prototype space. The clusters found can later refined by locally replacing each prototype by its constituent population and reanalyzing the cluster.

4 Techniques for dimensionality reduction, including principal component analysis and others can be used as preprocessing techniques to automatically derive the dimensions that contain most of the relevant information. See, *e.g.*, the singular-valued decomposition scheme to find the eigenvectors dominant in the data set in a related application involving finding small volcanos in Magellan images of Venus (Aubele et al. 1995).

5 Scientific verification and evaluation, testing, and follow-up on any of the newly discovered classes of objects, physical clusters discovered by these methods, and other astrophysical analysis of the results. This is essential in order to demonstrate the actual usefulness of these techniques for the NVO.

4.3.3 Scientific Verification. Testing of these techniques in a real-life data environment, on a set of representative science use cases, is essential to validate and improve their utility and functionality. Some of the specific scientific verification tests may include:

1 A novel and powerful form of the quality control for our data products, as multidimensional clustering can reveal subtle mismatch patterns between individual sky survey fields or strips, *e.g.*, due to otherwise imperceptible calibration variations. This would apply to virtually any other digital sky survey or other patch-wise collated data sets. Assured and quantified uniformity of digital sky

surveys data is essential for many prospective applications, *e.g.*, studies of the large-scale structure in the universe, *etc.*

2 A new, objective approach to star-galaxy separation could overcome the restrictions of the current accuracies of star-galaxy classifications that effectively limits the scientific applications of any sky survey catalog. Related to this is an objective, automated, multiwavelength approach to morphological classification of galaxies, *e.g.*, quantitative typing along the Hubble sequence, or one of the more modern, multidimensional classification schemes.

3 An automated search for rare, but known classes of objects, through their clustering in the parameter space. Examples include high-redshift quasars, brown dwarfs, or ultraluminous dusty galaxies.

4 An automated search for rare *and as yet unknown* classes of astronomical objects or phenomena, as outliers or sparse clusters in the parameter space, not corresponding to the known types of objects. They would be found in a systematic way, with fully quantifiable selection limits.

5 Objective discovery of clusters of stars or galaxies in the physical space, by utilizing full information available in the surveys. This should be superior to most of the simple density-enhancement type algorithms now commonly used in individual surveys.

6 A general, unbiased, multiwavelength search for AGN, and specifically a search for the long-sought population of Type 2 quasars. A discovery of such a population would be a major step in our understanding of the unification models of AGN, with consequences for many astrophysical problems, *e.g.*, the origins of the cosmic x-ray background.

This clustering analysis would be performed in the (reduced) measurement space of the catalogs. But suppose the clustering algorithm picks out a persistent pattern, *e.g.*, a set of objects, that for reasons not obvious to human from the measurements, are consistently clustered separately from the data. The next step is for the astronomer to examine actual survey images to study this class further to verify discovery or explain scientifically why the statistical algorithms find these objects different.

Some enhanced tools for image processing, in particular, probabilistic methods for segmentation (region growing) that are based both on pixel value, adjacency information, and the prior expectation of the scientist will need to be used to aid in analysis and possibly overcome some loss of information incurred when global image processing was performed.

There are also potential applications of interest for the searches for Earth-crossing asteroids, where a substantial portion of the sky would be covered a few times per night, every night. The addition of the time dimension in surveys with repeated observations such as these, would add a novel and interesting dimension to the problem. While variable objects obviously draw attention to themselves (*e.g.*, supernovae, gamma-ray bursts, classical pulsating variables, *etc.*), the truth is that we know very little about the variability of the deep sky in general, and a systematic search for variability in large and cross-wavelength digital sky archives is practically guaranteed to bring some new discoveries.

4.4. Astronomical Data Visualization

Effective and powerful data visualization would be an essential part of any virtual observatory. The human eye and brain are remarkably powerful in pattern recognition, and selection of interesting features. The technical challenge here is posed by the sheer size of the datasets (both in the image and catalog domains), and the need to move through them quickly and to interact with them "on the fly".

The more traditional aspect of this is the display of various large-format survey images. One of the new challenges is in streaming of the data volume itself, now already in the multi-TB range for any given survey. A user may need to shift quickly through different spatial scales (*i.e.* zoom in or out) on the display, from the entire sky down to the resolution limit of the data, spanning up to a factor of approximately 10^{11} (!) in solid angle coverage. Combining the image data from different surveys with widely different spatial resolutions poses additional challenges. So does the co-registration of images from different surveys where small, but always present systematic distortions in astrometric solutions must be corrected before the images are overlaid.

Another set of challenges is presented by displaying the information in the parameter spaces defined in the catalog domain, where each object may be represented by a data vector in tens or even hundreds of dimensions, but only a few can be displayed at any given time (*e.g.*, 3 spatial dimensions, color, shape, and intensity for displayed objects). Each of the object attributes, or any user defined mathematical combination of object attributes (*e.g.*, colors) should be encodeable on demand as any of the displayed dimensions. This approach will also need to be extended to enable the display of data from more than one survey at a time, and to combine object attributes from matched catalogs.

However, probably the most interesting and novel aspect is the combination and interaction between the image and catalog domains. This is only becoming possible now, due to the ability to store multi-TB data sets on line, and it opens a completely new territory. In the simplest approach this would involve marking or overplotting of sources detected in one survey, or selected in some manner, *e.g.*, in clustering analysis, on displayed images.

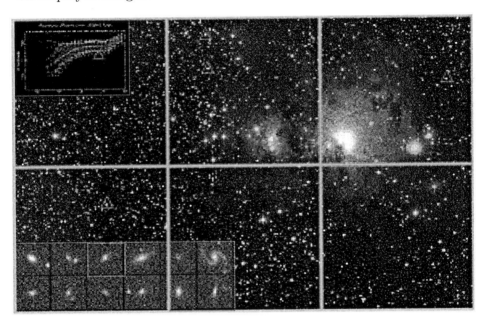

Figure 27.17. A prototype of the visualization services which would empower scientists to not only tackle current scientific challenges, but also to aid in the exploration of the, as yet unknown, challenges of the future. Note the intelligent combination of image and catalog visualizations to aid the scientist in exploring parameter space. Image Credit: Joe Jacob and the ALPHA group, (JPL).

In the next level of functionality, the user would be able to mark the individual sources or delineate areas on the display, and retrieve the catalog information for the contained sources from the catalog domain (see, *e.g.*, Figure 27.17 for a demonstration). Likewise, it may be necessary to remeasure object parameters in the pixel domain and update or create new catalog entries. An example may be measuring of low-level signals or upper limits at locations where no statistically significant source was cataloged originally, but where a source detection is made in some other survey, *e.g.*, faint optical counterparts of IR, radio, or x-ray sources. An even more sophisticated approach may involve development of new object classifiers through interaction of catalog and image domains, *e.g.*, using genetic algorithms.

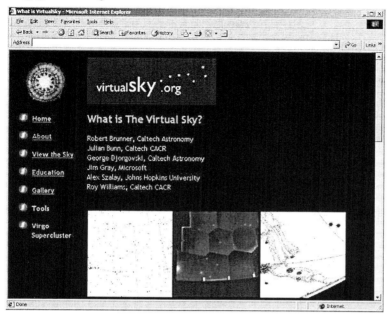

Figure 27.18. The Virtual Sky web-site which provides public access to several different data products, including images from the DPOSS survey.

Visualization of these large digital sky surveys is also a powerful education and public outreach tool. An example of this is the virtual sky project[62] (see Figure 27.18 for the project homepage).

5. Summary

We are at the start of a new era of information-rich astronomy. Numerous ongoing sky surveys over a range of wavelengths are already generating data sets measured in the tens of Terabytes. These surveys are creating catalogs of objects (stars, galaxies, quasars, *etc.*) numbering in billions, with tens or hundreds of measured numbers for each object. Yet, this is just a foretaste of the much larger data sets to come, with multi-Petabyte data sets already on the horizon. Large digital sky surveys and data archives are thus becoming the principal sources of data in astronomy. The very style of observational astronomy is changing: systematic sky surveys are now used both to answer some well-defined

---

[62] http://www.virtualsky.org

questions which require large samples of objects, and to discover and select interesting targets for follow-up studies with space-based or large ground-based telescopes.

This vast amount of new information about the universe will enable and stimulate a new way of doing astronomy. We will be able to tackle some major problems with an unprecedented accuracy, *e.g.*, mapping of the large-scale structure of the universe, the structure of our Galaxy, *etc.* The unprecedented size of the data sets will enable searches for extremely rare types of astronomical objects (*e.g.*, high-redshift quasars, brown dwarfs, *etc.*) and may well lead to surprising new discoveries of previously unknown types of objects or new astrophysical phenomena. Combining surveys done at different wavelengths, from radio and infrared, through visible light, ultraviolet, and x-rays, both from the ground-based telescopes and from space observatories, would provide a new, panchromatic picture of our universe, and lead to a better understanding of the objects in it. These are the types of scientific investigations which were not feasible with the more limited data sets of the past.

Many individual digital sky survey archives, servers, and digital libraries already exist, and represent essential tools of modern astronomy. We have reviewed some of them, and there are many others existing and still under development. However, in order to join or federate these valuable resources, and to enable a smooth inclusion of even greater data sets to come, a more powerful infrastructure and a set of tools are needed.

The concept of a virtual observatory thus emerged, including the incipient National Virtual Observatory (NVO), and its future global counterparts. A virtual observatory would be a set of federated, geographically distributed, major digital sky archives, with the software tools and infrastructure to combine them in an efficient and user-friendly manner, and to explore the resulting data sets whose sheer size and complexity are beyond the reach of traditional approaches. It would help solve the technical problems common to most large digital sky surveys, and optimize the use of our resources.

This systematic, panchromatic approach would enable new science, in addition to what can be done with individual surveys. It would enable meaningful, effective experiments within these vast data parameter spaces. It would also facilitate the inclusion of new massive data sets, and optimize the design of future surveys and space missions. Most importantly, the NVO would provide access to powerful new resources to scientists and students everywhere, who could do first-rate observational astronomy regardless of their access to large ground-based telescopes.

Finally, the NVO would be a powerful educational and public outreach tool.

Technological challenges inherent in the design and implementation of the NVO are similar to those which are now being encountered in other sciences, and offer great opportunities for multi-disciplinary collaborations. This is a part of the rapidly changing, information-driven scientific landscape of the new century.

Acknowledgments

We would like to acknowledge our many collaborators in the Virtual Observatory initiative. Particular thanks go to the members of the Digital Sky project and the NVO interim steering committee, where many of the ideas presented herein were initially discussed. We would also like to thank the editors for the immense patience with the authors. This work was made possible in part through the NPACI sponsored Digital Sky Project (NSF Cooperative Agreement ACI-96-19020) and generous grants from the SUN Microsystems Corporation and Microsoft Research. RJB would like to personally acknowledge financial support from the Fullam Award and NASA grant number NAG5-10885. SGD acknowledges support from NASA grants NAG5-9603 and NAG5-9482, the Norris foundation, and other private donors.

Bibliography

J.C̃. Aubele, L.S̃. Crumpler, U.M̃. Fayyad, P. Smyth, M.C̃. Burl, and P. Perona. Locating small volcanoes on venus using a scientist-trainable analysis system. In *Lunar and Planetary Science Conference*, volume 26, page 61, 1995.

R. J. Brunner, S. G. Djorgovski, and A. S. Szalay, editors. *Virtual Observatories of the Future*. Astronomical Society of the Pacific, San Francisco, 2001.

P. Cheeseman, J. Kelly, M. Self, J. Stutz, W. Taylor, and D. Freeman. AutoClass: A Bayesian Classification System. In *Proceedings of the Fifth International Conference on Machine Learning*, pages 54–64. Morgan Kaufmann Publishers, San Francisco, 1988.

R.R̃. de Carvalho, S.G̃. Djorgovski, N. Weir, U. Fayyad, K. Cherkauer, J. Roden, and A. Gray. Clustering analysis algorithms and their applications to digital poss-ii catalogs. In *Astronomical Data Analysis Software and Systems IV*, volume 77 of *ASP Conference Series*, pages 272–275. Astronomical Society of the Pacific, San Francisco, 1995.

R. Duda and P. Hart. *Pattern Classification and Scene Analysis.* John Wiley, & Sons, 1981.

I. Foster and C. Kesselman, editors. *The Grid: Blueprint for a New Computing Infrastructure.* Morgan Kaufmann, San Francisco, CA, 2001.

C. J. Lonsdale, T. Conrow, T. Evans, L. Fullmer, M. Moshir, T. Chester, D. Yentis, R. Wolstencroft, H. MacGillivray, and D. Egret. The OPTID Database: Deep Optical Identifications to the IRAS Faint Source Survey. In B. J. McLean, D. A. Golombek, J. J. E. Hayes, and H. E. Payne, editors, *New Horizons from Multi-Wavelength Sky Surveys*, volume 179 of *IAU Symposium Series*, pages 450–451. Kluwer Academic Publishers, Netherlands, 1998.

R.G. McMahon and M.J. Irwin. Apm surveys for high redshift quasars. In *ASSL Vol. 174: Digitised Optical Sky Surveys*, page 417, 1992.

D. Monet, A. Bird, B. Canzian, H. Harris, N. Reid, A. Rhodes, S. Sell, H. Ables, C. Dahn, H. Guetter, A. Henden, S. Leggett, H. Levison, C. Luginbuhl, J. Martini, A. Monet, J. Pier, B. Riepe, R. Stone, F. Vrba, and R. Walker. USNO-SA2.0. Technical report, U.S. Naval Observatory, Washington DC, 1996.

NVO Informal Steering Committee. Towards a national virtual observatory: Science goals, technical challenges, and implementation plan. In R.J. Brunner, S.G. Djorgovski, and A.S. Szalay, editors, *Virtual Observatories of the Future*, pages 353–372. Astronomical Society of the Pacific, San Francisco, 2001.

S.C. Odewahn, E.B. Stockwell, R.L. Pennington, R.M. Humphreys, and W.A. Zumach. Automated star/galaxy discrimination with neural networks. *Astronomical Journal*, 103:318–331, 1992.

R.L. Pennington, R.M. Humphreys, S.C. Odewahn, W. Zumach, and P.M. Thurmes. The automated plate scanner catalog of the palomar sky survey. i - scanning parameters and procedures. *Publications of the Astronomical Society of the Pacific*, 105:521–526, 1993.

I.N. Reid, C. Brewer, R.J. Brucato, W.R. McKinley, A. Maury, D. Mendenhall, J.R. Mould, J. Mueller, G. Neugebauer, J. Phinney, W. L.W. Sargent, J. Schombert, and R. Thicksten. The second palomar sky survey. *Publications of the Astronomical Society of the Pacific*, 103:661–674, 1991.

R.E. Rutledge, R.J. Brunner, T.A. Prince, and C. Lonsdale. XID: Cross-Association of ROSAT/Bright Source Catalog X-Ray Sources with USNO A-2 Optical Point Sources. *Astrophysical Journal, Supplement*, 131:335–353, 2000.

P. Smyth. *Model Selection for Probabilistic Clustering Using Cross-Validated Likelihood.* in press.

A. Szalay and R. J. Brunner. Astronomical Archives of the Future: A Virtual Observatory. *Future Generations of Computational Systems*, 16:63, 1999.

S. Urban, T. Corbin, and G. Wycoff. The act reference catalog. In *American Astronomical Society Meeting*, volume 191, page 5707, 1997.

W. Voges, B. Aschenbach, T. Boller, H. Bräuninger, U. Briel, W. Burkert, K. Dennerl, J. Englhauser, R. Gruber, F. Haberl, G. Hartner, G. Hasinger, M. Kürster, E. Pfeffermann, W. Pietsch, P. Predehl, C. Rosso, J. H. M.M. Schmitt, J. Trümper, and H.U. Zimmermann. The rosat all-sky survey bright source catalogue. *Astronomy and Astrophysics*, 349:389–405, 1999.

N. Weir, U. Fayyad, S. G. Djorgovski, and J. Roden. The SKICAT System for Processing and Analyzing Digital Imaging Sky Surveys. *Publications of the Astronomical Society of the Pacific*, 107:1243, 1995.

N.E. White, P. Giommi, and L. Angelini. Wgacat. *IAU Circular*, 6100: 1, 1994.

D.J. Yentis, R.G. Cruddace, H. Gursky, B.V. Stuart, J.F. Wallin, H.T. MacGillivray, and C.A. Collins. The cosmos/ukst catalog of the southern sky. In *ASSL Vol. 174: Digitised Optical Sky Surveys*, page 67, 1992.

Chapter 28

DATA MANAGEMENT IN ENVIRONMENTAL INFORMATION SYSTEMS

Oliver Günther
Institute of Information Systems, School of Economics and Business Administration, Humboldt-Universität zu Berlin, Germany
guenther@wiwi.hu-berlin.de

Abstract   This chapter describes the design and implementation of information systems to support decision-making in environmental management and protection. We employ a three-way object model: An environmental object (such as a lake) is described by one or more environmental data objects. Each environmental data object may in turn be described by some environmental metadata. The related data flow is structured into four phases. DATA CAPTURE concerns the collection and processing of environmental raw data, such as aerial photographs and measurement series. DATA STORAGE raises questions of suitable database designs and appropriate physical storage structures for the resulting massive amounts of data. DATA ANALYSIS techniques serve to prepare the available data for decision support purposes. METADATA is collected and aggregated throughout this data flow and serves to support search and browsing operations.

Keywords: Geographic and environmental information systems, Data management, Spatial databases, Statistical classification, R-trees, Simulation Models, Metadata.

1.   Introduction

The preservation of the environment has become an important public policy goal throughout the world. Citizens are taking a greater interest in the current and future state of the environment, and many are adapting their way of life accordingly. Companies are required to report on the environmental impact of their products and activities. Governments are concerned more than ever about their environmental resources and are establishing policies to control their consumption. To devise and implement such policies, administrators require detailed information about the current state of the environment and ongoing developments.

Moreover, an increasing number of governments are starting to recognize the right of their citizens to access the environmental information available. According to a recent directive of the European Union, for example, almost all environmental data that is stored at public agencies has to be made available to any citizen on demand (Council of the European Communities 1990). As the last few years have shown, the tendency to exert this right is rising steadily. Citizens, special-interest groups, and government agencies alike are requesting up-to-date information regarding air and water quality, soil composition, and so on.

As a result of these political and economic developments, there is a major demand for environmental information and appropriate tools to manage it. Given the amount and complexity of environmental data, these new information needs can only be served by using state-of-the-art computer technology.

Environmental information systems are concerned with the management of data about the soil, the water, the air, and the species in the world around us. The collection and administration of such data is an essential component of any efficient environmental protection strategy. Vast amounts of data need to be available to decision makers, mostly (but not always) in some kind of condensed format. The requirements regarding the currency and accuracy of this information are high. Details vary between applications (Stafford et al. 1994, Hilty et al. 1995). While earth scientists, for example, often emphasize the need to support system modeling (Eddy 1993), scientists from other disciplines may put their emphasis on powerful database systems (Gosz 1994) or systems integration (Strebel et al. 1994).

This paper describes the design and implementation of information systems to support decision-making in environmental management and protection. While the required information technology is rarely domain-specific, it is important to select and combine the right tools among those that are available in principle. This requires a thorough knowledge of related developments in computer science and a good understanding of the environmental management tasks at hand.

The first step in any kind of data processing, computer-based or not, concerns the mapping of real-world objects to entities that are somewhat more abstract and can be handled by computers or directly by the decision maker. In this paper we refer to the real-world objects of interest as *environmental objects*. Just about any real-world object can be regarded as an environmental object. Note in particular that the term is not restricted to natural entities (such as animals or lakes), but also includes human-made structures (such as houses or factories).

Each environmental object is described by, or mapped to, a collection of *environmental data objects*. These objects are abstract entities that can be handled by computers or by decision makers directly. A typical environmental data object would be a series of measurements that captures the concentration of a certain substance in a river (the corresponding environmental object).

While it is important to distinguish a data object from its physical and visual representations, the boundaries are sometimes fuzzy. An environmental data object can be available in analog or digital form. In the past, environmental information systems have been analog. Early attempts to understand and to manage the environment are as old as civilization itself. They are documented in ancient maps, collections of measurements and observations, hunting schedules, and so on – the environmental data objects of that period. The computer has opened up a whole new range of instruments to manage environmental data. As a result, environmental data objects are increasingly becoming available in digital form, with the trend toward digitized maps just being the most visible sign of this transition. As a consequence of these developments, we see that certain entities that would traditionally have been regarded as autonomous data objects are only visualizations of other data objects. A specialized map, for example, is often just a view or presentation tool for the underlying data.

In this paper we use the term *environmental data object* in a wide sense, oriented toward the users' perceptions. Whatever they perceive as a separate information entity is called an environmental data object. Nesting or overlaps between such objects are common.

Computerized environmental information systems are able to collect and process much greater amounts of data than anybody would have thought only a few years ago. Automatic data capture and measurement results in the processing of terabytes of data per day. Even in processed form, this kind of data is impossible to browse manually in order to find the information that is relevant for a given task. Modern information retrieval tools allow the automatic or semi-automatic filtering of the available data in order to find those data sets one is looking for. An important prerequisite for these tools is the availability of *metadata*, i.e., data about data.

In this paper, we call the metadata that refers to a particular environmental data object its *environmental metadata*. Each environmental data object is typically associated with one or more *metadata objects* that specify its format and contents. The documentation of the measuring series described above would be a typical example. It may include data about the spatial and temporal scale of the measurements, the main objectives of the project, the responsible agency, and so on. As soon as an environmental data object is created or updated, the change should be propagated to the metadata level. From there it may be forwarded to any application for which this modification may be relevant.

These considerations lead to a three-way object model (Fig. 28.1). An environmental object is described by one or more environmental data objects. Each environmental data object may in turn be described by some environmental metadata. The data flow that is typically associated with such an approach closely resembles the data flow in classical business applications. It can be structured into four phases: data capture, data storage, data analysis, and metadata management.

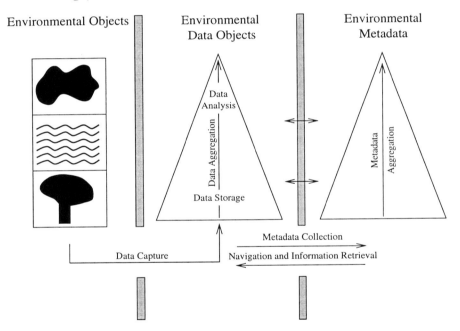

Figure 28.1. Data flow in environmental information systems.

1 The first phase, *data capture*, concerns the collection and processing of environmental raw data, such as measurement time series or aerial photographs. In this step the great variety of environmental objects is mapped onto a collection of environmental data objects, which usually have a structure that is much simpler and

more clearly defined than the original raw data. There are a variety of ways to perform such a mapping, including measurement and observation, but also value-based judgement. Data capture usually involves some considerable aggregation, where the raw data is condensed and enriched in order to extract entities that are semantically meaningful. In the case of image data, for example, this includes the recognition of geometric primitives (such as lines and vertices) in an array of pixels, the comparison of the resulting geometric objects with available maps, and the identification of geographic objects (such as cities or rivers) in the picture. The information can then be represented in a much more compact format (in this case, a vector-based data format, as opposed to the original raster data). Measurement time series also need to be aggregated and possibly evaluated by means of some standard statistical procedures. The aggregated data is then stored in a file or a database.

2 For *data storage*, one has to choose a suitable database design and appropriate physical storage structures that will optimize overall system performance. Because of the complexity and heterogeneity of environmental data, this often necessitates substantial extensions to classical database technology. Non-standard data types and operators need to be accommodated and supported efficiently. In particular the management of spatial data requires the integration of highly specialized data structures and algorithms.

3 In the *data analysis* phase, the available information is prepared for decision support purposes. This may require simultaneous access to data that is geographically distributed, stored on heterogenous hardware, and organized along a wide variety of data models. Data analysis is typically based on complex statistical methods, scenarios, simulation and visualization tools, and institutional knowledge (such as environmental legislation or user objectives). Only the synthesis of the input data and these kinds of model allows us to judge the state of the environment and the potential of certain actions, both planned and already implemented. With regard to aggregation, data analysis can be seen as a direct continuation of data capture. The main difference is that the aggregation is now more target-oriented, i.e., more specifically geared towards particular tasks and decision makers.

4 *Metadata* is collected and aggregated throughout the three phases described above. It is stored in appropriate data structures and

serves mostly in the data analysis phase to support search and browsing operations.

The overall objective of this complex aggregation process is to provide decision support at various levels of responsibility. Figure 28.1 uses the symbol of the pyramid to visualize this idea. Data aggregation corresponds to a bottom-up traversal of the central pyramid. Data can be used throughout that traversal for decision support purposes. While the data in the lower part of the pyramid tends to be used for local, tactical tasks, the upper part corresponds to strategic decision support for middle and upper management.

Given the fact that in most Western industrial nations one can recognize major bottlenecks especially in data aggregation and analysis, it is critical that information systems technology is put to much wider use in environmental management. While there are many parallels between the kind of sequential data processing sketched above and the data flow in traditional business applications and geographic information systems (GIS), environmental applications often combine several properties that are problematic from a data management point of view:

- The *amount of data* to be processed is unusually large. The amount of satellite imagery recorded per day, for example, is already in the terabyte range (Campbell and Cromp 1990). That is about two orders of magnitude beyond the typical size of large high-transaction databases in banking or airline reservation applications. The processing of such large amounts of data demands hardware and software tools that reflect the state of the art in computer technology. Classical storage technology and data management techniques are no longer sufficient.

- Data management is usually highly *distributed*. Environmental data is captured, processed and stored by a broad range of government agencies and other institutions.

- The data management is extremely *heterogenous*, in terms of both hardware and software platforms. Data is organized according to a wide variety of data models, depending on the primary objectives of the particular agency in charge.

- Environmental data objects frequently have a *complex internal structure*, i.e., they consist of subobjects. These components may be complex objects in turn and may be associated with heterogenous types and media (including sound or images).

- Environmental data objects are often *spatial-temporal*, i.e., they have a location and a spatial extension, and they change over time.

- Environmental data is frequently *uncertain*. Techniques from statistics and from artificial intelligence have to be employed to manage this uncertainty.

- Because environmental issues cut across traditional subject areas, the processing of user queries may require *complex logical connections and joins*. Data is often used for purposes that are very different from the context the original data providers had intended.

After a detailed discussion of these problem areas, this paper will present a variety of techniques that have been developed to handle these difficulties. The structure of the paper closely follows the data flow pictured in Fig. 28.1. One section each is devoted to data capture, data storage, data analysis, and metadata. Each section includes real-life examples taken from environmental information systems that are currently in use by government agencies and the general public.

Our treatment of the World Wide Web focuses on general principles how the Web can be used in the context of environmental information systems. It is not meant to provide an exhaustive list of relevant URLs, as the Web itself is a better place to generate and maintain such lists. However, we will provide pointers to indices, search engines, and metainformation systems that can be used to search for environmentally relevant sites (cf. Sect. 4.4.2 and Chap. 5). In addition, we provide an appendix with all relevant URLs cited in this paper.

2. Data Capture

Due to the political developments of the 1970s and 1980s, both the amount and the quality of the data that is being collected has increased considerably over the past 10 to 15 years. Environmental management and protection have become household terms. Many governments are encouraged by their constituencies to pay attention to environmental issues and to support efforts to increase the efficiency of environmental protection.

Governments in many Western industrial countries have reacted to these trends. Sensor networks have been installed and upgraded to monitor the quality of the water, the air, and the ground countrywide. Satellites are used increasingly to obtain environmental data, not only about remote areas. As a result, the availability of raw data about the environment is no longer a bottleneck, at least not for the Western industrial countries, Australia, New Zealand, Japan, and the Tiger states of South East Asia.[1] The question is rather how to process these large amounts of data in order to obtain efficient decision support.

Because of the political and administrative developments described above, the amounts of data that are collected about the environment have grown extremely large. The environmental data collected per day worldwide already exceeds the 10 terabyte mark, and we can expect this to grow by one or two orders of magnitude by the end of the century. Most of the collected data sets are unstructured raw data, in particular raster images. They require a considerable amount of processing before they can be used for decision support purposes. Given the size of the data sets, it is not realistic to rely solely on human expertise in order to aggregate and evaluate this data. In many areas there are simply not enough human experts available, and even if they were, it would not make economic sense to employ them for the often mundane task of raw data processing. Usually the government agency (or the company) in charge is neither willing nor able to pay for this service. As a result, there are already many cases where raw data sets are written to disk without anybody ever having had a look at them. Once again, we see that raw data is not the bottleneck – *evaluation* is!

2.1. Object Taxonomies

The term data capture denotes the process of deriving environmental data objects from environmental objects. As noted in the introduction, just about any real-world object can be regarded as an environmental object. Environmental objects can be grouped into a number of classes. The *atmosphere* includes all objects above the surface of the Earth, such as the air or most kinds of radiation. The *hydrosphere* contains water-related objects, such as lakes or rivers. For frozen water such as snow and ice, one sometimes uses the term *cryosphere*. The term *lithosphere* relates to soil and rocks. The *biosphere* is the collection of all living matter, i.e., animals and plants. The term *technosphere* is used to denote human-made objects, such as houses or factories.

Note that, while in common use, the taxonomy is far from being unambiguous. An artificial lake, for example, is part of both hydrosphere and technosphere. A given land parcel may be described both as a part of the soil (lithosphere) and as a piece of cultivated land.

A simpler version of this taxonomy structures the environment into just three *media: ground, water*, and *air*. This taxonomy, which is commonly used by government agencies, leads to a relatively straightforward mapping to the traditional scientific disciplines. Geology, mineralogy, and parts of the agricultural sciences are concerned with the ground.

Hydrology, oceanography, and limnology are about water, and the atmospheric sciences, including metereology, are concerned with the air. Many public administrations are still organized along this taxonomy. As we begin to understand the environment as an integrated and multiply connected complex system, however, this taxonomy is found increasingly insufficient. Many natural processes and effects cannot be assigned to one of those three media alone. Understanding the environment is an inherently interdisciplinary task that transcends those traditional boundaries. Academic programs in the environmental sciences often reflect this fact, drawing from the course offerings of a wide range of departments. Administrative structures are starting to change as well, albeit more slowly. While one still finds the classical departments, especially when it comes to data capture, interdisciplinary task forces and project-oriented matrix organizations are becoming increasingly common.

2.2. Mapping the Environment

2.2.1 Raw Data Processing.
Whatever taxonomy is used, the main question is which environmental objects should be monitored, and what data should be collected on them. There are many ways to obtain environmental data objects from environmental objects. One may use sensors to monitor, for example, the temperature of the air or the emissions of a factory. The results are archived as time series of measurements. More complex analytical techniques, such as chromatography or mass spectrometry, may be used to survey the chemical content of a lake. Air and satellite imagery is increasingly being used to monitor remote areas and to recognize long-term environmental developments. For that purpose, the raw imagery is usually processed and represented as a thematic map in order to visualize, for example, land utilization or temperature distribution. For forest and wildlife management, controlled observations and manual counts of animals or plants are often the most reliable source of data. For objects of the technosphere, it is frequently useful to study some written documentation in order to extract and condense the required environmental data objects.

In all of these cases, the incoming raw data first has to be subjected to some domain- and device-specific processing. Depending on the data source, this may include some optical rectification, noise suppression, filtering, or contrast enhancement.

Even at this early stage of data capture, it is not realistic to assume that this process can always be handled in an objective manner. Subjective influences, such as opinions and judgements of the people involved, often find their way into the data capture process. This may happen unknowingly or in full conscience (for instance, in order to judge the hazard potential of a factory).

2.2.2 Classification. If there is too much data for manual evaluation, the natural question is what parts of data capture could and should be automated. While the initial raw data processing, such as rectification and filtering, is usually performed directly by the sensor equipment, the following steps are somewhat more difficult to automate. A particularly important problem concerns the question how to classify a new observation with respect to a given taxonomy. The evaluation of satellite imagery provides a typical example. Given an unidentified pixel and a variety of possible land uses, the question is how to identify the most likely kind of land use for this pixel. Similar problems come up in other data capture situations as well.

The commonest technique for solving this problem is called *maximum likelihood*. This approach is based on the assumption that there exists a finite number of classes ω_i to which the new observation may belong. Each observation is represented as a point X in some n-dimensional feature space. X is sometimes referred to as the observation's *signature*. Furthermore, for each class ω_i there exists an n-dimensional probability distribution $p(X|\omega_i)$ that indicates the probability that a member of class ω_i assumes the observation value X. This probability distribution, however, is usually unknown. In most applications, one therefore assumes a normal distribution. As empirical studies have shown, slight violations of the normal distribution do not affect the classification accuracy in a major way (Swain and Davis 1978).

Once a normal distribution has been adapted, the problem remains how to estimate the parameters of the distribution. Both the mean μ_i and the covariance matrix Θ_i have to be estimated for each class ω_i. This is usually done using a suitable set of *training data*. The training data consists of observations whose class affiliation is known. In order to avoid singularities, one needs to take a representative sample of at least $n+1$ observations. More observations improve the classification performance; in some applications a sample size of $10n$ is not uncommon (Swain and Davis 1978). Given a sample X_{i1}, X_{i2}, ..., X_{iN} drawn from class ω_i ($N > n$), one obtains the following maximum likelihood estimators for μ_i and the elements $(\Theta_i)^{uv}$ of the covariance matrix ($1 \leq u, v \leq n$):

$$\hat{\mu}_i = \frac{\sum_{k=1}^{N} X_{ik}}{N}$$

$$(\hat{\Theta}_i)^{uv} = \frac{\sum_{k=1}^{N}(X_{ik}^u - \hat{\mu}_i^u)(X_{ik}^v - \hat{\mu}_i^v)}{N}$$

The distribution $p(X|\omega_i)$ takes the form

$$p(X|\omega_i) = \frac{1}{(2\pi)^{n/2} |\hat{\Theta}_i|^{1/2}} \, exp\left(-\frac{1}{2}(X - \hat{\mu}_i)^T \hat{\Theta}_i^{-1}(X - \hat{\mu}_i)\right)$$

Once probability distributions have been obtained for all classes ω_i, one can use Bayes' rule to compute the conditional probabilities that an observation with value X belongs to class ω_i:

$$p(\omega_i|X) = \frac{p(X|\omega_i) \, p(\omega_i)}{\sum_{k=1}^{M} p(X|\omega_k) \, p(\omega_k)}$$

Here, $p(\omega_i)$ denotes the a priori probability for the occurrence of class i. If those probabilities are not known, one often assumes that membership is equally distributed among the M classes, i.e., $p(\omega_i) = 1/M$. In that case we obtain:

$$p(\omega_i|X) = \frac{p(X|\omega_i)}{\sum_{k=1}^{M} p(X|\omega_k)}$$

Maximum likelihood now simply means that the observation is assigned to the class whose value $p(\omega_i|X)$ is maximal.

2.2.3 Validation and Interpretation.

Based on the results of the classification, one can now form an initial collection of environmental data objects that are more condensed and therefore easier to interpret than the raw measurements. Then one seeks to identify those environmental data objects that seem interesting or noteworthy according to a predefined set of criteria. This may in particular involve the extraction of data objects that seem to signify *unusual* events or developments.

Because of the uncertainties associated with the data acquisition process, validation procedures need to be employed to avoid undetected hazards, false alerts, and other mishaps (Pham and Wittig 1995). One usually starts with *in situ*, context-free validation procedures. Recent measurements are compared to previous measurements *(temporal validation)* and possibly to some reference data obtained under similar circumstances. Measurement values that do not fit the usual patterns are subjected to cross-validation with measurements from other sensors in the same area that measure the same parameter *(geographic validation)*. This may also include previous measurements from those sensors *(space-time validation)*. Those values that still do not fit the norm are forwarded to a cross-validation with sensors that measure different parameters *(inter-parameter validation)*. This last step is highly labor-intensive and requires considerable knowledge about the underlying analytical chemistry, about the land use that is typical for the site in question, the chemical substances that are common there, and so on. As we shall see later, however, any attempt to automate the measurement and validation process is futile if it does not take these various knowledge sources into account.

Obvious measurement errors and outliers are then removed from the output and not taken into account by later processing steps. Whenever one detects continual irregularities in the measurement process, one triggers an operation alert. This alert should eventually result in a closer examination of the sensor in question and, if necessary, appropriate maintenance. It is obvious that this validation procedure is somewhat biased towards avoiding false environmental alerts, rather than making sure no hazard remains undetected. For the latter case, special drills are necessary, where one induces a test substance into the system that is harmless but detectable.

2.3. Data Fusion and Uncertain Information

As we saw in the previous section, environmental data capture can be performed with techniques that are more or less standard in the areas of statistical classification, database management, and artificial intelligence. If one is now trying to combine the various pieces of information that are available at the beginning of the capture process, however, one requires more sophisticated concepts. In particular, this *data fusion* process is often based on techniques for the management of *uncertainty*.

When raw imagery and measurement data is aggregated and evaluated, the input data is only one of many sources of information. Other circumstantial information is also taken into account in order to extract those environmental data objects that the user is interested in. Examples of such circumstantial information include:

- experience with the methods and technical devices that are used for data capture

- knowledge about the chemical substances and products involved

- information about the circumstances of the sample (local agriculture, temperature, weather, date, time of day, etc.).

Human experts always take such information into account when evaluating a sample, albeit sometimes unconsciously. Any attempt to automate this process without including such circumstantial information is doomed to failure.

A promising strategy is to form a working hypothesis, and to support or challenge this hypothesis based on the information available. This technique somewhat resembles a lawsuit that is based on circumstantial evidence. The question is how to weigh the different information inputs, and how to combine the resulting evidence for and against the current working hypothesis. This has to include the possibility that the inputs may partly contradict each other.

One option would be to base this strategy on Bayesian probability theory. Each piece of evidence is weighed with a probability between zero and one. While this technique can be used in principle, it has been shown to have several disadvantages in this context. The main problems are:

- There is no way to consider the evidence *pro* and the evidence *contra* independently. Each piece of evidence is weighed with exactly one probability p that denotes the degree of support that this evidence gives to the working hypothesis. In the Bayesian model, however, this also means that it gives support of weight $1-p$ to the counterhypothesis. This is not always sufficient to model the actual situation, where the degrees of support *pro* and *contra* may be less dependent on each other.

- Bayesian probability theory has been proven to be sensitive to inaccuracies in the input probabilities. In particular, these inaccuracies may propagate through the model in a superlinear manner.

- The Bayesian model often requires that events are independent of each other. This assumption is rarely true in real life.

Recent research in applied mathematics and artificial intelligence has resulted in a number of alternative approaches to handling uncertainty. On the symbolic level, most notably there are *certainty factors* (Adams 1984), the *support intervals* (Dempster 1967; 1968, Shafer 1976), and *fuzzy sets* (Zadeh 1965). More recently, *neural networks* have attracted major interest as a broad class of techniques to represent uncertainty at the subsymbolic level (Minsky and Papert 1969, Rumelhart et al. 1985). Keller (1995) and Ultsch (1995) give several examples of how to apply neural networks to environmental problems.

There is a large number of books that discuss these and other techniques for handling uncertainty in much greater detail. Besides the book by Luger and Stubblefield (1989), the reader may consult, for example, the collections edited by Clarke et al. (1993), Riccia et al. (1995).

2.4. A Case Study: Evaluating Satellite Imagery

Environmental data is increasingly becoming available in digital form from advanced multispectral sensor systems or from scanned aerial photographs. While there has been a notable trend towards using imagery for many years, it has recently gained considerable momentum due to the "peace dividend" mentioned above. Military contractors are trying to recover some of the revenue that has been missing since the end of the cold war and are entering new markets. Satellite technology is one of the technologies that are most easily converted from military to civilian purposes.

There are a variety of satellite systems dedicated mainly to environmental data capture. Table 28.1 gives an overview of some of the best-known systems.

Table 28.1. Comparison of satellite systems (Belgian Federal Government 1995)

| Satellite | Landsat | Landsat | NOAA |
Sensor	TM	MSS	AVHRR
Main applications	land use	land use	meteorology, land use
Spatial resolution	30 m	80 m	1.1–4 km
Surface/scene [km]	185×172	185×172	2 000×1 000
Repeat coverage [d]	16	16	1–8
First launch	1982	1978	1986
Number of bands	7	4	5
Spectral range [μm]	0.45–12.4 blue-TIR	0.50–1.10 green-NIR	0.58–12.50 green-TIR
Min. recorded scale	1:75 000	1:100 000	1:3 000 000
Max. recorded scale	1:250 000	1:500 000	1:5 000 000
Price/scene [U.S.$]	4,500	800	15

(a)

| Satellite | SPOT | SPOT |
Sensor	HRV multispectral	HRV panchromatic
Main applications	land	land
Spatial resolution	20 m	10 m
Surface/scene [km]	60×60	60×60
Repeat coverage [d]	3–26	3–26
First launch	1986	1986
Number of bands	3	1
Spectral range [μm]	0.50–0.89 green–NIR	0.51–0.73 green–red
Min. recorded scale	1:50 000	1:25 000
Max. recorded scale	1:100 000	1:100 000
Price/scene [U.S.$]	3 000	4 000

(b)

The *Landsat* system (United States National Aeronautics and Space Administration 1997) provides periodic high resolution multispectral data of the Earth's surface on a global basis. Starting with its inception in 1972 (then called Earth Resources Technology Satellite – ERTS), the program was managed by the U.S. Federal Government. In 1985, there was a politically motivated privatization effort that resulted in the founding of the Earth Observation Satellite Company (EOSAT). The Clinton Administration decided to reverse this decision. Since 1993, Landsat has again been controlled by the U.S. Federal Government, represented by the National Oceanic and Atmospheric Administration (NOAA). The current operational satellite, Landsat 5, was launched in March 1985. Landsat 6 was lost shortly after its launch in October 1993.

Landsat orbits the Earth once every 16 days, taking pictures of each part of the Earth's surface. Each scene covers an area of 185 by 172 square kilometers. One TM scene now costs around \$4,500; partial scenes are also available. There are two sensors on board the Landsat system: the Thematic Mapper (TM) and the older Multispectral Scanner (MSS). TM captures data with a resolution of 30 meters and in seven channels (bands) for different parts of the electromagnetic spectrum. Four of the seven bands cover the visual spectrum (0.45–0.8 μm), the others are infrared (up to 12.4 μm). MSS has four spectral bands (0.50–1.10 μm, i.e., green to near infrared) and a resolution of 80 meters. Landsat images are used in a broad range of applications, including land use monitoring, map making, pollution monitoring, snow extent assessments, erosion detection, and forest fire monitoring.

SPOT (Satellite pour l'observation de la terre) (SpotImage 1997) is an Earth observing satellite coordinated by the French government; partners are Sweden and Belgium. There are three SPOT satellites currently in space, launched in 1986, 1990, and 1993. Each of the three satellites carries two High Resolution Visible (HRV) sensors that can work in two modes. The multispectral mode corresponds to a ground resolution of 20 meters and three channels or spectral bands (0.50–0.89 μm, i.e., green to near infrared). The panchromatic mode only covers the spectrum between 0.51 and 0.75 μm (green to red) but obtains a 10 meter ground resolution. SPOT features some sophisticated scanning technology to support stereoscopic imagery and other advanced viewing options. Main applications include environmental impact studies, geologic exploration, and thematic map making. Since 1997, the SPOT Web site also allows access to the DALI image archive, administered by the Centre National d'Etudes Spatiales. This browsable archive contains more than 4 500 000 SPOT images dating back to 1992. Given a location's coordinates, the server returns up to five images of the requested area, in either color or black and white, with less than 10% cloud coverage.

AVHRR, the *Advanced Very High Resolution Radiometer* (United States National Oceanic and Atmospheric Administration 1997, Loveland and Ohlen 1993) is another satellite system managed by the NOAA. Other than Landsat, however, AVHRR is mainly geared towards meteorological and environmental applications. Typical application areas are weather forecasting, pollution monitoring, toxic algal bloom detection, and sandstorm monitoring. There are two AVHRR satellites, one with four and one with five spectral bands. One band is in the visual range (V), one in the near infrared (NIR), the other two or three in the middle and far infrared. The quotient $(NIR-V)/(NIR+V)$ has been found empirically to be a good indicator of vegetation vigor. Under the name *Normalized Difference Vegetation Index (NDVI)* it is used widely to map global vegetation on a regular basis (Lillesand and Kiefer 1987). AVHRR ground resolution is only about 1 km. On the other hand, the two AVHRR satellites provide two images per day of any given area, i.e., daily global coverage.

The U.S. Geological Survey's EROS Data Center (EDC) has begun to develop AVHRR time series data sets (Loveland et al. 1991, Loveland and Ohlen 1993). EDC produces four levels of data products that are designed for both the biophysical and land cover data requirements of global change researchers:

- processed daily AVHRR scenes of most of North America and selected other areas

- composite images for specified time intervals for the conterminous U.S., Alaska, Mexico, and Eurasia

- time series sets for those areas

- a land cover characteristics prototype database for the conterminous U.S.

Aerial and satellite imagery is increasingly available through the World Wide Web. NASA, for example, is preparing a Web version of its Earth Observing System Data and Information System (EOSDIS). (United States National Aeronautics and Space Administration 1998, Doan et al. 1997) EOSDIS contains data from all satellite systems described above and others. The company Earth Observation Sciences (EOS) is running a Web-based Catalogue and Browse System (CBS) to provide instant access to aerial and satellite imagery (http://www.eos.co.uk). CBS allows users to search for data both temporally and spatially, possibly using additional parameters defined by the data archive (such as cloud cover or orbit number). The search functionality includes a global map interface with user-defined polygon searching and complex temporal searches such as searching for annual repeating time periods. CBS provides the ability to browse the results returned from a query before downloading, requesting or ordering data.

The digital raster data obtained from those satellite systems contains the desired environmental information in an implicit form only. Moreover, the same areas and phenomena are never recorded the same way because of the ever changing conditions of the atmosphere, the illumination, and the phenology of vegetation. Analyzing this kind of imagery is a well-developed branch of engineering. It involves techniques from a variety of disciplines, including signal processing, statistics and classification, and, increasingly, artificial intelligence. Problem areas include the classification of a given pixel, the grouping of adjacent pixels to form objects, and the recognition of features. Similar to the structure presented in Sect. 2.2, one can distinguish three phases: iconic image processing, classification, and symbolic image processing (Günther et al. 1993):

Iconic Image Processing. Iconic image processing essentially corresponds to the *raw data processing* described in Sect. 2.2.1. Here one applies a variety of image processing algorithms to the incoming raw data in order to produce an appropriate visualization. This typically includes (Swain and Davis 1978):

- rectifications of the image geometry to match the scale and coordinates of applicable maps

- corrections of the spectral intensities by contrast enhancement;

- elimination of sensor faults

- noise suppression

- edge improvement.

In order to emphasize certain features, one may combine selected spectral bands, often by simple arithmetic operations, and assign the resulting raster data to the RGB (red/green/blue) video signal to display a color image. The computation and visualization of the NDVI vegetation index, as described above, is a typical example.

In general, this first processing step is heuristic in nature and based on human experience and expertise. The overall goal is to produce a set of images that are visually appealing and easy to interpret with respect to the given task.

Classification. This second step serves to classify the imagery with respect to a given, application-specific taxonomy (such as land use). Here one usually applies techniques like the ones presented in Sect. 2.2.2. Features known to be typical for the class to be recognized are propagated over the whole image to find all regions with similar characteristics. The training areas, which are typically selected manually, determine the expected spectral reflectance of those classes. For each class ω_i, one computes the average signatures $\hat{\mu}_i$ of all pixels X_{ik} in all corresponding training areas. Remember that

$$\hat{\mu}_i = \frac{\sum_{k=1}^{N} X_{ik}}{N}$$

is a point in multidimensional feature space. The dimensions of this space are not restricted to the channels of the incoming signal but may also be used to encode secondary information. A pixel to be classified belongs to a class if its signature does not differ too much from the average signature of the corresponding training areas; see Sect. 2.2.2 for more details.

The procedure works well if the signatures of the different classes are sufficiently homogeneous and do not overlap in feature space. If that is not the case, the average signature may not be sufficient to characterize a given class, and one must take other criteria into account. Some classes may also be characterized by combinations of signatures. For an orchard, for example, one typically obtains a variety of spectral values that are mixtures of the characteristic signatures for trees and for grass. Other classes may be recognized based on their texture rather than on their spectral signature (Zamparoni 1988).

Once again, this step usually involves a considerable amount of human interaction. It is a heuristic process based on experience, technical expertise, and a thorough knowledge of the application domain and the requirements of those who use the imagery data for their decision making.

Symbolic Image Processing. Symbolic image processing corresponds to the *validation and interpretation* phase described in Sect. 2.2.3. The data on the resulting image is condensed, grouped into objects, interpreted, and associated with geographic entities. Usually this involves the conversion of the incoming raster data to a vector representation that is more compact and easier to process (Ehlers et al. 1991,

Riekert 1993). The geographic entities may be administrative units such as states, districts, counties, or land parcels. They may also be ecological units such as forests, agricultural areas, or residential areas.

For this processing step it is usually indispensable to interact with a GIS that contains data about the regions in question (Ehlers et al. 1991, Davis et al. 1991). This input data can help to improve the classification results, for example, by providing a priori probabilities (cf. Sect. 2.2.2). Symbolic image processing once again requires experience and knowledge that is not contained in the image itself. The data sets resulting from this step are often ready to be stored in a GIS again although problems of accuracy and format compatibility have to be taken into account (Lunetta et al. 1991).

Figure 28.2 illustrates the three processing steps described above. Figure 28.2a shows the original Landsat TM data. It is taken from the near infrared channel, which is appropriate to discriminate land and water-covered areas. Figure 28.2b shows a water mask generated from this data by a threshold classification. Figure 28.2c shows water objects derived from the water mask by vectorization and some cartographic generalization.

Most of the operations in this sequence are *unreliable* in the sense that they introduce uncertainties into the classification result. Error sources include atmospheric conditions, sensor operation, geometric inaccuracies, and human judgement. Many of these errors cannot be avoided completely. It is therefore necessary to monitor possible error sources and to assess their impact on the result of the image classification. This usually involves a cost/benefit analysis. Lunetta et al. (1991) give a comprehensive overview of error sources and possible remedies.

If one obtains processed remote sensing data without complete or reliable information about its history (the *lineage*), it may be desirable to verify the classification results by means of sampling. Moisen et al. (1994) give a good comparative overview of sampling methods that are common in remote sensing, including simple random sampling, systematic sampling, and cluster sampling. For an introduction to the underlying statistical concepts, see the textbook by Cochran (1977). More recent approaches also involve knowledge-based techniques to support the interpretation of remote sensing data (Goodenough et al. 1987, McKeown 1987, Desachy 1990, Ton et al. 1991).

3. Data Storage

The way environmental data is stored is currently changing at a rapid pace. A considerable fraction of relevant data is still only available in

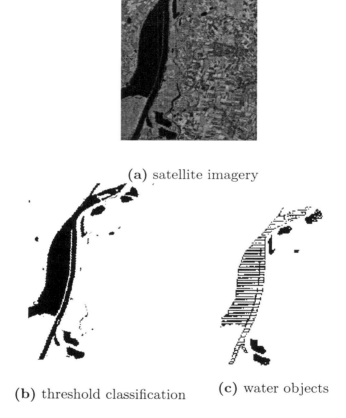

Figure 28.2. Generation of water objects from Landsat TM imagery [GHM+93, with kind permission from Kluwer Academic Publishers]

analog form. This concerns historical data but also a large number of more recent thematic maps, images, and documents. While this is a problem in environmental information management today, it is not going to be a major impediment in the near future. Those historical data sets that are of relevance in current and future applications are rapidly being digitized. This process is supported by the continuous progress in scanning technology as well as by the emergence of a market for digitizing historical data that is dominated by a handful of specialized software companies and publishers. *New* data is almost exclusively captured in some digital format, and it is mainly a question of logistics to make those digital versions available. The World Wide Web will help to speed up this process considerably.

In this paper we concentrate on software techniques for the management of *digital* environmental information. There are essentially two options for storing a given digital data set:

- a *database management system (DBMS)* with a well-defined data model, typically relational, object-relational, or object-oriented
- an application-specific *file system*, as it is still used by many geographic information systems (GIS).

It is important to view this choice independently of the kind of software tools that are used later to query and update the data. While in the past, storage and processing were tightly coupled, more recent systems make a clear distinction between those tasks. This trend is a direct result of the general tendency towards *open systems*. As users demand comfortable interfaces between different hardware and software tools across heterogenous platforms, vendors have been forced to decompose their products along the lines of more narrowly defined functionalities.

GIS in particular used to be monolithic systems, taking care of storage, querying, and visualization of geographic information in a tightly integrated manner. More recently, GIS products tend to follow a toolbox approach instead. Users are increasingly free to choose between a vendor-specific file system and a commercial database system for *storing* the geographic data. On top of the chosen storage system, GIS then provide advanced functionalities to *query, modify,* and *visualize* the data. Other software tools may be added to compute statistical aggregates, to perform fast full-text searches, and so on.

In this paper, we first discuss the strengths and weaknesses of commercial software systems for the management of very large environmental data sets. Section 3.1 starts with a presentation of geographic information systems and a discussion of the major problem spots. Section 3.2

continues with an analysis how conventional database technology supports the special needs of geographic and environmental data management. We then show how some recent developments in database research can be utilized for more efficient data management in this domain. In particular, we discuss the design and implementation of spatial data types and operators, abstract data types, and spatial query languages. Section 3.3 presents the state of the art in spatial access methods. Section 3.4 discusses the basic features of object-oriented databases and their potential for the management of environmental data. Section 3.5 concludes with an overview of likely future developments.

3.1. Data Storage in GIS

Geographic information systems are an essential tool for the management and visualization of large amounts of environmental information. Motivated by the potential of computers for cartographic tasks, research and development in this area started about 25 years ago. In the late 1970s, the first systems entered the market. Most of the early customers were from the public sector. Later on, GIS were also used in the context of management information systems and environmental information systems.

While GIS are a crucial component of environmental information management, they are not the main focus of this paper. In the following we will assume some basic knowledge of GIS and the underlying concepts. Readers not familiar with GIS are advised to consult the related literature. Common references include a two-volume treatment of the subject by Maguire et al. (1991), a collection of introductory readings edited by Peuquet and Marble (1990b), as well as textbooks by Aronoff (1989), Bill and Fritsch (1997), Bill (1996) (in German), Bonham-Carter (1994), Burrough (1986), Star and Estes (1990), and Tomlin (1990). For surveys on database issues in GIS, see the article by Medeiros and Pires (1994) and the collection edited by (Adam and Gangopadhyay 1997). The relationship between GIS and environmental information systems is discussed in a survey article by Bill (1995).

3.1.1 Traditional GIS Functionalities.
The original idea for GIS was to "computerize" the metaphor of a thematic map. As a result, GIS are designed to support the whole life cycle of map data. More generally, "GIS are computer-based tools to capture, manipulate, process, and display spatial or geo-referenced data" (Fedra 1993). This definition should remind the reader of the data flow that we observed for environmental data in Chap. 1. Geographic data is a special and important type of environmental information, and the data flow described

by Fedra can indeed be viewed as a special case of our scheme, although the details are somewhat different at each stage.

With regard to *data capture*, GIS provide software support for digitizing analog map data. This software is typically used in combination with digitizing tables that are equipped with mouse-like devices for tracing shapes on a given analog map. The movements are captured by the system, and whenever the user clicks on the digitizing device, its current location is translated into coordinates and stored in a file. Modern scanning technology has helped to facilitate this somewhat cumbersome task, but so far it has not been possible to automate the process completely. Maps are an important part of our culture, and as a result we have become used to very high standards for accuracy and presentation. Human understanding of maps is still essential to perform the conversion process from analog to digital in such a way that those high standards are met by digital maps as well.

For *data storage*, GIS used to rely exclusively on customized file system solutions. Since the mid-1980s, vendors have started to use commercial relational database systems. For the alphanumeric (non-spatial) part of the GIS data, these have quickly become the state of the art. The spatial data, however, is still mostly held in proprietary file systems. Most of the underlying data models are *layer-based*: information is encoded in a number of thematic maps, such as vegetation maps, soil maps, or topographic maps. With regard to geometry, each such map corresponds to a partition of the universe into disjoint polygons. Each polygon represents a region that is sufficiently homogenous with respect to the theme of the map. Maps may be enhanced by lines and points to represent specific features, such as roads or cities.

Data analysis then mainly consists in intersecting different map layers *(map overlay)* and aggregating the available data in a task-specific manner. Map overlay allows users to connect and interrelate different contents, and visualize the results in new, customized maps. The non-spatial data corresponding to the two input maps needs to be synthesized in a suitable manner, e.g., by creating a new relation whose columns are the union of the columns of the two input relations. To perform aggregations on the non-spatial data, one may use the aggregate functions of the database query language SQL. Additional software modules may be available for more complex statistical aggregation, environmental modeling, and other special functionalities; see Sect. 4.3 for an extensive discussion. Support for *spatial* aggregation, on the other hand is still rudimentary. The automatic aggregation of cartographic data has proven to be extremely difficult (McMaster 1987; 1988), and once again, manual intervention is often required.

3.1.2 Problem Areas.
Most commercial GIS are the result of evolutionary application-oriented developments based on early research activities of the geoscientific community. Database research has not focused on GIS-relevant questions until fairly recently. By the late 1980s, several conceptual deficiencies in the resulting GIS designs had become obvious, especially in comparison to modern relational database systems. Many of these problems can be traced back to their original orientation towards file systems, which provide only rudimentary database functionalities and do not scale up very well to large amounts of data.

Ad Hoc Query Facility. There are typically three types of interface to a relational database system: a programming language interface; an interactive interface that allows the user to state SQL queries in an ad hoc manner; and a graphical interface, whose expressive power is usually limited compared to the other two options. Many older GIS do not provide an equivalent to the interactive SQL interface. Whenever users want to retrieve or update a data item, they have to go through the graphical interface provided by the system. Most of these interfaces are menu-driven and do not have nearly the expressive power that an advanced database query language such as SQL can offer.

Persistence. An important aspect of databases is that they provide persistent storage, i.e., the stored data is available permanently to be used by different applications and ad hoc queries. For that purpose, a database system has to store not only some isolated pieces of raw data but also some structural information to aid the interpretation of the data records. In relational databases, for example, the system maintains catalog information about the different tables in the database in a special relation (the *relation* relation). GIS traditionally store their data on some proprietary file system, which is kept in persistent storage. The problem is that a lot of the related structural information is hidden inside different GIS *programs* that are not accessible to most users. Any attempts to use the data without those programs and without a full understanding of the underlying structure may lead to serious misinterpretations.

Concurrency. Because they operate in a multi-user environment, data-base systems provide mechanisms to facilitate the controlled sharing of data and resources. Certain applications may require that many users be able to read or update a subset of the database concurrently. At other times, one has to make sure that certain data items can only be read or updated by a selected group of individuals. Current database technology provides an array of sophisticated techniques for the management of complex transactions, concurrency, and security. Because of the complexity of these techniques and because of the significant overhead involved, most commercial GIS have originally been designed as single-user systems. Issues of security or concurrency have not been addressed until recently.

Recovery. Simple file systems, as used by many GIS, do not provide any facilities to ensure the consistency of the data after system crashes. If a system crash occurs after a user has manipulated certain data items in main memory, it is not immediately clear which updates have been written back to permanent memory before the crash, and which ones have not. This behavior of the system can lead to inconsistent states of the data. In database systems, on the other hand, it is ensured that a transaction (i.e., a sequence of steps that is just one operation from the user's point of view) is carried out either completely or not at all. If a system crash occurs, the system reestablishes a consistent state after coming back up. Ideally, the user does not even notice that some kind of recovery operation has occurred.

Distribution. In a distributed database, data can be stored at different locations without jeopardizing the integrity of the database. The distribution of the data is usually transparent to the user, and updates and retrievals can be performed exactly as in the non-distributed case. It is now possible, however, to store data where it is needed most in order to improve system performance, and to establish decentralized ownership privileges without affecting the logical consistency of the database. Distributed data management has made significant progress since the introduction of the relational data model, which greatly facilitates data distribution. This is also true for many object-oriented data models. Traditional GIS data models, on the other hand, do not lend themselves easily to distributed data management, and many GIS therefore do not provide any such features.

Data Modeling. Commercial GIS are often too static and inflexible with regard to new applications that may require customized data types and interfaces. Further problems include the modeling of structured objects, and the integration of efficient access methods for very large spatial databases.

Semantic Integrity. Most GIS do not offer functionalities to preserve semantic integrity. For example, it is not possible for a user to specify that a value must be included in a particular value range or that it is valid only in connection with certain other values. Modern DBMS, on the other hand, offer triggers and similar techniques to maintain consistency according to user-defined semantic constraints (Stonebraker 1990, Lohman et al. 1991).

3.1.3 Open GIS: GIS in a Client-Server Environment.

Many of these problems of commercial GIS are now disappearing. The reason is that more and more vendors give up on their traditional strategy of selling closed, proprietary systems, in favor of an open toolbox or "open GIS" approach (Voisard and Schweppe 1994; 1998).

In 1994, an international group of GIS users and vendors founded the Open GIS Consortium (OGC), which has quickly become a powerful interest group to promote open systems approaches to geoprocessing (OGC 1997). The OGC defines itself as a "membership organization dedicated to open systems approaches to geoprocessing." It promotes an *Open Geodata Interoperability Specification (OGIS)*, which is a computing framework and software specification to support interoperability in the distributed management of geographic data. OGC seeks to make geographic data and geoprocessing an integral part of enterprise information systems. Possible applications in the context of environmental information management have been discussed by Gardels (1997). More information about OGC including all of their technical documents are available at the consortium's Web site, http://www.opengis.org.

Due to strong customer pressure, the trend towards such open GIS has increased significantly ever since. Commercial database systems can be integrated into open architectures in a relatively simple manner. A GIS can thus gain directly from the traditional strengths of a modern database system, including an SQL query facility, persistence, transaction management, and distribution. Most GIS vendors have recognized these advantages and offer interfaces to one or more commercial database systems. Moreover, it is often possible to purchase the chosen DBMS bundled together with the GIS at a special price. Although object-oriented databases would often offer superior functionalities, most interfaces are to relational systems, due to customer demand.

Until recently, the database systems were used only for the *non-spatial* data, while the spatial data remained in proprietary file systems. Previous versions of ESRI's GIS ARC/INFO (Morehouse 1985, Peuquet and Marble 1990a, ESRI Inc. 1998) were a typical example of this approach. For the non-spatial data, the user could choose between a commercial relational database system or ESRI's file system solution (INFO). The spatial data was always stored under ESRI's proprietary system (ARC).

In those cases where the *spatial* data is stored in a relational database system as well, the database usually just provides *long fields* (also called *binary large objects (BLOBs)* or *memo fields*, cf. Sect. 3.2.2) that serve as containers for the geometric data structures. Those are in turn encoded in a proprietary spatial data format. The database system can not interpret the content; it can therefore not provide any data-specific support at the indexing and query optimization level. What the user gains, however, is increased data security and concurrency, as well as accessibility of the spatial data via the ad hoc query interface.

A typical representative of this approach is Siemens Nixdorf's GIS SICAD/ open (Siemens Nixdorf Informationssysteme AG 1998). For data storage, SICAD/open offers a component called GDB-X (Ladstätter 1997) that provides an interface to several commercial relational database systems (currently Oracle or Informix). As in previous SICAD versions, both spatial and non-spatial data are stored in the same database in an integrated manner. For the spatial data, the relations serve as containers that manage the geometries as unstructured BLOBs.

More recent versions of ARC/INFO also follow this paradigm. ESRI's Spatial Database Engine (SDE) provides the means to store and manage spatial data within a commercial relational database. Users can currently chose between Informix, Oracle, Sybase, and IBM's DB2 (ESRI Inc. 1997c).

Yet another direction is pursued by the GIS vendor Smallworld Systems. Their object-oriented Smallworld GIS stores both spatial and non-spatial data in a proprietary data manager called Version Managed Data Store (VMDS) (Batty and Newell 1997). VMDS provides efficient support for version management and long transactions – areas where commercial relational databases are still notoriously weak. External applications can access VMDS data either via an SQL server or via an application programming interface (API). Vice versa, it is possible to access commercial relational databases from Smallworld GIS. When creating a new table, users can choose whether to store it in VMDS or in some commercial DBMS linked to the GIS. Both kinds of tables can be accessed the same way, using the object-oriented concept of overloading. The main difference is that updates to the relational database are visible immediately to all users, whereas updates made to VMDS are only visible in the version in which they were made.

The ultimate open GIS would be a collection of specialized services in an electronic marketplace (Voisard and Schweppe 1994, Abel 1997, Voisard and Schweppe 1998). It is not clear at this point whether this is really what the customer wants. Advantages include vendor independence and scalable cost structures. In theory, users just buy the services they need without any unnecessary extras. On the other hand, this kind of toolbox approach requires considerable expertise on the users' part to select and assemble the right services for their purposes. In a sense, such an approach would be a reversal from the "turnkey" philosophy that says that users should have the full functionality of a system available immediately after installation (Theriault 1997).

3.2. Spatial Database Systems

For storing spatial data efficiently, database systems still have to overcome some of their notorious weaknesses. Spatial database systems represent an attempt by the database community to provide suitable data management tools to developers and users in application areas that deal with spatial data. Note that we consider a spatial database as an enabling technology that serves as a basis for the development of application systems – such as a CAD system or a GIS. As noted by Güting (1994), "it is not claimed that a spatial DBMS is directly usable as an application-oriented GIS."

The data management requirements of spatial applications differ substantially from those of traditional business applications. Business applications tend to have simply structured data records, each occupying just a few bytes of memory. There is only a small number of relationships between data items, and transactions are comparatively short. Relational database systems meet these requirements extremely well. Their data model is table-oriented, therefore providing a natural fit to business requirements. By means of the transaction concept, one can check integrity constraints and reject inconsistencies.

With regard to spatial applications, however, conventional DBMS concepts are not nearly as efficient. Spatial databases contain multi-dimensional data with explicit knowledge about objects, their extent, and their position in space. The objects are usually represented in some vector-based format, and their relative position may be explicit or implicit (i.e., derivable from the internal representation of their absolute positions). To obtain a better understanding of the requirements in spatial database systems, let us first discuss some basic properties of spatial data:

1 Spatial data objects often have a *complex structure*. A spatial data object may be composed of a single point or several thousands of polygons, arbitrarily distributed across space. It is usually not

possible to store collections of such objects in a single relational table with a fixed tuple size.

2. Spatial data is *dynamic*. Insertions and deletions are interleaved with updates, and data structures used in this context have to support this dynamic behavior without deteriorating over time.

3. Spatial databases tend to be *large*. Geographic maps, for example, typically occupy several gigabytes of storage. The seamless integration of secondary and tertiary memory is therefore essential for efficient processing (Chen et al. 1995).

4. There is *no standard algebra* defined on spatial data, although several proposals have been made in the past (Scholl and Voisard 1990, Güting 1989, Güting and Schneider 1993). This means in particular that there is no standardized set of base operators. The set of operators heavily depends on the given application domain.

5. Many spatial operators are *not closed*. The intersection of two polygons, for example, may return any number of single points, dangling edges, or disjoint polygons. This is particularly relevant when operators are applied consecutively.

6. Although computational costs vary, spatial operators are generally *more expensive* than standard relational operators.

7. Spatial data management suffers from an impedance mismatch between its theoretical requirements for *infinite accuracy* and the limited accuracy provided by computers.

Recent database research has helped to solve many related problems. This includes both extensions to the relational data model (Stonebraker and Rowe 1986, Kemper and Wallrath 1987, Haas et al. 1990, Stonebraker et al. 1990, Schek et al. 1990, Silberschatz et al. 1991, Gaffney 1996) and the development of flexible object-oriented approaches (Orenstein 1990, Worboys et al. 1990, Deux et al. 1990, Scholl and Voisard 1992, Bancilhon et al. 1992, Shekhar et al. 1997). We review some of the most relevant contributions in the remainder of this section. For more detailed information, the reader may consult the proceedings of the *Symposia on Spatial Databases (SSD)*, which have been held bi-annually since 1989 (Buchmann et al. 1990, Günther and Schek 1991, Abel and Ooi 1993, Egenhofer and Herring 1995, Scholl and Voisard 1997). There are also several survey articles (Günther and Buchmann 1990, Güting 1994, Medeiros and Pires 1994) and textbooks (Samet 1990b;a, Laurini and Thompson 1994, Scholl et al. 1996) on the subject.

3.2.1 Spatial Data Types and Operators.
An essential weakness in traditional commercial databases is that they do not provide any spatial data types. Following their orientation towards classical business applications, they may sometimes offer non-standard types such as *date* and *time* in addition to the classical data types *integer, real, character*, and *string*. Spatial data types, however, are not included in any of the standard commercial DBMS. On the other hand, such data types are a crucial requirement when it comes to processing geographic and environmental data.

For vector data, there have been several proposals on how to define a coherent and efficient spatial algebra (Güting 1989, Scholl and Voisard 1992, Egenhofer 1992; 1994, Güting and Schneider 1995, Schneider 1997). It is generally assumed that the data objects are embedded in d-dimensional Euclidean space E^d or a suitable subspace thereof. In this section, this space is also referred to as *universe* or *original space*. Any point object stored in a spatial database has a unique location in the universe, defined by its d coordinates. Unless the distinction is essential, we use the term *point* both for locations in space and for point objects stored in the database. Note, however, that any point in space can be occupied by several point objects stored in the database.

A *(convex) d-dimensional polytope* P in E^d is defined as the intersection of some finite number of closed halfspaces in E^d, such that the dimension of the smallest affine subspace containing P is d. If $a \in E^d - \{0\}$ and $c \in E^1$ then the $(d-1)$-dimensional set $H(a,c) = \{x \in E^d : x \cdot a = c\}$ defines a *hyperplane* in E^d. A hyperplane $H(a,c)$ defines two closed halfspaces, the *positive halfspace* $1 \cdot H(a,c) = \{x \in E^d : x \cdot a \geq c\}$, and the *negative halfspace* $-1 \cdot H(a,c) = \{x \in E^d : x \cdot a \leq c\}$. A hyperplane $H(a,c)$ *supports* a polytope P if $H(a,c) \cap P \neq \emptyset$ and $P \subseteq 1 \cdot H(a,c)$, i.e., if $H(a,c)$ embeds parts of P's boundary. If $H(a,c)$ is any hyperplane supporting P then $P \cap H(a,c)$ is a *face* of P. The faces of dimension 1 are called *edges*; those of dimension 0 *vertices*.

By forming the union of some finite number of polytopes $Q_1, ..., Q_n$, one obtains a *(d-dimensional) polyhedron* Q in E^d that is not necessarily convex. Following the intuitive understanding of polyhedra, one usually requires that the Q_i $(i = 1, ..., n)$ have to be connected. Note that this still allows for *polyhedra with holes*. Each face of Q is either the face of some Q_i, or a fraction thereof, or the result of the intersection of two or more Q_i. Each polyhedron P divides the points in space into three subsets that are mutually disjoint: its *interior*, its *boundary*, and its *exterior*.

One often uses the terms *line* and *polyline* to denote a one-dimensional polyhedron and the terms *polygon* and *region* to denote a two-dimensional polyhedron. If, for each k $(0 \leq k \leq d)$, one views the set of k-dimensional polyhedra as a data type, one obtains the common collection of spatial data types {*point, line, polygon, ...*}. Combined types sometimes also occur. Curved objects can be obtained by extending the definitions given above.

An object in a spatial database is usually defined by several non-spatial attributes and one attribute of some spatial data type. This spatial attribute describes the object's *spatial extent*. Other common terms for spatial extent include *geometry*, *shape*, and *spatial extension*. For the *description* of the spatial extent, one finds the terms *shape descriptor/description*, *shape information*, and *geometric description*, among others.

As there exists neither a standard spatial algebra nor a standard spatial query language, there is also no consensus on a canonical set of spatial operators. Different applications use different operators, although some operators (such as intersection) are generally more common than others. Queries are often expressed by some extension of SQL that allows abstract data types to represent spatial objects and their associated operators; see Sect. 3.2.3 for a discussion. The result of a query is usually a set of spatial objects.

In the following, we give an informal definition of several common spatial database operators. The operators are grouped into six classes, depending on their respective input and output behavior (their *signature*). At least one of the operators has to be of a spatial data type. The input behavior refers to whether it is a unary, binary, or (in rare cases) n-ary $(n > 2)$ operator, as well as to the type of its operands. The output behavior refers to the type of result.

Class U1: Unary operators with a Boolean result. These operators test a given spatial object for some property of interest. Examples include *triangle* or *convex*. They are often implemented ad hoc to serve some application-specific requirements. These operators are usually not part of the standard system architecture.

Class U2: Unary operators with a scalar result. These operators map a spatial object into a real or integer number. Examples include *dimension* or *volume*. Some operators in this class (such as *volume*) are an essential part of spatial information systems in general.

Class U3: Unary operators with a spatial result. The most important operators in this class are the *similarity operators* that map a spatial object into a similar object: *translation, rotation,* and *scaling*.

Class B1: Binary operators with a Boolean result. These operators are known as *spatial predicates* or *spatial relationships*. They take two spatial objects (or sets of spatial objects) as input and produce a Boolean as output. Several authors have tried to structure this range of operators (Pullar and Egenhofer 1988, Egenhofer 1990, Worboys 1992, Güting 1994). A common classification distinguishes between topological relationships, direction relationships, and metric relationships.

Topological relationships are invariant with respect to topological transformations, such as translation, rotation, and scaling. Examples are *intersects*, *contains*, *adjacent*, and *is_enclosed_by*.

Direction relationships refer to the current location of spatial objects and are therefore sensitive to some topological transformations, in particular rotations. Typical examples are *northwest_of* or *above*.

Metric relationships are sensitive to most topological transformations. They are in particular not invariant with respect to rotations and scalings. A typical example is *distance* < 10.

For the evaluation of spatial predicates (i.e., binary operators with a Boolean result) in a database context, the *spatial join* operator has been introduced. In analogy to the classical join operation, it denotes the combination of two classes of spatial objects based on some spatial predicate. More formally, a spatial join takes two sets of spatial objects as input and produces a *set of pairs* of spatial objects as output, such that each pair fulfills the given spatial predicate. Examples include:

- Find all houses that are less than 10 km from a lake.

- Find all buildings that are located within a biotope.

- Find all schools that are more than 5 km away from a firestation.

All these queries combine one class of objects (e.g., *houses*) with another class of objects (e.g., *lakes*) and select those mixed pairs that fulfill the given predicate (e.g., *are less than 10 km from*).

Class B2: Binary operators with a scalar result. These operators take two spatial objects (or sets of spatial objects) as input and compute an integer or real number as a result. The *distance* operator is a typical representative of this class.

Class B3: Binary operators with a spatial result. This set of operators comprises a wide range of different operators. Some of those most frequently used in practice include set operators and search operators.

The class of *set operators* comprises the union, difference, and intersection of two spatial objects or sets of objects.

To avoid dangling edges and similar anomalies, Tilove (1980) introduced a slight variation of the standard set operators called *regularized set operators*. These regularized operators first compute the standard union, intersection, or difference of the operands, then compute its interior, and finally add the boundary to the result. This way one always obtains a closed spatial object that has the same dimension as the two inputs. This approach, however, is not a panacea. There are numerous applications, especially in geographic and environmental information management, where the user is interested in all or at least some of the lower-dimensional parts of the result. It is therefore necessary to discuss those potential problems with the user before choosing a particular implementation.

Map overlays are an important application of set operators. A map overlay is nothing but an intersection operation performed on two sets of polygons (the *maps*) (Kriegel et al. 1991).

During the 1980s and 1990s, major progress has been made with regard to efficient algorithms for the computation of set operators. Especially the problem of computing polygon intersections has received a lot of attention in the computational geometry literature (Preparata and Shamos 1985, Edelsbrunner 1985, Günther and Wong 1991).

The class of *search operators* concerns spatial search in a possibly large set of spatial objects. Formally speaking, a search operator takes two inputs. One is a set of spatial objects (the database), the other one is a single spatial object (the search pattern). Figure 28.3 illustrates some of those queries. The two most common search operations are the *point query* (Fig. 28.3a) and the *range query* (Fig. 28.3b). The point query asks for all spatial objects in the database that contain a given search point. The range query requests those objects that overlap a given search interval. Sometimes the search interval is replaced by a search object of arbitrary polygonal or polyhedral shape (*region query,*). Common variations include the query for objects that are *near* a given search object, that are *adjacent* to a search object (*adjacency query,*), that are *contained* in a given interval (*containment query,*), or that *contain* a given interval (*enclosure query,*). There has been a considerable amount of research on access methods to support the computation of search queries. Section 3.3 gives a detailed overview.

3.2.2 Implementation Issues.

For the efficient computation of spatial operators one requires special implementations of the spatial data types mentioned above. Moreover, it is sometimes useful to repre-

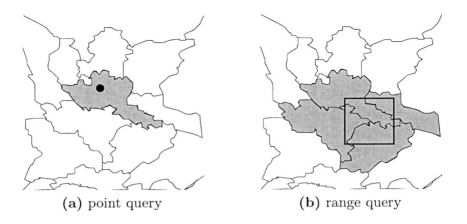

(a) point query (b) range query

Figure 28.3. Spatial search queries (Gaede and Günther 1998)

sent the same spatial object in more than one way in order to represent a broad variety of spatial operators. A polygon, for example, may be represented as a vertex list – one of the most common representations. A vertex list is a list of the polygon's vertices, such as [(1,1),(5,1),(4,4)]. The vertex list is well suited to support similarity operators. A translation, for example, corresponds to the addition of the translation vector to each of the coordinates. A scaling corresponds to a scalar multiplication, and a rotation to a matrix multiplication. The vertex list representation of a polygon is also reasonably well suited to support set operators; it is the input representation of choice for most common algorithms.

Problems with that particular representation include the fact that it is *not unique*. The triangle described above could equally be described by the lists [(5,1),(4,4),(1,1)] and [(4,4),(1,1),(5,1)]. If one drops the implicit assumption that all listed points have to be vertices, one obtains an infinite number of alternative representations, such as [(2.5,2.5), (1,1), (5,1), (4,4)]. Furthermore, there are *no invariants* with respect to set operations; a translation, rotation, or scaling changes each single element of the representation. This also means that it is not easily possible to recognize whether two vertex lists represent congruent or similar polygons. Finally, it is not possible to recognize easily whether a vertex list corresponds to a self-intersecting line or whether it represents a polygon with holes. Depending on the current definition of a polygon, such vertex lists may be *invalid* representations.

Regardless of which representation is chosen, one has to map it into the data model of the given database. In classical relational environments, there are two options. First, the geometric representation may be hidden in a long field. This means that one of the columns in the relation is declared to have variable (in theory, infinite) length. The geometric representation is then stored in such a long field in a way that the application programs can interpret. The database system itself is usually not able to decode the representation. It is therefore not possible, for example, to ask SQL queries against that column. In the case of a vertex list, a typical relation *Polygon* could look as in Table 28.2.

Table 28.2. Polygon: Relation with a long field

ID	Color	Shape
2	blue	[(1,1),(2,7),(3,9),(7,9)]
4711	red	[(2,1),(4,2),(1,5)]
...	...	...

An alternative to this approach is to exclude the geometry completely from the database system and just give the name of the *external file* where it is stored; see Table 28.3 for an example.

Table 28.3. Polygon: Relation with pointers to external files

ID	Color	Shape
2	blue	/usr/john/pol2
4711	red	/usr/john/pol4711
...	...	...

Both approaches are somewhat problematic because they rely on software modules external to the database in order to interpret the representation. The database system itself does not have enough information to evaluate it, and it therefore has no concept of the geometry and topology of the stored objects. There is no database support for geometric operations. Furthermore, it is difficult to control redundancy because common components (e.g., shared corner points) cannot be extracted by the database. They are hidden somewhere within the long field. In the second option, the spatial information is not even protected by the standard database techniques of recovery and concurrency. It is not subject to transaction management because it is stored outside the database.

In a GIS context, the first option corresponds to the solution where the spatial data is stored inside a database but remains essentially a black box for the database management system. This approach is now favored by most major GIS, including ESRI's ARC/INFO and Siemens Nixdorf's SICAD. It also complies with the OpenGIS Simple Features Specifications proposed by the Open GIS Consortium (cf. Sect. 3.1.3). Many vendors provide efficient implementations of this design although the basic conceptual drawbacks remain.

Abstract data types (ADTs) provide a more robust, if not more efficient, way to integrate complex types into a database system. The basic idea is to encapsulate the implementation of a data type in such a way that one can communicate with instances of the data type only through a set of well-defined operators. The (interior) implementation of the data type and the associated operators are hidden to the (exterior) users. They have no way to review or modify those interior features.

The ADT concept can easily be adapted to the implementation of spatial data types and operators. It may also be used to give experienced users the opportunity to customize a DBMS according to their particular requirements, and to define those data types and operators that are most specific and appropriate for a given application. No DBMS could foresee all those needs. In order to appeal to a broad range of users, DBMS have to be designed independently of any particular application. With ADTs the system needs to offer only a small number of base data types and operators directly. A user may perceive only those base data types plus the ADTs that have been declared specifically for the given application. This view-like approach simplifies using the system, especially for inexperienced users, and helps to reduce training time.

In an SQL-style database environment, the embedding of the ADT concept is possible with minor syntax extensions. To define an ADT *Polygon*, for example, one has to provide some basic information about the implementation, such as the internal space requirements, the default value, and the routines used to input and output objects of that type.

Abstract data types have been shown to greatly enhance data security and to facilitate application programming. They can adapt to user requirements in a flexible manner by encapsulating data structures and operators of arbitrary complexity. Disadvantages of this approach include the duality of the connected programming paradigms: one always has to switch back and forth between the database-internal mode, which typically involves a non-procedural language such as SQL, and the external procedures, which are usually written in a procedural programming language. Furthermore, the internal structure of the data type is lost for the outside application; there is no way to retrieve any structural information from the ADT. This is a problem in particular for the database query optimizer. Without special accommodation, it is impossible for the optimizer to obtain any information about the complexity of the

ADT operators that are included in a given query (Gaede and Günther 1995).

3.2.3 Spatial Query Languages.

As we saw in the previous sections, any serious attempt to manage spatial data in a relational database framework requires some significant extensions at the logical and the physical level. These kinds of extension need to be supported at the query language level as well. Besides an ability to deal with spatial data types and operators, this involves in particular concepts to support the interactive working mode that is typical for many GIS/EIS applications. Pointing to objects or drawing on the screen with the mouse are typical examples of these dynamic interactions. Further extensions at the user interface level include (Voisard 1995): the graphical display of query results, including legends and labels; the display of unrequested context to improve readability; and the possibility of stepwise refinement of the display (logical zooming).

For many years, the database market has been dominated by a single query language: the Structured Query Language SQL. There has been a long discussion in the literature as to whether SQL is suitable for querying spatial databases. It was recognized early on that relational algebra and SQL alone are not able to provide this kind of support (Frank 1982, Härder and Reuter 1985, Egenhofer and Frank 1988, Laurini and Milleret 1988).

In his 1992 paper "Why not SQL!" (Egenhofer 1992), Egenhofer gives numerous examples of SQL's lack of expressive power and limitations of the relational model in general. With regard to the user interface level, Egenhofer notes the difficulties one encounters when trying to combine retrieval and display aspects in a single SQL query. Besides requiring specialized operators, this kind of combination usually leads to long and complex queries. The integration of selection by pointing (to the screen) is also problematic. There is no support in SQL for the stepwise refinement of queries, which is particularly important in a spatial database context. The underlying problem is that SQL does not provide a notion of state maintenance that allows users to interrupt their dialogue at a given point and resume their work later on.

Moreover, SQL does not support the notion of object identity in the presence of value changes. An object is defined only by its values. Object-oriented databases solve this problem by maintaining immutable object identifiers (cf. Sect. 3.4.1).

Finally, the relational model does not provide much support for meta-queries, i.e., queries referring to column names and other database schema information. Partly as a result of that, queries such as "What is this data item?," "Which unit of measurement is used for this item?," "What is the relation between the widths of I-95 and Highway 1?," or "What are possible soil classifications?" are difficult to frame in a relational context.

In some sense, however, with the stellar success of SQL the discussion about its appropriateness has become a moot point. The question is not whether SQL should be used – SQL is and will be used to query spatial databases as well. The question is rather which kind of extensions are desirable to optimize user-friendliness and performance of the resulting spatial data management system.

Various extensions to SQL have been proposed to deal specifically with spatial data, including PSQL (Roussopoulos and Leifker 1984; 1985), Spatial SQL (Egenhofer 1991; 1994), GEOQL (Ooi et al. 1989, Ooi 1990), and the SQL-based GIS query languages for KGIS (Ingram and Phillips 1987) and TIGRIS (Herring et al. 1988). Egenhofer (1992) gives a detailed overview of this work. Table 28.4 summarizes the features provided by those systems.

Table 28.4. SQL extensions to handle spatial data (Egenhofer 1994).

Feature	GEO-QL	Ext'd SQL	PSQL	KGIS	TIG-RIS	Spatial SQL
Spatial ADT	+	+[1,2]	+	+[2]	+[2]	+
Graphical presentation	+	+	+	+	+	+
Result combination	-	-	-	+[3]	-	+
Context	-	-	-	+[3]	-	+
Content examination	-	-	-	-	-	+
Selection by pointing	+	-	-	+	-	+
Display manipulations	-	-	+[4]	-	-	+
Legend	-	-	-	-	-	+
Labels	-	-	+	-	-	+
Selection of map scale	-	-	+	-	-	+
Area of interest	-	-	-	+	-	+

[1] Only spatial relationships
[2] No data definition
[3] Only for context
[4] As part of the picture list in the on clause

3.3. Multidimensional Access Methods

An important class of spatial operators that needs special support at the physical level is the class of *search operators* described above. Retrieval and update of spatial data is usually based not only on the value of certain alphanumeric attributes, but also on the spatial location of a data object. A retrieval query on a spatial database often requires the fast execution of a geometric search operation such as a point or range query. Both operations require fast access to those data objects in the database that occupy a given location in space.

To support such search operations, one needs special *multidimensional access methods*. This section gives a brief overview of the main problem areas in this domain. For more details, we refer the reader to the extensive survey article by Gaede and Günther (1998).

3.3.1   Fundamentals.   The main problem for the design of multidimensional access methods is that *there exists no total ordering among spatial objects that preserves spatial proximity*. In other words, there is no mapping from two- or higher-dimensional space into one-dimensional space such that any two objects that are spatially close in the higher-dimensional space are also close to each other in the one-dimensional sorted sequence.

This makes the design of efficient access methods in the spatial domain much more difficult than in traditional databases, where a broad range of efficient and well-understood access methods is available. Classical examples of such *one-dimensional access methods* include linear hashing (Litwin 1980, Larson 1980), extendible hashing (Fagin et al. 1979), and the B-tree (Bayer and McCreight 1972, Comer 1979). These methods are an important foundation for almost all multidimensional access methods.

A natural approach to handle multidimensional search queries consists in the consecutive application of such single key structures, one per dimension. As Kriegel (1984) has pointed out, however, this approach can be quite inefficient. Since each index is traversed independently of the others, we cannot exploit the possibly high selectivity in one dimension for narrowing down the search in the remaining dimensions. Another interesting approach is to extend hashing simply by using a hash function that takes a d-dimensional vector as argument. A structure based on this idea is the grid file (Nievergelt et al. 1984). Unfortunately this approach sometimes suffers from superlinear directory growth.

As these few examples already demonstrate, however, there is no easy and obvious way to extend single key structures in order to handle multidimensional data. On the other hand, there is a great variety of requirements that multidimensional access methods should meet, based on the properties of spatial data and their applications (Robinson 1981, Lomet and Salzberg 1989):

1 *Dynamics.* As data objects are inserted and deleted from the database in any given order, access methods should continuously keep track of the changes.

2 *Secondary/tertiary storage management.* Despite growing main memories, it is often not possible to hold the complete database in main memory. Access methods therefore need to integrate secondary and tertiary storage in a seamless manner.

3 *Broad range of supported operations.* Access methods should not support just one particular type of operation (such as retrieval) at the expense of other tasks (such as deletion).

4 *Independence of the input data.* Access methods should maintain their efficiency even when the input data is highly skewed. This point is especially important for data that is distributed differently along the various dimensions.

5 *Simplicity.* Intricate access methods with special cases are often error-prone to implement and thus not sufficiently robust to be used in large-scale applications.

6 *Scalability.* Access methods should adapt well to growth in the underlying database.

7 *Time efficiency.* Spatial searches should be fast.

8 *Space efficiency.* An index should be small in size compared to the size of the data set.

9 *Concurrency and recovery.* In modern databases where multiple users concurrently update, retrieve, and insert data, access methods should provide robust techniques for transaction management without significant performance penalties.

10 *Minimum impact.* The integration of an access method into a database system should have minimum impact on existing parts of the system.

As for time efficiency, elapsed time is obviously what the user cares about, but one should keep in mind that the corresponding measurements depend greatly on implementation, hardware utilization, and other details. In the literature, one therefore often finds a seemingly more objective performance measure: the number of disk accesses performed during a search. This approach, which has become popular with the B-tree, is based on the assumption that most searches are I/O-bound rather than CPU-bound – an assumption that is not always true in spatial data management, however. In applications where objects have complex shapes, the refinement step can incur major CPU costs and change the balance with I/O (Gaede 1995, Hoel and Samet 1995). Of course, one should keep the minimization of disk accesses in mind as *one* design goal. Practical evaluations, however, should always give some information on elapsed times and the conditions under which they were achieved. A major design goal of multidimensional access methods is to meet the performance characteristics of one-dimensional B-trees: access methods should guarantee a logarithmic worst-case search performance for *all* possible input data distributions regardless of the insertion sequence.

A common approach to meet the requirements listed above consists in a two-step approximation approach (Fig. 28.4). The idea is to abstract from the actual shape of a spatial object before inserting it into an index. This can be achieved by approximating the original data object with a simpler shape, such as a bounding box or a sphere. Given the minimum bounding interval $I_i(o) = [l_i, u_i]$ ($l_i, u_i \in E^1$) describing the extent of the spatial object o along dimension i, the d-dimensional *minimum bounding box (MBB)* is defined by $I^d(o) = I_1(o) \times I_2(o) \times ... \times I_d(o)$.

An index may only administer the MBB of each object, together with a pointer to the description of the object's database entry (the *object ID* or *object reference*). With this design, the index only produces a set of *candidate solutions*. This step is therefore termed the *filter step*. For each element of that candidate set we have to decide whether the MBB is sufficient to decide that the actual object *must* indeed satisfy the search predicate. In those cases, the object can be added directly to the query result. However, there are often cases where the MBB does not prove to be sufficient. In a *refinement step* we then have to retrieve the exact shape information from secondary memory and test it against the predicate. If the predicate evaluates to true, the object is added to the query result as well, otherwise we have a *false drop*.

3.3.2 A Case Study: The R-Tree.

The *R-Tree* (Guttman 1984) is one of the most common spatial access methods. It corresponds to a hierarchy of nested d-dimensional intervals (boxes). Each node ν of the R-tree corresponds to a disk page and a d-dimensional interval $I^d(\nu)$. If ν is an interior node then the intervals corresponding to the descendants ν_i of ν are contained in $I^d(\nu)$. Intervals at the same tree level may overlap. If ν is a leaf node, $I^d(\nu)$ is the d-dimensional minimum bounding box (MBB) of the objects stored in ν. For each object in turn,

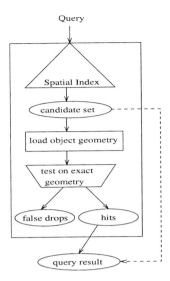

Figure 28.4. Multi-step spatial query processing (Brinkhoff et al. 1994)

ν only stores its MBB and a reference to the complete object description. Other properties of the R-tree include:

- Every node contains between m and M entries unless it is the root. The lower bound m prevents the degeneration of trees and ensures an efficient storage utilization. Whenever the number of a node's descendants drops below m, the node is deleted and its descendants are distributed among the sibling nodes (*tree condensation*). The upper bound M can be derived from the fact that each tree node corresponds to exactly one disk page.

- The root node has at least two entries unless it is a leaf.

- The R-tree is height-balanced, i.e., all leaves are at the same level. The height of an R-tree is at most $\lceil \log_m(N) \rceil$ for N index records ($N > 1$).

Searching in the R-tree is similar to the B-tree. At each index node ν, all index entries are tested for whether they intersect the search interval I_s. We then visit all child nodes ν_i with $I^d(\nu_i) \cap I_s \neq \emptyset$. Due to the overlapping region paradigm, there may be several intervals $I^d(\nu_i)$ that satisfy the search predicate. Thus, there exists no non-trivial worst-case bound for the number of pages we have to visit. Figure 28.5 gives an example R-tree. Each m_i is the MBB of a polygonal data object r_i. A point query with search point X results in two paths: $R_8 \to R_4 \to m_7$ and $R_7 \to R_3 \to m_5$.

Because the R-tree only manages MBBs, it cannot solve a given search problem completely unless, of course, the actual data objects are interval-shaped. Otherwise the result of an R-tree query is a set of candidate objects, whose actual spatial extent then has to be tested for intersection with the search space (cf. Fig. 28.4). This step, which may cause additional disk accesses and considerable computations, has not been taken into account in most published performance analyses (Guttman 1984, Greene 1989).

To insert an object o, we insert the minimum bounding interval $I^d(o)$ and an object reference into the tree. In contrast to searching, we traverse only a single path from the root to the leaf. At each level we choose the child node ν whose corresponding interval $I^d(\nu)$ needs the least enlargement to enclose the data object's interval $I^d(o)$. If several intervals satisfy this criterion, Guttman proposes to select the descendant associated with the smallest (d-dimensional) interval. As a result, we insert the object only once, i.e., the object is not dispersed over several buckets. Once we have reached the leaf level, we try to insert the object. If this requires an enlargement of the corresponding bucket region, we adjust it appropriately and propagate the change upwards. If there is not enough space left in the leaf, we split it and distribute the entries among the old and the new page. Once again, we adjust each of the new intervals accordingly and propagate the split up the tree.

As for deletion, we first perform an exact match query for the object in question. If we find it in the tree, we delete it. If the deletion causes no underflow we check whether the bounding interval can be reduced in size. If so, we perform this adjustment and propagate it upwards. On the other hand, if the deletion causes node occupation to drop below m, we copy the node content into a temporary node and remove it from the index. We then propagate the node removal up the tree, which typically results in the adjustment of several bounding intervals. Afterwards we reinsert all orphaned entries of the temporary node. Alternatively, we can merge the orphaned entries with sibling entries. In both cases, one may again have to adjust bounding intervals further up the tree.

In his original paper, Guttman (1984) discusses various policies to minimize overlap during insertion. For node splitting, for example, Guttman suggests several algorithms, including a simpler one with linear time complexity and a more elaborate one with quadratic complexity. Later work by other researchers led to the development of more sophisticated policies. The *packed R-tree* (Roussopoulos and Leifker 1985), for example, computes an optimal partitioning of the universe and a corre-

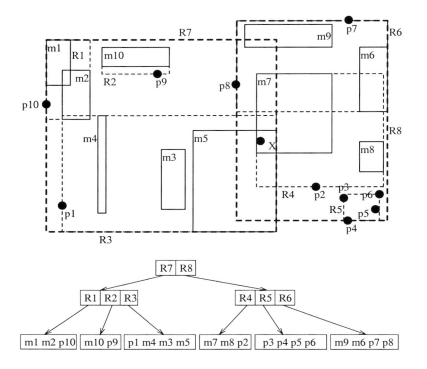

Figure 28.5. R-tree

sponding minimal R-tree for a given scenario. However, it requires all data to be known a priori. Ng and Kameda (1993; 1994) discuss how to support concurrency and recovery in R-trees.

Based on a careful study of the R-tree behavior under different data distributions, Beckmann et al. (1990) identified several weaknesses of the original algorithms. In particular, they confirmed the observation of Roussopoulos and Leifker (1985) that the insertion phase is critical for search performance. The design of the R^*-*tree* therefore introduces a policy called *forced reinsert*: if a node overflows, they do not split it right away. Rather, they first remove a certain number of entries from the node and reinsert them into the tree. Beckmann et al. report performance improvements of up to 50% compared to the basic R-tree.

3.3.3 Outlook. Research in spatial database systems has resulted in a multitude of multidimensional access methods. Even for experts it is becoming more and more difficult to recognize their merits and faults, since every new method seems to claim superiority to at least one access method that has been published previously. In this section we have not tried to resolve this problem but rather to give an overview of the pros and cons of a variety of structures. It will come as no surprise to the reader that at present no access method has proven itself to be superior to all its competitors in whatever sense. Even if one benchmark declares one structure to be the clear winner, another benchmark may prove the same structure to be inferior.

But why are such comparisons so difficult? Because there are so many different criteria for defining optimality, and so many parameters that determine performance. Both the time and space efficiency of an access method strongly depend on the data to be processed and the queries to be answered. An access method that performs reasonably well for iso-oriented rectangles may fail for arbitrarily oriented lines. Strongly correlated data may render an otherwise fast access method irrelevant for any practical application. An index that has been optimized for point queries may be highly inefficient for arbitrary region queries. Large numbers of insertions and deletions may deteriorate a structure that is efficient in a more static environment.

Initiatives to set up standardized testbeds for benchmarking and comparing access methods under different conditions are important steps in the right direction (Kriegel et al. 1990, Günther et al. 1997b) But note that clever programming can often make up for inherent deficiencies of a structure (and vice versa). Other factors of unpredictable impact are the programming language used, the hardware, buffer size, page size, data set, etc. Hence, it is far from easy to compare or rank different access methods. Experimental benchmarks need to be studied with care and can only be a first indicator for usability.

Another interesting direction for future research consists in recognizing common features of different access methods and using them to build configurable methods in a way that leads to modular and reusable implementations. The *Generalized Search Tree (GiST)* of Hellerstein et al. (1995) is such a generic method. A GiST is a balanced tree of variable fanout between kM and M ($2/M \leq k \leq 1/2$), with the exception of the root node, which may have fanout between 2 and M. It thereby unifies disparate structures such as B^+-trees and R-trees and supports an extensible set of queries and data types.

When it comes to technology transfer, i.e. to the use of access methods in commercial products, most vendors resort to structures that are easy to understand and implement. Quadtrees in SICAD (Siemens Nixdorf Informationssysteme AG 1998) and Smallworld GIS (Newell and Doe 1997), R-trees in the relational database system Informix (Informix Inc. 1997), and z-ordering in Oracle (Oracle Inc. 1995) are typical examples. Performance seems to be of secondary importance for the selection, which comes as no surprise given the relatively small differences among methods in virtually all published analyses. The tendency is rather to take a structure that is simple and robust, and to optimize its performance by highly tuned implementations and tight integration with other system components.

Nevertheless, the implementation and experimental evaluation of access methods is essential, as it often reveals deficiencies and problems that are not obvious from the design or a theoretical model. To make such comparative evaluations both easier to perform and easier to verify, it is essential to provide platform-independent access to the implementations of a broad variety of access methods. Some WWW-based approaches may provide the right technological base for such a paradigm change (Günther et al. 1997b;a). Once every published paper includes a URL, i.e., an Internet address that points to an implementation, possibly with a standardized user interface, transparency will increase substantially. Until then, most users will have to rely on general wisdom and their own experiments to select an access method that provides the best fit for their current application.

3.4. Object-Oriented Techniques

An object-oriented database management system (OODBMS) is a DBMS with an object-oriented data model (Dittrich 1986, Atkinson et al. 1989, Dittrich 1990). On the one hand, this simple definition entails the full functionality of a DBMS: persistence, secondary storage management, concurrency control, recovery, ad hoc query facility, and possibly data distribution and integrity constraints. We have already discussed the significance of these features for environmental data management. The second part of the definition refers to an *object-oriented data model*. This implies, among other things: object identity, classes and inheritance, complex object support, and user-defined data types. In geographic and environmental data management, these features are of varying significance. In this section, which is based on a survey article by Günther and Lamberts (1994), we explain those features and discuss their relevance for environmental information systems.

3.4.1 Object Identity. In an OODBMS the existence of an object is independent of its value. In contrast to the philosophy of the relational model, it is possible for two objects to have equal values and nevertheless to be identified unambiguously. Each object is labeled by some unique object identifier (OID) created by the system in order to guarantee systemwide, if not worldwide, uniqueness. It is not visible to the user and does not change during the lifetime of an object.

OIDs are an important concept for geographic and environmental information systems because they can be used to implement complex objects and shared subobjects, and to assign *multiple* representations (e.g., different scales, or raster versus vector) to a single object. Moreover, one can easily distinguish data objects that have been generated and managed at different locations, which is of particular importance in many environmental applications.

3.4.2   Classes and Inheritance.   Object-oriented systems use the concept of the abstract data type (Sect. 3.2.2) for defining the structure of object *classes*. A class is a collection of objects of the same (abstract) data type. They thus all have the same structure. Classes support the two basic concepts underlying abstract data types: *abstraction* and *encapsulation*. An object can only be accessed through the operators defined on its class, i.e., it is only characterized through its behavior. The user is prevented from applying unsuitable operators to the object, and its internal representation is hidden.

Operators (methods) and attributes are attached to a class, which means that they are valid for all objects that belong to it. Classes may form an inheritance hierarchy. This means that all attributes and methods of a class apply to its subclasses as well, unless they are explicitly overwritten.

Figure 28.6 gives an example of an inheritance hierarchy. Rectangles symbolize classes, and ovals represent attributes. The superclass *EnvObject* has two attributes: *Name* and the spatial attribute *Shape*. *EnvObject* has four subclasses: *Biotope*, *Road*, *River*, and *AdminUnit*. Both *AdminUnit* and *Biotope* add some attributes to the ones inherited from *EnvObject*. *AdminUnit* has two subclasses in turn: *City* and *District*. *City* extends the class definition some more, thus resulting in five attributes: *Name, Shape, Population, Mayor,* and *Districts*. *Districts* is a set-valued attribute: it contains all districts that make up the city in question. All five attributes are forwarded in turn to the descendants of *City*, i.e., to *UnivTown* (with the additional attribute *NoStudents*) and to *Spa* (with the additional attribute *Tax*). *Biotope*, on the other hand, adds four attributes to the ones inherited from *EnvObject*: *BiotopeID* is a special application-specific identifier, *ProtStatus* defines the current and intended protection status, *EndType* lists the various types of endangerment, and *ProtSpecies* lists the protected species living in the given biotope.

There may be classes that do not contain any objects because they only serve as a container for a number of subclasses. This allows one to factorize those attributes and methods that the subclasses have in common. Depending on the application, *EnvObject* could be defined as such an *abstract class*. In this case, any environmental object would have to be either a biotope, a road, a river, or an administrative unit.

An object may belong to multiple classes. For example, the city of *Stuttgart* is a university town and a spa, and therefore has attributes *NoStudents* and *Tax* (in addition to the attributes attached to *City*). *Berkeley* is only a university town, and therefore has just the attribute *NoStudents* (in addition to the attributes attached to *City*) and not the attribute *Tax*. The opposite is true for *Baden-Baden*.

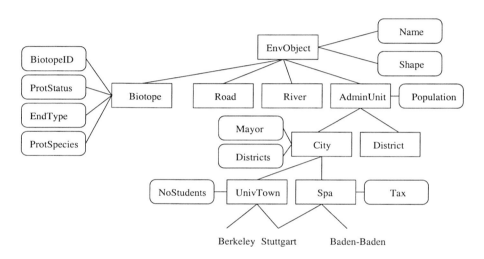

Figure 28.6. Inheritance hierarchy

In environmental information systems, there are many possible uses for inheritance hierarchies. For example, the extensive classification schemes in the natural sciences can easily be translated into inheritance hierarchies. Environmental object taxonomies, such as the ones presented in Sect. 2.1, can also be represented by inheritance hierarchies in a natural manner.

3.4.3 Structural Object-Orientation.

Real-world geographic and environmental objects are often structured in a hierarchical manner. Federal republics like the United States or Germany, for example, consist of several states, which are split up into counties, which are in turn divided into districts, and so on. Cities may consist of districts, streets, blocks, buildings, etc. Hardly any commercial GIS or conventional DBMS can provide sufficient support for these *complex* (also called *composite* or *structured*) objects. Instead the user typically has to split up objects into components until they are atomic, i.e., until the base type level of the GIS or DBMS is reached.

In a relational DBMS, the components may have to be stored in different relations. This kind of fragmentation considerably complicates the modeling and may have a negative impact on the system's performance. The user's ideas are not reflected appropriately in the underlying data structures, which complicates the interpretation of modelled objects. It is not possible to provide for spatial clustering, i.e., to make sure that component parts of the same larger structure are stored close together on the disk.

Object-oriented techniques provide solutions to some of these problems. By analogy to corresponding features in many typed programming languages, most object-oriented systems allow the user to build complex type structures. A data model is called *structurally object-oriented* if it supports the construction of complex objects. This means in particular that attributes of a tuple do not have to be atomic (as in the relational model), but may be composite in turn. To build application-specific complex object structures from atomic types (*integer, character*, etc.), users are offered special *type constructors*, such as *tuple, set*, or *list*.

3.4.4 Behavioral Object-Orientation.

A data model is called *behaviorally object-oriented* if it supports user-defined types, and operators (in this context usually called *methods*) that are applied to these types. Behavioral object-orientation is a direct outcome of the work on integrating abstract data types into databases (cf. Sect. 3.2.2). After specifying a type and its associated operators, users can treat the new type just like a system-defined type. Operators are defined by an interface or *signature* (name, input parameters, output parameters) and a program to compute it. The OODBMS guarantees that operators cannot be applied to inappropriate objects. To apply an operator, one only needs to know its interface; no information is required about its implementation.

Typically, users have little interest in the implementation details of the data types and operators. On the contrary, they usually prefer an encapsulation approach, where abstract operators are the only way to manipulate or communicate with the objects. Encapsulation allows one to specify whether an attribute or method is visible to the user. Moreover, one can define different ways these attributes and methods may be accessed.

3.5. Summary

In this section, we gave a comprehensive overview of storage techniques in geographic information systems and spatial databases. We began with an overview of data storage in GIS (Sect. 3.1), pointing out some of the notorious problem areas. Classical GIS, in which the data is administered in simple file systems, have major difficulties in efficiently managing the large amounts of data that are typical especially for environmental applications. Furthermore, increasing user requirements such as structured object modeling, concurrency, and recovery are difficult to realize in traditional GIS.

Classical database management systems, on the other hand, also seem unable to support complex geographic and environmental applications in an efficient manner. Due to recent research results, however, this is about to change in the near future. Spatial database research has developed powerful techniques to handle complex geometric data types and operators. Work on object-oriented and object-relational database systems provides the means to integrate spatial data management techniques into a commercial setting. This flexibility is a clear advantage compared to the fixed set of data types and operators that is typical for classical commercial DBMS. It can be used in a variety of ways to enhance the functionality and efficiency of an environmental information system.

Section 3.2 described the efforts of database researchers to integrate spatial data types and operators into a classical database framework. Section 3.3 focused on a specific problem area of spatial databases: the development and evaluation of efficient multidimensional access methods. Section 3.4 was devoted to object-oriented approaches to spatial data management.

We conclude that modern database technology is essential for the efficient handling of geographic and environmental data. For the necessary integration of GIS and modern database technology, there are essentially four options:

1 *Extension of an existing GIS by database functionalities.* Earlier versions of Siemens Nixdorf's GIS SICAD are a typical example of this approach. SICAD's proprietary data management system GDB offers advanced database functionalities and serves as integrated storage for both spatial and the non-spatial data.

2 *Coupling of a GIS with a commercial DBMS.* Such a coupling has been common for some time for storing the non-spatial data in a commercial relational database. ESRI's ARC/INFO, for example, has been offering this as an alternative to its proprietary INFO

component. Since the mid-1990s, many vendors started to store spatial data in a relational DBMS as well. In SICAD/open, for example, GDB has been replaced by a solution called GDB-X (Ladstätter 1997) that relies on commercial relational database systems. As in previous SICAD versions, both spatial and non-spatial data are stored in the same database. For the spatial data, however, the relations just serve as containers that manage the geometries as unstructured long fields. ESRI provides a similar solution with its Spatial Database Engine (SDE) (ESRI Inc. 1998; 1997c).

3 *Extension of an existing DBMS by geometric and geographic functionalities.* For this approach the use of an OODBMS seems to be an interesting option. Early prototypes were based on POSTGRES (Oosterom and Vijlbrief 1991), O_2 (Scholl and Voisard 1992), or ObjectStore (Günther and Riekert 1993). The object-relational database company Illustra, which was later bought by Informix, is a successful commercial example. Domain-specific extensions are packaged in specific modules called *DataBlades* (Gaffney 1996, Informix Inc. 1997).

4 *Open toolbox approaches* that see a GIS just as a collection of specialized services in an electronic marketplace (Voisard and Schweppe 1994; 1998). While there is currently no commercial system that strictly follows this architecture, many vendors are starting to integrate similar ideas into their products. GeoServe of Siemens Nixdorf (Ladstätter 1997) and ESRI's OpenGIS proposals (ESRI Inc. 1997a) represent first steps in this direction.

4. Data Analysis and Decision Support

In this section, we discuss how the stored environmental data can be prepared for decision support purposes. This usually involves some more aggregation and detailed analysis of the available data. Compared to the techniques presented in previous chapters, decision support is more target-oriented in the sense that it takes the particular requirements of a given decision-making task into account. As a result, the selection of data sets and analysis techniques is more application-specific than in the case of, say, raw data processing. Related work goes back into the late 1980s (Guariso and Werthner 1989). The topic is increasingly attracting attention among both environmental and management scientists.

We begin with a description of environmental monitoring, which is a particularly important application of data analysis and decision support in the environmental sciences. Section 4.2 gives an overview of simulation models for environmental applications. Section 4.3 lists some of the data analysis tools offered by commercial geographic information systems. Section 4.4 discusses online databases and the impact of the World Wide Web on environmental information management. Section 4.5 gives an overview of environmental information systems in the enterprise. We conclude with a case study: Section 4.6 presents UIS, the environmental information system of the Southern German state of Baden-Württemberg.

4.1. Environmental Monitoring

The collection and monitoring of environmental data is an essential component of any environmental management and protection strategy. Environmental monitoring consists of a continuous evaluation of the incoming data streams. Fritz (1992) and Günther et al. (1995) give an overview of related international activities. The purpose is to recognize unusual developments early on to avoid serious damage. Unusual developments may be event-oriented, such as a sudden increase in the concentration of a dangerous substance, or gradual, such as the slow change of a river's topology. Both kinds of developments are potentially dangerous. Once such a development has been spotted, one has to obtain more detailed information on the potential sources and, possibly, initiate countermeasures.

Most governmental agencies employ a multi-level approach for environmental monitoring. Management of the measuring stations and control of the primary data capture is usually performed by local government agencies, i.e., agencies at the city and county level. Those agencies use the data for tactical tasks, such as short-term resource distribution, quality assessment, pollution detection, and treatment monitoring. Aggregate data is forwarded to state, national, and possibly international agencies for middle- and long-term strategic planning, including policy assessment and legislation.

With regard to the data flow sketched in Chap. 1, this process forms a natural continuation of the data capture and storage phases. It corresponds to a further semantic aggregation, as symbolized by the pyramid in Fig. 28.1. At the technical level, this process can be supported by a modular system architecture. Pham and Wittig (1995), for example, propose a system design for river quality monitoring that is based on four functional entities (Fig. 28.7):

1 Sensor data acquisition

2 Validation procedures

3 Situation description

4 Situation assessment.

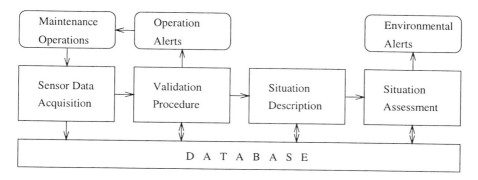

Figure 28.7. Functional modules of a river quality monitoring system (Pham and Wittig 1995)

The first two of those four entities are part of what we termed *data capture*. In the third stage, Pham and Wittig propose to go beyond measurement values in order to obtain a description of the overall situation. This step involves the computation of derived values based on multiple measurements, comparisons with historical data, and similar data fusion operations.

At some point, human judgement will be introduced with the objective of obtaining a coherent assessment of the environmental situation and ongoing developments. If necessary, an environmental alert will be triggered. This last step, termed *situation assessment*, is hard to separate from the situation description stage, as human judgement usually enters the analysis in a gradual manner. In most applications, human judgement is required before a major alert is triggered. This design reflects the current state of related software tools. While useful for the low-level acquisition and aggregation procedures, there rarely exists a situation where the high-level analysis and decision-making can be left to a computer. Given the complexity and inherent uncertainty of environmental data, as well as the political dimension of environmental decision-making, human judgement will remain indispensable for the foreseeable future.

This whole sequence of steps is supported by a comprehensive database or, more typically, a federation of several location- and task-specific databases. These databases contain all the measurement data, procedural data, and aggregate data that is used and generated as part of the process.

It is the last two functional modules that we want to emphasize in this section: situation description and assessment. Both tasks require some higher-level synthesis of the available data and, frequently, an evaluation by human experts. During those stages it is sometimes difficult to separate the data analysis from the decision making itself because the analysis is performed with a particular agenda in mind. If the various steps of the analysis are distributed among a hierarchically organized group of people, such as a government agency, this phenomenon becomes very obvious. Everybody in the hierarchy may have a certain bias on what is important, which may have an impact on the kind of information that is sent higher up. This is not necessarily a bad thing. After all, that is what expert counsel is all about: the selection and analysis of what the expert deems significant. It is important, however, that decision makers are aware of the various sources of error and bias that the information has been exposed to before being made available to them.

4.2. Simulation Models

4.2.1 Background. Models and simulation have long been an important analysis tool in a variety of disciplines. A *model* is an abstract description of a real-life phenomenon. The abstraction usually involves some simplification and results in a formal representation. To obtain insights about the underlying real-life phenomenon, one often performs experiments with the model, hoping that the results of such a *simulation* somehow correspond to the behavior of the system in real life. Many models have been developed specifically for performing simulations on them; such models are termed *simulation models*.

Simulation models have traditionally been one of the most important and demanding computer applications. Especially in the early age of the computer, these applications have been a driving force behind the development of ever more powerful hardware. Even today, simulations are one of the most common applications for supercomputers. To support such computations, there exists a variety of specialized programming languages (such as SIMULA) and commercial simulation software environments (such as Simulink (Scientific Computers GmbH 1997)). Figure 28.8 shows the various connections between modeling, simulation, software implementation, and validation.

Arguably the most notorious applications of simulation models are population dynamics and weather forecasting. Simulation models for population dynamics go back to the 1920s, when Lotka and Volterra formulated their model to simulate the interdependencies between two adverse populations (see, for example, Richter (1985)). Typical parame-

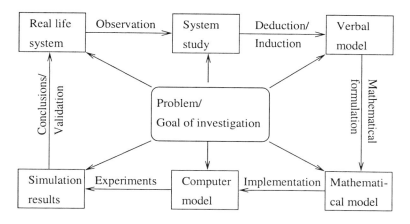

Figure 28.8. Modeling and simulation (Grützner et al. 1995)

ters entering the computation are food supply, or birth and death rates. Meteorological simulation models, usually based on the Navier–Stokes equations on the preservation of mass and impulse, are of more recent origin.

4.2.2 Environmental Applications. From those foundations, environmental information science has developed a great variety of models for different applications. Sydow (1996) lists, among others, the analysis and prevention of acid rain, the evaluation of environmental consequences of a nuclear strike ("nuclear winter"), and the analysis of carbon dioxide and ozone concentration in the stratosphere. Some of these applications help to explain general principles underlying our ecosystem. Others serve specifically to forecast the effect of certain intended actions. Especially the latter kind of models – also called *scenarios* – are of obvious importance to planners and decision makers.

The field of environmental modeling has a long history, and it is beyond the scope of this paper to cover this field in great depth. Good starting points for further reference include the textbook by Bossel (1994), two collections of articles edited by Goodchild et al. (1993; 1996), and the survey articles by Fedra (1993; 1994) and Hilty et al. (1995). Readers of German may also refer to a survey article by Grützner et al. (1995) and to a recent collection edited by Grützner (1997),

There also exists a comprehensive online listing of simulation models in environmental applications: the Register of Ecological Models (REM), maintained by a research group at the University of Kassel (Germany) (Benz and Voigt 1995; 1996). REM is available at http://dino.wiz.uni-kassel.de/ecobas.html. It allows users to search for models by name or by content. Content-oriented search can be structured according to the medium the model is concerned with (*air, terrestrial, freshwater*, or *marine*), and its main subject (*bio-/ geochemical, population dynamics, hydrology, (eco-)toxicology, meteorology, agriculture*, or *forestry*).

4.2.3   Problem Areas.   A key problem in environmental simulation models concerns the match between the data available and the data required by a given model. Environmental measurements are not standardized; they differ both in *what* they capture and *how* they capture it. As for the latter aspect, there are different measuring techniques but also more mundane differences that could easily be harmonized, such as scale and frequency of measurements. The United Nations Environmental Program (UNEP) has established a task force to work on this problem, called UNEP-HEM (Harmonization of Environmental Measurements). Keune et al. (1991) give an overview of the group's mission and concrete plans. While their discussion of data availability problems is somewhat obsolete by now, due to the rise of the Internet, their description of syntactic and semantic heterogeneities is as current now as it was back in 1991.

Another notorious problem area concerns the models' user interfaces and visualization features (Denzer 1993, Grützner et al. 1995, Denzer et al. 1995a). Many important models have been developed incrementally over several decades. They are typically programmed in Fortran and run in a mainframe-based computing environment. In addition, they are often badly documented – in summary, a typical legacy software dilemma. Making such models more user-friendly, easier to maintain, and compatible with open client/server architectures provides a long-term challenge for the modeling community.

The World Wide Web and related tools (such as the Common Gateway Interface (McCool 1994)) provide additional incentives to do so because they greatly facilitate the distribution of software tools. Models coupled with state-of-the-art user interfaces and visualization tools could be packaged as interactive services and made available to the community via the Internet, either as a free service or as a commercial product. For the older generation of Fortran- and mainframe-based simulation software, this will be difficult to achieve. Reimplementations will often be necessary to enter this new era of "Internet marketplaces" (Abel 1997).

4.3.   Data Analysis in GIS

While data analysis has always been an important part of GIS, the breadth and depth of related work has increased considerably since the early 1990s. By extending their functional spectrum beyond the traditional domains of data capture, storage, and visualization, GIS are gradually moving into the mainstream of computing. Rather than providing support just for the geosciences, GIS vendors are trying to position their products as spatial data management components that should be a part of just about any information system architecture – simply because just about any information has a spatial aspect. Interfaces to business software such as SAP's R/3 (SAP AG 1998) and the development of *spatial decision support systems* (Densham 1991, Abel et al. 1992) are among the most visible signs of this trend.

In order to achieve these ambitious goals, GIS vendors have to provide data analysis capabilities that go far beyond simple map overlays (cf. Sect. 3.1). Openshaw (1991) gives a systematic overview of possible analysis capabilities in GIS. He uses the term *spatial analysis method* in a broad sense to define a large class of complex methods to interpret spatial data and obtain decision-relevant information. Back in 1991, his assessment was as follows: "The geographical information revolution demands a new style of spatial analysis that is GIS appropriate and GIS proof. The existing spatial analytical toolbox is largely inadequate, consequently there is an urgent need to create more relevant methods and also to educate users not to expect the impossible when analyzing geographical data. The real challenge is the need to develop new, largely automated, spatial data exploratory techniques that can cope with the nature of both the geographical data created by GIS and the skill base of typical GIS users."

Since then, the situation has improved considerably, and most major GIS vendors have added advanced analysis tools to their products. They have done so in two different ways: either by integrating more complex functionalities directly into the GIS, or by offering interfaces to special-purpose analytical software from third-party vendors. Given the current trend towards "open" GIS (cf. Sect. 3.1.3), the second option is becoming increasingly relevant. Scripting languages, such as ARC/INFO's AML (ARC Macro Language) or ArcView's[2] Avenue, serve to access and combine the various functionalities. In some cases, it is also possible to combine GIS and analysis functionalities by calling the required modules from some standard programming environment (such as a C++ program) via application programming interfaces (APIs). ESRI's Open Development Environment (ODE), for example, provides such an API to selected ARC/INFO functionalities.

In the remainder of this section, we briefly describe some of those analysis features that seem most relevant for environmental applications.

4.3.1 Spatial Operators. An obvious way to improve GIS analysis capabilities is based on more complex spatial operators. In particular, the combination of the distance function with spatial searches (cf. Sect. 3.2.1) yields a powerful set of analysis tools. Typical related queries are:

- Find all yellow houses that are less than 2 km away from Lake Powell.

- Find all hospitals that are more than 3 km away from a bus route.

- Find the firestation closest to my house.

The first two of these queries involve the construction of a *buffer zone*, i.e., a polygon with a possibly complex shape that contains all points whose distance from the given polyline of polygon is less (or more) than a certain distance. The buffer zone then serves as input to a region query (cf. Sect. 3.3). The last query is a nearest-neighbor query that relates a single spatial object (here: the house) to a set of spatial objects (all firestations) to find the one with minimum distance. All queries involve computational geometry algorithms of considerable complexity. If they are applied in a database context, spatial join techniques may be used to optimize their performance.

A related class of analysis tools is based on the spatial modeling of three- or higher-dimensional objects. Three-dimensional modeling has traditionally been an important prerequisite for many applications in the geosciences. Set and similarity operators (cf. again Sect. 3.2.1) need to be computed efficiently in this context, as do more specific operators, such as the generation and manipulation of spline-based surfaces. Raper and Kelk (1991) give an overview of related techniques. Higher-dimensional modeling is also increasingly used to simulate environmental phenomena, e.g., by three-dimensional transport models, or by spatial-temporal models, where one of the dimensions is used to model time (Langran 1992).

4.3.2 Optimization. Other analysis features establish a connection between GIS and optimization and decision support software. Such connections are becoming increasingly important, as environmental management is taking greater advantage of modern decision support technologies (see, for example, Bhargava and Tettelbach (1997)).

Depending on historical reasons and marketing considerations, the decision support software is sometimes packaged as a separate system component, sometimes integrated into the GIS. ESRI, for example, offers a special network analysis package for its ArcView GIS. Their ArcView Network Analyst solves shortest-path and similar graph problems. Siemens Nixdorf, on the other hand, has long specialized in utilities applications, and many of its generic network analysis functionalities are consequently an integral part of its SICAD GIS. In addition, SICAD offers application-specific customizations for electricity networks (SICAD-UT-E), gas networks (SICAD-UT-G), and water networks (SICAD-UT-W).

4.3.3 Statistics and Visualization. While visualization is generally considered a core functionality of GIS, it should nevertheless be noted that most GIS visualization components concentrate on the classical map paradigm. What one sees on the screen is a map enhanced with some theme-specific labeling. The labels may be rather complex depending on the given application. The integration of business graphics, such as pie charts or histograms, for example, can be handled by most major commercial systems. However, if one compares these facilities to the visualization tools known to the statistics community, their deficiencies become clear.

Statistical packages provide advanced features to project and visualize multivariate data in a variety of ways that allow users to quickly recognize dependencies between variables. Certain dynamic techniques give users a sense of the overall shape of the data. Clusters and nonlinearities can easily be identified during such an *exploratory data analysis (EDA)*. GIS users could take advantage of such advanced analytical features in a large variety of applications that involve the interpretation of large spatial data sets. Nevertheless, classical GIS provide no such features.

Haining et al. (1996) describe the state of affairs as of 1996: "In order to carry out a program of SSA [Spatial Statistical Analysis], the user needs to access software that can store and manipulate spatially referenced information, can execute appropriate statistical analysis and finally allow good interactive visualization of the raw data or analysis outcomes. In each case there already exist packages which provide these facilities – in the order listed above GIS, statistical packages [...], and visualization packages. However, no current package provides all three."

In the meantime, more researchers and vendors have recognized the need for a consistent integration of GIS, statistics, and visualization. Most of the resulting system architectures leave the statistical computing and visualization to special-purpose software and construct interfaces between such packages and GIS to produce the desired results. This approach, which represents the current trend towards "open systems," should be seen in contrast to the assumption of some authors that GIS, statistical computing, and visualization are soon going to be integrated into comprehensive scientific computing environments. Rhyne, for example, predicts that the integration of GIS and visualization software will progress quickly along the following four phases: rudimentary (minimal data sharing), operational (consistency of data), functional (transparent communication), and merged (comprehensive toolkit) (Rhyne 1997).

Symanzik et al. (1997b), Cook et al. (1997) used the Remote Procedure Call (RPC) protocol to construct a bidirectional interface between ESRI's ArcView GIS and XGobi (Swayne et al. 1991), an interactive dynamic statistical graphics program. In the terminology of Rhyne, their solution would be somewhere between operational and functional. Both ArcView and XGobi are set up to act as both RPC clients and RPC servers. This allows ArcView to pass data to an XGobi process to perform specific operations, and vice versa. In extension of this work, Symanzik et al. (1997a) later constructed a link between their ArcView/XGobi environment and the statistical computing environment XploRe (Härdle et al. 1995). Symanzik et al. give an overview of the implementation (which is again RPC-based) and discuss an interesting application in satellite imagery interpretation.

Scott (1994) describes a link between ArcView and STATA, and Anselin and Bao (1996) report on an interface between ArcView and the spatial data analysis software SpaceStat. ARC/INFO has been linked to MathSoft's S-PLUS (MathSoft Inc. 1996) and other statistical packages (Haining et al. 1996).

4.3.4 Modeling. A last class of advanced analysis features is based on simulation models. Like statistical software, simulation software is rarely integrated into a GIS directly. One rather relies on interfaces that allow the simulation software to read and write geodata from the GIS. ARC/INFO, for example, maintains a variety of such interfaces, including one to FEFLOW, a simulation system for the modeling of transport phenomena. FEFLOW, which is marketed by a vendor called WASY (WASY GmbH 1997), has been developed independently of ARC/INFO over several years and provides advanced functionalities for the two- and three-dimensional modeling of groundwater flow. Like WASY and ESRI, many vendors of simulation software have close working relationships with GIS vendors to ensure the continued compatibility and efficiency of their interfaces.

Two collections of articles edited by Goodchild et al. (1993; 1996) give a broad overview of related system solutions and developments.

4.4. Environmental Information Online

As noted in Chap. 2, environmental information is increasingly available in digital form. Classical online databases have been available since the early 1980s. Typical contents include fact sheets about chemical substances, information about recycling options, environmentally relevant standards and legislation, or surveys about ongoing research projects and publications. The databases are accessed primarily through special providers and dial-up connections. Users are usually charged a combination of a monthly or yearly base fee and usage-dependent charges. The traditional user community of such databases consists of researchers and practitioners working on environmental issues. Usage by non-experts or the general public has been rare.

As in many other domains, the rise of the World Wide Web has changed this situation fundamentally. Online database providers offer Web-based access to their systems, which has resulted in a considerable enlargement of their traditional user communities. Moreover, the ease with which data can be made available on the Web has increased both the volume and the quality of the data that is available. Many agencies and commercial vendors now offer their data directly on the Web without going through a third-party provider. Advanced functionalities that support the automatic conversion of legacy data sets into Web-compatible formats accelerate this trend. This applies not only to data in traditional office formats (such as MS Word) but also to non-standard data – such as maps.

We continue with a more detailed treatment of online databases, followed by a discussion of the Web and its impact on environmental information management.

4.4.1 Online Databases.

Online databases have been a valuable source of up-to-date information on selected topics since the early 1980s. Environmental scientists have traditionally made heavy use of this medium, both for publishing and for retrieving information (Stoss 1991, Gayle Alston 1991, Gayle Alston and Stoss 1992, Hane 1992).

Access to online databases requires users to register with some service provider, in this context called *host*. The host bills its users periodically. Well-known hosts include Dialog (http://www.dialog.com), Data-Star (http://www.krinfo.com/dialog/publications/data-star-mini-catalogue.html), and STN International (http://www.cas.org/stn.html). Dialog and Data-Star belong to the American media company Knight-Ridder, and STN is run by a consortium consisting of the American

Chemical Society (ACS), the German host FIZ, and the Japan Science and Technology Corporation (JICST).

Note that the host is usually different from the actual information provider. Typical information providers are scientific or economic institutions, or service organizations close to them. The Institute for Scientific Organization (ISI), for example, (http://www.isinet.com) collects data from many sources, reorganizes them, supplies keywords and thesaurus descriptors, and offers the resulting online databases to different hosts for distribution. Their *Current Contents* cover a wide range of topics and are offered by numerous hosts.

Online databases store their data in a record structure with both textual and numeric attributes. Attribute names mostly consist of two letters. Dialog, for example, uses the following acronyms for its most common attributes:

AB: abstract
TI: title
AU: author
DT: document type
DE: descriptor
ID: identifier (for classification purposes)
LA: language
PY: year of publication

There have been several attempts to define a taxonomy of online databases (Ortmaier 1995, Voigt and Brüggemann 1995), depending on their formats and contents. One can basically distinguish three types of databases:

1 *Bibliographic databases* contain catalog entries and possibly abstracts of publications related to a given topic. They closely resemble library catalogs except that their search capacities have traditionally been more powerful. Figure 28.9 gives an example from the database Enviroline. The first two columns contain the attribute names. Enviroline is offered by Dialog, Data-Star, and several other hosts (see, for example, http://www.krinfo.com/dialog/databases/html2.0/bl031.html). It corresponds to the print medium Environment Abstracts and covers a broad range of environmentally relevant information. For more information about Enviroline, see Fig. 28.9.

2 *Literature databases*, also called *full-text databases*, go one step further by offering the complete text of selected publications. Reuters'

```
FN- DIALOG(R)File  40:Enviroline(R)|
CZ- (c) 1995 CIS, Inc. All rts. reserv.|
AN- 00247706
AA-(ENVIROLINE) 93-06159
TI- Toxicity Testing of Sediment Elutrates Based on Inhibition of
    Alpha-Glucosidase Biosynthesis in +i Bacillus licheniformis+r
AU- Campbell, Marjorie Univ of Florida, Gainesville; Bitton,
    Gabriel; Koopman, Ben
CS- Campbell, Marjorie Univ of Florida, Gainesville; Bitton,
    Gabriel; Koopman, Ben
JN- Arch Environ Contam Toxicol
PD- May 93
SO- v24, n4, P469(4)
LA- English
AV- Full text available from Congressional Information Service.
DT- research article
SF- 2 graph(s); 12 reference(s); 3 table(s)
AB- Elutriates were prepared from 66 sediments collected from nine
    hazardous-waste sites in Florida and screened by the Microtox
    and (gr)a-glucosidase biosynthesis (AGB) assays, using +i
    Photobacterium phosphoreum +r and +i Bacillus licheniformis+r,
    respectively. Total concentrations of lead, cadmium, zinc, and
    copper were determined for each elutrate. A linear relationship
    between AGTB and Microtox results was found. The percent
    agreement between the bwo bioassays was 85%. In terms of the
    heavy metals, AGB was more sensitive than Microtox results.
DE- (MAJOR) BIOASSAY; SEDIMENT; MEASUREMENTS & SENSING
DE- (MAJOR) METAL CONTAMINATION; LEAD; CADMIUM; ZINC; COPPER;
    TOXICOLOGY; BIOLOGICAL INDICATORS
DE- (MINOR) TOXIC SUBSTANCES
SH- 02
```

Figure 28.9. Enviroline data record (Ortmaier 1995)

Textline database, offered by Data-Star, is a typical example. Traditionally, full-text databases were restricted to ASCII formats and could therefore not include graphics. Due to advanced communication and multimedia facilities, however, this is changing rapidly.

3 *Factual databases* do not refer to publications like the two previous types of online databases but rather contain original data sets. Typical contents include measurement data, physical and chemical properties of substances, project data, economic indices, or information about companies and institutions. We give three concrete examples.

- The database PDLCOM offers manufacturer's and literature test data on the chemical compatibility and environmental stress crack resistance of plastics. PDLCOM is supplied by a company called Chemical Abstracts Service and offered by STN. PDLCOM can be reached on the Internet at http://info.cas.org/ONLINE/DBSS/pdlcomss.html.

- The research database UFORDAT is maintained by the German Federal Environmental Agency and offered by several hosts including STN (http://info.cas.org/ONLINE/DBSS/ufordatss.html). UFORDAT stores information about environmentally relevant research projects in Germany and beyond.

- CHEMTOX is a database containing information about substances, such as molecular formula and weight, chemical name, boiling and melting point, and toxicity. CHEMTOX is offered by a variety of hosts and can be reached, e.g., at http://www.krinfo.com/dialog/databases/html2.0/bl0337.html.

Requests to online databases are formulated in proprietary query languages that allows users to specify the desired database and to formulate conditions on the data sets to be retrieved. Similar to SQL, this includes the formulation of basic Boolean predicates on the attributes, using the two-letter acronyms discussed above. One can also specify the desired output format. Example query languages are Dialog, DSO, and Messenger. Note that both the query language and the user interface are host-specific.

Many databases offer special tools to support searches, such as indices or thesauri. The *basic index* of a database contains all terms occurring in selected text attributes. Depending on the kind of database, this may only concern special attributes (such as title or abstract) or the whole data set. Users can specify a number of terms to be matched against the basic index. The result of such a query consists of all records where any of the specified terms occurs in one of the text attributes. A *thesaurus* is a controlled domain-specific vocabulary that serves to capture the essential concepts of the chosen domain in a systematical manner. Each data set is indexed using one or more of the thesaurus terms (the *descriptors*). A thesaurus also maintains semantic associations between descriptors, such as synonym, generalization, and specialization relationships. These associations can often be used to improve search performance.

The contents of an online database are described in a structured description called a *bluesheet*. The bluesheet contains information about the database's thematic focus, the data sources, the period to which the data refers, the update cycle, and facts about the information provider.

It also gives an example data set, an explanation of the attributes, an overview of search facilities (indices, thesaurus, etc.), and a listing of possible output formats. To find those databases that are relevant for a given search problem, most hosts offer a metadatabase with bluesheets of all of their databases.

Metadatabases from non-profit hosts often refer to a broader spectrum of online databases. The German Research Center for Environment and Health (GSF), for example, edits a *Metadatabase of Online Databases (DADB)* (Voigt and Brüggemann 1993, Benz and Voigt 1995, Voigt and Brüggemann 1995). DADB concentrates on environmental chemicals and contains descriptions of more than 500 databases. It is complemented by metadatabases about Internet resources (DAIN), CD-ROMs (DACD), and printed documents (DALI).

Another important source of metainformation is the European Commission Host Organization (ECHO). Their database *IM Guide* contains a large number of references to online databases and CD-ROMs. ECHO can be used free of charge and is available on the Internet at http://www.echo.lu.

Traditionally, queries were forwarded to online database via a dial-up connection to the host. Since the mid-1980s, Internet-based connections have become more common, starting with simple protocols such as FTP, Telnet, and Gopher. By now, many online databases offer comfortable Web interfaces based on the Hypertext Transfer Protocol (HTTP). Online databases were thus able to take advantage of the great popularization of the Internet. While there were almost exclusively used by specialists in the past, they now have the potential to become information providers for the mass market.

4.4.2 Environmental Information on the Web. The rise of the World Wide Web has greatly enhanced both the quality and the quantity of environmental data available. Large parts of these enhancements are offered to users practically without cost. With regard to quality, the World Wide Web offers excellent facilities for managing multimedia data, which is of obvious relevance for environmental applications. Regarding quantity, it is mainly the ease with which one can post information that causes the current exponential growth of available data.

Public agencies, for example, are increasingly recognizing the Web as a simple and cheap medium to distribute their data to other government agencies and the general public. Environmental agencies are often leading this movement; they usually have large data sets that are publicly available in principle, and there is an increasing demand on the part of the public to review them. Other major providers of environmental data include GIS vendors and companies who sell raw data, such as satellite imagery or measurement series. Besides their regular products, they usually offer some promotional material free of charge. Finally there are universities and other institutions providing educational material, again mostly free of charge. The great majority of courses offered relates to geographic information systems.

This section is not meant to provide a comprehensive list of Web sites devoted to environmental information. The goal is rather to describe various uses of the Internet to manage and distribute environmental data, and to illustrate these concepts by a few examples. For more material, the reader may consult one of the many search engines available on the Internet.

The most obvious way to find environmentally relevant information on the Web is to use a general search engine, such as Altavista (http://altavista.digital.com) or Lycos (http://www.lycos.com). One will likely experience the notorious disadvantages of this type of search tool: depending on the specificity of the query, users are often overwhelmed by a large number of answers. Only a small number of the cited documents may actually correspond to the objectives of the user. Nevertheless, with a query that is reasonably specific one will at least obtain several starting points for a more targeted search.

Another approach is to use a catalog or metainformation system that is geared specifically to environmentally relevant material. Like generic search engines, many of these systems allow keyword-oriented access and the formulation of simple queries that include Boolean operators (AND, OR, etc.). Some also provide a thesaurus. Examples include (all URLs have to be prefixed by http://):

- AlfaWeb Hazardous Substance Information System (www.iai.fzk.de/~weidemann/lfu/lfu.htm)

- ASK – Global Change Directory of Information Services (ask.gcdis.usgcrp.gov:8080)

- Brown is Green Resource Conservation Program (www.brown.edu/Departments/Brown_Is_Green)

- Catalogue of Data Sources (www.mu.niedersachsen.de/cds/webcds)

- Central European Environmental Data Request Facility (www.cedar.univie.ac.at)

- Cygnus Group – Integration of Environmental and Business Concepts (www.cygnus-group.com)

- DAIN – Internet Resources on Environmental Protection (dino.wiz.uni-kassel.de/dain.html)
- Earth Observing System Data and Information System (spsosun.gsfc.nasa.gov/EOSDIS_main.html
- Earth Pages (starsky.hitc.com/earth/earth.html)
- EcoWeb (ecosys.drdr.virginia.edu/EcoWeb.html)
- Environmental News Network (www.enn.com)
- Envirolink (www.envirolink.org)
- European Commission Host Organisation (www.echo.lu)
- Global Change Master Directory (gcmd.gsfc.nasa.gov)
- Global Network for Environmental Technology (gnet.together.org)
- Government Information Locator Service (www.epa.gov/gils)
- Landsat Pathfinder (amazon.sr.unh.edu/pathfinder1/index.html)
- National Environmental Data Index Catalog (www.nedi.gov/NEDI-Catalog)
- National Environmental Satellite Data and Information Service (ns.noaa.gov/NESDIS/NESDIS_Home.html)
- Register of Ecological Models (dino.wiz.uni-kassel.de/ecobas.html)
- UDK Austria (udk.bmu.gv.at)
- WWW Virtual Library Environment (earthsystems.org/Environment.shtml).

While many of these systems have been developed specifically for the Web, others are Web versions of older metainformation systems, such as library catalogs. Most of the materials indexed by these older systems are either paper documents (books, articles, etc.) or material that is digital but not available on the Internet (such as CD-ROMs). As more and more data sources are available online, however, these systems gradually assume search engine functionalities. As a result, the differences between search engines, online catalogs, and metainformation systems are gradually disappearing. Section 5 discusses this trend in more detail and gives several examples of metainformation systems in practice.

More advanced search and presentation functionalities are available for specific types of data. As mentioned in Sect. 2.4, for example, a company called Earth Observation Sciences (EOS) offers a Catalogue and Browse System (CBS) to provide content-oriented access to selected aerial and satellite imagery (http://www.eos.co.uk). NASA will soon offer content-oriented Web access to its well-known Earth Observing System Data and Information System (EOSDIS), which contains large amounts of imagery from a variety of satellite systems (United States National Aeronautics and Space Administration 1998, Doan et al. 1997).

Many GIS vendors have started to sell special products to support their users in publishing maps on the Web. ESRI's MapObjects Internet Map Server, for example, helps Web developers to integrate spatial data and GIS functionalities into Web sites (ESRI Inc. 1997b). The ArcView Internet Map Server focuses on a wider audience by supporting users to publish any kind of ArcView GIS information on the Web. Similar products from other vendors include MapInfo's ProServer, or Genasys's Web Broker.

Note that map servers are usually much more powerful than simple image servers. On the one hand, map servers deliver not only static maps that have been produced ahead of time and stored at the server site. They rather allow users to specify the required information and possibly some display parameters, then generate the desired map at request time. The map thus serves as a customized visualization of some underlying (spatial or non-spatial) data set. On the other hand, the maps delivered are usually not just images but interactive entities that can be used to specify further queries. For example, users may be able to mark a point or rectangle on these maps, thereby requesting more detailed information about the selected location or region (*logical zoom*).

All these efforts clearly indicate the high potential GIS vendors see in Web-related business. ESRI estimates that by mid-1998 up to 20 percent of GIS computing may be provided through Internet and Intranet services (ESRI Inc. 1997b).

The formats delivered by tools from different vendors are not always compatible but users have started to work on conversion tools to improve interoperability. The Environment Protection Authority of the Australian state of New South Wales, for example, has developed Web server mapping software to overlay map layers from different Web sites into interactive maps for viewing on standard Web browsers (http://www.epa.nsw.gov.au/soe/maps). Both source map layers and output maps are held in the standard GIF format used throughout the Web. The approach could therefore be used to combine output from different proprietary Web map servers. With some further development of tools and standards, an organization would thus only need to publish a GIF map file and accompanying metadata file on their Web site and notify a map server of its existence, in order to link their map layer to the pool of existing layers for their geographic area. This would in particular require a standard format for the metadata files to accompany map images. Section 5 will discuss related problems in greater detail.

As for Web-based *courseware*, most of the available material is on GIS-related topics. There is a search engine available that allows specifically to search for online GIS courses (http://www.frw.ruu.nl/eurogi). Among the university offerings we find three particularly noteworthy:

- The U.S. National Center for Geographic Information and Analysis (NCGIA) offers a well-known GIS Core Curriculum (National Center for Geographic Information and Analysis (NCGIA) 1997) that is now entirely Web-based (http: //www.ncgia.ucsb.edu/education/ed.html) and provides a comprehensive coverage of the field.

- UNIGIS is an international consortium of universities led by Manchester Metropolitan University in the UK. It provides GIS courses and various degrees for distant learners. More information is available at http://www.unigis.org.

- A consortium led by U.C. Berkeley offers access to its public-domain GIS GRASSLinks (Berkeley 1998). GRASSLinks is a raster-based GIS that offers most standard GIS functionalities, including visualization, map overlay, buffering, and metadata management.

Most GIS vendors also offer access to tutorial materials on their Web sites. In addition to online introductions to their systems, this may include access to proceedings of user conferences or links to problem-specific discussion groups. ESRI's Web site (http://www.esri.com) is a typical example. There are also several commercial Web sites that offer GIS-related videos (see, for example, http: //www.amproductions.com/contentg.html)

4.5. Environmental Management Information Systems

Most applications presented in this paper are taken from *public* environmental information systems, i.e., from environmental information systems designed for and managed by public administrations. There is a simple reason for that: most environmental information systems in practice are run by public administrations. Gradually, however, the private sector is recognizing the need for collecting and managing environmentally relevant information as well. Books by Denton (1994) and

Steger (1993) (in German) discuss the possible benefits of environmental management at great length. Denton's book covers in particular the well-known Pollution Prevention Pays (3P) program of the American company 3M. According to 3M, the program saved over $500 million since its inception in 1975.

This trend towards environmental (data) management is reinforced by legislation efforts in many industrial countries that oblige companies to provide detailed reports on those activities that may have a significant environmental impact. The British Standard Institute, for example, published their specification for environmental management systems (BS 7750) in 1992 (British Standard Institution 1992). The European Union followed suit one year later with their regulation on eco-management and audit schemes (EMAS regulation) (Council of the European Communities 1993). Both standards require that companies collect and compile data about inputs and outputs and their possible impact on the environment. On the input side, the EMAS regulation lists in particular raw materials, energy, and water. On the output side, it mentions solid and liquid waste, as well as air emissions and noise.

But those legal regulations are not the only reason why companies contemplate stronger information system support in the environmental sector. Wicke et al. (1992) distinguish between internal and external uses of environmental information.

Internally, environmental information helps to make the environmental impact of a company's production facilities and its products more transparent. This is likely to have a direct impact on the departments concerned with production and materials. They can use this information to look for specific ways to minimize resource utilization, waste, and emissions, and to improve recyclability of their products.. Indirectly, other departments are concerned as well. Accounting and auditing may use the new insights to improve their bookkeeping procedures in order to provide a better documentation of material flow, marketing may be inspired to propose other ways of packaging a product, and so on.

Externally, the environmental data is primarily used to fulfill legal requirements. In addition, companies are likely to volunteer some of that information to their business partners and customers for a variety of reasons. Insurers, for example, have shown an increasing interest in environmental matters since courts tend to weigh related aspects more heavily in liability lawsuits. The liability problem also concerns investors who may often prefer companies that take possible environmental hazards explicitly into account. For example, the question whether a piece of land contains hazardous waste may well decide over the profitability of a related investment. Some investors also have strong ideological reasons to prefer companies that have shown concern for the environment. Investment companies are therefore offering "green funds" that include only shares of companies that fulfill certain environmental standards. Environmental information may also be useful to outside suppliers and corporate customers to optimize their own environmental strategies. Finally, environmental information about production processes and products – if favorable – is increasingly used for marketing purposes. In some countries and regions, such as Germany or Northern California, the "environmental correctness" of a product has become an important selling point.

Many companies are reacting to these insights and developments by building or purchasing specialized software. These *environmental management information systems (EMIS)* (also called *computer-aided environmental information and management (CAEM) systems* (Hilty 1995)) may be stand-alone programs, or they may be integrated with existing information systems in the enterprise. Unlike most public environmental information systems, the majority of such systems concentrate on the technosphere rather than the biosphere. Most of the data stored in EMIS is about human-made systems (such as plants, production processes, or waste) and on their impact on the environment.

Stand-alone systems are by their very nature limited in their functionalities. Most of them are simple report generators that collect the necessary input via form-based interfaces and produce printed reports to fulfill legal requirements. They are sometimes coupled with a dictionary component that gives access to legal texts, or to data about hazardous substances. More sophisticated systems also provide some database functionalities to manage data about the company's technical equipment, important deadlines, past measurements, and other kinds of information that may be environmentally relevant. While relatively cheap in terms of licensing fees, the operating expenses of such stand-alone systems may greatly exceed their purchase price. In particular, data input can become a time-consuming chore and therefore a major cost factor. In addition, employees may quickly become tired of typing in data that they know is already stored elsewhere, and therefore – consciously or subconsciously – boycott the system.

For those reasons, more sophisticated systems provide interfaces to the existing infrastructure within the enterprise. Most important are connections to accounting and auditing, and to production. Based on those two possible cornerstones of an environmental management information system, one can distinguish two lines of related research and development: *accounting-oriented* and *production-oriented* systems.

The current situation can be seen as analogous to the "end-of-pipe" solutions in other areas of environmental protection and management. Legislative support for more comprehensive solutions cannot be reasonably expected as long as many Western industrial economies are still in a fragile state. Until there are visible signs of a long-term recovery, legislators will be hesitant to impose stricter guidelines on companies. It is thus up to software vendors and researchers to show more convincingly that a comprehensive environmental (information) management has more to offer than marketing advantages and a slight drop in insurance premiums – that it can in fact save resources and improve the overall efficiency of production.

From a research point of view, the whole area of environmental management information systems is of very recent origin and still very much in flux. In addition, many related issues are of an institutional character and therefore depend on the given legislative and organizational framework. As a result, publications in this area are often specific to a particular country or branch of industry.

Readers interested in learning more about EMIS should consult some of the available survey articles and conference proceedings for more information. Hilty (1995), Hilty and Rautenstrauch (1995) give an overview of the field as of 1995. A later literature survey by the same authors (Hilty and Rautenstrauch 1997) covers more recent publications in the field and structures them depending on their level of abstraction. Arndt and Günther (1996) briefly describe the current legal and organizational situation in Europe and present possible system architectures that fit into this framework. The conference proceedings edited by Haasis et al. (1995b;a) give a good impression of ongoing projects. Except for Hilty (1995), all of these publications are in German. At least Hilty and Rautenstrauch (1997), however, should also be useful to readers who do not speak German, simply because it includes a large number of references to articles written in English. Some of the relevant work in North America is described in Denton's book on environmental management (Denton 1994).

4.6. UIS Baden-Württemberg: An Integrated Public EIS

Baden-Württemberg is a highly industrialized state in the southwest of Germany with a population of about 10 million. The State Ministry of the Environment was founded in 1987, which was also the starting point for a large-scale project called *UIS Baden-Württemberg*. UIS is an acronym for the German translation of *environmental information sys-*

tem. UIS was conceived as a showcase project for the state's environmental activities in the wake of environmental disasters like Chernobyl and the Sandoz Rhine pollution case. The state hired a major management consulting company to conceive a first system design in cooperation with various government agencies (Umweltministerium Baden-Württemberg and McKinsey and Company, Inc. 1987). This design has since been implemented and developed further under the leadership of a working group in the Ministry of the Environment (since 1996 the Ministry of Environment and Traffic). The annual budget of the UIS project group is between 30 and 40 million DM per year.

The main objectives of UIS are (Mayer-Föll 1993, Mayer-Föll et al. 1996, Mayer-Föll 1997):

- to support the administration in their environmental management and planning tasks;

- to implement an efficient environmental monitoring, including data capture, analysis, and forecasting;

- to support the management of environmental emergencies;

- to make environmental information available to the executive branch as well as to the general public;

- to protect past investments by coordinating existing system solutions and integrating them into a common system architecture.

From the start, UIS was not conceived as a monolithic system but as a network of subsystems and services. It was particularly important to be able to integrate existing systems, in order to avoid losing past investments and to ensure acceptance among other government agencies that are involved in environmentally relevant tasks. TCP/IP, relational databases, and client/server technology had just become popular at the time UIS was conceived, and have since become cornerstones of the overall system architecture. The World Wide Web was integrated into the design soon after it had become widely available. More recent system components follow middleware architectures with *object request brokers (ORBs)* negotiating between UIS service providers and consumers (Koschel et al. 1996b, Riekert et al. 1997).

While the UIS project has been a major force for the harmonization of terminologies and naming conventions throughout the state, constructing an all-encompassing data model (e.g., by means of an entity-relationship approach) was not attempted. This decision results from the lessons learnt in the 1970s and 1980s when many companies attempted – and failed – to build an enterprise-wide data model. UIS adopts a bottom-up approach instead, building on existing application-specific data models and database schemata. Thesauri and metainformation systems (such as the UDK, cf. Chap. 5) are used to establish cross-references and allow users to access subsystems with their particular terminologies.

The original UIS design (Umweltministerium Baden-Württemberg and McKinsey and Company, Inc. 1987) has gone through a natural evolution over the years. Some of the original objectives and modules have become more relevant than expected, and vice versa. Today one can distinguish five kinds of system components (Fig. 28.10):

1. *Generic base components* are systems that are not exclusively used for environmental tasks but for administrative tasks in general. These systems are part of the government's general computing infrastructure. Typical examples are official digital map collections, statistical databases, the cadastre, or the government intranet.

2. *UIS-specific base components* extend this basic infrastructure. Typical examples are data dictionaries, regulations for the formatting and exchange of spatial data sets, or operative rules for measuring networks.

3. *Task-specific components* are systems that have been designed for particular environmental applications. They are the most important source of environmental data and provide an essential foundation for UIS. Many task-specific components are concerned with capture and aggregation of environmental raw data, such as measuring series or aerial imagery. Others are devoted to more analytical tasks like water treatment monitoring.

4. *Integrative components* are intended for inter-agency and inter-domain aggregation of environmental data. The goal of these systems is to provide an integrated view of complex environmental phenomena that transcend the traditional boundaries defined by media and organizational hierarchies.

5. The *environment and traffic information service UVIS* is a strategic management information system for high-level decision support. It provides compact visualizations of complex, highly aggregated data from numerous connected subsystems.

A good overview of the complete system as of 1993 is given in a special chapter of (Jaeschke et al. 1993). Since then, the main focus has been to adapt UIS to open client/server architectures, WWW technology, and

ENVIRONMENTAL INFORMATION SYSTEMS

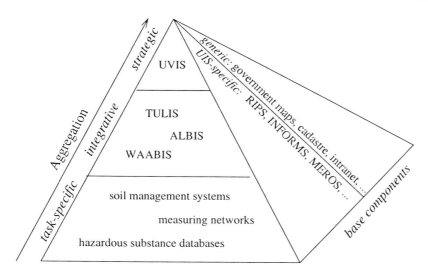

Figure 28.10. Pyramid architecture of the UIS Baden-Württemberg

middleware standards such as CORBA (Mayer-Föll et al. 1996, Mayer-Föll 1997). UIS can be accessed via the Internet at http://www.uis-extern.um.bwl.de.

5. Metadata Management

As we saw in the previous chapters, there is a major need for convenient navigation aids that help users to take advantage of network-based, distributed information, regardless of their computer literacy. Starting from an environmental query or problem formulation, such navigation aids should help users to localize the relevant data sets and to retrieve them quickly and in a user-friendly manner.

An essential prerequisite for both navigation and data transfer is the availability of appropriate *metadata*, i.e., data about the format and the contents of the data. The key idea is to enhance data sets by concise self-descriptions in order to improve both the speed and the accuracy of related search operations. The metadata serves as a kind of online documentation that can be read and utilized by appropriate tools as well as by human users. Note that there is no intrinsic distinction between data and metadata; it is rather a question of context whether a given data item represents metadata or not.

In this chapter we discuss the question of metadata in environmental data management in greater detail. Section 5.1 gives a more elaborate definition of metadata and shows how metadata can be integrated into a traditional data management architecture. Sections 5.2–5.4 describe several concrete approaches to metadata management. Section 5.2 presents the U.S. initiative to create a National Information Infrastructure (NII) and, within the NII framework, a National Spatial Data Infrastructure (NSDI). This includes discussions of the Government Information Locator Service (GILS), the Spatial Data Transfer Standard (SDTS), and the FGDC Content Standards for Digital Geospatial Metadata. Sections 5.3 and 5.4 continue with descriptions of two European systems: the Catalogue of Data Sources (CDS) developed by the European Environment Agency (EEA), and an environmental data catalog called UDK, whose development was coordinated by Austria and several German state governments. Section 5.5 concludes with a summary and an outlook on future work.

5.1. Metadata and Data Modeling

Our further discussion is based on the three-way data model described in Chap. 1. We distinguish between environmental objects, environmental data objects, and environmental metadata (Fig. 28.1). As we already discussed at length, the data flow in many environmental applications is usually associated with a complex aggregation process to provide decision support at various levels of responsibility. It closely resembles the data flow known from classical business applications: data capture, data storage, and data analysis.

Metadata may be collected throughout this aggregation process and built into the corresponding data structures. As Strebel et al. (1994) have pointed out, it is important to collect metadata in a timely manner – if possible, simultaneously with the collection of the original data. Figure 28.11 pictures their empirical results. It shows on the one hand (dotted line) how the effort to collect metadata increases dramatically if the collection effort is delayed with respect to the original data capture. On the other hand (solid line), the amount of metadata that can typically be recovered drops with increasing delay.

So far, metadata collection is a mostly manual process. The problem is that the most qualified people to perform this process are the data providers themselves. Those people, however, usually have little motivation to do so – they know how to obtain and interpret the data. What we see here is a variation of the classical software documentation problem. In analogy, there are two solutions: one can either establish an organizational framework that requires data providers to perform the necessary metadata capture, or one can search for ways to extract the metadata automatically.

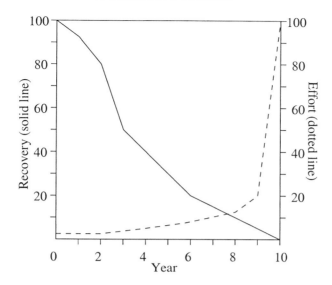

Figure 28.11. Effect of delay between data capture and metadata collection (Strebel et al. 1994)

The first option is typical for hierarchical organizations, such as companies and government agencies. To ensure the continuity and quality of the metadata, and in order to maintain motivation among employees, it is important to complement such a policy by software tools that facilitate the extraction process. Web browsers with their form-based graphical interfaces offer a solution that is both cost-efficient and simple to implement.

The automatic extraction of metadata is slowly becoming a serious alternative (Drew and Ying 1997). Riekert et al. (1997) present an approach where incoming documents are matched against a thesaurus to extract relevant terms. They intend to add a similar matching procedure for a gazetteer, i.e., a geographic index. In addition to matching geographic names, this requires a spatial component that determines the spatial extension the data set refers to.

Once the metadata has been collected, it can be used for a variety of purposes, especially during the data analysis phase:

- Computerized environmental information systems are able to collect and process much greater amounts of data than anybody would have imagined only a few years ago. Automatic data capture and measurement collects terabytes of new data per day (Campbell and Cromp 1990). Even in processed form, this kind of data is impossible to browse manually in order to find the information relevant for a given task. Modern information retrieval tools allow automatic or semi-automatic *filtering* of the available data in order to find quickly those data sets one is looking for. Metadata forms an important foundation for these tools by serving as a condensed representation of the underlying data. As such, it supports browsing, navigation, and content-oriented indexing.

- Environmental data management is extremely *heterogenous*, both in terms of hardware and software platforms. Data is organized according to a wide variety of data models, depending on the primary objectives of the particular agency in charge. Metadata can help to overcome these heterogeneities by specifying the platforms on which a given data item is located. This way, appropriate conversion routines can be introduced (semi-)automatically, wherever necessary.

- Environmental data is frequently *uncertain*. Metadata can be used to specify the accuracy of a data item, so users can judge from the metadata whether the corresponding environmental data objects are relevant for their current needs.

- Metadata can also help to *inventory* existing data holdings, to unify naming schemes, and to record relationships between different data items and data sets. This aspect of metadata has become very popular as one of the core functionalities of *data warehouses* (Hallmark 1995, Chaudhuri and Dayal 1997).

The concept of metadata is not new. Online documentation of programs and data sets has been in common use for many years. As we saw in Sect. 4.4.1, online database providers have long offered index databases that help users to find the data sources relevant for a given query. Machine-readable metadata has also been known for a long time, in particular in the context of relational databases, where the internal database structure (the *database schema*) is typically represented in a relational format itself. What is new, is the more systematic approach to providing machine-readable metadata, and the trend to standardize metadata in certain application areas.

For the subsequent discussion, it is useful to distinguish between two kinds of metadata (Melton et al. 1995). The term *denotative metadata* is used to refer to the kind of metadata that describes the *logical structure* of a data set; a relational schema would be a typical example. The term *annotative metadata*, on the other hand, is used to describe data that provides *content-oriented context information*, such as the documentation of the measuring series described above. Following Melton et al. (1995), further examples of annotative metadata include "information in scientific notebooks, instrument logs, manuals, and reports that document the platform and instrument conditions, the operational environment, interfering sources of noise, and that uniquely identify the software and computer platforms used for analysis, modeling and simulation." In the remainder of this chapter, we will concentrate on annotative metadata and use the term *metadata* in that sense, unless noted otherwise.

The relevance of metadata for the management and analysis of complex data sets was pointed out early on by McCarthy (1982) and pursued further in the area of statistical and scientific databases. Siegel and Madnick (1991) built on those ideas, concentrating on possible applications in financial data analysis. The IEEE Mass Storage Systems and Technology Committee has sponsored several metadata workshops whose results are available on the Web (URL http://www.llnl.gov/liv_comp/metadata/metadata.html).

The use of metadata in geographic and environmental information systems is of a more recent nature (Radermacher 1991). Lately, however, there has been broad agreement that metadata are a crucial factor to improve both the quality and the availability of geographic and environmental data. Several conferences on spatial databases and GIS have devoted parts of their program to metadata (Günther and Schek 1991, ESRI Inc. 1991; 1995b), and there has been a variety of workshops dedicated exclusively to metadata management in the geosciences and environmental sciences (Medykyj-Scott et al. 1991, Melton et al. 1995).

In terms of practical consequences, metadata technology is increasingly being integrated into commercial GIS. Most commercial systems have always maintained some basic metadata on the objects to be administered. ARC/INFO, for example, generates and maintains metadata on the spatial registration, projection, and tolerances of a coverage or grid (ESRI Inc. 1995a). Every time one creates a coverage, the system creates a set of metadata files, including the TIC file (containing data about the coverage's coordinate registration), the LOG file (tracking all ARC operations performed on the coverage), and the BND file (containing the coordinate values that denote the outer boundary or spatial extent of your coverage). There is also denotative metadata giving some schema information on the INFO tables that contain the non-spatial data components.

The practical use of metadata, however, extends far beyond this somewhat narrow scope. One trend is to collect more information about the detailed content of the data. Vendors typically choose some bibliography-style format to represent this information. In the case of spatial data, conformity with the FGDC Content Standards (see Sect. 5.2) is increasingly required. The ARC/INFO component DOCUMENT.AML (ESRI Inc. 1995a) is a typical example of such a tool.

Another trend is to describe the history and quality (also called *lineage*) of data sets and their sources in more detail. Geolineus from Geographic Designs is a common tool for this purpose (Geographic Designs Inc. 1995). Geolineus represents the data in a GIS by means of dataflow diagrams, where coverages and grids are shown as icons. Icons along the top of the diagram represent the *source data* on which the GIS is based. Icons further down represent data layers that were *derived* with spatial analysis operations like *buffer* or *intersect*. Finally, icons at the bottom of the diagram represent *products*, i.e., derived data items that represent the final steps in a GIS application. Geolineus shows the type of data in the corresponding layer for each icon and maintains command histories for each coverage. The system allows documentation about each layer to be stored in a frame-based format.

Drew and Ying describe a concrete approach to use metadata in order to provide uniform access to a heterogenous collection of GIS and spatial databases (Drew and Ying 1997). Based on metadata about those systems and their contents, their GeoChange system serves as a navigation and access tool. To a large extent, it is non-intrusive, i.e., it can be implemented on top of an existing collection of independent systems without major changes to the underlying architectures and implementations.

Other trends in metadata management include the inclusion of more spatial elements in the metadata itself (Seaborn 1995) and the use of metadata to describe and access not only other data sets, but also models and algorithms. We already described the Register of Ecological Models (REM) in Sect. 4.2.2. A related effort is described by Lenz et al. (1994).

Parallel to these application developments, metadata management has become a focus in an increasing number of government R&D projects. Besides the efforts described in the following sections, there has been a project by the European Space Agency (ESA) to develop an online geosciences metadata system, called the ESA Prototype International Directory (Walker 1991). At about the same time, the United Nations Environmental Program (UNEP) started its project on Harmonization of Environmental Measurements (HEM, cf. Sect. 4.2.3) (Keune et al. 1991). This was later followed by UNEP's Global Resource Information Database (GRID), which includes a Metadata Directory (GRID MdD).

Also actively involved in the harmonization of environmental data in research and monitoring is the International Council of Scientific Unions (ICSU), represented by its Scientific Committee for the Problems of the Environment (SCOPE) and its Committee on Data for Science and Technology (CODATA) (Bardinet et al. 1995). The Norwegian SAMPO project uses ESRI's ArcView to catalog its spatial data holdings (Mikkonen and Rainio 1995). The Austrian Ministry of the Environment has developed the Central European Environmental Data Request Facility (CEDAR) (Pillmann and Kahn 1992), which can be reached at http://www.cedar.univie.ac.at. Other efforts include the CIMI system of the Dutch Ministry of Transportation, Public Works and Water Management (Kuggeleijn 1995), the Australian FINDAR system (Johnson et al. 1991) and the New South Wales Department of Conservation and Land Management's Data Directory (Miller and Forner 1994, Miller and Bullock 1994).

Coordination between this great variety of efforts is difficult. As we will see in Sect. 5.3, the newly founded European Environment Agency will have an important role to play here. One promising effort concerns the development of a common European geodata standard. With strong support from the European Center of Normalization, Germany and Belgium's Geographic Data Files (GDF) are generally considered the frontrunner (Ostyn 1995). Further standardization is required, however. Environmental phenomena do not stop at national borders. In this domain, international cooperation on a broad scale is essential for making progress.

5.2. Metadata in the U.S. National Information Infrastructure

Since the early 1980s, the U.S. Government has been working intensively on creating a National Spatial Data Infrastructure (NSDI). A major motivation for this effort was to abolish the notorious incompatibilities among the internal formats used by various government agencies. Examples include DLG, TIGER/Line, and GRASS of the U.S. Geological Survey, DIGEST and the Vector Product Format (VPF) of the Defense Mapping Agency (DMA), and DX90 of the National Ocean Service. The parallel use of such a variety of standards led to considerable expenses to the taxpayer that could at least in part have been avoided.

Most of the early efforts on NSDI were coordinated by the U.S. Geological Survey, an agency under the supervision of the U.S. Department of the Interior. One of the first major results was the development of the Spatial Data Transfer Standard (SDTS), a Federal Information Processing Standard to facilitate the online exchange of spatial data (United States Geological Survey 1992). The goal is to accommodate different spatial data models, to preserve topologies, and to maintain even complex relationships, as data is transferred across different computer platforms and software systems. Other than many existing standards (such as VPF), the SDTS is not an exchange format. It rather provides guidelines that need to be translated into a native application-specific format before they can be used. Most GIS vendors provide interfaces and tools for that purpose (ESRI Inc. 1995c).

In the early 1990s, the NSDI was integrated into the National Information Infrastructure (NII). NII is a program coordinated by the White House to reorganize and renew the computer infrastructure throughout all levels of government. A middle-term goal of NII is to offer e-mail and Web access to every single government employee. Some of the key technical cornerstones of NII are:

- client/server technology
- TCP/IP-based intranets
- Internet connectivity
- relational databases.

NSDI has since been coordinated by a working group called the Federal Geographic Data Committee (FGDC), which is composed of representatives of the Departments of Agriculture, Commerce, Defense, Energy, Housing and Urban Development, the Interior, State, and Transportation; the Environmental Protection Agency; the Federal Emergency Management Agency; the Library of Congress; the National Aeronautics and Space Administration; the National Archives and Records Administration; and the Tennessee Valley Authority. The committee is chaired by the Department of the Interior, represented by the U.S. Geological Survey.

In May 1994, the FGDC published a draft for the new Content Standards for Digital Geospatial Metadata (United States Federal Geographic Data Committee 1994), which was later approved by the National Institute of Standards and Technology as a Federal Information Processing Standard. The implementation of the standard is based on the Executive Order 12906, "Coordinating Geographic Data Acquisition and Access: The National Spatial Data Infrastructure," which was signed on April 11, 1994, by President Clinton (United States Government 1994). In addition to providing a long-needed political foundation for the NSDI, the order requires all government agencies to use the FGDC Content Standards for documenting all new geospatial data it collects or produces as of April 11, 1995.

The FGDC Content Standards represent an impressive effort to establish a uniform way to document digital geospatial data sets. While mainly targeted at the description of geographic data, it also provides a solid basis for an environmental metadata system. Such an extension would entail a more detailed semantic framework, especially with regard to theme-related information

While both the SDTS and the FGDC Content Standards refer to metadata about spatial data, they have distinctly separate functions. The SDTS is a language for communicating spatial data across different platforms without losing any structural or topological information. The FGDC Content Standards, on the other hand, specify the kind of annotative metadata that federal agencies are required to collect on a spatial data set they maintain. The only two sections that both standards have in common concern data quality and the data dictionary information.

The objective of the *Government Information Locator Service (GILS)* was to build a national metainformation system to serve as a navigation and organization aid. GILS supports users for searching public databases and accessing official documents and files. Moreover, it is used as an internal documentation and organization tool.

GILS is organized as a decentralized cooperation of existing domain-specific information systems in ministries and subordinate government agencies. The integration of these systems is performed *ex post* and *bottom-up*, i.e., without compromising their autonomy. All federal agencies, however, are obliged to document new data sets in accordance with GILS guidelines, and gradually to provide documentation for older data sets as well. GILS is coordinated by the Office of Management and Budget (OMB), the National Archive and Record Association (NARA), and the National Institute of Standards (NIST). In summary, GILS serves to

- *document* public information resources held by government agencies;

- *describe* the information available;

- help users to query and retrieve the information *(navigation aid)*.

GILS uses the Z.39.50 protocol to access several distributed servers simultaneously. There exists a WWW Z.39.50 conversion routine to translate GILS queries that have been posed using a Web browser into a format compatible with Z.39.50. Conversely, the results of a query are reconverted to be displayed in the chosen Web browser.

5.3. The Catalogue of Data Sources

The European Union (EU) has been working on similar issues, especially since the 1994 foundation of its European Environment Agency (EEA), located in Copenhagen. In comparison to the American activities, the EEA efforts to build a Catalogue of Data Sources have a wider focus, concentrating not only on spatial data, but on environmental data in a more general sense. On the other hand, the results obtained so far are not quite as concrete as the FGDC recommendations described above.

The original goal of the EU activities was the implementation of an integrated European environmental information system. Based on the results of a previous project called CORINE CDS (1985–1989), in 1992 the EU commissioned a study entitled "Catalogue of Data Sources for the Environment – Analysis and Suggestions for a Meta-Data System and Service for The European Environment Agency" (European Environment Agency 1993). It became clear immediately that the construction of a European environmental information system from scratch is neither economically feasible nor politically viable. Many member countries already have some kind of national environmental information system. A European system should take advantage of these developments and attempt a bottom-up integration of the systems that are already functional. Devised as a *meta*-information system, such a CDS would only store descriptions of data sets that are locally available.

In the meantime, the European Environment Agency has intensified their related efforts and started a European Topic Centre on Catalogue of Data Sources (ETC/CDS), based in Hanover (Germany). In 1995, ETC/CDS commissioned a first CDS prototype that is focusing on the *locator* aspect of a metainformation system. The objective is to provide answers to the following questions:

- Who in Europe holds environmental information on a given topic?
- In which format is it stored?
- How can one access it?

Multilinguality is achieved by using a thesaurus. CDS is available on the Web at www.mu.niedersachsen.de/system/cds/.

5.4. UDK: Environmental Data Catalog

The *UDK (Umwelt-Datenkatalog = Environmental Data Catalog)* is a metainformation system and navigation tool that documents collections of environmental data from the government and other sources. These data sets may be available either online or by request to the responsible data administrator. Potential users of the system include government agencies, industry, and the general public. Similar to CDS, the UDK helps them to obtain answers to the following questions:

- Which relevant information is in principle available for a given problem?
- Where is this information stored?
- How can this information be retrieved?

The UDK design presented in this section is the result of several years of research and development (Lessing 1989, Lessing and Schütz 1994). In 1990, the Environmental Ministry of the State of Lower Saxony launched a research project with funding from the German Federal Environmental Agency. Two years later, an international working group was formed to oversee the UDK design and its further development into a practical software tool. In 1994, Austria passed an Environmental Information Law that introduced the UDK as the official navigation tool for all environmental information on record. In 1995, first versions of the UDK were made available in Austria and the German states of Baden-Württemberg and Lower Saxony. Other German states and several other European countries are considering to adapt the UDK as well. Several key concepts of the UDK have found their way into other systems, including the Catalogue of Data Sources described above.

The UDK is based on a three-way object model that is very similar to the data model described in the introduction (Fig. 28.1). The UDK distinguishes between environmental objects, environmental data objects, and UDK (meta-)objects. Each real-world environmental object is described by a collection of environmental data objects. Each environmental data object is in turn associated with exactly one *UDK object*, i.e., a metadata object that specifies its format and contents.

Although it is desirable to handle the creation (and deletion) of UDK objects with great flexibility, the decision to create a new object has to be based on a cost/benefit analysis, depending on the particular applications a user has in mind. The effort to create and maintain a UDK object is not negligible. Recent empirical data suggests that creation takes close to one person-day on average. Maintenance involves not only the occasional update of attributes but also the dynamic tracking of semantic associations between UDK objects and the corresponding environmental data objects.

At this time, most of the related work is performed by specialized personnel from higher-level government agencies or consulting firms, and is therefore relatively expensive. It is unlikely that the work can be delegated to less-qualified support staff in the near future. The idea of leaving the creation of UDK objects to local domain experts (biologists, chemists, etc.) is also unrealistic at the present time. The process is still too technical and time-consuming for someone who is not a UDK expert.

Austria and several German states have released WWW implementations of the UDK (see, for example, http: //udk.bmu.gv.at and http: //www.uis-extern.um.bwl.de). Access is mainly keyword-based. The result of a search is a list of relevant UDK objects. More details on a particular object are available by checking it and sending the marked-up form back to the server. A CGI script then retrieves the corresponding additional attributes. Further details of the current implementation and related issues have been described by Koschel et al. (1996a), Kramer and Quellenberg (1996), Kramer et al. (1997).

5.5. Summary

The goal of this chapter has been to show how metadata is becoming increasingly popular in environmental information systems, in order to improve both the availability and the quality of the information delivered. The growing popularity of Internet-based data servers has accelerated this trend even further. After a general discussion of the term *metadata* and of the possibility to integrate metadata into traditional information system architectures, we have discussed several case studies in detail. Particular emphasis has been put on the U.S. efforts to build a National Spatial Data Infrastructure, and on several European projects to integrate environmental information processing at the national and international level.

Despite the remaining heterogeneities and inefficiencies, the outlook seems positive. The ubiquitous trend towards open systems as well as the rise of the World Wide Web are two recent developments that will greatly improve the way we manage environmental information. Users will have faster and more comfortable access to ever greater amounts of information, and metadata will be an essential component of the underlying software architectures.

Finally, we envisage an increasing number of applications where metadata is used to administer not only simple data sets but also complex software tools, such as domain-specific aggregation methods or environmental simulation models. In those applications, the metadata will be used for two purposes: (i) to find the appropriate software tool for a given problem, and (ii) to to apply the tool to a given data set over the

Internet without having to port the software to a local machine. Our own MMM project (Günther et al. 1997a) is one example of a software architecture that supports this paradigm.

6. Conclusions

Environmental information systems are one of the most challenging applications areas for computer scientists. Due to the volume and complexity of environmental data, users require state-of-the-art techniques from a variety of disciplines. They are often willing to try out new approaches, and as a result, cooperations between users, researchers, and vendors are unusually intense.

Statistical classification, possibly enhanced by modern approaches to uncertainty management, is used to perform the initial filtering. Terabytes of seemingly unstructured raw data are transformed into formatted data sets that represent a semantic aggregation of the given inputs. Latest database technology is then needed to store and query those data sets. Recent advances in spatial databases and geographic information systems provide the means for doing so efficiently. To prepare the data for decision support purposes, one requires sophisticated analysis tools, such as simulation models, information retrieval techniques, and visualization packages. Systems for the management of associated metadata provide crucial support for searching and processing environmental information.

Throughout this information flow, data is aggregated and compressed to serve the needs of decision makers. In this paper we have tried to cover the base techniques underlying this complex aggregation process. Our treatment is structured into four major parts: data capture, data storage, data analysis, and metadata management. Because of the multitude of disciplines involved we were not able to discuss every technique in detail. References to more specialized literature are provided for the interested reader.

In addition to our discussion of the *concepts* underlying environmental information management, we presented several *case studies* of applied research projects and of environmental information systems in practice. As most of these systems are under continuous development, one should not be surprised to see minor differences between the descriptions given here and the functionalities at the time this paper appears in print. For more detailed and up-to-date documentation, readers are therefore advised to contact the responsible agencies.

Looking at the field of environmental informatics in a more general sense, the breadth and depth of related activities is still increasing.

Besides a large number of recent publications in journals and conference proceedings, there have been comprehensive anthologies in English (Avouris and Page 1995) and German (Page and Hilty 1995). In addition, the various communities involved in such projects are starting to meet and to conduct joint activities. In the past, the field has been somewhat fragmented along both national and subject borders.

Germany and Austria, for example, have a long history of related research and development activities. Since 1987, the German *Gesellschaft für Informatik* has organized an annual symposium on environmental informatics; see (Jaeschke et al. 1993, Hilty et al. 1994, Kremers and Pillmann 1995, Lessing and Lipeck 1996, Geiger et al. 1997) for recent proceedings. While the first conferences were mostly local, more recent events were advertised internationally and involved an increasing number of colleagues from the United States, France, and other countries. In 1997, the symposium was held for the first time in a non-German speaking country (France).

North American efforts do not suffer from the same language barriers. On the other hand, there have traditionally been few activities that transcend the traditional subject boundaries between the various disciplines that conduct environmentally relevant research. For example, researchers in the geosciences and environmental sciences have long been working on adequate computer support for their core activities. Commercial geographic information systems are among the most important outcomes of these efforts. Computer scientists have been involved only marginally in the development of such systems. More generally speaking, cooperations between computer scientists and the environmental science community have been rare until recently.

Since the early 1990s, however, the increasing number of contacts is bearing fruit. There have been several joint projects of high visibility, including the National Center for Geographic Information and Analysis (National Center for Geographic Information and Analysis (NCGIA) 1998) and the Sequoia 2000 project (Stonebraker 1993, Frew 1994). U.C. Berkeley's Digital Environmental Library (ELIB) project (University of California at Berkeley, Digital Library Project 1998, Wilensky 1996), and U.C. Santa Barbara's Alexandria project (Smith 1996) are pursuing similar goals; both projects are funded through the NSF/ARPA/NASA Digital Library Initiative. Since 1993, there has also been the bi-annual ISESS symposium series on environmental software systems (Denzer et al. 1995b; 1997).

These and related activities have resulted in numerous joint publications and conferences, as well as system solutions and commercial products. Besides the proceedings of the conferences listed above, good starting points into the related literature include the collections edited by Goodchild et al. (1993; 1996), Michener et al. (1994), Avouris and Page (1995), and Günther, editor (1997).

We certainly hope that this paper contributes to making the area of environmental information systems known to a broader audience, establishing the subject in the academic curriculum, and increasing the number of researchers and practitioners working on related problems. It is after all an exceptionally attractive opportunity for computer scientists and engineers to contribute directly to one of the great challenges of our time: to maintain and protect our natural environment in an era of unprecedented population and industrial growth. While technical progress has been a key factor in the developments that led to the current environmental crisis, it is now essential to turn to technology once again to help us solve those problems. Inventing and developing computer technology for environmental applications is one important cornerstone of this endeavor.

Bibliography

D. Abel and B. C. Ooi, editors. *Advances in Spatial Databases*. LNCS 692. Springer-Verlag, Berlin/Heidelberg/New York, 1993.

D. Abel, K. Yap, R. Ackland, M. Cameron, D. Smith, and G. Walker. Environmental decision support system project: An exploration of alternative architectures for geographic information systems. *Int. J. Geographical Information Systems*, 6(3):193–204, 1992.

D. J. Abel. Spatial Internet marketplaces: A grand challenge. In M. Scholl and A. Voisard, editors, *Advances in Spatial Databases*, LNCS 1262, Berlin/Heidelberg/New York, 1997. Springer-Verlag.

N. R. Adam and A. Gangopadhyay, editors. *Database Issues in Geographic Information Systems*. Kluwer Academic Publishers, Norwell, Mass., 1997.

J. Barclay Adams. Probabilistic reasoning and certainty factors. In B. Buchanan and E. Shortliffe, editors, *Rule Based Expert Systems*, pages 263–271. Addison-Wesley, Reading, Mass., 1984.

L. Anselin and S. Bao. Exploratory spatial data analysis linking SpaceStat and ArcView. Technical report, West Virginia University, Morgantown, WVa., 1996.

H.-K. Arndt and O. Günther. Betriebliche Umweltinformationssysteme. *UmweltWirtschaftsForum*, 4(1), 1996.

S. Aronoff. *Geographic Information Systems: A Management Perspective*. WDL Publications, Ottawa, 1989.

M. Atkinson, F. Bancilhon, D. DeWitt, K. Dittrich, D. Maier, and S. Zdonik. The object-oriented database system manifesto. In *Proc. First Int. Conf. on Deductive and Object-Oriented Databases*, 1989. Reprinted in Stonebraker (1994).

N. M. Avouris and B. Page, editors. *Environmental Informatics – Methodology and Applications of Environmental Information Processing*. Kluwer Academic Publishers, Norwell, Mass., 1995.

F. Bancilhon, C. Delobel, and P. Kannelakis, editors. *Building an Object-Oriented Database System*. Morgan Kaufmann, San Mateo, Calif., 1992.

C. Bardinet, J. E. Dubois, J. P. Caliste, J. J. Royer, and J. C. Oppeneau. Data processing for the environment analysis: A multiscale approach. In *Space and Time in Environmental Information Systems*, Marburg, Germany, 1995. Metropolis.

P. Batty and R. G. Newell. GIS databases are different, 1997. URL http://www.smallworld-us.com.

R. Bayer and E. M. McCreight. Organization and maintenance of large ordered indices. *Acta Informatica*, 1(3):173–189, 1972.

N. Beckmann, H.-P. Kriegel, R. Schneider, and B. Seeger. The R*-tree: An efficient and robust access method for points and rectangles. In *Proc. ACM SIGMOD Int. Conf. on Management of Data*, pages 322–331, 1990.

Belgian Federal Government. *The TELSAT Guide for Satellite Imagery*. Federal Office for Scientific, Technical and Cultural Affairs, 1995. URL http://www.belspo.be/telsat.

J. Benz and K. Voigt. Umwelt-Metadatenbanken im Internet. In B. Page and L. M. Hilty, editors, *Umweltinformatik – Informatikmethoden für Umweltschutz und Umweltforschung*. Oldenbourg, Munich/Vienna, 2nd edition, 1995.

J. Benz and K. Voigt. Aufbau eines Systems zur strukturierten Suche nach Informationsquellen für den Umweltschutz im Internet. In H. Lessing and U. Lipeck, editors, *Informatik im Umweltschutz*, Marburg, Germany, 1996. Metropolis.

U.C. Berkeley. About GRASSLinks: The public access GIS, 1998. University of California at Berkeley, Research Program in Environmental Planning and Geographic Information Systems, URL http://www.regis.berkeley.edu/grasslinks/about_gl.html.

H. K. Bhargava and C. G. Tettelbach. A Web-based decision support system for waste disposal and recycling. *Computers, Environment, and Urban Systems*, 21, 1997.

R. Bill. Spatial data processing in environmental information systems. In N. M. Avouris and B. Page, editors, *Environmental Informatics – Methodology and Applications of Environmental Information Processing*, chapter 5, pages 53–74. Kluwer Academic Publishers, Norwell, Mass., 1995.

R. Bill. *Grundlagen der Geoinformationssysteme II*. Wichmann, Karlsruhe, Germany, 2nd edition, 1996.

R. Bill and D. Fritsch. *Grundlagen der Geoinformationssysteme I*. Wichmann, Karlsruhe, Germany, 4th edition, 1997.

G. Bonham-Carter. *Geographic Information Systems: Modeling with GIS*. Pergamon, New York, 1994.

H. Bossel. *Modeling and Simulation*. A. K. Peters, Wellesley, Mass., 1994.

T. Brinkhoff, H.-P. Kriegel, R. Schneider, and B. Seeger. Multi-step processing of spatial joins. In *Proc. ACM SIGMOD Int. Conf. on Management of Data*, pages 197–208, 1994.

British Standard Institution. Specification for environmental management systems (BS 7750). London, 1992.

A. Buchmann, O. Günther, T. R. Smith, and Y.-F. Wang. *Design and Implementation of Large Spatial Databases*. LNCS 409. Springer-Verlag, Berlin/Heidelberg/New York, 1990.

P. A. Burrough. *Principles of Geographical Information Systems for Land Resources Assessment*. Clarendon Press, Oxford, 1986.

W. J. Campbell and R. F. Cromp. Evolution of an intelligent information fusion system. *Photogrammetric Engineering and Remote Sensing*, 56(6):867–870, 1990.

S. Chaudhuri and U. Dayal. An overview of data warehousing and OLAP technology. *SIGMOD Record*, 26(1):65–74, 1997.

L. T. Chen, R. Drach, M. Keating, S. Louis, D. Rotem, and A. Shoshani. Access to multidimensional datasets on tertiary storage systems. *Information Systems*, 20(2):155–183, 1995.

M. Clarke, R. Kruse, and S. Moral, editors. *Symbolic and Quantitative Approaches to Reasoning and Uncertainty*. LNCS 747. Springer-Verlag, Berlin/Heidelberg/New York, 1993.

W. G. Cochran. *Sampling Techniques*. John Wiley & Sons, New York, 1977.

D. Comer. The ubiquitous B-tree. *ACM Comp. Surv.*, 11(2):121–138, 1979.

D. Cook, J. Symanzik, J. J. Majure, and N. Cressie. Dynamic graphics in a GIS: More examples using linked software. *Computers and Geosciences*, 23(4), 1997.

Council of the European Communities. Council Directive (90/313/EEC) of 7 June 1990 on the freedom of access to information on the environment. *Official Journal of the European Communities*, L158:56–58, 1990.

Council of the European Communities. Council Regulation (93/1836/EEC) of 29 June 1993 allowing participation by companies in the industrial sector in a Community eco-management and audit scheme. *Official Journal of the European Communities*, L168:1–18, 1993.

F. W. Davis et al. Environmental analysis using integrated GIS and remotely sensed data. *Photogrammetric Engineering and Remote Sensing*, 57(6):689–697, 1991.

A. P. Dempster. Upper and lower probabilities induced by a multivalued mapping. *Annals of Mathematical Statistics*, 38, 1967.

A. P. Dempster. A generalization of Bayesian inference. *Journal of the Royal Statistical Society*, 30:205–247, 1968.

P. J. Densham. Spatial decision support systems. In D. J. Maguire, M. F. Goodchild, and D. W. Rhind, editors, *Geographical Information Systems: Principles and Applications*, volume 1, chapter 26, pages 403–412. Longman, Harlow, Great Britain, 1991.

D. K. Denton. *Enviro-Management – How Smart Companies Turn Environmental Costs Into Profits*. Prentice-Hall, Englewood Cliffs, N.J., 1994.

R. Denzer. Concepts for the visual presentation of environmental data. In A. Jaeschke, T. Kämpke, B. Page, and F. J. Radermacher, editors, *Informatik im Umweltschutz*, Informatik aktuell, Berlin/Heidelberg/New York, 1993. Springer-Verlag.

R. Denzer, H. F. Mayer, and W. Haas. Visualization of environmental data. In N. M. Avouris and B. Page, editors, *Environmental Informatics – Methodology and Applications of Environmental Information Processing*, chapter 6, pages 75–92. Kluwer Academic Publishers, 1995a.

R. Denzer, G. Schimak, and D. Russell, editors. *Environmental Software Systems*. Chapman and Hall, London, 1995b.

R. Denzer, D. A. Swayne, and G. Schimak, editors. *Environmental Software Systems*. Chapman and Hall, London, 1997.

J. Desachy. ICARE: An expert system for automatic mapping from satellite imagery. In L. F. Pau, editor, *Mapping and Spatial Modeling for Navigation*. Springer-Verlag, Berlin/Heidelberg/New York, 1990.

O. Deux et al. The story of O_2. *IEEE Trans. Knowledge and Data Eng.*, 2(1):91–108, 1990.

K. R. Dittrich. Object-oriented database systems: The notion and the issues. In *Proc. 1986 Int. Workshop on Object-Oriented Database Systems*, Washington, DC, 1986. IEEE Computer Society Press.

K. R. Dittrich. Object-oriented database systems: The next miles of the marathon. *Information Systems*, 15(1):161–167, 1990.

K. Doan, C. Plaisant, B. Shneiderman, and T. Bruns. Query previews for networked information systems: A case study with NASA environmental data. *SIGMOD Record*, 26(1):75–81, 1997.

P. Drew and J. Ying. Metadata management for geographic information discovery and exchange. In W. Klas and A. Sheth, editors, *Managing Multimedia Data: Using Metadata to Integrate and Apply Digital Data*. McGraw-Hill, New York, 1997.

J. A. Eddy. Environmental research: What we must do. In M. F. Goodchild, B. O. Parks, and L. T. Steyaert, editors, *Environmental Modeling With GIS*, chapter 1, pages 3–7. Oxford University Press, New York/Oxford, 1993.

H. Edelsbrunner. *Algorithms in Combinatorial Geometry*. Springer-Verlag, Berlin/Heidelberg/New York, 1985.

M. Egenhofer. A formal definition of binary topological relationships. In *Proc. 3rd Int. Conf. on Foundations of Data Organization and Algorithms*, LNCS 367, pages 322–338, Berlin/Heidelberg/New York, 1990. Springer-Verlag.

M. Egenhofer. Extending SQL for cartographic display. *Cartography and Geographic Information Systems*, 18:230–245, 1991.

M. Egenhofer. Why not SQL! *Int. J. Geographical Information Systems*, 6(2):71–85, 1992.

M. Egenhofer. Spatial SQL: A query and presentation language. *IEEE Trans. Knowledge and Data Eng.*, 6(1):86–95, 1994.

M. Egenhofer and A. Frank. Towards a spatial query language: User interface considerations. In *Proc. 14th Int. Conf. on Very Large Data Bases*, pages 124–133, 1988.

M. J. Egenhofer and J. R. Herring, editors. *Advances in Spatial Databases*. LNCS 951. Springer-Verlag, Berlin/Heidelberg/New York, 1995.

M. Ehlers, D. Greenlee, T. Smith, and J. Star. Integration of remote sensing and GIS: Data and data access. *Photogrammetric Engineering and Remote Sensing*, 57(6):669–675, 1991.

ESRI Inc., editor. *Proc. 11th ESRI User Conference*. ESRI Inc., Redlands, Calif., 1991.

ESRI Inc. Metadata management in GIS. Technical report, ESRI Inc., Redlands, Calif., 1995a. URL http://www.esri.com/resources/papers/papers.html.

ESRI Inc., editor. *Proc. 15th ESRI User Conference*. ESRI Inc., Redlands, Calif., 1995b.

ESRI Inc. SDTS – Supporting the Spatial Data Transfer Standard in ARC/INFO. Technical report, ESRI Inc., Redlands, Calif., 1995c. URL http://www.esri.com/resources/papers/papers.html.

ESRI Inc. ESRI proposes solutions to OpenGIS. *ARC News*, 18(4):21, 1997a.

ESRI Inc. GIS for everyone – now on Web! *ARC News*, 18(4):1–3, 1997b.

ESRI Inc. IBM and Informix select ESRI to spatially enable companies' respective technologies. *ARC News*, 18(4):9, 1997c.

ESRI Inc., 1998. URL http://www.esri.com.

European Environment Agency. Catalogue of datasources for the environment. Copenhagen, August 1993. Version 930831.

R. Fagin, J. Nievergelt, N. Pippenger, and R. Strong. Extendible hashing: A fast access method for dynamic files. *ACM Trans. Database Systems*, 4(3):315–344, 1979.

K. Fedra. GIS and environmental modeling. In M. F. Goodchild, B. O. Parks, and L. T. Steyaert, editors, *Environmental Modeling with GIS*. Oxford University Press, Oxford/New York/Toronto, 1993.

K. Fedra. Model-based environmental information and decision support systems. In L. M. Hilty, A. Jaeschke, B. Page, and A. Schwabl, editors, *Informatik für den Umweltschutz*, Marburg, Germany, 1994. Metropolis. Vol. 1.

A. Frank. Mapquery – database query language for retrieval of geometric data and its graphical representation. *ACM Computer Graphics*, 16: 199–207, 1982.

J. Frew. Bigfoot: An earth science computing environment for the Sequoia 2000 project. In W. K. Michener, J. W. Brunt, and S. G. Stafford, editors, *Environmental Information Management and Analysis: Ecosystem to Global Scales*. Taylor & Francis, London, 1994.

J.-S. Fritz. Environmental monitoring and information management programmes of international organizations. Technical report, United Nations Environmental Program, Harmonization of Environmental Measurement (HEM) Office, GSF Neuherberg, Germany, 1992.

V. Gaede. Optimal redundancy in spatial database systems. In M. J. Egenhofer and J. R. Herring, editors, *Advances in Spatial Databases*, LNCS 951, pages 132–151, Berlin/Heidelberg/New York, 1995. Springer-Verlag.

V. Gaede and O. Günther. Efficient processing of queries containing user-defined functions. In *Proc. 4th Int. Conf. on Deductive and Object-Oriented Databases*, LNCS 1013, pages 281–298, Berlin/Heidelberg/New York, 1995. Springer-Verlag.

V. Gaede and O. Günther. Multidimensional access methods. *ACM Comp. Surv.*, 30, 1998.

J. Gaffney. Illustra's Web DataBlade module. *SIGMOD Record*, 25(1): 105–112, 1996.

K. Gardels. Open GIS and on-line environmental libraries. *SIGMOD Record*, 26(1):32–38, 1997.

P. Gayle Alston. Environment online: The greening of databases. Part 2: Scientific and technical databases. *Database*, 14(5):34–52, 1991.

P. Gayle Alston and F. W. Stoss. Environment online: The greening of databases. Part 3: Business and regulatory information. *Database*, 14(6):17–35, 1992.

W. Geiger, A. Jaeschke, O. Rentz, E. Simon, Th. Spengler, L. Zilliox, and T. Zundel, editors. *Umweltinformatik '97 – Informatique pour l'Environnement '97*. Metropolis, Marburg, Germany, 1997. Two volumes.

Geographic Designs Inc. Online documentation of Geolineus 3.0. Santa Barbara, Calif., 1995.

M. F. Goodchild, B. O. Parks, and L. T. Steyaert, editors. *Environmental Modeling with GIS*. Oxford University Press, New York/Oxford, 1993.

M. F. Goodchild, L. T. Steyaert, and B. O. Parks, editors. *GIS and Environmental Modeling: Progress and Research Issues*. GIS World Books, 1996.

D. G. Goodenough et al. An expert system for remote sensing. *IEEE Trans. Geosci. Remote Sensing*, 25(3):349–359, 1987.

J. R. Gosz. Sustainable biosphere initiative: Data management challenges. In W. K. Michener, J. W. Brunt, and S. G. Stafford, editors, *Environmental Information Management and Analysis: Ecosystem to Global Scales*, chapter 3, pages 27–39. Taylor & Francis, London, 1994.

D. Greene. An implementation and performance analysis of spatial data access methods. In *Proc. 5th IEEE Int. Conf. on Data Eng.*, 1989.

R. Grützner, editor. *Modellierung und Simulation im Umweltbereich*. Vieweg, Braunschweig/Wiesbaden, 1997.

R. Grützner, A. Häuslein, and B. Page. Werkzeuge für die Umweltmodellierung und -simulation. In B. Page and L. M. Hilty, editors, *Umweltinformatik – Informatikmethoden für Umweltschutz und Umweltforschung*. Oldenbourg, Munich/Vienna, 2nd edition, 1995.

G. Guariso and H. Werthner, editors. *Environmental Decision Support Systems*. John Wiley & Sons, New York, 1989.

O. Günther. *Environmental Information Systems*. Springer-Verlag, Berlin/Heidelberg/New York, 1998.

O. Günther and A. Buchmann. Research issues in spatial databases. *SIGMOD Record*, 19(4):61–68, 1990.

O. Günther, G. Hess, M. Mutz, W.-F. Riekert, and T. Ruwwe. RESEDA: a knowledge-based advisory system for remote sensing. *Applied Intelligence*, 3(4), 1993.

O. Günther and J. Lamberts. Object-oriented techniques for the management of geographic and environmental data. *The Computer J.*, 37(1):16–25, 1994.

O. Günther, R. Müller, P. Schmidt, H. K. Bhargava, and R. Krishnan. MMM: A WWW-based approach for sharing statistical software modules. *IEEE Internet Computing*, 1(3), 1997a.

O. Günther, V. Oria, P. Picouet, J.-M. Saglio, and M. Scholl. Benchmarking spatial joins à la carte. In J. Ferrieé, editor, *Proc. 13e Journées Bases de Données Avancées*, Grenoble, 1997b.

O. Günther, F. J. Radermacher, and W.-F. Riekert. Environmental monitoring: Models, methods, and systems. In N. M. Avouris and B. Page, editors, *Environmental Informatics – Methodology and Applications of Environmental Information Processing*, chapter 3, pages 13–38. Kluwer Academic Publishers, Norwell, Mass., 1995.

O. Günther and W.-F. Riekert. The design of GODOT: An object-oriented geographic information system. *IEEE Data Engineering Bulletin*, 16(3), 1993.

O. Günther and H.-J. Schek, editors. *Advances in Spatial Databases*. LNCS 525. Springer-Verlag, Berlin/Heidelberg/New York, 1991.

O. Günther and E. Wong. A dual approach to detect polyhedral intersections in arbitrary dimensions. *BIT*, 31:2–14, 1991.

O. Günther, editor. Special section on environmental information systems. *SIGMOD Record*, 26(1):3–38, 1997.

R. H. Güting. Gral: An extendible relational database system for geometric applications. In *Proc. 15th Int. Conf. on Very Large Data Bases*, pages 33–44, 1989.

R. H. Güting. An introduction to spatial database systems. *The VLDB J.*, 3(4):357–399, 1994.

R. H. Güting and M. Schneider. Realms: A foundation for spatial data types in database systems. In D. Abel and B. C. Ooi, editors, *Advances in Spatial Databases*, LNCS 692, Berlin/Heidelberg/New York, 1993. Springer-Verlag.

R. H. Güting and M. Schneider. Realm-based spatial data types: The ROSE algebra. *The VLDB J.*, 4:100–143, 1995.

A. Guttman. R-trees: A dynamic index structure for spatial searching. In *Proc. ACM SIGMOD Int. Conf. on Management of Data*, pages 47–54, 1984.

L. Haas, W. Chang, G.M. Lohman, J. McPherson, P.F. Wilms, G. Lapis, B. Lindsay, H. Pirahesh, M. J. Carey, and E. Shekita. Starburst midflight: As the dust clears. *IEEE Trans. Knowledge and Data Eng.*, 2(1):143–161, 1990.

H.-D. Haasis, L. M. Hilty, J. Hunscheid, H. Kürzl, and C. Rautenstrauch, editors. *Umweltinformationssysteme in der Produktion*. Metropolis, Marburg, Germany, 1995a.

H.-D. Haasis, L. M. Hilty, H. Kürzl, and C. Rautenstrauch, editors. *Betriebliche Umweltinformationssysteme – Projekte und Perspektiven*. Metropolis, Marburg, Germany, 1995b.

R. Haining, J. Ma, and S. Wise. Design of a software system for interactive spatial statistical analysis linked to a GIS. *Computational Statistics*, 11(4):449–466, 1996.

G. Hallmark. The Oracle Warehouse. In *Proc. 21st Int. Conf. on Very Large Data Bases*, 1995.

P. Hane. *Environment Online: The Greening of Databases. The Complete Environmental Series from Database Magazine*. Eight Bit Books, Wilton, 1992.

T. Härder and A. Reuter. Architecture of database systems for non-standard applications. In A. Blaser and P. Pistor, editors, *Datenbanksysteme in Büro, Technik und Wissenschaft*, Informatik-Fachberichte 94, pages 253–286, Berlin/Heidelberg/New York, 1985. Springer-Verlag.

W. Härdle, S. Klinke, and B. A. Turlach. *XploRe: An Interactive Statistical Computing Environment*. Springer-Verlag, Berlin/Heidelberg/New York, 1995.

J. M. Hellerstein, J. F. Naughton, and A. Pfeffer. Generalized search trees for database systems. In *Proc. 21st Int. Conf. on Very Large Data Bases*, pages 562–573, 1995.

J. Herring, R. Larsen, and J. Shivakumar. Extensions to the SQL language to support spatial analysis in a topological data base. In *Proc. GIS/LIS Conf.*, pages 741–750, San Antonio, Tex., 1988.

L. M. Hilty. Information systems for industrial environmental management. In N. M. Avouris and B. Page, editors, *Environmental Informatics – Methodology and Applications of Environmental Information Processing*, chapter 22, pages 371–384. Kluwer Academic Publishers, Norwell, Mass., 1995.

L. M. Hilty, A. Jaeschke, B. Page, and A. Schwabl, editors. *Informatik für den Umweltschutz*. Metropolis, Marburg, Germany, 1994. Two volumes.

L. M. Hilty, B. Page, F. J. Radermacher, and W.-F. Riekert. Environmental informatics as a new discipline of applied computer science. In N. M. Avouris and B. Page, editors, *Environmental Informatics – Methodology and Applications of Environmental Information Processing*, chapter 1, pages 1–11. Kluwer Academic Publishers, Norwell, Mass., 1995.

L. M. Hilty and C. Rautenstrauch. Betriebliche Umweltinformatik. In B. Page and L. M. Hilty, editors, *Umweltinformatik – Informatikmethoden für Umweltschutz und Umweltforschung*. Oldenbourg, Munich/Vienna, 2nd edition, 1995.

L. M. Hilty and C. Rautenstrauch. Betriebliche Umweltinformationssysteme (BUIS) – eine Literaturanalyse. *Informatik-Spektrum*, 20 (3), 1997.

E. G. Hoel and H. Samet. Benchmarking spatial join operations with spatial output. In *Proc. 21st Int. Conf. on Very Large Data Bases*, pages 606–618, 1995.

Informix Inc. *Informix Spatial DataBlade Module – User's Guide (Version 2.2)*. Menlo Park, Calif., 1997. Part No. 000-3713.

K. Ingram and W. Phillips. Geographic information processing using a SQL-based query language. In N. Chrisman, editor, *Proc. AUTO-CARTO 8*, pages 326–335, Baltimore, Md., 1987.

A. Jaeschke, T. Kämpke, B. Page, and F. J. Radermacher, editors. *Informatik im Umweltschutz*. Informatik aktuell. Springer-Verlag, Berlin/Heidelberg/New York, 1993.

D. Johnson, P. Shelley, M. Taylor, and S. Callahan. The FINDAR directory system: a meta-model for metadata. In D. Medykyj-Scott, I. Newman, C. Ruggles, and D. Walker, editors, *Metadata in the Geosciences*, pages 123–137, Loughborough, UK, 1991.

H. B. Keller. Neural nets in environmental applications. In N. M. Avouris and B. Page, editors, *Environmental Informatics – Methodology and Applications of Environmental Information Processing*, chapter 9, pages 127–146. Kluwer Academic Publishers, Norwell, Mass., 1995.

A. Kemper and M. Wallrath. An analysis of geometric modeling in database systems. *ACM Comp. Surv.*, 19(1):47–91, 1987.

H. Keune, A. B. Murray, and H. Benking. Harmonization of environmental measurement. *GeoJournal*, 23(3):249–255, 1991.

A. Koschel, R. Kramer, R. Nikolai, W. Hagg, and J. Wiesel. A federation architecture for an environmental information system incorporating GIS, the World Wide Web, and CORBA. In *Proc. Third International Conf. on Integrating GIS and Environmental Modeling*, 1996a. URL http://www.ncgia.ucsb.edu/conf/santa_fe.html.

A. Koschel, R. Kramer, D. Theobald, G. von Bültzingsloewen, W. Hagg, J. Wiesel, and M. Müller. Evaluierung und Einsatzbeispiele von CORBA-Implementierungen für Umweltinformationssysteme. In H. Lessing and U. Lipeck, editors, *Informatik im Umweltschutz*, Marburg, Germany, 1996b. Metropolis.

R. Kramer, R. Nikolai, A. Koschel, C. Rolker, and P. Lockemann. WWW-UDK: A Web-based environmental meta-information system. *SIGMOD Record*, 26(1):16–21, 1997.

R. Kramer and T. Quellenberg. Global access to environmental information. In R. Denzer, G. Schimak, and D. Russell, editors, *Proc. 1995 International Symposium on Environmental Software Systems*, pages 209–218, New York, 1996. Chapman and Hall.

H. Kremers and W. Pillmann, editors. *Raum und Zeit in Umweltinformationssystemen*. Metropolis, Marburg, Germany, 1995. Two volumes.

H.-P. Kriegel. Performance comparison of index structures for multikey retrieval. In *Proc. ACM SIGMOD Int. Conf. on Management of Data*, pages 186–196, 1984.

H. P. Kriegel, T. Brinkhoff, and R. Schneider. An efficient map overlay algorithm based on spatial access methods and computational geometry. In G. Gambosi, M. Scholl, and H.-W. Six, editors, *Geographic Database Management Systems*, pages 194–211, Berlin/Heidelberg/New York, 1991. Springer-Verlag.

H.-P. Kriegel, M. Schiwietz, R. Schneider, and B. Seeger. Performance comparison of point and spatial access methods. In A. Buchmann, O. Günther, T. R. Smith, and Y.-F. Wang, editors, *Design and Implementation of Large Spatial Databases*, LNCS 409, pages 89–114, Berlin/Heidelberg/New York, 1990. Springer-Verlag.

R. Kuggeleijn. Managing data about data. *GIS Europe*, 4(3):32–33, 1995.

P. Ladstätter. Geodaten-Management mit SICAD/GDB-X: Neue Entwicklungen und Positionierung. In *Proc. 5th Int. User Forum*, Munich, 1997. Siemens Nixdorf.

G. Langran. *Time in Geographic Information Systems*. Taylor & Francis, London, 1992.

P. A. Larson. Linear hashing with partial expansions. In *Proc. 6th Int. Conf. on Very Large Data Bases*, pages 224–232, 1980.

R. Laurini and F. Milleret. Spatial data base queries: Relational algebra versus computational geometry. In M. Rafanelli, J. Klensin, and P. Svensson, editors, *Proc. 4th Int. Conf. on Statistical and Scientific Database Management*, LNCS 339, pages 291–313, Berlin/Heidelberg/New York, 1988. Springer-Verlag.

R. Laurini and D. Thompson. *Fundamentals of Spatial Information Systems*. Academic Press, 1994.

R. Lenz, M. Knorrenschild, C. Herderich, O. Springstobbe, E. Forster, J. Benz, W. Assoff, and W. Windhorst. An information system of ecological models. Technical Report 27/94, GSF, Oberschleissheim, Germany, 1994. URL http://www.gsf.de/UFIS/ufis.

H. Lessing. Umweltinformationssysteme – Anforderungen und Möglichkeiten am Beispiel Niedersachsens. In A. Jaeschke,

W. Geiger, and B. Page, editors, *Informatik im Umweltschutz*, Berlin/Heidelberg/New York, 1989. Springer-Verlag.

H. Lessing and U. Lipeck, editors. *Informatik im Umweltschutz*. Metropolis, Marburg, Germany, 1996.

H. Lessing and T. Schütz. Der Umwelt-Datenkatalog als Instrument zur Steuerung von Informationsflüssen. In L. M. Hilty, A. Jaeschke, B. Page, and A. Schwabl, editors, *Informatik für den Umweltschutz*, Marburg, Germany, 1994. Metropolis.

T. M. Lillesand and R. W. Kiefer. *Remote Sensing and Image Interpretation*. John Wiley & Sons, New York, 1987.

W. Litwin. Linear hashing: A new tool for file and table addressing. In *Proc. 6th Int. Conf. on Very Large Data Bases*, pages 212–223, 1980.

G. Lohman, B. Lindsay, H. Pirahesh, and K. B. Schiefer. Extensions to Starburst: Objects, types, functions and rules. *Comm. ACM*, 34(10): 94–109, 1991.

D. B. Lomet and B. Salzberg. The hB-tree: A robust multiattribute search structure. In *Proc. 5th IEEE Int. Conf. on Data Eng.*, pages 296–304, 1989.

T. R. Loveland, J. M. Merchant, D. O. Ohlen, and J. F. Brown. Development of a land-cover characteristics database for the conterminous U.S. *Photogrammetric Engineering and Remote Sensing*, 57(11):1453–1463, 1991.

T. R. Loveland and D. O. Ohlen. Experimental AVHRR land data sets for environmental monitoring and modeling. In M. F. Goodchild, B. O. Parks, and L. T. Steyaert, editors, *Environmental Modeling with GIS*, chapter 37, pages 379–385. Oxford University Press, New York/Oxford, 1993.

G. F. Luger and W. A. Stubblefield. *Artificial Intelligence and the Design of Expert Systems*. Benjamin/Cummings, 1989.

R. Lunetta, R. Congalton, L. Fenstermaker, J. Jensen, K. McTwire, and L. Tinney. Remote sensing and geographic information system data integration: Error sources and research issues. *Photogrammetric Engineering and Remote Sensing*, 57(6):677–687, 1991.

D. J. Maguire, M. F. Goodchild, and D. W. Rhind, editors. *Geographical Information Systems: Principles and Applications*. Longman, Harlow, UK, 1991. Two volumes.

MathSoft Inc. *S+Gislink*. Seattle, 1996.

R. Mayer-Föll. Das Umweltinformationssystem Baden-Württemberg. In A. Jaeschke, T. Kämpke, B. Page, and F. J. Radermacher, editors, *Informatik im Umweltschutz*, Informatik aktuell, Berlin/Heidelberg/New York, 1993. Springer-Verlag.

R. Mayer-Föll. Umweltinformationssystem Baden-Württemberg. In *Proc. InterGeo'97*, Karlsruhe, Germany, 1997.

R. Mayer-Föll, J. Strohm, and A. Schultze. Umweltinformationssystem Baden-Württemberg – Überblick Rahmenkonzeption. In H. Lessing and U. Lipeck, editors, *Informatik im Umweltschutz*, Marburg, Germany, 1996. Metropolis.

J. McCarthy. Metadata management for large statistical databases. In *Proc. 8th Int. Conf. on Very Large Data Bases*, 1982.

Rob McCool. *The Common Gateway Interface*.
URL http://hoohoo.ncsa.uiuc.edu/cgi/overview.html, 1994.

D. M. McKeown. The role of artificial intelligence in the integration of remotely sensed data with geographic information systems. *IEEE Trans. Geosci. Remote Sensing*, 25(3):330–347, 1987.

R. B. McMaster. Automated line generalization. *Cartographica*, 24(2): 74–111, 1987.

R. B. McMaster. Cartographic generalization in a digital enviroment: A framework for implementation in geographic information systems. In *Proc. GIS/LIS Conf.*, 1988.

C. Bauzer Medeiros and F. Pires. Databases for GIS. *SIGMOD Record*, 23:107–115, 1994.

D. Medykyj-Scott, I. Newman, C. Ruggles, and D. Walker, editors. *Metadata in the Geosciences*. Group D Publications, Loughborough, UK, 1991.

R. B. Melton, D. M. DeVaney, and J. C. French, editors. *The Role of metadata in managing large environmental science datasets (Proc. SDM-92)*, Richland, Wash., 1995. Pacific Northwest Laboratory. Technical Report No. PNL-SA-26092.

W. K. Michener, J. W. Brunt, and S. G. Stafford, editors. *Environmental Information Management and Analysis: Ecosystem to Global Scales*. Taylor & Francis, London, 1994.

K. Mikkonen and A. Rainio. Towards a societal GIS in Finland – ArcView application queries data from published geographical databases. In ESRI Inc., editor, *Proc. 15th ESRI User Conference*, Redlands, Calif., 1995. ESRI Inc.

D. Miller and K. Bullock. Metadata for land and geographic information – an Australia-wide framework. In *Proc. AURISA '94*, pages 391–398, Sydney, Australia, 1994.

D. Miller and B. Forner. Experience in developing a natural resource data directory for New South Wales. In *Proc. AURISA '94*, pages 391–398, Sydney, Australia, 1994.

M. Minsky and S. Papert. *Perceptrons*. MIT Press, Cambridge, Mass., 1969.

G. G. Moisen, T. C. Edwards, and D. R. Cutler. Spatial sampling to assess classification accuracy of remotely sensed data. In W. K. Michener, J. W. Brunt, and S. G. Stafford, editors, *Environmental Information Management and Analysis: Ecosystem to Global Scales*, chapter 11, pages 159–176. Taylor & Francis, London, 1994.

S. Morehouse. ARC/INFO: A geo-relational model for spatial information. In *Proc. 7th Int. Symp. on Computer Assisted Cartography*, 1985.

National Center for Geographic Information and Analysis (NCGIA). *Core Curriculum*. University of California, Santa Barbara, 1997. Three volumes. URL http://www.ncgia.ucsb.edu/education.

National Center for Geographic Information and Analysis (NCGIA), 1998. URL http://www.ncgia.ucsb.edu.

R. G. Newell and M. Doe. Discrete geometry with seamless topology in a GIS, 1997. URL http://www.smallworld-us.com.

V. Ng and T. Kameda. Concurrent accesses to R-trees. In D. Abel and B. C. Ooi, editors, *Advances in Spatial Databases*, LNCS 692, pages 142–161, Berlin/Heidelberg/New York, 1993. Springer-Verlag.

V. Ng and T. Kameda. The R-link tree: A recoverable index structure for spatial data. In D. Karagiannis, editor, *Proc. 5th Conf. on Database and Expert Systems Applications (DEXA '94)*, LNCS 856, pages 163–172, Berlin/Heidelberg/New York, 1994. Springer-Verlag.

J. Nievergelt, H. Hinterberger, and K. C. Sevcik. The grid file: An adaptable, symmetric multikey file structure. *ACM Trans. Database Systems*, 9(1):38–71, 1984.

The Open GIS Consortium and the OGIS Project, 1997. URL http://www.opengis.org.

B. C. Ooi. *Efficient Query Processing in Geographic Information Systems*. LNCS 471. Springer-Verlag, Berlin/Heidelberg/New York, 1990.

B. C. Ooi, K. J. McDonell, and R. Sacks-Davis. Extending a DBMS for geographic applications. In *Proc. 5th IEEE Int. Conf. on Data Eng.*, 1989.

P. van Oosterom and T. Vijlbrief. Building a GIS on top of the open DBMS "Postgres". In *Proc. European Conference on GIS*, pages 775–787, 1991.

S. Openshaw. Developing appropriate spatial analysis methods for GIS. In D. J. Maguire, M. F. Goodchild, and D. W. Rhind, editors, *Geographical Information Systems: Principles and Applications*, volume 1, chapter 25, pages 389–402. Longman, Harlow, Great Britain, 1991.

Oracle Inc. Oracle 7 multidimension: Advances in relational database technology for spatial data management, March 1995. White Paper.

J. Orenstein. An object-oriented approach to spatial data processing. In *Proc. 4th Int. Symp. on Spatial Data Handling*, Zürich, 1990.

A. Ortmaier. Online-Datenbanken zur Informationsgewinnung für Zwecke des Umweltschutzes – Eine praktische Anleitung zur sinnvollen Auswahl und effizienten Nutzung. Master's thesis, Humboldt-Universität, Berlin, 1995.

F. Ostyn. The EDRA – Fueling GIS applications with required geographical information. In ESRI Inc., editor, *Proc. 15th ESRI User Conference*, Redlands, Calif., 1995. ESRI Inc.

B. Page and L. M. Hilty, editors. *Umweltinformatik – Informatikmethoden für Umweltschutz und Umweltforschung*. Oldenbourg, Munich/Vienna, 2nd edition, 1995.

D. J. Peuquet and D. F. Marble. ARC/INFO: an example of a contemporary geographic information system. In D. J. Peuquet and D. F. Marble, editors, *Introductory Readings in Geographic Information Systems*, pages 90–99. Taylor & Francis, London, 1990a.

D. J. Peuquet and D. F. Marble, editors. *Introductory Readings in Geographic Information Systems.* Taylor & Francis, London, 1990b.

H.-N. Pham and T. Wittig. An adaptable architecture for river quality monitoring. In N. M. Avouris and B. Page, editors, *Environmental Informatics – Methodology and Applications of Environmental Information Processing*, chapter 14, pages 217–235. Kluwer Academic Publishers, Norwell, Mass., 1995.

W. Pillmann and D. J. Kahn. Distributed environmental data compendia. In A. M. Aiken, editor, *Proc. 12th IFIP World Computer Congress*, Amsterdam, 1992. Elsevier.

F. P. Preparata and M. I. Shamos. *Computational Geometry.* Springer-Verlag, Berlin/Heidelberg/New York, 1985.

D. Pullar and M. Egenhofer. Towards formal definitions of topological relations among spatial objects. In *Proc. 3rd Int. Symp. on Spatial Data Handling*, pages 225–242, 1988.

F. J. Radermacher. The importance of metaknowledge for environmental information systems. In O. Günther and H.-J. Schek, editors, *Advances in Spatial Databases*, LNCS 525, pages 35–44, Berlin/Heidelberg/New York, 1991. Springer-Verlag.

J. F. Raper and B. Kelk. Three-dimensional GIS. In D. J. Maguire, M. F. Goodchild, and D. W. Rhind, editors, *Geographical Information Systems: Principles and Applications*, volume 1, chapter 20, pages 299–317. Longman, Harlow, Great Britain, 1991.

T. M. Rhyne. Going virtual with geographic information and scientific visualization. *Computers and Geosciences*, 23(4), 1997.

G. Della Riccia, R. Kruse, and R. Viertl, editors. *Mathematical and Statistical Methods in Artificial Intelligence.* Springer-Verlag, Berlin/Heidelberg/New York, 1995.

O. Richter. *Simulation des Verhaltens ökologischer Systeme – Mathematische Methoden und Modelle.* VCH-Verlag, Weinheim, Germany, 1985.

W.-F. Riekert. Extracting area objects from raster image data. *IEEE Comp. Graphics and App.*, 13(2):68–73, 1993.

W.-F. Riekert, R. Mayer-Föll, and G. Wiest. Management of data and services in the Environmental Information System (UIS) of Baden-Württemberg. *SIGMOD Record*, 26(1):22–26, 1997.

J. T. Robinson. The K-D-B-tree: A search structure for large multidimensional dynamic indexes. In *Proc. ACM SIGMOD Int. Conf. on Management of Data*, pages 10–18, 1981.

N. Roussopoulos and D. Leifker. An introduction to PSQL: A pictoral structured query language. In *IEEE Workshop on Visual Languages*, 1984.

N. Roussopoulos and D. Leifker. Direct spatial search on pictorial databases using packed R-trees. In *Proc. ACM SIGMOD Int. Conf. on Management of Data*, 1985.

D. E. Rumelhart, J. L. McClelland, and the PDP Group, editors. *Parallel Distributed Processing: Explorations in the Microstructure of Cognition*. Bradford Books, Cambridge, Mass., 1985.

H. Samet. *Applications of Spatial Data Structures*. Addison-Wesley, Reading, Mass., 1990a.

H. Samet. *The Design and Analysis of Spatial Data Structures*. Addison-Wesley, Reading, Mass., 1990b.

SAP AG, 1998. URL http://www.sap.com.

H.-J. Schek, H.-B. Paul, M. H. Scholl, and G. Weikum. The DASDBS project: Objectives, experiences and future prospects. *IEEE Trans. Knowledge and Data Eng.*, 2(1):25–43, 1990.

M. Schneider. *Spatial Data Types for Database Systems*. LNCS 1288. Springer-Verlag, Berlin/Heidelberg/New York, 1997.

M. Scholl and A. Voisard. Thematic map modeling. In A. Buchmann, O. Günther, T. R. Smith, and Y.-F. Wang, editors, *Design and Implementation of Large Spatial Databases*, LNCS 409, Berlin/Heidelberg/New York, 1990. Springer-Verlag.

M. Scholl and A. Voisard. Object-oriented database systems for geographic applications: An experiment with O_2. In F. Bancilhon, C. Delobel, and P. Kanellakis, editors, *The O_2 Book*, pages 585–618. Morgan Kaufmann, San Mateo, Calif., 1992.

M. Scholl and A. Voisard, editors. *Advances in Spatial Databases*. LNCS 1262. Springer-Verlag, Berlin/Heidelberg/New York, 1997.

M. Scholl, A. Voisard, J.-P. Peloux, L. Raynal, and P. Rigaux. *SGBD Géographiques*. International Thomson Publishing France, Paris, 1996.

Scientific Computers GmbH, 1997. URL http://www.scientific.de.

L. M. Scott. Identification of a GIS attribute error using exploratory data analysis. *The Professional Geographer*, 46(3):378–386, 1994.

D. Seaborn. Database management in GIS: Is your system a poor relation? *GIS Europe*, 4(5):34–38, 1995.

G. Shafer. *A Mathematical Theory of Evidence*. Princeton University Press, Princeton, NJ, 1976.

S. Shekhar, M. Coyle, B. Goyal, D.-R. Liu, and S. Sarkar. Data models in geographic information systems. *Comm. ACM*, 40(4), 1997.

M. Siegel and S. Madnick. A metadata approach to resolving semantic conflicts. In *Proc. 17th Int. Conf. on Very Large Data Bases*, 1991.

Siemens Nixdorf Informationssysteme AG, 1998. URL http://www.sni.de.

A. Silberschatz, M. Stonebraker, and J. Ullman. Database systems: Achievements and opportunities. *Comm. ACM*, 34(10):110–120, 1991.

T. R. Smith. A digital library for geographically referenced materials. *IEEE Computer*, 29(5), 1996.

SpotImage. *The SPOT Satellite System*, 1997. URL http://www.spot.com.

S. G. Stafford, J. W. Brunt, and W. K. Michener. Integration of scientific information management and environmental research. In W. K. Michener, J. W. Brunt, and S. G. Stafford, editors, *Environmental Information Management and Analysis: Ecosystem to Global Scales*, chapter 1, pages 3–19. Taylor & Francis, London, 1994.

J. Star and J. Estes. *Geographic Information Systems: An Introduction*. Prentice-Hall, Englewood Cliffs, N.J., 1990.

U. Steger. *Umweltmanagement – Erfahrungen und Instrumente einer umweltorientierten Unternehmensstrategie*. Gabler, Frankfurt/Main, 1993.

M. Stonebraker. On rules, procedures, caching and views in data base systems. In *Proc. ACM SIGMOD Int. Conf. on Management of Data*, 1990.

M. Stonebraker. The Sequoia 2000 project. In D. Abel and B. C. Ooi, editors, *Advances in Spatial Databases*, LNCS 692, Berlin/Heidelberg/New York, 1993. Springer-Verlag.

M. Stonebraker, editor. *Readings in Database Systems*. Morgan Kaufmann, San Mateo, 1994.

M. Stonebraker and L. Rowe. The design of POSTGRES. In *Proc. ACM SIGMOD Int. Conf. on Management of Data*, 1986.

M. Stonebraker, L. A. Rowe, and M. Hirohama. The implementation of POSTGRES. *IEEE Trans. Knowledge and Data Eng.*, 2(1):125–142, 1990.

F. W. Stoss. Environment online: The greening of databases. Part 1: General interest databases. *Database*, 14(4):13–27, 1991.

D. E. Strebel, B. W. Meeson, and A. K. Nelson. Scientific information systems: A conceptual framework. In W. K. Michener, J. W. Brunt, and S. G. Stafford, editors, *Environmental Information Management and Analysis: Ecosystem to Global Scales*, chapter 5, pages 59–84. Taylor & Francis, London, 1994.

P. H. Swain and S. M. Davis. *Remote Sensing: The Quantitative Approach*. McGraw-Hill, New York, 1978.

D. Swayne, D. Cook, and A. Buja. XGobi: Interactive dynamic graphics in the X Window System with a link to S. In *ASA Proc. Section on Statistical Graphics*, pages 1–8, Alexandria, Va., 1991. American Statistical Association.

A. Sydow. Computersimulation – ein Schlüssel zum Verständnis der Umwelt. *Der GMD-Spiegel*, 26(4):16–18, 1996.

J. Symanzik, T. Kötter, S. Schmelzer, S. Klinke, D. Cook, and D. F. Swayne. Spatial data analysis in the dynamically linked ArcView/XGobi/XploRe environment. *Computing Science and Statistics*, 29, 1997a.

J. Symanzik, J. J. Majure, D. Cook, and I. Megretskaia. Linking ArcView 3.0 and XGobi: Insight behind the front end. Technical Report 97-10, Department of Statistics, Iowa State University, Ames, Iowa, 1997b.

D. G. Theriault. Smallworld GIS: An open system architecture for a workstation-based geographical information system, 1997. URL http://www.smallworld-us.com.

R. B. Tilove. Set Membership Classification: A Unified Approach to Geometric Intersection Problems. *IEEE Transactions on Computers*, C-29(10), 1980.

C. D. Tomlin. *Geographic Information Systems and Cartographic Modeling.* Prentice-Hall, Englewood Cliffs, N.J., 1990.

J. Ton, J. Sticklen, and A. Jain. Knowledge-based segmentation of LANDSAT images. *IEEE Trans. Geosci. Remote Sensing*, 29(2):222–232, 1991.

A. Ultsch. Einsatzmöglichkeiten von neuronalen Netzen im Umweltbereich. In B. Page and L. M. Hilty, editors, *Umweltinformatik – Informatikmethoden für Umweltschutz und Umweltforschung.* Oldenbourg, Munich/Vienna, 1995.

Umweltministerium Baden-Württemberg and McKinsey and Company, Inc. *Konzeption des ressortübergreifenden Umweltinformationssystems im Rahmen des Landessystemkonzepts Baden-Württemberg.* Umweltministerium Baden-Württemberg, Stuttgart, Germany, 1987. Five volumes.

United States Federal Geographic Data Committee. *Content Standards for Digital Geospatial Metadata.* U.S. Government, Federal Geographic Data Committee, Washington, DC, 1994. URL ftp://fgdc.er.usgs.gov.

United States Geological Survey. Spatial Data Transfer Standard (SDTS). Reston, Virg., 1992. URL http://mcmcweb.er.usgs.gov/sdts.

United States Government. Coordinating geographic data acquisition and access: The national spatial data infrastructure. Washington, DC, April 1994. U.S. Executive Order 12906.

United States National Aeronautics and Space Administration. *Landsat Pathfinder.* U.S. Government, National Aeronautics and Space Administration,Humid Tropical Forest Inventory Project, 1997. URL http://amazon.sr.unh.edu/pathfinder1/index.html.

United States National Aeronautics and Space Administration. Earth observing system data and information system (EOSDIS), 1998. URL http://spsosun.gsfc.nasa.gov/EOSDIS_main.html.

United States National Oceanic and Atmospheric Administration. *NESDIS Home Page.* U.S. Government, National Oceanic and Atmospheric Administration (NOAA), National Environmental Satellite, Data, and Information Service, Washington, DC, 1997. URL http://ns.noaa.gov/NESDIS/NESDIS_Home.html.

University of California at Berkeley, Digital Library Project, 1998. URL http://elib.cs.berkeley.edu.

K. Voigt and R. Brüggemann. Metadatenbank der Online Datenbanken. *Cogito*, 6:8–13, 1993.

K. Voigt and R. Brüggemann. Meta information systems for environmental chemicals. In N. M. Avouris and B. Page, editors, *Environmental Informatics – Methodology and Applications of Environmental Information Processing*, chapter 19, pages 315–336. Kluwer Academic Publishers, Norwell, Mass., 1995.

A. Voisard. Mapgets: A tool for visualizing and querying geographic information. *Journal of Visual Languages and Computing*, 6:367–384, 1995.

A. Voisard and H. Schweppe. A multilayer approach to the open GIS design problem. In N. Pissinou and K. Makki, editors, *Proc. 2nd ACM GIS Workshop*, pages 23–29, New York, 1994. ACM Press.

A. Voisard and H. Schweppe. Abstraction and decomposition in interoperable GIS. *Int. J. Geographical Information Systems*, 12, 1998.

D. R. F. Walker. Introduction to metadata in the geosciences. In D. Medykyj-Scott, I. Newman, C. Ruggles, and D. Walker, editors, *Metadata in the Geosciences*, Loughborough, UK, 1991. Group D Publications.

WASY GmbH, 1997. URL http://www.wasy.de.

L. Wicke, H.-D. Haasis, F. Schafhausen, and W. Schulz. *Betriebliche Umweltökonomie – Eine praxisorientierte Einführung*. Vahlen, Munich, 1992.

R. Wilensky. Toward work-centered digital information services. *IEEE Computer*, 29(5), 1996.

M. F. Worboys. A generic model for planar geographical objects. *Int. J. Geographical Information Systems*, 6:353–372, 1992.

M. F. Worboys, H. M. Hearnshaw, and D. J. Maguire. Object-oriented data and query modelling for geographical information systems. In *Proc. 4th Int. Symp. on Spatial Data Handling*, Zürich, 1990.

L. A. Zadeh. Fuzzy sets. *Information and Control*, 8:338–353, 1965.

P. Zamparoni. Feature extraction by rank-order filtering for image segmentation. *Int. J. Pattern Recognition Artif. Intelligence*, 2(2):301–319, 1988.

Chapter 29

MASSIVE DATA SETS ISSUES IN EARTH OBSERVING

Ruixin Yang
Center for Earth Observing & Space Research (CEOSR), School of Computational Sciences (SCS), George Mason University, Fairfax, VA 22030, USA
ryang@gmu.edu

Menas Kafatos
Center for Earth Observing & Space Research (CEOSR), School of Computational Sciences (SCS), George Mason University, Fairfax, VA 22030, USA
mkafatos@gmu.edu

Abstract Current and next decade global Earth observing, other remote sensing and related climate analysis data collected by space and operational U.S. agencies such as NASA and NOAA, the European ESA, the Japanese NASDA and other international agency missions of India, China, Russia, etc. will reach unprecedented data volumes, exceeding many petabytes. This is rushing in a new era in Earth system science. Along with the technology and data volumes afforded by remote sensing, there has been an unprecedented increase of capabilities in distributed data systems in the last few years. The existence of the Internet and the World Wide Web afford users access to data at diverse distributed sites that would had been very difficult in the past and only available to specialists. These data can be accessed by a variety of scientists, applications experts and the general public. Yet, the unprecedented large volumes of such missions are also presenting formidable challenges to wide user access and require both higher bandwidths of future Internet systems as well as more focused, user-centered data productions. Specialized data productions are best achieved in federated data systems. The large data volumes present a challenge of access, storage and distribution. What is most important is not just the volume of the data itself but the information they contain. A data system's usefulness is related to the ease that its users can search and access products and as such obtain information on the actual content of data before proceeding to order large volumes of data sets which may or may not serve their needs. Here we explore different functionalities in distributed Earth observing data systems and associated interoperability options. As an example, we examine options in the Earth Science Information Partners Federation (ESIPs) funded by NASA to extend the usage of NASA remote sensing data holdings. Results of several interoperability options applicable to federated systems are also presented. We also examine challenges presented by the use of regional remote sensing missions such as hyperspectral imaging.

Keywords: Massive data sets, Earth science.

1. Earth System Science Data in the Internet Era

The beginning of the 21st century will be marked by an unprecedented rate of collection of remote sensing data as U.S. and international agencies such as the National Aeronautics and Space Administration (NASA), the National Oceanic and Atmospheric Administration (NOAA), European and Japanese agencies such as the European Space Agency (ESA), the Japanese NASDA and other international agencies of China, India, Russia, etc. launch a series of Earth observing/remote sensing missions aimed at studying the Earth and its environs. The successful launches of *Landsat 7* and the first *Earth Observing System* (EOS) platform *Terra* by NASA in 1999 are ushering this new era. Large scientific database issues, such as interoperability, virtual environments, infrastructure such as networking, distributed databases, metadata, etc. (see EUU (2001)) cut across different fields and certainly apply to Earth system scientific databases. Geospatial databases have been established at many countries, principally in the U.S. and in E.U. countries (see Earth System Science URLs).

Current and future data will provide scientists with information on the physical and environmental processes operating on the Earth and assist scientists in their understanding of the Earth and its components. Current missions (e.g. the joint NASA/NASDA *Tropical Rainfall Measuring Mission*, or TRMM; ocean color missions such SeaWiFS; the NOAA operational polar orbiters, etc.), the recently launched *Terra* and *Landsat 7* missions, and future missions are collecting or will collect data exceeding many hundreds of terabytes or even petabytes in volume, some platforms producing data rates that will exceed $\sim$ terabyte/day. Specifically, the EOS program, a series of three main Earth observing platforms (*Terra*, *Aqua* and EOS-CHEM), *Landsat 7* and several related smaller missions, will produce the requisite Earth observations and associated massive data products (Asrar and Greenstone (1995)) extending to the early

2000's. NASA is also building an associated data information system termed Earth Observing System Data Information System (EOSDIS), designed for data production from these missions and distribution to users. The existing and future data products are or will be distributed among several science data centers termed Distributed Active Archive Centers (DAACs) (Asrar and Greenstone (1995)). The EOSDIS original centralized design can afford baseline products such as Level 0, Level 1 and higher products such as Level 2 as needed but due to cost considerations, design problems, etc. might not produce all data products as originally anticipated.

Of equal importance and competing in volume holdings, climate prediction model and analysis data sets can reach equally high volumes and rates. Daily rates of model output data from NOAA's different centers such as the NOAA/National Environmental Satellite, Data and Information Service's National Climatic Data Center (NCDC), the Climate Prediction Center (CPC), the Pacific Marine Environmental Laboratory (PMEL), etc. or from the European Center for Medium Weather Forecast (ECMWF) exceed tens to hundreds of Gigabytes with total volume holdings in excess of petabytes. Model output data are serving Earth system science and global change areas, in helping to produce integrated predictions and complementing Earth observing data.

Centralized solutions cannot easily provide more focused data products in support of different specific science and applications communities while at the same time flexible new information technology developments, which are themselves undergoing major changes every 2- 3 years, are hard to be integrated in a centralized system. Therefore, in its 1995 study of the U.S. Global Change Research Program (USGCRP) and NASA's Mission to Planet Earth (presently the Earth Science Enterprise or ESE), the National Research Council's Board on Sustainable Development recommended augmenting the current EOSDIS system through the formation of a federation of Earth Science Information Partners (ESIPs). This federation was envisaged to consist of several independent data information entities providing innovative solutions to Earth science data systems (BSD (1995)). The NRC recommended that "Responsibility for product generation and publication and for user services should be transferred to a federation of partners selected through a competitive process open to all". This should be taken as referring to functions beyond the baseline system (although where that dividing line might be is somewhat ambiguous). The intent was for NASA to share the processing and distribution tasks with universities and other research labs where additional talent and expertise could be brought to bear on the difficult problem of dealing with the large data sets and perhaps even more important, the information content of these data sets. Long-term evolutionary vision of NASA's data systems include the New Data Information System & Services (NewDISS) which will include distributed solutions and P.I.(Principal Investigator)-led data processing and distribution.

The USGCRP and NASA's ESE, focus on several interdisciplinary areas including seasonal to interannual (S-I) climate studies, land use/land cover, ozone depletion, climate change and natural hazards. On seasonal to interannual timescales extending to several years, satellite coverage and *in situ* measurements can help in our understanding of extreme events such as the 1997-98 warm or El Niño event, droughts, flood seasons and other global and regional climate events. The reason is that the timescales of data collection from existing NOAA polar orbiters have afforded scientists data spanning, in some cases, more than a decade or several El Niño/Southern Oscillation (ENSO) cycles, the latter being 3–7 years.

The last decade has witnessed a thousand-fold in computer speed, great innovations in remote sensing (RS) detectors and capabilities, a great reduction in the cost of computing and data storage and widespread access to the information highways (for similar arguments as presented here see Kafatos et al. (1999)). The existence of the Internet and the World Wide Web (WWW) has rushed in an era of wide access to information and data impossible even a few years ago. As the data volumes continue to grow, storage costs are dropping but not fast enough to accommodate the data increases. Moreover, a challenge remains for the long-term archiving of RS data, media degradation will occur faster than data will be able to be transferred to new media. We should also point out that the number of Earth scientists is remaining constant or slowly increasing at best.

Although general-purpose search engines are still limited in providing few step-specific, useful information to users and require instead several searches to yield the desired results, it is clear that users can now access data sets and information that before 1995 (when the National Science Foundation's NSFnet went public opening the door for today's Internet) was reserved for specialists at government labs and small number of academic institutions. Scientists, applications specialists, graduate and undergraduate students, high school students and even the general public can now access, order or even download data to their own systems for their own use. Along with the existence of the vast web information contained in the WWW, current Internet is, however, being stretched by precisely this volume of information and usage, mostly of commercial or private nature, limiting effective access to large data volumes by, ironically, the very specialists and scientists who were benefiting in the past and who were the reason the whole Internet revolution was started to begin with. Today, scientists, researchers and applications users need, however, not just access to information and data but *efficient* access. If a user requires some data sets for a specific application which are, say, hundreds of megabytes or even gigabytes, general purpose kilobit on-line access rates are, clearly, inadequate. The user will have to order the data sets in hard media and the advantage of fast, on-line access is clearly lost.

These are some of the reasons why the clogging of the conventional Internet highways has led to the need for a faster, more sophisticated network that is being created much the same way as the original Internet was created, i.e. through focused funding (Kafatos et al. (1999)). Two related initiatives, the university-led Internet2 (I2) and the Government's Next Generation Internet (NGI) will allow transfer of high-priority information at a high rate while passing lower- priority information (e.g. conventional e-mail) at lower frequencies (Finley (1998)). These initiatives will prove instrumental for future distributed Earth observing data systems. Whereas, Internet2 is the university community's response to the need to return to dedicated bandwidth for academic/research needs, the Next Generation Internet (NGI (1998)) is a Government multi- agency R&D initiative.

Both I2 and NGI are benefiting from the existence of the very-high-performance Backbone Network Service (vBNS), a joint project between the NSF and MCI Telecommunications. vBNS is a nationwide network connecting many of the 120 I2 institutions. The present vBNS backbone runs at 622 Mbps or (OC12) rates and most I2 members participating on the vBNS are connected at OC3 rates (155 Mbps), while some are connected at T3 and others up to OC12 (Finley (1998)). It is expected that the existence of vBNS will accelerate and assist in the implementation of both I2 and NGI. For example, both I2 and NGI are exploring ways to share in the establishment of GigaPoPs (gigabit-capacity points of presence), the main links in the future high-bandwidth Internet systems (Kafatos et al. (1999)).

In addition to the network challenge, we focus here our attention on global Earth observing systems and the associated Earth science data sets and the challenges they present to both distribution and "mining" of the information they contain. Although other science fields may not face identical problems as Earth science, many of the conclusions we will draw apply to any distributed science data system with large data holdings.

Applications for a variety of regional projects including environmental monitoring, inventories, forestry applications and transportation needs require data at much higher spatial and spectral resolutions than Earth observing missions. New technologies such as hyperspectral imaging and synthetic aperture radar (SAR) are becoming available. Universities in different countries can enter into cooperative programs with area industries and governments in support of user communities needing these data. Geographic Information Systems (GIS) tools integrated with remote sensing products, relevant algorithms and laboratory spectral measurements are all required. Moreover, innovative information technology products and techniques in computer science, statistics and visualization can enhance remote sensing tools and applications.

We here concentrate on issues related to, primarily, Earth observing for both global Earth system science and regional applications as opposed to model output data as the former are more diverse in distribution, size, and data types.

2. Issues in Federated Earth Science Data Systems

The existence of different data centers with their own data holdings of different formats and metadata, with different local information management systems, different database schema and serving different user communities, implies that achieving complete interoperability between different large databases will be difficult to achieve. What is more realistic is to form federation(s) of Earth system science data providers with open and flexible systems. Such federated systems would share resources with minimum procedures, make their data available to the appropriate scientific communities supported by each federated system, achieve some level of interoperability and present a common face to the outside world. The most crucial, top-level federation issues are the following:

- Ownership of federation information system
  The federation partners must have real ownership of the overall information system and the technology solutions associated with it. Otherwise the system will ultimately not serve the partners and, by implication, the users served by these partners.

- Long-term stewardship of federation data & data system
  Data holdings must ultimately be preserved. The Earth science data should last several decades to provide future generations of scientists and the societies which they serve with the environmentally rich information that prolonged observations of the Earth

and its environs can provide. This provides a challenge to federated systems as the partners of the federation might not be as permanent as Government data centers are. The data and the information products must outlast individual data and information providers, or the funding which supported their existence.

- Type and degree of interoperability in the federation (and associated information technology, or IT, issues)
  How can information technology solutions provide seamless access to data and information products across the entire federation(s)? Since the federation partners are, by definition, diverse, the IT solutions they bring to bear are also diverse. Interoperability must, however, provide common solutions to IT access which, hopefully, extend beyond the simplest common denominator. Here, the individual partners must do more to develop/accept IT solutions enhancing their current capabilities for the common good.

- How can success of the federation be assessed?
  A federation should be more than the sum of its parts or in more concrete terms, a federation should provide benefits that are more than the sum of the benefits provided by individual partners. Assessment of the success of such a partnership requires metrics that do not go counter to the individual metrics and mission of each partner and, at the same time, are well accepted by the community of scientists and data providers. The federation partners need to adhere to commonly developed standards and principles.

In terms of enabling information technology, the following principles ought to be adhered to as interoperability design options for the federation's data information system architecture:

- User-Centric and Custom-Design
  The highest goal of the federation is to serve the users of the various partners and provide custom-based solutions to their data and information needs. IT solutions are sought to provide user-centric services and, moreover, they are not an end in themselves. IT solutions should enable users' access and usage of federation data and information products. The data information system should be designed to provide easy access and usage of data and information products by the communities of users. No system should be built if it hampers these efforts. In summary, as such, IT solutions are to be designed to customers' needs.

- Autonomy of federation partners
  The federation should preserve the autonomy of its partners. No

federation standards or principles should be adopted that limit the autonomy of its partners. If such standards or principles were to be adopted, they would ultimately adversely affect the needs of the users and the entire system would not be user-centric. As such, no IT solutions that hamper the autonomy of the partners including the individual partners' IT solutions should be implemented. Related to this issue, IT designs that protect the independence and uniqueness of individual partners are to be sought and promoted.

- Enabling the federation
  IT design should enable the entire federation, furthering its goals and mission. IT federation options include both catalogue search protocols as well as data search and access solutions. Specific points are, of course, federation-related and can be decided by each federation or its management.

- Open
  Following an ever-changing information technology environment, an important IT principle is openness: Design of system architecture can change as needed, following users', federation partners' and federation's needs as well as new overall IT developments. Openness can be furthered by adopting existing IT standards, freely-available or low-cost commercial off-the-shelf (COTS) tools, etc.

- Incrementally Developed
  Design and implementation of system architecture and its IT components need to follow an incremental approach with some functionalities to be in place from the very beginning of the federation. This means that prototypes ought to be designed, tried and adopted as needed.

- Modular
  IT components are to be built, tried out and fit together. If an IT component does not work, it should not hinder overall design or goals, it should be bypassed and substituted by other component(s). Modules may also exist at different physical sites. A modular approach becomes an integrated approach as the modules are tested, integrated and adopted by more and more federation partners.

- Reuse
  Another IT principle that allows flexibility is to reuse existing IT solutions. As such, components can be assembled from other systems, including the federation partners' own designs or prototypes,

COTS, etc. Minimum new coding should be developed as needed and to assure linkage of existing modules. No central, large-scale coding should be followed.

- On-line
  The prevalence of the WWW implies that IT-enabled functionalities should be distributed. This includes platform-independent coding such as Java, etc. The overall design should make it at least as easy for users to do what they are already doing on their own systems and this generally means on-line capabilities. Many existing and future systems employ the client-server paradigm, particularly useful to federated systems. An example is the VDADC architecture which capitalized on the 1994 EOSDIS Independent Architecture Study-proposed concept of Domain Application Data Centers (DADCs) (Kafatos et al. (1994); Kerschberg et al. (1994)) as well as on the NASA's DAACs access, management and distribution of data mechanisms. The VDADC, on the other hand, relies on a client-server architecture utilizing the World Wide Web (WWW).

A proposed federated architecture that can apply to diverse Earth science data systems (Kerschberg et al. (1994)) consists of four layers: The top layer consists of different user communities. Instead of diverse user communities accessing data sets directly from the data centers where data are archived (bottom layer), the users access specialized DADCs in the *InfoMart* layer. The *InfoMarts* provide value-added data and information products to their constituent communities (e.g. S-I scientists). The various DADCs feed popular products to a common *Data Warehouse* layer. The *Data Warehouse* is not just an archive layer. It is the main federation data repository. Its data holdings change in time as new data and information products are archived by the federation partners and as user communities and their data needs themselves are changing.

3. Data for Regional Applications

Global Earth observing was primarily intended for scientific research, such as monitoring the ozone layer or the atmospheric warming, and there have been numerous efforts to extend the utility of these missions to applications and value-added products for regional applications. The business prospects and benefits of regional remote sensing can be many folds. From the pure business point of view, the global remote sensing market is estimated at $5.1 billion by 2005. The European Space Agency has provided remote sensing systems, such as the ERS series, intended

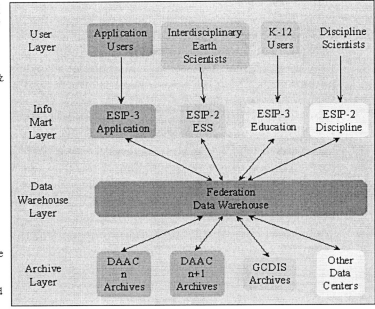

Figure 29.1. A Multi-Layer Federated Information Architecture.

for both scientific and applications usage. Along with global Earth observing missions, new RS missions and sensors with increased spatial and or spectral resolutions are becoming available. Here we give examples of a few RS capabilities and applications, concentrating on hyperspectral considerations, and data fusing/GIS applications. Missions providing regional application data can equal or even surpass global Earth observing data requirements. We give here two examples of regional RS application areas: Hyperspectral imaging and its many application areas; and RS for regional disasters. These serve to bring out the unique issues relevant to regional applications.

Hyperspectral Remote Sensing

Hyperspectral remote sensing or hyperspectral imaging (HSI) is an emerging technology for accurate remote sensing of the Earth for a variety of regional applications. Emerging capabilities in satellite technology, remote sensing technology and large database management afford the remote sensing industry and many stakeholders including agricultural concerns, environmental agencies, policy makers, oil industries as well as local government agencies with opportunities that did not exist even a few years ago. HSI platforms can be airborne (such as NASA's AVIRIS) or spaceborne (such as OrbImage's *OrbView 4*) and will be providing users and stakeholders with remote sensing information products that do not currently exist for a variety of applications, with image clarity, increased accuracy, ease of use, timeliness and, is hoped, at a competitive cost.

Usage of such sensors covers many different applications. HSI instruments, that will be mounted on low Earth-orbit satellites, and/or coupled with similar instruments aboard aircraft, can provide data and information products with sensing information that accesses seemingly dense surfaces to locate mineral and petroleum deposits. Other products can assess the productivity of large bodies of water, monitor soil before planting and analyze and monitor crops as they grow, assess road traffic flow, map routes for roads and pipelines, and help educate the next generation of research and applied scientists through modeling, simulations and algorithm developments. The data archiving and processing requirements of HSI can, however, be very challenging:

For example, consider a 210-channel imaging spectrometer (similar to AVIRIS or the future spaceborne systems *Near Earth Map Observer*, NEMO, and *OrbView 4*), having a 50 km swath, and, say, a 10 meter resolution. Such a system would have $(50 \times 10^3 m)/(10m) = 5,000$ pixels across track. The satellite moves $\sim 7.5 km/s$ i.e. $7.5 \times 10^3/10m$ or 750 pixels along track per second. The number of pixels imaged per second would then be $(5,000) \times (750) = 3.75 \times 10^6 pixels/sec$ and the number of data points generated would be $(210 channels) \times (3.75 \times 10^6 pixels/sec) = 7.875 \times 10^8 pixels/sec$. If one were to compare these data points with a spectral library containing, say, 500 spectra, then one would need a ~ 400 Gigaflop supercomputer!

To take advantage of these emerging opportunities, centers of excellence for remote sensing with a strong component in hyperspectral imaging which will be involved in applied and fundamental research in remote sensing can be set up, similarly to federal Earth observing and climate data centers. An HSI center can provide an environment for remote sensing research to thrive, concentrating on pilot projects carried out by universities. Furthermore, the hyperspectral and remote sensing research and applications at such a center can be coupled to other center activities allowing the center to:

- serve as a vehicle for government sponsored research

- develop talent pools upon which industry can call
- facilitate beneficial interactions among academia, industry, and government
- grant funds for tailored remote sensing research
- manage structured forums for resolving focussed issues, and
- manage the research life cycle: requirements to research data sources and research to applications

Regional centers' research activities can, among other things, emphasize:

- the detection, tracking, monitoring, assessing & leveraging spectral signatures
- the development of appropriate hyperspectral remote sensing algorithms
- the correlation of contaminants, tracers and spectral signatures

A regional center can also serve local and regional policy makers and regulators by providing a participation forum for the exchange and dissemination of appropriate information. Finally, a regional RS center can also build a hyperspectral science and technology infrastructure for the benefit of local and regional applications.

A regional RS center's initial research support areas, such as wetlands inventory, mineral exploration, pollution control, agricultural concerns and forestry inventories, can be selected to provide maximum immediate local and regional benefits. Researchers and data specialists can be involved in these areas of research that are going to be particularly relevant and beneficial to local and regional users.

For example, applying hyperspectral data to regional water control analysis is an important HSI application area. Developing or obtaining signatures detectable by hyperspectral instruments, e.g. chlorophyll alpha, phyto-plankton, dissolved organics, and amino and organic acids, can form the basis for relating remote sensed and *in situ* data, reducing reliance upon local data acquisition for monitoring purpose. Remote sensing then becomes the low cost alternative to provide the highly needed data independent of local weather conditions.

Obtaining continuous sets of water quality parameters, phytoplankton concentrations and algal abundance data sets can serve several functions:

1 As environmental model and simulation inputs and verification data to validate model outputs.

2 For public health and safety reasons such as the threat of algal concentrations exceeding allowable levels and requiring continual monitoring and tracking of any algal blooms that may appear.

3 Fishing and tourist constraints: knowledge of fishing 'Hot Spots' to avoid is essential from a health and safety viewpoint ("don't drink the water; don't eat the fish") as well as avoiding tourist visits to the polluted areas.

Other areas to be supported by HSI RS monitoring may include:

- Mineral/Petroleum Exploration
- Environmental Monitoring and Assessment
    - Fresh Water Watershed Runoffs: Pesticides, Fertilizers, Tire Residue, Nitrogenous Elements, etc.
- Agriculture & Other Renewable Resources
- Environmental Mapping
    - Coastal mapping
    - Bathymetry Maps
    - Civil, commercial, and military maps Program:

Figures 29.2 and 29.3 indicate the amount of detail that different spectral bands (Figures 29.3 is for a particular HSI band) can provide as contrasted with fewer bands. Generally, though, HSI sensors cannot achieve the high-resolution, meter-level panchromatic information that sensors like IKONOS can achieve. Combining HSI data with high-resolution, meter-level data and other remote sensing data such as Synthetic Aperture Radar (which are particularly useful for all-weather hydrology-related information such as floods), will result in more advanced fused data and information products.

Integrated Remote Sensing for Oil Spill Risk Mitigation

Remote sensing can be utilized for oil spill mitigation in environmentally-sensitive areas. For example, tankers traversing the Suez Canal are one of the most important sources of national revenue for the country of Egypt. An equally important revenue source is tourism in the Red Sea area, which has recently blossomed with the spread of world-class resorts due to its unique weather, coral reefs and sandy beaches. Lastly, oil exploration from platforms situated in the Red Sea also produce economic

Reagan National Airport/City of Alexandria Hyperspectral Image

Figure 29.2. Reagan National Airport/City of Alexandria Hyperspectral Image.

benefits, but risks as well. This proximity of the ecologically sensitive resort areas, the high traffic of commercial vessels and oil tankers, and oil platforms present a real need for timely detection, tracking and mitigation of the effects of potential oil spills using advanced remote sensing techniques.

Coupling remote sensing with advanced information technology tools including geographical information systems, (GIS), and data analysis and visualization tools has the potential for producing critically needed information products for mitigation response teams. Figure 29.4 shows an oil spill disaster, 10 days after it occurred, as captured by the European Space Agency's ERS-1. The very dark area is heavily polluted sea, while the gray area and dark streaks are older and more dispersed oil.

Information technology infrastructure for automating oil spill detection, tracking, and mitigation efforts can, in addition, be introduced. This IT can include GIS as well as innovations in the areas of Earth science data ingest, analysis, and visualization environments. In contrast

Figure 29.3. Reagan National Airport/City of Alexandria Hyperspectral Image, in a particular HSI band.

to the previous examples of HSI, oil spill mitigation and other natural and man-made disasters, would require real-time or near real-time data access, data manipulation, data fusion and information extracted for rapid response of relevant clean-up agencies and crews, law informant, etc. Although the data volumes may not be as large as the global Earth observing and regional HSI applications, the need for real-time information extraction (e.g. amount of spill, magnitude of disaster, etc.) presents additional IT challenges.

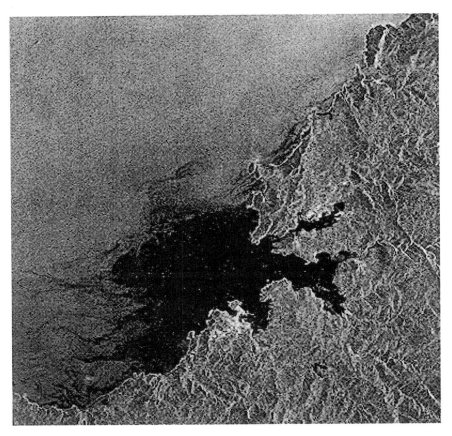

Figure 29.4. ERS-1 SAR image 10 days after an oil spill disaster.

Advances in remote sensing and information technology make it possible to develop systems that can detect, track, and support mitigation efforts for regional disasters such as oil spills in the sensitive Suez Canal and Red Sea areas, in a timely fashion through information products, utilizing Geographical Information Systems (GIS) techniques, and SAR data (such as ERS-1 and ERS-2), global Earth observing NASA and NOAA satellite data, and Earth science information technology systems in the areas of data access, selection, retrieval, analysis, and visualization. The need for real-time or near real-time response goes however, beyond the needs associated with less time-sensitive Earth sciences and regional applications production of data and information products.

4. Information Technology Support for Massive Data

Information technology (IT) can provide innovative solutions for many aspects associated with massive data sets in Earth sciences and the crosscutting, multidisciplinary research areas of Earth system science and global change. These include high performance databases; parallel and distributed data systems; Federated data systems; and data mining.

High Performance Databases

As we saw, data access to EOS and other Earth observing platforms is being provided by NASA's baseline system termed EOSDIS. Data rates from the EOS *Terra* platform are expected to reach terabytes per day. Even current relatively smaller missions like the *Tropical Rainfall Measuring Mission* (TRMM) have data rates in excess of 20 Gbytes/day. The overall capacity of current data centers like the Goddard Distributed Active Archive Center (DAAC) is measured in hundred of terabytes to be increased dramatically in the petabyte range in the EOS era. Whereas EOSDIS will provide basic functions for data access and ordering, scientists need focused searches for specific data products of interest to their field of study. Moreover, intelligent data querying needs to be implemented to allow focused searches.

The current approach in which a new definition of a data product requires a significant amount of software to be developed will not be adequate for data systems that are easily accessible and provide up-to-date data. Furthermore, a change in hardware or software platforms may introduce significant and non-trivial efforts in porting programs for the generation of the already implemented data products. Scientists are also often interested in data that are derived from heterogeneous data sources. High-performance database systems can be developed to support fast derivation of and efficient access to data and information products that are up-to-date and are probably derived from heterogeneous sources. Such a high-level system should have a narrow interface with a wide range of underlying hardware/software platforms, so that platform's change would only require changing a small number of software functions supporting the interface, while leaving unchanged the products' generation software.

Parallel and Distributed Data Systems

To meet the demands of accessing high volumes of Earth system science data, parallel and distributed computing capabilities can be utilized, including Beowulf clusters of commodity computers. In addition to the parallel access to data storage provided by the parallel computing infrastructure the key issue is distributed query processing, i.e., how queries can be processed in a way the parallel capabilities are utilized automatically (instead of manual adjustments). The idea is to allow scientific users/programmers to write applications queries that only focus on the required scientific processing, while being released from the burden of specifying what parts of code will run on what processor or computer.

Techniques to decompose queries into parallel sub-queries can be employed, utilizing high-level constraint-based languages as the basis for the above decomposing techniques. Results of such systems would include algorithms that efficiently use the parallel and distributed computers to process queries.

Federated Databases

As stated above, federated solutions can be brought to bear on the problem of massive data sets. It is often desirable to have specialized sub-databases created for the purpose of (1) providing an easy data acquisition environment for specialists, and (2) providing more efficient accesses to specialized data products. However, the use of a sub-database should not be a limitation to the user, i.e., the user should be able to easily access other parts of centralized systems such as EOSDIS. Also, software systems are desirable so that these sub-databases can be easily built, and various sub-databases combined to create larger sub-databases.

In a federated database approach, each database in the federation is considered as a sub-database that is built for or by some specialists of a research field. Data sets from sub-databases of other federation members can be accessible via federation data exchange, and query exchange protocols. The sub-databases in the federation maintain their own relative autonomy, i.e., each management of the sub-database may decide the structure of the database, user access methods and optimization techniques. One possible solution to support such a flexible environment, is to have metadata of the federation members represented using the XML standard, and a standard query facility on the XML metadata can be designed as the basis for the metadata query exchange protocol (see also below).

Other queries supported by the federation members may include so-called "content-based" or data mining types, i.e., queries that access the values in data sets themselves (in addition to the metadata).

Content-Based and Data Mining Systems

Current data mining systems that support content-based searches can be divided into information technology (IT)-focused data mining approaches and Earth science-based data mining or content-based approaches. This is because there is an increasing emergence of content-based searches with a primary focus being Earth science-centered approaches (utilizing, of course, existing IT approaches such as on-line data analysis and exploratory data analysis).

Statistical Data Mining and Visualization There are various statistical data mining techniques involving, among others, multi-dimensional visualization tools (such as parallel coordinates and multidimensional projection techniques); as well as 3-D density clustering projections. Moreover, statistical exploratory data analysis such as the "grand tours" (Wegman and Carr (1992)) allow the user to obtain a quick view of the data without the need of time-consuming computations.

Related technologies from the statistical community involve the visualization of massive data sets. Most scientific data are massive in size and effective visualization of such large databases is critical to scientific inquiry. For instance, Carr (Carr (1995)) suggested new methods to visualize large databases. New statistical graphics techniques have been introduced. Several powerful statistical graphics techniques for visualization of spatial data, including the micromaps (Carr and Pierson (1996)), have been developed. Some of these tools have already been implemented in the web environment to be used by any web users (Symanzik et al. (1998; 1999)). In addition, new models to represent and visualize global data effectively have been suggested (Carr et al. (1997)). Virtual reality technology has also been employed to explore and analyze high-dimension data (Wegman et al. (1999)).

Machine Learning and Discovery A variety of inductive learning tools currently exist in the machine learning and inference communities (also known as Artificial Intelligence or AI). These support scientific analyses and data mining and should possess the following properties: The discovered pattern (or the form of knowledge representation) should be relatively easy to interpret by humans, and; the data mining system should be able to benefit from the domain knowledge (which is easy for scientific work since the domain knowledge is well-documented).

Inductive learning systems that generate decision rules (Mikalski et al. 1986, Wnek et al. 1995) have been utilized for scientific queries related to El Niño events (Li et al. (1997)). Existing software allows rules to be represented in notations such as the Variable-valued Logic notation. When building a decision rule, machine learning tools perform a heuristic search through the space of logical expressions to determine those that account for all positive examples. Such tools are also able to handle inconsistent data and noisy data. Interesting hypotheses can be generated by applying inductive machine learning tools which can then be tested by the scientists.

Earth Science-based Content Based Systems The core of content-based search or on-line data mining, in our opinion, is to find a region in spatial and temporal dimensions on which parameter values are in given defined ranges. Instead of trying to find the exact solution for content-based queries, we define statistical range queries and propose methods to answer the queries quickly.

A simple statistical query is to find a set of regions, such that for each region, the probability of data values of a given parameter in certain range is no less than a specific confidence level. Connecting simple queries by boolean connectives will form more general statistical range queries. Simple statistical queries can also be combined with other kind of query restrictions. Most (if not all) content-based queries should be covered by the statistical range queries.

Statistical queries make sense for scientific data sets. The values of geophysical parameters are usually continuous in space and time. Therefore, the regions satisfying the content-based search conditions are usually of certain size. If we give a relatively high accuracy requirement, say 95%, most interesting regions should be selected by a statistical range query. Values in the given range but not selected are usually isolated or sparse in the distribution. Actually, it is good to not include single space points or tiny regions in the query result. To quickly answer the statistical range queries, a data pyramid model has been proposed (Li et al. (1998)). The data pyramid is built to store aggregation information of scientific data sets in a compact and approximate way. The problem space (temporal, spatial, and parameter) is divided into cells in different levels. In each level, all cells are disjoined and all cells together cover the whole study space. Each immediate high-level cell covers a space which is the union of a number of low level cells. Each low-level cell is associated with one and only one immediate upper-level cell. Mappings of values from the grouped low-level cells to the associated high-level cell are defined for constructing the data pyramid (see Figure 29.5). The

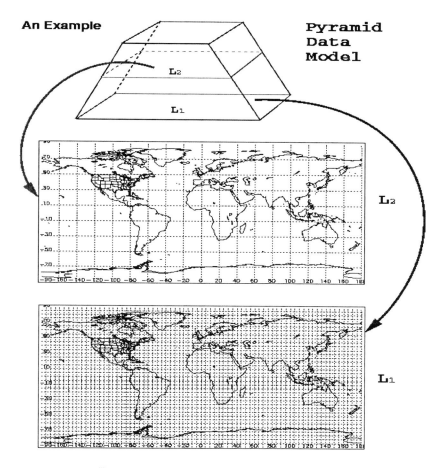

Figure 29.5. Data pyramid structure.

aggregation values of this data pyramid are number of data points, mean value, standard deviation, minimum value, maximum value, and probability density function based on data values in that cell (Figure 29.6). The density function can also be defined to be histograms (see below). With the data pyramid, a content-based query can be answered very quickly since the aggregation values are used to check the conditions. In addition, since the data pyramid provides data at different resolutions, one may specify the resolution for a query result (see Figure 29.5).

Some simple Earth science scenarios for content-based/data mining are given here:

1 Vegetation scenarios (for land, vegetation applications):

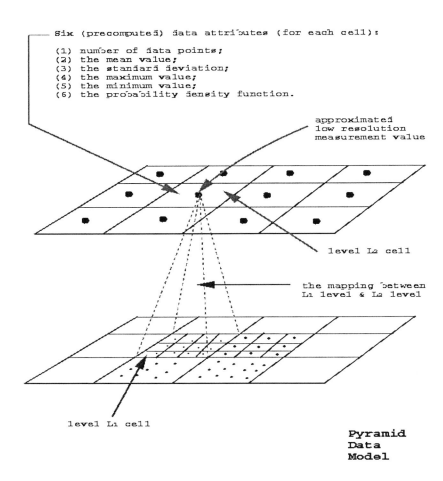

Figure 29.6. Data pyramid mapping relations.

- Find regions with, e.g., $\overline{NDVI} \geq 0.5$, where $\overline{NDVI}$ is the spatially-averaged NDVI value
- Find deciduous forests in a particular geographical region, e.g. NE U.S.
- Find the spatial regions in North America where the proportion of cultivated land is greater than 50% of the land coverage, etc.

2 Ocean scenarios

- Correlate Sea Surface Temperature (SST) Anomalies (SSTA) in the tropical Pacific with AVHRR Normalized Difference Vegetation Index (NDVI) in specific regions, such as S. Africa, continental U.S., etc.

- Correlate the Southern Oscillation Index (SOI–an index which provides information about the occurrence of El Niño phenomena) with agricultural production in S. Africa, S. America and Australia for drought/flood studies

Correlations are in time series of spatially averaged values of Sea Surface Temperature Anomaly (SSTA) and NDVI as well as SOI and NDVI.

Recently, we have studied a new method to enhance the data pyramid model for obtaining more accurate answers to statistical range queries (Yang et al. (2001)). In the new model, all values on higher levels are obtained by aggregation of values of highest resolution (at the lowest level), and only histograms are used to answer a query. Instead of using individual histogram on each cell, we group histograms into clusters and define a representative histogram for each group. When a query is answered, the system checks cells in groups instead of individual cells. In this way, the query speed should be higher than that working on individual cells. Moreover, from the representative histograms, the accuracy of the result can be assessed.

The above content-based browsing techniques could be utilized to rapidly search Level 3 (uniformly gridded and temporally spaced, e.g. monthly-mean) data and avoid time-consuming and expensive searches of the entire, large-volume databases (such as searching complete MODIS, etc. data sets).

Geographic Information Systems

Earth observing data are of interest not only to discipline Earth scientists, but also of broad-base interests to many scientists conducting research related to the environment. Environmental scientists in general, but more specifically ecologists, climatologists, oceanographers, and even emergency response personnel (such as the Federal Emergence Management Agency–FEMA) constitute some of these groups. These scientists and practitioners often include many Geographic Information Systems (GIS) users. They study various types of environmental phenomena and of various levels of geographic scale (global, regional, and local). Space observing or remote sensing data can be combined with other types of spatial data, such as population or Earth science data, to provide more comprehensive analyses. The former are likely to be massive, whereas the latter can be massive or not–depending on the level of resolution (e.g. city-level, county-level, country-level, etc.).

Web-based GIS functions can easily be incorporated into the system to allow users to conduct more sophisticated and more GIS-analysis in real time on the web. On the other hand, after browsing the data and completing preliminary analysis on the web, GIS users can elect to download the data for local usage. But most Earth observing data are stored in particular formats that cannot be directly imported to GIS. Conversion procedures to disseminate remote sensing in popular GIS formats need to be built.

Other related IT concepts for Earth system science and global change data include relational database management and object-relational database management systems. It is often advantageous to use the latter (as objects are more easily searchable and provide additional advantages) but most existing large databases at the big data centers are relational databases.

5. Federation and Interoperability Options

We now turn our attention to a concrete example of an Earth system science federation and the description of one of its nodes. What we will have to say here will be useful for other federated systems as the specific points will bring out general issues.

Specifically, we are discussing the Earth Science Information Partners (ESIP) Federation. In response to the NRC recommendations, NASA solicited proposals for a so-called working prototype (WP) ESIP federation in 1997. Since that time, the ESIP Federation (henceforth referred to as Federation) has become an entity with its own structure, ways of operation, interoperability options and serving communities beyond the original communities that were being served by the individual partners-ESIPs. The Federation consists of four types of partners, the so called ESIP-1, -2, -3 and -4 types. ESIP-1's are the NASA Distributed Active Archive Centers and other data centers serving the data needs of diverse communities. ESIP-2's are science-oriented and serve the Earth sciences and global change communities. ESIP-3's are value-added providers and tend to serve broader communities such as regional applications. ESIP-4's are expected to be funding partners such as federal agencies (currently, only NASA is an ESIP-4 member). Clearly the Federation presently has massive Earth science data holdings (in excess of hundreds of terabytes) and the addition of new members (such as NOAA data centers) as well as through the future massive increases in data provided by missions such as *Terra*, etc., will increase its holdings by many-fold, to several petabytes. It is expected that the Federation will serve as an experiment for new ways to hold and distribute data (such as the NASA concept of NewDISS). As such, interoperability options become particularly crucial in a federation of such diversity.

Degrees of Interoperability

Considering various modes or options of interoperability, the Federation could follow an option in a spectrum from total openness or no structure to total transparency or complete structure. Clearly a centralized system tends to concentrate more towards the latter type of structure but even a large, distributed architecture can achieve various degrees of transparency. Below we list several options. Clearly, each successive step requires additional coordination between the Federation nodes (or ESIPs) and is the price the Federation has to agree to for achieving a high degree of interoperable operations or an increasing degree of transparency.

- No Structure
  No effort to create the Federation is made. Clearly this is an unacceptable solution since it offers no common advantages.

- WWW
  There could be a Federation homepage centrally listing all the ESIPs and each ESIP could provide a link to that page. In fact, this has already been achieved, the Federation page is at http://www.esipfed.org/

- GCMD
  Each ESIP would submit a description of its data sets to the Global Change Master Directory (GCMD) and allow ESIP data to become compliant with the Federal Geospatial Data Committee (FGDC) requirements and searchable via the GCMD. It would thus make ESIP data much easier to find and provide a catalogue-level searching capability for the Federation. This step has also been achieved and it is a requirement of NASA and the Federation itself (see a schematic description shown in Figure 29.7).

The steps listed above could occur autonomously, i.e. they don't require much coordination among the members.

- Common front page
  There could be an HTML page that is the "front" to all ESIPs. It would explain the Federation and guide users to which ESIP

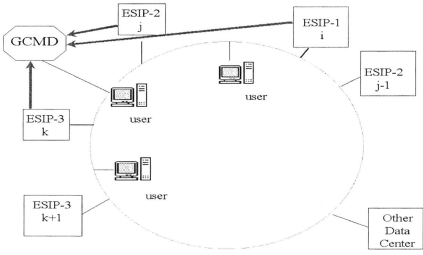

Figure 29.7. GCMD Metadata Compliance.

to visit. This option would require more communication between the ESIPs and the maintainer of this common page, because it would contain more information than the WWW level of federation. Moreover, it would require agreement as to the appearance of the front page. So far, the Federation has chosen not to follow this particular mode.

- Catalogue/Inventory
  This would allow users or another system to know what data are held at other ESIPs. There are several different levels of support for this type of interoperability.

  - Catalogue Search
    In general, if each Federation node indexed their site, then there could be keyword searches (via Z39.50 or other protocols) between different nodes to guide users (systems) to appropriate data sources. The Federation has agreed for an enhanced GCMD directory and catalogue search and a MERCURY system catalogue search. The latter (as the GCMD) allows keyword searches through metadata specifications that will be submitted to the GCMD, allows free text searches and in addition, allows HTTP access to data.

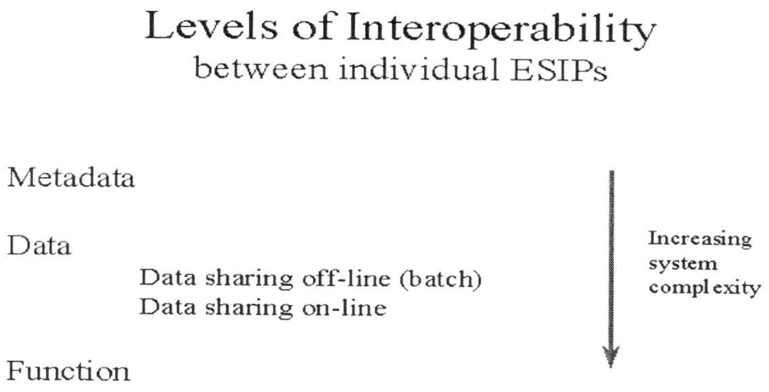

Levels of Interoperability
between individual ESIPs

Metadata

Data
        Data sharing off-line (batch)
        Data sharing on-line

Function

Increasing system complexity ↓

Figure 29.8. Levels of interoperability between individual ESIPs.

- Inventory Search
  A user (system) of one ESIP could search the inventory of other ESIPs. This means searching information provided by metadata at the inventory (or data file) level. However, no data exchanges would take place here.

- Place orders at other ESIPs
  The user (system) could place an order for data from within the interface of another ESIP. This could occur in batch mode (Figure 29.8). It is a partial data interoperability option.

- Import data from other ESIPs
  For complete data interoperability, data sharing on-line could be achieved. As more and more transparency is achieved (see Figures 29.8 and 29.9), members of the Federation become more and more transparent to outside users. Complete data interoperability could be implemented in a client-server mode (as is the case of establishing the client-server interoperability between two ESIPs, namely the DODS ESIP and SIESIP–see below). ESIPs could then freely import data from other ESIPs for interactive sessions with users (systems).

- Complete Transparency
  Each ESIP would mesh smoothly with each other, so the user or another system is not aware of any distinction between ESIPs. This would imply that support of a variety functions (as, for ex-

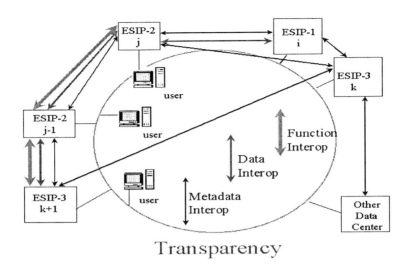

Figure 29.9. Levels of transparency in a federated data system.

ample, provided by CORBA), would be afforded to different user communities accessing different ESIPs (as is also the case of establishing a GrADS/DODS server implementation serving both the DODS and SIESIP user communities–see below).

System-Wide Interface Layer Interoperability Options

Any federation would implement an overall interoperability architecture. For the ESIP Federation, a system-wide interface layer (SWIL) for implementing interoperability is being adopted. Below are five SWIL interoperability options that were specified originally:

- Emerging Technologies
  This is not specified in any detail but it refers to technologies that would permit a Federation node (or ESIP) to be automatically searched and queried from remote clients as if part of a larger whole (i.e., constituting a "Federation").

- Version 0
  The Version 0 Information Management System (IMS) provides search and order tools for accessing more than 700 Earth science data & services held at the NASA DAACs and NOAA centers. It is going to be augmented by the EOSDIS Version 2 that will serve data sets collected by *Terra* and other future EOS missions.

- EOSDIS Core System (ECS)
  This is the Version 2 system serving access to *Terra* data (e.g. MODIS, etc.). A SWIL based on ECS (Version 2) EOSDIS infrastructure would provide scientists computing architecture for a variety of goals.

- CEOS Catalog Interoperability Protocol (CIP)
  Z39.50 1995 Version 3 is the base protocol for CIP (for search control, ordering & authentication). CIP provides distributed search, retrieval, order & other services with access to V0. It would align CIP to FGDC Clearinghouse systems.

- Federal Geographic Data Committee (FGDC) Clearinghouse
  FGDC is a Federal Government standard and indicates an activity for distributed search & retrieval of digital geospatial data from multiple sites using a common search vocabulary.

To enable these SWIL options, a variety of interoperability modes which themselves imply different implementations are clearly available. These include catalogue, inventory and data access options. It is clear from the above discussion that any federation of Earth science data providers and specifically the ESIP Federation has many options open. At the lowest end, Federation partners could choose minimum interoperability afforded by a data advertisement or directory services such as GCMD (already adopted) to more complete catalogue services as provided by the Oak Ridge National Laboratory's DAAC MERCURY system and the CEOS Interoperability Protocol (CIP). Advancing to higher levels of transparency, if data access and exchanges are needed, the Distributed Oceanographic Data System (DODS) or part of the Seasonal to Interannual ESIP (see below) could be utilized. If full functionality transparency is needed, tight interoperable solutions will need to be found (as, for example, provided by CORBA or a GrADS/DODS server implementation). As different Federation nodes cluster together, it is clear that different cluster requirements (driven by their combined user communities) will drive the interoperability designs. These will also determine whether the existing Internet can serve these needs (likely if only advertisement services and rudimentary catalogue services are needed) or whether I2 will be required (clearly needed for large data transfers and full functional interoperability). To a large extent, which options are going to be adopted is dependent on what the Federation and its clusters of ESIPs (or clusters of data providers) are attempting to accomplish. It is likely that more "tight" interoperability options will be pursued by individual clusters of data providers which are naturally working together, whereas minimal solutions such as Global Change Master Directory (GCMD) directory and advertisement search options (which require ESIPs to complete Data Interchange Format, DIFs, entries for their data holdings) have already been adopted by the Federation as a whole.

In summary, the overall ESIP Federation will be enabled by different modes of interoperability between individual nodes or clusters of nodes. At a simplified level, these modes could include metadata exchanges for catalogue and inventory search interoperability; data exchanges or sharing (at batch modes or even data sharing on-line); and full functional interoperability where on-line analysis, data exchanges, and full interoperable operations are enabled. It follows that different levels of transparency may exist between sub-partners of a specific Federation partner (such as SIESIP itself, see below) which might involve metadata, data exchanges and full functional interoperability. Or, interoperability between a particular ESIP and other ESIPs could take place at a lower level than inside the ESIP itself.

Metadata for Interoperability

Metadata provide important information about data. There are metadata for catalogue (i.e. information about locations of major data holdings, and providing general information about such data holdings) and advertisement services; metadata for directory (i.e. providing general information as well as approximate temporal and spatial coverage of each data set) data searches; metadata for inventory (i.e. providing even more information about the data files or data granules themselves). The role of metadata is to bring different data structures together by focusing on what is common among them. In Earth sciences, metadata follow conventions and standards often dictated by agreements of the scientific communities which themselves reflect to a large extent the scientific disciplines utilizing the data under consideration (e.g. netCDF following the Cooperative Ocean-Atmosphere Research Data Service or COARDS metadata convention, GRIB, HDF, etc.). Metadata structures may also reflect the information technology or machine structure where the data are located.

Clearly metadata structures (or schemas) are of the outmost importance in all scientific fields including Earth system science. For broad communities access, metadata can be less detailed. Most often, the details of the metadata schemas are invisible to the user. The challenge, however, is to allow diverse schemas applying to some data centers to be compatible with other schemas at different data centers. It is, of course, important for the user to quickly obtain information about where the data are. This in itself, though, will not allow the user to any data access. Further steps in the search process have to be undertaken linking the user to the data files themselves. For many scientists, location of the data might not be an issue (they may already know where the data are) but access to the data always is. It is important, therefore, that the overall framework in distributed Earth system science and global change databases allow the user to get down to at least the inventory-level information and services and hopefully provide access to the data themselves.

Bringing together broad communities of Earth science data users can lead to large metadata schemas and standards, for example the GEO-1 metadata schema has over 300 hundred concepts or fields (see EUU (2001)), the EOSDIS "core metadata" nearly 300 fields, etc. One approach would be to break the metadata standard into small pieces, dividing the metadata standards by discipline or community of users. This is almost imperative in a federation of different data providers and users where it is almost impossible to impose (or re-impose) standards from above. As such, the current trend to consider a community as a hierarchy of sub-communities (see EUU (2001)) is particularly natural in a federation environment. The metadata standards are then themselves hierarchical and represent varying levels of detail. The advantages here are that each sub-schema can be small and manageable and individual communities (or users associated with particular federation nodes, as in the ESIP Federation) can adopt their own standards while data providers or federation nodes (e.g. ESIPs) can choose which schemas to support based on the level of interoperability they want to or can afford to support. For example, Earth system science data providers can easily register their data with GCMD by filling in the appropriate DIFs and in this way be able to advertise to large numbers of users the type of data they are holding.

Structures of metadata can be built to reflect directory, inventory and data levels. The metadata at each level can be syntactic or semantic, the former referring to metadata required to define the organization of the data or syntax of the data set, and the latter referring to the contents of the data (Cornillon (2000)). For example, at the directory level, semantic metadata are "parameter", "range" and "descriptive"; the first is the description of the variables in the data set, the second the ranges of these variables while the third is information about the data set (Cornillon (2000)). Metadata definitions and data models need to be considered in light of the interoperable operations that are undertaken in a distributed data framework. For example, "parameter" and "range" metadata are required for system interoperability but "descriptive" metadata are not, etc.

A standard that is particularly useful for distributed data systems and takes advantage of the freeware and resources of the WWW is the eXtensible Markup Language (XML). XML is a markup language that is based on the same standard as HTML, namely the Standard Generalized Markup Language (SGML), and offers the stability and proven record of SGML. Thus, XML is gaining popularity as a standard for metadata representations and allows the definition of customized markup languages through the use of the so-called Data Type Definitions (DTDs). Specialized tags that have meaning for particular Earth science communities can be built for XML. As such, XML documents can, therefore, be automatically validated against the DTD to which they are applicable. This would allow scientists to search diverse data holdings via an XML-based engine, utilizing knowledge base information or the metadata which are appropriate for their discipline field. Moreover, specific XML files that have defined DTDs containing a variety of metadata can be developed and linked to diverse databases at different sites. They can even allow users to add their own free-text annotations and provide customized data descriptions with little or no need for IT knowledge, experience or maintenance requirements (which are not welcome by scientists). A great advantage of this approach is that the data providers don't have to follow a particular schema; they don't even have to build RDBMS or OODBMS. XML may become the standard for Earth science and global change research metadata searches.

6. Distributed Framework for Data Access, Querying and Analysis: An Example

We now give an example of a specific federation node in the ESIP Federation. The issues covered here go beyond the specifies of this node and should help bring out some general points applicable to distributed data centers.

In responding to the ESIP opportunity, George Mason University's (GMU) Center for Earth Observing and Space Research (CEOSR) and faculty from other GMU units, scientists from the Center for Ocean-Land-Atmosphere Studies (COLA), data specialists from the Goddard DAAC and faculty from the University of Delaware formed a collaborative team or mini- federation, serving a particular science area of the USGCRP and NASA's Earth Science Enterprise, namely the seasonal to interannual (S-I) science areas.

The GMU-led team is a distributed data and information technology consortium focusing on the seasonal-to-interannual science area (S-I) with a distribution of tasks as shown in Figure 29.10. The S-I ESIP (SIESIP) consortium http://www.siesip.gmu.edu/) is designed to be an innovative source of climate data and other products to S-I researchers, TRMM scientists, process studies such as hydrological processes and regional experiments in the tropical Pacific. As such, SIESIP forms itself a mini-federation and the issues it faces and lessons learned can apply to larger distributed Earth science data systems.

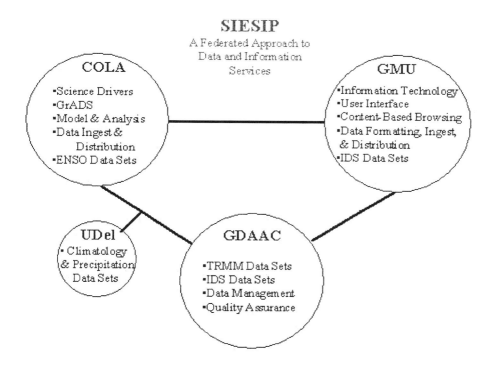

Figure 29.10. SIESIP mini-federation.

SIESIP is one of the 12 ESIP-2 (science-focused) Federation projects selected by NASA (NAS (2001)) (http://www.esipfed.org/). As we saw before, the Federation also consists of ESIP-1's (as the DAACs are known in the Federation) and ESIP-3's which are producing value-added products and services, focusing on more commercial approaches. All ESIPs are expected to conform to the Federation rules, interoperability protocols and interoperability that the Federation is establishing. SIESIP itself being a science-driven effort, is designed to address the science issues and needs of its community of users through new data and products, an information technology, distributed architecture and associated software and hardware (http://www.siesip.gmu.edu/science.html). As such, innovative IT methods are utilized to serve the data (often massive) and information needs of the relevant science communities.

In Figures 29.11 and 29.12, the main distributed tasks of the SIESIP mini-federation are shown. Tasks include data production, distribution of user access, analysis and data ordering. An innovative 3-phase data access is part of the SIESIP strategy (see below).

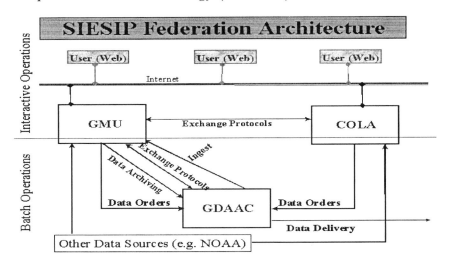

Figure 29.11. SIESIP mini-federation architecture.

SIESIP's goal and implementation strategy is to provide S-I climate scientists with both data and information solutions. The SIESIP mini-federation consisting of the three main distributed sites (GMU, COLA and GDAAC) is focusing on different tasks: George Mason University on information technology, data searches and analysis, and interdisciplinary Earth system science. GMU's contribution is in information technology design and implementation, leveraging existing prototypes such as the VDADC (http://www.ceosr.gmu.edu/~vdadcp/vdadc/vdadc.html) prototype project which resulted from GMU's 1994 study of alternative architectures for EOSDIS (Kafatos et al. (1994)). The VDADC focuses on global data sets, available via the WWW, by utilizing on-line tools allowing the user to quickly look of the data. The architecture and GUI of the VDADC are shown in Figures 29.13 and 29.14, respectively (Kafatos et al. (1997)).

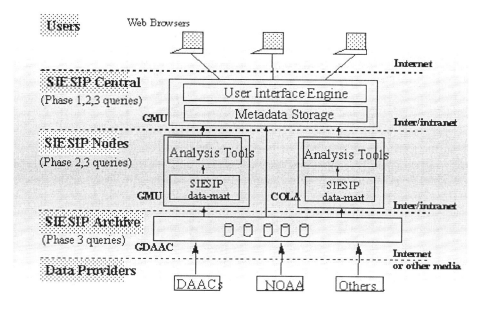

Figure 29.12. SIESIP mini-federation architecture (detailed).

GMU is also building the relational database and is porting data to the central SIESIP server, linking data to the on-line analysis tool GrADS, developing specialized TRMM products, as well as developing innovative data mining and search capabilities. GMU's communities include interdisciplinary Earth science users and graduate students. COLA is providing science scenarios, S-I data such as NOAA data sets and model output data, as well as enhancing the S-I popular graphical analysis tool GrADS developed at COLA (with several thousand users) for satellite (swath) data and other functions (http://grads.iges.org/grads/head.html). The Goddard DAAC is contributing data management, TRMM data products and will also assume some of the operational data distribution roles (http://daac.gsfc.nasa.gov/DAAC_DOCS/gdaac_home.html). GDAAC is supplying TRMM subsets and regional precipitation data in support of the South China Sea Monsoon Experiment, also known as SCSMEX. Finally, data are also being supplied by the University of Delaware researchers who have considerable expertise in climatology and station (rain gauge arrays) data.

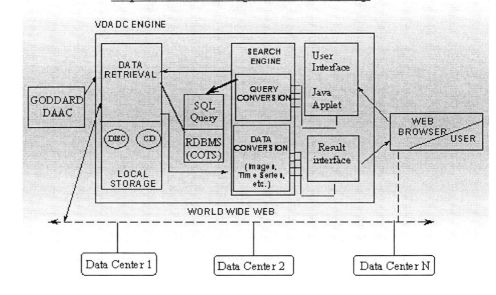

Figure 29.13. VDADC architecture.

Data

The SIESIP mini-federation produces a variety of data products (see Table 29.1). The GMU/COLA/DAAC/UDel consortium serves S-I and other related scientists who have requirements for specific data. Other user communities SIESIP serves are TRMM scientists, working on GCM intercomparison studies; hydrology and other interdisciplinary science investigators; other process studies such as SCSMEX (Table 29.1). For TRMM, data subsets are being assembled as described in Table 29.2.

The data products in the SIESIP system are chosen to facilitate research and data access for particular seasonal-to-interannual phenomena such as the El Niño/Southern Oscillation (ENSO), monsoons event-driven phenomena (hurricanes), etc. On-line preliminary analysis of ENSO teleconnections, interdisciplinary tools such as temporal lags among phenomena, etc. will be built-in as tools to the WWW search mechanism. Three-dimensional on-line visualization tools that are used to study specific tropical storms have also been built. The data sets are most often global and prepared on uniform grids (1 degree x 1 degree)

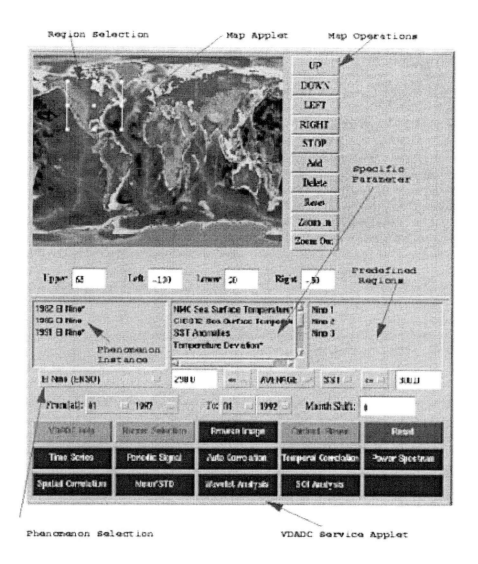

Figure 29.14. VDADC GUI.

and same temporal specifications (either monthly mean, 5-day or 10-day mean), i.e. so-called Level 3 (L3). Higher resolution data sets, e.g. SC-SMEX, are also being prepared for regional scales. The present GMU prototype VDADC has developed a suite of data products (Table 29.3)

Table 29.1. SIESIP Data.

S-I Data
NASA Data (TRMM subsets)
Specific Process Studies DATA (SCSMEX)
Station Data (UDel)
Pentads/Decade
Climatology Interdisciplinary Data Collection (CIDC)

Table 29.2. TRMM Data Subsets for SIESIP.

Coincidence Subsets (CSI):
collection of satellite scans and GV volume scans
when satellite passes over 4 GV primary radar sites.
(340 MB/day, 68MB/day compressed).
GDAAC to calculate percent of rain at ingest
Gridded subsets (G):
TRMM orbit data of TMI surface rainfall binned to 0.5 degree,
TRMM combined binned to 0.1 degree,
and VIRS radiance re-sampled to 0.25 degree grids
Regional Subset over SCSMEX region

---

in order to facilitate research by interdisciplinary Earth scientists and geoscientists.

The SIESIP project has developed the following products:

- El Niño and S-I data sets in GrADS format, primarily large volume model output and satellite data

- Pentads/decade (i.e. 5-day and 10-day) means of interdisciplinary science data sets

- Prototype TRMM gridded products and hurricane data

- SCSMEX data

- Climatology and precipitation data sets (CIDC)–a set of 4 CD's has been jointly developed between GDAAC and CEOSR (Kyle et al. (1998))

- Station data for the tropics (temperature and precipitation) developed by UDel.

The data sets can be accessed at the SIESIP home page, at http://www.siesip.gmu.edu/data.html. Similar enhanced and additional products will be developed, particularly bringing together model output (NOAA data sets provided through COLA) and RS data. The volumes here are very large, 100's of Terabytes, and even a bigger challenge is merging the data in meaningful scientific ways.

Table 29.3. VDADC Data Products.

Parameter	Period	coverage	Level
Total Precipitable Water	02/88–11/91	Ocean	L3
NMC Sea Surface Temperature	11/81–05/96	Ocean	L3
GISST2 Sea Surface Temperature	01/80–12/94	Ocean	L3
SST Anomalies	11/81–05/96	Ocean	L3
NDVI	07/81–08/94	Land	L3
Wind Fields($2^o \times 2^o$)	03/85–11/93	Land	L4
Precipitation	01/86–12/94	Land	L3
Temperature Deviation($5^o \times 5^o$)	01/84–12/94	Global	L3
Surface Skin Temperature	01/87–12/90	Global	L3

Information Technology

The overall information technology approach for the SIESIP distributed information system is:

- Development of science scenarios which drive the data access and query process to serve particular user communities and to minimize random searches of large databases

- Web accessibility of data and services

- Integration of tools accessibility with data set accessibility to allow meaningful, user-specified queries, and

- Integration of freely accessible visualization and analysis tools (GrADS) with relational data base management system

- Content-based browsing or human-assisted data mining

The SIESIP design is a multi-tiered client-server architecture, with three physical sites or nodes, distributing tasks in the areas of user services, access to data and information products, archiving as needed, ingest and interoperability options and other aspects. This architecture can serve as a model for many distributed Earth system science data.

This is enabled by a low cost, scalable information technology architecture. The information technology is built to allow on-line search of metadata, catalogue and data access, analysis and finally data ordering. To enable these functions, i.e. for easy access and WWW-enabled browsing and analysis capabilities, we have chosen as the central tool used by the SIESIP engine an enhanced version of GrADS (Doty and Kinter (1995); Doty et al. (1997)), presently under development, which is integrated into a database (Oracle) management system as well as an XML-based search capability as described above.

The SIESIP GUI or user interface allows a user to query the data system to search the data holdings (at GMU, COLA, GDAAC or UDel) in order to obtain an idea of what data sets are available and to allow a quick estimate of the content of data sets available. As we saw, this is termed content-based browsing and allows the existence of statistical summaries of data, which are themselves searchable. There are, correspondingly, three phases of user interaction with the data and information system (see also Figure 29.12). Each phase can be followed by other phase(s) or can be conducted independently (see Kafatos et al. (1998)).

Phase 1 or Metadata Access: Using the metadata and browse images provided by the SIESIP system, the user browses the data holdings. Metadata knowledge is incorporated in the system and queries explore this knowledge.

Phase 2 or Data Discovery/On-line Data Analysis: The user gets a quick estimate of the type and quality of data found in phase 1. Analytical tools are then applied as needed and these include statistical functions and visualization algorithms available via WWW. The SIESIP GUI also incorporates a spectrum of statistical data mining algorithms if applicable. We have also begun to implement tools for finding positive correlations among different data types, which provides a realistic, human-aided data mining capability. We have applied this approach and other data mining systems (Li et al. (1997)) to ENSO teleconnections with some positive results, e.g. in identifying anti-correlations with vegetation in tropical Africa and in the NE coastal U.S. (Li and Kafatos (2000)). These are being further explored.

Phase 3 or Data Order: After the user locates the data sets of interest she is now ready to order data sets. If the data are available through SIESIP, the information system will handle the data order; otherwise, an order will be issued to the appropriate data provider (e.g. GDAAC) on behalf of the user, or necessary information will be forwarded to the user for this task for further action.

These phases are designed to be transparent to the user and as such allow a mini-federation approach to a variety of functions. Phase 1 allows the user to search and discover what data are in the system. Phase 2 allows preliminary on-line analysis, with information obtained on the content of data, while Phase 3 allows the user to retrieve data for further analysis at the user's own site.

SIESIP's goal is ease of use for all the services by the user communities it serves (e.g. by deploying innovative products and information technology and allowing users to locate, browse and order data easily). SIESIP assists S-I scientists and its different communities by collecting relevant S-I data sets into a single point of access, at GMU, by integrating complementary data sets to enhance information, and by producing needed products. A single analysis tool (GrADS) is applied across diverse data sets, creating ease of use and compatible data interuse. As such, the SIESIP consortium will create an interdisciplinary Earth science source of data for S-I researchers in order to expand the usage and usefulness of NASA data for these communities and allow the merging of diverse model and satellite data sets. The SIESIP consortium under the guidance of an advisory board of science experts is selecting the most appropriate data from the GDAAC, NOAA, COLA, the University of Delaware as well as other data sources for S-I research applications. Metadata and summary statistics are also extracted and stored in databases at GMU for more efficient on-line searching (Li et al. (1998)).

The distributed information system implementation design is shown in Figure 29.15. Several functionalities can be supported at a variety of different physical nodes. The system allows the possibility of more physical nodes to join the mini-federation by providing new data sets and services. As such, the SIESIP consortium can provide useful lessons of implementing interoperability across diverse nodes for the entire federation. XML protocol development is presently under way between GMU and COLA and other protocols between GMU and GDAAC which include data ingest implementations.

To succeed in the above strategy, SIESIP is developing sets of metadata which can be used to implement phases 1, 2 and 3. For example, the principle guiding the first phase is to reveal, as quickly as possible, what data in the SIESIP inventory might be useful to the user. As such, besides the usual "description metadata" (which include catalogue and inventory metadata) as found in traditional data systems providing information such as scientific investigation, phenomenon under investigation, relevant parameter(s), holdings, etc., the system contains pre-computed "content-based metadata" which are pre-computed statistical properties

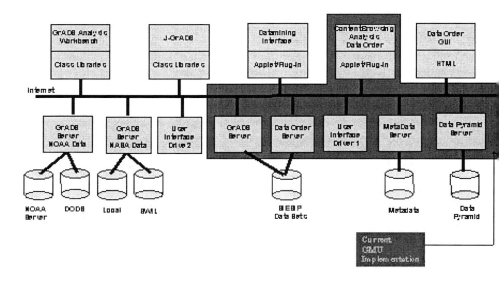

Figure 29.15. SIESIP implementation diagram.

of the underlying data. Content-based browsing and its associated Data Pyramid (Li et al. (1998)) have several advantages. The representations in the data pyramid allow the user to perform preliminary analysis on the data. Although the data are at reduced resolution and some errors are introduced, this is compensated for in speed of processing and the reduction of network traffic. Certain calculations, such as correlations between related parameters are optimized. We are presently exploring other methods such as clustering which would allow even more efficient searches. This phase is presently being implemented as a client-based JAVA GUI (see below, Figure 29.17), which interacts with the user by using the WWW. Phase 1 is distinguished as a search of the metadata, concluding with a system-generated list of files which the user has identified.

For the description metadata, SIESIP has developed an E-R diagram (Figure 29.16) which is a simplified version of corresponding data holdings at GDAAC. We are also developing ways to integrate COLA's data into a searchable metadata schema by exploring XML implementations. At present, data sets from UDel have been ingested into the database and we are in the process of ingesting SCSMEX and TRMM subsets as well.

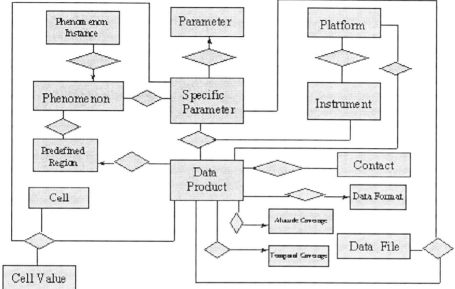

Figure 29.16. SIESIP ER diagram.

The third phase assumes the researcher is now ready to proceed with data ordering, following the preliminary results obtained in the second phase. The researcher may then place an order through the interface for access to the high-resolution data as needed. The requested data will either be staged for ftp pickup or the request will be submitted to the GDAAC if the data are not stored at the SIESIP GMU site.

Figure 29.17 above shows the JAVA-based WWW (based on the Swing technology) user interface (the particular figure shows the spatial section GUI) featuring three advances: 1) content-based searching of the summary metadata 2) exploratory analysis of on-line data and 3) phenomenon-based searching. The GUI allows the user to place in her "Workplace" the data sets that are needed. The GUI has metadata and description capabilities of the appropriate parameters stored at the SIESIP. At present, analysis functionalities include browsing of images in user-specified regions; wavelet decompositions; time series, etc.

Spatial Selection Interface

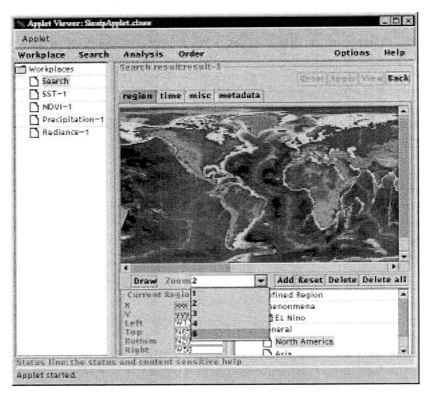

Figure 29.17. SIESIP GUI.

SIESIP has developed a rapid implementation of the system with various components developed modularly. The system description and associated results can be found at the URL http://www.siesip.gmu.edu/deliverables/deliverables.html. This URL gives all important developments including data, architecture, information about ingest and interoperability (which could be useful for the wider ESIP federation, http://www.esipfed.org/).

An interoperability implementation between two nodes of the Federation, that would be of importance to the general issue of data interoperability between Earth system science nodes, has been developed. Specifically, the data-level interoperability between SIESIP and DODS will result in substantially improved data access for the seasonal to interannual or climate (served by SIESIP) communities and the oceanographic (served by DODS) communities. The data interoperability allows XML capabilities for the treatment of metadata and interfacing GrADS to DODS both as a client and a server. The data interoperability allows for the installation of DODS servers for data sets that are of interest to one and/or the other of the two communities. As such, interoperability at the data level allows two diverse science communities to access each other's data sets; operate on the data with a diverse set of tools; and produce new, merged data and information products

7. Summary

In summary, lessons from the specific SIESIP federation-node implementation which have applicability to wider Earth system science and global change databases are

- Research projects and applications (with a specific science focus) are concrete manifestations of usage of interest to various communities
- On-line tools should be available to user communities; these are important to users and add value to analysis that can be performed on-the-fly
- Building the right interfaces with visualization and analysis tools is important for accessing data
- Interoperability between diverse nodes in a federated data system is crucial to the success of distributed large databases
- Support for different types of metadata is important for distributed databases; accessing by information content is as important as accessing data by description metadata, and
- Users can be information producers and add value to data.

Acknowledgments

We acknowledge partial prototype funding support from the NASA ESDIS Project (NAG5-8607), from the Goddard Global Change Data Center (NCC 5-143), and particularly from the Earth Science Enterprise WP-ESIP CAN Program (NCC 5-306), as well as from George Mason University.

*Other members of the SIESIP team include: B. Doty, J. Kinter & C. Steinmatz (COLA); C. Lynnes, L. Pham & G. Serafino (GDAAC); P. Chalermwat, L. Chiu, X. Deng, P. Hertz, Z. Li, Z. Liu, J. McManus, D-B. Shin, J. Vongsaard, C. Wang, X.S. Wang, H. Weir, & K-S. Yang (GMU);

K. Miyakoda, P. Schopf & J. Shukla (GMU & COLA); H. Wolf (IRMA & GMU), C. Willmott & K. Matsuura (UDel); and L. Jing (FIT).

Bibliography

G. Asrar and R. Greenstone, editors. *1995 MTPE EOS Reference Handbook*. NASA (Washington, DC), 1995.

BSD. *Board on Sustainable Development: A Review of the U.S. Global Change Research Program and NASA's Mission to Planet Earth/Earth Observing System*. National Academy Press, 1995.

D. B. Carr. Perspective on the Analysis of Massive Data Sets. In *Computing Science and Statistics, Proceeding of the 27th Symposium on the Interface*, volume 27, pages 410–419, 1995.

D. B. Carr, R. Kahn, K. Sahr, and A. R. Olsen. ISEA Discrete Global Grids. *Statistical Computing & Graphics Newsletter*, 8(2/3):31–39, 1997.

D. B. Carr and S. Pierson. Emphasizing Statistical Summaries and Showing Spatial Context with Micromaps. *Statistical Computing & Graphics Newsletter*, 7(3):16–23, 1996.

P. Cornillon. An Organizational Structure for Metadata. preprint, 2000.

B. E. Doty, J. L. Kinter III, M Fiorino, D. Hooper, R. Budich, K. Winger, U. Schulzweide, L. Calori, T. Holt, and K. Meier. The Grid Analysis and Display System (GrADS): An update for 1997. In *13th Conf. On Interactive Information and Processing Systems for Meteorology, Oceanography, and Hydrology*, pages 356–358. (American Meteorological Society, Boston, 1997.

B.E. Doty and J.K. III. Kinter. Geophysical Data Analysis and Visualization using GrADS. In E.P. Szuszczewicz and J.H. Bredekamp, editors, *Visualization Techniques in Space and Atmospheric Sciences*, pages 209–219. NASA, Washington, D.C., 1995.

Earth System Science URLs, 2001.

European Union-United States Joint Workshop on Large Scientific Databases, 2001.
http://www.cacr.caltech.edu/euus. Last visited on May 30, 2001.

A. Finley. Untangling the next Internet. *Sunworld*, 1998. April.

M. Kafatos, T. El-Ghazawi, X.S. Wang, and R. Yang. Earth Observing Data Systems in the Internet Era. *Photogrammetric Eng. & Remote Sensing*, 65:540–563, 1999.

M. Kafatos, Z. Li, R. Yang, and et al. The Virtual Domain Application Data Center: Serving Interdisciplinary Earth Scientists. In David Hansen and Yannis Ioannidis, editors, *Proceedings of the 9th International Conference on Scientific and Statistical Database Management*, pages 264–276. IEEE, Computer Society, 1997.

M. Kafatos, B. Moore, J. Kinter, and et al. The GMU ECS Federated Client-Server Architecture, report to NASA/Hughes, 1994. preprint.

M. Kafatos, X.S. Wang, Z. Li, R. Yang, and D. Ziskin. Information Technology Implementation for a Distributed Data System Serving Earth Scientists: Seasonal to Interannual ESIP. In Maurizio Rafanelli and Matthias Jarke, editors, *Proceedings of the 10th International Conference on Scientific and Statistical Database Management*, pages 210–215. IEEE, Computer Society, 1998.

L. Kerschberg, H. Gomaa, Daniel Menasce, George Micheals, Jong Pil Yoon, and Francis H. Carr. Chapter 4: Data and Information Management Architecture. In *The GMU ECS Federated Client-Server Architecture, Final Report*, 1994.

H.L. Kyle, J.M. McManus, S. Ahmad, and et al. *Climatology Interdisciplinary Data Collection, Volumes 1-4, Monthly Means for Climate Studies*. NASA Goddard DAAC Science Series, Earth Science Enterprise, National Aeronautics & Space Administration, NP-1998(06)-029-GSFC, 1998.

Z. Li and M Kafatos. Interannual Variability of Vegetation in the United States and Its Relation to El Niño/Southern Oscillation. *Remote Sensing of the Environment*, 71(3):239–247, 2000.

Z. Li, M Kafatos, and R. Michalski. Data Mining Application for El Niño Teleconnection Research. *GMU for Machine Learning Laboratory Report*, 1997.

Z. Li, X. S. Wang, M. Kafatos, and R. Yang. A Pyramid Data Model for Supporting Content-based Browsing and Knowledge Discovery. In Maurizio Rafanelli and Matthias Jarke, editors, *Proceedings of the 10th International Conference on Scientific and Statistical Database Management*, pages 170–179. IEEE, Computer Society, 1998.

R.S. Mikalski, I. Mozetic, J. Hong, and N. Lavrac. The AQ15 Inductive Learning System: An overview and Experiments. *ISG86-20, UIUCDCS-R-86-1260, Department of Computer Science, Univ. of Illinois, Urbana*, 1986.

NASA Selects Earth Science Information Partners, NASA Press release, Dec. 2, 1997, 2001.
ftp://ftp.hq.nasa.gov/pub/pao/pressrel/1997/97-277.txt.
Last visited on May 30, 2001.

NGI. NGI: Next Generation Internet–Implementation Plan, February 1998.

J. Symanzik, D. A. Axelrad, D. B. Carr, J. Wang, D. Wong, and T. J. Woodruff. HAPs, Micromaps and GPL–Visualization of Geographically Referenced Statistical Summaries on the World Wide Web. In *1999 American Congress on Surveying and Mapping, Annual Proceedings*, 1999.

J. Symanzik, D. Wong, J. Wang, D. B. Carr, D. A. Axelrad, and T. J. Woodruff. Web-based Access and Visualization of Hazardous Air Pollutants. In *Proceedings GIS in Public Health*, 1998.

University of Delaware: Udel Climate Station Data Products, 2001.
http://www.siesip.gmu.edu/presentations/climfields.html.
Last visited on May 30, 2001.

E. J. Wegman, J. Symanzik, J. P. Vandersluis, Q. Luo, and F. Camelli. The MiniCAVE - A Voice-Controlled IPT Environment. In H.-J. Bullinger and O. Riedel, editors, *International Immersive Projection Technology Workshop, 10/11 May 1999, Center of the Fraunhofer Society Stuttgart IZS*, pages 179–190. Springer, 1999.

E.J. Wegman and D.B. Carr. *Statistical Graphics and Visualization*. Center for Computational Statistics, George Mason University, 1992.

J. Wnek, K. Kaufman, E. Bloedorn, and R.S. Michalski. Inductive Learning System AQ15c: The Method and User's Guide. *Reports of the Machine Learning and Inference Laboratory*, MLI 95(4), 1995.

Ruixin Yang, Kwang-Su Yang, Menas Kafatos, and X. Sean Wang. Value Range Queries on Earth Science Data via Histogram Clustering. In J. F. Roddick and K. Hornsby, editors, *Proceedings of International Workshop on Temporal, Spatial and Spatio-Temporal Data Mining, TSDM2000, Lyon, France, Lecture Notes in Artificial Intelligence*, pages 62–76. Springer, Berlin, 2001.

Chapter 30

MINING BIOMOLECULAR DATA USING BACKGROUND KNOWLEDGE AND ARTIFICIAL NEURAL NETWORKS

Qicheng Ma
Department of Computer and Information Science, New Jersey Institute of Technology, Newark, NJ 07102, USA
qicheng@homer.njit.edu

Jason T. L. Wang
Department of Computer and Information Science, New Jersey Institute of Technology, Newark, NJ 07102, USA
jason@cis.njit.edu

James R. Gattiker
Los Alamos National Laboratory, Mail Stop E541, Los Alamos, NM 87544, USA
gatt@lanl.gov

Abstract   Biomolecular data mining is the activity of finding significant information in protein, DNA and RNA molecules. The significant information may refer to motifs, clusters, genes, protein signatures and classification rules. This chapter presents an example of biomolecular data mining: the recognition of promoters in DNA. We propose a two-level ensemble of classifiers to recognize E. Coli promoter sequences. The first-level classifiers include three Bayesian neural networks that learn from three different feature sets. The outputs of the first-level classifiers are combined in the second level to give the final result. To enhance the recognition rate, we use the background knowledge (i.e., the characteristics of the promoter sequences) and employ new techniques to extract high-level features from the sequences. We also use an expectation-maximization (EM) algorithm to locate the binding sites of the promoter sequences. Empirical study shows that a precision rate of 95% is achieved, indicating an excellent performance of the proposed approach.

Keywords: Biomolecular data mining, Bayesian neural networks, Background knowledge, Human genome project, Expectation maximization algorithm.

1. Introduction

As a result of the ongoing Human Genome Project (Frenkel 1991), DNA, RNA and protein data are accumulated at a speed growing at an exponential rate. Mining these biomolecular data to extract significant information becomes extremely important in accelerating genome processing (Wang et al. 1999b). Classification, or supervised learning, is one of the major data mining processes. Classification is to classify a set of data into two or more categories. When there are only two categories, it is called *binary classification*. In this chapter we focus on binary classification of DNA sequences.

In binary classification, we are given some training data including both positive and negative examples. The positive data belong to a target class, whereas the negative data belong to the non-target class. The goal is to assign unlabeled test data to either the target class or the non-target class. In our case, the test data are some unlabeled DNA sequences, the positive data are promoters and the negative data are non-promoters. Since our goal is to identify promoters in the unlabeled DNA sequences, we use the terms "classification" and "recognition" interchangeably in the chapter.

1.1. Related Work

Table 30.1 summarizes some work in biomolecular data mining. The first column indicates the "knowledge" to be mined in the biomolecular data, the second column shows the techniques used and the third column provides references. The knowledge to be mined includes DNA sequence signals such as splice sites and promoters, protein sequence classification rules and protein sequence motifs.

In this chapter we propose a two-level approach to recognizing E. Coli promoters in DNA sequences. The first-level classifiers include Bayesian neural networks (Mackay 1992b, Neal 1996) trained on different feature sets. The outputs of the first-level classifiers are combined in the second level to give the final classification result. Dietterich (1997) recently indicated that using an ensemble of classifiers can achieve a better recognition rate than using a single classifier when (i) the recognition rate

Table 30.1. A summary of work performed for biomolecular data mining.

Knowledge to be mined	Techniques used	References
Genes in DNA	Hidden Markov model	Kulp et al. (1996)
	Neural networks	Xu et al. (1996)
Splice sites in DNA	Pattern matching	Wang et al. (1999a)
	Markov chain	Salzberg (1997b)
Promoters in DNA	Neural networks	Opitz and Shavlik (1997)
	Decision tree	Hirsh and Noordewier (1994)
Protein classification rules	Hidden Markov model	Krogh et al. (1994)
	Neural networks	Wu et al. (1995)
Protein motifs	Minimum description length	Brazma et al. (1996)

of each individual classifier of the ensemble is greater than 0.5; and (ii) errors made by each individual classifier are uncorrelated. Our experimental results show that the proposed combined classifiers indeed outperform the individual classifiers made up solely by Bayesian neural networks.

Using an ensemble of classifiers to process biomolecular data has been studied by Brunak et al. (1991), Wang et al. (1996), and Zhang et al. (1992). In Brunak et al. (1991), Brunak *et al.* exploited the complementary relation between exon and splice sites to build a joint recognition system by allowing the exon signal to control the threshold used to assign splice sites. Specifically, a higher threshold was required to avoid false positives for the regions where there are only small changes in the exon activity. A lower threshold was used to detect the donor site for the regions where the exon activity decreases significantly. Similarly, a lower threshold was used to detect the acceptor site for the regions where the exon activity increases significantly. In Zhang et al. (1992), Zhang *et al.* developed a hybrid system, which included a neural network, a statistical classifier and a memory-based reasoning classifier, to predict the secondary structures of proteins. Initially, the three classifiers were trained separately. Then a neural network used as a combiner was trained to combine the outputs of the three classifiers by learning the weights for each classifier from the training data. The result of the classification was given by the combiner. In Wang et al. (1996), Wang *et al.* studied the complementarity of five classifiers for protein sequence recognition, and proposed an ensemble of the classifiers, which outperformed each individual classifier.

In contrast to the previous work, we apply Bayesian neural networks to recognizing promoters in DNA. The Bayesian neural networks make predictions by marginalization over the weight distribution. Furthermore, the Bayesian neural networks control the model complexity to avoid the overfitting problem (Mackay 1992a).

The rest of the chapter is organized as follows. Section 2 describes the characteristics of E. Coli promoters and our feature extraction methods. Section 3 and Section 4 present our two-level classification approach. Section 5 presents experimental results. Section 6 concludes the chapter.

2. Promoter Recognition

2.1. Encoding Methods

One important issue in applying neural networks to biosequence analysis is regarding how to encode the biosequences, i.e., how to represent the biosequences as the input of the neural networks. Good input representations make it easier for the neural networks to recognize the underlying regularities. Thus, good input representations are crucial to the success of the neural network learning (Hirsh and Noordewier 1994).

One of the encoding methods is *orthogonal encoding* (Opitz and Shavlik 1997). In orthogonal encoding, nucleotides or amino acids in a biosequence are viewed as unordered categorical values, and are represented by C dimensional orthogonal binary vectors, where C is the cardinality of the 4-letter DNA alphabet $\mathcal{D} = \{$A, T, G, C$\}$, or the cardinality of the 20-letter amino acid alphabet $\mathcal{A} = \{$A, C, D, E, F, G, H, I, K, L, M, N, P, Q, R, S, T, V, W, Y$\}$. That is, we use C binary (0/1) variables, among which only one binary variable is set to 1 to represent one of the C possible categorical values and the rest are all set to 0. For instance, we represent the nucleotide A by "1000", and amino acid Y by "00000000000000000001". The orthogonal encoding was frequently used in the early 1990s (Craven and Shavlik 1994, Hirst and Sternberg 1992). Figure 30.1 shows an example of the orthogonal encoding of a DNA sequence.

The orthogonal encoding requires that the biosequences be equal in length, or one must sample the biosequences of variable lengths by a window of fixed size. Another disadvantage is that it wastes a lot of input units in the input layer of a neural network. For instance, for a protein sequence of 100 amino acids, 2000 input units are required to represent the protein sequence. This requires many neural network weight parameters as well as many training data, making it difficult to train the neural network.

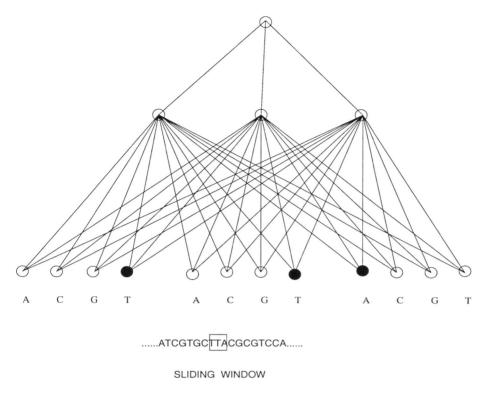

Figure 30.1. An example of the orthogonal encoding of a DNA sequence.

An alternative encoding method, as proposed in this chapter, is to use high-level features extracted from biosequences. The high-level features should be relevant and biologically meaningful. By "relevant", we mean that there should be high mutual information between the features and the output of the neural network, where the mutual information measures the average reduction in uncertainty about the output of the neural network given the values of the features. By "biologically meaningful", we mean that the features should reflect the biological characteristics of the sequences.

2.2. Characteristics of E. Coli Promoters

The E. Coli promoter is located immediately before the E. Coli gene. Thus, successfully locating the E. Coli promoter conduces to identifying the E. Coli gene. The uncertain characteristics of the E. Coli promoters contribute to the difficulty in the promoter recognition. The E. Coli promoters contain two binding sites to which the E. Coli RNA polymerase, a kind of protein, binds (Lisser and Margalit 1993). The two binding sites are the -35 hexamer box and the -10 hexamer box, respectively. Each binding site consists of 6 bases (nucleotides). The central nucleotides of the two binding sites are roughly 35 bases and 10 bases, respectively, upstream of the transcriptional start site. The transcriptional start site is the first nucleotide of a codon where the transcription begins; it serves as a reference point (position +1). The consensus sequences, i.e., the prototype sequences composed of the most frequently occurring nucleotide at each position, for the -35 binding site and the -10 binding site are **TTGACA** and **TATAAT**, respectively. But none of the promoters can exactly match the two consensus sequences. The average conservation is about 8 nucleotides, meaning that a promoter sequence can match, on average, 8 out of the 12 nucleotides in the two consensus sequences. Figure 30.2 shows an example promoter sequence with the -35 binding site being **TAGCGA** and the -10 binding site being **AAAGAT**. The conservation here includes only 6 nucleotides.

Figure 30.2. An example promoter sequence. The regions are highlighted by upper case letters. The -54 region, -44 region, -35 box, -29 region, -22 region, -10 box, and +1 region are CTTTGTAGC, CTTTCAC, TAGCGA, AACG, GAATGG, AAAGAT and CA, respectively.

The two binding sites are separated by a spacer. The length of the spacer has an effect on the relative orientation between the -35 region and the -10 region. A spacer of 17 nucleotides is most probable. The promoter sequence in Figure 30.2 has a spacer of 17 nucleotides. Another spacer between the -10 hexamer box and the transcriptional start site also has a variable length. The most probable length of this spacer is 7 nucleotides. The promoter sequence in Figure 30.2 has a spacer of 6 nucleotides.[1] Because of the variable spacing, it is not appropriate to use the orthogonal encoding to encode or view a promoter sequence as an n attribute tuple, where n is the length of the promoter sequence. Many promoter sequences have the pyrimidine (C or T) at the position -1 (one nucleotide upstream of the transcriptional start site), while the purine (A or G) is at the transcriptional start site (position +1). The +1 region includes the nucleotides at the position -1 and the transcriptional start site. The promoter sequence in Figure 30.2 has a nucleotide C at the position -1 and a nucleotide A at the transcriptional start site.

In addition to these salient characteristics in the two binding sites and the transcriptional start site, there are some non-salient characteristics in other regions. In Galas et al. (1985) and Mengeritsky and Smith (1987), a pattern matching method was applied to the characterization of E. Coli promoters. Some weak motifs were found around the -44 and the -22 regions. A weak motif is a subsequence, which occurs frequently in a region. We use the term "weak", since the frequency of a base of the motif is not as significant as the frequency of a base of the consensus sequences occurring in the binding sites. In Cardon and Stormo (1992), as many as 8 nucleotides (weak motifs) within the spacer region between the two binding sites were found to have contributions to the specificity of the promoter sequences. Recently, Pedersen and Engelbrecht (1995) adopted a neural network to characterize E. Coli promoters. The significance of a weak motif was measured by the decrease in the maximum correlation coefficient when all motifs except that weak motif were fed into the neural network. By using this method, the authors found some weak motifs in the +1, -22, and -44 regions. It is interesting to observe that these weak motifs are spaced regularly with a period of 10–11 nucleotides corresponding to one helical turn. This phenomenon suggests that the RNA polymerase makes contact with the promoter on one face of the DNA. Subsequently, the characterization of E. Coli promoter sequences was carried out by the hidden Markov model (Pedersen et al. 1996). It was observed that the position of the -35 box relative to the transcriptional start site is very flexible. More recently, a clustering analysis was carried out on a larger set of E. Coli promoter sequences containing 441 promoters (Ozoline et al. 1997). Some weak motifs were found in the -54 region.

These weak motifs were also revealed by the sequence logos described in Schneider and Stephens (1990). Figure 30.3 displays the sequence logos of 438 E. Coli promoters aligned according to the transcriptional start site.[2] Given a set of aligned sequences, the sequence logos measure the non-randomness of each position l independently by the Shannon entropy for that position:

$$R(l) = \log_2(|\mathcal{D}|) - (-\sum_{b \in \mathcal{D}} f(b,l) \log_2 f(b,l)), \qquad (30.1)$$

where $|\mathcal{D}|$ is the cardinality of the 4-letter DNA alphabet $\mathcal{D}$, $\log_2(|\mathcal{D}|) = 2$ is the maximum uncertainty at any given position, $f(b,l)$ is the frequency of base b at position l, and $-\sum_{b \in \mathcal{D}} f(b,l) \log_2 f(b,l)$ is the Shannon entropy of position l.

The height at each position represents the information content of that position. The higher the information content, the less random that posi-

1148   HANDBOOK OF MASSIVE DATA SETS

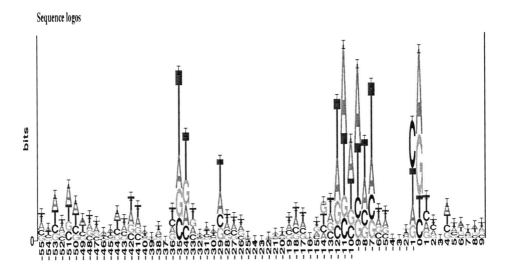

Figure 30.3. The sequence logos of 438 E. Coli promoter sequences. Position 0 in the figure is the transcriptional start site, which is equivalent to position +1 described in the text. The negative positions in the figure are consistent with those described in the text.

tion is. The size of each base at each position of the logos is proportional to the frequency of the base. Recall that a weak motif is a frequently occurring subsequence in a region. In the sequence logos, a weak motif consists of positions (bases) with non-zero information content. From Figure 30.3, it can be seen that some weak motifs exist in the +1, -22, -29, -44, and -54 regions.

Based on the characteristics of the E. Coli promoter sequences reported in the literature, we explore two methods for extracting high-level features in the following regions in the promoters: the -54 region (9 nucleotides long), the -44 region (7 nucleotides long), the -35 region (6 nucleotides long), the -22 region (6 nucleotides long), the -10 region (6 nucleotides long), and the +1 region (2 nucleotides long) (see Figure 30.2). The first method is the Maximal Dependence Decomposition (MDD) technique and the second one is a motif based method. Because the -29 region is 4 nucleotides long, and the 4 nucleotide long motif in the -29 region is not statistically significant, we apply only the MDD method to extracting features in the -29 region. In order to calculate these feature values, we must know, as precisely as possible, where these regions are. In the following subsection, we present an expectation-maximization (EM) algorithm for locating the binding sites of promoter sequences. Then we describe our feature extraction methods in detail.

2.3. Locating Binding Sites by the EM Algorithm

To align subsequences in the -35 region, the -29 region, the -22 region and the -10 region, we need to locate the two binding sites in the E. Coli promoters. Locating the -35 box and the -10 box may be done by the EM algorithm (Dempster et al. 1977). In general, the EM algorithm can be applied for the maximum likelihood estimation when data are incomplete. Locating the binding sites by the EM algorithm was first proposed by Lawrence and Reilly (1990). It was then generalized by Cardon and Stormo (1992) to allow for different spacers between the two binding sites. These published methods (Bailey 1995, Cardon and Stormo 1992, Lawrence and Reilly 1990) either assumed that the location of the binding sites is uniformly distributed or attempted to locate one "continuous region" that included the -35 box, the -10 box and a spacer of variable length between them. By contrast, our method does not make the assumption of uniform distribution and considers the binding sites separately from the spacer.

Let T represent the set of promoter sequences in the training set, i.e., T contains all positive training sequences. Let K denote the cardinality of T. For a promoter sequence $S_i \in T$, the length of the spacer between the -10 region and the transcriptional start site, denoted sp_{10}, and the length of the spacer between the -35 region and the -10 region, denoted sp_{35}, are unobserved, though S_i is observed. Specifically, we refer to the positive training sequences as "observed" data since they are given. These observed data are incomplete, because the lengths of the two spacers are not given (the lengths are referred to as "unobserved" or "missing" data). The proposed EM algorithm estimates the model parameters, defined later, from the incomplete data. Then based on the estimates of the model parameters, the algorithm determines the locations of the two binding sites for any DNA sequence.

In general, sp_{10} varies from 3 to 11 and sp_{35} varies from 15 to 21. Assume that the nucleotides at all positions are independent. Then one can use the Position Weight Matrix (PWM) described in Staden (1990) to model nucleotides at each position. Each binding site consists of 6 bases. Let $P_{10,j}(x), j = 1, \ldots, 6$, denote the probability of x, $x \in \mathcal{D}$, occurring at position j in the -10 region. Let $\mathbf{P_{10}}$ denote $(P_{10,1}, \ldots, P_{10,6})$. Let $P_{35,j}(x), j = 1, \ldots, 6$, denote the probability of x occurring at position j in the -35 region. Let $\mathbf{P_{35}}$ denote $(P_{35,1}, \ldots, P_{35,6})$. Let $P_o(x)$ denote the probability of x occurring in the regions outside the two binding sites. Let $P_s(Sp_{10} = m, Sp_{35} = n)$ denote the probability of $Sp_{10} = m$ and $Sp_{35} = n$ where Sp_{10} (Sp_{35}, respectively) is the random variable denoting the distance between the transcriptional start site (the -35 region, respectively) and the -10 region, $m \in \{3, \ldots, 11\}$ and $n \in \{15, \ldots, 21\}$. Let $\theta = (\mathbf{P_{10}}, \mathbf{P_{35}}, P_o)$.

The EM algorithm proceeds iteratively to converge. Each iteration consists of two steps: Expectation step (E step) and Maximization step (M step). The E step calculates the expected complete-data log likelihood, where the expectation is over the distribution of the missing data given the observed data and current estimates of θ. Assume that the promoter sequences in the training set T are independent. Then the E step calculates

$$E_{Sp_{10}, Sp_{35}|T, \theta} \log P(T, Sp_{10}, Sp_{35}|\theta)$$
$$= E_{Sp_{10}, Sp_{35}|T, \theta} \log(P(T|Sp_{10}, Sp_{35}, \theta) P_s(Sp_{10}, Sp_{35}))$$
$$= \sum_{i=1}^{K} \sum_{m=3}^{11} \sum_{n=15}^{21} P(Sp_{10} = m, Sp_{35} = n|S_i, \theta) \log($$
$$P(S_i|Sp_{10} = m, Sp_{35} = n, \theta) P_s(Sp_{10} = m, Sp_{35} = n)). \quad (30.2)$$

Suppose that all promoter sequences in the training set T are 65 nucleotides long and are aligned with respect to the transcriptional start site, which is at position 56. Let $S_{i,j}$ denote the nucleotide at position j of promoter sequence S_i. Define

$$I_{i,j,x} = \begin{cases} 1 & \text{if } S_{i,j} = x \\ 0 & \text{otherwise} \end{cases}. \quad (30.3)$$

Let $O_{i,m,n}(x)$ denote the number of occurrences of the nucleotide x outside the two binding sites of promoter sequence S_i given $Sp_{10} = m$ and $Sp_{35} = n$. Then

$$P(S_i|Sp_{10} = m, Sp_{35} = n, \theta) = \Pi_{j=1}^{6} P_{10,j}(S_{i,49-m+j}) *$$
$$\Pi_{j=1}^{6} P_{35,j}(S_{i,43-m-n+j}) *$$
$$\Pi_{x=A}^{T} P_o(x)^{O_{i,m,n}(x)}. \quad (30.4)$$

From (30.4), according to Bayes' rule, we have

$$P(Sp_{10} = m, Sp_{35} = n|S_i, \theta)$$

$$= \frac{P(S_i|Sp_{10}=m, Sp_{35}=n, \theta) P_s(Sp_{10}=m, Sp_{35}=n)}{P(S_i|\theta)}$$

$$= \frac{P(S_i|Sp_{10}=m, Sp_{35}=n, \theta) P_s(Sp_{10}=m, Sp_{35}=n)}{\sum_{m=3}^{11} \sum_{n=15}^{21} P(S_i|Sp_{10}=m, Sp_{35}=n, \theta) P_s(Sp_{10}=m, Sp_{35}=n)}. \quad (30.5)$$

where

$$P(S_i|Sp_{10}=m, Sp_{35}=n, \theta)$$
$$= \Pi_{j=1}^{6} P_{10,j}(S_{i,49-m+j}) \Pi_{j=1}^{6} P_{35,j}(S_{i,43-m-n+j})$$
$$P_s(Sp_{10}=m, Sp_{35}=n)$$
$$= \Pi_{x=A}^{T} P_o(x)^{O_{i,m,n}(x)} P_s(Sp_{10}=m, Sp_{35}=n)$$

Previous applications of the EM algorithm (Bailey 1995, Cardon and Stormo 1992, Lawrence and Reilly 1990) assumed $P_s(Sp_{10}, Sp_{35})$ to be uniformly distributed. The term $P_s(Sp_{10}=m, Sp_{35}=n)$ was deleted from (30.5). We do not assume $P_s(Sp_{10}, Sp_{35})$ to be uniformly distributed, as the most probable values of m are 6, 7, 8, and the most probable values of n are 16, 17, 18. Substituting (30.4) and (30.5) into (30.2), we have

$$K \sum_{j=1}^{6} \sum_{x=A}^{T} f_{10,j}(x) \log P_{10,j}(x) + K \sum_{j=1}^{6} \sum_{x=A}^{T} f_{35,j}(x) \log P_{35,j}(x) +$$

$$K(65-12) \sum_{x=A}^{T} f_o(x) \log P_o(x) + K \sum_{m=3}^{11} \sum_{n=15}^{21} f_s(m,n) \log P_s(m,n) \quad (30.6)$$

where

$$f_{10,j}(x) = \frac{1}{K} \sum_{i=1}^{K} \sum_{m=3}^{11} \sum_{n=15}^{21} I_{i,49-m+j,x} P(Sp_{10}=m, Sp_{35}=n|S_i, \theta),$$

$$f_{35,j}(x) = \frac{1}{K} \sum_{i=1}^{K} \sum_{m=3}^{11} \sum_{n=15}^{21} I_{i,43-m-n+j,x} P(Sp_{10}=m, Sp_{35}=n|S_i, \theta),$$

$$f_o(x) = \frac{1}{K(65-12)} \sum_{i=1}^{K} \sum_{m=3}^{11} \sum_{n=15}^{21} O_{i,m,n}(x) P(Sp_{10}=m, Sp_{35}=n|S_i, \theta),$$

$$f_s(m,n) = \frac{1}{K} \sum_{i=1}^{K} \sum_{m=3}^{11} \sum_{n=15}^{21} P(Sp_{10}=m, Sp_{35}=n|S_i, \theta).$$

Let θ^0 denote the value of θ at the beginning of the first iteration. θ^0 was initialized randomly so that the E step can proceed. In each iteration, we use the current estimate θ^t to calculate the expected complete data log likelihood.

The M step maximizes (30.6) with respect to θ. The maximum likelihood estimates of θ when the complete data were known are just sample frequencies $f_{10,j}$, $f_{35,j}$, f_o, and f_s, $j = 1, \ldots, 6$. That is,

$$P_{10,j}^{t+1}(x) = f_{10,j}(x), x \in \mathcal{D},$$
$$P_{35,j}^{t+1}(x) = f_{35,j}(x), x \in \mathcal{D},$$
$$P_o^{t+1}(x) = f_o(x), x \in \mathcal{D},$$
$$P_s^{t+1}(m,n) = f_s(m,n). \tag{30.7}$$

The new value of θ can be used in the next iteration. The process iterates to convergence. Given the model parameters calculated from the positive training sequences (i.e., the promoter sequences in the training data set T), we can determine the locations of the two binding sites of any DNA sequence S_n, which could be a training sequence or a test sequence, a positive sequence or a negative sequence, by choosing the two spacer lengths sp_{10} and sp_{35} that maximize $P(S_n, Sp_{10}, Sp_{35}|\theta) = P(S_n|Sp_{10}, Sp_{35}, \theta) P_s(Sp_{10}, Sp_{35})$. We then align the two binding sites of the training promoter sequences and extract features from the different regions using the MDD technique and the motif based method, described below.

2.4. The MDD Technique

The MDD technique was first proposed to detect the splice site in human genomic DNA in the gene prediction software GENSCAN (Burge and Karlin 1997). It was later adopted in the latest version of the gene prediction software MORGAN (Salzberg 1997a). MDD was derived from the PWM model to overcome the limitation of the consensus sequence by modeling the nucleotide distribution at each position. One disadvantage of PWM is that it assumes the positions are independent. This disadvantage was removed in the Weight Array Model (WAM) (Zhang and Marr 1993), which generalizes PWM by allowing for the dependencies between the adjacent positions.

WAM is essentially a first order Markov chain (conditional probability on the upstream adjacent nucleotide) which can be further generalized by the second-order Markov chain, third-order Markov chain, etc. However, the more dependencies one tries to model, the more free parameters the model has, thus requiring more training data to appropriately estimate the parameters in the model. In general, there is a danger when one tries to use more complex models, which have more free parameters, and does not have enough training data to estimate the free parameters.

For instance, suppose we have 438 promoter sequences available to estimate the parameters $P_i(x)$, $x \in \mathcal{D}$, in PWM where $P_i(x)$ represents the probability of nucleotide x occurring at position i in the sequences. Equivalently we have roughly $438/4 = 109$ promoter sequences available to estimate the parameters $P_i(x_i|x_{i-1})$ in WAM, where $P_i(x_i|x_{i-1})$ represents the conditional probability of x_i at position i given x_{i-1} at position $i-1$, which is the upstream neighbor of x_i at position i, x_{i-1}, $x_i \in \mathcal{D}$. 109 promoter sequences are too few to reliably estimate the free parameters.

MDD provides a flexible solution to the above problem by iteratively clustering the dataset based on the most significant adjacent or non-adjacent dependencies. It essentially models the first-order, second-order, third-order and even higher order dependencies depending on the amount of training data available. More specifically, MDD works as follows. Given a set U of aligned sequences, it first chooses the consensus nucleotide C_i at each position i. In our case, the set U includes subsequences in the same region of all the positive training sequences (i.e., promoter sequences). Then the χ^2 statistic χ_{ij} is calculated to measure the dependencies between C_i and the nucleotides at position j ($i \neq j$). If no significant dependencies are detected, then the simple PWM is used. If there are significant dependencies detected, but the dependencies exist only between adjacent positions, then WAM is used. Otherwise the MDD procedure is carried out.

The MDD procedure is an iterative process: calculate the sum $S_i = \sum_{i \neq j} \chi_{ij}$ for each i; select the position m such that S_m is maximal, and decompose the dataset U into two disjoint subsets, U_m (containing all sequences that have the consensus nucleotide C_m at position m) and $U - U_m$ (containing all sequences that do not have the consensus nucleotide C_m at position m).

The MDD procedure is then applied recursively to U_m and $U - U_m$ respectively until any one of the following conditions holds: no further decomposition is possible, no significant dependencies between positions exist in the resulting subsets, or the number of sequences in the resulting subsets is below a threshold, so that, reliable estimation of parameters is not possible after further decomposition.

We apply the MDD method to the -54 region, the -44 region, the -35 region, the -29 region, the -22 region, the -10 region, and the +1 region respectively, of the training promoter sequences. As a result, the -44 region and the +1 region are modeled by PWM, and one level decomposition is carried out in the other regions.

Given a set of sequences, the MDD feature values of each sequence are calculated as follows. First, the MDD technique is applied to all the positive training data (i.e., the E. Coli promoter sequences). The results are probability matrices for the -44 region and the +1 region as well as conditional probability matrices for the -54 region, the -35 region, the -29 region, the -22 region, and the -10 region. Secondly, for all the positive and negative sequences, these matrices are used to calculate the MDD feature values of each sequence. In particular, the feature value of a subsequence $X = x_1 x_2 x_3 \ldots x_n$, where $x_i \in \mathcal{D}$, $i = 1, 2, \ldots, n$, in the -44 region or the +1 region in a sequence S is calculated by

$$P(X) = p_1(x_1) p_2(x_2) p_3(x_3) \ldots p_n(x_n), \qquad (30.8)$$

where $p_i(x_i)$ is the probability of x_i at position i. For example, suppose the probability matrix in the +1 region of the positive training sequences is

$$\begin{array}{c} \\ \text{A} \\ \text{C} \\ \text{G} \\ \text{T} \end{array} \left(\begin{array}{cc} Position-1 & Position+1 \\ 0.168 & 0.481 \\ 0.392 & 0.100 \\ 0.110 & 0.271 \\ 0.330 & 0.148 \end{array} \right)$$

Then, for example, for the subsequence TG, $P(\text{TG}) = 0.330 \times 0.271 = 0.089$.

The feature value of a subsequence $X = x_1 x_2 x_3 \ldots x_n$ in the other regions of the sequence S is calculated by

$$P(X) = \begin{cases} p_m(C_m) \prod_{i=1}^{m-1} \{cp_i^m(x_i)\} \prod_{i=m+1}^n \{cp_i^m(x_i)\} & \text{if } C_m = x_m \\ p_m(x_m) \prod_{i=1}^{m-1} \{cp_i^{\overline{m}}(x_i)\} \prod_{i=m+1}^n \{cp_i^{\overline{m}}(x_i)\} & \text{otherwise} \end{cases},$$

(30.9)

where $p_m(C_m)$ ($p_m(x_m)$, respectively) represents the probability of C_m (x_m, respectively) at position m, $cp_i^m(x_i)$ ($cp_i^{\overline{m}}(x_i)$, respectively), $i = 1, 2, \ldots, m-1, m+1, m+2, \ldots, n$, represents the conditional probability of x_i at position i given $x_m = C_m$ ($x_m \neq C_m$, respectively).

2.5. The Motif Based Method

To calculate the motif feature values of each sequence, we first apply our pattern matching tool, **Sdiscover** (Wang et al. 1994), to the positive training data (i.e., the E. Coli promoter sequences) to find weak motifs in the -54 region, the -44 region, the -35 region, the -22 region and the -10 region respectively, in these sequences.

Let V be a set of sequences. We define the occurrence number of a motif (subsequence) to be the number of sequences in V that contain the motif. The **Sdiscover** tool can find all the motifs M where M is within the allowed Mut mutations of at least $Occur$ sequences in the given set

V and $|M| \geq Length$ where $|M|$ represents the number of nucleotides in M. $Mut, Occur, Length$ are user-specified parameters.

In our case, the set V includes all the subsequences in the same region (e.g., the -54 region) of the positive training sequences. The required length of the motifs is fixed for each region. In the study presented here, the length is 6 for the -54, -35 and -10 regions, the length is 5 for the -44 and -22 regions. The minimum occurrence number required is 2 and the allowed mutation number is 0. The occurrence number of a motif is assigned as the weight of the motif. Intuitively, the more frequently a motif occurs in a region of the positive training sequences, the higher weight it has.

Given a set of sequences, which could be training sequences or test sequences, positive sequences or negative sequences, the motif feature values in the -54, -44, -35, -22, -10 regions of each training sequence are calculated as follows. First, the motif based method described above is applied to the -54, -44, -35, -22, -10 regions of all the positive training data. The result is five sets of motifs in the -54, -44, -35, -22, -10 regions. Secondly, for each region of a sequence S, the subsequence in that region of S is matched against the motifs in that region. If there are matched motifs, the feature value of that region for S is the maximum weight of the matched motifs; otherwise the feature value is assigned to 0.

The motif feature value in the +1 region of any sequence is assigned to 1 if the nucleotide at position -1 is the pyrimidine (C or T) and the nucleotide at the transcriptional start site is the purine (A or G); otherwise it is assigned to 0.

3. Basic Classifiers

We have developed three basic classifiers: $Classifier\_0, Classifier\_1$ and $Classifier\_2$. Each of the classifiers is a Bayesian neural network. Each training sequence or test sequence is encoded by the high level features as described in the previous section and the feature values are used as the input of the neural networks.

The first classifier, $Classifier\_0$, is a Bayesian neural network with 5 hidden units and 9 input units including 7 MDD features and 2 distance features, which are the distance (i.e., the number of nucleotides) between the -35 box and the -10 box and the distance between the -10 box and the transcriptional start site. The second classifier, $Classifier\_1$, is a Bayesian neural network with 5 hidden units and 8 input units including 6 motif features and the above 2 distance features. The third classifier, $Classifier\_2$, is a Bayesian neural network with 6 hidden units and 15 input units including the above 7 MDD features, the above 6 motif features and the above 2 distance features. The number of hidden units is determined experimentally according to the evidence of the model.

The Bayesian neural network we use has one hidden layer with sigmoid activation functions. The output layer of the neural network has one output unit. The output value is bounded between 0 and 1 by the logistic activation function $f(a) = \frac{1}{1+e^{-a}}$. The neural network is fully connected between the adjacent layers. Figure 30.4 shows the Bayesian neural network architecture of $Classifier\_2$.

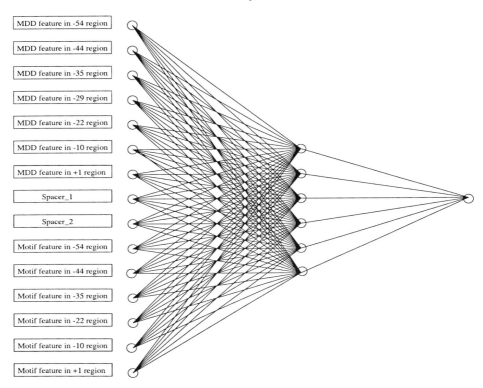

Figure 30.4. The Bayesian neural network architecture of $Classifier\_2$. Spacer_1 is the distance between the -35 box and the -10 box. Spacer_2 is the distance between the -10 region and the transcriptional start site.

3.1. Bayesian Neural Networks

The Bayesian neural network is the integration of Bayesian inference and the neural network. In Bayesian inference, a model (e.g. a neural network) M_i consists of a set of free parameters which are viewed as random variables. The *prior* of a model M_i is represented by $P(M_i)$. The *likelihood*, i.e., the probability of the data D given the model M_i, is specified by $P(D|M_i)$. The *posterior probability* of the model M_i is quantified by $P(M_i|D)$. From Bayes' rule, we have:

$$P(M_i|D) = \frac{P(D|M_i)P(M_i)}{P(D)}, \qquad (30.10)$$

where $P(D) = \int P(D|M_i)P(M_i)dM_i$ is a normalizing constant.

In our case, $D = \{\mathbf{x}^{(m)}, t_m\}, m = 1, 2, \ldots, N$, denotes the training dataset (including both positive and negative training data), where N is the total number of the training sequences in D, $\mathbf{x}^{(m)}$ is an input feature vector which contains 9 (8, 15, respectively) input values for $Classifier\_0$ ($Classifier\_1$, $Classifier\_2$, respectively), and t_m is the binary (0/1) target value for the output unit. That is, if $\mathbf{x}^{(m)}$ represents a promoter sequence, t_m is 1; otherwise, t_m is 0. Let $\mathbf{x}$ denote an input feature vector for a DNA sequence, which could be a training sequence or a test sequence. Given the architecture $\mathbf{A}$ and the weights $\mathbf{w}$ of the neural network, the output value y can be uniquely determined from the input vector $\mathbf{x}$. The output value $y(\mathbf{x}; \mathbf{w}, \mathbf{A})$ can be interpreted as $P(t = 1|\mathbf{x}, \mathbf{w}, \mathbf{A})$, i.e., the probability that $\mathbf{x}$ represents a promoter sequence given $\mathbf{x}, \mathbf{w}, \mathbf{A}$.

The likelihood, i.e., the probability of the data given the model, is calculated by:

$$P(D|\mathbf{w}, \mathbf{A}) = \Pi_{m=1}^{N} y^{t_m}(1-y)^{1-t_m} = exp(-G(D|\mathbf{w}, \mathbf{A})), \qquad (30.11)$$

where $G(D|\mathbf{w}, \mathbf{A})$ is the cross-entropy error function

$$G(D|\mathbf{w}, \mathbf{A}) = -(\sum_{m=1}^{N} t_m log y + (1-t_m) log(1-y)). \qquad (30.12)$$

$G(D|\mathbf{w}, \mathbf{A})$ is the objective function and is minimized in the non-Bayesian neural network training process, which is a maximum likelihood estimation based method, and assumes all possible weights are equally likely.

In the non-Bayesian neural network, the weight decay is often used to avoid overfitting on the training data and poor generalization on the test data by adding a term $\frac{\alpha}{2} \sum_{i=1}^{k} w_i^2$ to the objective function, where α is the weight decay parameter (hyperparameter), $\sum_{i=1}^{k} w_i^2$ is the sum of the square of all the weights of the neural network, and k is the number of weights. This objective function is minimized, to penalize the neural network with weights of large magnitudes, penalizing the over-complex model and favoring the simple model. However, there is no precise way to specify the appropriate value of α, which is often tuned offline.

In the Bayesian neural network, the hyperparameter α is interpreted as the parameter of the model, and is optimized online during the Bayesian learning process. The weight decay term $\frac{\alpha}{2}\sum_{i=1}^{k} w_i^2$ can be scaled to $(\frac{\alpha}{2\pi})^{\frac{k}{2}} exp(-\frac{\alpha}{2}\sum_{i=1}^{k} w_i^2)$ and interpreted as a prior probability of the weight vector $\mathbf{w}$ in the Gaussian distribution with zero mean and variance $\frac{1}{\alpha}$. Thus, larger neural network weights are less probable. The Bayesian neural network further generalizes the previous weight decay term by associating the weights in different layers with different variances. Thus, the hyperparameter becomes a vector α. Let $(\alpha_1, \alpha_2, \ldots, \alpha_n)$ represent the vector α, where α_i is associated with a group of weights w_j^i, $j = 1, 2, \ldots, q_i$, $i = 1, 2, \ldots, n$, q_i is the number of weights associated with α_i. Let E_W^c denote $\frac{1}{2}\sum_{j=1}^{q_c}(w_j^c)^2$, $c = 1, 2, \ldots, n$. Then the prior is:

$$P(\mathbf{w}|\alpha, \mathbf{A}) = \frac{exp(-\sum_{c=1}^{n} \alpha_c E_W^c)}{Z_W}, \qquad (30.13)$$

where $Z_W = \int exp(-\sum_{c=1}^{n} \alpha_c E_W^c) d\mathbf{w}$ is a Gaussian integral. From (30.12) and (30.13), we can get a posterior probability:

$$P(\mathbf{w}|D, \alpha, \mathbf{A}) = \frac{exp(-\sum_{c=1}^{n} \alpha_c E_W^c - G(D|\mathbf{w}, \mathbf{A}))}{Z_M}, \qquad (30.14)$$

where $Z_M = \int exp(-\sum_{c=1}^{n} \alpha_c E_W^c - G(D|\mathbf{w}, \mathbf{A})) d\mathbf{w}$ is also a normalizing constant.

3.2. The Training Phase

The Bayesian training of neural networks is an iterative procedure. In the implementation of the Bayesian neural network that we adopt,[3] each iteration involves two level inferences. Figure 30.5 illustrates the training process.

At the first level, given the value of hyperparameter α, which is initialized to the random value during the first iteration, we can infer the most probable value of the weight vector $\mathbf{w}^{mp}$ corresponding to the maximum of $P(\mathbf{w}|D, \alpha, \mathbf{A})$ by the neural network training, which minimizes $\sum_{c=1}^{n} \alpha_c E_W^c + G(D|\mathbf{w}, \mathbf{A})$. For the first level inference, Bayes' rule is

$$P(\mathbf{w}|D, \alpha, \mathbf{A}) = \frac{P(D|\mathbf{w}, \mathbf{A})P(\mathbf{w}|\alpha, \mathbf{A})}{P(D|\alpha, \mathbf{A})}, \qquad (30.15)$$

where $P(D|\mathbf{w}, \mathbf{A})$, $P(\mathbf{w}|\alpha, \mathbf{A})$, and $P(\mathbf{w}|D, \alpha, \mathbf{A})$ are given by (30.11), (30.13) and (30.14) respectively. The $P(\mathbf{w}|D, \alpha, \mathbf{A})$ can be approximated

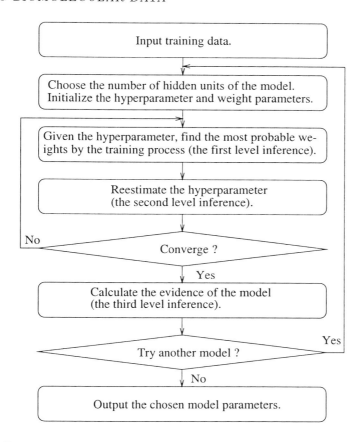

Figure 30.5. The training process of the Bayesian neural network.

by a Gaussian centered around $\mathbf{w}^{mp}$ (Mackay 1992a):

$$P(\mathbf{w}|D, \alpha, \mathbf{A}) \simeq P(\mathbf{w}^{mp}|D, \alpha, \mathbf{A})exp(-\frac{1}{2}(\mathbf{w} - \mathbf{w}^{mp})^T \mathbf{H}(\mathbf{w} - \mathbf{w}^{mp})), \qquad (30.16)$$

where $\mathbf{H} = -\bigtriangledown \bigtriangledown logP(\mathbf{w}|D, \alpha, \mathbf{A})|_{\mathbf{w}^{mp}}$ is the Hessian matrix evaluated at $\mathbf{w}^{mp}$.

At the second level, the hyperparameter α is optimized. For the second level inference, Bayes' rule is

$$P(\alpha|D, \mathbf{A}) = \frac{P(D|\alpha, \mathbf{A})P(\alpha|\mathbf{A})}{P(D|\mathbf{A})}. \qquad (30.17)$$

Because of the lack of the prior knowledge of $P(\alpha|\mathbf{A})$, we assume $P(\alpha|\mathbf{A})$ to be a constant and ignore it. Since the normalizing factor $P(D|\mathbf{A})$ is also a constant, the value of α maximizing the posterior $P(\alpha|D, \mathbf{A})$ can be inferred by maximizing the evidence of α, $P(D|\alpha, \mathbf{A})$, which is the normalizing factor in (30.15). So $P(D|\alpha, \mathbf{A}) = \int P(D|\mathbf{w}, \mathbf{A})P(\mathbf{w}|\alpha, \mathbf{A})d\mathbf{w}$. The evidence $P(D|\alpha, \mathbf{A})$ is maximized by differentiation with respect to α. We can find the new hyperparameter value α^{new} by setting the differentiation to zero,

$$\alpha_i^{new} = \frac{\gamma_i}{\sum_{j=1}^{q_i}(w_j^{mp})^2}, \qquad (30.18)$$

where γ_i is the number of "well-determined parameters" in the group i (Mackay 1992a). The new hyperparameter value α^{new} is then used in the next iteration. The process iterates until the convergence is reached.

The third level inference is the model comparison. This level is carried out manually. For the third level inference, Bayes' rule is

$$P(\mathbf{A_i}|D) = \frac{P(D|\mathbf{A_i})P(\mathbf{A_i})}{P(D)}, \qquad (30.19)$$

where $P(\mathbf{A_i})$ is the prior probability and is assumed to be a constant; $P(D)$ is also a constant. The posterior probability $P(\mathbf{A_i}|D)$ can be determined by the evidence $P(D|\mathbf{A_i})$, which is the normalizing factor at the second level inference. So $P(D|\mathbf{A_i}) = \int P(D|\alpha, \mathbf{A_i})P(\alpha|\mathbf{A_i})d\alpha$. Different models with a different number of hidden units were tested. The model with the largest evidence value was chosen.

3.3. The Testing Phase

In the testing phase, for the three basic classifiers we have developed, the output of a Bayesian neural network, y, is given by the marginalization of the network weight distribution. That is,

$$\begin{aligned} P(t=1|\mathbf{x}, D, \mathbf{A}) &= \int P(t=1|\mathbf{x}, \mathbf{w}, \mathbf{A})P(\mathbf{w}|D, \mathbf{A})d\mathbf{w} \\ &= \int y(\mathbf{x}; \mathbf{w}, \mathbf{A})P(\mathbf{w}|D, \mathbf{A})d\mathbf{w}. \end{aligned} \qquad (30.20)$$

The output of a Bayesian neural network, $P(t = 1|\mathbf{x}, D, \mathbf{A})$, is the probability that the unlabeled test sequence is a promoter. If it is greater than the decision boundary 0.5, the test sequence is classified as a promoter; if it is less than the decision boundary 0.5, the test sequence is classified as a non-promoter; otherwise the test sequence gets the "no-opinion" verdict.

4. Combination of Basic Classifiers

The three basic classifiers described in the previous section can be combined into one classifier in the second level (see Figure 30.6). We explore two methods for combining the three basic classifiers: $Combiner\_0$ and $Combiner\_1$.

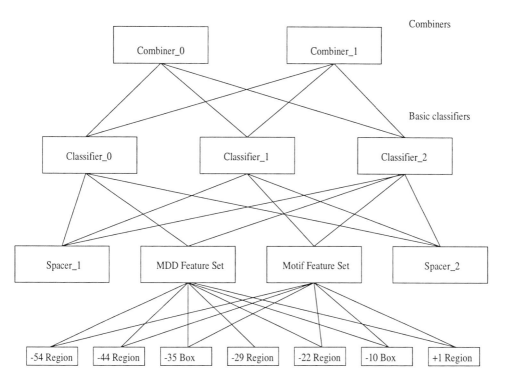

Figure 30.6. Our classification scheme.

$Combiner\_0$ employs an unweighted voter. Let $output_i$, $0 \leq i \leq 2$, be the output value of $Classifier\_i$. If the three basic classifiers agree on the classification results (Promoter, Non-promoter, No-opinion), the final result will be the same as the results of the three classifiers; if two classifiers agree on the classification results, the final result will be the same as the results of these two classifiers; if none of the classifiers agrees on the classification results, the final result will be the same as the result of a classifier whose $\min(1 - output_i, output_i)$ is minimal; otherwise the final result is "no-opinion".

$Combiner\_1$ employs a weighted voter. Its output is the weighted sum of the outputs of the three basic classifiers. That is, let w_i represent the weight of $Classifier\_i$, where the weight is the precision rate of $Classifier\_i$ in the training phase. The output of $Combiner\_1$ is given by

$$\frac{1}{w_0 + w_1 + w_2} \sum_{i=0}^{2} w_i \times output_i. \quad (30.21)$$

Note that if we assign equal weights to the three basic classifiers, then $Combiner\_1$ is reduced to $Combiner\_0$.

5. Experiments and Results

5.1. Data

In this study, we adopted E. Coli promoter sequences taken from the latest E. Coli promoter compilation (Ozoline et al. 1997). There were 441 E. Coli promoters aligned by the transcriptional start site. We trimmed each promoter sequence to a sequence of 65 nucleotides including nucleotides from -55 (55 nucleotides upstream of the transcriptional start site) to +10 (10 nucleotides downstream of the transcriptional start site). This gave us 438 promoter sequences.

The negative data (i.e., non-promoter sequences) was retrieved from Genbank.[4] The negative data are E. Coli genes with the preceding promoter region deleted. Each negative sequence is also 65 nucleotides long. There were 1,314 negative sequences.

5.2. Results

Table 30.2 gives the ten-fold cross validation results for the three basic classifiers and Table 30.3 gives the results for the combined classifiers. In ten-fold cross validation, the dataset containing both the positive data (promoters) and the negative data (non-promoters) was randomly split into ten mutually exclusive folds $D_1, D_2, \ldots, D_{10}$ of approximately equal size. Each Bayesian neural network was trained and tested ten times. During the ith run, the neural network was trained on $D - D_i$, and tested on D_i. We allocated the data in such a way that the training dataset $D - D_i$ (the test dataset D_i, respectively) has approximately $\frac{9}{10}$ ($\frac{1}{10}$, respectively) positive data and $\frac{9}{10}$ ($\frac{1}{10}$, respectively) negative data. The average over the ten tests was taken.

The machine used to conduct the experiments was a 350 MHz Pentium II PC running a Linux operating system. The time spent in extracting features by a basic classifier in a run was 0.69 seconds on average. The time spent in training the basic classifier in a run was 328.4 seconds on average. The time spent in the testing phase in a run was 0.35 seconds on average.

Table 30.2. Performance of the three basic classifiers.

	$Classifier\_0$	$Classifier\_1$	$Classifier\_2$
Precision rate	92.8%	92.9%	93.0%
Specificity	96.5%	95.5%	94.9%
Sensitivity	81.5%	85.2%	87.4%

We use the *precision rate* to measure the performance of the studied classifiers. The precision rate is defined as

$$\frac{C}{N} \times 100\%, \qquad (30.22)$$

where C is the number of test sequences classified correctly and N is the total number of test sequences. A *false positive* is a non-promoter test sequence that was misclassified as a promoter sequence. A *true positive* is a promoter test sequence that was also classified as a promoter sequence. The *specificity* is defined as

$$(1 - \frac{N_{fp}}{N_{ng}}) \times 100\%, \qquad (30.23)$$

where N_{fp} is the number of false positives and N_{ng} is the total number of negative test sequences. The *sensitivity* is defined as

$$\frac{N_{tp}}{N_{po}} \times 100\%, \qquad (30.24)$$

where N_{tp} is the number of true positives and N_{po} is the total number of positive test sequences.

Table 30.3. Performance of the two combined classifiers.

	$Combiner\_0$	$Combiner\_1$
Precision rate	93.9%	95.0%
Specificity	96.4%	97.6%
Sensitivity	86.3%	87.4%

From Table 30.3, we can see that $Combiner\_0$ and $Combiner\_1$ outperform the three basic classifiers. The $Combiner\_1$ gives the best precision rate 95%. The reason that $Combiner\_0$ has a higher precision rate than that of any one of the three basic classifiers can be explained by the Bernoulli model. For instance, assume that the three basic classifiers have the same precision rate of 93% and make classification errors completely independently. Then the $Combiner\_0$ makes a classification error when more than one classifier make errors at the same time. Thus, the precision rate of the unweighted voter of the three basic classifiers would be given by:

$$100\% - (\binom{3}{3}(1-93\%)^3 + \binom{3}{2}93\%(1-93\%)^2) = 98.5\%. \qquad (30.25)$$

The practical precision rate is a bit lower. The reason is that the $Classifier\_0$, $Classifier\_1$ and $Classifier\_2$ can not make errors completely independently.

Table 30.4. The complementarity of $Combiner\_0$ and $Combiner\_1$.

Classification results	Percentage of the test sequences
$Combiner\_0$ and $Combiner\_1$ agreed and both were correct	92.9%
$Combiner\_0$ and $Combiner\_1$ agreed and both were wrong	3.9%
$Combiner\_0$ and $Combiner\_1$ disagreed and $Combiner\_0$ was correct	1.0%
$Combiner\_0$ and $Combiner\_1$ disagreed and $Combiner\_1$ was correct	2.2%
$Combiner\_0$ and $Combiner\_1$ disagreed and both were wrong	0.0%

Table 30.4 illustrates the complementarity between $Combiner\_0$ and $Combiner\_1$. When $Combiner\_0$ and $Combiner\_1$ agree, the classification has a higher likelihood of being correct. From Table 30.4, when both agree, the probability that the classification is correct is given by 92.9%/(92.9%+3.9%) =95.9%. We can see that when $Combiner\_0$ and $Combiner\_1$ disagree, the probability that one is correct is 100%.

6. Conclusion

In this chapter we have proposed a two-level ensemble of classifiers to recognize E. Coli promoter sequences. The first-level classifiers include three Bayesian neural networks trained on three different feature sets. The outputs of the first-level classifiers are combined in the second-level to give the final result. A recognition rate of 95% was achieved. Currently we are extending the approach to classify protein sequences and to recognize the full gene structure.

Acknowledgments

The sequence logos software was developed by Dr. Thomas Schneider. We thank Dr. David Mackay for sharing the Bayesian neural network software with us. We also thank Dr. O. N. Ozoline for providing the promoter sequences used in the chapter. The anonymous reviewers gave useful comments, which helped to improve both the presentation and the quality of the chapter.

Bibliography

T. L. Bailey. Unsupervised learning of multiple motifs in biopolymers using expectation maximization. *Machine Learning*, 21:51–83, 1995.

A. Brazma, I. Jonassen, E. Ukkonen, and J. Viloi. Discovering patterns and subfamilies in biosequences. In *Proceedings of the Fourth International Conference on Intelligent Systems for Molecular Biology*, pages 34–43, 1996.

S. Brunak, J. Engelbrecht, and S. Knudsen. Prediction of human mrna donor and acceptor sites from the dna sequence. *Journal of Molecular Biology*, 220(1):49–65, 1991.

C. Burge and S. Karlin. Prediction of complete gene structures in human genomic dna. *Journal of Molecular Biology*, 268(1):78–94, 1997.

L. R. Cardon and G. D. Stormo. Expectation maximization algorithm for identifying protein-binding sites with variable lengths from unaligned dna fragments. *Journal of Molecular Biology*, 223(1):159–170, 1992.

M. W. Craven and J. W. Shavlik. Machine learning approaches to gene recognition. *IEEE Expert*, 9(2):2–10, 1994.

A. Dempster, N. Laird, and D. Rubin. Maximum likelihood from incomplete data via the em algorithm. *Journal of the Royal Statistical Society*, 39:1–38, 1977.

T. G. Dietterich. Machine learning research: Four current directions. *AI Magazine*, 18(4):97–136, 1997.

K. A Frenkel. The human genome project and informatics. *Communications of the ACM*, 34(11):41–51, 1991.

D. J. Galas, M. Eggert, and M. S. Waterman. Rigorous pattern-recognition methods for dna sequences: Analysis of promoter sequences from *escherichia coli*. *Journal of Molecular Biology*, 186(1): 117–128, 1985.

H. Hirsh and M. Noordewier. Using background knowledge to improve inductive learning of dna sequences. In *Proceedings of the Tenth Conference on Artificial Intelligence for Applications*, pages 351–357, 1994.

J. D. Hirst and M. J. E. Sternberg. Prediction of structural and functional features of protein and nucleic acid sequences by artificial neural networks. *Biochemistry*, 31(32):7211–7218, 1992.

A. Krogh, M. Brown, I. S. Mian, K. Sjolander, and D. Haussler. Hidden markov models in computational biology: Applications to protein modeling. *Journal of Molecular Biology*, 235(5):1501–1531, 1994.

D. Kulp, D. Haussler, M. G. Reese, and F. H. Eeckman. A generalized hidden markov model for the recognition of human genes in dna. In *Proceedings of the Fourth International Conference on Intelligent Systems for Molecular Biology*, pages 134–142, 1996.

C. E. Lawrence and A. A. Reilly. An expectation-maximization (em) algorithm for the identification and characterization of common sites in unaligned biopolymer sequences. *Proteins: Structure, Function, and Genetics*, 7:41–51, 1990.

S. Lisser and H. Margalit. Compilation of e. coli mrna promoter sequences. *Nucleic Acids Research*, 21(7):1507–1516, 1993.

D. J. C. Mackay. Bayesian interpolation. *Neural Computation*, 4(3): 415–447, 1992a.

D. J. C. Mackay. A practical bayesian framework for backprop networks. *Neural Computation*, 4(3):448–472, 1992b.

G. Mengeritsky and T. F. Smith. Recognition of characteristic patterns in sets of functionally equivalent dna sequences. *Computer Applications in the Biosciences*, 3(3):223–227, 1987.

R. M. Neal. *Bayesian Learning for Neural Networks*. Number 118 in Lecture Notes in Statistics. Springer-Verlag, 1996.

D. W. Opitz and J. W. Shavlik. Connectionist theory refinement: Genetically searching the space of network topologies. *Journal of Artificial Intelligence Research*, 6:177–209, 1997.

O. N. Ozoline, A. A. Deev, and M. V. Arkhipova. Non-canonical sequence elements in the promoter structure. cluster analysis of promoters recognized by *escherichia coli* rna polymerase. *Nucleic Acids Research*, 25(23):4703–4709, 1997.

A. G. Pedersen, P. Baldi, S. Brunak, and Y. Chauvin. Characterization of prokaryotic and eukaryotic promoters using hidden markov models. In *Proceedings of the Fourth International Conference on Intelligent Systems for Molecular Biology*, pages 182–191, 1996.

A. G. Pedersen and J. Engelbrecht. Investigations of *escherichia coli* promoter sequences with artificial neural networks: New signals discovered upstream of the transcriptional start point. In *Proceedings of the Third International Conference on Intelligent Systems for Molecular Biology*, pages 292–299, 1995.

S. Salzberg. A decision tree system for finding genes in dna. Technical Report CS-97-03, Department of Computer Science, Johns Hopkins University, 1997a.

S. Salzberg. A method for identifying splice sites and translational start sites in eukaryotic mrna. *Computer Applications in the Biosciences*, 13(4):365–376, 1997b.

T. D. Schneider and R. M. Stephens. Sequence logos: A new way to display consensus sequences. *Nucleic Acids Research*, 18(20):6097–6100, 1990.

R. Staden. Computer methods to locate signals in nucleic acid sequences. *Nucleic Acids Research*, 18(20):6097–6100, 1990.

J. T. L. Wang, T. G. Marr, D. Shasha, B. A. Shapiro, and G. Chirn. Discovering active motifs in sets of related protein sequences and using them for classification. *Nucleic Acids Research*, 22(14):2769–2775, 1994.

J. T. L. Wang, T. G. Marr, D. Shasha, B. A. Shapiro, G. Chirn, and T. Y. Lee. Complementary classification approaches for protein sequences. *Protein Engineering*, 9(5):381–386, 1996.

J. T. L. Wang, S. Rozen, B. A. Shapiro, D. Shasha, Z. Wang, and M. Yin. New techniques for dna sequence classification. *Journal of Computational Biology*, 6(2):209–218, 1999a.

J. T. L. Wang, B. A. Shapiro, and D. Shasha. *Pattern Discovery in Biomolecular Data: Tools, Techniques and Applications*. Oxford University Press, New York, 1999b.

C. H. Wu, M. Berry, Y. S. Fung, and J. McLarty. Neural networks for full-scale protein sequence classification: Sequence encoding with singular value decomposition. *Machine Learning*, 21:177–193, 1995.

Y. Xu, R. J. Mural, J. R. Einstein, M. B. Shah, and E. C. Uberbacher. Grail: A multi-agent neural network system for gene identification. *Proceedings of the IEEE*, 84(10):1544–1551, 1996.

M. O. Zhang and T. G. Marr. A weight array method for splicing signal analysis. *Computer Applications in the Biosciences*, 9(5):499–509, 1993.

X. Zhang, J. P. Mesirov, and D. L. Waltz. Hybrid system for protein secondary structure prediction. *Journal of Molecular Biology*, 225(4): 1049–1063, 1992.

Chapter 31

MASSIVE DATA SET ISSUES IN AIR POLLUTION MODELLING

Zahari Zlatev
National Environmental Research Institute Frederiksborgvej 399, P. O. Box 358 DK-4000 Roskilde, Denmark
zz@dmu.dk

Abstract   Air pollution, especially the reduction of the air pollution to some acceptable levels, is a highly relevant environmental problem, which is becoming more and more important. This problem can successfully be studied only when high-resolution comprehensive mathematical models are developed and used on a routine basis. However, such models are very time-consuming, even when modern high-speed computers are available. Indeed, if an air pollution model is to be applied on a large space domain by using fine grids, then its discretization will always lead to huge computational problems. Assume, for example, that the space domain is discretized by using a (480x480) grid and that the number of chemical species studied by the model is 35. Then ODE systems containing 8064000 equations have to be treated at every time-step (the number of time-steps being typically several thousand). If a three-dimensional version of the air pollution model is to be used, then the above quantity must be multiplied by the number of layers. Moreover, hundreds and even thousands of simulation runs have to be carried out in most of the studies related to policy making. Therefore, it is extremely difficult to treat such large computational problems. This is true even when the fastest computers that are available at present are used. The computing time needed to run such a model causes, of course, great difficulties. However, there is another difficulty which is at least as important as the problem with the computing time. The models need a great amount of input data (meteorological, chemical and emission data). Furthermore, the model produces huge files of output data, which have to be stored for future uses (for visualization and animation of the results). Finally, huge sets of measurement data (normally taken at many stations located in different countries) have to be used in the efforts to validate the model results. The necessity to handle efficiently huge data sets, containing input data, output data and measurement data, will be discussed in this paper.

Keywords: Air pollution modelling, Comprehensive mathematical models, Meteorological data, Emission data, Output model results, Validation of the model results, Measurements.

1. Why is Large Scale Air Pollution Modelling Needed?

The control of the pollution levels in different highly developed regions of Europe and North America (and also in any other highly industrialized regions of the world) is an important task for the modern society. Its relevance has been steadily increasing during the last two decades. The need to establish reliable control strategies for the air pollution levels will become even more important in the 21st century. Large scale air pollution models can successfully be used to design reliable control strategies. Many different tasks must be solved before starting to run operationally an air pollution model. The following tasks are most important:

1 describe in an adequate way all important physical and chemical processes,

2 apply fast and sufficiently accurate numerical methods in the different parts of the model,

3 ensure that the model runs efficiently on modern high-speed computers (and, first and foremost on different types of parallel computers),

4 use high quality input data (both meteorological and emission data) in the runs,

5 verify the model results by comparing them with reliable measurements taken in different parts of the space domain of the model,

6 carry out some sensitivity experiments to check the response of the model to changes of different key parameters, and

7 visualize and animate the output results to make them easily understandable also for non-specialists.

It is absolutely necessary to solve efficiently all these tasks. However, the treatment of such large scale air pollution models, in which all relevant physical and chemical processes are adequately described, leads to great computational difficulties related both to the CPU time needed to run the models and to the input and output data used during the runs and saved for future applications. Therefore it is necessary to give an answer to the question:

> *Why are large scale comprehensive mathematical models needed (and also used) in many environmental studies?*

A short, but also adequate, answer to this question can be given in the following way. It is well-known, also by the broad public in Europe, North America and all other parts of the world, that the air pollution levels in a given region depend not only on the emission sources located in it, but also on emission sources located outside the region under consideration, and even on sources that are far away from the studied region. This is due to the transboundary transport of air pollutants. The atmosphere is the major medium where pollutants can be transported over long distances. Harmful effects on plants, animals and humans can also occur in areas which are long away from big emission sources. Chemical reactions take place during the transport. This leads to the creation of secondary pollutants, which can also be harmful. The air pollution levels in densely populated and highly developed regions of the world, such as Europe and North America (but this will soon become true also for many other regions), must be studied carefully to find out how the air pollution can be reduced to safe levels and, moreover, to develop reliable control strategies by which air pollution can be kept under certain prescribed critical levels. The reduction of the air pollution is an expensive process. This is why one must try to optimize this process by solving two important tasks:

- the air pollution levels must be reduced to reliably determined critical levels (but no more if costs are high)

and (what is even more important)

- since the air pollution levels in different parts of a big region (as, for example, Europe or North America) are varying in a quite wide range, the optimal solution (a solution which is as cheap as possible) will require to reduce the emissions with different amounts in different parts of a big region.

Reliable mathematical models, in which all relevant physical and chemical processes are adequately described (according to the available at present knowledge about these processes), are needed in the efforts to solve successfully these two tasks. Large comprehensive models for studying air pollution phenomena were first developed in North America: see, for example, Binkowski and Shankar (1995), Carmichael and Peters (1984), Carmichael et al. (1986), Carmichael et al. (1996), Carmichael et al. (1999), Chang et al. (1987), Dabdub and Seinfeld (1994), Peters et al. (1995), Stockwell et al. (1990) and Venkatram et al. (1988). There are several such large models in Europe: see Ackerman et al. (1999), Aloyan (1999), Builtjes (1992), Builtjes (1999), D'Ambra et al. (1999), Elbern et al. (1999), Jonson et al. (1999), Jose et al. (1999), Simpson (1992), Simpson (1993), Tomlin et al. (1999) and Zlatev (1995).

The recent development of many of the existing large scale air pollution models is described in many of the papers of the proceedings of the NATO Advanced Research Workshop on *"Large Scale Computations in Air Pollution Modelling"*, see Zlatev et al. (1999). A comprehensive review of latest results achieved in the efforts to improve the performance of the large scale environmental models is given in a very well-written paper of Ebel, see Ebel (2000).

The use of the Danish Eulerian Model (fully described in Zlatev (1995); see also the web-site of the model Zlatev and Skjøth (1999)) in the solution of the tasks listed above will be discussed in this paper. However, the particular choice of this model is made only in order to facilitate the understanding of the statements. The ideas used in the development of all comprehensive mathematical models for studying air pollution phenomena are rather similar. Therefore, the consideration of any other large-scale air pollution model will lead to the same results and conclusions.

2. The Danish Eulerian Model

Some information about the last versions of the Danish Eulerian Model, developed at the National Environmental Research Institute and used in many environmental studies, will be presented in this and in the following sections.

2.1. Some Historical Information

The work on the development of the Danish Eulerian Model has been initiated in the beginning of 80's. Only two species, sulphur dioxide and sulphate were studied in the first version of the model. The space domain was rather large (the whole of Europe), but the discretization was very crude; the space domain was divided into (32×32) grid, which means, roughly speaking, that $(150\ km \times 150\ km)$ grid-squares were used. Both (a two-dimensional and a three-dimensional) versions of this model were developed. In both versions the chemical scheme was very simple. There was only one, **linear**, chemical reaction: it was assumed that a fixed part of the sulphur dioxide concentration is transformed, at each time step, into sulphate concentration. The two-dimensional version of this model is fully described in Zlatev (1985), while the three-dimensional version was discussed in Zlatev et al. (1983b).

This first version was gradually upgraded to much more advanced versions by carrying out successively the following actions:

1 Introducing better physical mechanisms.

2 Increasing the number of chemical compounds which can be studied by the model.

3 Improving the spatial resolution of the model by transition from a (32×32) grid to a (96×96) grid and a (480×480) grid, i.e. by reducing the $(150\ km \times 150\ km)$ grid-squares used in the original versions to $(50\ km \times 50\ km)$ and $(10\ km \times 10\ km)$ grid-squares.

4 Developing more advanced three-dimensional versions of the model with ten vertical layers.

5 Attaching new and faster numerical algorithms in the different parts of the model.

6 Preparing the model for efficient runs on different high-speed computers (both vector processors and parallel computer architectures).

The Danish Eulerian Model is well-structured and the above changes have been made by replacing only one module at a time. In this way the response of the model to different changes was also studied. Different versions of the model obtained during the upgrading procedure were described and discussed in many publications; see, for example, Zlatev (1995), Zlatev et al. (1994) and Zlatev et al. (1996a).

Results, obtained in comparisons of the concentrations and depositions calculated by different versions of the model with measurements taken at stations located in different European countries, were presented in Zlatev (1995) and Zlatev et al. (1993), Zlatev et al. (1992). Some comparisons of model results with measurements taken over sea were also carried out; see Harrison et al. (1994).

Different versions of the model were also used in long simulations connected with studies of the relationships between high ozone concentrations in Europe and related emissions. Many results, which were obtained in these simulations, were discussed in Skjøth et al. (2000), Birk et al. (1997) and Zlatev et al. (1996b).

The improvement of the physical and chemical mechanisms used in the model demanded an improvement also of the numerical algorithms used in the treatment of the model. Furthermore, high-speed computers became necessary. It is normally very difficult to achieve a good performance on the new modern computer architectures. A long and careful work, which is still carried out, is necessary in order to achieve a high computational speed when the model is run on such architectures. On the other hand, it will be impossible to carry out several hundreds (and in many cases even several thousands) of runs with the Danish Eulerian Model in the long simulation process when its code does not run very efficiently on the computers available. Different procedures used in the attempts to achieve high computational speeds on different computers are presented in Brown et al. (1995), Georgiev and Zlatev (1998), Georgiev and Zlatev (2000), Owczarz and Zlatev (2000a), Owczarz and Zlatev (2000b) and Zlatev (1988), Zlatev (1990), Zlatev (1995).

Some more details about the last versions of the Danish Eulerian Model will be given in the following sub-sections.

2.2. Chemical Species Treated in the Model

The pollution levels in Europe for all important chemical pollutants can be studied by using the Danish Eulerian Model. The chemical species involved in the model are:

- sulphur pollutants,
- nitrogen pollutants,
- ozone,
- ammonia-ammonium,
- several radicals

and

- a large number of relevant hydrocarbons.

2.3. Mathematical Formulation of the Model

The Danish Eulerian Model, as many other large-scale air pollution models, is described mathematically by a system of partial differential equations (**PDE's**):

$$\frac{\partial c_s}{\partial t} = -\frac{\partial(uc_s)}{\partial x} - \frac{\partial(vc_s)}{\partial y} - \frac{\partial(wc_s)}{\partial z} \quad (31.1)$$

$$+ \frac{\partial}{\partial x}\left(K_x \frac{\partial c_s}{\partial x}\right) + \frac{\partial}{\partial y}\left(K_y \frac{\partial c_s}{\partial y}\right) + \frac{\partial}{\partial z}\left(K_z \frac{\partial c_s}{\partial z}\right)$$

$$+ E_s + Q_s(c_1, c_2, \ldots, c_q)$$

$$- (\kappa_{1s} + \kappa_{2s}) c_s,$$

$$s = 1, 2, \ldots, q.$$

The different quantities that are involved in the mathematical model have the following meaning:

- the concentrations are denoted by c_s;
- u, v and w are wind velocities;
- K_x, K_y and K_z are diffusion coefficients;
- the emission sources in the space domain are described by the functions E_s;
- κ_{1s} and κ_{2s} are deposition coefficients;
- the chemical reactions used in the model are described by the non-linear functions $Q_s(c_1, c_2, \ldots, c_q)$.

The number of equations q is equal to the number of chemical species that are involved in the model and varies in different studies. Until now, the model has mainly been used with a chemical scheme containing 35 species (it may be necessary to involve more species in the future; experiments with chemical schemes containing 56 and 168 species have recently been carried out). The chemical scheme with 35 species is the well-known **CBM IV** scheme (described in Gery et al. (1989)) with a few enhancements which have been introduced in order to make it possible to use the model in studies concerning the distribution of ammonia-ammonium concentrations in Europe as well as the transport of ammonia-ammonium concentrations from Europe to Denmark. Recently, some other chemical schemes were developed and successfully tested in different models; see, for example, Stockwell et al. (1997). There are plans to implement some of these new chemical mechanisms in the Danish Eulerian Model in the near future.

The number of chemical species used in different large air pollution models varies from 20 to about 200 (see, for example, Borrell et al. (1990), Carmichael and Peters (1984), Carmichael et al. (1986), Chang et al. (1987), Ebel et al. (1997), Hass et al. (1993), Hass et al. (1995), Peters et al. (1995), Stockwell et al. (1990) and Venkatram et al. (1988)). The use of less than 20 chemical species will require crude parameterization of some chemical processes and, thus, the use of such chemical schemes is not advisable, when the accuracy requirements that are to be satisfied are not very crude. On the other hand, the use of more than 200 chemical species is connected with huge tasks that cannot be handled, at least when long-term simulations are to be carried out, on the computers available at present without imposing a sequence of simplifying assumptions. This leads to a crude parameterization of the other physical processes involved in the air pollution models (advection, diffusion, deposition and emission) and/or to discretizations of the models on coarse grids. Thus, 35 seems to be a good choice. Nevertheless, as mentioned above, experiments with two other chemical schemes, containing 56 and 168 chemical species, are at present carried out. It should be pointed out that although the use of more than 100 chemical species (but less than 200) leads at present, as mentioned above, to the necessity both to apply certain simplifying assumptions and/or to use coarse grids, the computers are becoming faster and faster and, therefore, such actions will probably not be necessary in the near future.

The non-linear functions Q_s from (31.1) can be rewritten in the following form:

$$Q_s(c_1, c_2, \ldots, c_q) = -\sum_{i=1}^{q} \alpha_{si} c_i + \sum_{i=1}^{q} \sum_{j=1}^{q} \beta_{sij} c_i c_j, \qquad s = 1, 2, \ldots, q.$$

(31.2)

This is a special kind of non-linearity, but it is not obvious how to exploit this fact during the numerical treatment of the model.

It follows from the above description of the quantities involved in the mathematical model that all five physical processes (advection, diffusion, emission, deposition and chemical reactions) can be studied by using the above system of **PDE's**. The most important processes are the advection (the transport) and the chemical reactions. Kernels for these two parts of the model must be treated numerically by using both fast and sufficiently accurate algorithms.

2.4. Space Domain

The space domain of the model (it contains the whole of Europe together with parts of Asia, Africa and the Atlantic Ocean) is discretized by using a (96 × 96) grid in the version of the Danish Eulerian Model which is operationally used at present. This means that

- Europe is divided into 9216 grid-squares

and

- the grid resolution is approximately (50 km × 50 km).

It has been mentioned above that some work is carried out at present to run the Danish Eulerian Model by using much higher resolution applying a (480 × 480) grid. If such a high resolution is used, then

- Europe is divided into 230400 grid-squares

and

- the grid resolution is approximately (10 km × 10 km).

It will be shown in the next sections that the high resolution version of the Danish Eulerian Model imposes severe storage requirements; both the input files and the output files that are needed in the runs become much larger than those used in the operational version. Of course, also the computing time needed to run this version is increased very considerably.

2.5. Initial and Boundary Conditions

Appropriate initial and boundary conditions are needed. If initial conditions are available (for example from a previous run of the model), then these are read from the file where they are stored. If initial conditions are not available, then a five day start-up period is used to obtain initial conditions (i.e. the computations are started five days before the desired starting date with some background concentrations and the concentrations found at the end of the fifth day are actually used as starting concentrations).

The choice of lateral boundary conditions is in general very important. This issue has been discussed in Brost (1988). However, if the space domain is very large, then the choice of lateral boundary conditions becomes less important; which is stated on p. 2386 in Brost (1988): "For large domains the importance of the boundary conditions may decline". The lateral boundary conditions are represented in the Danish Eulerian Model with typical background concentrations which are varied, both seasonally and diurnally. The use of background concentrations is justified by the facts that:

- the space domain is very large ($4800\ km \times 4800\ km$)

and

- the boundaries are located far away from the highly polluted regions (in the Atlantic Ocean, North Africa, Asia and the Arctic areas).

Nevertheless, it would perhaps be better to use values of the concentrations at the lateral boundaries that are calculated by a hemispheric or global model.

For some chemical species, as for example ozone, it is necessary to introduce some exchange with the free troposphere. Three such rules have been tested in Zlatev et al. (1993). The third rule described in Zlatev et al. (1993) is at present used in the Danish Eulerian Model, because the experiments indicate that this rule performs in general better than the other two rules.

2.6. Splitting the Model

It is difficult to treat the system of **PDE's** (31.1) directly. This is the reason for using different kinds of splitting. A simple splitting procedure, based on ideas discussed in Marchuk (1985) and McRae et al. (1984), can be defined, for $s = 1, 2, \ldots, q$, by the following sub-models:

$$\frac{\partial c_s^{(1)}}{\partial t} = -\frac{\partial (u c_s^{(1)})}{\partial x} - \frac{\partial (v c_s^{(1)})}{\partial y} \tag{31.3}$$

$$\frac{\partial c_s^{(2)}}{\partial t} = \frac{\partial}{\partial x}\left(K_x \frac{\partial c_s^{(2)}}{\partial x}\right) + \frac{\partial}{\partial y}\left(K_y \frac{\partial c_s^{(2)}}{\partial y}\right) \tag{31.4}$$

$$\frac{d c_s^{(3)}}{dt} = E_s + Q_s(c_1^{(3)}, c_2^{(3)}, \ldots, c_q^{(3)}) \tag{31.5}$$

$$\frac{d c_s^{(4)}}{dt} = -(\kappa_{1s} + \kappa_{2s}) c_s^{(4)} \tag{31.6}$$

$$\frac{\partial c_s^{(5)}}{\partial t} = -\frac{\partial (w c_s^{(5)})}{\partial z} + \frac{\partial}{\partial z}\left(K_z \frac{\partial c_s^{(5)}}{\partial z}\right) \tag{31.7}$$

The horizontal advection, the horizontal diffusion, the chemistry, the deposition and the vertical exchange are described with the systems (31.3)-(31.7). This is not the only way to split the model defined by

(31.1), but the particular splitting procedure (31.3)-(31.7) has three advantages:

1. The physical processes involved in the big model can be studied separately.
2. It is easier to find optimal (or, at least, good) methods for the simpler systems (31.3)-(31.7) than for the big system (31.1).
3. If the model is to be considered as a two-dimensional model (which often happens in practice), then one should just skip system (31.7).

It must be emphasized here that, while splitting facilitates the treatment of large air pollution models, it also introduces errors. It is difficult to control these errors. Normally, it is merely assumed that the errors caused by the splitting procedure used are sufficiently small and do not affect seriously the model results. It is necessary, however, both to derive some theoretical means which will allow us to judge the size of the splitting errors and develop some easily applicable algorithms for evaluating the magnitude of the errors. Some results in this direction have been reported in Lancer and Verwer (1999). These results are obtained by studying carefully the L-commutativity of the operators involved in the air pollution models. The concept of L-commutativity has recently been used also in Dimov et al. (1999) in an attempt:

- to improve in some sense the results obtained by Lancer and Verwer (Lancer and Verwer (1999))

and

- to obtain directly error estimations that are applicable to the Danish Eulerian Model.

Further research results in this direction are highly desirable, because it is necessary to know more about the nature of the splitting errors and to be able to develop constructive methods for evaluation of these errors.

2.7. Space Discretization

Assume that the space domain is a parallelepiped which is discretized by using a grid with $N_x \times N_y \times N_z$ grid-points, where N_x, N_y and N_z are the numbers of the grid-points along the grid-lines parallel to the Ox, Oy and Oz axes. Assume further that the number of chemical species involved in the model is $q = N_s$. Finally, assume that the spatial derivatives in (31.1) are discretized by some numerical algorithm. Then the system of PDE's (31.1) will be transformed into a system of ODE's (ordinary differential equations)

$$\frac{dg}{dt} = f(t,g), \qquad (31.8)$$

where $g(t)$ is a vector-function with $N_x \times N_y \times N_z \times N_s$ components. Moreover, the components of function $g(t)$ are the concentrations (at time t) at all grid-points and for all species. The right-hand side $f(t,g)$ of (31.8) is also a vector function with $N_x \times N_y \times N_z \times N_s$ components which depends both on the particular discretization method used and on the concentrations of the different chemical species at the grid-points. If the space discretization method is fixed and if the concentrations are calculated (at all grid-points and for all species), then the right-hand side vector in (31.8) can also be calculated.

As mentioned above, the large air pollution models are not discretized directly. Some kind of splitting is always used. If the model is split into sub-model as in the previous sub-section, then the application of discretization methods (normally different methods for the different sub-models) will lead to the following ODE systems:

$$\frac{dg^{(1)}}{dt} = f^{(1)}(t, g^{(1)}), \qquad (31.9)$$

$$\frac{dg^{(2)}}{dt} = f^{(2)}(t, g^{(2)}), \qquad (31.10)$$

$$\frac{dg^{(3)}}{dt} = f^{(3)}(t, g^{(3)}), \qquad (31.11)$$

$$\frac{dg^{(4)}}{dt} = f^{(4)}(t, g^{(4)}), \qquad (31.12)$$

$$\frac{dg^{(5)}}{dt} = f^{(5)}(t, g^{(5)}), \qquad (31.13)$$

where $g^{(i)}$ and $f^{(i)}$, $i = 1, 2, 3, 4, 5$, are again vector-functions with $N_x \times N_y \times N_z \times N_s$ components. The functions $f^{(i)}$, $i = 1, 2, 3, 4, 5$, depend on the discretization methods used. Some particular numerical methods that can be used in the discretization of the five sub-models (31.3)-(31.7) in order to obtain ODE systems of type (31.9)-(31.13) are listed below.

The discretization of the spatial derivatives in the sub-model describing the horizontal advection, the PDE system (31.3), can be carried out either by using pseudospectral expansions or by applying finite elements; see Zlatev (1995) and Zlatev et al. (1994), Zlatev et al. (1996a). The finite elements used at present in the Danish Eulerian Model have been applied in connection with air pollution models in Pepper and Baker (1979) and Pepper et al. (1979). Some more details about the implementation of this method in the advection part of the Danish Eulerian Model can be found in Zlatev (1995) and Georgiev and Zlatev (2000).

Other numerical methods can also be applied. It is important to emphasize that the resulting system (31.9) contains $q = N_s$ independent ODE systems (i.e. one such a system per each chemical species). This fact could easily be exploited on parallel computers.

The discretization of the second-order spatial derivatives in the horizontal diffusion sub-model, the PDE system (31.4), is quite similar. Both the pseudospectral algorithm and finite elements can be used. Again, other numerical algorithms can also be applied. The number of independent systems of ODE's in the resulting system (31.10) is again $q = N_s$; i.e. the same as in the previous case. It may be useful to apply different numerical algorithms in the advection and diffusion parts; see again Zlatev (1995) and Zlatev et al. (1994), Zlatev et al. (1996a).

The transition from (31.5) to (31.11) as well as the transition from (31.6) to (31.12) is trivial, because there are no spatial derivatives in (31.5) and (31.6). In the first case, the resulting ODE system (31.11) contains $N_x \times N_y \times N_z$ independent systems, each of them with N_s equations (because the chemical species at a given grid-point react with each other but not with chemical species at other grid-points). In the second case, the resulting system (31.12) consists of $N_x \times N_y \times N_z \times N_s$ independent ODE's (because the deposition of a given species at a given grid-point depends neither on the deposition of the other species nor on the deposition processes at the other grid-points). It is seen from the above discussion that a lot of parallel tasks arise in a natural way when the ODE systems (31.11) and (31.12) are to be handled.

Finite elements can be applied in the discretization of the spatial derivatives in the vertical exchange sub-model (31.7). The resulting ODE system (31.13) consists of $N_x \times N_y \times N_s$ independent ODE systems; each of them is defined on a vertical grid-line and, therefore, contains N_z equations.

2.8. Time Integration

It is necessary to couple the five ODE systems (31.9)-(31.13). The coupling procedure is connected with the time-integration of these systems. Assume that the values of the concentrations (for all species and at all grid-points) have been found for some $t = t_n$. According to the notation introduced in the previous sub-section, these values are components of the vector-function $g(t_n)$. The next time-step, time-step $n+1$ (at which the concentrations are found at $t_{n+1} = t_n + \triangle t$, where $\triangle t$ is some increment), can be performed by integrating successively the five systems (31.9)-(31.13). The values of $g(t_n)$ are used as an initial condition in the solution of (31.9). The solution of each of the systems (31.9)-(31.12) is used as an initial condition in the solution of the next system. The solution of the last system (31.13) is used as an approximation to $g(t_{n+1})$. In this way, everything is prepared to start the calculations in the next time-step, step $n+2$.

Predictor-corrector methods with several different correctors are used in the solution of the ODE systems (31.9) and (31.10). The correctors are carefully chosen so that the stability properties of the method are enhanced; see Zlatev (1984), Zlatev (1995) and Zlatev and Berkowicz (1988). The reliability of the algorithms used in the advection part is verified by using the well-known rotational test proposed simultaneously in 1968 by Crowley (1968) and Molenkampf (1968).

Consider now the solution of (31.11). Very often the QSSA method is used in this part of the model. The QSSA (quasi-steady-state approximation; see, for example, Hesstvedt et al. (1978) or Hov et al. (1988)) is simple and relatively stable but not very accurate (therefore it has to be run with a small time-stepsize). The QSSA method can be viewed as an attempt to transform dynamically, during the process of integration, the system of ordinary differential equation (31.11) into two systems: a system of ordinary differential equations and a system of non-linear algebraic equations. These two systems, which have to be treated simultaneously, can be written in the following generic form:

$$\frac{dg_1}{dt} = f_1(t, g_1, g_2), \tag{31.14}$$

$$0 = f_2(t, g_1, g_2). \tag{31.15}$$

In this way we arrive at a system of differential-algebraic equations (DAEs). There are special methods for treating such systems as, for example, the code DASSL (see Brenan et al. (1996)). Problem-solving environments (such as MATLAB or Simulink) can be used in the preparation stage (where a small chemical systems at one grid-point only is used in the tests). More details about the use of such problem solving environments can be found in Shampine et al. (1999). Solvers for systems of DAEs have not been used yet in the air pollution models.

The classical numerical methods for stiff ODE systems (such as the Backward Euler Method, the Trapezoidal Rule and Runge-Kutta algorithms) lead to the solution of non-linear systems of algebraic equations and, therefore, they are more expensive; see, for example, Hairer and Wanner (1991) or Lambert (1991). On the other hand, these methods can be incorporated with an error control and perhaps with larger time-steps. The extrapolation methods (see Deuflhard (1983), Deuflhard (1985), Deuflhard and Nowak (1986) or Deuflhard et al. (1987), Deuflhard et al. (1990) are also promising. It is easy to calculate an error estimation and to carry out the integration with large time-steps when these algorithms are used. However, it is difficult to implement such methods in an efficient way when all five systems (31.9)-(31.13) are to be treated successively. The experiments with different integration methods for the chemical sub-model (31.11) are continuing. The QSSA will be used in most of the experiments described here. However, it must be emphasized that many research groups in different countries are actively working to develop more accurate and faster numerical algorithms for the chemical part of a large-scale air pollution model. Other numerical methods for the chemical part of a large air pollution model are discussed in, for example, Alexandrov et al. (1997), Birk et al. (1998), Burrage (1995), Chock et al. (1994), Hertel et al. (1993), Jay et al. (1997), Odman et al. (1992), Sandu et al. (1996), Sandu et al. (1997b), Sandu et al. (1997a), Shieh et al. (1988), Skelboe and Zlatev (1997), Verwer et al. (1996), Verwer and van Loon (1994) and Verwer and Simpson (1995).

The matrices involved in the chemical part of the model are sparse. However, the direct utilization of general purpose sparse matrix algorithms, such as those proposed in Duff et al. (1986) or Zlatev (1991), is rather inefficient. In Alexandrov et al. (1997) it is explained why this is so. The application of several efficient algorithms, based on the use of an advanced sparse matrix scheme and adapted for the particular problem under consideration, on parallel computers is discussed in Georgiev and Zlatev (2000). Several ideas discussed in Willoughby (1970) are used in the development of the algorithm in Georgiev and Zlatev (2000). It should be mentioned here that similar ideas are also used in Swart and Blom (1996).

The next ODE system, (31.12), contains, see the previous sub-section, $N_x \times N_y \times N_z \times N_s$ independent ODE's. Moreover, all these ODE's are linear. Therefore they are solved exactly during the numerical treatment of the model.

The last ODE system, (31.13), can be solved by using many classical time-integration methods. The so-called θ-method (see, for example, Lambert (1991)) is currently used in the three-dimensional version of the Danish Eulerian Model (Zlatev (1995) and Zlatev et al. (1994), Zlatev et al. (1996a)).

2.9. Need for High Speed Computers

The size of the systems that arise after the space discretization and the splitting procedures used to treat (31.1) numerically is enormous. Consider the case where the model is two-dimensional, and let us assume that the model is discretized on a (96x96) grid (such a grid is used in the Danish Eulerian Model after 1993) and that $q = 35$. Then the number of equations in each of the four systems of ODE's (31.9)-(31.12) is 322560. The time-stepsize used in the advection step is 15 min. The chemical sub-model (31.11) cannot be treated with such a large time-stepsize (because it is very stiff; especially when photochemical reactions are involved). Therefore six small time-steps are carried out for each advection time-step (this means that the chemical time-stepsize is 2.5 min.).

From the above description it is clear that in fact nine systems of ODE's (each of them containing 322560 equations) are to be treated per advection step. Assume that a one-month run is to be carried with the model. This will result in 3456 advection time-steps (taking here into account that it is necessary to use five extra days in order to start up the model).

Consider now the case where the model is three-dimensional. Assume that ten layers are used in the vertical direction. Then the number of equations in every system of ODE's is 3225600 (i.e. ten times greater than in the previous case). The number of systems that are to be treated at every time-step is increased from four to five. The number of time-steps remains the same, 3456. The chemical sub-model must again be integrated by using smaller time-steps. If six chemical time-steps are to be carried out per an advection step, then the actual number of ODE systems that are to be handled at every advection time-step is ten.

More general, the number of equations in each of the five ODE systems is equal to the product of the number of the grid-points and the number of chemical species. Therefore, this number grows very quickly when the number of grid-points and/or the number of chemical species is increased. Some results, which are obtained for different numbers of grid-points and for different numbers of chemical species, are given in Table 31.1.

Table 31.1. The number of equations per system of ODE's that are to be treated at every time-step. The typical number of time-steps is 3456 (when meteorological data covering a period of one month + five days to start up the model is to be handled). The number of time-steps for the chemical sub-model is even larger, because smaller step-sizes have to be used in this sub-model.

Number of species	$(32 \times 32 \times 10)$	$(96 \times 96 \times 10)$	$(480 \times 480 \times 10)$
1	10240	92160	2304000
2	20480	184320	4608000
10	102400	921600	23040000
35	358400	3225600	80640000
56	573440	5160960	129024000
168	1720320	15482880	387072000

It is clear (and this has already been mentioned) that such large problems can be solved **only** if new and modern high speed computers are used. Moreover, it is necessary to select the right numerical algorithms (which are most suitable for the high speed computers available) and to perform the programming work very carefully in order to exploit fully the great potential power of the vector and/or parallel computers. Different versions of the Danish Eulerian Model have already been successfully run on several high-speed computers (see Brown et al. (1995), Georgiev and Zlatev (1998), Georgiev and Zlatev (2000), Owczarz and Zlatev (2000a), Owczarz and Zlatev (2000b), Zlatev (1995) and Zlatev et al. (1994)). Standard tools for achieving parallelism, such as OPEN MP (MP (1999)), the Parallel Virtual Machine (PVM, Geist et al. (1994)) and the Message Passing Interface (MPI, Gropp et al. (1994)), have been used to facilitate the transition of the code from one high-speed computer to another and, hopefully, to facilitate the implementation of the code on computer architectures which will appear in the near future.

However, we are still not able to solve all the problems listed in Table 31.1. Some of the problems are only treated as two-dimensional models. Therefore, it is still necessary to improve the performance of the different algorithms on different high speed computers. Therefore many more experiments are needed (and will be carried out in the future). High-quality algorithms and software for solving some standard mathematical problems are now available (see, for example, Anderson et al. (1992), Barrett et al. (1994), Demmel (1997), Dongarra et al. (1998) and

Trefethen and III (1997)) and one should try to use them extensively in the attempts to achieve high efficiency when large air pollution models are to be treated numerically on high performance computers.

Finally, it should be emphasized that there are many other activities in the efforts to utilize in a better way the great potential power of the modern high-speed computers; see Bruegge et al. (1995), D'Ambra et al. (1999), Dabdub and Manohar (1997), Dabdub and Seinfeld (1996), Elbern (1997), Elbern et al. (1999), Jacobson and Turco (1994).

3. Input Data

It is necessary to prepare large files of input data which are to be used when an air pollution model is run on the available computers. The problems connected with the necessity to handle huge input data sets will be discussed in this section.

3.1. General Information about the Input Data

Two types of input data are needed when a large scale air pollution model is to be handled numerically on a computer:

- meteorological data

and

- emission data.

The input data needed in the Danish Eulerian Model (both the meteorological data and the emission data) have been prepared within EMEP (the European Monitoring and Evaluation Programme; this is common European project in which nearly all European countries participate; the project has been initiated in the 70s and is supported financially by the United Nations Economic Commission for Europe, UN-ECE) and have been sent to us from DNMI (the Norwegian Meteorological Institute). Occasionally, input data received from other places have been used.

3.2. Meteorological Data

The meteorological data for the two-dimensional version of the Danish Eulerian Model contain horizontal wind velocity fields, vertical wind velocity fields (on the top of the boundary layer), temperatures (both surface temperatures and temperatures of the boundary layer), precipitation fields, mixing heights and pressures. The resolution of the meteorological fields is coarser than the resolution used in the model: the time resolution for all fields except the mixing heights fields is six hours (the

MASSIVE DATA SET ISSUES

resolution for the mixing heights fields is 12 hours), the spatial resolution is, roughly speaking, $150\ km \times 150\ km$. Simple linear interpolation rules are used both in time (because the time-steps used are 15 min. for the $50\ km \times 50\ km$ grid and 2.5 min. for $10\ km \times 10\ km$ grid) and in space (because we are not running the model with a grid resolution $150\ km \times 150\ km$ anymore).

The main meteorological fields which are used at present in the Danish Eulerian Model are given in Table 31.2.

Table 31.2. Meteorological fields used in the Danish Eulerian Model at present.

Meteorological input data	Level
Horizontal wind velocities	$\sigma = 0.925, z = 750m$
Vertical wind velocities	$\sigma = 0.850, z = 1500m$
Mixing height	Above ground
Average precipitation	Ground
Cloud covers	Free troposphere
Temperature in the mixing layer	$\sigma = 0.925$
Surface temperature	$z = 2m$
Relative humidity	$\sigma = 0.925$

The total amount of meteorological fields needed to run the model over a time period of one month is about 8 MBytes. It is clear that this figure is increased by a factor of 12 if a one-year run is to be carried out (which is the operational mode when the code is run on the $50\ km \times 50\ km$ grid) and by a factor of 120 when the code is run over a time period of ten years.

In the near future it is expected that the resolution of the meteorological data will be improved (in both space and time). If the meteorological data is prepared on a $50\ km \times 50\ km$ grid and the time resolution is three hours (instead of six hours), then the amount of meteorological data needed for a ten year run will be about 48 GBytes.

Let us consider the three-dimensional case now. The meteorological centres which prepare the data are normally calculating the data on layers which are not quite compatible with the layers used in the air pollution model. Therefore, every time when new data are received one should either adapt the model to the data used or interpolate the data (adapt the data to the model). Anyway, normally the amount of data is proportional to the number of layers. This means that the three-dimensional runs put severe storage requirements. The problems are tackled either by using some kind of secondary storage or by preparing a long series of runs (instead of running one job over a very long time-interval).

3.3. Emission Data

Four fields containing human-made (anthropogenic) emissions are currently used in the Danish Eulerian Model:

- SO_2 emissions,
- NO_x emissions,
- VOC emissions

and

- ammonia-ammonium emissions.

The emission data in all these four fields are available on a 50 km × 50 km grid. However, **only** the annual human-made emissions are given (i.e. only one number per grid-square per year). Simple rules must be used to get seasonal variations for the SO_2 emissions and the ammonia-ammonium emissions. Both seasonal variations and diurnal variations are simulated for the NO_x emissions and for the human-made VOC emissions.

While interpolation in time is needed when the model is discretized on a 50 km × 50 km grid, interpolation in both space and time is needed when the fine resolution version, discretized on a 10 km × 10 km grid, is run.

Only natural VOC emissions are at present used in the Danish Eulerian Model. These emissions are calculated on an hourly basis by using the mechanism proposed by Lübkert and Schöp (1989). Information about the forest distribution in Europe and about the surface temperatures is needed in this algorithm.

The amount of emission data is much lower than the amount of meteorological data, because the emissions are at present given on annual basis. The total amount of data needed to run the two-dimensional version of the model on a time-period of ten years (1989-1998) is more than 4 MBytes. However, improvement of the temporal resolution of the emissions is highly desirable. If the emission inventories are prepared on a monthly basis (even this is a rather crude approximation), then the amount of the emission data will be increased by a factor of twelve. If the emission inventories are prepared on a 10 km × 10 km grid, then the amount of the emission data will be further increased by a factor of 25.

The amount of emission data does not grow too much in the transition from two-dimensional computations to three-dimensional computations, because most of the emission sources are located on the surface. However, it may be necessary to distinguish between low sources and higher sources. The latter sources must be put in higher layers (in most of the cases, in the second layer).

Emissions from the aircraft traffic are normally not used in the large scale air pollution models, in spite of the fact that such emissions have been the central topic in several studies. However, the emissions of the aircraft traffic have been steadily growing during the last decade (and they will continue to grow in the near future). Therefore, the modellers must start to include such emissions in their models. Of course, it is necessary, first and foremost, to prepare high-quality emission inventories for the emissions from the aircraft traffic and to make them easily available for the modellers.

It should be mentioned that some emission fields with very high resolution (both a very high time resolution and a very high spatial resolution) are available for rather short time periods (from several days to several weeks). These fields are mainly used to study episodes (short periods with very high concentrations of certain pollutants).

3.4. Input Data for Air Pollution Prediction

The prediction of high air pollution levels (which may be harmful for plants, animals and humans) is becoming more and more important. In some extreme cases, when certain pollution levels are likely to become too high in the next two-three days, the population must be warned (see, for example, the Draft for the EU Ozone Directive (Commission 1998) and the Ozone Position Paper (Commission 1999)). This can be done if air pollution forecasts for the next two-three days become available on a routinely basis. If such an air pollution forecast is to be prepared one needs the following data:

- data from some weather forecast for the same period (the period for which an air pollution prediction is required),

- an initial field of the concentrations for the starting point of the air pollution forecast,

and

- reliable emissions for the period for which air pollution prediction is required.

Some meteorological input data are to be received in the first stage. Such data are normally calculated at meteorological offices and have to be received by using some fast and reliable tools (such as ftp, internet or others). Some limited area model (LAM) for weather forecasts might be selected and used in parallel with the model for preparing the air pollution forecast. The major advantage of such an approach is the possibility to prepare the needed meteorological data on precisely the same grid as that used in the air pollution model. It should be emphasized here, however, that the necessity of receiving data is not completely avoided when some weather forecast model is run in parallel with the air pollution model. The problem is that one still needs reliable starting meteorological data which can only be prepared at a meteorological office where one has access to many meteorological measurements (which are necessary for obtaining, by using data assimilation procedures, of reliable initial fields for the meteorological input data). Furthermore, some meteorological data are needed at the boundaries of the space domain used in the limited area weather forecast models. These data have also to be transmitted from some meteorological office. The amount of meteorological data which has to be received is reduced considerably when some limited area model for weather forecasts is run in parallel with the air pollution model. Interpolation rules have to be applied to attach the received data to the limited area model for weather forecasts. It is important to reiterate here that the limited area meteorological weather forecast model and the air pollution prediction model must be run on the same grid; if this is not done, then some interpolation of the meteorological data is necessary (at every time step) and this may affect seriously the air pollution forecast. The best solution is to merge totally the two models (i.e. at every time step to prepare first the meteorological data and then the air pollution forecast).

Some current measurements and data assimilation techniques (similar to the data assimilation technique used for the chemical scheme in air pollution models by Elbern and Schmidt (1999) and Elbern et al. (1997), Elbern et al. (1999), Elbern et al. (2000a), Elbern et al. (2000b)) can in principle be applied to prepare the initial concentration fields. It may be difficult to provide current measurement data. However, the situation is gradually improved. Some measurement data are already available on the internet. The use of assimilation techniques in this field is still quite new, but some progress has been made during the last two-three years; see again Elbern and Schmidt (1999) and Elbern et al. (1997), Elbern et al. (1999), Elbern et al. (2000a), Elbern et al. (2000b).

The third task is still not sufficiently well resolved. Some more efforts are needed here to prepare reliable and robust algorithms by which the needed emissions can be prepared by using given emission inventories. The major problem here is that the time resolution of the existing emission inventories is very poor.

Some work in this direction is carried out at present at the National Environmental Research Institute (see Brandt et al. (2000)). Such models for air pollution forecasts can be modified for usage in case of nuclear accidents, see Brandt et al. (1999).

4. Output Data

When a run with the model is completed, several very big data files containing digital information are prepared and moved from the high speed computers at the Danish Computing Centre (UNI-C) to several powerful work-stations at the National Environmental Research Institute. Thus, the output data are at present treated by using these work-stations. The main principles that are applied in the treatment of the output data will be discussed in this section.

4.1. Visualization of the Output Data

The digital data cannot directly be used in attempts to understand the relationships between different quantities. These data must be visualized by using powerful graphic tools. Four types of visualizations are currently used (according to the processes that are studied and to the requirements for presentation of the relationships of interest):

1. scatter plots where the mean values of the calculated concentrations and depositions are compared with corresponding measurements (mainly on monthly, seasonal and yearly basis),

2. time-series plots where temporal variations of the calculated concentrations and depositions at a given measurement station are compared with the corresponding observations,

3. colour plots presenting concentration and deposition levels of the calculated quantities in a given area (the whole of Europe or a prescribed part of Europe),

4. animations of the concentration and deposition levels in Europe or in a part of Europe in a given period of time (say, the ozone concentration levels during an episode with high ozone concentrations).

4.2. Types of Output Data

The output quantities, which are used in the visualization procedures, are given in Table 31.3. Average values of the concentrations and depositions are normally used during the visualization procedures. The different averaging lengths are given in Table 31.4.

Table 31.3. Output quantities used in the visualization programs.

Output quantities
Air concentrations
Concentrations in precipitation
Wet depositions
Dry depositions
Total depositions

Table 31.4. Time averages used in the visualization programs.

Averaging periods	Remark
Hourly mean values	Only for ozone
Daily mean values	
Monthly mean values	
Seasonal mean values	For pollutants with a strong seasonal variation
Yearly mean values	Mainly in connection with critical loads

4.3. Amount of the Output Data

The information given above shows clearly that the visualization of the results obtained by a large scale air pollution model is by no means an easy task. A large environmental model, in which all physical and chemical processes are adequately described, produces huge output files. The output data produced after a long simulation (as the simulations carried out in Birk et al. (1997) and Zlatev et al. (1996b)) contain normally several hundred MBytes or even several GBytes digital information. Fast end efficient visualization routines have to be used in order to represent the great amount of digital information in a form from which the main trends and relationships, which are hidden behind millions and millions of numbers, can be easily seen and understood (also by non-specialists).

The figures given above are for the two-dimensional case. In the three dimensional case the amount of the required data might become very large. However, one is very often interested only in the situation in the surface layer. If this is the case, then nearly the same amount of output data as in the two-dimensional case is required.

The amount of output data depends very essentially on the particular requirements. If, for example, the variation of the daily means of the concentrations has to be studied over a long time-period, then the amount of output data can become very large. However, this is true if we do not know in advance the points where data are needed (in such a case we have to store information at all grid-point in the space domain).

Normally, it is known where such information is needed (for example, at some measurement stations). The amount of output data can be reduced very considerably in such a case.

The main conclusion is that the amount of output data can vary from one run with the model to another. Therefore, one should be prepared to take into account the specific requirements for output data for the run under consideration by doing the necessary changes in the program in order to meet these requirements.

4.4. Need of Fast Graphical Tools

Modern graphic tools, both hardware and software, are needed in the solution of this task. Therefore, a powerful computer, an **ONYX2** from Silicon Graphics, has been bought by NERI and UNI-C together. This computer will give us the possibility to solve efficiently the tasks connected with the visualization and animation of the processes described in the output files.

It is not sufficient to have a fast and big computer in order to visualize and animate the output information. It is also necessary to use efficient software. A powerful program has been developed, which is based on new and modern ideas used at present in the visualization techniques. Some more efforts are needed to make this program both faster and more friendly to the users.

4.5. Requirement for Flexible Visualization Programs

Sometimes it is required to see the pollution levels in the whole space domain (in the whole of Europe). In other cases one need more detailed information in a part of the space domain (for example, in a part containing Denmark and some surroundings of Denmark). Therefore, the visualization program must allow the user to get such more detailed information when this is required by applying the following devices:

- using different zooming procedures,
- representing the grid on the plots,
- giving the numerical values of the changes in each grid-square,
- plotting the wind fields or the temperatures when the changes of the pollution levels are animated.

The visualization program developed at the National Environmental Research Institute can be used to get better information about the studied processes by applying the above devices.

5. Measurement Data

The reliability of the output data is an important issue. Measurement data are mainly used to verify the reliability of the model results. Measurement data can also be used in data assimilation algorithms (see Elbern and Schmidt (1999) and Elbern et al. (1997), Elbern et al. (1999), Elbern et al. (2000a), Elbern et al. (2000b)).

5.1. Measurements Used in the Verification Procedures

Mainly measurements collected at the EMEP network of measurement stations have been used to verify the reliability of the model output results. There are more than 100 such stations located in different European countries. The measurement data that are taken at the EMEP network of stations are collected at the Norwegian Air Research Institute (NILU, Oslo, Norway), where these are analyzed and then many modellers in Europe can obtain these data and use them in their studies. There are many NILU publications about the measurements collected there; see, for example Hjellbrekke (1995).

There are many chemical species which are measured at the stations of the EMEP network of measurement stations. The measurements, which are relevant for the validation of the results from the Danish Eulerian Model, can be divided into two groups:

- measurements of concentrations in air

and

- measurements of concentrations in precipitation.

The measurements from the **first group**, which have mainly been used until now in comparisons with results obtained by using the Danish Eulerian Model, are:

- nitrogen dioxide,
- nitrate,
- sulphur dioxide,
- sulphate,
- ozone

and

- ammonia + ammonium.

The measurements of the **second group**, which have mainly been used until now in comparisons with results obtained by using the Danish Eulerian Model, are:

- nitrate,
- sulphate

and

- ammonium.

It should be mentioned here that not all measurement stations measure all concentrations listed above. Moreover, there are periods (some days, some months and even some years) in which some measurements at some stations are missing.

Recently, concentrations of several hydrocarbons have been measured at some of the measurement stations in the EMEP network. A few comparisons of model results with the corresponding measurements of hydrocarbons have been carried out. This activity is still in a very preliminary stage.

Finally, it should be mentioned that not only measurements from the EMEP network have been used in the efforts to verify the reliability of the results obtained by the Danish Eulerian Model. Occasionally, measurements from some other sources have also been used in some studies.

5.2. Time and Space Resolution of the Measurement Data

Most of the measurement data are available as daily mean values of the concentrations over a long time-period (the measurement data used in the Danish Eulerian Model are from the period 1989-1998).

The ozone concentrations can also be obtained as hourly mean values (the ozone concentrations vary on a diurnal basis; the availability of hourly mean values of the ozone concentrations allows the modellers to study the diurnal variations of the ozone concentrations in different places in Europe in different seasons).

The space distribution of the available measurements depends very much on the measured compound and on the year of consideration. For some compounds (as, for example ozone and sulphate) the number of stations that measure regularly (at least 15 measurements per month) is about 60. For the remaining compounds the number of measurement stations is considerably smaller. The number of stations which measure regularly (at least 5 measurements per month) concentrations in precipitation is also rather small

In the ten-year period of consideration the number of measurement stations in the EMEP network tends to grow during the last years. However, most of the measurement stations are still in the countries of Western Europe and Scandinavia.

5.3. Amount of the Measurement Data

The files containing measurement data are, at present, smaller than the files containing input data and output data. Anyway, if the model results are to be compared with measurements over a long time-period, then the files of the measurement data needed are quite large; several hundreds of MBytes. The amount of these data tends to grow, because more and more stations have been opened during the last years and more and more of the existing stations are starting to measure more compounds.

5.4. Measurement Data for Real Time Computations

It has already been stated that measurement data are needed for preparation of initial fields of concentrations (by using data assimilation techniques) when air pollution forecasts are to be prepared. It is important to have these data as quickly as possible (say, one or two hours after performing the measurements). This was, of course, a very big problem only two-three years ago. Now the situation is rapidly changing due to the possibilities to use the internet and to carry out fast transition of large amounts of data. The needed data can be taken from many sites where such data are available. It should be mentioned here that also the National Environmental Research Institute is regularly exposing measurement data on the internet.

6. Some Results Obtained by the Danish Eulerian Model

The Danish Eulerian Model, together with the graphical tools associated by this model, can be considered as a powerful tool for

- treating very large sets of input data,

- preparing very large sets of output data,

- verifying the sets of output data by using very large sets of measurement data

and

- applying the verified sets of output data in different studies by also adding results from long sequences of scenarios performed by the model.

A few results will be presented in this section. Much more results obtained by this model have been described and discussed in Skjøth et al. (2000), Birk et al. (1997), Birk et al. (1998), Harrison et al. (1994), Zlatev (1995), Zlatev and Skjøth (1999) and Zlatev et al. (1993), Zlatev et al. (1992),Zlatev et al. (1994), Zlatev et al. (1996a) Zlatev et al. (1996b).

Only three features of the Danish Eulerian Model, but these are very important issues, will be discussed here and/or illustrated by some results which will be presented in the next sub-sections:

- the ability of the model to calculate output results by using very high resolution and to zoom at sub-domains that are of particular interest,

- the ability of the model to produce reliable results

and

- the use of the model with different scenarios.

6.1. Results Obtained with the High Resolution Version

In the two-dimensional high-resolution version of the Danish Eulerian Model the space domain (containing the whole of Europe together with parts of Asia, Africa and the Atlantic Ocean) is discretized by using a (480×480) grid into $(10\ km \times 10\ km)$ grid-squares. It is still not possible, as mentioned in Section 2, to run this version of the model on long time-intervals (a year or several years). Therefore, only some interesting situations, which occur on shorter time-intervals (the typical length of such a time-interval being a month), were studied by this version until now. Some results obtained when the model was run with meteorological data for July 1994 are shown in Fig. 31.1 and Fig. 31.2.

The distribution of the nitrogen dioxide pollution in Europe is shown in Fig. 31.1. It is seen that the most polluted areas in Europe are parts of England, the Netherlands, Belgium, Germany and parts of France, the Check Republic and Poland as well as the Northern part of Italy.

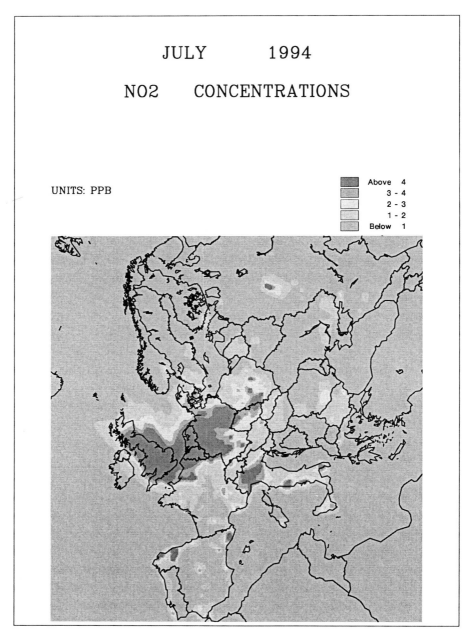

Figure 31.1. The distribution of the nitrogen dioxide concentrations in different parts of Europe and its surroundings

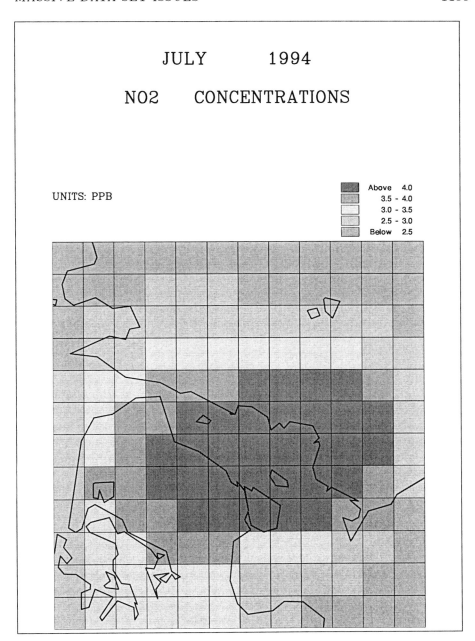

Figure 31.2. The nitrogen dioxide concentrations in area around the Danish capital Copenhagen and the Swedish city Malmö (the Øresund region)

Moreover, in the parts of Europe which are not very polluted one can locate large cities, such as Madrid, Rome, Oslo, Stockholm, Helsinki, Sct. Petersburg and Moscow, which are large sources of nitrogen pollution (the most important reason being the fact that pollution from the traffic is one of the major sources for nitrogen emissions).

The same output data as those used to draw Fig. 31.1 are also used to draw Fig. 31.2. Indeed, Fig. 31.2 could be viewed as a result from zooming in Fig. 31.1 onto the area around Copenhagen, the capital of Denmark, and the Swedish city of Malmö.

The grid lines are drawn in Fig. 31.2. This is impossible when the whole space domain is used. Indeed, Fig. 31.1 will become complete black if the option for drawing the grid-lines is not switched off when results on the whole space domain, containing 230 400 grid-squares, are plotted.

The number of grid-squares used in Fig. 31.2 is 144, obtained by using only a tiny part, a (12×12) sub-grid, of the whole (480×480) grid. At the same time, this area is nearly twice smaller than one grid-square of the (32×32) grids which were commonly used only a few years ago, and are still used in some models (see, for example, Amann et al. (1999) and the EMEP Report 1/98 (EMEP 1998)). While the size of only one grid-square is 22500 km^2 when a (32×32) grid is applied, the size of the area of 144 grid-squares used in Fig. 31.2 is 14400 km^2. This means that, while it will not be possible to see any difference in the area shown in Fig. 31.2 when a coarse (32×32) grid is used, the results from the high resolution model shown in Fig. 31.2 show clearly that there are several different levels of nitrogen pollution in this area. This illustrates the great potential power of the high resolution models. However, there is a price that is to be paid: very large sets of digital data are to be handled when such models are run.

It is also necessary to emphasize that Fig. 31.2 is only one example for zooming. Zooming could also be done for any other sub-domain of the space domain of the Danish Eulerian Model.

6.2. Checking the Reliability of the Model Results

The validation of the results calculated by a model is an important task. Many questions related to the validation procedures used are still open. Some results obtained in the validation of the results calculated by the Danish Eulerian Model will be discussed in this sub-section.

The first question which is to be answered during the validation process is connected to the reliability of the numerical algorithms used in the model. There are different tests which can be used to test the accuracy of the numerical algorithms. For the advection part, the so-called rotation test is well known. It has, as mentioned above, been introduced simultaneously by Crowley (1968) and Molenkampf (1968) in 1968. Both the classical rotation test and many modifications of this test have been used in several hundred publications in the field of air pollution modelling; see, for example, Chock (1985), Chock (1991), Chock and Dunker (1983), Fornberg (1975) and Fornberg (1996), Pepper and Baker (1979) and Pepper et al. (1979). A set of tests concerning both the advection and the diffusion part is discussed in Zlatev (1995) and Zlatev et al. (1983a). An extension of the rotation test for the combined advection-chemistry module of an air pollution model has been proposed in Hov et al. (1988); see also Zlatev (1995).

It should be emphasized that successful tests of the performance of the numerical method **do not** ensure the reliability of the results when the full model is run. The success of the numerical tests is only telling us that if there are errors, then these errors are not caused by the numerical methods and, thus, the source for the errors should be searched in other places (as, for example, uncertainties in the emission inventories, uncertainties in the meteorological data, uncertainties in parameters used in the description of some physical and/or chemical mechanisms, etc.). The conclusion is that one should not overestimate the importance of the numerical tests. On the other hand, one should not underestimate the importance of the numerical tests either. Successful numerical tests indicate that

- one of the major sources for errors in the model results has been removed (it should be mentioned here that this is crucial when **Eulerian models** are used)

and (what is even more important)

- there will be no interference of numerical errors when the sensitivity of the model results to variation of some key parameters is studied.

Comparison of the distribution of the concentrations in the different parts of the space domain with the distribution of related emissions can give us some qualitative estimation of the reliability of the results. Comparing the results in Fig.1 with the distribution of the NO_x emissions in Europe, we can conclude that

- the Danish Eulerian Model is producing highest levels of the nitrogen dioxide pollution precisely in the parts of the space domain where the NO_x emissions are highest and

- in the large European cities (which are in general also big sources of NO_x emissions), the levels of the nitrogen dioxide pollution is also high.

The qualitative correctness of the model results allows us to draw two conclusions. The first conclusion is that there are no crude errors in the description of the physical and chemical mechanisms. The second conclusion is that there are not too big uncertainties in the meteorological data used in the model. It should be noted here that both conclusions are drawn under an assumption that the numerical methods work sufficiently well (which emphasizes once again the importance of checking the accuracy of the numerical algorithms that are implemented in the model).

Of course, quantitative correctness of the model results is the most desirable property of an air pollution model. Most of the unresolved problems appear when one attempts to validate quantitatively the model results. The best which we can do at present (in the attempts to validate quantitatively the model results) is to compare model results with measurements taken at representative measurement stations located in different parts of the space domain.

There are two major ways to present the results which are obtained when calculated by the model concentrations are compared with corresponding measurements. The results from such comparisons can be represented:

- by scatter plots where the mean values of the calculated concentrations and depositions are compared with corresponding measurements (mainly on monthly, seasonal and yearly basis),

and

- by time-series plots where temporal variations of the calculated concentrations and depositions at a given measurement station are compared with the corresponding observations.

Both approaches have extensively been used when different versions of the Danish Eulerian Model have been developed. Results from validation procedures applied to different versions of the Danish Eulerian Model can be found in many publications; see, for example, Skjøth et al. (2000), Birk et al. (1997), Birk et al. (1998), Harrison et al. (1994), Zlatev (1985), Zlatev (1995), Zlatev and Skjøth (1999), Zlatev and Berkowicz (1988) and Zlatev et al. (1993), Zlatev et al. (1992), Zlatev et al. (1994), Zlatev et al. (1996a), Zlatev et al. (1996b). Nevertheless, much more comparisons are necessary.

6.3. Running the Model with Different Scenarios

Results from comprehensive mathematical models can successfully be used in different studies (assuming here that some validation process based on comparison of model results and measurements has successfully been completed). An example for such a study will be given here as an illustration. Other examples can be found in Skjøth et al. (2000), Birk et al. (1997), Birk et al. (1998), Zlatev (1995), Zlatev and Skjøth (1999) and Zlatev et al. (1996b).

Different critical levels of the concentrations of certain pollutants have been defined and used in many countries; especially in the European Union (EU), see, for example, Amann et al. (1999). Concerning ozone, one of the critical levels, which is under discussion now in the European Union, is the number of days in which the ozone concentrations were higher than 120 ppb in at least one hour (stated in the draft for the forthcoming EU directive for ozone and the "Ozone Position Paper" of the Commission of EU: Commission (1998), Commission (1999)). According to the draft for the forthcoming EU directive for ozone and the "Ozone Position Paper" of the Commission of EU (Commission (1998), Commission (1999)), the population must be warned if this is likely to happen in the next two-three days (this is one of the reasons for calculating air pollution forecasts - see also Section 4).

On the other hand, there are some predictions concerning the anthropogenic (human-made) emissions in Europe for year 2010. These emission inventories have been prepared by a large group of European specialists in this field. It is agreed that the European countries will work for achieving the predicted levels latest in year 2010. These emission levels are well-known (in Europe, at least) as Scenario 2010.

In connection with this critical level (number of days in which the ozone concentrations were higher than 120 ppb in at least one hour) and with the emissions from Scenario 2010, it is important to have the answers to the following two important questions:

- How frequently is the ozone level of 120 ppb exceeded?

and

- What will be the effect of using the emissions from Scenario 2010 on the number of days in which the ozone level of 120 ppb is exceeded?

The Danish Eulerian model, and more precisely the version of this model which is discretized by using ($50\ km \times 50\ km$) grid-squares, have been run twice over a time-interval of ten years (from 1989 to 1998).

For each year N, where $N = 1989, 1990, \ldots, 1998$, the existing emission inventories for year N were used in the first run. The predicted by Scenario 2010 emissions were used for all of the ten years in the second run. The meteorological data used in both runs were the same; for each year N, where $N = 1989, 1990, \ldots, 1998$, the meteorological data for year N were used. The first run will be called *Basic scenario*, while the name *Scenario 2010* will be used for the second run.

Results concerning Denmark and the region surrounding Denmark are given in Fig. 31.3 (the Basic scenario) and in Fig. 31.4 (Scenario 2010). The total numbers of days (for the whole period of ten years) in which the ozone level of 120 ppb has been exceeded in at least one hour are given in these two figures.

The results given in Fig. 31.3 show that the total numbers of days (for the all ten years) in which the ozone level of 120 ppb has been exceeded in at least one hour vary between 3 and 10 in the different parts of Denmark when the Basic scenario is used. This means that the answer to the first question for the different parts of Denmark is the following:

- *the exceedance of this critical level does not occur very often in Denmark, but may take place sometimes.*

The results given in Fig. 31.4 are the same as those in Fig. 31.3 (numbers of days in which the critical level of 120 ppb is exceeded), but here the emissions from Scenario 2010 are used. The results indicate that practically

- *no exceedance of the critical level for ozone of 120 ppb takes place in Denmark when the emissions from Scenario 2010 are used.*

Thus, the reduction of the emissions to the levels prescribed in Scenario 2010 will be very beneficial for Denmark in connection with the critical level of 120 ppb.

Only results related to Denmark were presented here. However, the huge output data sets can be used to zoom at any other sub-domain of the space domain of the Danish Eulerian Model. It should be mentioned here that the results indicate that, while it is true that the number of occurrences of days in which the critical level of 120 ppb is reduced considerably in the countries in Central and Western Europe when Scenario 2010 is used, it is also true that the reduction of the European emissions as planned in Scenario 2010 will not totally remove the exceedances of this critical level for ozone in these countries. This is why some scenarios for additional reduction of the anthropogenic emissions in the 15 EU countries are currently discussed (see, for example, Amann et al. (1999)).

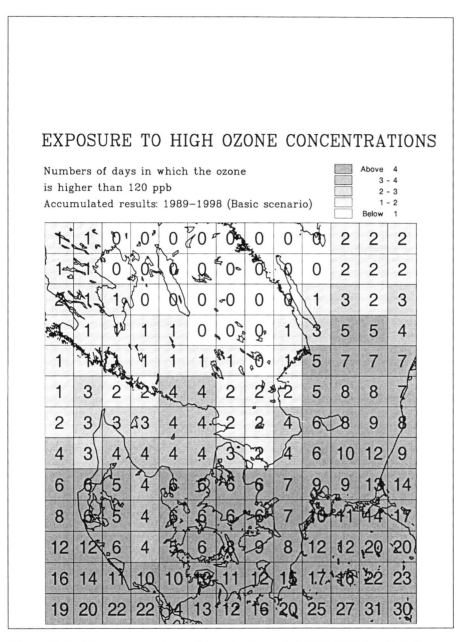

Figure 31.3. The total numbers of days in the period 1989-1998 in which the ozone concentrations in Denmark exceeded the critical level of 120 ppb - Basic Scenario

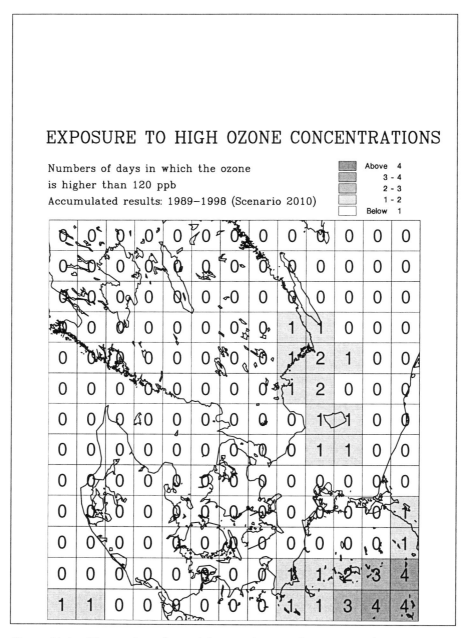

Figure 31.4. The total numbers of days in the period 1989-1998 in which the ozone concentrations in Denmark exceeded the critical level of 120 ppb - Scenario 2010

Most of the experiments are at present carried out by varying in a suitable way the anthropogenic emissions. However, it is also necessary to study the sensitivity of the concentrations and depositions also to changes of the biogenic (natural) emissions. This issue is becoming more and more important, because last investigations indicate that the biogenic emissions are strongly underestimated (by up to ten times: see, for example, Bouchet et al. (2000a), Bouchet et al. (2000b)).

6.4. Sensitivity Tests

Systematic sensitivity tests of the variations of the model results caused by the variations of certain physical parameters can give very useful results. If the results are very sensitive to the variations of the selected parameter, then this is an indication that it is necessary to use a very accurate value of the parameter under consideration. If we are not able to provide a sufficiently accurate value of the critical parameter, then it may be more profitable to replace the physical mechanism in which the critical parameter occurs with another mechanism which is perhaps not so accurate, but more robust (in the sense that the results are not very sensitive to the physical parameters involved in the new mechanism).

Some carefully chosen tests, appropriate to study the sensitivity of the model results to variations of the rate coefficients of the chemical reactions, were carried out in Dimov and Zlatev (1999). Monte Carlo techniques were used in Dimov and Zlatev (1999) in order

- to find the rate coefficient to the variation of which the model results are most sensitive

and

- to study in a more systematic way the variation of the model results caused by random variations in a prescribed range of the selected rate coefficient.

A simple box-model (one grid-point only and no other processes except the chemical reactions) is quite sufficient in the first part. The second part is much more complicated, because it requires a large number of runs of the whole air pollution model. It has been established in this study that the magnitude of the changes in the different parts of Europe was not the same. More details and other conclusions can be found in Dimov and Zlatev (1999).

7. Concluding Remarks

Remark 1. Different problems connected with the needs of using large data sets in air pollution modelling have been discussed by using a particular model, the Danish Eulerian Model. It has already been emphasized that this approach has been chosen only in order to simplify the exposition of the results. The same conclusions could be drawn also if any other large scale air pollution model is used. Indeed, the development of the Danish Eulerian Model is similar to the development of many other models:

- *starting with models defined on coarse grids with a few chemical compounds*

and

- *arriving gradually at very advanced mathematical models utilizing fine grids and involving much more chemical compounds.*

Remark 2. The question: *Are the mathematical models used at present sufficient for studying the most important environmental problems?* is relevant. The answer is, of course, no. While many problems can successfully be studied and **are** successfully studied by the models that are available at present, it is also true that many other problems cannot be handled with the models which are existing at present. The question of having models discretized on finer grids (say, by using $2\ km \times 2\ km$ grid-squares) is already a topic of discussion. Also the high-resolution three-dimensional models are still not tractable even on the largest computers existing at present (at least in the case where calculations over a long time-interval are required). This indicates clearly that

- *there is a lot of work to be done in this field*

and

- *the need of developing good tools for handling very large data sets (input data sets, output data sets and data sets containing measurements) is steadily increasing in the environmental studies which are carried out by using large scale mathematical models.*

Remark 3. The performance of the computers available is a central problem in this field. We are going to handle larger and larger environmental models on computers. The following two questions arise in connection with this development:

- Are the computers that are available at present large enough?
- Do we need bigger and faster computers for the treatment of the environmental problems which have to be resolved in the near future?

These questions can best be answered by using a quotation taken from a paper written by Jaffe (1984):

- *"Although the fastest computers can execute millions of operations in one second they are always too slow. This may seem a paradox, but the heart of the matter is: the bigger and better computers become, the larger are the problems scientists and engineers want to solve".*

This paper of Jaffe was published in 1984. At that time it was difficult to handle the version of the Danish Eulerian Model containing, after the discretization, 2048 equations. The biggest problems solved now by the Danish Eulerian Model contain, again after the discretization, more than 8000000 equations. One of the major reasons for being able to handle such huge problems, about 4000 times larger than the problems that were solved in 1984, is the availability of much bigger and much faster computers. There is no reason to assume that the development of bigger and faster computers will not continue. Therefore, the scientists could continue to plan the formulation and the treatment of bigger and bigger tasks, tasks which impose strong requirements of handling bigger and bigger data sets, because the solutions obtained in this way

- will be *closer* to the reality,
- will contain *more* details and *more useful* details

and

- will give *reliable* answers to questions which are at present open.

Acknowledgments

This research was supported by the NATO Scientific Programme under the projects ENVIR.CGR 930449 and OUTS.CGR.960312, by the EU ESPRIT Programme under projects WEPTEL and EUROAIR and by NMR (Nordic Council of Ministers) under a common project for performing sensitivity studies with large-scale air pollution models in which scientific groups from Denmark, Finland, Norway and Sweden are participating. Furthermore, a grant from the Danish Natural Sciences Research Council gave us access to all Danish supercomputers.

Several colleagues of mine have read the manuscript of this papers and made many useful suggestions for improvements. I should like to thank them very much.

Bibliography

I. Ackerman, H. Hass, B. Schell, and F. S. Binkowski. Regional modelling of particulate matter with made. *Environmental Management and Health*, 10:201–208, 1999.

V. Alexandrov, A. Sameh, Y. Siddique, and Z. Zlatev. Numerical integration of chemical ode problems arising in air pollution models. *Environmental Modelling and Assessment*, 2:365–377, 1997.

A. E. Aloyan. Modelling of global and regional transport and transformation of air pollutants. In Z. Zlatev, J. Brandt, P. J. H. Builtjes, G. Carmichael, I. Dimov, J. Dongarra, H. van Dop, K. Georgiev, H. Hass, and R. San Jose, editors, *Large Scale Computations in Air Pollution Modelling*, pages 15–24. Dordrecht-Boston-London: Kluwer Academic Publishers, 1999. NATO Science Series, 2. Environmental Security - Vol. 57.

M. Amann, I. Bertok, J. Cofala, F. Gyarfas, C. Heyes, Z. Klimont, M. Makowski, W. Schöp, and S. Syri. Cost-effective control of acidification and ground-level ozone. Technical report, IIASA (International Institute for Applied System Analysis), 1999.

E. Anderson, Z. Bai, C. Bischof, J. Demmel, J. Dongarra, J. Du Croz, A. Greenbaum, S. Hammarling, A. McKenney, S. Ostrouchov, and D. Sorensen. *LAPACK: Users' Guide*. SIAM, Philadelphia, 1992.

R. Barrett, M. Berry, T. F. Chan, J. Demmel, J. Donato, J. Dongarra, V. Eijkhout, R. Pozo, C. Romine, and H. van der Vorst. *Templates for the Solution of Linear Systems: Building Blocks for Iterative Methods*. SIAM, Philadelphia, 1994.

F. Binkowski and U. Shankar. The regional particulate model 1: Model description and preliminary results. *Journal of Geophysical Research*, 100:26191–26209, 1995.

A. B. Birk, J. Brandt, I. Uria, and Z. Zlatev. Studying cumulative ozone exposures in europe during a seven-year period. *Journal of Geophysical Research*, 102:23917–23935, 1997.

A. B. Birk, J. Brandt, and Z. Zlatev. Using partitioned ode solvers in large air pollution models. *Systems Analysis Modelling Simulation*, 32:3–17, 1998.

P. Borrell, P. M. Borrell, and W. Seiler. *Transport and Transformation of Pollutants in the Troposphere*. SPB Academic Publishing, The Hague, 1990.

V. S. Bouchet, R. Laprise, E. Torlaschi, and J. C. McConnell. Studying ozone climatology with the canadian regional climate model. part i: Model description and evaluation. *Journal of Geophysical Research*, 2000a. to appear.

V. S. Bouchet, R. Laprise, E. Torlaschi, and J. C. McConnell. Studying ozone climatology with the canadian regional climate model. part ii: Climatology. *Journal of Geophysical Research*, 2000b. to appear.

J. Brandt, J. Christensen, L. M. Frohn, R. Berkowicz, and Z. Zlatev. Optimization of operational air pollution forecast modelling from european to local scale. *Analli*, 2000. to appear.

J. Brandt, J. Christensen, and Z. Zlatev. Real time predictions of transport, dispersion and deposition from nuclear accidents. *Environmental Management and Health*, 10:216–223, 1999.

K. Brenan, S. Campbell, and L. Petzold. *Numerical solution of initial value problems in differential-algebraic equations*. SIAM, Philadelphia, 1996.

R. A. Brost. The sensitivity to input parameters of atmospheric concentrations simulated by a regional chemical model. *Journal of Geophysical Research*, 93:2371–2387, 1988.

J. Brown, J. Wasniewski, and Z. Zlatev. Running air pollution models on massively parallel machines. *Parallel Computing*, 21:971–991, 1995.

B. Bruegge, E. Riedel, A. Russell, E. Segall, and P. Steenkiste. Heterogeneous distributed environmental modelling, 1995. SIAM News, No. 28.

P. J. H. Builtjes. The lotos - long term ozone simulation project. Technical report, TNO (Institute of Environmental Sciences, Energy Research and Process Information), P. O. Box 342, 7300 AH Apeldoorn, The Netherlands, 1992. TNO Report No. R92/240.

P. J. H. Builtjes. The eurotrac contribution to the development of tools for the study of environmentally relevant trace constituents. In P. Borrell, P. Builtjes, P. Grennfelt, and H. Østein, editors, *Photo-Oxidants, Acidification and Tools: Policy Applications of EUROTRAC Results*, pages 143,194. Springer, Berlin, 1999.

K. Burrage. *Parallel and sequential methods for ordinary differential equations.* Oxford University Press, 1995.

G. R. Carmichael and L. P. Peters. An eulerian transport-transformation-removal model for so_2 and sulphate: I. model development. *Atmospheric Environment*, 18:937–952, 1984.

G. R. Carmichael, L. P. Peters, and T. Kitada. A second generation model for regional-scale transport-chemistry-deposition. *Atmospheric Environment*, 20:173–188, 1986.

G. R. Carmichael, A. Sandu, F. A. Potra, V. Damian, and M. Damian. The current state and the future directions in air quality modeling. *Systems Analysis Modelling Simulation*, 25:75–105, 1996.

G. R. Carmichael, A. Sandu, C. H. Song, S. He, M. J. Phadnis, V. D. Iordache, and F. A. Potra. Computational challenges of modelling interactions between aerosol and gas phase processes in large-scale air pollution models. In Z. Zlatev, J. Brandt, P. J. H. Builtjes, G. Carmichael, I. Dimov, J. Dongarra, H. van Dop, K. Georgiev, H. Hass, and R. San Jose, editors, *Large Scale Computations in Air Pollution Modelling*, pages 99–136. Dordrecht-Boston-London: Kluwer Academic Publishers, 1999. NATO Science Series, 2. Environmental Security - Vol. 57.

J. S. Chang, R. A. Brost, I. S. A. Isaksen, S. Madronich, P. Middleton, W. R. Stockwell, and C. J. Walcek. A three dimensional eulerian acid deposition model: Physical concepts and formulation. *Journal of Geophysical Research*, 92:14691–14700, 1987.

D. P. Chock. A comparison of numerical methods for solving advection equations - ii. *Atmospheric Environment*, 19:571–586, 1985.

D. P. Chock. A comparison of numerical methods for solving advection equations - iii. *Atmospheric Environment*, 25A:853–871, 1991.

D. P. Chock and A. M. Dunker. A comparison of numerical methods for solving advection equations - i. *J. Comp. Phys.*, 19:571–586, 1983.

D. P. Chock, S. L. Winkler, and P. Sun. Comparison of stiff chemistry solvers for air quality models. *Environ. Sci. Technol.*, 28:1882–1892, 1994.

European Commission. Amended draft of the daughter directive for ozone. Technical report, European Commission, 1998. European Commission, Directorate XI: "Environment, Nuclear Safety and Civil Protection".

European Commission. Ozone position paper. Technical report, European Commission, 1999. European Commission, Directorate XI: "Environment, Nuclear Safety and Civil Protection".

W. P. Crowley. Numerical advection experiments. *Monthly Weather Review*, 96:1–11, 1968.

D. Dabdub and R. Manohar. Performance and portability of an air pollution model. *Parallel Computing*, 23:2187–2200, 1997.

D. Dabdub and J. H. Seinfeld. Numerical advective schemes used in air quality models - sequential and parallel implementation. *Atmospheric Environment*, 29:403–410, 1994.

D. Dabdub and J. H. Seinfeld. Parallel computation in atmospheric chemical modelling. *Parallel Computing*, 22:111–130, 1996.

P. D'Ambra, G. Barone, D. di Serafino, G. Giunta, A. Murli, and A. Riccio. Pnam: parallel software for air quality simulations in the naples area. *Environmental Management and Health*, 10:209–215, 1999.

J. W. Demmel. *Applied Numerical Linear Algebra*. SIAM, Philadelphia, 1997.

P. Deuflhard. Order and stepsize control in extrapolation methods. *Numerische Mathematik*, 41:399–422, 1983.

P. Deuflhard. Recent progress in extrapolation methods for ordinary differential equations. *SIAM Review*, 27:505–535, 1985.

P. Deuflhard, E. Hairer, and M. Zugch. One step and extrapolation methods for differential-algebraic equations. *Numerische Mathematik*, 51:501–516, 1987.

P. Deuflhard and U. Nowak. Efficient numerical simulation and identification of large reaction systems. *Ber. Bensendes Phys. Chem.*, 90: 940–946, 1986.

P. Deuflhard, U. Nowak, and M. Wulkow. Recent development in chemical computing. *Computers Chem. Engng.*, 14:1249–1258, 1990.

I. Dimov, I. Farago, and Z. Zlatev. Commutativity of the operators in splitting methods for air pollution models. Technical report, Central Laboratory for Parallel Processing, Bulgarian Academy of Sciences, 1999. Report 04/99.

I. Dimov and Z. Zlatev. Testing the sensitivity of air pollution levels to variations of some chemical rate constants. In M. Griebel, O. P. Iliev,

S. D. Margenov, and P. S. Vassilevski, editors, *Large-Scale Scientific Computations and Environmental Problems*, pages 167–175. Vieweg, Braunschweig/Wiesbaden, 1999.

J. J. Dongarra, I. S. Duff, D. C. Sorensen, and H. A. van der Vorst. *Numerical Linear Algebra for High-Performance Computers*. SIAM, Philadelphia, 1998.

I. S. Duff, A. M. Erisman, and J. K. Reid. *Direct methods for sparse matrices*. Oxford University Press, Oxford-London, 1986.

A. Ebel. Chemical transfer and transport modelling. In P. Borrell and P. M. Borrell, editors, *Transport and Chemical Transformation of Pollutants in the Troposphere*, pages 85–128. Springer, Berlin, 2000.

A. Ebel, M. Memmesheimer, and Jacobs. Regional modelling of tropospheric ozone distributions and budgets. In C. Varotsos, editor, *Atmospheric Ozone Dynamics: I. Global Environmental Change*, pages 37–57. Springer, Berlin, 1997. NATO ASI Series, Vol. 53.

H. Elbern. Parallelization and load balancing of a comprehensive atmospheric chemistry model. *Atmospheric Environment*, 31:3561–3574, 1997.

H. Elbern and H. Schmidt. A four-dimensional variational chemistry data assimilation scheme for eulerian chemistry transport modelling. *Journal of Geophysical Research*, 104:18583–18598, 1999.

H. Elbern, H. Schmidt, and A. Ebel. Variational data assimilation for tropospheric chemistry modelling. *Journal of Geophysical Research*, 102:15967–15985, 1997.

H. Elbern, H. Schmidt, and A. Ebel. Implementation of a parallel 4d-variational chemistry data-assimilation scheme. *Environmental Management and Health*, 10:236–244, 1999.

H. Elbern, H. Schmidt, and A. Ebel. Improving chemical state analysis and ozone forecasts by four dimensional chemistry data assimilation. In WMO, editor, *Proceedings of the Third WMO International Symposium on Assimilation of Observations in Meteorology and Oceanography*. Geneva, 2000a. to appear.

H. Elbern, H. Schmidt, O. Talagrand, and A. Ebel. 4d-variational data assimilation with an adjoint air quality model for emission analysis. *Environmental Modelling and Software*, 2000b. to appear.

EMEP. Transboundary acidifying air pollution in europe: Part 1, estimated dispersion of acidifying and eutrofying compounds and comparison with observations. Technical report, EMEP MSC-W (Meteorlogical Synthesizing Centre - West), Norwegian Meteorological Institute, 1998. Status Report 1/98.

B. Fornberg. On a fourier method for the integration of hyperbolic equations. *SIAM J. Numer. Anal.*, 12:509–528, 1975.

B. Fornberg. *A practical guide to pseudospectral methods*. Cambridge University Press, Cambridge, 1996. Cambridge Monographs on Applied and Computational Mathematics.

A. Geist, A. Beguelin, J. Dongarra, W. Jiang, R. Manchek, and V. Sunderam. *PVM: Parallel Virtual Machine, A User's Guide and Tutorial for Networking Parallel Computing*. MIT Press, Cambridge, Massachusetts, 1994.

K. Georgiev and Z. Zlatev. Running an advection-chemistry code on message passing computers. In V. Alexandrov and J. Dongarra, editors, *Recent Advances in Parallel Virtual Machine and Message Passing Interface*, pages 354–363. Springer, Berlin, 1998.

K. Georgiev and Z. Zlatev. Parallel sparse matrix algorithms for air pollution models. *Parallel and Distributed Computing Practices*, 2000. to appear.

M. W. Gery, G. Z. Whitten, J. P. Killus, and M. C. Dodge. A photochemical kinetics mechanism for urban and regional modeling. *Journal of Geophysical Research*, 94:12925–12956, 1989.

W. Gropp, E. Lusk, and A. Skjellum. *Using MPI: Portable programming with the message passing interface*. MIT Press, Cambridge, Massachusetts, 1994.

E. Hairer and G. Wanner. *Solving Ordinary Differential Equations, II: Stiff and Differential-algebraic Problems*. Springer, Berlin-Heidelberg-New York-London, 1991.

R. M. Harrison, Z. Zlatev, and C. J. Ottley. A comparison of the predictions of an eulerian atmospheric transport chemistry model with measurements over the north sea. *Atmospheric Environment*, 28:497–516, 1994.

H. Hass, A. Ebel, H. Feldmann, H. J. Jacobs, and M. Memmesheimer. Evaluation studies with a regional chemical transport model (eurad) using air quality data from the emep monitoring network. *Atmospheric Environment*, 27A:867–887, 1993.

H. Hass, H. J. Jacobs, and M. Memmesheimer. Analysis of a regional model (eurad) near surface gas concentration predictions using observations from networks. *Atmos. Phys.*, 57:173–200, 1995.

O. Hertel, R. Berkowicz, J. Christensen, and Ø. Hov. Test of two numerical schemes for use in atmospheric transport-chemistry models. *Atmospheric Environment*, 27A:2591–2611, 1993.

E. Hesstvedt, Ø. Hov, and I. A. Isaksen. Quasi-steady-state approximations in air pollution modelling: Comparison of two numerical schemes for oxidant prediction. *International Journal of Chemical Kinetics*, 10: 971–994, 1978.

A. G. Hjellbrekke. Ozone measurements 1990-1992. Technical report, Norwegian Institute for Air Research (NILU), 1995. EMEP/CCC Report 4/95.

Ø. Hov, Z. Zlatev, R. Berkowicz, A. Eliassen, and L. P. Prahm. Comparison of numerical techniques for use in air pollution models with non-linear chemical reaction. *Atmospheric Environment*, 23:967–983, 1988.

M. Z. Jacobson and R. P. Turco. Smvgear: a sparse-matrix, vectorized gear code for atmospheric models. *Atmospheric Environment*, 28:273–284, 1994.

A. Jaffe. Ordering the universe: The role of mathematics. *SIAM Rev.*, 26:475–488, 1984.

L. O. Jay, A. Sandu, F. A. Potra, and G. R. Carmichael. Improved qssa methods for atmospheric chemistry integration. *SIAM J. Sci. Comput.*, 18:182–202, 1997.

J. E. Jonson, L. Tarrason, and J. Sundet. Calculation of ozone and other pollution for the summer, 1996. *Environmental Management and Health*, 10:245–257, 1999.

R. S. Jose, M. A. Rodriguez, E. Cortes, and R. M. Conzalez. Emma model: an advanced operational mesoscale air quality model for urban and regional environments. *Environmental Management and Health*, 10:258–266, 1999.

J. D. Lambert. *Numerical Methods for Ordinary Differential Equations*. Wiley, Chichester-New York-Brisbane-Toronto-Singapore, 1991.

D. Lancer and J. G. Verwer. Analysis of operators splitting in advection-diffusion-reaction problems in air pollution modelling. *Journal of Computational and Applied Mathematics*, 111:201–216, 1999.

B. Lübkert and W. Schöp. A model to calculate natural *voc* emissions from forests in europe. Technical report, International Institute for Applied System Analysis (IIASA), 1989. Report WP-89-082.

G. I. Marchuk. *Mathematical Modeling for the Problem of the Environment*. North-Holland, Amsterdam, 1985. Studies in Mathematics and Applications, No. 16.

G. J. McRae, W. R. Goodin, and J. H. Seinfeld. Numerical solution of the atmospheric diffusion equations for chemically reacting flows. *Journal of Computational Physics*, 45:1–42, 1984.

C. R. Molenkampf. Accuracy of finite-difference methods applied to the advection equation. *Journal of Applied Meteorology*, 7:160–167, 1968.

OPEN MP. Web-site for open mp tools, 1999. Available at: http://www.openmp.org.

M. T. Odman, N. Kumar, and A. G. Russell. A comparison of fast chemical kinetic solvers for air quality modeling. *Atmospheric Environment*, 26A:1783–1789, 1992.

W. Owczarz and Z. Zlatev. Running a large air pollution model on an ibm smp computer. *Analli*, 2000a. to appear.

W. Owczarz and Z. Zlatev. Templates for parallel runs of air pollution models. *International Journal of Computer Research*, 2000b. to appear.

D. W. Pepper and A. J. Baker. A simple one-dimensional finite element algorithm with multidimensional capabilities. *Numerical Heath Transfer*, 3:81–95, 1979.

D. W. Pepper, C. D. Kern, and Jr. P. E. Long. Modelling the dispersion of atmospheric pollution using cubic splines and chapeau functions. *Atmospheric Environment*, 13:223–237, 1979.

L. K. Peters, C. M. Berkowitz, G. R. Carmichael, R. C. Easter, G. Fairweather, S. J. Ghan, J. M. Hales, L. R. Leung, W. R. Pennell, F. A. Potra, R. D. Saylor, and T. T. Tsang. The current state and future direction of eulerian models in simulating tropospheric chemistry and transport of trace species: A review. *Atmospheric Environment*, 29: 189–222, 1995.

A. Sandu, F. A. Potra, G. R. Carmichael, and V. Damian. Efficient implementation of fully implicit methods for atmospheric chemical kinetics. *J. Comp. Phys.*, 129:101–110, 1996.

A. Sandu, J. G. Verwer, J. G. Bloom, E. J. Spee, and G. R. Carmichael. Benchmarking stiff ode systems for atmospheric chemistry problems: Ii. rosenbrock solvers. *Atmospheric Environment*, 31:3459–3472, 1997a.

A. Sandu, J. G. Verwer, M. van Loon, G. R. Carmichael, F. A. Potra, D. Dabdub, and J. H. Seinfeld. Benchmarking stiff ode systems for atmospheric chemistry problems: I. implicit versus explicit. *Atmospheric Environment*, 31:3151–3166, 1997b.

L. F. Shampine, M. W. Reichelt, and J. A. Kierzenka. Solving index-1 daes in matlab and simulink. *SIAM Rev.*, 41:538–552, 1999.

D. S. Shieh, Y. Chang, and G. R. Carmichael. The evaluation of numerical techniques for solution of stiff ordinary differential equations arising from chemical kinetic problems. *Environ. Software*, 3:28–38, 1988.

D. Simpson. Long-period modelling of photochemical oxidants in europe. model calculations for july 1985. *Atmospheric Environment*, 26A: 1609–1634, 1992.

D. Simpson. Photochemical model calculations over europe for two extended summer periods: 1985 and 1989. model results and comparisons with observations. *Atmospheric Environment*, 27A:921–943, 1993.

S. Skelboe and Z. Zlatev. Exploiting the natural partitioning in the numerical solution of ode systems arising in atmospheric chemistry. In L. Vulkov, J. Wasniewski, and P. Yalamov, editors, *Numerical Analysis and Its Applications*, pages 458–465. Springer, Berlin, 1997.

C. A. Skjøth, A. B. Birk, J. Brandt, and Z. Zlatev. Studying variations of pollution levels in a given region of europe during a long time-period. *Systems Analysis Modelling Simulation*, 2000. to appear.

W. R. Stockwell, F. Kirchner, and M. Kuhn. A new mechanism for regional atmospheric chemistry modelling. *Journal of Geophysical Research*, 102:25847–25879, 1997.

W. R. Stockwell, P. Middleton, J. S. Chang, and X. Tang. A second generation regional acid deposition model chemical mechanism for regional air quality modeling. *Journal of Geophysical Research*, 95: 16343–16367, 1990.

J. Swart and J. Blom. Experience with sparse matrix solvers in parallel ode software. *Comp. Math. Appl.*, 31:43–55, 1996.

A. S. Tomlin, S. Ghorai, G. Hart, and M. Berzins. 3d adaptive unstructured meshes for air pollution modelling. *Environmental Management and Health*, 10:267–274, 1999.

L. N. Trefethen and D. Bau III. *Numerical Linear Algebra*. SIAM, Philadelphia, 1997.

A. Venkatram, P. K. Karamchandani, and P. K. Misra. Testing a comprehensive acid deposition model. *Atmospheric Environment*, 22:737–747, 1988.

J. G. Verwer, J. G. Blom, M. van Loon, and E. J. Spee. A comparison of stiff ode solvers for atmospheric chemistry problems. *Atmospheric Environment*, 30:49–58, 1996.

J. G. Verwer and D. Simpson. Explicit methods for stiff ode's from atmospheric chemistry. *Appl. Numer. Math.*, 18:413–430, 1995.

J. G. Verwer and M. van Loon. An evaluation of explicit pseudo-steady state approximation for stiff ode systems from chemical kinetics. *J. Comput. Phys.*, 113:347–352, 1994.

R. A. Willoughby. Sparse matrix algorithms and their relation to problem classes and computer architecture. In J. K. Reid, editor, *Large Sparse Sets of Linear Equations*, pages 255–277. Academic Press, London-New York, 1970.

Z. Zlatev. Application of predictor-corrector schemes with several correctors in solving air pollution problems. *BIT*, 24:700–715, 1984.

Z. Zlatev. Mathematical model for studying the sulphur pollution in europe. *J. Comp. Appl. Math.*, 12:651–666, 1985.

Z. Zlatev. Treatment of some mathematical models describing long-range transport of air pollutants on vector processors. *Parallel Computing*, 6:87–99, 1988.

Z. Zlatev. Running large air pollution models on high speed computers. *Math. Comput. Modelling*, 14:737–740, 1990.

Z. Zlatev. *Computational methods for general sparse matrices*. Kluwer Academic Publishers, Dordrecht-Boston-London, 1991.

Z. Zlatev. *Computer treatment of large air pollution models.* Kluwer Academic Publishers, Dordrecht-Boston-London, 1995.

Z. Zlatev and R. Berkowicz. Numerical treatment of large-scale air pollution models. *Computers and Mathematics with Applications*, 15: 93–109, 1988.

Z. Zlatev, R. Berkowicz, and L. P. Prahm. Testing subroutines solving advection-diffusion equations in atmospheric environments. *Comput. Fluids*, 11:12–38, 1983a.

Z. Zlatev, R. Berkowicz, and L. P. Prahm. Three dimensional advection-diffusion modelling for regional scale. *Atmospheric Environment*, 17: 491–499, 1983b.

Z. Zlatev, J. Brandt, P. J. H. Builtjes, G. Carmichael, I. Dimov, J. Dongarra, H. van Dop, K. Georgiev, H. Hass, and R. S. Jose. *Large Scale Computations in Air Pollution Modelling.* Kluwer Academic Publishers, Dordrecht-Boston-London, 1999.

Z. Zlatev, J. Christensen, and A. Eliassen. Studying high ozone concentrations by using the danish eulerian model. *Atmospheric Environment*, 27A:845–865, 1993.

Z. Zlatev, J. Christensen, and Ø. Hov. An eulerian air pollution model for europe with nonlinear chemistry. *Journal of Atmospheric Chemistry*, 15:1–37, 1992.

Z. Zlatev, I. Dimov, and K. Georgiev. Studying long-range transport of air pollutants. *Computational Science and Engineering*, 1(3):45–52, 1994.

Z. Zlatev, I. Dimov, and K. Georgiev. Three-dimensional version of the danish eulerian model. *Zeitschrift für Angewandte Mathematik und Mechanik*, 76(S4):473–476, 1996a.

Z. Zlatev, J. Fenger, and L. Mortensen. Relationships between emission sources and excess ozone concentrations. *Computers and Mathematics with Applications*, 32(11):101–123, 1996b.

Z. Zlatev and C. A. Skjøth. The web-site of the danish eulerian model developed at the national environmental research institute (neri), 1999. Available at:
http://www.dmu.dk/AtmosphericEnvironment/DEM.

Index

Access gap, 362
Adjacency predicate model, 125
Aggregate views, 20, 712
Air pollution modelling, 1170
Antidictionaries, 180
Archival metadata, 964
Arithmetic coding, 260
Association rules, 186
Astronomical data, 474
Astronomy, 932
Attribute accuracy, 646
B-spline, 480
B-trees, 315
Background knowledge, 1141
Bandwidth, 48
Batched graph problems, 396
Batched sorting, 359
Bayesian neural networks, 1142
Billing, 895
Binning, 502
Biomolecular data mining, 1142
Biorthogonal wavelets, 479
Bipartite graphs, 126
Block addressing, 203
Boolean queries, 197
Bucketing, 502
Burrows-Wheeler transform, 178
CC-NUMA machines, 804
Checkpointing, 31
Clustering, 419
Clustering analysis, 962
Clusters, 595
Completeness, 645
Comprehensive mathematical models, 1171
Compression, 152
Compression entropy, 473
Concave function, 440
Convex polyhedral set, 424
Customer profiles, 912
Customer relationship management, 921
Data cleaning, 712
Data compression, 246
Data cube, 548
Data envelopment analysis, 419

Data gathering, 661
Data intensive computing, 854
Data layout, 867
Data maintenance, 662
Data management, 986
Data mining, 593, 966
Data mining astronomy, 932
Data organization, 662
Data quality, 644, 661
Data squashing, 579
Datastore, 743, 898
Data technology, 593
Data warehouse configuration problem, 726
Data warehouses, 594
Data warehousing, 548, 661, 711
Decision trees, 595
Decomposition algorithms, 429
Digital images, 248
Digital sky survey, 932
Digitized radiology, 870
Disk block, 362
Disk striping, 361
Distributed computing, 858
Distributed data, 854
Distributed data mining, 856
Distributed-memory MIMD machine, 802
Dynamic caching, 726
Dynamic databases, 593, 912
Earth science, 1095
Emission data, 1170, 1186
Encoding, 152
Entropy, 250
Environmental information systems, 982
Environmental metadata, 983
Evolving data, 605
Expectation-maximization algorithm, 1142
External memory, 362
Fingerprinting, 167
Flynn taxonomy, 793
Fourier transform, 375
Fraud, 912
Fraud detection, 914, 916
Frequent itemsets, 595

1221

Fuzziness, 651
Generalized support vector machine, 444
Geographic and environmental information systems, 1059
Geographic information systems, 643, 1115
Graph clustering, 516
High performance fortran, 794
HTTP, 29
Huffman coding, 253
Human genome project, 1142
Hyperspectral remote sensing, 1102
Hypertext, 5
Incidence function model, 125
Incremental maintenance, 912
Indexer, 10
Indexer process, 28
Indivisibility assumption, 388
Information retrieval, 3
Information theory, 249
Internet archive, 27
Internet traffic, 77
Interoperability, 1116
Inverted files, 196
Inverted list, 197
I/O, 856, 896
I/O complexity, 370
I/O model, 313
Java, 27
Kinetic data structures, 337
Knowledge discovery in databases, 594, 961
Lagrange multipliers, 444
Legacy formats, 867
Lineage, 645
Linear and quadratic programming, 439
Linear programming, 421
Linear separability, 439
Link extractor, 30
Locality and load balancing, 371
Logical consistency, 646
Lotka's law, 99
LZW coding, 178
Markov chains, 160
Massive data sets, 743, 895, 1093
Massive graphs, 99
Measurements, 1173
Mercator, 26
Message passing interface, 795
Metadata, 652, 983
Meteorological data, 1185
Metrics, 643
Moment matching, 582
Monitoring data characteristics, 593
Moore's law, 50
Multidimensional binary search, 504
Multiscale transforms, 475

Multi-threading, 834
Name resolution, 38
Nonlinear kernels, 447
Non-parametric estimation, 419
OLAP, 548
Online transaction processing, 661
Out-of-core algorithms, 360
Output model results, 1170
Parallel computing, 853
Parallel disk model, 360
Parallel disk systems, 876
Parallel disks, 889
Parallel virtual machine, 795
Particle physics, 869
Periodicities, 152
Planar point location, 325
Polyhedral norm, 439
Positional accuracy, 647
Power laws graphs, 108
Probabilistic automata, 183
Promoter recognition, 1144
Property testing, 125
Proximity queries, 315
Pyramidal transform, 478
Querying, 662
R-trees, 1024
RAID, 362
Random evolution, 97
Randomized algorithms, 123
Range queries, 550
Regular expressions, 151
Relational databases, 745
RISC processor, 801
Sampling, 580, 884
Scalable clusters, 853
Scale invariance, 106
Search engines, 9
Semistructured data model, 743
Sequential updating, 920
Shared and distributed memory machines, 792
Shared-memory SIMD machines, 795
Shingling, 15
Signature design, 918
Simulation models, 1032
SOR algorithm, 464
Sorting and selection algorithms, 876
Space science, 932
Space telescope, 939
Spatial data, 645
Spatial databases, 1005
Spider, 9
Splitting attribute, 602
Statistical classification, 981
String matching, 170
Substitute dataset, 580
Suffix arrays, 209

INDEX

Supraindex, 213
Telecommunications, 895
Telephone billing systems, 743, 896
Thresholding, 525, 914
Transaction data, 912
URL resolver process, 28
URL server process, 27
Validation of the model results, 1170

Vector machines, 442
Virtual observatory, 949
Visualization, 494, 654
Wavelets, 475
Web, 3
Web crawler, 25
Web crawling, 26
Zipf's law, 99